The Wild Mammals of Japan

Second Edition

Published by:
SHOUKADOH Book Sellers and the Mammal Society of Japan

Copyright © 2015 by the Mammal Society of Japan

All right reserved, but the right to each photo and figure belong to the person stated in the caption. No part of the book may be reproduced in any form by Photostat, microfilm, or any other means, without the written permission of the publisher.

ISBN978-4-87974-691-7　C0645

First edition: July 2009
Second printing: June 2010

Second edition: July 2015
Printed in Japan

SHOUKADOH Book Sellers
Shimodachiuri Ogawa Higashi, Kamigyo-ku, Kyoto 602-8048, Japan
E-mail: shoukadoh@nacos.com

Cover design: Hiromasa Igota

The Wild Mammals of Japan

Second Edition

Edited by

Satoshi D. OHDACHI

Yasuyuki ISHIBASHI

Masahiro A. IWASA

Dai FUKUI

Takashi SAITOH

SHOUKADOH Book Sellers

SHOUKADOH
KYOTO

Contents

Foreword vi
Preface vii
Contributors
    Core editors ix
    The Committee for "*the Wild Mammals of Japan, 2nd edition*" ix
    Authors xii
    Photographers xv
Acknowledgments xvi
Explanations for abbreviations and specific criteria in this book xvii
Species list with red list status xxii

● Soricomorpha
    Soricidae 2
    Talpidae 28

● Erinaceomorpha
    Erinaceidae 48

● Chiroptera
    Pteropodidae 52
    Rhinolophidae 58
    Hipposideridae 68
    Vespertilionidae 72
    Miniopteridae 126
    Molossidae 130

● Primates
    Cercopithecidae 134

● Rodentia
    Gliridae 148
    Muridae 150
    Myocastoridae 188
    Sciuridae 190

● Lagomorpha
    Ochotonidae 210
    Leporidae 212

● Carnivora
    Canidae 222
    Procyonidae 232
    Felidae 234
    Ursidae 240
    Mustelidae 248
    Herpestidae 272
    Viverridae 275
    Phocidae 280
    Otariidae 292

● Sirenia

Dugongidae 300

● Artiodactyla

Cervidae 304
Suidae 312
Bovidae 314

● Cetacea

Balaenidae 322
Eschrichtiidae 324
Balaenopteridae 328
Physeteridae 344
Kogiidae 346
Ziphiidae 350
Delphinidae 364
Phocoenidae 398

● Research Topics

1. Distribution patterns and zoogeography of Japanese mammals 44
2. Comparative phylogeography of sika deer, Japanese macaques, and black bears reveals unique population history of large mammals in Japan 142
3. Mosaic genome structure of multiple subspecies in the Japanese wild mouse *Mus musculus molossinus* 186
4. Acorns as a resource for forest-dwelling mammals in Japan 206
5. Wild mammal-borne zoonoses and mammalogists in Japan 228
6. The feral cat (*Felis catus* Linnaeus, 1758) as a free-living pet for humans and an effective predator, competitor and disease carrier for wildlife 238
7. Conservation and management of large mammals focusing on sika deer and bears in Japan 246
8. A molecular phylogenetic view of mammals in the "three-story museum" of Hokkaido, Honshu, and the Ryukyu Islands, Japan 269
9. Invasive alien mammal problems in Japan 277
10. Population dynamics of terrestrial mammals in Japan 309
11. History of mammalogy in Japan 404

● Reference list

408

● Indices

Common names in English 494
Common names in Japanese ("Kana" order) 496
Scientific names 498
Research Topics 504
Authors 506

● Appendix (Maps)

Foreword

"*The Wild Mammals of Japan*" was published in 2009 as a memorial project of the Ninth International Mammalogical Congress, which was held in 2005 in Sapporo. Thereafter, the first edition of the book received high praise from the world as the only book providing comprehensive information about Japanese mammals in English. Moreover, we are very honored that this book has been donated to awardees of the Albert R. and Alma Shadle Fellowship in Mammalogy and the ASM Fellowship at the request of the American Society of Mammalogists.

"*The Wild Mammals of Japan*" is an essential book not only for researchers and specialists of wildlife but also for amateur naturalists, because it contains purely academic descriptions with the articles written in nontechnical English. It is a good "concierge" for mammalogists, as this book clearly describes what is known and unknown for each species.

Until now, we have been expectantly waiting for the new edition of the book. Therefore, I am very glad to announce that "*the Wild Mammals of Japan, the Second Edition*" has been published as a project of the Mammal Society of Japan. I greatly appreciate the efforts of all contributors to this book.

I recommend this book with confidence to all persons who are interested in mammals and mammalogy in Japan. I hope "*the Wild Mammals of Japan, the Second Edition*" will contribute to the development of research of wildlife in the world.

Koichi Kaji

President of the Mammal Society of Japan

June, 2015
Tokyo

A male sika deer on the Shiretoko Penisula, Hokkaido (K. Kaji).

Preface for the first edition

The Ninth International Mammalogical Congress (IMC 9; chair: Noriyuki Ohtaishi) was held 31 July–5 August 2005 in Sapporo, Hokkaido, Japan (see *Mammal Study* vol. 30, *Supplement*, 2005 and *Mammalian Science* vol. 45, 2005, pp. 193–236). Soon after the congress it was proposed that we publish a book about wild mammals in Japan. It would be the most comprehensive and up to date guide available, and would be readable by anyone in the world who had an interest in Japanese mammals. The project was conceived by the core editors of this volume. Soon after, the committee of the publication for "*The Wild Mammals of Japan*" (*WMJ*) was organized as an official working group of the Committee for the Memorial Projects of IMC 9. The committee members for *WMJ* were appointed by the core editors. Contributing authors were determined by the committee and manuscript writing began in winter 2006.

There were three main goals for this book. First, it had to be readable to anyone in the World with an interest in mammals in Japan, and so would need to be written in English. There is a large amount of research published on Japanese mammals including local reports of basic information. Unfortunately, much of this is written in Japanese, thus unavailable to non-Japanese readers. In *WMJ*, articles are written in simple English (except for technical terms) so that even non-advanced readers of English will be able to read the text. Second, this book should explicitly demonstrate what is known and what is unknown for each species. Particularly, clear demonstration of "unknown" matters is important to motivate future investigation. In addition, the whole range of a species, including outside of Japan, is presented to illustrate its entire distribution. Third, data sources and citations for each species would be needed for authenticity and to assist further reading, especially for persons starting research on a particular species. There are several general books of Japanese mammals, such as "*Coloured Illustrations of the Mammals of Japan*" by Yoshinori Imaizumi (1960, Hoikusha, Osaka. In Japanese), "*The Encyclopaedia of Animals in Japan. Mammals I & II*" (1996, Heibonsha, Tokyo. In Japanese) and "*A Guide to the Mammals of Japan*" edited by Hisashi Abe (2005, Tokai Univ. Press, Hadano. In Japanese and English). These books list major references but lack citation lists or data sources for descriptions, and so the reader is less able to refer to the original sources. In *WMJ*, we present detailed references to allow referral to the original sources of descriptions.

Because of the limited number of pages in this volume, reluctantly we decided to omit species identification keys and synonym lists of scientific names for all the species. For identification keys of Japanese mammals, please consult the books by Y. Imaizumi and H. Abe mentioned previously and "*Illustrated Skulls of Japanese Mammals, Revised edition*" by H. Abe (2007, Hokkaido Univ. Press, Sapporo. In Japanese) for land mammals and pinnipeds, and "*Marine Mammals of the World. A Comprehensive Guide to their Identification*" by T. A. Jefferson et al. (2008, Academic Press, London) and "*Field Guide to Whales, Dolphins and Porpoises in the Western North Pacific and Adjacent Waters. New edition*" edited by Seiji Ohsumi (2009, University of Tokyo Press, Tokyo. In Japanese) for whales.

"*The Wild Mammals of Japan*" includes accounts for 169 species of mammals. Eleven research topics are included to introduce advances in mammalogical research in Japan. Fundamentally, the unit of the accounts is the species. However, the Iriomote and Tsushima leopard cats, which are subspecies of *Prionailurus bengalensis*, are treated as independent units because they long have been treated as different species and are distributed in unique and distant islands. Alternatively, Bryde's whale and Eden's whale are treated as one unit, i.e., the *Balaenoptera brydei/B. edeni* complex, because they are under taxomic revision and it is very difficult to distinguish them in past studies. Among the 169 included species of mammals, 14 species were introduced into Japan and 3 are commensal murids. Feral dogs and cats are omitted. Serious ecological problems caused by feral cats in Japan are described in Research Topic 6 in this volume. There may be more exotic species which have become established, but there is little information available for minor aliens such as these. See Research Topic 9 for details of exotic mammals in Japan. Three recently extirpated species, the gray wolf and 2 bat species, are included in *WMJ*. The river otter and the Japanese sea lion probably are extirpated in Japan, but governmental and some academic authorities have not declared their elimination yet; they are treated as critically endangered species by the Ministry of the Environment, Japan. Basically, we followed the taxonomic scheme of "*Mammal Species of the World. A Taxonomic and Geographic References. 3rd edition*" (*MSW3*) edited by D. E. Wilson and D. M. Reeder (2005, Johns Hopkins Univ. Press, Baltimore). However, when the authors of *WMJ* have different taxonomic opinions and the core editors and the taxonomic adviser accept their classification, scientific names differ from those used in *MSW3*.

Descriptions for each species include basic information such as red list status, distribution, fossil record, morphology, dental and mammae formulae, genetics, reproduction, lifespan, diet, habitat, home range, behavior, natural enemies and parasites. Taxonomic issues, relationship with humans, and other matters are documented in the remarks section if needed. Topic sections are included in some species accounts to describe unique characteristics. In addition to the English common name, Japanese, Chinese (both in mainland and Taiwan), Korean, and Russian common names are provided for the convenience of peoples in East Asia.

Species accounts for the 169 species of mammals and the 11 research topics were written by 63 authors. The manuscripts were compiled by the editors-in-charge. Next, they were reviewed by the core editors and then by the editors for English. After modification of the manuscripts by the authors, final drafts were produced. Galley proofs were examined by both the editors-in-charge and the core editors consulting the authors. Final responsibility for the articles rests with the authors, although the core editors also share responsibility for the layout, style and usage of English. Photos generously were contributed by many volunteers including professional photographers (except for those of 2 bat species offered by The Natural History Museum, London). Animal icons, the cover and animal drawings also were illustrated by volunteers. Hence, this book is the result of many voluntary actions. It is the most detailed compilation to date of research concerning Japanese mammals. In the second printing of this volume, careless mistakes in the first printing have been corrected.

We hope this book will be enjoyed as a standard text on Japanese mammals for amateurs, students, and professional researchers alike. We also hope that it may contribute to the advancement of the study of mammals around the World. There could be no better reward for the participants of IMC 9.

Core editors:
Satoshi D. OHDACHI
Yasuyuki ISHIBASHI
Masahiro A. IWASA
Takashi SAITOH

June 2009, Kyoto

Preface for the second edition

The first edition of "The Wild Mammals of Japan" (*WMJ1*) was published in July 2009, as a memorial project of the 9th International Mammalogical Congress, held in August 2005 in Sapporo, Japan. Two thousands copies were printed and sold out within a year. Such high sales were unexpected because this kind of book usually is not in much demand in Japan. Then, an additional 500 copies with some minor corrections were printed in July 2010 with the support of the Mammal Society of Japan (MSJ).

The first edition has gained a high reputation not only in Japan but also in the world, which owes to all contributors' high academic level and strong motivations to make it better. The main purpose of publishing *WMJ1*—namely, informing the world of the present status of Japanese knowledge of mammalogical study—was successful, and we believe *WMJ1* became a major landmark in the history of Japanese mammalogy. We are very proud of the book and would like to express our deep gratitude to the contributors and readers.

However, by January 2014, the second printing was sold out. Also, new knowledge has accumulated and some taxonomy and red data statuses changed since 2009. Hence, we thought that it was time for a new edition. Although the biggest hurdle was publication expense, after tough negotiations, the council of MSJ agreed in September 2014 to publish the second edition of *"the Wild Mammals of Japan"* (*WMJ2*).

After the decision of publication, the core editors immediately began to prepare for the new edition, adding Dai Fukui as a new core editor. We planned to publish the second edition by mid-June 2015, as the 5th International Wildlife Management Congress, co-hosted by MSJ and the Wildlife Society, will be held in July 2015 in Sapporo and many wildlife researchers from all over the world will gather in Japan. This is a golden opportunity to let them know of recent achievements in Japanese mammalogical study. Herein, we have successfully accomplished our task on time.

The style of *WMJ2* is basically the same as *WMJ1*, but the lists of references of each taxonomic group were merged into one list because of the increased number of references cited. In *WMJ2*, 170 species (or subspecies) are described, and, as in *WMJ1*, we fundamentally followed the taxonomic regime for order level taxa in *"Mammal species of the World. 3rd edition"* (2005, the Johns Hopkins Univ. Press) although orders Soricomopha and Artiodactyla/Cetacea are now not used by many researchers. In *WMJ2*, Miniopteridae is treated as independent from Vespertilionidae, unlike *WMJ1*. Additionally, taxonomic status and/or scientific names (including year) have been changed or corrected for some species. The authors are the same as in *WMJ1* although one author joined in *WMJ2*. However, very regrettably, two authors of *WMJ1*, Drs. Kazuo Moriwaki and Go Ogura, passed away, but they were included in the authors of this edition. The previous 11 research topics are included in *WMJ2* and slightly modified with the latest knowledge. In *WMJ2*, species were numbered and a table listing the species with their red list statuses as determined by 4 organizations is included for the convenience of readers. In addition, the new cover of *WMJ2* was drawn by Hiromasa Igota as was *WMJ1*.

Most striking difference of *WMJ2* from *WMJ1* is the decrease of "unknown" and "not recorded" descriptions, which means the development of mammalogical investigation since *WMJ1* has progressed rapidly. Thus, this book is the frontier of studies of Japanese mammals. Nonetheless, the contents of this book also will be outdated soon. We expect Japanese mammalogists of the next generation to publish a new and much better book than *WMJ2*.

Core editors:

Satoshi D. Ohdachi　　Yasuyuki Ishibashi　　Masahiro A. Iwasa　　Dai Fukui　　Takashi Saitoh

June 2015, Kyoto

Contributors

● Core editors

Satoshi D. OHDACHI (editor in chief)	Institute of Low Temperature Science, Hokkaido University, Kita-19 Nishi-8, Kita-ku, Sapporo, Hokkaido 060-0819, Japan
Yasuyuki ISHIBASHI	Hokkaido Research Center, Forestry and Forest Products Research Institute, Hitsujigaoka 7, Toyohira-ku, Sapporo, Hokkaido 062-8516, Japan
Masahiro A. IWASA	College of Bioresource Sciences, Nihon University, Kameino 1866, Fujisawa, Kanagawa 252-0880, Japan
Dai FUKUI	The University of Tokyo Hokkaido Forest, The University of Tokyo, Yamabe-Higashimachi 9-61, Furano, Hokkaido 079-1563, Japan
Takashi SAITOH	Field Science Center, Hokkaido University, Kita-11 Nishi-10, Kita-ku, Sapporo, Hokkaido 060-0811, Japan

● The Committee for "*the Wild Mammals of Japan, 2nd edition*"

Committee chair

T. SAITOH — Hokkaido University

Editors

Soricidae
S. D. OHDACHI — Hokkaido University

Talpidae & Erinaceidae
Shin-ichiro KAWADA — Division of Vertebrates, Department of Zoology, National Museum of Nature and Science, Amakubo 4-1-1, Tsukuba, Ibaraki 305-0005, Japan

Chiroptera
Akira SANO (chief) — Mie Prefecture Forestry Research Institute, Nihogi 3769-1, Hakusan, Tsu, Mie 515-2602, Japan

Kuniko KAWAI — Department of Biology, School of Biological Sciences, Tokai University, Minamisawa 5-1-1-1, Minami-ku, Sapporo, Hokkaido 005-8601, Japan

D. FUKUI — The University of Tokyo

Primates
Hideki ENDO — The University Museum, The University of Tokyo, Hongo 7-3-1, Bunkyo-ku, Tokyo 113-0033, Japan

Gliridae, Muridae & Myocastoridae
M. A. IWASA (chief) — Nihon University

Yukibumi KANEKO — Takaya-cho 502-4, Sakaide, Kagawa 762-0017, Japan

Sciuridae
Tatsuo OSHIDA — Laboratory of Wildlife Ecology, Obihiro University of Agriculture and Veterinary Medicine, Inada-cho, Obihiro, Hokkaido 080-8555, Japan

Lagomorpha
Fumio YAMADA — Wildlife Ecology, Forestry and Forest Products Research Institute, Matsunosato 1, Tsukuba, Ibaraki 305-8687, Japan

Contributors

Canidae & Procyonidae

Kohji URAGUCHI — Hokkaido Institute of Public Health, Kita-19 Nishi-12, Kita-ku, Sapporo, Hokkaido 060-0819, Japan

Felidae

Masako IZAWA (chief) — Faculty of Science, University of the Ryukyus, Senbaru 1, Nishihara-cho, Okinawa 903-0213, Japan

Nozomi NAKANISHI — Faculty of Science, University of the Ryukyus, Senbaru 1, Nishihara-cho, Okinawa 903-0213, Japan

Ursidae

Yoshikazu SATO — Laboratory of Wildlife Ecology, Department of Environmental Symbiotic Science, Rakuno Gakuen University, Bunkyo-dai Midori-machi 582, Ebetsu, Hokkaido 069-8501, Japan

Mustelidae, Herpestidae & Viverridae

Ryuichi MASUDA — Faculty of Science, Hokkaido University, Kita-10 Nishi-8, Kita-ku, Sapporo, Hokkaido 060-0810, Japan

Pinnipedia & Sirenia

Masatsugu SUZUKI — Department of Veterinary Medicine, Faculty of Applied Biological Sciences, Gifu University, Yanagido, Gifu, Gifu 501-1193, Japan

Artiodactyla

Junco NAGATA — Forestry and Forest Products Research Institute, Matsunosato 1, Tsukuba, Ibaraki 305-8687, Japan

Cetacea

Masao AMANO (chief) — Faculty of Fisheries, Nagasaki University, Bunkyo-machi 1-14, Nagasaki, Nagasaki 852-8521, Japan

Motoi YOSHIOKA — Department of Life Sciences, Mie University, Kurimamachiya-machi 1577, Tsu, Mie 514-8507, Japan

Research topics

Takuya SHIMADA — Tohoku Research Center, Forestry and Forest Products Research Institute, Nabeyashiki 92-25, Shimo-kuriyagawa, Morioka, Iwate 020-0123, Japan

Editors for English

Leslie CARRAWAY — Department of Fisheries and Wildlife, Oregon State University, Nash 104, Corvallis, OR 97331, USA

Chris J. CONROY — The Museum of Vertebrate Zoology at UC Berkeley, 3101 Valley Life Sciences Building, University of California, Berkeley, CA 94720, USA

Darrin LUNDE — Division of Mammals, National Museum of Natural History, Smithsonian Institution, 10th and Constitution Ave, NW, Washington D.C., USA

Richard P. SHEFFERSON — Department of General Systems Sciences, Graduate School of Arts and Sciences, The University of Tokyo, Komaba 3-8-1, Meguro-ku, Tokyo 153-8902, Japan

Edward DYSON — 41 Beaufort Road, Staple Hill, Bristol BS16 5JU, UK

Parasitological adviser

Yasushi YOKOHATA — Graduate School of Science and Engeneering, University of Toyama, Gofuku, Toyama, Toyama 930-8555, Japan

Paleontological adviser

Naoki KOHNO — Division of Biotic Evolution, Department of Geology and Paleontology, National Museum of Nature and Science, Amakubo 4-1-1, Tsukuba, Ibaraki 305-0005, Japan

Taxonomic adviser

Masaharu MOTOKAWA — The Kyoto University Museum, Kyoto University, Yoshida Honmachi, Sakyo-ku, Kyoto, Kyoto 606-8501, Japan

Korean adviser

Sang-Hoon HAN — National Institute of Biological Resources, Korea, Environmental Research Complex, Gyoungseo-dong, Seo-gu, Incheon 404-708, Republic of Korea

Russian adviser

Alexei V. ABRAMOV — Laboratory of Mammalogy, Zoological Institute Russian Academy of Sciences, Universitetskaya nab., 1, Saint-Petersburg, 199034, Russia

Mainland Chinese adviser

M. MOTOKAWA — Kyoto University

Mahmut HALIK — College of Life Sciences and Technology, Xinjiang University, Urumqi, Xinjiang 830046, China

Taiwanese Chinese adviser

M. MOTOKAWA — Kyoto University

Ya-Ju, CHEN

Red list adviser

Atsushi KAWAHARA — Hokkaido Regional Environment Office, Ministry of the Environment, Japan, Kita-8, Nishi-2, Kita-ku, Sapporo, Hokkaido 060-0808, Japan

Treasurer

Y. SATO — Rakuno Gakuen University

Animal icon design

Miki KAWASHIMA — Pinniped Research Group (Hireashi Kenkyu Kai)

Book cover design

Hiromasa IGOTA — Faculty of Environmental Systems, Rakuno Gakuen University, Bunkyodai, Ebetsu, Hokkaido 069-8501, Japan

Animal drawings

Makiko KASHIWAGI

Bibliography list

Mayuko TANIGAWA

Contributors

● Authors (alphabetical order)

Hisashi ABE	Katsuraoka-cho 26-17, Otaru, Hokkaido 047-0264, Japan
Shintaro ABE	Biodiversity Center of Japan, Ministry of the Environment, Kenmarubi 5597-1, Kamiyoshida, Fujiyoshida, Yamanashi 403-0005, Japan
Masao AMANO	Faculty of Fisheries, Nagasaki University, Bunkyo-machi 1-14, Nagasaki, Nagasaki 852-8521, Japan
Kyle N. ARMSTRONG	Australian Centre for Evolutionary Biology and Biodiversity, The University of Adelaide, North Terrace Campus, South Australia 5005, Australia
Lazaro M. ECHENIQUE-DIAZ	Institute of Arts and Sciences, Yamagata University, Kojirakawa-Machi 1-4-12, Yamagata, Yamagata 990-8560, Japan
Hideki ENDO	The University Museum, The University of Tokyo, Hongo 7-3-1, Bunkyo-ku, Tokyo 113-0033, Japan
Dai FUKUI	The University of Tokyo Hokkaido Forest, The University of Tokyo, Yamabe-Higashimachi 9-61, Furano, Hokkaido 079-1563, Japan
Kaoru HATTORI	Fisheries Management Division, Hokkaido National Fisheries Research Institute, Katsurakoi 116, Kushiro, Hokkaido 085-0802, Japan
Osamu HOSON	National Research Institute of Tohoku National Fisheries Institute, 25-259 Shimomekurakubo, Samemachi, Hachinohe, Aomori 031-0841, Japan
Tohru IKEDA	Regional Sciences, Human Sciences, Graduate School of Letters, Hokkaido University, Kita-10 Nishi-7, Kita-ku, Sapporo, Hokkaido 060-0810, Japan
Takao INOUÉ	Geriatric Health Services Facility, Utopia, Kawasaki 581-3, Yonago, Tottori 683-0852, Japan
Yasuyuki ISHIBASHI	Hokkaido Research Center, Forestry and Forest Products Research Institute, Hitsujigaoka 7, Toyohira-ku, Sapporo, Hokkaido 062-8516, Japan
Mari ISHIDA	Akiyoshi-dai Museum of Natural History, Akiyoshidai, Mine, Yamaguchi 754-0511, Japan
Nobuo ISHII	Tokyo Woman's Christian University, Zempukuji 2-6-1, Suginami-ku, Tokyo 167-8585, Japan
Hajime ISHIKAWA	Whale Laboratory, Shimonoseki Academy of Marine Science, Arcaport 6-1, Shimonoseki, Yamaguchi 750-0036, Japan
Tsuyoshi ISHINAZAKA	Shiretoko Nature Foundation, Shiretoko National Park Nature Center, 531 Iwaubetsu, Shari, Hokkaido 099-4356, Japan
Takeomi ISONO	Fisheries Management Division, Hokkaido National Fisheries Research Institute, Katsurakoi 116, Kushiro, Hokkaido 085-0802, Japan
Masahiro A. IWASA	College of Bioresource Sciences, Nihon University, Kameino 1866, Fujisawa, Kanagawa 252-0880, Japan
Masako IZAWA	Faculty of Science, University of the Ryukyus, Senbaru 1, Nishihara-cho, Okinawa 903-0213, Japan
Koichi KAJI	Laboratory of Wildlife Conservation, Tokyo University of Agriculture and Technology, Saiwaicho 3-5-8, Fuchu, Tokyo 183-8509, Tokyo
Yayoi KANEKO	Carnivore Ecology and Conservation Research Group, Tokyo University of Agriculture and Technology, Saiwaicho 3-5-8, Fuchu, Tokyo 183-8509, Japan

Yukibumi KANEKO	Takaya-cho 502-4, Sakaide, Kagawa 762-0017, Japan
Shin-ichiro KAWADA	Division of Vertebrates, Department of Zoology, National Museum of Nature and Science, Amakubo 4-1-1, Tsukuba, Ibaraki 305-0005, Japan
Atsushi KAWAHARA	Hokkaido Regional Environment Office, Ministry of the Environment, Japan, Kita-8, Nishi-2, Kita-ku, Sapporo, Hokkaido 060-0808, Japan
Kuniko KAWAI	Department of Biology, School of Biological Sciences, Tokai University, Minamisawa 5-1-1-1, Minami-ku, Sapporo, Hokkaido 005-8601, Japan
Kazumitsu KINJO	Okinawa International University, Ginowan 2-6-1, Ginowan, Okinawa 901-2701, Japan
Mari KOBAYASHI	Faculty of Bioindustry, Tokyo University of Aguriculture, Yasaka 196, Abashiri, Hokkaido 099-2493, Japan
Yuuji KODERA	Center for Weeds and Wildlife Management, Utsunomiya University, Mine-machi 350, Utsunomiya, Tochigi 321-8505, Japan
Asato KUROIWA	Faculty of Science, Hokkaido University, Kita-10 Nishi-8, Kita-ku, Sapporo, Hokkaido 060-0810, Japan
Ryuichi MASUDA	Faculty of Science, Hokkaido University, Kita-10 Nishi-8, Kita-ku, Sapporo, Hokkaido 060-0810, Japan
Kyoichi MORI	Department of Animal Sciences, Teikyo University of Science & Technology, Yatsusawa 2525, Uenohara, Yamanashi 409-0193, Japan
Junji MORIBE	Research Center for Wildlife Management, Gifu University, Yanagido 1-1, Gifu, Gifu 501-1193, Japan
Kazuo MORIWAKI (deceased)	RIKEN Tsukuba Institute
Masaharu MOTOKAWA	The Kyoto University Museum, Kyoto University, Yoshida Honmachi, Sakyo-ku, Kyoto, Kyoto 606-8501, Japan
Takahiro MURAKAMI	Shiretoko Museum, Honmachi 49, Shari-cho, Hokkaido 099-4113, Japan
Junco NAGATA	Forestry and Forest Products Research Institute, Matsunosato 1, Tsukuba, Ibaraki 305-8687, Japan
Atsushi NAKAMOTO	University Education Center, University of the Ryukyus, Senbaru 1, Nishihara-cho, Okinawa 903-0213, Japan
Nozomi NAKANISHI	Faculty of Science, University of the Ryukyus, Senbaru 1, Nishihara-cho, Okinawa 903-0213, Japan
Keisuke NAKATA	Forestry Research Institute, Hokkaido Research Organization, Koshunai, Bibai, Hokkaido 079-0198, Japan
Mitsuo NUNOME	Graduate School of Bioagricultural Sciences, Nagoya University, Furo-cho, Chikusa-ku, Nagoya, Aichi 464-8601, Japan
Keiji OCHIAI	6-62-1 Asumigaoka, Midori-ku, Chiba, Chiba 267-0066, Japan
Go OGURA (deceased)	Faculty of Agriculture, University of the Ryukyus
Satoshi D. OHDACHI	Institute of Low Temperature Science, Hokkaido University, Kita-19 Nishi-8, Kita-ku, Sapporo, Hokkaido 060-0819, Japan
Tatsuo OSHIDA	Laboratory of Wildlife Ecology, Obihiro University of Agriculture and Veterinary Medicine, Inada-cho, Obihiro, Hokkaido 080-8555, Japan

Contributors

Midori SAEKI	Wildlife Management Laboratory, National Agricultural Research Center, Kannondai 3-1-1, Tsukuba, Ibaraki 305-8666, Japan
Hiroaki SAITO	Department of Zoology, Graduate School of Science, Kyoto University, Kitashirakawa-Oiwakecho, Sakyo, Kyoto 606-8502, Japan
Takashi SAITOH	Field Science Center, Hokkaido University, Kita-11 Nishi-10, Kita-ku, Sapporo, Hokkaido 060-0811, Japan
Akira SANO	Mie Prefecture Forestry Research Institute, Nihogi 3769-1, Hakusan, Tsu, Mie 515-2602, Japan
Hiroshi SASAKI	Department of Contemporary Social Studies, Chikushi Jogakuen University, Ishizaka 2-12-1, Dazaifu, Fukuoka 818-0192, Japan
Yoshikazu SATO	Laboratory of Wildlife Ecology, Department of Environmental Symbiotic Science, Rakuno Gakuen University, Bunkyo-dai Midori-machi 582, Ebetsu, Hokkaido 069-8501, Japan
Takuya SHIMADA	Tohoku Research Center, Forestry and Forest Products Research Institute, Nabeyashiki 92-25, Shimo-kuriyagawa, Morioka, Iwate 020-0123, Japan
Miki SHIRAKIHARA	Faculty of Science, Toho University, Miyama 2-2-1, Funabashi, Chiba 274-8510, Japan
Hitoshi SUZUKI	Graduate School of Enviromental Earth Science, Hokkaido University, Kita-10 Nishi-5, Kita-ku, Sapporo, Hokkaido 060-0810, Japan
Masaaki TAKIGUCHI	Japan Wildlife Research Center, Kotobashi 3-3-7, Sumida-ku, Tokyo 130-8606, Japan
Hidetoshi TAMATE	Department of Biology, Faculty of Science, Yamagata University, Kojirakawa 1-4-12, Yamagata, Yamagata 990-8560, Japan
Noriko TAMURA	Tama Forest Science Garden, Forestry and Forest Products Research Institute, Todori 1833, Hachioji, Tokyo193-0843, Japan
Harumi TORII	Education Center for Natural Environment, Nara University of Education, Takabatake, Nara, Nara 630-8528, Japan
Kohji URAGUCHI	Hokkaido Institute of Public Health, Kita-19 Nishi-12, Kita-ku, Sapporo, Hokkaido 060-0819, Japan
Shigeki WATANABE	Seian University of Art and Design, Ohginosato-higashi, Otsu, Shiga 520-0248, Japan
Fumio YAMADA	Wildlife Ecology, Forestry and Forest Products Research Institute, Matsunosato 1, Tsukuba, Ibaraki 305-8687, Japan
Tadasu K. YAMADA	Division of Vertebrates, Department of Zoology, National Museum of Nature and Science, Amakubo 4-1-1, Tsukuba, Ibaraki 305-0005, Japan
Koji YAMAZAKI	Laboratory of Forest Ecology, Department of Forest Science, Faculty of Regional Environment Science, Tokyo University of Agriculture, Sakuragaoka 1-1-1, Setagaya-ku, Tokyo 156-8502, Japan
Yasushi YOKOHATA	Graduate School of Science and Engineering, University of Toyama, Gofuku, Toyama, Toyama 930-8555, Japan
Motoi YOSHIOKA	Graduate School of Bioresources, Mie University, Kurimamachiya-machi 1577, Tsu, Mie 514-8507, Japan

● Photographers (alphabetical order)

Persons (including those of the 1st edition but not including the 2nd edition)

Shintaro Abe, Masao Amano, Sylvia Brunner, Hideki Endo, Toshihiko Fujioto, Dai Fukui, Kimitake Funakoshi, Yoko Goto, Masashi Harada, Azusa Hayano, Kazuaki Higashi, Tatsuya Hiragi, Hirofumi Hirakawa, Lipke B. Holthuis, Michael. A. Huffman, Hidetaka Ichiyanagi, Junpei Igota, Shigeru Ikemura, Takao Inoué, Yasuyuki Ishibashi, Mari Ishida, Nobuo Ishii, Tsuyoshi Ishinazaka, Takeomi Isono, Motoki Iwai, Masahiro A. Iwasa, Takamichi Jogahara, Ikuko Kanda, Masataka Kanda, Masami Kaneko (Rakuno Gakuen University), Yayoi Kaneko, Hiromitsu Katsu, Ami Kato, Shin-ichiro Kawada, Makoto Kawaguchi, Atsushi Kawahara, Kuniko Kawai, Kazumitsu Kinjo, Gohta Kinoshita, Minoru Kinoshita, Tetsuo Kirihata, Maki Kishiro (National Research Institute of Far Seas Fisheries), Hiroyuki Kobayashi, Mari Kobayashi, Yuuji Kodera, Kazunobu Kogi, Naoki Kohno, Teruyuki Komiya, Norihisa Kondo, Kyoji Koyamagi, Nozomi Kurihara, Toshio Kurita, Tomoaki Kuwahara (Kon Photography & Research, Sapporo), Shigeru Matsuoka, Natsuki Miyazawa, Masahiko Mizuno, Akinobu Mochizuki, Masaharu Motokawa, Junji Moribe, Mitsuru Mukohyama (deceased), Takahiro Murakami, Kenji Murata, Susumu Murata, Hiroshi Nagata, Hiroaki Nakajima, Atsushi Nakamoto, Koji Nakamura, Katsushi Nakata, Keisuke Nakata, Yuri Nakayama, Tomoyuki Namba, Teruki Nishizawa, Yoshiyori Nitta, Tatsuya Noro, Keiji Ochiai, Kenji Oda, Kazuya Ogawa, Jun-ichi Ohara, Satoshi D. Ohdachi, Jun'ichi Ohki, Hideaki Okada, Hiroyuki Okazaki, Katsuki Oki, Takashi Ono (EnVision), Keiichi Onoyama, Yuichiro Ooya, Yushi Osawa, Nao Ozaki, Midori Saeki, Akira Sano, Junko Sano, Hiroshi Sasaki, Akiyoshi Sato, Fumihiro Sato, Haruka Sato, Haruko Sato (Sea Life Watch), Mitsunori Sato, Shuhei Sato, Masahiko Satô, Takuo Sawahata, Yasuko Segawa, Mayumi Shigeta, Junzo Shimogai, Norihito Shirakawa, Yoshifumi Sugiura, Kei Suzuki, Kinji Suzuki, Osamu Takahashi, Masae Takaishi, Masaaki Takiguchi, Hisao Tamura, Noriko Tamura, Hajime Taru, Shirow Tatsuzawa, Mitsuhiko Toda, Harumi Torii, Kouji Tokutake (Yokohama Hakkeijima Seaparadise/Okinawa Churaumi Aquarium), Kimiyuki Tsuchiya, Akinari Uemura, Takeshi Ueyama, Kohji Uraguchi, Morio Urano, Nobutaka Urano, Yuya Watari, Fumio Yamada, Tadasu K. Yamada, Seizo Yamamoto, Terumasa Yamamoto, Masami Yamanaka, Koji Yamazaki, Hisashi Yanagawa, Yasushi Yokohata, Hiroshi Yokota, Motoi Yoshioka, and Somyos Yossundara

Institutes

Akiyoshi-dai Museum of Natural History, Amami Wildlife Conservation Center of Ministry of the Environment, Ehime Prefectural Science Museum, Institute of Cetacean Research, Kagoshima City Aquarium, Kamogawa Sea World, Kanagawa Prefectural Museum of Natural History, Kyodo Senpaku Kaisha Ltd., Mammal Ecology Laboratory at University of the Ryukyus, Marine World "Umino-Nakamichi", Mikurashima Tourist Association, Ministry of the Environment of Japan, Morioka Zoological Park, National Museum of Nature and Science (Tokyo), National Research Institute of Far Seas Fisheries, Natural History Museum of Botanic Garden at Hokkaido University, Ogasawara Whale Watching Association, Shimonoseki Marine Science Museum "Kaikyokan", The Natural History Museum (London), and Toba Aquarium

A brown bear under a rainbow on the Shiretoko Pen., Hokkaido (T. Kuwahara, Kon Photography & Research).

Acknowledgments

We would like to express our deep gratitudes to the following persons (alphabetical order).
Committee members and authors were excluded from this list.

Go Abe, Yuji Abe, Kanjana Adulyanukosol, Katie Anderson (The Natural History Museum, London), Kazutoshi Arai (Kamogawa Sea World), Mitsuhiko Asakawa, Shih-Wei Chang, Yu-Cheng Chang, Hiroshi Dewa, Nikolai E. Dokchaev, Kimitake Funakoshi, Masami Furuta (Toba Aquarium), David L. Garshelis, Masashi Harada, Yukihiko Hashimoto, David A. Hill, Hirofumi Hirakawa, Hiroaki Homma, Chao-Lung Hsu, Hiroshi Iida, Makoto Inaba, Osamu Ishibashi, Takuya Ito, Paula Jenkins (The Natural History Museum, London), Masami Kaneko, Fumikazu Kikuchi (formerly of Tama Zoological Park, Tokyo), Toru Koizumi, Teruyuki Komiya, Norihisa Kondo, Yoichi Kondo, John L. Koprowski, Kyoji Koyanagi, Tadashi Kuramoto, Naoko Kurose, William Z. Lidicker, Jr. (former President of International Federation of Mammalogists), Karli Lawson, Kishio Maeda, Tsutomu Mano, Hayato Masuya, Sumiko Matsumura, Yukiko Matsuura, Akiko Mikasa, Shingo Minamikawa, Shusaku Moteki, Mitsuru Mukohyama (deceased), Ryohei Nakagawa, Kenichi Nakanishi (deceased), Chris Newman, Sen-ichi Oda (former President of MSJ), Seiji Ohsumi, Noriyuki Ohtaishi (former President of MSJ), Masayuki Oishi, Kimiko Okabe, Takashi Ono, Keiko Osawa, Tomowo Ozawa, Tae-Geon Park, Deborah A. Roach, Shinjiro Sasaki, Akiko Sato, Jun J. Sato, Masahiko Satô, Krzysztof Schmidt, Tomofumi Shimada, Akio Shinohara, Chris Smeenk, Keiko Sone, Yuko Tajima, Hiroshi Takahashi, Keiichi Takahashi, Osamu Takahashi, Yuji Takakuwa, Hisao Tamura, Masaya Tatara, Shirow Tatsuzawa, Kozue Tomomatsu, Morihiko Tomozawa, Yukitoshi Totake, Kimiyuki Tsuchiya, Tsunenori Tsujimoto, Mayumi Ueno, Daiki Usuda, Peter Vogel, Yuya Watari, Yuriko Yamaga, Terumasa Yamamoto, Hisashi Yanagawa, Sachiko Yasui, Keiichi Yokoyama, Satoko Yoshikura, Mizuko Yoshiyuki, Kazunori Yoshizawa, and Shuyi Zhang.

We also thank the following institutions for their support:
Akiyoshi-dai Museum of Natural History
Amami Wildlife Conservation Center, Ministry of the Environment
Ehime Prefectural Science Museum
Iriomote Wildlife Conservation Center, Ministry of the Environment
Okinawa District Forest Office
Tsushima Wildlife Conservation Center, Ministry of the Environment
Yamashina Institute for Ornithology

Special thanks to Enami Ishida (Nakanishi Printing – Shoukadoh, Kyoto)

Committee members of IMC 9 (Y. Ishibashi)

Explanation for abbreviations and specific criteria in this book

Japanese RDB Categories by Ministry of the Environment, Japan (2014)

- EX (Extinct): Thought to have died out in Japan
- EW (Extinct in the Wild): Known to exist only in captivity and in cultivation
- CR (Critically Endangered): Facing an extremely high risk of extinction in the very near future
- EN (Endangered): Facing a very high risk of extinction in the future, although the risk is not as serious as for CR taxa
- VU (Vulnerable): Facing an increasing risk of extinction
- NT (Near Threatened): Unlikely to go extinct at present, but may become qualified for CR, EN or VU status in the future, depending on change of conditions
- DD (Data Deficient): Inadequate information available to evaluate risk of extinction
- LP (Threatened Local Population): Isolated locally and facing high risk of extinction in certain localities

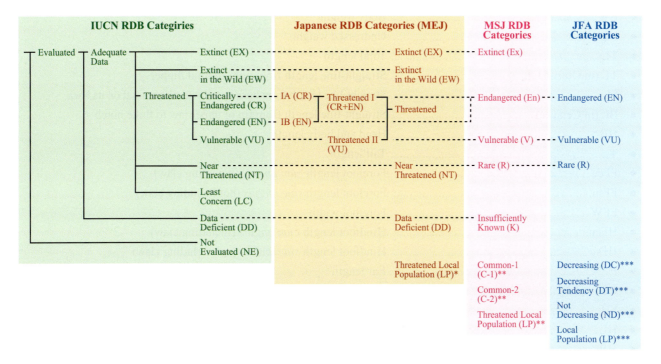

Fig. 1. Correspondences among the red list categories of three Japanese RDBs based on the IUCN RDB categories. One category with single asterisk, three categories with double asterisks and four categories with triple asterisks were created by MEJ, MSJ and JFA, respectively.

NOTE — Definition of categories

All the red list categories in this book fundamentally follow "the Japanese Red List (Japanese RDB Categories)" by MEJ (2014) (see http://www.biodic.go.jp/rdb/rdb_f.html for the latest version of red list about Mammalia). "MSJ", "JFA" and "IUCN" in parentheses following the red list category (or categories) in a species refer to whether the category was originally ranked by "the Mammalogical Society of Japan (renamed the Mammal Society of Japan)", "Japan Fisheries Agency" or "International Union for Conservation of Nature", respectively. The red list categories of MSJ (1997) are identical to the old categories of the Japanese RDB by the then Japan Environment Agency (1991). The red list categories of JFA were created by JFA (1998) for aquatic organisms.

Explanation for abbreviations and specific criteria in this book

Distribution map

- Green..Natural distribution
- Red...Artificially introduced distribution
- Purple..Extinct distribution
- Blue..Unknown whether natural or artificially introduced distribution

Statistics

- n...Number of samples
- SD...Standard deviation
- CV...Coefficient of variation

Morphology

- BW..Body weight
- UBW..Undressed weight
- TL..Total length
- TL (in Otariidae)..Straight-line length from nose to hind flipper
- SL..Straight-line length from nose to tail with the animal on its back
- BL (in Cetacea)...Straight-line length from tip of upper jaw to fluke notch
- HB...Head and body length
- T..Tail length
- FFcu..Forefoot length *cum unguis* (including claw)
- FFsu..Forefoot length *sine unguis* (not including claw)
- FFW..Forefoot width
- HFcu...Hindfoot length *cum unguis* (including claw)
- HFsu...Hindfoot length *sine unguis* (not including claw)
- E..Ear length
- CG...Chest girth
- SH...Shoulder height
- HL...Horn length
- HW..Height at withers
- FA..Forearm length
- Tib...Tibia length
- Tr..Tragus length

Vertebral formula

- Cx + Tx + Lx + Sx + Cdx..............................C: cervical vertebrae

  T: thoracic vertebrae

  L: lumbar vertebrae

  S: sacral vertebrae

  Cd: caudal vertebrae

  x: number of each kind of vertebra

Dental formula

- Ix/y, Cx/y, Px/y and Mx/y (PC$x/y$)..................I (or i*): incisor
  C (or c*): canine
  P (or p*): premolar
  M (or m*): molar
  PC: post canine teeth
  x: number of upper teeth
  y: number of lower teeth
  *Small letters are used for deciduous teeth.
- Superscript number.......................................Upper tooth
- Subscript number..Lower tooth

NOTE — Dental formula of Cetacea

Odontocetes are monophyodont (do not have deciduous teeth), and most of them have homodont (all teeth are similar shaped) dentition with variable number of teeth among individuals. Thus, dental formula of odontocetes indicates only the range of the number of teeth in each row of upper (above the slash) and lower jaw (under the slash). All extant mysticetes lack teeth but horny baleen plates in the upper jaw, and the range of the number of baleen plates are shown.

Mammae formula

- $x + y + z = w$ ....x: pectoral mammae
  y: abdominal mammae
  z: inguinal mammae
  w: total number of mammae

NOTE — Supplementary explanation for mammae formula

All bat species in Japan (Chiroptera) have a pair of teats in the anteriolateral pectoral position, with a single exception of *Vespertilio murinus* having two pairs. All macaque species in Japan (Primates) have a pair of teats on the chest. In whales (Cetacea), a pair of teats is present in mammary slits located at the both sides of genital slit. Therefore, there is fundamentally no mention of mammae formula for these animals.

Genetics

- $2n$...Diploid chromosome number
- FN ....Fundamental arm number
- A..acrocentric
- M..metacentric

NOTE — Two types of fundamental arm number description

FN refers to fundamental arm (brachial) number both including and excluding sex chromosomes. However, FNa refers to fundamental arm number of autosomes without sex chromosomes.

- mtDNA..mitochondrial DNA
- PCR...polymerase chain reaction

Home range

- MP...Minimum convex polygon method

Explanation for abbreviations and specific criteria in this book

Echolocation calls in Chiroptera

- FM .... Steep frequency-modulated components in the pulse
- CF .... Constant-frequency elements in the pulse
- QCF .... Quasi-constant-frequency: narrowband, shallow frequency-modulated components in the pulse
- EF .... End frequency: frequency at the end of the pulse
- PF .... Peak frequency: frequency of maximum energy of the pulse
- SF .... Start frequency: frequency at the start of the pulse
- FMAX .... Maximum frequency in the pulse
- FMIN .... Minimum frequency in the pulse

Organizations

- DDBJ .... DNA Data Bank of Japan
- EMBL .... European Molecular Biology Laboratory
- GenBank .... DNA Data Bank by National Center for Biotechnology Information
- IUCN .... International Union for Conservation of Nature
- IWC .... International Whaling Commission
- JFA .... Japan Fisheries Agency
- MEJ .... Ministry of the Environment, Japan
- MSJ .... Mammal Society of Japan (formerly, Mammalogical Society of Japan)
- USDI .... U. S. Department of the Interior

Natural Monument

"Natural Monument" is a rank that usually is enacted by the Japanese Government (sometimes prefectural governments) and maintained by the Agency for Cultural Affairs of Japan, for animals, plants, geological features and minerals. It is an indication of the marvelously natural conditions that exist under "Act on Protection of Cultural Properties" in Japan. Some "Natural Monuments" also are ranked as "Special Natural Monuments" which are considered as especially invaluable nature objects.

Appendix: Maps

- **Map 1:** World map (without polar areas) for major explanations of terrestrial and sea names.
- **Map 2:** Map of East Asia with major explanations of terrestrial and sea names.
- **Map 3:** Four mainlands (Hokkaido, Honshu, Shikoku and Kyushu), prefectures and the governmental district-level divisions in Japan.
- **Map 4:** Terrestrial and sea names of Japanese geography. Major names of islands and seas (strait, sea, bay, current, etc.) are described. If you need details for small islands and other geography, please see more specific maps.

    NOTE
    1) Some words are abbreviated due to lack of space (Isl., Island; Arch., Archipelago; Pen., Peninsula; Str., Strait).
    2) In island names, "-ohshima" is used for some island names and actually means a slightly larger island in Japanese. Usually, "⋯-oshima" is the correct spelling, but there is local name for "Oshima" Peninsula (in Hokkaido) in this map, and "⋯-oshima" and "Oshima" may be confused. Therefore, we chose "⋯-ohshima" to prevent any confusion here.
    3) In some species, "⋯ Island Group" is used in text and is identical to "⋯ Isls. (Arch.)" in Map 4.

References

IUCN. The IUCN Red List of Threatened Species (http://www.iucnredlist.org/).

Japan Environment Agency (1991) Threatened Wildlife of Japan, Red Data Book. Japan Wildlife Research Center, Tokyo (in Japanese).

水産庁 (Japan Fisheries Agency) (ed.) (1998) 日本の希少な野生水生生物に関するデータブック. 日本水産資源保護協会 (Japan Fisheries Resource Conservation Association), Tokyo (in Japanese).

Ministry of the Environment, Japan (2014) Threatened Wildlife of Japan, Red Data Book 2014. Vol.1 Mammalia. Gyosei Co., Tokyo (in Japanese with English list).

Mammalogical Society of Japan (1997) Red Data of Japanese Mammals (compiled by T. Kawamichi). Bun-ichi, Tokyo (in Japanese with English list).

An Iriomote cat taken by photo-trap in a subtropical forest on Iriomote-jima Isl., Okinawa (Mammal Ecology Laboratory, University of the Ryukyus). They mainly are nocturnal but sometimes active in day.

A Tokunoshima spiny rat, possessing the XO sex chromosome system, on Tokunoshima Isl., Okinawa (S. Ikemura).

A skull of a gray wolf, exterminated in Hokkaido in the late 19C, at Natural History Museum, Botanic Garden, Hokkaido Univ., Sapporo, Hokkaido (S. D. Ohdachi).

A roaring male Steller sea lion at Otaru Aquarium, Hokkaido (S. D. Ohdachi).

Species list with red list status

Species list with red list status by four organizations. Index Number (ID No) is indicated at the head of the first page for each species. "Hondo" means the island group of Honshu, Shikoku, Kyushu, and the neighboring small islands.

ID No.	Page	Scientific name	Red list category			
			MEJ (2014)	MSJ (1997)	IUCN (on line)	JFA (1998)
Soricomorpha						
001	2	*Sorex minutissimus*	VU as *S. m. hawkeri*	V as *S. m. hawkeri*	LC	—
002	4	*Sorex hosonoi*	NT	R	LC	—
003	6	*Sorex shinto*	NT as *S. c. shikokensis*	C-1 as *S. caecutiens shinto*, R as *S. c. shikokensis* and *S. sadonis*	LC	—
004	8	*Sorex caecutiens*	—	C-1 as *S. c. saevus*	LC	—
005	11	*Sorex unguiculatus*	—	C-1	LC	—
006	14	*Sorex gracillimus*	—	C-1	LC	—
007	16	*Chimarrogale platycephalus*	LP for the Kyushu population as *C. platycephala*	C-2 as *C. platycephala*, LP for the populations in Kyushu and Shikoku	LC	R for the population in Honshu, Shikoku and Kyushu as *C. himalayica*
008	19	*Crocidura shantungensis*	NT	R as *C. suaveolens shantungensis*	LC	—
009	21	*Crocidura watasei*	NT	R as *C. horsfieldii watasei*	LC	—
010	23	*Crocidura dsinezumi*	—	C-2 as *C. d. dsinezumi*, LP for the Hokkaido population as *C. d. dsinezumi*, R as *C. d. umbrina*	LC	—
011	25	*Crocidura orii*	EN	V	EN	—
012	26	*Suncus murinus*	—	C-1 as *S. m. temmincki*	LC	—
013	28	*Dymecodon pilirostris*	—	C-2, LP for the Shikoku, Kyushu and Kii Pen. populations	LC	—
014	30	*Urotrichus talpoides*	—	C-1	LC	—
015	32	*Euroscaptor mizura*	NT	R	LC	—
016	34	*Mogera imaizumii*	—	C-2, LP for the Shodo-shima Isl. population	LC	—
017	36	*Mogera wogura*	—	C-1	LC	—
018	38	*Mogera etigo*	EN	R	EN	—
019	40	*Mogera tokudae*	NT	R including *M. etigo*	NT	—
020	42	*Mogera uchidai*	CR	V	DD	—
Erinaceomorpha						
021	48	*Erinaceus amurensis*	—	—	LC	—
Chiroptera						
022	52	*Pteropus dasymallus*	CR as *P. d. daitoensis* and *P. d. dasymallus*	En as *P. d. dasymallus* and *P. d. daitoensis*, C-1 as *P. d. inopinatus* and *P. d. yayeyamae*	NT	—
023	54	*Pteropus loochoensis*	EX	Ex	DD	—
024	56	*Pteropus pselaphon*	EN	En, R for the Minami-iwo-to Isl. population	CR	—
025	58	*Rhinolophus ferrumequinum*	—	C-1	LC	—
026	61	*Rhinolophus cornutus*	EN as *R. c. orii*	C-2, K for the Amami Isls. population	LC	—
027	63	*Rhinolophus pumilus*	EN as *R. p. pumilus*, EX as *R. p. miyakonis*	En for the Miyako-jima Isl. population, V for the Ryukyu Isls. population	LC	—
028	65	*Rhinolophus perditus*	VU	En for the Ishigaki-jima Isl. population, V for the Iriomote-jima Isl. population	LC	—
029	68	*Hipposideros turpis*	LP for the Yonaguni-jima Isl. and Hateruma-jima Isl. populations	En for the Hateruma-jima, Ishigaki-jima and Yonaguni-jima Isls. populations, V for the Iriomote-jima population	NT	—
030	72	*Eptesicus japonensis*	VU	V	EN	—
031	74	*Eptesicus nilssonii*	—	R	LC	—
032	76	*Nyctalus aviator*	VU	R	NT	—
033	80	*Nyctalus furvus*	EN	V	VU	—
034	82	*Pipistrellus abramus*	—	C-2, K for the Okinawa Pref. population	LC	—
035	85	*Pipistrellus endoi*	VU	V	EN	—
036	87	*Pipistrellus sturdeei*	EX	Ex	DD	—
037	88	*Barbastella darjelingensis*	LP for the Honshu and Shikoku populations as *B. leucomelas*	V as *B. leucomelas*	LC as *B. leucomelas*	—
038	90	*Plecotus sacrimontis*	LP for populations in Kinki District and westward	R as *P. auritus*	LC	—
039	92	*Hypsugo alaschanicus*	DD	K as *Pipistrellus savii*	LC as *Pipistrellus savii*	—
040	94	*Vespertilio murinus*	DD	—	LC	—
041	96	*Vespertilio sinensis*	—	R as *V. superans*	LC	—
042	99	*Myotis rufoniger*	CR as *M. formosus*	K as *M. formosus*	LC as *M. formosus*	—
043	100	*Myotis frater*	—	R	DD	—

ID No.	Page	Scientific name	Red list category			
			MEJ (2014)	MSJ (1997)	IUCN (on line)	JFA (1998)
044	102	*Myotis gracilis*	VU	R as *M. mystacinus*	LC as *M. brandtii gracilis*	—
045	104	*Myotis ikonnikovi*	LP for the populations in the Kii Pen. and Chugoku District, as *M. i. hosonoi*	V for the Hokkaido population, R for the Honshu population	LC	—
046	107	*Myotis macrodactylus*	—	C-1	LC	—
047	110	*Myotis bombinus*	VU as *M. nattereri bombinus*	R as *M. nattereri*	NT	—
048	112	*Myotis petax*	—	R as *M. daubentonii*	LC as *M. daubentonii*	—
049	114	*Myotis pruinosus*	VU	V	EN	—
050	116	*Myotis yanbarensis*	CR	—	CR	—
051	117	*Murina hilgendorfi*	—	R as *M. leucogaster*	LC	—
052	120	*Murina ryukyuana*	EN	—	EN	—
053	122	*Murina tenebrosa*	DD	K	CR	—
054	123	*Murina ussuriensis*	—	R	LC	—
055	126	*Miniopterus fuliginosus*	—	C-1	LC	—
056	128	*Miniopterus fuscus*	EN	R	EN	—
057	130	*Tadarida insignis*	VU	K	DD	—
058	132	*Tadarida latouchei*	DD	—	DD	—
Primates						
059	134	*Macaca fuscata*	LP for populations in the Ouu and Kitakami Mountains, Kinkazan Isl.	R for *M. f. yakui* on Yaku-shima Isl., LP for *M. f. fuscata* on the Shimokita Pen. and Tohoku District	LC	—
060	137	*Macaca cyclopis*	—	—	LC	—
061	139	*Macaca mulatta*	—	—	LC	—
Rodentia						
062	148	*Glirulus japonicus*	—	V	LC	—
063	150	*Myodes rufocanus*	—	C-2 as *Clethrionomys r. bedfordiae*, LP for the Daikoku Isl. population	LC	—
064	154	*Myodes rex*	NT	R as *Clethrionomys rex*	LC	—
065	156	*Myodes rutilus*	—	C-2 as *Clethrionomys r. mikado*, LP for the Oshima Pen. population	LC	—
066	158	*Eothenomys andersoni*	—	C-2, LP for some populations in Honshu	—	—
067	160	*Eothenomys smithii*	—	R, LP for several populations in Kyusyu and Honshu	LC	—
068	162	*Ondatra zibethicus*	—	C-1	LC	—
069	163	*Microtus montebelli*	—	C-2, LP for the Noto-jima Isl. population	LC	—
070	165	*Tokudaia osimensis*	EN	En as *T. o. osimensis*	EN	—
071	168	*Tokudaia muenninki*	CR	En as *T. o. murnninki*	CR	—
072	169	*Tokudaia tokunoshimensis*	EN	—	EN	—
073	170	*Micromys minutus*	—	K, LP for the populations in Tokyo, Saitama and Hyogo Prefs.	LC	—
074	172	*Apodemus agrarius*	CR	R	LC	—
075	173	*Apodemus peninsulae*	—	C-1 as *A. p. giliacus*	LC	—
076	175	*Apodemus speciosus*	—	C-2, LP for the populations on Miyake-jima, Izu-ohshima and Nii-jima Isls.	LC	—
077	178	*Apodemus argenteus*	—	C-2, LP for the populations on Tsushima, Oki, Yaku-shima and Tane-gashima Isls.	LC	—
078	180	*Rattus norvegicus*	—	C-1	LC	—
079	181	*Rattus rattus*	—	C-1	LC	—
080	182	*Rattus exulans*	—	—	LC	—
081	183	*Diplothrix legata*	EN	En	EN	—
082	184	*Mus caroli*	—	C-1	LC	—
083	185	*Mus musculus*	—	C-1	LC	—
084	188	*Myocastor coypus*	—	C-1	LC	ND
085	190	*Sciurus vulgaris*	—	C-1 as *S. v. orientis*	LC	—
086	192	*Sciurus lis*	LP for the populations in Chugoku and Kyushu District	C-2, LP for several populations in western Honshu, and probably extinct in Kyushu	LC	—
087	196	*Callosciurus erythraeus*	—	C-1 as *C. e. thaiwanensis*	LC	—
088	198	*Tamias sibiricus*	DD	C-1 as *T. s. lineatus*	LC	—
089	200	*Petaurista leucogenys*	—	C-1	LC	—
090	202	*Pteromys momonga*	—	C-2, LP for the Kyushu population	LC	—
091	204	*Pteromys volans*	—	C-1 as *P. v. orii*	LC	—
Lagomorpha						
092	210	*Ochotona hyperborea*	NT as *O. h. yesoensis*	R as *O. h. yesoensis*, LP for the Yubari Mts. population	LC	—
093	212	*Pentalagus furnessi*	EN	En	EN	—
094	214	*Lepus timidus*	—	C-1 as *L. t. ainu*	LC	—

Species list with red list status

ID No.	Page	Scientific name	Red list category			
			MEJ (2014)	MSJ (1997)	IUCN (on line)	JFA (1998)
095	216	*Lepus brachyurus*	NT as *L. b. lyoni*	C-1	LC	—
096	218	*Oryctolagus cuniculus*	—	—	NT	—
Carnivora						
097	222	*Vulpes vulpes*	—	C-1 as *V. v. japonica* in "Hondo" and *V. v. schrencki* in Hokkaido	LC	—
098	224	*Nyctereutes procyonoides*	—	C-1 as *N. p. viverrinus* in "Hondo" and *N. p. albus* in Hokkaido	LC	—
099	226	*Canis lupus*	EX	EX	LC	—
100	232	*Procyon lotor*	—	C-1	LC	—
101	234	*Prionailurus bengalensis euptilurus*	CR	En as *Felis bengalensis euptilura*	LC as *P. bengalensis*	—
102	236	*Prionailurus bengalensis iriomotensis*	CR	En as *Felis iriomotensis*	CR	—
103	240	*Ursus arctos*	LP for the populations in the west side of Ishikari District and Teshio-Mashike Mts.	C-2, LP for the populations in south-western and northern Hokkaido and Konsen Plain	LC	—
104	243	*Ursus thibetanus*	LP for the populations in Shikoku, Chugoku District, Kii and Shimokita Pens.	C-2 as *Selenarctos thibetanus*, LP for some populations in Honshu, Shikoku, and Kyushu	VU	—
105	248	*Mustela itatsi*	—	C-1	LC	—
106	250	*Mustela sibirica*	NT as *M. s. coreana*	C-1 as *M. s. coreana*	LC	—
107	252	*Mustela nivalis*	NT as *M. n. namiyei*	V as *M. n. namiyei* in Honshu, C-1 as *M. n. nivalis* in Hokkaido	LC	—
108	254	*Mustela erminea*	NT as *M. e. nippon* and *M. e. orientalis*	R as *M. e. nippon* in Honshu, V as *M. e. orientalis* in Hokkaido	LC	—
109	256	*Neovison vison*	—	—	LC	—
110	258	*Martes melampus*	NT as *M. m. tsuensis*	V as *M. m. tsuensis* on the Tsushima Isls., C-1 as *M. m. melampus* in "Hondo"	LC	—
111	260	*Martes zibellina*	NT as *M. z. brachyura*	K as *M. z. brachyura*	LC	—
112	262	*Lutra lutra*	EX	En as *L. nippon*	NT	EN as *L. nippon*
113	264	*Enhydra lutris*	CR	—	EN	EN
114	266	*Meles anakuma*	—	C-1 as *M. meles anakuma*	LC	—
115	272	*Herpestes auropunctatus*	—	—	LC as *H. javanicus*	—
116	275	*Paguma larvata*	—	C-1	LC	—
117	280	*Phoca vitulina*	VU	En as *P. v. stejnegeri*	LC	VU as *P. v. stejnegeri*
118	282	*Phoca largha*	—	C-1	DD	ND
119	286	*Pusa hispida*	—	K as *Phoca h. ochotensis*	LC	DT
120	288	*Histriophoca fasciata*	—	C-1 as *Phoca fasciata*	DD	ND
121	290	*Erignathus barbatus*	—	K as *E. b. nauticus*	LC	DC
122	292	*Eumetopias jubatus*	NT	V	NT	R
123	295	*Callorhinus ursinus*	—	C-1	VU	DC
124	297	*Zalophus japonicus*	CR	En as *Z. californianus japonicus*	EX	EN as *Z. californianus japonicus*
Sirenia						
125	300	*Dugong dugon*	CR	En	VU	EN
Artiodactyla						
126	304	*Cervus nippon*	LP for the Mage-shima Isl. population	R for the population of the Kerama Isls., C-1 for the other populations	LC	—
127	307	*Muntiacus reevesi*	—	—	LC	—
128	312	*Sus scrofa*	LP for *S. s. riukiuanus* on Tokuno-shima Isl.	R for *S. s. riukiuanus* on the Ryukyu Isls., C-2 for *S. s. leucomystax* in Honshu, Shikoku, and Kyushu but LP for the populations on Nakadouri Isl. of the Goto Isls. and Awaji-shima Isl.	LC	—
129	314	*Capricornis crispus*	LP for Kyushu population	C-2, LP for Shikoku and Kyushu populations	LC	—
130	318	*Capra hircus*	—	—	—	—
Cetacea						
131	322	*Eubalaena japonica*	—	En as *E. glacialis* for the western North Pacific population	EN	VU as *E. glacialis*
132	324	*Eschrichtius robustus*	—	En for the Asian stock	CR for the western subpopulation	EN for the Asian stock
133	328	*Balaenoptera acutorostrata*	—	LP for the populations in the Sea of Japan, the Yellow Sea and the East China Sea, C-2 for the population in the Sea of Okhotsk and the western Pacific	LC	ND
134	331	*Balaenoptera borealis*	—	V	EN	DC

xxiv

ID No.	Page	Scientific name	Red list category			
			MEJ (2014)	MSJ (1997)	IUCN (on line)	JFA (1998)
135	333	*Balaenoptera brydei*	—	V as *B. edeni* for 1 or 2 populations from the Tosa Bay to the East China Sea, C-1 for the population in the western North Pacific	DD	R for the Eastern China Sea population as *B. edeni*, ND for the population in the western North Pacific as *B. edeni*
136	333	*Balaenoptera edeni*	—	V as *B. edeni* for 1 or 2 populations from the Tosa Bay to the East China Sea, C-1 for the population in the western North Pacific	DD	—
137	336	*Balaenoptera musculus*	—	En	EN	R for the western North Pacific population as *B. m. musculus*
138	338	*Balaenoptera omurai*	—	—	DD	—
139	340	*Balaenoptera physalus*	—	En for the population in the East China Sea, V for the population in the Sea of Japan, R for the population in the western North Pacific	EN	VU for the populations in the East China Sea and in the Sea of Japan, ND for the population in the western North Pacific
140	342	*Megaptera novaeangliae*	—	V for the population wintering off Ogasawara Isls. and Okinawa (Ryukyu) Isls.	LC	R for the population breeding off the Ogasawara Isls. the Okinawa Isls. and other North Pacific populations
141	344	*Physeter macrocephalus*	—	R for the two populations in the western North Pacific, K for the other populations in the North Pacific	VU	ND
142	346	*Kogia breviceps*	—	K	DD	—
143	348	*Kogia sima*	—	K as *K. simus*	DD	—
144	350	*Berardius bairdii*	—	R for the 3 populations of Japan	DD	DC
145	352	*Indopacetus pacificus*	—	R for the population in the tropical western North Pacific	DD	—
146	354	*Mesoplodon carlhubbsi*	—	R	DD	R
147	356	*Mesoplodon densirostris*	—	R	DD	R
148	358	*Mesoplodon ginkgodens*	—	R	DD	R
149	360	*Mesoplodon stejnegeri*	—	R	DD	R
150	362	*Ziphius cavirostris*	—	K	LC	R
151	364	*Delphinus capensis*	—	K	DD	R
152	366	*Delphinus delphis*	—	K	LC	ND
153	368	*Feresa attenuata*	—	R	DD	—
154	370	*Globicephala macrorhynchus*	—	R for the southern and northern forms, Common for the offshore population in the western North Pacific	DD	DC for the population north of Choshi to Hokkaido coasts, ND for south of Choshi
155	372	*Grampus griseus*	—	K	LC	ND
156	374	*Lagenodelphis hosei*	—	R	LC	ND
157	376	*Lagenorhynchus obliquidens*	—	C-1	LC	ND
158	378	*Lissodelphis borealis*	—	C-1	LC	—
159	380	*Orcinus orca*	—	R for the population in adjacent waters of Japan	DD	R for the population in adjacent waters of Japan
160	382	*Peponocephala electra*	—	R	LC	—
161	384	*Pseudorca crassidens*	—	LP for the population off the Pacific coast of Japan, C-2 for the other populations in the western North Pacific and adjacent seas	DD	DT for the population off the Pacific coast of Japan, DC for the populations off Iki-shima Isl. and in the East China Sea
162	386	*Stenella attenuata*	—	K	LC	DT for population off Japan in the western North Pacific
163	388	*Stenella coeruleoalba*	—	V for the populations off the Izu Pen. and Wakayama Pref. in Japanese waters, Common for the other populations in the western North Pacific	LC	R for the population in Pacific coasts of Japan, ND for populations in offshore North Pacific
164	390	*Stenella longirostris*	—	K	DD	—
165	392	*Steno bredanensis*	—	K	LC	—
166	394	*Tursiops aduncus*	—	—	DD	—
167	396	*Tursiops truncatus*	—	C-2, LP for the population in offshore waters of the western North Pacific	LC	ND
168	398	*Neophocaena asiaeorientalis*	—	En for the Omura Bay population as *N. phocaenoides*, R for the other populations off the Japanese coasts	VU	R as *N. phocaenoides*
169	400	*Phocoena phocoena*	—	R for population(s) inhabiting coasts of northern Japan, K for other populations in the Far East	LC	R
170	402	*Phocoenoides dalli*	—	K for the 2 populations in Japanese waters	LC	ND

A stone-carved guardian mouse at Ōkuni-sha, Ōtoyo Shrine, Kyoto (S. D. Ohdachi). The mouse is considered a messenger of Ōkuninushi, a noble deity of Shinto.

A stone-carved guardian wolf at Mitsumine Shrine, Saitama Pref. (S. D. Ohdachi). The wolf is considered a messenger at some Shinto shrines or even regarded as a noble deity. The Ainu, an indigenous people in Hokkaido, also regard the wolf as a deity.

A stone-carved guardian boar at Goō Shrine, Kyoto. The wild boar is a regarded as a messenger of Goō Shrine (S. D. Ohdachi).

A stone-carved guardian fox at Chanoki Inari Shrine in the Kamegaoka Hachimangū Shrine, Tokyo (S. D. Ohdachi). The red fox is considered a messenger of Inari Shrine, or even regarded as a deity itself.

A stone-carved sika deer behind the temizuya, an ablution pavilion, at Kasuga-taisha Shrine, Nara Pref. (Y. Nitta). The sika deer is regarded as a sacred beast at some Shinto shrines and Buddhist temples.

A stone-carved macaque at Hiyoshi-sha, Ōtoyo Shrine, Kyoto (S. D. Ohdachi). The Japanese macaque is considered a messenger of Sannō Gongen, a mountain deity of Shinto, or even regarded as a deity itself.

Soricomorpha

Soricomorpha   SORICIDAE

Red list status: VU as *S. m. hawkeri* (MEJ);
V as *S. m. hawkeri* (MSJ); LC (IUCN)

Sorex minutissimus Zimmermann, 1780

EN Eurasian least shrew   JP チビトガリネズミ (chibi togarinezumi)   CM 姬鼩鼱   CT 姬鼩鼱   KR 꼬마뒤쥐
RS крошечная бурозубка

An overwintered male eating a sea flea near a fruit of rugosa rose on a sand dune in Nemuro, Hokkaido (S. D. Ohdachi).

1. Distribution

Fennoscandia, Estonia, European part of Russia, Siberia, Mongolia, Chukotka, Kamchatka, Primorye, Shumushu Isl. (the northern Chishima = Kuril Isls.), Korean Pen., Northeast China (Manchuria), Sakhalin (Karafuto) and Hokkaido [2611]. Also distributed in Alaska, as the *Sorex minutissimus–S. yukonicus* complex [468, 939]. Among islands near the Hokkaido mainland, it is known from Kenbokki [1442, 1443] and Kunashiri Isls. [2611]. Thus, in Japan, its range is restricted to Hokkaido. In Hokkaido, it is recorded from the eastern, northern and central regions [1254, 1442, 1443, 2432].

2. Fossil record

A fossil referred to *S. minutissimus* was obtained from the Middle Pleistocene in western Honshu [1514, 2916] but it was reidentified as *Sorex hosonoi* [470]. In Northeastern Asia, no fossil records exist [2461] but some are known from western Europe from the middle Late Pleistocene [2915].

3. General characteristics

Morphology   One of the smallest mammals in the world, and the smallest shrew species in Hokkaido. Tail relatively shorter in relation to total body length, and more slender than that of *S. gracillimus*, a co-occurring small species of *Sorex* in Hokkaido. The external and cranial morphology is similar to that of *S. hosonoi*, but *S. minutissimus* is allopatric in Japan and smaller [8]. External measurements (mean ± SD) based on S. D. Ohdachi's database of 1988–2006 in millimeters or grams from Hokkaido were as follows (numbers in parentheses are sample size). In overwintered individuals (adult): BW, 2.4 ± 0.31 (12), TL, 80.3 ± 2.64 (12), T, 29.9 ± 0.68 (12: 2 males + 10 females), FFsu, 5.9 ± 0.39 (12), FFcu, 6.2 ± 0.44 (12), HFsu, 9.0 ± 1.56 (12), HFcu, 9.64 ± 1.08 (12). In young-of-the-year individuals: BW, 2.0 ± 0.24 (22: 14 males + 8 females), TL, 81.3 ± 2.99 (22), T, 32.5 ± 2.15 (22), FFsu, 6.0 ± 0.34 (22), FFcu, 6.6 ± 0.35 (15), HFsu, 9.0 ± 0.38. Skull measurements (in mm) were obtained only from young-of-the-year. Greatest length of skull, 13.69 ± 0.117 (5), width of braincase, 6.49 ± 0.154 (5), depth of braincase, 3.78 ± 0.131 (5), interorbital width, 2.79 ± 0.104 (5), rostral width between the 3rd unicuspids, 1.49 ± 0.065, length of upper tooth row, 4.91 ± 0.094 (5), length of upper unicuspid tooth row, 1.75 ± 0.085 (5). Total penis length of overwintered males 8.8 ± 1.57 SD mm ($n = 6$) in naturally stretched condition. Other measurements of this species in Japan were reported [8, 1011, 1252, 1444].
Dental formula   I 3/1 + C 1/1 + P 3/1 + M 3/3 = 32. Upper unicuspid tooth row consists of 5 teeth from I^2 to P^2 [29, 1011].
Mammae formula   0 + 0 + 3 = 6 or 0 + 1 + 2 = 6 [29, 1011].

Sorex minutissimus Zimmermann, 1780

Genetics Chromosomes: $2n = 42$ and $FN = 74$ in Hokkaido [2210]. On the continent, $2n = 42$ and $FN = 74$ in Siberia, but $2n = 38$ and $FN = 74$ in Finland [4253]. It is phylogenetically most closely related to *S. hosonoi* among extant species based on DNA analyses [478, 2619]. There exists great divergence of mitochondrial cytochrome *b* gene and the control region sequences between eastern and western Eurasian populations [939, 2611, 2623]. Local diversification of sequences of the mitochondrial control region are also prominent, but almost no prominent diversification within Hokkaido has been reported [2623]. A pair of PCR primers for sex determination is available [1995].

4. Ecology

Reproduction Mating system, gestation and lactation periods unknown. In July 25, 2006, the author (SDO) caught a late pregnant female with 4 embryos (ca. 13 mm total length) in Shiranuka, eastern Hokkaido.

Lifespan Under captivity, maximum life span is 303 days for a female kept in Hamanaka-cho, Hokkaido, and 704 days for a male kept in Tama Zoo, Tokyo [1444]. The average life span of captive shrews excluding those died in a month and of the longest records above, 355.7 ± 143.2 SD days ($n = 23$) in Hamanaka-cho and 401 ± 12.8 SD days ($n = 18$) in Tama Zoo [1444]. Lifespan in wild is unknown.

Diet Centipedes (Geophilomorpha and Lithobiomorpha), spiders, harvestmen, adult coleopterans, amphipodes and other arthropods were observed from Hokkaido [2432]. In Far East Russia, spiders, centipedes (Lithobiomorpha), adult coleopterans, and lepidopterans were main food items, and the dietary niche is closest to that of *S. gracillimus* among 7 sympatric shrew species [393]. Under captivity, the least shrew eats crickets (Gryllinae) and mealworms (*Tenebrio* sp.), ants, adult coleopterans, other insects, spiders, beach fleas (Talitridae), but never eats earthworms [1444].

Habitat It inhabits a wide variety of habitats such as coastal sand dunes, peat bog, shrubland, *Picea glehni* forest, deciduous broadleaved forests, mixed forest of coniferous and broad leaved trees in Japan [1254, 1442–1444]. It has been captured from 0 m to 890 m. On Kenbokki Isl. where *S. minutissimus* and *S. unguiculatus* co-occur, the former is captured closer to the seashore line than is the latter [1444].

Home range On Kenbokki Isl. in eastern Hokkaido, mean home range size is estimated to be 239.3 m^2 [1444].

Behavior It seems to be more active at night than daytime based on captive time and most active around 2:00 a.m. [1444]. Under captivity, least shrews frequently climb grass stalks and stay on them for long periods (sometimes for 30 minutes or more) [1444]. They move on grass stalks dexterously, and eat insects and spiders opportunistically. The shrew usually rests at arboreal position of grass. The nest is made of grass on the ground. A shrew makes 3 or 4 nests made of grass on the floor of a cage (40 cm × 30 cm area). The nest is ball-shaped and approximately 80 mm (long dimension) 40 mm (short dimension) × 50 mm (height) in size. When many crickets and beach fleas are available, shrews cache captured prey in holes that they dig or in grass on the ground under captivity. Severe aggressive behaviors against conspecific individuals have not been observed under captivity. In winter conditions, the shrews tend to stay under soil than above ground in a cage.

Natural enemies Not reported in Japan. Probably, there is no direct competition with conspecific *S. unguiculatus* on Kenbokki Isl., Hokkaido [1444].

Parasites Not reported in Japan.

5. Remarks

Rare species in Hokkaido, where populations might be fragmented. Protection of natural coastal sand dunes and wetlands in eastern and northern Hokkaido is important to maintain populations. The shrew in Alaska was once regarded as an independent species (*S. yukonicus*) from *S. minutissimus* [469], but it tentatively should be regarded as a subspecies or a local population of *S. minutissimus*, judging from molecular phylogeny based on mitochondrial DNA sequences [939, 2623].

S. D. OHDACHI & A. KAWAHARA

A Eurasian least shrew moves on slender stalks of mugwort (A. Kawahara).

A least shrew and a carabid beetle (*Carabus blaptoides*), captured in the same trap line in eastern Hokkaido (S. D. Ohdachi).

Soricomorpha SORICIDAE

002

Red list status: NT (MEJ); R (MSJ); LC (IUCN)

Sorex hosonoi Imaizumi, 1954

EN Azumi shrew   JP アズミトガリネズミ（azumi togarinezumi）   CM 本州鼩鼱   CT 本州尖鼠
KR 아즈미뒤쥐   RS азумийская бурозубка

A young individual in Mugikusa Pass, Chino, Nagano Pref. (J. Moribe).

1. Distribution

Distribution limited to central Honshu: Ishikawa, Gifu, Toyama, Nagano, Yamanashi, Gunma, Saitama and Shizuoka Prefs. [1738]. Type locality is Azumi-gun (900 m), Nagano Pref. Range of most habitats is from about 1,000 to 3,000 m.

2. Fossil record

Fossils referred to this species were obtained from the Middle Pleistocene in central and northeastern Honshu [470].

3. General characteristics

Morphology Small-sized *Sorex* species in Honshu. External measurements (mean ± SD) based on J. Moribe's database of 2002–2006 in millimeters or grams were as follows. BW, 3.6 ± 1.0 (range = 2.7–5.5, n = 6), TL, 104.9 ± 6.8 (96.6–114.0, n = 6), T, 49.2 ± 4.5 (42.0–55.7, n = 6), HFsu, 11.2 ± 0.8 (10.5–12.3, n = 6), E, 6.1 ± 0.7 (5.3–6.8, n = 6). Pelage of individuals are black bister, underparts wood brown [8]. Viewed from the side, first upper unicuspid clearly larger than second and third, which are nearly equal; second and third unicuspid clearly exceed fourth; fifth smaller than fourth in lateral view [8, 29]. Greatest length of skull is about 16 mm [29]. *Sorex hosonoi* is smaller than *S. shinto*, which is sympatric in Honshu. Similar dental and cranial characteristics of *S. hosonoi* indicate a close relationship with *S. minutissimus* [8, 29].

Subspecies *S. h. shiroumanus* has paler hair color compared to *S. h. hosonoi* [1004]. However, another researcher regarded the morphological difference as insignificant variation, as skull and teeth are quite similar between both subspecies [8].

Dental formula I 3/1 + C 1/1 + P 3/1 + M 3/3 = 32. Upper unicuspid consists of 5 teeth from I^2 to P^2 [8, 1004].

Mammae formula 0 + 0 + 3 = 6 or 0 + 1 + 2 = 6 [29, 1011].

Genetics Chromosome: $2n$ = 42 and FNa (fundamental numbers of autochromosomes) = 66 + XY sex chromosomes [2211]. The relationship between the karyotypes of *S. hosonoi* and *S. shinto* was explained by one pericentric inversion at chromosome no. 5 of *S. hosonoi* and chromosome no. 9 of *S. shinto*. A rearrangement in *S. shinto-hosonoi* differed from the rearrangements occurring on no. 5 of *S. shinto-caecutiens/unguiculatus* [2211]. Phylogenetically most closely related to *S. minutissimus*, based on mitochondrial cytochrome *b* gene sequences [2614]. Thus, *S. hosonoi* probably speciated from a common ancestor of *S. minutissimus* and *S. hosonoi* in proto-Honshu [2611]. A pair of PCR primers for sex determination is available [1995].

Sorex hosonoi Imaizumi, 1954

4. Ecology

Reproduction Mating system unknown. Newborn individuals appear from August to October. Gestation period unknown. A female with six embryos has been reported and breeding season in Mt. Hakusan, Ishikawa Pref., is from July to September [1738].

Lifespan Not reported. Certainly, no more than two years in the wild as in other *Sorex* species.

Diet Not reported. However, terrestrial arthropods such as spiders and small insects seem to be its main foods as in *S. caecutiens* [9].

Habitat Grasslands, shrublands and coniferous forests in subalpine to alpine zones and rarely found in either lower or higher mountain forests. The lowest habitats (approximately 1,000 m) are along streams or around waterfalls [2210].

Home range Not reported.

Behavior Not reported.

Natural enemies Ermines (*Mustela erminea*), Japanese martens (*Martes melampus*), and red foxes (*Vulpes vulpes*) have been observed eating soricid shrews [3720], but it has not been determined whether they were *S. hosonoi* or *S. shinto*.

Parasites *Neoskrjabinolepis singularis* is reported as a parasitic helminth [3128].

J. MORIBE

A habitat of the Azumi shrew at 2,700 m alt. in Mt. Tateyama, Toyama Pref. (J. Moribe). This species occurs in higher regions of central Honshu.

Morphological comparison between *S. shinto* (upper) and *S. hosonoi* (lower), both collected in Chino, Nagano Pref. (J. Moribe).

Soricomorpha   SORICIDAE

003

Red list status: NT as *S. caecutiens shikokensis* (MEJ); C-1 as *S. c. shinto*, R as *S. c. shikokensis* and *S. sadonis* (MSJ); LC (IUCN)

Sorex shinto Thomas, 1905

EN shinto shrew   JP シントウトガリネズミ (shinto togarinezumi)   CM 日本鼩鼱   CT 日本尖鼠
KR 일본뒤쥐   RS японская бурозубка

A young individual in Chino, Nagano Pref. (J. Moribe).

1. Distribution

Three subspecies of *Sorex shinto* are described: *S. s. shinto* (Honshu togarinezumi), *S. s. sadonis* (Sado togarinezumi), and *S. s. shikokensis* (Shikoku togarinezumi). *Sorex s. shinto* is distributed on Honshu. The ranges of *S. s. shinto* are limited to the high mountain areas in the central part of Honshu, but it is found at gradually lower elevations in the northern part of its range. *Sorex s. sadonis* is from the northern part of Sado(-gashima) Isl., Niigata Pref. This subspecies is found from lowlands (20 m) to near the top of the highest mountain (896 m). *Sorex s. shikokensis* is from Shikoku. Its distribution is limited to the Ishizuchi and Tsurugi-san mountain systems. No *Sorex* shrews are recorded from Kyushu. Distribution data based on author's database.

2. Fossil record

Fossils from the Middle to the Late Pleistocene are from Honshu, Shikoku and Kyushu [470, 819, 821, 1489, 1518, 2373, 3205].

3. General characteristics

Morphology   Medium size *Sorex* species in Honshu. No clear morphological demarcations were found among three subspecies, *S. s. shinto*, *shikokensis*, and *sadonis* [471]. However, *S. s. shikokensis* and *S. s. sadonis* tend to be larger in body and cranial size than *S. s. shinto*, and are morphologidally similar to *S. caecutiens* on Cheju(-do) Isl., South Korea [8, 471, 2618, 4214]. Comparing the three subspecies, *S. s. sadonis* is darker than *S. s. shinto*, and *S. s. shikokensis* is more light-brownish than *S. s. shinto* [8]. External measurements are as in Table 1.

Cranial size is greater in the populations on Sado and Shikoku than in Honshu [2618]. For example, greatest length of skull (GL) from Sado is 18.35 ± 0.22 SD mm ($n = 2$) and that from Shikoku is 18.00 ± 0.11 (8) while GL from Honshu is 17.54 ± 0.05 (33) [2618].

Dental formula   I 3/1 + C 1/1 + P 3/1 + M 3/3 = 32. Upper unicuspid consists of 5 teeth from I^2 to P^2 [29, 1011].

Mammae formula   0 + 0 + 3 = 6 or 0 + 1 + 2 = 6 [29, 1011].

Genetics   Chromosome: $2n = 42$ and FNa = 66 + XY sex chromosomes in Aomori, Nagano, and Yamanashi Prefs. [2211, 3392, 3637] and Sado and Shikoku [2210]. *Sorex s. sadonis* was described as a new species, *S. sadonis* [34, 4214]; however, many researchers now regard it as a subspecies or local population of *S.*

Sorex shinto Thomas, 1905

Table 1. External measurements in grams and mm (J. Moribe's database of 2002–2006). Mean ± SD (*n*).

subspecies	BW	TL	T	HFsu	HFcu	E
shinto	5.2 ± 1.08 (49)	110.0 ± 5.43 (49)	49.2 ± 2.85 (49)	11.9 ± 0.42 (49)	12.7 ± 0.45 (49)	6.6 ± 0.69 (49)
sadonis	6.2 ± 0.83 (8)	109.4 ± 3.42 (8)	45.8 ± 1.39 (8)	12.8 ± 0.35 (8)	13.6 ± 0.41 (8)	8.0 ± 0.20 (8)
shikokensis	6.2 ± 0 (2)	119 ± 2.23 (2)	52.0 ± 2.83 (2)	13.5 ± 0.14 (2)	14.4 ± 0.57 (2)	7.5 ± 0.71 (2)

shinto, according to molecular phylogenetic studies [2356, 2611, 2614].

4. Ecology

Reproduction   Mating system and gestation period unknown. Average number of embryos is 4.9 (ranging 3–6, *n* = 8) in May, 4.0 (2–5, *n* = 4) in June in Nagano pref., 3 (*n* = 3) in Yamanashi pref., and 2 (*n* = 1) in Ehime Pref. (*S. s. shikokensis*), including all stages of pregnancy [2135]. *Sorex s. sadonis* has 4–6 (*n* = 6) embryos in March and May [1741].

Lifespan   Not reported. Lifespan unlikely to exceed two years in the wild as in other *Sorex* species.

Diet   Not reported. However, terrestrial arthropods, such as spiders and small insects seem to be main foods as in *S. caecutiens* [9].

Habitat   Grasslands, shrublands and coniferous forests in subalpine to alpine zones. Rarely found in lower mountain forests, except for Sado Isl. (lowest point is at ca. 20 m) [4214]. In the lower elevations in Honshu (ca. 1,000 m) they were found along streams or waterfalls.

Home range   Not reported.

Behavior   Not reported.

Natural enemies   Ermines (*Mustela erminea*), Japanese martens (*Martes melampus*) and red foxes (*Vulpes vulpes*) ate *Sorex* shrews in Ishikawa Pref. [3720], but they were not identified as either *S. shinto* or *S. hosonoi*. Japanese mamushi pitvipers (*Gloydius blomhoffii*) also eat shinto shrews [2200, 3676].

Parasites   Several helminth species are recorded [1738, 3128]. The tick *Ixodes ovatus* [598] and several flea species [2968] are recorded as ectoparasites.

5. Remarks

Sorex chouei is a synonym of *S. shinto*. *Sorex chouei* was described on the basis of one specimen of an old individual with very worn teeth [1004], and its holotype represents an extreme point within the variation of *S. shinto* [471].

J. MORIBE

Habitats of the shinto shrew at 1,300 m alt. in Nagano Pref. (left) and at 300 m alt. on Sado Isl. (right) (J. Moribe). Usually, this species occurs at higher altitudes in central Honshu and Shikoku while it occurs at even lower altitudes on Sado Isl.

Soricomorpha SORICIDAE 004

Red list status: C-1 as *S. c. saevus* (MSJ); LC (IUCN)

Sorex caecutiens Laxmann, 1788

EN Laxmann's shrew, Eurasian common shrew
JP バイカルトガリネズミ（baikaru togarinezumi）, エゾトガリネズミ（ezo togarinezumi）for Hokkaido population
CM 中鼩鼱   CT 中尖鼠，北方尖鼠   KR 뒤쥐   RS средняя бурозубка

An overwintered male on the ground in Tomakomai, Hokkaido (T. Namba).

1. Distribution

Fennoscandia, northern part of Eurasia from eastern Europe via Siberia and northern Central Asia to Chukotka and Kamchatka, Far East Russia, Korean Pen., Cheju(-do) Isl. (Korea), Northeast China (Manchuria), Sakhalin, and Hokkaido [2617]. Records from Gansu, Henan and Tibet in China [968, 4249] should be re-examined [3278]. Among small islands near Hokkaido, it is recorded only from Kunashiri Isl. Thus, in Japan, its distribution is known only from Hokkaido.

2. Fossil record

A fossil referred to this species is known from the late Middle Pleistocene of Aomori Pref., Honshu [470]. In Far East Russia, fossils are recorded from the late Quaternary [2461]. Also it is found from the Middle to Late Pleistocene from eastern Europe [2915], but the fossils should be re-examined because species identification based on cranial and dental morphology is quite difficult in the *S. caecutiens/shinto* group [2618].

3. General characteristics

Morphology Medium size *Sorex* species in Hokkaido. Body size overlaps with overwintered (adult) *S. caecutiens* and young *S. unguiculatus* co-occurring in Hokkaido, but the former has a larger ratio of tail to total body length, and much shorter claws than *S. unguiculatus*. Body size overlaps with young *S. caecutiens* and overwintered *S. gracillimus* co-occurring in Hokkaido, but the former has a relatively longer tail and longer feet. Allopatric *S. shinto* is morphologically very similar to *S. caecutiens* in Hokkaido but the latter has a greater relative basal width of the upper premolar [471]. Among local populations of the *S. caecutiens/shinto* group in Northeastern Asia, *Sorex caecutiens* in Hokkaido is morphologically most similar to *S. caecutiens* in Sakhalin, while *S. caecutiens* from Cheju Isl. is most similar to *S. shinto* from Sado Isl. [2618]. Except for body weight in overwintered individuals, sexual difference in size is trivial or negligible. External and cranial measurements (mean ± SD) in millimeters or grams from the Hokkaido mainland population are given in Tables 1 and 2 (S. D. Ohdachi's database 1976–2006).

Total penis length of overwintered males 15.0 ± 1.91 SD mm ($n = 6$) in the naturally stretched condition (S. D. Ohdachi's database, 2005). More information from morphometric studies of *S. caecutiens* in Hokkaido is given elsewhere [5, 2618]. Summer pelage of young and adult individuals is brownish and winter pelage is more blackish [8]. Structure of hair has been examined under the scanning electron microscope comparing this species with *S. unguiculatus* and other shrew species in Japan [1686]. An

Sorex caecutiens Laxmann, 1788

Table 1. External measurements (in grams or mm) in Hokkaido. Mean ± SD (n).

	BW	TL	T	FFsu	FFcu	HFsu	HFcu
Overwintered							
male	7.7 ± 1.23 (107)	119.8 ± 4.51 (104)	47.8 ± 3.34 (103)	8.0 ± 0.45 (41)	8.8 ± 0.34 (15)	12.5 ± 0.47 (41)	13.7 ± 0.65 (14)
female	7.2 ± 1.37 (56)	120.3 ± 5.13 (54)	48.4 ± 4.34 (54)	7.7 ± 0.34 (26)	8.9 ± 0.48 (11)	12.4 ± 0.44 (25)	13.7 ± 0.50 (9)
Young-of-the-year							
male	5.0 ± 0.59 (282)	113.6 ± 4.26 (246)	48.4 ± 3.48 (246)	8.0 ± 0.43 (123)	9.3 ± 0.42 (56)	12.5 ± 0.60 (123)	13.8 ± 0.62 (55)
female	5.0 ± 0.63 (268)	112.9 ± 4.11 (233)	48.0 ± 3.11 (233)	7.9 ± 0.43 (147)	9.1 ± 0.45 (60)	12.3 ± 0.58 (148)	13.6 ± 0.72 (55)

Table 2. Cranial and dental measurements (in mm) in Hokkaido. Mean ± SD (n).

	GL	WB	DB	IOW	RW	UTL	UUL
Overwintered							
male	17.68 ± 0.321 (70)	8.88 ± 0.189 (70)	5.42 ± 0.166 (70)	3.77 ± 0.104 (70)	1.91 ± 0.061 (68)	6.65 ± 0.136 (70)	2.61 ± 0.100 (70)
female	17.62 ± 0.360 (33)	8.86 ± 0.237 (33)	5.30 ± 0.146 (33)	3.76 ± 0.130 (33)	1.94 ± 0.072 (32)	6.69 ± 0.155 (33)	2.59 ± 0.078 (33)
Young-of-the-year							
male	17.81 ± 0.359 (142)	8.98 ± 0.223 (143)	5.64 ± 0.260 (143)	3.73 ± 0.121 (143)	1.89 ± 0.059 (141)	6.77 ± 0.163 (143)	2.68 ± 0.089 (143)
female	17.70 ± 0.340 (110)	8.86 ± 0.201 (108)	5.61 ± 0.312 (109)	3.73 ± 0.111 (110)	1.90 ± 0.064 (110)	6.45 ± 0.169 (110)	2.66 ± 0.093 (110)

GL (greatest length of skull), WB (width of braincase), DB (depth of braincase), IOW (interorbital width), RW (rostral width between the 3rd unicuspid), UTL (length of upper tooth row from most anterior part of the first unicuspid to posterior part of the third molar), UUL (maxmum length of upper unicuspid tooth row).

individual with red eyes and covered with completely white fur (probably albino) was found in eastern Hokkaido in 2011 [2621].

Dental formula   I 3/1 + C 1/1 + P 3/1 + M 3/3 = 32. Upper unicuspid tooth row consists of 5 teeth from I^2 to P^2 [29, 1011].

Mammae formula   0 + 0 + 3 = 6 or 0 + 1 + 2 = 6 [29, 1011].

Genetics   Chromosome: $2n$ = 42 and FNa = 66 + XY sex chromosomes in Hokkaido [2760], whereas FNa varied from 66 to 68 depending on authors and local populations [1153, 2760, 4253]. Phylogenetically most closely related to *S. shinto* among extant species based on DNA analyses [478, 2619]. There is little diversification of sequences of the mitochondrial cytochrome *b* gene among localities throughout the Eurasian continent and Sakhalin, but the Hokkaido population was genetically unique from other local populations [2600, 2611, 2617]. Several PCR primers for microsatellite DNA are available [2357]. Based on microsatellite DNA analysis, it is suggested that most local populations in Hokkaido experienced recent bottleneck events [2357]. A pair of PCR primers for sex determination is available [1995].

4. Ecology

Reproduction   Mating system unknown. Newborn individuals appear from April to November with the peak of emergence in Hokkaido in July [2613]. Gestation period unknown. *S. caecutiens* most likely mates from late March to early November, judging from age structure and reproductive conditions [2613]. Average number of embryos recorded is 6.5 ± 1.15 SD (ranging 5–9, n = 13) including all stages of pregnancy in Hokkaido (S. D. Ohdachi's database, 1976–2006), whereas another researcher reported an average value of 7.1 for 8 individuals in Hokkaido [8]. On the Eurasian continent, the average litter size varied from 5.9 to 8.9 among populations [776]. All overwintered individuals are sexually mature in the spring in Hokkaido (S. D. Ohdachi's database, 1976–2006). Only 3 out of 187 young-of-the-year females were pregnant (1.6%) and another 3 were sexually mature but non-pregnant, (S. D. Ohdachi's database 1988–2006). Thus, 3.2% of females become sexually mature when they are young-of-the-year. There is no record of sexually mature males among young-of-the-year. Thus, most individuals do not become sexually mature before spring of the calendar year, although a few females become sexually mature before winter in Hokkaido. Testis and uterus size abruptly increases in spring [9, 2613].

Lifespan   In the wild, no individuals overwinter in the second winter of life [9, 2612]. Females tend to survive longer than males [9]. The maximum lifespan in captivity is 609 days [2606].

Diet   Main food items are epigeal small arthropods such as lepidopteran larvae, adult coleopterans, spiders (Araneae), harvestmen (Opiliones), and centipedes [1048, 2605]. In addition, they consume earthworms (Oligochaeta), slugs and snails [9], but earthworms are not highly preferred by *S. caecutiens* [2605]. Winter foods are unknown.

Habitat   *Sorex caecutiens* is a common species in central Hokkaido but less abundant in the southern and northern part. It occurs in many habitat types but is more abundant in habitats with volcanic ash and sandy soil, and less abundant in wetlands [2612]. When it occurs with other *Sorex* species in an area, it seems to prefer the more arid habitat [2616].

Home range Poorly known. There is one report of an average maximum home range size of 225.0 m² ($n = 3$) in Yufutsu meadow, central Hokkaido [1939, 2602, 2616].

Behavior *Sorex caecutiens* shows epigeal habits and is a ground surface wanderer [2604, 2607]. *Sorex caecutiens* can dig soil but cannot construct tunnel systems by itself [2604]. Basically it is nocturnal with two peaks of higher activity just after sunset and before sunrise [2603, 4180]. Even in the daytime, it becomes active every few hours [2603, 4180]. It may be found either above or below ground during resting periods [2607].

Natural enemies In Hokkaido, red foxes (*Vulpes vulpes*) [18] and sables (*Martes zibellina*) [2506] are known as mammal predators (or killers), and Ural owls (*Strix uralensis*) [4147] and Blakiston's fish owls (*Ketupa blakinstoni*) [40, 3982] are known as avian predators. Sables have been documented as predators of several species of *Sorex* in northern Hokkaido [2506], probably *S. unguiculatus*, *S. caecutiens*, and *S. gracillimus*. *Sorex unguiculatus* is a very small part of the prey of most of these predators, but for the Ural owl it is an important food item. Other carnivorous mammals, snakes [2200], other owls and raptors [2235] may also predate on *S. caecutiens*. The Sakhalin taimen (*Hucho perryi*) is an ichthyic predator in upstream rivers [2622]. In addition, an unidentified *Sorex* species (*S. unguiculatus* or *S. caecutiens*) was found in the stomach of a maruta dace (*Tribolodon brandti*) from a tributary of Teshio river in Horonobe, Hokkaido [2616].

Parasites A lot of helminth species are recorded [211, 742, 1543, 1738, 2595, 3128]. Parasitic acari and fleas are reported from Hokkaido [3, 2616, 2718, 2719, 2721, 2722, 2968].

5. Remarks

Subspecies of *S. caecutiens* in Hokkaido are classified as *S. c. saevus*, whose type locality is Korsakov (= Ohdomari) in Sakhalin (= Karafuto) in several papers [29, 1011]. However, Hokkaido and Sakhalin populations of *S. caecutiens* are phylogenetically rather distant [2356, 2611, 2617, 2618]. Thus, subspecies *saevus* is not appropriate for the population in Hokkaido. In addition, a new subspecies name is given for the population on Kunashiri Isl.: *S. c. kunashiriensis* Hutterer & Zaitsev, 2004 [969]. Further genetic and morphological comparisons between the Hokkaido mainland and Kunashiri populations are necessary to determine the appropriate subspecific name for the population on the Hokkaido mainland.

S. D. OHDACHI

An overwintered female captured in Sapporo, Hokkaido (M. A. Iwasa).

A Laxmann's shrew with white fur and reddish eyes, captured in Hamanaka-cho, Hokkaido (A. Kawahara).

Soricomorpha SORICIDAE

005

Red list status: C-1 (MSJ); LC (IUCN)

Sorex unguiculatus Dobson, 1890

EN long-clawed shrew   JP オオアシトガリネズミ (ooashi togarinezumi)   CM 长爪鼩鼱   CT 長爪尖鼠，粗腳尖鼠
KR 긴발톱첨서   RS когтистая бурозубка

A young individual in Sapporo, Hokkaido (M. A. Iwasa).

1. Distribution

Primorye and Khabarovsk regions of Far East Russia, North Korea, Northeastern China (Manchuria), Sakhalin, the southern Kuril Isls., Hokkaido and adjacent islands [2611]. Near the Hokkaido mainland, it is recorded from islands of Rebun, Rishiri, Teuri, Daikoku, Kenbokki [1442, 1443], Taraku (= Polonsky), Shibotsu (= Zelenyi), Akiyuki (= Anutchin), Yuri, Suisho (= Tanfilyev), Shikotan, and Kunashiri [*not* from Yagishiri and Etrofu Isls]. Thus, in Japan, the range is restricted to Hokkaido and some adjacent islands.

2. Fossil record

Some fossils of the late Middle Pleistocene of Yamaguchi Pref., Honshu are assigned to *S. unguiculatus* [470], and many fossils are known from Primorye in the late Quaternary [2461].

3. General characteristics

Morphology The biggest *Sorex* shrew in Japan. Claws, especially of forefoot, are very long. Tail relatively shorter in relation to total body length than in the co-occurring *S. caecutiens* in Hokkaido. *Sorex unguiculatus* is morphologically similar to *S. isodon* Turov, 1924, which co-occurs in Sakhalin and the Asian continent, but the latter has much shorter claws and relatively longer unicuspid tooth row, and is allopatric in Hokkaido [474]. Body size tends to be greater on small islands near Hokkaido than on the Hokkaido mainland. Sexual and age differences in size are obvious. External and cranial measurements (mean ± SD) in millimeters or grams from Hokkaido mainland and two islands are given in Tables 1 and 2 (S. D. Ohdachi's database from 1976–2006).

Penises of over-wintered (= sexually matured) males are huge relative to body size: total penis length 34.8 ± 3.57 SD mm ($n = 11$) in the naturally stretched condition (captured in Hokkaido, 2002–2005). Other morphometric studies were reported elsewhere [5, 8, 1011].

Molting of fur seems to happen at least three times during life although it is not well studied: juvenile pelage, summer young pelage, wintering adult pelage, and overwintered adult pelage (= summer adult pelage). Summer pelage brownish and winter pelage more blackish and the tip of belly fur is whitish gray [8, 1011]. Structure of hair has been observed under the scanning electron microscope compared with *Sorex caetutiens* and other shrew species in Japan [1686].

Dental formula I 3/1 + C 1/1 + P 3/1 + M 3/3 = 32. Upper unicuspid tooth row consists of 5 teeth from I^2 to P^2 [29].

SORICIDAE

Table 1. External measurements (in grams or mm) from Hokkaido mainland and two small islands. Mean ± SD (n).

	BW	TL	T	FFsu	FFcu	HFsu	HFcu
[Hokkaido mainland]							
Overwintered							
male	15.3 ± 1.66 (117)	133.7 ± 5.86 (117)	49.5 ± 2.83 (116)	9.9 ± 0.54 (116)	12.1 ± 0.58 (46)	14.1 ± 0.60 (115)	15.5 ± 0.82 (42)
female	12.1 ± 1.73 (142)	132.7 ± 5.35 (141)	49.6 ± 2.39 (140)	9.7 ± 0.53 (140)	12.1 ± 0.58 (63)	13.8 ± 0.55 (140)	15.3 ± 0.57 (57)
Young-of-the-year							
male	9.3 ± 1.06 (502)	125.9 ± 4.16 (473)	50.9 ± 2.42 (473)	9.9 ± 0.50 (475)	12.37 ± 0.571 (205)	14.1 ± 0.61 (474)	15.6 ± 0.74 (197)
female	8.8 ± 1.02 (445)	123.5 ± 4.19 (428)	49.5 ± 2.73 (429)	9.8 ± 0.46 (427)	12.2 ± 0.53 (197)	13.8 ± 0.56 (427)	15.3 ± 0.62 (183)
[Teuri Isl.]							
Overwintered							
male	20.5 ± 1.26 (9)	141.2 ± 4.02 (9)	50.9 ± 1.59 (8)	10.1 ± 0.17 (9)	–	14.9 ± 0.33 (9)	–
female	16.4 ± 1.76 (3)	135.2 ± 3.82 (3)	49.7 ± 1.53 (3)	9.9 ± 0.12 (3)	–	14.1 ± 0.27 (3)	–
Young-of-the-year							
male	10.8 ± 0.94 (11)	131.2 ± 2.93 (11)	52.0 ± 1.39 (11)	10.2 ± 0.42 (11)	–	14.8 ± 0.27 (11)	–
female	10.2 ± 0.92 (19)	127.8 ± 4.24 (19)	50.4 ± 2.13 (17)	10.0 ± 0.26 (19)	–	14.3 ± 0.33 (19)	–
[Rishiri Isl.]							
Overwintered							
male	17.0 ± 1.00 (7)	137.4 ± 2.35 (6)	48.2 ± 1.49 (6)	10.3 ± 0.27 (7)	–	14.7 ± 0.45 (7)	–
female	12.3 (1)	136.0 (1)	56.0 (1)	10.2 (10)	–	14.1 (1)	–
Young-of-the-year							
male	9.8 ± 0.54 (18)	128.1 ± 2.94 (18)	52.3 ± 1.67 (18)	10.4 ± 0.50 (18)	–	14.6 ± 0.53 (18)	–
female	9.6 ± 0.91 (22)	125.1 ± 3.54 (22)	50.3 ± 2.06 (22)	10.2 ± 0.39 (22)	–	14.2 ± 0.46 (22)	–

Table 2. Cranial and dental measurements (in mm) from Hokkaido mainland and two small islands. Mean ± SD (n).

	GL	WB	DB	IOW	RW	UTL	UUL
[Hokkaido mainland]							
Overwintered							
male	20.09 ± 0.390 (142)	10.36 ± 0.258 (141)	6.10 ± 0.236 (142)	6.10 ± 0.236 (142)	2.27 ± 0.076 (137)	7.57 ± 0.167 (141)	2.83 ± 0.012 (141)
female	19.70 ± 0.399 (105)	10.03 ± 0.243 (105)	5.78 ± 0.206 (105)	4.30 ± 0.119 (106)	2.23 ± 0.077 (105)	7.56 ± 0.167 (106)	2.85 ± 0.015 (106)
Young-of-the-year							
male	20.20 ± 0.416 (258)	10.24 ± 0.200 (258)	6.33 ± 0.223 (259)	4.32 ± 0.120 (259)	2.23 ± 0.079 (255)	7.68 ± 0.173 (258)	2.94 ± 0.170 (258)
female	20.05 ± 0.356 (207)	10.13 ± 0.231 (209)	6.33 ± 0.21 (208)	4.31 ± 0.118 (210)	2.22 ± 0.078 (209)	7.67 ± 0.158 (209)	2.92 ± 0.097 (210)
[Teuri Isl.]							
Overwintered							
male	21.30 ± 0.253 (4)	10.77 ± 0.172 (4)	6.45 ± 0.124 (4)	–	2.32 ± 0.033 (4)	7.99 ± 0.153 (4)	3.22 ± 0.125 (4)
female	20.99 (1)	10.34 (1)	5.90 (1)	–	2.34 (1)	8.10 (1)	3.26 (1)
Young-of-the-year							
male	21.41 ± 0.24 (8)	10.40 ± 0.093 (8)	6.55 ± 0.130 (8)	–	2.32 ± 0.037 (8)	8.17 ± 0.127 (8)	3.37 ± 0.094 (8)
female	21.20 ± 0.239 (6)	10.20 ± 0.238 (6)	6.48 ± 0.086 (6)	–	2.31 ± 0.069 (6)	8.10 ± 0.050 (6)	3.33 ± 0.051 (6)

GL (greatest length of skull), WB (width of braincase), DB (depth of braincase), IOW (interorbital width), RW (rostral width between the 3rd unicuspid), UTL (length of upper tooth row from most anterior part of the first unicuspid to posterior part of the third molar), UUL (maxmum length of upper unicuspid tooth row).

Mammae formula 0 + 0 + 3 = 6 or 0 + 1 + 2 = 6 [29, 1011].

Genetics Chromosome: $2n = 42$ and FNa = 68 + XY sex chromosomes in Hokkaido but FNa = 66 in Far East Russia [4253]. It is phylogenetically most closely related to *S. isodon* based on DNA analyses [478, 2619]. There is no local diversification of sequences of the mitochondrial cytochrome *b* gene throughout the whole range [2611]. A complete nucleotide sequence of the mitochondrial genome has been accomplished [2472]. Several PCR primers for microsatellite DNA are available [2357]. Based on microsatellite DNA analysis, it is suggested that most local populations in Hokkaido have experienced recent bottleneck events [2357]. Several sets of PCR primer pairs for sex determination are available [1286, 1995].

4. Ecology

Reproduction Mating system little known. Territory of an adult female overlapped with more than one adult male [1045], implying polyandry. Newborn individuals appear from April to

November with the peak of emergence in July in Hokkaido [2613]. Gestation period is unknown. *S. unguiculatus* likely mates from late March to early November, judging from age structure and reproductive conditions [2613]. Estimated lactating period is 27.6 ± 3.7 SD days (ranging 20–37, $n = 55$) in the wild [1039]. Average numbers of embryos was 5.6 ± 1.0 (ranging 4–8, $n = 20$) in early pregnant stage (embryo length < 8 mm) and 5.5 ± 0.5 (ranging 5–6, $n = 11$) in late stage (≥ 8 mm) [2601]. A female has at least two litters in a year (= lifetime reproduction) [1039]. The numbers of placental scars and corpora lutea are not good indicators of reproductive history because they decrease immediately after parturition [2601]. Postpartum estrus probably occurs [2601]. All overwintered individuals are sexually mature in spring in Hokkaido (S. D. Ohdachi's database, 1976–2006) and testis and uterus size abruptly increases [9, 2613]. Sexually mature individuals have not been recorded among the young-of-the-year. Thus *S. unguiculatus* does not become sexually mature before the next spring of the calendar year in Hokkaido. Under captivity, the onset of puberty for males is induced by low temperature and long day period [2680].

Lifespan In the wild, no individuals overwinter the second winter of their life [9, 2613]. Females tend to survive longer than males [9]. In the wild, the estimated lifespan of males is 408 ± 51.1 SD days ($n = 18$) while in females it is 459 ± 32.4 days ($n = 27$), although these estimates might be underestimated as observation ended before the death of some individuals [1039]. The maximum lifespan under captivity is 946 days [2606].

Diet Main food items in Hokkaido are earthworms (Oligochaeta), slugs & snails (Gastropoda), lepidopteran larvae, adult coleopterans, ants, spiders (Araneae), harvestmen (Opiliones), centipedes and other terrestrial small arthropods [1048, 2605]. Earthworms are the preferred food items for this species [2604]. Winter food habits are little known.

Habitat It is the most abundant soricid in Hokkaido, occurring in most habitat types including forest and open land, but less abundant in habitats with sandy soil and volcanic ash and in bogs [2612]. Interestingly, *S. unguiculatus* fundamentally occurs only in forests in Far East Russia and Sakhalin [467, 2462] but is ubiquitous in Hokkaido. *Sorex unguiculatus* prefers humus layer rich forest, compared with co-occurring *S. gracillimus* in Tokachi Region, Hokkaido [2511].

Home range Home range is rather territorial just after independence from parents [1044]. Females are more territorial than males [1045]. Home range is larger in males than females [1045]. Mean home range size of resident individuals is 188.3 m^2 ($n = 7$) [2602].

Behavior *Sorex unguiculatus* digs in the soil and makes tunnel systems [2604, 4180]. It is basically nocturnal with two peaks of higher activity just after sunset and before sunrise [2603, 4180]. During daytime, it is usually under ground but becomes active every few hours and sometimes emerges to the surface [2603, 2607]. Females are more natal philopatric than males [1039, 1045]. Postnatal development of behavior is reported [2460]. Some mating behaviors are observed under captivity; a male walks around a female shaking the posterior part of body and average copulating duration is 39.4 seconds for 51 bouts of copulation by a pair [2680].

Natural enemies In Hokkaido, red foxes (*Vulpes vulpes*) [18], introduced American minks (*Neovison vison*) [3741] and domestic cats (*Felis catus*) [2616] are known as mammal predators (or killers), and Ural owls (*Strix uralensis*) [2020, 4147], long-eared owls (*Asio otus*) [2019], and Blakiston's fish owl (*Ketupa blakinstoni*) [40, 3982] are known avian predators. Sables (*Martes zibellina*) are known to eat *Sorex* sp., certainly including *S. unguiculatus*, *S. caecutiens*, and *S. gracillimus* in northern Hokkaido [2506]. *Sorex unguiculatus* comprises a very small part of the prey of most predators, but for the Ural owl it is an important food item. Other mustelids, raccoon dogs (*Nyctereutes procyonoides*), raccoons (*Procyon lotor*), snakes [2200], other owls and raptors [2235] and bull-head shrikes (*Lanius buchephalus*) [3414] could be predators. The Sakhalin taimen (*Hucho perryi*) is an ichthyic predator in upstream rivers [2622]. In addition, unidentified *Sorex* species (*S. unguiculatus* or *S. caecutiens*) have been found in the stomach of a maruta dace (*Tribolodon brandti*) from a tributary of Teshio river in Horonobe, Hokkaido [2616]. For the red fox (*Vulpes vulpes*), *Sorex unguiculatus* seems to be less preferable as prey to the rodents (*Myodes rufocanus* and *Apodemus speciosus*), because most bodies of the shrew were not carried by the fox whereas most bodies of those rodents were carried away in a field experiment [1688].

Parasites A lot of helminth species are recorded [211, 1543, 1738, 2595, 3128]. Parasitic acari and fleas are reported from Hokkaido [2405, 2718, 2719, 2721, 2722, 2968].

5. Remarks

Ecological roles of *Sorex unguiculatus* in soil ecosystem in Hokkaido are examined based on field experiments. The presence of the shrews did not change the structure of microbial community in soil [3999]. Abundances of some soil arthropods and decomposition rate of litter were changed when the shrews are present although the effect of the shrew on the soil ecosystem is unclear [2430].

S. D. OHDACHI

Soricomorpha SORICIDAE

Red list status: C-1 (MSJ); LC (IUCN)

Sorex gracillimus Thomas, 1907

EN slender shrew   JP ヒメトガリネズミ (hime togarinezumi)   CM 细鼩鼱   CT 細尖鼠
KR 쇠뒤쥐   RS тонконосая бурозубка, дальневосточная бурозубка

A young individual captured in Kiritappu Moor, Hokkaido (M. A. Iwasa).

1. Distribution

Primorye, Khabarovsk, and Magadan regions of Far East Russia, North Korea, Northeast China (Manchuria), Sakhalin, Southern Kuril Isls., Hokkaido, and adjacent islands [2611]. Near Hokkaido mainland, it is recorded from islands of Rebun, Rishiri, Taraku (= Polonsky), Shibotsu (= Zelenyi), Suisho (= Tanfilyev), Shikotan, and Kunashiri. Thus, in Japan, the range is restricted to Hokkaido and some adjacent islands.

2. Fossil record

Fossils are not recorded from Japan, but they are recorded from Primorye [2461]. A fossil assigned to *S. gracillimus* was found in northern Honshu [2608] but it was reidentified as *S. hosonoi* [470].

3. General characteristics

Morphology   Small size *Sorex* species in Hokkaido. Body size overlaps between young *S. caecutiens* and overwintered (adult) *S. gracillimus* when co-occurring in Hokkaido, but the latter has a robust tail with sparse bristles and smaller fore- and hindfeet. Also body size overlaps in young *S. gracillimus* and overwintered *S. minutissimus* co-occurring in Hokkaido, but the latter has a short and delicate tail and smaller fore- and hindfeet. Skull and tooth morphology of allopatric *S. hosonoi*, which occurs only on Honshu, is quite different from *S. gracillimus* [29] although body size is similar. Sexual difference in size is trivial or negligible even in overwintered individuals. External and cranial measurements (mean ± SD) in millimeters or grams from Hokkaido mainland population are given in Tables 1 and 2 (S. D. Ohdachi's database 1976–2006).

On Rishiri Is., body measurements (in g or mm) are as follows. For overwintered male, BW = 4.6 ± 0.33 SD (n = 7), TL = 101.7 ± 3.55 (n = 7), T = 44.8 ± 1.55 (n = 7), FFsu = 6.7 ± 0.13 (n = 7), and HFsu = 11.3 ± 0.18 (n = 7). For overwintered female, BW = 5.5 ± 0.50 (n = 4), TL = 109.3 ± 1.71 (n = 4), T = 45.5 ± 1.00 (n = 4), FFsu = 6.9 ± 0.17 (n = 4), and HFsu = 10.9 ± 0.30 (n = 4). For young-of-the-year male, BW = 3.4 ± 0.10 (n = 3), TL = 101.7 ± 0.58 (n = 3), T = 44.7 ± 0.29 (n = 3), FFsu = 6.7 ± 0.21 (n = 3), and HFsu = 11.1 ± 0.12 (n = 3). For young female, BW = 3.3 ± 0.23 (n = 6), TL = 100.4 ± 2.29 (n = 6), T = 45.4 ± 1.64 (n = 6), FFsu = 7.1 ± 0.29 (n = 6), and HFsu = 11.0 ± 0.44 (n = 6). On Rebun Is., body measurements were obtained only for young-of-the-year male; BW = 3.5 ± 0.25 (n = 7), TL = 98.9.7 ± 3.56 (n = 7), T = 42.9 ± 1.99 (n = 7), FFsu = 7.0 ± 0.24 (n = 7), and HFsu = 11.4 ± 0.42 (n = 7).

Total penis length of overwintered males is 9.91 ± 1.16 SD mm (n = 11) in the naturally stretched condition (S. D. Ohdachi's database from Hokkaido mainland, 2002–06).

Summer pelage is somewhat tricolored and the tail bicolored [8]. Winter pelage is more blackish and bicolored.

Dental formula   I 3/1 + C 1/1 + P 3/1 + M 3/3 = 32. Upper unicuspid tooth row consists of 5 teeth from I^2 to P^2 [29, 1011].

Sorex gracillimus Thomas, 1907

Table 1. External measurements (in grams or mm) on the Hokkaido mainland. Mean ± SD (*n*).

	BW	TL	T	FFsu	FFcu	HFsu	HFcu
Overwintered							
male	4.5 ± 0.45 (96)	101.1 ± 4.42 (92)	43.3 ± 2.08 (92)	6.6 ± 0.42 (92)	7.4 ± 0.40 (30)	10.9 ± 0.53 (91)	11.7 ± 0.49 (29)
female	4.4 ± 0.79 (59)	103.5 ± 3.97 (54)	43.7 ± 2.18 (54)	6.6 ± 0.40 (53)	7.5 ± 0.44 (20)	10.9 ± 0.48 (53)	11.8 ± 0.53 (20)
Young-of-the-year							
male	3.3 ± 0.36 (172)	98.7 ± 2.83 (132)	44.0 ± 2.16 (133)	6.8 ± 0.41 (133)	7.5 ± 0.43 (47)	11.0 ± 0.50 (133)	11.9 ± 0.55 (47)
female	3.3 ± 0.41 (168)	99.8 ± 3.83 (134)	44.3 ± 2.07 (134)	6.7 ± 0.31 (134)	7.4 ± 0.42 (36)	10.9 ± 0.44 (133)	11.9 ± 0.47 (36)

Table 2. Cranial and dental measurements (in mm) on the Hokkaido mainland. Mean ± SD (*n*).

	GL	WB	DB	IOW	RW	UTL	UUL
Overwintered							
male	15.48 ± 0.229 (23)	7.32 ± 0.155 (23)	4.45 ± 0.138 (23)	3.16 ± 0.098 (23)	1.52 ± 0.120 (23)	5.89 ± 0.078 (23)	2.38 ± 0.057 (23)
female	15.37 ± 0.236 (20)	7.22 ± 0.202 (20)	4.30 ± 0.103 (20)	3.14 ± 0.063 (20)	1.54 ± 0.134 (20)	5.92 ± 0.093 (20)	2.39 ± 0.055 (20)
Young-of-the-year							
male	15.67 ± 0.269 (37)	7.37 ± 1.251 (37)	4.76 ± 0.228 (37)	3.11 ± 0.086 (38)	1.53 ± 0.091 (36)	5.96 ± 0.133 (38)	2.43 ± 0.088 (38)
female	15.67 ± 0.307 (33)	7.56 ± 0.183 (32)	4.83 ± 0.312 (33)	3.02 ± 0.540 (34)	1.48 ± 0.290 (34)	5.94 ± 0.118 (34)	2.44 ± 0.109 (34)

GL (greatest length of skull), WB (width of braincase), DB (depth of braincase), IOW (interorbital width), RW (rostral width between the 3rd unicuspid), UTL (length of upper tooth row from most anterior part of the first unicuspid to posterior part of the third molar), UUL (maximum length of upper unicuspid tooth row).

Mammae formula $0 + 0 + 3 = 6$ or $0 + 1 + 2 = 6$ [29, 1011].
Genetics Chromosome: $2n = 36$ and FNa $= 60 + XY$ sex chromosomes [4253]. There is local diversification of sequences of the mitochondrial cytochrome *b* gene among local populations [2611]. Phylogenetic position among *Sorex* (subgenus *Sorex*) was unclear due to the short diversification period among species of *Sorex* [478, 2619]. A pair of PCR primers for sex determination is available [1995].

4. Ecology

Reproduction Mating system unknown. Newborn individuals appear from April to November with the peak of emergence is in July in Hokkaido [2613]. Gestation period unknown. *Sorex gracillimus* likely mates from late March to early November, judging from age structure and reproductive conditions [2613]. Average number of embryos was $5.8 ± 1.21$ SD (ranging 4–8, $n = 6$) including all stages of pregnancy in Hokkaido (S. D. Ohdachi's database 1976–2006). All overwintered individuals are sexually mature in spring in Hokkaido and no reproductively active individuals have been found in the young-of-the-year (S. D. Ohdachi's database of 1976–2006). Thus, most individuals do not become sexually mature before the next spring of the calendar year.
Lifespan In the wild, there is no record of individuals overwintering in the second winter of life (S. D. Ohdachi's database 1976–2006). The maximum lifespan (estimation) in captivity is 419 days [2606].
Diet Main food items: epigeal small arthropods such as lepidopteran larvae, adult coleopterans, spiders (Araneae), harvestmen (Opiliones), and centipedes [2605]. *Sorex gracillimus* does not consume earthworms (Oligochaeta) [2605] and disfavors them [2604]. *Sorex minutissimus* has the most similar dietary composition with *S. gracillimus* among sympatric shrew species [393, 2432]. Thus, interspecific competition, if any, will be severe between the two species in Hokkaido. Winter foods are unknown.
Habitat *Sorex gracillimus* is common in eastern and northern Hokkaido but less abundant in the central and southern parts. It occurs in many habitat types but is more dominant in wetlands and higher places [2612]. *Sorex gracillimus* prefers litter rich forest, compared with co-occurring *S. unguiculatus* in Tokachi Region, Hokkaido [2511].
Home range Average home range of three individuals was 259.4 m^2 ($n = 3$) in northern Hokkaido [2602]. These three *S. gracillimus* had exclusive home ranges within species but overlapped with *S. unguiculatus* [2602].
Behavior *Sorex gracillimus* seems to have epigeal habits, although this is statistically not significant in laboratory experiments, and it may be a ground surface wanderer [2604, 2605, 2607]. *Sorex gracillimus* can dig soil but does not construct a tunnel system by itself [2604].

Basically nocturnal with two peaks of higher activity just after sunset and before sunrise [2603]. Even in the daytime, it becomes active every few hours [2603]. In laboratory match tests using *S. unguiculatus*, *S. caecutiens* and *S. gracillimus*, *S. gracillimus* was weakest in direct fighting and received much more attacks from *S. caecutiens* than conspecific individuals and *S. unguiculatus*.
Natural enemies In northern Hokkaido, sables (*Martes zibellina*) ate several species of *Sorex* [393], probably *S. unguiculatus*, *S. caecutiens*, or *S. gracillimus*. Other carnivorous mammals, snakes, owls and raptors could be predators for *S. gracillimus*. The Sakhalin taimen (*Hucho perryi*) is an ichthyic predator in upstream rivers [2622].
Parasites Several helminth species are recorded [211, 1738, 2595, 3128]. Parasitic acari and fleas are frequently observed in Hokkaido although they have not been identified [2616].

S. D. OHDACHI

Soricomorpha   SORICIDAE

007

Red list status: LP for the Kyushu population as *C. platycephala* (MEJ); C-2 as *C. platycephala*, LP for the populations in Kyushu and Shikoku (MSJ); LC (IUCN); R for the population in Honshu, Shikoku and Kyushu as *C. himalayica* (JFA)

Chimarrogale platycephalus (Temminck, 1842)

EN Japanese water shrew   JP カワネズミ (kawa nezumi)   CM 日本水駒   CT 日本水駒
KR 계류뒤쥐   RS японская водяная белозубка

A Japanese water shrew diving into a stream in Tsuru, Yamanashi Pref. (H. Ichiyanagi).

1. Distribution

Honshu and Kyushu, Japan, but no recent records from Shikoku [32, 1738]. Honshu: recorded from all 33 prefs. except Chiba Pref. [32, 377, 1612, 1738, 2678, 4039]. Kyushu: recorded from all seven prefs. [1738, 2390].

2. Fossil record

Fossils have been recorded from Honshu and Shikoku: from Yamaguchi Pref. in the Middle Pleistocene, from Hiroshima Pref. in the Late Pleistocene [1514], from Shiga Pref. in the Holocene [819] and from Ehime Pref. in the Late Pleistocene [819].

3. General characteristics

Morphology Very large shrew with a long tail. The body is well adapted to the aquatic mode of life. Tail, with long hairs on the under side, is shorter than the length of the head and body. Feet have a fringe of flattened, stiff hairs on both lateral edges of each toe, acting as a web in water. Ears are small and concealed in the fur, with a valvular antitragus closing the opening. Fur is soft and dense, the guard hairs with silvery tips scattered along the upper and lateral surface of body, especially dense on the rump. Summer pelage is grayish black or grayish drab above, deep olive buff or mouse gray (young) below; winter pelage is blackish brown above, whitish below. Tail is dark brown above, usually whitish below. The braincase is broad and flattened, the profile making a flat angle with that of the rostrum. Teeth are usually white, but the tips have a pale brownish pigmentation on some unworn teeth. The first upper incisor is slender, long and hook-like, with a lower posterior cusp; three upper unicuspids subequal in size to one another, about as high as the anterior cusp of premolar; lower incisor blade-like, with nearly straight cutting edge [8, 29]. Body and skull are larger in specimens from Honshu than in Kyushu and also larger in males than in females [176]. They are larger, as a whole, than those of the other species of *Chimarrogale* [176, 2268]. External and skull characters resemble, except for size, those of *C. himalayica* from Nepal, but the shape of the fourth upper premolar (Pm^4) is different: the concave posterior emargination of Pm^4 is evidently weaker (shallower) in *C. platycephalus* than in *C. himalayica* [12, 36]. This is the sole significant diagnostic character in these water shrew species. *C. platycephalus* is a member of the subfamily Soricinae according to tooth characters [2878] and sperm morphology [2206]. All the measurements used here are based on the author's database collected from 2000 to 2002 [32]. External measurements and GL of skull in millimeters or grams (mean ± SD) are given in Table 1.

Chimarrogale platycephalus (Temminck, 1842)

Table 1. External and cranial measures (in grams or mm). Mean ± SD (*n*).

	BW	HB	T	FFsu	HFsu	E	GL
[Honshu]							
overwintered							
male	49.29 ± 5.99 (39)	131.90 ± 4.69 (39)	102.81 ± 5.72 (36)	15.32 ± 0.65 (39)	27.27 ± 0.94 (39)	7.61 ± 0.52 (39)	27.90 ± 0.50 (39)
female	42.29 ± 5.42 (41)	125.17 ± 4.76 (41)	98.13 ± 4.26 (39)	14.55 ± 0.48 (41)	26.13 ± 0.78 (41)	7.27 ± 0.34 (41)	26.92 ± 0.45 (39)
Young-of-the-year							
male	38.78 ± 7.71 (25)	124.72 ± 6.73 (25)	104.52 ± 5.04 (25)	15.46 ± 0.53 (25)	27.75 ± 0.88 (25)	7.44 ± 0.42 (25)	27.36 ± 0.56 (25)
female	35.06 ± 6.54 (26)	120.00 ± 5.81 (26)	97.15 ± 5.42 (26)	14.52 ± 0.51 (26)	26.12 ± 0.72 (26)	7.22 ± 0.28 (26)	26.77 ± 0.33 (23)
[Kyushu]							
overwintered							
male	34.50 ± 6.74 (6)	118.17 ± 6.62 (6)	87.40 ± 3.13 (5)	13.85 ± 0.53 (6)	24.67 ± 0.83 (6)	6.97 ± 0.26 (6)	25.79 ± 0.93 (6)
female	25.00 (1)	112.00 (1)	82.00 (1)	13.00 (1)	23.00 (1)	8.00 (1)	25.14 (1)
Young-of-the-year							
male	28.50 (1)	111.00 (1)	86.00 (1)	13.50 (1)	24.50 (1)	6.80 (1)	25.04 (1)

GL: greatest length of skull.

Dental formula The dental formula of this shrew has been thought to be I 3/1 + C 1/1 + P1/1 + M 3/3 = 28, but revised to be I 3/1 + C 0/0 + P 2/2 + M 3/3 = 28 based on premaxillary suture and the homology of soricid teeth [1739].

Mammae formula 0 + 0 + 3 = 6 [8, 1011].

Genetics Chromosome: $2n = 52$ and $FNa = 100$; the sex chromosomes are of the XX (female)-XY (male) type [2550]. *C. platycephalus* in Japan forms a monophyletic group with *C. himalayica* in Taiwan while *C. himalayica* in Nepal and Vietnam makes another monophyletic group, based on mitochondrial cytochrome *b* (*Cytb*) gene sequences [2619]. However, there is considerable interspecific difference in *Cytb* sequence between *C. himalayica* in Taiwan and central and southern China (Fujian, Hubei, Shaanxi Province) and *C. platycephalus* in Japan, and both species diverge from *C. phaeura* in Borneo with more than 14% sequence difference [2268, 4229]. In *C. platycephalus*, four *Cytb* haplotype lineages are found in Japan, showing specific geographic distributions for each haplotype [1162].

4. Ecology

Reproduction Mating system unknown. Two breeding seasons are recorded: the main breeding season is in Spring, usually from February to June, and the other in Autumn from October to December [691, 695, 4231, 4232]. Average number of embryos is 4.2, ranging 1–6 (*n* = 10) [32]. Gestation and lactating periods unknown. A breeding nest constructed with broadleaves underground has been reported [4231].

Lifespan It has been estimated by a mark and release survey that the longevity is more than 3 years [4232]. The age composition of 131 specimens also supports this observation [4].

Diet Main food items are aquatic insects (Trichoptera, Ephemeroptera, Plecoptera, Corydalids), small fish, salamanders, frogs, snails, horsehair worms, crayfish, shrimp, and crabs. In addition, terrestrial insects (Diptera), earthworms, leeches, and spiders were reported as food item [37, 694, 1613, 4232].

A water shrew carrying a juvenile white-spotted char in Tsuru, Yamanshi Pref. (H. Ichiyanagi).

Habitat Water shrews inhabit swift streams or riversides in mountain forests. They prefer habitats with abundant shelters on the river bed such as boulders, large rocks or fallen logs with underside cavities or slits, and water-eroded cavities made along river banks [32]. Environmental factors were evaluated in relation to the capture rate [2944].

Home range Published data are very scarce. Range length is about 300 meters for females and about 600 m for males [975]. Territoriality was suggested from trapping data, but it has been supported only in females from radio tracking data [32, 975].

Behavior Water shrews are adept at swimming and diving. They have multiple activity cycles in a day, and activity time is longer at night than daytime [2279]. Trapping and rearing methods were evaluated [599, 600].

Natural enemies Japanese mamushi pit vipers (*Gloydius blomhoffii*) [1738], Japanese weasels (*Mustela itatsi*) [1612, 1738] and northern goshawks (*Accipiter gentilis*) [47] are known predators.

Parasites Several trematodes are reported [2595]. No fleas are reported from the water shrew [2967], but some unidentified fleas have been collected (H. Abe's collection at Hokkaido University Natural History Museum, Botanic Garden, Sapporo). An acarine mite, *Chimarrogalobia yoshiyukiae* is known as an ectoparasite [3700]. About 15 species of trematodes are reported from the water shrew [1543, 2941].

5. Remarks

They sometimes damage fish in fish-farming ponds in mountain regions.

Recently, it was proposed that *Chimarrogale platycephalus* is the correct scientific name for this species [968], instead of previously common used *Chimarrogale platycephala* [including *WMJ1*]. The reason for change has not been described in detail [968], and there has been debate which species name is the correct one.

This species was originally described as *Sorex platycephalus* by Temminck (1842) [3547], and then moved to the genus *Chimarrogale*. Concerning genus-group names, *Sorex* is masculine noun of Latin (shrew), while *Chimarrogale* is formed from "chimarro-" (Greek adjective, "winter-flowing") and "gale" (Greek feminine noun, a name given to various animals of the weasel-kind). Therefore, genders of *Sorex* and *Chimarrogale* are considered masculine and feminine, respectively. In contrast, species-group names "*platycephalus*" and "*platycephala*" are compounded of "platy-" (Greek prefix, "flat") and "-cephalus" (Greek suffix, making masculine noun, "head") or "-cephala" (Greek suffix, making feminine noun, "head"), respectively. In the International Code of Zoological Nomenclature [2513], a species-group name that is an adjective should agree in gender with the generic name (Article 31.2), while a species-group name that is a simple or compound noun in apposition need not agree in gender with the generic name with which it is combined, and the original spelling is to be retained (Articles 31.2.1, 34.2.1). The species-group name "*platycephalus*" is compound noun, and thus the original spelling should be retained. Species name, therefore, becomes "*Chimarrogale playtcephalus*" (feminine noun in genus-group name and masculine noun in species-group name) as recently proposed [968]. Previous use of the incorrect species name "*Chimarrogale platycephala*" may have been due to incorrect interpretation of the Code and / or misunderstanding of "*platycephalus*" to be adjective; and was the result of matching the gender (from masculine to feminine) of species-group name with the gender (feminine) of the generic group name.

H. ABE, H. SAITO & M. MOTOKAWA

A male water shrew captured from Yatarou Riv., Kiyokawa, Kanagawa Pref. (M. A. Iwasa).

Soricomorpha SORICIDAE

008

Red list status: NT (MEJ); R as *C. suaveolens shantungensis* (MSJ); LC (IUCN)

Crocidura shantungensis Miller, 1901

EN Asian lesser white-toothed shrew   JP アジアコジネズミ（ajia kojinezumi）   CM 亚洲小麝鼩   CT 小麝鼩，亞洲小麝鼩
KR 작은땃쥐   RS дальневосточная малая белозубка, манчжурская белозубка

An Asian lesser white-toothed shrew on Tsushima Isls. (K. Tsuchiya).

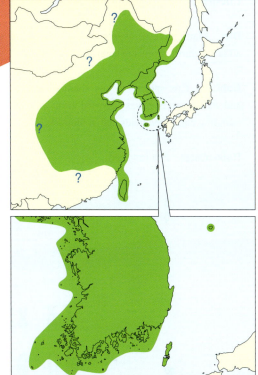

1. Distribution

Tsushima Isls. (Japan), Cheju(-do) Isl. (Korea), Korean Pen., Ullung Isl. (Korea), southern Primorye including small islands near Vladivostok (Russia), eastern mainland China, and Taiwan [2256, 3278]. Due to taxonomic inconsistency, distribution on the Asian Continent is still unclear [2620].

2. Fossil record

Not reported in Japan.

3. General characteristics

Morphology Small *Crocidura* species in East Asia. Color of the winter pelage is pale grayish bister above and pale mouse gray or dull white below, and summer pelage is seal brown above and dull white below [8]. Long bristles are abundant along most of the tail except for its tip [2256]. Tail ratio in relation to head and body length is 50–67% and is small among *Crocidura* species [2256]. No sexual differences were observed in Tsushima and Korean populations, but extensive sexual dimorphism in skulls was reported for Cheju and Taiwan shrews [2273]. External measurements for the Tsushima population are as follows. BW, 3.7–7.3 for both sexes and all age classes [8]. HB, 63.88 ± 4.21 SD mm (range, 59.0–69.0); T, 44.00 ± 2.48 (42.0–47.5); HFsu, 11.83 ± 0.64 (10.9–12.4); E, 8.38 ± 0.55 (7.6–8.9) for male: HB, 55.38 ± 2.29 (53.0–58.0); T, 37.70 ± 2.48 (34.0–40.5); HFsu, 11.50 ± 0.35 (11.0–12.0); E, 7.88 ± 1.23 (5.8–9.0) for female [2273]. Condylo-incisive length of skull measured from the condyles to the anterior tip of incisors is 17.59 ± 0.57 mm (16.60–18.02) for males and 17.24 ± 0.57 (16.58–18.00) for females [2273]. In a study of geographic variation in East Asia, the Tsushima population was most similar to the peninsular Korean population, but is distinct from the Cheju and Taiwan populations [2273].

Dental formula   I 3/1 + C 1/1 + P 1/1 + M 3/3 = 28. Upper unicuspid tooth row consists of 3 teeth from I^2 to C [2256, 2973].

Mammae formula   0 + 1 + 2 = 6 or 0 + 0 + 3 = 6 [8, 1011].

Genetics Chromosome: $2n$ = 39 or 40 and FNa = 46 + XY sex chromosomes in the Tsushima Isls. An individual was reported to have $2n$ = 39, FNa = 46 and it was interpreted as a Robertsonian polymorphism from $2n$ = 40, FNa = 46 karyotype [3640]. Phylogenetically close to *C. suaveolens* and *C. sibirica* possessing the same karyotype features; a detailed phylogenetic study of these species in Eurasia was conducted based on DNA analyses [254, 255, 477, 479, 2620]. Cytochrome *b* sequence phylogeography suggested three distinct clades of Tsushima clade, Jeju clade, and Taiwan–Ullung–continental East Asian clade [1665].

4. Ecology

Reproduction Mating system, reproductive season, and gestation period unknown in Japan. Litter size from 4 to 7 in wild [34, 2256].

Lifespan Not reported.

Diet Main food items are small insects and spiders [9]. In addition, earthworms are eaten [2256].

Habitat This species is less abundant on the Tsushima Isls., and its occurrence was reported from riverbanks, shrubs surrounding cultivated fields and in foothills [34].

Home range Not reported.

Behavior Not reported.

Natural enemies Not reported.

Parasites The cestode *Vampiroplepis tsushimaensis* has been reported [3128], as has the flea *Palaeopsylla nippon* [2968].

5. Remarks

Many previous studies considered *C. shantungensis* conspecific with *C. suaveolens* [8, 1014], and the population in Taiwan also was treated as *C. suaveolens* [535]. However, recent morphological and genetic studies show that *C. shantungensis* (including the Taiwan population) is a species distinct from *C. suaveolens* [254, 255, 477, 479, 923, 968, 1236, 2261, 2273, 2619, 2620]. Due to taxonomic problems, biological data for this species are scarce.

M. MOTOKAWA

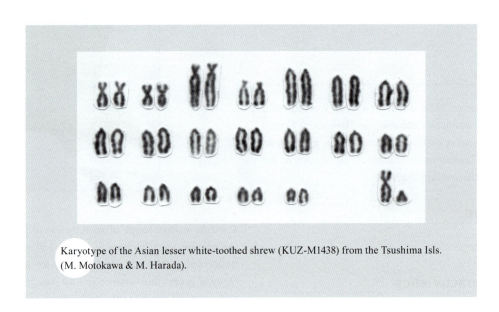

Karyotype of the Asian lesser white-toothed shrew (KUZ-M1438) from the Tsushima Isls. (M. Motokawa & M. Harada).

Soricomorpha   SORICIDAE

Red list status: NT (MEJ); R as *C. horsfieldii watasei* (MSJ); LC (IUCN)

Crocidura watasei Kuroda, 1924

EN Watase's shrew   JP ワタセジネズミ（watase jinezumi）   CM 琉球麝鼩   CT 渡瀬氏小麝鼩
KR 류큐땃쥐   RS белозубка Ватасэ

A Watase's shrew under fallen leaves on Okinawa-jima Isl. (T. Jogahara).

1. Distribution

Species endemic to the central Ryukyus, Japan: the Amami Isls. (Amami-ohshima, Uke-jima, Kikai-jima, Tokuno-shima, Okinoerabu-jima, Yoron-jima Isls.), the Okinawa Isls. (Ioutori-shima, Iheya-jima, Ie-jima, Okinawa-jima, Sesoko-jima, Hamahiga-jima, Tonaki-jima, Tokashiki-jima, Aka-jima, Yakabi-jima, Kume-jima Isls.) [2256, 2261]. Distribution on other islands, where this species is likely to occur, has not been confirmed.

2. Fossil record

Not reported.

3. General characteristics

Morphology   Small *Crocidura* species in the Ryukyu Arch. Color of winter pelage is a uniform fuscous or clove brown above and mouse gray below. Summer pelage chaetura drab above and hair brown below [8]. Tail long, with bristles sparsely distributed in the proximal half of the tail [2256]. Tail ratio greater than 70% [2256]. Body size ranges largely overlap between adult and subadult individuals, but the adults have larger mean value than subadults [2271]. No sexual differences were observed [2271]. In three islands of the Amami Isls., *C. watasei* co-occurs with *C. orii*, but the former is characterized by smaller body size without overlap with the latter. Body weight from Amami-ohshima Isl. ranged 3.7–7.3 g [8]. External measurements of adults from Okinawa-jima Isl.: HB, 66.0 ± 3.99 SD mm (range, 54.5–74.0), T, 54.08 ± 3.08 (47.0–60.0), HFsu, 11.73 ± 0.40 (10.85–12.25), FFsu, 7.25 ± 0.29 (6.70–7.70), E, 8.43 ± 0.53 (7.00–9.55) [2271]. Condylo-incisive length of skull measured from the condyles to the anterior tip of incisors from Okinawa-jima Isl. is 18.15 ± 0.34 mm (17.40–18.80) [2271]. Geographic differences in external and cranial measurements were reported among islands [847, 2271]. Some anatomical studies have been conducted [509, 846, 1802].

Dental formula   I 3/1 + C 1/1 + P 1/1 + M 3/3 = 28. Upper unicuspid tooth row consists of 3 teeth from I^2 to C [2256]. A case of dental abnormality with missing upper I^3 was reported from Tokashiki-jima Isl. [2259].

Mammae formula   0 + 1 + 2 = 6 or 0 + 0 + 3 = 6 [8, 1011].

Genetics   Chromosome: $2n = 26$ and FNa = 48 + XY sex chromosomes [797]. G- and C-differential staining techniques were applied to compare *C. watasei* and *C. dsinezumi* ($2n = 26$ and FNa = 52) karyotypes [797]. Phylogenetic position among the East Asian *Crocidura* was studied with cytochrome *b* gene sequences [2277, 2619, 2620]. Phylogenetically most closely related to *C. horsfieldii* but with divergence [2619]. No sequence differences were found between Amami-ohshima and Okinawa-jima Isls. samples [2277]. Evolutionary history in karyotypes and gene divergence among *Crocidura* species were reviewed [2278].

4. Ecology

Reproduction   Mating system is monogamy [837]. Reproduction occurs throughout the year [2256]. Gestation period is unknown. Litter size was reported to be from 2 to 4 in the wild and from 1 to 4 with an average of 2.5 in captivity [845, 2256]. Nests for reproduction are made from dead grass on the ground [2256].

Lifespan   Not reported.

Diet   Main food items are small insects (Coleoptera, Hymenoptera, Diptera, Plecoptera, Neuroptera, Orthoptera, Lepidoptera, Hemiptera), spiders, centipedes, and crustaceans [9, 34]. In addition, lizards and geckos were reported as food items [2239].

Habitat   Abundant species in the central Ryukyus, but trapping rates were different among islands and localities [2265]. This species occurs in bushes, grassland, cultivated fields, and among shrubs on low mountains [34, 2255, 2265].

Home range   Not reported.

Behavior   Caravanning behavior with mother and sibling(s) was observed in the wild, and occurred 5–20 days after birth in captivity [845].

Natural enemies   Habu pitvipers (*Protobothrops flavoviridis*), himehabu pitvipers (*Ovophis okinavensis*), Ryukyu odd-tooth snakes (*Dinodon semicarinatum*), introduced small Indian mongooses (*Herpestes auropunctatus*), introduced Japanese weasels (*Mustela itatsi*), and feral cats are known predators [54, 1242, 2200, 2256, 2588, 3163].

Parasites   *Vampirolepis amamiensis* and *Neogliphe hinoi* are reported as helminths [2595, 3128]. The flea (*Palaeopsylla nippon*) is reported [2968].

5. Remarks

There was an attempt to establish *C. watasei* as an experimental animal [845].

The holotype was destroyed by fire during World War II, but a paratype still exists in the Natural History Museum, London [2276]. *Crocidura watasei* had been a subspecies of *C. horsfieldii*, but is now considered to be a valid species based on karyotype differences [847, 2261, 2271]. It probably has been isolated in the central Ryukyus since the Pliocene from a zoogeographic view [2262]. Species taxonomy and natural history were reviewed in detail [2256, 2261].

M. MOTOKAWA

A habitat of the Watase's shrew in a subtropical evergreen forest on Amami-ohshima Isl. (Y. Watari).

Five Watese's shrews found in the stomach of a feral mongoose on Amami-ohshima Isl. (Y. Watari).

Soricomorpha SORICIDAE

010

Red list status: C-2 as *C. d. dsinezumi*, LP for the Hokkaido population as *C. d. dsinezumi*, R as *C. d. umbrina* (MSJ); LC (IUCN)

Crocidura dsinezumi (Temminck, 1842)

EN Japanese white-toothed shrew, dsinezumi shrew JP ニホンジネズミ (nihon jinezumi)
CM 日本麝鼩   CT 日本麝鼩   KR 일본땃쥐   RS японская белозубка

An adult shrew in Kitashitara-gun, Aichi Pref. (J. Moribe).

1. Distribution

Hokkaido, Honshu, Shikoku, Kyushu, Awaji-shima Isl., Tobi-shima Isl., Awa-shima Isl., Sado(-gashima) Isl., the Oki Isls. (Dogo, Nishino-shima, Chiburi-jima Isls.), the Izu Isls. (Toshima, Nii-jima, Shikine-jima Isls.), the Seto Inland Sea Isls. (Ohmi-shima Isl., Kamikamagari-jima Isl., Kurahashi-jima Isl., Nohmi-jima Isl., Oh-shima Isl., Yashiro-jima Isl.), Mishima Isl., Aishima Isl., Okino-shima Isl. (Fukuoka Pref.), Nakadohri-jima Isl., Fukue-jima Isl., Kamikoshiki-jima Isl., Nakakoshiki-jima Isl., Shimokoshiki-jima Isl., the Ohsumi Isls. (Tane-gashima, Yaku-shima, and Kuchinoerabu-jima Isls.), the Tokara Isls. (Kuchino-shima, Nakano-shima Isls.) in Japan [2256, 3412, 3413]. Distribution on some other small islands, where this species may be distributed, has not been confirmed. This species is also distributed in Cheju(-do) Isl., Korea [659, 765, 1167, 2256, 2270, 2273, 2620, 3411]. Populations in Hokkaido and Cheju Isls. may have been introduced by humans, as suggested from a cytochrome *b* gene sequence study [2620]. Elevational distribution is known to be between 0 m to 1,780 m above sea level [2256].

2. Fossil record

Fossils were recorded from the Middle and the Late Pleistocene in Honshu and Kyushu [1504, 1514, 2360].

3. General characteristics

Morphology Medium-sized *Crocidura* species in the Japanese Isls. Color of pelage variable geographically. Winter pelage is between grayish pale bister and clove brown on back, and between drab and brown on the belly; summer pelage is snuff brown or clove brown above, and buffy brown or hair brown below [8]. Tail long, with bristles sparsely distributed in the proximal part. Tail ratio is usually less than 70% [34]. External measurements: BW, 5.0–12.5 g; HB, 61.0–84.0 mm; T, 39.0–54.0; FFsu, 7.0–9.0; HFsu, 11.5–15.0; E, 7.1–9.1 [8]. Overall skull size is considerably variable among localities with condylo-incisive length measured from the condyles to the anterior tip of incisors of pooled samples as 16.70–21.00 mm [2263]. The overall size of the skull may follow a southwest-northeast cline of decreasing size, but there are some exceptions [2263]. Growth curve in body weight after birth was reported in captivity showing sexual dimorphism [2631]. Sometimes *C. dsinezumi* is sympatric with *Sorex* species, such as with *S. shinto* in Sado(-gashima) Isl. and Honshu, and with *S. caecutiens* in Hokkaido. They are similar in overall body size, but *C. dsinezumi* is distinguishable from *Sorex* species in having longer tail bristles and larger ears. Teeth of *C. dsinezumi* are white, while those of *Sorex* species have reddish pigmentation in their cusps [34].

Dental formula I 3/1 + C 1/1 + P 1/1 + M 3/3 = 28. Upper unicuspid tooth row consists of 3 teeth from I^2 to C [2256, 2973]. Juvenile tooth morphology was reported [2258].

Mammae formula 0 + 1 + 2 = 6 or 0 + 0 + 3 = 6 [8, 1011].

Genetics Chromosome: $2n = 40$ and FNa = 52 + XY sex chromosomes [797, 3391]. G- and C-banded chromosomes were compared with *C. watasei* and *Suncus murinus* [797, 3391]. Phylogenetically most closely related to *C. lasiura–C. kurodai* cluster among extant species based on DNA cytochrome *b* gene analyses [2269, 2277, 2619, 2620]. Within Japan, there is cytochrome *b* sequence divergence between eastern and western populations, and the Hokkaido and Izu Isl. populations were included within the eastern Honshu cluster [2257, 2620]. Evolutionary history in karyotypes and gene divergence among *Crocidura* species were reviewed [2278].

4. Ecology

Reproduction Mating system is unknown. Reproductive season appears to be from April to October [2256], when testis size is developed [9]. Gestation period is 28.9 days on average (26–33 in range) in captivity [2631]. Litter size was reported to be 1–5 in the wild [2256], and 2.6 on average (1–4 in range) in captivity [2631].

Lifespan In the wild, longevity is about one year [34]. The maximum lifespan in captivity was 28 months in females and 27 months in males [2631].

Diet Main food items are insects (Coleoptera, Lepidoptera, Hymenoptera), spiders and centipedes. In addition, crustaceans, gastropods, and earthworms were reported as food items [9].

Habitat Widely distributed in Honshu, Shikoku, and Kyushu; common but not abundant. Its occurrence was reported from river banks, waterfronts, bushes surrounding cultivated fields on lowlands and low mountain regions between 0 m to 1,780 m elevation [34, 2256]. Occurrences in Hokkaido and Cheju(-do) Isl. are considered to be the result of introductions by humans [2620].

Home range Not reported.

Behavior In newborns, caravanning behavior with mother and sibling(s) was observed in the wild and in captivity [2256].

Natural enemies Ural owls (*Strix uralensis*), Japanese mamushi pit vipers (*Gloydius blomhoffii*), and feral cats are reported as predators and natural enemies [2256].

Parasites Several helminth species are reported [3128]. The cestode *Aonchotheca crociduri* is specific to *C. dsinezumi* [211]. *Androlaelaps himizu* [3702] and *Ixodes ovatus* [598, 1088, 3408] are recorded as ticks.

5. Remarks

There was an attempt to establish *C. dsinezumi* as an experimental animal [2631]. Several subspecies of *C. dsinezumi* were previously proposed [1014, 2261], but no subspecies are currently recognized [2263]. The subspecies *C. d. umbrina* has been applied to the Yaku-shima Isl. population [8, 1014], but the type locality of this taxon is probably not from Yaku-shima Isl. Exact type locality has not yet been determined [2261]. The holotype of *C. d. quelpartis* from Cheju(-do) Isl. is not *C. dsinezumi*, but *C. shantungensis* [2273]. Populations in the northern Ryukyus (the Osumi and Tokara Isls.) are morphologically divergent and their taxonomic status should be reevaluated [2273]. Species taxonomy and natural history was reviewed in detail [2256, 2261].

M. MOTOKAWA

A typical habitat of the dsinezumi shrew in an unused rice terrace in a mountainous village in Kyoto (S. D. Ohdachi).

Soricomorpha   SORICIDAE

Red list status: EN (MEJ); V (MSJ); EN (IUCN)

Crocidura orii Kuroda, 1924

EN Orii's shrew   JP オリイジネズミ（orii jinezumi）   CM 折居氏麝鼩   CT 折居氏麝鼩   KR 오리이땃쥐
RS белозубка Ории

An Orii's shrew found on Amami-ohshima Isl. (T. Jogahara).

1. Distribution

Endemic to the Amami Isls. in the central part of the Ryukyu Arch., Japan: Amami-ohshima, Kakeroma-jima, Tokuno-shima Isls. [34].

2. Fossil record

Not reported.

3. General characteristics

Morphology   Larger *Crocidura* species in the Amami Isls. Color of pelage is dorsally light brown and ventrally light gray. Long bristles are sparsely distributed in the proximal half of tail [2260]. Body size is larger than *C. watasei*, which co-occurs in the Amami Isls. Tail ratio is usually less than 65% (50.2–65.4%) [2260]. External measurements in mm: HB, 79.3 ± 7.49 SD (range, 65.0–90.0); T, 44.4 ± 3.86 (41.0–51.0); HFsu, 14.7 ± 0.40 (14.0–15.0), Condylo-incisive length 21.3 ± 0.61 (20.7–21.9) [2260]. Sometimes considered to be a subspecies of allopatric *C. dsinezumi*, but *C. orii* is distinct from *C. dsinezumi* in having broad forefeet with long claws, long and narrow rostrum, posteriorly positioned anterorbital foramen, and rounded braincase [2260].
Dental formula   I 3/1 + C 1/1 + P 1/1 + M 3/3 = 28. Upper unicuspid tooth row consists of 3 teeth from I^2 to C [2256].
Mammae formula   Not reported.
Genetics   Chromosome: not reported. Mitochondrial cytochrome *b* gene analysis suggested that *C. orii* is distinct from other East Asian congeners [2619].

4. Ecology

Reproduction   Not reported.
Lifespan   Not reported.
Diet   Coleopterans are reported from stomachs [1729].
Habitat   Little known, but a broad-leaved natural forest is reported as a habitat [34, 2256].
Home range   Not reported.
Behavior   Not reported.
Natural enemies   The himehabu pit viper (*Ovophis okinavensis*) is reported as a predator [2256].
Parasites   Not reported.

5. Remarks

Reported *Crocidura orii* specimens are restricted to about 10 specimens [34, 1729, 2260]. Previously considered as a subspecies of *C. dsinezumi*, but detailed morphological comparison revealed that this is a distinct species [2260]. Recent genetic study also supports this view [2619]. This species may have been isolated in the Amami Isls. Group since the Pliocene from zoogeographic view [2262]. Taxonomic study has not been conducted for the Kakeroma-jima and Tokuno-shima populations, both known by one capture record each.

M. MOTOKAWA

Soricomorpha SORICIDAE

Red list status: C-1 as *S. m. temmincki* (MSJ); LC (IUCN)

Suncus murinus **(Linnaeus, 1766)**

EN musk shrew, house shrew   JP ジャコウネズミ（jakou nezumi）   CM 大臭鼩   CT 臭鼩，錢鼠
KR 사향땃쥐   RS домовая многозубка, гигантская белозубка

A musk shrew on Okinawa-jima Isl. (T. Jogahara).

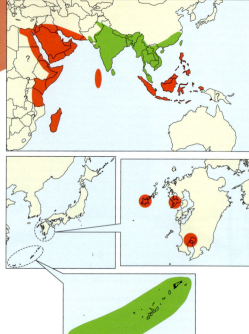

1. Distribution

Widely distributed in the Old World, but many populations were introduced by humans. Original range has been estimated to be Afghanistan, Pakistan, Bangladesh, India, Sri Lanka, Nepal, Bhutan, Myanmar (Burma), China, Taiwan, Japan, and continental and peninsular Indo-Malayan region. Likely introduced by humans into Guam, the Maldive Isls., the Philippines, and other Pacific islands, coastal Africa (Egypt to Tanzania), Madagascar, the Comores, Mauritius, Réunion, and coastal Arabia (Iraq, Bahrain, Oman, Yemen, Saudi Arabia) [968, 1240]. In Japan, it is probably naturally distributed in the Ryukyu Arch. from the Amami Isls. (Amami-ohshima, Tokuno-shima, Okinoerabu-jima, Yoron-jima Isls.), the Okinawa Isls. (Izena-jima, Iheya-jima, Okinawa-jima, Ie-jima, Minna-jima, Sesoko-jima, Kudaka-jima, Aguni-jima, Tokashiki-jima, Tonaki-jima, Zamami-jima, Aka-jima, Kume-jima), the Miyako Isls. (Miyako-jima, Irabu-jima), and the Yaeyama Isls. (Iriomote-jima, Ishigaki-jima, Tarama-jima, Yonaguni-jima). Populations in Nagasaki and Kagoshima in Kyushu and Fukue-jima Isl. were probably introduced by humans [2256, 2265]. Old written records in Japan were examined and evaluated [4044].

2. Fossil record

Not reported from Japan.

3. General characteristics

Morphology A large *Suncus* species. Color and body size are extensively variable among localities and populations. Those differences were sometimes considered to be the basis for semi-species or incipient species [968]. Pelage color from Kagoshima is brown above and mouse gray below. Tail is thick at the base and tapers off towards the end, and retains scattered long bristles along its entire length [34]. Scent glands are large and well developed in both sexes. Body size is much larger than co-occurring *Crocidura watasei*, *C. orii*, and *C. dsinezumi*. Sexual dimorphism is considerable and males are larger than females. Tail is usually thicker in males than in females [8]. External measurements from Kagoshima: BW, 45.0–78.0 g; HB, 116.0–157.0 mm; T, 61.0–77.0; FFsu, 12.5–15.0; HFsu, 18.5–22.0; E, 11.7–14.2 [8]. Greatest length of skull 27.0–33.0 mm [8]. Growth pattern, tooth eruption, and epiphyseal union were studied in captivity [3195] or in animals of laboratory strains derived from the Nagasaki wild population [1115, 1117]. The Okinawa-jima population is smaller in overall skull size than the Taiwan population [2450]. Morphological geographic divergence and its genetic basis were studied

using wild-derived experimental strains [769, 1116, 1118, 1119]. An anatomical study reported on this species' myology [3435].
Dental formula   I 3/1 + C 1/1 + P 2/1 + M 3/3 = 30 [771, 2256]. Dental anomalies were studied for the Ryukyu populations [1240, 1740, 2451].
Mammae formula   0 + 0 + 3 = 6 [8, 1011].
Genetics   Chromosome: $2n = 40$ and $FNa = 50$ in the Ryukyu populations [143, 2545, 3391, 4224]. Karyotypes were different between samples from Okinoerabu-jima and Okinawa-jima Isls. [143, 4224]. Polymorphism in X and Y sex chromosomes were reported [2545, 3391, 4223]. G-band karyotypes revealed 85% homology with *Crocidura dsinezumi* [3391]. Karyotype differences are extensive at the species level and $2n$ varies between 30 and 40 [4223]. Genetic geographic divergences were studied using wild-derived experimental strains [3958–3960]. Phylogenetic position in relation to other species of *Crocidura* was studied based on mitochondrial cytochrome *b* gene sequences [478, 2277, 2619]. *Suncus murinus* is phylogenetically closest to *Suncus montanus* (Kelaat, 1850) in Indonesia based on mitochondrial and nuclear gene sequences [478].

4. Ecology

Reproduction   Mating system unknown. Reproduction occurs throughout the year in Nagasaki with a peak from July to August. Gestation period is 31 days and lactation period is about 14 days in Nagasaki [2241, 2242, 2245]. Litter size is from 1 to 6 (3.21 in average) in Nagasaki, 4 to 6 in Kagoshima, and 1 to 6 in Okinawa-jima Isl. [9, 2241, 2242, 2568, 2569]. Reproductive biology in captivity was studied for the Nagasaki and Okinawa-jima populations [1295, 2243, 2244, 2246].
Lifespan   Lifespan in the wild unknown. In captivity, the mean longevity was reported as 556 ± 226 SD days in females and 662 ± 137 days in males [2567]. The maximum lifespan reported in captivity was 1,154 days in a female individual [2567].
Diet   Omnivorous. In the wild, food items include insects (Coleoptera, Diptera, Orthoptera, Lepidoptera), earthworms, leeches, frogs, peanuts, sweet potatoes, and garbage from humans [9, 2256].
Habitat   This species shows human-commensal habits. Thus, it is found under the floors of houses, in sewers, and along riverbanks in urban areas, as well as cultivated fields and grassland near human residences [34, 2256].
Home range   Average home range size is $2,556 \pm 909.5$ m^2 (mean ± SD) with range of 1,490–3,773 m^2 in Okinawa-jima Isl. [2386].
Behavior   Caravanning behavior with mother and her young occurs from 1 to 3 weeks after birth [2256], and detailed descriptions were made based on wild-derived experimental strains [3651, 3652]. Daily activity rhythm in captivity was basically nocturnal [2386, 3650]. Detailed activity pattern in the wild was studied using radio-tracking and actogram in Okinawa-jima Isl. [2386]. Behavioral development was reported in captivity or wild-derived experimental strains [2442].
Natural enemies   Habu pit vipers (*Protobothrops flavoviridis*), himehabu pit vipers (*Ovophis okinavensis*), and Iriomote cats (*Prionailurus bengalensis iriomotensis*) are known predators [2256].
Parasites   A lot of helminth species have been reported [1543, 2595, 3128]. Fleas and parasitic acari have been reported as ectoparasites [3404, 3699].

5. Remarks

History of introductions into Kyushu (Nagasaki and Kagoshima) and Fukue-jima Isl. with taxonomic discussion was reviewed [2256]. It is used as a laboratory animal, *Suncus*.

6. Topic

Experimental insectivorous animal for evolutionary studies

   Suncus murinus is now widely studied as an experimental animal that had been developed in Japan. It contributes not only to medical research, but also to evolutionary studies. The developmental history of *S. murinus* as an experimental animal and basic information on this species are summarized in a book compiled by Dr. Sen-ichi Oda and his colleagues [2567, 2570]. Besides the facts mentioned in this book, which were based on wild populations or laboratory colonies originating from Japan, many more studies using various experimental strains originally derived from Japan and other foreign countries have been conducted. Among the laboratory strains, several studies are also informative for understanding the natural history and evolutionary biology of wild populations of *S. murinus* such as embryonic development [2467, 2468, 4079, 4080], dental morphology and postnatal development [770, 772, 1707, 1708, 3009].

   Different strains have originated from wild populations in different localities with considerable variation in morphology and genetics. For example, the Nagasaki strain which originated from Kyushu, Japan, has a mean body mass of 52.9 g in males and 34.2 g in females at 120 days. The Bangladesh strain has a mass of 135.3 g in males and 82.0 g in females [1115], showing a greater than two-fold difference in body mass between the two. Considerable body size differences have been discussed concerning their breeding potential and genetic differences in relation to their taxonomic treatments, but the evaluation of breeding potential in wild allopatric populations is limited. In contrast, studies of breeding experiments between these strains led to a study of the influence of behavioral and genetic factors on breeding potential between the strains, as well as of the detection of genes relevant to body size [1115–1118]. These findings provided a new insight into the study of speciation especially in premating and postmating isolation mechanisms between allopatric populations.

   These various strains have also contributed to comparative and evolutionary studies of behavior, genetics, physiology, and morphology. *Suncus murinus* will become a candidate model species for speciation and diversity studies by integrating both field and laboratory experimental investigations.

M. MOTOKAWA

Soricomorpha TALPIDAE 013

Red list status: C-2, LP for the Shikoku, Kyushu and Kii Pen. populations (MSJ); LC (IUCN)

Dymecodon pilirostris True, 1886

EN True's shrew-mole, lesser Japanese shrew-mole　　JP ヒメヒミズ（hime himizu）　　CM 日本长尾鼩鼹
CT 日本長尾鼩鼴　　KR 쇠두더지사촌　　RS малый японский землеройковый крот

An individual collected in Aokigahara, Yamanashi Pref. (K. Tsuchiya).

1. Distribution

Endemic to Japan. Isolated in mountain areas of Honshu (except Chugoku District), Shikoku and Kyushu [35]. Broadly parapatric with *Urotrichus talpoides*, but generally *D. pilirostris* occurs at higher altitudes than *U. talpoides*. The lowest altitude is approximately 750 m in Kyushu [3682] and 900 m in Shikoku [9]. In Honshu the lowest altitude tends to be lower towards the north: 1,500–1,600 m in central Honshu [2143], 800–1,200 m in northern Honshu [9, 1576], and 100 m in northernmost Honshu (Shimokita Pen.) [1095]. However, *D. pilirostris* sometimes occurs at lower than usual altitudes in rocky areas where there is less soil [1028, 1576].

2. Fossil record

Recorded from the middle Middle Pleistocene outside of the current distribution area (Chugoku District) [1514].

3. General characteristics

Morphology　The smallest talpid in Japan. Snout is very long, slender and hairy. Auricles are lacking. Forefeet are only slightly broadened, the length being longer than the width; toes have slightly flattened and nearly straight claws. Tail is relatively long, with the tail ratio ranging 43–60% and with shorter hairs than those of *U. talpoides*, 5–7 mm at the middle, 8–15 mm at the tip. Fur on the back is grayish black to blackish brown, and the underside is paler than the back. It differs from *U. talpoides* by the very large first upper incisor with a broad and flat tip in front view and by the lower tooth row consisting of 9 teeth on each side [8, 35].

External measurements including both sexes from throughout its range are as follows: HB = 70–84 mm; T = 32–44 mm; HFsu = 12.8–15.2 mm; BW = 8–14.5 g (n = 36) [8].

Dental formula　I 3/2 + C 1/1 + P 3/3 + M 3/3 = 38 [4251] based on the trait that the first premolar is monophyodont. Another dental formula was proposed based on the position of Sutura maxilloincisiva in full awareness of the monophyodont fifth tooth: I 2/1 + C 1/1 + P 4/4 + M 3/3 = 38 [1029]. Possibly, the latter formula followed by the most recent Japanese researchers (e.g. [35]) is incorrect [1435].

Mammae formula　1 + 0 + 2 = 6 or 1 + 1 + 1 = 6 [8, 1011].

Genetics　Chromosome: $2n$ = 34 [1438, 3632]. Molecular phylogenetic data showed this species to be closely related to *U. talpoides* [3235].

Dymecodon pilirostris True, 1886

4. Ecology

Reproduction   Information is scarce. It seems to breed once in spring. Breeding males were obtained in April–June with testis lengths of 8 mm, and pregnant females in May–July with 3–5 embryos (average 3.7, $n = 6$) [2143]. Litter size (the number of embryos) is reported as 3–6 ($n = 3$) [1015] and 3 in one female [9].

Lifespan   Specimens collected in various locations consisted of 3 age classes I to III: the percentage of each class from young to old was 51, 41 and 8, respectively, and longevity was estimated at about 2 years [9].

Diet   Main food items are insects, earthworms (Oligochaeta), small centipedes, spiders (Araneae) and other small terrestrial invertebrates [9].

Habitat   Inhabits forests and grasslands in high mountains (generally subalpine and alpine zones) [9, 1015].

Home range   Range length was 25–26 m ($n = 3$) in a coniferous forest in central Honshu [1028].

Behavior   Semifossorial, but seems to be more active on the ground than *U. talpoides,* as can be speculated from the morphology.

Natural enemies   Not reported.

Parasites   Parasitic helminthes of *D. pilirostris* include: 2 species of trematodes, 1–3 of cestodes, 4 of nematodes and 1 of acanthocephalans [1545, 1557, 2594, 3132, 3137]. The parasitic helminth fauna of this species resembles that of *U. talpoides*. At least, 1 species of mite, 1 species of tick and 5 species of fleas have been recorded [598, 2968, 3702].

Morphological comparison between *Urotrichus talpoides* (upper) and *D. pilirostris* (lower), both collected at Mt. Takanosu, Tokyo (S. Kawada).

5. Remarks

Population density is estimated at 5–14/ha in a subalpine coniferous forest in Nagano Pref. [1015].

N. ISHII

A lesser Japanese shrew-mole in Shiojidaira, Iijima-machi, Nagano Pref. (T. Sawahata).

A typical habitat of the lesser Japanese shrew-mole in Shiojidaira, Iijima-machi, Nagano Pref. (T. Sawahata). The lesser Japanese shrew-mole prefer rockier places than the greater Japanese shrew-mole.

Soricomorpha TALPIDAE

014

Red list status: C-1 (MSJ); LC (IUCN)

Urotrichus talpoides Temminck, 1841

EN Japanese shrew-mole, greater Japanese shrew-mole   JP ヒミズ (himizu)   CM 日本鼩鼴   CT 日本鼩鼴
KR 두더지사촌   RS большой японский землеройковый крот

An adult female found in Akiruno, Tokyo (M. A. Iwasa).

1. Distribution

Endemic to Japan. Honshu, Shikoku, Kyushu, Awa-shima Isl., Noto-jima Isl., Awaji-shima Isl., Shodo-shima Isl., the Oki Isls. (Dogo Isl., and Nakano-shima and Nishino-shima Isls. of the Dozen Isls.), Mishima Isl., the Goto Isls. (Fukue Isl.), and Tsushima Isls. [35, 1737].

2. Fossil record

Recorded from the middle Middle Pleistocene [1514].

3. General characteristics

Morphology   Similar to True's shrew-mole but larger. Auricles are lacking. Tail is thick, clavate-like in form, with bottle-brush-like bristles. Fur on the back is black or blackish brown, and the underside is paler than the back. First upper incisor is very large, with a pointed tip. The Tsushima population has a longer tail (tail ratio of 37% in average) than the others (32–34%), and is thus recognized as a subspecies, *U. t. adversus* [8].

External measurements including both sexes from throughout its range are as follows: HB = 89–104 mm; T = 27–38 mm; HFsu = 13.8–16 mm; BW = 14.5–25.5 g (n = 272) [8].

Dental formula   I 3/2 + C 1/1 + P 3/2 + M 3/3 = 36 [4251] based on the trait that the first premolar is monophyodont. Another formula was proposed based on the position of Sutura maxillo-incisiva in full awareness of the monophyodont fifth tooth: I 2/1, C 1/1, P 4/3, M 3/3 = 36 [1029]. Probably, the latter formula followed by the most recent Japanese researchers (e.g. [35]) is incorrect [1435].

Mammae formula   1 + 0 + 2 = 6 or 1 + 1 + 1 = 6 [8, 1011].

Genetics   Chromosome: $2n$ = 34 [1438, 3632]. Two karyotypic forms (western and eastern, separated in central Honshu) are recognized [3641]. Molecular phylogenetic data suggests that these forms possibly retain genetic interchange [788].

4. Ecology

Reproduction   The breeding season is principally restricted to spring, and shows a tendency to be delayed in more northerly districts [9, 1096]. However, in Niigata and Hiroshima Prefs., a second smaller peak of breeding activity was observed in July–September [1024, 4233], suggesting that some individuals also breed in summer. The testis length of sexually active males was more than 5 mm [1096]. Litter size (the number of embryos) is 1–6, usually 3–4, and the average is 3.5 (n = 33) in Hiroshima [4233], 3.8

($n = 14$) in Niigata [1024], and 3.6 ($n = 8$) in Chiba [1096]. Both gestation and lactation periods were estimated at about 4 weeks [1096].

Lifespan Longevity was estimated to be about 3 years based on the age structure of specimens collected in various locations, which consisted of 4 age classes. The percentage of each class from younger to older was 47, 30, 20 and 3, respectively ($n = 294$) [9]. In a mark-recapture study, maximum longevity was estimated at 3 years and 1 month [1095].

Diet Main food items are insects (adults and larvae), earthworms, centipedes, spiders, and other terrestrial small animals, and plants such as seeds and fruits [9]. In Chiba Pref., earthworms and centipedes were found in the stomach during spring and summer more than in autumn and winter, while spiders and plants (berries and others) showed the opposite tendency [1095].

Habitat Inhabits forests, bushes and grasslands, mainly in the lower montane zone, but it is also found at higher altitudes [9, 35]. According to a study in northern Honshu [1095], they occur in various types of forests and grasslands, but are most abundant in deciduous forests and less so in coniferous plantations and grasslands.

Home range According to a mark-recapture study in Chiba Pref. [1097], during the non-breeding season, no significant differences in home range size were found between sexes. The range size (minimum convex polygon) was about 500–2,000 m^2, and the average was 1,533 m^2 ($n = 9$). The home ranges of breeding males were significantly larger than those of non-breeding males, while there were no significant differences in range size between non-breeding and breeding females. During the non-breeding season, home ranges tended to be mutually exclusive among individuals of the same sex, but overlapped extensively with those of the opposite sex. In the breeding season, male ranges largely overlapped each other, and female ranges also overlapped each other but to a lesser extent.

Behavior Semifossorial. Judging from their trapability, they are active both in daytime and at night, but are more active on the ground at night. The number of activity cycles is approximately 3 per day [4110].

Natural enemies Predators reported are as follows [4110]: mammals, *Prionailurus bengarensis* on Tsushima Isl., *Sus scrofa*, *Martes melampus*, and *Meles anakuma*; birds, *Spizaetus nipalensis*, *Falco tinnunculus*, *Strix uralensis*, and *Lanis bucephalus*; snakes, *Elaphe conspicillata* and *Gloydius blomhoffii*.

Parasites and pathogenic organisms Parasitic helminths of *U. talpoides* include: 5–6 species of trematodes, 3–4 of cestodes, 10–12 of nematodes and 1 of acanthocephalans [355, 1545, 1557, 1558, 1884, 2597–2599, 2941, 3131, 3133, 3978, 4110, 4120]. The parasitic helminth fauna of this species apparently resembles that of *D. pilirostris*. At least, 1 species of mite, 5 of ticks and 9 of fleas are recorded [598, 2140, 2968, 3408, 3702]. Hantavirus is detected from *U. talpoides* in Mie Pref. [177], although its pathogenicity for humans is unknown.

5. Remarks

According to a mark-recapture study [1095] in Chiba Pref., the population density was stable between 6.9 and 13.1 per ha during the non-breeding season (May–January), and did not fluctuate largely even in the breeding season (February–April). The survival rate was usually over 0.8/month, but fell in the breeding season (February–April) due to replacement and disappearance of breeding individuals. A rapid decrease in overwintered individuals occurred with the recruitment of the young from April to June. The proportion of young becoming adults was estimated at 0.48. The young settling down by June comprised over 70% of the population, while individuals passing 2 or more winters comprised less than 30% and those participating in breeding twice or more comprised about 20%.

N. ISHII

A typical habitat of the greater Japanese shrew-mole in Shiojidaira, Iijima-machi, Nagano Pref. (T. Noro). The greater Japanese shrew-mole prefer habitat with thick litter.

A greater Japanese shrew-mole in Shiojidaira, Iijima-machi, Nagano Pref. (T. Noro).

Soricomorpha   TALPIDAE

015

Red list status: NT (MEJ); R (MSJ); LC (IUCN)

Euroscaptor mizura (Günther, 1880)

EN Japanese mountain mole   JP ミズラモグラ (mizura mogura)   CM 本州高山鼴   CT 本州高山鼴鼠
KR 일본고산두더지   RS горный японский крот

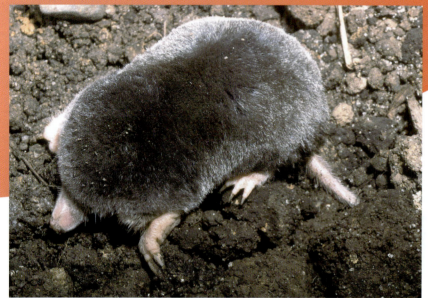

An adult mole collected at Mt. Iwaki, Aomori Pref. (M. A. Iwasa).

1. Distribution

Endemic to Japan. Known from several patches in the mountains of Honshu, its complete distribution is not fully known. Records of this species are concentrated in the mountain zone of Hiroshima to Aomori Prefs. [33]. In Kinki District, they occur in the mountains of Kii Pen. and the northern mountains of Kyoto Pref. Isolated populations are known from mountainous areas of Chugoku District and northern Tohoku District (Aomori and Iwate Prefs.). This species is usually collected in the high sub-alpine zone, but it also occurs at lower altitudes in Chugoku, Kinki and Tohoku Districts. The known distribution range of this species will widen with further research.

2. Fossil record

Recorded from Yamaguchi, Hiroshima and Gifu Prefs. from the Middle Pleistocene [1514].

3. General characteristics

Morphology   One of the smallest species of true mole in the world. Tail is long and rod-shaped with scattered hairs. Hind foot slightly longer relative to body length compared to *Mogera wogura* and *M. imaizumii*. Fur color is gray in southern Chubu and Kanto districts to dark slate black in northern populations. Snout is short and dorsal naked portion is triangular in shape with a longitudinal groove in the center. Penis is short and thick with many spines, similar to *Mogera*. Skull shaped like an isosceles triangle. Incisor row is conspicuously tapered as a sharp V-shape and first incisor is much larger than the others. Relative size of brain case occupies almost half of the greatest length of the skull.

The vertebral formula is C7 + T13 + L6 + S6 + Cd13. Pelvis has 2 sciatic foramina, but the ossifications between sacrum and both ischia are not fully developed [46, 4207].

External measurements including both sexes are as follows (sample size not shown): HB = 77.0–107.0 mm; T = 20.0–26.0 mm; FFsu = 11.0–16.0 mm, FFW = 8.0–15.3 mm, HFsu = 13.5–15.4 mm; BW = 26.0–35.5 g; greatest length of skull = 25.5–28.0 mm [35].

Dental formula   I 3/3 + C 1/1 + P 4/4 + M 3/3 = 44 [35, 1011].

Mammae formula   Not reported.

Genetics   Chromosome: $2n = 36$ and FNa (fundamental numbers of autosomes) = 54. *Euroscaptor mizura* shares G-banded karyotype of *M. wogura* [1437]. Molecular data for the mitochondrial cytochrome *b* gene showed that *E. mizura* is derived from the basal group of the talpine moles [3233]. Accumulation of high intraspecific molecular diversity is suggested among *E. mizura* populations [3234].

Euroscaptor mizura (Günther, 1880)

4. Ecology

Reproduction Reproductive behavior is unknown. Three newborns were found in a nest in May at Kyoto Pref. [2924]. Litter size is estimated at around 3. The dispersal season is thought to be from June to August, based on records of dead bodies [1303].

Lifespan Not reported. Possibly similar to other mole species in Japan.

Diet Stomach contents include mainly larval coleopterans [3611]. This corresponds to the nearly semifossorial behavior of this species. In captivity, *E. mizura* eats earthworms and mealworms.

Habitat This species has been considered as a subalpine species of Talpidae. Recently, collection records from relatively low mountains (500–1,000 m) have come to our attention. However, several dead bodies of this species have also been collected in alpine areas, for example in Tateyama in Toyama Pref. (2,300–2,400 m) [2306, 2307], so the species is apparently not rare in the alpine zone.

Habitat of this species is apparently limited in forested environments. In Kyoto Pref., they nest about 50 cm underground [2928]. Their nests are spherical, created by piling broad leaves into the nest rooms.

Home range Not reported.

Behavior Records of this species having been collected in pitfall traps set along the paths of shrews and shrew-moles suggest the species has semifossorial habits and that it often comes above ground.

Natural enemies A skull of *E. mizura* was found from the stomach of an unidentified owl [1011]. Some carnivorans, foxes, badgers and weasels, are also known to be predators of moles [4110]. This species is often found dead, and several kinds of carnivoran species are likely natural enemies.

Parasites Four species of parasitic nematodes are reported [4127]. The parasitic nematode fauna of this species apparently resembles to that of *Mogera* spp. Two types of parasitic protozoa (coccidia) are recorded [484].

5. Remarks

Euroscaptor mizura has been thought to be more primitive than other Japanese talpine species (*Mogera* spp.), because of its morphological features (principle dental formula of mammals, long tail, etc.), and fragmented distribution pattern, which suggest it is a relic species. However, this montane and forest species may occupy a slightly different niche from *Mogera* spp. and more biological and ecological information is required.

S. KAWADA & Y. YOKOHATA

A habitat of the Japanese mountain mole in a beech forest in Mt. Iwaki-san, Aomori Pref. (M. A. Iwasa).

A subadult Japanese mountain mole eating an earthworm (S. Kawada).

Soricomorpha TALPIDAE

016

Red list status: C-2, LP for the Shodo-shima Isl. population (MSJ); LC (IUCN)

Mogera imaizumii (Kuroda, 1957)

EN lesser Japanese mole, Japanese eastern mole   JP アズマモグラ（azuma mogura）   CM 东日本鼹   CT 東日本鼴鼠
KR 동일본두더지   RS восточная японская могера

An individual found in Fujisawa, Kanagawa Pref. (M. A. Iwasa).

1. Distribution

Endemic to Japan. Distributed in northeastern Honshu and the mountainous areas of southwestern Honshu, Shikoku and some surrounding islands (e.g. Awa-shima and Shodo-shima Isls.). The main habitat in southwestern Japan is steep mountains, where the distribution of *M. imaizumii* is usually surrounded by areas of *M. wogura* [30]. The distributions of these species are usually parapatric and sometimes sympatric. *Mogera imaizumii* is one of the most common wild mammals in its area of distribution.

2. Fossil record

Fossils referred to *M. imaizumii* are reported from Aomori Pref. from the Middle Pleistocene [1514].

3. General characteristics

Morphology Small species of Japanese *Mogera*. Body size variable, smaller in small mountainous populations, larger in large plain populations. Isolated southwestern populations have especially small bodies, with the skull size around 31 mm. Pelage color is also variable among populations, and darker with shorter hairs in the summer compared to winter. The tail is 10–15% of the head and body length, and rather hairy. The eyes are not opened during the whole lifespan. Skull shape is similar to *M. wogura*, but upper incisor row is arranged in a V-shape and projected forward (see the article on *M. wogura*). Individuals from mountain populations have a very slender rostrum, weak zygomatic arches and a rounded brain case. The vertebral formula is C7 + T14 + L5 + S6 + Cd12 and loss and addition of ribs are known [4207].

External measurements including both sexes are as follows (sample size not known): HB = 121.0–159.0 mm; T = 14.0–22.0 mm; FFsu = 15.5–22.5 mm; FFW = 16.5–23.0 mm; HFsu = 16.0–21.5 mm; BW = 48.0–127.0 g; greatest length of skull = 31.5–38.5 mm [35].

Dental formula I 3/2 + C 1/1 + P 4/4 + M 3/3 = 42 [35, 1011].

Mammae formula 2 + 1 + 1 = 8 or 1 + 2 + 1 = 8 [1011].

Genetics Chromosome: $2n = 36$ and FNa = 56 [1437]. The chromosomal pattern was estimated to be derived from that of *M. wogura* by a pericentric inversion in 1 pair of autosomes [1437]. Molecular data assigned by the mitochondrial COI gene showed the diversification of 4 groups: the northern and southern coasts of the main distribution, the isolated populations of the Kinki to Chubu Districts, and the Chugoku to Shikoku Districts [2675]. Similar results were obtained from examination of the

Mogera imaizumii (Kuroda, 1957)

mitochondrial cytochrome *b* gene [3643]. In eastern Tohoku District, 2 parapatrically distributed genetic populations are hypothesized to be the result of secondary contact between northern and southern groups of *M. imaizumii* [1165].

4. Ecology

Reproduction   These moles mate in spring but their reproductive behaviors are unknown. Pregnant females can be collected from April to July. Two to 6 offspring are born and both males and females can be reproductively active in the next spring [9, 953, 954]. Autumn reproduction of *M. imaizumii* is usually rare, but it is observed frequently in Echigo Plain, where this species is distributed parapatrically with *M. etigo* [829].

Lifespan   Based on tooth wear, the life span is estimated at around 3 years, though a few individuals survive longer [9, 953, 4114].

Diet   These moles eat various kinds of small invertebrate animals, i.e. annelids, arthropods, insects, etc. Major food items are earthworms (mainly, *Pheretima* spp. *sensu lato*). In captivity, they also eat frogs, mice and birds [4110], and it is thought that they can eat any animal they find in their tunnel systems.

Habitat   *Mogera imaizumii* ranges from near sea level to mountain regions of about 2,000 m in altitude, but rarely to the alpine meadows of relatively high mountains. This species is most abundant in low flat fields with deep soil having a fine texture and sufficient moisture [9]. These moles prefer soil with heavy leaf litter deposits.

Home range   In paddy fields, 3 individuals used 156–297 m lengths of levees and the sides of irrigation ditches as their exclusive territories [1462]. Territoriality is generally ubiquitous among *Mogera* spp. in Japan [4110]; *M. imaizumii* seems to be territorial.

Behavior   Two earthworm hunting tactics have been described: back-with-grip and bite-and-retreat [1025, 1026]. The former tactic is the biting of earthworms protruding from the wall of a tunnel and the pulling of the prey into the tunnel. The latter is repeated biting at struggling prey with advance and release with retreat, in order to immobilize them. They frequently eat their prey from head to tail, to prevent the escape of the earthworm through autotomy of the tail. This feeding behavior is observed also in *U. talpoides*. However, the frequency is higher in *U. talpoides* than in this species, probably to compensate for the slow eating speed of *U. talpoides*. Additionally, some experiments on the thigmotaxis of *M. imaizumii* have been performed [3558].

Natural enemies   Owls and raptors are the usual predators of this species. Some carnivores, foxes, badgers and weasels, are also known to prey on this species [4110], although moles are not their preferred food. Perhaps because of the bad smell of the moles, predators often abandon the moles after killing them. In the mole's dispersal season, people sometime find dead young moles with bite marks. Moles are easily captured above ground in this season, thus the mortality in young animals is high.

Parasites   As parasitic helminths of *M. imaizumii*, 1 species of trematode, 2 of cestodes and 13 of nematodes are reported [1676, 3977, 4118–4120, 4123, 4127]. The parasitic helminth fauna of this species apparently resembles that of other *Mogera* spp. Three types of parasitic protozoa (coccidia) are recorded [484].

5. Remarks

The specific name *imaizumii* first was given for a subspecies of the Japanese eastern mole that had been formerly named as *M. wogura* [1796]. After 1995, it was shown that the type specimen of *M. wogura*, collected by F. P. Siebold, corresponded to the Japanese western mole. Although a candidate of species name was *M. minor* (see [25]), this name was preoccupied by a fossil talpid species, *Talpa europaea* var. *minor* (see [1796]). Thus the valid name of this species is *M. imaizumii*.

Mogera imaizumii and other moles in Japan (*E. mizura* and *M. wogura*) make latrines near their nests, and are known to have symbiotic relationships with mushrooms (*Hebeloma* spp.) and trees of the genera *Quercus* and *Fagus*. In the relationships, the moles provide nitrogen to the mushrooms and trees, trees provide organic nutrient to the mushrooms and the mushrooms clean the moles' habitat near their nests. The mushrooms are looked at as a sign of a mole's nest, so that they are often called "mole-nest finders" [2925, 2927, 2928].

S. KAWADA & Y. YOKOHATA

Mole hills constructed along a road in Isumi-shi, Chiba Pref. (S. Kawada).

Soricomorpha TALPIDAE

017

Red list status: C-1 (MSJ); LC (IUCN)

Mogera wogura (Temminck, 1842)

EN large Japanese mole, Japanese western mole   JP コウベモグラ (koube mogura)   CM 西日本鼴   CT 西日本鼴鼠
KR 서일본두더지   RS западная японская могера

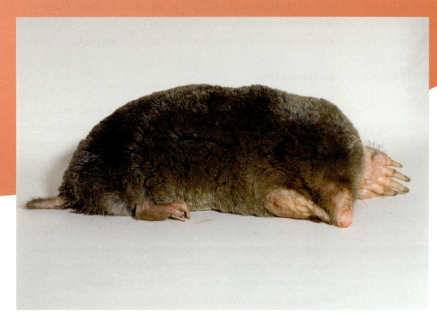

An individual collected in Minamikawachi, Osaka Pref. (K. Tsuchiya).

1. Distribution

Endemic to Japan. Distributed in southwestern Honshu, Shikoku, Kyushu and surrounding small islands, such as Shodo-shima Isl., Dogo Isl., Tsushima Isls, the Goto Isls., Tane-gashima Isl. and Yaku-shima Isl. In Honshu, the northeastern demarcation of its distribution is in Ishikawa, Gifu, Nagano and Kanagawa Prefs. In these prefectures, *M. wogura* borders the range of *M. imaizumii*, and these species usually have an obvious parapatric distribution [16, 20]. Within its distributional range, *M. wogura* is one of the most common wild mammal species in cities and towns.

2. Fossil record

Recorded from the Middle and middle Late Pleistocene in Yamaguchi and Shizuoka Prefs., respectively [1514], although species identification should be reconsidered (e.g. the latter record was thought to be *M. wogura* by [4113]).

3. General characters

Morphology A large species among Japanese *Mogera*. Body size is variable, ranging from small southern populations (skull length: 34 mm~) to large northern populations (~42 mm). The populations of Yaku-shima and Tane-gashima Isls. have the smallest body size [8], and were previously treated as a subspecies of *M. imaizumii* [1011]. Pelage color is lighter than *M. imaizumii* and shows a similar trend of variation. The tail is 10–15% of the head and body length, and rather hairy. The rostrum of *M. wogura* is wider than that of other *Mogera* species, with a U-shaped upper incisor row among adults [8, 25, 2267]. The vertebral formula is C7 + T14 + L5 + S6 + Cd13 [4207].

External measurements including both sexes are as follows (sample size not known): HB = 125.0–185.0 mm; T = 14.5–27.0 mm; FFsu = 16.8–25.5 mm; FFW = 16.0–25.0 mm; HFsu = 16.5–24.0 mm; BW = 48.5–175.0 g. Greatest length of skull = 33.0–42.1 mm [35].

Dental formula   I 3/2 + C 1/1 + P 4/4 + M 3/3 = 42 [35].

Mammae formula   2 + 1 + 1 = 8 or 1 + 2 + 1 = 8 [1011].

Genetics Chromosome: $2n$ = 36 and FNa = 54 [1437]. Continental *M. robusta*, previously considered as the same species, is different in FNa [1437]. Molecular data assigned by mitochondrial COI gene showed diversification to 3 groups, Kyushu and Yamaguchi Pref., Chugoku District and Hokuriku (= the Sea of Japan side of Chubu District) area, and the other Chubu District area [2675]. Similar results were obtained from examination of the mitochondrial cytochrome *b* gene [3643], and combined data suggested Kyushu and Yamaguchi population had larger genetic distance than continental *M. robusta* [1592].

Mogera wogura (Temminck, 1842)

4. Ecology

Reproduction Litter size is 3–6 [9, 613, 615]. The breeding season is usually limited to spring–summer [9, 613, 615], though it varies among regions. Mating and birth occur in late April–mid June and mid May–early July, respectively, in western Honshu [613, 615], whereas 3 young individuals were observed in their nest in October in central–western Honshu [2926] and a sexually active male and a lactating female were obtained in October and September, respectively, in Shikoku [1440].

Lifespan Based on the wear pattern of the upper molars, lifespan is estimated at 4 years maximum [4110]. Mortality rate is higher among younger animals [4111, 4112]. Mature individuals often show exceptionally long lifespans. For example, 5 annual rings were observed in the upper canine of an aged individual from western Honshu [4110].

Diet Earthworms (mainly *Pheretima* spp. *sensu lato*) are the main diet of *Mogera* spp. including *M. wogura*. Earthworms are frequently (65.5%) found from stomachs of this species collected from various localities in western Japan [9]. The amount of earthworms consumed shows little seasonal change [2706]. The diet includes larvae of coleopterans and lepidopterans, the mole cricket (*Gryllotalpa orientalis*) and chilopods in relatively higher frequency [9].

Habitat Moles of genus *Mogera*, including *M. wogura*, are distributed in various habitats in Japan, including forest, grassland, pasture, farm, and levees in paddy fields. They like soft and fertile soil such as wide pluvial plains, where their body size becomes larger [8, 26].

Home range No information is available in *M. wogura*, but each species of Japanese *Mogera* apparently uses its tunnel system as exclusive territory (e.g. [9]).

Behavior There is no precise study on the social system and activity patterns of *M. wogura*, but moles in Japan have been thought to be solitary and active both in day and night (e.g. [4110]). Some observations show thigmotaxis of *M. wogura* [1945].

Natural enemies *Mogera wogura* has been detected from stomach contents of Ural owls (*Strix uralensis*) and feces of Tsushima leopard cats (*Prionailurus bengalensis euptilurus*). Moreover, Japanese *Mogera* are predated by domestic cats, domestic dogs, red foxes (*Vulpes vulpes*), raccoon dogs (*Nyctereutes procyonoides*), and Japanese badgers (*Meles anakuma*). Several species of predatory birds, such as *Buteo buteo*, *Asio otus*, *Otus bakkamoena*, *Spizaetus nipalensis* and *Accipiter gularis*, also prey on this species [4110].

Parasites As parasitic helminths, 2 species of cestodes and 12 of nematodes have been reported [4126]. The parasitic helminth fauna of this species apparently resembles that of other *Mogera* spp. Three types of parasitic protozoa (coccidia) have been recorded [484].

5. Remarks

The type specimen of *M. wogura* was long misidentified and this name had been applied to the Japanese eastern mole, presently *M. imaizumii*, in the literature before 1995 [25]. The name *M. kobeae* was used previously but now is recognized as a junior synonym. In addition, the Japanese greater mole has been regarded as the same species of mole in Korea and Primorye, that usually is assigned to *M. robusta* or *M. wogura*. However, based on morphological and genetic relationships, the authors consider the Japanese greater mole (*M. wogura*) as a species distinct from the mole in Korea and Primorye. Thus, the only *Mogera* species in western Japan should be *M. wogura*.

Mogera wogura and other *Mogera* spp. often damage various crops physically and destroy agricultural structures, such as levees in paddy fields. Pest rodents invade farms to eat crops via the burrows of the moles. Moreover, many people believe that the construction of molehills and surface borrows spoils the beauty of parks and gardens. Usual countermeasures include trapping and evasive methods (e.g. oscillating apparatus and evasion chemicals). Modern structural improvements to paddy fields frequently prevent the habitation of moles.

S. KAWADA & Y. YOKOHATA

Mole hills constructed in a pasture in Shitara-cho, Aichi Pref. (S. Kawada).

Soricomorpha   TALPIDAE

Red list status: EN (MEJ); R (MSJ); EN (IUCN)

Mogera etigo Yoshiyuki & Imaizumi, 1991

EN Echigo mole   JP エチゴモグラ (echigo mogura)   CM 越后鼹   CT 越後鼴鼠
KR 에치고두더지   RS этигская могера

An individual collected in Maki-machi, Niigata Pref. (K. Tsuchiya). Note that the claws of this individual are elongated due to a long period of captivity.

1. Distribution

Endemic to Japan (Echigo Plain, Niigata Pref.). The distribution area is divided into 2 populations; one of them is located at an alluvial plain northern from Yahiko to Shibata, the other is a narrow belt-like area south of Mitsuke. These areas are surrounded by the distribution range of *M. imaizumii* [1027]. The division of the distribution range of *M. etigo* may have been caused by the range expansion of *M. imaizumii*.

2. Fossil record

Not reported.

3. General characters

Morphology   Largest mole species in Japan. Tail is rather long and rod-shaped with scattered hairs. Hind foot is slightly longer relative to body length compared to *M. wogura* and *M. imaizumii*. Skull shape is typical of the genus but the palate between the second upper molars is conspicuously broader than *M. wogura* and *M. imaizumii*. Posterior palatine foramina are located posterior of the line between protocones of the right and left second molars [35]. Infraorbital bridge is thick and connects to the proximal zygomatic arch above the mesostyle or fore part of the upper second molar [1436]. Incisor row is less protruded than in *M. tokudae*. The vertebral formula is C7 + T13 + L6 + S6 + Cd13, same as *E. mizura* [4215]. The pelvis has 2 well-developed sciatic foramina, and is thicker than *M. wogura* and *M. imaizumii*.

External measurements including both sexes are as follows (sample size not known): HB = 162.0–182.0 mm; T = 19.0–30.0 mm; FFsu = 21.0–24.5 mm; FFW = 21.0–25.5 mm; HFsu = 22.0–25.5 mm; BW = 112.0–168.0 g. Greatest length of skull = 39.65–43.25 mm [35].

Dental formula   I 3/2 + C 1/1 + P 4/4 + M 3/3 = 42 [4215].

Mammae formula   Only denoted as "10" in original description [4215].

Genetics   Chromosome: $2n$ = 36 and FNa = 56 [1437]. *Mogera etigo* shares the same G-banded karyotype of *M. imaizumii* and is derived from that of *M. wogura* by a single pericentric inversion based on comparative G-banding analysis [1437]. Molecular data of the mitochondrial cytochrome *b* gene showed that *M. etigo* is closely related to *M. tokudae* [3643].

4. Ecology

Reproduction   Pregnant females were mainly collected in March and April and rarely in July and August. Numbers of embryos and placental scars were 3.45 ± 0.82 SD and 3.21 ± 0.80 [829].

Lifespan   Not reported. Possibly similar to other mole species.

Diet   Stomach contents of 7 individuals examined had large amounts (64–100%) of earthworms (mainly, *Pheretima* spp. *sensu lato*) [4].

Habitat   This species lives in the Echigo Plain, which is a wide alluvial plain with soft and fertile soil [4110] and surrounding mountains [38].

Home range   Not reported.

Behavior   There are no studies on social system or activity patterns.

Natural enemies   Not reported.

Parasites   Eight species of parasitic nematodes have been reported [4118, 4123]. The parasitic nematode fauna of this species apparently resembles that of other *Mogera* spp.

5. Remarks

After the original description, this species had been treated as a geographic form of *M. tokudae*. The karyotypes of *M. etigo* and *M. tokudae* have different FNa caused by several pericentric inversions [1437]. It is possible that such a difference will result in reproductive isolation between these taxa. Thus *M. etigo* is considered a full species (also see [1434]).

Mogera etigo is distributed in agricultural areas, so this species often damages crops and farms (see the Remarks of *M. wogura*). Modern structural improvements of paddy fields, which prevent the habitation of moles, may accelerate the decline of *M. etigo* populations along with the expansion of *M. imaizumii* mentioned previously.

S. KAWADA & Y. YOKOHATA

A karyotype differentially stained by G-banding of a female Echigo mole collected in Maki-machi, Niigata Pref. (S. Kawada).

A skull and mandible specimen (NSMT-M29392) of the Echigo mole collected in Shirone, Niigata Pref. (S. Kawada). Bar = 10 mm.

019 Soricomorpha TALPIDAE

Red list status: NT (MEJ); R (MSJ) including *M. etigo*; NT (IUCN)

Mogera tokudae Kuroda, 1940

EN Sado mole, Tokuda's mole   JP サドモグラ (sado mogura)   CM 佐渡鼴   CT 佐渡鼴鼠
KR 사도두더지   RS могера Токуды

An individual collected on Sado Isl., Niigata Pref. (K. Tsuchiya).

1. Distribution

Endemic to Japan (Sado Isl., Niigata Pref.).

2. Fossil record

Not reported.

3. General characteristics

Morphology   A middle-sized Japanese mole. Tail is rather long and rod-shaped with scattered hairs. Hind foot is slightly longer relative to body length than *M. wogura* and *M. imaizumii*. Fur color of the back is brown in summer pelage, and dark brown in winter pelage. Skull shape is typical of the genus, but the breadth of the palate between the upper second molars is conspicuously broader than *M. wogura* and *M. imaizumii*. Posterior palatine foramina are located anterior of the line between protocones of right and left second molars [35]. Infraorbital bridge is slender and connects to the proximal zygomatic arch above the metastyle of the upper second molar [1436].

Incisor row is conspicuously protruded as a sharp V-shape and the first incisor is much larger than the others.

The vertebral formula is C7 + T13 + L6 + S6 + Cd13, same as *E. mizura* [4207]. The pelvis has 2 well-developed sciatic foramina, and is thicker than *M. wogura* and *M. imaizumii*.

External measurements including both sexes are as follows (sample size not known): HB = 149.0–167.0 mm; T = 22.0–28.0 mm; FFsu = 19.0–22.0 mm, FFW = 19.3–23.5 mm; HFsu = 19.5–22.5 mm; BW = 84.0–135.0 g. Greatest length of skull = 38.0–40.1 mm [35].

Dental formula   I 3/2 + C 1/1 + P 4/4 + M 3/3 = 42 [35].

Mammae formula   2 + 1 + 1 = 8 or 1 + 2 + 1 = 8 [1011].

Genetics   Chromosome: $2n = 36$ and FNa = 60. The karyotype of *M. tokudae* was derived from those of *M. imaizumii* and *M. etigo* by 3 pericentric inversions based on comparative G-banding analysis [1437]. Mitochondrial cytochrome *b* gene analysis shows that *M. tokudae* and *M. etigo* are closely related and the latter may represent local populations of *M. tokudae* [3643].

4. Ecology

Reproduction   Based on numbers of embryos and placental scars, litter size is 2 to 3 (mean: 2.8) [4].

Lifespan   Not reported. Probably similar to other mole species.

Diet Stomach contents ($n = 17$) were characterized by large volumes (40–100% of contents) of earthworms (mainly, *Pheretima* spp. *sensu lato*). Mole crickets (*Gryllotalpa orientalis*) were frequently detected (52.9%: 9/17) [4].

Habitat This species is limited to Sado Isl., which includes alluvial plains with soft soil, and steep mountains.

Home range No information is available in *M. tokudae*, but Japanese moles of the genus *Mogera* apparently use their tunnel system as their exclusive territory [9].

Behavior There is no precise study on the social system and activity patterns of *M. tokudae*, but they possibly resemble those of other mole species.

Natural enemies Possibly similar to other mole species.

Parasites As parasitic helminths of *M. tokudae*, 1 species of trematode and 5 of nematodes have been reported [2978, 4118, 4119, 4123]. The parasitic helminth fauna of this species apparently resembles that of other *Mogera* spp.

5. Remarks

Previously *M. etigo* was included in this species. Species recognition was confirmed by karyological study [1434].

S. KAWADA & Y. YOKOHATA

A karyotype differentially stained by G-banding of a male Sado mole collected on Sado Isl. (S. Kawada).

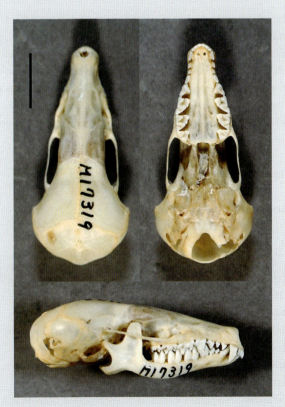

A skull and mandible specimen (NSMT-M17319) of a Sado mole collected in Sawada-machi, Sado Isl. (S. Kawada). Bar = 10 mm.

Soricomorpha TALPIDAE

020

Red list status: CR (MEJ); V (MSJ); DD (IUCN)

Mogera uchidai (Abe, Shiraishi & Arai, 1991)

EN Senkaku mole   JP センカクモグラ (senkaku mogura)   CM 琉球鼴   CT 琉球鼴鼠
KR 센가쿠섬두더지   RS сенкакуйская могера

The stuffed skin of the holotype deposited at Kyushu University (Y. Yokohata).

1. Distribution

Distributed only on Uotsuri-jima Isl. of the Senkaku Isls., Japan [27, 35, 46, 4117]. A single young female specimen has been obtained to date. It was captured near the northwestern seaside grassland.

2. Fossil record

Not reported.

3. General characteristics

Morphology   The Senkaku mole is the usual mole shape, but much smaller. The nostrils are directed outward and the median pad of the rhinarium is protruding and separated from the lower transverse pad by a horizontal groove. Dorso-posterior border of the rhinarium is strongly concave at the middle (V-shaped). Naked potion on the upper side of the muzzle is rather rectangular in outline, and has a longitudinal groove along the middle line. Manus is small, being 14.0 mm long (without claw) by 15.4 mm wide. Color of dorsal fur is close to dark grayish brown or dusky drab, slightly paler and dusky drab underside, with brown hair on the throat and chest [27, 46].

The vertebral formula is C7 + T14 + L5 + S7 + Cd4(+). Pelvis has 2 well-developed sciatic foramina, but the fusions of ischia and sacral bones are shallow. Width of ilia is thicker than Japanese *Mogera* and similar to Taiwanese *M. insularis*.

External measurements of the holotype as follows: HB = 129.9 mm; T = 12.0 mm; HFsu = 16.0 mm; BW = 42.7 g. Greatest length of skull = 31.82 mm [46].

Dental formula   I 3/2+ C 1/1 + P 3/3 + M 3/3 = 38 [46]. Upper and lower premolars are fewer than other Japanese moles in the genus *Mogera*. *Mogera insularis*, which is probably most closely related species to *M. uchidai*, frequently shows variations of dental number in Taiwan [2264, 2272]. Additionally, the lower third premolar of the holotype is an abnormal, connated tooth. It is possible that the dental formula of this species is not a specific character.

Mammae formula   Not reported.

Genetics   Not reported.

4. Ecology

Reproduction   Not reported.
Lifespan   Not reported.
Diet   Not reported.
Habitat   The sole specimen was obtained in grassland [46].
Home range   Not reported.
Behavior   Not reported.
Natural enemies   Not reported.
Parasites   No parasitic organisms are reported from *M. uchidai*.

Mogera uchidai (Abe, Shiraishi & Arai, 1991)

5. Remarks

This species was originally described as a monotypic taxon of a new genus, *Nesoscaptor* [46], but was later reconsidered [2272].

The general morphological characteristics of this species suggest that its ecological features resemble those of other fossorial *Mogera* spp., and probably also semifossorial *Urotrichus talpoides*. Uotsuri-jima Isl. is a very small island (only 3.8 km^2) and lacks other fossorial and semifossorial insectivores. Generally, fossorial talpids cannot become habituated to tropical and subtropical climates without much humus and organic soil, but there is a relatively cooler environment with well-developed cloud forests in the mountainous Uotsuri-jima, which rises to 362 m above sea level at the highest point, so *M. uchidai* might have been able to live on this island [27].

The ecosystem of Uotsuri-jima Isl. has probably been devastated by a drastic increase of goats (*Capra hircus*) since 1978, when they were deliberately introduced into the island by a political party in Japan. Because the Senkaku Isls. have been claimed as national territory by China, Taiwan, and Japan, researchers have been unable to land on Uotsuri-jima, nor have there been opportunities to eradicate the goats there [4113, 4116, 4129]. The effects of goats on the vegetation of the island were assessed using remote-sensing techniques, such as satellite images. The results showed that bare patches have emerged in several areas and occupied 13.6% of the island [4116, 4122]. Immediate eradication of the goats is necessary to protect the Senkaku mole and many other endemic animal and plant species on the island.

Y. YOKOHATA

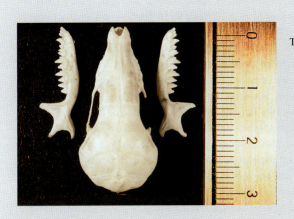

The skull of the holotype stored at Kyushu University (Y. Yokohata).

Three dimensional image of Uotsuri-jima Isl. in 1978 based on aerophotograph (left), showing no damage from introduced goats, and that in 2006 based on Quickbird satellite imaging (right), showing severe damage from introduced goats (M. Kaneko & T. Ono).

RESEARCH TOPIC 1

Distribution patterns and zoogeography of Japanese mammals
M. MOTOKAWA

1. Zoogeography of Japanese mammals

Japan is one of the most diverse zoogeographic regions in the world, ranging from 122°56'–153°59' E and 20°25'–45°33' N. It includes subtropical to cool-temperate zones, a maximum elevation of 3,776 m, and about 378,000 km² in land area (Fig. 1). Japan, except for the Ogasawara Island Group (I. G.), consists of continental islands located in the eastern coast of Eurasia. The Ogasawara I. G. is composed of oceanic islands in the Pacific Ocean located in the southeastern most of Japan, from which only two native species of bats are recorded. Geologically, Japan includes two regions: one is the so-called Japanese Isls. including the main islands (mainlands) of Japan (Hokkaido, Honshu, Shikoku, Kyushu), the other is the Ryukyu Arch., a chain of small islands in the southwestern part of Japan (7–14 in Fig. 1). With minor modification, these regions represent two major zoogeographic units for Japanese mammals. The boundary between the zoogeographic units is located in the Tokara Tectonic Str. between Akuseki-jima and Kodakara-jima Isls. in the Tokara I. G. of the northern Ryukyus. This boundary is known as Watase's line [2260, 2262, 2777].

There are 116 species of mammals in Japan excluding Order Cetacea, Order Sirenia and recently introduced species (see Research Topic 9); 50 of these species (43.1%) are endemic to Japan. Rates of endemic species are high in small mammals such as orders Soricomorpha (70.0%), Lagomorpha (50.0%), Rodentia (48.1%), and Chiroptera (43.2%), whereas rates are low for medium sized or larger mammals in orders Artiodactyla (33.3%) and Carnivora (12.5%). These differences in endemic rates among orders have been interpreted as differences in dispersal ability and longevity. Among 101 species, excluding five commensal species (*Suncus murinus*, *Rattus norvegicus*, *R. rattus*, *R. exulans*, and *Mus musculus*) two Ogasawara species and pinnipeds, 76 species are known from the Japanese Isls. and northern Ryukyu Arch. (Ohsumi/Tokara I. Gs.) north from the Tokara Tectonic Str. Seventeen species are known only from the central (Amami/Okinawa/Daito I. Gs.) and southern (Miyako/Yaeyama/Senkaku I. Gs.) Ryukyu Arch. Only eight species are distributed in both sides of the Tokara Tectonic Str., six of which are species of bats. The remaining two species (*Sus scrofa* and *Prionailurus bengalensis*) are also widely distributed in the continent.

2. Mammals in the central and southern Ryukyu Islands

Mammals distributed south of the Tokara Tectonic Str. are characterized by high endemic elements, and most of them are considered Oriental elements [2262]. It is known that most mammals in the Amami/Okinawa I. Gs. in the central Ryukyus, except for the several widespread species of bats such as the genus *Miniopterus* and commensal species, are endemic to this region. They include three endemic genera: two rat genera (*Tokudaia* and *Diplothrix*) and a rabbit (*Pentalagus*). The Amami/Okinawa I. Gs. are thought to have been separated from the north by the deep Tokara Tectonic Str. and from the south by the deep Kerama Gap at least from the Pliocene. Both straits have a sea depth of more than 1,000 m. Accordingly, it was hypothesized that many terrestrial mammals in the Amami/Okinawa I. Gs. have been isolated since the Pliocene [2262]. For several species, between–island differentiation also has been reported. The endemic rat genus *Tokudaia* has three species distributed in three islands: *T. osimensis* in Amami-ohshima Isl., *T. tokunoshimensis* in Tokuno-shima Isl., and *T. muenninki* in Okinawa-jima Isl. These three species are characterized by different karyotypes and morphology [180, 513, 1325, 1646, 2398]. Molecular phylogeny including many murid genera implies the split of these species from the other rat genera occurred several millions years ago, but the closest relative, which is probably distributed in East or Southeast Asia, has not yet been well detected [3052, 3356]. One mouse species *Mus caroli* is distributed only in Okinawa-jima Isl. in the Okinawa I. G. It was thought to have been introduced through human activity, but its zoogeographic origin is now enigmatic [2262]. Conspecific populations are distributed in Taiwan and the southeast Asian continent and islands, however, the Okinawa-jima *M. caroli* are morphologically and genetically divergent from these populations [2274, 3212, 3557]. Mammal species in the Yaeyama/Senkaku I. Gs. are limited (*Mogera uchidai*, *Apodemus agrarius*, *Prionailurus bengalensis iriomotensis*), and are thought to have close relationships with congeneric species or conspecific populations in the eastern continent and/or Taiwan [1988, 2272, 3346]. One species of fruit bat, *Pteropus dasymallus* in the Ryukyu Arch., has a distribution pattern that crosses the Tokara Tectonic Str. This species is distributed widely in the islands of the Ryukyu Arch. from the Tokara I. G. in the north to the Yaeyama I. G. and a small Taiwanese island, Liutao, in the south; but it is not distributed in the Ohsumi I. G. or Taiwan. The reason for this unique distribution pattern is uncertain, but probably is related to ecological factors relevant to island habitat.

3. Mammals in the Japanese mainlands

Mammals distributed north of the Tokara Tectonic Str. are thought to have a variety of colonization histories. They are considered the Palearctic elements among mammals of Japan [28, 463, 464]. There are four main-islands in the Japanese Isls.: Hokkaido (77,978 km² in area, 2,290 m in maximum elevation), Honshu (227,895 km², 3,776 m), Shikoku (18,292 km², 1,982 m), and Kyushu (36,716 km², 1,791 m).

These islands are not small and each has a complicated geomorphology. From a zoogeographic perspective, this area can be subdivided into two subareas: Hokkaido versus Honshu/Shikoku/Kyushu. Except for pinniped species between these two subareas, 19 species of mammals are known from Hokkaido and not found in Honshu/Shikoku/Kyushu, whereas 36 species are recorded from Honshu/Shikoku/Kyushu (including its adjacent islands) and not recorded from Hokkaido. Common between those two subareas are 21 species: 12 species of bats, six species of carnivores, one species of deer and two species of mice. Species of bats, carnivores and deer are considered to have high dispersal abilities compared with the small terrestrial mammals. Small non-volant mammals found both in Hokkaido and Honshu/Shikoku/Kyushu are limited to two endemic mouse species: *Apodemus speciosus* and *A. argenteus*. Tsugaru Str. between Honshu and Hokkaido has been considered an important zoogeographic boundary and sometimes is called Blakiston's line. It is about 140 m in maximum depth and thought to have last formed about 100,000 years ago [2632].

Species in Honshu/Shikoku/Kyushu include three endemic genera: two genera of shrew-moles (*Urotrichus* and *Dymecodon*) and a dormouse (*Glirulus*). They may have arrived in Japan during the Miocene. Many species in Honshu/Shikoku/Kyushu migrated from the continent in various ways by means of land bridge formations during the Pleistocene [28, 463, 464].

During the Miocene, the Japanese Isls. was a part of the eastern part of the continent. Thereafter, the Tokara Tectonic Str. were formed and the Japanese Isls. were separated from the continent. During Pleistocene glacial periods, the Japanese Arc. repeatedly was connected with and disconnected from the continent via the Korean Pen. and Sakhalin. Many species living in Honshu/Shikoku/Kyushu are considered to have migrated through the Korean Pen., but the timing and route has not been closely examined. Recent DNA phylogeographic studies are attempting to hypothesize detailed evolutionary histories for various species (see account for each species). Following these studies, more comprehensive study including conspecific and congeneric populations from East Asian countries, such as Korea, Russia, China, Vietnam, Mongolia, need to be conducted.

Zoogeographic studies of Japanese mammals until the end of the 20th century mainly focused on the problem of migration in relevance to the formations and collapses of land bridges in Tsushima & Korean Strs. and Tsugaru Str. during the Pleistocene. Recent studies, however, focused more on the zoogeographic events within Honshu. For several species, considerable genetic and cytogenetic divergence within Honshu have been found both for larger mammals such as *Cervus nippon* and *Macaca fuscata* [728, 1484, 2350, 3482] and small mammals such as *Crocidura dsinezumi*, *Chimarrogale platycephalus*, *Urotrichus talpoides*, *Mogera imaizumii*, *Apodemus speciosus* and *Glirulus japonicus* [788,

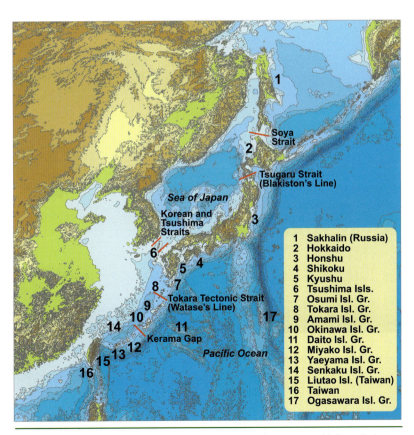

Fig. 1. Map of Japan illustrating geographic layout and zoogeographic boundaries.

1162, 1165, 2620, 3633, 4077]. Considerable divergence can be accounted for by the complicated zoogeographic histories of those species after migration into Japan across the land bridges. The high level of divergence probably was caused by environmental changes in the past associated with complicated weather and geomorphological features of Honshu, but detailed histories have not been interpreted for any species and is a task to be addressed in the future.

In contrast to Honshu, Hokkaido has no endemic species, and most species also are distributed in the continent. Because the Soya Str. between Hokkaido and Sakhalin is shallow and had formed a land bridge in the last glacial period, many Hokkaido species were thought to have migrated from the north during that period. Although many species are conspecific with the Sakhalin and continental Russian populations, recent studies showed considerable genetic divergence between them for several species and no difference in others [1164, 1173, 2611]. Genetic divergences within Hokkaido were found and multiple migrations from the north were suggested for several species such as *Ursus arctos* and *Sorex gracillimus* [1980, 1984, 2003, 2611]. These facts imply that the Hokkaido fauna also is as complicated as in Honshu and more phylogeographic studies for each species are necessary to reconstruct the real features of the zoogeography of mammals in Hokkaido.

Offshore small islands surrounding the Japanese main-islands also have been a focus for studies of zoogeographic history, especially in small mammals, because each island (or island group) has a different timing of connection with the main islands in the Pleistocene [3358]. The Tsushima Isls. are located between the Korean Pen. and Kyushu. Most species of Tsushima Isls. also are distributed in Kyushu, but several species are common with Korean mammals and not distributed in Honshu/Shikoku/Kyushu, such as *Mustela sibirica*, *Prionailurus bengalensis*, and *Crocidura shantungensis*. This mixed fauna in the Tsushima Isls. is hypothesized to have formed through different dispersal abilities among species [2273].

4. Conclusion

As previously discussed, the mammal fauna of Japan involves high species diversity and endemic elements, and a complicated zoogeographic background attained during several million years. It is obvious that this high diversity was obtained as a result of the variable geomorphology, climate, and habitat of Japan. Also, many extinct species of large and small mammals are recorded as Pleistocene fossils [1510]. This suggests considerable environmental changes during the Pleistocene. To understand the zoogeography of mammals from Japan, comprehensive reconstruction of the history of each species, in conjunction with paleoenvironmental and fossil evidence and detailed phylogeographic patterns of extant species, needs to be conducted.

A relief of Thomas W. Blakiston in Hakodate, Hokkaido (Y. Ishibashi).

The Iriomote cat.

The gray red-backed vole.

Erinaceomorpha

Erinaceomorpha ERINACEIDAE
021

Red list status: LC (IUCN); introduced

Erinaceus amurensis Schrenk, 1859

EN Amur hedgehog　　JP アムールハリネズミ（amuuru harinezumi）　　CM 东北刺猬　　CT 黑龍江刺猬
KR 고슴도치　　RS амурский ёж

An individual from the Izu Pen., Shizuoka Pref. (N. Ishii).

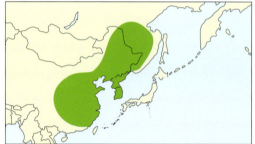

1. Distribution

Introduced species in Japan. Original distribution: Far East Russia, eastern to central China, and Korean Pen. In Japan, 2 populations originating from escaped or released animals have been established in the Odawara area (Kanagawa Pref.) and the eastern part of the Izu Pen. (Shizuoka Pref.), respectively [1099]. The former population was found in 1987 and occurs in a small area of about 1 km^2 and has a stable range. The latter was discovered in 1995, covered 58 km^2 in 2007 and has been expanding its range at a rate of 1 km/year [2122].

2. Fossil record

Fossils of *Erinaceus* sp. are found from the late Middle Pleistocene of Yamaguchi and Okayama Pref. Honshu [819, 1034, 1514].

3. General characteristics

Morphology Except for the face, legs and underside, the body is covered with dense, 2-cm long spines. The spines come in 2 different colors, one totally whitish and the other with one blackish-brown band [1099]. Individuals from the Odawara area are more whitish than those of the Izu area [2790]. The external character resembles that of *E. europaeus*, but the skull form is different (e.g. basisphenoid pit is V-shaped while U-shaped in other *Erinaceus* spp., and maxilla-nasal contact is long) [406], and thus the Japanese populations have been identified as *E. amurensis* [1099].

External measurements are as follows (only ranges are available) [1099]: Izu area (n = 8), HB = 239–269 mm, T = 32–40 mm, E = 23–30 mm, HFsu = 36–39 mm, BW = 540–920 g; Odawara area (n = 6), HB = 239–286 mm, T = 25–37 mm, E = 23–27 mm, HFsu = 37–41 mm, BW = 468–1,000 g. Individuals in the Izu area are slightly smaller than those in the Odawara area [1099].

Dental formula　I 3/2 + C 1/1 + P 3/2 + M 3/3 = 36 [406].

Mammae formula　0 + 2 + 2 = 8 or 0 + 3 + 2 = 10; it is sometimes observed in Japanese populations that the number of nipples is different between the two sides [2122].

Genetics　Chromosome: 2n = 48 [4060].

4. Ecology

Reproduction Occurrence of juveniles (individuals of body weight under 400 g) from June to November (mainly in July and November) suggests that the breeding season lasts from spring to autumn with 2 peaks [2122, 2790]. Two females with 3 embryos were captured in August and September in the Izu area [2122].

Lifespan　Not reported in Japan.

Erinaceus amurensis Schrenk, 1859

Diet Lepidopteran larvae, adults of ground beetles and orthopterans were the main food items found in the stomach contents of individuals captured in the Izu area [2122].

Habitat Found mainly in open areas such as cultivated fields, orchards, parks, golf courses and residential areas, but also in forests including coniferous plantations [2122].

Home range A study using radio transmitters in the Odawara area revealed that the range size of 1 male and 1 female were 1.9 and 1.8 ha, respectively [2005]. Territoriality is not known.

Behavior It is nocturnal and seems to hibernate in cold seasons, judging from its absence from December to March [2005].

Natural enemy Not reported in Japan.

Parasite Not reported in Japan.

5. Remarks

Hedgehogs may eat crops, especially strawberries, according to interviews with local residents [2790]. The mitochondrial DNA was analyzed for hedgehogs from the 2 populations established in Odawara and Isu Pen., and their identification was confirmed as *E. amurensis* [2663].

An individual in the Odawara area, Kanagawa Pref. (N. Ishii).

N. ISHII

A habitat of the Amur hedgehog in Odawara, Kanagawa Pref. (Y. Yokohata).

A rolled-up hedgehog in Shizuoka Pref. (N. Ishii).

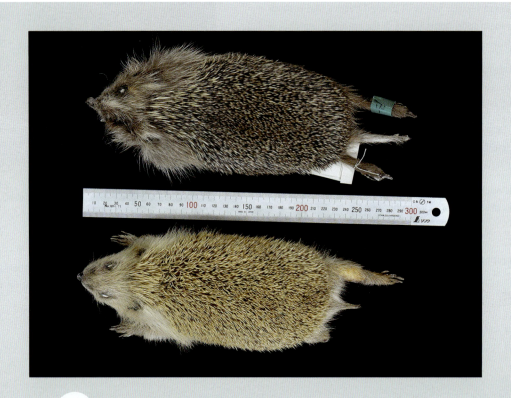

Two color types of the Amur hedgehog captured in Itoh, Shizuoka Pref. (S. D. Ohdachi).

A skeleton of the greater horseshoe bat, *Rhinolophus ferrumequinum* (M. Ishida).

Chiroptera

Chiroptera PTEROPODIDAE

Red list status: CR as *P. d. daitoensis*, *P. d. dasymallus* (MEJ); En as *P. d. dasymallus* and *P. d. daitoensis*, C-1 as *P. d. inopinatus* and *P. d. yayeyamae* (MSJ); NT (IUCN)

Pteropus dasymallus Temminck, 1825

EN Ryukyu flying fox　　JP クビワオオコウモリ (kubiwa ookoumori)　　CM 琉球狐蝠　　CT 琉球狐蝠
KR 류큐과일박쥐　　RS рюкюйская летучая лисица

A Ryukyu flying fox feeding on the fruits of a tropical almond at night on Okinawa-jima Isl., Okinawa Pref. (Y. Osawa).

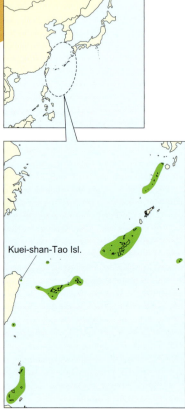

1. Distribution

Ranges from the southwestern islands of Japan and Taiwan [4208]. There is a sighting record from Uotsuri-jima isl. of the Senkaku Isls. [3448]. This species also occurs on the small northern islands of Batan, Dalupiri, and Fuga in the Philippines [874], but their taxonomic relationship is not clear. This species is divided into 5 subspecies in each island group [4208]: *P. d. dasymallus* (Erabu flying fox, erabu ookoumori): the Ohsumi Isls, Kuchinoerabu-jima and the Tokara Isls. (Nakano-shima, Taira-jima, Akuseki-jima, and Takara-jima) [656]. *P. d. inopinatus* (Orii's flying fox, orii ookoumori): Okinawa-jima Isl. and the adjacent islands (Okinoerabu-jima, Yoron-jima, Iheya-jima, Izena-jima, Ie-jima, Minna-jima, Kouri-jima, Yagaji-jima, Oo-jima (Nago-shi), Sesoko-jima, Kudaka-jima, Yabuchi-jima, Tsuken-jima, Henza-jima, Miyagi-jima, Hamahiga-jima, Ikei-jima, Oo-jima (Nanjo-shi), Senaga-jima, Tokashiki-jima, Aka-jima, Aguni-jima, and Kume-jima) [673, 674, 2387, 2742, 2744, 4141]. *P. d. daitoensis* (Daito flying fox, daito ookoumori): the Daito Isls. (Minami-daito and Kita-daito) [998]. *P. d. yayeyamae* (Yaeyama flying fox, yaeyama ookoumori): the Miyako I. G. (Tarama-jima, Irabu-jima, Shimoji-jima, Miyako-jima, Ikema-jima, Oogami-jima, and Kurima-jima) and the Yaeyama Isls. (Ishigaki-jima, Taketomi-jima, Kohama-jima, Iriomote-jima, Hateruma-jima, Kuro-shima, Hatoma-jima, and Yonaguni-jima) [1916, 2741]. *P. d. formosus* (Formosan flying fox): Taiwan, mainly on Lu-tao Isl. and Kuei-shan-Tao Isl., and occasionally on the east coast of the Taiwan mainland [1848]. Although *P. d. formosus* has been considered to be extinct since the early 1990s [1849], a few individuals were recently observed on Lu-tao Isl. [3285] and Kuei-shan-Tao Isl. [3286]. Distributional records on some islands does not mean that bats are always resident on that island because they often move among neighboring islands [2380].

2. Fossil record

Fossils of *Pteropus* sp. have been found from cave and fissure deposits of the Late Pleistocene to Holocene on Miyako Isl. and Ishigaki-jima Isl. [809, 1487, 1497, 2361], and semifossils of *P. dasymallus* are recorded from the stratum of ca. 2,500–3,000 BC in Okinoerabu-jima Isl. [2489].

Pteropus dasymallus Temminck, 1825

3. General characteristics

Morphology Medium-sized *Pteropus* species. No tail. Males are slightly larger than females [668]. Fur color is variable [4208]; usually dark to light brown with a mantle of light and yellowish brown hair on the head, neck, and shoulders. Hair on the neck is yellowish in adult males and paler in females. *P. d. daitoensis* is obviously different in coloration from other subspecies; a larger part of its body is whitish and yellowish, with some variation among individuals.

External measurements are given for each subspecies, including locality and sex if possible.

Measurements (in mm or g) are given in mean ± SD (range; sample size). *P. d. dasymallus*: BW = 550.8 ± 51.60 (449–662; $n = 9$), FA = 140.9 ± 2.23 (137.0–145.0; $n = 9$), HFsu = 43.8 ± 1.83 (41.5–47.3; $n = 14$), Tib = 71.0 ± 2.42 (66.3–74.3; $n = 14$), E = 19.1 ± 1.22 (18.8–21.0; $n = 5$) for males. BW = 550.8 ± 51.60 (460–560; $n = 5$), FA = 138.2 ± 2.48 (135.5–142.0; $n = 6$), HFsu = 43.8 ± 1.58 (41.2–46.2; $n = 11$), Tib = 68.0 ± 1.93 (64.5–70.3; $n = 11$), E = 20.2 ± 1.35 (17.3–20.9; $n = 7$) for females [668]. *P. d. inopinatus*: BW = 450.2 ± 46.27 (337–583; $n = 96$), FA = 138.6 ± 3.4 (130.6–146.7; $n = 96$), HFsu = 41.8 ± 1.84 (35.6–46.1; $n = 64$), Tib = 70.3 ± 1.85 (65.2–74.9; $n = 86$), E = 23.2 ± 1.89 (16.8–26.2; $n = 32$) for males. BW = 437.4 ± 61.96 (320–568; $n = 46$), FA = 135.89 ± 3.40 (126.4–142.6; $n = 47$), HFsu = 40.28 ± 2.54 (37.8–44.1; $n = 26$), Tib = 67.2 ± 2.03 (61.8–70.6; $n = 37$), E = 22.9 ± 1.48 (18.7–24.4; $n = 14$) for females (database of K. Kinjo, A. Nakamoto and M. Izawa in 1995–2007). *P. d. daitoensis*: BW = 492.3 ± 21.05 (455–510; $n = 7$), FA = 137.3 ± 1.63 (134.9–140.1; $n = 7$), E = 24.8 ± 1.77 (23.5–26.0; $n = 2$) for males. BW = 447.5 ± 51.30 (325–510; $n = 11$), FA = 133.9 ± 3.67 (123.9–137.6; $n = 14$), E = 24.1 ± 2.30 (22.0–28.0; $n = 5$) for females [2773]. *P. d. yayeyamae*: BW = 356.4 ± 34.6 (318–402; $n = 5$), FA = 130.3 ± 6.02 (124.0–136.0; $n = 3$) for males. BW = 388.5 ± 19.1 (375–402; $n = 2$), FA = 134.3 ± 4.55 (130.0–140.5; $n = 5$) for females [2773, 4208].

Dental formula I 2/2 + C 1/1 + P 3/3 + M 2/3 = 34 [4208].

Mammae formula Females of all Japanese bat species possess one pair of functional nipples which are located in the anteriolateral pectoral position (1 + 0 + 0 = 2), with a single exception of *Vespertilio murinus* having 2 pairs. Thus, hereafter, mammae formula is omitted in bat species.

Genetics Chromosome: $2n = 38$ and FN = 72 [3635]. The complete sequence of the mtDNA, 16,705 bp, is available [2471].

4. Ecology

Reproduction Mating season is mainly from September to December and females rear a single young from April to June. The offspring is born with complete milk teeth, and open eyes and ears. Birth weight is 50–70 g [1583]. The offspring becomes independent at the age of 3 to 4 months [3033].

Lifespan Little known. The greatest longevity record is 12 years in the wild [2385], and 24 years in captivity [3862].

Diet Feeds mainly on fruits and nectars of wild and cultivated plant species. Favorite food items are fruits of the Chinese banyan *Ficus microcarpa*, deciduous fig *Ficus superba*, tropical almond *Terminalia catappa* and common garcinia *Garcinia subelliptica*. Occasionally feeds on leaves and insects. *Pteropus d. dasymallus* especially utilizes leaves and insects in warm-temperate regions, and this is considered an adaptation to food shortage in winter [684]. Utilizes 106 plant species, including the fruits of 68 species, flowers of 37 species, and leaves of 30 species, as well as 11 insect species [2381].

Habitat Roosts in the canopies of trees in the daytime mainly individually or sometimes in small groups. In *P. d. daitoensis*, gregarious roosts of more than 100 individuals sometimes occur [1201]. Inhabits several types of habitats including forests, urban areas and plantations [2382].

Home range Daily home range size is 52.5 ha on average although the size is highly variable among individuals and seasons [2384]. Does not have a strict territory or fixed home range and shifts foraging areas and day roost sites according to spatial changes in food availability [2379]. Movements between islands have been observed in *P. d. daitoensis* [2962], *P. d. yayeyamae* [1586] and *P. d. inopinatus* [673, 2380].

Behavior Nocturnal [1585].

Natural enemies Natural enemies rarely reported. Potential predators are the Iriomote cats (*Prionailurus bengalensis iriomotensis*) on Iriomote Isl. [3846] and some introduced carnivores, feral cats and feral dogs [4059].

Parasites & pathogenic organisms

Ectoparasite
Acari: *Binuncus magnus* [3707].
Endoparasite
Not reported.
Pathogenic organisms: A bat adenovirus has been isolated from *P. d. yayeyamae* [1924, 1930].

5. Remarks

Although many species of flying foxes have been driven to extinction by over-harvesting for food in other countries, hunters have not targeted flying foxes in Japan. However, sometimes there are conflicts with farmers, who regard this species as a pest of commercial fruit crops. They may also be accidentally killed in orchard nets set to prevent birds or flying foxes from fruit crops. Population size may fluctuate greatly in relation to the frequency and intensity of striking typhoons [2388]. *P. d. dasymallus* and *P. d. daitoensis* were designated as Natural Monuments in 1975 and 1973, respectively. *P. d. daitoensis* were also designated as a National Endangered Species of Wild Fauna and Flora in 2004.

An important pollinator and seed disperser for some plant species in the Ryukyu Isls. [2383, 3627].

K. KINJO & A. NAKAMOTO

Chiroptera PTEROPODIDAE

Red list status: EX (MEJ); Ex (MSJ); DD (IUCN)

Pteropus loochoensis Gray, 1870

EN Okinawa flying fox   **JP** オキナワオオコウモリ (okinawa ookoumori)   **CM** 冲绳狐蝠   **CT** 沖繩狐蝠
KR 오키나와과일박쥐   **RS** окинавская летучая лисица

The type specimen stored in the Natural History Museum, London. © The Natural History Museum, London.

1. Distribution

Recorded only from Okinawa-jima Isl. in the 19th century [736]; however, it has not been found since this report. Now considered to be extinct [240].

2. Fossil record

Not reported.

3. General characteristics

Morphology   Known from only 2 specimens in the Natural History Museum, London [4208]. Dorsal surface of lower leg and humerus is naked, collar is remarkable. Fur is longer than in the allied species; approximate length: back 14–17 mm, mantle and belly 15–18 mm, least width of furred area of back is about 42 mm. Fur color of the type specimen (male, subadult, BM No. 49.1.5.2): back and rump are glossy blackish seal brown thinly sprinkled all over with grayish white. Breast, belly, and flanks are glossy seal brown sprinkled with silvery grayish and buffy hairs. The mantle is buffy, palest posteriorly, tinged with orange ocher in the middle, and passing into an ochraceous buff on the sides of the neck. Median line of foreneck is clouded with mars brown and seal brown. The concealed seal brown bases of buffy hairs are very short in the center of the mantle and somewhat longer posteriorly in the shoulder region, wanting on the sides of neck. Occiput is similar to the mantle or somewhat tawny; crown is mottled dark brown, russet, and buffy; the circumocular region is conspicuously paler (owing to predominance of buffy element) than the center of the crown. Cheeks are seal brown sprinkled with pale yellow. Chin and throat are blackish seal brown with some trace of a paler admixture. The dorsal surface of the leg is naked. No tail.

External measurements (in mm) of these 2 specimens are as follows; FA = 135.0, HFcu = 45.0, Tib = 60.0 in the subadult male, FA = 142.5, HFcu = 45.5, Tib = 61.5 in the adult female [4208].

Dental formula   I 2/2 + C 1/1 + P 3/3 + M 2/3 = 34 [4208].
Genetics   Not reported.

4. Ecology

Reproduction   Not reported.
Lifespan   Not reported.
Diet   Not reported.
Habitat   Not reported.
Home range   Not reported.
Behavior   Not reported.
Natural enemies   Not reported.
Parasites   Not reported.

5. Remarks

Some researchers treat this form as a synonym of *P. mariannus* [1712, 1887]. The taxonomic status of this species is uncertain and existing specimens are of unknown provenance.

K. KINJO

A Daito flying fox (*Pteropus dasymallus daitoensis*) (A. Nakamoto).

A view of Ie-jima Isl., Okinawa Pref. (A. Nakamoto). Traditionally many trees are planted around houses as windbreaks in Okinawa. Ryukyu flying foxes often visit these trees.

A Ryukyu flying fox feeding on fruits of the common garcinia (*Garcinia subelliptica*) planted in a private garden (A. Nakamoto).

Chiroptera PTEROPODIDAE

Red list status: EN (MEJ); En, R for the Minami-iwo-to Isl. population (MSJ); CR (IUCN)

Pteropus pselaphon Lay, 1829

EN Bonin flying fox   JP オガサワラオオコウモリ (ogasawara ookoumori)   CM 小笠原狐蝠   CT 小笠原狐蝠
KR 오가사와라과일박쥐   RS бонинская летучая лисица

A Bonin flyng fox that visited the American century plant to eat nectar at night in Chichi-jima Isl., Ogasawara Isls., Tokyo Pref. (Y. Osawa).

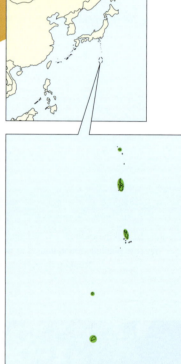

1. Distribution

Endemic to the Ogasawara Isls. (Chichi-jima and Haha-jima) and the Kazan Isls. (Iwo-to, Minami-iwo-to, and Kita-iwo-to) [4208]. Moreover, food remains of bats were found on Ani-jima Isl. [48], and also GPS telemetry surveys revealed use of both islands of Ani-jima and Otōto-jima [1052, 1053]. Additionary, there are some sighting records from Muko-jima Isl., 50 km apart from Chichi-jima Isl. [784].

2. Fossil record

Not reported.

3. General characteristics

Morphology Medium-sized *Pteropus* species. General form is larger than that of *P. dasymallus*. No tail. Fur color is almost black with a small scattering of silver and golden hairs. No collar. Fur is long and thick. Lower legs and hind feet are covered with hairs. The hair on the lower back is especially long [4208].

External measurements (mean ± SD in g or mm with range and sample size) are as follows: BW = 527.4 ± 46.02 (400–616; n = 54), FA = 138.6 ± 2.50 (132.1–145.0; n = 54), HFcu = 46.7 ± 2.87 (42.0–50.0; n = 6), Tib = 65.0 ± 1.72 (61.25–68.0; n = 38), E = 26.1 ± 2.66 (23.0–29.0; n = 4) for adult males. BW = 500.5 ± 65.75 (353–586; n = 23), FA = 138.5 ± 2.70 (131.2–143.1; n = 23), HFcu = 46.6 ± 3.44 (42.0–50.0; n = 5), Tib = 63.6 ± 2.06 (59.9–68.3; n = 16), E = 24.6 ± 1.70 (23.0–27.0; n = 4) for adult females [1051–1053, 3326, 4208].

Dental formula I 2/2 + C 1/1 + P 3/3 + M 2/3 = 34 [4208].

Genetics Chromosome: $2n$ = 38 and FN = 72 [786]. Genetic diversity and genetic structure among the inter-island populations, have been inferred from data on mtDNA control region sequences and microsatellite markers [2667].

4. Ecology

Reproduction Mating system may be female defense polygyny [3324, 3325] Mating season is mainly from December to April,

Pteropus pselaphon Lay, 1829

although they have an ability to breed year-round; Females give birth to a single young in June [1916, 3324].

Lifespan Little known but the greatest longevity record is 18.6 years in captivity [4174].

Diet Feeds mainly on fruits and nectar of wild and cultivated plant species. Utilizes 58 plant species, including the fruits of 31 species, flowers of 17 species, and the leaves of 18 species, as well as 1 insect species [1031, 2395, 3350]. Leaves of bird-nest fern (*Asplenium setoi*) are also used [2395, 3325]. A high proportion of the dietary composition consists of exotic agricultural fruits, such as bananas and citruses in Chichi-jima Isl. [1031]. Leaves are used more frequently by this species than by other flying foxes [1031].

Habitat Roosts in tree canopies in the daytime [48].

Home range Home range size is approximately 9 km^2 although the size is highly variable among individuals (range: 1.3–20.6) [1052, 1053]. Some GPS-collared individuals which captured on Chichi-jima Isl. also used neighboring Ani-jima Isl. and Ototo-jima Isl. [1052, 1053].

Behavior Fundamentally nocturnal, but diurnal activity has also been reported on Minami-iwo-to Isl. [1030, 1093, 3350]. Forms colonial roosts in winter and disperses in summer. On Chichi-jima Isl., over 100 individuals aggregate at a special roosting area, and almost all of these are individuals from the island; when this species roosts in winter, a unique behavior is observed; several dozen individuals form a cluster together by huddling. This huddling behavior is related to forming a harem for mating and for thermoregulation [3324, 3325]. On Chichi-jima Isl., the flying fox uses agricultural fields as feeding sites throughout the year.

Natural enemies Not recorded. However, feral cats are latent predators [1032].

Parasites
Ectoparasite
Acari: *Binuncus magnus* [3707].
Endoparasite
Not reported.

5. Remarks

The population on Chichi-jima Isl. was considered to be extinct in the 1970s, but they were rediscovered in the 1980's [1032]. The population size has slowly increased since then and is estimated at 100 to 150 individuals by recent surveys [3347]. On the other hand, the population on Haha-jima Isl. greatly decreased from more than 100 in the 1960's to a few individuals in the 2000's. The population on Minami-iwo-to Isl. was estimated at more than 100 in 1982 and 2007 [3347, 3350]. Consequently, the latest

A cluster of Bonin flying foxes on a winter colonial roost on Chichi-jima Isl. (K. Kinjo).

Hairs on the lower back are especially long (Y. Osawa).

total population size of this species is estimated at roughly 200–300 individuals [3347]. Threats for the species are habitat destruction such as deforestation, disturbance of roosting sites by tourism, and predation by feral cats. Notably, human disturbance to roosting sites is the most serious threat for this species, because the bats use a limited number of locations as colonial roosting sites [3324].

Additionally, there are conflicts with farmers of commercial fruits. Although farmers do not kill flying foxes directly, bats die when they are entangled in orchard nets which farmers set to prevent the animal from accessing fruit crops [1030, 1032]. This species was designated a "Natural Monument" in 1969, and as a "National Endangered Species of Wild Fauna and Flora" in 2009.

K. KINJO & M. IZAWA

Chioptera RHINOLOPHIDAE

025

Red list status: C-1 (MSJ); LC (IUCN)

Rhinolophus ferrumequinum (Schreber, 1774)

EN greater horseshoe bat   JP キクガシラコウモリ（kikugashira koumori）
CM 马铁菊头蝠   CT 馬鐵大蹄鼻蝠   KR 관박쥐   RS большой подковонос

An individual roosting on a cave ceiling (Y. Osawa).

1. Distribution

Widely distributed from southern Europe and north Africa to north India, China, Korea and Japan [423, 4208]. In Japan, occurs in Hokkaido, Honshu, Shikoku, Kyushu, and the islands of Izu-ohshima, Miyake-jima, Nii-jima, Sado-gashima, Oki, Tsushima, Iki-shima, Goto, Yaku-shima, Kuchino-shima, and Nakano-shima [31, 651, 656, 3115, 4208, 4220].

2. Fossil record

Many fossils have been found from cave and fissure deposits of the Middle Pleistocene, the Late Pleistocene and the Early Holocene ages in Honshu and Kyushu [1034, 1504, 1512–1516, 1518, 1519, 2360, 2687, 3602, 4158].

3. General characteristics

Morphology Fur is thick and glossy, pale brown or dark orange; nose-leaf is present, connecting process of the intermediate nose-leaf is broadly rounded above, posterior margin of posterior nose-leaf forms a wedge; lower labial plate has 1 groove and is divided into 2 parts; ears are large and lack a tragus; the second finger lacks phalanges, plagiopatagium is attached to base of metatarsal, ankle or lower portion of tibia [4208]. Wings are short and broad [631, 4208]. Females have a pair of functional pectorals and a pair of pseudomammillae anterior to the vulva. Largest rhinolophid in Japan.

Measurements (mean ± SD in mm with range and sample size) in various region in Japan are as follows. Hokkaido: FA = 58.9 ± 1.15 (56.2–61.7; n = 42), HB = 63.1 ± 5.83 (55.0–74.1; n = 14), T = 35.6 ± 2.06 (32.7–39.0; n = 14), HFcu = 12.9 ± 0.70 (11.5–14.1; n = 15), Tib = 25.8 ± 0.62 (25.0–27.5; n = 15), E = 23.0 ± 1.66 (20.0–26.0; n = 15) [630, 633, 840, 2343, 3058, 3079]. Honshu: FA = 59.9 ± 1.66 (n = 146), HB = 69.1 ± 4.54 (n = 96), T = 38.3 (30.4–43.2; n = 11), HFcu = 14.2 ± 0.99 (n = 112), Tib = 24.6 ± 1.66 (n = 106), E = 27.4 ± 1.49 (n = 112) [626, 1735, 1770, 2423, 3018, 3266, 4208]. Shikoku: FA = 59.9 ± 1.33 (n = 51), HB = 67.7 ± 4.51 (n = 41), HFcu = 14.3 ± 1.09 (n = 41), Tib = 24.5 ± 1.16 (n = 52), E = 27.7 ± 1.20 (n = 51), Kyushu: FA = 59.5 ± 1.38 (n = 112), HB = 67.8 ± 3.29 (n = 92), HFcu = 13.2 ± 0.76 (n = 54), Tib = 24.5 ± 1.36 (n = 73), E = 27.4 ± 1.61 (n = 83), and Sado Isl.: FA = 60.1 ± 1.30 (n = 13), HB = 70.2 ± 4.27 (n = 8), HFcu = 14.3 ± 0.70 (n = 8), Tib = 24.9 ± 0.82 (n = 8), E = 27.4 ± 1.94 (n = 8) [4208]. Tsushima Isls.: FA = 58.8 ± 1.21 (n = 18), HB = 65.6 ± 5.01 (n = 6), T = 34.4 ± 3.22 (n = 3), HFcu = 12.0 ± 0.15 (n = 6), Tib = 25.1 ± 0.93 (n =17), E = 27.2 ± 1.15 (n = 15) [1452, 4208].

Dental formula   I 1/2 + C 1/1 + P 2/3 + M 3/3 = 32 [4208].

Genetics Chromosome: $2n = 58$ and $FN = 62$ [147, 148]. Geographical variation of Japanese *R. ferrumequinum* has been revealed using complete sequences of the mitochondrial cytochrome *b* gene. Sequence divergence values in this species are very low [2974].

4. Ecology

Reproduction Reproductive pattern involves seasonal monoestry and delayed fertilization [2207, 2590, 2591, 3689]. Females form tightly packed maternity colonies to give birth and care for infants in summer. Sizes of maternity colonies are relatively small, averaging 85.0 (range 10–166) in Ishikawa Pref., central Honshu [2992], but 132.1 (50–200) in Yamaguchi Pref., western Honshu [1772]. Adult females, subadult females and males occupy 75.2%, 24.2% and 0.6% of a colony, respectively. Parturitions occur synchronously 2–3 weeks before and after mid-July in Ishikawa Pref. [2993, 2995]. Neonates are born at a relatively developed stage, with permanent teeth already beginning to erupt at birth [663, 664]. The proportion of neonatal to maternal size in FA and BW is 32.5% and 42.5% in Ishikawa Pref. [2992] and 30.0% and 46.3% in Kyushu, respectively [663]. Mothers recognize their own infants by vocal communication and selectively nurse them [2010, 2014, 2015]. Newborns grow rapidly and start self-feeding at about 32 days old, and are weaned at about 40 days old [2992]. Juvenile mortality is 3.6% [2992]. Fertility rate at 1, 2, 3 and 4 years of age averages 13.1%, 49.5%, 95.2% and 100%, respectively in Ishikawa Pref. [2995]. In Yamaguchi Pref., 1-year-old bats do not bear young, and fertility rate at 2 and 3 years of age averages 27.2% and 93.4%, respectively [1775]. Most females produce a single young annually after their first parturition, and delivery by a 23-year-old female has been observed [2990]. Adult females exhibit strong loyalty to their natal sites to give birth to young [2995]. Ecology of the mating season is little known.

Lifespan Greatest longevity recorded in Japan is 23 years and 8 months [1776].

Diet Prey includes Diptera, Lepidoptera, Coleoptera, Trichoptera, Plecoptera, Odonata, and Hemiptera [669, 676, 1085, 1770, 3607]. Feeds mainly on Lepidoptera, Coleoptera and Diptera measuring 8–45 mm in body length in open forests, woodland paths or forest edges [669, 676, 1085]. Hibernating bats often prey on troglophilic moths within caves [2998] and sometimes forage outside caves [2995].

Habitat Widely uses natural caves, abandoned mines, bomb shelters, unused tunnels, underground channels and buildings as day-roosts [2991, 3124]. Most common cave-dwelling bat species in Japan.

Home range Home range of adult females during summer averaged 1.5 ha [669].

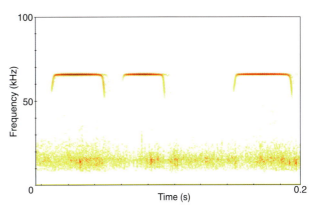

Fig. 1. Sonogram of perch call inside a mosquito net, recorded by D. Fukui in Hokkaido.

Behavior Nocturnal [1771, 2995]. Foraging style is flycatching (perch hunting) and aerial hawking, though flycatching is more common [669]. Often picks prey off the ground and from tree stems [1770, 1771].

Body weight increases by 25.8–28.2% in late autumn compared to summer in Kyushu [679]. Individuals tend to be scattered in winter, but dense clusters of mixed age and sex are formed when environmental temperatures fluctuate and fall; fidelity to hibernacula is relatively low, and size and member composition of the hibernating population changes from year to year [2995]. Long distance movement up to 130 km has been recorded in Kyushu [1064].

Echolocation and social calls Echolocation calls have an FM/CF/FM structure (Fig. 1). The CF values in Hokkaido, north Honshu, south Honshu, north Kyushu and south Kyushu populations average 65.0 kHz (62.8–66.6, $n = 12$) [624], 65.5 kHz (65.2–65.8, $n = 6$) [3527], 65.2 kHz ($n = 1$) [626], 68.5 kHz (68.3–68.8, $n = 10$) and 69.7 kHz (69.3–70.3, $n = 10$) [660], respectively. For prey selection, feeding bats are attracted by echoes with acoustical glints containing strong frequency and amplitude modulations [1590]. In mother-infant communication, they frequently emit supersonic calls and synchronize mutual calls fixing the range of emission timing; the interference sound generated by the overlaps of calls may serve as their recognition mark [2010, 2014, 2015].

Natural enemies Natural predation is rarely reported. Several were evidently eaten by feral cats, Ural owls (*Strix uralensis*), rat snakes (*Elaphe climacophora*) and striped snakes (*E. quadrivirgata*) [2990].

Parasites
Ectoparasite
Diptera: *Phthiridium hindlei, Penicillidia jenynsii, Brachytarsina kanoi* [653, 1760, 2183, 2187]. Acari: *Paraperiglischrus rhinolophinus, Eyndhoveni euryalis, Trombicula koomori, Sasatrombicula koomori, Speleognathopsis bastini, Opsonyssus zumpti, Ixodes vespertilionis* [1297, 1311, 2473, 3343, 3691, 3807, 4031].

Endoparasite
Trematoda: *Acanthatrium hitaense, A. isostomum, A. ovatum, A. papilligerum, A. setoense, Mesothatrium japonicum, Duboisitrema sawadai, Prosthodendrium yamizense, P. circulare, P. chilostomum, P. longiforme, P. macrorchis, P. miniopteri, P. parvouterus, P. postacetabulum, P. urna, P. radiatum, P. thomasi, P. radiatum, Lecithodendrium macrostomum, Pycnoporus rhinolophi, Plagiorchis corpulentus, Pl. koreanus, Pl. rhinolophi, Pl. vespertilionis, Pl. kyushuesis, Pl. minutifollicularis* [1541, 1542, 1546–1550, 1552, 1553]. Cestoda: *Insectivorolepis yoshidai, I. takasii, I. niimiensis, I. okamotoi, I. inuzensis, I. araii, I. osensis, I. ooyabui, Vampirolepis hidaensis, V. ikezakii, V. ogaensis, V. fujiensis, V. shirotanii, V. isensis, V. iwatensis, Hymenolepis nishidai, H. parva, H. rashomonensis, H. odaensis, H. subrostellata, H. iriei, H. tsuzurasensis, Rodentolepis macrotesticulata, Ro. hattorii, Oligorchis brevihamatus, Myotolepis grisea* [3106, 3109, 3110, 3116–3119, 3122, 3140]. Nematoda: *Strongylacantha pretoriensis, Molinostrongylus rhinolophi, Rictularia rhinolophi, Capillaria pipisterelli* [1257, 1260, 1262].

5. Remarks

The taxonomic status of *R. ferrumequinum* in Japan is currently under debate. It may be a subspecies (*R. f. nippon*) or potentially a distinct species based on molecular phylogenetic analyses [559, 2908, 3561].

A. SANO

View of a postpartum female of the greater horseshoe bat to show a mammilla (blue arrow) and pseudomammillae (red arrow) (A. Sano).

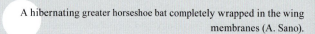

A hibernating greater horseshoe bat completely wrapped in the wing membranes (A. Sano).

A hibernating colony of the greater horseshoe bat formed on a mine ceiling (A. Sano).

Chiroptera RHINOLOPHIDAE

026

Red list status: EN as *R. c. orii* (MEJ); C-2, K for the Amami Isls. population (MSJ); LC (IUCN)

Rhinolophus cornutus Temminck, 1834

EN Japanese little horseshoe bat   JP コキクガシラコウモリ（kokikugashira koumori）
CM 日本小菊头蝠   CT 日本小蹄鼻蝠   KR 작은관박쥐   RS японский малый подковонос

Japanese little horseshoe bats hibernating in an abandoned mine (N. Urano).

1. Distribution

Endemic to Hokkaido, Honshu, Shikoku, Kyushu, the islands of Okushiri, Izu-ohshima, Miyake-jima, Nii-jima, Mikura-jima, Hachijo-jima, Sado-gashima, Tsushima, Iki-shima, Fukue, Yaku-shima, Tane-gashima, Kuchinoerabu-jima, Kuchino-shima, Nakano-shima, Amami-ohshima, Kakeroma, Tokuno-shima, and Okinoerabu-jima [31, 633, 651, 656, 659, 1919, 2217, 2778, 3124, 4208]. Once reported outside of Japan from Guanxi, southern China [3829], but there is little support for this identification [423].

2. Fossil record

Many fossils have been found in cave and fissure deposits of the Middle Pleistocene, the Late Pleistocene and the Early Holocene ages in Honshu and Kyushu [1504, 1513–1515, 1518, 1519, 2687, 3602, 4152].

3. General characteristics

Morphology General form is similar to *R. ferrumequinum*, but is remarkably small. Fur is thick and glossy, pale brown or dark orange; connecting process of the intermediate nose-leaf is sharply pointed above; lower labial plate has 3 grooves and is divided into 4 parts; ears are large and lack a tragus; wings are short and broad, the second finger lacks phalanges, plagiopatagium is attached to ankle or tibia [4208]. Wing aspect ratio is 5.00 [4136]. Females have a pair of pseudomammillae anterior to the vulva.

Measurements (mean ± SD in mm with range and sample size) from various regions in Japan are as follows. Hokkaido: FA = 40.5 ± 0.88 (38.3–42.5; n = 74), HB = 45.1 ± 5.05 (38.0–56.2; n = 17), T = 21.5 ± 3.24 (16.0–27.0; n = 16), HFcu = 8.2 ± 0.69 (7.0–9.1; n = 13), Tib = 18.8 ± 1.00 (17.0–21.1; n = 18), E = 15.8 ± 1.31 (14.0–17.5; n = 16) [450, 452, 630, 633, 840, 1697, 3058]. Honshu: FA = 40.2 ± 1.03 (n = 360), HB = 43.9 ± 2.35 (n = 106), T = 26.2 (25.3–29.5, n = 9), HFcu = 9.4 ± 0.62 (n = 282), Tib = 17.4 ± 0.80 (n = 273), E = 18.2 ± 0.96 (n = 285) [626, 1735, 1770, 3018, 3266, 4208]. Shikoku: FA = 39.5 ± 0.92 (n = 77), HB = 43.5 ± 2.41 (n = 37), HFcu = 9.0 ± 0.60 (n = 68), Tib = 17.3 ± 0.88 (n = 74), E = 18.0 ± 0.91 (n = 77) [4208]. Kyushu: FA = 38.8 ± 0.96 (n = 73), HB = 42.0 (39.8–43.2; n = 9), T = 22.6 (18.4–23.9; n = 9), HFcu = 9.3 ± 0.61 (n = 70), Tib = 17.3 ± 0.61 (n = 73), E = 17.9 ± 0.92 (n = 73) [1770, 4208]. Sado Isl.: FA = 39.7 ± 0.69 (n = 17), HB = 45.0 ± 1.93 (n = 17), HFcu = 9.6 ± 0.56 (n = 17), Tib = 17.5 ± 0.72 (n = 17), E = 19.3 ± 0.42 (n = 17), Hachijo-jima Isl.: FA = 37.6 ± 0.58 (n = 23), HB = 44.6 ± 2.20 (n = 22), HFcu = 9.6 ± 0.62 (n = 22), Tib = 15.4 ± 0.71 (n = 22), E = 17.8 ± 0.96 (n = 23). Fukue Isl.: FA = 38.9 ± 0.73 (n = 12), HFcu = 9.4 ± 0.31 (n = 12), Tib = 17.1 ± 0.56 (n = 12), E = 17.3 ± 0.40 (n = 12) [4208]. Tsushima Isls.: FA = 39.0 ± 0.64 (n = 127), HB = 38.2 ± 2.78 (n = 125), T = 15.5 ± 4.66 (n = 3), HFcu = 8.4 ± 0.43 (n = 127), Tib = 16.6 ± 0.53 (n = 127), E = 18.0 ± 0.70 (n = 123) [1452, 4208]. Amami-ohshima Isl.: FA = 37.7 ± 0.80 (n = 99), HB = 41.2 ± 1.83 (n = 12), T = 24.2 (20.8–26.7; n = 8), HFcu = 8.5 ± 0.60 (n = 97), Tib = 15.8 ± 0.79

($n = 88$), E = 17.3 ± 0.72 ($n = 93$) [1770, 4208].

Forearm length and cranium size are significantly greater in females than in males, and there is a clear decreasing cline in FA and skull size from north to south, with the smallest individuals representing the subspecies *R. c. orii* on the Amami Isls. [1906, 1913, 4208].

Dental formula   I 1/2 + C 1/1 + P 2/3 + M 3/3 = 32 [4208].

Genetics   Chromosome: $2n = 62$ and FN = 60 [148]. Geographical variation of Japanese *R. cornutus* has been revealed using complete sequences of the mitochondrial cytochrome *b* gene. Sequence divergence values in this species are very low [2974].

4. Ecology

Reproduction   Seasonal monoestry; males and females tend to be separated throughout the year except around the time of mating; adult females form maternity colonies of tens to thousands of bats in summer [672, 1770]. Mothers give birth to a single large-sized young [672, 4135]. The newborn is completely naked, with eyes closed, but the proportion of neonatal to maternal size in FA and BW reaches 42% and 43–44%, respectively [672, 4135]. Young start self-feeding at about 22 days and wean at about 45 days of age [4134]. Females generally achieve sexual maturity at 28 months and begin to reproduce in their third year [1778].

Lifespan   A 21-year-old male has been recorded [3554].

Diet   The prey includes Diptera, Lepidoptera, Coleoptera, Trichoptera, Orthoptera and Araneae [676, 679, 1770, 4266]. Feeds mainly on Lepidoptera and Diptera measuring 7–23 mm in body length [676]. During hibernation, often feeds on camel crickets (Orthoptera: Rhaphidophoridae) within caves and preys mainly on Diptera (e.g. Mycetophilidae, Tipulidae and Trichoceridae emerging in winter and early spring) outside the cave [679, 680].

Habitat   Roosts mainly in natural caves, abandoned mines, unused tunnels, bomb shelters and underground culverts [2991, 3124], and in abandoned houses on rare occasions [3098].

Home range   Not reported.

Behavior   Nocturnal [680]. Forages mainly in woodlands [676]. Large prey often taken to feeding perches. Body weight increases by 29.2–33.9% in late autumn compared to summer in Kyushu [679]. Hibernating bats prefer roosts with relatively high ambient temperature (9–15°C) and humidity (85–100%) and sleep lightly as compared with *R. ferrumequinum* [679, 680, 2728].

Echolocation and social calls   Echolocation calls are of relatively long duration and have an FM/CF/FM structure typical of *Rhinolophus* (Fig. 1). The CF values are generally similar across Honshu but are higher from Kyushu southwards (mean and range, in kHz): Aomori Pref. (northern Honshu) 104 (103–104); Yamaguchi Pref. (western Honshu) 106 (106–108); Wakayama Pref. (south Honshu) 106.3; in Kyushu: Fukuoka Pref. 107.7 (107.1–108.1), Ooita Pref. 108.3 (107.3–109.0), Kumamoto Pref. 108.2 (107.1–108.8), and Kagoshima Pref. 109.6 (108.7–110.8); and the range in frequency amongst the islands south of Kagoshima is similar to *R. pumilus* on northern Okinawa-jima: Yaku-shima Isl. 109 (108–109), Amami-ohshima Isl. 109 (109–111) and Tokuno-shima Isl. 112.8 [626, 660, 666, 2011, 4181,

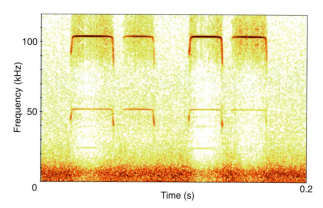

Fig. 1. Sonogram of perch call inside a laboratory, recorded by D. Fukui in Hokkaido.

4183]. The characteristic echolocation frequencies are generally negatively correlated with body size, with smaller bats at lower latitudes producing the higher frequencies [672]. No information is available on social calls.

Natural enemies   Natural predation is rarely reported. Several were evidently eaten by Japanese weasels (*Mustela itatsi*), rat snakes (*Elaphe climacophora*) [1782] and feral cats [2297].

Parasites

Ectoparasite

Diptera: *Nycteribia allotopa, N. parvula, Penicillidia jenynsii, Brachytarsina kanoi* [653, 2183, 2187, 3069]. Acari: *Paraperiglischrus rhinolophinus, Eyndhovenia euryalis, Trombicula alba, T. koomori, Speleognathopsis bastini, Opsonyssus zumpti, Ixodes vespertilionis* [1297, 1311, 3691, 3787, 4031].

Endoparasite

Trematoda: *Acanthatrium isostomum, A. setoense, A. ovatum, Mesothatrium japonicum, Prosthodendrium ascidia, P. chilostomum, P. miniopteri, P. urna, P. circulare, P. yamizense, Pycnoporus rhinolophi, Neoheterophyes sawadai, Plagiorchis rhinolophi, Pl. koreanus* [1549, 1550, 1553, 1556]. Cestoda: *Vampirolepis isensis, V. ikezakii, V. iwatensis* [3118, 3119, 3121, 3122]. Nematoda: *Strongylacantha pretoriensis, S. rhinolophi* [1259, 1261].

5. Remarks

The taxonomy of small *Rhinolophus* species (*cornutus, pumilus* and *perditus*) in Japan has not been resolved completely with various arrangements of species and subspecies given by different authors. The population amongst the Amami Isls. has been considered a subspecies, *R. c. orii* [4208]; and *R. pumilus* on Okinawa-jima Isl. and *R. perditus* on Ishigaki-jima Isl. have been considered synonyms of *R. cornutus* [423, 893, 3272]. According to a brief phylogenetic study based on analysis of mtDNA sequences, the East Asian small horseshoe bats, *R. monoceros, R. pusillus, R. cornutus* and *R. pumilus*, form a monophyletic group with low levels of sequence divergence [1845].

The construction of man-made tunnels has assisted in maintaining or promoting increased population size in this species, and the closure of such structures with inappropriate gates has the potential to deprive insular groups of valuable habitat, especially amongst the Izu Isls. and Tsushima Isls.

A. SANO & K. N. ARMSTRONG

Chiroptera RHINOLOPHIDAE

027

Red list status: EN as *R. p. pumilus*, EX as *R. p. miyakonis* (MEJ); En for the Miyako-jima Isl. population, V for the Ryukyu Isls. population (MSJ); LC (IUCN)

Rhinolophus pumilus Andersen, 1905

EN Okinawa little horseshoe bat   JP オキナワコキクガシラコウモリ（okinawa kokikugashira koumori）
CM 冲绳菊头蝠   CT 沖繩小蹄鼻蝠   KR 오키나와작은관박쥐   RS окинавский малый подковонос

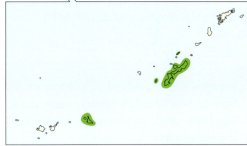

Okinawa little horseshoe bats roosting on a cave ceiling (H. Tamura).

1. Distribution

Endemic to the islands of Okinawa-jima, Iheya-jima, Tokashiki-jima, Kume-jima, Miyako-jima and Irabu-jima [1916, 1969, 3124, 4182]. Not recorded from Miyako-jima Isl. after 1971 [1895], the subspecies *R. p. miyakonis* is considered to be extinct. The population on Kume-jima Isl. has shown a marked decline [1969], and population sizes on Tokashiki-jima and Iheya-jima Isls. are very small [4182]. Distribution does not overlap with those of related 2 species (*R. cornutus* and *R. perditus*).

2. Fossil record

Not reported.

3. General characteristic

Morphology External characters are very similar to those of *R. cornutus* and *R. perditus*, but the noseleaf is smaller, skull length shorter and the length of the tibia is longer than that of *R. perditus* [1917, 4208].

Measurements (mean ± SD in mm with range and sample size) in northern Okinawa-jima Isl.: FA = 40.5 ± 0.73 (n = 266), HB = 43.3 ± 2.12 (n = 40), HFcu = 8.6 ± 0.49 (n = 40), Tib = 17.3 ± 0.65 (n = 40), E = 19.1 ± 0.60 (n = 40) [1972, 4208]. For subfossil specimens from the Miyako-jima Isl., HB = 36, 37 mm, FA = 38, 39 mm [1794]. Forearm lengths average 40.14 mm (n = 28) for males and 40.57 mm (n = 45) for females in Kume-jima Isl., 39.79 mm (n = 4) for males and 40.00 mm (n = 4) for females in Tokashiki-jima Isl. [4182], 40.5 mm (37.2–42.1; n = 108) for males and 40.9 mm (38.2–42.7; n = 109) for females in the northern Okinawa-jima Isl. population, but 39.2 mm (36.1–41.5; n = 124) for males and 39.8 mm (37.8–41.4; n = 138) for females in the southern Okinawa-jima Isl. [4183]. Mean FA is significantly greater in females compared to males in each population of Okinawa-jima Isl., and larger in the northern population than in the southern one in each sex [4183].

Dental formula   I 1/2 + C 1/1 + P 2/3 + M 3/3 = 32 [4208].

Genetics Chromosome: $2n$ = 62 and FN = 60 [3140, 3635]. A population genetic study with both nuclear and mitochondrial markers suggested that female bats did not move between the northern and southern regions of Okinawa-jima Isl., but gene flow was maintained by males. The parapatric condition of echolocation frequency differences between the 2 regions was suggested to be maintained by a process of vertical maternal transmission to developing young. Significant genetic differ-

A winter roost of the Okinawa little horseshoe bats (A. Sano).

Individuals in daily torpor (A. Sano).

ences were also noted between Okinawa-jima Isl. and Kume-jima Isl., with relatively low diversity on Kume-jima Isl. [4181].

The complete mtDNA sequence of this species is 16,869 bp, and mitochondrial markers have been used in various systematic and phylogeographic studies [1459, 2472].

4. Ecology

Reproduction Breeding cycle monoestrous; copulation observed between November and January; single infant born in May to June [1969, 1972]. No information on reproductive pattern, or age at sexual maturity.

Lifespan Not reported.

Diet Insectivorous but details unavailable, likely to be similar to *R. cornutus*.

Habitat Roosts in natural caves, abandoned mines and bomb shelters [1972, 3124, 4182, 4184].

Home range Not reported.

Behavior Nocturnal [1916, 4184]. Mainly forages in woodlands [4184], even in winter [1972].

Echolocation and social calls Echolocation calls have an FM/CF/FM structure as in *R. cornutus* (Fig. 1). The CF values average 108.2 kHz (105.6–110.9, $n = 108$) in males and 110.6 kHz (107.7–113.0, $n = 109$) in females for the northern Okinawa-jima Isl. population, but 116.0 kHz (112.8–119.3, $n = 124$) in males and 117.4 kHz (114.6–120.1, $n = 138$) in females for the southern Okinawa-jima Isl. population; the mean value is significantly higher in females than in males in each population, and higher in the northern population than in the southern one in each sex [4181, 4183]. Frequency range in *R. pumilus* does not overlap

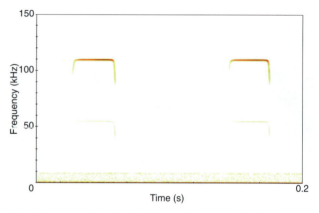

Fig. 1. Sonogram of perch call inside a small net, recorded by K. N. Armstrong on Okinawa-jima Isl.

with that of *R. perditus* [2011]. No information is available on social calls.

Natural enemies Not reported.

Parasites
Ectoparasite
Diptera: *Nycteribia allotopa* [2473].
Endoparasite
Trematoda: *Acanthatrium setoense* [1553]. Nematoda: *Capillaria pipistrelli* [3108].

5. Remarks

There is an opinion that the population on Okinawa-jima Isl. is a subspecies *R. p. pumilus* and that on Miyako-jima Isl. is *R. p. miyakonis* [4208]. Other researchers regard these taxa as synonyms of *R. cornutus* [423, 893, 3272]. Habitat destruction and tourism-related disturbance are key threats [1898].

A. SANO & K. N. ARMSTRONG

Chiroptera RHINOLOPHIDAE

028

Red list status: VU (MEJ); En for the Ishigaki-jima Isl. population,
V for the Iriomote-jima Isl. population (MSJ); LC (IUCN)

Rhinolophus perditus Andersen, 1918

EN Yaeyama little horseshoe bat　　JP ヤエヤマコキクガシラコウモリ（yaeyama kokikugashira koumori）
CM 八重山菊头蝠　　CT 八重山小蹄鼻蝠　　KR 야에야마작은관박쥐　　RS яеямский малый подковонос

An individual roosting on a cave ceiling (K. Koyanagi).

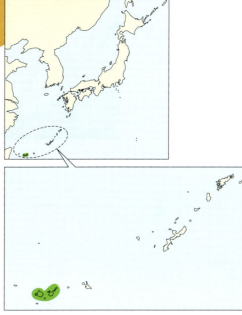

1. Distribution

Endemic to the Yaeyama Isls.: Ishigaki-jima, Iriomote-jima, Kohama and Taketomi-jima Isls. [1916].

2. Fossil record

Not reported.

3. General characteristic

Morphology External characters are very similar to those of *R. cornutus* and *R. pumilus*, but the noseleaf is larger, skull length greater and the length of the tibia is shorter than that of *R. pumilus* [1917, 4208].

Measurements (mean ± SD in mm with sample size) are as follows. Ishigaki-jima Isl.: FA = 40.5 ± 0.78 (n = 46), HB = 46.1 ± 2.62 (n = 26), HFcu = 8.8 ± 0.69 (n = 41), Tib = 16.4 ± 0.89 (n = 40), E = 18.5 ± 0.91 (n = 40). Iriomote-jima Isl.: FA = 41.6 ± 0.95 (n = 113), HB = 45.3 ± 2.07 (n = 32), HFcu = 9.4 ± 0.52 (n = 99), Tib = 16.9 ± 0.73 (n = 101), E = 18.8 ± 1.32 (n = 102) [4208]. Tail lengths for populations in islands of Ishigaki-jima and Iriomote-jima average 19.5 mm (18.1–21.7; n = 6) and 24.6 mm (21.2–26.1; n = 4), respectively [1770].

Dental formula I 1/2 + C 1/1 + P 2/3 + M 3/3 = 32 [4208].

Genetics Chromosome: $2n$ = 62 and FN = 60 [144, 3635]. Mitochondrial DNA markers have been used in systematic and phylogeographic studies [1459].

4. Ecology

Reproduction Seasonal monoestry. Females form maternity colonies of hundreds to over 1,000 individuals in May and give birth to a single young [1731, 2010]. Little is known of their breeding ecology.

Lifespan Not reported.

Diet Prey includes Lepidoptera, Coleoptera, Hymenoptera, Diptera, Hemiptera, Orthoptera, Neuroptera, Trichoptera and Araneae; feeds mainly on insects of the former 4 orders [637].

Habitat Roosts in natural caves, abandoned mines, underground culverts and bomb shelters [1731, 3124].

Home range Not reported.

An infant cluster in a maternity roost (K. Koyanagi).

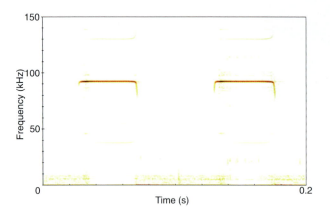

Fig. 1. Sonogram of perch call inside a small net, recorded by K. N. Armstrong on Iriomote-jima Isl.

Behavior Nocturnal [1918, 3441]. Forages mainly in woodlands [1923, 3441]. Active in all months of the year [679].

Echolocation and social calls Echolocation calls have an FM/CF/FM structure (Fig. 1). The CF values for the populations of Ishigaki-jima Isl. and Iriomote-jima Isl. are significantly lower than in the 2 related species *R. cornutus* and *R. pumilus*, being 97 kHz (96–98, $n = 10$) and 92 kHz (92–93, $n = 20$), respectively [2011]. No information is available on social calls.

Natural enemies Not reported.

Parasites
Ectoparasite
Diptera: *Brachytarsina amboinensis, B. suzukii* [2185, 2187, 3069].
Endoparasite
Trematoda: *Anchitrema sanguineum, Prosthodendrium urna* [1550]. Cestoda: *Vampirolepis isensis, V. iriomotensis* [3122, 3140]. Nematoda: *Capillaria pipistrelli* [3129].

5. Remarks

The population on Iriomote-jima Isl. is treated by some as a distinct species, *R. imaizumii* [423, 893, 3272]. Deforestation, roost destruction and tourism-related disturbance threaten this species [1898].

A. SANO & K. N. ARMSTRONG

Time-lapse photographs of flying Japanese bats (Y. Osawa)

Upper: A flying Japanese pipistrelle against the Tokyo Skytree (time-lapse 20 Hz).

Middle: A Japanese pipistrelle catching an insect (time-lapse 30 Hz).

Lower: A birdlike noctule emerged from the slit of an elevated railway (time-lapse 8 Hz).

Chiroptera   HIPPOSIDERIDAE

029

Red list status: LP for the Yonaguni-jima Isl. and Hateruma-jima Isl. populations (MEJ); En for the Hateruma-jima, Ishigaki-jima and Yonaguni-jima Isls. populations, V for the Iriomote-jima population (MSJ); NT (IUCN)

Hipposideros turpis Bangs, 1901

EN lesser leaf-nosed bat   JP カグラコウモリ（kagura koumori）   CM 琉球蹄蝠   CT 中葉鼻蝠
KR 나뭇잎코박쥐   RS японский листонос

A lesser leaf-nosed bat roosting on a cave ceiling (H. Tamura).

1. Distribution

Endemic to the Yaeyama Isls: Ishigaki-jima, Iriomote-jima, Yonaguni-jima and Hateruma-jima Isls. [1916, 3569].

2. Fossil record

Not reported, but many fossils of *Hipposideros* bats very similar to this species have been found from sediments in a cave on Miyako-jima Isl. [1497]. In another cave on Miyako-jima Isl., fossils of a *Hipposideros* bat species which is slightly different from extant *H. turpis* have been discovered [909].

3. General characteristics

Morphology   Fur color is brownish or drabby brown, but reddish orange individuals also exist; nose-leaf is broad, posterior margin of posterior nose-leaf is fairly roundish, intermediate nose-leaf is without a sella and connecting process; ears are large and lack a tragus; wings are short and broad, the second finger lacks phalanges, plagiopatagium is attached to ankle [4208]. Adult females have a pair of well-developed false teats (pseudomammillae) anterior to the vulva [2010]. Size is clearly smaller than that of a related species, *H. terasensis*, which occurs in Taiwan. Adult males have a well developed foetid gland on their forehead.

Measurements (mean ± SD in mm with range and sample size) are as follows [3569, 4208]. FA = 69.3 ± 1.51 (64.0–73.2; n = 323), HB = 79.6 ± 6.04 (69.0–88.0; n = 23), T = 46.9 ± 2.37 (42.0–51.0; n = 24), HFcu = 15.2 ± 1.00 (13.4–18.0; n = 43), Tib = 30.4 ± 1.50 (26.9–33.5; n = 50), E = 30.1 ± 2.16 (26.3–34.2; n = 51). Those by local population are as follows [4208]. Ishigaki-jima Isl.: FA = 68.1 ± 2.15 (64.5–70.3; n = 8), HB = 69.5 (n = 1), T = 45.5 (n = 1), HFcu = 14.9 ± 1.06 (13.4–16.0; n = 8), Tib = 28.6 ± 1.10 (26.9–30.0; n = 7), E = 28.5 ± 1.21 (26.5–30.5; n = 8). Iriomote-jima Isl.: FA = 69.2 ± 1.39 (66.0–71.4; n = 33), HB = 80.5 ± 5.44 (69.0–88.0; n = 21), T = 47.4 ± 2.11 (44.0–51.0; n = 21), HFcu = 15.3 ± 1.01 (13.5–18.0; n = 33), Tib = 30.5 ± 1.32 (28.0–32.7; n = 34), E = 30.9 ± 1.94 (26.3–34.2; n = 34). Yonaguni-jima Isl.: FA = 68.9 (68.8–68.9; n = 2), HB = 70.0 (n = 1), T = 43.0 (42.0–44.0; n = 2), HFcu = 14.8 (14.5–15.0; n = 2), Tib = 29.3 (29.0–29.5; n = 2), E = 30.8 (30.5–31.0; n = 2).

Dental formula   I 1/2 + C 1/1 + P 2/2 + M 3/3 = 30 [4208].

Genetics   Chromosome: $2n$ = 32 and FN = 60 [144, 3635]. There are 6 pairs of PCR primers for microsatellites available [490]. Genetic relationships among island populations based on microsatellite data suggest that the Yonaguni-jima Isl. population is isolated, having the lowest genetic variation, while the Iriomote and Ishigaki populations are connected by gene flow [491]. This pattern of genetic relationship could be either historical [489], or relate to current bat dispersal among the Yaeyama Isls. [491].

4. Ecology

Reproduction Maternity colonies comprise both sexes; single young are born from late May to June [2010]. Self-feeding starts at 5–6 weeks of age; most females first give birth at the end of their 3rd year [2013].

Lifespan The greatest recorded longevity is 13 years [2013].

Diet Prey includes Diptera, Lepidoptera, Coleoptera, Trichoptera, Odonata, Hemiptera, Hymenoptera, Orthoptera, Neuroptera, Blattaria and Araneae; feeds mainly on Coleoptera from spring to summer; *Anomala albopilosa* seems to be frequently consumed [637].

Habitat Roosts in natural caves, abandoned mines, and bomb shelters [1730, 1731, 1918, 1928, 3124, 3569]. In a cave on Iriomote-jima Isl., there is a large colony with more than 10,000 bats [1928].

Home range Not reported.

Behavior Nocturnal [1918, 3441]. Feeds mainly in woodlands [1923, 3441]. In the roost, hanging bats keep a certain distance by mutual exclusivity throughout the year [2010]. Hibernates even in roosts with high ambient temperatures of up to 20°C [679].

Echolocation and social calls Echolocation calls have a CF/FM structure (Fig. 1). The CF value averages at 81.3 ± 1.7 kHz (77.4–83.6; $n = 36$) [3569]. Those by local population are as follows. Iriomote-jima Isl.: 77–84 kHz (range, $n = 100$); Ishigaki-jima Isl.: 82 kHz (mean, $n = 12$); Yonaguni-jima Isl.: 78 kHz (mean, $n = 8$) [2011]. No information is available on social calls.

Natural enemies There are several accounts of predation by red banded odd-tooth snakes (*Dinodon rufozonatum*) [2010], feral cats, jungle crows (*Corvus macrorhynchos*), legant scops owls (*Otus elegans*), Sakishima habu pit vipers (*Protobothrops elegans*) and beauty rat snakes (*Elaphe taeniura*) [2013].

Parasites

Ectoparasite

Diptera: *Brachytarsina suzukii* [2187]. Acari: *Whartonia iwasakii* [3343].

Endoparasite

Trematoda: *Anchitrema sanguineum, Prosthodendrium parvouterus* [1550]. Nematoda: *Capillaria pipistrelli* [1262].

5. Remarks

Formerly, there was an opinion that a synonymous species was also distributed in southern Thailand and Vietnam [315, 410, 892, 3272, 3610], but extensive morphological and genetic analyses indicate that Vietnamese and Thailand populations are distinct from *H. turpis* [3569]. Main threats to this species are deforestation, roost destruction and tourism disturbance [1898].

A. SANO

Lesser leaf-nosed bats keeping a certain distance from each other in a day roost (K. Koyanagi).

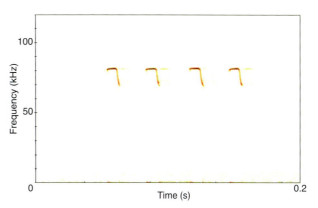

Fig. 1. Sonogram of a perch call inside a small net, recorded by K. N. Armstrong on Ishigaki-jima Isl.

6. Topic

Hipposideros turpis *in the Yaeyama Islands*

L. M. ECHENIQUE-DIAZ

The lesser leaf-nosed bat (*Hipposideros turpis* Bangs, 1901) is the only species of the family Hipposideridae in Japan, occurring on Ishigaki-jima, Iriomote-jima, Yonaguni-jima, and Haterumajima Isls. in the Yaeyama Isls., southern Okinawa Pref. [1916]. It is a strict cave-dwelling species that uses limestone caves, abandoned mines and war-time shelters as roosts, with the availability of these resources limiting the species distribution within particular islands. Data on its ecology regarding reproduction, development, behavior, foraging areas, population size estimates, activity patterns, dispersal, genetic variation and conservation status have been gathered during the last 25 years (see the description for *H. turpis* in this volume). This information has been sufficient to assign a conservation status to the species in Japan. However, most of the data remains unpublished and have been collected on only 2 islands (Iriomote-jima and Ishigaki-jima Isls.) over several years of surveys related to long-standing conflicts between conservationists and land developers. Recent studies suggest that significant differences in geological composition, forested area, and environmental history among the islands where the species has been confirmed to occur [488, 489, 491] affect ecological features such as roosting behavior, population size and genetic

HIPPOSIDERIDAE

Lesser leaf-nosed bats roosting in a limestone cave (T. Yamamoto).

Lesser leaf-nosed bats roosting in an air raid shelter (K. Koyanagi).

variation. These in turn result in different degrees of extinction risk for each island population. Therefore, ecological differences among islands should be taken into account in regional conservation plans of *H. turpis*. Some of these ecological differences are described here for 3 of the islands where the species is reported.

Iriomote-jima Isl. is mostly mountainous and large pristine forest covers most of its surface (262.48 km^2, 90.74%) [1931]. Typical Ryukyu limestone, an old formation where large caves are usually found, is scattered along the coast of the island and accounts for less than 10% of its surface [2789]. As a result, only 2 limestone cavities are known to hold year-round colonies of *H. turpis*. One of these caves, Ohtomi-daiichi-do, is known to hold the largest colony of the species with over 10,000 individuals [1928]. Demographic fluctuations in this colony have been recorded regularly since 1973 until the present by The Okinawa Environmental Research and Technology Center. The other 2 relatively large colonies (500–1,000 individuals) are found in abandoned mines. Bats rest in dark, humid and warm areas within these roosts. Basic studies on habitat use and dispersal ability within the island have shown that *H. turpis* moves through particular forest paths to its feeding areas, mainly broad-leaf semideciduous forest, although it also uses mangrove forest [1066]. High gene flow among colonies on the island makes them a single panmictic unit with the highest genetic variation among the Yaeyama Isls. Main threats to the species on Iriomote-jima are regular roost disturbances by so-called "ecotourists" and their local guides, and the development of tourism infrastructures near known roosts [491].

Ishigaki-jima Isl. has an area of 228 km^2 of which 42% is covered by different kinds of forest, including a few areas of broad-leaf semideciduous forest, secondary forest, and plantation [488]. Major patches of suitable habitat for *H. turpis* are mainly found in mountainous areas, given that topological conditions have made economic assimilation of these landscapes more difficult [488]. Flat parts of the island have undergone most of the transformation and are characterized by large gaps of pastures, cultivated areas, and human settlements. Typical Ryukyu limestone is exposed in almost all flat areas [2789], where a large number of caves are found. In the southern parts of the island extensive deforestation is prevalent, and a few partially isolated roosts of *H. turpis* are known. In the middle and northern parts of the island, bats roost in limestone caves scattered over flat, partially deforested areas and in war-time shelters found on forested hills [488]. Bats rest in dark, humid and warm areas within these roosts. Colony sizes range from 100–3,000 individuals, and the total population may be over 8,000 bats. There are at least 2 distinct populations of *H. turpis* on the island [491], and patterns of genetic variation among the known colonies appear to be related to patterns of habitat fragmentation [488]. Major threats to the species on this island are the construction of an airport, and the potential isolation of colonies because of extreme habitat fragmentation.

Yonaguni-jima Isl. is the most geographically isolated in the Yaeyama Isls. With 28.84 km^2 it has suffered a vast deforestation and natural vegetation degradation. Much of the remaining vegetation on the island (less than 50%) is a mixture of secondary formations and native forest. The island has a unique geological history among the Yaeyama Isls. A series of uplifting and subsiding movements coupled with sedimentation and different sea levels during past glaciations [1568, 2789], has resulted in different terrace-like levels. Some of these terraces are formed by typical Ryukyu limestone rocks, emerging on about 50% of the island. Bat roosts are mainly found on this rock formation, located at the base of terraces or in flat areas. The most common formations used by *H. turpis* on the island are crevices, dissolution holes, overhangs, and the vertical walls of terraces, shaded by thick vegetation. In these refuges bats are exposed to both daylight and eventual weather events, showing a roosting behavior not seen in any of the other islands where they occur. The population size has been estimated to be less than 1,000 bats [488]. Colonies on this island show the lowest genetic variation and are genetically isolated from other island populations [491]. Major threats to the species on this island are small population size, roost disturbances due to religious celebrations, and habitat lost.

Main day-roosts used by bats in Japan.
(1) Foliage: A Ryukyu flying fox roosting in tree foliage (A. Nakamoto).
(2) Cave (natural cave, abandoned mine, unused tunnel, air-raid shelter, underground channel, etc.): Eastern bent-winged bats flying around the entrance of an abandoned mine (A. Sano).
(3) Crevice: Rock crevice used by Oriental free-tailed bats as a day-roost (A. Sano).
(4) Building: An emerging Japanese pipistrelle from a day-roost in a house (J. Sano).
(5) Tree cavity: A birdlike noctule departing from a tree cavity (A. Sano).

Chiroptera　VESPERTILIONIDAE
030

Red list status: VU (MEJ); V (MSJ); EN (IUCN)

Eptesicus japonensis Imaizumi, 1953

EN Japanese northern bat　　JP クビワコウモリ（kubiwa koumori）　　CM 日本棕蝠　　CT 日本北棕蝠
KR 일본문둥이박쥐　　RS Японский кожанок

An individual from Fukushima Pref. (O. Takahashi).

1. Distribution

Endemic to Japan. Recorded from Fukushima [1577], Tochigi [1736], Saitama [4208], Ishikawa [4006], Yamanashi [1542, 4200, 4208], Nagano [1003, 1006, 3134, 3823, 3993, 4005, 4208], Toyama [4160] and Shizuoka [1262, 3098] Prefs. Red solid circles in the map denote sites where this species was observed in Japan.

2. Fossil record

Late Pleistocene fossils are recorded in western Honshu [4158].

3. General characteristics

Morphology　Dorsal fur color is dark blackish brown, basal portion of hair is not essentially different, hairs on back and behind shoulders have inconspicuous lighter (buffy) tips; ventral fur is light yellowish brown with deep golden brown collar behind the ears to anterior thorax; ears are comparatively short with extremities rounded off; tragus is short, with height less than half that of ear conch, anterior border is straight, posterior border is gently convex from narrowly rounded tip to small sharp basal lobe, widest region is at the base to a quarter of the anterior border; plagiopatagium is inserted at the base of the metatarsus of the first toe; tail is short, and its length is about 60% of HB [4208].

External measurements (mean ± SD in mm with range) are as follows [1003, 1006, 1577, 3823, 4005, 4006, 4200, 4208]. FA = 40.5 ± 0.64 (39.6–41.5; n = 12), HB = 62.5 ± 4.02 (56.5–68.0; n = 6), T = 37.7 ± 1.78 (35.0–39.5; n = 6), HFcu = 9.6 ± 1.15 (8.7–11.5; n = 6), Tib = 18.2 ± 0.76 (17.0–19.0; n = 6), E = 15.0 ± 0.95 (14.0–16.5; n = 6), Tr = 6.9 ± 0.42 (6.5–7.5; n = 5) for females. FA = 39.8 ± 1.20 (38.0–40.7; n = 4), HB = 58.0 ± 4.00 (54.0–62.0; n = 3), T = 38.2 ± 4.19 (35.5–43.0; n = 3), HFcu = 9.5 ± 0.71 (9.0–10.0; n = 2), Tib = 17.2 ± 1.08 (16.0–18.0; n = 3), E = 13.5 ± 2.29 (11.5–16.0; n = 3), Tr = 5.4 ± 0.85 (4.8–6.0; n = 2) for males.

Dental formula　I 2/3 + C 1/1 + P 1/2 + M 3/3 = 32 [4208].

Genetics　Chromosome: $2n$ = 50 and FN = 48 [142, 787].

4. Ecology

Reproduction　In Norikura Highland, Nagano Pref., up to 200 pregnant females assemble to form maternity colonies in June [3986]. Litter size, development, sexual maturity and breeding system are unknown.

Lifespan　Not reported.

Diet　Not reported.

Habitat Although this species characteristically uses tree cavities as roosts, some roosts have been reported in buildings [3986]. Hibernation sites and feeding habitats are unknown.

Home range Not reported.

Behavior Not reported.

Natural enemies Not reported.

Parasites
Ectoparasite
Not reported.
Endoparasite
Trematoda: *Plagiorchis* (*Plagiorchis*) *rhinolophi*, *Prosthodendrium* (*Prosthodendrium*) *postacetabulum*, *Pycnoporus rhinolophi* [1542]. Cestoda: *Vampirolepis balsaci* [3134]. Nematoda: *Molinostrongylus skrjabini longispicula* [1262].

Echolocation and social calls Not reported.

5. Remarks

Treated as a synonym of *E. nilssonii* or subspecies *E. nilssonii parvus* by some researchers [407–409, 1712, 2913, 3823]. However other researchers have treated the Japanese northern bat as an independent species in Japan, based on differences in cranial (brain case) and external (tail and tibia length) characters [1006, 3272, 4208].

In Norikura Highland, Nagano Pref., a bat house has been constructed to conserve this species.

D. FUKUI

A bat house built for the conservation of Japanese northern bats (A. Sano).

Chiroptera VESPERTILIONIDAE

Red list status: R (MSJ); LC (IUCN)

Eptesicus nilssonii (Keyserling & Blasius, 1839)

EN northern bat　　JP キタクビワコウモリ (kita kubiwa koumori)　　CM 北棕蝠　　CT 北方棕蝠，北首輪蝙蝠
KR 생박쥐　　RS северный кожанок

An individual captured in the Abashiri region, Hokkaido (H. Nakajima).

1. Distribution

Distributed from France and Norway through northern and central Europe and Asia east to the Pacific seaboard [2913, 3272]. In Japan, recorded from Hokkaido except for the southern part of the island [92, 267, 450–452, 455, 523, 624, 625, 1092, 1094, 1448, 1554, 1704, 1706, 1914, 1926, 1927, 1936, 2186, 2458, 2712, 2780, 2913, 3017, 3056, 3061, 3063, 3065, 3071, 3122, 3316, 3469, 3528, 3706, 3732, 3736, 3951–3954, 3956, 3957, 4051, 4053, 4055, 4211], Kunashiri (Kunashir) [1460] and Etorofu (Iturup) Isls. [1460]. Red solid circles in the map denote sites where this species was observed in Japan.

2. Fossil record

Late Pleistocene fossils are reported in western Honshu [4158].

3. General characteristics

Morphology　Ground color of upper parts ranges from prouts-brown to a light wood-brown; basal portion of hairs is not essentially different; hairs on back behind shoulders have conspicuous lighter (buffy) tips; underpart is slightly paler, sometimes approaching ochraceous-buff, but never sufficiently contrasting to produce a line of demarcation along the sides of the neck; ears are comparatively short, with extremities rounded off; tragus is short, less than half the height of the ear conch, its anterior border is straight, its posterior border is gently convex from narrowly rounded tip to upper edge of small but distinct basal lobe, its greatest width is at the level of the middle of the anterior border, equal to or slightly more than half the length of the anterior border; plagiopatagium is inserted at the base of the metatarsus of the first toe [4208].

External measurements (mean ± SD in mm with range) are as follows [267, 452, 625, 1006, 1914, 1926, 1927, 1936, 2458, 3017, 3063, 3065, 3071, 3732, 3736, 3953–3955, 4053, 4208, 4211]. FA = 40.9 ± 1.34 (36.6–43.4; n = 74), HB = 55.4 ± 6.14 (46.0–62.2; n = 18), T = 43.4 ± 3.63 (32.9–48.0; n = 18), HFcu = 10.5 ± 1.39 (7.6–12.0; n = 14), Tib = 16.8 ± 1.28 (15.2–19.5; n = 18), E = 14.5 ± 0.95 (12.0–16.0; n = 17), Tr = 6.1 ± 0.78 (4.5–7.0; n = 16) for females. FA = 39.7 ± 1.80 (36.3–44.0; n = 21), HB = 54.8 ± 4.66 (47.2–64.2; n = 12), T = 40.4 ± 3.69 (32.9–46.0; n = 12), HFcu = 9.8 ± 1.44 (7.2–11.5; n = 9), Tib = 17.1 ± 1.18 (15.0–19.0; n = 12), E = 13.9 ± 1.64 (9.7–15.2; n = 12), Tr = 5.8 ± 0.89 (4.0–6.8; n = 12) for males.

Dental formula　I 2/3 + C 1/1 + P 1/2 + M 3/3 = 32 [4208].

Genetics Chromosome: $2n = 50$ and $FN = 50$ [2712, 3635]. Molecular phylogenetic relationships among species of Vespertilionidae have been inferred from data on the ND1 gene of mtDNA, vWF gene of nuclear DNA and the SINE insertion [1459].

4. Ecology

Reproduction Pregnant females form maternity colonies from spring (May) to autumn (September) [3736], and colony size varies from a few dozen to hundreds of individuals [267, 450, 1936, 3736]. Parturition occurs from late June to mid-July [1094, 3955]. One newborn infant was observed in captivity; the newborn had no hair, closed eyes and deciduous teeth; body weight was 4 g at 8 days after birth, and reached 7 g at 25 days after birth; forearm length was 16 mm at 3 days after birth, and increased rapidly up to 4 weeks after birth [3955]. Colony size gradually decreases from mid-August to September owing to dispersal of mothers and their young. Sexual maturity and breeding system have not been investigated.

Lifespan Not reported in Japan, but a 12 year-old individual was discovered in France [1967].

Diet Not reported.

Habitat Roosts in tree cavities [1926] and buildings [267, 450, 1936, 3736]. Feeding habitats vary from forests to urban areas [3815], and forest and riparian habitats are preferred [1094]. Often forages around street-lamps [1685, 3065, 3953].

Home range Foraging ranges of reproductive females vary from 32.34 ha to 310.69 ha (mean: 209.36 ha, $n = 8$), and tend to increase as reproductive process (gestation and lactating) progresses [1094].

Behavior Nocturnal [1685]. Hibernation behavior has not been investigated.

Natural enemies The Eurasian hobby (*Falco subbuteo*) is reported as a predator [3268].

Parasites
Ectoparasite
Diptera: *Penicillidia monoceros* [2186]. Acari: *Spinturnix kolenatii* [3706].

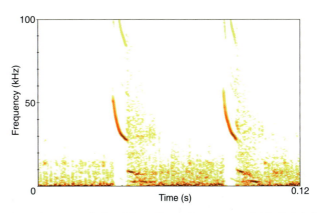

Fig. 1. Sonogram of a free-flying call around a roost, recorded by D. Fukui in Hokkaido

Endoparasite
Trematoda: *Plagiorchis minutifollicularis*, *Pl. koreanus*, *Prosthodendrium chilostomum*, *Pr. yamizense*, *Pr. ascidia*, Lecithodendliidae gen. spp. [1544, 1554, 3469]. Cestoda: *Vampirolepis ezoensis* [3122]. Nematoda: *Aonchotecha pipistrelli* [3469].

Echolocation and social calls Echolocation calls are FM-QCF calls of rather long duration in the search phase, and FM calls in the scanning phase or in back-cluttered space (Fig. 1). Mean values of EF and PF near the roost are 26.1 kHz (24.3–27.7, $n = 12$) and 30.5 kHz (28.3–31.6, $n = 12$), respectively [624]. Social calls have not been investigated.

5. Remarks

The populations in Korea, Sakhalin and Japan are treated as a subspecies *E. n. parvus* by some researchers because of the differences in the length of tibia and the width of inter-orbital constriction [4208].

Eptesicus nilssonii resembles *E. japonensis*. See the morphological description of *E. japonensis* for the differences between these species.

Conflicts occur with house owners because colonies are frequently formed in buildings.

D. FUKUI

Chiroptera **VESPERTILIONIDAE**

032

Red list status: **VU (MEJ); R (MSJ); NT (IUCN)**

Nyctalus aviator Thomas, 1911

EN birdlike noctule   JP ヤマコウモリ (yama koumori)   CM 大山蝠   CT 日本山蝠
KR 멧박쥐   RS вечерница-авиатор, восточная вечерница

An individual emerging from a tree hole in the Kamikawa region, Hokkaido (H. Nakajima).

1. Distribution

Distributed in eastern China, Korean Pen. and Japan, and possibly in Far East Russia [3272]. In Japan, recorded from Hokkaido [92, 267, 451, 452, 455, 630, 633, 840, 1092, 1888, 1902, 1903, 1905, 1910, 2715, 3017, 3528, 3542, 4024, 4046, 4051, 4053, 4208], Aomori [1553, 2286, 2300, 2301, 2711, 3401, 3703, 4208], Iwate [519, 794, 1549, 1551, 1553, 2986, 3111, 3823, 4208], Akita [590], Miyagi [96, 3389], Yamagata [794, 1542, 1723, 2731, 3242, 4208], Fukushima [1575, 4208], Tochigi [3036], Niigata [1001, 4208], Saitama [2931, 3036, 3823, 4208], Chiba [2966, 4208], Tokyo [2436, 3593, 3756, 4208], Kanagawa [168, 1551, 1790, 2553, 3125, 3534, 3969, 4218], Nagano [2134, 2141, 2179, 2253, 2352, 3135, 3224, 3706, 4208], Yamanashi, Shizuoka [1262, 4208], Ishikawa [3719, 3989], Gifu [1889], Aichi [3552], Mie [3000], Okayama [1910, 3942], Hiroshima [801, 3837], Tokushima [41], Ehime [2221], Kochi [2221, 4208], Nagasaki [3758, 3823, 4208], Kumamoto [178, 654], Kagoshima [660, 3823] and Okinawa [1584, 1971] Prefs. Red solid circles in the map denote sites where this species was observed in Japan.

2. Fossil record

Not reported.

3. General characteristics

Morphology The largest insectivorous bat in Japan. Fur is dense and velvety; color of dorsal hairs is dark yellowish brown, near wood-brown and cinnamon, and that of ventral hairs is slightly lighter than back; ears extend to about halfway from eyes to nostril when laid forward; anterior border of conch is abruptly convex below, then nearly straight to broadly rounded off at the extremity; posterior border is convex throughout; antitragus is long and low, well marked off posteriorly, its anterior border extends to just below the angle of mouth; tragus is very short, its greatest width is about equal to height; wings are long and slender; the fifth finger is shortest, and its distal point reaches the basal third of the first phalanx of the fourth finger; wing membranes are inserted at the basal portion of the metatarsus of the first toe [4208]. Mean wing aspect ratio and wingtip shape index are 6.73 and 1.37 ($n = 15$), respectively [631].

External measurements from 3 regions are as follows [1902, 4208]. For females, Hokkaido (mean in mm with range): FA = 62.3 (59.0–65.0; $n = 29$), HB = 100.4 (89.0–108.0; $n = 26$), T = 59.1 (55.0–65.0; $n = 26$), HFsu = 14.1 (12.0–15.0; $n = 28$), Tib = 24.5 (23.0–27.0; $n = 29$), E = 19.1 (17.5–20.5; $n = 27$), Tr = 8.6 (8.0–10.0; $n = 27$), Aomori Pref. (mean ± SD in mm with range): FA = 60.4 ± 1.43 (58.9–62.3; $n = 5$), HB = 91.0 ± 8.50 (82.5–103.5; $n = 6$), T = 53.8 ± 4.67 (45.0–59.0; $n = 6$), HFcu = 15.1 ± 1.34

(13.5–17.0; $n = 7$), Tib = 21.8 ± 1.81 (19.5–24.0; $n = 6$), E = 19.7 ± 2.02 (16.5–22.0; $n = 7$), Tr = 7.9 ± 0.80 (7.5–9.5; $n = 6$), and Tokyo, Kanagawa and Chiba Prefs. (mean ± SD in mm with range): FA = 61.1 ± 1.24 (59.2–64.0; $n = 15$), HB = 92.6 ± 7.22 (79.0–104.0; $n = 15$), T = 56.8 ± 3.31 (53.0–65.0; $n = 15$), HFcu = 14.8 ± 0.67 (13.5–16.0; $n = 14$), Tib = 21.3 ± 0.78 (20.0–23.0; $n = 12$), E = 20.1 ± 1.16 (18.5–22.5; $n = 15$), Tr = 9.8 ± 1.31 (7.0–11.0; $n = 14$). For males, Hokkaido (mean in mm with range): FA = 61.0 (57.0–64.0; $n = 22$), HB = 99.7 (90.5–106.0; $n = 21$), T = 59.0 (51.5–67.0; $n = 21$), HFsu = 14.4 (14.0–15.5; $n = 21$), Tib = 24.1 (21.5–26.0; $n = 22$), E = 18.8 (17.0–20.5; $n = 21$), Tr = 8.8 (8.0–10.0; $n = 20$), Aomori Pref. (mean ± SD in mm with range): FA = 61.3 ± 1.51 (60.0–63.2; $n = 4$), HB = 86.8 ± 4.47 (80.0–93.0; $n = 7$), T = 54.5 ± 2.40 (50.0–58.0; $n = 7$), HFcu = 14.1 ± 1.30 (12.5–16.0; $n = 8$), Tib = 23.6 ± 0.88 (22.5–25.0; $n = 8$), E = 19.2 ± 1.31 (18.0–21.5; $n = 8$), Tr = 7.4 ± 1.72 (5.0–10.5; $n = 8$), and Tokyo, Kanagawa and Chiba Prefs. (mean ± SD in mm with range): FA = 60.9 ± 1.12 (58.5–62.5; $n = 15$), HB = 96.5 ± 7.41 (83.0–106.0; $n = 15$), T = 56.5 ± 2.07 (54.0–61.0; $n = 15$), HFcu = 14.5 ± 0.73 (13.0–16.0; $n = 15$), Tib = 21.6 ± 1.01 (20.0–24.0; $n = 15$), E = 20.0 ± 0.95 (17.5–21.5; $n = 15$), Tr = 9.9 ± 1.32 (6.5–11.0; $n = 14$).

Dental formula I 2/3 + C 1/1 + P 2/2 + M 3/3 = 34 [4208].

Genetics Chromosome: $2n = 42$ and FN = 50 [142, 785, 794, 2711]. Molecular phylogenetic relationships among species of Vespertilionidae have been inferred from data on the ND1 gene of mtDNA, vWF gene of nuclear DNA and the SINE insertion [1459].

4. Ecology

Reproduction From June to August, females form maternity colonies [1903]. The size of a maternity colony varies from tens to more than a hundred individuals [455, 1903, 2253, 2300, 2352, 3135]. Parturition occurs during late June to early July in Iwate and Hokkaido [519, 1902]. Litter size is sometimes 1 but usually 2 [519, 1902, 2141]. Newborn infants have no or sparse hair, and closed eyes; eyes open at 4 to 11 days of age; forearm length at birth and its ratio to mother are 22.1 mm (range = 20.0–27.0; $n = 10$) and 0.36 ($n = 10$) on average respectively; forearm length increases rapidly up to 4 weeks after birth, and reaches nearly the same length as that of an adult up to 6 weeks of age; body weight at birth is 6.3 g on average (5.4–8.3; $n = 9$), and increases to 25 g up to 4 weeks of age; young begin to fly at about 40–45 days of age [1902]. Yearling females copulate in the first autumn but yearling males do not; the testes of adult males reach a maximum size in July and this state continues until the beginning of October; delayed ovulation and fertilization occur in the next spring [1903].

Lifespan Longevity is at least 6 years [1903].

Diet The stomach contents of bats caught in flight weighed 6.6 g on average (range = 3.1–9.9; $n = 4$) and the average ratio of stomach contents to BW was 15.1% [1903]. Mainly feeds on Coleoptela, Lepidoptera, Trichoptera, Ephemeroptera and Hemiptera [628, 1423]. In spring and from autumn to early winter, this species regularly preys on small birds of Passeriformes [628, 1423].

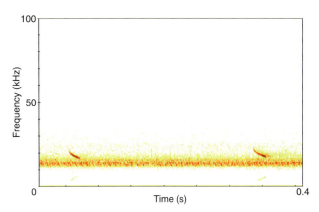

Fig. 1. Sonogram of a free-flying call around a roost recorded by D. Fukui in Hokkaido.

Habitat Roosts mainly in tree cavities [168, 455, 633, 1001, 1598, 1903, 2286, 2300, 2321, 2352, 3000, 3135, 3719, 3756, 4209, 4218], but occasionally uses slits under elevated bridges [3036], buildings [4209] and bird boxes [4046]. Environments around roosts vary from forest to urban areas. Feeding habitats are unknown.

Home range Not reported.

Behavior Nocturnal [1903], but often shows day-time feeding mainly in early winter [163, 2986]. A 74 km movement from a summer roost to a winter roost is known [3036].

During the period when females are involved in maternity colonies, males roost in solitary or form small colonies consisting of several to tens of individuals [1903]. Colonies consisting of several to tens of individuals of both sexes are formed from autumn (October) to spring (May) in tree cavities, buildings, and slits under elevated bridges [1903, 3036, 3719, 4218].

Natural enemies Owls (*Strix uralensis* and *Ninox scutulata*) [1903] and the Eurasian hobby (*Falco subbuteo*) [4149] are reported as predators.

Parasites
Ectoparasite
Hemiptera: *Cimex japonicus* [3025, 3761, 4024]. Acari: *Spinturnix acuminatus*, *Sp. vespertilionis*, *Liponyssus britanicus*, *Macronyssus flavus*, *Steatonyssus spinosus*, *St. superans*, *Ascoschoengastia narai*, *A. Mukohyamai* [1971, 2708, 3401, 3690, 3691, 3703, 3706, 4024]. Siphonaptera: *Ischnopsyllus elongatus*, *Nycteridopsylla galba* [2715, 2966, 4024].
Endoparasite
Trematoda: *Astiotrema* sp., *Acanthatrium rotundum*,

Lecithodendrium macrostomum, *Mesothatrium japonicum*, *Plagiorchis* (*Plagiorchis*) *koreanus*, *Pl.* (*Pl.*) *minutifollicularis*, *Prosthodendrium* (*Prosthodendrium*) *chilostomum*, *Pr. hurkovaae*, *Pr.* (*Pr.*) *yamizense*, *Pycnoporus rhinolophi* [1542, 1549, 1551, 1553, 3837, 4025]. Cestoda: *Staphylocystis bacillaris*, *Vampirolepis kawasakiensis*, *V. stenocephala*, *V. toohokuensis*, *V. multihamata* [3116, 3119, 3135, 4025]. Nematoda: *Capillaria pipistrelli*, *Molinostrongylus skrjabini longispicula* [1262, 3129, 4025].

Echolocation and social calls Echolocation calls are QCF calls of rather long duration in the search phase (Fig. 1), and FM calls in the scanning phase or in back-cluttered space. Mean values of EF and PF near the roost are 20.2 kHz (17.6–22.6, $n = 11$) and 21.1 kHz (20.2–23.3, $n = 11$), respectively (Hokkaido) [624].

Mean values of EF and PF in search phase are 17.7 kHz (16.1–18.9, $n = 10$) and 19.5 kHz (18.5–19.9, $n = 10$), respectively (Kagoshima Pref.) [660]. Social calls are unknown.

5. Remarks

Formerly, the birdlike noctule was treated as a subspecies of *N. lasiopterus*, *N. l. aviator* [499], but now it is treated as a valid species on the basis of morphological differences [404, 1910, 3272, 4208].

There is concern about population degradation by deforestation because roosts are usually found in large old trees.

D. FUKUI

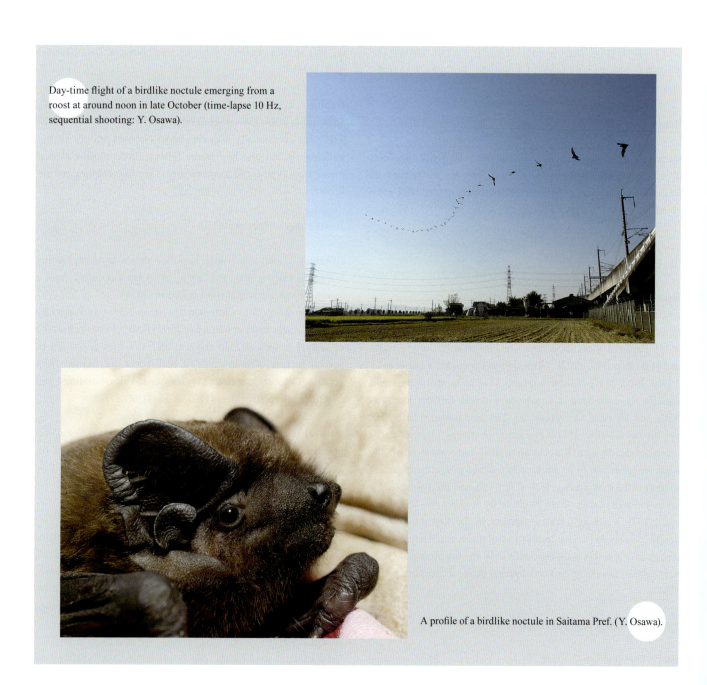

Day-time flight of a birdlike noctule emerging from a roost at around noon in late October (time-lapse 10 Hz, sequential shooting: Y. Osawa).

A profile of a birdlike noctule in Saitama Pref. (Y. Osawa).

Japanese bats and cherry blossoms (Y. Osawa).

Upper: A Ryukyu flying fox feeding on cherry blossoms, *Cerasus cerasoides*.

Lower: A Japanese pipistrelle flying around cherry blossoms.

Chiroptera VESPERTILIONIDAE 033

Red list status: EN (MEJ); V (MSJ); VU (IUCN)

Nyctalus furvus Imaizumi & Yoshiyuki, 1968

EN Japanese noctule JP コヤマコウモリ (ko yama koumori) CM 日本小山蝠 CT 日本小山蝠, 日本夜蝠
KR 작은멧박쥐 RS японская вечерница

An individual captured in Aomori Pref. (M. Mukohyama).

1. Distribution

Endemic to Japan. Recorded from Aomori [2290], Iwate [794, 1022, 1549, 1553, 3111, 4208], Fukushima [4194, 4204, 4208], Tochigi [4178] and Nagano Prefs. [4005]. Information is very limited; to date, the number of recorded individuals is less than 30. Red solid circles in the map denote sites where this species was observed in Japan.

2. Fossil record

Late Pleistocene fossils are reported in western Honshu [4158].

3. General characteristics

Morphology Middle-sized noctule bat. Color of dorsal hairs is dark brown; ears extend to about halfway from eyes to nostrils when laid forward; anterior border of conch is abruptly convex below, then nearly straight to broadly rounded off at the extremities; posterior border is convex throughout; antitragus is long and low, well marked off posteriorly, its anterior border extends to just below the angle of the mouth; tragus is very short, its greatest width is larger than its height, and widest at the basal third or quarter of the anterior border; wings are long and slender; the fifth finger is shortest, and its distal point reaches the basal third of the first phalanx of the fourth finger; wing membranes are inserted at the basal portion of the metatarsus of the first toe [4208].

External measurements (mean ± SD in mm with range) from Iwate Pref. are as follows [1022, 4208]. FA = 50.8 ± 0.29 (48.4–52.7; n = 11), HB = 79.8 ± 2.37 (77.1–84.2; n = 10), T = 50.5 ± 0.72 (47.5–54.0; n = 10), HFcu = 11.7 ± 0.25 (10.0–13.0; n = 10), Tib = 17.3 ± 0.15 (16.6–18.1; n = 11), E = 16.8 ± 0.40 (14.6–18.5; n = 10), Tr = 8.3 ± 0.19 (7.4–9.2; n = 10).

Dental formula   I 2/3 + C 1/1 + P 2/2 + M 3/3 = 34 [4208].

Genetics Chromosome: $2n$ = 44 and FN = 50–52 [142, 794, 2711].

4. Ecology

Reproduction Not reported.
Lifespan Not reported.
Diet Not reported.
Habitat Not reported.
Home range Not reported.
Behavior Not reported.
Natural enemies Not reported.

Parasites
Ectoparasite
Not reported.

Nyctalus furvus Imaizumi & Yoshiyuki, 1968

Endoparasite
Trematoda: *Prosthodendrium hurkovaae, Pr. (Prosthodendrium) urna, Pycnoporus rhinolophi* [1549, 1553]. Nematoda: *Molinostrongylus skrjabini* [3129].

Echolocation and social calls Only one release call was recorded. It consists of FM-QCF pulses (Fig. 1). Mean values of EF and PF at release are 27.5 kHz (25.9–29.5, $n = 8$) and 31.7 kHz (30.6–32.3, $n = 8$), respectively (Tochigi Pref.) [4178].

5. Remarks

In 1968, *N. furvus* was described as a new species based on specimens collected in 1961 from Iwate Pref., northeastern Honshu [1022]. In Japan, there were several reports of *N. noctula motoyoshii* before the description of *N. furvus* [1011, 1790, 1794, 2436, 2437]. However, none of these individuals were *N. noctula*, instead they were identified as *Vespertilio superans* (synonym of *V. sinensis*) when reconsidered after the description of *N. furvus* [1013].

Since morphological characteristics of *N. furvus* are similar to *N. noctula* and *N. plancyi*, some researchers formerly included

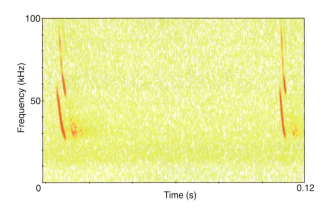

Fig. 1. Sonogram of a release call, recorded by S. Yoshikura in Tochigi Pref.

N. furvus in *N. noctula* [404, 409]. However, now it is treated as a valid species [3272] based on morphology [1022] and the difference in the numbers of chromosomes (*N. noctula*: $2n = 42$, *N. plancyi*: $2n = 36$ and *N. furvus*: $2n = 44$) [794, 1851, 2711, 2987].

D. FUKUI

Natural forest is a habitat of Japanese noctules (T. Yamamoto). A harp-trap is laid out on the river bank for a bat survey.

Chiroptera VESPERTILIONIDAE 034

Red list status: C-2, K for the Okinawa Pref. population (MSJ); LC (IUCN)

Pipistrellus abramus (Temminck, 1840)

EN Japanese pipistrelle　　JP アブラコウモリ（abura koumori）　　CM 东亚伏翼　　CT 東亞家蝠
KR 집박쥐　　RS восточный нетопырь

An individual hanging from a tree branch in Mie Pref. (A. Uemura).

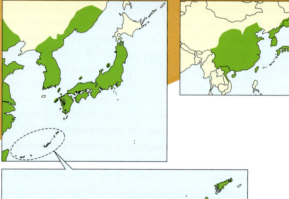

1. Distribution

Distributed in southern Ussuri area (Russia), China, Taiwan, Korea, Vietnam, Myanmar (Burma), India and Japan [3272]. In Japan, occurs in urban and suburban areas in all prefectures, including smaller islands; Tsushima Isls., Nagasaki Pref. [4199], Amami-ohshima Isl. [4208], Tokuno-shima Isl., Kagoshima Pref. [1919], Iriomote-jima Isl. [1202], Geruma-jima Isl., Miyako-jima Isl. [1971] and Okinawa-jima Isl., Okinawa Pref. [1202]. Although there had not been records from Hokkaido previously, 1 pregnant female was captured in Hakodate, in southernmost Hokkaido in 2001 [635].

2. Fossil record

Several fossils are reported from caves and fissure deposits of the Late Pleistocene in Honshu [4158].

3. General characteristics

Morphology The fur is not dense; color varies but in general the upper surface is grayish olive, slightly frosty; undersurface is light grayish brown; juveniles are darker than adults; ears are thin; the tragus is scarcely half as high as the ear; wings are rather narrow, with no special peculiarities of form; the wing membrane is inserted at the distal portion of the metatarsus of the first toe; the terminal tail vertebra is included in the uropatagium or slightly free from the membrane; the outer upper incisor is more than half as high as the inner incisor, with 2 cusps; the upper canine is strong but its secondary cusp weak; the anterior upper premolar (P^2) is visible from the outer side; its crown area is nearly equal to that of the outer incisor; the baculum is longer than that of *P. endoi*, and is curved remarkably, S-shaped [4208].

External measurements (in mm) are as follows. Hokkaido: FA = 35.1, HB = 52.4, T = 32.1, Tib = 14.7, E = 10.3, Tr = 5.2 for an adult female ($n = 1$) [635]. Saitama Pref. (mean ± SD): FA = 33.84 ± 0.82 ($n = 20$), HB = 50.02 ± 1.72 ($n = 20$), T = 39.53 ± 2.61 ($n = 20$), HFcu = 8.69 ± 0.61 ($n = 20$), Tib = 12.04 ± 0.66 ($n = 20$), E = 12.18 ± 0.50 ($n = 20$), Tr = 6.31 ± 0.62 ($n = 20$), Tokyo Metropolis: FA = 32.84 ± 1.08 ($n = 31$), HB = 48.31 ± 3.68 ($n = 30$), T = 36.36 ± 2.04 ($n = 30$), HFcu = 7.95 ± 0.75 ($n = 30$), Tib = 12.23 ± 0.94 ($n = 21$), E = 11.43 ± 0.94 ($n = 28$), Tr = 5.81 ± 0.67 ($n = 25$), Aichi Pref.: FA = 33.56 ± 0.97 ($n = 21$), HB = 54.26 ± 2.69 ($n = 21$), T = 36.11 ± 1.99 ($n = 21$), HFcu = 8.42 ± 0.64 ($n = 21$), Tib = 12.56 ± 0.95 ($n = 21$), E = 12.06 ± 0.61 ($n = 21$), Tr = 6.56 ± 0.35 ($n = 21$), Kagawa Pref.: FA = 32.90 ± 1.06 ($n = 14$), HB = 48.36 ± 3.61 ($n = 11$), T = 34.55 ± 1.46 ($n = 11$), HFcu = 8.14 ± 0.91 ($n = 14$), Tib = 13.04 ± 0.60 ($n = 14$), E = 11.25 ± 0.91 ($n = 14$), Tr = 6.21 ± 0.51 ($n = 14$), Ehime Pref.: FA = 32.55 ± 1.13 ($n = 11$), HB = 45.14 ± 3.41 ($n = 11$), T = 33.82 ± 2.59 ($n = 11$), HFcu = 7.45 ± 0.76 ($n = 11$), Tib = 11.82 ± 0.81 ($n = 11$), E = 10.23 ± 0.93 ($n = 11$), Tr = 4.55 ± 0.61 ($n = 11$) (no sex and no range given) [4208]. Tsushima Isls., Nagasaki Pref.: FA = 31.4 (*n* = 1), HB = 51.5 (*n* = 1), T = 37.0 (*n* = 1), HFcu = 11.0 (*n* = 1), Tib = 13.0 (*n* = 1), E = 11.0 (*n* = 1), Tr = 6.0 (*n* = 1) for a male [4199]. Iriomote-jima Isl., Okinawa Pref.: FA = 31.5, T = 35.0, HFcu = 8.0, Tib = 12.5, E = 11.0, Tr = 6.0 for a male (*n* = 1) [1202].

Dental formula I 2/2 + C 1/1 + P 2/2 + M 3/3 = 34 [4208]. Supernumerary mandibular incisors have been reported [2137]. Growth rings in the root dentine of the second lower molar indicate age in years [678].

Genetics Chromosome: $2n = 26$ and FN = 44 [142, 785, 2551, 2711, 3635]. Unlike other species in the genus *Pipistrellus*, the X chromosome is acrocentric [142]. The complete sequence of the mtDNA of this species is 16,976 bp [2472]. Molecular phylogenetic relationships with other species of Vespertilionidae have been inferred with the ND1 gene of mtDNA, vWF gene of nuclear DNA and the SINE insertion [1459]. Mitochondrial cytochrome *b* sequences and nuclear alpha-globin gene sequences have been determined [753, 2974].

4. Ecology

Reproduction The reproductive pattern is delayed fertilization type [3684]. After copulation in October, sperm are stored in the uterus and oviduct; the oocyte, containing a resting nucleus before and during the period of hibernation, enters the diffuse stage after awakening from hibernation; ovulation occurs at the end of April, and the number of ovulated ova, as well as the number of corpora lutea, is always less than 3 [3683, 3684]. Parturition occurs in early July in Kyushu [3683] and Kagawa Pref. [2215]. Gestational period is approximately 70 days [3683]. Litter size ranges between 2 and 4 with a mean of 2.6–3.0 [2141, 2214, 2508, 3683, 4213]. The mean body weight of neonatal infants is 0.86 g [2214], sex ratio (male/female) is 0.88 [678]. Parturition occurs mostly in the daytime [2214, 3678] and requires 4.5–5.0 hours [3678]. Ratios of neonatal to maternal values in FA, HB and HFcu are 29–31%, 44–51% and 80–90%, in Fukuoka Pref. [3678], and 30%, 55% and 80% in Kagawa Pref. [2223], respectively. Eyes open at 8–9 days of age [2215, 3678], pinnae are folded just after birth, and are flattened by 3 days of age in 80% of newborns [2215]. At 14 days of age, young are completely furred with black hair on the dorsum and long, dark fur over the scapulae; all young can fly freely at 25–30 days of age [2215, 3678]. The first curvature in the baculum develops at 10 days of age, the second occurs at 34 days of age, and the baculum is completely developed by 110 days of age when the testes become sexually mature [3666]. In late August, the size of infants reaches that of adults [678, 2216, 2225], but their body weight is significantly less until the autumn of their first year [2224]. In the same year, females enter estrus [3678] and mate within the natal colony [678]. The number of newborns decreases to about half over the duration of nursing [678, 682]. The disappearance rate from weaning to 1 year of age is 18–29% in females and 85–96% in males [682].

Two mothers and 2 infants in a narrow gap on the ceiling of an overpass in Kuki, Saitama Pref. (Y. Osawa). One of the mothers just left the roost.

In July, colonies consist of either all or mostly females, the number of adult males in a colony is less than 2 [156, 1574, 2225, 3678, 4082, 4213]. In late August, infants and adult females roost together [2225, 4082]. From May to September, adult males were found as solitary [4082]. In September, colonies comprise males and females; adiposity is observed earlier in females than in young or adult males [2225], and spermatogenesis is also observed [3678]. In late October, the mating season begins [3684], and the sex ratio of the colony is 1:1 [2225].

Colony size (mean ± SD with range) before the onset of flight in newborns is 5.1 ± 5.7 (1–26, *n* = 27) in Tokyo [4089] and 20 (4–36, *n* = 25) in Fukuoka Pref. [678]; colony size is sometimes over 100 individuals [3678]; after the onset of flight the colony size increases to 10.3 ± 13.1 (1–61, *n* = 22) in Tokyo [4089], and ranges from 11 to 27 in Fukuoka and Kagawa Prefs. [2222, 3678].

There is little exchange of individuals among colonies, even during the mating season in October [678].

Lifespan Most males die within 10 months [2225]. On the other hand, female disappearance rate from 1 to 3 years of age is relatively low, while that from 4 to 5 years of age is high; after weaning, male mortality always appears to be higher than in females [682]. The greatest age that has been observed is 5 years in females [682, 2225], and 3 years in males [682].

Diet Diet composition may be different between suburban areas and rice fields [895]; feeds mainly on Lepidoptera, Diptera and Hemiptera in Fukuoka urban areas [678], but on Diptera, Hemiptera and Hymenoptera above rice fields in Kyoto Pref. [895]. The diet composition varies seasonally, the main prey changing from Coleoptera in July to Diptera in October in Fukuoka Pref. [678], from Diptera in June to Hemiptera in September in Kyoto Pref. [895].

Habitat Roosts have been found in various narrow spaces of buildings under roof tiles, parapet caps, small roof covers above windows, in the spaces of aluminum frame windows, inside of sliding windows [2989, 4089] and in the spaces of bridges [649]. A few reports are from caves [2212, 4199, 4208] and in nests of the red-rumped swallow (*Hirundo daurica*) [910]. Roosts are not changed frequently [2225]. Forages in open areas such as parks, rice fields, reservoirs and rivers [678, 3199, 3626]. They have several feeding areas and change them one after another during the night [678, 3199].

Home range Not reported.

Behavior Foraging begins 10–30 minutes after sunset in summer, with a major peak in activity soon after sunset, and a minor peak just before sunrise from May to August, but in October feeding occurs only after sunset [678]. Emergence time is determined by a combination of several factors including energy consumption at each life stage, ambient temperature and insect abundance [2218, 2226]. Bats seldom emerge when the ambient temperature is lower than 15°C in Fukuoka, Fukuoka Pref. [678], and lower than 12°C in Kagawa Pref. [2218].

The increased ratio of body weight from summer to late autumn is 30.6% and 29.2% in adult females and males, respectively [678]. From late October to November, most individuals enter hibernation [2218, 3678]. Bats awaken from hibernation in March [678, 2218]. Weight loss during hibernation is measured as 21.3–25.3% in adult females, 20.0–26.0% in yearling females and 24.3–27.1% [2224] or 32% [678] in yearling males.

Natural enemies Long-eared owl (*Asio otus*) [375, 3363], Ural owl (*Strix uralensis*) [2227], peregrine falcon (*Falco peregrinus*) and jungle crow (*Corvus macrorhynchos*) [2213]. The impalement of an individual on a tree branch by a shrike (*Lanius bucephalus*) has been reported [164, 2213].

Parasites
Ectoparasite
Siphonaptera: *Ischnopsyllus elongatus*, *I. indicus* [2940, 2968]. Acari: *Steatonyssus longispinosus*, *Speleognathopsis bastini*, *Argas vespertilionis* [1311, 1759, 3703, 4031].
Endoparasite
Cestoda: *Vampirolepis urawaensis* [3121]. Trematoda: *Acanthatrium ovatum*, *Lecithodendrium macrostomum*, *Mesothatrium japonicum*, *Plagiorchis koreanus*, *Pl. minutifollicularis*, *Prosthodendrium*

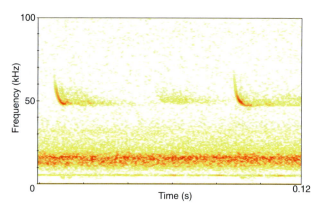

Fig. 1. Sonogram of a free-flying call, recorded by D. Fukui in Shizuoka Pref.

hurkovaae, *Pr. chilostomum*, *Pr. thomasi*, *Pr. parallelorchus*, *Pycnoporus heteroporus*, *Py. rhinolophi*, *Py. transversus* [1542, 1547–1551, 1553, 1554, 1559, 2219].

Echolocation and social call Echolocation calls are FM-QCF in the search phase (Fig. 1), and FM in the scanning phase or in back-cluttered space. Mean values ± SD of PF (FM-QCF) are 43.4 ± 0.50 kHz ($n = 8$) in Hokkaido and 42.9 ± 0.59 ($n = 8$) in Kanagawa Pref., respectively [635]. Mean values ± SD of PF and EF (FM) are 45.3 ± 1.23 (43.3–46.9; $n = 10$), 43.8 ± 0.99 (41.9–45.1; $n = 10$) in Kagoshima Pref. [660]. The echolocation behavior during exposure to artificial jamming sounds during flight was investigated [3419]. Social calls are unknown.

5. Remarks

Previously regarded as a subspecies of *P. javanicus* (e.g. [1713]), *P. abramus* is clearly separable from *P. javanicus*; the karyotype of *P. abramus* from Japan ($2n = 26$ and FN = 44) is different from those of *P. javanicus* from India ($2n = 36$) [480] and Malaysia ($2n = 34$) [3782], also the sutures of the cranium are different [4208]. There is no difference in the karyotype of *P. abramus* among Japan, Taiwan and Korea [1850, 4155].

This is the bat species most frequently encountered by humans in Japan. Because it roosts in buildings, conflicts sometimes arise due to fecal contamination and smell.

K. KAWAI

Chiroptera   VESPERTILIONIDAE

035

Red list status: VU (MEJ); V (MSJ); EN (IUCN)

Pipistrellus endoi Imaizumi, 1959

EN Endo's pipistrelle   JP モリアブラコウモリ (mori abura koumori)   CM 本州伏翼   CT 遠藤氏家蝠，本州伏翼
KR 삼림집박쥐   RS нетопырь Эндо

An individual captured in Ajigasawa, Aomori Pref. (M. Mukohyama).

1. Distribution

Endemic to Japan. The type locality is Horobe, Ashiro-cho, Ninohe, Iwate Pref. This species has been recorded from 15 prefectures, Aomori [1880, 2301], Iwate [3588, 4132, 4208], Miyagi [96, 1451], Tochigi [2070, 4085], Saitama [3359, 4208], Tokyo [1354, 3755], Kanagawa [4208], Niigata [4210], Ishikawa [4004, 4006], Yamanashi [1797, 1798, 4208], Nagano [4208], Gifu [1733, 1891, 4003], Shizuoka [3098], Nara [1643, 3904], Hiroshima [3714, 3716] and Ehime [41]. Red solid circles in the map denote the sites where this species was observed in Japan.

2. Fossil record

Fossils are reported from Late Pleistocene sediments in Honshu [4158].

3. General characteristics

Morphology   Like *P. abramus*, the ear and tragus are comparatively short and wide; the widest region of the ear is situated at the base of the anterior border of the tragus; the fur is glossy reddish brown; the ear and membrane are blackish brown; nostrils are naked, inter-narial septum broad; the plagiopatagium is inserted on the distal region of the metatarsus of the first toe; the calcar is long, but shorter than half of the lateral border of the uropatagium; the uropatagium has a keel and terminal lobe; the tail vertebra extends about 1 mm beyond the margin of the uropatagium [4208]. The baculum is nearly straight and slightly shorter than it of *P. abramus*, its straight-line length is 9–10 mm [1010].

External measurements (mean ± SD in mm with range) are as follows. FA = 31.98 ± 0.89 (29.8–33.4; n = 33), HB = 44.84 ± 2.08 (40.0–49.5; n = 35), T = 34.53 ± 2.52 (28.0–38.5; n = 35), HFcu = 7.41 ± 0.46 (6.5–8.2; n = 35), Tib = 11.51 ± 0.70 (10.0–12.5; n = 20), E = 11.46 ± 0.56 (10.0–12.5; n = 34), Tr = 5.85 ± 0.40 (4.9–6.5; n = 34) (sex and locality not designated) [1010]. Miyagi Pref.: FA = 32.6 ± 0.58 (32.1–33.7; n = 6) for females, FA = 32.0 ± 1.00 (30.8–34.2; n = 15) for males [1451].

Dental formula   I 2/2 + C 1/1 + P 2/2 + M 3/3 = 32 [4208].

Genetics   Chromosome: $2n$ = 36 and FN = 50 [142, 145, 2711]. The X chromosome is acrocentric like *P. abramus*, unlike others in the genus *Pipistrellus* [142].

4. Ecology

Reproduction   The reproductive pattern is the delayed fertilization type [3290]. Parturition may occur between mid-July to mid-August, as suggested by several observations: lactating females have been captured in late July in Ishikawa Pref. [4006] and in early August in Aomori Pref. [1880]. A pregnant female was captured in mid-August in Hiroshima Pref. [3716]. A

VESPERTILIONIDAE

Fur color of *P. endoi* (left) is darker than that of *P. abramus* (right) (M. Urano).

maternity colony was observed in a cavity in a large *Fagus crenata* [2287]. One female under captivity gave birth to 2 neonates [1451]. Development, sexual maturity and mating system are unknown.

Lifespan One female was kept for a total of 4 years and 4 months from just after birth until death [1451].

Diet Not reported.

Habitat Regarded as a forest dwelling species, this species is usually captured in natural forests, but there are 2 records from artificial forests in suburban Tokyo and Miyagi Pref. [1354, 1451]. Usually recorded from mixed forests with coniferous and broad-leaved trees at altitudes over 1,000 m [41, 1451, 1891, 3714, 3716, 4006, 4085]; it is sometimes recorded from altitudes lower than 500 m [1354, 1451, 1880, 3359, 3755]. Roosts in tree cavities [2287, 4210] and under tree bark [1354], solitary individuals have been found in buildings [1797, 1798, 4132] and rock crevices [1451].

A hibernating individual has been observed in a cavity in Japanese cedar (*Cryptomeria japonica*) in January [4210]. Individuals hibernating in solitary were observed from November to late March in rock crevices on Miyato Isl., Miyagi Pref. [1451].

Home range Not reported.

Behavior Not reported.

Natural enemies Not reported.

Parasites
Ectoparasite
Not reported.
Endoparasite
Trematoda: *Plagiorchis koreanus*, *Prosthodendrium hurkovaae*, *Pr. thomasi*, *Pr. urna*, *Pr. yamizense*, *Pycnoporus rhinolophi* [1549, 1550].

Echolocation and social call Not reported.

5. Remarks

This species is morphologically similar to *P. abramus*, but is slightly smaller in the length of the head and body, forearm and tibia. In suburban Tokyo, *P. abramus* and *P. endoi* exist sympatrically; however, they usually have an allopatric distribution, with *P. abramus* in urban areas, and *P. endoi* in mountainous regions. Careful examination is required in suburban areas, because it may be sympatric with *P. abramus*.

Reduction of natural forests poses a threat to this species.

K. KAWAI

Natural forest is a habitat of Endo's pipistrelles (T. Yamamoto).

Chiroptera VESPERTILIONIDAE

036

Red list status: EX (MEJ); Ex (MSJ); DD (IUCN)

Pipistrellus sturdeei Thomas, 1915

EN Sturdee's pipistrelle JP オガサワラアブラコウモリ (ogasawara abura koumori) CM 斯氏伏翼 CT 斯氏家蝠
KR 오가사와라집박쥐 RS бонинский нетопырь

The type specimen stored in the Natural History Museum, London.
© The Natural History Museum, London.

1. Distribution

Haha-jima Isl. (Hillsborough Isl.), the Ogasawara (Bonin) Isls., Japan. There has been no information since 1915 when the type specimen was collected.

2. Fossil record

Not reported.

3. General characteristics

Morphology Small size; ears are rather shorter and rounder than those of *P. abramus*, the inner margin is rounded at the base but not prominently convex, the tip is broadly rounded off, the outer margin is flattened above and slightly convex below with a well-marked anti-tragal lobule; the tragus resembles that of *P. abramus* but is shorter, its broadest point opposite the lower third of its inner margin; wings to the base of the toes; a narrow post-calcarial lobule; tail with the usual 7 vertebrae, only its extreme tip projecting; the fur color is blackish; wings are dark brown, without marked marginal lines; narrow delicate skull with a wide space between the canine and posterior premolar (P^4); incisors are short, the anterior prominently bicuspid, outer incisor surpassing the second cusp of the inner one [4208].

Type specimen: female, slightly mature; external measurements (in mm) are as follows [3567]. FA = 30, HB = 37, T = 31, Ear = 7.7, Tr = 3.

Dental formula As in *P. abramus* (I 2/2 + C 1/1 + P 2/2 + M 3/3 = 34), but smaller [3567].

Genetics Not reported.

4. Ecology

Reproduction Not reported.
Lifespan Not reported.
Diet Not reported.
Habitat Not reported.
Home range Not reported.
Behavior Not reported.
Natural enemies Not reported.
Parasites Not reported.
Echolocation and social call Not reported.

5. Remarks

This species is morphologically similar to *P. abramus*, but the forearm is shorter, skull is more slender, ears shorter, tragus broader and wing membrane darker [4208]. Based on skull characters, this species does not belong to the *P. javainicus* group, but to the *P. coromandra* group [3540].

K. KAWAI

Chiroptera VESPERTILIONIDAE

Red list status: LP for the Honshu and Shikoku populations as *B. leucomelas* (MEJ); V as *B. leucomelas* (MSJ); LC as *B. leucomelas* (IUCN)

Barbastella darjelingensis (Hodgson, 1855)

EN eastern barbastelle   JP チチブコウモリ (chichibu koumori)   CM 亚洲宽耳蝠   CT 寬耳蝠
KR 명주박쥐   RS азиатская широкоушка

An individual captured in Iwate Pref. (M. Mukohyama).

1. Distribution

Distributed from Egypt, Eriteria, northern Iran, and the Caucasus to Afghanistan, the Pamirs, India, Nepal, western China and Japan [3272]. In Japan, recorded from Hokkaido [93, 451, 455, 840, 1448, 1704, 1706, 1926, 1927, 1936, 2458, 2473, 2525, 2534, 2780, 3017, 3061, 3070–3072, 3075, 3078, 3086, 3166, 3732, 3736, 3951, 3953, 3954, 3956, 4058, 4208], Honshu (Iwate [1020, 1549, 3111, 3588, 3823, 4196, 4208], Fukushima [1564, 1577], Saitama [2435, 3823, 4208, 4221], Tokyo [1598], Kanagawa [3969], Nagano [1422, 3023], Gifu [1897, 3985], Shizuoka [915, 1794, 3098] Prefs.), Shikoku (Ehime [1, 3686] and Kochi [3912, 3913] Prefs.) and Kunashiri (Kunashir) Is. [1460]. In Tokyo, it was recorded only in the late 19th century. Red solid circles in the map denotes sites where this species was observed in Japan.

2. Fossil record

Late Pleistocene fossils are reported in the Akiyoshi-dai cave system, Yamaguchi Pref. [4157, 4158].

3. General characteristics

Morphology Color of hairs is dark blackish brown, those of upperparts tipped with light glossy ivory, those of under-parts tipped with a paler brown; ears are large, broad and triangular, joined at the front; tragus is large, somewhat triangular in outline, its width is widest above the middle of the anterior margin; posterior margin of tragus is concave on the upper quarter of the margin then convex in the widest region; wings are broad and inserted at the base of the first toe [4208].

External measurements (mean ± SD in mm with range) from Hokkaido are as follows [93, 451, 455, 840, 1448, 1936, 2458, 3061, 3070–3072, 3732, 3736, 3951, 3953, 3954, 3956, 4208]. For females, FA = 41.0 ± 0.79 (39.0–42.0; n = 19), HB = 55.3 ± 3.89 (50.9–61.6; n = 8), T = 48.2 ± 3.62 (41.0–52.0; n = 8), HFcu = 8.7 ± 1.40 (6.6–10.2; n = 8), Tib = 19.4 ± 0.68 (18.4–20.4; n = 9), E = 16.5 ± 1.58 (14.1–19.0; n = 8), Tr = 8.5 ± 1.58 (5.0–9.8; n = 9). For males, FA = 41.7 ± 0.83 (40.3–43.0; n = 11), HB = 49.7 ± 2.31 (47.0–51.0; n = 3), T = 44.8 ± 1.04 (44.0–46.0; n = 3), HFcu =

6.8 ± 0.25 (6.5–7.0; $n = 3$), Tib = 20.0 ± 0.91 (19.0–20.7; $n = 3$), E = 14.1 ± 1.71 (12.5–15.9; $n = 3$), Tr = 7.6 ± 1.37 (6.0–8.5; $n = 3$).

Dental formula   I 2/3 + C 1/1 + P 2/2 + M 3/3 = 34 [4208].

Genetics   Chromosome: $2n = 32$ and FN = 50 [142, 2711, 3686]. Molecular phylogenetic relationships among species of Vespertilionidae have been inferred from data on the ND1 gene of mtDNA and the SINE insertion [1459].

4. Ecology

Reproduction   A female with an embryo was observed once [4196].

Lifespan   Not reported.

Diet   Not reported.

Habitat   Roosts in artificial caves [1897, 3023, 3969, 3985], tunnels [3070, 3098] and rock crevices [4221]. Feeding habitats are unknown.

Home range   Not reported.

Behavior   Not reported.

Natural enemies   Not reported.

Parasites
Ectoparasite
Diptera: *Basilia truncata endoi* [2186, 2473]. Siphonaptera: *Nycteridopsylla* spp. [1422].
Endoparasite
Nematoda: *Plagiorchis* (*Plagiorchis*) *rhinolophi*, *Prosthodendrium* (*Prosthodendrium*) *urna* [1549, 1553].

Echolocation and social calls   Not reported.

5. Remarks

Formerly, the Japanese population was included in *B. leucomelas* [3272]. There is an opinion that the subspecies *B. l. darjelingensis* occurs from the Caucasus to China and Japan [1020, 4208]. Some researchers have also suggested that the Japanese population is distinctive at the species level because of the rather invariant dimensions of the ear [4208]. Furthermore, a recent molecular phylogenetic and morphological study suggested that the Asian population is a distinct species from *B. leucomelas* [274, 4246]. In this book, we adopt the species name *B. darjelingensis* for the Japanese population. However, further study is needed to reach a conclusion because *B. darjelingensis* may contain cryptic species as indicated by genetic data [4246].

D. FUKUI

Eastern barbastelles hibernating in a tunnel (Akiyoshi Sato).

Chiroptera VESPERTILIONIDAE

Red list status: LP for populations in Kinki District and westward (MEJ); R as *P. auritus* (MSJ); LC (IUCN)

Plecotus sacrimontis G. M. Allen, 1908

EN Japanese long-eared bat **JP** ニホンウサギコウモリ (nihon usagi koumori)
CM 日本大耳蝠 **CT** 日本大耳蝠, 長耳蝠, 兔耳蝠 **KR** 일본토끼박쥐 **RS** Японский ушан

A flying individual in a forest in the Kushiro region, Hokkaido (H. Nakajima).

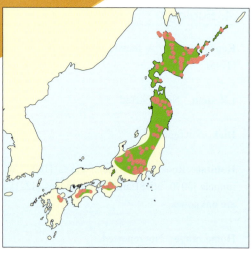

1. Distribution

Endemic to Japan [3300]. Recorded from Hokkaido [92, 93, 267, 450–452, 455, 625, 630, 633, 839, 840, 1084, 1092, 1446, 1448, 1704, 1706, 1896, 1926, 1927, 1935, 1936, 2458, 2473, 2525, 2711, 2780, 3017, 3061, 3063, 3064, 3070, 3072, 3075, 3076, 3086, 3123, 3166, 3191, 3316, 3469, 3528, 3542, 3732, 3953, 3954, 4047, 4055, 4056, 4058, 4208, 4263], Honshu (Aomori [2301, 2670, 2711, 3124, 3786, 4208], Iwate [3107, 3111, 3124, 3588, 3823, 4131, 4135, 4208], Miyagi [96], Akita [224, 2711, 3823], Yamagata [2731, 3622, 4208], Fukushima [1575, 1577], Tochigi [2836, 4083, 4093, 4208], Gunma [3018, 3110, 3124, 3221, 3823, 4203, 4204, 4208], Saitama [2932, 3124, 4208], Tokyo [1356, 1598, 4208], Kanagawa [1821], Niigata [605], Toyama [2305, 4208], Ishikawa [3005, 3984, 3987], Yamanashi [785, 3242, 3340, 4200, 4208], Nagano [380, 1287, 3698, 3993, 4208], Gifu [1595, 1889, 2092, 3988, 3990], Shizuoka [2434, 3098, 3266, 3745, 4177, 4200, 4208], Mie [2994, 3228], Nara [291, 960, 1643, 1892, 1894, 3107, 3110, 3124, 3136, 4208], Wakayama [1921, 3752], Okayama [3124] Prefs.), Shikoku (Tokushima [1630, 2220] and Ehime [41, 2228, 3126, 4208] Prefs.), Kyushu (Oita Pref. [665]), Kunashiri (Kunashir) [1460] and Etorofu (Iturup) [1460] Isls. Red solid circles in the map denote sites where this species was observed in Japan.

2. Fossil record

Pleistocene fossils are reported in Honshu [4157, 4158].

3. General characteristics

Morphology There are 2 types of fur color (pale and dark types); the back of the pale type is pale yellow gray, anterior of back is washed brown gray, underpart is ivory; the back of the dark type is dark brown gray or burnt umber, underpart is paler brown; ears are very large, their length much longer than the head, and they are joined across the forehead; tragus is simple and erect, its length about half as high as the conch, anterior border is straight below and slightly convex above, tip is narrowly rounded, posterior border is faintly concave above and distinctly convex below; wings are broad, and join the foot at the base of the outer toe [4208].

External measurements (mean ± SD in mm with range) from 5 localities are as follows [2220, 4208]. For females, Hokkaido: FA = 41.6 ± 2.49 (37.7–44.1; n = 5), HB = 52.9 ± 1.93 (50.0–54.0; n = 4), T = 49.8 ± 2.22 (47.0–52.0; n = 4), HFcu = 12.0 ± 0.56 (11.5–12.6; n = 3), Tib = 19.2 ± 0.85 (18.3–20.5; n = 5), E = 37.5 ± 0.45 (36.9–38.0; n = 4), Tr = 18.0 ± 0.82 (17.0–19.0; n = 4), Saitama Pref.: FA = 42.1 ± 0.72 (41.2–43.3; n = 7), HB = 53.4 ± 3.25 (48.0–57.5; n = 7), T = 49.1 ± 1.07 (48.0–51.0; n = 7),

HFcu = 10.9 ± 0.63 (10.0–12.0; *n* = 7), Tib = 19.1 ± 1.16 (17.7–21.0; *n* = 7), E = 41.7 ± 0.49 (41.0–42.0; *n* = 7), Tr = 18.4 ± 0.85 (17.5–19.5; *n* = 7), Yamanashi and Shizuoka Prefs.: FA = 42.6 ± 1.08 (40.7–44.1; *n* = 15), HB = 55.2 ± 4.56 (47.0–62.0; *n* = 17), T = 48.3 ± 2.89 (42.0–52.0; *n* = 17), HFcu = 11.5 ± 0.80 (9.8–12.5; *n* = 17), Tib = 20.7 ± 1.19 (19.5–24.5; *n* = 17), E = 41.4 ± 1.10 (39.0–43.5; *n* = 17), Tr = 19.8 ± 0.74 (18.5–21.0; *n* = 17), and Tokushima Pref.: FA = 41.4 ± 0.79 (40.0–42.8; *n* = 13), HB = 50.1 ± 2.61 (44.5–53.0; *n* = 13), T = 51.7 ± 2.30 (48.1–55.8; *n* = 13), HFcu = 10.4 ± 0.44 (9.9–11.1; *n* = 13), Tib = 20.8 ± 0.47 (20.2–21.8; *n* = 13), E = 36.7 ± 1.88 (33.2–39.5; *n* = 13), Tr = 18.8 ± 0.93 (17.0–20.0; *n* = 13). For males, Hokkaido: FA = 39.9 ± 1.13 (38.3–41.0; *n* = 4), HB = 49.9 ± 4.19 (45.5–53.0; *n* = 4), T = 46.8 ± 3.30 (43.0–50.0; *n* = 4), HFcu = 11.2 ± 0.76 (10.5–12.0; *n* = 3), Tib = 17.6 ± 1.31 (16.0–18.8; *n* = 4), E = 37.3 ± 0.81 (36.5–38.0; *n* = 4), Tr = 17.1 ± 0.70 (16.3–18.0; *n* = 4), Yamanashi and Shizuoka Prefs.: FA = 41.3 ± 1.36 (38.0–43.5; *n* = 16), HB = 51.6 ± 3.20 (44.5–55.0; *n* = 16), T = 48.1 ± 3.08 (42.0–55.5; *n* = 16), HFcu = 11.1 ± 0.87 (9.0–12.5; *n* = 16), Tib = 20.2 ± 1.85 (15.5–22.5; *n* = 16), E = 39.1 ± 1.58 (36.5–41.5; *n* = 16), Tr = 18.5 ± 1.24 (16.0–20.5; *n* = 16), and Nagano Pref.: FA = 40.4 ± 0.91 (39.0–41.5; *n* = 6), HB = 49.9 ± 3.80 (46.0–56.0; *n* = 6), T = 46.4 ± 1.56 (44.0–48.0; *n* = 6), HFcu = 10.9 ± 0.49 (10.0–11.5; *n* = 6), Tib = 21.3 ± 1.89 (19.8–24.5; *n* = 5), E = 38.9 ± 1.11 (38.0–41.0; *n* = 6), Tr = 18.0 ± 1.84 (15.0–19.5; *n* = 5).

Dental formula   I 2/3 + C 1/1 + P 2/3 + M 3/3 = 36 [4208].

Genetics   Chromosome: $2n$ = 32 and FN = 50–54 [142, 785, 2711, 3635]. Molecular phylogenetic relationships among species of Vespertilionidae have been inferred from data on the ND1 gene of mtDNA, vWF gene of nuclear DNA and the SINE insertion [1459]. Genetic distance among regions are relatively low [665]. Sequencing primers for the mitochondrial cytochrome *b*, control region and 16S rRNA gene are available [2974, 3300].

4. Ecology

Reproduction   Maternity colonies consisting of pregnant and immature females are formed in caves and buildings from April to August. The size of a maternity colony varies from several to several dozen individuals [4130, 4177]. In Yamanashi and Iwate Prefs., females give birth to a single newborn infant in July [3340, 4130]. The sex ratio (male/female) of young is 1.18 [4177]. Newborns are about 25% of adult mass, naked, and with closed eyes; they reach adult size and start feeding themselves at about 4 weeks after birth [4130]. Females give birth to their first offspring at the end of their first or second year [4177]. Age of sexual maturity is unknown in males.

Lifespan   Not reported.

Diet   Forages on Lepidoptera and Neuroptera, which are mainly 15–20 mm in length [4203]. Prey are carried to the night roost [4203].

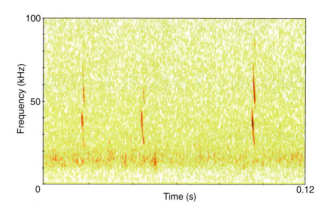

Fig. 1. Sonogram of release call, recorded by D. Fukui in Hokkaido.

Habitat   Roosts in caves [605, 960, 1356, 1630, 2220, 2228, 3136], tunnels [267, 2305, 3070, 3987, 3988], tree cavities [2321] and buildings [267, 1446, 2458, 3017, 3166, 3340, 3588]. They use caves and buildings as night roosts [3990, 4203]. Feeding habitats are unknown.

Home range   Not reported.

Behavior   Not reported.

Natural enemies   The Eurasian hobby (*Falco subbuteo*) is reported as a predator [4149].

Parasites
Ectoparasite
Diptera: *Basilia truncata endoi*, *Nycteribia pleuralis*, *Penicillidia jenynsii* [2187, 2473]. Acari: *Pteracarus submedianus*, *Spinturnix plecotinus* [3698, 3706].
Endoparasite
Trematoda: *Brachylaima* sp., *Plagiorchis* sp. [3469]. Cestoda: *Vampirolepis ozensis* [3110, 3123, 3126]. Nematoda: *Rictularia rhinolophi* [3469].

Echolocation and social calls   Echolocation calls are low intensity FM calls with a second harmonic [625, 1084] (Fig. 1). Mean values of EF and SF at release calls are 21.5 kHz (19.6–23.5, *n* = 5) and 47.0 kHz (45.0–51.3, *n* = 5), respectively. PF is indistinct (Hokkaido) [625]. Social calls are unknown.

5. Remarks

Formerly, the Japanese population was included in the European *P. auritus* [3272]. Some researchers have suggested that it is an endemic subspecies of *P. auritus*, *P. a. sacrimontis*, because of the differences in the length of the FA, E, Tr and condylobasal [4208]. Recent genetic and morphological analyses indicate that the Japanese long-eared bat is a distinct species, *P. sacrimontis* [3300].

When a roost is formed in buildings, there may be a conflict between bats and the building owner.

D. FUKUI

Chiroptera VESPERTILIONIDAE
039

Red list status: DD (MEJ); K as *Pipistrellus savii* (MSJ); LC as *Pipistrellus savii* (IUCN)

Hypsugo alaschanicus (Bobrinskii, 1926)

EN Alashanian pipistrelle   JP クロオオアブラコウモリ（kuro ooabura koumori）
CM 山伏翼   CT 沙維氏家蝠, 薩氏伏翼   KR 검은집박쥐   RS алашаньский кожановидный нетопырь

An individual captured in Otaru, Hokkaido (D. Fukui).

1. Distribution

Mongolia, China, Russian Far East, Korea and Japan [941]. In Japan, 10 individuals from 3 prefectures have been collected: Hokkaido [35, 636, 839, 1007, 1695], Aomori Pref. [2294, 4208] and Tsushima Isls., Nagasaki Pref. [4208]. Resident in Hokkaido [636]. Red solid circles in the map denote sites where this species was observed in Japan.

2. Fossil record

Not reported.

3. General characteristics

Morphology Color of basal hairs is dark reddish brown, nearly black; ears are broad, posterior border is slightly concave near the tip, anterior border is remarkably convex at the base; the antitragus is small; the tragus is half as high as the ear, and very wide; tubular nostrils protrude beyond the upper lip, large naris is naked, internarial septum is wide; the length of the calcar with keel is half of the lateral border of the uropatagium; the tibia length is about 36–37% of FA; terminal 1 or 2 tail vertebrae project from the margin of the uropatagium [4208].

External measures (mean ± SD in mm with range) are as follows. Hokkaido: FA = 36.3 ± 0.26 (36.0–36.5; n = 3), HB = 52.0 and 45.0, E = 13.0 and 13.0, Tr = 6.0 and 3.0, T = 39.0 and 36.0, Tib = 13.3 ± 0.52 (13.0–13.9; n = 3), HFcu = 8.0 and 9.0 for females; FA = 35.8 and 36.0 (n = 2), HB = 47.0 and 50.3 (n = 2), E = 12.9 (n = 1), Tr = 5.1 (n = 1), T = 37.6 (n = 1), Tib = 15.0 and 14.8 (n = 2), HFcu = 7.8 (n = 1) for males [636, 839, 1007, 1695]. Aomori Pref.: FA = 36.5 and 36.9 (n = 2), HB = 54.0 (n = 1), E = 13.1 (n = 1), Tr = 6.5 (n = 1), T = 39.0 (n = 1), HFcu = 8.0 (n = 1) for females [2294, 4208]. Tsushima Isls., Nagasaki Pref.: FA = 36.6, HB = 41.5, E = 14.0, Tr = 5.5, T = 40.5, Tib = 14.0, HFcu = 9.0 for a female, FA = 35.7, HB = 5, E = 13.0, Tr = 6.0, T = 38.0, Tib = 14.1, HFcu = 7.4 for a male [4208].

Dental formula I 2/3 + C 1/1 + P 2/2 + M 3/3 = 34. Presence of P^2 is unstable, often vestigial. If the premolars are absent, the total number of teeth is reduced by 2 [1695, 4208].

Genetics Chromosome: not investigated in Japanese specimens. The karyotype of those in Korea and Greece is $2n$ = 44 and FN = 50 [2802, 3782]. Apparently this species differs from "true" *Pipistrellus* species with $2n$ = 44 in the state of chromosome nos. 11 and 23 [3782].

4. Ecology

Reproduction In Aomori Pref., 2 individuals were observed in an artificial, concrete dome construction, and 1 of these was a lactating female [4217].

Lifespan Not reported.
Diet Not reported.

Habitat In Hokkaido, more than 12 individuals used a building as a day roost from summer to autumn. The top of the wall (about 7 m high) was utilized as a night roost [636]. Collected individuals of this species have been found in buildings, in both summer and winter, suggesting that this species might be present throughout the year in Japan.

Home range Not reported.
Behavior Not reported.
Natural enemies Not reported.
Parasites Not reported.

Echolocation and social call Echolocation calls are FM-QCF in the search phase (Fig. 1), and FM in the approach and terminal phases in the feeding buzz [636]. Mean values of frequency in consecutive pulses vary as follows (14 calls); PF: 33.9 to 36.4 kHz, FMAX: 41.0 to 62.0 kHz, FMIN: 32.0 to 34.4 kHz [636]. Social calls are unknown.

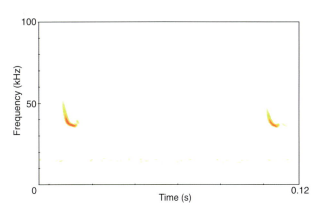

Fig. 1. Sonogram of an echolocation call, recorded by D. Fukui in Hokkaido.

5. Remarks

Formerly included as a subspecies of *H. (Pipistrellus) savii*. However, there is a strong argument for it to be recognized as a valid species of *Hypsugo* (*Pipistrellus*) [941]. *Pipistrellus coreensis* from Korea and Japan (Tsushima Isls.) was treated as a distinct species from *P. savii*, and that from Hokkaido was designated as *P. savii* [1007, 4208]. However, the Korean specimens did not differ substantially from other material of *H. alaschanicus*, and therefore *coreensis* has been regarded as *H. alaschanicus*, possibly as a separate subspecies [941].

K. KAWAI

A sediment mass including many Late Pleistocene fossils of vespertilionid bats found in the Akiyoshi-dai Plateau, Yamaguchi Pref. (M. Ishida).

Chiroptera VESPERTILIONIDAE

040

Red list status: DD (MEJ); LC (IUCN)

Vespertilio murinus Linnaeus, 1758

EN parti-colored bat   JP ヒメヒナコウモリ（hime hina koumori）   CM 双色蝙蝠   CT 欧州蝙蝠，姫雛蝙蝠
KR 북방애기박쥐   RS двухцветный кожан

An individual captured in Hokkaido (H. Nakajima).

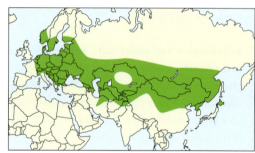

1. Distribution

Distributed from central, northern and eastern Europe and Siberia to the Pacific coast. The northern limit is above 60°N, and the southern limit passes through the Balkans, northern Iran, northern Pakistan, China and Korea. In Japan, known from 6 localities in Hokkaido, Aomori and Ishikawa Prefs. [1453, 1461, 1692, 3062]. Red solid circles in the map denote sites where this species was observed in Japan.

2. Fossil record

Not reported.

3. General characteristics

Morphology Slightly smaller than *V. sinensis*. Dorsal hairs are dark brown with golden tips, producing its "frosted" appearance. The only Japanese species having 2 pairs of teats [1692].

External measures (in mm) in Japan are as follows [1453, 1692, 3062]. Ozora, Hokkaido (8 adult females, mean ± SD): FA = 44.5 ± 0.79, Rebun Isl., Hokkaido (1 female): FA = 45.0, HB = 55.0, T = 46.0, Tib = 18.0, E = 15.0, Tr = 4.0, Chitose, Hokkaido (1 female): FA = 44.8, HB = 64.7, T = 36.9, HFcu = 9.3, Tib = 17.3, E = 16.0, Tr = 4.7, Haboro, Hokkaido (1 male): FA = 44.2, and Minmaya, Aomori Pref.: (1 male): FA = 46.0, HB = 62.0, T = 41.0, HFcu = 10.5, Tib = 18.0, E = 14.0, Tr = 5.5.

Dental formula   I 2/3 + C 1/1 + P 1/2 + M 3/3 = 32 [35].

Genetics Chromosome: $2n = 38$ and $FN = 50$ in Germany [3781]. Genetic distances among Japanese, European and Russian individuals are relatively low [1453]. The mitochondrial ND1 and cytochrome *b* genes from 7 Japanese individuals have been sequenced [1453, 1461] and are available in DDBJ.

4. Ecology

Reproduction A newly found maternity colony consists of 60 adult females [1692]. Parturition occurs during late June to early July [1692]. About 30 days after birth, new born young become independent, and adult females and young individuals emigrate from their colony in this order [1692].

Lifespan Not reported.

Diet Not reported.

Habitat The only known maternity colony in Japan is formed in a building (old gymnasium) surrounded by a rural area [1692]. Feeding habitats are unknown.

Home range Not reported.

Behavior Not reported.

Natural enemies Not reported.

Parasites
Ectoparasite
Siphonaptera: *Ischnopsyllus needhami* [3087].
Endoparasite
Trematoda: *Plagiorchis koreanus*, *Prosthodendrium chilostomum*, *Pr. ascidia* [1544].

Echolocation and social calls Echolocation calls are QCF calls of rather long duration in the search phase, and FM calls in the scanning phase or in back-cluttered space (Fig. 1). Mean values of FMIN and PF near the roost are 22.1 kHz (19.0–26.7, $n = 187$) and 26.1 kHz (22.6–35.2, $n = 187$), respectively (Hokkaido) [1692].

5. Remarks

Before a maternity roost was found in 2011, it was thought that this species occured in Japan as a result of migration, transportation by storms and accidental translocation from the continent [1453, 1692].

D. FUKUI

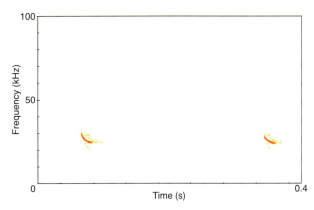

Fig. 1. Sonogram of a free-flying call near a roost, recorded by D. Fukui in Hokkaido.

A newly-discovered roost of *V. murinus*. Several tens individuals roost inside walls of an old gymnasium in Hokkaido (D. Fukui).

Chiroptera VESPERTILIONIDAE
041

Red list status: R as *V. superans* (MSJ); LC (IUCN)

Vespertilio sinensis (Peters, 1880)

EN Asian parti-colored bat   **JP** ヒナコウモリ (hina koumori)   **CM** 中华蝙蝠，东方蝙蝠   **CT** 霜毛蝠
KR 안주애기박쥐   **RS** восточный двухцветный кожан

An individual captured in the Tokachi region, Hokkaido (H. Nakajima).

1. Distribution

Distributed in China, Mongolia, Ussuri (Russia), the Korean Pen. and Taiwan [3272]. In Japan, recorded from Hokkaido [92, 267, 450–452, 455, 630, 633, 790, 839, 840, 1092, 2715, 2780, 3017, 3029, 3081, 3082, 3087, 3469, 3528, 4053, 4055, 4208], Aomori [1546, 1547, 1550, 1880, 2285, 2298–2300, 2302, 2670, 3110, 3401, 3703, 3823, 4208], Iwate [521, 790, 2295, 2302, 2985, 3588, 3823], Miyagi [96, 2295, 3323, 3823], Akita [1565, 2296], Yamagata [1413, 2423, 4197, 4208], Fukushima [1565, 1577, 2120, 3105, 4208], Ibaraki [4043, 4081], Tochigi [1011, 1559, 1735], Gunma [1353, 3018], Saitama [1013, 2437, 2740, 2743, 2931, 2932, 3098, 3124, 3823, 4208], Chiba [200], Tokyo [913, 1351, 1355, 1356, 1598, 3593, 3756], Kanagawa [169, 1134, 1414, 3969, 3970, 4208], Niigata [605, 617, 3142], Ishikawa [3718, 4006], Fukui [1556, 3134, 4208], Yamanashi [3746], Nagano [3023], Gifu [1891], Shizuoka [3098, 3266, 3823], Mie [2999, 3002], Shiga [1556, 2220, 3127, 3553], Osaka [2324, 3749], Hyogo [1556, 3754], Nara [291, 1643], Wakayama [790, 1550, 3111], Tottori [2672, 2673], Okayama [493], Hiroshima [1302, 3716], Ehime [41, 2521], Fukuoka [681, 790, 1550, 1554, 3111, 3822, 3823, 4208], and Kumamoto [4208] Prefs. Red solid circles in the map denote sites where this species was observed in Japan.

2. Fossil record

Late Pleistocene fossils are reported in western Honshu [4158].

3. General characteristics

Morphology Slightly larger than *V. murinus*. Color variation among individuals is remarkable. Dorsal hairs are reddish brown or blackish brown with whitish tips, producing a "frosted" appearance; ears are thick, their length and width are equal, the posterior border is straight; ear reaches its widest point at 1/3 of the length of anterior margin from the base; low antitragus developed from the basal region of the posterior margin and extends nearly to the angle of the mouth; tragus is short, its length is about half as long as the ear, its anterior margin is concave, posterior margin is convex from below tip to middle portion, then nearly straight to the base, tip is broadly rounded; wings are rather narrow, plagiopatagium is inserted at the distal end of the metatarsus [4208]. Mean wing aspect ratio and wingtip shape index are 6.68 and 1.25 ($n = 20$), respectively [631]. The structures in the nasal cavity suggests this species possesses a very keen sense of smell [3263].

External measurements (mean ± SD in mm with range) from 4 localities are as follows [4208]. Hokkaido: FA = 48.3 ± 2.27 (43.7–52.8; $n = 10$), HB = 72.6 ± 3.32 (65.9–79.2; $n = 10$), T =

Vespertilio sinensis (Peters, 1880)

43.2 ± 2.98 (37.2–49.1; $n = 10$), HFcu = 13.2 ± 0.53 (12.1–14.2; $n = 10$), Tib = 16.6 ± 2.29 (12.1–21.2; $n = 10$), E = 17.3 ± 1.03 (15.3–19.4; $n = 10$), Tr = 7.6 ± 0.34 (6.9–8.3; $n = 10$), Aomori Pref.: FA = 50.2 ± 1.30 (47.6–52.8; $n = 28$), HB = 70.1 ± 4.39 (61.3–78.9; $n = 25$), T = 43.7 ± 3.09 (37.5–49.9; $n = 25$), HFcu = 12.8 ± 1.13 (10.5–15.1; $n = 26$), Tib = 18.3 ± 1.16 (16.0–20.7; $n = 23$), E = 18.0 ± 0.98 (16.1–20.0; $n = 25$), Tr = 8.0 ± 0.85 (6.3–9.7; $n = 24$), Saitama Pref.: FA = 48.4 ± 1.24 (45.6–51.3; $n = 18$), HB = 67.8 ± 4.19 (59.4–76.1; $n = 14$), T = 43.3 ± 2.07 (39.2–47.5; $n = 14$), HFcu = 11.5 ± 1.28 (8.9–14.0; $n = 13$), Tib = 17.6 ± 0.63 (16.3–18.9; $n = 13$), E = 17.1 ± 1.42 (14.3–20.0; $n = 14$), Tr = 7.3 ± 1.02 (5.3–9.4; $n = 12$), and Fukui Pref.: FA = 48.7 ± 1.39 (45.9–51.5; $n = 10$), HB = 67.2 ± 3.05 (61.1–73.3; $n = 10$), T = 41.1 ± 3.28 (34.5–47.7; $n = 10$), HFcu = 12.5 ± 0.71 (11.1–13.9; $n = 10$), Tib = 16.2 ± 1.21 (13.8–18.6; $n = 10$), E = 17.3 ± 1.25 (14.8–19.8; $n = 10$), Tr = 6.8 ± 0.72 (5.3–8.2; $n = 10$).

Dental formula I 2/3 + C1/1 + P 1/2 + M 3/3 = 32 [4208]. The complete number of deciduous teeth in newborn young is i 2/3 + c 1/1 + pm 2/2 = 22 [681].

Genetics Chromosome: $2n = 38$ and FN = 50–54 [142, 790, 2546, 2711, 2713]. Molecular phylogenetic relationships among species of Vespertilionidae have been inferred from data on the ND1 gene of mtDNA, vWF gene of nuclear DNA and the SINE insertion [1459]. Sequencing primers for the mitochondrial cytochrome *b* gene are available [2974].

4. Ecology

Reproduction From spring (May) to summer (August), maternity colonies are formed only by pregnant females [627, 681, 839, 840, 2285]. The size of a colony ranges from tens to several thousands of individuals [627, 681, 839, 840, 1353, 2285, 2298, 2300, 2302, 2743, 2985]. In Fukuoka Pref., parturition occurs from late June to early July; litter size (mean ± SD) is 2.0 ± 0.30 (range = 1–3, $n = 34$), and the sex ratio (male/female) of young is 0.76 [681].

At birth, the newborn are completely naked, with closed eyes; hair fully covers the body within about 20 days of age; eyes open at 8 to 12 days of age; deciduous teeth are lost at 24 to 26 days of age [681]; forearm length at birth and its ratio between young and postpartum females are 14.5 mm ($n = 14$) and 0.30, respectively; the body mass of newborn young is about 12% of the mass of postpartum females; forearm growth is almost complete at the weaned stage at 5–6 weeks old, but is slightly shorter than that of adults; body weight rapidly increases until 24 days of age, the increase is retarded during the weaning period from 24 to 33 days of age, and then self-feeding recovers the retardation; after the young become independent, adult females, young males and young females emigrate from their colony in this order; individuals of both sexes reach sexual maturity in their first autumn [681]. Prolonged storage of spermatozoa occurs during the period of hibernation [3460].

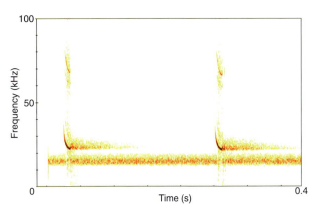

Fig. 1. Sonogram of a free-flying call near a roost, recorded by D. Fukui in Hokkaido.

Lifespan Not reported.

Diet Mainly feeds on Coleoptera, Lepidoptera and Diptera [623]. In August, they prey on Coleoptera (86%), Lepidoptera (5%) and others (9%), and the size of taken prey is about 20 mm in body length [681]. Daily food intake of pregnant and lactating females is estimated at about 6.0 g (ca. 33% of pre-feeding BW) and 6.4 g (ca. 36% of pre-feeding BW), respectively [681].

Habitat Roosts in tree cavities [267, 521, 627, 2300, 2302], buildings [267, 627, 839, 840, 1565, 2298, 2300, 2302], slits under elevated bridges [617, 1353], caves [1355, 2931, 2985, 3124], tunnels [1356] and rock crevices [681, 3822]. In winter, they hibernate in buildings and rock crevices in roosts of 1 to several tens of individuals [2740, 2745, 3023, 3970]. Non-reproductive individuals prefer warmer roosts in summer but cooler roosts in autumn [638]. Environments around roosts vary from forest to urban areas. Feeding habitats are unknown.

Home range Not reported.

Behavior Nocturnal [681]. Feeding activity of pregnant females begins about 10 minutes after sunset and continues until about 02:00 hr, and that of lactating females has 2 peaks with a major one soon after sunset and a minor one before sunrise [681].

During the maternity season, males appear to form small colonies consisting of a few individuals [521].

In captivity under near-natural conditions, body weight decreased from 27 to 19 g during hibernation [683].

Natural enemies The Eurasian hobby (*Falco subbuteo*) is reported as a predator [2298, 2985].

Parasites & Pathogenic organisms
Ectoparasite
Hemiptera: *Cimex japonicus* [3398, 4262]. Acari: *Androlaelaps casalis*, *Argas vespertilionis*, *Ascoschoengastia narai*, *A. Mukohyamai*, *Microtrombicula tenmai*, *M. vespertilionis*, *Speleognathopsis bastini*, *Steatonyssus spinosus*, *S. superans* [1311,

3398, 3401, 3703]. Siphonaptera: *Ischnopsyllus* (*Ischnopsyllus*) *needhami*; *I.* (*I.*) *obscurus* [2715, 3087, 3398].

Endoparasite

Trematoda: *Acanthatrium* (*Acanthatrium*) *hitaense*, Lecithodendliidae gen. spp., *Mesothatrium japonicum*, *Plagiorchis* (*Plagiorchis*) *latus*, *Pl.* (*Pl.*) *koreanus*, *Pl.* (*Pl.*) *minutifollicularis*, *Pl.* (*Pl.*) *rhinolophi*, *Prosthodendrium chilostomum*, *Pr. parallelorchus*, *Pr. postacetabulum*, *Pr. thomasi*, *Pycnoporus acetabulatus*, *Py. heteroporus* [1546, 1547, 1549, 1550, 1554, 1556, 1559, 3130, 3469]. Cestoda: *Vampiolepis multihamata* [3105, 3110, 3130, 3134]. Nematoda: *Molinostrongylus rhinolophi*, *Capillaria* sp. [3129, 3130].

Pathogenic organisms

Virus strains that are similar to Flaviviruses but distinct from Japanese encephalitis virus (JEV) have been isolated [2120].

Echolocation and social calls Echolocation calls are QCF calls of rather long duration in the search phase, and FM calls in the scanning phase or in back-cluttered space (Fig. 1). Mean values of EF and PF near the roost are 21.8 kHz (18.1–23.2, $n = 21$) and 24.2 kHz (21.8–26.5, $n = 21$), respectively (Hokkaido) [624]. Mean values of EF and PF in search phase are 21.9 kHz (18.4–24.8, $n = 10$) and 24.5 kHz (21.2–26.5, $n = 10$), respectively (Fukuoka Pref.) [660]. Social calls unknown.

5. Remarks

Formerly, 2 species, *V. superans* and *V. orientalis*, were recognized in Japan based on morphological differences of the baculum [3823]. However, *V. orientalis* was found to be a synonym of *V. sinensis* (formerly *V. superans*) because this difference appears only with age [4156]. The specific name *superans* was commonly applied to this species until it was demonstrated that *sinensis* has priority over *superans* [940]. There are reports of 6 cases of middle-distance migration (130–230 km) [1565, 2295, 2296].

Because of conflicts with house owners, colonies formed in buildings are occasionally exterminated by the owners. There are reports of successful transplant of colonies to a "bat house" [2298, 2299].

D. FUKUI

Asian parti-colored bats roosting in a slit under elevated Shinkansen railway. Reddish brown bats are mothers, and others with frosty pale brown fur are infants (about 30 days after birth) (Y. Osawa).

Chiroptera   VESPERTILIONIDAE

042

Red list status: CR as *M. formosus* (MEJ); K as *M. formosus* (MSJ); LC as *M. formosus* (IUCN)

Myotis rufoniger (Tomes, 1858)

EN red and black Myotis   JP クロアカコウモリ (kuroaka koumori)   CM 緋鼠耳蝠   CT 緋鼠耳蝠，黒赤蝙蝠
KR 붉은박쥐   RS красно-чёрная ночница, восточноазиатская ночница

An individual captured on the Tsushima Isls., Nagasaki Pref. (O. Takahashi).

1. Distribution

Distributed in North Korea, South Korea, the Tsushima Isls. (Japan), China (Fujian, Jiangxi, Jilin, Shanghai, and Sichuan), Taiwan, Laos, and Vietnam [422]. In Japan, only 11 individuals have been recorded in the Tsushima Isls. [1452, 3758]. Red solid circles in the map denote sites where this species was observed in Japan.

2. Fossil record

Not reported in Japan.

3. General characteristics

Morphology   Fur is thick, wooly; the region from the top of the head to the back of the neck is reddish orange; the posterior portion of the back is grayish brown, while the ventral surface is vivid yellowish orange; the basal zone of the dorsal and ventral fur is dark gray; ears are reddish orange, both on frontal and posterior surfaces, the margins black; the plagiopatagium and uropatagium are reddish orange both above and below, the spaces between the third and fourth, and between the fourth and fifth fingers have large, triangular, blackish patches, the inner side of the fifth finger with a large quadrangular blackish patch; upper margin of the propatagium, thumb and hind foot are black; the dorsal wing membrane, upper arm, basal half of propatagium, basal plagiopatagium, and basal half of uropatagium are furred sparsely; on ventral surface, the area from the knee to elbow on the plagiopatagium and basal half of uropatagium are furred; the thumb of the upper leg is comparatively long, about 11 mm; the wing membrane is wide, the plagiopatagium is inserted in the basal portion of the first toe; the calcar is more than half of the free margin of the uropatagium, without a keel and terminal lobe; the tail is as long as about 79% of HB, its terminal vertebra is free from the margin of the uropatagium [4208].

External measurements for sex-unknown adults (mean ± SD in mm with range) are as follows [4208]. FA = 48.8 ± 1.75 (46.3–50.0; $n = 4$), HB = 62.8 ± 6.49 (57.4–70.0; $n = 3$), T = 47.2 ± 4.25 (43.0–51.5; $n = 3$), HFcu = 12.0 ± 0.00 (12.0–12.0; $n = 4$), Tib = 23.4 ± 0.48 (23.0–24.0; $n = 4$), E = 16.4 ± 0.53 (16.0–17.0; $n = 3$), Tr = 8.5 and 9.0.

Dental formula   I 2/3 + C 1/1 + P 3/3 + M 3/3 = 38 [4208].
Genetics   Chromosome: not reported for Japanese populations. In Korea, $2n = 44$ and FN = 50 [4155].

4. Ecology

Reproduction   Parturitions may occur during June to July [3758]. A female with a juvenile was captured in August 1968 [3758]. Sexual maturation and mating season are unknown.
Lifespan   Not reported.
Diet   Not reported in Japan.
Habitat   All individuals have been recorded in forests, not in caves. On the Korean Pen., this species concentrates in caves during autumn, preliminary to winter hibernation, and disperses to other habitats such as forests and shrubby areas for summer [3888]. However, in Japan there is no information on seasonal changes in habitats. Day roosts are also unknown in Japan.
Home range   Not reported.
Behavior   Not reported.
Natural enemies   Not reported.
Parasites   Not reported.
Echolocation and social call   Not reported.

5. Remarks

In Japan, there had not been any information on this species since 1968, however, 4 individuals were captured in September 2006 [1452]. Formerly included in *M. formosus*. However, Csorba et al. (2014) determined that *M. formosus* contained several morphologically distinct forms, and dealt "*M. formosus*" in Tsuhima Isl. with *M. rufoniger* as a member of subgenus *Chrysopteron* [422].

K. KAWAI

Chiroptera VESPERTILIONIDAE

Red list status: R (MSJ); DD (IUCN)

Myotis frater Allen, 1923

EN fraternal Myotis   JP カグヤコウモリ (kaguya koumori)   CM 长尾鼠耳蝠   CT 華南鼠耳蝠
KR 긴꼬리수염박쥐   RS длиннохвостая ночница

A hanging individual in the Kushiro region, Hokkaido (H. Nakajima).

1. Distribution

Distributed in eastern Siberia, Ussuri, Krasnoyarsk (Russia) to the northern part of the Korean Pen., Heilungjiang (China), south-eastern China and Japan [3272, 3888]. In Japan, reported from 16 prefectures: Hokkaido [267, 326, 451–455, 630, 633, 840, 1092, 1445, 1448, 1449, 1700, 1704, 1706, 1881, 1896, 1922, 1926, 1927, 1935, 2359, 2458, 3017, 3061, 3063, 3072, 3073, 3077, 3086, 3166, 4047, 4051, 4053–4056, 4208, 4211], Aomori [1008, 1880, 2288, 2290, 2301, 3789, 4195, 4208], Iwate [520, 2303, 3588, 3981, 4208], Miyagi [914], Akita [2301], Yamagata [2423], Tochigi [1734, 1735, 4084, 4086], Gunma [3018, 4208], Saitama [3359], Niigata [603–605], Toyama [3117], Ishikawa [4004, 4006], Yamanashi [4208], Nagano [3993, 4208], Gifu [1733, 1890, 3991, 4003] and Shizuoka [3266, 4208]. In Hokkaido, this species is common in forests from low to high altitude, but in Honshu, it is limited to forests at high altitude and it is a relatively rare species. Red solid circles in the map denote sites where this species was observed in Japan.

2. Fossil record

Fossils are reported from fissure deposits from the Late Pleistocene in Yamaguchi Pref. [4158].

3. General characteristics

Morphology Fur is soft and woolly without gloss, slightly curled, fur color varies with age; in juveniles, the basal coloration of hairs on the dorsal and ventral surfaces is dark grayish brown, tips of the hairs on the former are pale brown, those of the latter are pale gray; in subadults and adults, hair tips on the dorsal surface are dark brown, with a darker base, hairs on the ventral side are dark grayish brown at the base, and yellowish brown to pale beige at the tip; ears are rather short, not extending to the anterior end of the muzzle when laid forward; the tragus is about 56% of ear length with a posterior basal small lobe; the membrane is almost naked, base of uropatagium is sparsely covered with hairs on both the dorsal and ventral surfaces; the plagiopatagium is inserted at the distal portion of the metatarsus of the first toe [4208]. An observation suggested the possibility that hair colour in females changes from dark to lighter brown during lactation, and returns to the darker colour afterwards [604]. A female with reduced pigmentation was captured in the Rishiri Isl. [3057].

External measurements (mean ± SD in mm with range) are as follows. Hokkaido: FA = 39.2 ± 0.97 (37.8–40.5; n = 19), HB = 46.5 ± 1.14 (44.0–49.0; n = 19), T = 42.6 ± 2.16 (38.0–46.0; n = 19), HFcu = 10.8 ± 0.43 (10.0–11.8; n = 19), Tib = 20.1 ± 0.45 (19.0–21.0; n = 19), E = 12.4 ± 0.46 (11.5–13.5; n = 19), Tr = 6.6 ± 0.36 (6.0–7.5; n = 19) for females; FA = 38.2 ± 0.94 (36.9–39.9; n = 10), HB = 48.0 and 52.0 (n = 2), T = 44.0, HFcu = 10.0 and 11.0 (n = 2), Tib = 18.5 ± 0.94 (17.5–20.0; n = 4), E = 12.0 and 13.5 (n = 2), Tr = 5.0 and 7.5 (n = 2) for males [452, 3077, 4208]. Honshu (Aomori, Iwate, Tochigi, Niigata, Ishikawa, Yamanashi, Nagano, Gifu and Shizuoka Prefs.): FA = 39.4 ± 1.24 (36.7–41.4; n = 20), HB =

Myotis frater Allen, 1923

45.9 ± 3.04 (43.5–49.5; n = 3), T = 39.6 ± 3.91 (36.8–44.1; n = 3), HFcu = 8.0 and 10.5 (n = 2), Tib =20.1 ± 1.59 (15.1–22.0; n = 19), E = 11.7 ± 1.36 (10.0–13.3; n = 4), Tr = 6.5 ± 1.23 (5.0–8.0; n =4) for females; FA = 38.8 ± 1.14 (35.7–40.6; n = 21), HB = 51.7 ± 2.95 (46.5–56.5; n = 10), T = 42.1 ± 1.66 (39.0–45.0; n = 11), HFcu = 10.3 ± 0.73 (8.6–11.5; n = 11), Tib = 19.2 ± 1.43 (16.0–22.2; n = 19), E = 13.3 ± 0.44 (12.5–14.0; n = 11), Tr = 7.3 ± 0.58 (6.5–8.5; n = 9) for males [604, 1733, 1735, 3789, 4003, 4006, 4084, 4208].

Dental formula I 2/3 + C 1/1 + P 3/3 + M 3/3 = 38 [35]. Deciduous teeth are i 2/3 + c 1/1 + pm 2/2. All deciduous teeth except the posterior one of the upper premolar erupt over the mucosa at birth [1922].

Genetics Chromosome: $2n$ = 44 and FN = 52 [142, 796]. Molecular phylogenetic relationships within Vespertilionidae have been inferred with the ND1 gene of mtDNA [1459]. The status of Japanese and East Asian bats of the genus *Myotis* has been discussed based on ND1 and cytochrome *b* sequences of mtDNA [1458].

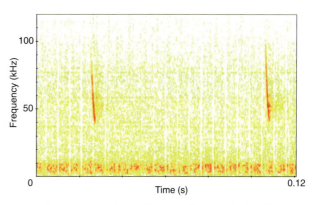

Fig. 1. Sonogram of a release call, recorded by M. Ishida in Rishiri Isl.

4. Ecology

Reproduction Parturitions occur from mid-June to mid-July in Hokkaido [1922, 3474]. In other regions of Hokkaido, parturition may occur between mid-July and mid-August, as suggested by several observations: lactating females and an adult male in a tree cavity have been captured in late July, and lactating females and juveniles have been captured in middle August [4051].

Maternity colonies consisting of more than 100 females have been found in buildings in Hokkaido [1922, 3474]. The largest maternity colony was estimated to consist of at least 500 individuals of adult females and juveniles in a barn in Hokkaido [3474]. Sexual maturation in females is estimated to occur between 15–16 months of age [1922]. Litter size is 1 [1922]. Sex ratio (male/female) of neonates is 0.83; ratios of neonates to maternal size in FA, HB and T are 43.4%, 64.6% and 42.2%, respectively [1922].

Lifespan One male has been observed in the same tunnel from 1994 to 2005 in summer; this observation suggests that the maximum longevity is over 11 years and the individual has used the same roost continuously [4006].

Diet Not reported.

Habitat As day roosts, they use tree cavities [448, 2321, 4051], caves (colonies of 15 to 25 individuals) [605, 4195, 4208], tunnels [3077, 4006], slits of bridges [92], buildings [450, 840, 1922, 2178], bat boxes [3528] and broken nests of the Asian house martin (*Delichon dasypus*) [3993]. A night roost was observed in the wall of a building [2423].

Hibernation sites are unknown.

Home range Not reported.

Behavior Feeding behavior has been observed in an open area along a street, several individuals flying back and forth covering 30–40 m along the street at heights of 1–2 m from the ground [520]. In Hokkaido, the number of individuals in summer roosts begins to reduce from early August, with none remaining by mid-September; emergence time from maternity roosts is 20–45 minutes after sunset [3474]. In summer, roost sites of females and males are segregated from each other. Colonized adult males in a test adit or a tunnel have been observed in August, 12 individuals in Hokkaido [3077], 4 individuals in Ishikawa Pref. [4006] and 10–12 individuals from early July to early August in Niigata Pref. [604]. At the end of August, the same number of females and males (4–5 individuals each) congregated in the same place in a tunnel [604]. Behavior from mid-September to spring is unknown because there are very few records during this period [457]. Observations of drinking behaviors on the wing under captivity was reported [458].

Natural enemies Predation by the Eurasian hobby (*Falco subbuteo*) are reported [4149].

Parasites

Ectoparasite

Diptera: *Basilia rybini japonica*, *Nycteribia pleuralis*, *Penicillidia monoceros* [3054, 3068, 3069, 3084, 3469]. Hemiptera: *Cimex* cf. *japonicus* [3028].

Endoparasite

Cestoda: *Vampirolepis kaguyae*, *V. yoshiyukiae* [3119, 3121, 3122]. Nematoda: *Molinostrongylus skrjabini*, *M. tsuchiyai* [3129, 3469]. Trematoda: *Acanthatrium* (*Acanthatrium*) *isostomum*, *A. mukooyamai*, *Mesothatrium japonicum*, *Palalecithodendrium ovimagnosum*, *Plagiorchis* (*Plagiorchis*) *koreanus*, *Pl. rhinolophi*, *Pl. vespertilionis*, *Prosthodendrium* (*Prosthodendrium*) *chilostomum*, *Pr. thomasi*, *Pr. urna* [1542, 1544, 1549, 1550, 1552, 1554, 1556].

Echolocation and social call Echolocation calls are FM calls (Fig. 1). The mean values of EF and PF from multiple pulses (release calls of 2 individuals) are 43.8 ± 2.11 kHz (40.8–50.3) and 53.8 ± 3.20 kHz (47.7–58.4), respectively (Rishiri Isl.) [1084]. Social calls unknown.

5. Remarks

Four subspecies are recognized, *M. f. frater* (southeastern China), *M. f. longicaudatus* (eastern Siberia and southwestern Tadjikistan), *M. f. bucharensis* (central China), *M. f. kaguyae* (Japan) and another subspecies may occur in Russia [941]. *Myotis kaguyae*, which was described in Japan [1008], is reclassified as a subspecies of *M. frater*, with a slightly shorter tail and tibia than those of the type specimen of *M. frater* [4195].

K. KAWAI

Chiroptera VESPERTILIONIDAE

Red list status: VU (MEJ); R as *M. mystacinus* (MSJ); LC as *M. brandtii gracilis* (IUCN)

Myotis gracilis Ognev, 1927

EN Ussuri whiskered bat   JP ウスリホオヒゲコウモリ (usuri hohige koumori)
CM 乌苏里鼠耳蝠   CT 須鼠耳蝠   KR 우수리큰수염박쥐   RS дальневосточная ночница Брандта

An individual captured in Shibetsu, Hokkaido (K. Kawai).

1. Distribution

Distributed in eastern Siberia, Sakhalin, Kamchatka, the Ussuri area Amur, Trans-Baikalia, Hokkaido, Kunashiri (Kunashir) Isl. and Etorofu (Itrup) Isl. Distribution area in Japan is limited to northern and eastern Hokkaido [454, 1449, 1455, 1460, 1704, 1926, 2359, 2780, 3017, 3063, 3071, 3072, 3077, 3316, 3952, 4208]. Sometimes occurs sympatrically with the sibling species *M. ikonnikovi* [454, 1449, 1455, 1704, 2780, 3071, 4206]. Red solid circles in the map denote sites where this species was observed in Japan.

2. Fossil record

Not reported in Japan.

3. General characteristics

Morphology   General form is similar to *M. ikonnikovi*, but the venation pattern of the tail membrane is different and of the "straight type" (Fig. 1) [1457, 1699]. Ears are rounded off at the top, not truncated, and with the anterior margins turned outward above center; the tragus is straight or slightly turned outward and its length is under half of ear length; the dorsal pelage is unicoloured gray-brown, ventral pelage is pale black; the anteorbital ridge is distinct, the lacrymal foramen is not exposed; depth of braincase is low, less than 73% of width; the basicochlear fissure is long, the distance between I^3 and C^1 is longer than the diameter of C^1; P^3 and P_3 perfectly on tooth row; upper molars with protoconules; the plagiopatagium is inserted at the distal portion of the metatarsus of the first toe [4208].

External measurements (mean ± SD in mm with range) are as follows. Hokkaido: FA = 35.2 ± 0.91 (33.1–37.0; n = 19), HB = 46.3 ± 3.39 (38.0–51.0; n = 15), T = 35.4 ± 1.96 (32.0–38.0; n = 15), HFcu = 8.9 ± 0.69 (8.0–9.9; n = 15), Tib = 15.2 ± 0.78 (13.7–16.0;

Fig. 1. Venation pattern of the tail membrane, a "straight type" (N. Kondo).

$n = 15$), E = 14.3 ± 0.88 (12.0–15.5; $n = 15$), Tr = 6.9 ± 1.04 (5.0–8.6; $n = 15$) for females [454, 4208]; FA = 33.7 ± 1.24 (32.0–35.5; $n = 6$), HB = 44.0 and 38.0 ($n = 2$), T = 36.0 and 35.0 ($n = 2$), HFcu = 9.0 ($n = 1$), Tib = 13.8 ($n = 1$), E = 13.1 ($n = 1$), Tr = 7.3 ($n = 1$) for males. Kunashiri (Kunashir) Isl.: FA = 34.82 and 34.85 ($n = 2$) for females, FA = 33.43 (31.9–34.18; $n = 4$) for males. Etorofu (Itrup) Isl.: FA = 34.55 (33.58–35.27; $n = 6$), HB = 47.0 and 42.0 ($n = 2$), T = 37.0 and 41.0 ($n = 2$), HFcu = 8.0 and 7.0 ($n = 2$), Tib=15.2 and 15.3 ($n = 2$), E = 15.0 and 15.0 ($n = 2$), Tr = 7.0 and 8.0 ($n = 2$) for females, FA = 34.89 (32.75–35.77; $n = 6$) for males [1460, 3077, 3952, 4208].

Dental formula   I 2/3 + C 1/1 + P3/3 + M 3/3 = 38 [4208].

Genetics   Chromosome: not reported. The status of the Japanese and East Asian bats of the genus *Myotis* has been discussed with this species based on ND1 and cytochrome *b* sequences of mtDNA [1458; described as Japanese *M. mystacinus*] (see Remarks). Several haplotypes are recognized [1457]. However, genetic distances among *M. gracilis* are very small.

4. Ecology

Reproduction   Maternity colonies have been found in old wooden buildings [839; described as *Myotis* sp., 1449, 1690, 3059]. The maternity colonies consist almost entirely of adult females but they sometimes included non-pregnant females, and the number of females in the colonies are more than 50 [1690, 3059]. Parturition may occur in early July [839, 3059], lactating females and juveniles are observed from late July to early August in Hokkaido [1449, 3059, 3077, 4056]. Litter size is 1 [839]. Development, sexual maturity and mating system are unknown in Japan.

Lifespan   Not reported.

Diet   Not reported.

Habitat   This species has been considered a forest dwelling species. Roosts, except for maternity colonies in old wooden houses [839, 1449, 1690, 3059], are unknown. Capture points can be divided into 2 types: agricultural and forest areas at low altitudes [1449, 1704, 1926, 2359, 3017, 3063, 3071, 3072, 3077, 3952, 4056, 4208], and forests at high altitudes [454, 1455, 2780, 4208]. Hibernation sites are unknown.

Home range   Not reported.

Behavior   Not reported.

Natural enemies   Not reported.

Parasites
Ectoparasite
Siphonaptera: *Ischnopsyllus obscurus*, *Myodopsylla* sp. [3087].
Hemiptera: *Cimex* cf. *japonicus* [3028].
Endoparasite
Trematoda: *Plagiorchis* (*Plagiorchis*) *latus* [1553].

Echolocation and social call   Not reported.

5. Remarks

Three cryptic species, *M. ikonnikovi*, *M. yesoensis* and *M. gracilis*, were recognized in Hokkaido [4208]. On the other hand, *M. yesoensis* was described as a synonym of *M. ikonnikovi*, and *M. gracilis* as a synonym of *M. mystacinus* [1893]. A reexamination of this systematic treatment of the *M. mystacinus* group treated *M. gracilis* from Hokkaido as a subspecies of *M. brandtii*, *M. b. gracilis* [275]. Molecular phylogenetic analyses suggest that *M. gracilis* and *M. brandtii* are far from *M. mystacinus* and closely related to each other. All are clearly distinguished from each other by genetic distance [1458]. In response to recent studies, usage of "*M. mystacinus*" in Japan was dropped in favor of *M. gracilis*, following Yoshiyuki (1989) [4208] and based on information that the taxon was distinct from *M. mystacinus* [1916]. The genetic distance between *M. gracilis* and *M. brandtii* suggests that *M. gracilis* should be considered the valid species and not a subspecies of *M. brandtii* (as suggested by [275, 1713, 3272]), conforming to Yoshiyuki (1989) [1457]. In this volume, we follow this latest revision of Japanese bat taxonomy. Additionally, there are differences between populations on either side of the Ob River in the *M. brandtii* group. The animals captured east side of the Ob are genetically identical to Far East *M. gracilis* and differ from *M. brandtii*. The type locality of *M. sibiricus* Kastschenko, 1905 appears to be within the range of *M. gracilis* under consideration. On this basis, there is an opinion that the animals captured east side of the Ob river must be regarded as valid for *M. sibiricus* pending subsequent examination of topotypes [1748].

K. KAWAI

Chiroptera VESPERTILIONIDAE

045

Red list status: LP for the populations in the Kii Pen. and Chugoku District, as *M. i. hosonoi* (MEJ); V for the Hokkaido population, R for the Honshu population (MSJ); LC (IUCN)

Myotis ikonnikovi Ognev, 1912

EN Ikonnikov's Myotis   JP ヒメホオヒゲコウモリ (hime hohige koumori)   CM 伊氏鼠耳蝠   CT 伊氏鼠耳蝠
KR 쇠큰수염박쥐   RS ночница Иконникова

An individual captured in the Kamikawa region, Hokkaido (H. Nakajima).

1. Distribution

Distributed in Primorye and North Korea to Lake Baikal, the Altai Mts., and Mongolia, Northeast China, Sakhalin, Kunashiri (Kunashir) Isl. and Honshu and Hokkaido, Japan [1460, 1720, 3272]. In Japan, reported from 20 prefectures: Hokkaido [42, 93, 267, 451–455, 624, 625, 630, 632, 633, 1447–1449, 1455, 1687, 1699, 1701, 1704, 1706, 1881, 1935, 1936, 2343, 2359, 2458, 2780, 3017, 3061, 3063, 3064, 3067, 3071–3075, 3077–3082, 3085, 3086, 3166, 3316, 3372, 3732, 3736, 3954, 3956, 3957, 4053, 4055, 4058, 4206, 4208], Aomori [1880, 2288, 2290, 2291, 2301], Iwate [518, 2293, 3588, 4133, 4208], Akita [2301], Yamagata [1723, 2423], Fukushima [1564, 1577, 4204], Tochigi [1734, 1735, 2070, 4084, 4086–4088, 4092, 4093], Gunma [3018, 4204], Saitama [3359], Niigata [4259], Toyama [3117], Ishikawa [4006], Yamanashi [3035, 3242, 4208], Nagano [2066, 2178, 3035, 3993, 4204, 4208], Gifu [3991, 4003], Shizuoka [3035, 3098, 3266], Mie [3227, 3228], Hyogo [1591], Nara [1643] and Hiroshima [3714, 3716]. Sympatric distributions with similar-sized *Myotis* species have been reported: *M. gracilis* in Hokkaido [454, 1449, 1455, 1704, 2780, 3071, 4206] and *M. pruinosus* in Honshu (Tochigi [1734, 4084], Ishikawa [3991] and Shizuoka [3099] Prefs.). Red solid circles in the map denote sites where this species was observed in Japan.

2. Fossil record

One fossil is reported from fissure deposits of the Late Pleistocene in Yamaguchi Pref. [4158].

3. General characteristics

Morphology General form is similar to *M. gracilis*, but tail membrane venation pattern is different and of the "dog-leg type" (Fig. 1) [1457, 1699]. A metallic luster on the dorsal surface of the fur is unclear or not presented; the plagiopatagium is inserted at the base of the first toe; the rostrum is short; in lateral view, the outline of the skull is curved at the superior orbital margin and rises up at the neurocranium; the zygomatic arch is narrow, nearly straight in lateral view; the canine and premolars are relatively large, molars are relatively small; the mental foramen of the mandible is posterior to canine [35]. Wing has intermediate aspect ratio and wingtip shape index [631].

External measurements (mean ± SD in mm with range) are as

Myotis ikonnikovi Ognev, 1912

Fig. 1. Venation pattern of the tail membrane, a "dog-leg type" (N. Kondo).

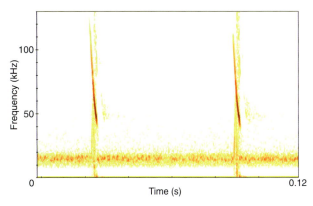

Fig. 2. Sonogram of a release call, recorded by D. Fukui in Hokkaido.

follows. Hokkaido: FA = 33.4 ± 0.99 (31.6–35.0; n = 18), HB = 44.5 ± 3.28 (41.0–50.5; n = 9), T = 34.7 ± 2.87 (30.5–39.0; n = 9), HFcu = 7.0 ± 0.55 (6.4–7.5; n = 3), Tib = 15.3 ± 1.25 (12.5–16.6; n = 15), E = 12.3 ± 0.65 (11.0–13.2; n = 10), Tr = 6.9 ± 0.37 (6.5–7.5; n = 9) for females; FA = 33.7 ± 1.07 (31.5–35.5; n = 37), HB = 45.6 ± 3.20 (41.0–51.5; n = 17), T = 35.9 ± 2.95 (31.5–41.0; n = 17), HFcu = 8.4 ± 0.78 (7.0–9.0; n = 15), Tib = 15.2 ± 1.54 (11.4–17.5; n = 29), E = 12.7 ± 0.63 (11.4–13.7; n = 17), Tr = 6.9 ± 0.49 (6.0–7.9; n = 17) for males [93, 452, 455, 633, 1936, 3072, 3077, 3736, 3956, 3957, 4053, 4208]. Honshu (Iwate, Yamagata, Fukushima, Tochigi, Ishikawa, Yamanashi, Nagano, Gifu, Mie and Hiroshima Prefs.): FA = 34.4 ± 1.19 (31.1–35.6; n = 26), HB = 47.6 ± 3.41 (39.0–51.5; n = 11), T = 35.0 ± 2.06 (31.5–38.0; n = 12), HFcu = 7.9 ± 1.18 (6.0–9.5; n = 18), Tib = 14.7 ± 0.75 (13.5–15.9; n = 18), E = 11.8 ± 1.95 (8.0–14.0; n = 18), Tr = 6.4 ± 1.56 (3.5–8.0; n = 17) for females; FA = 34.3 ± 0.85 (32.0–35.4; n = 16), HB = 48.7 ± 4.76 (43.0–55.0; n = 6), T = 33.4 ± 2.77 (29.0–37.0; n = 6), HFcu = 7.6 ± 1.21 (6.0–9.2; n = 7), Tib = 15.9 ± 0.88 (14.4–17.1; n = 16), E = 12.4 ± 1.79 (10.2–15.0; n = 8), Tr = 6.4 ± 1.32 (4.8–8.0; n = 6) for males [518, 1577, 1723, 1734, 3227, 3716, 3991, 3993, 4006, 4084, 4208].

Dental formula I 2/3 + C 1/1 + P 3/3 + M 3/3 = 38 [4208].

Genetics Chromosome: $2n$ = 44 and FN = 52 [142, 785; described as *M. hosonoi*]. The status of Japanese and East Asian bats of the genus *Myotis* has been discussed with this species based on ND1 and cytochrome b sequences of mtDNA [1458]. Several haplotypes of cytochrome b sequences are recognized in Japanese populations [1457]. However, the genetic distances among *M. ikonnikovi* are very small.

4. Ecology

Reproduction Parturition may occur from mid-June to late July [459, 620, 1455, 2178, 3166, 3242, 4196]. Lactating females and juveniles have been observed from late July to early August [1701, 1734, 1880, 1936, 3166, 3716, 3736, 3956, 4087]. Litter size is 1 [2178, 3242, 4196]. A maternity colony consisting of at most 100 individuals was formed in a barn in Hokkaido from mid-May to mid-September [459]. Presumed maternity colonies have been found in an old wooden building (approximately 100 individuals) [3993], in the roof of a farm tent (30 adults plus an infant each) [2178] and under tree bark (19–31 individuals) [620]. Development, sexual maturity and breeding system are unknown.

Lifespan A male was recaptured in the same tunnel after 5 years and 3 months [3099].

Diet Not reported.

Habitat Forest dwelling bats. In Hokkaido, they are distributed widely, from agricultural land to high mountain forest independent of forest type. In Honshu, found in forest at relatively high altitude (from 600 m to more than 2,000 m) except for the northern part of the island, such as Aomori Pref. (200–300 m [1880]; 500–600 m [2290]). In Tochigi Pref., the distribution of this species depends on the distribution of natural forest [4088]. Day roosts are located under bark, in small cavities of broken branches, in narrow spaces between climbing vines (liana) and tree trunks, in old wooden buildings and the roofs of farm tents [448, 457, 620, 2178, 3993, 4087], caves and tunnels [3035, 3099, 3227, 4133]. A night roost has been observed in a tunnel [3990]. Feeding habitats are unknown.

Home range Not reported.

Behavior Observation of drinking behaviors on the wing under captivity were reported [458].

Natural enemies Not reported.

Parasites
Ectoparasite
Acari: *Penicillidia monoceros* [3084]. Diptera: *Nycteribia pygmaea*, *Basilia truncata endoi* [3069, 3082], Siphonaptera: *Ischnopsyllus indicus* [3087].
Endoparasite
Cestoda: *Vampirolepis brevihamata*, *V. rikuchuensis*, *V.*

uchimakiensis V. sp. [3119, 3121, 3122, 3469]. Trematoda: *Acanthatrium* (*Acanthatrium*) *mukooyamai*, *A.* (*A.*) *isostomum*, *Palalecithodendrium ovimagnosum*, *Plagiorchis* (*Plagiorchis*) *koreanus*, *Pl.* (*Pl.*) *minutifollicularis*, *Pl.* (*Pl.*) *chilostomum*, *Pl.* (*Pl.*) *rhinolophi*, *Pl. sp. Prosthodendrium macrorchis*, *Pr.* (*Prosthodendrium*) *parvouterus* [1544, 1547, 1549, 1552, 1553, 1556, 1559, 3117, 3469], Nematoda: *Aonchotecha pipistrelli, Molinostrongylus tsuchiyai, Riouxgolvania kapapkamui* [814, 3469].

Echolocation and social call

Echolocation calls are FM calls (Fig. 2). The values of EF and PF at hand release or inside mosquito nets average 43.2 kHz (37.8–50.1) and 50.6 kHz (47.3–55.9), respectively, for the Hokkaido population ($n = 32$) [624]. Social calls are unknown.

5. Remarks

The systematics of Japanese *Myotis* is complicated. Several species and/or subspecies had been recognized from regional populations in Japan; for example, *M. fujiensis* (Mt. Fuji and northeastern Honshu), *M. hosonoi* (Nagano Pref.), *M. ozensis* (Oze marshland), *M. ikonnikovi* (Lake Tôya, Hokkaido), *M. yesoensis* (Hokkaido) and *M. gracilis* (Hokkaido) (cf. [4208]).

On the other hand, *M. fujiensis*, *M. hosonoi*, *M. ozensis* and *M. yesoensis* were synonyms of *M. ikonnikovi*, and *M. gracilis* a synonym of *M. mystacinus* [1893]. Although there are several variations in the size of external characters among those species, they can be treated as geographical variations [1915]. Two individuals of *M. ozensis* were characterized as having 2 pairs of upper premolars compared to 3 pairs in all other species of *Myotis* [1005]; these individuals came to be considered as variant individuals of *M. ikonnikovi* [1915]. On the other hand, irrespective of their classification as species or subspecies, "*fujiensis*" should be recognized as a junior synonym of "*ikonnikovi*" based on morphological characteristics [3670]. In addition, a molecular phylogenetic study also came to the conclusion that *M. fujiensis* should be included in *M. ikonnikovi* [1458]. Furthermore, extensive field sampling in Hokkaido suggested that there is only 1 species, *M. ikonnikovi*, and that the occurrence of *M. yesoensis* is unlikely [1457]. In this volume, *M. fujiensis*, *M. hosonoi*, *M. ozensis* and *M. yesoensis* are treated as synonyms of *M. ikonnikovi*, and measurements from previous studies were recalculated together on a regional basis. Subspecies of those taxa would necessitate a detailed study.

K. KAWAI

An Ikonnikov's Myotis roosting under the exfoliating bark of a snag (M. Mizuno).

Chiroptera VESPERTILIONIDAE

046

Red list status: VU as *M. nattereri bombinus* (MEJ); R as *M. nattereri* (MSJ); NT (IUCN)

Myotis macrodactylus (Temminck, 1840)

EN Japanese large-footed bat   JP モモジロコウモリ (momojiro koumori)   CM 大足鼠耳蝠   CT 日本大足蝠
KR 큰발윗수염박쥐   RS длиннопалая ночница

An individual captured on Kunashiri (Kunasir) Isl. (K. Kawai).

1. Distribution

Occurs in eastern Siberia, southern Sakhalin, Korea and Japan. In Japan, reported from Hokkaido, Honshu, Shikoku, Kyushu, and the islands of Kunashiri (Kunashir), Sado-gashima, Oki, Tsushima, Iki-shima, Fukue-jima (the Goto Isls.), Tane-gashima, Amami-ohshima and Tokuno-shima [1456, 1460, 1694, 1720, 1895, 1919, 3124, 4208].

2. Fossil record

Fossils are reported from cave and fissure deposits of the Late Pleistocene and Early Holocene ages in Honshu [1519, 4158].

3. General characteristics

Morphology Dorsal fur is grayish or blackish brown, ventral fur is paler grayish brown, white hairs are found on lower abdomen and surface of femur; ears are fairly narrow and long; tragus is long and pointed, but shorter than 9 mm; plagiopatagium attaches to ankle or lower part of tibia; posterior edge of interfemoral membrane lacks a fringe of fine hairs; hind foot is large, HFcu occupies 70–86% of lower leg [4208]. Wing aspect ratio is 7.91 [1770].

External measurements (mean in mm with ± SD or range) from 5 localities are as follows. Hokkaido: FA = 37.7 (33.6–42.0; n = 373), HB = 51.8 (40.0–71.0; n = 187), T = 37.6 (29.0–49.0; n = 192), HFcu = 10.4 (9.0–12.7; n = 24), Tib = 16.8 (15.2–20.7; n = 194), E = 15.0 (12.0–16.2; n = 194) [93, 326, 450, 452, 455, 456, 625, 630, 633, 840, 1446, 1904, 2343, 3058, 3061, 3076–3079, 3081, 3732, 3736, 3953, 3954, 3956, 3957, 4051, 4054]. Honshu: FA = 37.6 ± 1.13 (n = 230), HB = 48.4 ± 3.47 (n = 80), T = 38.7 (34.7–43.0; n = 35), HFcu = 10.3 (8.0–11.0; n = 49), Tib = 15.8 ± 0.99 (n = 101), E = 14.4 ± 1.46 (n = 84) [626, 1356, 1735, 1770, 2423, 3018, 3266, 4003, 4006, 4208]. Shikoku: FA = 37.7 ± 0.93 (n = 63), HB = 47.5 ± 2.60 (n = 63), T = 34.6 ± 1.79 (n = 43), HFcu = 10.8 ± 0.63 (n = 43), Tib = 15.4 ± 0.77 (n = 63), E = 15.0 ± 0.60 (n = 63), and Kyushu: FA = 36.5 ± 0.66 (n = 11), HB = 49.1 ± 1.78 (n = 11), Tib = 14.5 ± 0.75 (n = 11), E = 15.3 ± 0.57 (n = 11) [4208]. Kunashiri (Kunashir) Isl.: FA = 37.8 (32.7–40.3, n = 109), HB = 45.8 (44.8–46.8, n = 2), T = 42.7 (40.2–45.2; n = 2), HFcu = 13.1 (13.0–13.2; n = 2), Tib = 17.4 (17.3–17.4, n = 2), E = 15.2 (14.2–16.2, n = 2) [1460], Tsushima Isls.: FA = 37.0 ± 0.88 (n = 14), HB = 48.2 ± 4.40 (n = 13), T = 32.8 (32.8–32.9; n = 2), HFcu = 9.6 (9.3–9.9; n = 2), Tib = 15.9 ± 0.68 (n = 11), E = 14.7 ± 0.65 (n = 8) [1452, 4208].

Dental formula I 2/3 + C 1/1 + P 3/3 + M 3/3 = 38 [4208].

Genetics Chromosome: $2n$ = 44 and FN = 52 [142, 2551]. The status of Japanese and East Asian bats of the genus *Myotis* has been discussed on the basis of ND1 and cytochrome *b* sequences of mtDNA [1458]. These sequences are available [1458, 2974].

4. Ecology

Reproduction Reproductive pattern involves seasonal monoestry and delayed fertilization [2591, 3688]. Females produce a single infant per litter. Ratios of neonatal to maternal values in FA and BW reach 40.0% and 21.1%, respectively [1770]. Most females produce their first infant at the end of their second year; pregnancy rates in adult females are extremely high; some females breed up to 15 years of age [1775]. In summer, adults and subadults of both sexes form maternity colonies of tens to hundreds of bats [1773]. Often forms clusters consisting of different bat species, such as *Rhinolophus ferrumequinum*, *R. cornutus*, *Miniopterus fuliginosus*, *Myotis bombinus* and *My. petax*; During spring to autumn, they also give birth and rear infants in mixed-species colonies [92, 625, 1783]. Copulation occurs in autumn [1770, 3688], but reproductive ecology during the mating season is unknown.

Lifespan The greatest recorded longevity is 19 years [1776, 1777].

Diet Preys mainly on Diptera, Trichoptera, and Lepidoptera measuring 7–20 mm in body length; many spiders (Araneae) are also found in the diet [676].

Habitat Mainly roosts in natural caves, abandoned mines, unused tunnels and bomb shelters [2991, 3124].

Home range Not reported.

Behavior Nocturnal [1916]. Forages typically above water surface [630, 3316], Crawls well using 4 limbs, frequently gets into and roosts in the crevices of cave ceilings [1773, 1783, 3687]. In winter, usually hibernates solitarily or in small groups in caves [1773]. This species can burrow with its forelimbs, and is occasionally found roosting under stones along riverbanks [3001].

A long distance migration of 37 km has been recorded in Hokkaido [3066].

Natural enemies Not reported.

Parasites & Pathogenic organisms

Ectoparasite

Diptera: *Nycteribia allotopa*, *N. parvula*, *N. pleuralis*, *N. pygmaea*, *N. uenoi*, *Phthiridium hindlei*, *Penicillidia jenynsii*, *Pe. monoceros*, *Brachytarsina kanoi* [2186, 2187, 2473, 3054, 3068, 3069, 3076, 3078, 3079, 3081, 3084, 3086]. Acari: *Spinturnix myoti*, *Macronyssus granulosus*, *Ichoronyssus scutatus*, *Amblyomma testudinarium*, *Haemaphysalis flava*, *H. formosensis*, *Ixodes simplex*, *I. vespertilionis* [1760, 2708, 3691, 3706, 4031].

Endoparasite

Trematoda: *Acanthatrium mukooyamai*, *A. isostomum*, *Duboisitrema sawadai*, *Mesothatrium japonicum*, *Prosthodendrium postacetabulum*, *Pr. chilostomum*, *Pr. longiforme*, *Pr. miniopteri*, *Pr. parvouterus*, *Pr. ascidia*, *Pr. hurkovaae*, *Pr. thomasi*, *Pr. parallerorchus*, *Pr. circulare*, *Pr. parvouterus*, *Pr. thomasi*, *Pr. yamizense*, *Neoheterophyes sawadai*, *Plagiorchis koreanus*, *Pl. kyushuensis*, *Pl. vespertilionis*, *Pl. minutifollicularis*, *Pl. rhinolophi*, *Pl. magnacotylus* [1541, 1546–1551, 1554, 1555, 1559]. Cestoda: *Vampirolepis wakasensis*, *V. tanegashimensis* [3122]. Nematoda: *Molinostrongylus rhinolophi*, *M. skjrjabini*, *Rictularia rhinolophi*, *Riouxgolvania kapapkamui*, *Aonchoteca pipistrelli* [814, 3113, 3129, 3136, 3469].

Pathogenic organisms

Virus strains that are similar to Flaviviruses but distinct from Japanese encephalitis virus (JEV) have been isolated [2120].

Echolocation and social calls Echolocation calls are FM calls (Fig. 1). The EF and PF values (in kHz) average 35.8 (28.3–47.7) and 53.8 (43.8–68.8), respectively, for the Hokkaido population ($n = 564$) [624, 625, 4018], 37.4 and 60.3, respectively, for the Wakayama Pref. (south Honshu) population ($n = 1$) [626], and 38.4 (35.4–45.3) and 49.5 (47.4–53.7), respectively, for the Kyushu population ($n = 10$) [660]. Mean PF value of the free flying call is 49.1 kHz for the Tokuno-shima Isl. population ($n = 17$) [666]. No information is available on social calls.

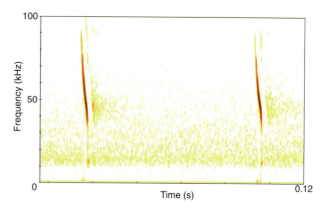

Fig. 1. Sonogram of a release call, recorded by D. Fukui in Hokkaido.

A. SANO

5. Topic

Life history traits and roosting ecology of the Japanese large-footed bat revealed by long-term banding studies at Akiyoshi-dai Plateau

M. ISHIDA

Akiyoshi-dai Plateau, located in Yamaguchi Pref., western Honshu, is the largest limestone plateau in Japan (Fig. 2). About 450 limestone caves are scattered throughout this area. In the

Fig. 2. Akiyoshi-dai Plateau (Akiyoshi-dai Museum of Natural History).

Myotis macrodactylus (Temminck, 1840)

Table 1. Survival records, ages of first parturition and long distance movement of 6 cave bat species

	Survival record		Sex	Reference	Age of first parturition	Reference	Long distance movement (km)	Reference
	Year	Month						
Myotis macrodactylus	18 (19)		F	[1776]	2	[1775]	30	[1778]
Myotis bombinus	14 (15)	10 (3)	F	[1775]	1	[1774]	–	
Miniopterus fuliginosus	14	7	F	[1775]	3	[1779]	130	[1776]
Murina hilgendorfi	7 (8)	11 (8)	F	[1083, 1086]	–		–	
Rhinolophus ferrumequinum	23 (23)	(8)	F	[1776]	2–3	[1775]	51	[1774]
Rhinolophus cornutus	14	11	M	[1774]	2–3	[1778]	20	[1778]

Numbers in parentheses are minimum longevities calculated from the date when bats were captured first.

Fig. 3. A mixed-species colony of Hilgendorf's tube-nosed bats and Japanese large-footed bats (N. Urano).

Fig. 4. A maternity colony consisting of male and female Japanese large-footed bats (A. Sato).

caves on this plateau, a long-term banding investigation for 6 bat species, *Rhinolophus ferrumequinum*, *R. cornutus*, *Myotis macrodactylus*, *My. bombinus*, *Miniopterus fuliginosus* and *Murina hilgendorfi*, has been continuously conducted by staff of the Akiyoshi-dai Museum of Natural History since 1966 [1083, 1086, 1774–1779, 1783]. As a result, 42,857 bats have been banded, and the total number of recaptures has reached 18,868 at the time of writing (Table 1). On this topic, I introduce the results of the long-term study of *My. macrodactylus*, as well as the other species captured along with it.

Female *My. macrodactylus* produce their first infant at 2 years of age, and become pregnant almost every year after sexual maturity [1775]. A 15-year-old pup-bearing female and a 19-year-old survivor have been confirmed [1775–1777]. This species is iteroparous, maintaining a high pregnancy rate during the long pubertal period.

Myotis macrodactylus has a unique roosting ecology, forming dense clusters with species belonging to different genera (*Miniopterus* and *Murina*) and to a different family (Rhinolophidae) (Fig. 3) [1773, 1783, 3687]. Mixed-species colonies of various sizes have been observed throughout the year in crevices and the flat ceilings of caves. A small number of females frequently join maternity colonies of other species, where they give birth and care for their own infants.

They also form single-species maternity colonies (Fig. 4). The participation of males in these colonies is notable. In spring and autumn, males and females tend to separate. In summer, however, adults and subadults of both sexes gather in the maternity colony [1773]. This is an exceptional type of chiropteran society, as classified by Bradbury (1977) [325].

Myotis macrodactylus may take advantage of the huddling effect produced by clusters of species, and by the joining of males to the colony to increase thermoregulating potential.

Chiroptera VESPERTILIONIDAE

Red list status: VU as *M. nattereri bombinus* (MEJ); R as *M. nattereri* (MSJ); NT (IUCN)

Myotis bombinus Thomas, 1906

EN Far Eastern Myotis   JP ノレンコウモリ (noren koumori)   CM 远东鼠耳蝠   CT 遠東鼠耳蝠
KR 아무르박쥐   RS амурская ночница

A profile of the Far Eastern Myotis with relatively long ears and pointed tragus (A. Sano).

1. Distribution

Distributed in northeastern China, southeastern Russian, Korea and Japan [3272]. In Japan, found in Hokkaido, Honshu, Shikoku, Kyushu and the islands of Kunashiri (Kunashir) and Kuchinoerabu-jima (the Ohsumi Isls.) [659, 1456, 1460, 4208, 4259]. Red solid circles in the map denote sites where this species was observed in Japan.

2. Fossil record

Late Pleistocene fossils are reported from caves in Akiyoshi-dai Plateau, Yamaguchi Pref. [4158].

3. General characteristics

Morphology Dorsal fur is grayish brown or dark brown, ventral fur is paler; interfemoral membrane has a fringe of fine hairs along the posterior edge (Fig. 1); plagiopatagium attaches to the base of the outer toe; tragus is conspicuously long, its length over 9 mm (11.2 ± 0.82 SD mm, 9.0–12.0 mm; $n = 23$) [4208].

External measurements (mean ± SD in mm with range) are as follows. Hokkaido: FA = 39.0 ± 1.65 (37.5–41.3; $n = 5$), HB = 45.2 (43.0–47.3; $n = 2$), HFcu = 8.9 (8.7–9.0; $n = 2$), Tib = 18.0 (17.5–18.4; $n = 2$), E = 15.7 (15.0–16.3; $n = 2$) [840, 1703, 2458]; Honshu: FA = 40.0 ± 1.60 (37.0–42.1; $n = 10$), HB = 46.4 ± 1.99 (44.0–49.0; $n = 7$), T = 41.1 ± 2.41 (38.0–45.0; $n = 7$), HFcu = 10.0 ± 0.91 (8.5–11.5; $n = 7$), Tib = 17.3 ± 1.41 (15.5–20.3; $n = 9$), E = 16.4 ± 1.12 (15.0–18.0; $n = 9$) [626, 3991, 4006, 4208], and Kyushu: FA = 40.0 ± 1.08 (38.15–41.85; $n = 18$), HB = 47.8 ± 2.54 (41.5–52.0; $n = 18$), T = 42.3 ± 1.62 (40.0–45.0; $n = 18$), HFcu = 11.0 ± 0.42 (10.2–12.0; $n = 17$), Tib = 17.2 ± 0.78 (16.0–19.2; $n = 17$), E = 17.4 ± 0.88 (14.5–18.5; $n = 18$) [4208]. Kunashiri (Kunashir) Isl.: FA = 39.6 ± 0.93 (38.6–40.5; $n = 4$) [1460].

Fig. 1. Fringe of fine hairs along the posterior edge of the tail membrane (H. Nakajima).

Dental formula   I 2/3 + C 1/1 + P 3/3 + M 3/3 = 38 [4208].

Genetics   Chromosome: $2n = 44$ and FN = 50 [142, 796]. The status of Japanese and East Asian bats of the genus *Myotis* has been discussed on the basis of ND1 and cytochrome *b* sequences of mtDNA [1458]. These *b* sequences are available [1458, 2974].

4. Ecology

Reproduction   Reproductive pattern involves seasonal monoestry and delayed fertilization [2591, 3688]. In summer, pregnant females and some adult males congregate and form maternity colonies of several to ca. 200 bats; a single young is born in early June in Kyushu [655, 657]. Ratios of neonatal to maternal values of FA and BW are 34.4% and 16.1%, respectively [1770]. Females attain sexual maturity by their first autumn and give birth for the first time at 1 year of age; the pregnancy ratio in adult females is extremely high [657, 1775]. Parturition by 15-year-old females has been confirmed [1775]. Copulation occurs in autumn [3688], but reproductive ecology during mating season is little known.

Lifespan   Greatest longevity recorded in Japan is 15 years [1775].

Diet   Feeds mainly on Lepidoptera, Coleoptera, Diptera, Trichoptera and Araneae in and around woodlands [676].

Habitat   Roosts in natural caves, abandoned mines, unused tunnels, and sometimes in buildings and on the undersurfaces of bridges [655, 1703, 2286, 2996, 3098, 3124]. Rarely roosts in tree cavities [3721].

Home range   Not reported.

Behavior   Nocturnal [1916]. Hibernation ecology has not been reported.

Natural enemies   Not reported.

Parasites
Ectoparasite
Diptera: *Nycteribia pleuralis*, *Penicillidia monoceros*, *Basilia rybini* [2187, 2473, 3068, 3084].

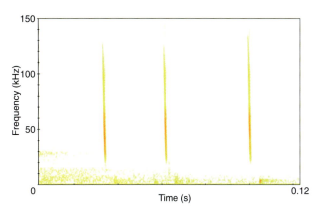

Fig. 2. Sonogram of a release call, recorded by M. Ishida in Chugoku District.

Endoparasite
Trematoda: *Acanthatrium isostomum*, *Mesothatrium japonicum*, *Prosthodendrium klausrohdei*, *P. parvouterus*, *P. yamizence*, *Plagiorchis corpulentus*, *Neoheterophyes sawadai* [1548, 1550, 1551, 1554, 1559]. Cestoda: *Vampirolepis brevihamata*, *Hymenolepis rashomonensis*, *Insectivorolepis mukooyamai* [3116, 3119, 3122].

Echolocation and social calls   Echolocation calls are broadband FM calls (Fig. 2). The EF and PF values (mean ± SD in kHz) average 28.2 and 60.2, respectively, for the Wakayama Pref. (south Honshu) population ($n = 1$) [626], and 29.3 ± 0.63 (28.2–30.3) and 36.4 ± 2.34 (32.6–40.1), respectively, for the Kagoshima Pref. (south Kyushu) population ($n = 10$) [660].

In communicating, infants and mothers use characteristic isolation and directive calls, respectively [4019].

5. Remarks

Formerly this species from Japan was included in *M. nattereri* which was widely distributed in Europe and East Asia, and treated as a subspecies *M. n. bombinus* [4208]. Recently, molecular phylogenetic analyses revealed that *M. bombinus* is distinct from *M. nattereri* and is widely distributed in east Asia [1458, 1749, 3272].

A. SANO

Chiroptera VESPERTILIONIDAE

048

Red list status: R as *M. daubentonii* (MSJ); LC as *M. daubentonii* (IUCN)

Myotis petax Hollister, 1912

EN eastern water bat, Daubenton's bat   JP ドーベントンコウモリ（dobenton koumori）
CM 水鼠耳蝠   CT 水鼠耳蝠，道氏鼠耳蝠   KR 우수리박쥐   RS восточная водяная ночница

A flying individual in the Kushiro region, Hokkaido (H. Nakajima).

1. Distribution

Distributed in southern and eastern Siberia, Transbaikalia, northern China, Mongolia, Far East Russia, Korea, Sakhalin, Kunashiri (Kunashir) Isl., Etorofu (Itrup) Isl. and Hokkaido, Japan [1460, 1746, 2032]. In Japan, occurs only in Hokkaido [92, 453, 601, 625, 840, 1092, 1449, 1687, 1701, 1704, 1706, 1911, 1912, 1926, 1927, 2187, 2322, 2343, 2359, 2369, 3017, 3054, 3061, 3063, 3064, 3068, 3069, 3071, 3072, 3074, 3084, 3166, 3316, 3952, 3954, 4051, 4053, 4208]. Red solid circles in the map denote sites where this species was observed in Japan.

2. Fossil record

Fossils are reported from fissure deposits of the Late Pleistocene in Yamaguchi Pref. [4158].

3. General characteristics

Morphology Fur is silky and short; dorsal color is grayish brown; ventral color is paler than dorsal color; the basal portion is the same as the dorsal color; ears and membranes are grayish brown and transparent; ears reach tip of muzzle or extend about 1.0 mm beyond muzzle when laid forward; the tragus is slender, straight, about half as high as ears; wing membranes are wide; the fifth finger is comparatively long, about 79.5% of the third finger in length; the length of the thumb is about 19.5% of FA; the claw is long and its distal portion strongly curved; the third metacarpal is longest, the fifth the shortest, the third terminating about 2.3 mm short of elbow; the patagium is narrow; the hind foot is large and its length is about 69.1% of tibia length; the lower leg is comparatively short; the calcar is slender, its length about 2/3 of the length of the lateral border of the uropatagium, lacking a keel and with a terminal lobe; terminal vertebra free of the margin of the uropatagium; the plagiopatagium is inserted on the metatarsus of the first toe [4208; described as *M. daubentonii ussuriensis*].

External measurements (mean ± SD in mm with range) are as follows [1460, 2369, 3077, 4208]. Hokkaido: FA = 36.4 ± 0.88 (35.1–38.3; n = 33), HB = 49.2 ± 2.45 (46.0–54.2; n = 10), T = 34.3 ± 4.10 (28.0–41.0; n = 10), HFcu = 10.2 ± 1.19 (8.3–12.0; n = 10), Tib = 16.0 ± 1.02 (14.5–17.2; n = 10), E = 13.6 ± 0.99 (11.5–15.3; n = 10), Tr = 5.7 ± 1.05 (4.0–7.0; n = 10) for females; FA = 36.2 ± 0.93 (34.1–38.0; n = 17), HB = 47.0 ± 4.11 (40.4–51.5; n = 7), T = 34.9 ± 5.09 (27.0–41.0; n = 7), HFcu = 10.5 ± 1.26 (9.0–12.0; n = 6), Tib = 16.2 ± 0.72 (15.2–17.0; n = 7), E = 13.3 ± 0.75 (12.0–15.3; n = 7), Tr = 5.6 ± 1.05 (4.0–7.1; n = 7) for males. Kunashiri (Kunashir) Isl.: FA = 37.0 (35.6–38.0; n = 4), HB = 46.4 (n = 1), T = 43.2 (28.0–41.0; n = 10), HFcu = 13.2 (n = 1), Tib = 17.4 (n = 1), E = 16.2 (n = 1), Tr = 5.2 (n = 1) for females; FA = 36.6 (35.6–38.0; n = 4) for males, Etorofu (Itrup) Isl.: FA = 36.8 (35.3–37.9; n = 14) for females, FA = 36.97 (36.2–37.7; n = 14), HB = 45.0 and 49.0, T = 40.0 and 37.0, HFcu = 10.0 and 10.0, Tib = 18.4 and 18.1, E = 16.0 and 14.0, Tr = 6.0 and 6.5

for males. No sexual differences in external measurements have been recognized, but there were geographic variations in the length dimensions of the skull and the height of the braincase [1912; described as *M. daubentonii ussuriensis*].

Dental formula I 2/3 + C 1/1 + P 3/3 + M 3/3 = 38 [4208].

Genetics Chromosome: not reported in Japan. $2n = 44$ and $FN = 52$ in Korea (described as *M. daubentonii*), $2n = 44$ and $FN = 50$ in Far East Russia (described as *M. daubentonii ussuriensis*) [1715]. The status of Japanese and East Asian bats of the genus *Myotis* has been discussed based on ND1 and cytochrome *b* sequences of mtDNA [1458]. See also Remarks.

4. Ecology

Reproduction Parturition may occur in July. Lactating females are observed early July to early August [1449, 1912, 2369, 3072, 4056]. Developed testes are observed in early August [2369]. A maternity colony has been found under a bridge [92]. A colony of approximately 30 individuals of mixed ages and sexes has been observed under a bridge in early August [1449]. Male-biased colonies in tunnels were reported in summer [3066, 3316]. Litter size, development, sexual maturity and mating system are unknown in Japan.

Lifespan Not reported.

Diet Not reported.

Habitat Common near aquatic environments, at relatively low altitudes except for central Hokkaido, such as records in Daisetuzan National Park (500–900 m) [1449, 1706, 1912, 2458]. In summer, day roosts have been found in tree cavities [601, 1450, 3316], under a bridge [92, 1449, 3316], in an old shelter [3066], in a bomb tunnel [3316] and in a sluiceway [453, 625, 1912, 2322]. Sometimes mixed colonies with *M. macrodactylus* have been observed [92, 625, 1912, 2322, 3061]. A night roost has been observed under a bridge [3017]. Hibernation sites are unknown because all past records were obtained from spring to autumn.

Home range Not reported.

Behavior The recapture ratio of males in summer is relatively high compared to *M. macrodacylus* using the same roosting site [3066]. Mean distance between sites of capture and recapture were greater for females than males (2.5 km and 0.8 km, respectively) [2322].

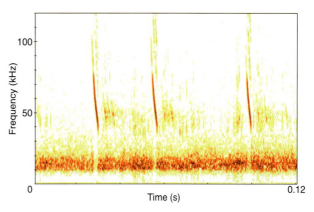

Fig. 1. Sonogram of a release call, recorded by D. Fukui in Hokkaido.

Natural enemies Not reported.

Parasites
Ectoparasite
Hemiptera: *Cimex japonicus* [2421], *C.* cf. *japonicus* [3028]. Diptera: *Penicillidia monóceros*, *Nycteribia pygmaea*, *Penicillidia monoceros* [2186, 2187, 3054, 3068, 3069, 3084].
Endoparasite
Nematoda: *Aonchotheca pipistrelli*, *Rictularia rhinolophi* [3469]. Trematoda: *Brachilaima* sp. *Plagiochis* sp. [3469].

Echolocation and social calls Echolocation calls are FM calls (Fig. 1). The values of EF and PF at hand release average 37.8 ± 1.6 kHz and 49.5 ± 1.9 kHz, respectively ($n = 4$) [625]. Social calls are unknown.

5. Remarks

Formerly included in *M. daubentonii* as a subspecies. However, according to morphological data, DNA sequences and Inter-SINE PCR results, the species has been divided into 2 groups, the "Western" and "Eastern" groups [1458, 1746, 2032]. The oldest name for the latter is *M. d. petax* from Altai, south of western Siberia. Based on these results, *M. petax* has been proposed as the valid name for the "Eastern" group [1746, 2032]. Three subspecies, *M. p. ussuriensis*, *M. p. loukashkini* and *M. p. chosanensis*, are included in the "Eastern" group [2032], and a junior synonym, *M. abei*, is included in this group [3671]. The Hokkaido population was not significantly different from *M. p. ussuriensis* from Sakhalin, Korea and Etorofu (Itrup) Isl., except for the height of the braincase [1912].

K. KAWAI

Chiroptera VESPERTILIONIDAE

049

Red list status: VU (MEJ); V (MSJ); EN (IUCN)

Myotis pruinosus Yoshiyuki, 1971

EN frosted Myotis   JP クロホオヒゲコウモリ (kuro hohige koumori)   CM 霜白鼠耳蝠   CT 霜白鼠耳蝠
KR 검은큰수염박쥐   RS седая ночница

An individual captured in Akita Pref. (M. Mukohyama).

1. Distribution

Endemic to Japan. Distribution range is limited to Honshu, Shikoku, and Kyushu from 18 prefectures; Aomori [2289, 2300], Iwate [522, 793, 4201], Akita [522, 4208], Fukushima [1565], Tochigi [1734, 1735, 4084], Niigata [3648], Ishikawa [1895, 4004, 4006], Gifu [3991], Shizuoka [3098, 3266], Shiga [1933], Kyoto [522, 3238], Nara [1892, 1934, 3904], Wakayama [1934], Hiroshima [3716], Tokushima [1], Ehime [41, 1895, 4193], Kumamoto [675] and Miyazaki [675, 1895]. Sometimes sympatric with *M. ikonnikovi* but usually allopatorically distributed (e.g. [3266, 4006]). Red solid circles in the map denote sites where this species was observed in Japan.

2. Fossil record

Fossils of Late Pleistocene age are reported from Akiyoshi-dai Plateau, Yamaguchi Pref. [4158].

3. General characteristics

Morphology   Color is dark, dorsal surface of body dark grayish brown and conspicuously frosted by ashy or whitish hair tips; ears are narrow and short, the tragus is relatively short; the hind foot cum unguis is large; the lower leg is slender; the plagiopatagium is inserted in the base of the first toe; tail and calcar are long [4208].

External measurements (mean ± SD in mm with range) are as follows [1565, 2289, 3716, 4084, 4208], Honshu (Aomori, Iwate, Akita, Tochigi, Ishikawa, Hiroshima Prefs.): FA = 32.0 ± 1.03 (30.6–33.3; n = 12), HB = 42.4 ± 1.20 (40.1–43.5; n = 6), T = 34.9 ± 1.90 (33.4–38.0; n = 6), HFcu = 8.3 ± 0.27 (8.0–8.6; n = 6), Tib = 14.6 ± 1.30 (13.0–16.6; n = 12), E = 12.8 ± 0.38 (12.4–13.3; n = 6), Tr = 6.3 ± 0.36 (5.7–6.8; n = 6) for females; FA = 32.2 ± 1.00 (30.1–33.8; n = 22), HB = 41.0 and 42.0, T = 37.5 and 37.5, HFcu = 9.1 and 9.0, Tib = 14.6 ± 0.75 (12.9–15.9; n = 21), E = 12.6 and 12.1, Tr = 6.1 and 6.5 for males, Kyushu (Kumamoto and Miyazaki Prefs.): FA = 33.4 ± 0.98 (32.3–35.0; n = 7), HB = 40.9 ± 4.78 (34.0–46.9; n = 7), T = 35.6 ± 1.99 (32.0–38.0; n = 7), Tib = 15.4 ± 0.39 (14.8–16.0; n = 7), E = 11.5 ± 0.56 (10.7–12.5; n = 7), Tr = 5.7 ± 0.42 (5.0–6.3; n = 7) for females; FA = 32.5 ± 0.49 (31.9–33.1; n = 4), HB = 40.1 ± 1.59 (38.1–41.6; n = 4), T = 32.9 ± 1.16 (31.8–43.2; n = 4), Tib = 15.2 ± 0.40 (15.0–15.7; n = 3), E = 12.2 ± 0.78 (11.5–13.3; n = 4), Tr = 5.8 ± 0.36 (5.3–6.1; n = 4) for males. No sexual differences in external measurements have been recognized, but there were geographical variation in the length dimensions of the skull and the wide of the braincase [675].

Dental formula   I 2/3 + C 1/1 + P 3/3 + M 3/3 = 38 [4208].

Genetics   Chromosome: $2n$ = 44 and FN = 52 [793]. The status of Japanese and East Asian bats of the genus *Myotis* has been discussed based on ND1 and cytochrome *b* sequences of mtDNA (see Remarks) [1458]. Other molecular phylogenies of Japanese bats have been investigated based on cytochrome *b* sequences, and this species has been used as an outgroup for these analysis [2974]. Geographical variations of the cytochrome *b* sequences were suggested [675].

4. Ecology

Reproduction Parturition occurs in the middle of July; litter size is 1 [522]. Sexual maturation and mating season are unknown.

Lifespan Not reported.

Diet Not reported.

Habitat Day roosts have not been found. This species has been regarded a forest dweller. However several individuals were recorded in caves from Ehime Pref. in March and Shiga Pref. in August [1933, 4193]. It has been observed flying along the edges of forests, trails and streams and above grasslands [522]. One hibernating individual has been observed in a crack on the roof of a cave [4193].

Home range Not reported.

Behavior Not reported.

Natural enemies Not reported.

Parasites
Ectoparasite
Diptera: *Basilia rybini* [3068], *B. r. japonica* [3069].

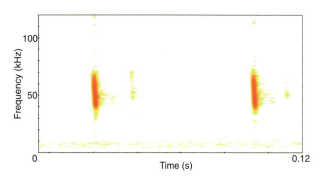

Fig. 1. Sonogram of an echolocation call in a mosquito net, recorded by K. Funakoshi in Miyazaki Pref.

Endoparasite
Trematoda: *Acanthatrium mukooyamai*, *Mesothatrium japonicum*, *Paralecithodendrium ovimagnosum*, *Prosthodendrium chilostomum*, *Pr. circulare*, *Pr. urna*, *Neoheterophyes yamato* [1552–1554, 1556, 1559].

Echolocation and social call Echolocation calls are short and broadband FM (Fig. 1). The mean values (± SD) of EF and PF at hand release or inside a mosquito net are 38.7 ± 2.47 kHz ($n = 10$) and 52.4 ± 1.73 kHz ($n = 10$), respectively, for the Kyushu population [660].

5. Remarks

Molecular phylogenetic studies show that this species is grouped with *M. yanbarensis* (distributed in evergreen forest in Okinawa Pref.), *M. montivaus* from Malaysia and *M. annectans* from Thailand [287, 1458]. Ancestral *M. pruinosus* and *M. yanbarensis* may have adapted to warm temperatures, with a probable origin in south East Asia [1458].

K. KAWAI

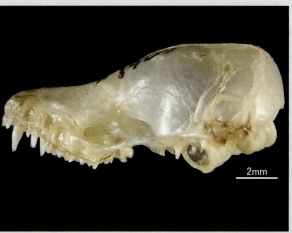

Lateral views of the skull and lower jaw of the frosted Myotis (type specimen: NSMT-M 14842) (K. Kawai).

Chiroptera VESPERTILIONIDAE

Red list status: CR (MEJ); CR (IUCN)

Myotis yanbarensis Maeda & Matsumura, 1998

EN Yanbaru Myotis   JP ヤンバルホオヒゲコウモリ (yanbaru hohige koumori)   CM 琉球鼠耳蝠   CT 山原鼠耳蝠
KR 오키나와큰수염박쥐   RS янбарская ночница

An individual captured on Tokuno-shima Isl., Kagoshima Pref. (K. Funakoshi).

1. Distribution

Two individuals, the holotype and paratype, were recorded from forests in northern Okinawa-jima Isl., Okinawa Pref. in 1996 [1929]. After these first records, 2 individuals were found on Tokuno-shima Isl., Kagoshima Pref. in 1999 [1919] and 2 individuals on Amami-ohshima Isl., Kagoshima Pref. in 2000 [1932]. Red solid circles in the map denote sites where this species was observed in Japan.

2. Fossil record

Not reported.

3. General characteristics

Morphology Dorsal hairs and membrane are blackish; the tips of the guard hairs are dull silver; wing membranes attach to the base of the first toe; the anterior border of the ear is turned outward, and the posterior border concave at the center; the tragus is slender and tapers gradually towards the pointed tip, and its distal third bends externally very slightly if at all; the braincase is very small in length and also in width; the rostrum is long [1929].

External measurements (in mm) are as follows. Okinawa-jima Isl.: FA = 37.5 and 36.5 (n = 2), HB = 43.0 and 41.5 (n = 2), T = 46.0 (n = 1), HFcu = 10.0 and 9.5 (n = 2); Tib = 17.0 and 16.5 (n = 2); E = 14.0 and 14.5 (n = 2); Tr = 7.0 and 7.0 (n = 2) for adult males [1929]. Amami-ohshima Isl.: FA = 35.6, HB = 38.0, T = 39.0, E = 14.0, Tr = 7.0 for a female (n = 1); FA = 35.3, HB = 36.5, T = 40.0, E = 13.5, Tr = 7.0 for a male (n = 1) [1932]. Tokuno-shima Isl.: FA = 38.2 mm for a female; FA = 35.2 for a male (n = 1) [1919].

Dental formula I 2/3 + C 1/1 + P 3/3 + M 3/3 = 38 [35].

Genetics Chromosomes: not reported. The status of Japanese and East Asian bats of the genus *Myotis* has been discussed based on ND1 and cytochrome *b* sequences of mtDNA [1458]. See also Remarks of *M. pruinosus*.

4. Ecology

Reproduction Not reported.
Lifespan Not reported.
Diet Not reported.
Habitat All individuals have been captured in evergreen broadleaf forest. Day roosts and feeding habits are unknown.
Home range Not reported.
Behavior Not reported.
Natural enemies Predation by a spider (*Nephila pilipes*) has been reported [1033].
Parasites
Endoparasite
Trematoda: *Prosthodendrium thomasi*, *Mesothatrium japonicum* [1559].
Echolocation and social call Not reported.

5. Remarks

See Remarks of *M. pruinosus*.

K. KAWAI

Chiroptera   VESPERTILIONIDAE

051

Red list status: R as *M. leucogaster* (MSJ); LC (IUCN)

Murina hilgendorfi (Peters, 1880)

EN Hilgendorf's tube-nosed bat   JP テングコウモリ（tengu koumori）   CM 白腹管鼻蝠   CT 白腹管鼻蝠
KR 관코박쥐   RS большой трубконос

An individual captured in Akita Pref. (O. Takahashi).

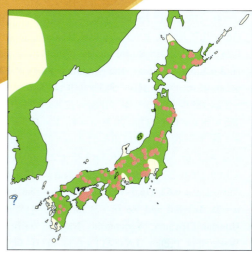

1. Distribution

Distributed in northern China, upper Yenisei River (Russia), the Altai Mts. (Russia, Kazakhstan and Mongolia), Korea, Primorye (Russia), Sakhalin, Kunashiri (Kunashir) Isl. and Japan [1460, 3272]. In Japan, reported from 39 prefectures: Hokkaido [267, 452, 624, 625, 632, 1445, 1448, 1701, 1704, 1706, 1936, 2343, 2780, 3017, 3166, 3316, 3528, 3736, 3954, 4046, 4053, 4058], Aomori [2288, 2291, 2301, 3788], Iwate [436, 516, 2292, 3981, 4133, 4208], Miyagi [96, 2396], Akita [2301], Fukushima [373, 374, 1577], Tochigi [1734, 1735, 2070, 3914, 4083, 4093, 4208], Gunma [3018], Saitama [4208], Tokyo [1649], Kanagawa [3967–3969], Niigata [603, 605], Toyama [4208], Ishikawa [4004, 4006], Fukui [948, 2190, 3114, 3264], Yamanashi [4208], Nagano [3993, 4208], Gifu [3991, 3992, 4208], Shizuoka [3098, 3266, 4208], Aichi [3552], Mie [2997, 3003, 3229], Shiga [60], Kyoto [791, 3751], Osaka [3103, 3747, 3757], Hyogo [3103, 3753], Nara [98, 1925, 3748], Wakayama [1920, 3750], Tottori [2673], Shimane [1818, 2592, 2728, 2729], Okayama [3107], Hiroshima [612, 614, 2100, 3715, 4232], Yamaguchi [1083, 1086, 1769, 1774–1777, 1779, 1780, 2012, 4020, 4021], Kagawa [2221], Ehime [2228, 3107, 3112, 3980, 4002, 4208], Kochi [3088, 3912, 4208], Fukuoka [1806, 3685], Kumamoto [648], Oita [658] and Miyazaki [1120]. Red solid circles in the map denote sites where this species was observed in Japan.

2. Fossil record

Many fossils have been found from cave and fissure deposits of the Middle Pleistocene, the Late Pleistocene and the Early Holocene in Honshu and Kyushu [1512, 1514, 1515, 1518, 1519, 1522, 2360, 2687, 4158].

3. General characteristics

Morphology   Face is slender; tubular nostrils protrude 1.0 mm externally; the inter-narial septum is shallow and wide; ears have 2 concavities in the external margin and the shape is ovate and rounded; the tragus is turned outward and more than half of the ear in height; thumbs are very large in length, about 29% of FA; the claw is slender; fur consists of silky straight hairs and soft curly hairs; guard hairs are silvery and glossy; the dorsal surface of the uropatagium, plagiopatagium and thumb are thickly covered with fur; the plagiopatagium is attached to the base of the first toe of the hind foot [4208]. Wing has low wing aspect ratio and high wingtip shape index [631].

External measurements (mean in mm with range) are as follows.

Hokkaido: FA = 41.6 (40.0–43.4; n = 18) for females [633]; FA = 41.1 (39.5–43.9; n = 13), HB = 55.0 (n = 1), T = 45 (n = 1), Tib = 21.7 (n = 1), E = 19.0 (n = 1), Tr = 10.0 (n = 1) for males [452, 633]. Iwate Pref.: FA = 42.3 ± 1.18 SD (40.7–44.6; n = 9), HB = 55.8 ± 3.49 (54.0–60.0; n = 5), T = 34.6 ± 1.67 (33.0–37.0; n = 5), HFcu = 11.9 ± 1.26 (10.5–13.8; n = 9), Tib = 17.1 ± 1.55 (15.0–18.9; n = 7), E = 16.1 ± 1.57 (13.8–18.5; n = 9), Tr = 10.3 ± 1.01 (8.5–11.8 ; n = 9) for females; FA = 42.1 ± 0.98 (40.6–43.5; n = 8), HB = 55.4 ± 2.57 (53.0–58.2; n = 4), T = 37.7 ± 2.50 (35.5–41.3; n = 4), HFcu = 12.7 ± 1.15 (10.5–13.8; n = 7), Tib = 18.6 ± 2.53 (16.1–21.7; n = 4), E = 17.2 ± 1.02 (15.0–18.2; n = 8), Tr = 10.1 ± 0.78 (9.0–11.1; n = 8) for males [516, 4208]. Eastern Honshu (Fukushima, Tochigi and Saitama Prefs.): FA = 43.3 ± 1.26 (42.0–45.0; n = 4), HB = 59.3 ± 9.19 (46.5–67.5; n = 4), T = 37.5 ± 2.84 (35.0–41.4; n = 4), HFcu = 12.8 ± 0.25 (12.5–13.0; n = 3), Tib = 18.9 ± 1.38 (17.5–20.5; n = 4), E = 17.3 ± 0.69 (16.6–18.2; n = 4), Tr = 10.5 ± 0.50 (10.0–11.0; n = 3) for females; FA = 42.0 and 42.7 (n = 2), HB = 56.2 ± 3.75 (52.5–60.0; n = 3), T = 37.0 ± 1.00 (36.0–38.0; n = 3), HFcu = 11.8 ± 1.17 (10.5–12.8; n = 3), Tib = 17.0 ± 0.62 (16.5–17.7; n = 3), E = 15.6 ± 0.55 (15.0–16.0; n = 3), Tr = 8.0 and 9.6 (n = 2) for males [1577, 4208]. Central Honshu (Toyama, Yamanashi, Nagano, Gifu, Shizuoka and Wakayama Prefs.): FA = 43.1 ± 0.74 (42.0–44.0; n = 5), HB = 59.4 ± 4.29 (53.5–64.0; n = 5), T = 38.7 ± 4.18 (33.5–45.0; n = 5), HFcu = 13.5 ± 0.68 (13.0–14.5; n = 5), Tib = 17.6 ± 1.25 (16.0–19.0; n = 4), E = 18.7 ± 1.01 (18.0–20.3; n = 5), Tr = 9.5 ± 2.00 (7.0–12.0; n = 5) for females; FA = 42.1 ± 1.28 (40.0–44.3; n = 15), HB = 60.4 ± 5.99 (50.0–69.5; n = 14), T = 37.5 ± 3.31 (32.0–43.0; n = 14), HFcu = 12.5 ± 1.30 (10.5–15.0; n = 14), Tib = 18.8 ± 1.74 (15.0–21.0; n = 13), E = 17.8 ± 1.00 (16.5–20.0; n = 14), Tr = 9.9 ± 1.45 (8.2–12.0; n = 12) for males [1577, 1920, 4006, 4208]. Western Honshu (Hiroshima and Yamaguchi Prefs.): FA = 44.0 ± 1.44 (42.0–45.8; n = 8), HB = 55.4 ± 4.24 (50.0–64.1; n = 8), T = 43.5 ± 2.41 (39.0–47.0; n = 8), HFcu = 11.6 ± 1.65 (9.5–14.6; n = 8), E = 15.8 ± 2.81 (11.5–19.0; n = 8), Tr = 9.7 and 9.9 (n = 2) for adult females; FA = 42.7 ± 0.75 (42.0–43.5; n = 3), HB = 52.2 ± 5.25 (47.0–57.5; n = 3), T = 41.4 ± 3.49 (37.2–44.0; n = 3), HFcu = 10.8 ± 1.42 (9.5–12.3; n = 3), E = 14.3 ± 4.04 (12.0–19.0; n = 3), Tr = 9.3 (n = 1) for adult males [1769, 2100]. Shikoku (Ehime and Kochi Prefs.): FA = 44.0 ± 1.43 (41.0–45.1; n = 7), HB = 62.0 ± 3.92 (58.0–67.0; n = 4), T = 37.8 ± 3.40 (35.0–42.0; n = 4), HFcu = 12.1 ± 1.03 (11.0–13.5; n = 4), Tib = 19.9 ± 0.66 (19.0–20.4; n = 4), E = 18.9 ± 0.95 (17.5–19.5; n = 4), Tr = 10.1 ± 0.48 (9.5–10.5; n = 4) for females; FA = 43.8 ± 0.90 (42.5–46.2; n = 23), HB = 61.2 and 67.4 (n = 2), T = 34.6 and 36.7 (n = 2), HFcu = 12.6 and 13.1 (n = 2), E = 17.1 (n = 1), Tr = 10.2 (n = 1) for males [3088, 3912, 3980, 4208]. Kyushu (Fukuoka Pref.): FA = 44.5, T = 41.5, E = 15.0, Tr = 9.5 for a male (n = 1) [3685].

Dental formula I 2/3 + C 1/1 + P 2/2 + M 3/3 = 34 [4208]. Deciduous teeth: i 3/2 + c 1/1 + pm 2/2 = 22 [1780].

Genetics Chromosome: 2n = 44 and FN = 50–58 [142, 789]. ND1 and cytochrome b genes of mtDNA have been sequenced [1458; described as *M. leucogaster*, 2974].

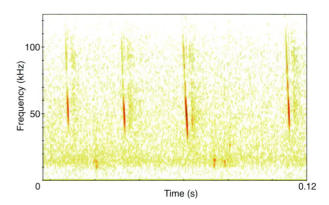

Fig. 1. Sonogram of a release call, recorded by D. Fukui in Hokkaido.

4. Ecology

Reproduction Parturition occurs in July [1780, 1806, 3024, 4022]. Litter size varies from 1 to 3 [1780, 1806, 4022]. Several observations of their reproductive traits have been reported; 6 adults with 8 infants have been found in a tree canopy [1806], and several colonies consisting of adults and infants have been observed in a tunnel in July [3024]. Ratios of neonatal to maternal values of body weight are 14–17%; all the deciduous teeth erupt at birth; ears stand erect at 3–4 days of age, eyes open at 8–10 days, and hairs cover fully at 12–14 days; replacement of deciduous teeth with permanent ones is complete at 23–24 days of age [1780]. Sexual maturity and breeding system are unknown.

Lifespan The longest survivors in a long-term banding study were estimated to be approximately 5 years and 9 months in the case of one male, and 8 years and 8 months for a female [1086].

Diet Diets under natural conditions are not reported. Under captive conditions, consumes 4–10 g mealworm *Tenebrio molitor* larvae per day but does not consume basket worms [3685].

Habitat Roosting individuals have been found in various structural objects, foliage [2367, 2396, 3992], tree branches [1083, 1806, 4020], bat boxes [3528], nest boxes of Russian flying squirrels [4046], houses [267, 1083, 2070, 2997, 3229, 3685, 4232], abandoned mines, caves and tunnels [60, 373, 374, 603, 1083, 1086, 1577, 1704, 1769, 1776, 1777, 2012, 2190, 2228, 2592, 2728, 2729, 2997, 3024, 3117, 3229, 3264, 3550, 3748, 3750, 3751, 3753, 3757, 3967]. The individuals in the caves were more commonly observed in narrow crevices rather than on open ceilings [1086]. This species exhibits hovering flight [516]. The occurrence probability of feeding individuals decreases with increasing canopy gap size [631].

Home range Not reported.

Behavior Excepting maternity colonies, 1 or a few individuals of both sexes roost in caves from early summer to early autumn. One observation reported that more than 10 males formed a colony in caves in October [2220]. In one instance, 10 to more than 100 individuals aggregated in a cave between early February and late June [60, 374, 1083, 2012, 3753]. The number of individuals in an aggregation reaches a maximum in late April to early May, and aggregations disappear by early July [60, 1083, 2012]. The sex ratio in an aggregation colony is almost 1:1 [2012]. The role of such aggregation is still unclear.

Natural enemies Predation by domestic cats has been reported [3993].

Parasites
Ectoparasite
Diptera: *Basilia truncata* [2187]. Acari: *Spinturnix maedai* [3706].
Endoparasite
Not reported.

Echolocation and social call Echolocation calls are short and broadband FM (Fig. 1). Mean values of EF and PF at release or inside mosquito nets are 43.6 kHz (range = 35.6–50.1; $n = 33$) and 51.2 kHz (47.0–56.5; $n = 33$) for Hokkaido population; 27.4 ± 1.38 (SD) kHz (range = 25.3–28.4; $n = 10$) and 49.6 ± 4.13 kHz (44.0–54.7; $n = 10$) for Miyazaki population, respectively [624, 660].

Sounds emitted by infants are of 2 types, FM signals and isolation calls; the frequency range of FM signals extends rapidly to high frequency for echolocation just before the age when juveniles began to fly [4022].

5. Remarks

Formerly included in *M. leucogaster* and treated as a subspecies, which was widely used for larger tube-nosed bats from Siberia and East Asia including Japan (e.g. [1713]). In Japan, the subspecific rank was not generally used, and therefore this bat was simply referred as *M. leucogaster* (e.g. [1893]). Alternatively, there was an opinion that *M. l. hilgendorfi* from Japan should be treated as a valid species called *M. hilgendorfi*, particularly in comparisons of specimens from Korea and Sakhalin [4208]. A study pointed out that the variability of forms within the genus *Murina* might reflect the low migration activity of those bats [1747]. A cline in morphology from south to north along the Japanese Isls. may be present, in which case a reassessment of the taxa may be needed.

K. KAWAI

Hilgendorf's tube-nosed bats hibernating in a cave (N. Urano).

Chiroptera VESPERTILIONIDAE

Red list status: EN (MEJ); EN (IUCN)

Murina ryukyuana Maeda & Matsumura, 1998

EN Ryukyu tube-nosed bat   JP リュウキュウテングコウモリ（ryukyu tengu koumori）   CM 琉球管鼻蝠   CT 琉球管鼻蝠
KR 류큐관코박쥐   RS рюкюйский трубконос

An individual captured on Tokuno-shima Isl., Kagoshima Pref. (K. Funakoshi).

1. Distribution

Endemic to Japan. Five individuals, the holotype and paratypes, were captured in a natural forest on the northern part of Okinawa-jima Isl., Okinawa Pref. in 1996, and described as a new species [1929]. After the first description, more than 25 individuals were recorded from Tokuno-shima Isl., Kagoshima Pref., and more than 2 from Amami-ohshima Isl., Kagoshima Pref. [660, 666, 1919, 1932]. Red solid circles in the map denote sites where this species was observed in Japan.

2. Fossil record

Not reported.

3. General characteristics

Morphology The dorsal surface is generally pale brown with the bottom 4 mm of dark brown clearly demarcated from the upper part of pale brown; on the ventral side, hairs are 7–8 mm long, the top 3–4 mm is pale brown and the bottom 4 mm is dark brown; lengths of ear and tragus are conspicuously long; tubular nostrils are conspicuously projecting outwards about 1.5 mm from the base; wings are comparatively wide; the ratio of the length of the third digit to that of the fifth digit is about 1.2; wing membranes attach to the second phalange of the first toe, or the base of the phalange; the tail is shorter than the head and body length, about 1 mm of its terminal portion is free from the interfemoral membrane [1929].

External measurements (in mm) are as follows. Okinawa-jima Isl., Okinawa Pref.: FA = 37.0 and 35.5, HB = 47.0 and 47.0, T = 45.0 and 45.0, HFcu = 10.5 and 10.5, Tib = 19.0 and 19.0, E = 18.5 and 18.5, Tr = 10.5 and 10.5 for 2 adult males [1929]. Amami-ohshima Isl., Kagoshima Pref.: FA = 32.3, HB = 46, T = 40.5, E = 18.5, Tr = 10.0 for a male [1932]. Tokuno-shima Isl., Kagoshima Pref. (mean ± SD): FA = 37.1 ± 0.54 (range = 36.3–38.1; n = 11) for females; FA = 35.1 ± 0.74 (range = 33.8–36.3; n = 17), HB = 44.0 (n = 1), T = 34.0 (n = 1), HFcu = 11.0 (n = 1), Tib 18.5 (n = 1), E = 18.5 (n = 1), Tr = 10.5 (n = 1) for male [666, 1919]. FA of females is significantly longer than that of males [666].

Dental formula   I 2/3 + C 1/1 + P 2/2 + M 3/3 = 34 [35].

Genetics Chromosome: not reported. ND1 gene of mtDNA has been sequenced [1458].

4. Ecology

Reproduction Parturition may occur from May to July in Tokuno-shima Isl. [666]. Several observation of their reproductive traits have been reported in Tokuno-shima Isl.; a juvenile was

Murina ryukyuana Maeda & Matsumura, 1998

captured in beginning of June, 5 of 6 females were lactating with the rest of the group likely comprising subadults in July and August, females had atrophied mammary gland in middle of August [666, 1919]. Testis is growing from beginning of September and its size becomes maximum in November [666]. Sexual maturation may occur after 1 year old [1919].

Lifespan Not reported.

Diet Not reported.

Habitat All individuals were captured in evergreen broadleaf forests and their surroundings. Solitary males roosting in dead-leaf foliage of 3 tree species (*Idesia polycarpa, Schefflera heptaphylla* and *Symplocos cochinchinensis*) were observed in September, in Tokuno-shima Isl. [666, 3852]. Artificial roosts made by tied dead-leaves of *Mallotus japonicus* were used by solitary bats in August, September and November [666].

Home range Not reported.

Behavior Not reported.

Natural enemies Not reported.

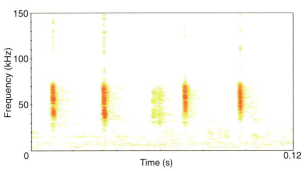

Fig. 1. Sonogram of an echolocation calls in a mosquito net, recorded by K. Funakoshi in Tokuno-shima Isl., Kagoshima Pref.

Parasites
Ectoparasite
Not reported.
Endoparasite
Trematoda: *Prosthodendrium klausrohdei, P. radiatum, Lutziella microacetabularis, Macroorchis himizu, Zonorchis hokkaidensis* [1559].

Echolocation and social call Echolocation calls are short and broadband FM (Fig. 1). The mean values (± SD) of EF and PF at hand release or inside a mosquito net are 43.6 ± 1.52 kHz ($n = 6$) and 58.3 ± 3.05 kHz ($n = 6$), respectively [660].

K. KAWAI

Dead leaves of *Idesia polycarpa* were used as day-roosts by the Ryukyu tube-nosed bat (K. Funakoshi).

An individual roosting in artificial roosts made by tied dead-leaves of *Mallotus japonicus* (K. Funakoshi).

Chiroptera VESPERTILIONIDAE

Red list status: DD (MEJ); K (MSJ); CR (IUCN)

Murina tenebrosa Yoshiyuki, 1970

EN gloomy tube-nosed bat   JP クチバテングコウモリ (kuchiba tengu koumori)   CM 对马管鼻蝠   CT 對馬管鼻蝠
KR 대마도관코박쥐   RS цусимский трубконос

The type specimen captured on Tsushima Isls., Nagasaki Pref., and stored in the National Science Museum, Tokyo (K. Kawai).

1. Distribution

Endemic to Japan. Occurred on Tsushima Isls., Nagasaki Pref. [4208]. There has been no information since 1962 when the type specimen was captured. The red solid circle in the map denotes the site where the type specimen was captured in Japan.

2. Fossil record

Not reported.

3. General characteristics

Morphology The dorsal surface of the uropatagium is almost naked; the plagiopataium is attached to the base of the claw of the first toe of the hind foot; the calcar is stout with a posterior lobe. The skull is larger than *M. ussuriensis*; the rostrum is deep and massive; the temporal ridge is evident; the anterior narial emargination is shallow, cochleae small; the upper anterior premolar (P^2) is comparatively large; tubular nostrils are long, projecting outwards, shorter than that of the other Japanese *Murina* [4208]. This species is different from *M. ussuriensis* in that the dorsal surface of the uropatagium is almost naked.

External measurements (in mm) are as follows [4198]. FA = 34.4, HB = 50.5, T = 34.5, HFcu = 8.8, Tib = 14.4, E = 16.0, Tr = 8.5 for the holotype (female).
Dental formula   I 2/3 + C 1/1 + P 2/2 + M 3/3 = 34 [4208].

Genetics Not reported.

4. Ecology

Reproduction Not reported.
Lifespan Not reported.
Diet Not reported.
Habitat The holotype was collected in an abandoned mine on the northern part of Tsushima Isls. on 15th February 1962 [4198].
Home range Not reported.
Behavior The holotype was presumed to be hibernating in the abandoned mine where it was collected [4198].
Natural enemies Not reported.
Parasites Not reported.
Echolocation and social call Not reported.

5. Remarks

Measurements of the skull are slightly larger than those in *M. ussuriensis*. For this reason, a specimen from Yaku-shima Isl. had been included in *M. tenebrosa* [4208]. However, demonstrating a sort of mixture of features among specimens from Honshu and Hokkaido indicates that the Yaku-shima Isl. population should be classified as *M. ussuriensis*, not as *M. tenebrosa* [633; described as *M. silvatica*].

K. KAWAI

Chiroptera VESPERTILIONIDAE

054

Red list status: R (MSJ); LC (IUCN)

Murina ussuriensis Ognev, 1913

EN Ussurian tube-nosed bat   JP コテングコウモリ (kotengu koumori)   CM 乌苏里管鼻蝠   CT 烏蘇里管鼻蝠
KR 작은관코박쥐   RS уссурийский трубконос

An individual captured in Miyagi Pref. (O. Takahashi).

1. Distribution

Distributed in the Korean Pen., Primorye, Sakhalin, Kunashiri (Kunashir) Isl. and Japan [35, 1460]. In Japan, reported from 34 prefectures: Hokkaido [93, 451–455, 625, 627, 630, 633, 839, 840, 900, 903, 935, 1447, 1448, 1700, 1706, 1896, 1935, 1936, 2021, 2343, 2715, 2780, 3054, 3055, 3060, 3061, 3063, 3064, 3067, 3070, 3072–3077, 3079–3082, 3085, 3086, 3316, 3372, 3732, 3736, 3951–3954, 3956, 4055, 4058, 4208], Aomori [2288, 2290, 2301, 3788, 4208], Iwate [436, 515, 2293, 4208], Miyagi [96, 4208], Akita [2301], Yamagata [2423, 3623, 4208], Fukushima [1577, 3338, 4194, 4204, 4208], Ibaraki [4081, 4091, 4264], Tochigi [1734, 1735, 2132, 4086, 4092], Gunma [3018], Saitama [3359, 4208], Kanagawa [3969], Niigata [603, 605, 2577, 3141], Ishikawa [3992, 4006], Yamanashi [1787, 3242], Nagano [162, 2066, 2362, 3915], Gifu [3991, 3992, 4216], Shizuoka [3034, 3098], Aichi [3555], Mie [3228, 3603], Kyoto [188], Hyogo [1591], Nara [3228, 3904], Tottori [2673], Oki Isl., Shimane [1421, 4208], Okayama [3107, 3939], Hiroshima [1301, 3715, 3716, 4232, 4235], Tokushima [1], Ehime [3107, 4002], Kochi [3912, 3913], Tsushima Isls., Nagasaki [1452, 3758, 4199], Kumamoto [1732], Miyazaki [658], Kagoshima [650, 658, 659] and Yaku-shima Isl., Kagoshima Pref. [634, 659]. Red solid circles in the map denote sites where this species was observed in Japan.

2. Fossil record

Many fossils have been found from cave and fissure deposits of the Middle Pleistocene, the Late Pleistocene and the Early Holocene in Honshu [1519, 1522, 4158].

3. General characteristics

Morphology Ears are short, tips rounded broadly, anterior and posterior margin is convex; the tragus is more than half as high as the conch, anterior and posterior borders are almost straight; the muzzle is tubular, nostrils protrude 1–3 mm externally, internarial septum is deep; thumbs are long about 10 mm, 35.3% of forearm length; the propatagium attaches to the second phalanx of the first toe; the fur is very soft with a silky texture and glossy appearance; upper part of guard hairs extend densely to the plagiopatagium and uropatagium; general color of upper parts is golden rufous or light grayish brown, these 2 color phases do not relate to sex, age, season and locality; ears and membranes are brown [4208]. Wing has low aspect ratio and high wingtip shape index [631].

External measurements (mean ± SD in mm with range) are as follows. Hokkaido: FA = 31.38 ± 1.07 (29.00–34.00; n = 27), Tib = 15.39 ± 0.67 (14.30–17.00; n = 26), E = 13.82 ± 1.03 (11.50–16.00; n = 26), Tr = 7.84 ± 0.69 (6.00–9.00; n = 26) for females;

FA = 29.74 ± 0.85 (28.10–31.40; n = 26), Tib = 15.13 ± 0.64 (14.00–16.20; n = 24), E = 13.80 ± 1.03 (12.0–16.00; n = 24), Tr = 7.37 ± 0.63 (6.00–8.00; n = 24) for males, Kunashiri (Kunashir) Isl.: FA=30.9 for a female, FA = 28.0 and 29.4 (n = 2), Tib=14.6 (n = 1), E = 4.2 (n = 1) for males, northern Honshu (Aomori, Akita and Iwate Prefs.): FA = 30.66 ± 1.41 (28.30–32.30; n = 9), Tib = 13.63 ± 0.94 (12.00–14.50; n = 9), E = 15.19 ± 0.99 (14.00–16.50; n = 9), Tr = 8.28 ± 0.76 (7.00–9.50; n = 9) for females; FA = 30.05 ± 0.65 (28.80–31.00; n = 9), Tib = 13.70 ± 0.84 (12.50–15.00; n = 9), E = 14.70 ± 0.49 (14.0–15.50; n = 9), Tr = 7.94 ± 0.68 (6.50–9.00; n = 9) for males, central Honshu (Nagano, Gifu and Yamanashi Prefs.): FA = 31.81 ± 0.76 (30.50–33.00; n = 9), Tib = 15.53 ± 0.59 (14.80–17.00; n = 25), E = 13.81 ± 0.97 (12.00–15.50; n = 25), Tr = 7.37 ± 0.70 (6.00–9.00; n = 25) for females; FA = 30.70 ± 0.83 (29.40–31.80; n = 12), Tib = 15.49 ± 0.57 (14.60–16.40; n = 12), E = 14.00 ± 0.95 (12.0–15.00; n = 12), Tr = 7.67 ± 0.58 (6.50–9.00; n = 12) for males, and Yaku-shima Isl., Kagoshima Pref.: FA = 33.82 ± 1.03 (31.20–35.80; n = 22), Tib = 17.0 (n = 2), E = 16.0, 17.0 (n = 2), Tr = 9.0 and 9.5 (n = 2) for females; FA = 32.82 ± 0.76 (31.30–34.00; n = 17), Tib = 17.0 and 17.0 (n = 2), E = 15.5 and 17.0, Tr = 9.00 and 9.50 (n = 2) for males [634; described as *M. silvatica*]. The Oki Isls., Shimane Pref.: FA = 31.50, HB = 43.0, T = 33.00, HFcu = 10.00, Tib = 15.00, E = 15.00, Tr = 7.50 for an old male [4208; described as *M. silvatica*]. Tsushima Isls., Nagasaki Pref.: FA = 32.77 (32.30–33.10; n = 3) for females; FA = 31.61 (30.70–32.90; n = 8) for males; FA = 32.83 and 30.87, HB = 49.38 and 42.50, T = 26.93 and 38.50, HFcu = 8.85 and 8.00, Tib = 16.30 and 14.00, E = 15.00 and 15.00 for 2 sex-unknown individuals [1452, 1460]. FA of females are significantly larger than it of males in Kagoshima Pref.; FA = 32.6 ± 0.89 (n = 10) for adult females, FA = 30.7 ± 0.64 (n = 18) for adult males [671].

There is no distinct cline in skull morphology among Hokkaido, northern and central Honshu populations; however, all measurements are significantly greater for the Yaku-shima population than for the others [634; described as *M. silvatica*]. Measurements of the skull and external characters in the Tsushima population are also similar to those of the Yaku-shima population [1452].

Dental formula I 2/3 + C 1/1 + P 2/2 + M 3/3 = 34 [4208].

Genetics Chromosome: $2n$ = 44 and FN = 56–60 [142; described as *M. aurata*, 789; described as *M. aurata ussuriensis*]. Molecular phylogenetic relationships among vespertilionids have been inferred with the ND1 gene of mtDNA, vWF gene of nuclear DNA and the SINE insertion [1459].

4. Ecology

Reproduction Litter size varies from 1 to 2 [517, 839, 1907, 3762, 4235]. Parturition occurs from late May [671, 3762] to late July [517, 839, 1907, 4235]. Several observations of their reproductive traits have been reported; 8 infants of several developmental stages have been found in a tree cavity [4235]. Maternity roosts were observed in warm temperate rain forest of in Yaku-shima Isl. included 4 types: vertically suspended clusters of dead leaves, tree hollows, a crevice in a root plate and a completely exposed site under a branch. The number of individuals recorded roosting together ranged from 2 (a mother with infant) to 22 individuals. The maternity colonies show very frequent roost-switching [629]. One mature male and 4 adult females have been found separately in dead furled leaves in autumn, within a 40 m radius; this case might describe the mating system of this species [4216]. Females enter estrus in their first autumn, but it is not clear when the first spermatogenesis occurs in males [1907]. Size of testis grows up from the end of August and become it in maximum size in September; maximum size of epididymis is in October in Kagoshima Pref. [671].

Lifespan Not reported.

Diet Diet under natural condition is not reported. In captivity, eats the house fly (*Musca domestica*), Coleoptera and Orthoptera, consuming more than 3 g per day with the aid of the uropatagium [515].

Habitat Individuals have been found in various structural objects, in tree cavities [517, 629, 902, 1732, 1799, 2425, 4235], under bark [4235], in foliage [629, 650, 671, 903, 934, 1735, 2021, 2132, 2343, 2367, 2425, 3372, 3939, 3991, 3992, 4216], on the ground [451, 1577, 3338, 4208], under leaf litter [839, 1896], in tree canopies [900], in a crevice in a root plate [629], under a branch [629] in abandoned mines and tunnels [3098, 3107, 3603, 3623], in houses [517, 605, 612, 614, 3623, 4208, 4232], under frost cover on a chrysanthemum [4235], inside a fallen tree [162], inside an agricultural barn [3951, 3956], and on the snow [96, 839, 904, 1301, 1787, 2366, 2577, 3546, 3715, 4058]. Day roosts in foliage are used by 1 to 5 individuals from summer to autumn [650, 671, 903, 934, 1735, 2021, 2132, 2343, 2367, 2425, 3372, 3939, 3991, 3992, 4216]. Usually, the bats select larger leaves, and height from ground is not related to roost selection in spring and summer in Hokkaido [2021]. Females roost in tree canopies and change roosts everyday in summer [900]. A night roost has been observed in a tunnel [4002]. More than 30 cases of these bats straying into houses at night have been reported (e.g. [521]). The occurrence probability of feeding individual decreases with increasing canopy gap size [631].

Home range Not reported.

Behavior Foraging behavior has been observed 1 hour after sunset over the surface of a spring in forest, and 40 minutes before sunrise over trails [517]. This species exhibits hovering flight [515].

Hibernating traits are unknown but there have been several observations in winter; the 3 of 5 cases where bats have been found in tree cavities [517, 1732, 1799, 2425, 4235], 1–8 individuals were found in pits or crevices inside tunnels from December to March [3034]. Bats have been found during the day on snow remaining in spring, typically in April and May [96, 839, 1301,

Murina ussuriensis Ognev, 1913

Fig. 1. A bat sleeping on a patch of snow in Hokkaido (H. Nakajima).

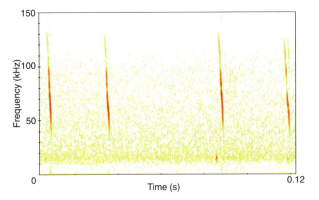

Fig. 2. Sonogram of a release call, recorded by D. Fukui in Hokkaido.

2366, 2577, 3546, 4058], with the exception of 2 records in November [904] and January [3715]. Individuals were also observed in snow pockets less than 10 cm in depth (Fig. 1) [96, 839, 2366, 2577, 3546, 3715, 4058], except for 2 observations 25 cm and 50–80 cm under the snow surface, respectively [904, 1301]. It remains unknown whether those individuals hibernated in snow or entered torpor after activities in spring.

Natural enemies Predation by the Japanese rat snake (*Elaphe climacophora*) [901] and domestic cats [1452]. Impaling on a barbed wire by the brown shrike (*Laninus cristatus*) has been reported [3915].

Parasites
Ectoparasite
Diptera: *Nycteribia pleuralis*, *N. pygmaea* [2187, 3069]. Acari: *Mycteridopsylla nipopo*, *Spinturnix maedai* [2715, 2968, 3706].
Endoparasite
Trematoda: *Mesothatrium japonicum*, *Plagiorchis* (*Plagiorchis*) *minutifollicularis*, *P. latus*, *Plagiorchis* sp. *Prosthodendrium* (*Prosthodendrium*) *thomasi* [215, 1550, 1554, 1559, 3469]. Cestoda: *Vampirolepis yakushimaensis*, *Vampirlepis* sp. [215, 3122, 3469].

Echolocation and social call Echolocation calls are short and broadband FM (Fig. 2). PF varies with individuals because there is no clear PF. Mean values ($\pm$ SD) of EF and PF in kHz: 46.3 ± 6.8 ($n = 12$) and 69.8 ± 9.4 ($n = 12$), respectively in the Hokkaido population [625]; 51.6 ± 5.65 (range = 40.4–63.8; $n = 6$) and 61.9 ± 2.75 (52.0–69.0; $n = 6$), respectively in Miyazaki Pref. [660]. Various types of social calls were recorded at maternity roosts. The pulses of social calls were lower frequency (typically within the range 14–80 kHz), higher intensity, longer duration (6–30 ms) and given at irregular intervals. These included curved sweeps, stepped sweeps, complex FM-CF combinations and trills [629].

5. Remarks

The systematics of the Ussurian tube-nosed bat in Far East Russia, Sakhalin and East Asia including Japan has been disputed. In Japan, *M. ussuriensis* was recorded for the first time on Yakushima Isl., Kagoshima Pref. in 1920 [100]. After the description, most Japanese mammalogists used this name for the small-sized *Murina* collected from Hokkaido, Honshu and Tsushima Isls. Provisionally it was included in *M. aurata* or treated as a subspecies of *M. aurata* that is distributed in southwest China and Myanmar [500]; however, the small-sized *Murina* in Japan and Far East Russia is clearly different from *M. aurata* on characteristics of the skull and baculum [1908]. In 1983, *M. silvatica* was described in Japan discriminating from syntypes of *M. ussuriensis* [4205], and several mammalogists followed this description (e.g. [634, 1713, 4208]). However, there is the opinion that "*M. silvatica*" is a subspecies of *M. ussuriensis* [1747, 1916]. An examination of features of the mainland Russia and Japanese forms of the small-sized *Murina* indicated that the Japanese specimens need to be morphologically reassessed, and that any species or subspecies reclassification should be validated by genetic studies [1747].

K. KAWAI

Chiroptera MINIOPTERIDAE

055

Red list status: C-1 (MSJ); LC (IUCN)

Miniopterus fuliginosus (Hodgson, 1835)

EN eastern bent-winged bat   JP ユビナガコウモリ (yubinaga koumori)   CM 亚洲长翼蝠   CT 亞洲長翼蝠
KR 긴가락박쥐   RS тёмный длиннокрыл

A small cluster of eastern bent-winged bats hibernating in an abandoned mine (N. Urano).

1. Distribution

Ranges from Afghanistan to India, China and Japan [1909]. In Japan, known from Honshu, Shikoku, Kyushu, and the islands of Sado-gashima, Izu-ohshima (the Izu Isls.), Tsushima, Oki, Fukue-jima (the Goto Isls.) and Yaku-shima [106, 3124, 4208].

2. Fossil record

Many fossils are reported in cave and fissure deposits of the Middle Pleistocene, the Late Pleistocene and the Early Holocene in Honshu and Kyushu [1514, 1515, 1518, 2360, 4158, 4159].

3. General characteristics

Morphology   Fur is short, velvety, and colored dark brown; wings are narrow, second phalanx of third finger is nearly 3 times as long as the first, membrane is inserted on the tibia or slightly above the ankle; ears are short and rounded, a relatively long tragus has a blunt tip [4208]. Wing aspect ratio is 7.42 [4136].

External measurements (mean ± SD in mm) from 5 localities are as follows. Honshu: FA = 47.3 ± 1.07 ($n = 190$), HB = 59.8 ± 3.64 ($n = 99$), T = 53.1 ± 1.69 ($n = 7$), HFcu = 11.3 ± 0.44 ($n = 36$), Tib = 19.5 ± 1.12 ($n = 159$), E = 11.5 ± 0.87 ($n = 168$) [1564, 1909, 3018, 3266, 4094, 4208]. Shikoku: FA = 47.4 ± 0.92 ($n = 97$), HB = 62.1 ± 4.90 ($n = 41$), HFcu = 11.6 ± 0.17 ($n = 39$), Tib = 19.8 ± 0.97 ($n = 89$), E = 12.2 ± 0.70 ($n = 84$) [1340, 1909, 4208], and Kyushu: FA = 47.2 ± 0.98 ($n = 117$), HB = 60.3 ± 3.44 ($n = 83$), and HFcu = 11.1 ± 0.40 ($n = 21$), Tib = 18.1 ± 1.27 ($n = 94$), E = 11.8 ± 0.80 ($n = 104$) [1909, 4208]. Sado Isl.: FA = 46.7 ± 0.80 ($n = 41$), HB = 65.0 ± 2.77 ($n = 17$), Tib = 19.3 ± 0.66 ($n = 41$), E = 11.3 ± 0.70 ($n = 41$) [4208]. Tsushima Isls.: FA = 47.3 (45.6–49.4; $n = 42$), HB = 64.3 (59.3–69.7; $n = 3$), T = 41.1 (38.3–43.7; $n = 3$), Tib = 20.8 (20.2–21.2; $n = 3$), HFcu = 8.9 (8.6–9.6; $n = 3$) [1452].

Dental formula   I 2/3 + C 1/1 + P 2/3 + M 3/3 = 36 [4208].

Genetics   Chromosome: $2n = 46$ and FN = 52 [142]. Molecular phylogenetic relationships with Vespertilionidae have been inferred with the ND1 gene of mtDNA, vWF gene of nuclear DNA and the SINE insertion, and these sequences are available [1459]. cytochrome *b* of mtDNA sequence is also available in Genbank [2974].

4. Ecology

Reproduction   Reproductive pattern involves seasonal monoestry and delayed implantation; copulation, ovulation and fertilization occur in quick succession in autumn, but placentation

Miniopterus fuliginosus (Hodgson, 1835)

occurs after mid-March [1566, 1567, 2208, 2209].

Adult and subadult bats of both sexes coexist in a cluster from autumn to spring, but maternity colonies consist almost entirely of adult females [654, 1770]. The sizes of maternity colonies sometimes exceed 10,000, and the huge, dense clusters promote postnatal growth in infants [654]. Parturition occurs synchronously from late June to early July in Kyushu [654], with a single infant per litter. Neonates are completely naked with their eyes closed [654, 661]. Ratios of neonatal to maternal values in FA and BW are 36.0% and 23.0%, respectively [1770]. Mothers nurse their own young exclusively; juvenile mortality is 9% [654]. Most females give birth for the first time at the end of their second year [1781].

Lifespan The greatest longevity record is 15 years [1776, 3551].

Diet Prey includes Diptera, Lepidoptera, Coleoptera, Trichoptera, Ephemeroptera and Plecoptera [676, 677, 1770]. Forages typically in open spaces over grasslands, woodlands and open water, and feeds mainly on Lepidoptera, Diptera and Trichoptera measuring 5–25 mm in body length [676, 677].

Habitat Roosts in natural caves, abandoned mines, bomb shelters, unused tunnels and underground culverts [1909, 2991, 3124, 4090].

Home range Not reported.

Behavior Nocturnal [677, 4090]. A migratory species [3903]. Roosts shift seasonally, and are usually formed of dense clusters of more than ten bats in all seasons [652, 677, 1770, 2592, 2729]. Sizes of hibernating clusters often number in thousands, and a cluster up to 83,000 bats was recorded in central Honshu [3200]. Huge clusters decrease body weight loss in hibernating bats [679].

The BW ratio increases from late autumn to summer from 18.3 to 22.7% in Kyushu. Arousal frequencies during hibernation depend upon environmental temperature; the ratio of bats that forage out rapidly increases when outside air temperatures rise above 7°C [679].

A long distance migration of over 200 km from the natal place has been recorded in Honshu [3903].

Natural enemies Not reported.

Parasites & Pathogenic organisms
Ectoparasite
Diptera: *Nycteribia allotopa, N. parvula, N. pygmaea, Penicillidia jenynsii, Ascodipteron speiserianum* [653, 1339, 1760, 2183, 2187, 2473, 3069]. Acari: *Alabidocarpus fujii, Cheyletus malaccensis, Spinturnix psi, Pteracarus faini, Calcarmyobia japonica, Trombicula uchidai, Microtrombicula uchidai, Macronyssus coreanus, Ichoronyssus miniopteri, Speleognathopsis bastini, Argas vespertilionis, Ixodes simplex* [1297, 1311, 1760, 3691, 3698, 3706, 3806, 4031].

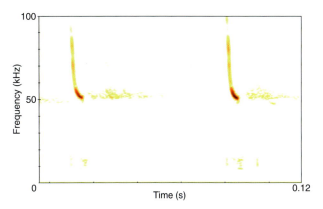

Fig. 1. Sonogram of a free-flying call near a roost, recorded by D. Fukui in Aomori Pref.

Endoparasite
Trematoda: *Duboisitrema sawadai, Mesotharium japonicum, Prosthodendrium chilostomum, Pr. parvouterus, Pr. longiforme, Pr. miniopteri, Pr. postacetabulum, Pr. urna, Pr. radiatum, Pr. thomasi, Pr. circulare, Pr. yamizense, Acanthatrium isostomum, A. ovatum, A. setoense, Paralecithodendrium ovimagnosum, Lecithodendrium macrostomum, Plagiorchis corpulentus, Pl. koreanus, Pl. rhinolophi, Pl. vespertilionis* [1541, 1542, 1546–1550, 1554–1556]. Cestoda: *Vampirolepis hidaensis, V. minatoi, Hymenolepis odaensis, Pseudodiorchis clavatus* [3116, 3119–3121]. Nematoda: *Strongylacantha rhinolophi, Molinostrongylus rhinolophi* [1257, 1258].

Pathogenic organisms
Virus strains that are similar to Flaviviruses but distinct from Japanese encephalitis virus (JEV) have been isolated [2120]. Bat coronaviruses have been also detected [3260].

Echolocation and social calls Echolocation calls are FM-QCF calls in the search phase, and FM calls in the scanning phase or back-cluttered space (Fig. 1). EF and PF values of search phase calls for the south Kyushu population ($n = 10$) average 47.1 kHz (45.9–48.0) and 48.3 kHz (47.2–49.8), respectively [660]. Newborn infants emit low frequency calls with a long duration, but the vocalization development is completed by weaning [661]. No information on social calls is available.

5. Remarks

Traditionally, the genus *Miniopterus* was placed in Vespertilionidae as a subfamily Miniopterinae. However, the recent genetic analysis showed that the *Miniopterus* represents its own well-defined family [2073]. One researcher treats *M. fuliginosus* as a subspecies of *M. schreibersi* ranging widely from Europe, Asia, Africa and Australia, namely, *M. s. fuliginosus* [4208]. Another regards this taxon as a synonym of *M. schreibersii* [3272]. MtDNA analyses support the recognition of Asian *M. schreibersii* as a valid species, *M. fuliginosus* [172, 3573].

A. SANO

Chiroptera MINIOPTERIDAE

Red list status: EN (MEJ); R (MSJ); EN (IUCN)

Miniopterus fuscus Bonhote, 1902

EN East-Asian little bent-winged bat   JP リュウキュウユビナガコウモリ (ryukyu yubinaga koumori)
CM 东南亚长翼蝠   CT 東南亞長翼蝠   KR 류큐긴가락박쥐   RS восточноазиатский длиннокрыл

A small cluster of East-Asian little bent-winged bats on a cave ceiling (H. Tamura).

1. Distribution

Endemic to the islands of Amami-ohshima, Tokuno-shima, Okinoerabu-jima, Okinawa-jima, Kume-jima, Ishigaki-jima and Iriomote-jima [1919, 4182, 4208].

2. Fossil record

Not reported.

3. General characteristics

Morphology General forms are very similar to *M. fuliginosus*, size is distinctly small [4208]. Pelage color is individually variable from reddish brown to blackish brown (Fig. 1).

External measurements (mean ± SD in mm) from 3 islands are as follows [666, 1909, 4202, 4208]. Amami-ohshima Isl.: (in mm) FA = 45.1 ± 1.08 (n = 5), Tib = 18.5 ± 0.41 (n = 5), HFcu = 9.4 ± 0.12 (n = 5), E = 10.9 ± 0.20 (n = 5), Tokuno-shima Isl.: FA = 43.2 (n = 1); Okinoerabu-jima Isl.: FA = 43.8 ± 0.73 (n = 24), HB = 53.0 ± 6.26 (n = 15), HFcu = 9.4 ± 0.18 (n = 5), Tib = 18.0 ± 0.83 (n = 24), E = 10.9 ± 0.64 (n = 24), and Iriomote-jima Isl.: FA = 44.1 ± 0.85 (n = 134), HB = 54.1 ± 3.29 (n = 65), HFcu = 9.7 ± 0.34 (n = 34), Tib = 17.9 ± 0.97 (n = 97), E = 11.0 ± 0.57 (n = 99). Tail lengths average 54.0 ± 0.30 mm (range 52.5–55.0, n = 4) for the populations on Ishigaki-jima and Iriomote-jima Isls. [1770].

Dental formula I 2/3 + C 1/1 + P 2/3 + M 3/3 = 36 [4208].

Genetics Chromosome: $2n$ = 46 and FN = 50–52 [787, 3140]. Molecular phylogenetic relationships with Vespertilionidae have been inferred with the ND1 gene of mtDNA, vWF gene of nuclear DNA and the SINE insertion [1459].

4. Ecology

Reproduction Mating season and female reproductive pattern unknown. Thousands of females congregate to form maternity colonies from June to August, with females bearing a single young in early June on Okinawa-jima Isl.; most females give birth beginning at 2 years of age, but a few do so at 1 year of age; only 2 maternity roosts are presently known from the islands of Okinawa-jima and Iriomote-jima [3487].

Lifespan Not reported.

Diet Prey includes Diptera, Lepidoptera, Coleoptera, Hemiptera, Hymenoptera, Orthoptera, Neuroptera and Psocodea; feeds mainly on Lepidoptera, Hymenoptera and Diptera [637].

Habitat Roosts in natural caves, abandoned mines and bomb shelters [1731, 1909, 1919, 1972, 3124, 4182].

Miniopterus fuscus Bonhote, 1902

Fig. 1. A large cluster of East-Asian little bent-winged bats on a cave ceiling (H. Tamura).

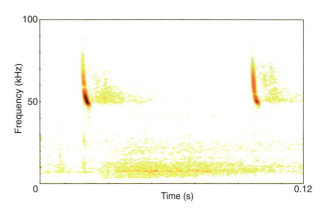

Fig. 2. Sonogram of a free-flying call, recorded by K. Funakoshi on Ishigaki-jima Isl.

Home range Not reported.

Behavior Nocturnal [1918, 3441]. Mainly forages over woodlands [1923, 3441]. Active even in winter [679].

Natural enemies Not reported.

Parasites
Ectoparasite
Diptera: *Nycteribia allotopa, N. parvula, Ascodipteron speiserianum, Brachytarsina amboinensis* [2185, 2187, 2473, 3069].
Endoparasite
Trematoda: *Anchitrema sanguineum, Mesothatrium japonicum, Paralecithodendrium ovimagnosum, Prosthodendrium parvouterus, Pr. thomasi, Pr. urna Pr. ascidia* [1550, 1554]. Cestoda: *Vampirolepis hidaensis* [3140]. Nematoda: *Molinostrongylus rhinolophi* [1262].

Echolocation and social calls Echolocation calls are FM-QCF calls in the search phase (Fig. 1). Mean PF value of the free flying call is 54.8 kHz ($n = 14$) [666].

5. Remarks

Traditionally, the genus *Miniopterus* was placed in Vespertilionidae as a subfamily Miniopterinae. However, recent genetic analysis shows that the *Miniopterus* represents its own well-defined family [2073]. Habitat destruction by development is the main threat [1898]. The concentration of maternity colonies to a small number of caves makes widespread local extinction a real possibility.

A. SANO

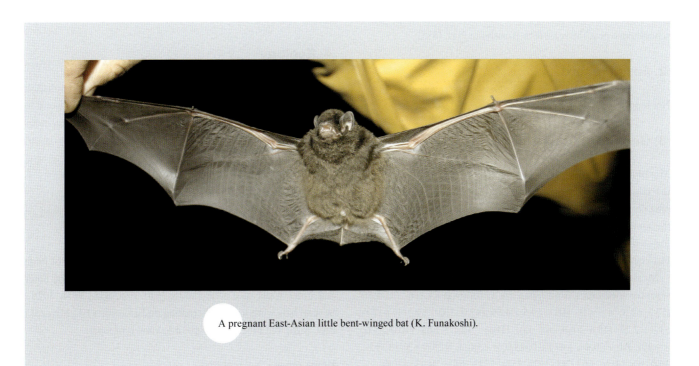

A pregnant East-Asian little bent-winged bat (K. Funakoshi).

Chiroptera MOLOSSIDAE

057

Red list status: VU (MEJ); K (MSJ); DD (IUCN)

Tadarida insignis (Blyth, 1861)

EN Oriental free-tailed bat   JP オヒキコウモリ（ohiki koumori）   CM 东方犬吻蝠   CT 游離尾蝠
KR 큰귀박쥐   RS восточный складчатогуб

Oriental free-tailed bats roosting in a rock crevice on a desert island (A. Sano).

1. Distribution

Known from China, Korea, the Ussuri area (Russia), Taiwan and Japan [4208]. In Japan, there are more than 30 distributional records from Hokkaido, Honshu, Shikoku and Kyushu [4208]. Colonies are known from only 6 places, 4 desert islands in Kyoto [1916], Mie [3004], Kochi [685] and Miyazaki Prefs. [670], a rock cliff projecting into the sea in Shizuoka Pref. [3026], and a schoolhouse near the seaside in Hiroshima Pref. [835].

2. Fossil record

Not reported.

3. General characteristics

Morphology   Tail projects by at least 1/3 of its length beyond the edge of tail membrane (Fig. 1). The head is flat and the muzzle is broad; ears are very large and rounded, arising from the same point on forehead; wings are narrow, plagiopatagium is inserted to the basal 1/3 or 1/2 portion of the tibia; pelage color is blackish brown [4208].

Measurements (mean ± SD in mm with range and sample size) for specimens from Japan are as follows; FA = 62.6 ± 1.50 (57.7–65.3; n = 87), HB = 89.5 ± 4.89 (82.0–102.1; n = 17), T = 53.7 ± 2.77 (48.0–60.0; n = 63), HFcu = 12.1 ± 1.63 (10.0–15.0; n = 13), Tib = 20.4 ± 1.47 (17.0–23.0; n = 58), E = 29.6 ± 3.11 (23.5–34.0; n = 14) [6, 667, 761, 1021, 2514, 2733, 3004, 3603, 4208].

Dental formula   I 2/3 + C 1/1 + P 2/2 + M 3/3 = 34 [4208].

Genetics   Chromosome: $2n$ = 48 and FN = 54 [2710]. Molecular phylogenetic relationships with Vespertilionidae have been inferred with the ND1 gene of mtDNA, vWF gene of nuclear DNA and the SINE insertion [1459].

4. Ecology

Reproduction   Mating season is unknown. Adult and subadult females and subadult males form a maternity colony of several to hundreds bats from May to September [670, 685]. Parturitions occur in July and August [670]. Age at sexual maturity is unknown.

Lifespan   Not reported.

Diet   Insectivorous; feeds mainly on Lepidptera [835].

Habitat   Roosts have been recorded from the crevices of huge rocks facing the sea and in the narrow spaces between the outer and inner walls of a schoolhouse [670, 685, 835, 3004, 3026]. Several bats were found hibernating on Birou-jima Isl., Miyazaki Pref. [670], but hibernacula are little known in Japan.

Home range   Not reported.

Tadarida insignis (Blyth, 1861)

Fig. 1. An individual captured in Mie Pref. (A. Sano).

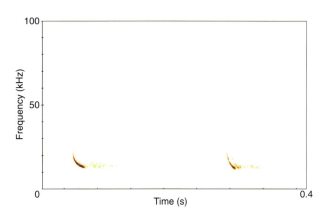

Fig. 2. Sonogram of a free-flying call around a roost, recorded by K. Funakoshi in Miyazaki Pref.

Behavior Flies fast, mainly forages in wide space over woodlands and open water at night [2990]. Although body mass greatly increases with fat deposition in November [670], feeding bats often are observed even in winter [2990].

Echolocation and social calls Emitting audible calls. Long QCF calls in the search phase, and FM calls in the scanning phase (Fig. 2). EF and PF values of search phase calls for the Miyazaki Pref. population ($n = 10$) average 12.6 kHz (12.2–13.1) and 14.2 kHz (13.5–15.3), respectively [660]. No information is available on social calls.

Natural enemies Not reported.

Parasites Not reported.

5. Remarks

There is a taxonomic opinion that Japanese *T. insignis* is a subspecies of *T. teniotis* [409, 500]. A large population of more than 500 bats had roosted at a schoolhouse in Hiroshima Pref., but they disappeared after reconstruction work [835].

A. SANO

A habitat of Oriental free-tailed bats on Mimiana-jima Isl., Mie Pref. (A. Sano).

Chiroptera MOLOSSIDAE

058

Red list status: DD (MEJ); DD (IUCN)

Tadarida latouchei Thomas, 1920

EN Oriental little free-tailed bat   JP スミイロオヒキコウモリ（sumiiro ohiki koumori）
CM 宽耳犬吻蝠   CT 黑色游離尾蝠，小游離尾蝠   KR 쇠큰귀박쥐   RS малый восточный складчатогуб

A subadult male captured on Kuchinoerabu-jima Isl., the Ohsumi Isls. (K. Funakoshi).

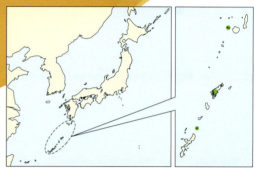

1. Distribution

Known from China, Laos, Thailand and Japan [878, 1650, 3568, 4212]. In Japan, a single specimen has been collected since 1969 from each of Yoron-jima, Amami-ohshima and Kuchinoerabu-jima Isls., respectively [667, 4212, 4219].

2. Fossil record

Not reported.

3. General characteristics

Morphology General form is very similar to *T. insignis*, but is clearly smaller; plagiopatagium is inserted to the basal 1/3 portion of the tibia; compared with *T. insignis*, ears are smaller and thinner, pelage color is darker and dorsal hairs are shorter; fur contains guard hairs with a hoary tip [667, 4212].

Measurements (in mm) for 3 specimens from Japan are as follows; FA = 55.65, 53.6, 53.0, HB = 81.50, 67.3, 69.0, T = 45.00, 41.2, 44.0, HFcu = 12.11, 12.5, 13.0, Tib = 16.89, 17.2, 16.0, E = 29.0, 22.2, 25.0 [667, 4212, 4219].

Dental formula   I 1/3 + C 1/1 + P 2/2 + M 3/3 = 32 [4212].

Genetics   Not reported.

4. Ecology

Reproduction   Not reported.
Lifespan   Not reported.
Diet   Not reported.
Habitat   Not reported.
Home range   Not reported.
Behavior   Not reported.
Echolocation and social calls   Not reported.
Natural enemies   Not reported.
Parasites   Not reported.

5. Remarks

Some workers have treated *T. latouchei* as a synonym of *T. teniotis insignis* [404, 409].

A. SANO

Dorsal views of *T. latouchei* (right) and *T. insignis* (left) (K. Funakoshi).

Primates

Primates CERCOPITHECIDAE

059

Red list status: LP for populations in the Ouu and Kitakami Mountains, Kinkazan Isl. (MEJ); R for *M. f. yakui* on Yaku-shima Isl., LP for *M. f. fuscata* on the Shimokita Pen. and Tohoku District (MSJ); LC (IUCN)

Macaca fuscata (Blyth, 1875)

EN Japanese macaque, Japanese monkey JP ニホンザル (nihon zaru) CM 日本猴 CT 日本獼猴
KR 일본원숭이 RS Японский макак

A Japanese macaque in a snowy forest on the Shimokita Pen., Aomori Pref, the northernmost part of the range (M. Kinoshita).

1. Distribution

The Japanese macaque is endemic to Japan, and distributed in Honshu, Shikoku and Kyushu, and the islands of Kinkazan, Awaji-shima, Shodo-shima, Ko-jima, Tane-gashima and Yaku-shima. Its distribution is fragmented in some regions in Tohoku District (northern Honshu), whereas it is widely distributed in central and western Japan [1100, 1661]. The macaque in Aomori Pref. is the northernmost non-human primate population in the world.

2. Fossil record

Fossil records suggest that the ancestor of the Japanese macaque immigrated from the Asian Continent between 0.63 MYA and 0.43 MYA [570, 1514]. One of the oldest macaque fossils was discovered in a fissure at the Ando Quarry in Yamaguchi Pref. [1155]. Another example is a right humerus found in the late Middle Pleistocene of Chiba Pref., Honshu [1156].

3. General characteristics

Morphology The species is a relatively large-sized macaque. The pelage is dark brown in the dorsal region, light brown or whitish in the ventral region and the medial area of the limbs [757]. Hairless face and buttocks are red. The body shape is robust, especially in the shoulder and hip regions in comparison with other macaques. The hair is long in the dorsal part of the head and trunk; the Japanese macaque also has long whiskers and dense fur adapted to cold temperatures [1035, 1036]. HB is about 500–600 mm in male and 480–550 mm in female [570]. T is shorter (60–130 mm in adult) than in any other species of macaques. Other external measurement data including BW have been detailed in each geographical population [758]. Geographical variation in this species have been morphologically confirmed among various localities [570, 758, 759, 1800, 3994, 3995]. Morphological differences between basic (core) and marginal (peripheral) populations can be recognized [2283, 3994]. Morphological characteristics of the teeth have been detailed [236, 2929, 2930, 3526]. Detailed morphological investigations have clarified the growth pattern of the Japanese

Macaca fuscata (Blyth, 1875)

macaque [570, 754, 755, 759, 2282, 2363, 3994]. Seasonal changes in BW have also been examined [756, 1785].

The Japanese, rhesus (*M. mulatta*), long-tailed (*M. fascicularis*) and Taiwan (*M. cyclopis*) macaques form the *M. fascicularis* taxonomical group [565, 566]. Morphological similarities in the skull are more conspicuous between the Japanese and rhesus macaques in this group [2282].

Sexual dimorphism in the skull is weaker in this species than in the rhesus macaque [2282]. Pelage of Yaku-shima Isl. population is distinguishable from other populations because of its long, coarse and blackish-gray hair. Since the body size of this population is much smaller than that of the others, in addition to the pelage characteristics, the subspecies *M. f. yakui* on Yaku-shima Isl. can be considered as a different subspecies from *M. f. fuscata* in other parts of the range [1794].

Dental formula   I 2/2 + C 1/1 + P 2/2 + M 3/3 = 32 [29].

Genetics   Chromosome: $2n = 42$ [371, 3303]. The final divergence among the Japanese, Taiwan and rhesus macaques in the *fascicularis* group is estimated at 0.7 MYA from mtDNA sequence data [857]. Other data, however, suggest that the common ancestral population (or species) of the Japanese and Taiwan macaques colonized Japan and Taiwan at 0.38–0.44 MYA, and that the age of isolation of *mulatta-fuscata-cyclopis* may be about 0.17 MYA [390]. The variability of blood proteins in the Japanese macaque is less than that of many other macaques; however, various groups can be geographically distinguished [859, 2526, 2527]. Blood protein data has demonstrated that the groups in Honshu and Kyushu are genetically different from those of the peripheral peninsular populations in the Izu, Boso, and Shimokita Pens. or insular populations in Shodo-shima and Yaku-shima Isls. Haplotypes found in the Japanese macaque from various localities can be separated into 2 major clusters by mtDNA analysis [1484]. One cluster consists of haplotypes from Tohoku, Kanto, Chubu and Kinki Districts, and the other from Chugoku, Shikoku and Kyushu Districts, and Yaku-shima Isl. It has also been demonstrated that individuals from various isolated localities in the Tohoku District (northern Honshu) have a single haplotype, suggesting the recent expansion of these populations by small numbers of immigrants from southern Honshu after post-glacial warming. MtDNA sequences did not show the subspecies *M. f. yakui* on Yaku-shima Isl. to be genetically separated from the other Japanese macaque populations [1484, 1951]. Thus, phenotypic analyses of the morphological and blood protein variations are not consistent with phylogenetic relationships based on mtDNA sequences. Hybrids of the Japanese macaque and introduced rhesus macaque are found in the Boso Pen. [1480], while those of Japanese and Taiwan macaques are found in Wakayama and Aomori Prefs. [1481–1483]. This suggests that the genetic and morphological features of the native Japanese macaque might be disturbed by hybridization with other introduced species of macaques.

Japanese macaques on a subtropical forest floor on Yaku-shima Isl., Kagoshima Pref., the southernmost part of the range (M. Kinoshita).

4. Ecology

Reproduction   The age of sexual maturation is about 6–7 years in males, although mounting behavior, ejaculation and descent of the testis can be observed at younger ages [1042, 2466, 3437]. Females attain sexual maturity at 4–6 years when para-anal pubertal swellings are observed [755, 2199, 2655, 3437]. Reproductive activity can be mainly seen in individuals of 6 to 18 years old. Females remain in their native troops, while males leave the native troop when older [570, 3437]. The estrous cycle is typically 26–32 days in the mating season [136, 755, 2465, 3437], while the duration of estrous periods is highly variable. One to 6 obvious estruses are observed during the mating season, and even after conception females still show additional estrus [3437]. The litter size is 1 and rarely 2. Females reproduce every 2 or 3 years. The average gestation period is approximately 170 days [2524, 2771, 3223, 3616]. The lactation period is about 7–10 months, although it continues to 1 year in some cases [817, 2782]. Reproductive activity is seasonally restricted, although geographical variations mainly according to latitude are clearly observed [569, 705, 1463, 1727, 2523, 3844]. In most areas, newborn are produced from April to July, although variation in the mating season has been confirmed [569, 570]. Mating occurs in multimale and multifemale groups. Mean composition of groups is 18% adult males, 32% adult females, 35% juveniles and 15% infants [570, 1136]. A dominance hierarchy among males occurs in each group.

Lifespan   In a wild group provisioned by humans, the longest recorded lifespan is 33 years old in females and 28 years old in males [1727, 2378, 2810]. Most adult macaques die before the age of 25 in the provisioned group [2810, 3439], although the lifespan of non-provisioned individuals, which totally depend on wild foods, may be much shorter.

Diet   The natural foods of the Japanese macaque are seeds, nuts, fruits, young leaves, flowers, buds and shoots [72, 779, 782, 1158, 3851]. Seasonal and interannual changes in food plants have been observed [503, 3658]. Additionally, the diet may include roots, grasses, herbs, fungi, insects, crabs, spiders, mollusks, fish and the eggs of birds [81, 780, 3842]. They also occasionally eat frogs, lizards, and dead shrike [3371].

Habitat Unlike the other macaques, the Japanese macaque is also distributed in regions of lower temperature. Deciduous broadleaved forest is the typical habitat for the species in the northern part and at higher altitudes. In the south, macaques inhabit evergreen broadleaved forests [3339, 3713, 3843]. This species shows a wide elevational range of distribution from sea level to higher than 3,000 m. Some populations in Honshu move down to lower locations during snowy winters.

Home range The home range of the Japanese macaque differs according to forest type and group density [3450]. The home range size of a group varies from 1.4 to 6.4 ha in evergreen broadleaved forests, and from 9 to 79 ha in deciduous broadleaved forests. Aggressive behavior between adjacent groups frequently occurs in warmer areas such as Yaku-shima Isl., but only occasionally or not at all in the colder parts [1192, 2937]. Variations in group density have been examined [781, 782].

Behavior Japanese macaques generally live in groups of 10 to more than 150 individuals; however, groups provisioned by humans are much larger, having more than 1,000 individuals [1205, 3330]. Groups consist of females, their offspring and some adult males [1136, 1728, 3451]. In addition, solitary males are observed around groups. The density of males ranging alone was estimated to be 1.2–5.7 individuals/km^2 in Yaku-shima Isl. [2786]. The species is diurnal and because it is active on the ground and on the trees, it is classified as semiterrestrial, whereas the crab-eating macaque is arboreal and the pig-tailed macaque (*M. nemestrina*) is terrestrial [366]. Japanese macaques sleep on trees [3793], although they sleep on the ground on small islands without predators and enemies [3420, 3657]. The Japanese macaque sometimes swims.

Natural enemies Predators are mainly hawk eagles (*Spizaetus nipalensis*) [980], which tend to take middle-sized individuals. Other enemies are feral dogs and introduced raccoon dogs [1157]. On Yaku-shima Isl., the Japanese weasel is a potential predator of young Japanese macaques.

Parasites Parasitic helminth and protozoan species are reported [141, 1143, 3100]. Several flea species are also reported [3100].

5. Remarks

Agricultural damage and other nuisances by Japanese macaques are reported in many districts [2330]. Various practices to allow coexistence between humans and Japanese macaques are used in Japan. More than 5,000–10,000 Japanese macaques are exterminated to prevent agricultural damage each year [1100, 1968, 2656]. Six areas are protected as natural habitats.

H. ENDO

Primates CERCOPITHECIDAE

060

Red list status: LC (IUCN); introduced

Macaca cyclopis Swinhoe, 1863

EN Taiwan macaque, Taiwanese macaque   JP タイワンザル（taiwan zaru）   CM 台湾猴   CT 臺灣獼猴
KR 대만원숭이   RS тайваньский резус

A Taiwan macaque at Chang-shou shan (Mt. Longevity), Gao-xiong Xian, Taiwan (M. A. Huffman).

1. Distribution

The Taiwan macaque is an introduced species in Japan and endemic to Taiwan. In Taiwan it is widely distributed, but human activity has restricted its distribution to the hills and mountains [1761, 1836]. Habitats before World War II have been mostly destroyed and urbanized. In Japan, feral populations originating from escaped individuals have been found in the Shimokita Pen., Aomori Pref., the Kii Pen., Wakayama Pref. [1482, 1483] and Izu-ohshima Isl. [404]. However, all individuals of the Shimokita Pen. were removed in 2004 [1100, 1481].

2. Fossil record

Not reported in Japan.

3. General characteristics

Morphology   A middle-sized macaque. The pelage color is gray and brown in the dorsal part, and paler in the ventral part and the medial area of the limbs. The blackish extremity of the limbs is distinguished from the lighter skin color of the Japanese macaque. Hairless face and buttocks are red. HB is about 400–550 mm in males and 350–450 mm in females. T is obviously longer (250–350 mm in adult) than the Japanese macaque. BW is about 6–11 kg and males are a little heavier than females [573, 1100]. Variation in relative tail length and the number of caudal vertebrae have been examined in hybrid populations between Taiwanese and Japanese macaques [760, 3580].

Dental formula   I 2/2 + C 1/1 + P 2/2 + M 3/3 = 32 [29].

Genetics   Chromosome: $2n = 42$ [369]. Blood protein analyses shows that the species is closely-related to *M. mulatta* [572, 573, 2528]. The Taiwan, Japanese (*M. fuscata*), and crab-eating (*M. fascicularis*) macaques form the *M. fascicularis* group that is widely distributed in Asia. Restriction-enzyme and sequence analyses of mtDNA show that within the genus *Macaca*, the Taiwan, rhesus (*M. mulatta*), and Japanese macaques form a single cluster separated from the other species [857, 858, 2059]. The final divergence of the 3 species is estimated at 0.7 MYA by mtDNA sequence analysis [857], although another study suggests that the ancestral population (or species) of the Taiwan and Japanese macaques colonized Taiwan and Japan at 0.38–0.44 MYA, and that populations of *M. mulatta-fuscata-cyclopis* might have separated about 0.17 MYA [390]. The Taiwan macaque shows a distinctive lineage related to China or Myanmar populations of the rhesus macaque [3625]. Serum proteins and mtDNA analyses indicate that hybrids with Japanese macaques have been established in Wakayama Pref. since 1955 [1482, 1483]. Taiwan macaques have existed in Aomori Pref. since the 1950s. Hybrids with Japanese macaques have been present in the Shimokita Pen. [1481], and all individuals related to the Taiwan macaques were removed from the wild population in 2004. Geographical variation within Taiwan revealed by mtDNA data have not yet been published, but suggest that 4 regional groups can be identified by sequence divergence [573].

4. Ecology

Reproduction The mating season is from November to January and births are from April to June. Females produce their first young at 4–5 years, whereas sexual maturation is 5–6 years in males [962, 3896]. Females enter estrus several times in the reproductive season showing para-anal pubertal swellings [2517, 3896]. The estrous cycle has been determined to be about 27–30 days [964, 2814]. Gestation period is approximately 162 days [962, 2814]. The usual litter size is 1, however twins have been reported [964]. The body weight of a newborn is about 400 g. Females of 5 to 9 years old give birth every 2 years, whereas older females have offspring every year [2910]. As in other macaque species, nursing lasts for about 1 year [573].

Lifespan Not well known. In wild individuals, however, the lifespan seems to be generally shorter than 25 years [573].

Diet Taiwan macaques eat a wide variety of foods including fruits, leaves, berries, seeds, insects, and other small invertebrates [573]. Diet and seasonal variation of foods have been evaluated by fecal analyses [3312].

Habitat Not reported in Japan. Wild populations were once seen from the seashore to the highlands in Taiwan, but are not observed in warm-temperate and lower altitude zones today.

Home range Not reported in Japan. Troop range is 0.6–2 km^2 in area and population density is 15–17 per km^2 in Taiwan [99].

Behavior Not reported in Japan. The Taiwan macaque forms multimale-multifemale social groups containing 10–200 individuals. Solitary male and multiple-male groups are also seen in Taiwan. Solitary males occasionally mate with females of other groups [963, 3896]. The species is diurnal, and generally ground-dwelling, but also arboreal [1872]. Social structure and mating patterns have been examined in a wild group [295].

Natural enemies Potential natural enemies are wild carnivores including the clouded leopard in the macaque's native range. The hawk eagle *Spizaetus nipalensis* can also take macaques [2852].

Parasites The cestode *Streptopharagus pigmentatus* has been collected in feral Taiwanese macaques in Japan [3100].

5. Remarks

Agricultural and forestry damage and other conflicts with human activity have been reported in Taiwan and Izu-ohshima Isl. They have been taken for use as food, companion and experimental animals in their native range. It was suggested that 1,000–2,000 Taiwan macaques were annually captured for such purposes, and their habitats have been largely destroyed in their native range [2851]. The Taiwan macaque is listed as threatened by USDI and vulnerable by IUCN.

Hybridization between Taiwan and Japanese macaques threatens the genetic identity of endemic Japanese macaque populations.

H. ENDO

Primates   CERCOPITHECIDAE

061

Red list status: LC (IUCN); introduced

Macaca mulatta (Zimmermann, 1780)

EN rhesus macaque, rhesus monkey   JP アカゲザル（akage zaru）   CM 猕猴   CT 普通獼猴，恆河獼猴
KR 붉은털원숭이   RS макак-резус

Rhesus macaques at Primate Research Institute of Kyoto University, Aichi Pref. (H. Endo).

1. Distribution

In Japan feral individuals have been observed in the southern region of the Boso Pen., Chiba Pref. The rhesus macaque is widely distributed in Afghanistan, India, the southern part of China, and the northern part of the Indochinese Pen. Since the long-tailed macaque (*M. fascicularis*) is distributed in the southern part of the Indochinese Pen., hybrids between rhesus and long-tailed macaques have been observed in the central regions of Vietnam, Laos, and Thailand [508, 1206, 2737].

2. Fossil record

Not reported in Japan.

3. General characteristics

Morphology   The species is a relatively large-sized macaque [318]. HB is 450–620 mm in males and 400–550 mm in females. BW is about 4–10 kg. The tail length ratio shows a wide variation among localities. From Pakistan to Myanmar the ratio is about 45%, but only 30% in populations from Thailand, Laos, Vietnam and China [568, 571]. The pelage color is gray-brown in the dorsal region and whitish in the ventral region and the medial area of the limbs. The tips of its 4 limbs are blackish. The typical color pattern of the trunk of those observed in southeastern Asian countries is grayish in the anterior part and reddish in the posterior part, although the dorsal pelage color is much lighter in individuals of Pakistan and India. In addition, entirely whitish trunks or dark blue-brown anteriors are seen in some individuals from the Indochinese Pen. The hair has whorls in the vertex of the head in Chinese and Indian populations; in contrast, those from northern Thailand have all back hair. The hair in the cheek shows wide variation among localities. Morphometrical differences in the skull are weaker between the rhesus and Japanese macaques than between the long-tailed and Japanese macaques [2282].

Dental formula   I 2/2 + C 1/1 + P 2/2 + M 3/3 = 32.

Genetics   Chromosome: $2n = 42$ [371, 485, 3303, 3310]. Restriction-enzyme and sequence analyses of mtDNA show that within the genus *Macaca*, the rhesus, Japanese, and Taiwan macaques form a single cluster separated from the other species [859, 2059]. MtDNA sequence data demonstrates 4 major clusters in the rhesus macaque: 1) populations from India, 2) populations from India and Myanmar, 3) Eastern China and Vietnam, 4) Western China [3279]. DNA sequence analysis indicates that the feral population in Japan is closely related to the population from China [752]. There are hybrids between the rhesus and Japanese

139

macaques in the Boso Pen. Hybrid macaques show characteristics of both species in the mtDNA and TSPY genes [1480, 3624].

4. Ecology

Reproduction   Reproduction is clearly seasonal in wild populations [2897]. In South Central Asia, births are confirmed from February to May and from September to October. The estrous cycle is typically 28 days in the mating season [2897]. Litter size is usually 1 every 2 years in the wild. The gestation period is 135–194 (mean of 166) days [747, 1660, 2897, 3770]. Newborns weigh 400–500 g and are nursed for about 1 year. The age of sexual maturation is about 2–3 years in males and 2.5–4 years in females [2897, 3273]. Most reproductive activity is over before 25 years old.

Lifespan   The lifespan is usually 20–25 years in the wild [571, 2897].

Diet   Not reported in Japan. The rhesus macaque is omnivorous, feeding on fruits, berries, grains, leaves, seeds and flowers, insects, other small invertebrates, crabs and shellfish [1852, 2897].

Habitat   Not reported in Japan. The rhesus macaque is adapted to various habitats including subtropical and temperate forests, semi-desert and snowed areas [567, 2517]. It can be observed from the seashore to elevations of 2,500 m in its original range [1852]. Typically rhesus macaques are rare in broad-leaved evergreen forests but often observed in secondary, deciduous, coniferous and mangrove forests. They also live in cities and towns as in the northern Indian State of Uttar Predesh [3296]. In Calcutta, groups live in houses and buildings [3297].

Home range   Not reported in Japan. Group home range has been recorded as 16 km^2 in sub-Himalayan forests, 1–3 km^2 in other low forests and 0.05 km^2 in city areas [1852, 2897]. Population density is 5–15 individuals per km^2 in high forests, 57 per km^2 in low forests and up to 753 per km^2 in city areas [2897].

Behavior   In the native range, the species seems terrestrial rather than arboreal. They often swim. Rhesus monkeys live in groups of 8–180 individuals. The group essentially consists of from 2 to 4 times as many adult females as adult males [1841, 1852, 2897, 3296]. A dominance hierarchy among males is observed in each group [3273].

Natural enemy   Not reported in Japan. Various carnivores, raptors, snakes and large lizards are potential enemies in their original range.

Parasite   Not reported in Japan.

5. Remarks

In their native range populations have been reduced due to hunting for food, commercial trade and the destruction of habitat. In the late 1950s, 200,000 rhesus monkeys were exported to the USA [2188]. Damage to agriculture and forestry has fueled overhunting in many parts of the Asian continent. Hybridization between rhesus and Japanese monkeys may cause extinction of the original populations of Japanese macaques. The species is also used in laboratory research all over the world.

H. ENDO

A Japanese macaque eating a twig on a tree on the Shimokita Pen., Aomori Pref. (T. Kuwahara, Kon Photography & Research).

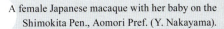

A female Japanese macaque with her baby on the Shimokita Pen., Aomori Pref. (Y. Nakayama).

A female Japanese macaque in snow on the Shimokita Pen., Aomori Pref. (Y. Nakayama).

RESEARCH TOPIC ❷

Comparative phylogeography of sika deer, Japanese macaques, and black bears reveals unique population history of large mammals in Japan
H. TAMATE

1. Introduction

Sika deer (*Cervus nippon*), Japanese macaques (*Macaca fuscata*) and Japanese black bears (*Ursus thibetanus*) are common species in Japan. They inhabit forests with various floras, from warm-temperate evergreen forests to temperate or cool-temperate forests. At present, natural populations of sika deer are distributed widely throughout the Japanese Isls. except in the high mountains. Japanese macaques are distributed from the northernmost part of Honshu to Shikoku and Kyushu, but not in Hokkaido. Japanese black bears are distributed similarly, although local populations have been extirpated or are seriously endangered in southern Japan.

Like other large mammals, these 3 species each occupy an irreplaceable niche in the forests, and have been important game animals since the pre-historic period of Japan. Therefore, the distribution, population dynamics and population structures of these species have been influenced not only by ecological factors, but also by anthropological activities. Recent phylogeographical studies using DNA markers unveiled hidden population structures of these species, giving deep insights into the history of the mammalian fauna in Japan.

2. An enigmatic bifurcation in mitochondrial DNA phylogeography

Geographical distribution of the mitochondrial DNA (mtDNA) variations in sika deer, macaques and black bears show striking similarities: for each species mtDNA haplotypes are grouped into 2 major clades in phylogenetic trees (Fig. 1). Haplotype lineages of the mtDNA control region and the cytochrome *b* gene from sika deer both are divided into 2 major clades, i.e., northern and southern [2350, 3482]. Haplotypes in the northern clade are observed among populations from Hokkaido and eastern Honshu, whereas those in the southern clade are found among populations from western Honshu, Shikoku, Kyushu, and other southern islands. In macaques, a phylogenetic analysis of sequence variations at the mtDNA control region showed 2 major clades, of which one is distributed in eastern Honshu and the other in western Honshu, Shikoku and Kyushu [1484]. Phylogeographical analyses of black bear mtDNA also revealed the presence of 3 distinct monophyletic lineages: eastern, western and southern clades. A border between the eastern and western clades lies in Kinki District of Honshu, as observed in sika deer and macaques [1071, 2630, 4095]. Therefore, for the 3 species, phylogenetic trees based on mtDNA variation show the same bifurcating pattern that corresponds to geographically separated—northern vs. southern, or eastern vs. western—distributions (Fig. 2). It also should be noted that, in these species, a geographical boundary between the northern and southern (or eastern and western) clades lies in the middle of Honshu where there are no apparent impediments to the animals' dispersal.

Similarity also is observed among the 3 species in the level of genetic diversity: there is a general trend that the haplotype diversity is greater in populations from southern (western) Japan than in those from northern (eastern) Japan. In macaques, only 1 haplotype was found in a large area of eastern (northern) Honshu, whereas a number of haplotypes exist in western (or southern) Honshu [1484]. In sika deer, northern populations from Hokkaido to eastern and central

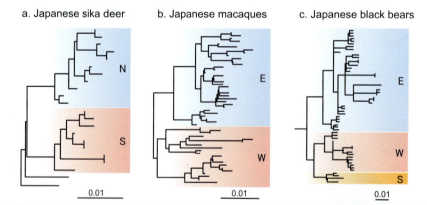

Fig. 1. Phylogenetic trees of the mtDNA control region sequences observed among sika deer (*Cervus nippon*) [2350], Japanese macaques (*Macaca fuscata*) [1484] and Japanese black bears (*Ursus thibetanus*) [2630]. (a) In sika deer, 2 major haplotype lineages, N and S, correspond geographically to the northern and southern groups, respectively. (b) In Japanese macaques, 2 major haplotype lineages, E and W, correspond geographically to the eastern and western groups, respectively. (c) In Japanese black bears, 2 major haplotype lineages, E and W, are geographically demarcated by a border in the northern Kinki District. Haplotypes in Shikoku and the Kii Peninsula are clustered into a separate clade S (see also Fig. 2). Bars represent 0.01 nucleotide substitutions/site. Figures were reconstructed from the data in the papers (a [2350]; b [1484]; c [2630]).

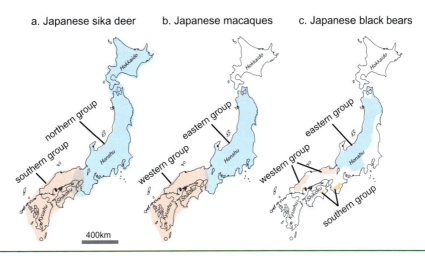

Fig. 2. Geographical distribution of the major mtDNA clades in (a) sika deer (*Cervus nippon*), (b) Japanese macaques (*Macaca fuscata*), and (c) Japanese black bears (*Ursus thibetanus*).

Honshu possess only 2 haplotypes of mitochondrial cytochrome *b* sequences, whereas southern populations from western Honshu and Kyushu retain more than 9 haplotypes [3482]. The reduced diversity in the northern (eastern) populations of sika deer and macaques indicates a bottlenecking event followed by rapid population expansion in northern Japan. Furthermore, the southern (western) populations were supposed to have consisted of large stable populations.

The results of the above-mentioned studies clearly indicate that each of the 3 species—sika deer, macaques and black bears—consist of 2 genetically distinct groups in northern and southern Japan that probably have different population histories. It is not, however, clear whether such bifurcating phylogeography is a common feature for other mammals in Japan. In the case of Japanese wild boars, the geographical boundary of different mtDNA haplotypes was suggested to lie between Kinki and Chugoku Districts, although the phylogenetic relationship among the haplotypes was not as clear as that shown in sika deer and macaques [3848]. The Japanese large field mouse (*Apodemus speciosus*) is known to have dimorphic variation in karyotypes: populations with $2n = 48$ and $2n = 46$ are located in eastern and western Honshu, respectively, and are demarcated by a clear border in central Honshu with a narrow hybrid zone [3642]. In mtDNA lineages, however, field mice (*A. speciosus* and *A. argenteus*) represent 2 major clades that correspond to central and peripheral populations, respectively, and not to northern and southern localities [3358]. In Japanese dormice (*Glirulus japonicus*), phylogeography based on mtDNA shows a more complex pattern [4077], and does not fit the northern vs. southern scheme. To date, 18 mammalian species have been studied for phylogeography across the Japanese Isls. [3477]. Among them, Japanese hare (*Lepus brachyurus*) and Japanese white-toothed shrews (*Crocidura dsinezumi*) show similar bifurcating pattern as sika deer, macaques and black bears [2532, 2620].

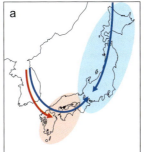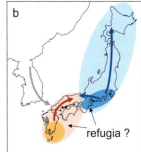

Fig. 3. Multiple-colonization (a) and refugia (b) hypotheses. Possible routes for immigration and expansion are shown for the northern (blue arrows) and southern (red arrows) groups. The multiple-colonization hypothesis assumes that the 2 groups colonized Japan at different times and via different routes: 2 alternative immigration routes are considered for the northern group. However, the refugia hypothesis assumes that after the immigration of the ancestors (gray arrow) the 2 lineages were confined separately to different refugia during the last glacial maximum period and expanded their distributions thereafter.

3. "Multiple-colonization" and "refugia" hypotheses

Why do we see remarkable phylogeographical similarities especially among sika deer, macaques and black bears? The bifurcation of mtDNA lineages clearly indicates that a big split of an ancestral population occurred in the past. Two alternative hypotheses are proposed to account for the cause of the split: multiple-colonization and refugia (Fig. 3).

The multiple-colonization hypothesis is based on the molecular data that the split time of the northern and southern groups of the Japanese sika deer was estimated to be about 0.3–0.5 MYA (million years ago). According to paleontological studies, immigration of extant species of large mammals to the Japanese Isls. occurred at least twice during the last half-million years [805]. The first immigration took place in the Middle Pleistocene (about 0.43 MYA), with a group of large mammals that had adapted to temperate forests, such as extinct elephants (*Palaeoloxodon naumanni*). They were supposed to have migrated to Honshu and Kyushu via the Korean Peninsula. During the last glacial period (about 0.01–0.02 MYA), another group of large mammals—such as moose (*Alces alces*) and extinct bison (*Bison priscus*)—from the northern part of the continent reached the Japanese Isls. via a land bridge that was formed between the continent and Hokkaido [1509]. Since the divergence time between the northern and southern haplotype groups is as old as the time of the first immigration, it is hypothesized that one of the groups colonized first, during the Middle Pleistocene, in southern Honshu and Kyushu, and subsequently another immigrated to Japan and spread over Hokkaido and northern Honshu in the late Late Pleistocene (about 0.01–0.02 MYA).

The underlying assumption for the multiple-colonization hypothesis is vicariance: closely related taxa isolated geographically from one another by a natural barrier. Morphological characteristics as well as chromosome numbers differ remarkably between southern and northern sika deer populations in Japan. The mean body mass of adult males, for example, varies from about 40 kg in the southern subspecies *C. n. yakushimae*, to over 100 kg in the northern subspecies *C. n. yesoensis*, from Hokkaido. This led to the assumption that ancestors of the northern and southern sika deer were separated in different environments for long enough to achieve their differentiation of the characteristics. Comparative morphometry revealed significant differences in mandibular and dental morphology between the northern and southern mtDNA lineages of sika deer [2794]. As the differences do not necessarily correspond to current habitat vegetation or feeding types, it was suggested that the 2 lineages were selected for under different conditions. Japanese sika deer, therefore, may have multiple origins that can be traced back to different ancestral populations that diverged in the Asian continent. It should be noted, however, that the multiple-colonization hypothesis is not supported by paleontological studies: Kawamura (2009) who examined fossil antlers of sika deer concluded that they could not have immigrated to Japan via the land bridge between the continent and Hokkaido [1511].

Alternatively, the refugia hypothesis explains the northern vs. southern splitting pattern as a consequence of recent allopatric fragmentation of populations within the Japanese Isls. during the last glacial maximum period (LGM, 15,000–10,000 years ago). Palynological studies showed that during the LGM temperate deciduous broad-leaved forests and warm-temperate evergreen forests, which provide favorable habitat for sika deer, macaques and black bears, retreated to the southwestern Pacific coasts of Honshu, Shikoku and Kyushu. During the same period, the rest of the Japanese Isls. mostly was covered with boreal coniferous or mixed forest, which is a rather harsh environment for the 3 species. Considering the past changes in vegetation, it was hypothesized that western and eastern haplotype groups of macaques survived separately in different refugia—somewhere in southwestern Japan—during LGM and expanded their distribution over the Japanese Isls. [1484]. Thus, the bifurcation in mtDNA phylogeography is explained by the post-LGM expansion of haplotypes that had been segregated in 2 different refugia. The absence of macaques and black bears in Hokkaido is well explained by the refugia hypothesis as they might have been unable to cross over the strait between Honshu and Hokkaido.

Supporting evidence for the refugia hypothesis has been provided by recent phylogeographical studies on major tree species of temperate forests, Japanese beech (*Fagus crenata*), Japanese cedar (*Cryptomeria japonica*) and 4 species of oaks (*Quercus* spp.). In Japanese beech, chloroplast DNA (cpDNA) haplotypes among populations are divided into 2 major lineages (clade I, and clade II/III) [588], which roughly overlap the distribution of the 2 major haplotype-groups observed in sika deer, macaques and black bears. The genetic structures of natural populations of Japanese cedar show a clear geographical trend that corresponds to putative refugia of the trees [3663]. The localities of some haplotypes observed among the macaques were close to the refugia of Japanese cedar [1484]. Of the 2 cpDNA haplotypes observed among 4 species of oaks, one is restricted to eastern Japan, whereas another is widespread from Korea to western Japan [1341], implying that after LGM they expanded their distributions from at least 2 refugia in Honshu. As the phylogeographical patterns observed among sika deer, macaques and black bears coincides with those of the tree species, the refugia hypothesis which was proposed for the Quaternary flora now is considered to explain the history of the mammalian fauna in Japan.

Basically, the 2 hypotheses differ in the chronology of allopatric differentiation within a species, but are not necessarily mutually exclusive. It is possible that the genetic differentiation between 2 ancestral populations that had colonized Japan at different times were reinforced by the isolation in refugia during LGM. Also, it should be emphasized that the same hypothesis may not always hold in all species. The validity of the hypotheses should be tested by further studies on the phylogeography, karyology, and paleontology of closely related species in the Asian continent.

4. Recent population history revealed by genetic analyses

Until the 19th century, sika deer, macaques and black bears were abundant throughout Japan. However, rapid modernization that started in the 1860s led to the destruction of forest habitats, which resulted in population fragmentation of wildlife. Uncontrolled hunting, combined with severe winters in 1879 and 1903, hastened the extirpation of local populations. Consequently, populations of sika deer, macaques, and black bears were confined to small areas during the 20th century. Then, in the late 1980s, some local populations of sika deer and macaques started to expand rapidly in population size and distribution area. They now are over-

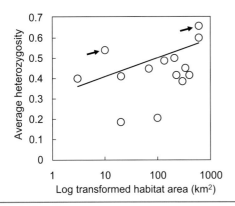

Fig. 4. Correlation between the average heterozygosity and habitat size of sika deer (*Cervus nippon*) populations in Honshu and Kyushu ($r = 0.1321$, $P < 0.05$; based on the data from [728]). Arrows indicate local populations that have been under the protection of shrines.

abundant. Little is known about recent demographic changes of black bear populations, although the number of bears culled for nuisance control is increasing. Population-genetic studies have focused on the impact of the changing environments in which these species live, and provided empirical data to establish a management strategy for each population.

Of the 3 species, black bears have been the most favored game animal because their gallbladders are economically valuable for Chinese medicine. Due to the probable high hunting pressure and loss of habitats, black bear populations have been reduced, especially in the western part of Honshu. Microsatellite analysis showed there was significant genetic differentiation among 4 local populations, from Chugoku to Kinki Districts [2955]. Genetic diversity measures such as allelic richness and heterozygosity in 3 of those populations were significantly lower than expected for large continuous populations [2629]. Both mtDNA and microsatellite data showed that gene flow is extremely restricted between some populations in Chugoku and Kinki Districts [1071, 2629, 2955]. These results imply that the restoration of corridors that connect fragmented habitats will be necessary to prevent further loss of genetic diversity within the threatened bear populations.

Significant differences in genetic diversity among populations also were documented for sika deer at a nationwide scale [728]. Despite the large population size and habitat area, sika deer in Hokkaido show a low level of genetic diversity. Studies on genetic structures of current and ancient populations of sika deer in Hokkaido suggested that sika deer in Hokkaido experienced severe bottlenecking in the past [2338, 2348]. Such reduction in genetic diversity also was more or less observed in local populations of Honshu and Kyushu [728, 3480]. There was no sign of isolation-by-distance in population-genetic measures such as F_{ST} among the populations, and the effective population size [728] and genetic diversity measure generally were correlated with the current habitat area of each population (Fig. 4). It is, therefore, concluded that most current genetic differentiation and the differences in genetic diversity probably resulted from genetic drift under bottlenecking [728]. However, genetic diversity is relatively high in populations that have been protected for religious reasons by Shintoism (Fig. 4).

Population genetic structures of sika deer were studied further at hierarchical levels. At a regional scale, in the southern Kanto region of Honshu, gene flow among subpopulations has been interrupted [4230] probably because of recent anthropogenic disturbance, such as the development of road networks and urbanization. At a local scale, a biased distribution of microsatellite alleles was observed within a small population on Kinkazan Isl. [2669], suggesting that the range of dispersal of individuals is quite limited on this island.

Low genetic diversity and biased distribution of genetic variation within populations also are evident in macaques on Yaku-shima Isl. Mismatched distribution analysis of mtDNA haplotypes suggested that the Yaku-shima population experienced a bottleneck event which was followed by rapid expansion [850]. The geographical distribution of the haplotypes was not uniform on the island; one of the haplotypes was distributed widely, whereas the others were observed only in the lowland forests. The authors hypothesized that environmental heterogeneity during the expansion could have determined such spatial genetic structure within the population. Microsatellite analysis also revealed that the northernmost population in Tohoku District of Honshu was genetically differentiated from others and possessed a reduced level of genetic diversity [1485]. Unlike sika deer and black bears, the reduced genetic diversity in the populations of macaques was attributed to a consequence of demographic changes over several thousand years as the genetic signatures of recent bottlenecking were not observed.

In conclusion, phylogeography tells us that sika deer, macaques and black bears immigrated to the Japanese islands through a complex process of lineage sorting, and that more recent demographic events—fragmentation, bottlenecking, and rapid expansion—further shaped current population structures at both regional and local scales. Genetic diversity and population structures of these species are influenced greatly by human activities, which means that we humans are fully responsible for the fate of sika deer, macaques and black bears living in this country.

A female Japanese macaque on the Shimokita Pen., Aomori Pref. (Y. Nakayama).

A Japanese black bear.

A sika deer in summer pelage in front of a gate of Todaiji temple, Nara Pref. (H. Torii).

Rodentia

Rodentia   GLIRIDAE

062

Red list status: V (MSJ); LC (IUCN)

Glirulus japonicus (Schinz, 1845)

EN Japanese dormouse   JP ヤマネ (yamane)   CM 日本睡鼠   CT 日本睡鼠
KR 일본겨울잠쥐   RS японская соня

An adult from Tokushima Pref., Shikoku (K. Tsuchiya).

1. Distribution

Honshu, Shikoku, Kyushu and Dogo Isl. of the Oki Isls. Occurs in mountain ranges from 700 m to 3,000 m in elevation in central Honshu [1327, 2364].

2. Fossil record

Fossils in the Middle and the Late Pleistocene, and the Holocene have been discovered from Honshu [1499–1501, 1503, 1508, 1513, 1521, 1725, 2688].

3. General characteristics

Morphology   TL = 68–84 mm, T = 44–54 mm, HFsu = 15–17 mm, BW = 14–20 g in summer or after hibernation and 34–40 g before hibernation. Coat color is pale brown with dark brown stripe on back and paler belly and dark brown around eyes. Coat color of individuals from eastern Honshu more grayish than those from western Japan (Fig. 1). Tail is covered with long hairs, approximately 20 mm long [1011, 1327, 2364, 2365].

Dental formula   I 1/1 + C 0/0 + P 1/1 + M 3/3 = 20 [29].

Mammae formula   2 + 1 + 1 = 8 [1327, 2364].

Genetics   Chromosome: $2n = 46$ (XX and XY) and FN = 56 [2765, 3635]. G-, C- and NOR (nucleolar organizer regions) banded patterns are present [2765]. Molecular phylogenetic data are available [3351, 3635, 4076, 4077].

4. Ecology

Reproduction   Litter size 3–5, occasionally up to 7. Breeds from spring to autumn. Females give birth twice a year [1327, 2364, 2365].

Lifespan   Three years longevity under wild conditions, but up to 8 years under captive conditions [1327, 2364, 2365].

Fig. 1.   Coat colors of western type (upper, specimen No.: KT3063, Shikoku) and eastern type (lower, specimen No.: KT3733, central Honshu).

Glirulus japonicus (Schinz, 1845)

An adult from Yamanashi Pref., Honshu (K. Tsuchiya).

Diet Eats small insects including larvae and pupae, fruits, small animals and eggs of small birds [1327, 2364].

Habitat Inhabits mature forests in lower and subalpine regions. Makes spherical nests with leaves, bark and mosses in tree hollows or between branches [1327, 2364].

Home range Large home range for its body size, with nearly 1 ha for females and 2 ha for males based on a study in central Honshu [2078, 2364, 2365].

Behavior Fundamentally nocturnal and arboreal. Nocturnally feeding. Occasionally, active during daytime before hibernation [2, 13]. Hibernation behavior of this species is well studied [3183].

Natural enemies Snakes, martens and owls are reported as predators. If hibernation occurs under or near the ground, dormice are occasionally attacked by moles, foxes, raccoon dogs, weasels and martens [2364, 2365].

Parasites Eight species of parasitic mites are known as ectoparasites [2365]. One of them is specific to *G. japonicus*: *Steatonyssus nakazimai* (Acarina, Macronyssidae) [222]. Another three species of mites are observed from the nests [2364, 2365].

5. Remarks

In 1975, this species was designated as a Natural Monument [1327].

M. A. IWASA

An adult Japanese dormouse from Inabu, Aichi Pref., Honshu (S. Kawada).

Rodentia MURIDAE

063

Red list status: C-2 as *Clethrionomys r. bedfordiae*, LP for the Daikoku Isl. population (MSJ); LC (IUCN)

Myodes rufocanus (Sundevall, 1846)

EN gray red-backed vole   **JP** タイリクヤチネズミ（tairiku yachi nezumi）
CM 棕背䶄   **CT** 棕背鼠，棕背䶄，亞洲紅背田鼠   **KR** 대륙밭쥐   **RS** красно-серая полёвка

An adult male from Hokkaido (M. A. Iwasa).

1. Distribution

Widely distributed in Eurasia: Scandinavia through Siberia to Kamchatka, Sakhalin, southern Urals, Altai and Korea [13, 19, 1327, 1335, 2334, 3173]. In Japan, Hokkaido, Rishiri, Rebun, Teuri, Yagishiri and Daikoku Isls., and neighboring islands including Shibotsu, Shikotan and Kunashiri Isls. [1327, 3173].

2. Fossil record

Fossils of a species, *M. japonicus*, related to *M. rufocanus* were discovered at Ikumo and Ando Quarries of the Middle Pleistocene in Yamaguchi Pref., western Honshu, and some fossils of the *Myodes–Phaulomys* transitional form were discovered at Kumaishi-do and Sugi-ana Caves of Gifu Pref., central Honshu [1507].

3. General characteristics

Morphology Small-sized rodent with a short tail, small eyes and short ears for semi-fossorial lifestyle. Coat color is dark brown on back and ivory white or murky white on belly [1327]. Sometimes variant coat colors observed [82, 594, 1161, 2155, 2320, 2420, 2422, 3422, 3461, 3710, 3712]. Dorsum has wide reddish-brown stripe. Enamel pattern of upper third molar is simple with 2 inner re-entrant angles and 3 inner salient angles in adults (Fig. 1). This dental characteristic is important to identify this species among 3 *Myodes* species in Hokkaido. External measurements (range) are as follows: HB = 110.0–142.0 mm, T = 39.0–55.0 mm, HFsu = 19.0–22.5 mm, BW = 27–50 g [14, 15, 19, 80, 1327, 1335].

Dental formula I 1/1 + C 0/0 + P 0/0 + M 3/3 = 16 [19, 29, 1327].

Mammae formula 2 + 0 + 2 = 8 [19, 1327].

Genetics Chromosome: $2n = 56$ (XX and XY) and FN = 56 [3635, 3636]. Sequence data of mitochondrial cytochrome *b* gene, *Sry* and *G6pd* gene intron and restriction fragment length polymorphism data for nuclear ribosomal genes are available [1170, 1173, 3348]. Several microsatellite markers are also available [1074, 1075].

4. Ecology

Reproduction Length of breeding season varies greatly; it usually starts in April and ends by October, although winter breeding is known to occur. Reproductive activity may be affected by many factors including season, population density, fluctuation phase and habitat [1335, 2415]. Litter size (number of embryos) is 1–12 and on average 5.3 [593]. Gestation period is 18–19 days under laboratory conditions [10]. Average weight at birth is 2.0 g. Pups are naked and blind. Eyes open at 12–13 days. Nursing period is 15–20 days [10]. Spring-born voles may reach sexual maturity at 30 days under mild conditions. Females may attain maturity after reaching 16 g and males after reaching 24 g in the wild [2418]. Autumn-born voles usually do not reproduce until the following spring, at which time they constitute the major part of the breeding population. Mating system is promiscuous and multiple paternity is observed (see Topic).

Lifespan In the wild, survival to 14 months has been recorded [593]; longer in captivity.

Diet Mostly herbivorous. Range of food habits is wide, including many kinds of herbs, roots, seeds and insects. The preferred food markedly varies with the seasonal variation of resources in the habitat [3434]. Under low temperature food intake increases, and kidney, heart and small intestine are hypertrophied [1817].

Habitat Grasslands, scrublands, plantations and forests; inhabits open habitats as well as forests in Hokkaido, possibly owing to the absence of *Microtus* [1335, 2774, 2781]. Dominant in habitats having dense vegetation near the ground surface and with thick litter layers [13, 22].

Home range Breeding males have overlapping home ranges, whereas females' ranges are more mutually exclusive (i.e. territorial; see Topic) [2948]. Such exclusiveness is relaxed during the non-breeding season [1077].

Behavior Data gathered from trapping suggest that the species is more nocturnal during the summer and diurnal during winter [449]. In summer, activity peaks at dusk and dawn. In winter nests are placed on the ground under the snow. Females are philopatric, whereas males are prone to disperse (see Topic). Male-biased natal dispersal has a function to avoid inbreeding (see Topic).

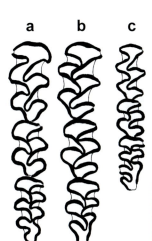

Fig. 1. Schematic enamel patterns of left upper molars of adult *M. rex* (a), *M. rufocanus* (b) and *M. rutilus* (c) specimens. Bar = 2.0 mm (Specimen No.: a, HEG147; b, HEG119; c, MAI-123).

Natural enemies Preyed upon by many birds, mammals reptiles, and fishes, including owls [2019, 2020, 4147], foxes [18, 1688, 1705, 2094], martens [2314], weasels [4137], American minks [3741], Sakhalin taimens [2622] and probably stoats and snakes. Predators tend to prefer this species to other small mammals [1335, 1688, 4143].

Parasites Parasites and diseases have been thoroughly reviewed [4140]: of ectoparasites, 26 genera of Acari (mites and ticks), 15 genera of Siphonaptera (fleas), and 2 genera of Anoplura (lice) are listed; *Radfordia* (*Microtimyobia*) *lemnina rufocani*, a polyxenic subspecies of myobiid mite, is currently parasitic on 2 host genera *Myodes* and *Eothenomys* [3701, 3704]. Helminths found in Hokkaido are reviewed [207, 812]; *Mastophorus muris*, a nematode collected from species of *Eothenomys* and *Apodemus*, is additionally listed [2714]; of nematodes, *Heligmosomum* (*Paraheligmosomum*) *yamagutii* shows strict a host-relationship [208]. Apicomplexa, a parasitic Protozoa, are reported [1041, 2596, 4243].

5. Remarks

Known as a major forestry pest, causing considerable damage to plantations of fruit trees, woody-stemmed ornamental plants and forestry trees [1335, 2410]. Have been found to be favorable intermediate host of the cestode *Echinococcus multilocularis*, an endoparasite which causes serious disease (alveolar echinococcosis) for humans in Hokkaido [3426]. The vole in Hokkaido and neighboring islands is ranked as a distinct subspecies, *M. rufocanus bedfordiae* [3563] based on morphological criteria [19, 1335]. This subspecies is slightly larger than *M. rufocanus rufocanus* [13]. *Myodes rex* and *M. rutilus* also inhabit Hokkaido, but *M. rex* and *M. rutilus* are larger and smaller than *M. rufocanus*, respectively, in body size. Recently, the generic name of vole taxa in Hokkaido was revised from *Clethrionomys* to *Myodes* [349, 1313]. Island populations are morphologically differentiated and have been confused taxonomically. For example, populations of Rishiri Isl. and Daikoku Isl. were classified into valid subspecies [1000, 1011, 2779] but we have no evidence that these populations are classified into valid subspecies based on recent findings [14, 15, 440, 2954].

K. NAKATA, T. SAITOH & M. A. IWASA

6. Topic

Social organization and mating system

T. SAITOH & Y. ISHIBASHI

Social organization has been studied, focusing in particular on spacing and dispersal behavior. Female territoriality has been observed during the breeding season in this as well as other species of this genus [1078, 2415, 2947, 2948, 2951, 3521]. Females maintain breeding territories with a size ranging from 200 to 600 m^2 [2951]. Males have larger home ranges covering several females' territories (400–1,300 m^2) [2776], and their home ranges frequently overlap [2948].

MURIDAE

Sexual maturation of young females is inhibited when they fail to establish a territory, even though they have reproductive potential [2947, 2950]. Consequently, the number of breeding females per unit area is limited [2415, 2951], although the reduction of home range (territory) size has been observed at high density [2951]. Females whose home ranges overlap with other females fail to become pregnant [1526]. Females are better able to acquire a territory when the neighboring female is a sister than when the neighbor is not a sister [1526].

The gray red-backed vole aggregates with common use of shelters during cold periods in the non-breeding season [2949]. Huddling or communal nesting is widely known in arvicoline rodents and the relatively low mortality in winter has been thought to be related to the huddling and the change in social system from being territorial during the breeding season to aggregation during winter [1078]. Kin-biased overwintering groups are commonly formed [1077]. Winter groups consist of individuals which had neighboring home ranges at the end of the breeding season. Since most juveniles are philopatric in the autumn, most winter groups consist of maternal relatives.

Males disperse further than females from the birth place to the location of the initial breeding attempt [2952]; average dispersal distances (SD) are 64.9 (51.1) m and 35.3 (45.6) m for males and females, respectively. In the studied population, 51.2% of females settled within 1 home range length from the natal site, and 22.0% settled further than 2 range lengths; only 24.8% of males settled within 1 home range length from the natal site, and 51.2% dispersed further than 2 range lengths [2952].

As a consequence of male-biased dispersal and female philopatry, spatial genetic structure differs between sexes. At a fine spatial scale, allozyme alleles are distributed more heterogeneously in females than that in males [1525] and kin-related clusters by breeding female relatives are formed [1076]. At a spatial scale of 2 km, genetic differentiation of mtDNA haplotypes is not apparent between 0.5-ha plots apart 1.0 km or less in both sexes, whereas all combinations between plots apart more than 1.0 km show significant differentiation for females, but not for males [1079].

Promiscuous mating, as have been suggested previously [1524, 1527], has been documented in a gray red-backed vole population, in which parentage was established by genotyping at several microsatellite DNA loci [1072]. Litters sired by multiple males are observed throughout the breeding season [1072]. The proportion of multiple paternity is positively correlated with local male density around estrous females. Male mating success is related to male body mass. Multi-male mating may occur because dominant males cannot deter subordinates from access to their mates when local male density is high [1072]. Inbreeding between relatives (relatedness $r \geq 0.25$) is rare [1073]. Male-biased dispersal is partly effective to avoid incestuous mating, but it does not provide complete separation of male and female close relatives [1073]. Additional mechanisms such as kin discrimination based on familiarity may work in inbreeding avoidance of the vole.

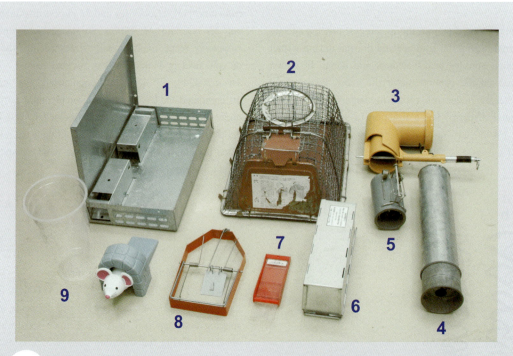

Traps for small insectivores and rodents.
1, live trap, "Tin Cat"; 2, mesh-type live trap, "Amikan"; 3, harpoon-type mole trap; 4, tunnel-type mole trap; 5, Nishi-type mole trap (with a slight modification); 6, live trap, "Sherman"; 7, snap-type trap, "Pan-chu"; 8, snap trap; 9, pit fall trap (M. A. Iwasa).

A subnivean nest of the gray red-backed vole in April when the snow melted at Bibai, central Hokkaido (Keisuke Nakata).

Larch trunk gnawed by the gray red-backed vole in a plantation at Teshikaga, eastern Hokkaido, showing distinct grooves made by the incisors and scattered fragments of the rhytidome left uneaten on the ground (Keisuke Nakata).

Rodentia MURIDAE

064 Red list status: NT (MEJ); R as *Clethrionomys rex* (MSJ); LC (IUCN)

Myodes rex (Imaizumi, 1971)

EN dark red-backed vole   JP ムクゲネズミ (mukuge nezumi)
CM 黑背䶄   CT 黃墨背鼠, 黃墨背䶄, 暗背田鼠   KR 일본대륙밭쥐   RS японская рыжая полёвка

An adult male from Rishiri Isl., northern Hokkaido (M. A. Iwasa).

1. Distribution

Distributed in Hokkaido (mountain ranges of Hidaka, Daisetsu, Kitami, Teshio, Mashike, Yubari, Shiribeshi, Youtei and Oshima), Rishiri, Rebun Isls., and neighboring Isls. including Shibotsu, Shikotan and Kunashiri Isls. [1327, 1335, 1719, 2266, 2413, 2419, 3173]. Occurs also in Sakhalin [19, 68, 1327, 2419, 3173].

2. Fossil record

See text in *M. rufocanus*.

3. General characteristics

Morphology Small-sized rodent with a short tail, small eyes and short ears for semi-fossorial lifestyle. Appearance is very similar to that of the gray red-backed vole. Body size is larger than that of *M. rufocanus*. Coat color is darkish yellow brown on back lacking dorsal red zone, and grayish white on belly. Enamel patterns of upper third molars showing 3 inner re-entrant angles and 4 inner salient angles in adults are diagnostic for this species (Fig. 1 and see *M. rufocanus* for details) [1017, 1018]. External measurements (range) are as follows: HB = 112–149 mm, T = 44–65 mm, HFsu = 19.5–22.7 mm, BW = 33–78 g for adults [19, 1168, 1336]. Sexual size dimorphism unclear [19, 1017, 1018, 1168, 1336, 2410]. This species closely resembles *M. rufocanus* in ratios of the large intestinal length to small intestinal length and the caecum length to small intestinal length [2726].

Dental formula   I 1/1 + C 0/0 + P 0/0 + M 3/3 = 16 [29].

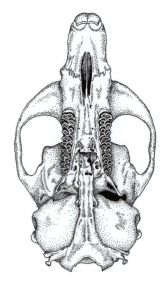

Fig. 1. Ventral view of the skull of the holotype (Specimen No.: NSMT M10823).

Myodes rex (Imaizumi, 1971)

Mammae formula   2 + 0 + 2 = 8 [19, 1017, 1327].

Genetics   Chromosome: $2n = 56$ (XX and XY) and FN=56. G-banded patterns are similar to those of *M. rufocanus* [1363, 3636]. Mitochondrial DNA sequences, *G6pd* intron 1 sequences and restriction fragment length polymorphism data for nuclear ribosomal genes are available [1166, 1170, 1173, 1454, 3348, 3816]. Some microsatellite markers also are available [1075].

4. Ecology

Reproduction   Average number of placental scars is 5.3–6.8 and that of corpora lutea is 16.8–18.7; the large number of ovulations may be due to elevation [593]. Reproductive activity varies seasonally and multiannually [2412].

Lifespan   Not reported.

Diet   Mainly plant materials (seeds and berries) and invertebrates, taking a wide variety of foods [748]. Under laboratory conditions, they show a higher preference for seeds of herbs and trees in comparison with *M. rufocanus*; mealworms were also eaten [3639]. Also known to debark stems and twigs of trees [2416] in late autumn and winter.

Habitat   Forests, plantations, scrublands and grassy fields. Occurs in wooded, mountain areas from 20 m to 1,900 m in elevation on the Hokkaido mainland. It also inhabits grasslands and abandoned farmlands on Rishiri and Rebun Isls. Prefers mixed undergrowth of *Sasa* bamboos and herbs, and moist areas from bottoms to rather moderate slopes of small valleys [2417, 2419].

Home range   Home range size varies depending on habitat, season and sex [2412]. Similar autumnal sizes to *M. rufocanus* are recorded [2413].

Behavior   Not reported.

Natural enemies   No data available. Possibly similar to *M. rufocanus*.

Parasites
Ectoparasite
Acari (mites): *Neotrombicula nagayoi* and *Radfordia* (*Microtimyobia*) *lemnina rufocani* [3143, 3701, 3704], *Hirstionyssus isabellinus* and *Laelaps* (*Laelaps*) *clethrionomydis* [2720]. Siphonaptera (fleas): *Peromyscopsylla hamifer takahasii* and *Megabothris sokolovi* are reported [4140].
Endoparasite
Cestoda and Nematoda: of the former, *Echinococcus multilocularis* is recorded [207, 208, 217, 3424].

5. Remarks

In 1971, *M. rex* was first described as a new species, *Clethrionomys rex*, for individuals from Rishiri Isl. [1017]. Subsequently, *M. montanus* (formerly *C. montanus*) was described for individuals from Hidaka Mountains in central Hokkaido in 1972 [1018]. However, it was suggested that *M. montanus* is the junior synonym of *M. rex* [1312]. Some mammalogists, especially Russian mammalogists, used *M. sikotanensis* for this species but this specific name "*sikotanensis*" has been considered as a local variation of *M. rufocanus* in southern part of the Kuril Isls. [1168], considering the description of *Neoaschizomys sikotanensis* for samples from Shikotan Isl. [3591]. It was concluded in 2008 that *N. sikotanensis* is a variant of *M. rufocanus* based on the re-examination of type specimens [2266]. Accordingly, *M. sikotanensis* from Sakhalin including the so-called "*microtinus*" form should be reidentified as *M. rex* [68]. The generic rank of vole taxa in Hokkaido was revised from *Clethrionomys* to *Myodes* [349, 1313].

K. NAKATA & M. A. IWASA

An example of the habitat of *Myodes rex* in Teshio, northern Hokkaido, showing *Sasa* bamboos, grasses and broad-leaved trees in summer (M. A. Iwasa).

Rodentia MURIDAE

065

Red list status: C-2 as *Clethrionomys r. mikado*, LP for the Oshima Pen. population (MSJ); LC (IUCN)

Myodes rutilus (Pallas, 1779)

EN northern red-backed vole   JP ヒメヤチネズミ（hime yachi nezumi）
CM 紅背䶄   CT 紅背鼠，紅背䶄，北方紅背田鼠   KR 숲들쥐   RS красная полёвка

An adult male from Hokkaido (M. A. Iwasa).

1. Distribution

Widely distributed from Scandinavia through Siberia and northeastern Asia to Alaska and northwestern Canada [19, 1327, 2334, 3173]. In Japan, this species occurs only in Hokkaido [19].

2. Fossil record

Not reported in Japan.

3. General characteristics

Morphology Small-sized rodent with a short tail, small eyes and short ears for semi-fossorial life style. Coat has an apparent red zone in the center of brown on back, light brown on sides and ivory white or cream on belly [19, 1327]. Tail densely haired and apparently bicolored with brown in upper half and ivory white in lower half [19, 1327, 2410]. Among the 3 *Myodes* species of Hokkaido, *M. rutilus* is the smallest based on the external and skull dimensions (see Fig. 1 in *M. rufocanus*). External measurements (range) are as follows: HB = 74.2–114.0 mm, T = 28.6–41.2 mm, HFsu = 15.9–19 mm, E = 10.2–13.0 mm, BW = 15.0–27.7 g [19, 1327].

Dental formula   I 1/1 + C 0/0 + P 0/0 + M 3/3 = 16 [29].

Mammae formula   2 + 0 + 2 = 8 [19, 1327].

Genetics Chromosome: $2n = 56$ (XX and XY) and FN = 56 [2543, 3636]. Autosomal G-band patterns of *M. rutilus* are not identical to those of *M. rufocanus* but are to *M. glareolus* [2182, 2543]. Sequence data of mitochondrial cytochrome *b* gene and *G6pd* gene intron are available [1170]. A few microsatellite markers also are available [1075].

4. Ecology

Reproduction Litter size (number of embryos) is 2–7 and the mode is 5 or 6 in the wild [593]. Average weight at birth is 2.0 g; newborns are naked and blind; eyes open at 14 days; females mature by 32 days under laboratory conditions [593]. Breeding season is probably from April to September, with an unimodal pattern in litter production; most of the current year's individuals do not mature until the next season [593].

Lifespan Few data are available. In captivity they live to at least 600 days [593].

Diet Relatively confined to fewer items such as a few kinds of herbs, seeds and insects, and less variable seasonally [3434].

Myodes rutilus (Pallas, 1779)

Habitat Occurs from lowlands to alpine zones. Inhabits forests, plantations, scrublands and grasslands with dense cover and humus, especially dominant in coniferous forests with highly closed canopies or in relatively dry habitats often associated with sand dunes [7, 22]. When sympatric with *M. rufocanus*, microhabitat segregation is observed [2774].

Home range Home ranges vary in size according to habitat and season and may decrease in non-breeding season; 22–73 m in range length and 150–600 m² in range area [2727].

Behavior Mainly crepuscular or nocturnal, similar to *M. rufocanus*.

Natural enemies Taken by carnivores and birds of prey [18, 1705, 2019, 2020, 3741, 4143, 4147].

Parasites
Ectoparasite
Acari (mites and ticks): genera *Leptotrombidium, Neotrombicula, Gahrliepia, Radfordia, Speleorodens, Hirstionyssus, Haemogamasus, Androlaelaps, Laealps, Macrocheles, Ixodes, Listrophorus* and *Myocoptes* [3545, 3704, 4140]. Of these, *Radfordia* (*Microtimyobia*) *lemnina mikado* infests only *Myodes rutilus mikado* in Hokkaido, and it differs from *R.* (*M.*) *lemnina rutila* parasitizing *Myodes rutilus* in Europe [3701, 3704]. Siphonaptera (fleas): genera *Ctenophtalmus, Corrodopsylla, Neopsylla, Catallagia, Stenoponia, Hystrichopsylla, Rhadinopsylla, Nosopsyllus, Ceratophyllus* and *Peromyscopsylla*. Anoplura (louse): genera *Hoplopleura* and *Polyplax* [4140].
Endoparasite
Four species of Cestoda and 8 species of Nematoda [206–208]; 2 species of nematodes, *Heligmosomum* (*Paraheligmosomum*) *mixtum* and *Syphacia petrusewiczi*, are host-specific. Apicomplexa, a parasitic Protozoa, are reported [4243]. Have been found to be one of the favorable intermediate hosts of the cestode *Echinococcus multilocularis*, a parasitic organism causing the disease alveolar echinococcosis in humans [3426].

5. Remarks

Similar species: See text of *M. rufocanus*. Taxonomic remarks: Hokkaido population is ranked as a subspecies, *M. rutilus mikado* [3563]. No record of damage in man-made plantations.

K. NAKATA & M. A. IWASA

An example of the habitat of *Myodes rutilus* in Kiritappu Marsh, eastern Hokkaido, showing grasses and shrubs in summer (M. A. Iwasa).

Rodentia   MURIDAE

066

Red list status: C-2, LP for some populations in Honshu (MSJ)

Eothenomys andersoni (Thomas, 1905)

EN Anderson's red-backed vole   JP ヤチネズミ（yachi nezumi）   CM 本州絨鼠
CT 安德森氏紅背絨鼠，安德森氏天鵝絨鼠   KR 일본비단털들쥐   RS южноазиатская полёвка Андерсона

An adult female from Miyagi Pref., Honshu (M. A. Iwasa).

1. Distribution

Endemic to Japan. Roughly central and northern areas, and southern Kii Pen. of Honshu [1208, 1327, 2334] but exact distribution range is still unknown. The Kii Pen. population is considered as relictual [1011, 1208, 1327]. Vertical distribution varies from 170 m (Aomori Pref. of northernmost of Honshu) to 3,420 m (Mt Yarigatake of central Honshu) [1786]. In Honshu, *E. andersoni* and *E. smithii* occur parapatrically and are distributed at higher and lower elevation, respectively [1334, 1563].

2. Fossil record

Fossils of *Eothenomys* (=*Phaulomys*) sp. and *Aschizomys* (= *Eothenomys*) *andersoni* are recorded from the Late Pleistocene in central and western Honshu [1507, 1512, 1513, 1518, 1519].

3. General characteristics

Morphology   Small-sized rodent with a short tail, small eyes and short ears for semi-fossorial life style. Appearance of this species similar to that of *E. smithii* and *Microtus montebelli*. Body size is larger than that of *E. smithii* but similar to *M. montebelli*. Coat color is red-brown or buff-brown on back and yellowish gray-brown on belly [1327]. External measurements (range) are as follows; HB = 79–118 mm, T = 40–63 mm, HFsu = 16.5–19.2 mm, BW = 11–40 g in northeastern Honshu; HB = 79–118 mm, T = 50–77 mm, HFsu = 18.5–21.5 mm, BW = 19–42 g in central Honshu; HB = 79–127 mm, T = 45–78 mm, HFsu = 18.9–22.3 mm, BW = 18–60 g in Kii Pen. Tail ratio ranges greatly from 40 to 80% in Kii Pen. (n = 42) [1011, 1208, 1327, 3226]. As noted above, this species shows a great deal of morphological size variation (see Remarks).

Dental formula   I 1/1 + C 0/0 + P 0/0 + M 3/3 = 16 [29].

Mammae formula   2 + 0 + 2 = 8 [29, 1327].

Genetics   Chromosome: $2n$ = 56 (XX and XY) and FN = 56–62 [1172, 1617, 3636, 4169]. Autosomal G-band patterns of *E. andersoni* are not identical to those of *Myodes glareolus* but are to *M. rufocanus* [1172, 1617]. Sequence data of mitochondrial cytochrome *b* gene, nuclear *Sry*, *G6pd* gene intron and 18S ribosomal RNA gene, and restriction fragment length polymorphism data for nuclear ribosomal genes are available [597, 1169–1171, 3348]. A few microsatellite markers are also available [1075].

4. Ecology

Reproduction   Litter size (number of embryos) is 1–5 and the average is 3.2 [1327]. In the Kii Pen. population, the mean litter size of the wild-caught 1st generation was 3.88 ± 0.21 (n = 25, range 2–6) [1615]. Breeding seasons in northeastern and central Honshu has a mode in summer, and from winter to spring in the Kii Pen.

Lifespan Under captive condition, 796 days recorded [1616].

Diet Under captive condition, mainly eats plant matter. Sometimes eats bark at young plantations (Japanese cypress and Hinoki cypress) and causes serious damage for wood growth [1614].

Habitat Rocky areas and streamsides without sunshine, artificial stone dykes or walls, and cultivated areas [1327, 3226]. Prefers wet conditions throughout range.

Home range Not reported.

Behavior Not reported.

Natural enemies Potential predators are owls and mustelids [2787].

Parasites
Ectoparasite
Acari (mites): *Leptotrombidium, Radfordia, Lealaps, Eulaelaps, Haemolaelaps, Haemogamasus, Hirstionyssus,* and *Ixodes* [3407, 3693, 3696, 3701], Siphonaptera (fleas) [3694, 3695] and Anoplura (lice) [3696].
Endoparasite
Several taxa of Cestoda and Nematoda [206, 208, 210, 3007].

5. Remarks

In central Honshu, *E. andersoni* and *E. smithii* occur parapatrically (with narrow sympatric zones) along vertical distributions [1334, 1563]; the former and the latter are confirmed in higher and lower elevations, respectively. However, *E. andersoni* sometimes occurs at lower elevation without the occurrence of *E. smithii* [1160]. Such specimens have smaller body sizes and are similar to *E. smithii* [1160, 1334, 1563]. Morphological identification for *E. andersoni* and *E. smithii* is possible based on the relationship between tail length and hind foot length in areas of sympatry [1160, 1334, 1563]. Some taxonomists classified Anderson's red-backed vole into 3 valid species: *E. andersoni* in northeastern Honshu, *E. niigatae* in central Honshu and *E. imaizumii* on the Kii Pen. based on the morphological and genetic data [1011, 1208, 2334]. However, consistent morphological characters for the classification above have not been established. In particular, the former 2 are distributed continuously and the morphological discrimination depends on only statistic state of morphological dimensions among their clines. Further, crossbred offspring among the 3 species are fertile and viable [1618]. Namely, we have no diagnoses to positively recognize the 3 species. Thus, it is appropriate to recognize *E. niigatae* and *E. imaizumii* as junior synonyms of *E. andersoni* [80]. The Anderson's red-backed vole has been ranked within the genera *Evotomys, Craseomys, Aschizomys, Myodes, Eothenomys, Clethrionomys* and *Phaulomys* [2334]. At present, the genus *Eothenomys* is appropriate because of the presence of unrooted molars into consideration [1328].

M. A. IWASA

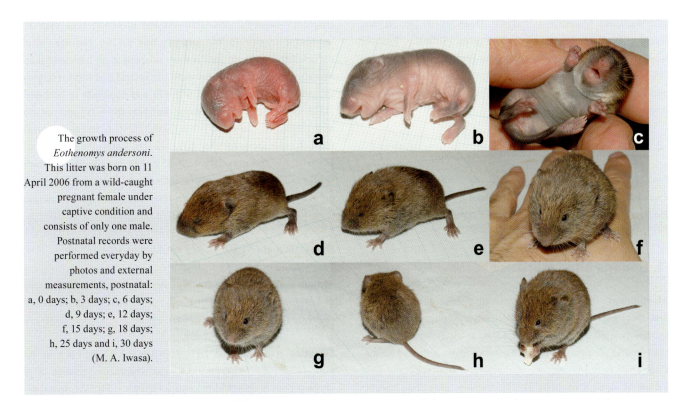

The growth process of *Eothenomys andersoni*. This litter was born on 11 April 2006 from a wild-caught pregnant female under captive condition and consists of only one male. Postnatal records were performed everyday by photos and external measurements, postnatal: a, 0 days; b, 3 days; c, 6 days; d, 9 days; e, 12 days; f, 15 days; g, 18 days; h, 25 days and i, 30 days (M. A. Iwasa).

Rodentia MURIDAE

067

Red list status: R, LP for several populations in Kyusyu and Honshu (MSJ); LC (IUCN)

Eothenomys smithii (Thomas, 1905)

EN Smith's red-backed vole   JP スミスネズミ (sumisu nezumi)   CM 日本绒鼠   CT 史密氏紅背絨鼠，日本天鵝絨鼠
KR 스미스비단털들쥐   RS южноазиатская полёвка Смита

An adult male from Tokyo Metropolis, Honshu (M. A. Iwasa).

1. Distribution

Distributed in Honshu (except for northernmost parts, Boso Pen., Miura Pen., Mt. Yahiko-yama, Chita Pen., Shimane Pen. and Wanizuka Mountains), Shikoku, Dogo Isl. of Oki Isl. and Kyushu. Vertical range is between 60 m (Shimane Pref. of western Honshu) and 2,400 m (Nagano Pref. of central Honshu) [1327, 1334]. In Honshu, *E. andersoni* and *E. smithii* occur parapatrically and are distributed at higher and lower elevations, respectively [1334, 1563].

2. Fossil record

Fossils of *E. smithii* (including *E. kageus*, *Eothenomys* sp., and *Phaulomys* cf. *smithii*) are recorded from the Late Pleistocene–Holocene layers in central and western Honshu [1500, 1501, 1507, 1512, 1513, 1518, 1519, 1725, 2139, 2145].

3. General characteristics

Morphology   Small-sized rodent with a short tail, small eyes and short ears for semi-fossorial lifestyle. Appearance of this species is similar to that of *E. andersoni*. Body size is smaller than that of *E. andersoni* and *Microtus montebelli* in Honshu. Coat color is red-brown or yellowish-brown on back and light yellow to orange on belly [1327]. Usually, arvicoline rodents carry simple, intermediate and complex types of enamel patterns of upper third molar, and this species shows lower frequency (right = 63.2%, left = 65.8%, $n = 30$) of intermediate type than *E. andersoni* (right = 80.0%, left = 80.9%, $n = 38$) and higher frequency (right = 36.9%, left = 34.2%) of simple type than *E. andersoni* (right = 16.7%, left = 16.7%) [2136]. Namely, it is expected that *E. andersoni* is more adapted as an herbivore than this species. The enamel between the second and third triangles (T2 and T3), and fourth and fifth (T4 and T5) tends to be fused [1322, 2136]. Baculum also shows variation and is not appropriate as a diagnostic character [1208, 1318, 1321]. External measurements (range) are as follows; HB = 70–116.3 mm, T = 30–50 mm, HFsu = 15.5–18.2 mm, BW = 20–35 g with geographic variation [1321, 1327]. Measurements of individuals from southern areas are larger than those of northern ones. Tail to body ratio ranges greatly, from 35 to 60% [1011, 1321, 1327, 4162].

Dental formula   I 1/1 + C 0/0 + P 0/0 + M 3/3 = 16 [29].

Mammae formula   0 + 0 + 2 = 4 and 1 + 0 + 2 = 6. Both types and asymmetrical type, 1/0 + 0 + 2 = 5, occur within a population [1009, 1318, 1321, 1327].

Genetics   Chromosome: $2n = 56$ (XX and XY) and FN = 56–62 [137, 1172, 1617, 3636, 4169]. A heterochromatic variation of X chromosome is observed in central Honshu [138]. Autosomal G-

Eothenomys smithii (Thomas, 1905)

band patterns of *E. smithii* are not identical to those of *M. glareolus*, but are to *M. rufocanus* [1172]. A biochemical analysis revealed isozyme differentiation [4169]. Sequence data of mitochondrial cytochrome *b* gene, nuclear *Sry*, *G6pd* gene intron and 18S ribosomal RNA gene, restriction fragment length polymorphism data for nuclear ribosomal genes are available [597, 1169–1171, 3348]. A few microsatellite markers are also available for molecular ecological studies [1075].

4. Ecology

Reproduction Average litter sizes are 2.3 ($n = 11$ at alt. 1,900–2,400 m) in Nagano Pref. of central Honshu, 2.6 ($n = 13$ at alt. 470–650 m) in Hiroshima Pref. of western Honshu, 2.3 ($n = 4$ at alt. 65–140 m) in Kagawa Pref. of Shikoku, 2.23–2.78 ($n = 35$–44 at alt. 700–1,300 m) in Kochi Pref. in Shikoku and 4.1 ($n = 13$ at alt. 40–200 m) in Fukuoka Pref. of Kyushu [1320, 2144, 3523, 4161, 4234, 4236, 4237]. Breeding seasons vary geographically: May–October in Nagano Pref., September–April in Hiroshima Pref., October–April in Kagawa Pref., September–June at alt. 1,000 m and October–May at alt. 600 m in Kochi Pref., and November–March in Fukuoka Pref. [976, 1320, 2144, 4161, 4234, 4236, 4237]. Females become sexually mature at 23–28 days of age and males at 31–34 days of age based on the vaginal opening and epididymis maturation [139].

Lifespan Over 3 years under captive conditions [139, 1321].

Diet Mainly green parts of plants, seeds and fruits, particularly those of oak, chestnut, mulberry, Japanese bird cherry and giant dogwood [4238]. Small pieces of broadleaves and grasses have been observed in their stomach contents [773].

Habitat Forests, coniferous plantations, agricultural areas, rocky terrain from lower elevations to alpine zones. Particularly prefers wet areas [1321, 1327, 3385, 3517].

Home range Range length is 23.1–25.7 m and range size is 0.057–0.101 ha [3522].

Behavior Daily activity shows 2 or 3 peaks under dark-light condition by an experiment using an exercise wheel [44].

Natural enemies Copperhead, forest rat snake, owls and mustelids [1321, 4238].

Parasites
Ectoparasite
Acari (mites): Trombidiformes (myobiid mites): *Radfordia lemnina rufocani* [3701]; Trombiculidae (chigger mites): *Leptotrombidium miyazakii* [3431]. Siphonaptera (fleas) [2963, 3694, 3695]. Anoplura (louse): *Hoplopleura acanthopus* [1315].
Endoparasite
Trematoda, Cestoda and Nematoda [206, 208, 210, 3007].

5. Remarks

Similar species: See text in *E. andersoni*. Taxonomic rank: Some taxonomists divide this species into 2 species: *E. kageus* in northeastern and central Honshu and *E. smithii* in western part based on the morphological criteria, particularly differences in mammae formula, baculum morphology and genetic data [1009, 1011, 3636]. However, number of mammae and baculum morphology are variable and these criteria do not seem to be diagnostic. Further, it is thought that the Y chromosomal criteria do not lead to reproductive isolation. Thus, it is appropriate to recognize them as conspecific. Smith's red-backed vole taxon has been ranked within the genera *Evotomys*, *Myodes*, *Anteliomys*, *Eothenomys*, *Clethrionomys* and *Phaulomys* [1208, 1507, 2334, 3563]. At present, genus *Eothenomys* is appropriate because of the presence of unrooted molars [1328].

Yukibumi KANEKO & M. A. IWASA

An example of the habitat of *Eothenomys smithii* at the foot of Mt. Fuji, Shizuoka Pref., Honshu, showing shrubs and rocks covered with bryophytes in early spring. *Eothenomys smithii* is a typical rodent inhabiting the foot of Mt. Fuji (M. A. Iwasa).

Rodentia   MURIDAE

068

Red list status: C-1 (MSJ); LC (IUCN), introduced

Ondatra zibethicus (Linnaeus, 1766)

EN muskrat   JP マスクラット（masukuratto）   CM 麝鼠   CT 麝鼠，麝田鼠   KR 사향쥐   RS ондатра

A swimming muskrat in eastern Saitama Pref., Honshu (Kinji Suzuki).

1. Distribution

Introduced from North America (probably 1943) as a fur animal for clothes of airplane pilots. First feralization was confirmed in 1947. Information on occurrence is obtained from Ichikawa (Chiba Pref.), Satte, Kasukabe, Koshigaya (Saitama Pref.) and Katsushika Ward (Tokyo Metro.), along Edogawa River [1327, 3334].

2. Fossil record

Not reported in Japan.

3. General characteristics

Morphology   Coat color is entirely brown, dark brown or black. Pelage short, dense and soft. Ears and eyes small. Whitish long sharp claws in both fore and hind feet. Tail vertically flat and long, over 200 mm. External measurements are as follows; HB = approx. 300 mm, HF = approx. 70 mm, BW = approx. 800 g. Condylobasal length over 50 mm. Adults show molar roots. Presence of interdigital hard hairs as paddles contributing to swimming [1327, 2933, 3334].
Dental formula   I 1/1 + C 0/0 + P 0/0 + M 3/3 = 16 [29]
Mammae formula   1 + 0 + 2 = 6 [1327].
Genetics   Not reported for Japanese populations.

4. Ecology

Reproduction   In North America, breeds actively throughout year, 5 or 6 times per year. Gestation period is 25–30 days and litter size (number of embryos) is 1–11 [1327].
Lifespan   Not reported in Japan.
Diet   Eats stems, subterranean stems and roots of poaceous and iridaceous grasses at watersides. Sometimes eats mollusks and crayfish [1327, 2933, 3334].
Habitat   Nests in grasses within tunnel systems in waterways or along rivers. Entrances of the tunnel are open under the water. In the tunnels, there are spaces for eating, excreting, and ventilation [1327].
Home range   Does not go far from nests [3334].
Behavior   Semi-aquatic and tough against the cold. Generally active in day and night. Sleeps during morning and active in afternoon. Before evening, returns to nest and after sunset, active again. Swims well but terrestrial locomotion is weak. Secretions with musk mixed in urine [1327, 2933, 3334].
Natural enemies   Dogs, raccoon dogs, cats, raccoons, mustelids, raptors and snakeheads [2933, 3334].
Parasites   Not reported in Japan.

5. Remarks

Recently, loss of wetlands related to urbanization has caused populations to decrease [1327, 2933]. No serious agricultural damage by this species in Japan [3334].

M. A. IWASA

Rodentia   MURIDAE

069

Red list status: C-2, LP for the Noto-jima Isl. population (MSJ); LC (IUCN)

Microtus montebelli (Milne-Edwards, 1872)

EN Japanese field vole   JP ハタネズミ (hata nezumi)   CM 日本田鼠   CT 日本田鼠   KR 일본밭쥐
RS японская полёвка

An adult female from Shizuoka Pref., Honshu (M. A. Iwasa).

1. Distribution

Endemic to Japan. Distributed in Honshu, Kyushu, Sado Isl. and Noto-jima Isl. [1317, 1327].

2. Fossil record

Fossils in the Middle and the Late Pleistocene layers are recorded from Honshu and in Holocene layers from Honshu and Kyushu [821, 1499–1503, 1505, 1507, 1512, 1513, 1518, 1519, 1521, 1725, 2145, 2375, 2440, 2688, 3201, 3203, 3205, 3206, 3354, 3442, 3602]. Although the population in Shikoku became extinct, the fossil record is known from the Late Pleistocene of Ehime Pref. [1520].

3. General characteristics

Morphology   Small-sized rodent with a short tail, small eyes and short ears for semi-fossorial lifestyle. Body size is larger than that of *Eothenomys smithii* but similar to *E. andersoni*. Tail of this species is shorter than that of *E. andersoni*. Coat color is non-reddish brown or yellow-grayish brown on back and gray on belly [1317, 1327]. The typical number of pads on the hind foot is 5, but 6 pads may be observed in *M. montebelli* or *Eothenomys* species [1317, 1327]. To identify *M. montebelli* exactly, observation of the structure of the posterior edge of the palate is essential. If the shape of the palatine posterior terminal is complex, it is genus *Microtus* (Fig. 1). External measurements (range) are as follows; HB = 95–136 mm, T = 29–50 mm, HFsu = 16.5–20.4 mm, BW = 22–62 g with a geographic cline; southern populations are larger than northern ones [1317, 1319, 1327].

Dental formula   I 1/1 + C 0/0 + P 0/0 + M 3/3 = 16 [29].

Mammae formula   2 + 0 + 2 = 8 [1317, 1327].

Genetics   Chromosome: $2n = 30$ (XX and XY) and FN = 56 [3635, 3979]. G-, C- and NOR band patterns are revealed [3979]. Some microsatellite DNA markers are available [1075]. A few chromosomal aberrations known from an illegal dumpsite are likely due to influence by chemical contaminations [1820, 2544].

4. Ecology

Reproduction   Breeding occurs spring through autumn in eastern Honshu and spring and autumn in western Honshu and Kyushu [57, 1317, 2142, 3248]. Gestation period is about 21 days. Average litter size (number of embryos) is 4.3 ($n = 88$) in Iwate Pref., 3.5 in Nagano Pref., 5.1 ($n = 40$) in Kyoto, western Honshu, and 5.0 ($n = 8$) in Fukuoka Pref. [57, 1317, 2142, 3248]. Vaginal opening can be observed at 29–43 days [57].

Lifespan   783 days for a female and 825 days for a male under captive conditions [3245]. Aging available by a lens weight analysis

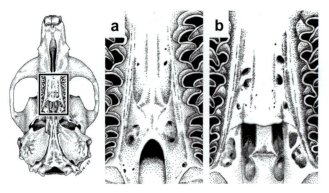

Fig. 1. Shapes of the posterior edge of the palate in the Japanese arvicolids, *Microtus*, *Myodes* and *Eothenomys*: (a) complex type in *Microtus* and (b) simple type in *Myodes* and *Eothenomys*.

[3909].

Diet Mainly eats plants: Asteraceae, Cucurbitaceae, Caprifoliaceae, Rubiaceae, Malvaceae, Lamiaceae, Convolvulaceae, Solanaceae, Primulaceae, Apiaceae, Vitaceae, Euphorbiaceae, Araliaceae, Violaceae, Geraniaceae, Fabaceae, Rosaceae, Brassicaceae, Basellaceae, Portulacaceae, Acanthaceae, Chenopodiaceae, Amaranthaceae, Polygonaceae, Urticaceae, Moraceae, Ulmaceae, Fagaceae, Salicaceae, Liliaceae, Plantaginaceae, Dioscoreaceae, Commelinaceae, Araceae, Cyperaceae, Poaceae, Pinaceae, Equisetaceae. Sometimes eats animal matter, gastropods, larvae of Lepidoptera and Orthoptera [1317, 3841].

Habitat Occurs from low to high elevation. Mainly inhabits cultivated fields, coniferous plantations, river banks and pastures, and also natural forests. In cultivated fields, uses leaf fibers of *Alopecurus aequalis*, *Poa annua*, *Imperata cylindrical*, *Veronica didyma*, *Capsella bursa-pastoris*, *Cyperus microiria*, wheat and rice for nesting underground [1317].

Home range Maximum range length is 25 m in females and 30 m in males at standard density of about 50 individuals per ha [1305].

Behavior Usually nocturnal but diurnally active under cloudy or rainy conditions [3841]. Under captive conditions, active periods are confirmed at evening–midnight, dawn and daytime [2181]. Further, under wild conditions, active in nighttime during spring–autumn, and active in both night and daytime or completely active in daytime during winter [1127]. Digs tunnels with diameter 3–4 cm.

Natural enemies Forest rat snake, Japanese striped snake, kites, dogs and mustelids [1488, 2787, 3841].

Parasites
Ectoparasite
Acari (mites): genera *Gahrliepia*, *Trombicula*, *Neotrombicula* and *Acomatacarus* [3400]. Siphonaptera (fleas): genera *Hystricopsylla*, *Atyphloceras*, *Stenoponia*, *Rhadinopsylla*, *Ctenophathalmus*, *Neopsylla*, *Peromyscopsylla*, *Frontopsylla*, *Malaraeous* and *Megabothris* [2963]. Anoplura (louse): genera *Polyplax* and *Hoplopleura* [1315].
Endoparasite
Several taxa of Trematoda, Cestoda and Acanthocephala [206].

5. Remarks

Recent changes in natural forests to artificial ones have caused the expansion of its distribution to areas of higher elevation. This species has been extirpated from former typical distribution areas, such as rice fields and farms. External appearance of *M. montebelli* is similar to *Eothenomys andersoni* and *E. smithii*.

M. A. IWASA

An example of the habitat of the Japanese field vole in Hirosaki, Aomori Pref., Honshu, showing rice fields and reeds in autumn (M. A. Iwasa).

Rodentia   MURIDAE

070

Red list status: EN (MEJ); En as *T. o. osimensis* (MSJ); EN (IUCN)

Tokudaia osimensis (Abe, 1933)

| EN Amami spiny rat | JP アマミトゲネズミ（amami toge nezumi） | CM 琉球刺鼠 | CT 裔鼠，琉球刺鼠 |
| KR 아마미가시털쥐 | RS амамийская колючая мышь | | |

An adult from Amami-ohshima Isl., the Satsunan Isls. (K. Tsuchiya).

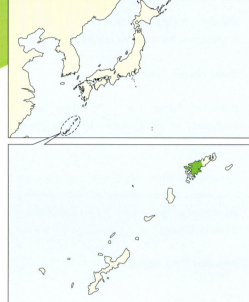

1. Distribution

Endemic to Amami-ohshima Isl. of the Satsunan Isls. [1327].

2. Fossil record

Fossils of *Tokudaia* sp. were discovered in the Late Pleistocene and Holocene layers from Okinawa-jima Isl. and Ie-jima Isl. of the Ryukyu Isls. [822, 824, 1508, 1725, 3443].

3. General characteristics

Morphology   Coat color is orangish dark-brown on back and sides, and gray or white on belly. Brown on the upper and white on the lower on tail. Tactile hairs 30 mm in length around the mouth and eyes. Body hairs reach 20 mm in length with light, reddish and yellowish orange in color with light gray at the proximal end. Spinous hairs reach 20 mm in length with black at the terminus and light gray at the proximal end (Fig. 1). Spinous hairs are absent around the mouth, ears, feet and tail [1327]. Tail ratio ranges 60–100% as found in *T. muenninki*. External measurements (range) are as follows; HB = 103–160 mm, T = 83.5–135 mm, HFsu = 28.8–34 mm [1325, 1327].

Dental formula   I 1/1 + C 0/0 + P 0/0 + M 3/3 = 16 [29].

Mammae formula   0 + 0 + 2 = 4 [1327].

Genetics   Chromosome: $2n = 25$ (XO in both sexes) and FN = 46 [937, 1805, 3644]. A few gene localizations on chromosomes have been made [180]. Mitochondrial cytochrome *b* gene sequence data and restriction fragment length polymorphism data for nuclear ribosomal genes, and some nuclear gene sequence data are available [1569, 1804, 3052, 3349].

Fig. 1.   A typical view of spinous hairs.

4. Ecology

Reproduction Litter size ranges from 1 to 7. Breeding season is during October–December [1327].

Lifespan Not reported.

Diet Seeds of chinquapins, ants, ant larvae and sweet potatoes [1327].

Habitat Prefers broadleaved secondary forests, particularly chinquapins, because they mainly eat seeds of chinquapins [1327].

Home range Not reported.

Behavior Jumps up to 30–40 cm in height with strong hindlimbs, to avoid attacks by habu pitvipers (*Trimeresurus flavoviridis*) [1327].

Natural enemies Feral cats and dogs, exotic small Indian mongooses, and native habu pitvipers (*Trimeresurus flavoviridis*) are known predators [1327]. This fact causes social and environmental problems.

Parasites Not reported.

5. Remarks

Described in 1933 and designated as a "Natural Monument" in 1972.

M. A. IWASA

6. Topic

Unique and interesting sex chromosome evolution in **Tokudaia**

A. KUROIWA

Variations of karyotypes and unique sex chromosomes

Variation in karyotypes has been observed across the 3 species of *Tokudaia*. *Tokudaia muenninki* on Okinawa-jima Isl. has XX/XY sex chromosome constitution as is typical for almost all eutherian mammals, and the chromosome number is $2n = 44$ [3644]. *Tokudaia tokunoshimensis* and *T. osimensis*, however, have XO/XO sex chromosome constitution with lack of the Y chromosome, and the chromosome numbers of *T. tokunoshimensis* and *T. osimensis* are $2n = 45$ and $2n = 25$, respectively (Fig. 2) [936, 937]. The evolutionary process of speciation in *Tokudaia* was suggested (Fig. 3) [3333] based on comparison of these karyotypes. The genus *Tokudaia* probably evolved on an island that included present-day Okinawa-jima, Amami-ohshima, and Tokuno-shima Isls. Okinawa-jima Isl. was first separated from the other 2 islands with a rise in sea level. The Ryukyu spiny rat that inhabited Okinawa-jima Isl. became *T. muenninki* ($2n = 44$, XX/XY). The Y chromosome loss event occurred in the populations that remained on the island. When Tokuno-shima and Amami-ohshima Isls. were still connected, the ancestral species of *T. tokunoshimensis* and *T. osimensis* might have developed a new sex-determining system without the Y chromosome. After separation of Tokuno-shima and Amami-ohshima Isls., the chromosome number of the Amami-ohshima populations might have decreased to $2n = 25$. Comparison of chromosome painting map with chromosome-specific DNA probes of the laboratory mouse (*Mus musculus*) between *T. tokunoshimensis* and *T. osimensis* indicated that the ancestral karyotype of the 2 species is $2n = 48$, XX/XY. This karyotype was similar to the one of *T. tokunoshimensis*, and the

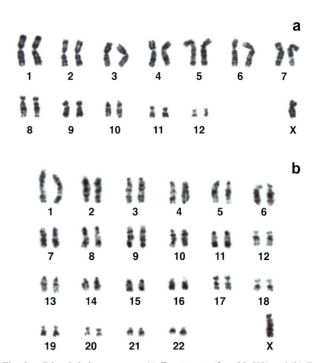

Fig. 2. G-banded chromosomes: (a) *T. osimensis* ($2n = 25$, XO) and (b) *T. tokunoshimensis* ($2n = 45$, XO).

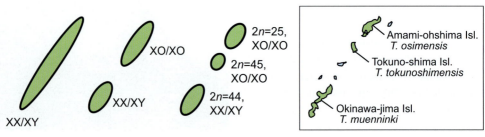

Fig. 3. The evolutionary process of speciation in *Tokudaia*.

karyotype of *T. osimensis* would then have been established through at least 14 chromosomal changes, mainly centric fusions and tandem fusions, from the ancestral karyotype [2398].

The single X chromosomes of *T. osimensis* and *T. tokunoshimensis* are submetacentric and subtelocentric, respectively. Comparative mapping of the mouse cDNA clones of the X-linked genes between the 2 *Tokudaia* species revealed that the chromosome rearrangement, which occurred in either of the X chromosomes after the 2 species diverged from a common ancestor, is not a pericentric inversion but the centromere repositioning (CR). It was also suggested that the CR event might also occur in X chromosomes of *T. osimensis* lineage (Fig. 4) [1647].

Y chromosome loss event

Most of mammalian Y-linked genes are specialized for sex, such as testis differentiation and spermatogenesis. These functions are essential for males. Although the Y chromosome is absent in *T. tokunoshimensis* and *T. osimensis*, the 2 genes, that linked to human and mouse Y chromosomes, were located on the distal part of the long arm of the X chromosome in both species [180]. Furthermore, another 3 Y-linked genes of human and mouse were also mapped on the X chromosome and/or autosomes in *T. osimensis* [1803]. This result suggests that the X chromosome carries a translocated Y segment in these species, and translocations occurred several times in the ancestral Y-linked genes of the 2 species of *Tokudaia* during Y chromosome loss event.

Novel sex determination mechanism in Tokudaia

Ryukyu spiny rats are unique and interesting mammals for studying the mechanism of sex determination. The *Sry* gene, which is the Y-linked male dominant sex-determining gene in mammals, has not been detected in the genome of *T. tokunoshimensis* and *T. osimensis* [3295, 3333]. This indicates that the 2 species have acquired a novel sex-determining system that does not depend on the *Sry* gene in the evolutionary process. No sex-specific chromosome regions were observed between sexes in the 2 species by comparing the fine G-banding patterns and the CGH (comparative genomic hybridization) method [1646]. The sex specific chromosome region therefore might be very minute, and an evolutionary event might have occurred in a small chromosome region with the disappearance of the Y chromosome in the 2 species. It has been suggested that a single sex differentiation-related gene evolved into a new sex-determining gene, which acquired a novel function and consequently superseded the Y-linked *Sry* gene. The species of *Tokudaia* are scientifically important as source materials for understanding the evolution of sex chromosome and the mechanism of sex determination.

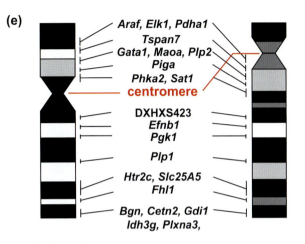

Fig. 4. Chromosome mapping of the *Fhl1* gene on *T. osimensis* (a, b) and *T. tokunoshimensis* (c, d) chromosomes. The hybridization signals are indicated by arrows. The *Fhl1* gene was localized to *T. osimensis* Xq23.2 and *T. tokunoshimensis* Xq24.2, respectively. R- and Hoechst G-banding patterns are demonstrated in a, c and b, d, respectively. Comparative cytogenetic maps of 22 X-linked genes in *T. osimensis* and *T. tokunoshimensis* (e). The locations of genes and gene orders are shown on the side of each X chromosome.

An Okinawa spiny rat in a forest on Okinawa-jima Isl. (T. Jogahara).

Rodentia   MURIDAE

071

Red list status: CR (MEJ); En as *T. o. muenninki* (MSJ); CR (IUCN)

Tokudaia muenninki (Johnson, 1946)

EN Okinawa spiny rat   JP オキナワトゲネズミ (okinawa toge nezumi)   CM 冲绳刺鼠   CT 琉球刺鼠，沖繩刺鼠
KR 오키나와가시털쥐   RS окинавская колючая мышь

An adult from Okinawa-jima Isl., the Ryukyu Isls. (K. Tsuchiya).

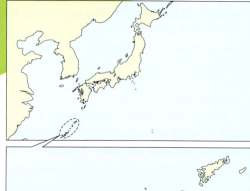

1. Distribution

Endemic to Okinawa-jima Isl. of the Ryukyu Isls. [1327].

2. Fossils

See *T. osimensis*.

3. General characteristic

Morphology   Coat color is black and yellowish brown with reddish yellow on back, and gray with reddish yellow on belly. Dorsal spines are black or amber and darker than *T. osimensis* at the tips [1327]. Tail ratio ranges 60–100% as well as *T. osimensis*. External measurements (range) are as follows; HB = 112–175 mm, T = 92–132 mm, HFsu = 29.8–35 mm [1327].

Dental formula   I 1/1 + C 0/0 + P 0/0 + M 3/3 = 16 [29].

Mammae formula   0 + 0 + 2 = 4 [1327].

Genetics   Chromosome: $2n = 44$ (XX in female and XY in male) and FN = 50 [937, 2327, 3644]. Mitochondrial cytochrome *b* gene sequence data and restriction fragment length polymorphism data for nuclear ribosomal genes, and some nuclear gene sequence data are available [2326, 3349].

4. Ecology

Reproduction   Litter size ranges from 5 to 10. Breeding season is estimated to occur between October and December [1327].

Lifespan   Not reported.

Diet   Seeds of chinquapins, mollusks and river crabs [1327].

Habitat   Forests of chinquapins with higher coverage of tree layer and undergrowth herbs at higher than 300 m elevation. In the northern distribution area of Okinawa-jima Isl., mainly inhabits forests consisting of *Castanopsis sieboldii*, *Lithocarpus edulis*, *Distylium racemeosum* and *Schima wallichii*. Nests are made on slopes of forest floors with entrances dug horizontally, changing angle at 0.5–1.0 m intervals [1327, 3931].

Home range   Not reported.

Behavior   Completely nocturnal, with activity and resting periods in 1–2 hour intervals [1327]. Most active period is just after beginning of activity. Hops quickly [1327].

Natural enemies   Feral cats [1327].

Parasites   Not reported.

5. Remarks

Discovered in 1943 [1243] and designated as a "Natural Monument" in 1972. Habitats of this species have been reduced by logging.

M. A. IWASA

Rodentia   MURIDAE

Red list status: EN (MEJ); EN (IUCN)

Tokudaia tokunoshimensis
Endo & Tsuchiya, 2006

EN Tokunoshima spiny rat   JP トクノシマトゲネズミ (tokunoshima toge nezumi)
CM 德之島刺鼠   CT 德之島刺鼠   KR 가시털쥐   RS токуношимская колючая мышь

An adult from Tokuno-shima Isl., the Satsunan Isls. (K. Tsuchiya).

1. Distribution

Endemic to Tokuno-shima Isl. of the Satsunan Isls. [1327].

2. Fossil record

See *T. osimensis*.

3. General characteristics

Morphology   Coat color is dark-brown on back and light gray and buff on belly. Tail is blackish on the upper and whitish on the lower half. Spinous hairs are hard and reach 15–20 mm in length. They are absent around the mouth, ears, feet and tail [513]. External measurements (mean ± SD) are as follows; HB = 155 ± 14.53 mm (n = 3), T = 117.5 mm (n = 1), HFsu = 37.5 ± 0.71 mm (n = 3) for adults. Sexual size dimorphism unknown. The skull is the largest among all the *Tokudaia* species [1325].

Dental formula   I 1/1 + C 0/0 + P 0/0 + M 3/3 = 16 [29, 513].

Mammae formula   2 + 0 + 2 = 8 [1327].

Genetics   Chromosome: $2n$ = 45 (XO in both sexes) and FN = 58 [936, 937, 3644]. Mitochondrial cytochrome b gene sequence data and restriction fragment length polymorphism data for nuclear ribosomal genes, and male specific gene data are available [1804, 3349].

4. Ecology

Reproduction   Not reported.
Lifespan   Not reported.
Diet   Not reported.
Habitat   Broadleaved secondary forests to natural forests [1327].
Home range   Not reported.
Behavior   Not reported.
Natural enemies   Not reported.
Parasites   Not reported.

5. Remarks

In 2006, *T. tokunoshimensis* was first described as a new species based on 3 individuals from Tokuno-shima Isl., separated from *T. osimensis* on Amami-ohshima Isl. [513]. Designated as a "Natural Monument".

M. A. IWASA

Rodentia MURIDAE
073

Red list status: K, LP for the populations in Tokyo, Saitama and Hyogo Prefs. (MSJ); LC (IUCN)

Micromys minutus (Pallas, 1771)

EN harvest mouse   JP カヤネズミ (kaya nezumi)   CM 巢鼠   CT 巢鼠   KR 멧밭쥐   RS мышь-малютка

An adult male from Kanagawa Pref., Honshu (M. A. Iwasa).

1. Distribution

Widely distributed from Europe to East Asia. In Japan, Honshu except for northern areas, Shikoku and Kyushu, and small peripheral islands: the Oki Isls. (Dogo, Nishino-shima and Nakano-shima), Shimane Pref.; Awaji-shima Isl., Hyogo Pref.; Teshima Isl., Kagawa Pref.; Innoshima Isl. and Osakikamishima Isl., Hiroshima Pref.; the Tsushima Isls. and the Amakusa Isls. (Shimo-shima), and the Goto Isls. (Fukue), Nagasaki Pref.; the Osumi Isls. (Kuchinoerabu-jima), Kagoshima Pref., and found in lowlands—up to 1,200 m elevation. Only a few examples reveal collection records at 600–1,000 m elevation in Shikoku [1327, 3243]. Information about distributions on small islands is lacking.

2. Fossil record

Not reported in Japan.

3. General characteristics

Morphology Quite small. Coat color is dark brown or yellowish brown on back and white on belly. Tip of tail lacks hairs. Appearaence is similar to *Mus musculus* but incisor morphology is different [1]. External measurements (mean ± SD) are as follows; in females from Honshu ($n = 8$): HB = 65.9 ± 1.78 mm, T = 76.0 ± 2.59 mm, HFsu = 15.8 ± 0.21 mm. In males from Honshu ($n = 12$): HB = 63.2 ± 1.75 mm, T = 70.6 ± 1.84 mm, HFsu = 15.5 ± 0.11 mm. In females from Shikoku and Kyushu ($n = 45$): HB = 59.9 ± 0.84 mm, T = 64.7 ± 0.87 mm, HFsu = 15.4 ± 0.09. In males from Shikoku and Kyushu ($n = 50$): HB = 61.6 ± 0.77 mm, T = 65.6 ± 0.77 mm, HFsu = 15.2 ± 0.08 mm. In both sexes from Tsushima Isl. (range): HB = 50.0–59.0 mm ($n = 4$), T = 53.0–59.0 mm ($n = 6$), HFsu = 14.0–15.7 mm ($n = 6$) [166, 1000, 1011, 1594, 1792, 3243, 3562, 3592]. BW = 7–14 g [1327].

Dental formula   I 1/1 + C 0/0 + P 0/0 + M 3/3 = 16 [29].

Fig. 1. A spherical nest (arrow).

Micromys minutus (Pallas, 1771)

Mammae formula 2 + 0 + 2 = 8 [1327, 3243].

Genetics Chromosome: $2n = 68$ (XX and XY) [3635, 3636]. Sequence data of mitochondrial cytochrome *b* gene are available [4078].

4. Ecology

Reproduction Gestation period is unknown. Mating season is May–December in Honshu, and May–June and September–December in Kyushu. Litter sizes vary as 4.0 ± 0.34 during spring to autumn ($n = 17$) and 6.0 ± 0.23 in winter [3243]. Most populations show breeding peaks in spring and autumn but sometimes in summer [1327].

Lifespan In the wild, lifespan is estimated to be around 22–24 months based on age compositions [3243]. A maximum lifespan is recorded as 31 months under captive conditions [3247].

Diet Mainly seeds of millet, wheat, rice and sunflower, fruits of watermelon, cucumber and melon, and grasshoppers and beetles [3243].

Habitat Grasslands, paddy fields, cultivated fields, abandoned fields and wetlands with dense plants [1327, 3243]. During summer–autumn, spherical nests are built using materials: *Miscanthus*, Japanese blood grass, foxtail grass and others, at 70–110 cm from the ground (Fig. 1) [1327]. Nests are built higher above the ground in summer–autumn than in late spring and early winter [1327]. In winter, tunnels are burrowed in litter on surface or under ground [1327, 3243]. Although *Phragmites communis* is dominant in habitats of this mouse species, this species uses *Miscanthus* spp. in summer and *Carex confertiflora* after October for nesting [1801].

Home range Not reported.

Behavior Climbs stalks and moves well from stalk to stalk using tail as hanger and balancer [1327, 3243].

Natural enemies Not reported.

Parasites Only a few species of Trematoda and Nematoda [810, 3244].

5. Remarks

Recently, the environmental condition appropriate to nesting has diminished. One must pay attention to identify this species because of external similarities with *Mus musculus*.

M. A. IWASA

An example of the habitat of *Micromys minutus* in Fujisawa, Kanagawa Pref., Honshu, showing rice fields and reeds in spring (M. A. Iwasa).

Rodentia MURIDAE

074

Red list status: CR (MEJ); R (MSJ); LC (IUCN)

Apodemus agrarius (Pallas, 1771)

EN striped field mouse　　JP セスジネズミ（sesuji nezumi）　　CM 黑线姬鼠　　CT 赤背條鼠, 黑線姬鼠
KR 등줄쥐　　RS полевая мышь

An adult male from Uotsuri-jima Isl. of the Senkaku Isls. (K. Tsuchiya).

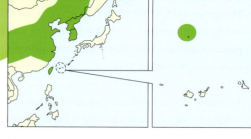

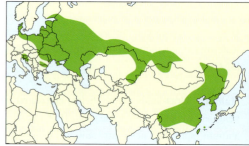

1. Distribution

Widely distributed in Eurasia: in the west from Europe to Baikal Lake and in the east Amur River to China and Korea, and Taiwan. In Japan, Uotsuri-jima Isl. of the Senkaku Isls., the Nansei Isls. [1327, 2334, 3246].

2. Fossil record

Not reported in Japan.

3. General characteristics

Morphology　Small-sized mouse-like appearance. Coat color is yellowish brown with apparent black stripe on back and white on belly [1327]. External measurements are as follows; HBL = 130.0 mm, T = 118.5 mm, HFLsu = 24.6 mm, BW = 56.4 g (n = 1, collected in 1979) [1327]. Length of tooth row and breadth of first upper molar of 2 specimens are slightly longer and wider than those on Taiwan and on Cheju-do Isl. of Korea. However, following characteristics: length of diastema, length of nasals, palatilar length, breadth of brain case, width of rostrum, interorbital width, and skull height at molar part of the Uotsuri-jima Isl. specimens were similar to those on Taiwan and on Cheju-do Isl. of Korea.
Dental formula　I 1/1 + C 0/0 + P 0/0 + M 3/3 = 16 [29].
Mammae formula　2 + 0 + 2 = 8 [1327].
Genetics　Chromosome: $2n$ = 49 with a supernumerary chromosome (n = 1) [3246].

4. Ecology

Reproduction　Not reported in Japan.
Lifespan　Not reported in Japan.
Diet　Not reported in Japan.
Habitat　Two specimens were collected in open grassland on Mt. Narahara in the western side of Uotsuri-jima Isl. [3246].
Home range　Not reported in Japan.
Behavior　Not reported in Japan.
Natural enemies　Not reported in Japan.
Parasites　Not reported in Japan.

5. Remarks

Recently, goats which were introduced to Uotsuri-jima Isl. have destroyed vegetation and bare ground has increased. This has threatened the survival of *A. agrarius*. Up to the present, only 3 specimens have been collected. By a mating experiment of individuals from Uotsuri-jima Isl. with ones from Taiwan, offspring were obtained. Therefore, the individuals from Uotsuri-jima Isl. were recognized as conspecific to *A. agrarius* [3246]. Now, we are not able to easily visit Uotsuri-jima Isl. because of international political problems, and additional research is impossible at present.

M. A. IWASA

Rodentia MURIDAE

Red list status: C-1 as *A. p. giliacus* (MSJ); LC (IUCN)

Apodemus peninsulae (Thomas, 1907)

EN East Asian field mouse   JP ハントウアカネズミ (hantou aka nezumi)   CM 大林姬鼠   CT 朝鮮姬鼠，大林姬鼠
KR 흰넓적다리붉은쥐   RS восточноазиатская мышь

An adult male from Hokkaido (K. Tsuchiya).

1. Distribution

Widely distributed in Siberia, northeastern China, Shansei, Kansu, Korea and Sakhalin. In Japan, Hokkaido [1327, 2334].

2. Fossil record

Not reported in Japan.

3. General characteristics

Morphology Medium-sized mouse-like appearance. Long tail, large eyes and long ears. Coat color is dark yellowish brown on back and white or grayish white on belly. Dorsal pelage soft and without spines. Entirely more grayish than *A. speciosus*. Appearance is quite similar to *A. speciosus*, especially young individuals of *A. speciosus* with which some confusion may be experienced. This species shows small granules between plantar pads on hind feet (Figs. 1 and 2) [2410] and this hind foot characteristic is useful to identify the 3 *Apodemus* species, *A. peninsulae*, *A. speciosus* and *A. argenteus* in Hokkaido. External measurements (range) are as follows; HB = 72–112 mm, T = 72–104 mm, HFsu = 21–24 mm, BW = 19–27.6 g for adults [1327, 2410].

Fig. 1. A typical plantar view of right hind foot of *Apodemus peninsulae*.

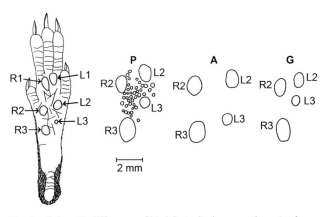

Fig. 2. Schematic differences of hind foot criteria among three *Apodemus* species, *A. peninsulae* (P), *A. speciosus* (A) and *A. argenteus* (G), in Hokkaido [2410].

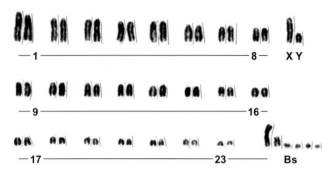

Fig. 3. A conventionally stained karyotype with B chromosomes (Bs) in a male individual (Specimen No.: HEG83).

Dental formula   I 1/1 + C 0/0 + P 0/0 + M 3/3 = 16 [29].

Mammae formula   2 + 0 + 2 = 8 [19, 1327].

Genetics   Chromosome: $2n$ = 48 (XX and XY) and FN = 46 [53, 868, 1644]. Supernumerary chromosomes (B chromosomes) with variable morphologies are observed: number of B's ranges 3–13 in Naganuma (Fig. 3) [868, 870], 2–6 in Sapporo and Hayakita [53], and 4–10 in Tomakomai [2902] on Hokkaido (Fig. 2) [1348, 1349, 1644]. Sequence data of mitochondrial cytochrome b gene are available [3165].

4. Ecology

Reproduction   Litter size (number of embryos) is 1–7 and the average is 5.0 [593]. Under laboratory conditions, number of newborns is 1–8; gestation period 20 days [3638]. Maturation may commence at 20 g or more. The main breeding season extends from April to August, showing a unimodal pattern [593].

Lifespan   Few data are available, but one study of a wild population found that marked mice survived 13 months [593]. In captivity they live to at least 63 months [3638].

Diet   Large or middle-sized seeds and small invertebrates, but is unable to crack walnuts [1900].

Habitat   Mainly scrublands, grasslands and secondary vegetation including plantations and farmlands. Also occurs in forest edges and mainly in man-made habitats such as cultivated field and shelterbelts, especially when *A. speciosus* sympatrically occur [595, 1636, 1689]. A rather rare species in Hokkaido.

Home range   Most of summer ranges in secondary forests and plantations are less than 30 m in length [2412].

Behavior   Chiefly crepuscular or nocturnal in activity, being less nocturnal in activity than *A. speciosus* and *A. argenteus* [4140]. A ground dweller [2775]. They build nests in a simple underground burrow as well as on the ground, using plants of Gramineae on the outside and lining with leaves of deciduous trees [1636].

Natural enemies   Probably similar to those of *A. speciosus*. Taken by the red fox and the Ural owl [4143, 4147].

Parasites
Ectoparasite
Acari (mites): genera *Leptotrombidium*, *Neotrombicula*, *Gahrliepia*, *Myobia*, *Hirstionyssus*, *Haemogamasus*, *Euleaps*, *Androlaelaps*, *Laealps* and *Myocoptes*. Siphonaptera (fleas): genera *Ctenophtalmus*, *Neopsylla* and *Nosopsyllus* [4140].
Endoparasite
Several taxa of Nematoda from Japan [207, 4140].

5. Remarks

Similar species: 3 *Apodemus* species inhabit Hokkaido mainland. These species can be distinguished by hind foot characteristics [2410]. Dimensional relationships of pads are specific to each species as shown in Fig. 2. Before the 1970s, this species was considered as a variant (or young individual) of *A. speciosus* based on the morphological similarities. However, karyological studies revealed the occurrence of *A. peninsulae* in Hokkaido [868]. Taxonomically, the Hokkaido population is ranked as a subspecies, *A. peninsulae giliacus*.

K. NAKATA & M. A. IWASA

An example of the habitat of the East Asian field mouse in Tomakomai, central Hokkaido, showing grasses and shrubs in summer (M. A. Iwasa)

Rodentia MURIDAE

076

Red list status: C-2, LP for the populations on Miyake-jima, Izu-ohshima and Nii-jima Isls. (MSJ); LC (IUCN)

Apodemus speciosus (Temminck, 1844)

EN large Japanese field mouse   JP アカネズミ（aka nezumi）   CM 日本大姬鼠   CT 日本大林姬鼠，日本紅森鼠
KR 큰흰배숲쥐   RS большая японская мышь, красная мышь

An adult male from Kanagawa Pref., Honshu (M. A. Iwasa).

1. Distribution

Hokkaido, Honshu, Shikoku and Kyushu, and small peripheral islands larger than about 10 km^2: Kunashiri Isl., Rishiri Isl., Okushiri Isl. (Hokkaido); Sado Isl. (Niigata Pref.); Noto Isl. (Ishikawa Pref.); the Izu Isls. (Izu-ohshima, Nii-jima, Shikine-jima, Kozu-shima and Miyake-jima: Tokyo); Toshi-jima Isl. (Mie Pref.); the Oki Isls. (Dogo, Nishinoshima, Nakanoshima and Chiburi-jima) (Shimane Pref.); Oshima Isl. (Wakayama Pref.); Awaji-shima Isl. (Hyogo Pref.); Shodo-shima Isl., Te-shima Isl. (Kagawa Pref.); Miya-jima Isl., Kurahashi-jima Isl., Nouke-jima Isl., Eda-jima Isl. (Hiroshima Pref.); the Geiyo Isls. (Ohshima, Hakata-jima, Ohmi-shima, Ikuchi-shima, Innoshima, Mukai-jima, Takane-jima and Osaki-Kamishima (Hiroshima and Ehime Prefs.), Yashiro-jima Isl., Ohshima Isl., Naga-shima Isl. (Yamaguchi Pref.); Iki Isl., Hirato-jima Isl., the Goto Isls. (Fukuejima and Nakadorijima), the Tsushima Isls. and Otoko-jima Isl. (Nagasaki Pref.); the Amakusa Isls. (Kami-shima, Shimo-jima and Oyano-jima) (Kumamoto Pref.), Yaku-shima Isl., Tane-gashima Isl., Shimokoshiki-jima Isl. and the Tokara Isls. (Kuchinoerabu-jima, Kuchino-shima and Nakano-shima) (Kagoshima Pref.) [1327].

2. Fossil record

Fossils have been recorded from Middle Pleistocene (Honshu and Kyushu), Late Pleistocene (Honshu and Shikoku), and Holocene layers (Honshu and Kyushu) [807, 819, 821, 1499–1503, 1505, 1506, 1508, 1512, 1513, 1518, 1519, 1521, 1725, 2139, 2145, 2373, 2440, 2441, 2688, 3202–3205, 3602].

An adult from Hokkaido (K. Tsuchiya). Mice in Hokkaido are slightly larger than those in Honshu.

Fig. 1. Toyama-Hamamatsu Line in central Honshu.

3. General characteristics

Morphology Medium or large sized mouse-like appearance. Long tail, large eyes and long ears. Coat color is brown or orange brown on back and snow-white belly. This species is a typical rodent endemic to Japan showing beautiful coat color. Unclear blackish-brown zone sometimes appears on back in old age. Adults of this species are consistently larger than those of *A. peninsulae* and *A. argenteus*. In many areas, this species and *A. argenteus* co-occur and it is possible to roughly identify both species by the hind foot length, which is longer in *A. speciosus* than in *A. argenteus*. External measurements (range) are as follows; HB = 80–140 mm, T = 70–130 mm, HFsu = 22–28 mm (22–26 mm in Honshu, Shikoku and Kyushu; 24–28 mm in Hokkaido), BW = 20–60 g with geographic variations [1327, 1689, 2308, 2410]. Tail length is similar or shorter than HB. Because young of this species are similar to *A. argenteus* in external morphology in Honshu and southward, it is necessary to identify them with critical examination of characters such as the shape of the anterorbital zygomatic plate (see Fig. 1 in *A. argenteus*). Sexual dimorphism has been found in wild populations [3709].

Dental formula I 1/1 + C 0/0 + P 0/0 + M 3/3 = 16 [29].

Mammae formula 2 + 0 + 2 = 8 [1327].

Genetics Chromosomes: $2n = 46$, 47 and 48 (XX and XY); there are 3 chromosomal races (see Topic). Mice of $2n = 46$ and 48 live in western and eastern Japan, respectively. The boundary of both races coincides with Toyama-Hamamatsu Line (Fig. 1). In the boundary area, a hybrid-type, $2n = 47$, occurs [2945, 3633, 3961]. Sequence data of mitochondrial cytochrome *b* gene and restriction fragment length polymorphism data for nuclear ribosomal genes are available [3051, 3352, 3355, 3358, 3609]. A sex chromosomal anomaly found from at an illegal dumpsite was likely influenced by chemical contamination [2544].

4. Ecology

Reproduction Litter size (number of embryos) is 2–12, and the average is 4.0–6.2 from various geographical locations in the wild [1691]. Under laboratory conditions, number of newborns is 1–8; sex ratio at birth is approximately equal (males/females ratio, 1.02–1.07); gestation period is 19–26 days with continuous parturition occurring at 25–30 day intervals [2313, 2589, 3638]. The breeding season varies geographically; the reproductive period ranges from April to September in Sapporo, Hokkaido, in March–April and September–November in Kyoto Pref., Honshu, and from October to March in Nagasaki Pref., Kyushu [1691, 2313].

Lifespan On the basis of live trapping, maximum lifespan in the wild is estimated to be 15 months in Hokkaido [593] and 26 months in Honshu [890].

Diet Root and stems (with no chloroplasts) of herbaceous plants, seeds, berries, and insects. Main foods are insects in summer and plants from autumn to spring in Honshu, varying among habitats and years [3543]. Caching of acorns and seeds is well known [43, 979, 1306, 1562, 2064, 2125, 3801]. An acclimation to acorn defensive chemicals, tannin-binding salivary proteins and tannase-producing bacteria are found in this species [3215].

Habitat Forests, plantations, riverside fields with dense grasses, paddy fields and cultivated fields. In Hokkaido, they are usually found in broad-leaved forests which produce nuts [22]. In Honshu and Kyushu, has a strong preference for ground vegetation dominated by gramineous plants [2481, 3240].

Home range Home ranges of both sexes in the breeding season are significant larger than those in the non-breeding season; mean ranges as 304–1,853 m^2 [2662]. Ranges of males overlap rather randomly, whereas those of females tends to be mutually exclusive. No evidence of pair bonds found [1710, 2662]. Interspecific interactions between 2 congeneric species appears to be one-way action from *A. speciosus* to *A. argenteus* [3164], although their habitat use may vary seasonally [2976, 3053].

Behavior Primarily nocturnal. A ground dweller. Less tendency to climb trees.

Natural enemies Taken by a wide variety of carnivores and birds of prey; foxes [18, 4143], owls [2019, 2020, 3373, 4147], weasels [589], probably including sables and martens. Freshwater salmon are potential predators [2622]. Predators prefer the species of *Myodes* to those of *Apodemus* [1688, 4140].

Parasites
Ectoparasite
Acari (mites and ticks): genera *Leptotrombidium*, *Neotrombicula*,

Apodemus speciosus (Temminck, 1844)

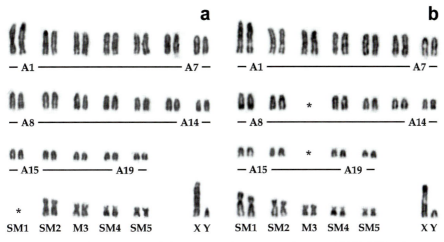

Fig. 2. Conventionally stained karyotypes of *A. speciosus*, typical eastern type, $2n = 48$ (a, specimen No.: NUWL-M91) and western type, $2n = 46$ (b, specimen No.: MAI-445).

Miyatrombicula, Eltonella, Cheladonta, Walchia, Gahrliepia, Myobia, Paraspeleognathopsis, Eucheyletia, Neocunaxoides, Pygmephorus, Tydeus, Lophioglyphus, Dermacarus, Xenoryces, Afrolistrophorus, Hypoaspis, Cosmolaelaps, Haemogamasus, Laealps, Euleaps, Hirstionyssus, Ornithonyssus and *Ixodes* [611, 1874, 2026, 2372, 2716, 3403, 3427, 3484, 3693, 3695, 3705, 4001, 4140]. Siphonaptera (fleas): genera *Stivalius, Atyphroceras, Hystricopsylla, Palaeopsylla, Ctenophtalmus, Neopsylla, Stenoponia, Hystrichopsylla, Rhadinopsylla, Nosopsyllus* and *Peromyscopsylla, Monopsyllus, Frontopsylla* [2724, 2725, 3693, 3695–3697, 3705, 4140]. Anoplura (lice): genera *Hoplopleura* and *Polyplax* [2717, 2723, 3697, 4140].

Endoparasite

Apicomplexa, Metamonada, Euglenozoa, Trematoda, Cestoda, Acanthocephala and Nematoda [206, 207, 1041, 1139, 2714, 2979, 3239, 3653].

5. Remarks

Damages agricultural field crops [2180]. Beneficial in preying upon harmful insects in woodlands [3388]. Both inhibits and aids forest regeneration by feeding on and burying tree seeds. Can be bred in captivity [2589, 2975, 3638].

K. NAKATA, T. SAITOH & M. A. IWASA

6. Topic

Chromosomal races in Apodemus speciosus

M. A. IWASA

In most areas of the Japanese Isls, *A. speciosus* is the dominant wild mouse in natural forests, plantations and farms. This species has 3 chromosomal constitutions, with diploid sets: $2n = 46$, 47 and 48. The karyotype of $2n = 46$ is found in the western part from Toyama-Hamamatsu Line (Fig. 2a) and the karyotype of $2n = 48$ in eastern part from the line (Fig. 2b). Furthermore, the karyotype of $2n = 47$ is also observed at parapatric zone (contact zone) of both karyotypes. The numbers of chromosomes, 46 and 48, are clearly distinguishable [2945, 3633]. Similar to *A. speciosus*, the greater Japanese shrew-mole, *Urotrichus talpoides* [788] also carries 2 types of karyotypes and they are recognized as chromosomal races.

According to previous studies, the difference between karyotypes, $2n = 46$ and 48 is explained by Robertsonian rearrangement between autosomes: Nos. 10 and 17 (Fig. 2) [2945, 3633]. In addition, it is considered that the hybrid type, $2n = 47$, maintains relatively normal fertility based on the histological observation of spermatogenesis [2946]. Therefore, between both chromosomal races, $2n = 46$ and 48, it is thought that post-mating isolation system is not established cytogenetically. Surprisingly, patterns of mitochondrial and nuclear DNA differentiation are not related to the chromosomal distribution patterns in mainlands: Hokkaido, Honshu, Shikoku and Kyushu [3352, 3609]. Considering these facts, in *A. speciosus*, chromosomal changes are not related to speciation at present. Nevertheless, such chromosomal rearrangements may cause chromosomal non-disjunction at meiotic stages as a post-mating isolation cue in future. Thus, *A. speciosus* is an appropriate model for elucidation of chromosomal speciation, as is the greater Japanese shrew-mole in Japan.

Rodentia MURIDAE

077

Red list status: C-2, LP for the populations on Tsushima, Oki, Yaku-shima and Tane-gashima Isls. (MSJ); LC (IUCN)

Apodemus argenteus (Temminck, 1844)

EN small Japanese field mouse   JP ヒメネズミ（hime nezumi）   CM 日本小姫鼠   CT 日本姬鼠，日本森鼠
KR 작은흰배숲쥐   RS малая японская мышь

An adult male from Shizuoka Pref., Honshu (M. A. Iwasa).

1. Distribution

Endemic to Japan: Hokkaido, Honshu, Shikoku and Kysuhu, and small peripheral islands larger than about 150 km^2: Kinkazan Isl., Miyagi Pref.; Awa-shima Isl. and Sado Isl., Niigata Pref.; the Oki Isls. (Dogo and Nishinoshima), Shimane Pref.; Awaji-shima Isl., Hyogo Pref.; Shodo-shima Isl., Kagawa Pref.; Miya-jima Isl., Hiroshima Pref.; the Goto Isls. (Fukue-jima and Miya-jima) and the Tsushima Isls., Nagasaki Pref.; Amakusa Shimo-jima Isl., Kumamoto Pref., Yaku-shima Isl. and Tane-gashima Isl., Kagoshima Pref. [1327].

2. Fossil record

Fossils have been recorded from Middle and Late Pleistocene (Honshu), Holocene layers (Shikoku and Kyushu) [94, 768, 807, 819, 821, 1499–1503, 1505, 1508, 1512, 1513, 1518, 1519, 1521, 1725, 2139, 2145, 2373, 2439–2441, 2688, 3202–3205, 3602].

3. General characteristics

Morphology Small-sized mouse-like appearance. Long tail, large eyes and long ears. Coat color is chestnut brown on back and grayish-white or ivory on belly. Sometimes partial whiting observed on the skin. Body size of this species is apparently smaller than that of *A. speciosus* and *A. peninsulae,* but similar to that of *Mus musculus*. However, the coat color of this species differs from that of *M. musculus* and it is easy to discriminate them. External measurements (range) are as follows; HB = 65–100 mm, T = 70–110 mm, HFsu = 17–21 mm, BW = 10–20 g [19,

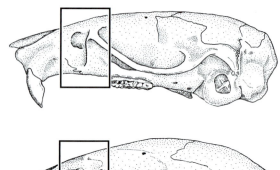

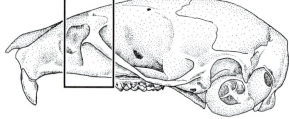

Fig. 1. Comparison between the shapes of anterorbital plates, indicated by rectangles, of an adult male of *A. argenteus* (upper, specimen No.: MAI-532, condylobasal length = 21.3 mm) and a young female of *A. speciosus* (lower, specimen No.: MAI-359, condylobasal length = 22.5 mm) from Honshu.

Apodemus argenteus (Temminck, 1844)

591, 1327, 2410]. Tail length is slightly longer than HB and HFsu is shorter than that of *A. speciosus*. This species and young of *A. speciosus* are similar in external appearances and it is possible to identify them with the anterorbital plate (Fig. 1) [1327] and hind foot criteria (see *A. peninsulae*).

Dental formula   I 1/1 + C 0/0 + P 0/0 + M 3/3 = 16 [29].

Mammae formula   2 + 0 + 2 = 8 [1327].

Genetics   Chromosome: $2n$ = 46 (XX and XY) and FN = 48. Supernumerary chromosomes (B chromosomes) are sometimes observed [750, 2547, 2548, 3635]. The X chromosome carries an apparently large heterochromatic block characterized by delayed fluorescent brightness of Q-banding [646, 2548]. Sequence data of mitochondrial cytochrome *b* gene are available [3165, 3358].

4. Ecology

Reproduction   Trapping studies reveal that most females have 1 or 2 litters in their life. Litter size ranges from 1 to 9, and averages 3.3–4.9 among geographical locations [591, 3541]. Litter size varies with density, population phase and body size; litter size of overwintered females is lower at high than at low population density, whereas litter size of the current year's females increases with body size [2414]. The mean sex ratio (proportion of males) is 0.45–0.51 in the wild [3186]. Young are born naked with the eyes closed. Females may attain maturity after reaching 8 g and males after reaching 10 g in the wild. The proportion of breeding females varies within and among years, and is suppressed at higher densities [2418]; similar responses are found for males. Breeding season varies geographically, suggesting that it is timed by changes in temperature [17, 2480]; it extends from April until October or even November in cold temperate regions, whereas it lasts from October to March in warmer temperate regions. Breeding is however found to be density- and phase-related [2418] and the effect of temperature on breeding has been discredited in populations exhibiting multi-annual fluctuations.

Lifespan   Based on live-trapping data, maximum lifespan in the wild is thought to be 27 months [591]. Few mice survive more than 1 year.

Diet   Takes seeds, green plants, fruits, and invertebrates (especially insects) [1900]. Acorn caching occurs [4179].

Habitat   Occurs in a variety of habitats, all of them wooded, from lowlands to alpine zones; also inhabits scrublands and plantations. Prefers late-successional forests consisting mainly of mature trees [591, 1327, 2774]. Nests are usually below ground, occasionally above ground in holes in trees and bird nesting boxes; commonly made of broad-leaves [2775]. Microhabitat segregation is observed between congeneric *Apodemus* species [45, 2481, 3240].

Home range   Home ranges vary in size according to habitat and season and may decrease in non-breeding season; 200–1,325 m^2 in ranges [2776]; smaller than those of *A. speciosus* in shared habitats. Home ranges overlap between and within sexes [2628, 2662]. Observations of male-female pairing pattern suggests a monogamous mating system [2482, 2662], though other studies have not found this and DNA analysis revealed polygynous mating [864, 2628, 3169].

Behavior   Mainly nocturnal. A ground dweller, but arboreal movements are common [2775]. Male-biased natal dispersal has been reported [2628].

Natural enemies   Various carnivores and birds of prey; foxes [18, 4143], owls [2019, 2020, 4147], probably including weasels, sables and martens. Freshwater salmon (*Hucho perryi*) are potential predators [2622]. Little information available on losses to avian and mammalian predators.

Parasites

Ectoparasite

Acari (mites and ticks): genera *Leptotrombidium*, *Neotrombicula*, *Gahrliepia*, *Myobia*, *Demacarus*, *Hypoaspis*, *Laelaps*, *Eulealaps*, *Haemogamasus*, *Hirstionyssus* and *Ixodes* [97, 611, 3484, 3692, 3693, 4140]. Siphonaptera (fleas): genera *Atyphloceras*, *Stenoponia*, *Ctenophthalmus*, *Neopsylla*, *Catallagia*, *Hystrichopsylla*, *Neoarctopsylla*, *Monopsylla* and *Peromyscopsylla* [2724, 2963, 4140]. Anoplura (lice): genera *Polyplax* and *Hoplopleura* [2717, 2723, 3697, 4140].

Endoparasite

Several taxa of Trematoda, Cestoda and Nematoda [97, 206, 207]; Protozoan parasites, Apicomplexa and Metamonada, are reported [1041, 1139, 3239].

5. Remarks

Damage to seedlings of Todo-fir [1899] and catkins of *Betula* trees [3192]. Predation of sawfly pupae and other lepidopterous larvae in larch stands [3037], possibly suppressing their outbreaks. It often enters cottages and homes in the country, particularly in the autumn.

K. NAKATA, T. SAITOH & M. A. IWASA

Rodentia MURIDAE

078

Red list status: C-1 (MSJ); LC (IUCN), introduced

Rattus norvegicus (Berkenhout, 1769)

EN Norway rat   JP ドブネズミ (dobu nezumi)   CM 褐家鼠   CT 溝鼠（挪威鼠，褐鼠）   KR 집쥐   RS серая крыса, пасюк

An adult from Miyazaki Pref., Kyushu (K. Tsuchiya).

1. Distribution

Cosmopolitan and commensal [1327].

2. Fossil record

Middle and Late Pleistocene-Holocene fossils of *R. norvegicus* and *Rattus* sp. are recorded from Honshu, and Honshu and Kyushu, respectively [1508, 1725].

3. General characteristics

Morphology A typical rat. Slightly larger than *R. rattus* in body size, however, ears are smaller than those of *R. rattus* and do not reach the eyes when folded forward. Coat color is yellowish dark on back side and grayish or yellowish white on belly [1327]. External measurements (range) are as follows; HB = 110–280 mm, T = 175–220 mm, HFsu = 27–42 mm, BW = 40–500 g [1327].

Dental formula   I 1/1 + C 0/0 + P 0/0 + M 3/3 = 16 [29, 1327].

Mammal formula   Varies: 1 + 1 + 2, 2 + 1 + 2, 2 + 2 + 2, 3 + 0 + 3, or 3 + 1 + 2 = 8–12 [1327].

Genetics   Chromosome: $2n$ = 42 (XX and XY) and FN = 64 with slight modification by chromosomal rearrangements [1843, 4222]. Mitochondrial DNA data available [3292].

4. Ecology

Reproduction   Litter size (number of embryos) ranges between 1–18 with the average of 8–9. Breeding is active in spring in Tokyo [1327].

Lifespan   Not reported in Japan.

Diet   Mainly animal matter (26–86%) with variation depending on habitat conditions [1327, 3907].

Habitat   Concentrated in cities, towns and farms. During summer they may move out into the fields, but the majority returns to the shelter of buildings during winter. Occupies barns, warehouses, dock areas, and stores. Prefers areas close to water. Inhabits moist conditions such as drains, kitchen sinks, garbage pits, underground markets, food storage, and paddy fields. Nesting in small spaces in houses or buildings, in broken drains and concrete cracks in the ground, and cavities under the concrete surface [1327, 3907].

Home range   Not reported in Japan.

Behavior   Mainly nocturnal but active diurnally without interfered condition [1327, 3907].

Natural enemies   Cats, mustelids, eagles, owls and crows [1159].

Parasites   Ectoparasites, such as fleas and mites, are possibly vectors causing serious diseases [3907].

5. Remarks

Recently, a "super rat" bearing resistance for poisonous baits was described [3907]. In underground urban areas, *R. norvegicus* actively depends on leftover foods [3907].

M. A. IWASA

Rodentia MURIDAE

Red list status: C-1 (MSJ); LC (IUCN), introduced

Rattus rattus (Linnaeus, 1758)

EN black rat   JP クマネズミ (kuma nezumi)   CM 家鼠   CT 玄鼠, 黑鼠   KR 곰쥐   RS чёрная крыса

A young from Miyazaki Pref., Kyushu (K. Tsuchiya).

1. Distribution

Cosmopolitan and commensal [1327].

2. Fossil record

Fossils of *R. rattus* and *Rattus* sp. are recorded from Honshu (Middle Pleistocene), and Honshu and Kyushu (Late Pleistocene and Holocene) [1508, 1725].

3. General characteristics

Morphology   Slightly smaller than *R. norvegicus* in body size. Coat color is black to grayish brown on back and gray or white with slate on belly [1327, 3907]. Sometimes white spots and melanism are observed in individuals from the Ogasawara and the Ryukyu Isls. [1282, 3907]. Ears are larger than those of *R. norvegicus* and reach the eyes when folded forward. External measurements (range) are as follows: HB = 150–240 mm, T = 150–260 mm, HFsu = 22–35 mm, BW = 150–200 g [1327].

Dental formula   I 1/1 + C 0/0 + P 0/0 + M 3/3 = 16 [1327].

Mammae formula   2 + 0 + 3 = 10 or 2 + 1 + 3 = 12 [1327].

Genetics   Chromosome: $2n = 38$ (Oceania and Europe origin) and 42 (Asia origin) and FN = 54 with slight modification by chromosomal rearrangements [353, 4222]. Mitochondrial and nuclear DNA data are available [171, 382, 1283, 1284].

4. Ecology

Reproduction   Litter size (number of embryos) ranges between 2–10 with the average of 5.6–6.3. Breeding is active in summer in Tokyo Metropolis [1327].

Lifespan   Not reported in Japan.

Diet   Mainly seeds (51–59%) and small amount of animal matter (5–11%) [1327, 3907].

Habitat   Inhabits drier conditions than *R. norvegicus*: higher places of buildings and roofs. Appears in grasslands and forests on Iriomote Isl., Nakanokami-shima Isl. and Yokoate-jima Isl. of the Ryukyu Isls. [1327, 3907].

Home range   Range of activity 50–100 m [3907].

Behavior   Able to climb vertical pillars or walls and walk thin wires. Mainly nocturnal [1327, 3907].

Natural enemies   Not reported in Japan.

Parasites   Ectoparasites, such as fleas and mites, are possibly vectors causing serious diseases [3907].

5. Remarks

Recently, occurs in urban office and residential areas, and sometimes gnaws electric wires causing serious electric problems [3907]. *Rattus rattus* shows genetic and morphological variation, and some taxonomists suggest that Asian black rats should be ranked as *R. tanezumi* [2334].

M. A. IWASA

Rodentia   MURIDAE

080

Red list status: LC (IUCN), introduced

Rattus exulans (Peale, 1848)

EN Polynesian rat   JP ポリネシアネズミ（porineshia nezumi）   CM 緬鼠   CT 緬甸小鼠   KR 남양쥐   RS малая крыса

A young from Thailand (K. Tsuchiya).

1. Distribution

Widely distributed in Southeastern Asia, New Guinea, and throughout the Pacific [2334, 4261]. In Japan, introduced onto Miyako-jima Isl. of the Ryukyu Isls. [1509].

2. Fossil record

Not reported in Japan.

3. General characteristics

Morphology   Coat color is light brown with slightly darker back and light gray belly [1327]. External measurements are as follows; HB = 115 mm, T = 115 mm, HFsu = 22–22.5 mm ($n = 2$ females) [1327, 2275]. Condylobasal length 25.4–26.5 mm, length of upper teeth row = 4.2–4.3 mm, molar width (maximum distance between the lateral borders of M^1) = 5.7 mm, maximum width of the upper first molar = 1.3–1.4 mm ($n = 2$ females) [1327, 2275].
Dental formula   I 1/1 + C 0/0 + P 0/0 + M 3/3 = 16 [29, 1327].
Mammae formula   2 + 0 + 2 = 8 [1327].
Genetics   Chromosome: $2n = 42$ (XX and XY) and FN = 36 [2275].

4. Ecology

Reproduction   Not reported in Japan.
Lifespan   Not reported in Japan.
Diet   Not reported in Japan.
Habitat   Houses, grain storages, cultivated fields, scrubs and the edge of forests [1327, 2275].
Home range   Not reported in Japan.
Behavior   Not reported in Japan.
Natural enemies   Not reported in Japan.
Parasites   Not reported in Japan.

5. Remarks

The first record of occurrence of this species in Japan was reported in 2001 but the specimens were collected before 1955 [1327]. A similar species, *R. rattus,* shows similar size of the skull but there are differences between *R. rattus* and *R. exulans* in braincase, molar width, length of upper teeth row and foramen magnum [2275].

M. A. IWASA

Rodentia MURIDAE

081

Red list status: EN (MEJ); En (MSJ); EN (IUCN)

Diplothrix legata (Thomas, 1906)

EN Ryukyu long-furred rat   JP ケナガネズミ（kenaga nezumi）   CM 琉球长毛鼠   CT 琉球鼠，琉球長毛鼠
KR 류큐긴털쥐   RS рюкюйская крыса

An adult from the Yanbaru forest, Okinawa-jima Isl., the Ryukyu Isls. (T. Hiragi).

1. Distribution

Endemic to Japan; western mountains of Amami-ohshima Isl., Mts. Amagi, Inokawa and Inutabu of Tokuno-shima Isl. and northern mountains of Okinawa-jima Isl. of the Nansei Isls. [1327].

2. Fossil record

Fossil records are known from south of Okinawa-jima Isl. and Miyako Isl. [1441, 1508–1510].

3. General characteristics

Morphology Coat color is yellowish brown on back and dark brown on belly. Ordinary hairs are about 29 mm long on ventral side and quite long bristle-hairs are about 50–60 mm. Spinous hairs are about 25 mm long and 0.5 mm wide, and slender and flattened. Proximal 3/5 of tail is blackish and terminal 2/5 is white with hairs about 3–4 mm long (Fig. 1) [1327]. External measurements (range) are as follows; HB = 220–330 mm, T = 240–372 mm, HFsu = 49–60 mm [1327].
Dental formula I 1/1 + C 0/0 + P 0/0 + M 3/3 = 16 [29].
Mammae formula 2 + 0 + 2 = 8 [1327].
Genetics Chromosome: $2n = 44$ (XX and XY) and FN = 54 [3635, 3636, 3645]. Sequence data of mitochondrial cytochrome *b* gene and a nuclear gene are available [3356].

4. Ecology

Reproduction Litter size is 2–5 [866].
Lifespan Not reported.
Diet Eats seeds of chinquapins, sweet potatoes, insects; an omnivorous diet [866, 1327, 2681].
Habitat Natural forests of chinquapins and oaks, and logging sites at 300–400 m elevation on Amami-ohshima Isl. Areas with forests cover of 60% at 300–400 m elevation on Tokuno-shima Isl. Natural forests of chinquapins and oaks from middle to top of mountains on Okinawa-jima Isl. Active on trees and builds nests in tree hollows using dry leaves and twigs [866, 1327, 2681].
Home range Not reported.
Behavior Nocturnal, based on observations under natural conditions [2681].
Natural enemies Feral cats and dogs, and exotic small Indian mongooses are known as predators [1327, 2681].
Parasites A nematode is reported [1135].

5. Remarks

Predation by introduced species threatens this species. Reduction of natural forests caused population decline of this species on 3 islands. This species was designated as a "Natural Monument" in 1972.

M. A. IWASA

Fig. 1. A typical view of hairs and colors of the tail.

183

Rodentia MURIDAE

Red list status: C-1 (MSJ); LC (IUCN)

Mus caroli Bonhote, 1902

EN Ryukyu mouse   JP オキナワハツカネズミ（okinawa hatsuka nezumi）   CM 琉球小家鼠   CT 月鼠，田鼷鼠
KR 오키나와생쥐   RS рюкюйская мышь

An adult from Okinawa-jima Isl., the Ryukyu Isls. (K. Tsuchiya).

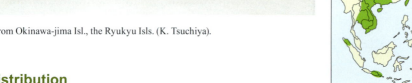

1. Distribution

Widely distributed in southeastern Asia: Taiwan, Hainan, southern China, Malaysia, Vietnam and Sumatra Isl. of Indonesia. In Japan, only Okinawa-jima Isl. of the Ryukyu Isls., the Nansei Isls. [1327].

2. Fossil record

Not reported in Japan.

3. General characteristics

Morphology External fur color: dark grayish-brown on back and white on belly [179]. Tail showing bicolor with light-brown in upper half and white in lower half [179]. TL is consistently longer than HB, and average tail:body ratio is over 110% and greater than that of *M. musculus* [1327]. Anterior edge of anterorbital plate is curved forward from base of zygomatic arch of premaxilla (Fig. 1). Curving form shows slightly backward or "S" shape. Anterior edge of upper incisor is positioned rather anterior to anterior edge of nasal (Fig. 1) [1327]. External measurements (range) are as follows; HB = 63–89 mm, T = 61–95 mm, HFsu = 14–19 mm, BW = 8–19 g [1327].
Dental formula   I 1/1 + C 0/0 + P 0/0 + M 3/3 = 16 [29].
Mammae formula   1 + 2 + 2 = 10 [1327].
Genetics Only a little genomic information available for Japanese populations about chromosomes [2000] and DNAs [3212, 3353, 3557].

4. Ecology

Reproduction Gestation period is about 20 days. Litter size

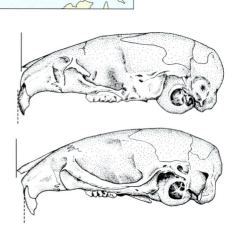

Fig. 1. Lateral view of skulls (upper: *M. caroli*, Specimen No.: NSMT-M2308, Okinawa-jima Isl.; lower: *M. musculus*, Specimen No.: MAI-372, central Honshu). Solid lines and dotted lines indicate anterior edge of nasal and anterior edge of incisor, respectively.

(number of embryos) is 3–5. Weaning is observed from about 17 days. Sexual maturity approached at about 40 days [179, 1327].
Lifespan   Not reported in Japan.
Diet   Not reported in Japan.
Habitat   Underground tunnels in sugarcane fields, abandoned fields, paddy fields and grasslands [179, 1327].
Home range   Not reported in Japan.
Behavior   Not reported in Japan.
Natural enemies   Not reported in Japan.
Parasites   A helminth is reported [813].

M. A. IWASA

Rodentia MURIDAE

083

Red list status: C-1 (MSJ); LC (IUCN), introduced

Mus musculus Linnaeus, 1758

EN house mouse   JP ハツカネズミ (hatsuka nezumi)   CM 小家鼠   CT 家鼷鼠, 小家鼠
KR 생쥐   RS домовая мышь

An adult female from Kanagawa Pref., Honshu (M. A. Iwasa).

An adult male of JF1 strain derived from the Japanese population.

1. Distribution

In Eurasia, Palaearctic region, as a commensal with humans. In Japan, most of the mainlands and peripheral islands [762, 1011, 1327]. Recent urbanization has lead to reduction of farms and thus the distribution range has declined.

2. Fossil record

Not reported in Japan.

3. General characteristics

Morphology   Smaller than the European mouse. Coat color is gray on back and white or buff on belly with white feet. Tail showing bicolor with brown in upper half and ivory white in lower half [762, 1327]. External measurements (range) are as follows; HB = 57–91 mm, T = 42–80 mm, HFsu = 13–17 mm, BW = 9–23 g [1327].
Dental formula   I 1/1 + C 0/0 + P 0/0 + M 3/3 = 16 [29].
Mammal formula   1 + 2 + 2 = 10 [762, 1009, 1318, 1321, 1327].
Genetics   Chromosome: $2n = 40$ (XX and XY) and FN = 38 [3636]. Genetic details have been published [2251, 3556, 3647, 4152].

4. Ecology

Reproduction   Pregnancy is observed in spring (March–May) and autumn (September–November) in Kyushu. Average litter size (numbers of embryos) is 5.32 ($n = 37$). Litter size is larger in autumn than in spring. Average estrous cycle is 6.001 ± 0.079 days ($n = 419$). Average age at vaginal opening is 64.18 ± 0.445 days ($n = 87$) [762].
Lifespan 705.85 ± 16.212 days ($n = 40$) under captive conditions [896].
Diet   Grass seeds, dicotyledonous seeds, plant leaves and stems and invertebrates [762, 1327, 3409].

Habitat   Inhabits houses, paddy fields, rice fields, riverbanks, grass lands and abandoned fields. Burrows in underground tunnels [762, 1327].
Home range   Not reported for Japanese populations.
Behavior   Considered nocturnally active, but details for Japanese populations unknown.
Natural enemies   Not reported for Japanese populations.
Parasites
Ectoparasite
Acari (gamasid mites and myobiid mites): *Laelaps nuttalli*, *L. algericus*, *Radfordia affinis*, and *Myobia musculi* [223, 762, 763]. Anoplura (lice): *Hoplopleura* sp.
Endoparasite
Several taxa of Diplomonadida and Cestoda [762, 764, 2665].

5. Remarks

Some taxonomists have ranked the Japanese populations of the house mouse as a valid species, *M. molossinus* [762] or a subspecies *M. musculus molossinus* [1011, 3592]. Actually, the Japanese wild mouse is slightly smaller than laboratory mouse (*M. musculus domesticus*) in body size [1961, 1963, 3375]. However, the type specimen of "*molossinus*" (deposited in Museum Volkenkunde: R.M.N.H.18824) is thought to be a hybrid between *M. castaneus castaneus* and *M. musculus manchu* morphologically [1962]. Namely, "*molossinus*" is considered to be an unassigned name based on the International Code of Zoological Nomenclature (article 23) and the Japanese wild mouse is recognized as *M. musculus manchu* [1962, 2334]. Some of the laboratory strains, such as JF1 and MSM, are derived from the Japanese populations [1669, 2248]. Because of recent urbanization, populations of this species are in decline.

M. A. IWASA

RESEARCH TOPIC ❸

Mosaic genome structure of multiple subspecies in the Japanese wild mouse *Mus musculus molossinus*

M. NUNOME, H. SUZUKI & K. MORIWAKI

House mice (*Mus musculus*) have a short coat of light brown to black fur, and a body length of 10–20 cm. They inhabit grasslands, fields, and sandy areas, and as their name suggests, in more recent times, human dwellings. In Japan, house mice, due to their docility and small size, were bred as companion animals in the Edo era (1600–1867). They feature in many folk songs, folklore tales, "netsuke" (miniature sculptures), and paintings (Fig. 1). Today, they are essential resources in the life sciences, and as biological and experimental models, they have provided essential information on the genetics, physiology, and disease etiology of humans. Since development of the laboratory mouse as a research subject, wild mice have begun to attract researchers' attention as unique study animals because of their greater genetic variation as emphasized previously in a book published in 1994 [2251].

Despite their worldwide distribution and association with humans (commensalism), the taxonomy and evolutionary history of *M. musculus* remained unresolved until the late 20th century. Based on morphological traits, *M. musculus* was classified into more than ten subspecies [3154]. However, biochemical and molecular studies assigned the many geographic variants to three major subspecies groups, i.e., the *domesticus* group (DOM) in western Europe, Africa, and America; the *musculus* group (MUS) in northern Eurasia; and the *castaneus* group (CAS) in southern Asia [313, 2247, 2249]. Molecular data [320, 321, 461, 2250, 3353, 4150, 4153] further identified western Iran, northern Iran and northern India as the likely source areas of DOM, MUS, and CAS, respectively, with initial divergence in ancient times, around 0.5–1.0 MYA. The subsequent range expansion across the whole of Eurasia including the Japanese Isls. occurred in comparatively recent times, perhaps associated with the spread of human populations and their agricultural practices [3357].

Japanese wild mice have some unique morphological features and were traditionally distinguished as a local subspecies, *M. m. molossinus* [1794, 2251]. Since the mid-20th century, studies of their morphology, cytogenetics, and biochemistry focused on determining their phylogenetic position [313, 1794, 1964, 2079, 2252]. Pioneering work on variation in mitochondrial DNA (mtDNA) demonstrated that 2 distinct haplotype groups occurred among Japanese wild mice. MUS haplotypes were detected across much of the Japanese Isls., whereas CAS haplotypes were confined to parts of northern Japan [4151]. From these data, it was inferred that mice with CAS mtDNA reached Japan in ancient times, whereas mice with MUS mtDNA colonized later through the northern part of Asian continent. As a result, the CAS mtDNA came to be restricted to the northern regions. This pattern of events mirrors the population history of the Japanese people who themselves seem to have arrived through 2 ancient migrations, an earlier one from southern Asia and a later one from northern Asia [775, 3516]. This concordance suggests that ancient dispersal of the mice was directly related to the movements of people. Notably, more detailed mtDNA analyses conducted recently have narrowed down the likely source areas of mice (and possibly, of the people). Japanese CAS haplotypes suggest an origin in South China, especially along the Yangtze River, whereas Japanese MUS haplotypes point to the Korean Pen. and neighboring areas (Fig. 2) [3357].

Sequence data from nuclear genes provides new opportunities to explore the evolutionary view of Japanese and Eurasian mice. Initial studies identified further geographic subdivisions in each of the subspecies groups [2529]. MUS can be divided into two geographic groups with considerable genetic differentiation: one group inhabits northern regions (eastern Europe, Siberia, Primorye, and Sakhalin), while the other inhabits southern regions (Uzbekistan, northern China, Korea, and Japan). Although the exact processes and routes of dispersal across Eurasia still are unknown, it is tempting to think that at least 2 different dispersal routes were followed during the eastward spread of MUS. From assessments of associated mtDNA diversity, the southern expansion route probably was followed earlier and in quite ancient times, allowing for the development of local genetic differentiation among several regions of East Asia including the Korean Pen., from whence Japanese MUS mice subsequently came, as previously mentioned [3357].

Nuclear gene sequences of Japanese wild mice also re-

Fig. 1. "Gakushu Nezumi": mice learning Japanese calligraphy. This picture was drawn in the 19th century by Kyosai Kawanabe, a Japanese "Ukiyoe" artist. This painting suggests that people of the Edo era possessed amicable feelings toward mice.

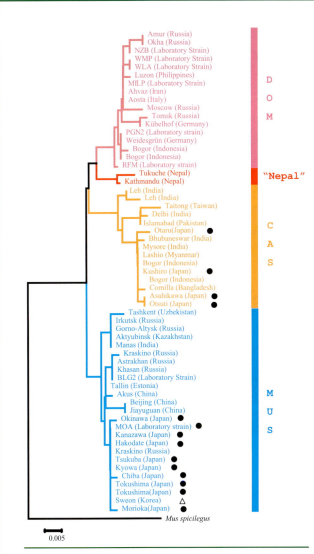

Fig. 2. Phylogenetic relationships among 3 subspecies groups (DOM, MUS, and CAS) of *Mus musculus* based on mitochondrial cyt *b* gene sequences. Two haplotypes from Nepal form a poorly sampled fourth clade ("Nepal"). In the MUS clade, Japanese mice (filled circle) converge with Korean mice (open triangle). A smaller subset of Japanese mice are found in the CAS clade.

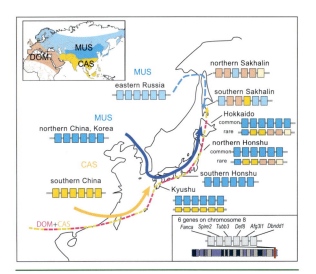

Fig. 3. Geographic ranges of three major subspecies groups as inferred from mtDNA and nuclear gene sequences (left upper box), and a schematic representation of the evolutionary history of Japanese wild mice as inferred from a haplotype analysis of linked nuclear genes (central main box). The gene array used for the phylogeographic inference is shown in the right lower box. It is inferred that CAS mice with introgressed DOM components initially migrated to the Japanese Isls. through a southern route via Kyushu (dashed arrow "DOM+CAS"), followed by a second introduction of CAS from southern China or Southeast Asia (solid arrow "CAS"). Genetic data indicate two subdivisions within the subspecies groups of MUS — northern (pale blue) and southern (dark blue) types. Recombinant haplotypes containing both northern and southern MUS components were recovered from Sakhalin and northern Japan, suggesting a different northern (dashed arrow) and southern (solid arrow) migration routes of MUS to Japan. The latest prehistoric colonizer of Japan — MUS via the Korean Pen. — is inferred as a result of the broad distribution of intact southern MUS haplotypes of in the Japanese Isls. (solid arrow "MUS").

veal a rather complicated history in the eastward movements of CAS (Fig. 3). Some of the most northerly Japanese mice, from Hokkaido and northern Honshu, have short segments of both CAS and western European-type DOM nuclear DNA embedded in a MUS genomic background [2529]. Furthermore, because the DOM segments have engaged in recombinant interactions with CAS segments, it would appear that the DOM elements entered a CAS genomic background during ancient introgression events among mice living on the Indian subcontinent, before the eastward dispersal that carried them to the Japanese Isls. [1651]. Even more remarkably, the DOM–CAS recombinants are found among mice in the Japanese Isls., on the Sakhalin, and on the Indian subcontinent but they are absent in Southeast Asian and South China [1651, 2529]. This points to 2 or more different waves of long range dispersals of CAS mice, and the possibility that parts to India are now occupied by secondarily dispersed CAS mice that reached as far as the Ryukyu Isls. and Kyushu, whereas more remote parts of the Japanese Isls. retained their original genetic structure, free from the influence of the second wave of dispersal of CAS.

Recent developments in genomics, including analysis of entire genome nucleotide sequences, have made *M. musculus* one of the most accessible organisms, not only for biomedical sciences, but also for evolutionary genetics and molecular taxonomy. As previously described, the analysis of variations in nucleotide sequence (single nucleotide polymorphisms, SNPs) across multiple gene loci represents a very powerful tool for tracing the evolutionary history of wild mice and of their human agents of dispersal. As SNP data from more gene loci in each chromosome become available in the future, more details concerning dispersal and genetic differentiation in the house mouse will occur.

Rodentia MYOCASTORIDAE

084

Red list status: C-1 (MSJ); LC (IUCN); ND (JFA), introduced

Myocastor coypus (Molina, 1782)

EN nutria　　JP ヌートリア（nutoria）　　CM 海狸鼠　　CT 河狸鼠，美洲巨水鼠　　KR 뉴트리아　　RS нутрия

An adult in Koyaike Park, Hyogo Pref. (K. Tsuchiya).

1. Distribution

Introduced from South America into Ueno Zoological Gardens (Tokyo Metropolis) in 1907. During 1939–1945, raised for fur production industry at many places across Japan. After World War II, this species escaped and became feral [1327, 2309]. To date, recorded in the following prefectures: Tokyo, Kanagawa, Yamanashi, Ishikawa, Nara, Wakayama, Kochi, Tokushima and Ehime. Now, recorded in following prefectures: Okayama, Hiroshima, Tottori, Shimane, Kagawa, Kyoto, Osaka, Hyogo, Fukui, Gifu, Aichi, Mie and Saitama, and several Isls of Seto Inland Sea [1327, 1441, 2106, 2309, 2666, 3334, 3840].

2. Fossil record

Not reported in Japan.

3. General characteristics

Morphology　Large nostrils and small eyes at upper part of head. Coat entirely grayish-brown. Whitish hairs appear at nose. HB = 50–70 cm, T = 35–50 cm, BW = 6–9 kg; the largest species among rodents in Japan [2309]. Small ears and webbed feet adaptive for aquatic lifestyle [2103, 2309]. Presence of interdigital webbing from 1st to 4th fingers of hind foot [3334].

Dental formula　I 1/1 + C 0/0 + P 1/1 + M 3/3 = 20 [29].

Mammae formula　2 + 2 + 0 = 8 [1327].

Genetics　Not reported in Japan.

4. Ecology

Reproduction　Breeds two or three times per year. Postpartum estrus 1–2 days after confinement and estrous period is 24–26 days. Gestation period is about 130 days. Average number of embryos is 5 (range: 1–12). Sexually mature 6–7 months after birth [1327, 2103].

Lifespan　Over 10 years under captive condition. Maximum lifespan is 8 years in male and 11 years in female in wild population of Okayama Pref. by an age determination using annual rings of molars [2103].

Diet　Mainly seeds of water oats (*Zizania latifolia*), water hyacinths (*Eichornia*), cattails (*Typha latifolia*), reeds (*Phragmites* sp.), water caltrops (*Trapa* sp.), leaves of Prickly water lily (*Euryale ferox*) and mizuaoi (*Monochoria korsakowii*), and mollusks [1327, 2103, 2309].

Habitat　Inhabits watersides such as rivers, lakes and pools.

Myocastor coypus (Molina, 1782)

Nesting on banks of the watersides. Number of entrances of nests ranges 1–3, and are located in water. Total length of nest tunnels over 10 m. Constructs a platform as a floating nest using water plants in winter [1327].

Home range Adult females are sedentary but adult males disperse. Home range (mean ± SD) of females (860 ± 430 m in adult and 430 ± 50 m in young) is smaller than that of males (1,300 ± 650 m). Telemetry researches show they move less than 10 m from water [1327, 2103].

Behavior Semi-aquatic. Active in early morning (3:00–9:00) and evening (19:00–22:00) for feeding [1327, 2103].

Natural enemies Not reported in Japan.

Parasites *Strongyloides myopotami* and *Calodium hepaticum* (Nematoda), and *Fasciola* sp. (Trematoda) are found in Japan [218, 2002].

5. Remarks

A game species. Designated as an invasive alien species in 2005 in Japan, because of agricultural damages for rice, wheat and potatoes [3334].

M. A. IWASA

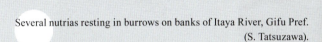
Several nutrias resting in burrows on banks of Itaya River, Gifu Pref. (S. Tatsuzawa).

Leakage of water through a burrow of nutria in Kasai, Hyogo Pref. (S. Tatsuzawa).

Vegetation destroyed by nutrias in Kasai, Hyogo Pref. (S. Tatsuzawa). In ponds and rivers, where nutrias have invaded, disturbance of habitats causes ecological problems.

Rodentia   SCIURIDAE

085

Red list status: C-1 as *S. v. orientis* (MSJ); LC (IUCN)

Sciurus vulgaris Linnaeus, 1758

EN Eurasian red squirrel   JP キタリス (kita risu), エゾリス (ezo risu) for Hokkaido population   CM 北松鼠   CT 北松鼠
KR 청설모   RS обыкновенная белка

A Eurasian red squirrel in winter pelage at Nopporo Forest Park, Ebetsu, Hokkaido (T. Ueyama).

1. Distribution

From Iberia and Great Britain, east to Kamchatka and Sakhalin (= Karafuto), and south to Mediterranean and Black Seas, northern Mongolia, Korea and Northeast China [3570]. In Japan, *S. vulgaris* is distributed in lowland to mountain forests (1,650 m) of Hokkaido [3449]. The population in Hokkaido is described as a subspecies *S. v. orientis*.

2. Fossil record

There are no exact Pleistocene fossil records in Japan. A Late Pleistocene fossil is known from the northern part of North Korea [3287].

3. General characteristics

Morphology   HB = 220–270 mm, T = 160–200 mm, E = 26–34 mm, HFsu = 50–65 mm [1101, 3449]. Mean BW of males is 350–375 g and that of females is 377–456 g [1837]. In the summer pelage, dorsum is reddish brown or dark brown and belly is pure white. In the winter pelage, dorsum is grayish brown and belly is pure white. Ear tufts in winter are ca. 40 mm in length [1101].

Dental formula   I 1/1 + C 0/0 + P 2/1 + M 3/3 = 22 [29, 1011].

Mammae formula   1 + 2 + 1 = 8 [29, 1011].

Genetics   Chromosome: $2n = 40$ (XX and XY) and FN = 74 [3016]. The G-, C-, and Q-banded karyotypes have been described [2757]. Nucleotide sequence data of mitochondrial control region [257, 1666] and cytochrome *b* [2754] and 12S ribosomal RNA [2756] genes are available. The 18S and 28S ribosomal RNA genes are localized on chromosomes [2770].

4. Ecology

Reproduction   There are 1 or 2 mating peaks a year, one of which is from February to March, and the next is in May. Three to 7 males chase the estrous female and the dominant male successfully mates after aggressive interactions among males [1837]. Females bear young at about 1 year old [3449]. Litter size is 1–7 [3449]. The mean number of weaned litters is 1.3 (range 0–2) [3449]. Gestation period is 38–39 days [3449].

Lifespan   A quarter of young individuals is alive until at least 1 year old [3449]. The longest record in captivity is 16 years [3449].

Diet   Sap (2%: e.g. Aceraseae, Beturaseae), buds (14%: e.g. *Picea* sp.), flowers (2%: e.g. Aceraseae), seeds (63%: e.g. *Juglans ailanthifolia* and *Pinus koraiensis*), fruits (4%: e.g. *Morus*

Sciurus vulgaris Linnaeus, 1758

A Eurasian red squirrel in winter pelage emitting an alarm call in February in Sapporo, Hokkaido (T. Kuwahara, Kon Photography & Research).

bombycis and *Prunus* sp.), mushrooms (1%), insects (2%: e.g. Hemiptera, Curculionidae), bird eggs, and carcasses [1837, 3449]. Hoarding behavior is important to survive winter [867]. Scatter-hoarding by *S. vulgaris* plays an important role for regeneration of walnut, *Juglans ailanthifolia* [2006]. Buds of *Abies sachalinensis* and *Picea jezoensis* are heavily eaten in winter [11].

Habitat They preferentially use evergreen coniferous forests as opposed to deciduous coniferous, deciduous broad-leaved and mixed forests [1203, 1838]. Evergreen coniferous trees (Pinaceae) provide good sources of food and suitable sites for building drays.

Home range Home range sizes vary with environments; 1 ha to 4 ha [3449], 72 ha for males and 11 ha for females [1837].

Behavior They are considered to be arboreal but are often observed caching foods on the ground in autumn. In winter, they descend to the ground and dig under the snow to retrieve the caches. Traffic accidents are major problems for survival in Obihiro, Hokkaido, because this species moves across the fragmented forests [4050]. The warning sign posts and eco-bridges over roads are effective measures to prevent accidents [4048, 4050]. They are active in early morning and evening, but active for a short time in midday in winter [3449]. The mean height of dreys is 6.8 m (3.5–9.8 m) ($n = 29$) in Ishikari [8], and 8.4 m in Obihiro [3972]. Squirrels select the large evergreen conifers as nest trees and select the high density stands as nest sites [3972].

Natural enemies Red foxes (*Vulpes vulpes*), sables (*Martes zibellina*), domestic cats, hawks (e.g. *Accipiter gentilis*) and owls (e.g. *Strix uralensis*) are known predators [3449].

Parasites Four species of fleas (*Aenigmopsylla grodekovi*, *Ceratophyllus indages*, *Rhadinopsylla japonica*, and *Tarsopsylla octodecimdentata*) and the louse (*Enderleinellus nitzschi*) are reported [1342, 2968]. Infection by a *Babesia microti*-like parasite is reported [3654].

5. Remarks

Delisted from a game species in 1994. An individual escaped from captivity was found in Sayama, Saitama Pref. [3198]. Eradication from Sayama started in 2014.

N. TAMURA

A female Eurasian red squirrel in summer pelage eating a walnut on a tree in October in Sapporo, Hokkaido (T. Kuwahara, Kon Photography & Research).

Rodentia SCIURIDAE

086

Red list status: LP for the populations in Chugoku and Kyushu District (MEJ); C-2, LP for several populations in western Honshu, and probably extinct in Kyushu (MSJ); LC (IUCN)

Sciurus lis Temminck, 1844

EN Japanese squirrel   JP ニホンリス (nihon risu)   CM 日本松鼠   CT 日本松鼠   KR 일본청설모
RS японская белка

A Japanese squirrel in summer pelage in Chino, Nagano Pref. (H. Nagata).

1. Distribution

Endemic to Japan (Honshu, Shikoku, Kyushu, and Awaji-shima Isl.). Rare in Chugoku District of western Honshu and urban areas of central Honshu. In Kyushu and Awaji-shima Isl., there are no recent observations [1102].

2. Fossil record

A Middle Pleistocene fossil of this species was reported from the southern part of Honshu (Yamaguchi Pref.); Late Pleistocene fossils are known from southern to central Honshu and Shikoku; Holocene fossils are found in northern to central Honshu [1507].

3. General characteristics

Morphology   HB = 160–220 mm, T = 130–170 mm, E = 22–31 mm, HFsu = 48–58 mm, BW = 250–310 g [1102]. This species is smaller than *S. vulgaris*. In the summer pelage, dorsum is reddish brown with orange parts on shoulder and thigh, and belly is pure white. In the winter pelage, dorsum is grayish brown and belly is pure white. Hair tufts on the ears occur in winter. *S. lis* has faint white rims on eyelids, but *S. vulgaris* lacks the white rims.

Dental formula   I 1/1 + C 0/0 + P 2/1 + M 3/3 = 22 [29, 1011].

Mammae formula   1 + 2 + 1 = 8 or 2 + 1 + 1 = 8 [1011].

Genetics   Chromosome: $2n = 40$ (XX and XY) and FN = 74 [2751]. The C- and Q-banded karyotypes are shown [2751, 2769]. Nucleotide sequence data of mitochondrial cytochrome *b* [2754] and 12S ribosomal RNA [2756] genes are available. The 18S and 28S ribosomal RNA genes are localized on chromosomes [2770].

4. Ecology

Reproduction   Breeds once or twice a year. Mating bouts are observed from January to March and June in Chiba Pref., January, July, and September in Tokyo, from February to March and from May to June in Nagano Pref. [4099]. In Nagano Pref., 68% of 16 females reproduced once a year, 19% did twice and 13% did not [2479]. Up to five females assemble in the home range of estrous females and each female copulates with 1 or 2 males within a single day of estrus [2478]. In captive condition, litter size is from 1 to 6 with a mean of 3.4 and gestation periods are 39 or 40 days [3020]. Young first appear from the natal nest at about 40 days age [2478]. The number of weaned litter is from 1 to 3 with a mean of 2.0 ($n = 6$) in Tokyo [4099] and from 1 to 5 with a

Sciurus lis Temminck, 1844

mean of 3.8 (*n* = 4) in Nagano Pref. [2478].

Lifespan The longest duration of stay of marked individuals is 36 months both for females and males in Tokyo [3497]. 63% of 63 marked individuals had disappeared within 1 year, 29% were lost within 2 years, and 8% were lost within 3 years [3497].

Diet Buds (e.g. *Larix leptolepis* and *Betula platyphylla*), flowers (e.g. *Juglans ailanthifolia* and *Prunus* sp.), fruits (*Akebia quinata* and *Morus bombycis*), seeds (e.g. *J. ailanthifolia*, *Pinus densiflora*, and *P. parviflora*), mushrooms, and insects [1420, 4100]. *Sciurus lis* needs various species of plants to cover a year-round diet, although walnuts and pines are the most important as a main food to sustain for a long period [1420, 4100]. They transport and cache seeds to consume during winter and/or food shortage [1420, 3505]. The mean transport distance was 20.7 m (up to 62.0 m) by tracing walnuts with radio-transmitters [3505], but it changed with food availability and competition [3495]. As 10% of cached walnuts remain for 6 months, the caching behavior plays an important role for regeneration of plants [3505]. The feeding remains of Japanese walnuts and pine cones are unique in shape, so that they are important field sign to know their distribution [3494, 3501, 4101].

Habitat *Sciurus lis* requires continuous forests supplying enough food, nests sites, and cover from predators [3506, 4103]. This species selectively uses evergreen forests including various tree species, but seldom use artificial conifer plantation and deciduous forests in lowland suburban forests of western Tokyo [3490]. For nest sites, they select forests with large evergreen trees in upper layer [3506]. The existence of squirrels is affected by woodlot size and distance from the large continuous forests in isolated woodlots in western Tokyo [1407]. In northern Honshu, however, squirrels selectively use walnut forests and survive in small isolated woodlots, if there are abundant walnuts [800].

A Japanese squirrel in winter pelage in Chino, Nagano Pref. (H. Nagata).

Sciurus lis are distributed in sub-alpine forests of Mt. Fuji up to 2,500 m in elevation and they preferentially use *Pinus parviflora* forests supplying both nest sites and food resources [1640]. In lowland forests from central to western Honshu, squirrels are often observed in *Pinus densiflora* forests which are suitable for making nests and getting foods [3494, 4100]. However, the population of this area is declining due to pine wilt disease [3501, 4101].

Home range Home range size varies among localities and/or habitat types. The mean home range size is generally larger in males than females. The mean home range size calculated by 100% convex polygon method is 12.0 ha (female) and 30.0 ha (male) at Tateshina, Nagano Pref., 4.5 ha (female) and 12.0 ha (male) at Karuizawa, Nagano Pref., 3.1 ha (female) and 9.7 (male) at City Park in Chiba Pref., 7.0 ha (female) and 18.9 ha (male) in mixed forests at Mt. Takao, western Tokyo, 8.2 ha (female) and 17.6 ha (male) in the red pine forests at Kawaguchiko-machi,

Typical food remains by the Japanese squirrel. *Juglans ailanthifolia* (left) and *Pinus densiflora* (right) (N. Tamura)

Yamanashi Pref., 15.3 ha (female) and 19.6 ha (male) in sub-alpine forests at Mt. Fuji, Yamanashi Pref., and 1.5 ha (female) and 3.7 ha (male) in the walnuts forests at Morioka, Iwate Pref. [1404, 1640, 2476, 3492, 4103]. Home range sizes increase with decreasing preferable habitats in their ranges at western Tokyo [3492]. Home ranges among males overlap each other while those among mature females usually do not overlap [4103], however, females also have overlapping home ranges in Morioka, Iwate Pref. with abundant walnuts [2476].

Behavior Diurnal activity is bimodal, early morning and evening, in spring-autumn [2478]. A unimodal pattern occurs in winter with a midday peak [2478]. Squirrels most frequently use tree crown and middle layer for locomotion, resting, and feeding. They often move on the ground to search and cache food, but seldom use the shrub layer [4100]. *Sciurus lis* uses cavities (tree hollows) and dreys (leaf nests) to breed young, to rest at night, and to avoid severe weather. They usually select arboreal dreys and cavities but occasionally use underground dens formed on the lava flows of Mt. Fuji [1405]. The drey is a spherical structure about 40 × 25 × 20 cm in size and made of two layers; an outer husk of branches and a central core of inner bark [4098]. The squirrels often use branches of the nearest tree regardless of species, but select inner barks of *Cryptomeria japonica* and *Chamaecyparis obtuse* in Chiba and Yamagata Prefs. [4098]. The mean height of dreys is 10.1 m (4.5–16.0 m, $n = 38$) above ground [4098]. Most of the dreys (85%) are set near the trunk at the base of horizontal branches [4098]. One individual has 3–20 nests in the home range and changes nests every 2.4 days by males and 3.4 days by females [2477]. *Sciurus lis* sometimes strip the bark of *Cryptomeria japonica* and *Chamaecyparis obtuse* to get nest materials, but the damage is trivial. In Shizuoka and Kochi Prefs., the log cultivation of shiitake mushrooms is damaged by *S. lis* that gnaw the log to get larvae of *Fornax* [609].

Natural enemies Known predators are red foxes (*Vulpes vulpes*), Japanese martens (*Martes melampus*), domestic cats, mountain hawk eagles (*Spizaetus nipalensis*), goshawks (*Accipiter gentilis*), sparrow hawks (*Accipiter nisus*), common buzzards (*Buteo buteo*), and crows (*Corvus* spp.) [4103].

Parasites The flea *Ceratophyllus indages* and the ticks *Haemaphysalis flava* (nymphs) and *Ixodes persulcatus* (larvae, nymphs and adult females) parasitize this squirrel [2968, 3496, 3973]. Another flea *Ceratophyllus argus* is reported from nests [2968].

5. Remarks

Recent studies indicate that *S. lis* is vulnerable to forest fragmentation and habitat change by human activities [1407]. Huge areas of pine forests were damaged by pine wilt disease since the 1950s from southwestern to northeastern Japan, so that habitats of the Japanese squirrel extensively decreased [3493, 4101]. Population decline has been reported in several areas, especially western Japan [3501, 4067]. This species was delisted from game species in 1994.

N. TAMURA

Red pine (*Pinus densiflora*) forest, which is a main habitat of the Japanese squirrel in western Honshu, has been largely damaged by pine wilt disease (N. Tamura).

A Japanese squirrel caching walnuts at the foot of Mt. Yatsugatake, Nagano Pref. (H. Nagata).

A Siberian chipmunk near Lake Utonai, Tomakomai, Hokkaido (S. D. Ohdachi).

Sucklings of the Siberian flying squirrel in a nest box in Furano, Hokkaido (T. Oshida).

Nest materials of the Siberian flying squirrel found in nest box in Furano, Hokkaido (T. Oshida).

A female Eurasian red squirrel with noticeable nipples in Sapporo, Hokkaido (S. D. Ohdachi).

Rodentia SCIURIDAE
087

Red list status: C-1 as *C. e. thaiwanensis* (MSJ); LC (IUCN), introduced

Callosciurus erythraeus (Pallas, 1779)

EN Pallas's squirrel   JP クリハラリス (kurihara risu)   CM 赤腹松鼠   CT 赤腹松鼠   KR 대만청설모
RS Палласова белка

A Pallas's squirrel in Kamariya, Yokohama, Kanagawa Pref. (S. Yamamoto).

1. Distribution

West of Irrawaddy River in India, Myanmar (Burma), and southeastern China. East of Irrawaddy River in Myanmar, Thailand, Malaysia Pen., Indochina, southern China and Taiwan [3570]. In Japan, introduced populations are naturalized on Izu-ohshima Isl. and Akiruno, Tokyo, Bando, Ibaraki Pref., Iruma, Saitama Pref., southeastern part of Kanagawa Pref., western Izu Pen. and Hamamatsu, Shizuoka Pref., Kinkazan, Gifu Pref., Osaka-jo Park, Osaka Pref., Himeyama Park, Hyogo Pref., Tomo-gashima Isl. and Wakayama-jo Park, Wakayama Pref., Fukue, Iki Isls., and Shimabara Pen, Nagasaki Pref., Takashima Isls., Ohita Pref., the Uto Pen, Kumamoto Pref, and Kirishima, Miyazaki Pref. [121, 1352, 3485, 4068, 4071].

2. Fossil record

Not reported from Japan.

3. General characteristics

Morphology   HB = 200–260 mm, T = 170–200 mm, E = 19–24 mm, HFsu = 45–54 mm [1103]. BW = 309–435 g for adult males [3507]. Geographic variation is considerable with different color forms. Ventral color varies from entirely agouti to maroon with central agouti stripe or without stripe on Taiwan and introduced populations in Japan [410].

Dental formula   I 1/1 + C 0/0 + P 2/1 + M 3/3 = 22 [29, 1011].

Mammae formula   0 + 1 + 1 = 4 [1011].

Genetics   Chromosome: $2n = 40$ (XX and XY) and FN = 70 [2764]. The G- and C-banded karyotypes are shown [2764]. Nucleotide sequence data of mitochondrial control region [2752, 2762] and cytochrome *b* gene [2766] are available. The 18S and 28S ribosomal RNA genes are localized on chromosomes [2770].

4. Ecology

Reproduction   Breeding occurs throughout the year both in Taiwan and Japan [3499]. Nine to 17 males gather near an estrous female and 4–11 males mate according to their dominance rank [3498]. Individual females breed 0–3 times a year with a mean of 1.9 times in Taiwan and 1.2 times in Japan [3488]. The mean

Callosciurus erythraeus (Pallas, 1779)

A group of Pallas's squirrels feeding on a tree in Yokohama, Kanagawa (S. Yamamoto). Their tolerant social system often causes high population density.

number of embryos is 1.8 in Taiwan [3384] and 2.3 (range 0–4) on Izu-ohshima Isl., and 2.0 (range 0–4) on the Uto Pen., Japan [3491, 4070]. The number of weaned litters is 1.1 in Taiwan and 1.3 in Japan [3499]. First copulations are observed at 11 months after birth in captive conditions [3507]. Gestation period ranges from 47 to 49 days in captivity [3491].

Lifespan Among adult females, 81% live beyond 1 year, 66% more than 2 years, and 11% for 4 years, but none more than 5 years in Japan [3499]. In contrast, 34% of adult females survive more than 1 year, and only 20% survive more than 2 years in Taiwan [3499].

Diet Various parts (leaves, buds, flowers, fruits, seeds, and barks) and species of vegetation materials, mushrooms, and animal materials, such as insects, bird eggs and snails [389, 2793, 3499]. Debarking in conifer plantations causes serious economic damage in Taiwan and Japan [1762, 3620]. *Callosciurus erythraeus* strip hardwood bark horizontally and lick the flowing sap, especially in early spring [3504]. Squirrels established in Japan damage orchard products, such as oranges, grapes, and loquats [121]. Squirrels damage flowers, seeds, and bark of *Camellia japonica* whose oil is an important product on Izu-ohshima Isl. [3197].

Habitat They are distributed in various types of forest (natural forests, conifer plantations, orchards, bushes and city parks), but they prefer mixed species broad-leaved evergreen forests [2683]. In Kanagawa Pref., Japan, squirrels settle in small isolated woodlots such as gardens, parks, and temple forests, with a mean of 6.6 ha in size, and expanded distribution in urban areas [3502].

Home range Mean home range size of females 0.3 ha in Taiwan and 0.5 ha in Kanagawa, Japan and that of males is 1.4 ha in Taiwan and 2.2 ha in Kanagawa [3499]. Females occupy exclusive home range in Kanagawa, but have overlapping home ranges in Taiwan, while home ranges among males and between sexes overlap extensively in both study sites [3499]. Because of overlapping small home ranges, the density of this species is usually high; 6.5–6.8/ha in Taiwan and 5.6–6.8/ha in Kanagawa [3499].

Behavior They are regarded as arboreal and often use electric wires and hedgerows for locomotion in urban areas of Japan. They are active in early morning and evening in Taiwan [389] but patterns are often affected by artificial feeding [3503]. Social interactions are frequently observed and dominance hierarchies are apparent in mating bouts and at feeding situations [3498]. This species emits different alarm vocalization in response to 3 types of predators, carnivores, snakes and hawks [3489]. Repetitive barks towards terrestrial carnivores make other squirrels escape from the ground to the tree, squirrels stop moving by single bark towards flying birds, and screams towards snakes gather squirrels to attend the mobbing bouts [3489]. This species often makes dreys, but sometimes use cavities, nest boxes, and door pockets as nest site. The dreys are built at 5.9 m in height (1.5–12.0 m) of broadleaved evergreen trees (62%), such as *Quercus myrsinifolia* and *Castanopsis sieboldii*, deciduous trees (26%), such as *Swida controversa* and *Mallotus japonicus*, and evergreen conifers (12%), such as *Chamaecyparis obtuse* and *Cryptomeria japonica* in Kanagawa [2684]. Inner bark of conifers is frequently used as nest materials [2684].

Natural enemies Known predators are carnivores (e.g. *Martes flavigula*, *Felis bengalensis*), snakes (e.g. *Trimeresurus* sp., *Elaphe* sp.), and hawks (e.g. *Spilornis cheela* and *Butastur indicus*) in China and southeast Asia, and domestic cats, snakes (e.g. *E. climacophora*), and hawks (e.g. *Buteo buteo*) in Japan [3488].

Parasites This introduced squirrel is parasitized in Japan by a flea (*Ceratophyllus anisus*), lice (*Enderleinellus kumadai* and *Neohaematopinus callosciuri*), and a tick (*Haemaphysalis flava*) [1342, 3236, 3237, 3496]. Two parasitic helminths *Brevistriata callosciuri* and *Strongyloides* sp. are also recorded from this species in Japan [2002].

5. Remarks

Designated as an invasive alien species in 2005 [3485]. Game species. Attempted eradications being conducted in Iruma, Saitama Pref., Akiruno, Tokyo, Bando, Ibaraki Pref., and the Uto Pen., Kumamoto Pref. are at an early stage [1352, 3487, 4069]. On Izu-ohshima Isl., Hukue Isl., and in Kanagwa Pref., population control is conducted to protect agriculture, forestry, and human life. Origin of introduced populations in Japan has been estimated by mitochondrial DNA analysis, and it was found that six populations (Izu-ohshima, the Izu Pen., Hamamatsu, Miyazaki, Kumamoto, and Fukue Isl.) may be derived from the Taiwan population [991], while the Hamamatsu population includes individuals with a different genetic pattern regarded as *C. finlaysonii* or hybrid between *C. erythraeus* and *C. finlaysonii* [2762].

N. TAMURA

Rodentia SCIURIDAE

088

Red list status: DD (MEJ); C-1 as *T. s. lineatus* (MSJ); LC (IUCN)

Tamias sibiricus (Laxmann, 1769)

EN Siberian chipmunk   **JP** シマリス (shima risu), エゾシマリス(ezo shima risu) for Hokkaido population   **CM** 花鼠
CT 亞洲花栗鼠, 金花鼠   **KR** 다람쥐   **RS** азиатский бурундук, сибирский бурундук

A Siberian chipmunk foraging on a tree in Hokkaido (H. Yanagawa).

1. Distribution

European and Siberian Russia to Sakhalin (= Karafuto), the Kuril Isls; eastern Kazakhstan to northern Mongolia, China, Korea, and Hokkaido of Japan [3570]. In and around Japan, this species occurs in Sakhalin, Hokkaido, Kunashiri, Rishiri, Rebun, Teuri, Yagishiri, and Etorofu (= Iturup) Isls. [1104]. The detailed distribution of this species is still unclear in Japan, so the map tentatively shows the possible distribution. Also, there are introduced populations in Austria, Belgium, France, Germany, Italy, the Netherlands and Switzerland [127, 3570] (these localities are not shown in the worldwide distribution map).

2. Fossil record

There is no fossil record, although ambiguous *Tamias* remains are reported from northern China [441].

3. General characteristics

Morphology In Hokkaido, HB = 124–165 mm, T = 105–133 mm, HFsu = 33–38 mm, E = 15–17 mm, BW = 71–116 g (n = 15, T. Oshida's database based on specimens captured in Hokkaido, 1957–1996). External fur color: back and tail are light brown, with five longitudinal stripes on the back, with cream-color between them. Belly and ear tips are white. A black stripe runs from nose to cheek.

Dental formula I 1/1 + C 0/0 + P 2/1 + M 3/3 = 22 [1011].

Mammae formula 1 + 1 + 2 = 8 or 1 + 2 + 1 = 8 [1011].

Genetics Chromosome: $2n$ = 38 (XX and XY) and FN = 58 [2339, 2767, 3016]. The 18S and 28S ribosomal RNA genes are localized on chromosomes [2770]. The C- and Q-banded karyotypes are reported [2767]. Segmental chromosome homology between *T. sibiricus* and human is reported [271]. Nucleotide sequence data of mitochondrial D-loop region [2909], cytochrome *b* [2552, 2841], cytochrome *c* oxidase II [2841] and 12S rRNA genes [2756], hibernation specific proteins [2426, 2709], proto-oncogene c-myc [3306] and interphotoreceptor binding protein [2060] genes and recombination activating gene 1 (RAG1) [3306] are available.

4. Ecology

Reproduction Breeding season is from spring to summer. Mating behavior is observed from middle April to late May [1477]. The gestation period is 30 days [1477]. In captivity, litter size is 2–7 (mean 4.7), and in the wild, it is estimated to be 3–6 (mean 4.8)

Tamias sibiricus (Laxmann, 1769)

[1477]. Juveniles begin to be active outside the nest from day 35, and all siblings remain in the same nest for 45 days, after their birth [1469].

Lifespan About 5–6 years [1104].

Diet Various plant items (acorns, seeds, fruits, buds, leaves, flowers, and sap) are used in different seasons, and also several insect species including larva and adult, spider, snail, and egg and nestling of birds are recognized as its food items. The diet consists of as many as 41 plant and 16 animal species [1466]. Seasonal changes of food habits are reported [1466]. In spring, various seeds such as acorns, cherry, maple, and bitter-dock, which were produced in the previous year, are utilized. This chipmunk consumes buds, young leaves, and flowers from early May to early June: buds and young leaves of maple group are the most favored. Its second favored food, buds of bamboo grass, is eaten from middle May to middle June. Third favored food, chickweed, is used from late April to early November. From late May, seeds of various plants are commonly eaten: from late May to early June, sedge seeds are used, and cherry seeds become important resources from late June. The seeds of *Prunus* are also eaten from late June to late August. The grass seeds such as touch-me-nots are mainly used from August to the early half of September. Acorns make up the main portion of the diet in October and November.

Habitat Inhabits various types of forest from coast to alpine zone, and prefers open microhabitats surrounded with forests [1104].

Home range The range and territory have been described [1465]. This chipmunk is not territorial. The range of male and female were recorded as 6,830 m^2 in autumn ($n = 22$) and 3,934 m^2 ($n = 39$). Both males and females have overlapping ranges, but the overlapping of males is wider than that of females. In the range of 1 male, 1–8 females make a nest for hibernation.

Behavior Nests in underground burrows and tree hollows. The Siberian chipmunk transports many foods in its cheek pouches to other places for scatter and larder (nest) hoarding: the hoarded food items are acorns, cherry seeds, and the seeds of *Kalopanax pictus*, touch-me-nots, wake-robins and bitter-docks; acorns are a major hoarding item [1466]. The scatter hoarding behavior is associated with increased cell proliferation in the hippocampus, suggesting that new cells in the hippocampus are affected by learning and memory tasks [2799].

From late October or early November to late April or early May, this chipmunk hibernates. Females hibernate significantly earlier than males [1467]. The hibernation behavior is reported in detail [1465, 1467, 1468]. The hibernation stage starts 7–95 days after selection of hibernation burrow. In the first stage, the entrance of the hibernation burrow is plugged by soil. The basic structure of burrows consists of one entrance (diameter: 4.0–6.0 cm, $n = 3$), one tunnel (length: 134.0–209.0 cm, $n = 3$) and one nest chamber (height: 20 cm and depth: 17 cm, $n = 17$) at the end of the tunnel. The deepest point of the burrow is at the bottom of the nest chamber or in a depression near the entrance of the nest chamber (68.0–107.0 cm, $n = 4$). The nest chamber is dry and clean, and is filled with foods and nest materials placed over the foods. In the second stage (around late November), the entrance of the burrow is opened, and the chipmunk mounds soil at the burrow entrance. Soil is tightly packed into the main tunnel, as a thick plug. Under ground, one or more new tunnels (subtunnels) are made from the nest chamber to a downward portion (length: 132.0–267.5 cm, $n = 5$). A toilet is prepared in a depression in the tunnel or at the end of subtunnel. In this stage, the deepest point of the burrow is at the end of subtunnel (84.0–144.0 cm, $n = 4$). In spring, the chipmunk makes a new tunnel to the surface. Position of the entrance of spring burrows is 0–410.0 cm ($n = 99$) from that of hibernation burrows made in the previous autumn. Commencement of hibernation is significantly earlier for adults than for juveniles.

The Siberian chipmunk shows precise circannual rhythms in the blood titer of hibernation-specific protein complex (HPc) [1702], which consist of HP55 (a homolog to α1-antitrypsin) combined with the HP20c [1696]. The HPc titers decrease during hibernation, and rise with its termination [1702].

Natural enemies In the feces and stomach contents of the Japanese sable, Siberian chipmunks are frequently found (19.3% of total materials) [2314].

Parasites The nematode (*Brevistriata skrjabini*) is reported [209].

5. Remarks

The smallest game mammal species in Japan. In Honshu, there are alien populations from the Eurasian Continent [2746]. In Hokkaido, hybridization and/or competition between native individuals and introduced continental individuals are suspected [1104, 2746, 2748]. In a worldwide biogeographical study based on skull morphology and external characteristics, 3 major groups were found: the Korean Pen., central China, and northern Eurasia groups. Hokkaido population is included in the northern Eurasian group [2552]. Also, in phylogeographical study, 2 major mtDNA lineages (the Korean Pen. and northern Eurasia) were recognized, showing that Hokkaido population is included in the latter [2552].

T. OSHIDA

Rodentia SCIURIDAE
089
Red list status: C-1 (MSJ); LC (IUCN)

Petaurista leucogenys **(Temminck, 1827)**

EN Japanese giant flying squirrel   JP ムササビ (musasabi), ホオジロムササビ (hoojiro musasabi)
CM 日本大鼯鼠   CT 日本飛鼠   KR 큰날다람쥐   RS японская гигантская летяга

A Japanese giant flying squirrrel at Tsuru, Yamanashi Pref. (H. Okazaki).

1. Distribution

Endemic to Honshu, Shikoku, and Kyushu of Japan [3570]. The exact distribution of this species is unknown in Japan, so that the distribution map tentatively shows the possible distribution area.

2. Fossil record

Middle Pleistocene fossils of this species are reported from the southern part of Honshu; Late Pleistocene fossils are known from southern to central Honshu, Shikoku and Kyushu; Holocene fossils are abundant in Honshu, Shikoku and Kyushu [1507].

3. General characteristics

Morphology HB = 272–485 mm, T = 280–414 mm, HFsu = 50–73 mm, E = 28–43 mm [1011], BW = 495–1,250 g. External fur color: back and tail are brownish or grayish with geographic variations; grayish individuals are known from central to northern parts of Honshu. Belly is white. White line runs between the eye and ear to the cheek.

Dental formula   I 1/1 + C 0/0 + P 2/1 + M 3/3 = 22 [1011].

Mammae formula   1 + 2 + 0 = 6 [2679].

Genetics Chromosome: $2n = 38$ (XX and XY) and FN = 72 [2758]. The G-, C-, Q, and R-banded karyotypes are reported [2758]. Geographic variation of C-heterochromatin is recognized [2759]. Nucleotide sequence data of mitochondrial cytochrome *b* [2753, 2755, 2761] and 12S ribosomal RNA [2756] genes and control region [2750] are available. The 18S and 28S ribosomal RNA genes were localized on chromosomes [2770].

4. Ecology

Reproduction There are 2 mating seasons: from middle November to late January (range 76 days) and from middle May to middle June (range 36 days) [1476, 1478]. Each reproductive bout requires 4–6 months, and mature females are engaged in reproduction all year [1476]. Seasonal change in testis size associated with mating seasons is reported [1475]. Gestation period is 74 days on average [1105]. Litter size is 1–4 (usually 2) [1105]. Young individuals begin foraging independently (80 days old for the earliest one) from late April and from middle October onward [1474].

Lifespan In captivity, its lifespan is over 10 years, and in the wild, it is 10 years [1105].

Diet Japanese giant flying squirrels feed entirely on plant material. As their food items, 45 species of trees belonging to 21 families have been identified: the families Fagaceae, Pinaceae, and Rosaceae are important resources [1474]. Their foods change seasonally. However, main food items are seeds, leaves, buds, and staminate cones of conifers; also, they utilize fruits, flowers, and bark of woods [152, 1474].

Habitat The Japanese giant flying squirrel inhabits warm-temperate evergreen forest and temperate evergreen mixed deciduous forest: these are primary or mature secondary forests, from the lowlands to sub-alpine zone [1105]. This flying squirrel is also found in coniferous plantation forest [1105].

Home range Females have non-overlapping home ranges of about 1.0–1.5 ha in size, and males have overlapping ranges of about 2.0–3.0 ha [1105]. The range size, however, varies from 0.46 to 2.68 ha in females ($n = 5$), and from 0.78 to 5.16 ha in males ($n = 4$) [237].

Behavior Japanese giant flying squirrels are nocturnal and arboreal. By using flying membranes, this flying squirrel glides from tree to tree; mean horizontal glide distance is 19.46 ± 10.75 m ($n = 57$) [3301]. This flying squirrel nests usually in tree hollows and sometimes on artificial buildings [151]. Shredded bark of *Cryptomeria japonica* is usually utilized as nest material [153]. Nests on branches (dreys) are sometimes observed [153].

Natural enemies There are no exact records on predators for this species. However, carnivores occurring in Japan such as foxes, weasels, and martens could prey on Japanese giant flying squirrels. Interestingly, Japanese macaques have been twice observed to attack solitary Japanese flying squirrels, although the behavioral meaning has not been resolved [1362].

Parasites The flea (*Monopsyllus argus*) was reported [4030]. The helminth, *Aprostatandrya petauristae* [3139] and *Sypharista kamegaii* [209] are reported.

A gliding Japanese giant flying squirrel individual at Mt. Takao, Tokyo (H. Okazaki).

5. Remarks

The Japanese giant flying squirrel was formerly treated as a game species, but it was excluded in 1994. Its fur is of poor quality. This species sometimes causes bark damage in coniferous plantations, and was controlled as a pest animal [981]. Ancient bones of this squirrel have been found frequently from the remains of the Jomon Period (10,000–400 B.C.), suggesting that it could have been popularly eaten by Jomon people [149]. There are at least 5 phylogroups (Northern, Central, Southeastern, Southwestern, and Southern lineages), which may have originated from glacial refugia during the Late Pleistocene; after the last glaciation, the Northern phylogroup, widely distributed in eastern Japan, had extensively expanded northward from its refugia, while, in western Japan, population expansion was restricted to western Japan [2755]. A breathing rate ranges from 24 to 44 per minute ($n = 5$, T. Oshida's database), and a heartbeat rate is 98–100 per minute ($n = 1$, T. Oshida's database).

T. OSHIDA

A habitat (*Cryptomeria japonica* dominated mixed forest) of the Japanese giant flying squirrel in Mt. Takao, Tokyo (M. Shigeta).

Rodentia SCIURIDAE

090 Red list status: C-2, LP for the Kyushu population (MSJ); LC (IUCN)

Pteromys momonga Temminck, 1844

EN Japanese flying squirrel JP ニホンモモンガ (nihon momonga) CM 日本飞鼠 CT 日本小飛鼠
KR 일본하늘다람쥐 RS японская малая летяга

A Japanese flying squirrel in Aomori Pref. (N. Shirakawa).

1. Distribution

The Japanese flying squirrel is endemic to Honshu, Shikoku and Kyushu of Japan [1106]. The exact distribution of this species in Japan is unclear, so the map shows the possible distribution.

2. Fossil record

Middle Pleistocene and Late Pleistocene fossils are reported from southern Honshu; Holocene fossils are also known from southern Honshu [1507]. Although a Pleistocene fossil of *Pteromys* was found in Aomori Pref., its taxonomic status is unclear [825].

3. General characteristics

Morphology HB = 139–200 mm, T = 95–140 mm, HFsu = 32–39 mm, E = 15–25 mm, BW = 150–220 g [1106]. External fur color: Back and tail are dark brown in summer and gray brown in winter with variations. Belly is white. Dark brown around the eyes.
Dental formula I 1/1 + C 0/0 + P 2/1 + M 3/3 = 22 [1011].
Mammae formula 2 + 2 + 1 = 10 [1011].
Genetics Chromosome: $2n = 38$ (XX and XY) and FN = 68 [2763]. The G- and C-banded karyotypes are reported [2763]. Nucleotide sequence data of mitochondrial cytochrome *b* gene are available [2749]. The 18S and 28S ribosomal RNA genes are localized on chromosomes [2770].

4. Ecology

Reproduction Not well studied. There are several case reports. A litter consisting of 2 young was observed in Shizuoka in late March [3611]. The author observed that at least 2 young were present in a nest cavity in Aomori Pref. in June. In November, a litter consisting of 3 young was observed in Fukui Pref. [4052]. This species may breed twice a year [4052].
Lifespan Not reported.
Diet Not well known. Japanese flying squirrels may eat mainly leaves, buds, bark, nuts, seeds, fruits, and mushrooms [1106].
Habitat Inhabits mixed-temperature forests from montane to subalpine zone [1106]. This species is also found in the plantation forest of Japanese cedar [4052].
Home range Not reported.
Behavior The Japanese flying squirrel is nocturnal and arboreal. This species usually nests in tree hollows [1106], and prefers tree hollows of coniferous trees such as *Cryptomeria japonica* and *Chamaecyparis obtuse* to that of broad-leaved trees [3361]. Shredded bark of *Cryptomeria japonica* is usually utilized as nest material [150].
Natural enemies Snakes attack their nest cavities and owls are potential predators [2747].
Parasites Not reported.

T. OSHIDA

Pteromys momonga Temminck, 1844

A Japanese flying squirrel carrying leaves in Kiyokawa, Kanagawa Pref. (Kei Suzuki).

A habitat of the Japanese flying squirrel in the mixed forest composed of Japanese cedars (*Cryptomeria japonica*) and several hardwood trees in Kiyokawa, Kanagawa Pref. (Kei Suzuki).

A habitat of the Japanese flying squirrel in the Japanese cedar-dominated mixed forest in Kiyokawa, Kanagawa Pref. (Kei Suzuki).

Rodentia   SCIURIDAE

091

Red list status: C-1 as *P. v. orii* (MSJ); LC (IUCN)

Pteromys volans (Linnaeus, 1758)

EN Siberian flying squirrel, Russian flying squirrel   JP タイリクモモンガ (tairikumomonga), エゾモモンガ (ezo momonga) for Hokkaido population   CM 小飞鼠   CT 西伯利亞小鼯鼠   KR 하늘다람쥐   RS обыкновенная летяга

A Siberian flying squirrel in Obihiro, Hokkaido (H. Yanagawa).

1. Distribution

The entire coniferous forest zone of Eurasia from Finland and Baltic Sea to eastern Siberia, the Ural and the Altai Mts., Mongolia, northern China, and Korea; Sakhalin (= Karafuto) and Hokkaido [3570]. Hokkaido population is regarded as an endemic subspecies, *P. v. orii* [1791].

2. Fossil record

Not reported in Japan.

3. General characteristics

Morphology   The following measurements (range) are from the Hokkaido population of Japan. HB = 130–167 mm, T = 92–118 mm, HFsu = 32.0–36.5 mm, E = 17.5–21.5 mm, BW = 62–123 g (n = 22, T. Oshida's database based on specimens captured in Hokkaido, 1960–2000). External fur color: back and tail are dark brown in summer and gray brown in winter with variations. Belly is white. The color around the eyes is dark brown.

Dental formula   I 1/1 + C 0/0 + P 2/1 + M 3/3 = 22 [1011].

Mammae formula   2 + 1 + 1 = 8 [1011].

Genetics   Chromosome: $2n$ = 38 (XX and XY) and FN = 68 [2768, 2866]. The G- and C-banded karyotypes are reported [2763]. Nucleotide sequence data of mitochondrial cytochrome *b* and 12S ribosomal RNA genes of *P. v. orii* are available [2749, 2756]. Complete mitochondrial genome sequence of Korean population is reported [2914]. The 18S and 28S ribosomal RNA genes are localized on chromosomes [2768].

4. Ecology

Reproduction   There are 2 mating seasons: from late February to early March and in June [4049]. This species gives birth in middle April–early May and late July–middle August [4049]. Litter size is 2–6 (usually 3) [1107, 4049]. In 1 case, young left the nest cavity at 60 days of age [4049].

Lifespan   In captivity, its lifespan is 4–5 years, and in the wild, it is rare over 3 years [1107].

Diet   Siberian flying squirrels feed entirely on plant material such as leaves, nuts, seeds, flowers, and buds [1107]. Their preferred foods change seasonally: young leaves, buds, and catkins of *Salix* spp. and young leaves of *Betula platyphylla* in April; leaves and buds of *Salix* spp., buds of *Populus maximowiczii*, catkins of *B. platyphylla*, leaves of *Quercus dentata*, and buds of

Ulmus davidiana and *Larix leptolepis* in May; buds, flowers, and young leaves of *Q. dentate, buds, leaves,* and seeds *of U. davidiana*, leaves of *Acer mono*, and seeds of *Picea glehnii* in June; leaves of *Populus nigra*, leaves and seeds of *P. maximowiczii*, catkins of *B. platyphylla*, seeds of *U. davidiana*, fruits of *Morus bombycis*, and seeds of *L. leptolepis* in July; catkins of *B. platyphylla*, seeds of *Q. dentate,* leaves of *Acer ginnala*, and seeds of *L. leptolepis* and *P. glehnii* in August; seeds of *Alnus japonica*, seeds of *Quercus crispula* and *P. glehnii*, and buds and catkins of *B. platyphylla* in September; leaves of *P. nigra*, catkins of *B. platyphylla*, and seeds of *Q. crispula* and *P. glehnii* in October; buds of *L. leptolepis*, leaves of *P. nigra*, seeds of *Pinus koraiensis* and *P. glehnii*, and buds of *L. leptolepis* in November; buds of *B. platyphylla* in January [226, 4049]. Also, buds of *A. japonica* and *U. davidiana* are eaten during winter [4049].

Habitat Inhabits coniferous, broad-leaved deciduous, and mixed forests from lowland to montane zone [1107]. In Obihiro of Hokkaido, this species occurs even in small groves [227].

Home range Home ranges of males and females are 2.2 ha and 1.7 ha, respectively [4049]. Also, by using GIS, ranges of males and females are estimated to be 4.84 ha and 1.69 ha, respectively in Obihiro, Hokkaido [3655]. Females have a territory in the reproductive season, while males do not have clear territory, showing overlapped range with other male and female individuals [4049]. In natural forest of Furano, Hokkaido, the population density is estimated to be 2 individuals/ha [3366].

Behavior Nocturnal and arboreal. The Siberian flying squirrel glides with gliding membranes: in Hokkaido, its mean launch height, landing height, and horizontal glide distance are 14.4, 2.7, and 21.4 m, respectively [3360]. Activity varies in seasons [3971, 4057]. In May–June and September–October, activity is bimodal, while unimodal activity is found in July–August. In November–December, 2 or 3 short activity periods alternate with long rest periods. Also, the activity pattern of lactating females is different from those of females during non-breeding season and males: lactating females return to the nest several times for lactation during nocturnal activity. Siberian flying squirrels usually nest in tree hollows [4049]. The Siberian flying squirrel is the most dominant user of cavities excavated by great spotted woodpeckers (*Dendrocopos major*) [1721]. In Obihiro, Hokkaido, average nest cavity height above ground is 5.4 m [225]. In natural forests of Hokkaido, nest cavities are most frequently found in *Abies sachalinensis* [1255, 1989]. Bark of vines such as *Vitis coignetiae* is usually utilized as nest material [4049]. During winter (from November to early March), 3–5 individuals may use a single nest cavity together [4049].

Natural enemies Owls, sables, foxes, and domestic cats prey on Siberian flying squirrels [4049]. Siberian flying squirrels are sometimes found in feces and stomach contents of sables [2314].

Parasites The parasitic nematode *Citellina petrovi* is reported [209]. *Echinococcus multilocularis* infection is recorded, although Siberian flying squirrel is not a common intermediary host [2659]. In Finland, 2 flea species (*Tarsopsylla octodecimdentata* and *Ceratophyllus indages*) are reported [848].

5. Remarks

In Siberia, this species was hunted for commercial use of fur [741]. This species has never been used for game hunting in Japan. In the legend of Ainu, the native people of Hokkaido, this species plays a role of guardian deity for children. Ainu people call this species "*A-kamui*".

In a worldwide phylogeographical study, 3 major mtDNA lineages were recognized: "Hokkaido", "Far Eastern" (= northeastern Asia) and "northern Eurasian" phylogroups [2749]. The divergence data among them indicate that the Hokkaido group separated earliest from the other groups.

T. OSHIDA

A Siberian flying squirrel returning to its nest at dawn in February in Sapporo, Hokkaido (T. Kuwahara, Kon Photography & Research).

Checking a nest box set for investigating a Siberian flying squirrel on a trunk of *Abies sachalinensis* in Furano, Hokkaido (A. Kato).

RESEARCH TOPIC 4

Acorns as a resource for forest-dwelling mammals in Japan
T. SHIMADA

1. Introduction

Acorns, the seeds of oaks (genus *Quercus*), are considered a staple resource for forest-dwelling mammals. Generally, they are relatively large and potentially digestible, thereby providing high levels of energy per seed [3773]. In addition, they are less subject to decay and degradation than seeds with a high water content such as berries [2046], thus are advantageous for consumers because they are available for extended periods. However, acorns are not a reliable food source, as acorn crops fluctuate greatly from year to year [952, 2045, 2956]. The intermittent production of such large seed crops is known as *masting*.

Population ecologists have focused on fluctuations in acorn crops as a key factor of population dynamics of mammals, particularly rodents. In some instances, they have found synchrony between acorn production and population levels [419, 2045, 2956]. This synchrony has been considered to demonstrate the importance of acorns as a regulating factor of population dynamics. In North America and Europe, the accumulation of long-term data has provided insights into the relationship between acorn crops and mammalian population dynamics. Until recently, such analyses have not provided clear results in Japan. Nonetheless, the idea that acorns have an important role in the population dynamics of mammalian consumers is accepted *a priori* in Japan.

However, recent research demonstrated that significant differences in nutritional characteristics exist between acorns in Japan and those in other regions, especially North America [3214]. Thus, it is not appropriate to assume that interrelationships between acorn crops and the population dynamics of consumers will be the same in Japan as in other regions.

In this topic, I will review the significance of acorns as a resource for mammals in Japan from the following viewpoints of the ecological and nutritional characteristics of acorns and the responses of mammals to fluctuations in the acorn crop.

2. Ecological and nutritional traits of acorns in Japan

The genus *Quercus* includes approximately 500 species worldwide, that are classified into three subgenera: *Lepidobalanus*, *Erythrobalanus*, and *Cyclobalanopsis* [1750]. Species in *Lepidobalanus* are distributed in North America, Europe, and Asia, whereas species in *Erythrobalanus* are restricted to North and South America. *Lepidobalanus* and *Erythrobalanus* in North America are known as the white and red oaks, respectively. *Cyclobalanopsis* is the group known as evergreen oaks, which are native to eastern and southeastern Asia.

More than 20 species of *Quercus* (subgenus *Lepidobalanus* and *Cyclobalanopsis*) are distributed widely in Japan, from subtropical to cool temperate zones [2674]. *Quercus* is one of the dominant genera of trees in Japanese temperate forests, and species in this genus are found in various stages of forest succession.

Acorns are characterized primarily by their large size. In most species, the average acorn mass exceeds 1 g, and acorns of some species reach almost 5 g [3773]. The gross energy content of acorns per unit mass is not particularly high (ca. 18 kJ/g) [3214]; for example, their gross energy content is slightly more than half that of beechnuts (from *Fagus crenata*), another staple mast species in Japan (27.2 kJ) [3317]. However, the gross energy per seed is much higher in acorns than in beechnuts because of their large seed size (e.g., *Q. crispula*, 1.7 g dry mass, 24.2 kJ per seed; *F. crenata*, 0.13 g dry mass, 2.8 kJ per seed) [951]. Thus, the nutritional value of acorns for mammals partly can be ascribed to their large size.

Carbohydrates are the primary nutrient in acorns and account for an average of more than 70% by dry weight (Table 1). Fat and protein contents vary widely among species. However, most species within a subgenus have similar values that differ from those of other subgenera. In addition to the nutritional elements, acorns of some species of oak accumulate high levels of tannins. Tannins, a diverse group of water-soluble polyphenolics with a high affinity for proteins, are widely distributed in various plant parts [3856]. They have several detrimental effects on consumers of these plant parts: reduction in protein digestibility [392, 2888], damage to the gastrointestinal mucosa and epithelia [278], kidney or liver failure [580], and endogenous nitrogen loss [278, 305, 3213]. The differences in the nutritional compositions between white and red oaks are well documented in North America [1593]: the acorns of white oaks (*Lepidobalanus*) contain lower levels of fat and tannins than are found in red oaks (*Erythrobalanus*).

As a result of these contradictory characteristics (acorns contain high levels of both nutrients and potentially noxious defensive chemicals), the value of acorns as a food source is not as easy to determine as has been believed. Recently, a detailed review of the chemical compositions of acorns of 27 species of oak from Japan, Europe, and North America indicated that acorns could be divided into three types [3214]: Type 1 acorns (exclusively in *Erythrobalanus*) are high in tannins but also high in fat and proteins (consequently, they are rich in included energy per unit); Type 2 acorns (found in *Lepidobalanus* and *Cyclobalanopsis*) are high in tannins but low in fat and proteins; and Type 3 acorns (mostly found in *Lepidobalanus*) are low in tannins and have intermediate levels of fat and proteins. Type 2 acorns must have the lowest nutritional value for consumers, because they are characterized not only by low fat and protein contents but also by high levels of defensive chemicals. Type 3 acorns

Table 1. Nutritional composition and contents of phenolic compounds in acorns of *Quercus* spp.

Region	Subgenus	Latin name	Type of acorns[1]	Protein (% DW)	Fat (% DW)	Fiber (% DW)	Ash (% DW)	Carbohydrate (% DW)	Total phenol[2] (% DW)	Gross energy[3] (kJ/g seed mass)	Reference
Japan	Lepidobalanus	*Q. acutissima*	3	5.2	4.6	1.9	1.8	86.5	6.0	17.9	[3214]
		Q. aliena	2	5.3	2.8	1.6	1.5	88.8	6.7	17.7	[3214]
		Q. crispula	2	4.4	1.7	2.1	1.5	90.3	11.7	17.3	[3213]
		Q. serrata	2	4.5	2.5	1.9	2.8	88.3	7.3	17.3	[3213]
		Q. variabilis	3	5.1	5.3	1.8	2.3	85.5	3.2	18.0	[3214]
	Cyclobalanopsis	*Q. gilva*	3	2.6	3.4	1.3	1.9	90.8	1.9	17.6	[2025] (transformed to % of dry weight)
		Q. glauca	2	3.1	3.2	1.5	2.7	89.5	7.5	17.4	[2025] (transformed to % of dry weight)
		Q. myrsinaefolia	2	3.0	3.4	1.9	2.9	88.8	7.6	17.3	[2025] (transformed to % of dry weight)
North America	Lepidobalanus	*Q. alba*	3	6.3	6.3	2.5	2.6	82.3	5.6	18.2	[3810]
		Q. prinoides	3	7.6	6.3	2.4	2.0	81.7	4.4	18.4	[3810]
		Q. prinus	2	6.9	5.1	2.6	2.2	83.2	10.4	18.0	[3810]
	Erythrobalanus	*Q. ilicifolia*	1	10.3	20.0	3.0	2.1	64.6	11.3	21.5	[3810]
		Q. rubra	1	6.6	20.8	3.1	2.4	67.1	9.8	21.3	[3810]
Europe	Lepidobalanus	*Q. petraea*	3	5.9	4.9	1.9	2.1	85.2	5.3	18.0	[3214]
		Q. robur	3	6.2	4.9	1.8	1.6	85.5	5.2	18.1	[3214]

1: See text for details.
2: Total phenols, an index of tannin content, was measured using the Folin-Denis method and related techniques.
3: Gross energy was estimated from the contents of protein, fat, and carbohydrates.

would seem to be more palatable, because they contain a lower quantity of tannins than Type 2 acorns, although their nutrient levels are similar. Type 1 acorns contain as much tannins as Type 2 acorns, but they also contain a higher level of energy than Type 2 acorns owing to their high contents of fat and proteins. Animals tend to follow a high-fat diet before they overwinter because they must store sufficient body fat as energy reserves. Body fat also can be synthesized from carbohydrates, but the cost of fat synthesis can be reduced by directly storing the metabolites of dietary fat [2888]. In this sense, fat-rich acorns likely are a precious food for wildlife. As a consequence of the different balances of nutrients and defensive chemicals, these three acorn types are not nutritionally equivalent; that is, the food value for consumers likely differs among acorn types.

European oaks (*Q. robur* and *Q. petraea*) are classified as having Type 3 acorns, whereas North American oaks generally belong to Type 1 (all *Erythrobalanus*) or Type 3 (all *Lepidobalanus*), with the exception of *Q. prinus* (Type 2). It should be noted that except for *Q. prinus*, Type 2 acorns are limited to Japan. Five of eight Japanese species of oak produce Type 2 acorns; the remainder have Type 3 acorns. As the Type 2 group includes *Q. serrata* and *Q. crispula*, which are the dominant species in Japanese temperate forests, acorns in Japan generally are characterized both by high levels of tannins and by low fat and protein levels. This suggests that a large proportion of Japanese oaks produce acorns with lower nutritional qualities for forest-dwelling mammals than is true for European and North American oaks.

3. Responses to the fluctuation of acorn crops

The above discussion suggests that consumers of acorns in Japan may respond differently than those in other regions to the fluctuation of acorn crops. Many mammalian species in Japan consume acorns: most rodents (e.g., field mice in the genus *Apodemus* [3210]; voles in the genus *Myodes* (formerly *Clethrionomys*) [2943]; squirrels in the genera *Sciurus* and *Callosciurus* [3486, 3500]; chipmunks in the genus *Tamias* [1466]; giant flying squirrels in the genus *Petaurista* [1474]), Japanese macaques (*Macaca fuscata*) [4239], sika deer (*Cervus nippon*) [3421, 3860], wild boars (*Sus scrofa*) [1654], and some Carnivora species, such as Japanese black bears (*Ursus thibetanus*) [830, 2368] and brown bears (*U. arctos*) [3091]. However, there have been few quantitative studies of the impact of acorns on mammalian behavior and population dynamics in Japan, and these studies have been limited mostly to rodents [952, 2956, 2960].

A recent study reviewed the responses of rodent populations to acorn masting considering the nutritional types of acorns previously mentioned, although the analyzed data were restricted to five species because of the limited published literature: *Q. robur* and *Q. petraea* (Europe, Type 3), *Q. alba* (North America, Type 3), *Q. rubra* (North America, Type 1), and *Q. crispula* (Japan, Type 2) [3214]. They found that masting of Type 1 and 3 acorns generated positive responses in rodent populations in most instances, such as improved overwinter survival, increased reproduction during the winter, and increased population density during the following year, whereas positive responses were not observed after the masting of Type 2 acorns. The traits of Type 2 acorns, namely their high tannin content and low fat and protein contents, may make it difficult to generate positive responses in rodent populations even when masting provides a plentiful resource.

More recently, however, significant positive relationships between Type 2 acorns and rodent population density have been reported [2956, 2960]: a population of field mouse *A. speciosus* evidently followed acorn dynamics of *Q. crispula* with one year lag, but populations of other sympatric species (*A. argenteus* and *M. rufocanus*) did not. Saitoh and coworkers hypothesized these differences might result from different tolerances for dietary tannins in the three species of rodents. Feeding on the acorns of *Q. crispula* potentially can cause detrimental effects for the consumers [3213]. However, *A. speciosus* can overcome the damage caused by acorn tannins through acclimation mediated by an increase in levels of tannin-binding salivary proteins and of tannase-producing enterobacteria [3216]. Thus, *A. speciosus* can consume acorns efficiently due to its ability to counteract the adverse effects of tannins. In contrast, *A. argenteus* and *M. rufocanus* may have weaker abilities to counteract the effects of tannins, although the tannin tolerance of these species is unknown. This study further supports the idea that Type 2 acorns may not be a palatable resource for rodent consumers, especially those without specific abilities to counteract the effects of tannins.

The situation for other mammals is less clear. Large mammals also consume large quantities of acorns. Positive impacts of large acorn crops on their behavior and population dynamics reported in Europe and North America were increased survival of offspring and an increased probability of successful reproduction (white-tailed deer [*Odocoileus virginianus*] [541]; American black bear [*U. americanus*] [3774]). The observation that acorns form a large part of the autumn food supply also has been reported for large Japanese mammals: Japanese black bears [2664], brown bears [3091], wild boars [1654] and sika deer [3421]. However, studies on ecological responses of those mammals are limited in Japan.

In autumn, Japanese black bears appear to rely heavily on acorns and other kinds of hard masts, such as beechnuts and chestnuts [831, 1742, 2169, 2368]. Indeed, black bears were demonstrated to change their home range use according to the abundance of *Q. crispula* acorns; they tended to expand home ranges to lower elevations in poor-mast years [1742]. Similarly, the damage to agricultural crops caused by brown bears native to Hokkaido increases in the year following a crop failure of *Q. crispula* acorns [3091]. This may indicate that acorns of *Q. crispula* form a staple food also for brown bears. In spite of these observations, the relationship between acorn crop and bear population dynamics has not been clarified.

4. Conclusions

The published literature suggests that the value of acorns as a food resource varies markedly among species of *Quercus* and among consumer species. Consequently, the responses of mammals to the fluctuation of acorn crops also may differ among species of oak. This emphasizes that findings from other regions should not be applied directly to the situation in Japan, and vice versa. In addition, a larger body of reliable data comparable with data from other regions must be accumulated to supplement our knowledge of this aspect of Japanese mammalogy.

The Japanese field mouse eating acorns.

Lagomorpha

Lagomorpha   OCHOTONIDAE

Red list status: NT as *O. h. yesoensis* (MEJ); R as *O. h. yesoensis*, LP for the Yubari Mts. population (MSJ); LC (IUCN)

Ochotona hyperborea (Pallas, 1811)

EN northern pika   JP キタナキウサギ（kita nakiusagi）   CM 东北鼠兔   CT 極北鼠兔，亞洲鼠兔
KR 우는토끼   RS северная пищуха

A northern pika in Shikaoi-cho, Hokkaido (K. Onoyama).

1. Distribution

Ural, Putorana, the Sayan Mts., east of Lena River to the Koryak area (Chukotka), Koryatsk and Kamchatka, upper Yenesei, Transbaikalia, Amur regions, and Sakhalin, Russia; north-central Mongolia; north-eastern China; North Korea; and Hokkaido, Japan [924]. In Hokkaido, the northern pika occurs in the Daisetsu Mts. and the Hidaka Mts. [1108]. The Hokkaido population is regarded as a subspecies, *O. h. yesoensis*.

2. Fossil record

Not reported in Japan. In the Late Pleistocene of Eurasia, the northern pika inhabited the area around Denisova Cave in the Altai Mts., Russia, but went extinct at the Pleistocene–Holocene boundary, because the taiga communities were replaced with steppe communities [1854].

3. General characteristics

Morphology   Fur color is reddish brown in summer and grayish to dark brown in winter. Ears are short and round. External measurements (range) are as follows: HB = 130–190 mm, T = 5–12 mm, HFsu = 24–27 mm, E = 15–20 mm, BW = approximately 150 g [1108].

Dental formula   I 2/1 + C 0/0 + P 3/2 + M 2/3 = 26 [29].

Mammae formula   1 + 0 + 1 = 4 or 2 + 0 + 1 = 6 [1011].

Genetics   Chromosome: $2n = 40$ (XX and XY) [869, 3785]. Nucleotide sequence data of mitochondrial control region [576], cytochrome *b* gene [1853, 2507], and cytochrome *c* oxidase gene subunit I [1853] are reported.

4. Ecology

Reproduction   In Hokkaido, breeding season is from spring to summer, and females reproduce once a year [1108]. Litter size ranges from 1 to 9 [3277]. In Hokkaido, the average litter size is 3.1 (range 1–5) [1472].

Lifespan   About 1 to 3 years [3277].

Diet   In Hokkaido, northern pikas eat leaves, stems, flowers, ferns, mosses, and mushrooms [1058, 1108, 2961]. From summer to autumn, they store large amounts of vegetation (stems and leaves) in the spaces between rocks or under tree roots in preparation for winter [749, 1058, 1470].

Ochotona hyperborea (Pallas, 1811)

Habitat In Hokkaido, the northern pika occurs from 50 m to near the top of Mt. Hakuundake (2,230 m) in elevation [1431] as long as there is suitable habitat: mainly in the alpine zone over 800 m in elevation [1108]. They inhabit rocky terrain in forests with moss laced rocky slopes [1062, 1470]. Detailed information on habitat of the northern pikas occurring in Hokkaido is recently presented [1430, 1432].

Home range An adult male and an adult female occupy a joint territory in which they remain sedentary [1470, 1471]. Each individual tends to stay in a fixed area for much of the day [1471]. Both sexes defend territories from individuals of the same sex. Following the death of territory's occupants, the territory's boundaries tend to remain the same even after resettlement by other individuals [3277].

Behavior It is reported that northern pikas are active day and night with peaks in the morning and evening [1473], however, a recent study [2660] suggests that they are more active during nighttime than daytime without bimodal activity peaks in the day, and with nocturnal activity highly conspicuous in autumn. Males utter a long call, and all individuals, irrespective of age and sex, produce a short call and a trill [1473]. The short call is used for communication between males and females: in spring, it is used only by females, while in autumn, both sexes use the short call. The spectrum of male long calls is individually specific [1678]. This individual variation doesn't differ with aging and through the reproductive and non-reproductive seasons [1678]. The spectrum of short calls varies geographically [577].

Natural enemies Not reported in Japan. In northeastern Yakutia of Russia, sables (*Martes zibellina*) consume significant numbers of northern pikas [2923].

A northern pika eating leaves in a forest of Shikaoi-cho, Hokkaido (T. Kuwahara, Kon Photography & Research).

Parasites In Japan, larva of the parasitic dipteran *Oestromyia leporina* have been reported [3315]. Outside Japan, protozoans *Eimeria banffensis*, *E. circumborealis*, *E. calentinei*, *E. klondikensis*, *E. princepsis* and *E. worleyi* are reported [917].

5. Remarks

In the 1930s, the Hokkaido population was considered a forest pest and was controlled [1062]. In the 1990s, a heated controversy about a road plan in a major habitat of the Hokkaido population continued for several years until the plan was finally discontinued to conserve the natural habitat that the northern pika occupied [932].

T. OSHIDA

A rocky bank habitat of the northern pika in Shikaoi-cho, Hokkaido (S. Sato).

Lagomorpha   LEPORIDAE

Red list status: EN (MEJ); En (MSJ); EN (IUCN)

Pentalagus furnessi (Stone, 1900)

EN Amami rabbit   JP アマミノクロウサギ(amamino kuro usagi)   CM 琉球兔   CT 奄美短耳兔，琉球兔，奄美黑兔
KR 아마미검은토끼   RS японский древесный заяц, лазающий заяц

A young Amami rabbit in subtropical forest on Amami-ohshima Isl. (H. Katsu).

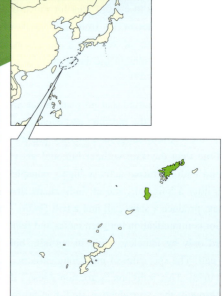

1. Distribution

The Amami rabbit is endemic to Amami-ohshima Isl. (712 km^2) and Tokuno-shima Isl. (248 km^2) in the Ryukyu Arch., in southern Japan [1011, 1479, 3675]. Both islands are located in the subtropical zone. Its vertical range is from sea level to 694 m (the highest point of the island) on Amami-ohshima Isl. and to 645 m (the highest point of the island) on Tokuno-shima Isl.

2. Fossil record

Two upper molariform teeth (right M^1 and left P^3) reported from Tokuno-shima Isl. constitute the first fossil record and date to the Late Pleistocene [3605]. Recently, fossils of *Pentalagus* spp. were reported from Okinawa-jima Isl. and these date to the Early Pleistocene (1.7–1.5 million years ago) [2795].

Fossils of *Pliopentalagus,* which is ancestral to the current Amami rabbit, are found in Huainan, Anhui Province, China, and date to the Late Miocene (ca. 6 mya) and Late Pliocene (ca. 3 mya) [3604]. Fossils of *Pliopentalagus* have also been recorded from Moldavia and Slovakia in Europe [439, 743].

3. General characteristics

Morphology   The Amami rabbit has been reported to have the most primitive characteristics in the family Leporidae [405]. The hair is very dark in color, with a soft plumbeous under-fur. The long hair is coarse and hispid, brownish black, with buff annulations or tips, becoming mahogany on the rump and brighter yellowish brown on the feet, except about the base of claws and tail [3311]. The heavy, curved claws are unusually long for rabbits (10–20 mm). Mean external measurements for 4 male adults and 3 female adults from Amami-ohshima Isl. are as follows [3928, 3930]: TL = 451 mm (range 430–470), 452 mm (397–530); E = 44 mm (40–50), 45 mm (42–49); HFsu = 86 mm (80–92), 89 mm (83–92); T = 27 mm (20–35), 30 mm (25–33); BW = 2,226 g (2,030–2,675), 2,477 g (2,000–2,880). Measurements of 1 male adult from Tokuno-shima Isl., are: TL = 470 mm; E = 44 mm; HFsu = 85 mm; T = 25 mm; BW = 2,240 g [3928, 3930].

Dental formula   I 2/1 + C 0/0 + P 3/3 + M 3(2)/3 = 30 (28) [29, 1014, 3930].

Mammae formula   1 + 1 + 1 = 6 [2027, 3923].

Genetics   Chromosome: $2n = 46$ and FN (including the X chromosome) = 80 [795]. No other leporid has the same diploid number. Nucleotide sequences of mitochondrial 12S ribosomal RNA and cytochrome *b* genes appear in the DDBJ/EMBL/GenBank with the accession numbers AB058603–AB058606, AB058608, AB058609, and AB058614 [3937]. A phylogeny of molecular data shows that *P. furnessi* appeared during the first generic radiation of the leporids in the Middle Miocene along with the other leporid genera [3937].

4. Ecology

Reproduction Litter size is generally 1, and parturition takes place during late March to May and September to December. Gestation period and timing of sexual maturation are unknown. A female may deliver 2 or 3 times a year [2980]. Each mother digs a burrow for nursing juveniles ca. 1 week before parturition. The length of the burrow is 150 cm, with a chamber (30 cm in diameter) full of leaves in the back. The Amami rabbit is altricial, and juveniles stay alone in the chamber during the daytime. Mothers visit their nursing burrow after 20:00–21:00 h and remove the soil on the entrance for suckling the juvenile. After suckling, the entrance (15 cm in diameter) is covered with soil again and camouflaged with twigs and leaves, which takes only 30 sec. When a juvenile grows to some degree, the mother stops covering the entrance with soil. There are 2 observations on the period of covering; one continued for 27 days after parturition, and the other for 47 days [2980].

A neonate 2 days after birth in captivity weighs 100 g, has short brown hair in the body, closed eyes and ears, erupted incisors, 3 pairs of nipples, and grown nails, the tips of which are white, filamentous and curled [2027]. The 3–4 month juvenile is 25–35 cm long, and the mother aggressively forces it from both the nursing burrow and mother's burrow for independence [2980].

Lifespan Not reported.

Diet The Amami rabbit mainly eats sprouts, young parts and cambiums of trees, and a wide range of plant species [3928, 3930]. It feeds on more than 29 species of plants including 12 species of herbaceous plants, *Adenostemma lavenia*, *Carex* sp., *Miscanthus sinensis*, *Peucedanum japonicum*, *Mosla dianthera*, etc., and 17 species shrub plants, *Castanopsis sieboldii*, *Melastoma candidum*, *Rubus sieboldii*, *Styrax japonica*, *Zanthoxylum ailanthoides*, etc. It also feeds on acorns of *Castanopsis seiboldii* from autumn to winter [3928, 3930].

Habitat When the Islands were originally covered by dense primary forests, the Amami rabbit lived mainly in primary forests, but after deforestation in the 1970s and 1980s, when 70–90% of primary forests were lost, the rabbit began to inhabit cut-over areas and forest edges covered by *Miscanthus sinensis* [3318] in addition to what remained of the primary forest. According to radio-telemetry surveys, the Amami rabbit is active mainly at night, moving around to feed and to drop fecal pellets in open places, such as forest roads where food plants are rich. Most activity occurs 100–200 m away from their burrows, which are usually located in small valleys covered by dense forests [3936].

Home range The average size revealed by radio-telemetry surveys is 1.3 ha for 4 males and 1.0 ha for 3 females [3936]. The home ranges of the females do not overlap with each other, whereas those of males overlap with those of other males as well as with those of females.

Behavior The Amami rabbit has a unique communication system involving vocalizations like pikas (*Ochotona*) and the use of their hind limbs for ground beating [3928, 3930]. At dusk, rabbits appear at the entrances of their burrows before they become active, and make calls that can be heard loud and clear in small valleys. A mother also vocalizes to attract her offspring when approaching her nursing burrow.

Natural enemies The only native predator is the habu pit viper (*Protobothrops flavoviridis*). Originally, there were no natural mammalian predators on the islands. However, invasive predators such as feral dogs and cats now occur on both Islands. In addition, the small Indian mongoose (*Herpestes auropunctatus*) was introduced in 1979 into Amami-ohshima Isl., in an attempt to control the habu pitviper [3927].

Parasites Amami rabbits host 13 species of 8 genera of trombiculid mites and 5 species of 3 genera of ticks [1625, 3930]. *Cordiseta nakayamai* and *Walchia pentalagi* are host specific to the Amami rabbit and can be used to distinguish active burrows from inactive ones [3342]. The trematode *Ogmocotyle* sp. and a kind of cestode (Anoplocephalidae gen. sp.) are reported as endoparasites [1288]. Systemic protozoal infection (probably toxoplasmosis), purulent bronchopneumonia due to Gram-negative bacilli infection and fibrinous pericarditis, accumulations of foamy macrophages (suspected endogenous lipid pneumonia), focal fungal pneumonia, focal pyogranulomatous pneumonia, pulmonary abscess, and renal abscess are found [1754].

5. Remarks

The size of its distributional range, as estimated by fecal pellet counting during 1992–1994, is 370.28 km^2 (52% of the Island) on Amami-ohshima Isl. and 32.97 km^2 (13%) on Tokuno-shima Isl. [3318, 3319]. The size of its range on Amami-ohshima Isl. in 1992–1994 was 20–40% smaller than those estimated in 1974 and 1977. In addition, 1 fragmented population was recorded therein (Fig. 1).

Impact of habitat loss is quite severe for the Amami rabbit, because extensive logging operations on the 2 islands have resulted in the area of old forests being reduced to less than 10–30% of its 1980 extent [3318]. Furthermore, the impact of invasive predators such as feral dogs, cats and mongoose is severe, and road kills often occur [3927].

A new program of mongoose eradication was restarted 2005 in Amami-ohshima Isl. by the Ministry of the Environment to protect the ecosystem of the island including the Amami rabbit [3927, 3928]. Consequently, recent mongoose populations have been very low, and native species, including endangered species, have recovered [618, 3322, 3854].

F. YAMADA

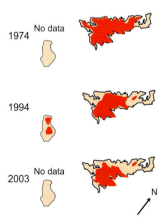

Fig. 1. Decrease of distribution of *Pentalagus furnessi* from 1974 to 2003 on Amami-ohshima Isl. (right), and distribution on Tokuno-shima Isl. in 1994.

Lagomorpha   LEPORIDAE

094

Red list status: C-1 as *L. t. ainu* (MSJ); LC (IUCN)

Lepus timidus Linnaeus, 1758

EN mountain hare   JP ユキウサギ(yuki usagi)   CM 雪兔   CT 雪兔   KR 눈토끼   RS заяц-беляк

A mountain hare in winter coat standing with alert posture in Sarobetsu Moor, Hokkaido (T. Fujimoto).

1. Distribution

Palearctic from Scandinavia to eastern Siberia except eastern Chukot (Russia), south to Sakahalin and Sikhote-Alin Mountains (Russia); Hokkaido (Japan); Heilungjiang, North Xinjiang (China); North Mongolia; Altai, North Tien Shan Mountains; North Ukraine, eastern Poland, and Baltics; isolated populations in the Alps, Scotland, Wales and Ireland. Introduced into England, Faeroes and Scottish Isls. [924]. In addition to the Hokkaido mainland, it is distributed on the following related islands: Kunashiri and Etorofu Isls. The Hokkaido population is regarded as a subspecies, *L. t. ainu* [258, 1479, 1589]. It has been introduced into the Faeroes in Denmark, England and various Scottish Islands, Germany, Switzerland, etc. [561, 924, 1859].

2. Fossil record

Not reported in Japan. The genus *Lepus* arose in the early Middle Pleistocene. Fossils of *L. timidus* appear in the Pleistocene in Hungary and from Pyrenees to the southeastern Europe [158].

3. General characteristics

Morphology   The mountain hare is a large leporid, but it is smaller than *L. europaeus*, with shorter forelimbs, ears, and tail, but with longer hind feet. Pelage is pale brown over the back and white below. Summer fur is softer, grayer and in winter most individuals turn white [2580]. *Lepus t. ainu* is a larger hare than *L. brachyurus* in southern Japan. External measurements (range):
HB = 50–58 cm; HFsu = 16–17 cm; T = 5–8 cm; E = 7–8 cm; BW = 2,000–3,950 g, and skull length (males) = 97.2 and 98 mm [1011].

Dental formula   I 2/1 + C 0/0 + P 3/2 + M 3/3 = 28 [29, 1011, 1014].

Mammae formula   2 + 2 + 0 = 8 [1011].

Genetics   Chromosome: $2n = 48$ [744]. Chromosomes of *L. t. ainu*: $2n = 48$ and FN = 72. M = 10, SM = 6, ST = 10, A = 20, X = M and Y = A [437, 3635, 3645]. Data on nucleotide sequences of mitochondrial 12S ribosomal RNA and cytochrome *b* genes of *L. t. ainu* are available in the DDBJ/EMBL/GenBank with the accession numbers AB058607, AB058610, and AB058611 [3937].

A mountain hare molting from the winter coat to the summer coat in May 2008, Sapporo, Hokkaido (H. Hirakawa).

Lepus timidus Linnaeus, 1758

4. Ecology

Reproduction In Hokkaido, the breeding season extends from February to July with young born between April and August. Gestation period of *L. t. ainu* is 49.7 days (range 48–52 days), and litter size is 3.0 (1–8) [3189]. BW of the newborns is 118 g (77–157 g) [1681]. Females and males become sexually mature at 8 and 10 months, respectively [3189]. Pregnancies occur during February–August in Sweden, and during January–September in Scotland, juvenile females do not breed in the year of their birth, and gestation period is 50.3 days (47–55) with no difference between first (50.8) and second (50.0) litters in Sweden [158]. One litter of 6.4–6.9 young each year in northern Russia, and 3 litters of 5–6 young a year in southern Norway and Scotland [561]. At birth leverets weight ca. 100 g (61–182 g) [561]. Postpartum estrus with copulation a few hours after parturition is the rule, and superfetation is not common [158]. The mountain hare is precocial; at birth the young are furred, their eyes are open and they start suckling at once [158].

Lifespan In Hokkaido, average lifespan is 1.14 years, maximum is less than 4 years and 25% of newborn survive over 1 year after birth [2653, 3189, 3190]. Juvenile survival from birth to the following spring is very low, e.g. 20% in Sweden [158].

Diet The food eaten varies with the habitat. The food in Hokkaido includes grasses, twigs and the bark of trees from many species [596].

Habitat In Hokkaido, *L. t. ainu* is common in grassy fields, scrublands, and open forests from sea level up to the mountains [596]. In Eurasia, mountain hares occupy tundra and open forest, particularly of pine, birch and juniper, and tend to be replaced by the European hare *L. europaeus* on agricultural land [561].

Home range In Hokkaido, distance moved in 1 night is 1,364 m (1,129–1,599 m) and density of the hares estimated by tracking in snow is 0.059/ha (0.041–0.072/ha) [862, 3189]. Home range size in Hokkaido is not reported. Home range sizes vary from 10–30 ha in Scotland to 72–305 ha in Finland, and densities are also variable, such as 0.01/ha in Sweden, 0.02–0.06/ha in Finland [158]. There is a 3–4 year cycle in abundance in Fennoscandia and an 8–12 year cycle in Scotland and Russia [561].

Behavior In Hokkaido, *L. t. ainu* is solitary and nocturnal, and the hares confuse their tracks by jumping to the side before resting for the day [596]. The hare is primarily nocturnal, but in Europe shows increased daylight activity in summer when the nights are short or in winter when food is scarce [561]. In snow, hares dig to reach vegetation and also make short tunnels for protection from adverse weather or aerial predators.

Natural enemies In Hokkaido, the red fox (*Vulpes vulpes*) is a major predator, and decline of hare populations from the 1980s to the 1990s was due to red foxes [3189]. In Scotland, the most important mammalian predator of the hare is the red fox, 75% of all hares are taken by the fox [158]. Other predators in the wild are wildcats (*Felis silvestris*), dogs, stoats (*Mustela erminea*), golden eagles (*Aquila chrysaetos*), buzzards (*Buteo buteo*), and hen harriers (*Circus cyaneus*). In other countries, predators include pine martens (*Martes martes*), wolves (*Canis lupus*), goshawks (*Accipiter gentiles*), and eagle owls (*Bubo bubo*) [158].

Parasites In Hokkaido, the cestode (*Multiceps* (= *Coenurus*) *serialis*) is a reported endoparasite [4026]. *Toxoplasma gondii* is mortal and a zoonosis [3189]. The hare can be infected by *Francisella tularensis* causing Tularemia, and it is an important zoonosis for the hare and humans [610]. In Europe, viral haemorrhagic disease (VHD) is highly contagious to *L. europaeus* and *L. timidus* and is known as European brown hare syndrome (EBHS), first described in the early 1980s in northern Europe [585].

5. Remarks

The mountain hare in Hokkaido was an important game species, with more than 100,000 hares captured in the 1970s, while less than 1,000 were captured in the 1990s and zero in the 2000s [2089]. Damage to forest plantations and farm products was significant, especially during the 1950s–1970s.

Other common names for "mountain hare" in English include varying hare, arctic hare and snow hare [158].

F. YAMADA

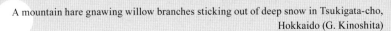
A mountain hare gnawing willow branches sticking out of deep snow in Tsukigata-cho, Hokkaido (G. Kinoshita)

Lagomorpha LEPORIDAE
095
Red list status: NT as *L. b. lyoni* (MEJ); C-1 (MSJ); LC (IUCN)

Lepus brachyurus Temminck, 1844

EN Japanese hare   JP ニホンノウサギ(nihon nousagi)   CM 日本兔   CT 日本野兔
KR 일본멧토끼   RS японский кустарниковый заяц

A Japanese hare in February in southern Kyoto Pref., where it does not turn white in winter (Y. Segawa).

1. Distribution

Endemic to Japan; Honshu, Shikoku, Kyushu and their islands including Sado Isl., the Oki Isls., the Goto Isls., the Amakusa Isls. and Shimokoshiki-jima Isl., except Hokkaido [1011, 1109, 1479].

2. Fossil record

Reported from the Middle Pleistocene in western Honshu [819, 1514].

3. General characteristics

Morphology The Japanese hare is a relatively small species of leporid [561]. Several color forms are recognized, varying from dark brownish-gray to reddish-brown, with variable amounts of white on the head and legs [1011]. In populations in snow zones on the Sea of Japan side of Honshu, and on Sado Isl., the fur color turns white in winter. The body color begins to change white from the middle of September to the end of November (except the tips of the ears which remain black) and begin to change brown from the end of January to May depending on photoperiod [2530, 2532, 2788]. HB = 45–54 cm, HFsu = 12–15 cm, T = 2–5 cm, E = 7.6–8.3 cm and BW = 2,100–2,600 g [3933]. Skull measurement: condylobasal length = 80.85 mm and zygomatic width = 44.04 mm [3933]. Variations in body size and cranial measurements are recognized among populations from islands and the mainland in Shimane Pref. [906].

Dental formula   I 2/1 + C 0/0 + P 3/2 + M 3/3 = 28 [29, 1011, 1014].

Mammae formula   1 + 2 + 1 = 8 [1682].

Genetics Chromosome: $2n$ = 48 and FN = 72; M = 10, SM = 6, ST = 10, A = 20, X = M and Y = A [3634, 3635, 3645]. Data on nucleotide sequences of mitochondrial 12S ribosomal RNA and cytochrome *b* genes are available in the DDBJ/EMBL/GenBank with the accession numbers AB058612, AB058613, AB058615, and AB058616 [3937]. Partial sequences of mitochondrial control (D-loop) region are also available (AB098026–AB098061 and AB158505–AB158509).

4. Ecology

Reproduction Breeding season and litter size vary between northern and southern populations of *L. brachyurus*. In the north (Yamagata Pref.), the breeding season extends from February to July with young born between April and August, but mostly in

Lepus brachyurus Temminck, 1844

May and June [2788], while in the south (Kagoshima Pref.) the breeding season extends all year [3525]. The average litter size is 1.86 (range 1–4) in the north and 1.16 (1–3) in the south [2788, 3525]. Gestation period is 42–43 days in the north and 45–48 days in the south [1681, 2788, 3525]. The weight at birth is 77–165 g in the north and 125–150 g in the south [1681, 3934]. Growth rate is faster and adult body weight is 20% heavier in the north, perhaps for surviving in winter [3932]. Females become sexually mature at 8–10 months after birth [2788, 3525]. The hare is an induced ovulator, and ovulation occurs 12–15 hr after stimulation of copulation [3932]. Parturition takes place in a shallow den (typically 5 cm in depth and 20 cm in diameter) dug by the mother. Parturition occurs speedily (2 mins), and a newborn young can run ca. 1 hr after delivery owing to precocity [3934]. Young are suckled at midnight only once a day for ca. 2 min [3934]. Young begin to feed on plants at day 8, but are nursed until 1 month after birth [3934].

Lifespan Average lifespan is 1.06 year, maximum is less than 4 years and 30% of newborns survive over 1 year after birth (Akita Pref.) [3187].

Diet Hares eat many plant species. More than 150 species of plants are recorded as food items [2788]. Hares eat over 64–75% of all plant species (53–59 species) which are found in forest habitats [3925]. Almost all grasses and forbs browsed by hares are eaten, but ca. 30% of twigs and shoots which are clipped by hares are not eaten because they are too thick and woody [3925]. Hares debark twigs more than 7 mm in diameter [3925]. Hares eat 200–500 g of plants per day, which is equivalent to 10–20% of their body weight [2788]. Hares also eat their soft feces to recover nitrogen content in the cecum (coprophagy) and some even eat hard feces if food is scarce [899].

Habitat Habitat varies greatly from seaside forests to mountainous areas and includes agricultural lands, forests, and meadows. They prefer open fields, forest edges and young forests which supply undergrowth as food and cover [3225]. However, the area of preferable habitat in mountainous areas has become smaller recently, because 40–50 year-old artificial coniferous forests where food and cover are not supplied sufficiently occupy most of their distribution range [3965, 4102].

Home range Home range size is 10–30 ha and distance of movement in a night is 1,176 m (841–1,729 m) [3611, 3619].

Behavior Solitary and nocturnal. The hare becomes active at about 19:00 h and returns to a resting place called a form at 07:00 to 08:00 h after it confuses its track by jumping to the side [3390]. Mating system is promiscuous. Males chase females and box to repel rivals. The time of copulation including mating behavior is very short (1–2 minutes) [3924].

A captive Japanese hare in winter phase in December in Nagano Pref. (T. Komiya). Fur of the hare turns white except for tips of the ears which remain black.

Natural enemies Red foxes (*Vulpes vulpes*), martens (*Martes melampus*), weasels (*Mustela itatsi*), and raptors (*Accipiter gentilis, Aquila chrysaetos, Spizaetus nipalensis*, etc.) are main predators; 36% of fox pellets and 13.8% of marten pellets contained bones and hair of the hare [3611].

Parasites & pathogenic organisms Cestodes *Mosgovoyia pectinata* and *Taenia pisiformis* are reported as helminths [3974]. Ticks *Haemaphysalis flava, Ixodes nipponensis* and *I. persulcatus* are reported as ectoparasites [598, 3408]. The hare can be infected by *Francisella tularensis* causing Tularemia, and it is an important zoonosis from the hare to humans [610].

5. Remarks

The Japanese hare is an important game animal, with about 0.6–1.2 million hares captured in the 1970s–1980s, and 30,000–200,000 hares in the 1990s–2000s [2089]. Damage to forest plantations and farm products has been significant especially in the 1950s–1970s.

Four subspecies are recognized; *L. b. angustidens* in the northern part of Honshu and on the Sea of Japan side of southern Honshu, *L. b. brachyurus* on the Pacific side of southern Honshu, and in Shikoku and Kyushu, *L. b. lyoni* on Sado Isl. and *L. b. okiensis* on the Oki Isls [1011]. *L. b. lyoni* is listed as the Near Threaten in the Japanese Red List (2007) [2086] because of predation of invasive Japanese marten *Martes melampus* introduced from mainland in 1960s and habitat loss [3225, 3929].

F. YAMADA

Lagomorpha   LEPORIDAE

Red list status: NT (IUCN); introduced

Oryctolagus cuniculus (Linnaeus, 1758)

EN European rabbit   JP アナウサギ（ana usagi）   CM 穴兔   CT 穴兔
KR 굴토끼   RS европейский дикий кролик

A feral rabbit on Nanatsushima-ohshima Isl. in Ishikawa Pref., which is a descendant of 2–3 pairs of a domestic breed released in 1984 (F. Yamada).

1. Distribution

In Japan, feral rabbits live on 13 islands (numbers correspond with those on the distribution map): 1) Oshima-ohshima and 2) Oshima-kojima in Hokkaido, 3) Nanatsushima-ohshima in Ishikawa Pref., 4) Jinai-to in Tokyo, 5) Mae-jima in Aichi Pref., 6) Matsu-shima of Ie-jima Isls. in Hyogo Pref., 7) Motoko-jima in Okayama Pref., 8) Hasa-jima in Kagawa Pref., 9) Ohkuno-jima in Hiroshima Pref., 10) Okino-shima of the Oki Isls. in Shimane Pref., 11) Ushibuka-ohshima in Kumamoto Pref., 12) Ie-jima of the Uji Isls. in Kagoshima Pref. and 13) Yanaha-jima and 14) Kayama-jima in Okinawa Pref. [861, 2977, 3921, 3926].

The European rabbit has spread all over the world except Antarctica and occurs in both wild and domestic forms (Fig. 1). They are present in most of Europe, North Africa, parts of South America, Australia and New Zealand, as well as on more than 800 islands, where they occupy a huge variety of ecosystems [560, 1859]. The original distribution was thought to be West and South Europe through the Mediterranean region to Morocco and North Algeria [924, 4265]. However, for most of its history, the European rabbit was confined to the Iberian Pen. where the species is supposed to have emerged in the Middle Pleistocene [549, 1863].

2. Fossil record

Not reported in Japan.

3. General characteristics

Morphology   Externally, the European rabbit is distinguished from most of the other sympatric leporids in Europe by its smaller size, relatively shorter ears lacking black tips and the white underside to its tail; also, unlike *Lepus*, the rabbit is altricial and young rabbits are usually born underground and are naked, blind and helpless [724]. In Japan, BW = 2,500 and 2,100 g for 2 males and 2,400 g for 1 female, HB = 448 and 687, and 532 mm, HFsu = 95 and 91.5, and 103 mm, T = 73 and 73, and 65 mm, E = 90 and 96, and 103 mm, respectively, in Ushibuka-ohshima Isl. [2977]. Body weights in native areas are 923–1,274 g in Spain and 1,017 g for males and 1,023 g for females in Portugal [2895]. In wild populations, HB = 350–450 mm, T = 40–70 mm, BW = 1,350–2,250 g [2515]. Not only has *O. cuniculus* become smaller in the course of its evolution over the last 15 millennia, it also exhibits a clear reduction in body size over its present distribution range from north to south in Europe [1863].

Dental formula   I 2/1 + C 0/0 + P 3/2 + M 3/3 = 28 [29, 1014].

Oryctolagus cuniculus (Linnaeus, 1758)

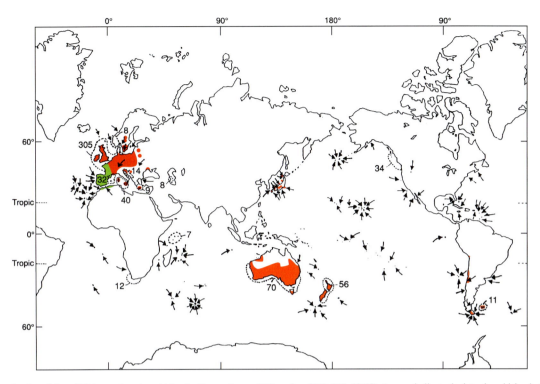

Fig. 1. Introduction of the rabbit to continents and islands all over the world (based on [560, 562, 4265]). Arrows indicate the introduced islands. Numbers for the areas enclosed by dotted lines are the total numbers of islands within area.

Mammae formula $1 + 2 + 1 = 8$ [751, 2977].

Genetics   Chromosome: $2n = 44$ for laboratory rabbits; M, SM or ST = 34, A or T = 8, X = SM and Y = SM [966]. Nucleotide sequences of mitochondrial 12S ribosomal RNA and cytochrome *b* genes are reported [3332].

4. Ecology

Reproduction   In Japan there are no observations for the rabbit in the wild. The European rabbit is typically a burrowing animal living communally in warrens. Birth peaks occur in spring when food production is at its maximum, and litter size is 3–9 young with a mean of 4–6 in mid-season [724]. Winter breeding is common in Mediterranean climates, where breeding ceases in late spring to mid-summer due to the dry season. The longest breeding seasons recorded are in the warm temperate zone of New Zealand where more than 50% of the adult females are pregnant for 9 months of the year [724]. Females show postpartum estrus and it lasts for 7 days. Gestation period is 28–30 days. Body weight is ca. 40–50 g at birth, and they remain in the nest for 20–21 days after which they are weaned. While young are in the nest, the mother (doe) visits her young to suckle only once a night for a short period (e.g. 5 mins). The number of young born per female per year ranges from 15 to 45. Males and females first become fertile at 3–4 months of age.

Lifespan   In Hokkaido, average lifespan is 0.99 years (maximum 3.5 years) and 28% of newborns survive more than 1 year after birth [861, 3188]. Survival and mortality show distinct regional differences. The survival rate of young in Spain (0.16) is much lower than in France (0.25–0.34), but adult survival rate in Spain (0.85) is much higher than in France (0.42–0.58) [2895]. More than 80% of the young are killed by predators in Australia [724]. In England, average lifespan is 1.45 years (maximum 4.5 years) and 42% of newborn survive over 1 year after birth [1857].

Diet   On Oshima-ohshima Isl., 14 out of 100 total plant species are selectively eaten, including *Aruncus dioicus*, *Maianthemum dilataum* and *Miscanthus sinensis* [861]. On Ushibuka-ohshima Isl., 19 out of 91 total plant species are selectively eaten [2977]. The rabbit eats mainly grasses and forbs when available; in Europe, 14–43 plant species are eaten, and they prefer gramineae to dicots and shrubs [2895]. Rabbits in Australia require a high-quality diet of less than 40% fiber with 10–12% protein for reproduction [3879].

Habitat   In Japan, feral rabbits mainly inhabit small offshore islands [3921, 3926]. Ideal rabbit habitat resembles conditions typical of the Iberian Pen.: a Mediterranean climate with a rainfall of less than about 1,000 mm per year, short herbage and well drained, loosely compacted soils that are easily dug, or with secure refuge areas in scrub adjacent to feeding grounds [724].

Home range   Not reported in Japan. Rabbits are strongly territorial and form social groups with a strict linear hierarchy of dominance [724]. Defended territories are usually small, often less than 1 ha; but if food is limited or distributed unevenly, rabbits may feed communally at night at distances of up to at least 500 m away from where they spend the day [724]. In Australia, rabbits spread over unoccupied territory at rates of up to 300 km a year, and in New Zealand, 15 km a year [724].

Behavior In Japan, rabbits are likely to be active in the early morning (5:00–7:00 h) and evening (15:00–19:00 h) [2977]. The dominant male (buck) serves several females (does), while some younger subordinates may live peripherally to the group and delay achieving a degree of dominance until they get older. At high density social groups may defend group territories, but at low density rabbits frequently live alone or in small groups of 2 or 3 animals. Neighboring groups may coalesce and feed together at night, or be joined by bucks visiting from other groups [724].

Natural enemies Not reported in Japan. In Europe all medium and large predators except the wolf depend on rabbits for food, sometimes exclusively, during the breeding season [2895]. The most important predators of young rabbits are mustelids and those of older rabbits are foxes in Europe.

Parasites & pathogenic organisms Not reported in Japan. Myxomatosis was introduced from South America, where the *Myxoma* virus is endemic and benign in local *Sylvilagus* sp., to Europe and Australia in 1950–1953 and transmission of the virus via flea and mosquitoes resulted in a tremendous reduction in the number of rabbits [543, 724]. Mortality from myxomatosis may be 10–50% according to year and season [2895]. Rabbit haemorrhagic disease (RHD) is a highly contagious and acute fatal disease of the European rabbit caused by a calicivirus (RHDV) [1828].

5. Remarks

The European rabbit divides into 3 groups: 1) an ancestral population in southern Spain, 2) separated populations in France and northern Spain from at least 50,000 years ago, and 3) populations from domestic rabbits and introduced rabbits in northern Europe [2895]. The European rabbit has one of the oldest fossil records, dating back to the Middle-Late Pliocene (3.5 million years ago) in the Iberian Pen. [1863]. *Alilepus* or *Trischizolagus*, which is a fossil genus, could be the ancestor of the European rabbit and the 4 parapatric species spread in Western Europe. The living species *O. cuniculus* is first recorded in the Middle Pleistocene (around 0.6 mya) in southern Spain, and is the only survivor.

The European rabbit has long been highly popular as a game animal and for use as food. Deliberate introduction of the species began in the Mediterranean region during Roman times, and in England and Ireland in the 12th century, and in the islands on the southern oceans in the late 18th and early 19th centuries by whalers eager to supplement stocks of food [560, 724]. It was also successfully introduced to Australia in 1859 and to New Zealand a few years later [560]. More than 150 domestic breeds of *Oryctolagus* probably originated in French monasteries between A.D. 600 and A.D. 1,000 are of considerable economic importance for meat, skin and fur, as laboratory animal, and as pets [560]. Because introduced rabbits are a major environmental and agricultural pest, a lot of strategic management has been carried out in their colonized areas (e.g. [2520]).

F. YAMADA

Soil erosion after overgrazing by feral rabbits on Nanatsushima-ohshima Isl., Ishikawa Pref. (F. Yamada).

Carnivora

Carnivora   CANIDAE

Red list status: C-1 as *V. v. japonica* in "Hondo" and *V. v. schrencki* in Hokkaido (MSJ); LC (IUCN)

Vulpes vulpes (Linnaeus, 1758)

EN red fox   JP アカギツネ (aka-gitsune)   CM 赤狐   CT 赤狐，紅狐   KR 여우   RS обыкновенная лисица

A red fox in pastureland in spring in Nemuro, Hokkaido (K. Uraguchi).

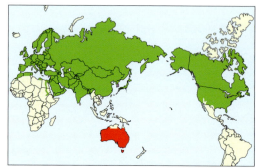

1. Distribution

Distributed throughout most of the northern hemisphere. The red fox has the widest distribution of all terrestrial carnivores. It has been introduced in Australia [1879]. In Japan, there are 2 subspecies: *V. v. schrencki* in Hokkaido, and *V. v. japonica* in Honshu, Shikoku and Kyushu [526].

2. Fossil record

Reported from the Middle and Late Pleistocene in Honshu [808, 819, 825, 1507, 2375, 3203].

3. General characteristics

Morphology   A medium-sized canid. Slender and pointed muzzle. Ears large, erect and pointed. Long and slender legs. Long, bushy and white-tipped tail. Fur is red-brown above and white in the underparts, chin and throat. Lower legs, feet and back of ears are black, but black on lower legs may be much reduced. Caudal scent gland usually marked by a conspicuous black patch. Guard hairs composed of white, yellow-brown and black bands, in various proportions depending on the part of the body. Underfur is gray. Melanic forms, called silver and cross foxes, are observed infrequently in Hokkaido. There are 2 moulting periods, but only that of spring is clearly visible. Body measurements are given in Table 1. Body sizes of the 2 subspecies in Japan are not significantly different. Males slightly larger than females but with considerable overlap. The width and the length of the cranium reach the maximum at about 4 and 9 months of age, respectively [3008]. All permanent teeth erupt by 7 months of age, corresponding to the age at which cranial growth to adult size and independence of offspring from parents are reached [3008].

Dental formula   I 3/3 + C 1/1 + P 4/4 + M 2/3 = 42 [1594].

Mammae formula   Normally 1 + 2 + 1 = 8 [3739].

Genetics   Chromosome: $2n = 34 + 3$–5 m [336, 2443, 2865]. The red fox demonstrates variation in chromosome number both between and within individuals. A few data for the mtDNA cytochrome *b* gene and control region are available [1050]. There are 2 lineages of mtDNA in Japanese foxes. One lineage is only found in Hokkaido, whereas the other lineage comprises haplotypes from Hokkaido, Honshu and Kyushu. Haplotypes from Hokkaido in the latter lineage appear to differ from those from Honshu and Kyushu [1050].

4. Ecology

Reproduction   Few data available in Japan. Monoestrous. Mating December–February [2429, 3661]. The onset of breeding is retarded in the higher latitudes [1856]. Gestation period is 52–53 days. Births occur March–May in Hokkaido, February–April in Kyushu [2429, 3661, 3739]. Vixens bear an average of 3 to 5 cubs in their den [1856, 2429, 3661, 3739].

Vulpes vulpes (Linnaeus, 1758)

Table 1. Body measurements of red foxes in Japan.

	*Vulpes vulpes schrencki**						*Vulpes vulpes japonica***					
	Males			Females			Males			Females		
	Mean	Range	n	Mean	Range	n	Mean	Range	n	Mean	Range	n
HB (mm)	652	568–705	150	623	388–690	109	648	555–700	60	614	515–670	46
T (mm)	376	313–440	150	356	250–413	109	374	314–420	58	357	285–420	45
HFsu (mm)	156	134–175	149	146	127–165	109	149	135–160	23	141	125–149	16
E (mm)	79	56–89	149	76	63–88	108	79	72–88	18	79	75–90	16
BW (kg)	5.1	3.5–6.7	150	4.2	2.9–5.7	108	5.1	4.0–6.6	17	4.4	1.9–6.1	17

* specimen collected from January to March in 1988–1991 in Hokkaido [3739].
** specimen collected in Tochigi Pref., central Honshu [3467].

Lifespan Proportion of animals less than a year old is about 70% and that of animals more than 4 years old is less than 5% in about 3,000 foxes shot in Hokkaido [3743]. The oldest age recorded in the wild is 14 years in Hokkaido [1940]. The mean longevity is 1.2–1.5 years in eastern Hokkaido, because of high mortality in the first year of life [4148].

Diet Seasonally and regionally varying. Red foxes are adaptable and opportunistic omnivores, with a diet ranging from invertebrates to mammals and birds, and vegetables. In Japan, they mainly eat small mammals, insects and fruits [18, 1705, 2094, 4142, 4143]. They also scavenge in rural areas on deer carcasses [1663]. In and around urban areas, garbage is also an important food source [1705, 2094, 3739].

Habitat Globally, red foxes inhabit tundra, deserts, sand dunes, forests, farmlands, grasslands and city centers, from coastal to alpine zones [1856, 1878, 3468]. Lack of specific habitat requirements is one of the reasons for their wide distribution. In Japan, they generally prefer mosaic patchworks of scrub, woodland and farmland. Such habitat offers a wide variety of cover and food. Foxes have intruded into urban areas in Hokkaido since the 1990s.

Home range Home range size varies, depending on habitat, from 100 to 800 ha in Japan [496, 2095, 3468, 3739], and is probably affected by food resources. They defend their home ranges as territories at least from spring to summer [3660]. In rural areas home ranges overlap during periods when food is abundant [3660].

Behavior Primarily nocturnal, but can be active during daytime. Vixens rearing cubs are active in the daytime and at night in Hokkaido [495, 496, 3468, 3659, 3739]. Red foxes can dig their own dens [2427, 2428, 3742], normally in banks, tree root systems, rocky crevices and even under buildings. In rural areas in Hokkaido, foxes prefer to den on slopes in woodland near grassland [3742]. Fox dens are mainly for breeding and rearing cubs.

Juveniles disperse in the autumn and winter of their first year, but some older animals may disperse, as well. Males are more likely to disperse, and disperse further than females [3628, 3739]. Dispersal distances are generally less than 6 km in the Nemuro Pen., eastern Hokkaido, but distances up to 50 km have been recorded in inland Hokkaido [3739]. Female cubs may remain on the parental range. These females, described as "helpers", play a role in rearing their mother's cubs in the next year [1876].

Natural enemies Few natural predators, although feral dogs may kill cubs.

Parasites & pathogenic organisms

Ectoparasite

Haemaphysalis flava [3973], *Ixodes persulcatus* [1597], *Sarcoptes scabiei* [3662]. Sarcoptic mange caused by *S. scabiei* is the most important and potentially fatal disease of foxes. This disease has been prevalent among foxes in Hokkaido since the late 1990s, and has apparently reduced some local fox populations [3425, 3662, 3744].

Endoparasite

Massaliatrema yamashitai, *Alaria alata*, *Cryptocotyle lingua* [1294, 1622, 1882, 3917], *Concinnum ten*, *Echinochasmus japonicus*, *Echinostoma hortense*, *Metagonimus miyatai*, *Paragonimus westermanii*, *Pricetrema* sp., *Pseudotroglotrema* sp., *Dipylidium caninum*, *Mesocestoides paucitesticulus*, *Mosocestoides* sp., *Spirometra erinacei*, *Taenia pisiformis*, *T. polyacantha*, *T. crassiceps*, *Corynosoma* sp., *Ancylostoma kusimaense*, *Arthrostoma miyazakiense*, *Capillaria* sp., *Crenosoma* sp., *Dirofilaria immitis*, *Heterakis spumosa*, *Molineus legerae*, *M. patens*, *Strongyloides planiceps*, *Toxascaris leonine*, *Toxocara canis*, *Trichinella spiralis*, *Trichuris vulpis* [2682], *Capillaria aerophila* [2593], *Crenosoma vulpis* [1294], *Uncinaria stenocephala* [1285], *Echinococcus multilocularis* [4023].

Pathogenic organism

The red fox is susceptible to some infectious diseases (e.g. distemper) [1856], but few data are available in Japan. The red fox is a major vector of rabies in Europe [1877]. There has been no rabies among Japanese animals since 1957 [3458].

5. Remarks

Much folklore surrounds the fox. It is also a good furbearer and game species. The harvest of foxes increased in the 1970s and reached a peak of about 20,000 in 1981 in Japan; about half of which were shot in Hokkaido. After that, the harvest has consistently declined in Japan. These fluctuations may depend on the value of pelts [2093]. In Hokkaido, while red foxes are very popular with tourists, many foxes are controlled as agricultural pests and definitive hosts for *Echinococcus multilocularis*, which causes alveolar hydatidosis.

K. URAGUCHI

Carnivora CANIDAE

Red list status: C-1 as *N. p. viverrinus* in "Hondo" and *N. p. albus* in Hokkaido (MSJ); LC (IUCN)

Nyctereutes procyonoides (Gray, 1834)

EN raccoon dog　　JP タヌキ (tanuki)　　CM 貉　　CT 狸, 貉　　KR 너구리　　RS енотовидная собака

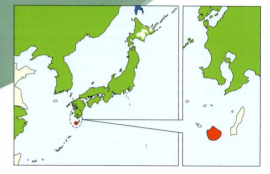

A raccoon dog carrying dead leaves in its mouth in deciduous forest, Ibaraki Pref. (M. Saeki).

1. Distribution

Naturally distributed in East Asia [101, 293, 306, 395, 404, 410, 735, 1429, 1793, 2058, 3302] and introduced into Europe [1429, 1844, 2099]. In Japan, there are 2 subspecies. *Nyctereutes p. albus* inhabits Hokkaido, including Okushiri Isl. [1214]. *Nyctereutes p. viverrinus* was once confined to Honshu, Shikoku, and Kyushu, including major islands between Tsugaru Str. (Blakiston's line) and Miyake (Ohsumi) Str. [1214]. It was introduced to Chiburi-jima Isl., Shimane Pref. in 1941 and Yaku-shima Isl., Kagoshima Pref. in the 1980s.

2. Fossil record

In Japan, fossils of *N. viverrinus nipponicus* (the middle Late Pleistocene), *N. v. genitor*, and *N. v. okuensis* (Neolithic) have been recorded [3203].

3. General characteristics

Morphology　External measurements (mean ± SD) are as follows: BW = 4,130 ± 900 g (n = 79) with seasonal variation (maximum in late autumn to winter); TL = 745 ± 61 mm (n = 67); T = 177 ± 22 mm (n = 68); HFsu = 109 ± 7 mm (n = 71); E = 48 ± 9 mm (n = 69) [2919]. Sexual dimorphism is insignificant. A defining feature is the black facial mask, similar to the common raccoon *Procyon lotor*, but unlike the raccoon the tail has no rings. The fur color varies from pale to reddish and dark gray, with black shoulders and legs.

Dental formula　Basically I 3/3 + C 1/1 + P 4/4 + M 2/3 = 42 [534]. M_3 sometimes missing [201]. i 3/3 + c 1/1 + p 3/3 = 28 (deciduous) [534].

Mammae formula　Normally 1 + 2 + 1 = 8 [2920].

Genetics　Chromosomes are highly polymorphic [3579], and there is variation in the number of B chromosomes inter- and intra-individually [3901]. For *N. p. viverrinus*, 2n = 38 + Bs [3798, 4226], 2n = 39 + Bs [3798, 4227], 2n = 41 + Bs [3798], FN = 66 [1943], FN = 68 [1941] and FN = 70 [4170]. For *N. p. albus* in Hokkaido, 2n = 38 + Bs [3800]. For continental subspecies, basically 2n = 54 + Bs [1429, 1942, 3799, 3832, 4225].

Many nucleotide sequence data on *N. procyonoides* are available on the DDBJ/EMBL/GenBank databases. For *N. p. viverrinus*, partial sequences of the mtDNA control region are deposited in the databases with the accession numbers D83614 and D83615.

4. Ecology

Reproduction　Basically monogamous [1426, 4066]. Testosterone levels in males peak in February–March in Japan, and progesterone levels in females are synchronous among females, even in the absence of males, suggesting that the raccoon dog is a monoestrous, seasonal and spontaneous ovulator [4190]. Raccoon dogs reach sexual maturity at 9 to 11 months [1429]. They show a back-to-back copulatory posture due to a copulatory tie like most Canidae [986]. Mating (lasting a few days) takes place between February and April, followed by parturition 9 weeks later [987]. Litter size is usually 4–6, and birth weight is around 100 g [987]. Both males and females exhibit parental care [2691, 2920, 3997]. Parents take turns attending the den site for 30 to 50 days, and they move together with the pups for another month or so after the pups emerge from the den site [987].

Lifespan 7–8 years (exceptionally 10 years; a record of 13 years in captivity) [1429].

Diet Raccoon dogs are opportunistic omnivores, and mainly eat fruit, invertebrates and small vertebrates. Plants are eaten throughout the year, while animals such as insects and earthworms are mainly eaten from summer to autumn [4008, 4010]. Persimmon and gingko fruits are important autumn foods in the countryside and suburbs [908]. They sometimes catch fish, birds, and small mammals, and eat carrion especially in winter [990, 2917, 3013, 4012].

Habitat Raccoon dogs use a wide range of habitats, including deciduous forests, broad-leaved evergreen forests, mixed forests, farmland and urban areas, from the coastal to sub-alpine zones throughout Japan [1214, 1429]. In their typical landscape of *satoyama* (a type of area where local people have influenced habitat through agriculture and forestry), some individuals occupy predominantly semi-natural habitat, whereas others use more managed farmlands, resulting in 2 types of habitat users in the same area [2917, 2921]. Understory vegetation is important for resting and foraging [3293].

Home range Home range size varies greatly from 10 to 600 ha in Japan [986, 4013], probably affected by food resources. Dispersing subadults and males without mates may have home ranges greater than 1,000 ha [2920, 4013]. Offspring start to move independently away from the parents within the same range in the late summer to autumn of the year of birth [2920, 3910]. Some juveniles start to disperse in the same autumn while others remain in the same area [2920, 3910]. Juvenile dispersal ranges from 1 to 10 km [2920, 4013].

Behavior Primarily nocturnal, but can be active during daytime. They do not bark like domestic dogs but growl or groan [986]; their vocalizations are higher in tone than those of domestic dogs and more or less resemble those of domestic cats [986]. The threatening posture is similar to that of felids, with an arched back, but the submissive posture is a lowered body consistent with that of canids [986], and, at least in captivity, they may show their belly [2919]. The basic social unit is the breeding pair, and some non-paired adult individuals may stay in a group or within the same area [1425, 2920]. Pairs share home ranges and travel together throughout years, whereas non-paired adults usually travel alone [1427, 1429, 2920]. Latrine sites are shared by family members [988, 3996], and they can distinguish feces of unknown conspecifics from their own [3996]. However, they probably do not use latrines as territorial boundary markings [3998]. They have sharp, curved nails, which probably enable them to climb trees and fences. They also 'play possum' (feign death when threatened). In regions with severe winter, the continental subspecies undergoes winter sleep for 3 to 5 months with a lowered metabolism [253, 1428, 2335], but the degree of reduction in metabolism is much less than that of true hibernators. Raccoon dogs in Japan lower their activity levels in cold periods [2921], and *N. p. albus* in Hokkaido shows body temperature reduction of 1–2°C [1623], similar to the Finnish subspecies [2335].

Natural enemies Feral dogs may kill raccoon dogs [1429, 2920].

Parasites & pathogenic organisms *Sarcoptes scabiei* (Acari: Sarcoptidae) may cause prominent hyperkeratosis, acanthosis and papillomatosis in raccoon dogs and may affect local populations with complications [1659, 2561, 3185, 3430, 3876, 3983]. Some other ectoparasites have been reported, such as *Haemaphysalis flava*, *Ixodes nipponensis*, *I. tanuki* and *I. ovatus* [598, 3408]. Many endoparasites, including Conoidasida [221, 1997, 2325, 2453], Trematoda [1082, 2128, 2317, 2318, 2377, 2730, 2941, 3040, 3041, 3046, 3182, 3378, 3679], Cestoda [1758, 1993, 3040, 3041, 3138, 4107], Secernentea [619, 945, 1993, 3040, 3041, 3045], Adenophorea [1300, 1645, 2938, 3044, 3045] and Palaeacanthocephala [3040], have been reported [221, 619, 860, 885, 945, 1082, 1263, 1300, 1645, 1758, 1993, 1997, 2128, 2317, 2318, 2325, 2358, 2377, 2453, 2510, 2730, 2938, 2941, 3040, 3041, 3044–3046, 3138, 3182, 3378, 3679, 4107, 4172].

An outbreak of canine distemper virus may cause local extinction [1886, 2453]. The raccoon dog can be the main vector and victim of rabies in its epidemic range, although rabies has not found in Japan since 1957 [933, 3458].

5. Remarks

Game species. Harvest has dropped from a peak of 76,000 in 1981 to 2,450 in 2011 [2089]. Legal culling as a crop-damage countermeasure has increased after the 1970s to about 10,600 in 2011 [2089]. They cause damage to various crops throughout Japan, especially corn, watermelons, and fruits [2918, 3364], but both the area of damage and the economic loss comprise less than 2% of crop damage by wild mammals [2917]. They used to be raised for fur and this was exported mostly to the USA before World War II [1620]. The fur is still used in brushes for calligraphy or in stuffed animals. Mortality in vehicle collisions is significant, and conservative estimates of road-kills are 110,000–370,000 per year [2922]. The raccoon dog has often appeared in Japanese folklore [2389], and Shigaraki pottery of the raccoon dog is popular although it is very different in shape from the actual animals.

M. SAEKI

Carnivora CANIDAE

Red list status: EX (MEJ, MSJ); LC (IUCN)

Canis lupus Linnaeus, 1758

EN gray wolf   JP オオカミ (ookami)   CM 狼   CT 狼，灰狼   KR 늑대   RS волк

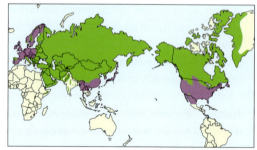

The extinct gray wolf in Hokkaido: a male (behind) captured in Sapporo, 1880 and a female (front) captured in Sapporo, 1882 (Natural History Museum, Botanic Garden, Hokkaido University).

1. Distribution

Widely distributed in northern and central Eurasia and North America [2516]. In Japan, the gray wolf was observed at least in Hokkaido, Honshu, Shikoku and Kyushu. However, the Hokkaido population became extinct in the late 19th century. The last recorded individual in Japan was killed in Nara Pref. in 1905.

2. Fossil record

Reported from the Middle and Late Pleistocene layers [504, 821, 1490, 1514].

3. General characteristics

Morphology Head and body length (HB) is about 950–1150 mm and T 300 mm from Honshu specimens, and HB 1,200–1,300 mm and T 250–400 mm from Hokkaido specimens [4145]. Since the animals in Japan are much smaller than the general continental ones, the populations from Hokkaido, and from Honshu, Shikoku and Kyushu have been considered as subspecies, *C. l. hattai* Kishida, 1931 and *C. l. hodophilax* Temminck, 1939, respectively. The latter has often been regarded as an independent species, *C. hodophilax*, since various morphological characteristics of the skeleton are distinguishable from those of the other populations. In continental populations, HB is 1,000–1,600 mm and T 350–560 mm [2516]. Southern populations are smaller in size within the species, according to Bergmann's rule. In the subspecies *C. l. arabs* still present in the Arabian Pen., HB is recorded as 820–1,440 mm [545, 2516]. The BW range of the gray wolf is thought to be 20–80 kg in males and 18–55 kg in females [2050, 2051]. The long pelage is gray or light brown, sometimes blackish in living populations. White individuals are often confirmed in tundra regions. Only about 10 specimens or fragments of skin have been kept from Japanese populations, and the color of these skins is brownish in the upper parts and light-grayish on the ventral side. Dark-brownish patches are observed in the dorsal part of the extremities in the forelimbs. The osteological and external characteristics of Japanese populations have been detailed from museum specimens and remains [510, 511, 1015, 1016, 2126, 2127, 2536–2538, 2849, 3314, 3535]. Three-dimensional image analyses were also undertaken in both Japanese and living continental populations [510, 512, 4171].

Dental formula I 3/3 + C 1/1 + P 4/4 + M 2/3 = 42 [35].

Mammae formula Not reported in Japan.

Genetics Karyotype of Japanese gray wolves is not recorded, but extant gray wolf is $2n = 78$, as in the coyote and the golden jackal of the same genus [3382]. Mitochondrial and nuclear DNA sequences were analyzed from extinct Japanese populations using skin and skeleton specimens [1087, 3386]. 600 bp sequence data of the control region indicated that individuals from Hokkaido (*C. l. hattai*) should be included within the cluster of the Canadian Yukon, whereas populations of *C. l. hodophilax* are phylogenetically distinguished from continental populations [1087, 1090]. MtDNA sequence data also pointed out that the populations of Honshu, Shikoku and Kyushu were genetically separated ca. 25,000–125,000 years ago from the other populations, whereas the population of Ezo wolf arrived more recently at Hokkaido from the Asian continent via the land bridge with Sakhalin Isl. that existed up to 10,000 years ago [2016]. Haplotypes of mtDNA have been analyzed from 162 gray wolves of 27 populations from throughout Europe, Asia and North America [3776]. The results demonstrated that the gray wolf populations could be separated into some major geographical clusters, for example, Alberta, Yukon, Israel, China, Iran, Saudi-Arabia, and Greece and Romania. However, data suggest recent hybridization between wolves and dogs, since populations of wild gray wolves and the domesticated dog cannot be genetically distinguished as separated clusters [3646].

Multiple origins of domesticated dogs have been confirmed in various clusters of wild gray wolves [3776], and gene flows of at least mtDNA have been confirmed among various populations of *C. lupus*, *C. familiaris* and *C. latrans* [70, 3777]. Since dogs from Eastern Asia contain higher genetic variability in both mtDNA [3102] and nuclear genomes [2803], it has been suggested that the center of the domestication of the dog from the gray wolf might be Eastern Asia. The mtDNA sequence data suggest that the Australian dingo branched from populations of *C. lupus-familiaris* from Indonesia, Malaysia and the Philippines about 5,000 year ago [3101]. Hypervariable tandem repeats in mtDNA have also been compared between wolves and dogs [3102].

4. Ecology

Reproduction   Not reported in Japan.

In extant continental populations, reproductive seasonality is obvious [2054, 4254]. In low latitudes mating occurs from January, whereas it is seen in April in high latitudes. The gestation period is about 62 days, and litter size is 1–11. Newborns weigh 400–500 g. Juveniles are weaned at about 5 weeks old, after which pack members hunt and bring food to the young. The age of sexual maturation is almost 22 months. Most reproductive activity finishes before 12–13 years [2050, 2051, 2053, 3764, 3765].

Lifespan   Longevity may be greater than 15 years [2050, 2051, 2516].

Diet   Not reported in Japan.

We believe that Japanese populations may have fed on Sika deer (*Cervus nippon*). In continental populations the gray wolf is one of the strongest predators, hunting wapiti, moose, bison, muskox and mountain sheep [2050–2052, 2516].

Habitat   Not reported in Japan.

Extant populations of the gray wolf have adapted to many habitats, except for tropical rainforests and arid deserts in the Eurasian and North American Continents [2516, 2804].

Home range   Not reported in Japan.

In living populations, home range varies widely. The largest range of a pack is 13,000 km^2, the smallest 18 km^2 [2050, 2843]. The range is recorded as 52–555 km^2 in Minnesota [584, 803, 3764], and 195–629 km^2 in summer and 357–1,779 km^2 in winter in Alberta [647], compared to 30 km^2 in areas with abundant food and 1,000 km^2 in tundra and deserts in Russia and Kazakhstan [286]. Population densities also widely vary in each district and natural environment. Densities of North American wolves were estimated to range from 1 individual per 25 km^2 to 1 per 500 km^2 [2050, 2843].

A specimen captured in Fukushima Pref., stored at the National Museum of Nature and Science, Tokyo (S. Kawada).

Behavior   No scientific records for Japanese populations. The gray wolf lives in "packs" that consist of the adult pair and their young of 1 or more years [2050, 2051]. Since the gray wolf is the most social of carnivores, a pack containing about 5–10 individuals acts as a highly-organized hunting team. The males and females of a pack have dominance hierarchies in both sexes. The leader of the pack is usually an alpha male. The most dominant pair generally mates, and they inhibit the sexual activity of the other members.

Natural enemies   There were basically no natural enemies.

Parasites   Not reported in Japan.

5. Remarks

The wolf has been both traditionally worshipped as an object of faith and detested as an ill-omened beast. Skeletal remains of the wolves have been widely discovered in archaeological sites from the Jomon Period (about 10,000 years ago) to the 19th century. The reason why the Japanese populations became extinct remains unclear. It has been suggested that wolves were excessively hunted for a long time. After the Meiji Restoration, strychnine was also used to kill wolves in various districts of Japan for the protection of livestock animals. The pressures of excessive hunting and long-term habitat disturbances have caused decline and regional extinctions around the world. In the Soviet Union after World War II, the number of the gray wolves was estimated to be 150,000–200,000, having decreased to 50,000 by the mid-1970s [2516]. In Alaska, numbers are estimated to be 4,000–7,000 and in Canada 30,000–60,000 [2516, 3559]. Only some hundreds survive in India, while populations in the British Isles disappeared in the 18th century because of hunting pressure [2516]. A decline in genetic variability has also occurred in small separated populations in many districts [3859], whereas hybridization with domestic dogs has altered the genetic make-up of other wild populations of the gray wolf [2516].

H. ENDO

RESEARCH TOPIC 5

Wild mammal-borne zoonoses and mammalogists in Japan
Y. YOKOHATA

1. Introduction

The word, "zoonosis" (or zoonoses, pl.), means an infectious disease transmitted between or among human and non-human animals. Drafting proposals and conducting the practice of countermeasures for zoonoses are one of the responsibilities of mammalogists. Some guidelines explicitly state the need for adequate care for the prevention of zoonoses in our daily activities (e.g., USA: [701], Japan: [397]).

More than 50 zoonoses are known from Japan, but many of them are attributable mainly to domestic and/or non-mammalian animals. Strict evaluation of the contribution of wild mammals to the occurrence of zoonoses often is difficult. At least 27 infectious diseases are known to occur in Japan as wild mammal-borne zoonoses, of which causative organisms, at least in part, use wild mammals as their hosts or vectors (Table 1). This table does not list all zoonoses involving wild mammals in Japan, but does include almost all that cause problems in this country.

Following are discussions concerning 2 representative zoonoses from wild mammals in Japan, alveolar echinococcosis and Lyme disease. They are provided to illustrate the role of mammalogists in dealing with the problem of zoonoses in Japan.

2. Alveolar echinococcosis in Hokkaido and mammalogical contributions

This disease is attributable to a species of cestode, *Echinococcus multilocularis*, in the family Taeniidae whose main intermediate and final hosts in Japan are the gray red-backed vole (*Myodes rufocanus*) and the red fox (*Vulpes vulpes*), respectively (Table 1-a). The major endemic region of this disease in Japan is Hokkaido, where more than 460 human patients were reported up to 2003 [3426] and 5 to 19 new patients have been reported every year since 1982 [1292]. They were infected orally by ingestion of the egg of this worm and parasitized by the larval worms in their livers and other viscera. This cestode is distributed widely in the Northern Hemisphere, and numerous species of wild and domestic mammals are known as its common or occasional intermediate hosts. In Hokkaido, other wild intermediate hosts include *Sorex unguiculatus*, *S. caecutiens*, *Myodes rutilus*, *M. rex*, *Rattus norvegicus*, *Apodemus argenteus* and *Mus musculus*. Domestic swine (*Sus scrofa*) and horses (*Equus caballus*) are known as eventual intermediate hosts, as are humans [1154, 1290, 3426]. Until the mid 1960s, this disease was limited to Rebun Isl., a small island near the mainland of Hokkaido. On Rebun Isl., foreign red foxes with *E. multilocularis* were introduced and released in the 1920s. In 1965, 2 human cases were diagnosed in Nemuro, the easternmost area of the mainland of Hokkaido. An ongoing intensive survey, based on necropses, of the prevalence of the worm in foxes and other animals has been conducted across most of Hokkaido. This disease and the worm have spread westward over Hokkaido [1292, 3426]. The mean prevalence of this worm in red foxes in Hokkaido has increased dramatically over the past 2 decades, from less than 10% in 1985 to 58.4% in 1998, although in 2002 the prevalence fell to just over 30% [1292]. The regional prevalence in foxes is positively and negatively influenced by the density of voles and snow depth, respectively, in Hokkaido [4124]. Based on data of detection of coproantigen by use of sandwich enzyme-linked immunosolvent assay (sandwich ELISA) and detection of taeniid eggs (whose morphological identification is difficult even at the generic level) from feces of foxes, a trial vaccination program of the wild fox population was performed in eastern Hokkaido. As a result, the prevalence of alveolar echinococcosis in foxes decreased [1292, 2238].

These advanced studies and countermeasures against this disease are based on some mammalogical contributions. For example, it was pointed out that an increase of livestock farming wastes such as cattle placentae provided winter food for foxes and may have helped to increase their population density in Hokkaido [18]. Predation pressure of foxes on vole populations decreases under dense ground cover, such as deep snow [4144]. Also, high dispersal ability of foxes may have contributed to the range expansion of this disease in Hokkaido [21]. Further, prevalence of the cestode species in foxes is affected by the delayed density-effect of vole populations in eastern Hokkaido [2959]. Findings from these studies explain some important phenomena concerning alveolar echinococcosis in Hokkaido. These contributions could not be achieved without many intensive ecological studies of small mammals and carnivores conducted and accumulated in this region over an extended period of time.

3. Lyme disease and some other *Borrelia* infections in small mammals in Japan

Lyme disease is a well-known tick-transmitted zoonotic bacteriosis carried by some wild animals, such as rodents and birds in Europe, North America, Japan, Western Asia and North Africa. Human patients of this disease show characteristic migrating skin erythema (erythema migrans) and meningitis with fever, headache, arthralgia, myalgia and extreme pain as first symptoms [1207]. Causative bacteria (spirochetes) are at least 3 molecular species of the genus *Borrelia*: *B. burgdorferi* sensu stricto, and *B. garinii* and *B. afzelii*, belonging to *B. burgdorferi* sensu lato. They had been indistinguishable, until some molecular biological methods became available during the mid 1990s [2129, 3402]. These species, vector tick species (mainly *Ixodes* spp.) and mammalian hosts of the spirochetes differ from each other among areas in the world, but the mammalian reservoirs apparently are limited to rodents. Larger mammals, such as cervids, were discussed as potential reservoirs of this disease, but such a possibility now is recognized as negligible [1207].

In Japan, *B. burgdorferi* sensu stricto is absent, and the causal *Borrelia* species of this disease largely are limited to *B. garinii* and *B. afzelii* (Table 1-b). They and some other species of *Borrelia* were detected in various Japanese wild

Table 1. Wild mammal-borne zoonoses in Japan

Disease	Causative organism	Wild mammalian host or vector (*direct infection source to human)	Occurrence in human	Human-human transmission	Reference
a. Parasitoses (sensu stricto)					
Helminthiasis					
Alveolar echinococcosis	*Echinococcus multilocularis*	Carnivores (mainly canids)* and small mammals (mainly *Myodes* spp.)	Endemic (mainly Hokkaido)	–	See text.
Angiostrongyliasis	*Angiostrongylus cantonensis*	*Rattus* spp., *Suncus murinus* (rare)	Endemic (Nansei Isls.)	–	[2340, 2341]
Clonorchiosis	*Clonorchis chinensis*	Various mammals (mainly carnivores and rodents)	Rare	–	[2002, 3524]
Echinostomasis	*Echinostoma* spp. (mainly *E. hortense*)	Various mammals (mainly carnivores and rodents)	Rare	–	[146]
Fasciolasis	*Fasciola hepatica* and *F. gigantica*	Various mammals	Rare	–	[140]
Gnathostomiasis	*Gnathostoma nipponicum*	Carnivores (mainly *Mustela* spp.) and boars (*Sus scrofa*)	Rare	–	[2128, 2792, 3182]
Paragonimiasis	*Paragonimus* spp.	Carnivores (mainly *Mustela* spp.) and boars (*S. scrofa*)	Sporadic	–	[2936]
Metagonimiasis	*Metagonimus* spp.	Various mammals (mainly carnivores)	Sporadic	–	[4244]
Trichinellosis	*Trichinella spiralis*	Bears (*Ursus* spp.)* and boars (*S. scrofa*)*	Sporadic	–	[1645]
Protozoosis					
Human babesiosis	*Babesia microti*	Shrews (*Sorex* spp.) and various murids	Rare	–	[1433, 2694, 3681, 4243]
Toxoplasmosis	*Toxoplasma gondii*	Various mammals (felids*, etc.)	Sporadic	–	[1067, 1991]
b. Bacterioses					
Japan spotted fever	*Richettsia japonica*	Murids and sika deer (*Cervus nippon*)	Sporadic	–	[1241]
Leptospirosis	*Leptospira* spp. (mainly *L. interrogans* sensu lato)	Various mammals (mainly murids)*	Sporadic (relatively more in southwestern Japan)	–	[3515, 3966]
Listeriosis	*Lysteria monocytogenes*	*Rattus* spp. (mainly *R. rattus*)*	Sporadic	–	[2965]
Lyme disease (Lyme borreliosis)	*Borrelia* spp. (*B. garinii* and *B. afzelii* in Japan)	*Apodemus* spp. (mainly *A. speciosus*) and *Myodes rufocanus*	Endemic (mainly northern Japan or high mountains)	–	See text.
Murine salmonellosis	*Salmonella enterica* serover. *typhimurinum*	Various murids*	Sporadic	+	[3838]
Murine typhus	*Richettsia mooseri*	*Rattus* spp.*	Rare	–	[4096]
Q fever	*Coxiella burnetii*	Various mammals (e.g. *C. nippon*)	Sporadic	–	[2837, 2965, 3408]
Rat-bite fever	*Streptobacillus moniliformis* and *Spirillum minus*	Murids (mainly *Rattus* spp.)*	Rare (?, precisely unknown)	–	[3838]
Scrub typhus (Tsutsuga-mushi disease)	*Rickettsia tsutsugamushi* and *R. orientalis*	Various mammals (mainly murids, e.g. *Microtus montebelli*)	Endemic in northern Honshu and sporadic in others	–	[187]
Tularemia	*Francisella tularensis*	Various mammals (mainly hares, *Lepus* spp. and rodents)*	Rare	–	[186]
Yersiniasis	*Yersinia enterocolitica*	Various mammals (mainly rodents)*	Sporadic	–	[2965]
c. Viroses					
Hantavirus infection	Hantavirus (Bunyaviridae)	Murids (mainly *Rattus norvegicus*)	Rare	–	[1846, 2240, 2965]
Hepatitis E	HEV (Hepeviridae)	Boar (*S. scrofa*)* and sika deer (*C. nippon*)*	Rare	–	[328]
Lymphocytic choriomeningitis	LCMV (Arenaviridae)	Murids	Unknown, but not many	–	[2446]
Severe fever with thrombocytopenia syndrome	SFTSV (Bunyaviridae)	Murids (mainly) and various mammals	Rare (but with high mortality)	–	[3097, 4027]
Tick-borne encephalitis	TBEV (Flaviviridae)	Murids	Rare	–	[4176]

Table 2. Small mammals from which *Borrelia* spp. were detected in Japan, geographic regions, assumed vector tick species, and species of *Borrelia*

Order Genus Species	Region	Vector tick species*	*Borrelia* species	Reference
Soricomorpha				
Sorex				
S. unguiculatus	Hokkaido	Io	B. japonica	[2405]
S. caectiens	Hokkaido	Io	B. japonica	[2405]
S. shinto	Gifu (central Honshu)	Unknown	Unknown	[3402]
Crocidura				
C. dsinezumi	Yamagata (northern Honshu)	Unknown	Unknown	[3402]
C. dsinezumi	Niigata, Fukui (central Honshu) and Kuju Mountains (Kyushu)	Io	B. japonica	[3402, 3406]
Rodentia				
Apodemus				
A. speciosus	Hokkaido to Kyushu	Io	B. japonica	[1088, 2405, 2406, 3402, 3406]
A. speciosus	Hokkaido to northern Honshu	Ip	B. afzelii and/or B. garinii	[2405, 2406, 3402]
A. speciosus	Fukushima (northern Honshu)	Ip	B. garinii	[1089]
A. speciosus	Fukui and Tokushima (Shikoku)	It	B. tanuki (probably)	[1088, 3402]
A. speciosus	Tsushima Isls. (Kyushu)	It	B. tanuki	[3406]
A. argenteus	Hokkaido to central Honshu	Io	B. japonica	[2406, 3402, 3406]
A. argenteus	Hokkaido	Ip	B. afzelii and/or B. garinii	[2406]
A. argenteus	Gifu	Unknown	Unknown	[2406, 3402]
Microtus				
M. montebelli	Aomori (northern Honshu)	Io	B. japonica	[3402]
M. montebelli	Kyoto (western Honshu)	Unknown	Unknown	[3402]
Eothenomys				
E. smithii	Ishikawa (central Honshu) and Fukui	Io	B. japonica	[3402]
E. smithii	Fukui	Ip	B. afzelii and B. garinii	[1089]
E. smithii	Fukui	It	B. tanuki (probably)	[3402]
E. smithii	Gifu and Kyoto	Unknown	Unknown	[3402]
E. andersoni	Aomori and Yamagata	Io	B. japonica	[3402]
Myodes				
M. rufocanus	Hokkaido	Io	B. japonica	[2406]
M. rufocanus	Hokkaido	Ip	B. afzelii, B. garinii and/or B. spp.	[3402]
M. rutilus	Hokkaido	Io	B. japonica	[2406]
Mus				
M. caroli	Okinawa-jima Isl. (Nansei Isls.)	Ig	B. valaisiana	[3404]

*Io: *Ixodes ovatus*; Ip: *I. persulcatus*; It: *I. tanuki*; Ig: *I grannulatus*

small mammals (Table 2). Human patients with Lyme disease have been reported in this country since the late 1980s [2129], and primarily have occurred in areas of northern Japan, such as Hokkaido.

In Hokkaido, *B. garinii* and *B. afzelii* have been detected from the large Japanese field mouse (*A. speciosus*), the small Japanese field mouse (*A. argenteus*) and 2 species of the red-backed voles (*Myodes* spp.) [2129, 2406]. These rodents are parasitized by at least 2 species of the genus *Ixodes*, *I. persulcatus* and *I. ovatus*. *Ixodes ovatus* was suspected as one of the vectors of the causative *Borrelia* species [2131], but the causative *Borrelia* actually are transmitted only by *I. persulcatus* [2129, 2407]. Spirochaetes of the genus *Borrelia* also were detected in the long-clawed shrew (*S. unguiculatus*) and Laxmann's shrew (*S. caecutiens*) in Hokkaido, but it was transmitted by *I. ovatus* [2405] and apparently was a non-causative species, *B. japonicus* [2129].

Not only occurring in Hokkaido, Lyme disease also has often occurred in mountainous regions in northern to central Honshu and rarely in higher areas in Kyushu [1088, 1089, 3402, 3405, 3406], because *I. persulcatus* usually occurs in cooler climates. In Kyushu, *B. garinii* was detected only from *I. persulcatus* without any detection from mammalian reservoirs, whereas *B. japonica* was detected not only from *I. ovatus*, but also from *A. speciosus* and the dsinezumi shrew (*C. dsinezumi*) [3406]. Moreover, *B. valaisiana* is known from the Ryukyu mouse (*M. caroli*) on Okinawa-jima Isl. in southern Japan [3404]. *Borrelia valaisiana* belongs to *B. burgdorferi* sensu lato, so special attention may be necessary to detect this species of *Borrelia* in the future in southern Japan.

Various species of small mammals are involved in the life

cycles of these *Borrelia* spp. Nevertheless, mammalogists in Japan apparently have not contributed to the knowledge and understanding of this disease differing from the case of alveolar echinococcosis in Hokkaido. Mammalogists may be less motivated to study this disease, because of fewer patients and reduced clinical problems with zero mortality. Additionally, 2 problems may inhibit development of mammalogical research of the disease caused by species of *Borrelia*. First, the ecological relationship between small mammals and *Borrelia* spp. is less specific than those between the ticks and the *Borrelia* spp. Infection by *B. burgdorferi* sensu lato hardly affects growth and reproductive parameters of *I. persulcatus* [2404], so that the relationships between them apparently are commensal, which suggests a deep host-parasite bond. Also, their distributions highly overlap, whereas the distributions of *B. garinii* and *B. afzelii* apparently are independent of the distribution of the primary mammalian reservoir in Japan, *A. speciosus*. This is because the 2 species of *Borrelia* dominate in northern Japan and are limited to higher mountains in western Japan. Second, in the case of echinococcosis, the main intermediate and final hosts, i.e., voles and foxes, have a strong prey-predator interrelationship, whereas the host-parasite interrelationship between the small mammals and ticks is more facultative with weaker host specificity. These 2 facts obscure the ecological relationships both between small mammals and *Borrelia* spp. and between these mammals and ticks.

4. Risk of foreign zoonosis invasion into Japan —raccoon roundworm infection

Fatal wild mammal-borne zoonoses, such as rabies and plague, currently are absent in Japan. However, this region has been under the risk of invasion by many foreign infectious diseases. There are many opportunities with various routes via which parasitic organisms could invade Japan. As previously mentioned, alveolar echinococcosis on Rebun Isl. was caused by intentionally introduced foreign foxes and infections among human patients resulted in high mortality. In the past, rabies and plague invaded Japan and created severe problems with numerous human patients and deaths [1296]. A researcher reviewed information on parasitic helminthes of 24 species and 2 subspecies of alien mammals in Japan [4115]. Of these helminthes, at least 2 cestode and 5 nematode species from them were alien parasites; however, most are assumed to show no or low pathogenicity for any animals (also see [212]).

As a relatively recent example, invasion of raccoon roundworm infection into captive raccoons in Japan has been noted since the 1990s (e.g. [2150, 3327]). This disease is caused by the raccoon roundworm, *Baylisascaris procyonis*, a parasitic nematode of the raccoon *Procyon lotor*. The raccoon was imported into Japan in recent years because of a brisk demand as a pet. Larvae of this nematode often randomly migrate in human bodies when their eggs are ingested orally and hatch in alimentary tracts, resulting in severe clinical troubles with occasional death (larva migrans [1528–1531, 3327]). In Japan, a high prevalence of this nematode in raccoons in zoos (39.9%) was reported [2150], and an outbreak of the larva migrans has occurred in zoo animals [3039, 3042]. Intensive investigations of numerous wild individuals of introduced raccoons in Japan have been conducted [213, 216, 1433, 1645, 1991, 1992, 1994, 3043, 3047], but *B. procyonis*-positive animals have never been found. In Japan, quarantine for 180 days is required when some animals listed as reservoirs of rabies are imported into Japan from countries infected by this disease (including all areas within the natural distribution of raccoons). Raccoons legally were added to the list of reservoir animals in 2000, so that the import of pet raccoons from these countries into Japan actually became impossible. These legal measures preceded establishment of "Invasive Alien Species (IAS) Act" of Japan in 2004, which prohibits the import of IAS including raccoons.

Social actions against the invasion of foreign zoonoses include scientific investigation, legal measures, and public education. As illustrated by the example of raccoon roundworm infection, such systems already have been established in Japan. In this instance, it is important that both countermeasures against damage to ecosystems and biodiversity by raccoons and against the invasion of a foreign zoonosis with this animal have advanced cooperatively. Such cooperation is essential in our present and future work.

5. Concluding remarks

For infectious diseases with human to human transmission, such as rabies and plague, countermeasures are very difficult. In Japan, very few wild mammal-borne zoonoses can be transmitted from human to human (Table 1). This fact has made the problem of dealing with wild mammal-borne zoonoses in Japan easier. We must maintain this situation in the future, although anthropogenic disturbances of ecosystems and invasions by foreign zoonoses may change this situation. Repeated emergence of new zoonoses in the world (e.g., [2817]) may be the result of such disturbances and invasions. Many people in Japan apparently have tolerated introductions of various foreign and inland organisms. This tolerance may be attributable partly to the lack of severe outbreaks of zoonotic diseases, but the tolerance is not appropriate under the changed situation.

On the other hand, at least 48 patients of severe fever with thrombocytopenia syndrome (SFTS), which was found in China in 2011 for the first time, have been found in Japan since 2012 [2446]. SFTS virus strain isolated in Japan shows a different molecular pattern from those isolated in China, so that it is endemic in Japan [4176]. This virus may transmit in not only mites–mites cycle, but also in mites–mammals cycle, which includes various wild mammals (Table 1-c). In Japan, the unexpected emerging of this disease with high human mortality (17/48 till 2013 in Japan, [2446]) may be attributable, at least partly, to increasing numbers of sika deer (*C. nippon*) and wild boars (*S. scrofa*) in many regions in Japan.

People in Japan have developed well-managed social systems that include various forms of information and technology transfered from other advanced countries. In this social system, Mammalogists in Japan have a great task against the wild mammal-borne zoonoses.

Carnivora PROCYONIDAE

100

Red list status: C-1 (MSJ); LC (IUCN); introduced

Procyon lotor (Linnaeus, 1758)

EN raccoon, North American raccoon   JP アライグマ（araiguma）   CM 浣熊   CT 浣熊
KR 미국너구리   RS енот-полоскун

Two raccoons climbing a tree in Tomakomai, Hokkaido (T. Namba).

1. Distribution

Introduced species in Japan. The original distribution is North, Central and South America, from southern Quebec, central Ontario, the prairie provinces and southern British Columbia to Colombia [1860].

The first report of naturalization in Japan was from Inuyama, Aichi Pref. in 1962 [73, 135]; the second from Eniwa, Hokkaido in 1979, and the third from Kamakura, Kanagawa Pref. in 1988 [292, 984, 993, 996, 997]. The raccoon has already been confirmed in 36 prefectures [292], and has been reported in all 47 prefectures in Japan [984]. The raccoon was also introduced into France, the Netherlands, Germany, Switzerland, Russia (and adjacent independent republics) and the West Indies [1860].

2. Fossil record

Not reported in Japan.

3. General characteristics

Morphology   The pelage color can range from gray-black to red-brown. The most characteristic physical features of the raccoon are the black facial "bandit" mask across the cheeks and eyes, and the bushy black ringed tail. Ears are rounded with white hair on the inside. Both fore and hind feet have five fingers/toes and the sense of touch in the forepaws is well developed. External measurements are as follows: BW = 3.6–9.0 kg, TL = 63.4–105.0 cm (males)/60.0–90.9 cm (females), T = 20.0–40.5 cm (males)/19.2–34.0 cm (females) in North America, place of origin [4245]. Data from Japan remain approximately within this range so far [219, 1217, 3335]. In rare cases, however, raccoons over 10 kg have been captured at areas abundant in food resources in Japan [992]. The largest records in North America are 25.4 and 28.3 kg [3157, 3889]. Generally males are larger than females.

Dental formula   I 3/3 + C 1/1 + P 4/4 + M 2/2 = 40. Two pairs of molars on both the upper and lower jaws are unique among carnivores [33].

Mammae formula   Usually 1 + 1 + 1 = 6 [714, 992]. However, 8 mammae have been observed in some individuals in North America [2988].

Genetics   Chromosomal constitutions have not been investigated in Japan. In North America, $2n = 38$ [1424], and for some raccoons $2n = 42$ have been reported [1865]. The nucleotide sequence at the 199-bp region of the D-loop of mitochondrial nucleotide has been determined [2029, 3410] and the MHC class II *DRB1* gene is used in polymorphism analysis [2028, 3399].

Procyon lotor (Linnaeus, 1758)

4. Ecology

Reproduction Mating season is usually from January to March [997]. Gestation period is 63 days on average. Mean pregnancy rate is 0.66 in yearlings and 0.96 in adults in west-central Hokkaido [220, 997], 0.88 in adults in Kanagawa Pref. [1944], 0.75 in adults in Nagasaki Pref. [1217], and 0.72 in adults in Wakayama Pref. [3336]. Mean litter size is 3.6 in yearlings and 3.9 in adults in west-central Hokkaido [220, 997]. The rate of population growth in west-central Hokkaido is estimated to range from 0.20 to 0.25 without harvest mortality [220].

Lifespan Most wild raccoons live less than 5 years, but up to 13–16 years has been reported in North America [1865]. In Hokkaido, an 8-year-old raccoon has been recorded [219].

Diet The raccoon in Japan shows opportunistic and omnivorous feeding habits, with a wide variety of foods such as insects, reptiles, crustaceans, birds, small mammals, fruits, crops and garbage [984, 997]. In general, diet is predominantly animal in summer and vegetable in autumn. In Hokkaido, a preference for bower actinidia (*Actinidia arguta*) in autumn has been observed [4241].

Habitat In Japan, the raccoon inhabits forests near water, often preferring the forest edge, but it is so adaptive that it can inhabit agricultural, suburban and urban areas [73, 983, 995]. Tree hollows and rock crevices are preferred as dens, but burrows of other animals and abandoned houses or roof spaces are also used. Drainpipes are used as travel paths in urban areas [997]. In Hokkaido, the raccoon uses mainly snowless riversides to travel in the winter season [997].

Home range Home range size varies depending on habitat, season and sex. Home range size in urban areas is smaller than in suburban and forested areas [984, 997]. They are up to 2,219 ha for males in forested areas, and as small as 35 ha for females in urban areas [984, 997]. Density also varies depending on habitat and is usually high in urban and agricultural areas. In North America, mean density is 1–27/km^2 [714], and densities up to 66.7–333.3/km^2 (mean 125/km^2) have been reported. In Japan, the highest record is 13.8/km^2 in Kanagawa Pref. [3877] and is normally under 5.0/km^2 in Hokkaido.

Behavior The raccoon is usually nocturnal, and has a polygamous mating system. The male tends to keep more than 1 female in its home range, and will come and go among several females in the mating season [997]. The raccoon is good at climbing up trees and swimming [984]. The raccoon does not hibernate, but in Hokkaido it mostly stays in its den and avoids energy expenditure in winter [2, 984].

Natural enemies In Japan the raccoon has no natural enemies, except for birds of prey, which prey on juveniles [984, 995]. Predators in North America include the bobcat, coyote, lynx, wolf, wolverine and great horned owl.

Parasites Cestodes *Taenia taeniaeformis* [1992] and *Taenia hydatigena* [1992], nematodes *Toxocara tanuki* [1994], *Molineus legerae* [205], and *Trichinella* T9 [1645], and protozoans *Eimeria* and *Isospora* [1991], *Babesia microti*-like parasites [1433] are reported as ectopasites in Japan. The parasitic mite *Sarcoptes scabiei* has also been reported [205].

5. Remarks

The raccoon was mainly imported as a pet animal in Japan, but it is so fierce in nature that many owners released them into the wild. In addition, the raccoon is clever with its hands and can escape from cages easily. Such irresponsible abandonment and escapes are the main causes for the establishment of the raccoon in Japan [2, 73, 984]. It is possible that the crab-eating raccoon (*P. cancrivorus*) was also imported and released into the wild, but this has not been confirmed yet.

Agricultural damage is serious [2, 984, 994]. The raccoon damages crops and fruits such as corn, melons, watermelons, strawberries, paddy rice, soybeans, potatoes, beets, oats, and so on in summer and autumn [2, 984, 997]; agricultural damage amounts to 340 million yen in Japan [2088]. Raccoons finding their way into houses are also a problem in urban areas [984]. A game species.

6. Topic

Impacts of the invasive alien raccoon and countermeasures
In addition to agricultural damage, the raccoon has various impacts on native species and ecosystems. Resource competition with the red fox (*Vulpes vulpes*) and the raccoon dog (*Nyctereutes procyonoides*) occurs and usually the red fox and the raccoon dog disappear after the raccoon invasion [3, 852, 994]. Another example of resource competition is the take-over of breeding sites from owls [997]. Direct predation on native species is also serious for the Japanese crayfish (*Cambaroides japonicus*), the Ezo salamander (*Hynobius retardatus*) and some species of frogs. Impacts on the Tokyo hynobiid salamander (*Hynobius tokyoensis*), the red clawed crab (*Chiromantes haematocheir*) and shore crab (*Helice tridens*) are also a cause for concern. Nest abandonment by gray herons (*Ardea cinerea*) due to raccoons attacking breeding colonies has been reported [943, 944, 982, 994]. The raccoon also poses the threat of infectious diseases such as raccoon roundworm (*Baylisascaris procyonis*) infection [2150, 3038] which has not yet been detected in naturalized raccoons, but has been detected in raccoons and other animals in captivity. In eastern North America, the raccoon is one of the major transmitters of rabies.

The raccoon is designated as an invasive alien species in "Invasive Alien Species Act" and is also listed in the Invasive Alien Species Lists of Japan [2085, 2090]. Countermeasures are taken throughout Japan in accordance with "Invasive Alien Species Control Action Plan" [2091].

T. IKEDA

Carnivora FELIDAE

Red list status: CR (MEJ); En as *Felis b. euptilura* (MSJ); LC as *P. bengalensis* (IUCN)

Prionailurus bengalensis euptilurus (Elliot, 1871)

EN Tsushima leopard cat   JP ツシマヤマネコ (tsushima yamaneko) for Tsushima population
CM 豹猫   CT 豹貓，黑龍江豹貓   KR 삵   RS дальневосточный лесной кот

A Tsushima leopard cat taken by photo-trap (Mammal Ecology Laboratory, University of the Ryukyus).

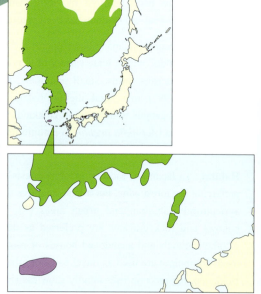

1. Distribution

In Japan, this subspecies of the leopard cat (*Prionailurus bengalensis*) occurs only on the Tsushima Isls., located between mainland Japan and the Korean Pen. *Prionailurus b. euptilurus* also occurs on the Korean Pen. and possibly part of east China [882]. The leopard cat has a wide distribution ranging from Siberia, China, Korea, Southeast Asia to Indonesia and is divided into 10 or more subspecies, though detailed information on the range of each subspecies is limited [2519].

2. Fossil record

Fossils of *Felis* cf. *microtis* Milne-Edwards (= *P. bengalensis*) have been excavated from limestone fissure deposits of the Late Pleistocene in Tochigi Pref., Honshu [3203].

3. General characteristics

Morphology A small-sized Felidae. The pelage color is light brown with dark spots throughout. Clear white and black stripes are on the forehead. Fur color pattern is clearer than the Iriomote cat (*P. b. iriomotensis*). Ears are rounded with a white spot on the back. Tail is thick to the tip. Individuals can be identified by their forehead stripe patterns.

External measurements (mean ± SD) are as follows: BW = 3.55 ± 0.43 kg ($n = 7$) for males and 3.15 ± 0.40 kg ($n = 11$) for females; HB = 569.00 ± 41.09 mm ($n = 7$) for males and 523.27 ± 32.46 mm ($n = 11$) for females; TL = 245.00 ± 20.14 mm ($n = 7$) for males and 223.70 ± 21.62 mm ($n = 10$) for females; HFsu = 111.16 ± 6.83 mm ($n = 7$) for males and 105.82 ± 4.28 mm ($n = 11$) for females; E = 42.01 ± 4.05 mm ($n = 6$) for males and 42.36 ± 2.72 mm ($n = 11$) for females [1200].

Dental formula   I 3/3 + C 1/1 + P 3/2 + M 1/1 = 30 [29].

Mammae formula   0 + 1 + 1 = 4 [1200].

Genetics Chromosome: $2n = 38$ [3900]. The Tsushima leopard cat is classified as a subspecies of the leopard cat, *P. bengalensis*. Mitochondrial cytochrome *b* region analysis reveals that this subspecies has the same mtDNA lineage as the leopard cat [1987]. Mitochondrial and Y-choromosomal DNA sequences have been analyzed to show the genetic diversity and phylogeography of the Asian leopard cat, including this subspecies [3476].

Prionailurus bengalensis euptilurus (Elliot, 1871)

4. Ecology

Reproduction Mating season ranges from January to March and 1–3 kittens are born in April and May [1200, 3963].

Lifespan Unknown.

Diet Mainly small mammals such as *Apodemus speciosus*, *A. argenteus*, *Micromys minutus* and *Crocidura shantungensis*. Also, birds, amphibians and insects are eaten and the proportion of each food item is different among seasons and areas [1043, 1215, 3539].

Habitat Main habitat is forest, but sometimes agricultural areas are used for foraging. The Tsushima Isls. belong to the temperate zone and are mostly covered with secondary forest and cedar-cypress plantation. A large part of the cat population is on the north island, Kami-shima [1197, 2081]. On the south island, Shimo-shima, the cat had not been recorded since 1984, until an individual was recorded by photo-trap in 2007 [1197, 2081]. Thereafter, the presence of the leopard cats has been confirmed by scat and photo in some areas in Shimo-shima, so probably a small population is present [1197, 2081].

Home range Home range size varies depending on season, breeding activity and habitat condition, ranging from 0.5 to 16.5 km^2 for males and 0.2 to 2.5 km^2 for females [1200, 1215]. Females have relatively small and stable home ranges, while males have larger home ranges and expand them in the mating season [1200].

Behavior Solitary except for the mother–kitten relationship. Active mainly at night. Peaks of activity are observed at dusk and dawn [1215].

Natural enemies Two other native carnivorous competitors, martens *Martes melampus tsuensis* and Siberian weasels *Mustela sibirica*, are distributed on the Tsushima Isls. Recently, introduced mammals threaten the existence of the Tsushima leopard cats; the domestic dog as a predator, the domestic cat as a competitor and a carrier of infectious disease, and the wild boar as a potential predator on kittens.

Parasites & pathogenic organisms Trematode, Cestode, Nematode, Acanthocephala and Apicomplexan parasites have been detected [1753, 4072–4074]. Feline immunodeficiency virus (FIV) has been reported to be transmitted from domestic cats [2488].

5. Remarks

The Tsushima leopard cat is protected by law as a "Natural Monument" (1971) and a "National Endangered Species" (1994). Major threats are habitat loss by agricultural development and road construction, deforestation, road kill, introduced animals, infectious disease and traps for prevention of poultry loss [466, 1197].

M. IZAWA & N. NAKANISHI

A skull specimen (Mn-06) of the Tsushima leopard cat stored at Tsushima Wildlife Conservation Center (Mammal Ecology Laboratory, University of the Ryukyus). Note that the second upper premolars are lost from tooth sockets in this specimen, but were originally present.

Carnivora FELIDAE

Red list status: CR (MEJ); En as *Felis iriomotensis* (MSJ); CR (IUCN)

Prionailurus bengalensis iriomotensis (Imaizumi, 1967)

EN Iriomote cat   JP イリオモテヤマネコ（iriomote yamaneko）for Iriomote-jima population
CM 西表猫   CT 西表山貓   KR 이리오모테삵   RSириомотейский лесной кот

An Iriomote cat taken by photo-trap (Mammal Ecology Laboratory, University of the Ryukyus). Fur color is darker than the Tsushima leopard cat.

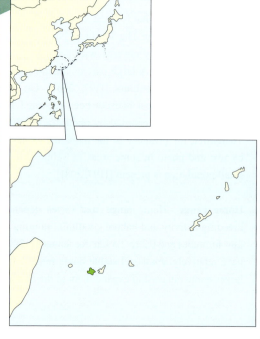

1. Distribution

Occurs only on Iriomote-jima Isl., part of the Yaeyama Islands in the Ryukyu Arch., Japan.

2. Fossil record

Fossils of *Felis* cf. *microtis* Milne-Edwards (= *P. bengalensis*) have been excavated from limestone fissure deposits of the Late Pleistocene in Tochigi Pref., Honshu [3203].

3. General characteristics

Morphology A small-sized Felidae. The pelage color is dusky brown with darker spots throughout. Black lines run from the forehead to the back of the neck. Two white lines run from the inner corner of the eyes onto the forehead. The ears are rounded with a white spot on the back. Tail is thick to the tip.

External measurements (mean ± SD) are as follows: BW = 4.01 ± 0.48 kg (n = 71) for males and 3.03 ± 0.25 kg (n = 42) for females; HB = 580.13 ± 26.73 mm (n = 70) for males and 527.56 ± 24.10 mm (n = 39) for females; TL = 249.15 ± 17.16 mm (n = 71) for males and 236.77 ± 17.03 mm (n = 39) for females; HFsu = 118.75 ± 3.45 mm (n = 71) for males and 112.04 ± 2.73 mm (n = 38) for females; E = 44.87 ± 2.72 mm (n = 69) for males and 43.85 ± 2.35 mm (n = 34) for females [2403].

Dental formula I 3/3 + C 1/1 + P 2/2 + M 1/1 = 28 [29, 1012].

Mammae formula 0 + 1 + 1 = 4 [1200].

Genetics Chromosome: $2n$ = 38 [3900]. The taxonomic status of the Iriomote cat has long been debated. The phylogenetic status of the Iriomote cat has been revealed by ribosomal DNA analysis [3346] and mtDNA sequence analysis [1987, 3476].

4. Ecology

Reproduction Mating season ranges from February to May. The kittens are born from April to July and become independent from the mother between August and December [2676]. The litter size is recorded as 2, based on the number of fetuses of an adult female killed by a car [99]. Descent of the testis is observed at 7 months old in captivity [887].

Prionailurus bengalensis iriomotensis (Imaizumi, 1967)

Lifespan The maximum life span is recorded as 13 years for a female in the wild [2401]. In captivity, maxima of 13 and 14 years have been recorded for males [886, 1065].

Diet A variety of animals including crustaceans (e.g. *Macrobrachium* spp.), insects (e.g. *Cardiodactylus novaeguineae* and *Rhabdoblatta yayeyamana*), amphibians (e.g. *Fejervarya sakishimensis* and *Rhacophorus owstoni*), reptiles (e.g. *Japalura polygonata ishigakiensis* and *Plestiodon kishinouyei*), birds (e.g. *Amaurornis phoenicurus* and *Turdus pallidus*), and mammals (e.g. *Rattus rattus* and *Pteropus dasymallus yayeyamae*) [2964, 3846].

Habitat Iriomote cats inhabit the whole area of Iriomote-jima Isl. except for residential areas [2399]. The island has an oceanic subtropical climate, and 80% of the area is covered with subtropical evergreen broad-leafed forest. High precipitation (> 2,000 mm/year) maintains many rivers, streams, swampy forests and mangrove forests.

Home range Home range size is 4.47 ± 2.24 km^2 (mean $\pm$ SD, $n = 20$, range 1.26–9.65) for males and 2.80 ± 1.08 km^2 ($n = 11$, 1.17–5.00) for females. Females have stable home ranges. Males maintain relatively exclusive home ranges among their own sex and enlarge their home ranges in the mating season. Male home ranges overlap with those of 1 or 2 females [2402, 2403, 3148]. Some males roam without fixed ranges as transients [2403].

Behavior Solitary except for the mother–kitten relationship. The activity pattern of this species is basically crepuscular, showing 2 peaks of higher activity at dusk and dawn [1195].

Natural enemies The Iriomote cat has no native predators on the island. The domestic cat and the dog are a potential competitor and predator, respectively [1197, 3846].

Parasites With respect to endoparasites, 8 species of helminth and hepatozoonosis have been reported [1753, 4074]. Three ixodid, 3 trombiculid and 1 mallophagan species have been recorded [815].

5. Remarks

The Iriomote cat is protected by law as a "Special Natural Monument" (1977) and as a "National Endangered Species" (1994). A paved road along the northeastern coast cuts across the home ranges of the cats. In recent years, road kills are the most serious threat.

Some researchers regard the Iriomote cat as a subspecies of the leopard cat *P. bengalensis*, based on genetic studies [1987, 3346]. Meanwhile others treat the Iriomote cat as an independent species *P. iriomotensis* [2518]. Thus its taxonomic status is still controversial and the best taxonomic treatment for the Iriomote cat is as *incertae sedis* [3894].

N. NAKANISHI & M. IZAWA

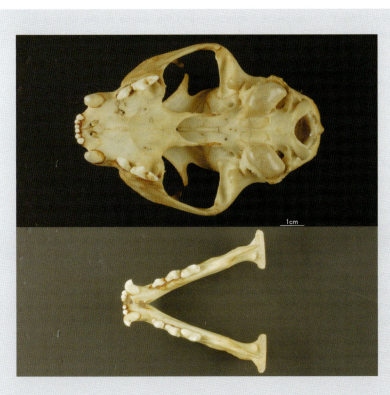

A skull specimen (RUMF-ZZ-00041) of the Iriomote cat stored at Fujyukan, University of the Ryukyus (Mammal Ecology Laboratory, University of the Ryukyus). The Iriomote cat does not have second premolars in its upper jaw although the Tsushima leopard cat does.

RESEARCH TOPIC 6

The feral cat (*Felis catus* Linnaeus, 1758) as a free-living pet for humans and an effective predator, competitor and disease carrier for wildlife
M. IZAWA

1. History and ecology of the domestic cat

"Feral cats" are domestic cats (*Felis catus*) that live independently from humans and reproduce in the wild. They are considered to have been domesticated in ancient Egypt from the small felid, *F. silvestris lybica*. Domestication of the cat followed a different process and purpose from those of other domestic animals such as cattle and dogs. The original purpose of domestication of the cat was for pest control. In return for protection of grain stores from rodents, cats were provided with shelter and food. Thus, their behavior, movement and reproduction were never controlled and they were expected to retain their natural hunting ability. Since the first domestication, the cat has moved to many countries along with the spread of human beings. In the history of the relationship between cats and humans, the domestic cat has filled the roles of grain store keeper, god, witch, and companion animal.

Domestic cats, including house cats belonging to fixed owners and feral cats, show high flexibility in their food habits, social system and habitat utilization [558, 1196, 1198, 1199, 1847]. The basic social system of most felids, including domestic cats, is a solitary way of life. Most likely, this is an adaptation to their hunting style in forests. However, cats inhabiting urban areas and depending upon humans for their food (feeding in garbage sites and direct feeding by humans) often live in groups; as a consequence, they reach higher population densities. The social system and diet of domestic cats can vary with the abundance and distribution of food resources. Despite their reliance on food obtained from humans, domestic cats are highly carnivorous and skillful hunters, and they maintain their hunting ability even in urban areas. Feral cats have been reported to hunt mainly small mammals in many areas throughout the world (e.g. [173, 558, 1626]).

2. Impact on wildlife

Such a history of domestication and high ecological flexibility allowed the domestic cat to easily become feral and disperse to a wide range of environments from urban to natural habitats. Now, the feral cat is regarded as an alien species having the broadest distribution, and as such is listed in the "100 of the World's Worst Invasive Alien Species" by the IUCN.

The feral cat may affect ecosystems in various ways: predation on native animals in lower levels of the food chain, competition with native animals occupying similar ecological niches, and hybridization with and disease transmission to wild felids.

The biggest problem observed in many areas is predation on native animals, which is particularly serious on small islands. Islands frequently lack native carnivorous mammals due to limited size and isolated geographic locations. As a consequence, resident animals have not evolved effective anti-predatory strategies against carnivorous mammals. Many examples of such animals can be found even in the Ryukyu Isls.: the Okinawa rail *Gallirallus okinawae*, the Okinawa spiny rat *Tokudaia muenninki*, the Ryukyu robin *Luscinia komadori*, and the Daito flying fox *Pteropus dasymallus daitoensis*. Many endemic animals on the Ogasawara Isls. suffer from the same predation pressure (e.g. [1032]). Serious threats from feral cats also have been observed in breeding colonies of seabirds located on small islets and rocks. This is the main reason that feral cats are listed in the "100 of the World's Worst Invasive Alien Species".

A particular threat related to domestic cats concerns Iriomote-jima Isl. and the Tsushima Isls. Two endangered wild felids inhabit these islands: the Iriomote cat *Prionailurus bengalensis iriomotensis* and the Tsushima leopard cat *P. b. euptilurus*. Possible threats include competition for food and habitat [3846] and hybridization. However, an unexpected problem occurred on the Tsushima Isls. One Tsushima leopard cat was found to be a carrier of Feline Immunodeficiency Virus (FIV) in 1996. DNA analysis of the virus revealed that it most likely was transmitted from domestic cats [1194, 2488]. Since then, two more individuals were confirmed to be carriers of this virus [3667]. There are important problems related to this disease: FIV is lethal, there is no effective vaccination against it, and the source of the virus is the domestic cat. No virus-carrier has been found in the Iriomote cat population. However, considering the population sizes of the two wildcats are about 100 individuals each and they occur in closed and insular habitats, an infectious lethal-disease may be a critical spark leading to their extinction.

3. Ways to solve feral cat problems

When dealing with feral cat problems, the following issues should be considered. 1) It is difficult to distinguish feral cats from house cats. Moreover, house cats easily become feral. 2) Free-ranging house cats may sometimes cause the same damage to wildlife as feral cats. 3) There have been no legal regulations concerning domestic cats in Japan because they do not cause any direct damage to humans and livestock. In contrast, free-ranging domestic dogs have been controlled strongly by law because of their ability to transmit rabies and their potential danger to humans and livestock. 4) If feral cats have formed an enduring population for many years, they may also have occupied a certain niche at the top of the food chain in the ecosystem. Therefore a sudden eradication of domestic cats may result in unpredictable ecological consequences for the local structure of an ecosystem. 5) Killing cats is opposed strongly, because they are considered pet animals.

There are two sensible approaches that aim at solving cat-related problems. The first is to restore natural wildlife habitat of wildlife to a cat-free condition. The second is to stop the release of more domestic cats into the wild. These two approaches are most effective when implemented concurrently.

The first approach (restoration of a cat-free habitat) has been started by local governments and the Ministry of the Environment, partly with the joint effort of NPOs (e.g. in Yanbaru [northern part of Okinawa-jima Isl.; Kunigami-son, Ohgimi-son and Higashi-son, Okinawa], Iriomote-jima Isl. [Taketomi-cho, Okinawa], the Ogasawara Isls. [Tokyo]). The life of feral cats is closely related not only to natural habitats but also to human life. The cat-removal program conducted on Iriomote-jima Isl. was triggered by the commencement of garbage dump management. Disposal of garbage was a public concern on the island. As the garbage sites had sustained feral cat populations, after they were cleared, the cats likely dispersed into the forests. This would cause negative interactions with wildlife including the Iriomote cat. Regarding wildlife conservation, it was clear that garbage management should be done simultaneously with cat control procedures.

Before commencement of the feral cat removal program, some local ordinances needed to be established: Ogasawara-son (Tokyo) in 1999, Taketomi-cho (Okinawa) in 2001, and Kunigami-son, Ohgimi-son and Higashi-son (Okinawa) in 2005. These ordinances outlined the obligations of cat owners concerning their pets' management; e.g. registration of house cats, identification of each cat in some manner such as using microchips, and control of breeding and free-ranging. Now many activities are involved in the removal of feral cats. The Government office checks the ownership of captured cats by registration, and the NPO keeps them in a shelter while looking for a new owner.

The second approach (preventing the increase of feral cats) also is conducted by the Government, NPOs and local people. An educational program includes descriptions of cat husbandry, promotion of birth control and castration, and garbage control to reduce food resources for feral cats.

Feral cat problems are attributed mostly to individual cat owners. However, other people also are causing the increase of feral cat populations through the garbage supply and supplementary feeding. Despite these problems, the situation has improved in recent decades and the goal of reducing the impact of feral cats on wildlife appears more realistic.

An Amami rabbit attacked by a feral cat on Amami-ohshima Isl. (Amami Wildlife Conservation Center, Ministry of the Environment).

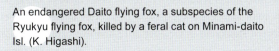

An endangered Daito flying fox, a subspecies of the Ryukyu flying fox, killed by a feral cat on Minami-daito Isl. (K. Higashi).

Carnivora   URSIDAE

103

Red list status: LP for the populations in the west side of Ishikari District and Teshio-Mashike Mts. (MEJ); C-2, LP for the populations in south-western and northern Hokkaido and Konsen Plain (MSJ); LC (IUCN)

Ursus arctos Linnaeus, 1758

EN brown bear   JP ヒグマ (higuma)   CM 棕熊   CT 棕熊   KR 불곰   RS бурый медведь

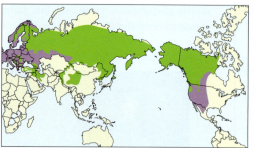

A female brown bear with two cubs on a seashore path in Shiretoko National Park, Hokkaido (M. Yamanaka).

1. Distribution

Distributed in Europe, Asia, and North America [3167]. In Asia, the species can be found in parts of Turkey, Syria, Lebanon, Iraq, Iran, the former Soviet Union, Mongolia, North Korea, China, India, Pakistan, and Japan. In Japan, the brown bear occurs only in Hokkaido. Their distribution decreased from the late 1900s, but has increased since 2003. They now cover approximately 60% of Hokkaido, a proportion that roughly corresponds to the forested area [1210, 1212], whereas originally they were distributed throughout the island [1948].

2. Fossil record

Some fossils of brown bears or related species dating to the middle and late Pleistocene have been found on Honshu [3446]. Brown bears might have become extinct on Honshu in about 17,000 y BP [1517].

3. General characteristics

Morphology   The largest terrestrial animal in Japan. The skulls of Hokkaido bears are, however, generally smaller than those of some of the other subspecies of northeastern Asia [265]. The pelage of most of the body is dark brown, but there is usually black fur with a golden portion in its ends from the face to their back. Occasionally bears have a white patch on the chest. Ininkari bear, having white fur only on the upper half of the body, are observed in Kunashiri (Kunashir) and Etorofu (Iturup) Isls. [3096]. The species has long spinous processes of the vertebrae and a hump on its shoulders. There have been only a few published reports on external measurements of wild bears in Hokkaido. For live-trapped adult bears from the Oshima Pen. in southern Hokkaido, BW (mean ± SD) is $81.7 ± 17.2$ kg ($n = 17$) for females and $127.6 ± 33.9$ kg ($n = 8$) for males [931]. BW of three adult females live-trapped in Urahoro, in central-eastern Hokkaido, was $104.7 ± 6.4$ kg ($n = 3$) [3031]. Among adult bears live-trapped on the Shiretoko Pen. in eastern Hokkaido, BW (mean, no SD given) was 102.9 kg ($n = 31$) for females and 192.4 kg ($n = 7$) for males [1667]. A credible maximum recorded BW for an adult male is 465 kg and for an adult female 175 kg; the two bears were hunted at Onbetsu, in eastern Hokkaido in October 2012 and live-captured at Urahoro in November 2012, respectively [3089]. Other measurements are as follows: HB = $180.0 ± 18.0$ cm ($n = 8$), Front Paw Width = $14.7 ± 1.4$ cm ($n = 8$) for adult males, $156.1 ± 11.4$ cm ($n = 17$), Front Paw Width = $11.2 ± 0.6$ cm ($n = 17$) for adult females on the Oshima

Pen. [931]. For captive brown bears at Noboribetsu Bear Park in Hokkaido, the maximum BW is 440 kg for males and 221 kg for females [1937]. The mean BW of 4- to 6-year-old bears at this Bear Park is 219.0 kg for males and 115.0 kg for females [1937]. CBL increases from southwest to northeast in Hokkaido [2610, 4146]. There are 3 lineages of the mtDNA control region in brown bears (see the genetics section below) [2003]. One lineage is specific to each of the southern, central, and eastern regions. Comparison of cranial and dental morphology among bears of the three lineages revealed that the skull is smallest in the southern haplotype group and largest in the eastern group [265]. In particular, bears from eastern Hokkaido, including Kunashiri and Etorofu Isls., have significantly larger skulls, smaller cheek teeth, and broader faces.

Dental formula   I 3/3 + C 1/1 + P 4/4 + M 2/3 = 42. First to third pairs of premolars are small and some of these teeth are often missing [29].

Mammae formula   2 + 0 + 1 = 6 [29].

Genetics   Chromosome: $2n = 74$, similar to that of the other ursids except for the giant panda (*Ailuropoda melanoleuca*: $2n = 42$) and spectacled bear (*Tremarctos ornatus*: $2n = 52$) [2444, 2445]. Phylogenetically *Ursus arctos* is closest to the polar bear (*U. maritimus*) and the cave bear (*U. spelaeus*), the latter being an extinct species [1864, 3475]. mtDNA analysis of Hokkaido brown bears has revealed 3 genetically distinct lineages located allopatrically in north-central (east to Ishikari Depression and north to Kushiro region), eastern (Akan-, Shiretoko-areas, Kunashiri and Etorofu Isls.), and southern (south to Ishikari Depression) Hokkaido [912, 2003]. Divergences between southern lineage and the other 2 lineages, and between the north-central and the eastern lineage has been estimated to have occurred approximately 268 ky BP and 165 ky BP, respectively, indicating that the lineages were divided not in Hokkaido but on the Asian continent [912]. A phylogeographic study of brown bear populations worldwide indicates that the north-central lineage is close to the continental Eurasia and western Alaska lineage, the eastern lineage is close to the eastern Alaskan lineage, and the southern lineage is close to the North America lineage [911, 912]. Immigration of brown bears to Hokkaido from the continent could have occurred at least 3 times through landbridges that formed around the Japanese Isls. in the Pleistocene [912]. The genetic structure includes 3 subpopulations and 3 admixed groups has been confirmed in a consecutive population in eastern Hokkaido [1145].

4. Ecology

Reproduction   Observations under captive conditions have revealed that the breeding season of the brown bear is from May to July [3629]. Fetuses start growing after delayed implantation and grow for about 2 months [3631]. Cubs are born in dens between late January and early February at about 400 g in BW [1901]; the cub is accompanied by its mother for the first year and experiences its first hibernation with her. The mating system is

A male brown bear (277 kg) captured in a barrel trap for satellite tracking in the Shiretoko National Park, Hokkaido (M. Yamanaka).

polygamous or promiscuous [418, 3308]. A case study of bears harvested on the Oshima Pen., in southwestern Hokkaido, found that the minimum age of parturition is 4 years, but reproductive success of females younger than 6 years old is low [1949]. On the Oshima Pen., the mean interval between births was 2.3 to 3.0 years, and the mean litter size was 1 for females younger than 7 years old and 1.8 for females of 7 years or older (maximum 3) [1949]. In the Rusha area of the Shiretoko Pen., located in eastern Hokkaido, where food resources may the be richest in Hokkaido, the fertility rate is estimated to range from 0.709 to 0.960 cubs per female per year [1667].

Lifespan   In the wild, there are records of 2 female bears giving birth to young at 26 years of age [160]. One female bear was killed at age 34 years [160], and 1 male was killed at age 30 [929]. The maximum lifespan in captivity is 38 years for a female bear raised at Higashiyama Zoo [228]. However, 98% of 823 bears killed for nuisance control or sport hunting were 16 years old or younger [930]. The mean age of male bears killed is substantially lower than that of females; this may be because young males (1 to 4 years old) are likely to disperse more widely and are therefore more likely to come into contact with humans [930, 1946].

Diet   Omnivorous but feeds mainly on plant material. Herbaceous plants are the dominant food in spring and summer, whereas fruit is dominant in autumn [39, 159, 2609, 4016]. Animal materials in the form of insects such as ants of the Formicidae and wasps of the Vespidae are also consumed in summer. The number of plant species eaten by brown bears is 50 in the Daisetsu Mts. of central Hokkaido [1142], and 56 on the Shiretoko Pen. [4017]. In the 1990s, some studies showed extensive use of agricultural crops in late summer (August and September) and an increase in the consumption of sika deer (*Cervus nippon*) as a result of the increasesing sika deer population in eastern Hokkaido [1641, 3090, 3095]. Brown bears eat various items in late summer, such as premature herbaceous plants, spawning fish, early-maturing fruits, and crops. In most regions of Hokkaido, bears use crops because of shortages of alternative natural foods in this season. Fruits are important in autumn. In years of low mast production by their major food sources, brown bears increase their use of crops as alternative food sources [3091].

Habitat Brown bears are found over a broad range of habitats in the Northern Hemisphere from the northern arctic tundra to dry desert [3167]. In Hokkaido, the bears use various habitats from forested mountains to the boundaries of cultivated lands or human residential areas. Habitat also ranges from sea level to an alpine zone around 2,000 m above sea level. However, the most important habitat is deciduous broad-leaved forest and/or mixed forest of coniferous and deciduous broad-leaved trees [931, 3094].

Home range The species has no territoriality. Adult females have a smaller annual home range than adult males. Females use almost the same areas for years, and their home ranges overlap with those of neighboring females. The annual home range size (mean ± SD) of adult females (100% MCP) was 13.4 ± 9.9 km^2 on the Shiretoko Pen. ($n = 10$) [4015], 15.1 ± 3.49 km^2 on the Oshima Pen. ($n = 19$) [931], and 43.0 ± 9.52 km^2 in the Urahoro region ($n = 5$) [3094]. These sizes are smaller than those reported from North America and Europe [4015]. There are few reports of the home range size of adult males because of the difficulty in tracking males over their very broad areas. The annual home range size of adult males (100% MCP) is 199 to 462 km^2 on the Shiretoko Pen. ($n = 2$) [4015] and 277 to 496 km^2 in Tomakomai region ($n = 3$) [3833].

Behavior Brown bears live solitarily, except for a dependent offspring with its mother, siblings after separation from their mother, and temporary couples in the mating season. In some places, called "ecocenters," such as along streams where spawning salmon are running, in crop fields, and on grasslands, brown bears aggregate and have some social hierarchy among individuals [417]. Separation of offspring from their mothers is considered to occur at age 15 to 27 months [1949]. After the separation, young males disperse from their natal places [300, 726], although there have been only a few illustrative cases in Hokkaido [1144, 1667, 3032]. During the mating season, tree-rubbing behavior is observed frequently in the wild; it is believed to have some communicative function among individuals [3093]. Field observations reveal that Hokkaido brown bears enter their dens to hibernate during late November to mid-December and emerge from hibernation during late March to late April. After emerging from the den, females with cubs start their movement later than lone adults and subadults [161, 1947]. Hokkaido brown bears hibernate in self-dug dens. In most cases dens are dug under tree roots spread over a slope [1060, 2627].

Natural enemies Although wolves (*Canis lupus*) can compete with brown bears [709], they were eliminated from Hokkaido in the late 19th Century. There are no predators of the species,

A young brown bear in Shiretoko, Hokkaido (T. Kuwahara, Kon Photography & Research).

except as occurs in intraspecific infanticide. Although direct observations of intraspecific killings are rare, there are a few reports of the killing of dependent offspring by adult males in Europe and North America [273]. These infanticides by males are considered to be a strategy to increase reproductive success [338, 3381]. However, in Hokkaido there is as yet no record of infanticide among brown bears.

Parasites Several helminth, protozoa, and tick species have been reported from brown bears in Hokkaido [214, 812, 1239, 1752, 2615, 3331].

5. Remarks

Game animal in Japan. Killed for sport hunting and nuisance-control purposes. The average annual number of bear kills in the last decade (2004–2013) was 82 for sport hunting and 476 for nuisance control [928]. Major conflicts between humans and brown bears causing human injuries, agricultural crop depredation, and latent fear of attacks, are dominant causes for nuisance control. In the last decade (2004–2013) 10 people were killed and 19 injured by bears in Hokkaido [928]. Most of these incidents occurred in forested areas and were caused by accidental encounters with bears. The amount of agricultural damage by brown bears is increasing gradually [928]. Although various practices such as non-lethal damage management and public education about coexistence between humans and brown bears are being implemented in some areas of Hokkaido [931, 3834], the major option for resolution of human-bear conflicts in most areas of Hokkaido is still the nuisance-control kill. Human-bear conflicts have never decreased, whereas the number of bear kills for nuisance-control in Hokkaido is increasing [3030, 3092]. Small and isolated populations on the west side of the Ishikari and Teshio-Mashike districts are listed in the Red List for Japan and the Red Data Book for Hokkaido [927, 2083].

Y. SATO

Carnivora   URSIDAE

(104)

Red list status: LP for the populations in Shikoku, Chugoku District, Kii and Shimokita Pens. (MEJ); C-2 as *Selenarctos thibetanus*, LP for some populations in Honshu, Shikoku, and Kyushu (MSJ); VU (IUCN)

Ursus thibetanus G. Cuvier, 1823

EN Japanese black bear, Asiatic black bear, Asian black bear, moon bear   JP ツキノワグマ (tsukinowaguma)
CM 黑熊   CT 亞洲黑熊   KR 반달가슴곰   RS гималайский медведь, белогрудый медведь

An adult Japanese black bear on a willow tree in Nikko National Park, Tochigi Pref. (H. Yokota).

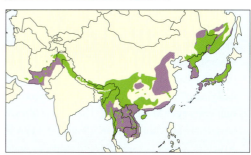

1. Distribution

Although fossil remains of the species have been found as far west as Germany and France [708], their distribution is now limited to Asia, from Japan in the east to Iran in the west [3168]; Iran, Afghanistan, Pakistan, India, Nepal, Bhutan, China, Bangladesh, Myanmar, Thailand, Laos, Cambodia, Vietnam, North and South Korea, Russia, Taiwan, and Japan [1211]. In Japan, they were originally widespread over Honshu, Shikoku and Kyushu; however, the population of Kyushu is thought to have gone extinct during the 1940s [1212, 2083]. The distribution area has been increasing over the last decade except Shikoku [1210, 4042].

2. Fossil record

A fossil of *U. thibetanus*-like bear is known from the late Middle or the Late Pleistocene of Tochigi Pref., although it was originally described as an extinct species, *U. tanakai* [3203]. Another fossil of the same specific status is known from the late Middle Pleistocene of Aomori Pref. [3897]. Several archaeological bear-remains dating from about 3,000 years ago have been excavated in western Japan. This suggests that bear habitat in western Japan has been limited since very early times, because rice paddy fields are dominant features on the plains in that area [1670].

3. General characteristics

Morphology   Medium-sized black colored bear, with a light muzzle and relatively large ears. A white "crescent moon" on the chest, which is sometimes in a V shape [3168, 3309] is distinct. Occasionally reddish-brown phases (*U. t. gedrosianus*) are also known [1361], and recently a blond color phase was discovered in Cambodia and Thailand [696]. The front half of the body is heaver than the rear; the front legs are stronger and longer than the hind legs, for tree climbing [3763]. In Russia, BW (mean ± SD) of sub-adult and adult is 101 ± 43.4 kg ($n = 13$) for males, and 70 ± 15.4 kg ($n = 11$) for females [329]. Mature males may weigh up to 250 kg after accumulating fat in autumn [182]. The Japanese black bear, *U. t. japonicus*, is a relatively small subspecies (like the other island subspecies, *U. t. formosanus*); unlike some other subspecies, it has a dark colored muzzle and lacks prominent bushy cheeks [2848, 3155]. The white chest marking is sometimes small or absent. External measurements (mean ± SD) for adults (≥ 4 years old) are as follows: BW = 62 ± 22.0 kg (n = 17) for males, and BW = 36 ± 7.5 kg (n = 13) for females in Okutama area of Chichibu-Tama-Kai National Park (NP), BW = 71 ± 23.7 kg (n = 8) for males and BW = 42 ± 5.8 kg (n = 6) for females in Ashio area of Nikko NP; HB = 129.0 ± 10.4 cm (n = 17) for males and HB = 115.0 ± 6.0 cm (n = 13) for females in Okutama area; HB = 129.7 ± 10.8 cm (n = 8) for males and HB = 124.8 ± 73.4 mm (n = 6) for females in Ashio area. Front Paw Width = 97 ± 5.2 mm (n = 17) for males and Front Paw Width = 81 ± 3.7 mm (n = 13) for females in Okutama area; Front Paw Width = 98 ± 5.3 mm (n = 8) for males and Front Paw Width = 85 ± 5.2 mm (n = 6) for females in the Ashio area (all measurements based on the author's database). Males are larger than females in skull size [118, 2522]. Most skull measurements cease to increase after 5 years old in males and 4 years old in females [118]. There are significant differences in skull morphology (skull width and masticatory organs) in Iwate Pref., where a river divides the bears into two populations, suggesting that gene flow between these two populations has been restricted by the river basin [79].

Dental formula I 3/3 + C 1/1 + P 4/4 + M 2/3 = 42 (some of P1-3 often missing) [29].

Mammae formula 2 + 0 + 1 = 6 [29].

Genetics Chromosome: $2n$ = 74, similar to that of the most of the Ursidae (all of the Ursinae) [2445, 2445]. MtDNA analysis suggests that the Asiatic black bear, the American black bear (*U. americanus*), the brown bear (*U. arctos*), and the sun bear (*U. malayanus*) diverged from a common ancestor about 2–3.5 million years ago (MYA) [3811]. The Asiatic black bear is thought to have come to Japan between 0.3 and 0.5 MYA [464], and genetic structure shows considerable divergences among local populations [1071, 2955]. Microsatellite DNA analysis has shown that populations to the west of Yura River have lower genetic diversity in comparison with larger populations in central Honshu [2629, 2955]. Such loss of genetic diversity may be associated with an increased risk of extinction; it is also a concern for other isolated populations, such as those in Shikoku, the Shimokita Pen. and the Kii Pen. [2658].

4. Ecology

Reproduction Japanese black bears attain physiological puberty at 2–4 years old in males [1680] and 4 years old in females [1408]. In the wild, breeding may occur later. The breeding season spans from June to August [4000]. This is followed by a period of delayed implantation lasting several months. Females give birth to young between January and February (as observed in captivity) and nurture cubs during the rest of hibernation [3083]. The mean litter size in the wild is 1.86 [1408]. The nutritional condition of the mother affects reproductive success, including implantation, fetal growth, parturition, litter size, and cub growth and survival [3630].

Lifespan Although there are no data on maximum lifespan in the wild, harvest records include a 23 year old female and 25 year old male bear killed on Honshu [1213]. The mean age of nuisance-killed bears in Tochigi Pref. is 5.8 years old for males (n = 63) and 6.2 years old for females (n = 26) (excluding those under 3 years old, because the number of young nuisance bears varies dramatically among years) [1976]. The mean age appears to be declining due to high numbers of nuisance kills [1976].

Diet Although omnivorous, Japanese black bears feed more on vegetation than meat, and food habits vary seasonally [832]. In spring, bears eat herbs, young leaves and buds of trees, and if available, nuts that fell in the previous autumn. In summer, bears turn to colonial insects (e.g. bees, ants), and also eat herbs and berries. In autumn, hard mast such as beech masts (*Fagus crenata*, *F. japonica*) and acorns (*Quercus crispula*, *Q. serrata*) are the staple food in northen and central Japan [616, 832, 971, 4035, 4041]. Southern bears had the same food item selection as northern and central bears, but tended to consume more *Swida* spp. fruits in autumn [2657]. During June to August, bears sometimes feed on the cambium layer of conifers (such as hinoki cypress *Chamaecyparis obtusa* and Japanese cedar *Cryptomeria japonica*) [686, 3839, 4037, 4173]; de-barking tends to occur when other food availability is poor in the area [4173]. The species occasionally hunts newborn sika deer (*Cervus nippon*) in early summer (from the author's original data in Nikko NP). It has been reported that Japanese black bears utilize 90 species of fruits, both soft and hard mast. In autumn they concentrate on lipid-rich fruits (nuts), unlike other omnivores such as Japanese martens (*Martes melampus*) and raccoon dogs (*Nyctereutes procyonoides*), which do not hibernate [1673]. Recently, stable isotope ratios (e.g. carbon and nitrogen) of hair have been used as a tool to investigate feeding history, particularly the relative use of meat and corn [2172, 2409].

Habitat The distribution of the species roughly coincides with forest distribution in southern and eastern Asia. In Japan, the black bear uses a variety of forested habitats from foothills to an elevation of 3,000 m [1204]. There is a close association between the geographic distribution of bears and that of broad-leaved deciduous forest dominated by beech and oak [767]. However, the total area of planted conifer forests has reached about one half of the total forest area after World War II, and thus habitat quality has been diminishing for the Japanese black bear [873, 2658].

Home range In the Nikko-Ashio area of Nikko NP, where bears shift their range according to the seasons, home range size for consecutive years (100% MCP) are large, as investigated by GPS telemetry: 256 km^2 (226.8 and 284.6, n = 2) for adult males and 205 km^2 (161.8 and 247.8, n = 2) for adult females [4036]. Home ranges (mean ± SD) for consecutive years (100% MCP) investigated by VHF telemetry are 46 ± 32.0 km^2 (n = 4) for adult males and 23 ± 9.7 km^2 (n = 5) for adult females in the Okutama area of Chichibu-Tama-Kai NP, where bears have a high fidelity to annual home ranges [4035]; and 93 ± 34.3 km^2 (n = 3) for adult males and 55 ± 25.0 km^2 (n = 4) for adult females in the Northern Japanese Alps [1204]. Thus, there are home range size differences between areas, and adult females have a smaller home range size than adult males. In the Chichibu area of Chichibu-Tama-Kai NP, the female home range size expanded in summer and shrank in autumn [830]. In the alpine zone, there is a remarkable change in elevational use with the seasons [1204]. In low mast production years, home range sizes became much larger than normal years [1742]. However, the influence of movement distance of females was larger than that of males, thus masting influenced the behavior of females more strongly than males [1672].

Behavior In northern latitudes, where food becomes unavailable in winter, both sexes hibernate. In the tropics, however, the species does not hibernate, except females giving birth to young during winter [972]. In Japan, the hibernation period lasts 5 or 6 months, from November to April. They hibernate in hollow trees, under large rocks, or in holes they excavate in the ground [873]. Emergence from dens varies with individual conditions. Lone females emerge about one month earlier than females that gave birth during the winter [2658]. These bears are solitary except for dependent cubs with their mother, siblings sometimes remaining together for a short time after separation from their mother, and temporary pairing in the mating season. Territoriality has not been observed, however males may socially exclude females from rich stands of hard mast [970]. The species is basically diurnal with activity peaks in early morning and at twilight [4040]. However, their nocturnal activity becomes higher near human settlements [189]. The level of their activity time budget gradually increased in spring and reached a peak in July, then decreased and reached a trough in late August, and increased and reached a peak again in October (from the author's GPS study data in Nikko NP) [1743]. All sexes and ages readily climb trees, for feeding or resting [3763]. When Japanese black bears feed in hard mast trees, they often break branches inward to reach the nuts, and then pile the branches up in the canopy, forming a seat (day bed) called "kumadana". They also sometimes make a ground nest with grass such as *Sasa* bamboos. They may rest and eat while in such nests [873]. Recently, it has been suggested that Japanese black bears are potentially effective seed-dispersers of the yamazakura cherry (*Prunus jamasakura*), transporting a large number of the seeds, and causing no damage to their germination [1671, 1674].

Natural enemies The Japanese black bear is the only large carnivore on Honshu and Shikoku. Their only natural enemies are other bears (and people). There is one record of infanticide by a bear in Ashio area of Nikko NP in 2006 [4036].

Parasites Many helminth species have been reported [3649, 3725, 3726, 4121]. Several tick species are known [3408, 3665].

5. Remarks

Every year, approximately 500 bears are harvested by sport hunters and 1,000 to 2,000 bears are killed as nuisance bears on Honshu [2658]. Conflicts between humans and bears can be categorized as one of two types: those in which the bears caused human injuries or death [157, 1213], and those in which the bears caused damage to agricultural crops [2658], livestock [3911] or planted conifer trees [686, 3839, 4037, 4173]. In the last decade, mass intrusions of Japanese black bears have occurred regularly, and have caused serious damage both to bears and humans. Mass intrusions refer to episodes when bears range into human-settled areas, and the numbers of nuisance kills are 2–3 times higher than normal years. In such years, over 2,000 to 4,000 bears are killed and around 100 people are injured [1213, 4042, 4139]. Bear food shortages in the mountains, the changing structure of *satoyama* (a traditional agricultural area between the mountains and town), and the lack of a scientific management system could all be causes for the increasing appearance of bears near human settlements in recent years [4038]. In central Japan, bears tend to select areas with high human access in summer [3436]. Because bear distribution has been rapidly expanding on some parts of Honshu [1210, 1212, 4042], there is an urgent need to formulate better management plans.

K. YAMAZAKI

A cub of the Japanese black bear in the Ashio Mts., Tochigi Pref. (K. Yamazaki).

RESEARCH TOPIC 7

Conservation and management of large mammals focusing on sika deer and bears in Japan

K. KAJI

1. Current status of sika deer and bears

Sika deer (*Cervus nippon*) populations remained at low density in the mid-1950s due to overharvesting, including illegal hunting. They started rebounding in the 1980s such that today they are widespread and abundant. Nationwide surveys of wildlife distribution were conducted by the Ministry of the Environment (MOE) in 1978 and 2003, revealing that the range of the sika deer has expanded by 1.7 times over that interval [3583]. The number of harvested deer remained about 30,000 in the 1970s and increased from about 100,000 in the early 1990s to about 410,000 in 2011. The increased harvest of deer, despite a drastic decline in the number of hunters, is a reflection of the rapid increase of the deer population. Associated with the significant population increase of deer, damage to agriculture and forestry by deer also increased. Comparing damage levels (damaged area) in 2012 to the beginning of the 1990s, damage to agriculture increased 3-fold and forestry 2-fold. Feeding by sika deer heavily affected vegetation on the floor of natural forests, leading to erosion in some areas. Associated with sika deer range expansion, the deer have influenced the natural vegetation of nearly half of the 83 national parks and quasi-national parks [1212]. The greatest concerns with deer damage are in Shiretoko National Park [1270, 3587] and Yaku-shima National Park [3918], both registered as World Natural Heritage Sites, where some endemic and rare plant species are threatened with extinction.

Originally, the Japanese black bear (*Ursus thibetanus*) was distributed widely over Honshu, Shikoku and Kyushu. Its numbers have declined especially since the 1950s due to forest cutting and nuisance kills [1211]. Brown bears (*U. arctos*) inhabited all of Hokkaido before the beginning of modern development in the late 19th century, and decreased due to habitat changes particularly from forest cutting and over-harvesting from the 1950s to 1970s [1211]. Black and brown bears slightly expanded their distributions between 1978 and 2003 [1212]. As of 2003, most of the areas they inhabited were relatively stable. The exceptions were on the Shimokita Pen. and in western parts of Japan, where habitat degradation has caused some bear populations to become isolated; they are listed as threatened local populations in the Red Data Book [1211]. Japanese black bears were culled mainly in eastern Japan and the annual harvest number was about 2,000 between the late 1960s and 1970s. The harvest number decreased to about 1,500 animals by the mid-1990s due to hunting regulation. Beginning in 2000, harvest levels began to increase because unusually high numbers of black bears began appearing in mountain villages and other residential areas in 2004, 2006, 2010 and 2014, and 2,221, 4,335, 3,053 and 3,369 bears were killed in those years, respectively. The cause of the mass occurrence of black bears in villages and surrounding areas was suspected to be related to bears searching for food because of reduced production of forest foods such as beechnut mast [1211, 2664].

2. Large mammals and severe damage to agriculture, forestry and ecosystems

Conflicts between wildlife and humans increased during the last two decades. A contributing reason may be the range expansion and population increases of large mammals such as sika deer, brown and Japanese black bears, wild boars (*Sus scrofa*) and Japanese macaques (*Macaca fuscata*). At least 5 factors affect the range expansion of these species. 1) The initial recovery of wildlife populations was brought about by legal protection and enforcement of hunting regulations. 2) Large scale forest cutting and conversion of natural forests to plantations threatened wildlife habitat between the mid-1950s and mid-1960s. These changes contributed to the decrease in bear habitat and increase in deer habitat. 3) The area of abandoned farm land increased three fold during the last 25 years, between 1985 and 2010, and reached 396,000 ha due to depopulation and acceleration of an aging society in farm and mountain villages [2080]. The abandoned farm land in mountainous and nearby areas provided good habitat for large mammals. 4) The hunter population reached a peak of 530,000 people in the 1970s, thereafter it drastically decreased to 181,000 people by 2012. The age structure of hunters shifted toward older age classes and aged hunters (60 years old and older) accounted for more than 60% of the hunter population. This decreasing hunting population and the increasing age of hunters resulted in the decline of overall hunting effort. 5) Recent shallow snowfall and mild winters combined with high nutrition from agricultural crops might be contributing to increasing wildlife population growth rates.

3. Responsibility for wildlife management in Japan

Originally, wildlife management in Japan was controlled by prefectural governments through sports hunting and nuisance kills under the "Wildlife Protection and Hunting Law" (WPHL) [2082]. In the changing socio-economical and natural environments of Japan, overabundant large mammals such as sika deer and wild boars have caused serious damage to agriculture, forestry and ecosystems, whereas some local populations of Japanese black bears have decreased to threatened levels. To solve these problems, the WPHL was revised and "Specified Wildlife Conservation and Management Plan" (SWCMP) by MOE [2087] was established under the WPHL in 1999. The SWCMP is drafted optionally by the Governor of a prefecture. When an SWCMP is drafted, the prefectural Governor can relax hunting regulation and expand hunting seasons. The goal of SWCMPs is the coexistence of wildlife with people through proper conservation and management measures, including population control of wildlife. The aims of institutionalizing SWCMPs are 1) setting management goals based on scientific information, 2) comprehensive management through habitat management, wildlife damage control, and population control, 3) inspection of management measures through monitoring and feedback of the results to the next management plan, 4) developing scientific procedures from goal setting to enforcement measures, 5) clarity of procedures for drafting SWCMPs, including consensus building, and 6) registration of adequate participation of the nation in SWCMPs when necessary. After revising the WPHL, the responsibility of wildlife management was changed from the nation to prefectures. For damage control, "Act on Special Measures for Prevention of Damage Related to Agriculture, Forestry and Fisheries Caused by Wildlife" was established by the Ministry of Agriculture, Forestry and Fisheries in 2007. The role of national wildlife conservation and management is restricted to endangered species and wildlife protection areas as designated by the national government. Without SWCMPs, the prefecture can conduct nuisance kills as usual under the WHPL. Thus, population control is conducted in three ways: sports hunting, control kills under SWCMPs, and nuisance kills. Because overabundant deer caused not only damage to agriculture and forestry but also to ecosystems and biodiversity in national parks, wildlife management should be considered as ecosystem management. The central role of the National Parks in Japan was traditionally to protect the landscape and provide for recreational opportunities, not for protection of wildlife habitat or ecosystem conservation. In 2002, the Natural Park law was revised to belatedly include the concept of conservation of ecosystems and biodiversity. Both the nation and the prefectures are responsible for the measure. Integrated management systems based on sound science, however, are still lacking. This might result from Zoning-Systems in natural parks, included private lands, Forestry Agency, and various kinds of land ownership, which separately manage with their own priorities. Wildlife management in Natural Parks as ecosystem management has presented a challenge for the government to come to a consensus. A trial of integrated management has just started in Shiretoko National Park with the cooperation of the Scientific Committee for Shiretoko Natural World Heritage Site since 2006. Thereafter, a culling operation for overabundant deer has been conducted under the sika deer management plan.

4. Feedback management and problems to be solved in SWCMP

Feedback management was started for a sika deer population in the eastern part of Hokkaido in 1998. The goal of management was to avoid population irruptions that led to severe damage to agricultural crops and forestry, and extinction of the sika deer population, by controlling hunting pressure [1998]. In this plan, 3 population levels were set: an irruption level, optimal level, and critical level for extinction. One of four management responses would be applied annually according to population status: an emergency culling, gradual population reduction, gradual population increase, or a ban on hunting. The hunting pressure on females is set to a higher level as population size increases. The concept of feedback management, based on uncertain information about the deer population [1998], was introduced into the main framework of the SWCMP, which for sika deer were drawn up in 40 prefectures from 2002 to 2014. Because of either underestimated population sizes or the expansion of deer distribution, population control via SWCMP has not reached its goal in most areas [3737]. Evaluation systems, cooperation between scientists and managers, and large scale management of deer populations covering multiple prefectures are essential for successful management [3737].

Of the 33 prefectures that bears inhabit, 21 have drawn up SWCMPs for bears. The problems for bear management are a lack of efficient monitoring methods due to their large home ranges and low density, and a lack of management systems within each prefecture to prevent further population reduction [3459]. Wildlife management systems have not worked well at the municipality level, where administrators in charge of wildlife and hunting are coping with wildlife management issues without any special knowledge regarding wildlife. Further, hunting as resource management and culling for ecosystem management should be synergistically combined under adaptive management [1272].

We are entering into a shrinking society where human populations are decreasing and wildlife populations are increasing on farmland. Because the hunter population is decreasing rapidly to a threatened level, clarifying the roles of the nation, prefecture, and municipality for wildlife management, personnel training for wildlife managers, and the establishment of genuine management systems urgently are needed in Japan. Thus "Wildlife Protection and Proper Hunting Act" was revised again as "Wildlife Protection and Management, and Proper Hunting Act" on March 11, 2014 (effective as from May 29, 2015), in order to further emphasize the importance of the management of wildlife. Under the revised law MOE can designate a wildlife population as "Specified Wildlife" that should be managed intensively and extensively, and the national or prefectural governments can relax existing regulations for hunting and capture of wildlife under a plan authorized by the law.

Carnivora   MUSTELIDAE

Red list status: C-1 (MSJ); LC (IUCN)

Mustela itatsi Temminck, 1844

EN Japanese weasel   JP ニホンイタチ（nihon itachi）   CM 日本鼬   CT 日本鼬   KR 일본족제비
RS японский колонок, итатси

A female Japanese weasel in Morioka, Iwate Pref. (Morioka Zoological Park).

1. Distribution

Endemic to Japan and distributed on the mainlands of Honshu, Kyushu, and Shikoku (except Hokkaido) as well as small islands such as Yaku-shima, Tane-gashima, Sado, Iki and Izu-ohshima [1011, 1794]. There are some reports of distribution in local areas such as Shodo-shima Isl., Kagawa Pref. [1718] and Gifu Pref. [154]. As described in "Remarks" below, this animal has been introduced to a lot of areas (the Ryukyu Isls., some islands of Nagasaki Pref., Rishiri and Rebun Isls. of Hokkaido, and Sakhalin etc) for rat control [3250] and it is difficult to precisely identify the islands where this species is native.

2. Fossil record

Fossils of *Mustela* sp. have been recorded from the Middle and Late Pleistocene layers in Honshu [1514] although it was unclear whether or not they were of *M. itatsi*.

3. General characteristics

Morphology   A medium-sized weasel. The summer coat color is dark brown, while winter color is reddish brown. Fur color around the eye is gray. Although *M. itatsi* is morphologically very similar to *M. sibirica*, the 2 species can be identified by their tail length ratios: around 40% of body length for *itatsi* and more than 50% for *sibirica* [1011]. The Japanese weasels have sexual dimorphism (Table 1), with males being much larger than females [1011]. Although the Japanese weasel "*itatsi*" has often been classified as a subspecies of *M. sibirica* [409, 3893], molecular phylogenetic data suggests that *itatsi* should be positioned at the distinct species level [1985]. Additional information on classification is described in the Genetics section.

Dental formula   I 3/3 + C 1/1 + P 3/3 + M 1/2 = 34 [29].

Mammae formula   3–4 pairs [2068].

Genetics   Chromosome: $2n = 38$ and $FN = 64$ [2535, 2540, 2542].

Molecular phylogenetic studies of mtDNA show that the genetic differences between the 2 taxa correspond with those among other mustelid species and that *itatsi* should be classified to a distinct species level [1985, 1986]. In addition, although chromosome numbers are identical, there are large differences in G- and C-banding patterns between them [1811]. MtDNA phylogeography suggests co-occurrence of the 2 lineages, the Honshu and the Shikoku/Kyoshu clades [1982].

Mustela itatsi Temminck, 1844

Table 1. Mean body measurements of *M. itatsi*. Ranges are shown in parentheses.

	Shiga Pref.		Osaka Pref.		Gifu Pref.
	Male ($n = 14$)	Female ($n = 2$)	Male ($n = 10$)	Female ($n = 5$)	Male ($n = 12$)
BW (g)	450 (306–575)	150 (125–175)	471 (270–600)	142 (110–180)	NA
HB (mm)	329 (295–365)	255 (245–265)	336 (310–355)	239 (229–248)	311 (278–369)
T (mm)	134 (120–145)	90 (90)	138 (120–150)	97 (92–105)	134 (112–151)
T/HB (%)	40.7 (36.9–46.0)	35.4 (34.0–36.7)	41.6 (35.3–47.7)	40.2 (37.1–42.9)	43 (37–47)

Measurements for Shiga Pref. are based on [3836] including unpublished data [3845].
Measurements for Osaka Pref. are based on [3835] including unpublished data [3845].
Measurements for Gifu Pref. are cited from [154].
NA, not available.

4. Ecology

Reproduction The mating season is March to May [2068]. Several males mate with 1 female [2068]. The gestation period is 37.3 days (range 36–40) in captivity. Litter size is 4.4 (1–8), and sex ratio is about 1:1 [2068].

Lifespan Less than 9 years in captivity [2068].

Diet The Japanese weasel is omnivorous; it eats rodents, small birds, fishes, crustaceans, other arthropods, amphibians, fruits, and other plants [589, 888, 2787]. Females eat more invertebrates, whereas males eat more vertebrates, probably because of the difference in body size [2677].

Habitat *Mustela itatsi* prefers to live near rivers [2671]. It inhabited lowlands in rural areas in western Japan before the alien species *M. sibirica* invaded. As a result, *M. itatsi* now inhabits mountainous areas in western Japan. In eastern Honshu, *M. itatsi* still inhabits lowlands.

Home range A telemetry survey for 26 days in spring showed that a male had a home range of about 35 ha around the Tama River in Tokyo [888].

Behavior Usually nocturnal, but active through the day in the mating season [2068]. They move by jumping and sometimes stand on their hind legs when they are interested in something [2068]. They generally swim to catch fish, even in winter [2068].

Natural enemies Not recorded, but large carnivores such as the red fox (*Vulpes vulpes*) and raptors may prey on *M. itatsi*.

Parasites Three nematode species, *Filaroides martis*, *Capillaria putorii* and *Skrjabingylus nasicola*, have been recorded [1289, 1883]. The acanthocephalan *Centrorhynchus elongates* has been recorded in Hokkaido [181].

5. Remarks

Mustela itatsi was once cultured in farms by the Japanese Government [2068], and released to many islands to control rat populations [3250]. On those islands, animals have become naturalized and cause deleterious effects on native ecosystems such as eating native small animals [806, 1056, 1059, 3415, 3717]. They sometimes live near houses, and attack chickens. Male Japanese weasels are hunted as game but females are not.

R. MASUDA & S. WATANABE

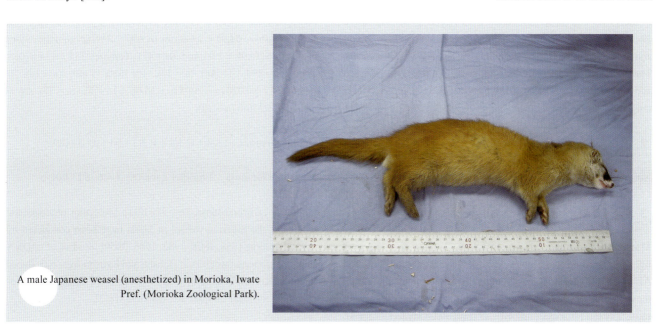

A male Japanese weasel (anesthetized) in Morioka, Iwate Pref. (Morioka Zoological Park).

Carnivora MUSTELIDAE

106

Red list status: NT as *M. s. coreana* (MEJ); C-1 as *M. s. coreana* (MSJ); LC (IUCN)

Mustela sibirica Pallas, 1773

EN Siberian weasel　　JP シベリアイタチ (siberia itachi)　　CM 黄鼬　　CT 黄鼬，黄鼠狼　　KR 족제비
RS колонок

A male Siberian weasel in Fukuoka Pref. (H. Sasaki).

1. Distribution

Distributed in Russia (from the western side of the Ural Mts. to Far East Russia), Mongolia, Pakistan, Kashmir, Himalayan region of India, Nepal, Bhutan, Myanmar (Burma), Thailand, Laos, Vietnam, China, Taiwan and Japan. In Japan, Siberian weasels are endemic only on Tsushima Isls., but were introduced in western Japan. Individuals escaped from fur farms in Hyogo in the 1930s and those introduced from Korea to Kyushu in the 1940s seem to be the origin of the population in western Japan, and these have spread in western parts of Japan [958, 2146, 3589]. The eastern fronts of their distribution are Fukui, Gifu, Nagano and Aichi Prefs. [155, 2133, 3014]. Recently the expansion of their range may have slowed down.

2. Fossil record

Not reported in Japan.

3. General characteristics

Morphology　A medium-sized weasel. Winter coat color is light brown and coat color changes to brown in summer. Although this species is morphologically very similar to *M. itatsi*, these 2 species can generally be identified by their tail length ratios: the tail ratio is around 40% of head and body length for *M. itatsi* and more than 50% for *M. sibirica* [1011]. The fur color around the eye of *M. sibirica* is the same as that of the body but in *M. itatsi* it is grey. Siberian weasels have conspicuous sexual dimorphism in body size: the body weight of males is about twice that of females.

External measurements (mean ± SD) are as follows [2646]. Male: BW = 717 ± 176 g (n = 35), HB = 358 ± 21 mm (n = 40), T = 196 ± 16 mm (n = 40), T/HB = 0.548 ± 0.047 (n = 40). Cranial measurements (mean ± SD) are as follows [2646]. GL (greatest skull length) = 64.95 ± 2.65 mm (n = 47), ZW (zygomatic width of skull) = 34.96 ± 2.04 mm (n = 47). Female: 355 ± 134 g (n = 18), HB = 294 ± 14 mm (n = 20), T = 161 ± 11 mm (n = 20), T/HB = 0.549 ± 0.020 (n = 20), GL = 55.08 ± 1.79 mm (n = 18), ZW = 28.92 ± 1.07 mm (n = 19).

Dental formula　I 3/3 + C 1/1 + P 3/3 + M 1/2 =34 [29].

Mammae formula　1 (or 0) + 2 (or 1) + 1 = 6–8 [3010].

Genetics　Chromosome: $2n$ = 38 [1811]. Molecular phylogenetic studies of the Siberian weasel in Japan have been conducted by several authors [958, 1808, 1982, 1986, 3050]. Studies of mtDNA show that Siberian weasels in Honshu and Kyushu were introduced from the Korean Pen. [958, 1982]. PCR primers for mtDNA cytochrome *b* for fecal samples are available [1813].

Mustela sibirica Pallas, 1773

4. Ecology

Reproduction Siberian weasels are solitary. Females become estrous in April and nurse their young from June to August. The volume of the male testis is high from March to May in Kyushu [3011]. Spermiogenesis is very active from March to June in the western parts of Japan [3723]. Litter sizes range from 4 to 6 in Japan [2069] and average litter size is 5.06 in China [3174].

Lifespan Lifespan in the wild is estimated at 2.1 years [2121].

Diet Siberian weasels show a wide variety of foods in their diet. Their main foods are rodents and insects [3015, 3539], but they will also eat sugar-cake, mayonnaise and other human foods.

Habitat Because Siberian weasels are tolerant to human activity and eat a wide range of foods, they are able to inhabit city centers as well as mountainous areas. Their suitable habitat might be plain areas because their body weight is heavier in plains than in mountain areas in China [3175]. Siberian weasels have expanded their range to urban habitat after Japanese weasels abandoned them. In mountain areas, Siberian weasels cannot invade new habitats that lack nearby villages with paddy and cultivated fields and also natural habitat where Japanese weasels occur. The 2 weasel ranges meet in the plains of Aichi Pref. The presence of Japanese weasels may prevent the expansion of the Siberian weasel range [3014].

Home range They have intrasexual territory, and female ranges overlap with male ranges, but males give up their territories at high density [3011]. In the breeding season, the home range size of males (mean ± SD) is 13.5 ± 8.4 ha in the Tsushima Isls. [687]. In the non-breeding season, home range size of males and females has been recorded as (mean ± SD) 1.42 ± 1.62 ha and 1.31 ± 1.12 ha respectively at a small village and 4.37 ± 3.87 ha and 1.67 ± 0.91 ha respectively in a grassland on a small island [3015].

Behavior *Mustela sibirica* does not appear to have specific daily activity patterns [687].

Natural enemies Not reported in Japan, but bigger carnivores and raptors may hunt them.

Parasites Ticks (*Haemaphysalis flava* and *Ixodes nippobensis*) are reported from the Tsushima Isls. [1277]. Trematodes (*Paragonimus ohirai* and *Clonorchis sinensis*) and nematodes (*Gnathostoma nipponicum*, *G. spinigerum* and *Dioctophyma renale* var. *yoshidai*) are reported as endoparasites [4115].

Sexual dimorphism is apparent in the Siberian weasel (H. Sasaki). A female (upper) from Fukuoka Pref. and a male (lower) from Saga Pref.

5. Remarks

Furs of weasels were important as an export product until the 1960s. In 1961, 40% of harvested furs of weasels in northern Kyushu were from Siberian weasels [2146]. However, the fur quality of Siberian weasels in Japan is not so good and the hunting of the 2 weasel species has lost its appeal with the availability of cheaper mink furs. In recent years, just a few hundred weasels have been hunted annually.

Male Siberian weasels are a game animal, whereas female Siberian weasels are protected by law.

The Siberian weasel was introduced to small islands for rat control. This method of rat control has been abandoned because the weasels damage the endemic fauna. In towns, Siberian weasels prefer to nest in insulation in the ceiling space under the roof. Their movement is noisy, and feces and urine near the nest cause a smell and damage houses. In rural districts, Siberian weasels occasionally attack chickens. Both legal and illegal trapping is conducted by people to resolve these problems. Only Siberian weasels in the Tsushima Isls. are classed as Near Threatened by the Ministry of the Environment because of their limited range in Japan.

H. SASAKI

Carnivora MUSTELIDAE
107
Red list status: NT as *M. n. namiyei* (MEJ); V as *M. n. namiyei* in Honshu, C-1 as *M. n. nivalis* in Hokkaido (MSJ); LC (IUCN)

Mustela nivalis Linnaeus, 1766

EN least weasel   JP イイズナ (iizuna)   CM 伶鼬   CT 伶鼬, 小黄鼠狼   KR 쇠족제비   RS ласка

A least weasel in November in Hirosaki, Aomori Pref. (M. Mukohyama).

1. Distribution

Widely distributed in circumboreal zones in the Holarctic regions, including most parts of Asia, Europe, North America, and North Africa [404, 2515, 3894]. In Japan, they occur in Hokkaido and northern Honshu including Aomori and Yamagata Prefs. [1011]. The distribution of this species in Aomori Pref. has been reviewed using data from the literature, specimen collections, and photo records [2549].

2. Fossil record

Fossils of this species have been recorded from the Late Pleistocene layers in Honshu [1514]. No fossil records from Hokkaido have been reported.

3. General characteristics

Morphology   The smallest *Mustela* species in the world. Coat color changes between summer and winter. In summer, the dorsal side is darker brown, while the ventral is white. In winter, the whole body becomes white. The tip of the short tail has no black hairs. Due to its smaller body size and shorter tail without black hairs, *M. nivalis* is easily distinguished from *M. erminea*, which is allopatrically distributed. Adult males are generally larger than adult females.

Few detailed data on body size of individuals are available from Japan. In 1 adult in summer coat color from Aomori Pref., HB = 190 mm, HFsu = 18.5 mm, and E = 11 mm [1794]. In 1 adult in winter coat color from Hokkaido, HB = 200 mm, T = 6 mm, HFsu = 20.5 mm, and E = 11 mm [1794]. Other data on body size are available without sampling localities: HB = 145–182 mm in 7 males and 112–200 mm in 6 females; T = 19–27 mm in 7 males and 21–32 mm in 5 females, HFsu = 21.4–26.2 mm in 6 males and 18.5–20.5 mm in 6 females; E = 8.5–12.5 mm in 7 males and 7.5–13.5 mm in 6 females [1011]. There are differences in skull morphology between the Aomori and Hokkaido populations [2535]. In addition, a high variability in morphology among populations around the world has been reported [66], in line with the genetic variation mentioned below.

Dental formula   I 3/3 + C 1/1 + P 3/3 + M 1/2 = 34 [29].

Mammae formula   4 pairs [3172].

Genetics   Karyological differences within Japanese *M. nivalis* have been reported: $2n = 38$ (FN = 66) for the Honshu population and $2n = 42$ (FN = 74) for the Hokkaido population and continental populations [2535, 2539, 2542]. Molecular phylogenetic studies of mtDNA show genetic differentiations between the 2 populations [1814, 1985]. In addition, the Hokkaido lineage appears closely related to North American lineages, whereas the Honshu lineage is more closely related to Eurasian continental lineages including eastern Siberia, European Russia, Altai, and Ukraine [1809]. Thus, genetic variation among populations of *M. nivalis* among populations are much larger than those of *M. erminea*, although the 2 species have similar distribution patterns in the northern hemisphere.

4. Ecology

Reproduction Gestation period is 34–36 days without delayed implantation [3172]. Few data are available for Japanese populations. A litter usually has 3 to 7 cubs [3466].

Lifespan Average lifespan is approximately 1 year [3172]. Not reported for the Japanese populations.

Diet Small rodents, small birds, amphibians, and insects [1794, 3466].

Habitat In Honshu, this species inhabits mountainous areas [1794], while they are relatively widespread in Hokkaido: mountains, coastal meadows, and even farmhouses [3466].

Home range Not reported in Japan.

Behavior This active hunter moves rapidly, investigating every hole in the field. Solitary and primarily nocturnal [3172], although it can be active during the day [3466]. This species does not hibernate.

Natural enemies Large carnivores such as the red fox (*Vulpes vulpes*) and raptors may prey on *M. nivalis*.

Parasites Two helminth species, *Filaroides martis* and *Centrorhynchus elongatus*, are recorded [1289].

5. Remarks

The least weasels of Hokkaido and Honshu were first classified in a subspecies (*M. n. namiyei*) [1011]. Based on the karyological characteristics and differences between the Honshu ($2n = 38$) and Hokkaido ($2n = 42$) populations, it was then proposed that the former should be recognized as a species *M. namiyei* and the latter should be a subspecies (*M. n. nivalis*) [66, 2535, 2539, 2542]. Molecular phylogenetic studies show that the genetic differences between the 2 populations correspond with differentiations at the subspecies level for other mustelids [1809, 1814, 1985].

R. MASUDA

A least weasel killed by a domestic cat in Shiretoko, Hokkaido (T. Murakami).

Carnivora   MUSTELIDAE

Red list status: NT as *M. e. nippon* and *M. e. orientalis* (MEJ); R as *M. e. nippon* in Honshu, V as *M. e. orientalis* in Hokkaido (MSJ); LC (IUCN)

Mustela erminea Linnaeus, 1758

EN ermine, stoat   JP オコジョ（okojo）   CM 白鼬   CT 白鼬   KR 검은꼬리털족제비   RS горностай

An ermine in winter coat color in Nagano Pref. (T. Nishizawa).

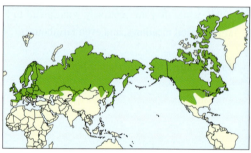

1. Distribution

Widely distributed in the tundra, boreal forest and deciduous forest zones in the Palaearctic and Nearctic regions [404, 1579, 2515, 2861, 3894]. Few data on the distribution in Japan are available. This species inhabits highlands from central to northern Honshu (subspecies *M. e. nippon*) and Hokkaido (subspecies *M. e. orientalis*) [1011].

2. Fossil record

Fossils of this species have been recorded from the Middle and Late Pleistocene layers of Honshu [1514]. In addition, ermine-like mandibles have been excavated from the Late Pleistocene layer of Kyushu [2360]. No fossil records from Hokkaido have been reported.

3. General characteristics

Morphology   Coat color changes between summer and winter. In summer, the dorsal side is darker brown, while the ventral is white. In winter, the entire coat turns white. However, the tip of the long tail is black through the year. Adult males are generally larger than adult females.

Few detailed data on body size are available. For the Honshu population; HB = 182–198 mm in males and 140–170 mm in females; T = 48–67 mm in males and 54–61 mm in females; HFsu = 30.5–34.2 mm in males and 26.0–28.0 in females; E = 14.5–15.5 mm in males and 11.8–12.5 mm in females [1011]. In the Hokkaido population; HB = 235–240 mm in males and 225 mm for a female; T = 70–88 mm in males and 70 mm in a female; HFsu = 39–42 mm in males and 36.5 mm in a female; E = 13.0–20.5 mm in males and 13.0 mm in a female [1011]. Data on morphological variation within Japan are not available. Geographic variation in ermine morphological characters are very small throughout the entire range [486, 2861], in agreement with molecular phylogenetic data as mentioned below.

Dental formula   I 3/3 + C 1/1 + P 3/3 + M 1/2 = 34 [29, 1794].

Mammae formula   Not reported.

Genetics   Chromosome: $2n = 44$ and $FN = 64$ [2541, 2542]. MtDNA phylogenetic studies show that *M. erminea* first speciated from the ancestral species among extant *Mustela* species [1808, 1985]. Karyological investigation also suggests that the ermine has the ancestral karyotype [2542]. Genetic variation of the mtDNA control region is very small among populations in the Japanese Isls. and even among populations from various regions of the Nearctic and Palaearctic including Japan [1809, 1814].

Mustela erminea Linnaeus, 1758

4. Ecology

Reproduction *Mustela erminea* usually mate once a year. Implantation is delayed 9–12 months, followed by a 4 week gestation [1578]. In Japan, they seem to mate in winter and have several cubs in spring [1794, 1979]. In the Shiga Plateau of Nagano Pref., they mate throughout the year and have on average 4–6 cubs per litter in April or May [4065]. They nest in underground burrows.

Lifespan Not reported in Japan. Average lifespan is 1–1.5 years in other regions [1578].

Diet Few data are available in Japan. The ermine appears to eat small rodents, small birds, bird eggs and insects, and sometimes attacks hares [1794, 1979]. It has been reported that *M. erminea* in the Shiga Plateau of Nagano Pref. eats hares, rodents, shrews, reptiles, amphibians, snails, insects and bird eggs [4065].

Habitat Inhabits forests and rocky areas in the high mountains in Honshu, and ranges from lower to higher land in Hokkaido. In Honshu, some individuals move to lower lands from high mountains in winter [1011, 1794].

Home range Home ranges (except winter) are reported to be about 40–83 ha for males and about 18–50 ha for females in Shiga Plateau of Nagano Pref. [4065].

Behavior They do not hibernate. They have quick movements, and jump between rocks [1979]. In winter, they make small burrows underground for nesting and mating [1794].

Natural enemies Few data on natural enemies are available. Large carnivores such as the red fox (*Vulpes vulpes*) and raptors may prey on *M. erminea*.

Parasites Fleas *Ctenophthalmus congener congeneroides* and *Ctenophyllus armatus* have been recorded from Nagano Pref. and Hokkaido, respectively [74].

5. Remarks

In the Shiga Plateau of Nagano Pref., the number of roadkills increases from spring to summer. *Mustela erminea* has been categorized as Near Threatened (NT) by the Ministry of the Environment, Japan, and is a non-game species. However, because of the high quality of fur, it is an important game species in some other countries such as Russia.

R. MASUDA

An ermine in summer coat color in Nagano Pref. (T. Nishizawa).

Carnivora MUSTELIDAE

109

Red list status: LC (IUCN); introduced

Neovison vison (Schreber, 1777)

EN American mink　　JP ミンク（minku）　　CM 美洲水貂　　CT 美洲水貂　　KR 밍크　　RS американская норка

An American mink in Tsurui, Hokkaido (Mitsunori Sato).

1. Distribution

Introduced species in Japan. Native range covers most of North America. Translocations associated with fur farming have led to the establishment of feral populations in several areas in the Palaearctic, including Britain, the Scandinavian countries, Ireland, Iceland, France, Spain, Germany, Russia [294, 310] and Japan. In Japan it is widespread in Hokkaido [926], and along the Chikuma River in Nagano Pref. [1604] and the Abukuma River in Fukushima Pref. [978]. Some hunting records exest from other prefectures in Honshu and Kyushu [2089].

2. Fossil record

Not reported in Japan.

3. General characteristics

Morphology　A medium-sized mustelid. Although their appearance resembles *Mustela itatsi*, this species is significantly larger than *M. itatsi*. Males are significantly larger than females. The sexual dimorphism ratio (male/female) of adults is 1.56 in BW and 1.13 in HB [1698]. In Japan the ratio in BW is slightly lower than those given in other countries (about 1.6–1.9) [2193]. Fur color is dark-brown, appearing almost black in the wild type. White spots, patches and sometimes only a few white hairs on the chin, throat and ventral surface. Mutation types, which may be white, pale brown, pale silvery gray etc., comprise 10–20% in Hokkaido [926].

External measurements in Hokkaido are given in Table 1 [1698].

Dental formula　I 3/3 + C 1/1 + P 3/3 + M 1/2 = 34 [29].

Mammae formula　1–4 pairs although the standard mammae formula was unknown [294, 357].

Genetics　Chromosome: $2n = 30$ and FNa = 54 [4252]. Microsatellite analyses suggest there are still polymorphisms in introduced populations in the Kushiro Wetland [3219] and Nagano Pref. [3217].

Table 1. Body measurements of American minks (adults) in Hokkaido (mean ± SD).

	Males ($n = 26$)	Females ($n = 20$)	Sexual difference
HB (mm)	406.0 ± 35.4	359.6 ± 11.6	***
T (mm)	195.3 ± 17.5	170.1 ± 12.6	***
HFsu (mm)	64.8 ± 4.4	56.8 ± 3.7	***
E (mm)	23.2 ± 2.5	20.9 ± 2.1	**
BW (g)	1063 ± 245.0	681 ± 123.2	***

Asterisks indicate the significant level of sexual difference in *t*-test; **: 1%, ***: 0.1%.

Neovison vison (Schreber, 1777)

An American mink attacking a mallard in Sapporo, Hokkaido (K. Murata).

4. Ecology

Reproduction Few data are available in Japan. The number of placental scars in Hokkaido in June and July were 6, 6 and 5 ($n = 3$). Litter size was 7 ($n = 1$) [926].

Lifespan Few data are available. In Hokkaido 63.2% of 76 minks collected throughout the year except for March and April were juvenile. The oldest age was five years in 45 males, and 6 years in 31 females [1698].

Diet Opportunistic predator taking a wide variety of mammal, fish, bird, amphibian and invertebrate prey. Composition of the diet shows spatial and seasonal variations [3741].

Habitat Inhabits a wide range of aquatic habitats, i.e. river, lake, pond, seashore, marsh and channel [926].

Home range Few data available in Japan. One male's range was linear in shape along a river, and about 700 m in length. There were several dens in the range [3740]. No data on territoriality.

Population density Few data available in Japan. Estimated density using the removal method were 2.8–4.8 per km along rivers [978].

Behavior Few data available in Japan. One male showed a diurnal pattern in continuous tracking for 23 hours [3740].

Natural enemies Red foxes (*Vulpes vulpes*) may prey on minks, since some fox stomachs include mink remains [1705].

Parasites Reported endoparasites are *Echinostoma hortense*, *Spirometra erinaceieuropaei*, and *Molineus patens* [2130].

5. Remarks

Originally included in *Mustela*. There are significant differences between the American mink and *Mustela* according to biochemical and cytogenetic data [64]. The level of the differences is higher than that among other *Mustela* species [1808]. Some analyses support significant divergence of *vison* from the *Mustela* lineage [1808].

A game species, but regarded as a pest by fish-farmers and poultry-keepers, though its commercial impact is unclear [926]. Modern mink breeding in Japan started in 1953, mostly in Hokkaido, and escaped minks may have become feral in Hokkaido in the 1960s [3741]. Density in relation to that of the Japanese weasel (*M. itatsi*) varies regionally. The mink dominates the Japanese weasel in almost all regions of Hokkaido except the southernmost area (the Oshima Pen.) [2942, 3740].

K. URAGUCHI

Footprints of minks on snow in Nemuro, Hokkaido (K. Uraguchi).

Carnivora MUSTELIDAE

Red list status: NT as *M. m. tsuensis* (MEJ); V as *M. m. tsuensis* on the Tsushima Isls., C-1 as *M. m. melampus* in "Hondo" (MSJ); LC (IUCN)

Martes melampus (Wagner, 1840)

EN Japanese marten JP ニホンテン (nihon ten) CM 日本貂 CT 日本貂
KR 노란담비 RS японский соболь

A Japanese marten in summer coat in the Tsushima Isls. (Y. Ooya).

1. Distribution

Endemic to the main islands of Honshu, Kyushu, and Shikoku, and the Tsushima Isls. [1011]. The populations on the main islands are classified as subspecies *M. m. melampus*, and the population on the Tsushima Isls. is a different subspecies *M. m. tsuensis*. This species was introduced to Hokkaido for fur farming, and naturalized populations currently occupy southern parts of Hokkaido [2319].

2. Fossil record

Fossils have been recorded from the Late Pleistocene layers of northern Honshu [1492, 1493]. Bones of *Martes* sp. have been recorded from the Middle and Late Pleistocene layers in Honshu [1514] although it is unclear whether or not they are of *M. melampus*.

3. General characteristics

Morphology A medium-sized mustelid. Narrow and slender shape, which is the common characteristic of mustelids. Tail long, bushy; legs are longer in proportion than *Mustela*. Adult males are generally larger than adult females. Body measurements of captured and killed animals of the Tsushima Isls. are shown in Table 1 [3538]. Statistically informative data on body sizes of animals from the main islands of Japan (Honshu/Kyushu/Shikoku) are not available. Measurements of skulls from Honshu indicate that animals of southern areas are smaller than those of northern areas [2007]. Coat color changes between summer and winter: summer coat color is brownish, whereas the winter coat is very yellowish. There are variations in winter coat colors from yellow (called "ki-ten") to darker brown ("susu-ten"), and animals with the darker brown coat color ("susu-ten") are restricted to western Japan including the Kii Pen. and Shikoku [956].

Dental formula I 3/3 + C 1/1 + P 4/4 + M 1/2 = 38 [29].

Mammae formula Not reported.

Table 1. Body measurements of captured and killed martens from the Tsushima Isls. (mean ± SD) [3538].

	Male ($n = 50$)	Female ($n = 33$)	Sexual difference (t-test)
BW (g)	1,558.3 ± 268.7	1,010.5 ± 186.4	$P < 0.001$
HB (mm)	468.5 ± 24.2	425.6 ± 21.2	$0.01 < P < 0.05$
T (mm)	189.6 ± 17.5	183.9 ± 15.3	ns ($P > 0.05$)
HFsu (mm)	81.9 + 3.0	76.4 ± 2.5	$0.01 < P < 0.05$
E (mm)	40.2 ± 2.1	38.6 ± 1.9	ns ($P > 0.05$)

Martes melampus (Wagner, 1840)

Genetics Chromosome: $2n = 38$ and $FN = 66$ [2541, 2542]. Molecular phylogenetic status in the Mustelidae and genetic variations have been described using mtDNA data [1812, 1985, 2316], nuclear ribosomal DNA data and others. Variation in the gene related to coat color has been reported [957]. A microsatellite study showed genetic differentiation between the North and South subpopulations on the Tsushima Isls. [1280]. The introduced populations on Hokkaido were reported to consist of at least 2 genetic lineages [1281].

4. Ecology

Reproduction In the Tsushima population, mating occurs mainly in summer, but young are born the next spring due to delayed implantation [3537]. Litter size is usually 2 [3537]. In captivity, a female had 3 cubs in April [1679].

Lifespan Not reported.

Diet Analyses of scats and stomach contents showed that this species prefers fruits and seeds, insects, small birds and small mammals such as rodents [2397, 2732, 2785, 3261, 3536, 3538, 3539].

Habitat Prefers natural broad-leaved forests to artificial forests [955]. On the Tsushima Isls., *M. melampus* sometimes has nests in human houses near forests [3537].

Home range The average home range size throughout the year is about 0.8 km^2 for males and about 0.7 km^2 for females on the Tsushima Isls. [955, 3536, 3538]. In the Kinki District of Honshu, a territorial home range of about 2.3 km^2 has been recorded [955].

Behavior Usually nocturnal, although on the Tsushima Isls., active animals are often observed even in daytime [3538]. Behavior varies greatly from day to day, and the rhythm within a day is not stable. The average ratio of the active time per day is 43% for females, with 72% for males in summer and 49% for males in winter [955, 3538]. The higher value for males in summer is probably related to mating behavior. The Japanese marten plays a role as seed disperser in broad-leaved forests [2785].

Natural enemies Not reported, but large carnivores such as the red fox (*Vulpes vulpes*) and large raptors may prey on *M. melampus*.

Parasites Two nematode species, *Filaroides martis* and *Capillaria putorii*, have been recorded [1289, 3917].

5. Remarks

The Japanese marten is a game species, but the population of the Tsushima Isls. is conserved as a "Natural Monument". Recently, genetic methods for species identification using fecal DNA were developed for future ecological studies of allopatric carnivores including *M. melampus* [1813, 2431, 3218].

The high quality fur produced in snowy regions was once highly prized by humans. This species prefers to eat fruits, including agricultural fruits, but the amount of agricultural damage is unknown. The species is frequently involved in collisions with cars.

R. MASUDA

A Japanese marten in winter coat on the Tsushima Isls. (M. Kawaguchi).

Carnivora   MUSTELIDAE

111   Red list status: NT as *M. z. brachyura* (MEJ); K as *M. z. brachyura* (MSJ); LC (IUCN)

Martes zibellina (Linnaeus, 1758)

EN sable, Japanese sable for population in Hokkaido
JP クロテン (kuro ten), エゾクロテン (ezo kuroten) for the Hokkaido population
CM 紫貂, 黑貂   CT 紫貂, 黑貂   KR 잘, 흑초   RS соболь

A Japanese sable in the winter coat in Maruseppu, eastern Hokkaido (M. Iwai).

1. Distribution

Distributed in eastern Russia, Sakhalin, North Korea, Mongolia [63, 249], northern China [3176, 3278] and Hokkaido [918, 1081, 2319]. The Japanese sable (*M. z. brachyura*) is a subspecies of the sable and distributed only in Hokkaido, mainly in the eastern and northern parts [2319]. In addition, in central Hokkaido, there may be still isolated populations [897, 898]. In Fennoscandia and western Russia, sables were extirpated approximately 300 years ago [881].

2. Fossil record

Not reported in Japan.

3. General characteristics

Morphology   The Japanese sable has a narrow slender shape, which is a common characteristic of mustelids. However the ratio of tail to body size is smaller than in other mustelids except for the least weasel, *Mustela nivalis*. Compared to the Japanese marten, both body length and tail/body length ratio of the Japanese sable are smaller. Measurements from mainland Hokkaido are shown in Table 1 [2311]. Fur color changes from whitish yellow in winter to black in summer. Most individuals have a yellowish mark around the throat. Color of the tail tip of Japanese sables is black to dark brown, while the Japanese martens' tail tip is white or yellow. Some individuals have a relatively correct color but without dark parts [905].

Dental formula   I 3/3 + C 1/1 + P 4/4 + M 1/2 = 38 [29].

Genetics   Chromosome: $2n = 38$ and $FN = 66$ [1163]. Compared to the Russian Far Eastern sable, the Japanese sable population has little mtDNA sequence variation [959, 1812, 3048]. However, analysis of the mitochondrial DNA control region suggested that there are 3 populations in eastern Hokkaido [1049]. Based on analyses of microsatellite DNA, 2 lineages were detected in eastern Hokkaido [2342]. Analysis of other nuclear genes relating to coat color revealed that the Hokkaido population has 2 lineages [1081]. A phylogenetic analysis of 2 nuclear genes, melanocortin-1 receptor (*Mc1r*) and transcription factor 25 (*Tcf25*), showed that there were 2 haplotype groups in the Hokkaido population: one was closely related to the Japanese marten (*M. melampus*) and the American marten (*M. americana*), and the other was closely related to the continental sable [1081].

Table 1. External measurements in millimeter or gram (mean ± SD) of the Japanese sable in Hokkaido.

	BW	HB	T	T/HB (%)	FFW
Male	1021.2 ± 136.5 (n = 45)	414.9 ± 22.8 (n = 47)	135.0 ± 10.4 (n = 47)	32.6 ± 2.6 (n = 47)	27.8 ± 3.7 (n = 44)
Female	662.2 ± 72.0 (n = 9)	366.7 ± 13.4 (n = 9)	121.2 ± 7.9 (n = 9)	33.1 ± 2.4 (n = 9)	24.1 ± 4.1 (n = 9)

FFW (Left foot).

4. Ecology

Reproduction The mating system is little known. As testis size reaches its maximum from June to August [2312], mating may occur around this period. This is similar to the Russian and Chinese sable, as their breeding season ranges from June to August [2811, 3176]. In Far East Russia, litter size ranges from 1 to 7 and is usually 3 to 4 [1757]. Litter size of the Japanese sable has not been reported yet.

Lifespan Not reported in Japan.

Diet The Japanese sable chiefly depends on small mammals such as voles (*Myodes* spp.), Siberian chipmunks (*Tamias sibiricus*) and field mice (*Apodemus* spp.) [2167, 2314]. In addition, insects, chiefly beetles, are one of their major foods in summer and berries are one of their major foods in autumn and winter [2167, 2314].

Habitat Japanese sables prefer areas with dense-tree cover or debris-rich microhabitats for resting and foraging [2168, 2315]. Japanese sables may avoid predators and strong wind by using dense canopy covers [2168].

Home range As home ranges of Japanese sables overlap with each other extensively, they do not seem to have strong territoriality [2168]. Radio-collared Japanese sables had 0.50–1.78 km^2 home ranges, but other sables were nomadic without establishing home ranges [2168].

Behavior Japanese sables adapt to tree climbing and often use trees. Over short distances, they can jump from tree to tree. In winter, they often use habitat underneath the snow for resting or feeding.

Natural enemies Not reported in Japan.

Parasites The flea *Stenopomia montana* is reported [2968].

A radio-collared male sable in Shiretoko, Hokkaido (T. Murakami).

5. Remarks

Japanese sable fur is of high quality, although its commercial value is considered lower than that of the Russian sable. By the early 20th century, many Japanese sables had been trapped for fur production. Since 1920, trapping Japanese sables has been prohibited. Generally, they do not appear around human habitation, although they sometimes damage hen houses or farm barns. Occasionally, chickens are killed by the Japanese sable [2310].

A closely related species, the Japanese marten, *M. melampus*, was introduced into Hokkaido in the 1940s, and currently its distribution overlaps with the Japanese sable around Sapporo-shi [897, 898, 2319]. There may be a risk of local extinction of the Japanese sable in areas overlapping with the Japanese marten [897]. More attention should be paid to isolated populations of the Japanese sable.

T. MURAKAMI

Carnivora MUSTELIDAE

Red list status: EX (MEJ); En as *L. nippon* (MSJ); NT (IUCN); EN as *L. nippon* (JFA)

Lutra lutra (Linnaeus, 1758)

EN Eurasian otter, river otter   JP カワウソ（kawauso）   CM 水獺   CT 水獺，歐亞水獺   KR 수달
RS речная выдра

A river otter specimen in Ehime Prefectural Science Museum (H. Sasaki). This individual died after being accidentally trapped in a fixed shore fishing net in 1966 in Ehime Pref.

1. Distribution

Distributed in Europe, Asia and the north of Africa. River otters were once a common species in Japan and occurred in Hokkaido, Honshu, Shikoku, Kyushu and Tsushima. Though the hunting of otters was prohibited in 1928 (when they still occurred through the country) because of decreasing population size, their range continued to shrink due to illegal hunting, habitat destruction and pollution [498, 693, 999, 2229]. The last record was confirmed in Honshu in 1954 and in Hokkaido in 1955. In Shikoku, the last otters were confirmed in Kagawa Pref. in 1944, in Ehime Pref. in 1975 and in Kochi Pref. in 1979. The otter in Japan was designated as extinct by the Environmental Ministry in 2012 and their extinction may have occurred in the 1990s [134].

2. Fossil record

Fossil records of *Lutra* sp. are reported from the Middle Pleistocene layers in Honshu [3602], and from the Late Pleistocene layer in Kyushu [818, 826]. Semifossils are reported from the Jomon Period layer in Honshu [820, 1023, 2008].

3. General characteristics

Morphology   Large-sized mustelid. Body elongate, legs short, head flat with small ears and broad muzzle with whiskers. Tail long and tapering evenly from a thick base. Feet webbed, five digits on each. Their fur color is dark brown, and color variation from dark brown to white is observed on the ventral side.

External measurements (mean ± SD) are as follows. For male, HB = 64.0 ± 13.1 cm ($n = 6$), T = 41.5 ± 7.1 cm ($n = 6$), BW = 6.5 ± 2.1 kg ($n = 2$). For female, HB = 66.3 ± 9.1 cm ($n = 5$), T = 41.5 ± 6.4 cm ($n = 4$), BW = 4.2 kg ($n = 1$) ([498, 3220]; measured from [372]).

Cranial measurements (mean ± SD) are as follows. For male, CBL (condylobasal length) = 116.6 mm ($n = 1$), ZW (zygomatic width) = 74.1 mm ($n = 1$). Female CBL = 112.6 ± 2.3 mm ($n = 6$), ZW = 70.7 ± 2.1 mm ($n = 6$) (measured from [1023]).

Dental formula   I 3/3 + C 1/1 + P 4/3 + M 1/2 = 36 [3222].

Mammae formula   0 + 1 + 1 = 4 [3220].

Genetics   Chromosome: $2n = 38$ [3771]. Chromosome number not reported in Japan. A 224-bp sequence of mtDNA cytochrome *b* gene from a river otter in Japan was sequenced and PCR primers for that gene are available [3374].

4. Ecology

Reproduction Litter size is 2. They are generally solitary, but the mother makes a small family group with her cubs [498].

Lifespan Not reported in Japan.

Diet Fish, crabs and prawns are their main food [692]. Flathead silverside *Hypoatherina valenciennei*, Common sea bass *Lateolabrax japonicus*, Sevenband grouper *Epinephelus septemfaciatus*, Japanese parrot fish *Oplegnathus fasciatus*, Barface cardinalfish *Apogon semilineatus*, and Flag fish *Goniistius zonatus* were identified from spraints, and also birds, carbs, barnacles, coronate moon turbans, oysters and cuttlefishes were reported as prey in the sea in Ehime Pref. [3916]. Carp *Cyprinus carpio*, crucian *Carassius auratus*, eel *Angoilla japonicus*, prawns *Macrobrachium nipponensis*, Japanese mitten crab *Eriocheir japonicus*, and bayberry *Myrica rubra* in rivers and along riversides are also eaten [3104].

Habitat The otter's main habitat was the river, but after the 1930s they had been observed mainly along the seashore in Shikoku and the ria coasts offered food and refuge from human activity for otters. They used rivers and the seashore as their habitat in Ehime and Kochi Prefs.

Home range Estimated range is 12–16 km along seacoasts [1002].

Behavior Nocturnal [498].

Natural enemies There is a record of a young otter being attacked by a crow [498].

Parasites Not reported in Japan.

5. Remarks

Otters in Japan, except for Hokkaido, are sometimes considered as a species, *L. nippon*, distinct from *L. lutra* [1023]. DNA [3374] and morphological [514] studies support this taxonomic status. Although some researchers list them as an independent species (e.g. [3894]), they are included in *L. lutra* in the IUCN Red List 2008 because of small sample size and numbers of analyzed base pairs used for the studies. A recent DNA survey suggests the Japanese otter was a local population of Eurasian otters [3817].

There were once many river otters in Japan. Thirty-five river otter specimens are stored in Ehime Prefectural Science Museum (H. Sasaki).

In the Meiji (1968–1912) and Taisho Eras (1912–1926), the hunting of otters became popular for skins and livers which were used in Chinese medicine, and they were subsequently excluded from the game list by law because of their decreasing population size [2105]. Before otters attracted people's attention in the 1950s, illegal hunting had gradually reduced them to small isolated populations and local extinction occurred in many places. After an otter was found in Ehime Pref. in 1954, many conservation measures (captive breeding, establishment of protected areas and so on) were taken. In 1965 otters were designated as a "National Natural Monument". However, ineffective conservation techniques and failure to gain the support of public opinion led to the extinction of the otter. From 1945 to 1979, 125 otters died in Shikoku: 39 drowned by fishing nets, 24 by failed live capture, 14 by intended killing, 37 by unknown reasons, 11 by other reasons (modified from [3012]). This strongly suggests that humans forced otters to extinction.

H. SASAKI

Carnivora MUSTELIDAE

Red list status: CR (MEJ); EN (IUCN, JFA)

Enhydra lutris (Linnaeus, 1758)

EN sea otter   JP ラッコ (rakko)   CM 海獺   CT 海獺   KR 해달   RS калан, морская выдра

A sea otter hauled out on a rock at Cape Nosappu, Hokkaido.

Point of sea otters observed since 1970s

1. Distribution

Occurs in coastal areas of the northern Pacific Ocean from the eastern part of Hokkaido to California [3772]. Three subspecies of sea otters are recognized: *E. l. lutris* (in the Commander Isls., the Kamchatka Pen., and the Kuril Isls., and eastern Hokkaido), *E. l. kenyoni* (in Alaska), and *E. l. nereis* (in California) [3881].

Some breeding groups are established at Harukarimoshiri and Todo Isls. in Hokkaido [838, 1693, 2736]. Some sea otters are recorded from the islands of Kunashiri, Kanakuso, Kabuto, Akiyuri, and Odoke of the southern Kuril Isls. [838]. In addition, they are observed along the eastern coast of Hokkaido from Erimo to Nemuro, and on the Shiretoko Pen. [844, 1114]. At Cape Erimo, 1 individual has been observed since 2002 [1114].

2. Fossil record

Although the historical distribution of sea otters is unknown, some remains (semifossils) such as skulls have been excavated from 6 archeological sites dating from before the Jomon Era to about the 9–14th century in Hokkaido [844].

3. General characteristics

Morphology   The sea otter is the largest mustelid species and the smallest of the marine mammals in Japan. Sea otters have large, flipper-like hind feet, no scent glands, retractile claws on their front feet, loose skin, and a horizontally flattened tail. For thermoregulation, sea otters have dense, water-resistant fur instead of blubber. Moderately sexually dimorphic, males tend to have a heavier head and neck than females [531, 1536].

Measurements of adults are as follows: TL, 126–145 cm in males and 107–140 cm in females; BW, 21.8–44.9 kg in males and 14.5–32.7 kg in females [1536, 3144]. From mainland Hokkaido, measurements of 2 males have been obtained; BW = 42 kg ($n = 1$), TL = 133.5 cm ($n = 2$), HB = 101.4 cm ($n = 2$), baculum length = 15.3 cm ($n = 2$) [1693].

Dental formula   I 3/2 + C 1/1 + P 3/3 + M 1/2 = 32 [29]. The sea otter is the only species of fissiped carnivore with 2 pairs of lower incisors [531]. The sea otter's dentition is adapted for crushing hard-shelled macroinvertebrates; molars are broad and flattened, and canines are rounded and blunt [1536, 2885].

Mammae formula   0 + 1 + 0 = 2 [1536].

Genetics   Chromosome: $2n = 38$ [531]. MtDNA analyses indicate low-level genetic diversity [307, 420, 838, 1827, 3158, 3672]. Several PCR primers for microsatellite DNA are available [78, 1827]. Some pairs of PCR primers for sex determination are also available [841, 2328, 3159].

Enhydra lutris (Linnaeus, 1758)

4. Ecology

Reproduction Sea otters exhibit a variable degree of polygyny, although many aspects of their mating system remain unclear [531]. There is no definite mating season, and seasonal peaks of pupping vary with region (in May/June in Russia) [3673]. Sea otters have delayed implantation [3275]. Their gestation period has been estimated to last 4 to 8 months [1209, 1536, 1869]. Sea otters give birth to a single pup, twinning occurs infrequently [3149]. Birth may occur on land or in the water [1536]. Males reach sexual maturity by age 5 to 6 years [3150], but males under 6 years are not successful breeders [706]. Most females have their first estrus at 4 to 5 years of age [308, 706, 1536]. Reproductive behavior has not been recorded around Hokkaido.

Lifespan Life expectancy of females appears to be about 15 to 20 years, males about 10 to 15 years in Alaska [347].

Diet Because of their small body size and lack of blubber, adult sea otters ingest about 20–33% of their body weight in food per day [411]. Sea otters generally feed on benthic macroinvertebrates, but can turn to fish if the invertebrate supply is depleted. On Etorofu Isl., sea otters feed mainly on sea urchins, bivalves and crabs [1717]. Around Hokkaido, feeding behavior has been recorded at Kiritappu, Moyururi Isl., Cape Nosappu and Cape Erimo [844, 1114]. Sea urchins and various kinds of bivalves were eaten at Cape Nosappu and Moyururi Isl., respectively.

Habitat Sea otters inhabit shallow coastal waters within the 25–40 m depth curve. They are keystone species in the nearshore environment, and they promote the growth of kelp, which in turn has a variety of community and ecosystem-level consequences [532]. The southern limit of sea otters coincides with the 20 to 22°C isotherm, which is also the approximate southern limit of cool water upwelling and the distribution of giant kelp (*Macrocystis*) [531]. Sea otters are mainly observed in rocky-bottom habitat along the eastern coast of Hokkaido, and low-relief rocks exposed at low tide are used as haul-out sites [844, 1114].

Home range Sea otters segregate by sex and age [1536, 3150]. In the USA studies, male areas were located at the front of the expanding population, while females inhabited areas that had been occupied for longer periods [707]. Some adult males can keep territory for reproduction in female areas [1536, 2886]. The dimensions of home ranges may vary in space and time. Around Hokkaido, home range size is unknown, although the sea otter at Cape Erimo is mainly observed around offshore reefs.

Behavior Diurnal activity cycles are generally characterized by crepuscular peaks in foraging activity [1536]. Sea otters occasionally prey on seabirds in California and Alaska [1536, 2885, 3769]. In addition, around Hokkaido, it was frequently observed that sea otters attacked seabirds such as long tailed ducks (*Clangula hyemalis*) and Japanese cormorants (*Phalacrocorax capillatus*) at Cape Nosappu [844, 1523], although predation has not been confirmed. A rock or other object is often used to break the prey's shell [555, 1114].

Natural enemies Around Hokkaido, there is no record of predation on sea otters. Predation by killer whales is having severe effects on the population of sea otters in western Alaska, resulting in an up to 70% decline [476]. White sharks, eagles, coyotes, and brown bears have also been reported to take sea otters [125, 2886, 3180].

Parasites Not reported in Japan.

5. Remarks

Historically, aboriginal hunting of sea otters took place throughout the range, including the Uruppu (Urup) and Etorofu (Iturup) Isls. (the southern Chishima = the Kuril Isls. Group) [1057, 1677, 2469]. Due to the excessive harvest of sea otters during the fur trade (1742–1911), sea otters were extirpated from most of their historical range [1536]. In southern Chishima, sea otters were hunted commercially by Russians from the late 18th century [3283, 3794]. In 1873, the government of Japan began to hunt around Etorofu Isl. [3724, 3794], and it continued until 1945 [3724]. Under protection by "International Fur Seal Treaty" established in 1911, during the 1970s, the sea otter population rapidly recolonized much of its former habitat. In Etorofu Isl., the population was reestablished by the early 1990s [1716], and population range expansion is likely to be occurring in the southern part of the archipelago [1717]. Nowadays, although the hunting of otters is prohibited (by a domestic law established in 1912), incidental harvest by fishing net has been reported around Hokkaido [844, 1114, 1693]. Some conflicts with the local shellfish industry have occurred at Cape Erimo, where 1 sea otter had been observed for more than 2 years [844, 1114]. On July 2014, one pup held by a female was observed at Moyururi Isl. [938]. This is the first reported mother–pup pair in existence on Hokkaido. Because reestablishment is in progress in the Habomai Isls., expansion into Hokkaido will occur in the near future. However, incidental harvest by fishing net could prevent the establishment of sea otter populations around Hokkaido, or else the habitation of sea otters could damage shellfisheries.

K. HATTORI

Carnivora MUSTELIDAE

114

Red list status: C-1 as *M. meles anakuma* (MSJ); LC (IUCN); see Remarks

Meles anakuma Temminck, 1842

EN Japanese badger　　JP ニホンアナグマ（nihon anaguma）　　CM 日本狗獾　　CT 日本獾　　KR 일본오소리
RS японский барсук

An adult female badger trapped at Hinode-machi, Tokyo (Yayoi Kaneko).

1. Distribution

Endemic to Japan, at the eastern extreme of the *Meles* distribution: Honshu, Kyushu, and Shikoku [3894].

2. Fossil record

Fossils of the Japanese badger have been excavated from the late Middle Pleistocene strata [1509, 1514]. Its absence from Hokkaido, the northernmost island of Japan, and the Ryukyu Isls., in the southern islands of Japan, suggests that the species migrated via the Korean Pen.

3. General characteristics

Morphology　A large-sized mustelid. It has a rather small head, a thick, short neck, a rather long and wedge-shaped body and a very short tail. The feet are armed with very strong claws, and are powerful for digging. The eyes are small in comparison with the size of the head. The face color pattern is distinguished by a conspicuous dark vertical stripe from the inner side of the ears surrounding the eyes on the white head. Ears are short and tipped with white.

External measurements (mean ± SE) for adults (2 years old or older) in Hinode, a suburban area of Tokyo, are as follows: BW = 7.7 ± 1.3 kg, TL = 78.7 ± 4.9 cm, and T = 13.9 ± 1.5 cm in males, and BW = 5.4 ± 0.8 kg, TL = 73.1 ± 3.5 cm, and T = 13.9 ± 1.9 cm in females [1324, 1331]. Badgers in Yamaguchi Pref. are smaller: BW = 5.7 ± 0.4 kg and TL = 66.8 ± 2.7 cm in males, and 4.4 ± 0.6 kg and 60.4 ± 2.4 cm in females [3514]. Among *Meles* species including continental badgers, the Japanese badger has the smallest skull compared to the other 2 species (*M. leucurus* and *M. meles*), except that the size of the canine teeth are similar in all three species [67].

Dental formula　I 3/3 + C 1/1 + P3–4/3–4 + M 1/2 = 34–38 [29, 266].

Mammae formula　1 + 1 + (1) + 1 = 6–7 [1316].

Genetics　Chromosome: $2n = 44$ and FNa = $64 + XY$ sex chromosomes [2542]. Mitochondrial [1810, 1952, 3532] and nuclear [3050, 3531, 3533] phylogenetic analyses suggest substantial differences between other continental species (*M. leucurus* and *M. meles*) and the Japanese badger.

4. Ecology

Reproduction　Females mature at 2 years old and males at 1 year old. Birth is given from mid-March (Yamaguchi Pref.) to mid-April (Tokyo). Average litter size is 2.5 ± 1.2 SE cubs (1–4 cubs) in suburban Tokyo [1324] and 2.3 cubs in Yamaguchi Pref. [3511]. Mating occurs just after birth, and the mating season lasts until October–November, with delayed implantation of blasto-

Meles anakuma Temminck, 1842

Table 1. Food items and home range size of the Japanese badger.

Study area	Landscape	Forest cover (%)	Altitude (m)	Main food (faecal analysis)				Home range size in ha		Reference
				Spring	Summer	Autumn	Winter	Male (n)	Female (n)	
Mt. Nyugasa, Subalpine zone of Honshu	Larch plantation, evergreen conifer, deciduous forest, pine	90~	1,500–2,000	earthworm, beetle	earthworm, beetle, berry, leftover	earthworm, beetle, fruits	—	407 (1)	200–269 (2)	[205, 206]
Yamaguchi, western Honshu	Pine forest, residential area, oak/chinquapin forest, cedar/cypress plantation	72	37–496	earthworm, beetle, berry	earthworm, beetle	earthworm, beetle	—	108–253 (1)	22–39 (3)	[198]
Hinode-machi, suburban Tokyo	Mozaic of residential area, agricultural area, and cedar/cypress plantation	69	200–500	earthworm, leftover	earthworm, beetle, berry, leftover	tree fruit (persimmon), leftover	—	22–72 (7)	5–19 (4)	[203, 209]

cysts until February [1324].

Lifespan The longest recorded lifespan is 7 years in the wild [1316].

Diet Predominantly earthworms from spring to autumn (Table 1) [1314, 1333, 3511, 4009, 4011]. They prefer berries and beetles in summer. Fruits of the persimmon (*Diospyros kaki*) are an important food in autumn in suburban Tokyo [1333].

Habitat In general, badgers (*Meles*) are forest dwellers and are adapted to a fossorial life-style, spending a long time in extensive burrow systems termed "setts" [2454]. Sett types vary from plain (no cover), rock crevices to the bases of (tree) roots [1333, 4007]. Setts are often located on hills/ridges to avoid chamber humidity. In suburban Tokyo, west-facing slopes, which are prone to landslides during heavy rainfall in summer, are avoided.

In suburban Tokyo (Hinode-machi) [1333], forest edge and agricultural areas have an important role in the badgers' diet in 3 categories: staple foods such as earthworms; temporary supplements such as beetles, berries and persimmon; and artificially provisioned food in Buddhist cemeteries. First, earthworm availability is affected not only by season but also by land use: in early summer, earthworm growth is slow in deciduous forest compared to the edges dividing grassland and coniferous forest, and earthworm biomass in grasslands is greater than in forests. During autumn in Hinode, persimmon trees, which are found only in agricultural areas accessible from forests, provide a patchy and super-abundant food source, while earthworm availability is diminished. Second, badgers eat insects and berries seasonally, mainly in early summer. These principally become available in deciduous forest and on the boundaries between forest and agricultural areas, both of which are rare patchy habitats in suburban Tokyo. Third, another highly patchy, intermittent, but artificial, food source is the cemetery adjoining the forest, where traditional Buddhist ceremonies frequently leave confectionaries and fruits behind.

Home range Home range size with reference to kinds of habitat is shown in Table 1. The badger's home range can be explained by food choice, driven by nutritional requirements and access to food [1333]. Yamaguchi and Hinode (Tokyo) appear to be more food rich habitats compared to the sub-alpine zone of Mt. Nyugasa.

In terms of population, adult males and yearlings (which usually have larger home ranges) utilize many small temporary setts and resting sites on the ground (couches) [1326]. In suburban Tokyo, a typical inter-sett distance is 630 m with activity focused within 300 m of setts [1333]. In Yamaguchi Pref., badgers utilize 13.5 setts on average per year (8–71 setts per individual home range), and male offspring share the natal sett with their mother for up to 26 months, while females remain with their mother for only 14 months [3511].

Density tends to be lower than that of other endemic medium-sized carnivores such as raccoon dogs (*Nyctereutes procnoides viverrinus*) and red foxes (*Vulpes vulpes*) [1214]. A density of 4 adults/km^2 in suburban Tokyo was estimated from 7 years (1990–1997) of mark-recapture data [1316].

Behavior Nocturnal [3511–3513]. Japanese badgers form mother-cub groups as a social unit [1329]. Male badgers are solitary but bond temporarily with one or several females in the mating season [1329]. Territoriality may exist within sexes [3514] or only among females [1329]. Scent marking at border latrines to exchange information and to advertise estrus status is observed at low density, where few opportunities for mutual encounter exist [1337].

Apparent hibernation from January to February, when little activity is observed and when they stay in their setts; badgers enter a state of torpor, dropping core body-temperature by 3°C [3511]. The badger's body-shape allows it to function with very variable levels of body-fat [2454], and Japanese badgers can add more than 50% to their summer mass before hibernation [1332].

Digging behavior is thought to have evolved in the Miocene, with ancestors, such as the Melodon, that dug up ground-dwelling prey [2454]. Evolutionary adaptations for digging come at the expense of agility and the arboreality of their primitive "marten-like" ancestors from the Paleocene onwards. With the opportunistic switch to earthworms as their main food, the present function of digging is to provide shelter and protection. Their usual strategy to escape aggressive encounters is to retreat underground, although they are tenacious and can be extremely fierce if cornered [2464].

Natural enemies Potentially, introduced feral raccoons (*Procyon lotor*); raccoons can exclude female badgers from breeding setts [1316].

Parasites Trematodes *Concinnum ten* (Syn. *Eurytrema ten*) [3679], *Isthmiophora meles* [1291], *Paragonimus miyazakii* [746], and *P. ohirai* [3378] are reported as endoparasites. The ticks *Haemaphysalis flava*, *Ixodes ovatus*, and *I. tanuki* [598] and the flea *Chaetopsylla mikado* [2968] are reported as ectoparasites.

5. Remarks

The Japanese badger was considered a subspecies *M. meles anakuma*, but is now considered a distinct species independent from its continental Asian and European congeners, *M. leucurus* and *M. meles* [3894]. This taxonomic status is supported by mtDNA [1810], head color pattern [65], baculum structure [266, 3514], and craniological characters [67].

The Japanese badger is currently listed as a game species in Japan. However, the annual hunting bag declined from 7,000 individuals per year in the 1970s to less than 2,000 by the late 1980s. This may mainly reflect a decreased interest in badgers as game animals among hunters. However, the badger's geographic range is shrinking demonstrably in 45 prefectures. In 2003 the Ministry of the Environment reported that the distribution area of badgers (about 126,000 km^2) is 29% of the land area of Japan including Hokkaido [1212]. Compared with survey data from 1978, a 7% reduction occurred over the past 25 years following 1978, especially evident in Nara and Chiba Prefs. Badgers are locally designated as a red data list species in 11 prefectures. Vulnerable (VU): Hyogo Pref. Near Threatened (NT or CR): Chiba, Kagawa, Tokyo, Oita, Okayama, Osaka and Yamaguchi Prefs. Data Deficient (DD): Aichi, Gunma, and Tochigi Prefs.

Yayoi KANEKO

Japanese badgers.

A sett of the Japanese badger in Hinode-machi, Tokyo (Yayoi Kaneko).

RESEARCH TOPIC 8

A molecular phylogenetic view of mammals in the "three-story museum" of Hokkaido, Honshu, and the Ryukyu Islands, Japan

H. SUZUKI

The Japanese Isls. comprise a "biodiversity hotspot" that harbors more than 100 native species of mammals, half of which occur only in Japan [24, 33]. The origin of these mammals is of great interest and has been the subject of intense debate from paleontological, zoogeographic, phylogenetic, and taxonomic perspectives (e.g., [1793, 3590, 3592]). Recent molecular phylogenetic studies of many groups of Japanese mammals, such as rodents, moles, mustelids, and leporids, have suggested exciting new hypotheses regarding the evolutionary history of the mammals of the Japanese Isls.

The Cenozoic, which began approximately 65 million years ago (MYA), was the most prominent period of mammalian evolution. However, it is the last 10–15 million years of evolutionary activity that largely has shaped the present-day worldwide distribution of mammals, including those of the Japanese Isls. In this study, the evolutionary histories of the terrestrial mammals of Japan are examined along with the development of continental species, with special emphasis on murids, leporids, mustelids, and talpids.

The murine rodents of Eurasia provide an excellent example of a "punctuated" evolutionary mode in mammals. The diversification of murines began around 12 MYA in Southeast Asia, resulting in a modern tally of more than 120 genera including *Mus*, *Micromys*, *Apodemus*, and *Rattus* (Fig. 1). Final species numbers range from a single species of *Micromys* to more than 60 species of *Rattus*. Nevertheless, these genera appear to have similar evolutionary histories, as exemplified by the patterns of *Rattus* [383], *Apodemus* [3165, 3345], and *Mus* [3211], each of which shows 2 periods of diversification that generated subgeneric and species lineages, respectively (Fig. 1). The genus *Mus*, for example, comprises 4 subgenera that emerged around 5–6 MYA, and the subgenus *Mus* has 4 distinct "species groups" that emerged around 2–3 MYA (Fig. 1) [3344, 3353].

Recent molecular phylogenetic studies on rabbits and hares identify 3 radiation-like events after initial divergence from the ancestral pika lineage [2030, 3895]. Assuming a general divergence of pikas and rabbits approximately 30–40 MYA, evolutionary divergence times are estimated to be 9–15 MYA for leporine genera, 4–7 MYA for species of *Lepus*, and 1–3 MYA for intraspecific diversity within *L. timidus*. Notably, 3 leporid species from Japan—the Amami rabbit (*Pentalagus furnessi*) from the Ryukyu Isls., the Japanese hare (*Lepus brachyurus*) from Honshu, and a Japanese form of the mountain hare (*L. timidus ainu*) from Hokkaido—are associated with each of these radiation events (Fig. 2).

Assuming an evolutionary split between raccoons and weasels 28.5 MYA, we can speculate that most lineages of weasels and martens differentiated over the last 10 my (million years) [3050] in 2 major radiation phases. In weasels and martens, the initial radiation events likely occurred 5–10 MYA in a temperate area of Eurasia [3050]. In the last

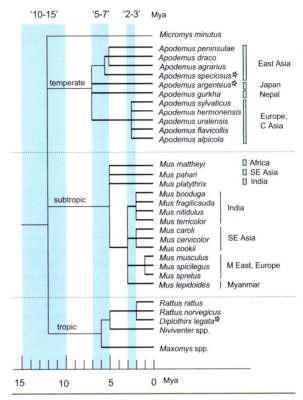

Fig. 1. Phylogenetic relationships of the three major groups of the subfamily Murinae—*Apodemus*, *Mus*, and *Rattus* (and its related genera)—and the monospecific lineage *Micromys*, based on nuclear gene sequences. Note that although these three groups now occupy three different geographic areas of Eurasia (temperate, subtropical, and tropical regions), their evolutionary patterns show coincidental patterns within the three time periods of 10–15, 5–7, and 2–3 MYA. Stars indicate species endemic to Japan.

2 my, a second radiation event generated the species of weasels and martens that now occupy the subarctic area of the Palearctic zone [958, 3049]. In mustelids, we can discern a similar trend to that seen in rabbits and hares; in both groups, evolutionary diversification probably occurred in a stepwise manner in the context of the tropical, temperate, and cool-temperate climates of the Middle Miocene, Early Pliocene, and Early Pleistocene, respectively. The Japanese lineages of the weasel (*Mustela itatsi*) and the marten (*Martes melampus*) emerged during the final radiation phases within their respective genera (Fig. 2).

Although temporal aspects of talpid evolution have not been described, the evolutionary patterns are likely to be similar to those of murids, mustelids, and leporids. Studies of the shrew-like moles (*Uropsilus*), desmans (*Desmana*), shrew moles (*Neurotrichus*, *Dymecodon*, *Urotrichus*), star-nosed moles (*Condylura*), and moles (*Euroscaptor*, *Mogera*, *Scalopus*,

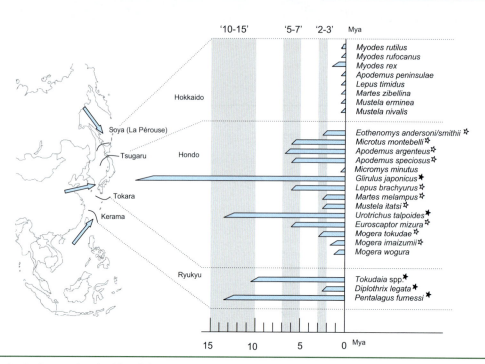

Fig. 2. Approximate degree of endemicity of Japanese terrestrial mammals such as muroids, mustelids, and leporids represented by their estimated divergence times. These times were obtained by comparing Japanese species with their nearest congeners on the continent by the use of published data sets of mitochondrial and nuclear gene sequences. Three geographic regions (subtropic, temperate, and subarctic) are possible source areas of the mammals of Japan (arrows). They are similar to the three geographic areas of Ryukyu, Hondo, and Hokkaido, respectively, which are separated by the Tsugaru and Tokara Straits. Most of the species lineages are presumed to have diverged at 3 periods of time, i.e., 10–15, 5–7, and 2–3 MYA. Stars indicate taxa endemic to each of the 3 areas at their genus (black) and species (white) levels. The Hokkaido taxon *Myodes* (=*Clethrionomys*) *rex* was reported to inhabit the southern part of Sakhalin as well [33]. The 2 endemic wood mice *Apodemus argenteus* and *A. speciosus* occur on Hokkaido and Hondo (=Honshu + Shikoku + Kyushu + their adjacent small islands).

Scapanus, Talpa) of Eurasia and North America suggest that the lineages were generated within a short period of evolutionary time [3231, 3232], probably in the mid-Miocene approximately 14–15 MYA [1871]. Subsequent divergence events in each of the specious genera *Mogera* and *Talpa* occurred in each of East Asia and Europe [3235, 3643], respectively, around the Miocene–Pliocene boundary (5–6 MYA) [1871]. Further radiation events likely date to around 2–3 MYA, the Pliocene-Pleistocene boundary [1592].

Examination of the phylogenetic tree suggests that several ancestral talpid species colonized the Japanese Isls. [3235, 3643]. The first colonization event involved shrew-moles (common ancestor of *Urotrichus talpoides* and *Dymecodon pilirostris*); the second event was a member of the genus *Euroscaptor* (*E. mizura*), which is rather small in size; the third event concerned the ancestor of *Mogera tokudae* that now inhabits Sado Isl.; a fourth event produced the Eastern Japanese mole (*Mogera imaizumii*); and a final colonization event involved the Western Japanese mole (*Mogera wogura*), which remains conspecific with moles from Korea and Russian Primorye (Fig. 2). Periodical radiation events in each of these mammalian groups involved both long-range dispersal (sometimes across continents) and subsequent regional differentiation (formation of spatial partitions), and these processes in combination likely shaped the current species diversity in the Northern Hemisphere from the mid-Miocene to the present. It is very tempting to further link these events to global or regional environmental changes that occurred in the last 15 my. According to geological evidence, 12–15 MYA was a time during which significant global cooling occurred, with a further rapid fall in global temperature at approximately 7 MYA [354]. In addition, the Pliocene–Pleistocene boundary period (ca. 2–3 MYA) also was an important time for global floral and faunal changes, related to the onset of cyclic glacial episodes. These global environmental changes might have promoted the radiation of lineages that were able to adapt to newly-developed habitats and/or undertake long-range dispersal.

Clear evidence of rapid, long-range dispersal comes from microevolutionary studies of the harvest mouse, *Micromys minutus*. In this species, mitochondrial genomes intermingled across the entire Eurasian continent from the British Isles in the west to the Japanese Isls. in the east within the last few tens of thousands of years [4078]. Mammals other than mice probably underwent similarly rapid dispersal across the Eurasian continent. Comparable long-range dispersal events in periods of the late Tertiary, including intercontinental exchange of lineages, are implied from the phylogenetic patterns of rabbits/hares [2030] and moles [3231, 3232].

Geographic partitioning is necessary to promote the sudden development of local-specific lineages. Spatial partitioning across the Eurasian Continent is an example that has encouraged evolution within a variety of mammalian groups. For example, divergence of subgenera and species groups within the genus *Mus* likely was promoted by spatial barriers between the Indian subcontinent, Myanmar, and Southeast Asia (Fig. 1) [3211, 3344]. Similar geographic subdivisions can be observed from a microevolutionary perspective. In the rice-field mouse, *Mus caroli*, genetic structure and geographic pattern analysis show country-specific mitochondrial clades [3212, 3557]. In *Apodemus*, 4 Eurasian geographic areas, including the Japanese Isls., are thought to have hosted the earliest phase of lineage differentiation (Fig. 1) [3165, 3352]. Perhaps in other mammals as well, the Japanese Isls. served as an important geographic region for the generation of region-specific evolutionary lineages.

This view of a long mammalian history for Japan stands in contrast with its very poor Tertiary fossil mammal record, and with the associated belief that terrestrial mammals only migrated to Japan rather recently, during the last half million years. By comparing endemic Japanese taxa with their most closely related continental counterparts, we were able to quantify the level of genetic endemism of Japanese taxa and estimate their times of divergence (Fig. 2). Our results indicated that levels of endemicity were high in the Hondo (Honshu, Shikoku, and Kyushu) and Ryukyu blocks, and low in Hokkaido. Extant mammalian lineages in Japan can be characterized by their level of endemicity, which mostly goes back to 2–3 MYA and 5–6 MYA in the Tertiary. As noted above, available fossil data recovered from the Japanese Isls. suggests that these anciently diverged lineages might have co-migrated from the continent to Japan in the recent geological past, such as during the Middle Pleistocene (0.5–0.6 MYA). However, this hypothesis requires the extinction on the continent of each of the Japanese endemic lineages, all within a short period of evolutionary time. Furthermore, why the continental counterpart species did not migrate first to Japan during that period rather than their Japanese ancestral species is a difficult question to address and even more difficult to explain. Because the striped field mouse *Apodemus agrarius* has a high level of genetic diversity that likely developed during the last half million years [766], it is conceivable that the Japanese fauna has been dominated by taxa that had similar niches compared with their continental counterparts. Moreover, fossil evidence from mainland East Asia does not support the preoccupation of the Japanese lineages on the continent, and some Japanese endemic species possess substantially large degrees of intraspecies genetic divergences. The Japanese dormouse (*Glirulus japonicus*), for example, has eight or more distinct local populations with divergent mtDNA that is estimated to have diverged several million years ago [4075, 4077]. It has been noted that global eustatic changes in sea level during the last 10 my had a significant impact on the faunal composition of Japanese mammals, assisting intermittent migration events as is predicted in the case of moles endemic to Japan and Taiwan [1592].

Whatever the evolutionary histories for colonization events, the Japanese Isls. are now preserving anciently divergent lineages. The region is in fact a "museum" that contains and maintains important lineages and species with high levels of diversity that migrated from the continent from ancient to more recent times. This museum has three floors—Hokkaido, Hondo, and Ryukyu islands—for lineages that migrated recently, in moderately ancient times, and in ancient times, respectively. The central and southern Japanese domains are each characterized by several endemic terrestrial mammalian species.

The Ryukyu domain spans the boundary of the Palearctic and Oriental regions and is home to several deeply endemic mammal species of obscure genetic origin. The Ryukyus, otherwise known (though somewhat ambiguously) as the Ryukyu Isls., contain many endemic species of plants, insects, birds, as well as mammals, and long have been noted for their biological and geological significance. An understanding of the phylogenetic relationships of the mammals of the Ryukyus helps explain numerous aspects of the geology, zoogeography, taxonomy, and conservation biology of these islands [3356]. Geographic variation within the mammal species of the Ryukyu domain provides another key to understanding the region's mammalian fauna. For example, considerable differences exist among regions of the Ryukyus, as defined by the Kerama and Tokara Straits. From taxonomic and genetic perspectives, outstanding levels of endemism are observed among some mammals in the central region (Fig. 2). In addition, as is observed in the Ryukyu spiny rat (genus *Tokudaia*), genetic differentiation has occurred on each island. Species from Okinawa-jima, Tokuno-shima, and Amami-ohshima Isls. exhibit karyotypic differences [i.e., $2n = 44$, $2n = 45$ (XO, both female and male) and $2n = 25$ (XO, both female and male)]. In XO species, the Y chromosome is absent and portions of it are translocated to a distal part of the X chromosome ([180]; see also Topic for *Tokudaia* species).

The central Japanese Isls. of Hondo host a high diversity of mammalian species. Typical examples from this domain are the Japanese dormouse (*G. japonicus*) [2531] and species of moles (*Mogera*, *Euroscaptor*) [3235, 3643]. The Hondo dormouse shows an extremely high level of endemicity; its divergence from the continental species of dormouse is estimated to have occurred 23 MYA [2531]. The large Japanese wood mouse (*A. speciosus*) exhibits an intriguing "doughnut-like" evolutionary pattern in which central (Hokkaido, Honshu, Shikoku, Kyushu) and peripheral (e.g., Hokkaido, Sado, Oki, Tsushima) populations show divergent genetic structures [3358] and distinctive morphological characters [1019].

In contrast, Japan's northern island of Hokkaido is home to only 1 phylogenetically distinct species of rodent, *Myodes* (=*Clethrionomys*) *rex*. This species shows a considerable degree of sequence divergence in the cytochrome *b* gene from its sister species, *M. rufocanus* [1173, 3348]. The Japanese population of the sable (*Martes zibellina*) on Hokkaido differs substantially in coat color compared with its continental counterpart [957]. However, single nucleotide substitutions are proposed to have caused the phenotypic variation and this implies rapid evolution within the lineage that dispersed from the continent into the new environment of Hokkaido, probably involving evolutionary modulation. Only now are we beginning to address this issue to elucidate the ecological meaning and genetic basis of such modulation.

Overall, the Japanese mammal fauna comprises a unique set of species with high endemicity and a variety of evolutionary histories. This fauna provides abundant opportunities for studying mammalian evolution over both long and short time frames—it should be both conserved and studied.

Carnivora   HERPESTIDAE

Red list status: LC as *H. javanicus* (IUCN); introduced

Herpestes auropunctatus (Hodgson, 1836)

EN small Indian mongoose   JP フイリマングース（fuiri manguusu）   CM 红颊獴   CT 紅頰獴
KR 회색몽구스   RS малый мангуст, малый индийский мангуст

A small Indian mongoose at Mt. Yuwan on Amami-ohshima Isl. (S. Abe).

1. Distribution

The mongoose is feral on Okinawa-jima Isl. (1,206 km^2; introduced in 1910) in Okinawa Pref. and on Amami-ohshima Isl. (712 km^2; introduced in 1979) in Kagoshima Pref., and at Kiire in Kagoshima and at Takae in Satsumasendai in Kagoshima Pref. in Japan [55, 1264, 1596, 2376, 3853]. The species is native to Iraq, Iran, Afghanistan, Pakistan, India, Nepal, Bhutan, Bangladesh, Myanmar (Burma), and southern China including Hainan Isl. [725, 2456, 2518]. After their first introduction from India to Jamaica in 1872, the species has been introduced to many countries and regions to control rats and snakes: Ambon (Indonesia), Antigua, Barbados (introduced in 1877), Beef Isl., Buck Isl. (1910), Cuba (after 1870), Fiji (1883), French Guiana, Grenada (1882), Guadeloupe, Gyuana, Hawaii (1883), Hispaniaola (1895), Jost Van Dyke, La Desirade, Lavango, Mafia (Tanzania), Marie-Galante, Martinique, Maui (1883), Molokai (1883), Nevis, Oahu, Puerto Rico (1887), St. Croix (1884), St. Vincent, Suriname (1900), Tortola, Trinidad (1870), Vieques, and Water Isls. [264, 725, 1859, 2456].

2. Fossil record

Not reported in Japan.

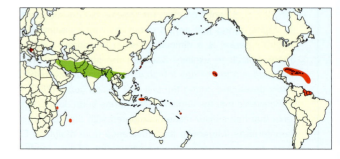

3. General characteristics

Morphology   The overall coloring, except for a portion of the ventral surface, is mottled with black and yellowish-white speckles. The contour hairs on the dorsal neck are 12 to 20 mm long, and each hair is alternately ringed with 5 or 6 dark and light bands [2587]. The body is slender with short legs [2456]. The head is elongated with a pointed muzzle. The tail is robustly muscular at the base and tapers gradually throughout its length. The ears are short and round, and project only slightly beyond the pelage. There are 5 digits on each limb, the hind foot is naked to the heel, and the foreclaws are sharp and curved. Small scent glands situated near the anus eject a vile-smelling secretion in both sexes. The external characteristics show sexual dimorphism as follows. Mean HB, 287 mm for females and 325 mm for males; T, 225 mm for females and 256 mm for males; BW, 376 g for females and 611 g for males on Okinawa-jima Isl. [2587]. On Amami-ohshima Isl., mean HB, 302 mm for females and 336 mm for males; T, 223 mm for females and 241 mm for males; BW, 456 g for females and 673 g for males [55].

Dental formula I 3/3 + C 1/1 + P 4/4 + M 2/2–3 = 40–42 [2456], the third molar in the lower jaw is often missing [2456], and is absent on Okinawa-jima Isl. [2584] and Amami-ohshima Isl. [49].

Mammae formula 1 + 1 + 1 = 6, and only the posterior 4 mammae are regularly used by cubs [2456].

Genetics Chromosome: $2n = 36$ in females and 35 in males with no apparent Y chromosome due to the translocation of the Y chromosome to an autosome [2456, 3143]. According to mtDNA analysis, there is almost no difference in cytochrome *b* between populations on Okinawa-jima and Amami-ohshima Isls., and the mongooses on Amami-ohshima Isl. are thought to have originated from the population established on Okinawa-jima Isl. [3162]. Microsatellite analysis supports that both populations have a recent common history with near equal allocation [3572].

4. Ecology

Reproduction Mating system is polygamous [2456]. Ovulation occurs by stimulation of copulation [1816, 2456]. Gestation period is ca. 7 weeks [56, 2456]. Females begin to copulate from February, and give birth to young from April to September and to nurse until November or December on Okinawa-jima Isl. [2586]. Mean litter size is 2.15 and mean number of placenta is 2.54 [2586]. On Amami-ohshima Isl., length of the breeding period varies from 7 months (from March to September) to 12 months, and mean litter size also varies from 2.26 to 2.93 depending on population density and other conditions [56]. Body weight on day 2 after birth is 27 g for a female and 24 g for a male on Okinawa-jima Isl. [2586]. All deciduous teeth erupt at 4 weeks and all permanent teeth erupt at 8 weeks after birth [2586]. Young undertake their first activity outside of their den at ca. 4 weeks of age and follow their mother on their first hunting trip at ca. 6 weeks of age [2456]. Time of weaning is ca. 7 weeks after birth and sexual maturity is reached at HB 255 mm for females, and more than 270 mm for males on Okinawa-jima Isl. [2584]. In the Caribbean, females past 6 months of age are normally sexually mature [2456].

Lifespan On Amami-ohshima Isl., mean lifespan is 0.99 years for females and 1.04 years for males, and approximately 90% of individuals within a population for both sexes are 2 years old or less [52].

Diet Omnivorous. Insects (71%), reptiles (18%), oligochaetes (earthworms) + molluscs (12%) and birds (10%) and mammals (4%) are observed in 83 digestive tracts of the mongoose in the northern part of Okinawa-jima Isl. [2588]. On Amami-ohshima

A small Indian mongoose in Nago-shi on Okinawa-jima Isl. (Katsushi Nakata).

A small Indian mongoose captured on Okinawa-jima Isl. (Katsushi Nakata).

Isl., insects (in 80% of case) and other invertebrates (65%), amphibians + reptiles (35%), mammals (25%) and birds (10%) were observed in 146 stomachs of the mongoose collected in the original release area from 1990–1992 [50, 54]. Insects (40%) and other invertebrate animals (20%), amphibians and reptiles (60%), mammals (20%) and birds (15%) were observed in 89 pellets of mongoose collected in the habitat of the Amami rabbit (*Pentalagus furnessi*) from 1993–1997 [3936]. Eight percent of pellets contained the Amami rabbit. Although the mongoose chiefly preyed on insects and birds in all seasons, it tended to prey more frequently on amphibians and reptiles in summer and on mammals in winter. However, there was little evidence of predation of the habu pit viper (*Protobothrops flavoviridis*) because the mongoose is diurnal, while the pit viper nocturnal [54, 527, 3936]. Indirect effects caused by mongoose feeding are also found on Amami-ohshima Isl. as follows; smaller species increase in abundance through top-down cascades, i.e. decreases in native predators such as frogs and lizards caused by the mongoose result in increases in the abundance of smaller animals [3855]. Accumulated amount of total mercury in the mongoose on both Okinawa-jima Isl. and Amami-ohshima Isl. is very high (1.75–55.5 µg/g wet wt.) and is similar to level found in marine mammals, indicating that the mongoose has an efficient metabolic potential for mercury in liver as in marine mammals [942].

Habitat Inhabits both farmlands and dense forests on both islands in Japan [51, 55]. Xeric habitats are preferred because of aversion to rain and water [872].

Home range On Amami-ohshima Isl., home range size is 12.9 ha for 8 adult females and 18.4 ha for 7 adult males estimated by 95% convex polygon method of radio-tracking investigation [528, 1098]. On Caribbean Isls., home range size is 2.2 ha for females and 4.2 ha for males, and home ranges of males overlap both male and female ranges as do home ranges of females [2456].

Behavior Investigation by trapping and remote camera set-ups, each with infrared sensors on both Okinawa-jima Isl. and Amami-ohshima Isl., have revealed that the mongoose is diurnal [1722, 3923]. In the Caribbean, the mongoose is also diurnal with most activity between 10:00 and 16:00 [2456].

Natural enemies No natural enemies on the Okinawa-jima and Amami-ohshima Isls. [56].

Parasites Introduced mongooses do not seem to have carried any parasites or diseases from their original sites to new habitats [2456]. On Okinawa-jima Isl., 9.3% of 2,406 mongooses carried 370 cat fleas (*Ctenocepahlides felis*) [1069], and 19.1% of 2,618 mongooses carried 1,317 ticks including 3 genus and 5 species; *Ixodes granulatus* (83% of total ticks), *Amblyomma testudinarium* (7%), *Haemaphysalis hystricis* (5%), *H. formosensis* (4%) and *H. flava* (1%) [1068]. Rabies is the most important disease transmitted by the mongoose in the Caribbean, and Leptospirosis is probably the second most important human disease [2456]. On Okinawa-jima Isl., the mongoose also carries *Leptospira interrogans* (30.1% of 133 mongooses) [1067], and *Salmonella enteritidis* (1.5% of 132 mongooses) [1070]. Antibody prevalence rate for *Toxoplasma gondii* in the mongoose on Amami-ohshima Isl. is 10.0% out of 150 mongooses [1137].

5. Remarks

In order to control rats and the poisonous habu pit viper, which is feared by local residents because of the severe consequences of its bite, on Okinawa-jima Isl., 13–17 mongooses from Bangladesh were released in Naha by Dr. Watase in 1910 [1596], and their descendants went north and eventually invaded the northern part of the island after 1990, where many endangered species live (Fig. 1) [51]. On Tonaki-jima Isl. (4 km^2), 4 mongooses were released in 1910, but they did not colonize [1596]. On Amami-ohshima Isl., 30 mongoose were released in Naze, Amami around 1979, and they have since expanded their distribution [55, 527].

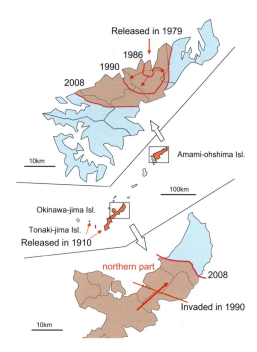

Fig. 1. Releases of the mongoose on Okinawa-jima, Tonaki-jima and Amami-ohshima Isls., and expansions of their distribution into the northern part of Okinawa-jima Isl. and on Amami-ohshima Isl.

However, the mongoose did not become a predator of the habu pit viper as expected. Since introduction, it has had serious negative impacts on to important native species on these islands of unique and high biodiversity. In addition, the mongoose was designated as one of the worst invasive alien species by IUCN [1063]. Eradication programs have been implemented by the Japanese government as a model for the conservation of biodiversity on subtropical islands of the Nansei Isls. since 2000 [622, 985, 2096, 2581, 3927]. Other eradication actions were begun by local governments in Kiire, Kagoshima and Takae, Satsumasendai in Kyushu in 2009. Consequently, recent mongoose populations, especially on Amami-ohshima Isl., have been becoming very low and native species including endangered species have been recovered [618, 3322, 3854]. A total of 115 mongooses (68 females and 47 males) along the coastal area (15 km in length × 4 km in width) at Kiire, Kagoshima, which are thought to have resulted from introduction in the 1980s, were eliminated during 2009–2010, and the population has likely been eradicated [662, 1264, 2581]. On the other hand, only 1 male mongoose at Takae in Satsumasendai, which is ca. 70 km northwest apart from Kiire, Kagoshima, was captured in 2011 and no more mongooses have been recorded so far [1264].

Some authors consider the small Indian mongoose and the Javan mongoose (*H. javanicus*) conspecific under the name *H. javanicus* or *H. auropunctatus*. Recent molecular studies suggest that they should be treated as separate species [725, 985, 2808, 3775].

F. YAMADA, G. OGURA & S. ABE

Carnivora   VIVERRIDAE

116

Red list status: C-1 (MSJ); LC (IUCN); introduced

Paguma larvata (Smith, 1827)

EN masked palm civet   JP ハクビシン（hakubishin）
CM 花面狸，果子狸   CT 白鼻心，果子狸   KR 사향고양이   RS гималайская цивета

Masked palm civets in an abandoned mine in Gifu Pref. (A. Sano).

1. Distribution

Bangladesh, Myanmar (Burma), Cambodia, China (south to Taiwan, Hainan Isl. and north to Hopei, Shanxi, and the vicinity of Beijing), India (southern Andaman Isls.), Indonesia (Borneo and Sumatra), Laos, Malaysia, Nepal, Pakistan, Singapore, Thailand, Vietnam, and Japan [3893]. This animal is regarded an introduced mammal in Japan; however, the original place where they were introduced and the period of introduction are unknown. They are now found across most of Japan [4138].

2. Fossil record

Not reported in Japan.

3. General characteristics

Morphology   The masked palm civet is a medium-sized mammal in Japan, and similar to any other Viverridae in shape, with a long body and tail, short-limbs, wide ears and big eyes. External measurements from Shizuoka Pref., central Japan, are shown in Table 1 (Torii's database 1976–1996). External fur color: dark brown dorsally, lighter ventrally. Underparts of the legs and tail black. The face is black with white patches from brow to nose and under both eyes, giving rise to its "masked" appearance. The tail is often tipped dirty white. The color of the body, and especially the face mask, varies among individuals [3611]. Fur color is also variable among localities [2892], and body and skull sizes vary among local populations or subspecies [1840].

Dental formula   I 3/3 + C 1/1 + P 4/4 + M 2/2 = 40 [1011].

Mammae formula $0 + 1 + 1 = 4$ ($n = 13$), $0 + 0 + 2 = 4$ ($n = 1$), $0 + 0 + 1 = 2$ ($n = 3$) in Shizuoka and Tokyo Prefs. [3617].

Genetics   Chromosome: $2n = 44$ for specimens from the National Zoological Park, Washington, D.C., USA [3899] and from Shizuoka Pref. [792]. FN = 68 [3899] and 66 [792]. Difference in the morphology of the Y chromosome has been found between specimens from China and Shizuoka Pref., Japan. The

Table 1.   External measurements in mm or kg (mean ± SD) of the older than 2-year-old masked palm civet in Shizuoka Pref.

	BW	HB	T	HFcu	E
Male	3.1 ± 1.2 ($n = 15$)	515.9 ± 45.5 ($n = 12$)	404.9 ± 35.0 ($n = 14$)	79.7 ± 4.8 ($n = 9$)	44.1 ± 6.1 ($n = 12$)
Female	2.9 ± 1.1 ($n = 8$)	515.2 ± 29.9 ($n = 6$)	419.7 ± 11.1 ($n = 6$)	83.0 ± 453 ($n = 4$)	45.2 ± 142 ($n = 4$)

former was A-element [3830], and the latter was M-element [792]. This indicates that there is geographical variation in the morphology of Y chromosomes. A phylogenetic study of mtDNA [1981] shows that in Japanese populations there are some mtDNA types that are different from those of Southeast Asian populations. Further phylogeographic study indicates that there are at least 2 mtDNA lineages in the Japanese population, and that Taiwan is one of the origins for the Japanese population [1983]. Microsatellite analysis of civets in Japan and Taiwan suggests the occurrence of 4 genetic clusters [1047]. The population in the Shikoku district had the large differentiation from other areas, namely the Kanto and Chubu districts, which were the locations where they were first found in Japan in the 1940s [203]. These results suggest 2 or more introduction routes into Japan.

4. Ecology

Reproduction Little is known about the mating system, but the occurrence of copulatory plugs [1235] suggests that they have a habit of promiscuity. According to 2 observations of captive civets, the gestation period was estimated to be 57–59 days [2400, 3617]. Embryo number (mean ± SD) is 2.2 ± 0.9 (1–4, $n = 6$) [3617]. Litter size (mean ± SD) is 3.0 ± 0.8 (2–4, $n = 7$), and the sex ratio is 1:1 in Shizuoka Pref. [3621]. A litter with 4 young was recorded in Nepal [2855]. Under captivity, litter size (mean ± SD) is 2.1 ± 0.8 (1–3, $n = 15$) in Asa Zoological Park, Hiroshima Pref. [356]. It is also recorded as 2 in Regent's Park Zoo [4257] and 3 in London Zoo [2850]. According to a questionnaire to a civet breeder in Taiwan, litter size is 2–4, while some females give birth to 6 cubs, and many females breed twice or more a year [370]. Newly born cubs are observed or caught throughout the year in Shizuoka Pref., but they may have 2 peaks of delivery per year, spring and fall [3611]. Eyes of cubs open in 9 days and body weight reaches that of adults at 3 months of age [2892]. According to other observations, eyes open in 4 days and body weight rapidly increases and reaches adult size in about 6 months [2400]. About 2 years are needed for full eruption of the teeth [2400]. Spermatogenesis starts at about 18 months of age, but first copulation occurs at about 21 months. First estrus occurs at about 20 months of age, with the last copulation confirmed 1 month later, and the first birth at about 23 months [1235].

Lifespan Not reported in the wild. Under captivity, a 19 year-old female (which died due to an injury) and an 8 year-old female were recorded in Nihondaira Zoo, Shizuoka Pref., Japan [2123]. According to information from outside Japan, captive animals generally live for approximately 15 years [2055].

Diet Stomach content analysis in Japan shows that they are omnivorous [3618]. They mainly eat fruits, although they make use of a wide variety of food materials such as fruit, arthropods, mollusks, fishes, reptiles, birds, mammals, and garbage [3618]. Civets living by the seaside also eat sea animals, such as the acorn barnacle *Balanus eburneus* and the giant pacific oyster *Crassostrea gigas* [3615]. Their food selection depends on the availability of food resources including human garbage.

Habitat Civets live at altitudes between 50 and 2,000 m in Taiwan [370]. They live from the seaside to snowy mountains in Japan. They can live in and around human dwellings, farmland and forest among other habitats [3612]; however, their most preferred habitat is near and around human dwellings.

Home range The home range of a radio-tagged masked palm civet was not large in the short term (ca. 30–120 ha), but they often shifted their range in long distance movements so that the total range throughout the year was considerably large [3614]. Range shifts may depend on the occurrence of their preferred foods. The ranges overlap between individuals regardless of sex [3614]. This indicates that they are probably not territorial.

Behavior The civet is fundamentally nocturnal and arboreal [1840]. Their behavior is not well studied in Japan. They live in families composed of a mother and her cubs [3611]. Based on observations of captive civets in the Taipei Zoo, they live in social groups or pairs, and show tolerant behaviors to other members of group [370].

Natural enemies Captures of cubs by house cats have been observed [3617].

Parasites & pathogenic organisms Canine distemper virus has been reported [907, 1885]. As endoparasite, *Metagonimus yokogawa*, and *Sporometra erinaceieuropaei* has been reported [3680]. Five species of ticks have been recorded on the masked palm civet in Shizuoka Pref.; *Haemaphysalis flava*, *H. longicornis*, *Ixodes nipponensis*, *I. ovatus*, and *I. tanuki* [3617]. The flea *Ctenocephalides felis* has been observed on the civet in Shizuoka Pref. [3617]. The occurrence of scabies *Sareoptes scabiei* has been reported [2475] and seems to be common.

5. Remarks

Being frugivorous, many kinds of agricultural crops, such as mandarin orange, persimmon, loquat, peach, sweet corn and so on, are subject to damage [3611]. They often settle in buildings, especially temples or shrines, where they make disagreeable noise and deposit urine and dung inside; so, many civets are culled as pests [3613].

A game species.

H. TORII

RESEARCH TOPIC 9

Invasive alien mammal problems in Japan
T. Ikeda

1. Alien mammals in Japan

Problems with invasive alien species are a major concern throughout the world. In Japan, rodents had already been introduced during the Jomon period approximately 6,000–7,000 years ago [203]. However, few alien mammals had been introduced until the end of the Edo era (1603–1867), when Japan restricted trade due to a policy of national seclusion. Subsequently, a large number of organisms were brought into Japan via international exchange operations after the opening of Japan's trade industry to the world in the Meiji period (1868–1912) and during the post-World War II (after 1945) rapid economic growth period.

The number of alien mammals established in Japan to date has reached 39. The number would be 46 if species whose levels of establishment have yet to be confirmed were included. Supposedly, these 7 species either escaped or were temporarily released (Table 1). The number of existing native species of terrestrial mammals in Japan is 107 [33]. Thus, alien species of mammals account for more than a quarter of the present day mammalian fauna of Japan.

Although the distribution of some species, such as small Indian mongooses (*Herpestes auropunctatus*) and Taiwan macaques (Formosan rock macaques, *Macaca cyclopis*), is restricted to a few islands or some areas, a number of alien mammals are already widely distributed and expected to further expand their distributions. For example, raccoons (*Procyon lotor*) occur throughout Japan if temporal invasions are included [984], nutrias (*Myocastor coypus*) are extending their distribution in the western part of Honshu and Shikoku [2309], and Siberian weasels (*Mustela sibirica*) have established themselves in most of the southwestern part of Honshu, Shikoku and Kyushu [174]. Similarly, feral domestic animals, such as cats (*Felis catus*) and crossbreeds of pig/wild boar (*Sus scrofa*), are causing serious problems in many parts of the country [1193, 1344].

2. Routes of introduction

The routes of introduction of alien mammals to Japan can be classified as follows:

1) Unintentional introduction

These species such as brown rats (Norway rats, *Rattus norvegicus*) and house mice (*Mus musculus*) were unintentionally introduced through cargo transport or human migration.

2) Intentional introduction as a natural enemy

This category includes small Indian mongooses, intentionally introduced to control the habu pit viper (*Trimeresurus flavoviridis*), a poisonous snake in Okinawa-jima and Amami-ohshima Isls [3922, 3935], and Japanese weasels (*Mustela itatsi*) introduced to control rats in some islands [3249].

3) Abandoned or escaped species originally introduced for the purpose of breeding, farming and exhibition

This includes pet breeding (such as raccoons and Amur hedgehogs, *Erinaceus amurensis*), animals bred for the purpose of producing meat, fur, skins, etc. (such as crossbreeds of pig/wild boar and nutrias), and animals introduced for exhibition (such as Taiwan macaques and Reeves's muntjacs, *Muntiacus reevesi*). These animals became established after escaping due to incomplete management or abandonment.

4) Animals transported and released into the food supply

This includes goats (*Capra hircus*) released on the Ogasawara Isls. for sailors on whaling ships and other vessels to procure food.

There is a growing trend to breed rare creatures and the use of animals has diversified nowadays. Escape and abandonment of reared animals consequently are the major causes of recent establishment of alien mammals in Japan.

3. Impacts of alien mammals

Alien mammals are causing various problems across Japan with damage to agriculture, forestry and fisheries industries noted first. Indeed, damage to agriculture by raccoons [994, 997], to forestry by Pallas's squirrels (*Callosciurus erythraeus*) [3485] and to cultured fish by American minks (*Neovison vison*) [3738] has been well documented.

Another problem is the transmission of zoonotic and parasite infections (see Research Topic 5), such as raccoon roundworm infection by raccoons [2151] and echinococcosis by red foxes (*Vulpes vulpes*) [3370].

A further alarming problem is their impact on native species and ecosystems. Small Indian mongooses introduced in the Okinawa-jima and Amami-ohshima Isls. prey on endangered species such as Watase's shrews (*Crocidura watasei*), mountain agamas (*Japalura polygonata*), Amami rabbits (*Pentalagus furnessi*) and Amami spiny rats (*Tokudaia osimensis*) [2585, 2588, 3935]. Their impact on Okinawa rail (*Gallirallus okinawae*) also is of concern. Goats on the Ogasawara Isls. and European rabbits (*Oryctolagus cuniculus*) transported to several small islands have been causing regression of vegetation, soil erosion, etc., due to excessive grazing [2520, 3417, 3582, 3584, 3922]. Moreover, hybridization of native Japanese macaques (*Macaca fuscata*) with alien Taiwan macaques in Wakayama and Aomori Prefs., and with alien rhesus macaques (*M. mulatta*) in Chiba Pref. has been confirmed, leading to significant genetic disturbance [1480, 1481, 1483].

Native species are unlikely to adjust or adapt quickly enough in the presence of invasive alien species to prevent catastrophic declines. Consequently, generally they are defenseless and vulnerable to the impact of invasive species. This might, in the worst situation, result in the extinction of the native species over a short period of time. Invasion of alien mammals at a superior trophic level may have a particularly significant impact. In particular, small islands with rich endemic faunas will suffer from even more severe effects.

4. Countermeasures against invasive alien mammals in Japan

In recent years, public awareness of the impact of invasive alien mammals in Japan has grown. Accordingly, countermeasures against them have been taken in many parts of the country. "Invasive Alien Species Act", enacted in 2005, designated 25 kinds of mammals as invasive alien species, and 41 kinds of mammals are listed on the Invasive Alien Species Lists of Japan, published in 2015 [2090] (Table 1 & Table 2). The Ministry of Environment, in cooperation with other ministries, also published the Invasive Alien Species Control Action Plan and invasive alien species in Japan will be

Table 1. Alien mammals in Japan

Common name	Scientific name	Purpose of introduction	Distribution	Remarks
Alien mammals from abroad				
Amur hedgehog	*Erinaceus amurensis*	pet	Kanagawa, Shizuoka, etc.	1, 2
Asian house shrew	*Suncus murinus*	unintentional	Nagasaki, Kagoshima	
Taiwan macaque	*Macaca cyclopis*	exhibition	Aomori, Wakayama, Izu-ohshima Isl. (Tokyo)	1, 2, 3
Rhesus macaque	*Macaca mulatta*	exhibition	Chiba	1, 2
Squirrel monkey	*Saimiri sciureus*	exhibition	Izu Pen. (Shizuoka)	2
European rabbit	*Oryctolagus cuniculus*	farming	islands around the country	2
Eurasian red squirrel	*Sciurus vulgaris*	pet	Sayama Hills (Saitama)	1 (excluding *Sciurus vulgaris orientis*), 2
Pallas's squirrel	*Callosciurus erythraeus*	pet	Kanagawa, Shizuoka, etc.	1, 2
Prairie dog	*Cynomys* sp.	pet	Hokkaido, Nagano	
Siberian chipmunk	*Tamias sibiricus*	pet	Hokkaido, Niigata, Yamanashi, Gifu, etc.	2
Muskrat	*Ondatra zibethicus*	farming	Chiba, Tokyo, Saitama	1, 2
Polynesian rat	*Rattus exulans*	unintentional	Miyako Isl. (Okinawa)	
Brown rat	*Rattus norvegicus*	unintentional	across the country	2
Roof rat	*Rattus rattus*	unintentional	across the country	2
House mouse	*Mus musculus*	unintentional	across the country	2
Nutria	*Myocastor coypus*	farming	Kinki, Chugoku and Shikoku Districts	1, 2
Raccoon	*Procyon lotor*	pet, farming	across the country	1, 2
Red fox	*Vulpes vulpes*	natural enemy	Rebun Isl. (Hokkaido)	eradicated
Dog	*Canis familiaris*	pet	across the country	2
American mink	*Neovison vison*	farming	Hokkaido, Nagano	1, 2
Masked palm civet	*Paguma larvata*	farming? Exhibition?	Honshu, Shikoku, Kyushu	2
Small Indian mongoose	*Herpestes auropunctatus*	natural enemy	Okinawa-jima Isl. (Okinawa), Amami-ohshima Isl. (Kagoshima)	1, 2
Cat	*Felis catus*	pet	across the country	2
Horse	*Equus caballus*	farming	Cape Toi (Miyazaki), Yururi Isl. (Hokkaido)	under control now
Pig/Boar	*Sus scrofa*	farming	across the country	2
Reeves's muntjac	*Muntiacus reevesi*	exhibition	Chiba, Izu-ohshima Isl. (Tokyo)	1, 2
Philippine brown deer	*Cervus mariannus*	unknown	(Chichi-jima Isl. (Tokyo) at one time)	2, extinct
Formosan sika deer	*Cervus nippon taiouanus*	exhibition	Tomo-gashima Isl. (Wakayama)	1, 2
Cattle	*Bos taurus*	farming	Kuchinoshima Isl. (Kagoshima), Iriomote-jima Isl. (Okinawa)	
Goat	*Capra hircus*	released for food, farming	Ogasawara Isls. (Tokyo), Uotsuri-jima Isl. (Okinawa)	2
Intranational alien mammals				
Dsinezumi shrew	*Crocidula dsinezumi*	unintentional	Oshima Pen. (Hokkaido)	
Japanese pipistrelle	*Pipistrellus abramus*	unintentional?	southern part of Hokkaido	native?
Red fox	*Vulpes vulpes schrencki*	natural enemy	Honshu	
Raccoon dog	*Nyctereutes procyonoides*	natural enemy	islands around the country	2
Japanese weasel	*Mustela itatsi*	natural enemy	Hokkaido, islands around the country	2
Siberian weasel	*Mustela sibirica*	natural enemy	southwestern part of Honshu, Shikoku and Kyushu	2
Japanese marten	*Martes melampus*	farming	Hokkaido, Sado Isl. (Niigata)	2
Sika deer	*Cervus nippon*	farming		2
Black sika deer	*Cervus nippon keramae*	exhibition	Kerama Isls. (Okinawa)	2
Alien mammals confirmed temporarily				
Ferret	*Mustela furo*	pet	not established?	2
Opossum	genus, species unknown	pet?	not established	
Crab-eating macaque	*Macaca fascicularis*	unknown	not established?	2
Yakushima macaque	*Macaca fuscata yakui*	unknown	not established?	
Capybara	*Hydrochoerus hydrochaeris*	unknown	not established?	
Skunk	genus, species unknown	unknown	not established?	
Mage sika deer	*Cervus nippon mageshimae*	unknown	not established?	2

Remarks:
1: Alien mammals designated as invasive alien species in Invasive Alien Species Act
2: Alien mammals listed on the Invasive Alien Species Lists of Japan
3: Eradicated in Aomori Pref.

Table 2. Alien mammals designated as invasive alien species in Invasive Alien Species Act

Order	Family	Genus	Species
Marsupialia	Phalangeridae	*Trichosurus*	Bushtail possum (*T. vulpecula*)
Soricomorpha	Erinaceidae	*Erinaceus*	Any species of the genus *Erinaceus*
Primates	Cercopithecidae	*Macaca*	Taiwan macaque (*M. cyclopis*)
			Crab-eating macaque (*M. fascicularis*)
			Rhesus macaque (*M. mulatta*)
			Hybrid between Taiwan macaque and Japanese Macaque (*M. fuscata*)
			Hybrid between Rhesus macaque and Japanese Macaque (*M. fuscata*)
Rodentia	Myicastoridae	*Myocastor*	Nutria (*M. coypus*)
	Sciuridae	*Callosciurus*	Pallas's squirrel (*C. erythraeus*)
			Finlayson's squirrel (*C. finlaysonii*)
		Pteromys	Russian (or Siberian) flying squirrel (*P. volans*) excluding Japanese subspecies (*P. volans orii*)
		Sciurus	Gray squirrel (*S. carolinensis*)
			Eurasian red squirrel (*S. vulgaris*) excluding Japanese subspecies (*S. vulgaris orientis*)
	Muridae	*Ondratra*	Muskrat (*O. zibethicus*)
Carnivora	Procyonidae	*Procyon*	Raccoon (*P. lotor*)
			Crab-eating raccoon (*P. cancrivorus*)
	Mustelidae	*Neovison*	American mink (*N. vison*)
	Herpestidae	*Herpestes*	Small Indian mongoose (*H. auropunctatus*)
			Small Asian (Javan) mongoose (*H. javanicus*)
		Mungos	Banded mongoose (*M. mungo*)
Artiodactyla	Cervidae	*Axis*	All species of the genus *Axis*
		Cervus	All species of the genus *Cervus* excluding
			C. nippon centralis
			C. nippon keramae
			C. nippon mageshimae
			C. nippon nippon
			C. nippon pulchellus
			C. nippon yakushimae
			C. nippon yesoensis
		Dama	All species of the genus *Dama*
		Elaphurus	Pere David's deer (*E. davidianus*)
		Muntiacus	Reeves's muntjac (*M. reevesi*)

controlled in accordance with this action plan basically from now on [2091]. But some countermeasures were promoted by the governments or residents, and some actions already have started exhibiting effects. The mongoose project in Amami-ohshima Isl. has been successful in decreasing the number of mongoose and aiding the recovery of affected native species [1098], and the raccoon project in Hokkaido and the nutria project in Hyogo Pref. also have begun to exert their effects [984]. Moreover, eradication of goats and roof rats (house rats, ship rats; *Rattus rattus*) has been successfully achieved on some islands of the Ogasawara Isls. [3584]. The population of Taiwan macaques in Wakayama Pref. has been reduced successfully and soon will be eradicated.

5. Challenges and future prospects

Because of growing public awareness, some of the current countermeasures against invasive alien mammals in Japan have succeeded in eradicating or suppressing their populations to some extent. However, there are a number of remaining problems, including clarification of ecological impacts of alien mammals; development of risk assessment and management protocols; development of methods for confirming surviving alien animals in low-density areas; and development of new, efficient and effective methods and devices for individual animals that have developed resistance to existing measures, etc.

Moreover, although existing countermeasures were developed to deal with single species, ecosystem management covering multiple species must be taken into consideration in the future [1080]. In addition, it is necessary to simultaneously work to restore affected native species in their natural habitats.

Problems concerning legal restrictions also remain. Although the current "Invasive Alien Species Act" is expected to be effective for preventing the emergence of new alien species by imposing import restrictions on new organisms, the law only covers species originating from abroad and does not cover domestic introductions beyond natural distribution limits. We must pay attention not only to invasive alien mammals from abroad, but also the inter-regional transport within the country of origin of native species such as the Japanese weasel and the Siberian weasel.

Human social factors also must be taken into account to resolve invasive alien species problems. Promotion of comprehensive countermeasures, rooted in biodiversity conservation, through the integration of natural scientific knowledge and human social aspects of issues (i.e., public awareness raising activities and consensus building) is necessary to develop successful countermeasures for invasive alien species.

Carnivora   PHOCIDAE

Red list status: VU (MEJ); En as *P. v. stejnegeri* (MSJ); LC (IUCN); VU as *P. v. stejnegeri* (JFA)

Phoca vitulina Linnaeus, 1758

EN (Kuril) harbor seal, harbour seal   JP ゼニガタアザラシ（zenigata azarashi）   CM 港海豹   CT 港海豹，麻斑海豹
KR 엽전무늬물범   RS обыкновенный тюлень

A dark-phased harbor seal on a haul-out site (H. Kobayashi).

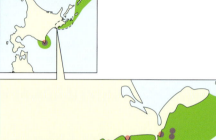

1. Distribution

The harbor seal has the widest distribution range in pinnipeds from about 30°N to 80°N in the eastern Atlantic region and about 28°N to 62°N in the eastern Pacific region [289]. The distribution of the harbor seal (*P. v. stejnegeri*) around Hokkaido is limited to the eastern Pacific coast affected by the Oyashio Current (cold water) from Yururi and Moyururi Isls. to Cape Erimo. Eight haul-out sites are known breeding areas in Hokkaido: Yururi and Moyururi Isls. in Nemuro, 2 sites in Hamanaka, Dikoku Isl. and 2 sites in Akkeshi and Cape Erimo [1639]. Eleven haul-out sites around Hokkaido are indicated as red solid circles on the maps.

2. Fossil record

Fossils (canine of the lower jawbone and molar tooth) referred to *Phoca* cf. *vitulina* were obtained from the end of the Middle Pleistocene deposit in northeastern Honshu (Cape Shiriyazaki, Aomori Pref.) [825].

3. General characteristics

Morphology   Each individual has a unique pattern of fine spots on the body and as the spot pattern does not change through life, the pattern can be used for individual identification [2470]. They vary in color from yellowish or yellowish-gray (light type) to blackish (dark type). Light-type seals are usually paler on the flanks and belly than on the back, and are covered with small black spots. The spots often have small pale rings usually on the slightly darker dorsum. Dark type seals also have dark spots, though the spots are largely masked by the background coloration. Usually darker seals have obvious light rings, especially on the dorsum [343]. In Hokkaido, almost all the seals are the dark type.

TL and BW of adult males range between 174 to 186 cm and 87 to 170 kg; those of adult females from 160 to 169 cm and 60 to 142 kg. Those of newborn pups are 98 cm and 19 kg [343]. Skulls of the harbor seal are generally larger than that those of the spotted seal. In particular the relative sizes of the palatal parts are much larger in pups of the former than those of the latter [3730]. The harbor seal might have apomorphic morphological characters, compared to the spotted seal.

Dental formula   I 3/2 + C 1/1 + P 4/4 + M1/1 = 34 [29] (I 3/2 + C 1/1 + PC 5/5 = 34 [3369]).

Phoca vitulina Linnaeus, 1758

Mammae formula   0 + 0 + 1 = 2 [29].

Genetics   Chromosome: $2n = 32$ [191–193]. MtDNA studies suggest that there are at least 2 groups in Hokkaido: the Erimo group and the eastern group (eastern area from Akkeshi) [3847].

4. Ecology

Reproduction   Pupping extends for about 10 weeks in spring, with a 2-week peak in general. Females bear a single pup on rocky shore. Newborn pups can enter the water, often being forced to do so by tidal inundation or by disturbance from birds scavenging afterbirth. Mother–pup bonding within the first hour of birth is critical for mutual recognition. A young pup often clings to its mother's back in the water [343]. Mothers nurse their pups for approximately 4-weeks [2470]. Pups start to forage by themselves during the late stages of the nursing period. Mating occurs in the water around the time when pups are weaned, although the first mating, or mating by females that have not given birth may occur outside the peak period of the post-parturient animals [343].

Mating system: polygyny with low reproductive skew [312]. There is inter-male competition for receptive females and no obvious social organization during the breeding season [343]. Breeding season: from mid-April through June in Hokkaido [1642]. Gestation period: like other pinnipeds, fertilization is followed by a prolonged period of delayed implantation (embryonic diapause) that lasts about 2.5 months. The actual gestation period, from fertilization to birth, is about 10.5 months [312]. Birth season: from late-April to late-May in Hokkaido with a peak in mid-May [1639]. Recently, pupping period appears to be elongated on Daikoku Isl., Hokkaido [1642]. Litter size: most sexually mature females bear a pup every year [312]. Age of sexual maturity: in general, females reach sexual maturity at 3 to 4 years of age and physical maturity by 6 to 7 years of age. For males 4 to 5 years of age and 6 to 9 years of age, respectively [3337].

Lifespan   The maximum recorded lifespan is around 34 years [2654] although there are few animals that live so long. In Daikoku Isl., it is reported that a 32 year-old female was pregnant in the wild [1642].

Diet   They are opportunistic predators that feed mainly on abundant and easily available prey, and thus the diet varies by season and region. Primary foods are small to medium size fishes, such as various members of the codfish family, herring, sardines, capelin, sculpins, a variety of flatfish, greenling, salmonids and many others. Their propensity for cod, salmon and other commercially important species has resulted in long-standing conflicts with fishermen in Hokkaido. Cephalopods (squid and octopus) are usually reported as secondarily important foods after fishes [1726, 2408].

Habitat   The habitats of harbor seals are different from those of other pinnipeds. Cape Erimo is the largest and most southern haul-out site in Hokkaido. There are many rookeries, spread from the tip of the cape to 1.5 km southeast and they also use some of them for rest and reproduction. By DNA analysis and marked investigation, the population of Cape Erimo is differentiated from populations of the eastern area of Hokkaido [3847]. The second largest haul-out site is Daikoku Isl. The number of individuals in Cape Erimo and Daikoku Isl. account for 70% of the entire Hokkaido population [384, 2935].

Home range   In Hokkaido, the harbor seal uses the same haul-out sites all year, so this subspecies is considered well established [851].

Behavior   In Hokkaido, harbor seals mainly haul out on land (i.e. rockeries) that is infrequently visited by humans. They use haul-out sites throughout the year, although the largest numbers are observed during the pupping and molting season. Regardless of season, haul-out activity is strongly affected by the situation of the tide, air temperature, wind speed, precipitation, and time of day. They lay close to the water when hauling out and usually flee to the water when disturbed, although habituation is not uncommon near large human population centers, if they are not harassed unduly [343].

Natural enemies   Harbor seals are preyed on by killer whales (*Orcinus orca*), sharks, Steller sea lions (*Eumetopias jubatus*), and pups are also taken by eagles, ravens and gulls [343].

Parasites   Anisakid nematodes are found in the stomach and the intestine. *Corynosoma strumosum*, *Phocitrema fusiforme* and *Diplogonoporus tetrapterus* are also found in the intestine. These species are common in other pinnipeds in Japan [1763, 2344].

5. Remarks

Five subspecies have been recognized: *P. v. vitulina* in the eastern Atlantic; *P. v. concolor* in the western Atlantic; *P. v. mellonae* in Ungava Pen. in eastern Canada; *P. v. richardii* in the eastern North Pacific; and *P. v. stejnegeri* in the western Pacific. Recent reports, however, show no significant genetic difference between *P. v. richardii* and *P. v. stejnegeri* [339, 3871]. The distribution of *P. v. stejnegeri* extends from the Aleutian and Commander Isls. to the Kamchatka Pen. of eastern Russia, southward in the Chishima (Kuril) Isls. and beyond to Hokkaido.

Fisheries have been damaged by the seals. In the Erimo area, 2% of the fixed nets for salmon are damaged on average, while in Akkeshi and Nosappu areas the rates of damage are 0.2% and 0.5%, respectively [587, 1638, 2934]. The yearly numbers of by-catch seals by fixed fish nets are about 60, 20 and 120 individuals in Erimo, Akkeshi and Nosappu, respectively.

M. KOBAYASHI

Carnivora   PHOCIDAE

Red list status: C-1 (MSJ); DD (IUCN); ND (JFA)

Phoca largha Pallas, 1811

EN spotted seal, larga seal   JP ゴマフアザラシ（gomafu azarashi）   CM 斑海豹   CT 斑海豹，堪察加麻斑海豹
KR 물범   RS ларга

Spotted seals resting on drift ice at the Sea of Okhotsk (H. Kobayashi).

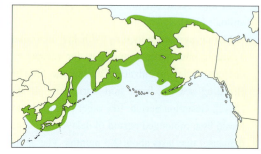

1. Distribution

Endemic to the North Pacific. They occur in Bering Sea, Chukot (Chukchu or Chukuchi) Sea (in summer), Beaufort Sea (in summer), and the Sea of Okhotsk, Tartar (Mamiya) Str., the Sea of Japan and northern Yellow Sea (Huang Hai Sea), and adjacent embayments. The southernmost breeding populations (about 38°N) are found in the Sea of Japan and Yellow Sea [343].

2. Fossil record

Fossils (molar teeth) referred to *Phoca* cf. *largha* were obtained from the late Middle Pleistocene deposits in northeastern Honshu (Cape Shiriyazaki, Aomori Pref.) [825].

3. General characteristics

Morphology   Coloration is generally pale, silver-gray above and below, with a darker mantle dominated by dark oval spots of fairly uniform size (1–2 cm) and generally oriented parallel to the long axis of the body. There may be light rings around some spots, or large irregular spots car blotches. Spotting tends to be of fairly even distribution and darkness overall. In harbor seals, spots are more faded and sparse on the underside. The face and muzzle are darker than in the harbor seal. Pups are born with a long, woolly, whitish lanugo, which is shed 2 to 4 weeks after birth. Body size of spotted seals is similar to that of the harbor seal [343]. In samples from the Bering Sea, TL and BW are 161–176 cm and 85–110 kg, respectively, for adult males, while 151–169 cm and 65–115 kg for adult females. In near-term fetuses and newborn pups from the Sea of Okhotsk body length and body weight are 78–92 cm and 7–12 kg [343].

Dental formula   I $3/2$ + C $1/1$ + P $4/4$ + M$1/1$ = 34 [29] (I $3/2$ + C $1/1$ + PC $5/5$ = 34 [3369]).

Mammae formula $0 + 0 + 1 = 2$ [29].

Genetics   Chromosome: $2n = 32$ [191–193]. MtDNA studies suggest no genetic differentiation in populations found around the Hokkaido coasts [2176].

4. Ecology

Reproduction   The use of sea ice as a platform on which to bear and nurture pups is central to the ecology of spotted seals. Pups spend most of the time on ice floes until weaned. The reproductive events (birth, nursing, wearing, and early independence) are directly related to sea ice conditions. The timing of birth coincides with the average period of greatest extent and stability of the sea ice cover and varies by region. Weaning, which occurs abruptly, coincides with the onset of ameliorating spring weather and disintegration of ice packs. Pups are born earlier in the more southerly parts of the distribution range. Mothers forage by themselves in the sea during the nursing period while pups remain on ice floes. Mating occurs at about the time when pups are weaned, and most females breed annually [343].

Mating system: spotted seals begin to form pairs early during the pupping season. They are considered to be annually monogamous and territorial. Triads consisting of a female, her pup, and an attending male can be seen on the ice. Pairs consisting of an adult female and male without a pup are also formed. In the Bering and Okhotsk Seas, such pairs are seldom seen on the ice in early April (prior to the molt of adults) [343]. Breeding season: in Okhotsk, birth season is between mid-March and mid-April [2177]. The total breeding season including birth and fertilization is between mid-March and May [324]. Gestation period: like other pinnipeds, fertilization is followed by a prolonged period of delayed implantation (embryonic diapause; 2 months). The actual gestation period, from fertilization to birth, is about 11 months including embryonic diapause [343]. Birth season: in the Yellow Sea the peak of the birth season occurs in late January; in the Sea of Japan it occurs between February and March; and in both the Okhotsk and Bering Seas the peak occurs in the first half of April. Healthy pups usually double or sometimes triple their weight during the first 3- to 4-weeks after birth [343]. Pups begin to forage for themselves about 10–15 days after they are weaned. During fasting and early in the independent feeding period they live on their accumulated fat reserves and lose 18–25% (sometimes up to 30%) of their weight. Litter size: like most pinnipeds, spotted seals give birth to a single pup each year. Age at sexual maturity: 2–5 years for females and 3–6 years for males [343].

Lifespan Not reported.

Diet The first food of pups is usually small amphipods or euphausiids. Schooling fishes are the main foods of older seals and, in the Sea of Okhotsk, account for 89% of foods for seals 1–4 years old and 70% for older seals. Cephalopods are next in importance, followed by decapods. Amphipods are still consumed by 1–4 year old seals but are not found in diets of older animals. The frequency of occurrence of cephalopods is higher in older animals. Spotted seals are reported to feed more in the morning and the evening than at other times of the day [343].

In Hokkaido, primary food items are small to medium sized fishes, such as various species of the codfish family, herring, smelt, capelin, lance, saury, mackerel, sculpins, a variety of flatfish, greenling, salmonids and many others [2408].

Habitat The spotted seal is well adapted to use the "front" and broken ice zones of seasonal sea ice that overlies continental shelves during winter and spring. Spotted seals resort to haul outs on the land during ice-free seasons of the year [343].

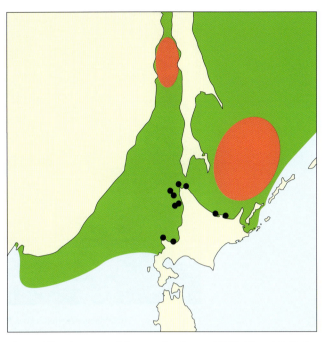

Fig. 1. Wintering range of the spotted seal around Hokkaido and Sakhalin. Black solid circles on Hokkaido coasts indicate haul-out sites. Large red ovals denote the breeding areas on sea ices.

Fig. 2. Summer range of the spotted seal around Hokkaido, Sakhalin and Primorye. Black solid circles on the Hokkaido coast indicate haul-out sites.

Home range Spotted seals migrate from northward into Sakhalin or the Chishima (Kuril) Isls. during ice-free months and from southward into the coast of Hokkaido during ice-cover months (the breeding season). There are great seasonal expansions and contractions of range, commensurately with the annual cycle of sea ice advance and retreat [343].

Behavior Spotted seals start to use ice with its formation. They often come together for the early ice that forms near river mouths and estuaries where they feed on spawning fishes. As the ice

An immature spotted seal on Rebun Isl., Hokkaido (M. Kobayashi).

thickens, becomes attached to land, and extends further from shore; spotted seals move seaward following drifting ice. Spotted seals migrate to maintain the association with marginal areas of the sea ice that are highly labile. During winter they rarely haul out. Peak haul out on the ice is found during the pupping and molting season [343]. When the sea ice cover retreats and disintegrates in late spring–early summer, spotted seals again move to their summer habitat. Large aggregations can often be seen close to shore and haul-out sites [343].

Spotted seals migrate to the Sea of Japan and the Sea of Okhotsk around Hokkaido, from Sakhalin or the Chishima (Kuril) Isls. from November to May, while they are observed on the Pacific side from February to May (Fig. 1) [1637]. They also have 2 summer habitats at Odai marsh and Lake Furen, where they can be observed from May to October (Fig. 2) [1637].

Natural enemies Spotted seals are preyed on by killer whales, sharks, Steller sea lions and brown bears, and pups are also taken by eagles, ravens and gulls [343].

Parasites Anisakid nematodes are found in the stomach and the intestine. *Corynosoma strumosum*, *Phocitrema fusiforme* and *Diplogonoporus tetrapterus* are also found in the intestine. These species are common in other pinnipeds in Japan [1763, 2344].

5. Remarks

The spotted seal was considered a subspecies of the harbor seal. However, recent studies on morphological, biochemical, and behavioral differences warrant its reclassification as a full species [343, 3171]. Conflicts with fisheries are on the rise in Hokkaido, especially along the coasts of the Sea of Japan. Fishing activities can affect this species adversely by causing incidental mortalities and by competing for fish the seals depend on for food [1642].

M. KOBAYASHI

A spotted seal.

A haul-out site of harbor seals on Dikoku Isl., Hokkaido (M. Kobayashi).

A haul-out site of spotted seals on a sandbank in Bakkai Port, Hokkaido (M. Kobayashi).

Harbor seal mother and pup on Dikoku Isl., Hokkaido (M. Kobayashi).

Carnivora   PHOCIDAE
119

Red list status: K as *Phoca h. ochotensis* (MSJ); LC (IUCN); DT (JFA)

Pusa hispida (Schreber, 1775)

EN ringed seal   JP ワモンアザラシ（wamon azarashi）   CM 环斑海豹   CT 環斑海豹
KR 고리무늬물범   RS кольчатая нерпа

A ringed seal at the Aquatic Wildlife Breeding Center, Monbetsu, Hokkaido (N. Miyazawa).

1. Distribution

Distributed in circumpolar Arctic coasts. They are found wherever there is open water in fast ice, even as far as the North Pole [1580]. In Japan, these are some reports of ringed seal pups in the Sea of Okhotsk, and also reports of pups in Honshu, Shikoku and Kyushu [2354].

2. Fossil record

Not reported in Japan.

3. General characteristics

Morphology   Gray-white rings are found on the generally gray backs, and the belly is usually silver and lacking dark spots. The rings are usually separate from each other but sometimes fuse together. Pups are born with a white woolly natal lanugo [586]. Average adult TL varies from 121 cm in Chukot Sea, 128.5 cm in Bering Sea [538] to 135 cm in the Canadian Arctic [2043]. Pups are 65 cm in body length and about 4.5 kg in BW in average. Average TL of adults around Japan is 130 cm for males and 120 cm for females [1642].

Dental formula   I 3/2 + C 1/1 + P 4/4 + M1/1 = 34 [29] (I 3/2 + C 1/1 + PC 5/5 = 34 [586]).

Mammae formula   0 + 0 + 1 = 2 [29].

Genetics   Chromosome: $2n = 32$ [191–193].

4. Ecology

Reproduction   Pregnancy rates vary geographically: 91–92% in the Baffin Isl. area [2043, 3280], 86% in the southern Chukot Sea [1246], and 53% in Alaska water in 1975–1977 [586].

Mating system is polygynous with low reproductive skew. Mating probably occurs in mid-April, shortly after parturition and while the female is still lactating [1580]. The lactation period is nearly 2 months. Gestation period is similar to other pinnipeds, fertilization is followed by a prolonged period of delayed implantation (embryonic diapause) that lasts about 2.5 months. The actual gestation period, from fertilization to birth, is about 9.1 months [412, 2044, 2884]. Parturition occurs between the middle of March and the middle of April [1532, 3281]. Like most pinnipeds, ringed seals give birth to a single pup each year. Age at sexual maturity is 6–7 years for both sexes with geographic variation; 3–5 years of age for *P. h. botnica*, 6–10 years of age for *P. h. hispida* [586].

Pusa hispida (Schreber, 1775)

Lifespan The maximum recorded age is 43 years old [2043].

Diet Ringed seals feed on small fish and also on a wide variety of small pelagic amphipods, euphausiids, and other crustaceans. Seventy-two food species were identified in stomach samples from the eastern Canadian Arctic. In shallow, inshore waters, the seals feed near the bottom, chiefly on the polar cod (*Boreogadus saida*) and on the small crustacean *Mysis*, whereas in deeper offshore waters they catch the planktonic amphipod *Themisto libellula* [1580].

Habitat Ringed seals are strongly associated with drifting or fast ice. Their habitat is limited at least to seasonal ice-covered areas [1222].

Home range It is thought that males have territories [1222].

Behavior Females make their lairs on bergs of ice or uneven glacial features covered with snow, where they can go in and out to sea and hide from polar bears [1222].

Natural enemies Ringed seals are preyed on by polar bears (*Ursus maritimus*) and hunted by humans [1222].

Parasites Anisakid nematodes are found in the stomach and the intestine. *Corynosoma strumosum*, *Phocitrema fusiforme* and *Diplogonoporus tetrapterus* are also found in the intestine. These species are common in other pinnipeds in Japan [1763, 2344].

5. Remarks

Although many local populations and/or subspecies have been reported, 5 subspecies are usually accepted [586]: *Pusa h. hispida* in the Arctic Ocean and the confluent Bering Sea, *P. h. ochotensis* in the Sea of Okhotsk, *P. h. saimensis* in Lake Saimaa (Finland), *P. h. ladogensis* in Lake Ladoga (Russia) and *P. h. botnica* in the Baltic Sea.

Many thousands of ringed seals are caught by humans in all areas where they occur, mostly for their skins, for their decorative fur, and also for blubber [1580]. A mass killing for human use was recorded in 1960 in the Baltic and North Seas.

M. KOBAYASHI

A skull & mandible specimen (NMNS-CA 166) of an adult ringed seal (sex unknown) captured off Abashiri, Hokkaido, stored at the National Museum of Nature and Science (N. Kohno).

Carnivora PHOCIDAE

120

Red list status: C-1 as *Phoca fasciata* (MSJ); DD (IUCN); ND (JFA)

Histriophoca fasciata (Zimmermann, 1783)

EN ribbon seal   JP クラカケアザラシ（kurakake azarashi）   CM 环海豹   CT 環海豹
KR 띠무늬물범   RS полосатый тюлень, крылатка

A mature male on drift ice in the Sea of Okhotsk (H. Kobayashi).

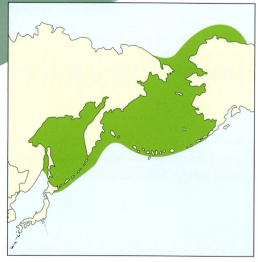

1. Distribution

Endemic to the Northern Pacific Ocean. Three populations are known: 2 in the Sea of Okhotsk and 1 in Bering Sea. A part of the latter population migrates to the southern region of Chukot Sea in spring and summer. There is, however, no difference in morphology between populations in these 2 sea areas [344, 537]. They migrate to north-eastern parts of Hokkaido from the end of February to the beginning of May; particularly at the Kitami–Yamato Bank, offshore of Abashiri Bay and Nemuro Str. They appear on the closer ice in 200–1,000 m deep sea areas [2649]. There are differences in age structure and sex ratio between the sea areas [2355, 3576].

2. Fossil record

Not reported in Japan.

3. General characteristics

Morphology There are 4 light stripes against a black or brown background: 1 stripe goes round the neck, another encircles the body at the bottom of the flippers and 2 more symmetrically underline the base of the pectoral flippers. The coloration is fairly dull at 1 year old, but becomes brighter as they grow. The skull is short, the cranium braincase and cheekbones are wide, and the face is short and narrow [722]. TL is 165–175 cm and BW is 72–90 kg for adults. TL and BW of newborn pups are 73–98 cm and 6–10 kg, respectively [536].

Dental formula I 3/2 + C 1/1 + P 4/4 + M1/1 = 34 [29] (I 3/2 + C 1/1 + PC 5/5 = 34 [3369]). Teeth are small; their number varies from 32 to 36 [342, 722].

Mammae formula 0 + 0 + 1 = 2 [29].

Genetics Chromosome: $2n = 32$ [191–193].

4. Ecology

Reproduction Reproduction is strongly influenced by ice formation. Ice conditions (its thickness, shape, the hummocks, the amount of snow on it, its location, and the speed of its decomposition) determine the time of rearing and pup growth [536].

The mating system is not reported. Breeding occurs from March to May [1534, 3281]. Gestation period is not reported. In the southern part of the Sea of Okhotsk, ribbon seals start delivering pups in the middle of March. In the northwestern part of the Sea of Okhotsk and in Bering Sea it takes place in April, with the peak in the middle of the month. Like other pinnipeds, ribbon seals give birth to a single pup every year. Age at sexual maturity is 2–4 years for females and 3–5 years for males [412, 2044, 2884].

Lifespan Studies of age as defined by horn covers on the claws and layers of cement on the tusk apexes show that the ribbon seals can reach 30 years of age and beyond [536].

Diet Yearling ribbon seals feed mostly on euphausiids, individuals of 1–2 years feed mostly on shrimps, and adult animals eat

Histriophoca fasciata (Zimmermann, 1783)

cephalopods and fish. In the Sea of Okhotsk, adults eat mostly Alaska pollack (65%), whereas in the Bering Sea they mostly feed on squid and octopus (67%). The ribbon seal's daily consumption is 8–10 kg, including invertebrates and fish [539].

Habitat The ribbon seal belongs to the ice seals, whose life is closely connected with ice in winter. Since they are little observed in summer, this species is thought to live in the open sea in summer [341].

Home range Not reported.

Behavior During the breeding season (March–April), adults dwell on the ice to give birth to pups, while immature seals are also there for molting. In May and June when adults start molting, ribbon seals gather together due to the melting and decomposition of the ice. In June, when little ice remains, one can see a mass of ribbon seals from several different populations [536].

Natural enemies Not reported.

Parasites Anisakid nematodes are found in the stomach and the intestine. *Corynosoma strumosum*, *Phocitrema fusiforme* and *Diplogonoporus tetrapterus* are also found in the intestine. These species are common in other pinnipeds in Japan [1763, 2344].

5. Remarks

Between 1956 and 1992, the ribbon seals were subjected to commercial hunting. Up until 1969 the hunt was not limited. The average annual catch was 11,000 animals in the Sea of Okhotsk and 9,000 in the Bering Sea. Between 1969 and 1992 the annual catches were reduced to 5,000 to 6,000 animals in the Sea of Okhotsk and 3,000 to 4,000 in the Bering Sea.

M. KOBAYASHI

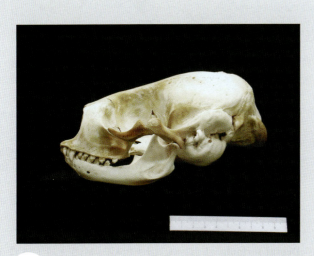

A skull & mandible specimen (NSMT-M 14988) of an adult male ribbon seal captured off Rausu-cho, Hokkaido, stored at the National Museum of Nature and Science (N. Kohno).

Carnivora PHOCIDAE

Red list status: K as *E. b. nauticus* (MSJ); LC (IUCN); DC (JFA)

Erignathus barbatus (Erxleben, 1777)

EN bearded seal　　JP アゴヒゲアザラシ（agohige azarashi）　　CM 髯海豹　　CT 髯海豹，太平洋鬚海豹
KR 턱수염물범　　RS лахтак, морской заяц

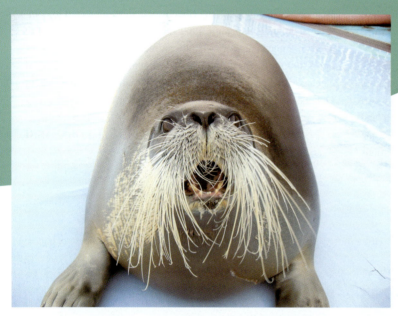

A bearded seal at the Aquatic Wildlife Breeding Center, Monbetsu, Hokkaido (M. Takaishi).

1. Distribution

Bearded seals have a patchy distribution throughout the Arctic and sub-Arctic region. In the Pacific, they migrate in the Sea of Okhotsk, but there are very few near the coast of Hokkaido; some pups have been caught in fishery nets in Nemuro Pen. in May and June [2354]. In Honshu, there are records of vagrant juveniles in Tokyo, Akita and Niigata Prefs. [2354].

2. Fossil record

No reported in Japan.

3. General characteristics

Morphology　The bearded seal is a large, arctic phocine. Adults are 2–2.5 m long (TL) and are gray-brown. Body weight vary markedly on an annual cycle, but an average would be 250–300 kg. Females, which are somewhat larger than males, can weigh in excess of 425 kg in the spring. Bearded seals often have irregular light-colored patches and can, in some geographic regions, also have rust-red faces and fore flippers. Pups are approximately 1.3 m long (TL) at birth and weigh an average of 33 kg. They are born partially molted but usually still bear a significant quantity of fuzzy gray-blue lanugo in combination with a second darker-gray smooth coat. Their faces have white cheek patches and white eyebrow spots [1724].

Their body shape is very rectangular. Their heads appear small compared to their body size. They have very square-shaped fore flippers that bear very strong claws. They also have extremely elaborate and smooth facial whiskers that tend to curl when dry [1724].

Dental formula　I 3/2 + C 1/1 + P 4/4 + M1/1 = 34 [29] (I 3/2 + C 1/1 + PC 5/5 = 34 [341]).

Mammae formula　0 + 1 + 1 = 4 [29].

Genetics　Chromosome: $2n = 34$ [191–193].

4. Ecology

Reproduction　Mating takes place around the time that females leave their offspring. Males sing underwater to attract females and they also fight with other males during the breeding season. Their songs are a series of complex, downwardly spiraling trills that can be heard over tens of kilometers in quiet weather. Little is known regarding the specifics of mating behavior although pairing takes place in the water [1724].

Mating system is polygynous with low reproductive skew

Erignathus barbatus (Erxleben, 1777)

[340, 341, 1533, 3560]. Breeding occurs between late-March and mid-May, depending on the locality. Gestation period is similar to other pinnipeds, fertilization is followed by a prolonged period of delayed implantation (embryonic diapause) that lasts about 2 months. The actual gestation period, from fertilization to birth, is about 9.5 months [340, 341, 1533, 3560]. Birth season: April in the Sea of Okhotsk [1724]. Litter size is similar to other pinnipeds, bearded seals give birth to a single pup each year. Age at sexual maturity: about 5 years old for females and normally 6 to 7 years of age for males [133, 412, 2884]. Parturition occurs on drift ice [1724].

Lifespan Bearded seals live 20–25 years [1724].

Diet They are for the most part benthic feeders. They likely use their whiskers to find foods in soft substrates. They feed on polar cods (*Boreogadus saida*), sculpins (*Cottidae*), rough dabs (*Hippoglossoides platessoides*) and eelblennies (*Lumpenus medius*). Bearded seals also eat a variety of invertebrates, including spider crab (*Hyas araneus*), shrimps, a variety of molluscs, cephalopods, polychaete worms and amphipods [1724].

Habitat Their preferred habitat is drifting pack ice in areas over shallow water shelves. They are often found in coastal areas. Some populations are thought to be resident throughout the year, whereas others follow the retraction of the pack ice northward during the summer and advance southward once again in the late fall and winter [1724].

Home range Juvenile animals wander quite broadly and can be found far south of the normal adult range [1724].

Behavior Bearded seals are largely solitary, although it is not unusual to see them hauled out together in small groups along leads or to holes in the spring or early summer [1724].

Natural enemies Polar bears (*Ursus maritimus*) and walruses (*Odobenus rosmarus*) are possible predators. Killer whales (*Orcinus orca*) and Greenland sharks (*Somniosus microcephalus*) may also take bearded seals, particularly pups and juveniles [1724].

Parasites Not reported in Japan.

5. Remarks

Bearded seals are an important subsistence resource in the Arctic. A few thousand animals are harvested annually for use as food, dog food, and for their thick leather, which is important for various traditional articles of clothing in Alaska, the Canadian Arctic and Greenland. Russia is the only country that has had a commercial scale harvest of bearded seals with catches that exceeded 10,000 animals in some years during the 1950s and 1960s. Quotas were introduced to limit the harvests of the declining populations in the Okhotsk and Bering Seas, and the catch dropped to a few thousand seals annually through the 1970s and 1980s [1724].

M. KOBAYASHI

 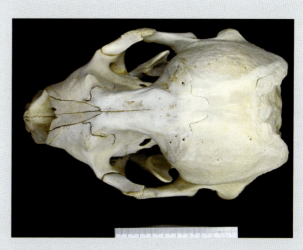

A skull & mandible specimen (NMNS-CA 165) of an adult female bearded seal captured in Hamatonbetsu-cho, Hokkaido, stored at the National Museum of Nature and Science (N. Kohno).

Carnivora OTARIIDAE

Red list status: NT (MEJ); V (MSJ); NT (IUCN); R (JFA)

Eumetopias jubatus (Schreber, 1776)

EN Steller sea lion, Steller's sea lion, northern sea lion   JP トド (todo)
CM 北海狮   CT 北海獅, 海驢, 北方海獅   KR 큰바다사자   RS сивуч

Young Steller sea lions at a resting site at Shiretoko Pen., Hokkaido (T. Ishinazaka).

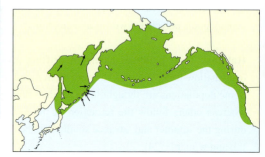

1. Distribution

Occurs through the North Pacific Ocean rim from Japan to southern California [1868]. The subspecies that was newly named as the western Steller sea lion (*E. j. jubatus*) occupies the area west of 144°W longitude [2839, 2840]. In Japan, they occur in winter mainly around Hokkaido, although some of them are seen around the Shimokita Pen., Aomori Pref. There is no rookery (breeding site) in Japan; therefore, few sea lions are found around Hokkaido in summer.

2. Fossil record

Some fossils referred to *E. jubatus* were obtained from the Middle Pleistocene deposit in northeastern Honshu [825].

3. General characteristics

Morphology   The largest otariid in the world. Adult fur color varies between a light buff to reddish brown. Immature animals are darker. Both sexes become blonder with age. Pups are born with wavy, chocolate brown fur that molts after about 3 months of age [1868]. Northern fur seals are darker in color and smaller in size. They also have relatively longer ears and hind flippers than Steller sea lions'.

They show marked sexual dimorphism, males being larger than females [1152]. Adult males have distinctive manes [1581]. Around Hokkaido, BW of 5+ year old males averages 452.9 ± 168.2 kg ($n = 17$, mean $\pm$ SD) and females 260.5 ± 47.9 kg ($n = 129$) (maximum of about 780 kg and 410 kg, respectively) [1130]. A maximum male BW of about 1,120 kg was recorded [1868]. Pup BW at birth is 16–23 kg [1868]. The average SL of males is 256.4 ± 30.5 cm ($n = 27$) and of females is 227.2 ± 11.7 cm ($n = 145$) (maximum of about 314 cm and 258 cm, respectively) [1130].

Dental formula   I 3/2 + C 1/1 + PC 5/5 = 34 with some individual variations [1581, 1868].

Mammae formula   0 + 1 + 1 = 4. Steller sea lions have 4 mammary teats, 1 pair in front of, the other pair behind and closer to, the umbilicus [1581].

Genetics   Chromosome: $2n = 36$ [1581]. MtDNA studies suggested that at least 2 stocks exist: an eastern stock (California through southeastern Alaska) and a western stock (Prince William Sound, Alaska, and westward) [288, 1867]. Further study indicates another subdivision within the western stock, and resulted in the recognition of an Asian stock [246]. Based on genetic data combined with morphological data, 2 subspecies have been recognized. One is the western Steller sea lion *E. j. jubatus* (western and Asian stocks), and the other is the Loughlin's northern sea lion *E. j. monteriensis* (eastern stock) [398, 2839, 2840].

4. Ecology

Reproduction Mating system is polygynous [2061]. Estrus occurs 10–14 days after parturition mostly in June, and copulation takes place at this time [2844]. Copulation usually takes place on land, though sometimes in shallow water [1581]. Gestation period is 1 year, including a delayed implantation period of about 3.5 months [1581]. Viable births begin in late May and continue through early July [1868]. The peak of birth on the Kuril Isls. is mid-June [272]. Like most pinnipeds, Steller sea lions give birth to a single pup each year; twins are rare [1868]. The pregnancy rate for mature females shot by local hunters was 90.5% ($n = 63$) in Nemuro Str. lying east to Hokkaido between 1995 and 1999 [1124].

The age of sexual maturity in females culled in Nemuro Str. was 4–5 years old ($n = 83$ [3731]; $n = 29$ [1125]; $n = 66$ [1124]). Generally, females reach sexual maturity between 3 to 8 years of age and may breed into their early 20s [1868]. Females can have a pup every year but may skip years as they get older or when stressed nutritionally [1123, 1868].

Males reach sexual maturity at 3 years of age [1123, 2844], but do not attain the physical size or social status to obtain and keep a breeding territory until 9 years of age or older [1868, 2818, 3571].

Russian scientists have branded pups at the main rookeries in the Russian Far East waters since 1989, and over 50 branded sea lions were re-sighted and/or recaptured around Hokkaido [949, 1132]. This re-sighting data indicates that the main birthplaces of the population wintering around Hokkaido are the Kuril Isls. and Iony and Yamsky Isls. in the northern part of the Sea of Okhotsk [949, 1131, 1132]. Animals born in Tuleny Isl. (southern part of the Sea of Okhotsk) also occurred around Hokkaido in winter [557, 842]. Known breeding rookeries of the Hokkaido wintering population are indicated by arrows on the map [1868].

Lifespan Males rarely live beyond their mid-teens, whereas females may live as long as 30 years [1868].

Diet Steller sea lions eat a variety of fishes and invertebrates. In the northern part of Nemuro Str., just east to the Shiretoko Pen., principal prey include walleye pollocks (*Theragra chalcogramma*), Pacific cods (*Gadus macrocephalus*), schoolmaster gonate squids (*Berryteuthis magister*), saffron cods (*Eleginus gracilis*), and flounders. In the Shakotan Pen. (western part of Hokkaido), octopuses, arabesque greenlings (*Pleurogrammus azonus*), Pacific cods, walleye pollocks, and flounders were the main items from stomach contents [733, 734].

Habitat Steller sea lions tend to avoid people and prefer isolated offshore rocks and islands to breed and rest. Rookeries and rest sites are specific and change little from year to year [1868]. Steller sea lions tend to make relatively shallow dives. They occur mainly in coastal areas or on the continental shelf. There are several specific rest sites of females along the coast of Rausu-cho (northern Nemuro Str.). These rest sites are not rocks for hauling-out (going ashore) but waters. Therefore, female Steller sea lions float and swim slowly in the rest sites during the daytime [1121]. Sea lion hunters of Rausu-cho call these rest sites "Tsukiba". A few coastal rocks in the Sea of Japan (Isoya, Cape Kamui, Cape Ofuyu, and Benten Isl. near from Cape Soya) are used as haul-out sites by sea lions in winter [557, 949, 950].

Home range Based on USA data, estimated home ranges are 320 km^2 for adult females in summer, and in winter approximately 47,600 km^2 for adult females and 9,200 km^2 for yearlings [1868].

In Hokkaido, little is known about foraging range size because only a few satellite telemetry researches have been conducted for this species in Japan. A young female tracked by satellite telemetry moved 440.4 km in total during 52 days (Nov. 25, 1993–Jan. 16, 1994) between Hokkaido and Sakhalin Isl. [239]. Some branded sea lions were found at both Cape Kamui and Cape Ofuyu in the winter of 2002/2003 [2983]. The distance between these capes is about 85 km.

Behavior Mostly nocturnal. They forage at night or in early morning in Nemuro Str., based on the digestive rank of the stomach contents for each shooting hour of the day [732]. During the daytime in winter, many sea lions rest at specific sites in coastal waters (Tsukiba) in Nemuro Str. [1121, 1126]. Along the coast of the northern Sea of Japan however, sea lions haul-out and rest on specific rocks [949]. During the breeding season, females with pups generally feed at night, however, males do not take foraging trips when defending their breeding territories [1868].

While swimming, Steller sea lions use the fore flippers primarily for movement and the rear flippers for braking and turning [1868]. Steller sea lions sometimes jump out of the water surface like fur seals and dolphins when they are active just after sunset and when they are chased by hunting boats [1126].

Groups wintering in Nemuro Str. usually consist of females of all ages and subadult males [1133], whereas groups wintering around the Shakotan Pen., in the northern part of the Sea of Japan, mainly consist of adult males [1133]. Based on observations of branded animals at the rocks of Cape Kamui and Cape Ofuyu, sea lions derived from different rookeries occur together in a large aggregation on a single rock during winter [950, 1122].

Natural enemies Predators include sharks and killer whales (*Orcinus orca*) [1152].

Parasites Anisakid nematodes (*Pseudoterranova decipiens*, *Anisakis simplex*, and *Contracaecum osculatum*) are usually found in the stomachs of Steller sea lions. A total of 83% of Steller sea lions ($n = 40$) were infested by adult halarachnid mites (*Orthohalarachne attenuata*) in their pharynges [1711].

5. Remarks

Fisheries around Hokkaido are damaged by Steller sea lions. Gill-net fisheries especially are damaged. They often break the gill-nets when the nets are full of fish.

Non game species. However, a total of 116 animals per year around Hokkaido had been shot under the fisheries damage control activities between 1993 and 2006. In 2007, Japan Fisheries Agency (JFA) calculated new numbers based on potential biological removal (PBR) from aerial survey data obtained in 2004 and 2005 [843]. A total of 227 animals per year have been allowed to be culled under human-derived factors (shooting and by-catch) in Japan since 2007 [836]. The quota for shooting was restricted to under 120 animals per year in 2007/08. PBR and quota for Steller sea lions around Hokkaido is often revised, and the quota for 2014/15 is 516 individuals [556]. A portion of the culled sea lions were utilized for local food and for canned meat in Hokkaido. The cans of sea lion meat are on sale as souvenirs for tourists visiting Hokkaido, but they are not very popular.

T. ISHINAZAKA

A haul-out site of Steller sea lions in May, 2009 in Ishikari, Hokkaido (T. Namba).

Carnivora   OTARIIDAE
123

Red list status: C-1 (MSJ); VU (IUCN); DC (JFA)

Callorhinus ursinus (Linnaeus, 1758)

EN northern fur seal   JP キタオットセイ（kita ottosei），オットセイ（ottosei）
CM 海狗   CT 海狗，北方海狗   KR 물개   RS северный морской котик

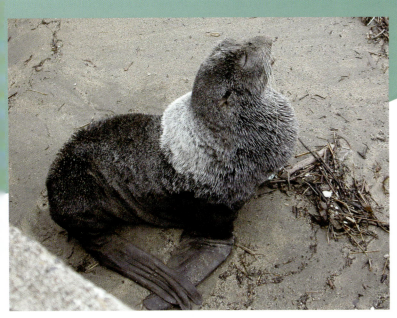

Incidental landing of a northern fur seal at Kushiro Port in Hokkaido (Y. Goto).

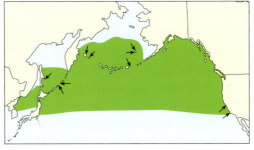

1. Distribution

Distributed in the North Pacific Ocean, Bering Sea, the Sea of Okhotsk and the Sea of Japan. Distribution differs depending on sex and age class; on the Pacific side adult males are mainly distributed from Tsugaru Str. to the north, and adult females and young are from the strait to the south [3792], on the Sea of Japan side mature males are more dominant compared to the Pacific side [946, 1826, 2800, 3792]. There is little information about the distribution of fur seals in the Sea of Okhotsk. Breeding sites are indicated by arrows.

2. Fossil record

Some fossils from the Middle Pleistocene are from northern Honshu [825]. A fossil referred to the extinct fur seal, *C. gilmorei*, was obtained from the Dainenji Formation of the Pliocene in the central part of Honshu [1668].

3. General characteristics

Morphology   Northern fur seals differ from other otariids, they have a noticeably shorter, down-curved rostrum, longer ear pinnae, and longer rear flippers. Longer ear pinnae, and longer rear flippers, dense fur and black coloration are characteristic as compared with Steller sea lions. Their large, bare flippers aid in regulating their body temperature. The fur is a mix of a permanent dense underfur and longer guard hairs which are molted once a year.

Sexual dimorphism is marked; males may be about 4.5 times heavier than females [3147]. Measurements of adults are as follows: SL is 213 cm for mature males and 150 cm for females; BW ranges between 182 and 272 kg for mature males, and between 43 and 50 kg for females [1581]. Both sexes are born black with a light underfur, then molt to silver to gray while young. Coloration of adult males ranges from reddish brown to black; females from brown to gray.

Dental formula   I 3/2 + C 1/1 + P 4/4 + M 2/1 = 36 [29]. (I 3/2 + C 1/1 + PC 6/5 = 36 [3146]).

Mammae formula   0 + 1 + 1 = 4 [2884].

Genetics   Chromosome: $2n = 36$ [192]. MtDNA analysis indicates that the northern fur seal has a basal relationship to the rest of the family Otariidae [3902]. A fast and non-hazardous method of DNA fingerprinting was developed for northern fur seals [1634]. Several PCR primers for microsatellite DNA developed for gray seals and northern elephant seals are applicable to northern fur seals [460, 1633].

4. Ecology

Reproduction Mating system is extreme polygyny (resource-defense polygyny) [311]; males have the opportunity to mate with 15–20 females during the breeding season, often more [261]. The age of sexual maturity is 4 years old in both sexes, although males do not become socially mature until 8–9 years old. At the rookeries, adult males set up territories in May; females arrive in June and give birth to 1 pup a few days later [721]. Females copulate 5–14 days after giving birth; peak mating season is late June–late July. Most pups are weaned at about 4 months old in November [720, 2838], begin their southward migration, and may remain at sea nearly 2 or 3 years before returning to their rookery of birth [290, 1538]. About 70–80% of mature females off the Sanriku coast (Sanriku Oki) are pregnant [235].

Fur seals migrate south to the open ocean or coastal waters during winter and spring (from about October to May), and return to their breeding grounds in summer [1538, 2871].

Approximately 50% of population breeds at the Pribilof Isls. (Alaska); the remainder breeds at the Commander Isls., the middle part of the Kuril Isls. and Tuleny Isl. (Russia), Bogoslof Isl. (Alaska) and Farallone and San Miguel Isls. (California) [717, 719]. Based on re-sighting data, birthplaces of the population wintering off the Sanriku coast are Tuleny Isl. (63%), the Commander Isls. (30%), the Pribirof Isls. (6%) and the Kuril Isls. (1%); almost all individuals wintering in the Sea of Japan are derived from the Tuleny Isl. population [235].

Lifespan Average lifespan is 16 years in males and 21–23 years in females [719, 1537].

Diet Fur seals are opportunistic feeders, taking a wide variety of prey nocturnally, and switching the diet according to the availability of prey species [1273, 4154]. They primarily prey on fishes and squids such as walleye pollocks (*Theragra chalcogramma*), Japanese sardine (*Sardinops melanostictus*), lantern fish (*Diaphus watasei*), and Sparkling enope squids (*Watasenia scintillans*) in the waters off the Sanriku coast [3791, 4154]. Most foraging dives are 20–130 m in depth; the deepest recorded dive is 190 m [1714]. It is often observed that fur seals worry and eat large fish such as cods at the surface.

Habitat Northern fur seals are more pelagic than other northern hemisphere otariids. Although seals can occur in the central northern Pacific, while at sea they concentrate in areas of upwelling over seamounts and along the continental slopes [719]. Off the Sanriku coast, dense populations are located at the boundary of cold (Oyashio) and warm (Kuroshio) currents, and along the continental slopes; in the Sea of Japan, located on the bank and along the continental slopes [1826, 2800, 3792]. They are also more likely to be found in waters with surface temperatures between 12–15°C, which reflect local abundances of fish [3790].

Home range Natal site fidelity among fur seals is strong and increases with age [248, 719]. In a study of migratory behavior, 1 subadult and 2 adult females were tracked for approximately 2 months from the Commander Isls., Russia, to a broad area of the western and central North Pacific [238]. From off Sanriku coast, adult females returned to Tuleny Isl. via Etorofu Strait, or to the Commander Isls. along the Kuril Isls., and they migrated approximately 3,000 km in total with an average speed of 7.1 km/h [235]. They migrate south with cold currents (Oyashio), and to the north avoiding warm currents (Kuroshio).

Behavior In the water, most animals sleep floating on one side with the upper front flipper held between the 2 rear flippers in an arch above the surface. This posture apparently serves a thermoregulatory function [1581]. Most fur seals are solitary or occur in pairs; but groups of 3 occasionally occur [3792]. Outside the reproductive season only sick individuals come to shore [718]. From Hokkaido to central Honshu, some incidental landings of individuals have been reported [1631].

In Japan, wintering seals occur in the Sea of Okhotsk, the Sea of Japan to the middle of Korea and Niigata Pref. and in the western Pacific Ocean to Ibaraki Pref. In the western Pacific, fur seals are first seen off Hokkaido in late November, and then move south off the Joban coast (Joban Oki) in Honshu and stay in large numbers off Sanriku coast till April [235, 1581].

Natural enemies Killer whales and white sharks (*Lamnidae* sp.) prey on fur seals [1581]. Occasionally young fur seals are taken by Steller sea lions.

Parasites The hookworm (*Uncinaria lucasi*) is one of the major causes of pup death during the first year of life [1540]. On Bering Isl., 84.6% of dead pups were infected by *U. lucasi* [2175].

5. Remarks

After their discovery in the late 17th century, the worldwide population has declined sharply several times by commercial sealing for their fur. In 1911 an agreement was reached between UK, Russia, USA and Japan to prohibit pelagic sealing in the north Pacific. Although this agreement was broken in 1941, an interim treaty for the management of fur seals was in effect from 1957 to 1985 [719]. Nowadays, the hunting of fur seals is prohibited by a law established in 1912 in Japan. Thousands of fur seals were incidentally caught and drowned by Japanese drift nets operated in the high seas till 1992. Some fur seals are taken incidentally in set net and gill nets in northern Japan [946, 1631, 1648, 3814]. Many animals are affected, sometimes lethally, by man-made debris such as portions of fishing nets and plastic bands [1581, 1632]. Around Hokkaido, fisheries are damaged seriously by fur seals as with Steller sea lions.

K. HATTORI

Carnivora OTARIIDAE

124

Red list status: CR (MEJ); En as *Z. californianus japonicus* (MSJ); EX (IUCN); EN as *Z. c. japonicus* (JFA)

Zalophus japonicus (Peter, 1866)

EN Japanese sea lion   JP ニホンアシカ（nihon ashika）
CM 日本海狮   CT 日本海獅   KR 바다사자   RS японский морской лев

A stuffed specimen (subadult male) in Tennoji Zoo Osaka, Osaka Pref. This individual was caught on Takeshima Isl. in 1930s (T. Inoué).

1. Distribution

Although lots of records and specimens from the Jomon to Modern eras were found, there are few recent sighting and capture records. In the 1950s, hundreds of Japanese sea lions still inhabited the vicinity of Takeshima Isl. [3145]. The latest record was that of a couple of the sea lions seen in December 1975 on this island [3892]. After the 1980s, no Japanese sea lions have been sighted.

By careful examination of distribution records, 28 places were confirmed as past habitats (from 1880 to the present) (see the above map). The following places were confirmed as breeding places (indicated as arrows on the map): 1 Takeshima Isl., Shimane Pref., 2 Kyuroku-jima Isl., Aomori Pref., 3 Shikine-jima and 4 Onbase-jima Isls., Tokyo (numbers correspond with those in the map) [1149]. These places overlapped with the southern part of the distribution of Steller sea lions and the Northern fur seals in the North West Pacific Ocean.

2. Fossil record

Fossils of sea lions are obtained from the middle Late Pleistocene in north and central Honshu [825, 3207].

3. General characteristics

Morphology Sexual dimorphism is evident in body size and morphology as in other otariids. In adult males, an upheaval typically develops in the parietal region, which distinguishes Japanese sea lions from other otariids. The body color is bright tan to dark brown [1147]. Available external measurements are limited. Reconstructed sizes in adult males [1150]: SL, average 239.2 cm (228.4–249.4 cm, $n = 4$). Total length, 287.9 cm ($n = 1$) and BW, 493.7 kg (473.7–513.7 kg, $n = 2$). Adult males are approximately 10% longer in body length and 30% heavier than the California sea lion *Z. californianus* [1150]. In subadult females, only one measurement has been available [3145]: SL, 164.0 cm ($n = 1$).

Sagittal crest is well developed in adult male *Zalophus* [1581]. The skull of the Japanese sea lion is larger and relatively wider in several measurements than the California and Galapagos sea lions *Z. wollebaeki* [335, 1148]. The skulls of adult males excavated from a shell mound estimated to be 500–800 years ago in Hokkaido were measured as follows [1148]: condylobasal length, average 313.8 mm (307.0–323.0 mm, SD = 5.42, n = 7); snout width at canine, average 71.7 mm (66.0–77.6 mm, SD = 4.22, n = 6); zygomatic width, average 187.5 mm (174.0–199.2 mm, SD = 7.52, n = 7); mastoid width, average 174.8 mm (161.0–186.4 mm, SD = 7.71, n = 7).

Dental formula   I 3/2 + C 1/1 + P 4/4 + M 1–2/1 = 34–36 [29] (I 3/2 + C 1/1 + PC 5-6/5 = 34–36 [1581]).

Mammae formula   Not reported.

Genetics   According to an ancient mtDNA study from skeletal remains [2969], the average amount of nucleotide substitution between the Japanese and California sea lions was 7.02%. These 2 sea lions were estimated to have diverged 2.2 million years ago (late Pliocene Epoch) [2969].

4. Ecology

Reproduction   The breeding season was from April to July and females gave birth to a single pup. Pups were black in color, 65 cm in SL, and 9 kg in BW [1150].

Lifespan   Not reported.

Diet   The diet of Japanese sea lions has been unclear. Japanese sea lions had been seen in fixed nets and fed on yellowtails (*Seriola lalandi*). According to an oral tradition of fishermen, they also fed on sardines and squids.

Habitat   Not reported.

Home range   Not reported.

Behavior   Not reported.

Natural enemies   Not reported.

Parasites   Not reported.

5. Remarks

Genus *Zalophus* had been treated as a single species comprising 3 distinct populations; *Z. c. californianus*, *Z. c. japonicus* and *Z. c. wollebaeki* according to inadequate specimens preserved in museums [1581, 3145]. However, it was recently reclassified as a separate species [3894] by means of morphological studies [335, 1148].

Ancient people from the Jomon period hunted Japanese sea lions to get the pelage and flesh. In the modern period, the skin was used for leather goods and the subcutaneous fat was rendered down for lamp oil. From the Meiji Era (1868–1912) to World War II, sea lions were hunted using firearms and nets. On Takeshima Isl., large-scale hunting was conducted from 1904 to 1911 with no consideration of sex and age: approximately 14,000 animals were killed in this period [1046].

Although IUCN regards this species as extinct in 1994, governmental and some academic authorities have not accepted the extinction yet, based on some unconfirmed sighting records in some areas (see the map), and the fact that 50 years have not passed since the last confirmed record at Takeshima Isl. in 1975 [3892].

T. ISONO & T. INOUÉ

Skull and mandible specimens of an adult male (upper) and an adult female (lower) of the Japanese sea lion in the same scale, stored at Historical Museum of Japanese History, Japan and Zoological Museum of Berlin, Germany, respectively (S. Brunner). There is a great sexual dimorphism in morphology.

Sirenia

Sirenia DUGONGIDAE

Red list status: CR (MEJ); En (MSJ); VU (IUCN); EN (JFA)

Dugong dugon (Müller, 1776)

EN dugong   JP ジュゴン (jugon)   CM 儒艮   CT 儒艮，南海牛   KR 듀공   RS дюгонь

Two dugongs in Kin Bay, Okinawa-jima Isl. (Ministry of the Environment, Japan).

1. Distribution

Tropical and subtropical coastal areas of more than 40 countries, from Indian Ocean to Pacific, lying between approximately 30°N and 30°S of the equator. The largest dugong populations are in the waters of northern Australia [2501]. In Japanese waters, this species was recorded throughout the Ryukyu Isls. during the 19th century [2084]. However, since the 1980s, dugongs have been confirmed only in the northern part of Okinawa-jima Isl. [1372, 1373, 2084, 2582]. The Ryukyu Isls. is the most northern dugong habitat in the world [1372].

2. Fossil record

The dugong fossil record in Japan comprises approximately 100 remains from the Ryukyu Isls. dating from the Jomon period (ca. 16,000–3,000 years ago) to the middle ages (The 12–14th century) [2233]. A fossilized tusk of *Dugong* sp. was obtained from a 1,500 BC kitchen midden or "kaizuka" on Iriomote Isl. [823].

3. General characteristics

Morphology The dugong has flipper-like limbs, a fluked tail, and smooth gray skin. The body shape of the dugong is more streamlined than that of a manatee, and resembles that of a cetacean. The snout bends downward, and hard sensory hairs are arranged around the rostral disk. Specimens of average size have been reported only in Australian waters, and there is little information about this species in other areas. TL of adults averages approximately 240–300 cm, and BW ranges from 200 to 900 kg [876]. The dugong exhibits a unique tooth-replacement system: its premolar and molar teeth progress from aboral to oral and fall out at the tip. Mature individuals have 2 or 3 molars, and the upper incisors are tusk-like and larger in males than in females. The incisors are erupted in adult males and in many older females [1956].

Dental formula Dental formula of functional teeth has been recognized as I 1/0 + C 0/0 + P 3/3 + M 3/3 = 26 [883]. However, the dugong has vestigial teeth under the horny pad of the mandible, and juveniles have 2 pairs of upper incisors [2098]. The complete dental formula is I 2/3 + C 0/1 + P 3/3 + M 3/3 = 36.

Dugong dugon (Müller, 1776)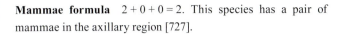

Mammae formula 2 + 0 + 0 = 2. This species has a pair of mammae in the axillary region [727].

Genetics Chromosome: $2n = 50$ [3267]. Phylogenetic research has indicated that the order Sirenia is closely related to the orders Proboscidea and Hyracoidea [194], but few sequence data of the mtDNA control region are available [2041, 2084, 3575]. Based on mtDNA analysis, sub-populations may reside off western and eastern Australia as well as in Asian waters [2041]. Populations in Japanese waters are thought to share a common ancestor or have genetic exchange with populations from Philippine waters [2084].

4. Ecology

Reproduction Not reported in Japan. Based on records from Australian waters, breeding occurs throughout the year. The gestation period has been estimated to be approximately 13–14 months, and litter size is usually a single calf. The TL of neonates ranges from 100 to 150 cm, and BW is approximately 20 kg [1956]. The estimated lactation period is at least 18 months, and the breeding interval is estimated to be approximately 3–7 years. Sexual maturity is attained at 250 cm TL and at least 9–10 years of age in both males and females [1954, 1956].

Lifespan Not reported in Japan. Dugongs are long-lived animals, and their estimated life span is more than 50 years. The age of this species can be estimated by counting the growth layers on a section of the tusk [1953]. The oldest recorded dugong was approximately 73 years old when it died [1954].

Diet The primary food item is seagrass [529, 876, 1247, 1955]. Stomach content analyses have indicated that the dugong consumes 18 different species of seagrass [95].

Habitat Not reported in Japan. Dugongs generally frequent coastal waters and tend to occur in wide shallow protected bays, wide shallow mangrove channels, and large inshore islands [877], which all harbor seagrass beds. Time-depth recordings have indicated that dugongs conduct 47% of their daily activities within 1.5 m of the sea surface [381], which may be a strategy to minimize the risk of shark predation [129].

Home range Dugong movements have been tracked using VHF or satellite transmitters in Australian waters [1960, 2856, 3179]. Many dugongs regularly make round trips of 15 to 40 km between foraging grounds, and some individuals travel even longer distances. For example, an adult female traveled 600 km in approximately 5 days [2856], while a male traveled a straight-line distance of 140 km 3 times in 6 weeks [1960].

Behavior Not reported in Japan. Feeding is the dominant activity of the dugong [130]. After feeding, a distinct feeding trail approximately 20 cm wide and 1 to 5 m long remains in the seagrass beds. The time of peak activity is unknown. Some studies have reported that dugongs feed at night, whereas others have indicated that tides are more important than photoperiod for the regulation of activity [1219, 2084, 2284]. Dugong diving behavior has been documented using time-depth recorders: the maximum recorded dive depth was 20.5 m, the mean maximum dive depth was 4.7 m, the mean dive duration was 2.6 min, and the average number of dives was 11.1 per hour [381].

Natural enemies Not reported in Japan. In Australia, killer whales and tiger sharks have been reported to prey on dugongs [131, 3274]; however, definitive evidence of predation pressure by animals other than humans is lacking.

Parasites *Labicola elongate* is found in abscesses in the upper lip, *Parachochkeotrema indicum* live in nasal mucosae, two species of *Opisthotrema* occur in the middle ear and Eustachian tube, and *Taprobanella bicaudata*, *Rhabdiopoeus taylori*, *Haerator capertaus* and *Indosolenorchis hirudinaceus* are found in the viscera [299, 492]. *Paradujardinia halicors* occur in large numbers in the cardiac gland of the stomach. This species is not only found in Australian waters but also the Asian region, including Japan [299, 492, 811]. In addition, *Cryptosporidium parvum* is found in the small intestine. 18S ribosomal RNA and acetyl CoA synthethase genes suggest that *Cryptosporidium parvum* infecting dugons is the variant that infects humans [891, 2194].

5. Remarks

Dugongs were trapped for meat, oil, medicine and other products [1959]. In addition, the by-catch of dugongs, the destruction of seagrass beds, and marine environmental pollution pose serious threats to this species [2807, 2857]. In some areas, dugong populations are locally extinct or nearly so [1959].

O. HOSON & G. OGURA

A skeleton of a male dugong (body was collected at Ginoza-son, Okinawa-jima Isl.) stored at Nago Museum, Nago, Okinawa Pref. (G. Ogura).

Census of seagrasses (*Halophila ovalis* and *Halodule uninervis*), which are the primary foods of dugongs, in Kasari Bay, Amami-ohshima Isl., Kagoshima Pref. (G. Ogura).

Artiodactyla

Artiodactyla   CERVIDAE

Red list status: LP for the Mage-shima Isl. population (MEJ); R for the population of the Kerama Isls., C-1 for the other populations (MSJ); LC (IUCN)

Cervus nippon Temminck, 1836

EN sika deer   JP ニホンジカ（nihon jika）, ニホンシカ（nihon shika）   CM 梅花鹿   CT 梅花鹿   KR 사슴
RS пятнистый олень

Adult and 1-year old males in the Shiretoko National Park, Hokkaido (H. Okada).

1. Distribution

In Japan, Hokkaido, Honshu, Kyushu, Shikoku, the Tsushima Isls., the Goto Isls., Mage-shima Isl., Tane-gashima Isl., Yaku-shima Isl., Kuchinoerabu-jima Isl., and the Kerama Isls. [1212, 2651]. The sika deer occupies over 40% of the area of Japan, where 70% of the distribution has been colonized during the last 3 decades [1212]. This species was originally widespread throughout northeastern Asia, from the Ussuri area (Russian) to northern Vietnam and Taiwan [183, 2037–2039, 2651, 2652, 2813, 3783, 3872]. Sika deer in China, the Korean Pen., Taiwan, and Vietnam are either extinct or have been reduced to population fragments [716, 965, 2037–2039, 2651, 2652, 2863, 3178, 3888, 3905]. Introductions have resulted in established populations on the British Isls., several countries of mainland Europe, USA, and New Zealand (map not shown) [256, 262, 542, 3380, 3872].

2. Fossil record

Sika deer fossil records in Japan have been summarized in [1511]. Fossil antlers were recorded from 2 Middle Pleistocene localities: Chiba Pref. [3445] and Shizuoka Pref. [3416]. Fossils were recorded from 4 late Pleistocene localities: Aomori Pref. [3444], Iwate Pref. [2009], Nagano Pref. [2512] and Ishikawa Pref. [1439]. Fossils of Holocene age have been recorded from eleven sites: 3 sites in Osaka Pref. [2738, 3530, 3760], 1 site in Nagasaki Pref. [621], 3 sites in Chiba Pref. [424, 1310, 4106], 1 site in Ishikawa Pref. [2156], 1 site in Miyagi Pref. [2008], 1 site in Fukui Pref. [3194] and 1 site in Shiga Pref. [3193, 3708]. Fossil records in China have been summarized in [1511] indicating that sika deer continuously inhabited northern China from the early Middle Pleistocene to Holocene. Sika deer remains have been found in Lesser Kuril Ridge, dating to about 100 years ago [3784].

3. General characteristics

Morphology   Sika deer in Japan show striking variation in size from north to south [2102, 2651, 3872]. 50–130 kg, 90–190 cm in HB, and 70–130 cm in shoulder height for adult males, 25–80 kg, 90–150 cm in HB, 60–110 cm in shoulder height for adult females [2102]. Weight is sexually dimorphic, with males larger than females [540, 2102]. Effects of environmental variables on body size have been reported [1755, 3548].

Two molts occur annually in sika deer [540]. In the summer pelage, the general color varies with locality from yellowish-red to reddish-brown with numerous white spots on both sides. In winter, white spots disappear and the pelage turns a dark reddish brown to a blackish brown [3872]. Males are slightly darker than females in summer and more markedly so in winter [350]. Sika deer on the Kerama Isls. tend to melanism and occasionally all-black animals occur [3262, 3872]. Antlers are only found among males. Antlers are usually 8-tined (4 on each side) and vary in length from about 30 cm to 80 cm depending on locality [2102, 2651, 3872]. However, it is rare for the numbers of antler tines on Yaku-shima and Kerama Isls. males to exceed 6 (3 on each side) [2651, 3872]. Adult males cast antlers in early March and new antlers are fully developed by August [2111]. Their velvet is shed by early September and hard antlers predominate by early September, just in time for intra-sexual selection activities [2111].

Dental formula I 0/3 + C 1/1 + P 3/3 + M 3/3 = 34 [29, 540, 1011].
Mammae formula 0 + 0 + 2 = 4 [1011].
Genetics Chromosome: $2n = 64–68$ (XX and XY) and FN = 70. Karyotype variations exist in Japan [2705, 3241, 4109, 4247]. Data on mtDNA; complete sequences, cytochrome b, cytochrome c oxidase, 12S rRNA, COX II, NADH, ATP synthase genes, control region sequences and restriction fragment length polymorphism, have been reported [61, 400, 639, 2337, 2338, 2346, 2348, 2350, 2351, 2791, 3470, 3479, 3481–3483, 3549, 3795–3797, 3940, 3941, 4185, 4230]. There are 2 major mitochondrial lineages in Japan: the Northern Japan group (Hokkaido Isl., most of the Honshu and a part of Shikoku Isl.) and the Southern Japan group (a part of the southern Honshu mainland, a part of Shikoku, Kyushu and small islands around Kyushu) [2346, 2350, 3482, 3941]. Divergence between the 2 lineages is estimated to have occurred approximately 0.35 million years before present [2350]. A few sex chromosomal gene makers are available for sex identification [640, 641, 3428, 4028]. Microsatellite DNA analyses have resolved the genetic structure and gene flow among Japanese sika deer populations [61, 728, 2349, 2668, 2669, 3276, 3480]. A few allozyme polymorphisms have been reported [61, 642, 643]. The basic genetic characteristics of sika deer have been summarized in [3478].

4. Ecology

Reproduction Sika deer are polygamous [350]. Breeding (rutting) usually occurs in autumn [506, 2111]. Conception date among females varies largely (39–100 days) [197, 1675, 2024, 3365]. Gestation period has been estimated to be 216–260 days [2024, 2111]. Fawning takes place from mid-May until November with a peak in June in Nara Park [2111], and in May and July in Hyogo Pref. [1675]. The fawn is usually single [1675, 3368] weighing 4.5–7.9 kg [540, 2518, 3368]. Large numbers of females attain sexual maturity as yearlings, and the pregnancy rate of yearlings and older females is over 90% [3367, 3368]. Fertility in the high density population is lower [1268].

Lifespan The maximum recorded longevity of wild males is 14 and that of females is 21 years [1267, 1675]. The average lifespan of wild females has been estimated to be 5.5 to 6.1 years [1267]. In the protected population at Nara Park, the maximum recorded longevity of males is 21 years and that of females is 24 years, and the mean longevities of males and females are 4.0 and 5.9 years, respectively [2650]. One captive deer was recorded to have lived 26 years and 3 months [3862].

Diet This species has a highly adaptable diet, and can either

An adult female and its fawn in Utoro, Hokkaido (H. Okada).

graze or browse. There is geographic variation in the food habits of Japanese sika deer. Northern deer eat graminoids, particularly dwarf bamboos (*Sasa* spp.), while southern deer browse leaves and fruits [3452]. The boundary between the two lies at a latitude of approximately 35ºN [1221]. Sika deer drastically shift their diets and exploit alternative foods under food limitation. They begin feeding on low quality foods such as fallen leaves, bark, twigs, and even unpalatable plants [2124, 3421, 3454].

Habitat Sika deer prefer the edges of forested areas with an understory but adapt well to a variety of other habitats [2518]. They are found at a variety of elevations from sea level to about 3,000 m: 1,000–1,900 m in Tochigi Pref. [1973, 1975]; above 1,300 m in Tanzawa, Kanagawa Pref. [316]; up to about 3,000 m in Minami Alps, central Honshu [391], sea level to 1,800 m on Yaku-shima Isl. [3284]. Sika deer avoid areas with deep snow [1973, 1975, 3453].

Home range Summer ranges are generally higher and larger than winter ranges [1973, 2981, 3457]. Sika deer in northern Japan tend to have larger home ranges than in southern Japan: the annual home ranges of resident females in Hokkaido were estimated to be 325.2 ha [3908]; the monthly home range sizes for both sexes in Nikko, Tochigi Pref. were found to be 21.0 to 284 ha [1973]; the annual home range of females in Chiba Pref. was 46.1–246.3 ha [3196]; the mean annual home ranges for males and females on Mt. Ohdaigahara in Kii Pen. were 211.3 ± 152.4 (SE) ha and 162.2 ± 106.7 ha [1938, 3920]; the mean summer home range in Nara Pref. was 11.7 ha [2107]; the annual home range of females in Miyazaki Pref. was 48.7–58 ha [3906]; the mean home range size in Nozaki Isl. in Nagasaki Pref. was 3.0–3.6 ha [505]. Large sika deer males are territorial during the rutting season. During the summer adult males begin to establish territories that average 4.76 ha (range 2.69–7.70) in Nara Pref. Non-territorial males have larger home ranges of 11.74 ha on average (5.56–17.55) [2113].

Behavior Activity occurs primarily from dusk to dawn but may also occur at times during daylight [2518]. Males tend to gather into single-sex groups throughout the year, except during the rutting season [1973, 2110, 2113]. Male groups begin to break up as older males wander off to their rutting areas. Then, each

A dwarf male approaching a female group in late rut (late November) in a subtropical habitat on Mage-shima Isl., Kagoshima Pref. (S. Tatsuzawa & K. Ogawa).

dominant male establishes a mating territory and tries to attract females to his harem. After the rut, males gather into single-sex groups again [2110, 2113]. The behavior and rank of each male is directly related to the configuration of his antlers [2113]. Males mark the boundaries of their territories by rubbing, pawing and thrashing, and urinating also serves as an olfactory signal to mark spots [2113]. Sexual segregation occurs in and around the calving and rearing seasons [1141, 1973]. Seasonal migrations have been reported especially in northern Japan with summer ranges being generally higher and larger than winter ranges, depending on such factors as snow cover, bamboo grass, coniferous cover, calving season, etc. [316, 1973, 1975]. Northern sika deer migrates longer distances than southern sika deer. For example, Hokkaido females migrate 35.1 km on average (range 7.2–100) [977], while in Kyushu females migrate 2 to 7.5 km [3906]. Females have high fidelity to seasonal ranges, especially winter home ranges and migratory routes [977, 2982, 3733].

Natural enemies The brown bear (*Ursus arctos*) is a predator in Hokkaido [1253].

Parasites

Ectoparasite

Ticks: *Haemaphysalis megaspinosa, H. flava, H. kitaokai, H. longicornis, H. cornigeraias, H. japonica, H. jezoensis, H. yeni, H. phasiana, Ixodes acatitarsus, I. nipponensis, I. monospinosus, I. ovatus, I. persulcatus, Amblyomma testudinarium, Haemaphysalis mageshimaensis,* and *Boophilus* spp. [1037, 1138, 1277, 1619, 1624, 2205, 2583, 3423, 3973, 4029].

Louseflies: *Lipoptena fortisetosa, L. japonica* and *L. sikae* [606, 644, 1276, 2184, 2187, 4032–4034].

Louse: *Solenopotes* sp. [1619], *Solenopotes* sp. cf. *binipilosus* [1276], Trichodectidae gen. sp. [1619], *Damalinia sika* [2173].

Endoparasite

Many endoparasites, including Trematoda [1619, 2056, 2174, 2579, 3962]. Nematoda [1619, 2056, 2174, 3962, 4128] and Castoda [1619, 2056] have been reported.

5. Remarks

An important big game species, especially in Hokkaido. Because of damage to agricultural crops, forests, and traffic and railway accidents, dense concentrations of sika deer have become a severe social problem [2647, 2648, 3891]. In addition to ordinary hunting, a number of them are killed as nuisance animals to protect agriculture and forestry. The velvet antlers of sika deer are used in Chinese traditional medicine. Because sika are religiously important for Japan's native religion, Shinto, they are sometimes kept on the grounds of Shinto shrines, such as Koganeyama shrine in Miyagi Pref., Nara Park in Nara Pref., and Itsukushima shrine in Hiroshima Pref.

The sika deer is divided into 14 subspecies [2651]. This classification is widely accepted at present. While *C. n. hortulorum* and *C. n. sichuanicus* remain, *C. n. manchuricus, C. n. mandarinus, C. n. grassianus, C. n. kopschi, C. n. taiouanus,* and *C. n. pseudaxis* are extinct or have been reduced to population fragments in the wild in China, the Korean Pen., Taiwan, and Vietnam [716, 965, 2651, 2652, 2863, 3178, 3888]. Six subspecies exist in Japan [2651]: *C. n. yesoensis* (Hokkaido population), *C. n. centralis* (Honshu mainland and Tsushima Isls. populations), *C. n. nippon* (Kyushu, Shikoku, and the Goto Isls. populations), *C. n. mageshimae* (Mageshima Isl. and Tane-gashima Isl. populations), *C. n. yakushimae* (Yaku-shima Isl. and Kuchinoerabujima Isl. populations) and *C. n. keramae* (the Ryukyu Isls. population). However, there are some conflicts regarding the taxonomy of the Tsushima Isls. population [507].

Hybridization between sika deer and red deer (*C. elaphus*) has occurred in several parts of the world where human introductions have brought the 2 species together, for example, in Europe [61, 262, 263, 804, 3380], USA [542] and New Zealand [256, 438].

J. NAGATA

Artiodactyla   CERVIDAE

Red list status: LC (IUCN), introduced

Muntiacus reevesi (Ogilby, 1839)

EN Reeves's muntjac, Chinese muntjac   JP キョン（kyon）   CM 小麂   CT 山羌   KR 애기사슴
RS китайский мунтжак

A male muntjac with velvet antlers on the Boso Pen., Chiba Pref. (Y. Sugiura).

1. Distribution

The native range is southeastern China and Taiwan. The muntjac was introduced as an exhibition animal and naturalized in the southern part of Chiba Pref. between the 1960s and 1980s [198], and onto Izu-ohshima Isl. in 1970 [204]. It was also introduced into Great Britain [363].

2. Fossil record

Not reported in Japan, but related *Dicrocerus* sp. (Ryukyu-mukashi-kyon or Ryukyu-paleo-muntjac) was found in the Ryukyu Isls. [822, 2772].

3. General characteristics

Morphology   Body color is deep to reddish brown. Fawns are born with off-white spots that disappear by 6 to 8 weeks [362]. The white underside of the raised tail of an alarmed animal is clearly visible. Males have V-shaped black lines running up each pedicle on their faces. Females have a diamond-shaped black patch on the forehead. The upper canines of the male are apparent and protrude from the side of the mouth. The preorbital and forehead glands are easily visible in both sexes. The following morphological data come from Chiba [444].

External measurements (mean ± SD) of adults (≥ 2 years old): males: BW = 10.0 ± 1.4 kg (n = 21), HB = 75.1 ± 4.9 cm (n = 22), T = 9.3 ± 1.4 cm (n = 18), CG (chest girth) = 50.4 ± 2.7 cm (n = 22), SH (shoulder height) = 38.8 ± 2.5 cm (n = 22), HFcu = 17.9 ± 1.1 cm (n = 22); females: BW = 8.9 ± 0.9 kg (n = 19), HB = 73.1 ± 3.2 cm (n = 19), T = 9.2 ± 1.0 cm (n = 15), CG = 47.8 ± 2.6 cm (n = 19), SH = 36.0 ± 2.0 cm (n = 19), HFcu = 17.6 ± 1.0 cm (n = 19). Significant sexual differences in size are apparent only in body weight, chest girth, and shoulder height. Significant differences are seldom found in external measurements among fawns 6 to 12 months old, yearlings, and adults 2 years old or more, although body size increases until 2 years. Bucks cast their antlers in spring, and velvet grows soon after [359, 444, 3177, 4175]. In Chiba, yearling males had 1-pointed antlers (n = 9), and males of 2 years or more (n = 24) had 1-pointed (33%) or 2-pointed antlers (67%). Antler length was 3.3 ± 1.4 cm (n = 9) in yearlings and 6.3 ± 1.5 cm (n = 24) in males 2 years or older. Body measurements and mandible sizes indicate that the population introduced into Chiba is closer to *M. r. micrurus* in Taiwan than to *M. r. reevesi* in China [442, 444].

Dental formula   I 0/3 + C 1/1 + P 3/3 + M 3/3 = 34 [29, 360].

Mammae formula   0 + 0 + 2 = 4 [2556].

Genetics   Chromosome: $2n$ = 46 [3181, 3898].

4. Ecology

Reproduction Birthing occurs at any time of the year, with no seasonal peak in China or Great Britain [361, 362, 3177], but with a seasonal peak of May to October in Chiba Pref. [444], and a large peak in April to June and a small peak in November–December in captivity in Japan [4175]. Single fawns are born after a gestation of 210 days [362]. Twins are rare (1.2% of 82 pregnant females in Chiba) [444]. Sexual maturity in females occurs at 6 (5–7) months of age [362, 3177], and pregnancy in a female of 7 months was observed in Chiba Pref. [444]. The pregnancy rate in winter in Chiba Pref. was 25% ($n = 20$) of fawns of 6 to 12 months, 59% ($n = 22$) of yearlings, and 78% ($n = 76$) of adults of 2 years old or older [445]. Muntjac have the potential to produce fawns at intervals of 7 months on account of post-partum estrus [362, 364].

Lifespan Maximum longevity in the wild has been recorded as 9 years [3177]. The maximum recorded longevity of captive muntjac is 23 years and 2 months [3862].

Diet Muntjac are primarily browsers. They feed mainly on leaves of broadleaved trees, and partly graminoids and fruits in Chiba Pref. [444]. They selectively consume higher-nutrient foods than sympatric sika deer (*Cervus nippon*) in Chiba Pref. [2555].

Habitat Muntjac in Chiba Pref. inhabit broadleaved forests and coniferous plantations. They also use open areas up to 150 m from forest edges at night [444].

Home range The mean winter home range size of females was 5.3 ha (range 3.4–7.7; $n = 3$) in Chiba Pref. [444].

Males had larger ranges than females in Great Britain [365], but there was no sexual difference in home range size in Taiwan [2040]. In Great Britain, female ranges overlapped, but core areas were largely exclusive. Male ranges were exclusive with each other, but overlapped with one or more female ranges. Males are territorial all year round. There is no evidence of antagonistic interactions between females [365].

Behavior There are peaks of activity at dawn and dusk, and lower levels of activity during the day and at night [365, 2040]. Males in captivity often interact aggressively with each other, and dominant males defend small areas by aggressive behavior and scent marking [2112]. Social behavior, including scent-marking and sexual behavior, in captivity has been reported [259, 260, 1822, 2108, 2109, 2112, 3919]. Suckling was observed until the age of 17 weeks in captivity [364]. Both sexes bark [3282], and the voice of an alarmed animal resembles a dog's bark.

Muntjac are usually solitary, but sometimes occur in groups of up to 3. In Chiba Pref., 92% of sightings were single animals; 7% were twos; 1% were threes ($n = 238$) [444].

Natural enemies Possibly not at risk of predation, but feral dogs could be predators in Japan.

Parasites Five species of ticks, *Haemaphysalis megaspinosa*, *H. flava*, *H. longicornis*, *H. cornigera* and *Ixodes ovatus* are collected from muntjacs in Chiba Pref. [3664].

5. Remarks

Damage to crops has been reported in Chiba Pref. and Izu-ohshima Isl. [198, 442–444, 3595]. A total of 4,348 muntjacs were culled during 1983–2011 in Chiba Pref. [378].

The Chiba population was estimated at 350–1,900 animals in 2002, 1,400–5,400 in 2007, and 14,500–26,400 in 2012, in a distribution range of 507 km^2 in 2002, 570 km^2 in 2007, and 1,377 km^2 in 2010 [196, 199, 443, 444]. The Izu-ohshima population was estimated at 1,894–2,435 animals in 2007 and 2,309–4,218 in 2011 [3594, 3595].

Impacts on vegetation have been reported in Izu-ohshima Isl. [3594, 3595]. Impacts on vegetation [401, 402, 2853] and interactions with native roe deer (*Capreolus capreolus*) [365, 879, 880] have been reported in Great Britain. Muntjacs were designated an invasive alien species under "Invasive Alien Species Act" of Japan in 2005.

K. OCHIAI

An adult male photographed using an infrared-triggered camera on the Boso Pen., Chiba Pref. (J. Ohki).

RESEARCH TOPIC 10

Population dynamics of terrestrial mammals in Japan
T. SAITOH

1. Historical background

Charles S. Elton learned of the cyclical population fluctuations of lemmings from a book about mammals in Norway, and in 1924 he published the first ecological paper about population fluctuation [501]. The phenomenon that animal populations fluctuate was new for zoologists [502] and Elton helped establish population ecology as a basic science in Europe. Also in the 1920s, a paper about rodent populations was published in Japan by a forester, Eiichiro Kinoshita [1588]. His main interest was the biology of rodents that damaged tree plantations. He focused on how to control rodent populations in order to reduce the damage, thus in Japan the population ecology of rodents began as an applied science.

Generally mammal populations have been studied in order to solve management problems in Japan. In this section I will introduce some significant studies of mammal populations in Japan; those on rodents, macaques and deer. Each combines a unique historical background with relatively rich information.

Recently, many mammalogists have been concerned by pest-damage and conservation. Sika deer, wild boars, bears (brown and black bears) and Japanese macaques have all caused serious agricultural and/or forestry damage. At the same time, some bear and macaque populations are isolated and considered to be facing a high risk of local extinction. Thus, complex management programs, combining conservation and damage management, are required for those mammals (see [2332]). Conservation based on surveys of abundance has been conducted for endangered species (e.g., the Amami rabbit, the Iriomote cat, the Tushima cat and others). Eradication programs for alien species (the mongoose on Amami-ohshima Isl. and Okinawa-jima Isl., and the raccoon) also have provided data on abundance. I will briefly discuss these recent research activities in Japan.

2. Rodent dynamics in Hokkaido

Forestry has been a major industry of Hokkaido. Rodents have often seriously damaged tree plantation through barking. The Forestry Agency of the Japanese Government in Hokkaido and the Hokkaido Government initiated a management-focused census program of small rodents in 1954 using standardized methods [592, 1335, 2957, 2958]. The data obtained have been analyzed by a research group consisting of the Forestry and Forest Products Research Institute, Hokkaido University and the University of Oslo [3305].

The target species of the rodent census is the gray red-backed vole (*Myodes* (= *Clethrionomys*) *rufocanus*), as a major cause of damage to tree plantations. Based on 225 sets of time series data of various periods (12 to 31 years), which together cover the whole island of Hokkaido, the following features of population dynamics have been observed:

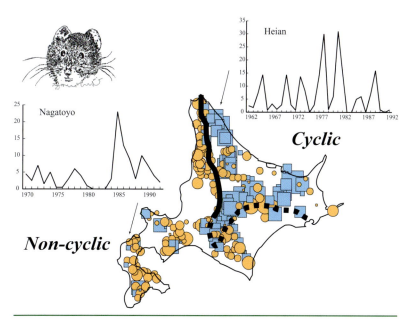

Fig. 1. Geographic gradient of cyclicity in gray red-backed vole populations of Hokkaido, Japan. Symbols represent index of cyclicity; squares: negative, circles: positive (see [298]). Highly negative values indicate clear cyclicity; i.e., squares show populations that showed evidence of 3–4 year cyclicity. Populations shown by circles exhibit irregular fluctuation (Illustration by Y. Ishibashi).

abundance (expressed by the average number of voles captured over three nights with 50 traps) ranges between 1.1 and 21.1 individuals with an average of 7.1, and variability (represented by the standard deviation of logarithmic abundance (base 10)) ranges from 0.20 to 0.63 with an average of 0.39. In comparison with vole populations (*Microtus* and *Myodes*) in Fennoscandia, where rodent populations have been studied intensively, vole populations in Hokkaido are less variable with higher average abundance. Cyclic populations (periodic large-scale fluctuations with high amplitude) are found in the central and north-eastern areas of Hokkaido. The typical interval between peak abundance is 3–5 years. Synchronism is observed on a limited spatial scale (less than 50 km) [297, 871].

A clear geographic gradient is found in variability and cyclicity, resembling populations in Fennoscandia. These population attributes become more pronounced in cooler areas (north-eastern parts; Fig. 1). Density dependence, in which population growth rate has a negative relationship with preceding densities, also exhibits a clear geographic gradient. Direct (one year lag) and delayed (two year lag) density dependence is stronger in cooler areas (north-eastern parts), whereas in Fennoscandia direct density dependence becomes weaker in colder areas which instead are characterized by constantly, strongly delayed density-dependence. Although generalist and specialist predators are postulated as agents of direct and delayed density dependence in Fennoscandia, respectively [778], the mechanisms that generate the geographic gradient in Hokkaido are still unknown.

Field (wood) mice (genus *Apodemus*) appear to have dynamics dominated by irregular irruption related to years of abundant seed production, and direct density dependence dominates over delayed density dependence [2953]. The effects of acorn (*Quercus crispula*) abundance on population dynamics differ between three common rodent species (*Apodemus speciosus*, *A. argenteus*, and *Myodes rufocanus*) in Hokkaido. *Apodemus speciosus* generally increases its abundance following acorn masting [2956]. However, such effects are not detected for the other two species of rodents. Acorns of *Quercus crispula* accumulate a considerable amount of tannins, which potentially have severe detrimental effects on herbivores ([3213, 3214, 3216], see Research Topic 4). *Apodemus speciosus* may reduce the damage caused by acorn tannins with tannin-binding salivary proteins and tannase-producing bacteria, whereas such physiological tolerance to tannins has not been discovered in the other two species.

Acorn abundance influences the strength of density dependence (intraspecific competition) of *A. speciosus*. When the abundance of acorns is high, density dependence is relaxed and as a result the equilibrium density at which the population growth rate decreases to zero becomes higher [2960]. These effects of acorn abundance are regarded as a nonlinear perturbation effect [2911]. Nonlinear density dependence has also been detected; higher densities have stronger effects on population growth rates [2960].

Effects of beechnuts on rodent populations have also been demonstrated in northern Honshu [952, 2063, 2065]. Abundance of rodents, especially for *A. speciosus*, increases dramatically every spring after masting events in forests dominated by *Fagus crenata*, whereas rodent populations fluctuate less in mixed forests.

3. The Japanese macaque

Another unique origin of population ecology in Japan is found in primatology. A charismatic biologist, Kinji Imanishi, established primatology in the mid-1950s to shed light on human evolution and began to study the Japanese macaque in various parts of Japan [3387]. Although the main interest was the evolution of social behavior, the studies generated a considerable amount of data on dynamics of populations (or troops) (see [462, 2330] for review).

Among non-human primate species, while most populations are in jeopardy [462], the situation of Japanese macaques is complex. The populations seem to have declined due to hunting for meat and traditional medicine and habitat destruction by logging until World War II, and in 1948 they were protected from hunting [2330]. Since then, some populations (mainly in central Japan) have expanded, whereas others are isolated and considered to be facing a high risk of local extinction, particularly in western parts of Japan [2330]. Some fed populations with low mortality have increased in abundance and cause agricultural damage. In areas with a low human population people counteract the damage by fencing fields and chasing macaques away, but in some cases they have abandoned the fields due to heavy damage by macaques [2330, 2331].

Demographic analyses in natural populations show the following features ([3329, 3341, 3440]; see [2330] for review). Japanese macaques become reproductive at 6–7 years of age and fertility (births/female/year) is generally 0.27–0.38. Average interbirth intervals are 1.5–1.6 years following the death of infants. These are shorter than those following surviving infants (2.2–2.4 years). Infant annual mortality is 0.25–0.53. In comparison with other macaque species, Japanese macaques seem to have a late reproductive onset and relatively long interbirth intervals [745].

In fed populations they become reproductive earlier (5–6 years of age), fertility is higher (0.50–0.62) and interbirth intervals following surviving infants are shorter (1.46–1.58 years). Infant mortality in fed populations is very low (0.10–0.19 [1146, 1727, 3438]; see [2330] for review).

4. Dynamics of deer populations in Hokkaido

Studies on deer populations in Hokkaido are a good example of management oriented data on abundance. Agricultural and forestry products have been seriously damaged by sika deer particularly since the mid-1980s, and forests in national parks have been damaged in Hokkaido as well as in other areas in Japan. The Hokkaido Government has administrated a deer management program based on feedback management theory since 1998 [1998]. Various problems related to population size estimation have been studied intensively [1999, 3735, 4014].

Hunting records indicate that following modern development associated with Japanese colonization of Hokkaido (1870s), overexploitation of sika deer (for meat and hides) combined with habitat destruction and severe winters caused populations to decline to a threatened level [1998]. Thereafter the deer population recovered because of the extinction of wolves by 1890 [1055], bans on hunting between 1890 and 1900 and between 1920 and 1956, bucks-only hunting between 1957 and 1993 [3734], and land use changes (expansion of pasturage has provided new habitats for deer). Since 1994 female hunting has been allowed, because of severe damages to agriculture and forestry. A new management program was introduced in 1998 based on feedback management theory [1998]. Effects of the feedback management program have been evaluated using population size indices (e.g., spot light survey data) [1998, 3735]. Estimates based on spot light survey data indicate that the target population of the feedback management program in eastern Hokkaido remains at high abundance, whereas the population in western Hokkaido is expanding at a high rate to approach an abundance level similar to that of the eastern population [4014]. The total number of sika deer in Hokkaido is estimated at 330,000–680,000 in 2004 [4014].

In the process of expansion deer have colonized vacant areas where deer populations had been locally extinct [1269]. On the Shiretoko Pen. (one of the recently colonized areas) the deer population exponentially increased and crashed (like in other cases of deer colonization), but then recovered to a level as high as before the crash and oscillated thereafter (Fig. 2) [1270, 1271]. A similar dynamic pattern (build up and crash) also was observed on the islet in lake Toya [1268]; following the crash, the population recovered to a higher peak than that of the first irruption [1266]. Post crash oscillation at high abundance levels has been rarely observed in ungulate populations, and these observations are good case studies to reconsider the traditional irruptive paradigm (which holds that after colonization, populations increase to peak abundance, crash, and then increase to lower peak abundance) [578].

5. Conservation oriented researches and Conclusions

The Japanese Government has been conserving four mammals species (or subspecies; the Iriomote cat, the Tsushima leopard cat, the Amami rabbit, and the Daito flying-fox *Pteropus dasymallus daitoensis*) based on "Law for the Conservation of Endangered Species of Wild Fauna and Flora". Based on monitoring data, the abundance of the Iriomote cat is considered stable, whereas the Tsushima leopard cat population appears to be decreasing, although details are unknown. The abundance of the Amami rabbit has been decreasing in spite of eradication programs for an alien predator (mongoose) [3928]. There is no reliable information about the Daito flying-fox.

Eradication programs for alien species (mongooses on Amami-ohshima Isl. and Okinawa-jima Isl., and raccoons in Hokkaido) are on going in order to remove their impact on native animals. Although the programs monitor abundance, reports on population dynamics have not been published yet.

As we have seen above, studies of mammal populations have mainly focused on pest control or conservation problems in Japan. No studies have been attempted to understand the common features of population dynamics in Japanese mammals. Terrestrial mammalian fauna in Japan is biased towards forest species; 70% of the species mainly inhabit forests [2104]. There may be some common features of population dynamics reflecting habitats and fauna. Comparative studies on population dynamics and studies aiming to understand interactive relationships among mammalian populations in Japan are needed, in addition to studies focused on practical solutions to specific problems.

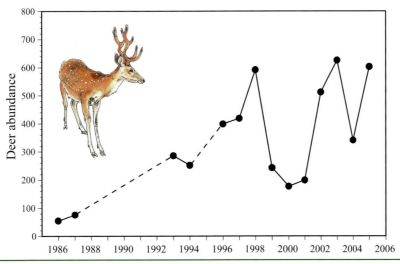

Fig. 2. Change in sika deer abundance after colonization on Cape Shiretoko, Hokkaido, Japan based on photographic censuses. (redrawn based on the data in [1270, 1271]; illustration by H. Ohtaishi).

Artiodactyla SUIDAE

Red list status: LP for *S. s. riukiuanus* on Tokuno-shima Isl. (MEJ); R for *S. s. riukiuanus* on the Ryukyu Isls., C-2 for *S. s. leucomystax* in Honshu, Shikoku, and Kyushu but LP for the populations on Nakadouri Isl. of the Goto Isls. and Awaji-shima Isl. (MSJ); LC (IUCN)

Sus scrofa Linnaeus, 1758

EN wild boar   JP イノシシ（inoshishi）   CM 野猪   CT 野豬   KR 멧돼지   RS кабан

A female and its daughter in Mt. Rokko, Hyogo Pref. (Y. Kodera).

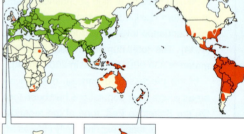

1. Distribution

Native wild boars inhabit a wide area of Europe, Northern Africa, the Near and Middle East, Indian Pen., Southeast Asia, East China, parts of Central Asia, Far East Russia, and Indonesia [426]. There are 2 subspecies in Japan [3432]. One is *S. s. leucomystax* which occurs on Honshu (except Aomori, Iwate, Akita, and Yamagata Prefs.), Shikoku, Kyushu, and Awaji-shima Isl., and the other is *S. s. riukiuanus* which occurs on the Nansei Isls. [1212].

Feral pigs (*S. s. domesticus*), cross-bred pigs, and wild boars have been introduced to North and South America, Australia and many Pacific isls. [426, 777, 3433]. In Japan there are cross-bred feral pigs in parts of Hokkaido [1212, 3432]. Feral pigs on Otōto-jima Isl. (the Ogasawara Isls.) were exterminated in 2007 [1218].

2. Fossil record

Fossils have been found throughout the current distribution range, as well as in Hokkaido [2004].

3. General characteristics

Morphology   Biometric studies with detailed statistics for wild boars are scarce in Japan. A simple description of body measurements for *S. s. leucomystax* is as follows [1345]. Females reach maxima of 50 kg in body weight (undressed weight; UBW), 125 cm in head and trunk length, 75 cm in height at withers (HW), and 95 cm in chest girth (CG) after 3.5 years old. Males continue to grow with age and reach 100 kg in UBW, 145 cm in head and trunk length, 120 cm in CG at the age of 5.5 years old. In general, the body weight of *S. s. leucomystax* males is greater than 100 kg while that of *S. s. riukiuanus* is not more than 40–50 kg, even in males [1807].

Dental formula   I 3/3 + C 1/1 + P 4/4 + M 3/3 = 44 [29, 865, 1011].

Mammae formula   4–5 pairs: usually 5 pairs (the standard mammae formula can not be used in this species) [1807].

Genetics   Chromosome: $2n = 36$ for the western European subspecies, $2n = 38$ for the eastern European population, and $2n = 36$ to 38 for Central and East Asia [1807, 2862]. Data on mtDNA have been reported by many authors [1091, 2347, 2689, 2690, 3848–3850].

4. Ecology

Reproduction In a wild population of Shimane Pref., western Japan, the average number of embryos was 5.13 ± 1.55 SD ($n = 15$) in the early stages of pregnancy (embryo weight was under 160 g) and 5.00 ± 1.05 SD ($n = 10$) in the late stages (160 g and up) (the author's database of 1997–2003).

Wild boars in French forests have 2 types of farrowing schedules [2033]. One is unimodal with a peak in April and May, while the other is bimodal (the first peak occurs in January and February, the second one in August and September). In Japan, the bimodal type (the first peak in April and May, the second one in September) has been observed on a single ranch [1338], but the unimodal schedule (1 peak in May) has been observed in a wild population [1346, 1658, 3656]. The average litter size is between 3.05 and 4.60 in Europe [547, 548, 2033] and is influenced by weight and age [2033]. In Japan, the average litter size reported on ranches is 4.19 ± 0.16 SE in 1 case [1338] and 4.4 in another [497]. The gestation period is about 120 days (119 days, $n = 18$ [2033], 117.1 ± 0.85 SE days [1338]).

Lifespan For Japanese wild boars, the average lifespan is reported as 13.8 months in males and 16.7 months in females in a population of Kanmuriyama, Chugoku District [1345]. In Europe the average male lifespan is reported as 21 months, and the female lifespan as 24 months in the wild without hunting [1234]. Wild boars less than 3 years old are characterized by high mortality [1234]. The maximum recorded longevities of captive males and females are 20 years and 11 months, and 20 years and 7 months, respectively [3862].

Diet Although omnivorous, wild boars are primarily vegetarian, usually eating roots, herbs, fruits, grains, and berries [426]. The seasonal food habits of wild boars in Japan have been recorded as follows: Kinki District region population: subterranean stem-roots and rhizomes in winter [202]; Chugoku District region population: acorns in winter [1346]; Shimane Pref. population: bamboos in spring, dicotyledons in summer, acorns in autumn, and tubers in winter [1654, 1655].

Habitat Wild boars occupy an extremely wide range of habitats: mixed forests, marshlands, steppes, deserts, mountains and croplands [426]. Wild boars in Shimane Pref. prefer abandoned paddy fields, broad-leaved forests, and bamboo thickets, but avoid coniferous plantations [1656].

Home range In Japan, information on home range is scarce. In Shimane Pref., home range in summer is recorded as 81.4–132.4 ha ($n = 3$) [1657]. In general, the home range of the wild boar is influenced by age, sex, the availability of food resources, population density, human interference, and large predators [1978, 3298]. Generally infants and their mothers have small home ranges (10 to 20 ha), and the juvenile home range is still smaller than the adult home range of a virgin female [3298]. Adult males settle in their home ranges in areas overlapping several female home ranges [3298].

A wild boar walking on a farm road in Hokuto, Yamanashi Pref. (J. Shimogai).

A wild boar eating acorns in a broad leaf forest on Mt. Rokko, Hyogo Pref. (Y. Kodera).

Behavior Almost unknown for wild boars in Japan.

In Europe, it is known that activity of the wild boar is influenced by temperature, time of sunset and/or sunrise, circadian rhythms, and ultradian rhythms that are close to a 3-hour interval and to multiples of this (6 hours, 9 hours and so on) [302, 415, 2912].

Natural enemies Before its extinction, the wolf (*Canis lupus*) was a major predator in Japan [1347].

Parasites The tick *Haemaphysalis flava* has been collected from wild boars [598]. An acanthocephalan (*Macracanthorhynchus hirudinaceus*) is recorded as an endoparasite [1293].

5. Remarks

A major big game species in southern Japan. They damage agricultural crops in Japan [494].

Y. KODERA

Artiodactyla   BOVIDAE

Red list status: LP for Kyushu population (MEJ); C-2, LP for Shikoku and Kyushu populations (MSJ); LC (IUCN)

Capricornis crispus (Temminck, 1836)

EN Japanese serow **JP** ニホンカモシカ (nihon kamoshika), カモシカ (kamoshika)
CM 日本髭羚 **CT** 日本長鬚山羊 **KR** 일본산양 **RS** японский сероу

An adult female in early spring on the Shimokita Pen., Aomori Pref. (K. Ochiai).

1. Distribution

Endemic to Japan: Honshu, Shikoku, and Kyushu.

2. Fossil record

Fossils of *C. crispus* or *Capricornis* sp. are recorded from the Late Pleistocene in Aomori, Iwate, and Ehime Prefs. [1491, 2375, 3204].

3. General characteristics

Morphology   Goat-sized with sturdy legs and a short bushy tail. Variable in coloration from whitish to grayish to darkish brown. Northern populations generally have light colored coats. Since sexual dimorphism in body size is minimal [2115], sex is determined from visible external genitalia in the field [1600]. The body stops growing by 3 years, but horns grow continuously throughout life [2115]. Horn rings are useful for age determination [2114]. Both sexes have large preorbital glands. The glands are a little heavier in males, and a sexual difference in sebaceous gland type has been reported [1653, 4125]. Serows also have interdigital and preputial glands [230–232].

External measurements (mean ± SD) of adults (≥ 3 years old) from central Japan culled during November to March were as follows: males: BW = 35.9 ± 4.5 kg (n = 292), CG = 80.8 ± 5.8 cm (n = 281), SH = 73.1 ± 5.0 cm (n = 290), HL = 13.2 ± 1.2 cm (n = 281); females: BW = 38.4 ± 5.0 kg (n = 279), CG = 82.6 ± 7.0 cm (n = 268), SH = 74.0 ± 6.0 cm (n = 275), HL = 13.0 ± 1.3 cm (n = 262) [2115]. Winter weight loss of both sexes ranged from 5% to 23% [2119]. Mandible sizes were also reported [2115].

Dental formula   I 0/3 + C 0/1 + P 3/3 + M 3/3 = 32 [29, 1011].

Mammae formula   0 + 0 + 2 = 4 [29, 1011].

Genetics   Chromosome: $2n$ = 50, with 5 pairs of metacentrics and submetacentrics [3289]. A complete sequence of the mtDNA control region has been reported [2686]. Five mtDNA haplotypes and 6 serum protein types have been found [1996, 2074, 2485]. A sex identification method using the amelogenin gene from feces is reported [2487]. A microsatellite marker assay for individual identification has been developed [2486].

4. Ecology

Reproduction   Peak conception occurs from late October to early November [1611, 3321]. Based on behavioral studies, the rutting season is September to November [1602, 2557]. Gestation lasts about 210–213 days [1140, 1683]. Single kids are usually born in May–June: 95% (n = 37) in Akita Pref. [1602] and 96% (n = 26) in Aomori Pref. [2559], northern Honshu. Twins are rare:

Capricornis crispus (Temminck, 1836)

Table 1. Home range size (ha) of resident adults (Mean ± SD).

Area (elevation)	Male	Female	Method*	Reference
Akita Pref. (150–574 m)	15.2 ± 1.1 ($n = 53$)	10.4 ± 0.6 ($n = 62$)	100% MCP	[1606]
Shimokita Pen., Aomori Pref. (0–240 m)	16.6 ± 6.2 ($n = 16$)	10.5 ± 3.6 ($n = 22$)	100% MCP	[2564]
Nagano Pref. (1,000–2,127 m)	18.4 ± 9.1 ($n = 6$)	11.9 ± 2.5 ($n = 7$)	100% MCP	[2336]
Asahi Mts. (500–1,000 m)	24.4 ± 8.1 ($n = 6$)	15.7 ± 10.2 ($n = 4$)	100% MCP	[1629]
Yamagata Pref. (200–1,000 m)	99 ($n = 16$)	87 ($n = 15$)	90% MCP	[2685]
Kamikouchi, Nagano Pref. (1,500–2,000 m)	109.2, 88.7 ($n = 1$)	—	100% MCP, 95% MCP	[3529]

* MCP: Minimum Convex Polygon

0.8% in 259 pregnant females [1611]. A relationship has been observed between horn growth and reproductive history in females [2118]. The mean fertility or reproductive rate of adult females in 2 studies of natural populations was 72% [2118] and 81% [1602]. Both males and females become sexually mature at about 2.5 to 3 years [3320, 3574], although ovulation and pregnancy were observed in about 10% and 1.5%, respectively, of 68 yearling females [1610]. Age at first reproduction ranged from 2 to 5 (4 on average) years [2118].

The serow is fundamentally monogamous. Home ranges of adult females and males overlap greatly, and mating units can be recognized by patterns of intersexual range overlap. In 2 studies pairs composed of 1 male and 1 female accounted for 81% and 71% of the total mating units; units of 1 male with 2 females accounted for 19% and 25%; and units of 1 male with 3 females accounted for 0% and 4% [1606, 2564]. The mean range size of polygynous males with 2 females was nearly twice that of monogamously paired males, because the ranges of the males covered 2 separate female ranges [1606]. While polygynous units were not as stable as monogamous pairs in Akita Pref. [1606], there was no significant difference in mating-unit retention between mating-unit types in Aomori Pref.: the mean duration of pairs was 3.3 ± 1.4 SD years ($n = 15$) and that of polygynous units was 2.6 ± 1.9 SD years ($n = 8$) [2564]. Changes in mating units occurred following the disappearance of territory holders or the transformation of home ranges [1603, 1606, 2559].

Lifespan The reported maximum longevity is 20 to 24 years, and the average lifespan is 5 to 6 years for both sexes in the wild; there are no sexual differences in age-specific survival rates [2101, 2116, 3586]. The mortality rate of kids within the 1st year was 0.56 ($n = 86$) in Akita Pref. [1602] and 0.37 ($n = 38$) in Aomori Pref. [2559].

Diet Serows are primarily browsers. They feed mainly on leaves and twigs of deciduous broadleaved trees and forbs [2560, 3456]. Conifers, evergreen broadleaved shrubs, and dwarf bamboos (*Sasa* spp.) are also important in the winter diet depending on habitat [89, 376, 1662, 2138, 3455, 3456, 3964]. There is little sex-based difference in diet [3362]. A comparative analysis of feeding ecology and digestive systems in the Japanese serow and the Mongolian gazelle (*Procapra gutturosa*) has been reported [1237].

Habitat The serow inhabits mainly deciduous broadleaved forests, as well as some broadleaved-coniferous mixed forests and coniferous forests, from low mountains to the subalpine zone, includings regions with heavy snow. The distribution largely overlaps that of the cool temperate forest zone in Japan. Although heavy hunting drove Japanese serows into the high mountains (1,500–2,700 m) in the past, their distribution is expanding once again into lower mountains.

Home range All territorial aggressive chases occur between animals of the same sex [91, 1606, 1628, 1629, 2558, 2564]. Adults of both sexes maintain intrasexually exclusive territories, defending most of the home range throughout the year [1606, 2558, 2564]. The mean duration of territory retention was 11.7 years for females ($n = 15$) and 12.4 years for males ($n = 11$) [2564]. Home range size varies depending on habitat, and is larger for males than for females (Table 1).

Behavior Serows alternate activity and rest every few hours, day and night [3230]. They have been observed to use 40% to 50% of their home range area in a 24-h period, and over 80% in a 72-h period [1603]. Territory holders frequently rub their preorbital gland against leaves, branches, trunks, rocks, and other projections. This marking behavior has been observed throughout the year, with higher frequencies in the rutting season, and among males [2554]. Offspring or wanderers begin to scent-mark frequently when they obtain their own territories [1605]. There is no evidence that serows scent-mark along territory boundaries with preorbital glands, horn rubbing, fecal pellets, or urine [2336, 2558]. Early mother and kid behavior [1601], suckling [1602, 2559], interactions between adults and offspring [1602, 2565], social behavior [1602, 1628, 1629, 1990, 2336, 2559], and feeding behavior in cultivated fields [446, 447] have been reported.

The serow is usually solitary, but sometimes occurs in groups of up to 4: 68–79% were single animals, 18–26% in twos, 2–5% in threes, and 0–2% in fours [90, 774, 1602, 2452, 2559, 2984]. In Aomori Pref. ($n = 2,397$), groups composed of 1 adult female and her offspring (0–2 years) accounted for 74% of groups numbering 2 to 4 animals. When groups of mother-offspring-adult male and groups of adult male-adult female are added to the above, the total accounts for 92% [2559].

Kids usually accompany their mothers. Yearlings begin to become independent, but they still remain within the mothers' territories. Both sexes are predominantly dispersers, although a little female-biased philopatry has been found [1599, 1602, 1603, 2557, 2565]. In Aomori Pref., 3 offspring (9%; 1 male and 2 females) remained philopatric following the disappearance of territory holders, and 30 offspring (91%; 16 males, 14 females) dispersed from the natal home ranges [2565]. In Akita Pref., 26 offspring (81%; 20 males, 6 females) dispersed, and 6 (19%) female offspring established territories within their natal home ranges following the disappearance of their mothers or by taking over part of their mothers' ranges [1599, 1602, 1603]. Mean age ± SD at dispersal was $2.9 ± 0.8$ years ($n = 16$, range 2.00–4.50) for males and $2.8 ± 0.9$ years ($n = 14$, range 2.00–4.25) for females [2565]. Natal dispersal can be explained by competition rather than by inbreeding avoidance [2565]. The longest dispersal distance in a study of 4 juveniles was 4 km by a male [2685].

Natural enemies Probably not at risk of predation, although feral dogs and Asian black bears (*Ursus thibetanus*) could be predators.

Parasites & pathogenic organisms
Ectoparasite
Sarcoptic and chorioptic mange are reported [2573, 3184, 3429]. Ticks: *Ixodes acutitarsus, Haemaphysalis kitaokai, H. japonica, H. megaspinosa, H. flava* [3728, 3973].
Endoparasite
Trematoda: *Ogmocotyle capricorni, Dicrocoelium dendriticum* [3376]. Cestoda: *Moniezia monardi* [3376]. Nematoda: *Okapinema japonica, Skrjabinema* sp., *Trichuris discolor, Protostrongylus shiozawai*, 2 species of *Onchocerca* [3376], *Cercopithifilaria bulboidea, C. minuta, C. multicauda, C. shohoi, C. tumidicervicata* [3729], *Loxodontofilaria caprini* [3727]. Conoidasida: *Eimeria capricornis, E. kamoshika, E. naganoensis, E. nihonis, E. serowi, E. crispus* [413, 1040].
pathogenic organisms
Parapoxvirus infection is occasionally severe [989, 1038, 1974, 3376, 3377]. *Toxoplasma gondii, Leptospira interrogans, Chlamydia psittaci*, Rotavirus, Akabane virus [1587], *Lyme borreliosis* [3328], Bovine viral diarrhea virus [799].

5. Remarks

The serow is no longer considered game, but was an important game species in the past. After continuous poaching, it has been strictly protected as a Special Natural Monument since 1955, and has rapidly recovered its numbers in most of Honshu. Culling for control under special license began in 1978, because damage to conifer plantations and crops became a problem in the 1970s. Around a thousand individuals are culled in some areas each year. By 1989, 13 protection areas for the serows had been established, ranging in size from 141 km^2 to 2,152 km^2 [2707].

The total distribution range in Japan has been estimated at about 34,500 km^2, with total numbers estimated at between 60,000 to 120,000 animals in the late 1970s [1970]. The population size in 1980–1984 was estimated at about 100,000 [3585]. The number of inhabited grids (5 km × 5 km) increased from 2,947 in 1978 to 5,010 in 2003 [1212]. A population decrease resulting from competition with sika deer (*Cervus nippon*) has been reported in Tochigi Pref. [1662]. Population density is related to food supply and territoriality [2562–2564, 3291]. The mean population density in 174 census areas in 10 prefectures was 2.6 individuals/km^2 [1970]. The density seldom exceeds 20 individuals/km^2.

The genus *Capricornis* includes 3 species: *C. sumatraensis* (mainland serow), *C. crispus* (Japanese serow), and *C. swinhoei* (Formosan serow) [2518]. According to Grubb (2005) [740], there are 6 *Capricornis* species; mainland species include *C. milneedwardsii, C. rubidus, C. sumatraensis*, and *C. thar*. Molecular phylogenetics shows that the Formosan serow is an independent species, and that the mainland serow and Formosan serow have the closest relationship; the Japanese serow is distant from the other 2 serows [379, 1855, 2074, 2686]. Serows and gorals are usually assigned to separate genera, *Capricornis* and *Nemorhaedus* (or *Naemorhedus*). A morphological study suggested that the 2 groups should be assigned to the genus *Nemorhaedus* [737], and some authors have followed that suggestion [410, 715, 739]. However, molecular studies suggest that serows and gorals belong to distinct genera [713, 738, 833, 2074].

K. OCHIAI

Mother and her 1-month old kid on the Shimokita Pen., Aomori Pref. (K. Ochiai).

Mother and her 10-months old kid on the Shimokita Pen., Aomori Pref. (K. Ochiai).

A subadult male in winter on the Shimokita Pen., Aomori Pref. (K. Ochiai).

Artiodactyla   BOVIDAE

Red list status: introduced

Capra hircus Linnaeus, 1758

EN feral goat   JP ヤギ (yagi), ノヤギ (no yagi)   CM 山羊   CT 山羊   KR 염소   RS домашняя коза

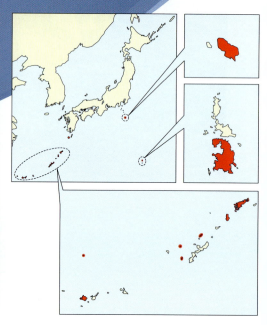

A male feral goat on Nishi-jima Isl., the Ogasawara Isls. (M. Takiguchi).

1. Distribution

Introduced species in Japan. Chichi-jima Isl. (the Ogasawara Isls.); Hachijo-jima Isl. (the Izu Isls.); Kata-shima Isl. (Yamaguchi Pref.); Madara-jima Isl. (Saga Pref.); Yaneo-jima Isl. (the Goto Isls.) [3418]; Ujimukae-jima Isl. (the Uji Isls.) [3472, 3608]; Kuro-shima Isl. (Kagoshima Pref.) [3473] Yaku-shima Isl. (the Ohsumi Isls.); the Tokara Isls.; islands of Amami-ohshima, Kakeroma-jima, Yoro-jima, Uke-jima, Edateku-jima, Eniyabanare-jima, Hanmya-jima, Tobira-jima and Tokuno-shima (the Amami Isls.) [104, 3418]; islands of Iheiya-jima, Yanaha-jima, Aguni-jima, and the Kerama Isls. (the Okinawa Isls.); Iriomote-jima (the Yaeyama Isls.), and Uotsuri-jima (the Senkaku Isls.) [3418].

2. Fossil record

Not reported in Japan.

3. General characteristics

Morphology   HB: $1{,}117 \pm 83$ mm SD ($n = 74$) adult males, $1{,}004 \pm 66$ mm ($n = 103$) adult females. T: 102 ± 17 mm ($n = 74$) adult males, 97 ± 16 mm ($n = 103$) adutl females. HFcu: 251 ± 14 mm ($n = 74$) adult males, 236 ± 12 ($n = 103$) adult females. SH: 625 ± 38 mm ($n = 74$) adult males, 583 ± 29 mm ($n = 103$) adult females. HL: 329 ± 75 mm ($n = 69$) adult males, 121 ± 39 mm ($n = 85$) adult females. BW: 30.9 ± 6.6 kg ($n = 73$) adult males, 23.1 ± 5.3 kg ($n = 100$) adult females. All data were derived from Ani-jima Isl. (the Ogasawara Isls.) [3596–3599]. A wide variety of coat colorations have been observed on the Ogasawara Isls., including white, black, dark brown, dark gray or mixed [2117].

Dental formula   I 0/3 + C 0/1 + P 3/3 + M 3/3 = 32 [2805].

Mammae formula $0 + 0 + 1 = 2$ [4258].

Genetics   Chromosome: $2n = 60$ [2805].

4. Ecology

Reproduction   On the Ogasawara Isls., goats likely breed throughout the year, but most goats give birth between November and April [3208, 3596]. The gestation period is 144–158 days [3581]. Litter size is 1–2 [3208].

Lifespan   12–15 years in captivity [3581].

Diet   Goats are browsers rather than grazers. In New Zealand, they eat a wide variety of plants but have strong preference for *Melicytus ramiflorus* and *Griselinia littorails* [2805]. They readily browse on seedlings, saplings, or epicormic shoots within reach, and litter-fall [2805].

Capra hircus Linnaeus, 1758

Habitat On the Ogasawara Isls., goats live in grasslands, scrub, and forest, but prefer rocky stretches rather than flat land.

Home range On Chichi-jima Isl., female goats exhibit strong fidelity to established home ranges. The typical territory area is from 0.4 to 0.5 km^2. The home range of males is 0.7 km^2 on average. Measured over the long term (more than 1 year), male home ranges are several times larger than those of females [3208].

Behavior Goats are gregarious mammals and avoid isolation. Group size and composition change with population density, disturbance, time of day, and time of year. The basic social unit is the female and her offspring for that year [2805].

Natural enemies Not reported in Japan.

Parasites Three species of ticks, *Boophilus micropus*, *Haemaphysalis flava*, and *H. hystricis* are collected from feral goat [3600].

5. Remarks

Feral goats severely depleted the native vegetation cover on Nakohdo-jima Isl. of the Ogasawara Isls. through overuse and trampling until 1999, when they were eradicated. Depleted vegetation has resulted in massive soil erosion [1216].

On Chichi-jima Isl., goats have caused damage to lemon, papaya, potato and other crops [4260].

M. TAKIGUCHI

A herd of feral goats on Ototo-jima Isl, the Ogasawara Isls. (M. Takiguchi).

Destroyed vegetation and soil erosion caused by feral goats on Nakohdo-jima Isl., the Ogasawara Isls. (M. Takiguchi).

Wild boars are popular big games and nuisance animals affecting agriculture in Japan, except in Hokkaido. Left: 4 hunted wild boars gathered to a meat broker in Hamada, Shimane Pref. (Y. Kodera). Right: a 1.5-year old male boar entering a cage trap for nuisance animals near a potato field in Minami-shimabara, Nagasaki Pref. (Y. Kodera).

Overpopulated sika deer are responsible for the destruction of vegetation in many regions in Japan. Herds of sika deer in Shiretoko, Hokkaido (M. Yamanaka). Left: a herd of females in June, 2009 in the Rusha area. Right: aero-census of deer in February, 2005 on Cape Shiretoko: more than 100 dots on snow are deer.

Body size of sika deer varies greatly according to latitude, showing "Bergmann's rule". Left: a group of sika deer in Nara Park, Nara Pref. (J. Moribe). They are medium sized sika deer in Japan. Right: a 120 kg male (medium size for an adult male in Hokkaido) hunted in Nishiokoppe, Hokkaido (J. Igota). Body size of sika deer in Hokkaido is largest in Japan and it is a popular big game. For scale, the person in the photos is 171 cm tall.

Cetacea

Cetacea BALAENIDAE

131

Red list status: En as *E. glacialis* for the western North Pacific population (MSJ); EN (IUCN); VU as *E. glacialis* (JFA)

Eubalaena japonica (Lacépède, 1818)

EN North Pacific right whale　　JP セミクジラ（semi kujira）　　CM 北太平洋露脊鯨　　CT 北太平洋露脊鯨
KR 북방긴수염고래　　RS японский кит

A surfacing North Pacific right whale in the western North Pacific (Institute of Cetacean Research, Tokyo). V-shaped blowing has been used for species identification from a distance.

1. Distribution

Distributed in the North Pacific, from Taiwan and northern Mexico to the Yellow Sea, the East China Sea, the Sea of Japan, the Sea of Okhotsk, and Bering Sea [1410, 3288].

2. Fossil record

Fossil of a right whale skull from the Holocene was discovered in Akita Pref. 1966 [2496]. A fossil of *Eubalaena* (late Miocene–early Pliocene) found in Nagano Pref. in 1938 was defined as the new species, *E. shinshuensis* [1573].

3. General characteristics

Morphology　Mean body length of physically mature North Pacific right whales is 16 m for females and 15.2 m for males [1411]. Body color is black or slate-black. Most of them have a white irregular shaped patch on the ventral side. Some individuals have a large white pattern which extends to the lateral side of the body. Compared to the balaenopterids, right whales are very large in girth relative to their length. They have a very rotund body and no dorsal fin. The Japanese name of "Semi-kujira" (kujira means whale) is derived from the beautiful dorsal shape of this species. The large head comprises about one-third of the total body length. The jaws are extremely arched with a bumpy upper edge on the lower jaw. Although the head is wide at the eyes, the width of the rostrum is quite narrow. The pectoral fins are short and broad. The tail flukes are broad and isosceles-shaped with a width of 25–30% of the body length [425, 1410]. Right whales lack the ventral grooves that are characteristics of the balaenopterids. The head and rostrum are decorated with callosities which are a characteristic feature of right whales. These are irregular-shaped patches of thickened keratinized tissue, which usually appear white or cream colored. The callosity at the tip of the snout is called the bonnet. Innumerable cyamids, amphipod crustaceans, inhabit the callosities. As the callosity patterns are congenital and unique to individuals, they are useful for individual identification [1535].

Dental formula　There are no teeth; instead there are 205–270 baleen plates on each side of the upper jaw [425]. These are black or dark bluish gray, and seem to darken with age. The baleen plates are narrow and long with a recorded maximum length of 277 cm in a 16.4 m male [2704]. There are very fine fringing hairs on the inner side of the baleen plates that make it easy to strain small prey species from sea water.

Eubalaena japonica (Lacépède, 1818)

Genetics Chromosome: $2n = 42$ [3577].

4. Ecology

Reproduction Females attain sexual maturity at 9–10 years when their body length is 15–16 m, and they give birth in winter [1535, 2704]. Gestation period is roughly 1 year. Calves are 5–6 m long at birth and are weaned by 6 to 7 months [425, 2494]. Although calving interval of this species is not known, long term observation of closely related southern right whales *E. australis* revealed that it ranged from 2 to 7 years with a 3-years interval being the most common [2812].

Lifespan There are few data on the longevity of right whales because an aging method for this species has not yet been found. Photo identification records of a female North Atlantic right whale, *E. glacialis*, suggested she was more than 70 years, however, recent research on closely related bowhead whales *Balaena mysticetus* suggests that they may live even longer [1535].

Diet North Pacific right whales feed primarily on calanoid copepods and sometimes small euphausiids [2704]. Different from rorquals, right whales are typical skimming feeders. They feed by swimming forward with their mouth open. Water flows into the opening at the front, and out through the baleen, straining their prey from the water [1535].

Habitat They migrate between the temperate waters for breeding in winter and subpolar cold waters for feeding in summer season. Historically, two populations of right whales migrated in the waters around Japan, at least prior to the 20th century. One migrated through the waters off the coasts of Mie, Wakayama and Kochi Prefs. The other migrated through waters off the coasts of Kyoto to Yamaguchi Prefs. and northern Kyushu. Both populations moved southwards in winter and northwards in spring [2700]. The breeding and calving grounds are not known, but in the western Pacific they are thought to be around the Ryukyu Isls., in the Yellow Sea, the Sea of Japan, or far offshore areas [1185]. Japanese research in the western North Pacific (JARPN and JARPN II, 1994–2007) recorded relatively small numbers of sightings, with an incidence of 1/10 of the sightings of blue whales [2018]. In recent years, there has been roughly 1 or 2 stranding and incidental catch records a year in Japanese coastal waters. Most of these occurred on the Pacific coast of western Japan in spring [1113].

Behavior Right whales are slow swimmers. They swim at 3.7 to 5.6 km/h during migration and even when startled they swim no faster than 9.3 km/h [2494]. As the blow from each nostril is visible separately, it looks V-shaped when observed from behind [2704]. Right whales, as well as humpback whales, often show aerial behavior above the surface of the water, including breaching, lobtailing and flipper slapping [1535]. The mating system of right whales is promiscuous. Females often mate with more than one male in succession, or even two simultaneously.

Natural enemies Killer whales and large sharks are probably natural enemies. Although there are only a few records of killer whales attacking adult right whales [425], predators may be more likely to attack calves or juveniles.

Parasites No endoparasites are reported from Japan, but heavy infections of cyamids, *Cymus ovalis* and *C. erraticus*, on the callocites are reported [2704]. Diatom *Coconeis ceticola* film on the fluke of a whale is also reported [2704].

5. Remarks

Right whales have long been treated as two species, Northern right whale *E. glacialis* and Southern right whale *E. australis*. The Scientific Committee of the International Whaling Commission (IWC/SC) reviewed right whale taxonomy in 2000. Although morphological differences are nearly absent, the most recent genetic evidence demonstrates that the right whales of the North Atlantic, North Pacific, and Southern Oceans possess fixed, unique genetic patterns indicating complete and long-established isolation. Based on the evidence, the committee recommended that the three populations be considered as separate species [1535]. According to IWC/SC recommendation, this article treats the North Pacific right whale as a full species *E. japonica*.

Whaling of right whales around Japan has a very long history. Annual catches of right whales by net whaling, which used many folds of nets and harpoons, in the first half of the 19th century were estimated at about 50, but they greatly decreased in the second half of the century, presumably by the impact of American whaling around Japan. Like other right whales worldwide, Japanese modern whaling, beginning near the end of the 19th century, severely decreased populations of North Pacific right whales, especially in the Sea of Japan [2700]. All right whale species were protected from commercial whaling in the early 1930s. Today, ship strikes and incidental catches in fishing gear are the most significant human-related sources of mortality, especially among western North Atlantic right whales [1185, 1535].

H. ISHIKAWA

Cetacea ESCHRICHTIIDAE

Red list status: En for the Asian stock (MSJ); CR for the western subpopulation (IUCN); EN for the Asian stock (JFA)

Eschrichtius robustus (Lilljeborg, 1861)

EN gray whale   JP コククジラ (koku kujira)   CM 灰鯨   CT 灰鯨   KR 귀신고래   RS серый кит

A gray whale feeding at the bottom of the sea off Ohshima Isl., Japan (34°47'N, 139°24'E) (K. Nakamura, Japan Underwater Films Co., Ltd.).

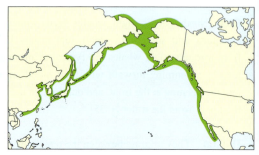

1. Distribution

Found only in the North Pacific Ocean and adjacent seas [170, 1248, 3863]. Although gray whales previously occurred in the North Atlantic, they were extinct by the 17th or the early 18th century [1248]. There are two stocks of gray whales in the North Pacific, one is the eastern (California) stock and another is the western (Korean or Asian) stock. Historic sightings and whaling records indicate that western gray whales occurred in areas of the northern Sea of Okhotsk, off the Korean Pen., the Yellow Sea and the Pacific side of Japan. Although it was suggested that there are two migration routes from the northern Sea of Okhotsk to the south of China, one is along the continental coastline and another is along the Pacific side of Japan [2698, 2699, 2701], recent study revealed the existence of another migration route along the western (the Sea of Japan) side of Japan [170, 1151, 2433, 3950]. Recent records of western gray whales are limited to individuals aggregating to feed along the shallow water shelf of northeastern Sakhalin Isl. and southeastern Kamchatka. Although there are occasional stranding or sightings along the coast of Japan, records on the continental coast are very few [1113, 1151, 1188, 2433, 3864].

2. Fossil record

A specimen from the late Pliocene (1.8–3.5 Ma) discovered in Hokkaido was assigned the oldest record of *Eschrichtius* sp. [974].

3. General characteristics

Morphology The body is more slender than that of the right whale and stockier compared to balaenopterids, which have a more streamlined body. Head is narrow and relatively small in proportion to total body length with two to four short ventral grooves on the throat. There is no dorsal fin but they have a low hump followed by a series of 8–14 small bumps (knuckles) along the top of the peduncle. Widely spaced tactile hairs sprout from small dimples on the upper and lower jaws. The body color is gray with white mottling, which is easily identified on the sea. The body surface is colonized by a lot of barnacles and three species of whale lice (cyamids).

Mean body length at physical maturity is 13.0 m for males and 14.1 m for females at ca. 40 years [2883]. There is a cyst-like fluid-filled structure (10–25 cm in diameter) on the ventral surface of the caudal peduncle. It is assumed to be the sebaceous

Eschrichtius robustus (Lilljeborg, 1861)

gland of land mammals and to function as a track-laying scent gland, although its exact function is unknown [1248, 3887].

Dental formula There are no teeth, but 140–180 yellowish baleen plates per side under the upper jaw [3887]. The number of baleen plates is the smallest in the baleen whales. The left and right baleen rows are separated in the front of the snout. Each baleen plate is short and thick and the fringes are very coarse, which is an adaptation to the unique ecology of bottom feeding in the species (see Diets) [2457, 3887]. The length of the baleen plates in the anterior part of the series is not bilaterally symmetrical and the plates on the right side are shorter than those in the corresponding positions on the left side. It is thought that feeding gray whales scoop the sediment of the bottom mostly with the right anterior region of the baleen plate rows [1400].

Genetics Chromosome: $2n = 44$ [1815].

4. Ecology

Reproduction Several biological parameters are well documented for the eastern stock of gray whales. Sexual maturity is attained at 6–12 years with a mean of 8 years and mean body length of 11.7 m for females and 11.1 m for males. Reproductive cycle of females lasts 2 years. Females come into estrus during about a 3-weeks period in late November and early January. Pregnancy lasts 11 to 13.5 months. Parturition occurs within a period of 5 to 6 weeks from late December to end of February. Newborn calves are about 4.6 m long and weigh 500 kg. The calf is nursed for about 6 months and weaned by July with a body length of 7 m. After weaning the calf, females are in anoestrus for 3 to 4 months until November or December, when most of them go into estrus and commence a new pregnancy. A few that either fail to ovulate, or ovulate but fail to conceive, remain in anoestrus for another year [1179, 3887]. Gray whales are thought to have a promiscuous mating system. Males and females do not form long-term pair associations and both sexes may copulate with several partners during the same breeding season [1248].

Lifespan Although neither maximum nor average lifespan is known, it was estimated to be 40–80 years of age [1249].

Diet The gray whale is unique among large cetaceans in that it feeds primarily upon benthic small crustaceans and other invertebrates. They forage on the ocean floor in shallow waters over continental shelves (4–120 m deep) during summer and autumn when they are in higher latitudes. For eastern gray whales, over 80 prey species have been identified, reflecting their opportunistic approach to feeding and nonselective nature of their feeding mechanism. On the northern summer feeding ground, they primary consume gammaridean amphipods ranging from 13 to 27 mm in length. In some areas, polychaete worms are their main food. Mysids, crab larvae, red crab, mobile amphipods, herring eggs and larvae, squid, megalops and baitfish are also eaten in the peripheral feeding areas south of the main feeding grounds.

When feeding, gray whales roll to one side bringing the head parallel with the seafloor. They then sweep the side of the mouth close over the bottom a few centimeters above it, and open their jaws slightly to suck sediment containing prey into the mouth. Then they strain water, sand and mud through the baleen, leaving the food trapped on the inner margin. They move slowly along the bottom, sucking up infauna in pulses, and surface with mud plumes streaming from the mouth. Although gray whales often use the right side of the head to suck sediment with prey items, the left side is also used. It is also observed that they suck sediment with the tip of snout moving forward. Bottom feeding leaves mouth-sized depressions called "feeding pits" in the sea floor. It is thought that removing the top layers of sediment promotes re-colonization, succession and maturing of the prey community, and hastens the growth and diversity of life on the seafloor. In addition to its bottom feeding abilities, the gray whale is also capable of feeding on pelagic prey by surface skimming and engulfing [1248, 2459].

Habitat The gray whale is the most coastal of all the large whales, and they inhabit primarily inshore or shallow, offshore continental-shelf waters. They make a long migration up to 20,000 km round trip every year largely without feeding, traveling along a nearshore route between a summer feeding zone of high productivity in the Arctic or subarctic waters, and a winter breeding zone in temperate or subtropical southern waters [1248, 3887]. The eastern stock of gray whales feeds mostly in the northern and western Bering Sea, Chukot Sea, and the western Beaufort Sea from the May through September. From November through December, they move south along the coastline of North America to the calving area of lagoons of Baja California. The western stock of gray whales feeds mostly off the northeastern coast of Sakhalin and southeastern Kamchatka in the summer season [1151]. Although there is no information on the breeding ground of this stock, it is presumed to be in the waters of southern China, and might possibly extend to Hainan Island [1151, 1417, 3828]. Resent study of satellite tagging and photo-identification revealed that some of the western gray whales that summered in the Sea of Okhotsk migrated to the wintering ground of the eastern stock of gray whales, raising necessity of a more comprehensive examination of movement patterns and population structure of this species [1191].

Behavior The gray whale is not a fast swimmer. Standard swimming speed of migrating gray whales is 7 to 9 km/h. They develop a speed up to 16 km/h when they are frightened. Individuals usually submerge for 3 to 5 min, then surface and blow five or six times. The bushy blow is 3 to 4.5 m high. They usually make shallow dives of 15 to 50 m, but can dive to depths of up to 170 m [1248, 3887]. During migration and in calving areas, gray whales prefer shallow waters and sometimes swim into the surf zone to rub themselves on beaches. They are active at the surface, and spy hopping, breaching, lobtailing, flipper slapping and boltering (lying on the side while waving a flipper in the air) are commonly performed.

Natural enemies Killer whales are the only known predators for gray whales, although several species of sharks scavenge on carcasses and might kill a small number of calves. Pods of killer whales cooperatively pursue gray whales, especially calves and juveniles, and seem to attack by repeatedly ramming along their sides, grasping the flukes and flippers to immobilize and drown them, and trying to open their mouth to bite into the tongue [1248].

Parasites From western gray whales stranded along the Japanese coast, an ectoparasite *Cyamus scammoni* (whale lice) and an epizoite *Cryptolepas rachianecti* (barnacles) are reported [2323, 3462]. Three helminth parasites, *Ogmogaster antarcticus* (Trematoda), *Diphyllobothrium macroovatum* (Cestoda) and *Priapocephalus eschrichtii* (Cestoda) are also reported [1766].

5. Remarks

The eastern stock of gray whales was severely overexploited to near extinction in the later half of the 19th century and early 20th century. However, following protection from commercial whaling, the eastern stock increased to 26,300 [1183] which is thought to be a nearly pre-exploitation population level. The western stock of gray whales experienced 20–30 annual takes in Japanese coastal waters by hand-harpooning and net whaling until the end of the 19th century [2699], and they were also severely hunted from 1900 to 1930 by modern whalers. As the initial population before 1900s was not so high, which was estimated at 1,000 to 1,500 [2883], it decreased rapidly and was thought to be extinct in the early 1970s [323], but is known to survive today as a remnant population [303, 331, 3863]. The current population size was reported as 112–130 [403]. Of concern is the fact that large-scale oil and gas development projects in the Sakhalin area may threaten a restricted feeding area of the western stock. Incidental catches by fisheries activities, especially in set nets along the coastal line of the migration route are also potential threats. The Government of Japan amended ministerial ordinances to prohibit the sale or possession of products of gray whales in 2008, with associated administrative guidance leading to promote the release of any gray whale bycaught in a set net or stranded alive [1151].

H. ISHIKAWA

A gray whale strayed into Tokyo Bay in May 2005 (K. Tokutake, Yokohama Hakkeijima Seaparadise/Okinawa Churaumi Aquarium). Note a lot of barnacles on the head. This animal died by a set net later and was an immature female with 7.81-m body length.

A replica of blue whale skeleton at Taiji Whale Museum, Wakayama Pref. (M. Yoshioka).

Cetacea BALAENOPTERIDAE

Red list status: LP for the populations in the Sea of Japan, the Yellow Sea and the East China Sea, C-2 for the population in the Sea of Okhotsk and the western Pacific (MSJ); LC (IUCN); ND (JFA)

Balaenoptera acutorostrata Lacépède, 1804

EN common minke whale   JP ミンククジラ，コイワシクジラ (minku kujira, koiwashi kujira)
CM 小须鲸   CT 小鬚鯨   KR 밍크고래   RS малый полосатик, остромордый полосатик

A minke whale is gulping prey in Nemuro Str. (Haruko Sato, Sea Life Watch). This species feeds on swarming zooplanktons (mainly krills) and schooling fish.

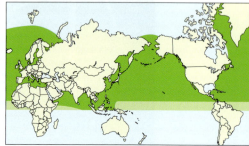

range of "O" stock

range of "J" stock

possible range

1. Distribution

Widely distributed in all northern oceans and neighboring seas. In waters around Japan, they are distributed in the western North Pacific from coastal to pelagic, the Sea of Okhotsk, the Sea of Japan, Yellow Sea and East China Sea. They are also found in Seto Inland Sea (Setonai-kai) but it is uncommon. Studies of genetics, morphology and conception data suggest that there are at least 2 different stocks ('O': Okhotsk Sea–West Pacific stock and 'J': the Sea of Japan–Yellow Sea–East China Sea stock) largely to the east and west of Japan [731, 1186] (see ''Habitat'' in detail).

2. Fossil record

A skull from the Holocene was discovered in the Osaka Plain [2739].

3. General characteristics

Morphology   Smallest among the rorquals and second smallest of the baleen whales. Mean body length of physically mature individuals is 8.0 m for females and 7.5 m for males, which is smaller than the Antarctic minke whale *B. bonaerensis* (8.9 m for females and 8.4 m for males) [1415]. In the North Pacific, recent biological records of the Japanese Whale Research Program under the Special Permit (JARPNII) showed that maximum length of common minke whales was 8.8 m for females [3510] and 8.6 m for males [608]. Body shape is streamlined. The dorsal fin is relatively large and sickle or triangle shaped, and is situated 25–30% of the body length from the notch in the tail flukes. The shape of the head is triangular with a narrow and pointed rostrum. There is a single ridge on the head. Body color is black on the dorsal side and white on the ventral side. In some individuals, the black dorsal color extends onto the left part of the ventral grooves. There is a remarkable white band on each pectoral fin, which is a diagnostic of this species, because the other baleen whales (including the Antarctic minke whale) have monochrome pectoral fins except for the dwarf minke whale in the Southern Hemisphere. Pale streaks often occur from the shoulder to the top of the dorsal side. Ventral side of tail flukes is white with black fringe. There are 40–60 parallel ventral grooves that do not reach the position of the navel.

Dental formula   There are no teeth but 260–300 baleen plates

Balaenoptera acutorostrata Lacépède, 1804

per side under the upper jaw [2457]. They are yellowish white and some individuals have a narrow black stripe on the outer side of the plate as is also seen in the Antarctic minke whales.

Genetics Chromosome: $2n = 44$ [3577].

4. Ecology

Reproduction Conception season of the common minke whales in the North Pacific is different between the 2 stocks in the 'O' stock and 'J' stock of Japan (see Distribution), former is winter and later is autumn [1415]. Gestation period is estimated at 10 months and body length at birth is 2.4–2.7 m [1415, 2823]. Lactation lasts 4–5 months. Calving interval is estimated at 1.2 years. The mean body length at sexual maturity is 6.3 m for males and 7.1 m for females [1415]. The age at sexual maturity is estimated at 6 years for males and 7.1 years for females [3307].

Lifespan Although 62 years of age was recorded as the maximum age for the Antarctic minke whale, about 50 years of age is a good estimate for the average lifespan of the common minke whale [1409].

Diet Minke whales are known as catholic feeders. They feed opportunistically on the most convenient and abundant food in the area at the time [1359]. Although the Antarctic minke whale mainly feeds on krill (*Euphausia superba*), the common minke whale feeds on various kinds of fish, squids, copepods and krill. Japanese research (JARPN 1994–1999) found geographical and temporal changes in feeding behavior of common minke whales in the western North Pacific. In the east of Japan, Japanese anchovy (*Engraulis japonicus*) was the most important prey species in May and June, while Pacific saury (*Cololabis saira*) in June and August and krill in September was the most important species. In the southern area of the Sea of Okhotsk, krill was the most important prey species in July and August [1186]. Walleye pollock (*Theragra chalcogramma*) and sand lance (*Ammodytes personatus*) are also important prey species in coastal waters. Feeding of minke whales is categorized as "swallowing" type. They feed on swarming zooplankton and schooling fish, pursuing single prey species aggregations [2457, 3508]. Analysis of stomach contents of common minke whales in the western North Pacific suggested that there was little diurnal change of feeding activity, whereas Antarctic minke whales showed diurnal changes in feeding activity depending on dispersion and/or vertical migration of krill [3508].

Habitat As in most of other large baleen whales, minke whales migrate between the temperate waters for breeding in winter and polar waters for feeding in summer season. It is known that the distribution of minke whales on feeding grounds is segregated by sex and maturity status; pregnant females tend to gather in higher latitude waters to the ice edge [1182, 3803]. Little is known about the breeding area, but they are probably located in low latitudinal offshore waters. Estimated migration route of 'O' stock is as follows; 1) immature males and females migrate into the northern coastal area of Japan (Miyagi and Iwate Prefs.) in April, then disperse to the coastal waters of the Pacific and the Sea of Okhotsk side of Hokkaido in summer, 2) mature males migrate through coastal and offshore waters from May to September, 3) mature females enter the Sea of Okhotsk in April and May, after which they move further north to the middle and northern area of the Sea of Okhotsk [834]. On the Pacific side of Japan, immature 'O' stock animals were found mainly in coastal areas whereas mature animals were found mainly in offshore areas. On the other hand, estimated migration route of 'J' stock is as follows; 1) northward migration begins in January–February, 2) pregnant females migrate into the southern part of Okhotsk Sea in April, and they start southward migration in July, 3) adult animals are distributed in offshore waters in the Sea of Japan, whereas juveniles stay close to the coasts of Japan and Korea almost year around [2806]. Recent analysis of bycatch individuals suggested 'J' stock animals are widely distributed on the southwestern Pacific coastal side of Japan [1190].

Behavior Common minke whales do not form close social schools. Although they commonly form a relatively large group in high latitudinal feeding areas, most sightings around Japan are of solitary individuals. Compared to other baleen whales, common minke whales are difficult to spot at a distance because the blow is small and inconspicuous and surfacing behavior is brief. They sometimes approach drifting vessels. Common minke whales are thought to be more tolerant of ship noise than Antarctic minke whales, presumably because of the greater volume of ship traffic in the North Pacific [1111]. Breaching is observed occasionally in feeding areas where large groups are formed. Common minke whale sounds include low frequency grunts (80–140 Hz, 165–320 ms duration), thumps (100–200 Hz, 50–70 ms duration), ratchets (850 Hz, 1–6 ms duration) and pinglike sounds and clicks at various frequencies extending to over 20 kHz [3884].

Natural enemies Killer whales are the natural enemy. In the Antarctic, it was observed that 85% of the killer whales diet was Antarctic minke whales [337]. Large sharks might attack a calf of minke whales.

Parasites An acanthocephalan *Bulbosoma nipponicum*, cestodes *Diphyllobothrium macroovatum*, *Diplogonoporus balaenopterae*, *Tetrabothrius* spp., trematode *Lecithodesmus goliath* are reported from Japan [1298, 1763, 3976]. Heavy infestation of *Anisakis simplex* is common in the western North Pacific minke whales' stomach and this species is considered a final host of this nematode [1768].

5. Remarks

Although minke whales had been classified as a single species *B. acutorostrata*, recent morphological and genetic studies suggest that they should be divided into 2 species. IWC recognizes the common minke whale *B. acutorostrata* in the northern hemisphere and the Antarctic minke whale *B. bonaerensis* in the southern hemisphere in 2000. IWC also recognizes an unnamed subspecies of the common minke whale based on morphology, genetics and distribution, which occurs in the southern hemisphere and is termed the "dwarf minke whale" [1184]. Some scientists recognize 3 subspecies of the common minke whale: the North Atlantic minke whale (*B. a. acutorostrata*), the North Pacific minke whale (*B. a. scammoni*), and the dwarf minke whale [2882].

Until the introduction and spreading of modern whaling in the 1920s, common minke whales were taken in small numbers by local people on both the Pacific and Atlantic coasts. Even in the modern whaling era, they were not the primary species pursued. Exploitation of common minke whales expanded as abundance and availability of larger baleen whales declined, as was true for the Antarctic minke whales in Southern Ocean [3307]. In Japanese waters, substantial commercial whaling of common minke whales started after World War II. Almost all Japanese minke whaling was operated by land-based whaling within 50–80 nautical miles of land stations. Annual catch of 300–400 had continued until 1987 when the Japanese government accepted the moratorium of commercial whaling decided by IWC [834, 1409]. Since 1994 the Japanese government has conducted the Japanese Whale Research Program under the Special Permit in the western North Pacific (JARPN and JARPNII) in which 100–220 common minke whales have been taken annually. Common minke whales are often caught incidentally by set net and other fishing gears in coastal water. In Japan, most incidental catches occur in set nets and more than 100 cases are reported every year. Relatively large numbers of catches are caused by other fishing gear such as gill nets and trawls in Korea (e.g. [1113, 1189]).

H. ISHIKAWA

A minke whale swimming with numerous seabirds in Nemuro Str., Hokkaido (Haruko Sato, Sea Life Watch).

Cetacea BALAENOPTERIDAE

134

Red list status: V (MSJ); EN (IUCN); DC (JFA)

Balaenoptera borealis Lesson, 1828

EN sei whale   JP イワシクジラ（iwashi kujira）   CM 鰛鯨   CT 鰮鯨   KR 보리고래
RS сейвал, ивасёвый кит, сайдяной кит

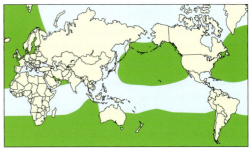

A surfacing sei whale in the western North Pacific (Institute of Cetacean Research/Kyodo Senpaku Kaisha, Ltd.). Laterally angled head and tall dorsal fin are characteristics of this species among the rorquals.

1. Distribution

Widely distributed from temperate to subpolar waters, though they are rarely found in neighboring seas. As the seasons are opposite in each Hemisphere, there is little opportunity for contact between the northern and southern populations. Consequently it is believed that sei whales are genetically isolated between Northern and Southern Hemispheres [3804]. In the North Pacific in summer, sei whales can be found from California to the Gulf of Alaska and across the Bering Sea and down to off Japan. In winter, the centers of abundance move southward to around 20°N [700]. Sei whales pass through offshore water during seasonal migration. Japanese research (JARPN and JARPNII, 1994–2002) shows that sei whales in the western Pacific are found north of 38°N from May to June and north of 40°N from July to August, which suggests the southern limit of distribution in this season [607]. They rarely occur in the Sea of Japan, East China Sea, Yellow Sea and the Sea of Okhotsk.

2. Fossil record

Some fossils of this species were discovered from Hokkaido, Fukushima and Ibaraki Prefs. Estimated geological epoch of those fossils was from the late Pliocene to the late Pleistocene [2661].

3. General characteristics

Morphology The third largest whale in the balaenopteriids. Like other large baleen whales, sei whales in the Northern Hemisphere are smaller than those in the Southern Hemisphere. In the North Pacific, mean body length of physically mature individuals is 15.2 m for females and 14.3 m for males [1977]. Recent biological records of the Japanese Whale Research Program under Special Permit (JARPNII) showed that the maximum length of sei whales was 16.3 m for females [3509] and 15.1 m for males [3510]; the maximum body weight was 31.0 t for females and 24.4 t for males [607]. Body shape is streamlined, with a relatively large dorsal fin that was located 28% of the body length from the notch in the tail flukes. The wide and tall dorsal fin is one of characteristics of the sei whale, with a height of 3.1–4.6% (mean 3.8%) of the body length. The head is large and long and occupies 22.7–25.7% (mean 23.9%) of the body length [607]. As the median crest of the head rises toward the position of the blow holes, the upper jaw of the head inclines more forward and laterally compared to other rorquals that have a relatively flat head. Body color is dark gray on the dorsal side and white on the ventral side. Ventral side of pectoral fins and tail flukes is gray. There are 32–60 parallel ventral grooves that which do not reach the position of the navel.

Dental formula There are no teeth but 320–380 (mean 340) baleen plates per side under the upper jaw of the North Pacific sei whale [2457]. They are black colored and relatively narrow and triangular in shape with fine inner fringe hairs.

Genetics Chromosome: $2n = 44$ [1374]. Analysis of microsatellite DNA indicated a single population inhabiting the western North Pacific [1190, 1308].

4. Ecology

Reproduction Gestation period was estimated at 10.5 months in the North Pacific [1977], and 12 months in the Antarctic [698]. In the North Pacific, parturition may occur in November and body length of calves at birth is 4.4 m. They are weaned after 6–7 months of age with 9 m of body length. As the resting phase of ovulation cycle after parturition is estimated at 6.5 months [1977], females may give birth every 2 or 3 years. The breeding ground is probably in low latitudinal offshore waters. Body length at sexual maturity is estimated at 12.7–13.1 m in males and 13.3–13.7 m in females [607]. The age at sexual maturity is thought to be shortened from 10 years in 1925 to 7 years in 1960 because the abundance of larger whales decreased by heavy exploitation and thus food availability was improved for this species [700, 1977].

Lifespan The maximum age determined from annual growth layers in the earplug is 60 years old [700].

Diet In the North Pacific, sei whales feed mainly on copepods (small crustaceans), followed by euphausiids (krill) [1495]. They also feed on schooling fish such as Japanese anchovy, Pacific saury and chub mackerel and squids. Prey species of North Pacific sei whales vary with area and season. In general, they prefer copepods and euphausiids north of 40°N and fish to the south of 40°N from spring to summer. In fall, they prefer squids north of 40°N and fish to the south of 40°N. Although the dominant prey species changes through the years, Japanese Research (JARPNII, 2002–2003) showed that the Japanese anchovy was the most important fish species for North Pacific sei whales [607]. Recent surveys indicated mackerel is also an important fish species for them [252]. Feeding of sei whales is categorized as "swallowing" and "skimming" type [2457]. They use both these feeding methods and are not greatly specialized to favor one method over the other.

Habitat Like most of the other large baleen whales, sei whales migrate between temperate waters for breeding in winter and high latitudinal waters for feeding in summer, although they do not enter so far into icy water like other rorquals such as blue, humpback and minke whales. Sei whales around Japan are distributed far off the Pacific coast and pass through offshore water during seasonal migrations. There are scarcely stranding or incidental catch records of sei whales in the coastal waters of Japan. Only 3 reliable records exist until 2013, in Mie Pref. in January 1964, in Nagasaki Pref. in June 2003, in Tokyo bay in October 2013 [1112, 1113]. Although there are many local records of the sei whale in the Sea of Japan, most of them are considered misidentifications of minke whales because local fishermen often call minke whales (Japanese name, "koiwashi kujira") sei whales ("iwashi kujira").

Behavior Sei whales do not form socially close schools and tend to travel alone or in a pairs, although they commonly form a larger group in the feeding area. The blow is columnar and 3–4 m tall. Sei whales are thought to be fast swimmers attaining speeds of 55 km/h over a short time [1486]. Sei whales tend to be shallow swimmers, with their head seldom emerging, and with no positive arching on diving [607, 947].

Natural enemies Although large sharks might attack a calf of the sei whale, killer whales *Orcinus orca* are the only natural enemy [947].

Parasites An acanthocephalan *Bolbosoma turbinella* and a cestode *Diplogonoporus balaenopterae* are reported from Japan [181, 1763].

5. Remarks

The sei whale was not exploited until the era of modern whaling because it swam fast and its dead body sank immediately after being harpooned. Since the end of the 1800s, large numbers of sei whales were caught worldwide by both land-based and pelagic whaling. The largest catches were made by the Antarctic pelagic fleets in the late 1960s, after the numbers of blue and fin whales had been reduced. In Japanese waters, the catch of sei whales started from the 1890s by land-based whaling, followed by pelagic whaling from 1940. In Japan, sei whales were confused with Bryde's whales until 1954 [607, 2696]. Commercial catch of sei whales in the North Pacific had been regulated since 1968, and has been banned since in 1976 by IWC [2637]. Since 2002, Japan has taken 50–100 sei whale a year in the Second Phase of Japanese Whale Research Program under the Special Permit in the western North Pacific (JARPN II).

H. ISHIKAWA

Cetacea BALAENOPTERIDAE

135
136

Red list status: V as *B. edeni* for 1 or 2 populations from the Tosa Bay to the East China Sea, C-1 for the population in the western North Pacific (MSJ); DD (IUCN); R for the Eastern China Sea population as *B. edeni*, ND for the population in the western North Pacific as *B. edeni* (JFA)

Balaenoptera brydei Olsen, 1913

EN Bryde's whale JP ニタリクジラ (nitari kujira) CM 布氏鲸 CT 布氏鯨
KR 브라이드고래 RS полосатик Брайда

Balaenoptera edeni Anderson, 1878

EN Eden's whale JP カツオクジラ (katsuo kujira) CM 鳀鲸 CT 鯷鯨
KR 에덴고래 RS полосатик Эдена

A Bryde's whale (cf. *Balaenotera edeni* = katsuo kujira) of inshore North Pacific population (A. Mochizuki). Three ridges on the head are clearly seen. Note that the location is close to the shore, off Kochi Pref.

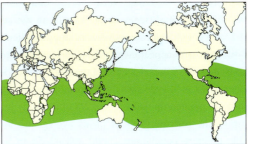

B. edeni
B. brydei

1. Taxonomic remark

The authors are of the opinion that 2 taxa, nominal *B. brydei* and *B. edeni*, are distinct at a certain taxonomic (possibly specific) level; however, the relationships among each of the 2 entities defined by the specific descriptions and biological characters of the known populations (at least 4) are still unresolved (see 6. Remarks). Thus, the exact classification of these 2 taxa following the code of nomenclature is hardly possible at this moment and hence they are described in a single chapter here. Although provisional expressions *B. e. edeni* and *B. e. brydei* were introduced [398], in the following descriptions these groups are treated as *B. edeni*/*B. brydei* complex ("Bryde's whales"), but whenever discrimination is necessary or possible, the population identity is indicated.

2. Distribution

North and South Pacific, North and South Atlantic and Indian Oceans within the area between 40°N and 40°S [1418]. Note that the distribution in the world map is an approximate range of "Bryde's whales".

3. Fossil record

Not reported in Japan.

4. General characteristics

Morphology "Bryde's whales", consisted of *B. edeni* and *B. brydei*, have a typical rorqual appearance. External morphology

A North Pacific offshore Bryde's whale surfacing (Institute of Cetacean Research/Kyodo Senpaku Kaisha Ltd.). Note the three distinct sagittal ridges and the bushy (and V-shaped) blow.

is quite similar to the sei whale, and they had not been distinguished in many areas including North Pacific until the late 1950s [2696]. Three sagittal ridges on the head, 1 in the center plus lateral 2, and longer ventral grooves reaching umbilicus are significant differences in appearance. In both species, the head is pointed and triangular in dorsal view, but the lateral borders are slightly more convex in "Bryde's whales". Dorsal fin is triangular with a posteriorly pointed tip that is about 2/3 of the body length from the tip of the rostrum. Ventral pleats average 62 for 14 North Pacific offshore populations [1498] and 42–54 in South African populations [283]. Baleen plates of offshore North Pacific Bryde's whales are also very different from those of the sei whale, having much coarser bristles and a lighter color. No sufficient baleen information is available for inshore populations of Bryde's whales of the North Pacific.

Body length of physically mature Bryde's whales in the North Pacific is 13.4 m for females and 12.8 m for males [1419], and in the South African Region, adult Bryde's whales of inshore populations are 13.7–14.0 m in females and 12.8–13.1 m in males, whereas 14.3–14.6 m and 13.7 m, respectively, for offshore populations [283]. Body color is dark gray on the back and white on the belly.

Dental formula No teeth, but 255–365 baleen plates for North Pacific offshore Bryde's whales [2703] and 276–289 plates per side of upper jaw for South African Bryde's whales [283]. They are dark gray in color with very coarse bristles. No baleen plate information is available for inshore populations of Bryde's whales.

Genetics Not reported.

5. Ecology

Reproduction Females of an offshore North Pacific population attain sexual maturity at a body length of 11.8–12 m and some of them are 9–10 years old [2697], whereas South African offshore females reach sexual maturity at a body length of 12.8–13.1 m [283]. That for males is 12.3 m in a North Pacific offshore population [2497], around 12 m in inshore and around 13 m in offshore populations in South Africa [282]. Gestation period is roughly 1 year and the body length of a newborn is estimated to be 4.08 m [282]. In the North Pacific offshore population, weaning may start sometime when the body length becomes close to 9 m [2642].

Lifespan Among 103 males and 171 females of the North Pacific offshore individuals captured through 1959–1965, the oldest male and female were 55 and 52 years old, respectively [2641].

Diet In the North Pacific, 345 stomachs of offshore Bryde's whales were examined and in summary a gonostomatid fish *Vinciguerria nimbaria* occupied more than 55%, followed by *Euphausia similis* (25.2%) and *Nematoscelis difficilis* (10.8%) [1496]. It is pointed out that possible major prey species shift to *Engraulis*, *Sardinops* and euphausiids, although this may have resulted from the change of sampling area from mainly around 30°N in commercial whaling to 35°N upward in JARPANII [607]. In South African waters, an inshore population fed mainly on *Engraulis*, *Trachurus* and *Sardinops*, whereas an offshore population consumed euphausiids [282].

Habitat Various populations of "Bryde's whales" are found both inshore and offshore waters within the range of 40°N and 40°S of all oceans. They prefer water temperature higher than 18°C [2641].

Behavior Swimming speed of an inshore population in South Africa is estimated as 6.5–8.3 km/h [284], however they can swim as fast as 20–25 km/h and dive up to 300 m in the North Pacific [1418]. Average school size of the South African population is 1.4 (range 1–8) [283]. The blow of Bryde's whale in the North Pacific is not clearly visible as those of other larger whales [2642], partly because of the higher water and atmosphere temperature of their habitat. The normal surfacing behavior is observed in South Africa and it consists of 1–6 short dives, followed by a longer dive of maximum nearly 10 min [283]. Low moaning sounds produced by "Bryde's whales" are recorded and analyzed in various areas [283].

Natural enemies Killer whales and large sharks could be the natural enemy. There are several publications [3270] and video resources on killer whales attacking North Pacific Bryde's whales.

Parasites Unpublished lists on parasites found in North Pacific offshore populations of Bryde's whales cited by [607] include internal parasites such as, Nematodes, Cestodes and Trematodes, and some barnacles, Copepodes as external parasites. In South African offshore populations *Pennella* and *Conchoderma* are reported, but no cyamids or diatom are confirmed [283].

6. Remarks

The species identification of "Bryde's whale" is chaotic. In 1879, Anderson described *B. edeni* with some description on the skeletal characteristics of the type specimen collected in Myanmar [128]. In 1912–1913, Olsen described *B. brydei* with no type specimen [2692]. A thorough osteological description of an individual supposed to be a *B. brydei* collected from the type locality of the species [1862] had functioned as a species description of *B. brydei*. In 1950, however, based on a specimen from Pulau Sugi, Singapore, Junge concluded that the 2 species are identical, and that *B. brydei* became a junior synonym of *B. edeni* [1251]. Although there were counter-opinions to encourage further careful examinations on more specimens [385, 3294], Junge's opinion was widely accepted [281, 2696]. However, the presence of 2 different eco-forms of "Bryde's whales" in the type locality of *B. brydei* has made the situation more complicated [281]. One of the causes for the confusion is the English vernacular name "Bryde's whale". At least, it should have been "Eden's whale", if *B. edeni* was supposed to be a distinct species. Summarizing the above, "Bryde's whales" were recognized as a monotypic species represented only by *B. edeni*, with possible reservation on the presence of *B. brydei*.

In the late 1970s in the Southern oceans, a small form of "Bryde's whales" was found among the whales captured under the special permit [2642, 2644] and their specific identity was argued based on the findings on isozyme analyses [3804]. In 1998, a baleen whale was killed by a collision with a fishing boat near Tsuno-shima Isl., Yamaguchi Pref. in western part of Japan. External morphology, body color, baleen characteristics and osteology did not correspond with those of any known baleen whale species. The complete control region sequence of mtDNA proved that this individual and 8 small-form Bryde's whales mentioned above were conspecific, and a new species *B. omurai* was described [3805]. During the processes of describing *B. omurai*, "Bryde's whale" specimens were extensively compared in morphology and it is concluded that *B. omurai* is distinct from Bryde's whales both morphologically and genetically [3805]. Later, analyses of complete mtDNA sequences and SINE insertion patterns revealed that *B. omurai* is far different from any other balaenopterid species [3022].

The taxonomy of "Bryde's whales" remains unclear even after *B. omurai* was sorted out. The *omurai* issue was a kind of noise, which had nothing to do with "Bryde's whales" taxonomy but made the scope complicated for a while. The specimens of "Bryde's whales" can be subdivided into several groups that can be defined by several morphological characteristic states. There are: Group 1, *B. edeni sensu stricto* is well defined with the type specimen. Those specimens that are morphologically very similar to the type specimen of *B. edeni*. (katsuo kujira: Eden's whale) can be grouped together based on genetic features, except for the type specimen itself, for which no genetic data is known. Group 2, those specimens that are morphologically very similar to the specimens described by Lönnberg are defined as offshore Bryde's whales of the North Pacific (nitari kujira). Group 3 (Indo-Pacific Bryde's whale), those specimens preserved in the Philippines, Indonesia and South Africa, are very close to Group 2 in skull morphology, but have consistent small differences in the morphology of the parietal. Group 4, those specimens from other waters, require further morphological and genetic examination. Also, these discussions have not paid proper attention to the 2 eco-types of South Africa presented by Best [281, 282]. At least 1 individual from either of the eco-types should genetically be grouped with Group 2 [4164].

Whales of Group 1 can be clearly defined in the Indian and North Pacific Oceans. A population known from coastal areas around Kyushu and Shikoku represents Group 1., and is *B. edeni* (katsuo kujira). Among those of Group 2, North Pacific specimens are recognizable both in morphology and genetics. These are recognized here as North Pacific Bryde's whales and is identical with *B. brydei* (nitari kujira). Individuals of Group 3 are found from the Indian and western North Pacific Oceans. The rest of Bryde's whales are included into Group 4, which requires further investigations to make their taxonomic status clear.

As mentioned above, "Bryde's whale" is one of the least known baleen whale species. Taxonomy of "Bryde's whales" requires further strenuous world-wide comparative works from various aspects including morphology and molecular genetics. Listing and cross-checking of various characteristics are needed and lumping superficially similar groups is harmful to reach a sound conclusion. Geographical and/or ecological differentiation of populations would be revealed, because Bryde's whales remain at lower latitudes and do not migrate extensively. The final conclusion would be that although "Bryde's whales" are classified in a single species now, they actually consist of a number of geographically and/or ecologically different populations. Most of those populations would be assigned to different species. After the first edition of this book was published several significant articles were published [2816], examined both inshore and offshore Bryde's whales of South Africa and concluded both of these 2 eco-forms were identified as *B. brydei*. Based on mtDNA sequence data. Genetic analyses on Bryde's whales of the Gulf of Mexico revealed that the population is consisted of a distinct genetic entity within *B. edeni* claster [2901]. An extensive genetic work on Bryde's whales of Indian and Pacific Oceans confirm the distinctiveness of *B. edeni* and *B. drydei* and also the close affinity of these taxon with *B. borealis* [1539].

T. K. YAMADA & H. ISHIKAWA

Cetacea BALAENOPTERIDAE

Red list status: En (MSJ); EN (IUCN); R for the western North Pacific population as *B. m. musculus* (JFA)

Balaenoptera musculus (Linnaeus, 1758)

EN blue whale JP シロナガスクジラ (shironagasu kujira) CM 蓝鲸 CT 藍鯨
KR 대왕고래 RS кит синий, блювал, голубой кит

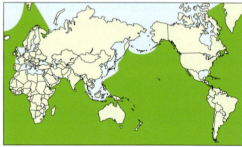

A surfacing blue whale is ready to blow in the western North Pacific (Institute of Cetacean Research/Kyodo Senpaku Kaisha, Ltd.). The shape of flat and wide head is likened to a Gothic arch.

1. Distribution

Widely distributed from tropical to polar waters. They are currently divided into 3 subspecies. In the Northern Hemisphere, *B. m. musculus* is distributed mainly in the North Pacific and Atlantic Oceans. *Balaenoptera m. intermedia* and *B. m. brevicauda*, known as a pygmy blue whale, are found in the Southern Hemisphere. In the Antarctic Sea, *B. m. intermedia* is distributed farther south than pygmy blue whales that are found mainly in the subantarctic waters of Indian Ocean and southeast Atlantic [4108]. Although blue whales are also distributed in the northern Indian Ocean, it is not clear whether these form a distinct population [3160]. In the North Pacific, there are at least 3 stocks on both the eastern and western side [2638, 3160]. North Pacific blue whales around Japan are distributed far off the Pacific coast. Although they were found in the southern Sea of Japan, Yellow Sea and East China Sea until late 1930's [1350], they are not distributed in neighboring seas including the Sea of Okhotsk at present [2635, 2638, 2695]. They are observed along the Kamchatka coast, the Commander Isls. and in the Anadyr Gulf in small numbers [3288], however, they do not extend into Bering Sea [2494, 2635]. As they are generally distributed in offshore waters, there have been no reliable stranding or incidental catch records from the coast of Japan [1113].

2. Fossil record

Not reported in Japan.

3. General characteristics

Morphology The blue whale is the largest cetacean and is also the largest animal known to have existed on Earth. The longest record is a 33.58 m female from the Southern Hemisphere [4108]. The heaviest weight is 190 t recorded in a female from the Southern Hemisphere [3288]. Blue whales in the Northern Hemisphere are smaller than those in the Southern Hemisphere. Mean body length of physically mature whales from the North Pacific is 23.4 m for males and 24.8 m for females [2634, 2645]. Body color is bluish gray with lighter color mottling as if it were painted with brush. Differing from other rorquals, the color of the ventral side is nearly the same as the dorsal region of the body. The ventral side of pectoral fins is white, whereas the ventral side of broad tail flukes is uniformly colored bluish gray. The body is streamlined. The head is flat, broad and U-shaped with a cusp when viewed from above (it is often likened to a Gothic arch). Along the center of the rostrum, there is a single prominent ridge, which ends in a fleshy crest surrounding 2 large blowholes anteriorly and laterally. A small dorsal fin is located about 25% of the body length from the notch in the tail flukes. There are 55–80 parallel ventral grooves that extend from the chin to near or the rear of the navel.

Dental formula There are no teeth but 300–400 baleen plates per side under the upper jaw [2457]. Maximum length of the baleen plate generally does not exceed 1 m. They are black in color and broadly triangular in shape with a coarse inner fringe [1495, 2638].

Genetics Chromosome: $2n = 44$ [3577]. Hybridization between blue and fin whales in nature has been reported in more than 10 cases [280].

4. Ecology

Reproduction In the Northern Hemisphere, blue whales calve and mate in late autumn and winter. Gestation period is roughly 11 months. Females give birth to a single calf every 2 or 3 years, probably in low latitudinal offshore waters. Calves are 6–7 m long at birth. They are weaned after 6–8 months with a body length approximately 16 m [3160]. In the North Pacific blue whale, it is assumed that the age at sexual maturity in both males and females decreased from 12 years of age in a virgin stock to 9 years in the 1960s because of depletion by the heavy exploitation. It is believed that pregnancy rate also changed from 25% before the 1950s to 35% in the 1960s [2638, 2645].

Lifespan Blue whales are thought to live for at least 80–90 years and probably longer [2494, 3160]. Estimated longevity is 120 years [2638].

Diet Blue whales feed almost exclusively on euphausiids (krill). In the North Pacific, only 1.3% of Copepoda (smaller crustaceans) were recorded as the stomach contents compared with 97.6% of euphausiids [1495]. Feeding of blue whales is characterized as "swallowing" or "gulping" type [2457]. A feeding blue whale swims into an area of heavy prey concentration and scoops up a mouthful of water and prey, then the whale closes its mouth, forcing the water out through the baleen plates and trapping the prey, which is moved back with a motion of the tongue [4108]. Although surface feeding has often been observed during the day, blue whales dive to at least 100 m into layers of euphausiid concentrations during daylight hours and rise to feed near the surface in the evening, following the ascent of their prey in the water column [3160].

Habitat They migrate northward to feeding grounds in spring and summer from subtropical and tropical wintering grounds. Although little is known about the breeding ground of this species, mother and calves are sighted regularly in the Gulf of California in late winter and spring in the eastern side of the Pacific [3160]. In the western side of the Pacific, they pass through offshore waters off Japan during seasonal migration. Although there are past coastal catch records of presumable calves (less than 15.2 m of body length) of this species in late winter and spring in the southeast side of Japan, nothing is known about the breeding area [2023].

A spraying blue whale (Institute of Cetacean Research/Kyodo Senpaku Kaisha, Ltd.). Unlike other rorquals, body color of blue whale is bluish gray with lighter color mottling. Some animals have yellowish or mustard coloration caused by the presence of a diatom film, which explains the alternative name "sulphur-bottom" whale.

Behavior Blue whales tend to travel alone or in small groups. They are one of the fastest swimming cetaceans. Swimming speed of blue whales ranges from 2 to 6 km/h while feeding, 5 to 33 km/h while cruising or migrating, to a maximum of 20 to 48 km/h when being chased or alarmed [1858]. Diving time of blue whales is generally 8–15 min when they are foraging or feeding at depth. The longest dive recorded was of 36 min, however, dives more than 30 min are rare [3160]. The blow is tall and vertical with 6–12 m in height. They vocalize regularly throughout the year with peaks from midsummer to winter. Its voice is one of the loudest (188 db) and the lowest (17–20 Hz) made by any animal [3160].

Natural enemies Although large sharks might attack a calf of the blue whale, killer whales are the only natural enemy [4108].

Parasites 18 endohelminth species and 8 ectoparasites or epizoa are reported from blue whales in the world. Of these parasites, *Tetrabothrius schaeferi* and *Crassicauda tortilis* are restricted to blue whales [2049]. It is suggested infection of *Crassicauda boopis* is potentially a major impediment to population recovery of this species [1825].

5. Remarks

Blue whales were heavily exploited after the 1860s when modern whaling with harpoon guns and steam driven vessels were developed. Because of its great size and the commercial value, a great number of blue whales were hunted, especially in the 1930s, and they were severely depleted in all areas. In the waters around Japan, they were targeted from the 1890s by both land based and pelagic whaling. The whaling of this species was banned by IWC in 1966 and blue whales are now protected worldwide.

H. ISHIKAWA

Cetacea BALAENOPTERIDAE

Red list status: DD (IUCN)

Balaenoptera omurai
Wada, Oishi & Yamada, 2003

EN Omura's whale   JP ツノシマクジラ (tsunoshima kujira)   CM 大村鯨   CT 大村鯨
KR 작은브라이드고래   RS полосатик Омуры

An Omura's whale sighted off Racha Isl. close to Phuket City, Thailand (S. Yossundara). Note that right lower jaw is white.

1. Distribution

Offshore and coastal waters of Indo-Pacific oceans between 40°N and 40°S, and 90°E and 150°E. Sightings, strandings, incidental and research captures occurred in the North Pacific Ocean, the Sea of Japan, Yellow Sea, East China Sea, South China Sea, Bohol Sea, Andaman Sea, Solomon Sea, Tasman Sea and the seas around Indonesia [1265, 2642, 2644, 2824, 3805, 3944-3946]. Sighting data, probably of *B. omurai*, are known from the southern Pacific near the Cook Isls. [849], Andaman Sea [71] and in Komodo National Park [1265]. Red solid circles in the map indicate stranding sites.

2. Fossil record

Not reported.

3. General characteristics

Morphology External morphology is that of a typical rorqual with a sleek streamlined profile; it is extremely difficult to identify this species. The head occupies about 1/4 of the body length, which may appear thicker than that of "Bryde's whales" (see 5. Remarks). The lateral contour line of the head in dorsal view is more convex than in other rorquals. There is a single median ridge on the dorsal aspect of the head starting from the tip of the rostrum to the blow holes, unlike "Bryde's whales", which normally have 2 additional lateral ridges. A relatively small falcate dorsal fin, that is much smaller than those of "Bryde's" and sei whales, is located about 2/3 of the body length from the tip of rostrum. The ventral pleats, reaching the umbilicus, are 80 to 90 between the flippers. Known body length of adults are between 10 to 12 m, and could be a little larger in females. Body color is dark gray to black on the dorsal surface but gradually changes lighter, almost to white on the ventral side. Lower jaw, throat, and the anterior thoracic regions are usually asymmetrically colored, black on the left side and white on the right, which is similar to the fin whale. At least some individuals hither to observed had a whiter patch at the angle of gape [3805].

Dental formula Baleen plate count is about 200, which is a very small number for a balaenopterid species. Plates are fairly wider in basal length relative to the height, compared to those of other rorquals, giving specific appearance. Baleen color of the holotype specimen is: all black in the posterior 1/6 to 1/5 of the plates, anterior 1/2 to 2/3 are black outside and yellowish white inside and anterior 1/3 could be yellowish white. Coloration of the plates is asymmetrical [3805].

Genetics Chromosome is not reported. Analyses of whole mtDNA sequence and SINE insertion patterns revealed that the species diverged first from the rest of balanopterids [3022], indicating significant distances between the Omura's whale and other balaenopterid species including "Bryde's whales".

4. Ecology

Reproduction Only available data are 3, 4, 2 and 11 corpora lutea and albicantia count for females with 10.1, 10.4, 10.3 and 11.5 m body length and 18, 19, 23 and 29 growth layer count in the earplug [3805]. A physically mature male with 1.9 and 1.7 kg testes was 9.6 m in body length with 38 growth layers in the earplug [3805]. A neonate of about 3 m in length was recorded in Miyazaki Pref. in the last week of August [3943].

Lifespan Valid age count was made only on 5 individuals (4 females and 1 male). Physically mature females with body length of 10.3 to 11.5 m in body length had 19 to 29 growth layers in the earplug, whereas physically mature single male with 9.6 m body length had 38 layers [3805].

Diet No published stomach contents data is available. We found highly digested crustacean and piscine remains from a 7-m female stranded in Seto Inland Sea. Three species of euphausiids were confirmed from the Southern Bryde's whales [1494], which possibly included *B. omurai*.

Habitat Hitherto known localities are both inshore and offshore waters of mainly tropical to warm areas [3945]. Seasonal migration is not reported for this species.

Behavior Because there are very few sighting records, little is known about their behavior. Based on very few sightings of whales highly suspected as *B. omurai*, they tend not to dive with their flukes up even when they start big dives [849].

Natural enemies Not reported.

Parasites No published papers for internal parasites. However, a 7-m female found in the Seto Inland Sea had several *Pennella* sp. [3943].

5. Remarks

The species was first noticed by scientists after the 1970s, and the distinction from "Bryde's whales" has been seriously argued [1834, 2882, 3804, 4164]. Several characteristics, such as skull and rib morphology, external head character, coloration and dorsal fin shape, however, are distinct from those of other balenopterid species including Bryde's whales. The only similarity shared with the Bryde's whales could be the shorter body length of about 12 m. In addition to the morphological differences, clear differences in genetic characteristics are also shown [3022]. In order to establish a good understanding on these species further examinations on external morphology including body color, baleen color, number and shape, skeletal morphology and genetic characteristics are essentially needed. Vernacular names suggesting affiliation to Bryde's whales should be avoided for this species, because it is misleading especially to non-specialists.

T. K. YAMADA

The skull of the holotype specimen (NSMT32505) of the Omura's whale (National Museum of Nature and Science).

Cetacea BALAENOPTERIDAE

Red list status: En for the population in the East China Sea, V for the population in the Sea of Japan, R for the population in the western North Pacific (MSJ); EN (IUCN); VU for the populations in the East China Sea and in the Sea of Japan, ND for the population in the western North Pacific (JFA)

Balaenoptera physalus (Linnaeus, 1758)

EN fin whale　　JP ナガスクジラ (nagasu kujira)　　CM 长须鲸　　CT 長鬚鯨
KR 참고래　　RS финвал, сельдяной кит

A surfacing fin whale in the western North Pacific (Institute of Cetacean Research/Kyodo Senpaku Kaisha, Ltd.). The asymmetrical pigmentation of the lower jaw is characteristic and the white color of right lower jaw is an important key to identify this species on the sea.

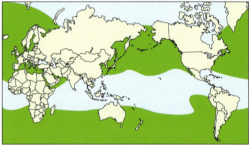

range of Asian stock

range of Sea of Japan stock

range of East China Sea stock

1. Distribution

Widely distributed from tropical to polar waters. Unlike blue and sei whales, fin whales are distributed in neighboring seas such as the Mediterranean Sea, Red Sea and the Sea of Japan but are absent near the ice limit in polar seas [3288]. As the seasons are opposite in each Hemisphere, there is little opportunity for contact between the northern and southern populations, which are recognized as different subspecies, *B. p. physalus* for the Northern Hemisphere and *B. p. quoyi* for the Southern Hemisphere [75, 699]. It is suggested there are at least 3 stocks of North Pacific fin whales around Japan; the Asian stock distributed off the Pacific coast and the Sea of Okhotsk, the Sea of Japan stock, and East China Sea stock [2636, 2639]. Blood typing, morphological and marking studies indicate differences among the East China Sea stock and the 2 Asian stocks (west and east side of the Aleutian Isls.) [602, 973], though genetic differences in mtDNA are not clear among these suggested stocks [730].

2. Fossil record

Not reported in Japan.

3. General characteristics

Morphology　The second largest species of cetacean next to the blue whale. Fin whales from the Northern Hemisphere are smaller than those from the Southern Hemisphere. Physical maturity is attained about 25 years of age, with mean body length of 18.8 m for males and 20.0 m for females in the Northern Hemisphere [2636]. Body shape is streamlined, with a sharp pointed or falcate dorsal fin that is located about 1/3 of the body length from the notch in the tail flukes. The upper ridge of the caudal peduncle forms a sharp ridge. The head is large and flat. There are 55–100 parallel ventral grooves that extend from the chin to the rear of the navel [2639]. Body color is black on the dorsal side and white on the ventral side. Ventral side of pectoral fins and tail flukes are also white. A V-shaped gray chevron pattern crosses the dorsal side behind the head. The asymmetrical pigmentation of the

Balaenoptera physalus (Linnaeus, 1758)

lower jaw is most characteristic of this species; while the left side is black, the right side is white.

Dental formula There are no teeth but 300–400 baleen plates per side under the upper jaw [2457]. They are olive green with a black band on the outer edge in some plates, but those of the front third on the right side are white or cream colored.

Genetics Chromosome: $2n = 44$ [75]. Hybridization between the blue whale and fin whale has been reported in more than 11 cases, with an estimated rate of 0.1–0.2% of fin whales [75, 280].

4. Ecology

Reproduction Gestation period of fin whales is roughly 11 months. Females give birth to a single calf every 2 or 3 years in winter, probably in low latitudinal offshore waters. Calves are 6.3 m long at birth. They are weaned after 6–7 months when they are about 11 m long. In the North Pacific fin whales, the age at sexual maturity is thought to be shortened from 12 years of age to 5 years of age because the population decreased by heavy exploitation [2636, 2639].

Lifespan Fin whales are thought to live for at least 80–90 years and a North Pacific fin whale of 101 years of age was recorded [75, 2636].

Diet Fin whales feed on a variety of food species depending on their availability. In the North Pacific, they feed mainly on euphausiids (krill), followed by Copepoda (small crustaceans) [1495]. They also feed on schooling fish such as capelin, herring, walleye pollack, arabesque greenling and sometimes squid [2457, 2636]. Feeding of fin whales is characterized as the "swallowing" or "gulping" type, the same as for blue whales [2457].

Habitat Most fin whales migrate between temperate waters for breeding in winter and polar cold waters for feeding in summer. However, some whales remain in higher latitudes during the coldest months of the year, whereas some may remain at lower latitudes year-round if food is available [75]. Nothing is known about the breeding area of this species around Japan [2636]. There are several stranding and incidental catch records of this species in both Pacific and the Sea of Japan side of Japan. Most of the stranding and incidental catches occur in the winter and spring seasons, which seems to reflect the migration of this species [1113].

Behavior Fin whales do not form socially close schools and tend to travel alone or in pairs. However, they commonly form larger groups and sometimes form a group of more than 100 individuals in a feeding area. The blow is columnar and 4–6 m tall. A series of shallow dives lasting 10–20 seconds may be followed by a longer dive for 15 min or more, down to a depth of 230 m [699]. Fin whales are known as one of the fastest swimmers among cetaceans, exceeding 40 km/h [1858]. Activity of fin whales is categorized: (1) blowing at the surface as the most visible behavior; (2) a short time (2–6 min) dive as the most common behavior; (3) a long time (6–14 min) dive relating to feeding behavior; (4) near surface slow swimming as presumable resting behavior; (5) rapid travel near the surface in transit; (6) surface feeding on schooled fish [3857]. A radio-tagging survey revealed that fin whales traveled 2,095 km over the course of 10 days. They produce higher frequency sounds (under 100 Hz); 20 Hz pulses, ragged broadband low frequency pulses and low frequency rumbles as well as non-vocal sharp impulsive sounds. The higher frequency sounds appear to be for communication with nearby fin whales, the 20 Hz single pulse seems to be used for longer distance communication, the patterned seasonal 20 Hz pulse appears to be for courtship displays, the low frequency rumble seems to have agonistic significance, and non-vocal sharp impulsive sounds are related to surface feeding [3857].

Natural enemies Although large sharks might attack a fin whale calf, killer whales are the only natural enemy.

Parasites 2 Trematodes, 1 Cestodes and 2 Nematodes are reported from the North Pacific fin whales [1950]. It is suggested that infection with *Crassicauda boopis* is potentially a major impediment to the population recovery of this species [1824, 1825].

5. Remarks

Fin whales were heavily exploited after the 1860s when modern whaling with harpoon guns and steam driven vessels were introduced, as was the case with most other large rorquals. The whaling of this species in the North Pacific was operated by Japan, Korea, Russia, Canada and USA mainly in the middle of the 20th century. Japan started whaling of fin whales in the late 19th century in the coastal waters of the Korean Pen. and it extended to the north as far as the Kuril Isls. and south as far as Taiwan. Commercial whaling for fin whales was banned by IWC in 1976. Today, a small number of fin whales are taken in the North Atlantic in both commercial and "aboriginal" whalings. Though the results of recent sighting surveys off the western North Pacific suggest recovery of this species, populations in the Sea of Japan and the East China Sea still seems to be at low levels [2018, 2639].

H. ISHIKAWA

Cetacea BALAENOPTERIDAE

Red list status: V for the population wintering off Ogasawara Isls. and Okinawa (Ryukyu) Isls. (MSJ); LC (IUCN); R for the population breeding off the Ogasawara Isls. the Okinawa Isls. and other North Pacific populations (JFA)

Megaptera novaeangliae (Borowski, 1781)

EN humpback whale   JP ザトウクジラ (zatou kujira)   CM 大翅鯨   CT 大翅鯨, 座頭鯨
KR 혹등고래   RS горбач, горбатый кит

Humpback whales migrate to low latitude winter breeding grounds, which are shallower and near the Ogasawara Isls. (Ogasawara Whale Watching Association).

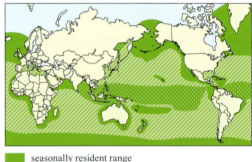

- seasonally resident range
- transient range
- possible range

1. Distribution

Cosmopolitan. They are highly migratory, spending summer through autumn on feeding grounds in mid or high latitude waters and wintering on breeding grounds in low latitude waters. On the map, the seasonally resident range denotes the current main winter breeding and summer feeding grounds. Their migrations often take them through oceanic waters. In the North Pacific, feeding grounds are wide from California to Alaska and along the Aleutian Chain into the western North Pacific, and breeding grounds are located in the eastern North Pacific along the coast of Mexico and near the offshore the Revillagigedo Isls., in the central North Pacific around the Hawaiian Isls., and in the western North Pacific from southern Japan to the Philippines [345].

Around Japan, humpback whales are regularly seen in the Ogasawara Isls. and the Okinawa and Kerama Isls. (Okinawa waters). Waters shallower than 200 m serve as winter breeding grounds in Japan [435, 889, 2204, 2572].

2. Fossil record

Fossils assigned to *Megaptera* were found from Aomori, Akita, Yamagata, Miyagi, Chiba and Kanagawa Prefs. in Japan [2345]. Among them, one find from the upper Middle Pleistocene in Chiba Pref. was identified as *M. novaeangliae* [2345].

3. General characteristics

Morphology   The body is robust. Body color shows black or dark gray on the back, variable from black to white or mottled on belly. The flippers are extremely long, up to approximately 1/3 of the body length. Ventral sides of the tail flukes have individually different black and white coloration, which is used for individual identification by scientists. The dorsal fin is low and wider-based like a hump. Maximum reliably recorded adult lengths are 16–17 m range, although 14–15 m is more typical [394].

Dental formula   300–370 (mean 330) baleen plates on each side on the upper jaw in North Pacific [1495].

Megaptera novaeangliae (Borowski, 1781)

Genetics Chromosome: $2n = 44$ [1823]. Along with photo-identification studies, genetic studies revealed differentiation among feeding grounds, suggesting persistent site fidelity to feeding ground in this species [247, 2798]. This may be a result of 'cultural' transmission of migration destination [247, 2798, 3886].

4. Ecology

Reproduction Humpback whales off the Okinawa and Kerama Isls. reach sexual maturity at about 11.9–12.1 m in females and 11.3–11.9 m in males in the 1950s [2491]. Gestation period lasts 10–11 months (or slightly longer) and length at birth is around 3.9–4.3 m in the Okinawa [2491]. Based on photo-ID studies in Ogasawara waters in the 1990's, calving intervals range from 1 to several years and are most commonly 2 years [3027].

Lifespan The oldest known humpback whale in the North Pacific is 77 years [2643].

Diet Humpback whales feed on euphausiids or krills (*Euphausia pacifica, Thysanoessa longipes, T. raschii* and *T. inermis*), mackerels, sand lances, capelins and herrings in the northern North Pacific and Bering Sea [1495, 3885]. Humpback whales feed on these prey in summer feeding grounds in mid- and high-latitude waters. They do not take any food in the winter breeding ground, but *Euphausia similes, Pseudoeuphausia latifrons*, small squid and small mackerel were found in the stomach as a rare case in Okinawa waters [2491].

Habitat Humpback whales feed and breed mostly in coastal waters over continental shelves or around islands.

Behavior Male humpback whales sing long and complex songs in the winter breeding ground [434]. Humpback whales often do aerial behavior: breaching, flipper slapping, tail slapping and so on. Humpback whales show unique foraging behavior. They are swallowing-type feeders and sometimes use air bubbles (nets, clouds or curtains) to corral or trap schooling fish (bubble net feeding).

Migration pattern is complex and migration route is poorly known. However, whales that feed in the eastern and western coasts of North Pacific tend to winter in the lower latitude breeding ground on the same side of Pacific (Asia in the west and mainland Mexico and Central America in the east), and those that feed in the central and higher latitude North Pacific tend to migrate to breeding grounds off the Hawaii or the Revillagigedo Arch. [345]. In Japanese waters, movements within Ogasawara or Okinawa waters are common and some interchanges are found

Mother and calf humpback whales are observed in winter and spring in the Ogasawara Isls. (Ogasawara Whale Watching Association).

between Ogasawara and Okinawa [345, 889, 2204]. A humpback whale was found twice off Kagoshima Pref. and off Shizuoka Pref. during its northbound migration [1177], although the exact migration route of this whale was not known.

Natural enemies Killer whales attack humpback whales, and sharks usually attack only dead or weakened animals. Crescentic scars may have been caused by the teeth of *Isistius brasiliensis*, a squaloid shark [3885].

Parasites Patches of external parasites, whale lice (order Amphipoda), barnacles (*Coronula* and *Conchoderma* spp.) are found on their bodies [3885].

5. Remarks

Humpback whale populations were severely depleted due to past commercial exploitation in all the oceans. The pre-1905 humpback population of the North Pacific was probably about 15,000 [2880], and approximate numbers following the cessation of commercial whaling have been estimated at 1,200 [1244]. IWC has protected humpback whales in the North Pacific since 1966. Humpback whales were taken around Japan by traditional whaling in the 15th–20th century and modern costal pelagic whaling in the 20th century [1387]. As a result, the western North Pacific population was considered endangered. Currently the population is showing evidence of recovery [345, 2195, 2572]. Humpback whales have been target species for whale watching off the Ogasawara, Okinawa and Kerama Isls. since the end of the 1980s.

K. MORI

Cetacea PHYSETERIDAE

141

Red list status: R for the two populations in the western North Pacific, K for the other populations in the North Pacific (MSJ); VU (IUCN); ND (JFA)

Physeter macrocephalus Linnaeus, 1758

EN sperm whale　　JP マッコウクジラ (makko kujira)　　CM 抹香鯨　　CT 抹香鯨　　KR 말향고래　　RS кашалот

A sperm whale spyhopping off the Kii Pen. (M. Yoshioka).

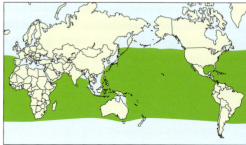

1. Distribution

Distributed in deep oceanic waters throughout the world, from the equator to the edges of the polar pack ice [2881]. The range is one of the widest of all mammals except humans and killer whales. The distributional range is segregated by sex and growth stage. Females, calves and juveniles inhabit warm deep waters at latitudes less than 40–50° year-round [3874]. Males gradually move to higher latitudes with growth, and large and old mature males may be found near the edge of pack ice. Sperm whales are found in the deep waters off Japan, mainly in the Pacific and are rare in the East China Sea, the Sea of Japan and the Sea of Okhotsk [1395, 2881]. Two populations or stocks are suggested in the western North Pacific; female groups of each population range to the north and south of the Kuroshio–Oyashio front, respectively [1395]. The northwestern North Pacific population winters off Choshi (Chiba Pref.) through Hokkaido, and females migrate northward to Attu Isl. (the Aleutian Isls.) in summer. The whales of the southwestern North Pacific population winters off Choshi (Chiba Pref.) through the Nansei Isls. (Kagoshima and Okinawa Prefs.) and migrates to the sea off the Sanriku coast in summer [1395]. At least 2 different vocal clans exist in the southwestern North Pacific population, off the Kumano coast and off the Ogasawara Isls. [108].

2. Fossil record

Not reported in Japan.

3. General characteristics

Morphology　Sperm whales show remarkable sexual dimorphism in body length and weight. Adult females reach about 11 m in length and 15–20 tons, and physically mature males are about 16 m and 45 tons. Newborn sperm whales are 3.5–4.5 m long. The head is squarish and huge, especially in large males, and occupies a quarter to a third of the total length. The blowhole is single and located at the front of the head, on the left of the upper surface. Lower jaw is rod-shaped and fits into the concave upper jaw when it closes. Body surface behind the head is wrinkled. Dorsal fin is low and round or obtuse. On the dorsal fin roughened skin or calluses are found in most mature females and a few immature males [1399]. Small bumps are lined on the back behind the dorsal fin. Flippers are relatively small and broad. Flukes are triangular with a straight trailing edge, but become convex in mature males [2504]. Body is dark brown or dark gray in color, sometimes black, with white areas around the mouth and often on the belly. Females reach physical maturity at 25–45 years old and 10.4–11.0 m long and males at 35–60 years and about 15.2–16.1 m [2881].

Dental formula　0/20–26. As many as 11 vestigial teeth, which rarely erupt, are present in the gum of each row of the upper jaw [2881].

Genetics Chromosome: $2n = 42$ [2881]. Modern genetic studies have been unable to find any distinct difference at geographical scales of less than an ocean basin [3874].

4. Ecology

Reproduction Females reach sexual maturity at about 9 years when 8–9 m long and give birth roughly every 4–6 years [285, 2881]. Female reproductive rates decline with age. About 25% of females are pregnant at ages 10–14 years, but only about 7% at older than 40 years [285]. Males reach puberty approximately between 7–11 years and then testes grow slowly until 18–21 years when they attain sexual maturity, after that growth of testes accelerates as well as body size [2881]. Males may not breed until their late twenties. Breeding season is prolonged between late spring and early summer, and conceptions may occur from January through August, with a peak between March and June in the northwestern Pacific [2640]. Gestation period is 14–15 months. Lactation period lengthens with age from 1.6 years for mothers younger than 10 years to 3.5 years for those older than 40 years [285]. Sporadic nursing may last much longer sometimes for over 10 years after calves start to eat solid food [285].

Lifespan The oldest recorded sperm whale is 75 years old for a female and 77 years old for a male [1416] and lifespan does not seem to be different between sexes.

Diet The most important food items are mesopelagic and bathypelagic cephalopods, mostly squids, followed by fishes. Diets of sperm whales are diverse and show considerable geographic difference. Histioteuthidae, Gonatidae, Onychoteuthidae and Octopoteuthidae are important cephalopod species, and rock fish and cod are important fish species in Japanese waters [1464]. Large fragments of cephalopod (i.e. *Architeuthis* sp., *Moroteuthis* sp. and *Haliphron atlanticus*) have sometimes been seen floating at the surface, where sperm whales were sighted off the Ogasawara Isls. [2201]. Sperm whales eat solid food before reaching 1 year old [3874].

Habitat Sperm whales are usually found in waters deeper than 1,000 m. In some areas, large males are found in waters less than 300 m deep, while female groups rarely enter the waters on the continental shelf. Sperm whales have a wide distributional range in the pelagic oceans, but there are waters where whales concentrate, referred to by whalers as "grounds". These areas are usually associated with high productivity and steep drop-off or strong oceanographic features [3875]. The range of female groups is limited to areas with surface water temperature higher than 15°C. Sperm whales may conduct seasonal migration in many areas, but this is not as regular or as well understood as for most baleen whales [3874].

Behavior Female sperm whales live in a matrilineal group called a "unit" for tens of years and often throughout their lives. A group observed at the surface is a temporal assemblage of animals from different units. Each unit has a unique repertory of pulsed sounds called "coda" used for mutual communication, which may be transmitted among unit members by social learn-

A sperm whale with a data logger tag on its back off the Ogasawara Isls. (M. Amano).

ing [2877, 3152]. Males leave their natal unit at around puberty and make a group with 10–15 similar sized individuals, a so-called "bachelor group". The bachelor group gradually becomes smaller in size and moves to higher latitudes as the males grow. Fully matured males are usually found alone in high latitudes and occasionally move to lower latitudes to copulate. Males rove among female groups and stay only a few hours in each group [3873]. Sperm whales are one of the deepest and longest divers among marine mammals. They routinely dive to depths more than 400 m for foraging and spend 60–80% of their time in foraging dives. They actively search for and pursue prey by means of echolocation [122, 2072, 3858]. Diel patterns of diving behavior were found off the Ogasawara Isls., where the whales dived deeper and swam faster during the day than at night, but such diel rhythms were not found off Kumano Coast [167]. Sperm whales spend a few hours at the surface usually once a day for socializing and resting. Sperm whales rest drifting in shallow depths in a vertical posture with their head upward [2071].

Natural enemies Killer whales have been observed attacking sperm whales and large sharks are also potential predators [3874].

Parasites 24 species of parasitic helminths and 24 ectoparasites and epizoites are reported from all over the world [2881]. A cestode *Phyllobothrium delphini* is reported from Japan [1763].

5. Remarks

Sperm whales were taken near Japan by Yankee whalers in the 19th century and modern coastal and pelagic whaling in the 20th century. Sperm whales off Japan were severely exploited and the northwestern Pacific stock was depleted by the taking of about 180,000 whales after World War II [1383]. Commercial whaling of this species ceased with an International Whaling Commission moratorium in 1988. Small numbers of sperm whales had been caught off Japan under the so-called "scientific whaling program" until 2013. In some years up to a dozen sperm whales are stranded on the beach or wander in the bay along the coast of Japan [4192]. The sperm whale is a main target species at several whale watching sites in Japan: Rausu, Hokkaido; Choshi, Chiba Pref.; Ogasawara Isls., Tokyo; the southern Kii Pen., Wakayama Pref., Muroto, Kochi Pref. and so on.

K. MORI, M. YOSHIOKA & M. AMANO

Cetacea   KOGIIDAE

Red list status: K (MSJ); DD (IUCN)

Kogia breviceps (de Blainville, 1838)

EN pygmy sperm whale   JP コマッコウ (komakko)   CM 小抹香鯨   CT 小抹香鯨
KR 꼬마향고래   RS карликовый кашалот

A pygmy sperm whale in captivity of Kamogawa Sea World, Chiba, Japan (M. Yoshioka). The whale is characterized by very short rostrum, small dorsal fin and "gill slit" behind the eye.

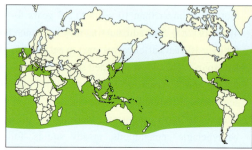

1. Distribution

Widely distributed in warm temperate and tropical waters of the world [1230]. In Japan, the species is found south of Hokkaido in both the Pacific and the Sea of Japan [2448].

2. Fossil record

Not reported in Japan. Fossil *Praekogia cedrosensis* from early Pliocene Mexico is reported to be clearly ancestral to extant *Kogia* [2036].

3. General characteristics

Morphology   The head is shark-like and has a narrow underslung lower jaw. It also has a spermaceti organ like sperm whales. Small and falcate dorsal fin is located well behind the center of the back. Body is counter-shaded with dark grayish back and white belly. It has a bracket-shaped pattern referred to as a "false gill" that extends from behind the eye to the flipper. The skull is markedly asymmetrical. Zygomatic arches are complete, unlike the dwarf sperm whale (*K. sima*). Maximum body length is up to 3.8 m [1230]. Females are somewhat larger than males.

Dental formula   0/12 (sometimes 10 or 11) –16 [1230]. Upper jaws lack teeth and those in the lower jaw are fanglike and fit into sockets in the upper.

Genetics   Chromosome: $2n = 44$ [190].
Some gene flow is suggested between the Atlantic and Indian Oceans [387].

4. Ecology

Reproduction   Information on the life history of pygmy sperm whales is limited. Gestation period is 11 months and birth size is about 120 cm [346]. Length at sexual maturity is 2.7–3.0 m [346] and age at sexual maturity is between 2.5 and 5 years [1230]. Births usually occur during spring and summer. Parturition in a stranded pygmy sperm whale was observed in captivity [967].

Lifespan   Although maximum known longevity is 23 years [2903], details are unknown.

Diet   Stomach contents of captured and stranded specimens included squids (primary item of diet), benthic (deep-sea) fishes, crabs and shrimps [270, 710, 1230, 3006]. Feeding habits were suggested to resemble those of dwarf sperm whales [3304].

Kogia breviceps (de Blainville, 1838)

Habitat Deep offshore warm waters, especially common over and near the continental slope [1225]. In comparison with dwarf sperm whales, pygmy sperm whales appear to occur more often in slightly colder seas [346].

Behavior Most sightings of pygmy sperm whales are of one to several individuals. They occasionally jump vertically out of the water and do not approach vessels [346]. The species spends considerable time lying motionless at the sea surface with the back of the head exposed and tail hanging down loosely (logging behavior). When startled, the species may emit a reddish brown intestinal fluid and escape [2872, 3938]. Sound production control related to the spermaceti organ is reported [396].

Natural enemies Not reported. Likely killer whales and large sharks.

Parasites Not reported in Japan, but heavy parasite infections are reported in other areas [304].

5. Remarks

It is unlikely that pygmy sperm whales are directly affected by humans. No significant levels of direct and incidental catches are reported. A few strandings are reported on Japanese coasts annually [1110]. Live-stranded individuals have been transported to aquariums in Japan.

M. YOSHIOKA

A pygmy sperm whale at Kamogawa Sea World, Chiba Pref. (M. Yoshioka). The pygmy sperm whale has a shark-like head with a narrow under-slung lower jaw.

Cetacea KOGIIDAE

Red list status: K as *K. simus* (MSJ); DD (IUCN)

Kogia sima (Owen, 1866)

EN dwarf sperm whale   JP オガワコマッコウ (ogawa komakko)   CM 侏儒抹香鯨   CT 侏儒抹香鯨
KR 쇠향고래   RS малый карликовый кашалот

A group of two dwarf sperm whales surface and swim slowly off Wakayama Pref., Pacific coast of Japan (K. Oda).

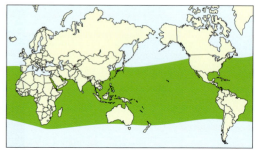

1. Distribution

Mainly over the continental shelf and slope off temperate and tropical coasts of all oceans. In Japan, strandings have been recorded from the south of Hokkaido on the Pacific side, and the south of Niigata Pref. in the Sea of Japan [1113], but usual range should be more south. Off Wakayama, dwarf sperm whales are occasionally sighted by whale watchers in summer [4188].

A female (BL = 2.2 m) stranded in Nachi-katsuura, Wakayama Pref., July 23, 2007 (T. Kirihata, Taiji Whale Museum). Dwarf sperm whales occasionally strand on Japanese beaches.

2. Fossil record

Not reported in Japan. Fossil *Praekogia cedrosensis* from early Pliocene Mexico is reported to be clearly ancestral to extant *Kogia* [2036].

3. General characteristics

Morphology As in pygmy sperm whales, the head is shark-like, with a narrow underslung lower jaw, but the head is smaller, and when viewed from above, the several short grooves on the throat and rostrum more squarish. Spermaceti organ is in its head. The dolphin-like falcate dorsal fin is located near the mid-point of the back and higher than that of pygmy sperm whales. Zygomatic arches are incomplete unlike pygmy sperm whales. Body is counter-shaded with dark grayish back and white belly. It has a light colored bracket-shaped pattern called a "false gill" behind each eye and extending to the flipper. The skull is markedly asymmetrical. Maximum body length is up to 2.7 m, smaller than pygmy sperm whales [1230]. Males may be slightly larger than females.

Dental formula 0–3/7–12. Teeth in upper and lower jaws are tiny and fanglike (30 mm in length and less than 4.5 mm in diameter) [1230].

Kogia sima (Owen, 1866)

Genetics Chromosome not investigated. Recent molecular genetic studies suggest the existence of two separate (one in the Atlantic and one in the Indo-Pacific) species [387].

4. Ecology

Reproduction Information on life history of dwarf sperm whales is limited. Length at birth is about 1 m [346]. Gestation period is 9 to 11 months [2077]. Length at sexual maturity is 2.2 m [346] and age at sexual maturity is between 2.5 and 5 years [1230]. Births may peak in summer.

Lifespan Maximum known longevity is 22 years, not appearing to be long [1230].

Diet Dwarf sperm whales primarily feed on deep-water cephalopods [1230]. About 40 different prey species including cephalopods, fish and crustaceans were found from the stomach contents of individuals from South African waters [2903].

Habitat Deep offshore warm waters, especially common over the continental slope [1225]. In comparison with pygmy sperm whales, dwarf sperm whales appear to occur more often in slightly warmer seas [346].

Behavior Most sightings of dwarf sperm whales are of small groups of less than 6 individuals (up to ten individuals per group) [2353]. Seasonal variation in group size and habitat use is reported [483]. Aerial behavior is rare and they do not approach vessels [346]. The species spends considerable time lying motionless at the sea surface with the back of the head exposed and tail hanging down loosely (logging behavior). When startled, the species may emit a reddish brown intestinal fluid and escape [2872, 3938].

Natural enemies An attack on a dwarf sperm whale by a pod of killer whales was observed in Kumano Sea, off Wakayama, Japan [4188]. Also, probably large sharks.

Parasites Not reported in Japan.

5. Remarks

There are no known human impacts on dwarf sperm whales including commercial catch and significant numbers of bycatches. A few strandings are reported in Japanese coasts annually [1113].

M. YOSHIOKA

A stranded dwarf sperm whale (M. Amano). The body shape resembles that of shark, which is thought to be mimicry.

Cetacea ZIPHIIDAE

Red list status: R for the 3 populations of Japan (MSJ); DD (IUCN); DC (JFA)

Berardius bairdii Stejneger, 1883

EN Baird's beaked whale   JP ツチクジラ (tsuchi kujira)   CM 贝氏喙鲸   CT 貝氏喙鯨
KR 큰부리고래   RS северный плавун

A breaching Baird's beaked whale off the Pacific coast of Japan (M. Kishiro and National Research Institute of Far Seas Fisheries).

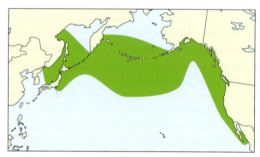

1. Distribution

Distributed in warm to cold temperate areas of the North Pacific Ocean, mainly in deep waters over the continental slope. Along the Pacific coast of Japan they appear during the summer from May to August. They are found on the Pacific side north of 34°N and in the Sea of Japan north of 36°N [250, 1396, 2484, 2502]. In the eastern North Pacific, they are sighted north of the California Pen. [1386]. Around Japan, the existence of 3 different populations, the Pacific, the Sea of Japan and the Okhotsk populations, have been suggested based on differences in population density and seasonal movements [1382].

2. Fossil record

Not reported.

3. General characteristics

Morphology The largest species of the Ziphiidae. Long trunk with a small head. Mean body length of individuals 15 years or older (when they are considered physically mature) is 10.45 m (SD = 0.31, $n = 22$) for females and 10.10 m (SD = 0.35, $n = 66$) for males [1388]. Sexual dimorphism is smallest in this species among all the Ziphiidae. Body weight can be as heavy as 12 t. The beak is longer and is demarcated more distinctively from the melon than it is in most of other beaked whales. There is, however, no crease between the beak and the melon. The blowhole forms a posteriorly opened crescent, which is unusual in odontocetes. There is a pair of reversed V-shaped grooves, typical of beaked whales, at the throat. The relatively short flippers fit into the "flipper pockets". The small round-tipped triangular dorsal fin is at about 2/3 of the body length from the tip of the rostrum. Most individuals have a slight depression instead of a notch at the middle of the flukes. Adult males and females are dark brown to black all over, sometimes with white patches on the ventral portion. The body is covered with scars made by the teeth of other individuals. The scars are more conspicuous in larger males, but are also found in females, which is unusual for Ziphiid species.

Dental formula Two pairs of teeth in the lower jaw, the front pair are larger. They erupt around the time when they become sexually mature in both sexes (exceptionally as a ziphiid species).

It should be noted here that ziphiid whales generally have 1 or 2 pairs of teeth in the mandible, which only matured males possess, and have no teeth in the maxilla. These teeth are non-functional in feeding behavior, but have an important role in social displays for males.

Berardius bairdii Stejneger, 1883

Genetics Existence of distinct two populations has been suggested [1621].

4. Ecology

Reproduction Sexual maturation has been investigated in individuals of 6–11 years and 9.1–9.8 m in body length for males by histological examination of the testes. First ovulation of females is observed at age 10 to 15 years, when they are 9.8–10.7 m. Gestation period is estimated as 10–17 months [2702]. They become physically mature before 15 years old. Mean neonatal body length is estimated at 4.56 m. Peak seasons for mating could be October–November and for calving March–April [1377, 2702]. Heavy linear scars on males, mostly consisting of parallel double lines, are believed to form in the struggle with other adult males during the mating season. Similar scars are also seen in females, which is very peculiar for beaked whales. It is suspected that females may also get involved in mating rituals, if any occur.

Lifespan The maximum recorded lifespan of the Baird's beaked whale is 54 years in females, and 84 years in males. It is remarkable among mammalian species that the male's lifespan is 30 years longer than that of females [1388].

Diet Ziphiid species generally feed on squids, but for Baird's beaked whales deep sea fishes such as rat tails and hake are also important prey species [2624, 3820].

Habitat The species appears to prefer the cold temperate waters of the North Pacific Ocean. They are found on the outer margin of the continental shelf to the continental slope where ocean depths are 1,000–3,000 m.

Behavior Group size ranges from 1–30, and is usually 2–9 [1386]. Dive time ranges from 1–67 minutes with a mean of 19.8 minutes, while the time spent at the surface ranges from 1 to 14 minutes [1382]. More males are found in adult populations because of the longer lifespan of males. Adult males seem to play a role in rearing calves, for example in deep-dive training [1388]. Diving times and depths have been recorded in detail for 1 individual (estimated body length of 9–10 m) off Chiba Pref. using a specially designed data logger. In 81 dives totalling more than 29 hours the average depths were 1,565.9 m for 5 deep dives, 379.2 m for 29 intermediate dives and 20.3 m for 46 shallow dives. Mean diving time records for the 3 dive types were 2,746, 1,504 and 609 seconds, respectively. There was no significant difference between descent and ascent rates, unlike results published for other beaked whale species [245, 2075].

Natural enemies Killer whales and large sharks could be natural enemies of this species. Scars caused by killer whale teeth are common especially on flukes and flippers [1386].

Parasites *Crassicauda* infestation of the kidney is very common and could be fairly severe. A large number of Anisakid nematodes are frequently found in the stomach. External parasites, *Conchoderma* sp. are found especially on the apical pair of teeth. Cyamid whale lice are commonly found [1386].

5. Remarks

There is a long history of whaling this species in Japan. At present the annual quota for this species is 66 for whalers based on land stations in Abashiri and Hakodate, Hokkaido, Ayukawa, Miyagi Pref. and Wadaura, Chiba Pref. Strong anthropogenic sound emissions could exert serious influences on this species, e.g. lethal tissue damage, such as gas bubble disease or periotic hemorrhage [416]. The possible existence of 2 distinct populations, a black group and a slate-colored group have been mentioned [2702]. Anecdotal information of "Kurotsuchi (Black *Berardius*)" has been reported [1368, 2490, 2493]. Further comments on the whaler's recogniton of three types of *Berardius*, the "Akatsuchi (red *Berardius*)", the "Kurotsuchi (black *Berardius*)" and the "Karasu (crow)" have been made [645]. The black *Berardius* [1621] may have some relationship with "black *Berardius*" in the above mentioned references.

T. K. YAMADA

A skull specimen (NSMT M30129) of Baird's beaked whale (National Museum of Nature and Science).

Cetacea ZIPHIIDAE

145

Red list status: R for the population in the tropical western North Pacific (MSJ); DD (IUCN)

Indopacetus pacificus (Longman, 1926)

EN Longman's beaked whale **JP** タイヘイヨウアカボウモドキ (taiheiyou akaboumodoki)
CM 朗氏喙鯨 **CT** 朗氏喙鯨 **KR** 태평양민부리사촌, 태평양민부리고래 **RS** австралийский ремнезуб

Head of an adult female Longman's beaked whale stranded in Kagoshima Pref. (Kagoshima City Aquarium).

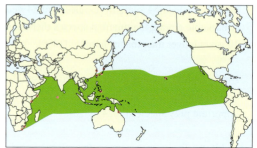

1. Distribution

Sightings of the Longman's beaked whale and/or individuals which appear to be this species are reported from tropical to subtropical waters of the Indian and Pacific Oceans, suggesting that it is fairly widespread and may be more abundant than previously thought. Sightings and strandings of this species are more common in the Indian Ocean and in the western Pacific than in the eastern Pacific Ocean. The waters around the Maldives are where individuals of this species are sighted most frequently [132, 697, 2254, 2845, 2847]. Red solid circles denote stranding sites. In Japan, stranding records are from Kagoshima, Hokkaido and Okinawa [1113].

2. Fossil record

Not reported.

3. General characteristics

Morphology For many years the species was recognized by only 2 skulls [234, 1861], and other information including the external morphology was not known. The species was originally described as a *Mesoplodon* species, but based on the morphology of several skulls it has been assigned to an independent genus *Indopacetus* [2191, 2882]. A little more than a dozen individuals were confirmed in Australia, Somalia, South Africa, Kenya, Maldives, Japan, Taiwan, Myanmar and the United States (Hawai'i) [69, 234, 431, 1861, 2815, 3826, 3868, 3947, 4063]. Body length ranges from 291 to 648 cm. The juvenile body form somewhat resembles that of the southern bottlenose whale (*Hyperoodon planifrons*) of similar size [431, 2904], in that the melon is bulbous and the beak is stout and relatively short. In larger individuals, however, the beak is much longer and more slender, and is clearly demarcated from the melon. Although there is no crease along the boundary between the beak and melon, the head appears more like that of the rough-toothed dolphin, so that the Philippines individual was first reported as "a gigantic bottlenose dolphin". Judging from the Philippines specimen, which was a subadult male of 576 cm, the melon of males could be more bulbous than that of females [69]. Still, the species seems to be less sexually dimorphic than most other beaked whale species. Although several authors have assigned sighting records of whales with very bulbous heads and stout beaks as adult males of this species, the characteristics of the subadult male specimen from the Philippines were suggestive of less developed secondary sexual characteristics. The adult body form is something between smaller beaked whale species (fusiform) and large beaked whales (cylindrical). At the throat is a pair of grooves forming a reversed V-shape, typical of beaked whales. The flippers are relatively short and fit into the 'flipper pockets". The triangular dorsal fin stands at about 2/3 of the body length from the tip of the head, and is somewhat falcate and proportionately taller than those of other beaked whales. There is no median notch at the middle of the flukes. Adult females are generally dark brownish gray all over but the melon and beak are light brown. Many of them show fairly dense cookie cutter shark bites.

Dental formula They have a pair of conical teeth at the tip of the lower jaw. As in the most of other ziphiid species, these teeth erupt only in males when they become matured, but the exact timing of the eruption is unknown.

Genetics Not reported.

4. Ecology

Reproduction A 596-cm female stranded in the Maldives in January had a fetus of 104 cm [431]. A stranding event of a possible mother–calf pair, specifically a lactating female with a calf, was reported from Taiwan in July [3826, 4063].

Lifespan Not reported.

Diet The only stomach contents hitherto identified include middle to deep squid species, such as *Taonius pavo*, *Moroteuthis loennbergi* and *Chiroteuthis imperator* [4097].

Habitat The Longman's beaked whale primarily inhabits waters off the continental shelf with depths of 250–3,500 m, and individuals traversing across the continental slope have been observed [132].

Behavior Average group sizes are reported as 4.2–7.2 for the western Indian Ocean, 8.6 for the eastern Pacific, 29.2 for the western Pacific, and 18.5 for the entire Indo-Pacific [132, 2847]. In one study, the mean dive time of 24 dives was 23 min (range = 11–33 minutes). A presumed mother and calf pair dove together for 23–25 minutes [132]. The blow has been expressed as conspicuous or low, bushy and usually visible, and obliquely forwarded, like that of the sperm whale [132, 2847].

Natural enemies Killer whales and large sharks could be natural enemies of this species. No further information is available at present.

Parasites Fairly large anisakid nematodes have been collected from the stomach. External parasites, at least one *Pennella* sp. and fairly numerous cyamid whale lice are found around the blowhole, throat groove, and ano-genital region.

5. Remarks

Since the 1960s, sightings of unidentified tropical bottlenose whales and *Indopacetus* sp. have been reported [132, 697, 1831, 2152, 2254, 2847, 3759, 3808]. There have been very few morphological descriptions. However, more morphological and molecular data have become available recently [69, 431, 3826, 3947, 4063], and further comparative investigations based on these new findings are needed for a better understanding of the taxonomy, phylogeny and ecology of this species. Morvillivirus infection, which is similar to the infections observed in pilot whales, was detected in a juvenile male individual stranded in Hawaii [3868].

T. K. YAMADA

A skull specimen (KGA M14) of an adult female Longman's beaked whale (Kagoshima City Aquarium).

Cetacea ZIPHIIDAE

Red list status: R (MSJ); DD (IUCN); R (JFA)

Mesoplodon carlhubbsi Moore, 1963

EN Hubbs' beaked whale JP ハップスオウギハクジラ (hubbusu ougiha kujira) CM 哈氏中喙鯨 CT 哈氏中喙鯨
KR 허브큰이빨부리고래 RS ремнезуб Хаббса

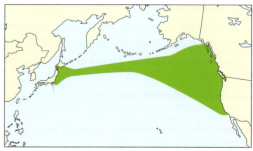

Head of a dead adult Hubbs' beaked whale live-stranded in Oiso, Kanagawa Pref. (H. Taru).

1. Distribution

Endemic to the North Pacific Ocean. All 17 strandings recorded in Japan are from the Pacific side of Honshu and Hokkaido (red solid circles). Other stranding data are from British Columbia to Southen California [2048].

2. Fossil record

Not reported.

3. General characteristics

Morphology A lateral view of this species shows a relatively large thorax with a small head and tail, which is somewhat similar to the other *Mesoplodon* species. Adult Hubbs' beaked whales have a very deep body but are less flattened, or more robust than Stejneger's beaked whales. The largest individuals are around 5.3 m and 1,500 kg, and no significant sexual differences are known [2048]. The beak continues smoothly to the melon. The melon, just anterior to the blowhole, is conspicuously raised in both sexes. In adult males a pair of tusks erupt just behind the true angle of the gape, which is located much anterior to the superficial angle of the gape, and the medial surfaces of the tusks are covered with oral mucus so that the medial surfaces are not exposed to the oral cavity. This would account for the lack of incision on the leading edge of the tusks at the gum line level. There is a pair of reversed V-shaped grooves, typical of beaked whales, at the throat. The flipper is shorter and wider than other beaked whales. Flippers fit into the "flipper pockets". The relatively small dorsal fin is at about 2/3 of the body length from the tip of the rostrum. There is no median notch between the flukes. Adult males and females are almost black all over, except for the beak, the top of the melon and dorsal (outer) surface of the flippers, which are characteristically white. Younger animals are counter-shaded with darker brownish gray on the dorsal side and almost white on the ventral side. Whiter symmetrical patterns of various shapes are seen on the ventral surface of the flukes. The body surface, especially around the ano-genital region, is heavily scarred with cookie cutter shark bites. Sexually mature males are heavily covered with linear scars mostly consisting of pairs of parallel lines.

Dental formula Hubbs' beaked whales have a single pair of teeth in the lower jaw almost at the angle of gape. Teeth do not erupt in immature males and females. In males, probably when they become sexually mature, the teeth break the gum and grow to a height almost halfway to the upper surface of the upper jaw.

Genetics Chromosome: $2n = 42$ [195].

Mesoplodon carlhubbsi Moore, 1963

4. Ecology

Reproduction   The mating system, gestation and lactation periods are unknown. Average length at birth is estimated as 2.5 m [2048]. Neonates of 2.3 m and 2.47 m in body length with fetal folds have been found in Iwate and Ibaraki Prefs. collected in June and April. The calving season for the eastern North Pacific population was estimated as 'summer' [2048]. Heavy linear scars, mostly consisting of parallel double lines, are believed to be incurred during struggles with other adult males during the mating season.

Lifespan   Not reported.

Diet   Squid beaks are often found in the stomachs of this species. Identified prey species include deep sea squids (*Gonatus* sp., *Onychoteuthis borealijaponicus*, *Octopoteuthis deletron*). Among the bones and otoliths of several fishes, *Chauliodus macouni* (Pacific viperfish) has been observed as a primary and significant species in stomach contents [2048]. Considering the prey species habitat, Hubbs' beaked whales are assumed to forage mainly in the middle to deep layers of the sea.

Habitat   The species appears to be primarily a mild to cold temperate resident, and seems to prefer slightly warmer water compared to Stejneger's beaked whale.

Behavior   Almost all the information on the species is from stranded animals and very little is known about its behavior.

Natural enemies   Killer whales and large sharks could be natural enemies of this species.

Parasites   *Crassicauda* infestation, which is severe in many of the beaked whales, is rare in this species. Anisakid nematodes are found in the stomach. External parasites, *Conchoderma* sp. are found especially on adult male tusks.

5. Remarks

This is one of the least reported species in Japan, with a total of 17 records [2499].

T. K. YAMADA

A skull specimen (KPM-NF1003632) of an adult male Hubbs' beaked whale. (Kanagawa Prefectural Museum of Natural History).

Cetacea ZIPHIIDAE

Red list status: R (MSJ); DD (IUCN); R (JFA)

Mesoplodon densirostris (de Blainville, 1817)

EN Blainville's beaked whale　　JP コブハクジラ (kobuha kujira)　　CM 柏氏中喙鯨　　CT 柏氏中喙鯨
KR 혹부리고래　　RS тупорылый ремнезуб

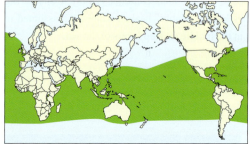

Head of an adult male Blainville's beaked whale which strayed into Hakata Bay, Fukuoka Pref. (Marine World Umino-Nakamichi). Two tusks are covered with numerous *Chonchoderma* barnacles forming tufts.

1. Distribution

Widely distributed from tropical to temperate waters of all oceans and often referred to as "cosmopolitan". The first record of this species and most of the stranding information from Japanese waters are from the coasts of the Okinawa Isls. [3677]. The northernmost record of the species in Japan is in Hitachi-shi, Ibaraki Pref. (36°30'N), where a young male stranded [1113]. There is an anecdotal report of this species captured in Toyama Bay, but details are not clear. Red solid circles denote stranding sites in Japan.

2. Fossil record

Not reported.

3. General characteristics

Morphology Adult males have a rather bizarre appearance. After the first astonishment, one will realize that the basic body characteristics except for those of the head are not so different from other species of the genus *Mesoplodon*. In adult males the alveolar portion of the mandibles are extremely elevated and the tusks erupt well above eye level. Females have a much more modest elevation of the alveoli, but this is more prominent than in females of other *Mesoplodon* species. They have a pair of throat grooves forming a reversed V-shape, typical of beaked whales. They have a median dorsal ridge starting at the posterior end of the head, which diminishes as it reaches the dorsal fin. The maximum body length is around 470 cm [1230], but measurements of several specimens collected in Japan have been generally less than 420 cm in males and around 450 cm in females. The smallest individual measured is a 181.5-cm neonate. A 4.39-m long individual was 809 kg [2867]. Adult males are heavily covered with linear scars believed to result from conflicts with conspecific adult males. Most of these scars are in parallel pairs. The flippers are relatively short and fit into the "flipper pockets". The triangular dorsal fin is at about 2/3 of the body length from the tip of the head. Usually they have no fluke notch. Most of them are covered with fairly dense cookie cutter shark bites, especially on the ventral side and around the ano-genital region. The body color is basically brownish dark gray with a whiter belly. Sometimes the body has a yellowish tint, possibly caused by diatom films. Neonates have a typical baby color of countershaded gray with a very dark patch around the eyes and over the head.

Dental formula They have a pair of teeth at the middle of the mandibles. As in the most of other ziphiid species, these teeth

Mesoplodon densirostris (de Blainville, 1817)

erupt only in males when they become matured, but the exact timing of the eruption is unknown. The tusks of adult males are not so flat as those of other *Mesoplodon* species, but are somewhat like those of the killer whales and often carry stalked barnacles *Conchoderma* sp.

Genetics Not reported.

4. Ecology

Reproduction A pair, probably a mother and calf, stranded in July in Numazu, Shizuoka Pref.; the larger female was 423 cm and the female calf 181.5 cm. In another possible mother and calf pair that stranded in March in Aichi Pref., body lengths were 449 cm and 261 cm. A neonate stranded in August in Hamamatsushi, Shizuoka Pref. was 208 cm in body length. These very limited data suggest that the calving season could be around July–August, and a yearing calf could be around 2.5 m in body length. A 471-cm female had milk in her mammary gland in South Africa. A 190-cm near term fetus that was found in a 433-cm mother was expected to be born within a month or two in March or April in South Africa [2904]. Low numbers of corpora lutea/albicantia counted in physically mature females in South Africa suggest that they might reach physical maturity soon after they attain sexual maturity [2904]. Milk sample was collected from a 423-cm female stranded in Shizuoka Pref., and consisted of 48.4% water, 34.8% fat, 15.8% protein, 0.2% carbohydrate and 0.8% ash [3019].

Lifespan Age estimation by growth layer group (GLG) count on 10 Blainville's beaked whales from South Africa indicated that the oldest female of 4.33 m had 12 or 13 GLGs, and the largest physically immature male of 4.15 m had 7 GLGs [2904]. Any time scale for GLG has not yet been established.

Diet Among 5 stomachs examined in South Africa, only one had stomach contents composed of fish otoliths of *Cepola* sp., *Scopelogadus* sp. and *Lampanyctus* sp. [2904].

Habitat This species prefers waters between 200 and 1,000 m in depth [1230].

Behavior Investigations using sound-and-orientation tags have revealed that 3 Blainville's beaked whales dove at depths between 640 and 1,251 m to feed, and the average foraging dive was 835 m and 47 minutes. These deep foraging dives occurred on average every 92 minutes, interspersed with several non-foraging shallower dives [3674]. Another 4 Blainville's whales tagged for dive data collection dove regularly for 48–68 minutes to depths greater than 800 m (median depth of 922 m and maximum 1,408 m). Ascent rates for long/deep dives were substantially slower than descent rates, while for shorter dives there were no consistent differences, and they spent prolonged periods of time in the upper 50 m of the water column. They appeared to prepare for long dives by spending extended periods of time near the surface [245]. Acoustic research has revealed that they selectively emit 2 different clicks to search for and capture prey [1245].

Natural enemies Killer whales and large sharks could be natural enemies of this species. A live stranded neonate had several fresh shark bites.

Parasites In some cases anisakid nematodes are collected from the stomach. Nematodes and cestodes are found in the intestine. *Crassicauda* sp. is found in the kidneys, but not as many as in some of the other beaked whales, such as *Mesoplodon stejnegeri*. *Chonchoderma* sp. is often found attached to adult male tusks.

5. Remarks

The most significant issue related to this species is the sonar problem, including decompression syndrome, which leads the animals to death [251, 546, 2896]. Further investigation on decompression syndrome and its physiological considerations are needed. Because of the species' extremely wide distribution, geographical variation in many of its traits should also be examined.

T. K. YAMADA

A skull specimen (NSMTM34021) of an adult male Blainville's beaked whale (National Museum of Nature and Science). Note the extremely elevated alveolar portion.

Cetacea   ZIPHIIDAE

148

Red list status: R (MSJ); DD (IUCN); R (JFA)

Mesoplodon ginkgodens Nishiawaki & Kamiya, 1958

EN ginkgo-toothed beaked whale   JP イチョウハクジラ (ichouha kujira)   CM 銀杏齒中喙鯨   CT 銀杏齿中喙鲸
KR 은행이빨부리고래   RS японский ремнезуб

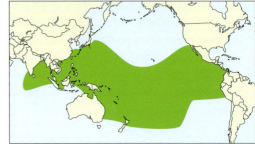

Head of an adult male ginkgo-toothed beaked whale stranded in Izu, Shizuoka Pref. (National Museum of Nature and Science).

1. Distribution

Distributed in tropical and temperate waters of Pacific and Indian Oceans. They are sometimes stranded along the Pacific side of the Japanese Isls. from Okinawa to Ibaraki Prefs., with an exceptional record in Muroran, Hokkaido. Records are more frequent in Shizuoka, Kanagawa and Chiba Prefs. [1110, 1299, 2371, 2393, 2498, 2500, 2574]. A red solid circle denotes the sampling site of the type locality of this species.

2. Fossil record

Not reported.

3. General characteristics

Morphology   The spindle shaped body consists of a relatively large thorax with small head and tail, similar to the other *Mesoplodon* species. Males of this species complete physical maturity when they attain a body length of around 4.7 m [2500]. Physically mature individuals are up to 5.3 m and 1–1.5 tons. There is no clear demarcation between the beak and the melon. Even in adult males, only the tips of the tusks break the gums. There is a pair of reversed V-shaped grooves, typical of beaked whales, at the throat. The small flippers fit into the "flipper pockets". The relatively small dorsal fin is at about 2/3 of the body length from the tip of the rostrum. The flukes are relatively wide and no median notch is present. Adult males and females are dark gray to almost black, with some whiter portions around the beak. One 3.12 m juvenile was reported to be "gray" all over [2371]. The body surface, especially around the ano-genital region, is heavily scarred with cookie cutter shark bites. It is of note that linear scars in parallel pairs, usually found in adult males of most of the beaked whales, are scarcely found in this species. This suggests that the species could be a "peaceful beaked whale" [432, 2500, 2872], or that the tusks are too small to leave scars. Healed fractures of the ribs are reported in some cases, which might indicate that they take to ramming each other but without a sharp weapon.

Dental formula   The ginkgo-toothed beaked whale has only 1 pair of teeth in the lower jaw. The teeth erupt only in adult males but even then only the tips of the teeth are visible.

Mesoplodon ginkgodens Nishiawaki & Kamiya, 1958

Genetics Analyses of mtDNA resurrected *Mesoplodon hotaula*, which was once treated as a synonym of *M. ginkgodens* [2192], being independent from *M. ginkgodens* [430].

4. Ecology

Reproduction Almost nothing is known about their mating system, gestation and lactation periods. The absence of heavy linear scars suggests that struggle between adult males during the mating season could be less fierce or that there is a different mechanism for male-to-male competition. In a 477-cm physically mature whale, the testes were 0.13 kg (R) and 0.14 kg (L). In this animal, the peripheral tissue in the testis was at the puberty stage; mature tissue and sperm being found only around the "ampulla of the vas deferens" [2500]. A 312-cm female whale had a fringed tongue, suggesting that she was still suckling [2371]. In Taiwan, a milk sample was collected from a lactating 489-cm mother with a 284-cm male calf. The fatty acid component was higher than those of terrestrial mammals but lower than those published for Stejneger's and Blainville's beaked whales [3021].

Lifespan Not reported.

Diet Squid beaks have been found in stomachs. It has been suggested that the major prey may be similar to other beaked whales.

Habitat Not reported.

Behavior Not reported.

Natural enemies Killer whales and large sharks could be natural enemies of this species.

Parasites Anisakid nematodes are found in the alimentary tract, especially in the stomach, Acanthocephalan thorny-headed worms in the intestine and *Crassicauda* sp. in the kidneys also occur in this species [2371]. External parasites, *Conchoderma* sp. are found especially on adult male tusks.

5. Remarks

This species was originally described based on a specimen stranded in Kanagawa Pref. [2498]. The earliest known specimen is, however, an individual landed at a fishing port in Miyazaki Pref. in 1935 [2574, 2576].

T. K. YAMADA

The skull of the holotype specimen (NSMTM8744) of the ginkgo-toothed beaked whale (National Museum of Nature and Science).

Cetacea ZIPHIIDAE

(149)

Red list status: R (MSJ); DD (IUCN); R (JFA)

Mesoplodon stejnegeri True, 1885

EN Stejneger's beaked whale   JP オウギハクジラ (ougiha kujira)   CM 史氏中喙鯨   CT 史氏中喙鯨
KR 큰이빨부리고래   RS командорский ремнезуб

A breaching juvenile of probably Stejneger's beaked whale. Note the juvenile coloration and well-defined flipper pocket (J. Ohara).

stranding is frequent
stranding is occasional
stranding is rare

1. Distribution

Endemic to the North Pacific Ocean. In Japan, stranding reports are mainly from the Sea of Japan side, Fukuoka in the southwest, up to Wakkanai in the northeast. Strandings are also reported from several localities along the Sea of Okhotsk, and the Pacific coasts of Hokkaido and northern Honshu. Records from areas other than Japan are from coasts of the Sea of Japan, the Sea of Okhotsk, the Bering Sea, the Aleutian Isls. and the West Coast of Alaska, Canada and the United States [2047].

2. Fossil record

Not reported.

3. General characteristics

Morphology   The lateral view of this species is similar to the other 3 *Mesoplodon* species known from the seas around Japan. Adult Stejneger's beaked whales are, however, extremely flattened laterally and the body is very deep [1866, 2047]. Physically mature individuals are between 4.6 and 5.3 m in body length and around 1,000 kg in BW. No significant sexual differences in body measurements are known. Neonates are about 2.2 m in length. The beak continues smooth to the melon, and in adult males a pair of triangular tusks erupt at the angle of gape. This pair of tusks in adult males pinches the upper jaw and grows to the level of the upper contour of the upper jaw. As they grow older the leading edge of the tusk at the gum line becomes worn to form a notch-like abrasion. There is a pair of reversed V-shaped grooves at the throat. The spatular flippers are about 10% of the body length, and fit into the "flipper pockets". The relatively small dorsal fin is at about 2/3 of the body length from the tip of the rostrum. Usually there is no median notch at the center of the flukes. Adult males and females are almost black all over, whereas younger animals are brownish gray, though darker on the dorsal and lighter on the ventral sides. Whiter symmetrical patterns of various shapes are seen on the ventral surface of the flukes. When males reach sexual maturity, the tusks break the skin, and the body surface is gradually covered with linear scars mostly consisting of 2 parallel lines. Numerous cookie cutter shark bites are reported in individuals of the eastern North Pacific [3818]. In contrast, they are extremely rare in individuals examined in Japan.

Dental formula   The species has a single pair of teeth in the lower jaw almost at the angle of the gape. No erupted teeth have been confirmed in females, where teeth remain unerupted and do not break the gum although the size may become larger as they grow. In males, probably in line with sexual maturity, teeth grow

Mesoplodon stejnegeri True, 1885

tall to almost reach the level of the upper surface of the upper beak and pinch the upper jaw. In some older males tusks are broken off at the level of the gum line.

Genetics Chromosome: not reported. In full sequences (about 1,000 bp) of the mtDNA control region obtained from 26 individuals of Stejneger's beaked whales collected from Japanese waters, only 3 haplotypes were found and thus genetic diversity was extremely low among these 26 individuals [1279].

4. Ecology

Reproduction The mating system, gestation and lactation periods are unknown. Neonate strandings have been reported from Ishikawa to Akita Prefs. mostly during the period between late April and early June [3943]. A fetus was found from a female stranded in Sado Isl. in October, 1988 [3943]. Fetuses with body lengths of 1.4 m, 1.7 m and 2 m were collected from mothers stranded in February, March and April in Yamagata and Niigata Prefs. [3943]. A 503-cm lactating female was stranded in Ishikawa Pref. in late April [1684]. Milk collected from the individual was examined [1684]. The fatty acid and amino acid components were significantly higher than those reported for the blue-green milk collected from another female investigated in Alaska [3722].

Lifespan The oldest animal recorded was aged to 34.5 GLGs. This was an adult male of about 5 m in body length [175].

Diet Squid beaks are often found in stomachs. Identified prey species include deep sea squids such as *Berryteuthis magister* or *Gonatopsis borealis* [3948]. Otoliths and bone remains of fishes have also been found. Based on frequency, however, fish seem to be a less important source of food. Nearly 70% of stomachs examined had various anthropogenic debris, mostly plastic bags.

Habitat The main habitat appears to be cold temperate and sub-arctic. According to very limited sources this species might spend more time in areas of the deep sea with significant slope [3209]. Stranding records are much more frequent during winter along the west coast of Honshu. Winter migration from the north to the central area of the Sea of Japan has been argued, but the apparent migration might simply be due to the extremely fierce northwesterly seasonal winds during winter.

Behavior The pod size ranges between 5 and 15 for 52 sighting records of what were probably Stejneger's beaked whales in waters along the Aleutian Isls. in June and July. The pods observed consisted of different sexes and ages and group cohesion seemed tight [1870]. According to limited sighting information from Japanese waters, they are seen in groups of less than 3, and are mostly solitary. The species is believed to be extremely shy and sighting data are very scarce; however, they are sometimes observed as very active. The parallel double linear scars mentioned above have an interval about 70 mm or smaller. Because the distance between the tips of both matured male tusks is

Dynamic breach of a female identified as a Stejneger's beaked whale. She was accompanied by a juvenile (see the left page) and they breached several times together (J. Ohara).

about the same, these linear scars are interpreted as the result of fighting between adult males during the mating season. Synchronization of sexual maturity, teeth (tusk) eruption and linear scars on the male body also support this hypothesis.

Natural enemies Killer whales and large sharks could be natural enemies of this species. Some stranded individuals were heavily injured probably because of attacks from larger sharks, but most injuries are considered to be the result of scavenging after death.

Parasites Most notorious is *Crassicauda* sp. found in abundance in the kidneys. They often cause glanuromoutous nephritis. Other nematode infestation is rare but sometimes found in the stomach. External parasites, *Conchoderma* sp. are very frequently found especially on adult male tusks often in clusters. In the eastern North Pacific, fairly heavy infestations of *Pennella* are reported [3818], but in the seas around Japan no record of *Pennella* is known.

5. Remarks

Since the first specimen of this species in Japan was collected in Akita Pref. [2492], the species had been understood as extremely rare. However, after active data and sample collections started in the end of the 1980s, fairly frequent strandings have been recorded and more than 100 specimens have been collected to date. The results of pathological investigations have found some diseases such as amyloidosis among the necropsied stranded individuals [3397].

T. K. YAMADA

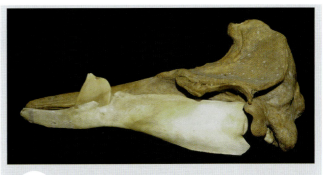

A skull specimen of an adult male Stejneger's beaked whale (National Museum of Nature and Science).

Cetacea ZIPHIIDAE

Red list status: K (MSJ); LC (IUCN); R (JFA)

Ziphius cavirostris G. Cuvier, 1823

EN Cuvier's beaked whale, goose-beaked whale JP アカボウクジラ (akabou kujira) CM 柯氏喙鯨 CT 柯氏喙鯨
KR 민부리고래 RS настоящий клюворыл

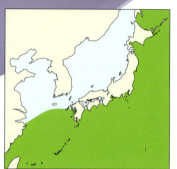
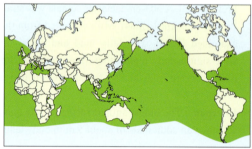

Head of an adult female Cuvier's beaked whale stranded in Shiraoi, Hokkaido (National Museum of Nature and Science).

1. Distribution

One of the most extensively distributed species in the Family Ziphiidae, being recorded in all oceans except for polar regions. In Japan, strandings are reported mainly from the Pacific side. Based on records from 1948–1952 ($n = 85$) and 1965–1970 ($n = 189$), most catches were done by whalers based in Chiba Pref., then Miyagi Pref., followed by Wakayama Pref. A few individuals were caught in the Sea of Okhotsk. Those whales were caught by whalers primarily hunting Baird's beaked whales [2503, 2702].

2. Fossil record

Not reported.

3. General characteristics

Morphology Cuvier's beaked whale is long and stocky, with a relatively small head and tail. The beak is inconspicuous compared to other beaked whale species. The basic proportions of the skull, however, are not so different from those of other beaked whales, suggesting the specific soft anatomy, at least of the relative size, of the melon is more massive than that in most other *Mesoplodon* beaked whales. A pair of throat grooves is in reversed V-shape. Relatively small flippers can be retracted to fit into the "flipper pocket", which could be suggestive of a deep diver. The small dorsal fin is almost opposite to the ano-genital region. The flukes are proportionately large with no median notch. The body is brownish or bluish dark gray, but the anterior portion of the head is white to very light brownish gray. The body is densely covered with cookie cutter shark bites especially on the ventral side.

Dental formula They have a pair of small conical teeth at the tip of the lower jaw. They often have rows of "vestigial" smaller teeth in the gum tissue of both jaws [581, 583, 2575].

Genetics Not reported.

4. Ecology

Reproduction Male Cuvier's beaked whales reach sexual maturity at 5.5–5.8 m. One 5.5-m female had a fetus, but pregnant females were mostly longer than 6 m. Pregnant females are 5.5–6.9 m [2503, 2702].

Lifespan The cementum layer count in a 5.79-m female was 36+ growth layer groups (GLGs) and in a 5.88-m male 30 GLGs [2904], although there is no confirmation that these GLGs are produced annually.

Ziphius cavirostris G. Cuvier, 1823

Diet   Generally they are said to forage mainly for squids [2097, 3606]. The predominant stomach contents of whales captured in the Kanto area were deep sea fishes, but those off the Sanriku coast and Wakayama were squids [2503]. Squid beaks (84 upper and 101 lower) of nearly 20 species and more than 40 morid fish otoliths were found from the stomach of a 5.88-m male [2904].

Habitat   In the whaling seasons of 1965–1970, Cuvier's beaked whales were captured along or near the 1,000 m isobath [2503].

Behavior   Seven Cuvier's beaked whales were examined with sound-and-orientation tags. They were estimated to forage mainly at depths between 613 and 1,297 m, and the average foraging dives were 1,070 m and 60 minutes. The average interval of these deep foraging dives was 63 minutes, interspersed with several non-foraging shallower dives [3674]. In Hawaii, 2 Cuvier's whales were studied using time-depth recorders attached by suction-cup. They regularly dove for 48–68 minutes to depths greater than 800 m (maximum 1,450 m). Ascent rates for long/deep dives were substantially slower than descent rates, while for shorter dives there were no consistent differences, and they spent prolonged periods of time (155 minutes) in the upper 50 m of the water column. Based on time intervals between dives, such long dives were likely aerobic. The species appeared to prepare for long dives by spending extended periods of time near the surface [245]. Record breaking diving depth of 2,992 m and duration for 137.5 min has been confirmed [3151]. This is the deepest dive recorded for a mammalian species.

Natural enemies   Killer whales and large sharks could be natural enemies of this species. One Cuvier's beaked whale was reported to have fairly large ante mortem shark bites [251]. No further information is available at present.

Parasites   Anisakid nematodes are found in the alimentary tract, especially the stomach, the Atlantic examples of which are identified as *Anisakis ziphidarum*, known to be specific to beaked whales [1635]. Numerous nematodes (e.g. *Crassicauda giliakiana*) are often found in the kidneys [1560]. External parasites, at least *Pennella* sp. and fairly numerous cyamid whale lice, are found around the blowhole, throat groove and ano-genital region.

5. Remarks

Cuvier's beaked whales are frequently referred to as possible serious victims of naval sonar. In Japanese waters, during the period 1960–1990, more than 46 animals were stranded in 8 unusual mass-stranding events in Sagami and Suruga Bays [1110], and the possible influence of military sonar has been argued [333]. Further physiological and anatomical considerations are needed, in relation to the above-mentioned diving profile, to confirm possible decompression syndrome [245, 251, 546, 2896, 3674]. Because of their extensive distribution, geographical variation in various characteristics should be examined.

T. K. YAMADA

A skull specimen (NSMTM34256) of an adult female Cuvier's beaked whale (National Museum of Nature and Science).

Cetacea DELPHINIDAE

151

Red list status: K (MSJ); DD (IUCN); R (JFA)

Delphinus capensis Gray, 1828

EN long-beaked common dolphin   JP ハセイルカ（hase iruka）   CM 长吻真海豚   CT 長吻眞海豚
KR 긴부리참돌고래   RS длинномордый дельфин

Long-beaked common dolphins are regularly observed in Tosa Bay, Kochi Pref. (M. Amano).

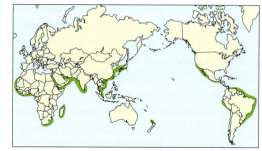

1. Distribution

Distributed in tropical to temperate waters in the Atlantic, Indian and Pacific oceans. It most commonly occurs in coastal and warm waters compared with short-beaked common dolphins (*D. delphis*), thus the distribution ranges of the 2 species are different. Off the Pacific coast of Japan, this species occurs south of the Izu Pen. and is relatively common in Tosa Bay. Common dolphins, identified as *D. capensis* also occur in the Sea of Japan [103]. It also occurs in the East China Sea [103].

2. Fossil record

Not reported in Japan.

3. General characteristics

Morphology Beak is distinct and longer than that of the short-beaked common dolphin. It shows clear hour-glass pattern coloration, but the contrast is weaker than that of the short-beaked common dolphin. The thoracic patch is especially paler. Gray flare in the center of dorsal fin is also weak even in older specimens. Flipper stripe gradually narrows and usually is less distinct than that of the short-beaked common dolphin. Although it is reported in North American specimens that flipper stripe fuses with lip patch at the 1/3 length anterior to the gape or runs in parallel with the gape, it fuses more anteriorly in most Japanese specimens [103]. Adult BL of specimens from Japanese waters is about 230–260 cm for males and 230 cm for females, which is greater than that of the short-beaked common dolphin [103]. Relative rostral length to zygomatic width is > 1.52 [103, 884]. *Delphinus* in the Indian Ocean possesses much longer rostrum and is considered a different subspecies *D. c. tropicalis* [1228]. Cranial measurements were reported [103, 1228].

Dental formula Largest number of teeth in a row is 50–60 off Japan (mean ± SD = 53.8 ± 2.6, *n* = 15) [103]; but is 54–67 for the *tropicalis*-form (mean ± SD = 59.5 ± 3.17, *n* = 58) [1228]; and 47–60/47–57 (*n* = 53/51) for individuals off California [2820].

Genetics Chromosome: karyotype is not reported but number of chromosomes is presumed to be the same as for other delphinids, 2*n* = 44. Very high genetic differentiation, even larger than that between *D. delphis* and *D. capensis*, was reported among populations in the Atlantic and eastern Pacific [2449]. No genetic information is available for western Pacific populations. Hybrids between *D. capensis* and *Tursiops truncatus* were reported and 1 female among them was fertile [4255].

Delphinus capensis Gray, 1828

A cow–calf pair of the long-beaked common dolphin found in Ise Bay, Mie Pref. (N. Ozaki).

4. Ecology

Reproduction Information on life history of *D. capensis* is scarce. In the population off South Africa, females attain sexual maturity at approximately 210–220 cm in body length, and give birth any time of the year with a peak in summer [2904].

Lifespan Not reported. The oldest individual among 15 specimens from the western North Pacific was 22 years old [103].

Diet Prey of long-beaked common dolphins was reported only in southern African and Californian waters. Off the coast of South Africa, long-beaked common dolphins feed on various schooling fishes (sardines, pilchards, herrings and anchovies) and cephalopods [124, 2904, 3153, 3161, 4228]. It is reported to feed on mesopelagic lantern fishes at night and in the early morning off South Africa [3161]. But they mostly depend on limited dominant prey species [124]. Long-beaked common dolphins in the Gulf of California change feeding habits toward higher trophic levels with age [2474].

Habitat Long-beaked common dolphins occur in more inshore (mainly < 100 miles off California) [533] and warmer waters [2820] than short-beaked common dolphins, although their sightings in similar locations suggest complex habitat partitioning in the area where the 2 species occur sympatrically [269].

Behavior Long-beaked common dolphins make schools of various sizes up to several thousands. School structure seems to be similar to that of short-beaked common dolphins, but no accurate study of the structure has taken place.

Natural enemies There is an observation that a pod of killer whales presumably attacked a group of long-beaked common dolphins [330]. Besides killer whales, larger sharks are possible predators.

Parasites A nematode *Anisakis simplex* was reported in the stomach of *D. capensis* stranded in Japan [1765]. Infestation of *Crassicauda* sp. in the cranium was reported from Peru [3768].

5. Remarks

In recent times, long-beaked common dolphins have not been taken in Japan. Former catches by drive or harpoon fisheries in western Kyushu during the 20th century should have included this species [1381]. Several by-catches in set nets and gill nets were reported off the southern Pacific and the Sea of Japan coasts.

M. AMANO

A school of long-beaked common dolphins in Ise Bay, Mie Pref. (N. Ozaki).

Cetacea   DELPHINIDAE

Red list status: K (MSJ); LC (IUCN); ND (JFA)

Delphinus delphis Linnaeus, 1758

EN short-beaked common dolphin   JP マイルカ（ma iruka）   CM 真海豚   CT 真海豚
KR 짧은부리참돌고래   RS дельфин-белобочка

Short-beaked common dolphins frequently jump out of water in schools (Haruka Sato).

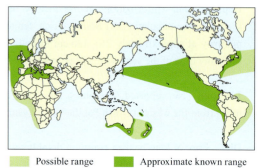

Possible range   Approximate known range

1. Distribution

Widely distributed in tropical to temperate pelagic waters of the Atlantic and Pacific. In the North Pacific, northern limit of distribution is near British Columbia to the east and at northern Honshu to the west. Although information is limited off Japan, short-beaked common dolphins occur mainly off the Pacific coast of Japan north of the Izu Pen. and seem parapatric to long-beaked common dolphins (*D. capensis*). But there is a possibility of sympatry around the Ryukyu Isls. It usually occurs in offshore waters and less commonly in coastal waters, compared with long-beaked common dolphins and other warm water dolphins such as striped (*Stenella coeruleoalba*) and pantropical spotted (*S. attenuata*) dolphins [1357].

2. Fossil record

Not reported in Japan.

3. General characteristics

Morphology   Beak is distinct and long but shorter than that of long-beaked common dolphins. Both *Delphinus* species have a V-shaped cape making an hour-glass pattern on the side and a yellowish thoracic patch. The contrast between the thoracic patch and cape is sharp. Falcate dorsal fin is black but a gray flare develops in the center with age. Flipper stripe distinctly narrows anterior to the eye, and fuses with lip patch at 1/3–1/2 length of mouth from the gape. There seems to be considerable individual and geographic variation in coloration. Mean body length (± SD) of adult animals (> 9 years) is 177.2 ± 6.22 cm ($n = 8$) for females and 183.2 ± 7.88 cm ($n = 11$) for males [103]. In the cranium, as shared characteristics for *Delphinus*, deep palatal grooves are separated by distinct mesial ridge and both maxillae are fused in the middle of the rostrum on the dorsal side. Relative rostral length is different between *D. delphis* and *D. capensis*; mean relative rostral length to zygomatic width is about 1.4 for *D. delphis*, but 1.6 for *D. capensis* [103, 884]. Cranial measurements were reported [103].

Dental formula   Largest number of teeth in a row is 46–56 (mean ± SD = 50.5 ± 2.14, $n = 39$) in the specimens from the western North Pacific [103] and 42–54/41–53 ($n = 49/47$) in individuals off California [2820].

Genetics   Chromosome: $2n = 44$ [190]. The short-beaked common dolphin is reported to be genetically distinct from the long-beaked common dolphin [1582, 2899]. Relatively lower genetic differentiation among populations in the Atlantic suggests significant gene flow [123, 1873, 2280, 2449]. Off eastern Australia,

Delphinus delphis Linnaeus, 1758

population differentiation is reported to coincide with oceanographic structure [2189]. There is no information on genetic variation within *Delphinus* in western Asia.

4. Ecology

Reproduction Information on life history of short-beaked common dolphins around Japan is less available. For the offshore North Pacific population, females attain sexual maturity at 7.2–8.5 years old and 182–189 cm in body length, and males 10.5 years old and 182–189 cm [553]. Gestation lasts about 11–12 months, lactation 14–19 months, and calving interval is about 1.3–2.6 years [433, 553, 2832, 3869]. Reproductive seasonality shows considerable geographic variation. Adult males have extremely large testes (up to 4 kg each), suggesting promiscuous mating and sperm competition [2333].

Lifespan The oldest individual among 44 specimens from the western North Pacific was 18 years of age [103], but individuals may live longer. Estimates of 28 years of age were reported in the Black Sea [103] and eastern North Atlantic [2333], and 25 years of age in the eastern tropical Pacific [433].

Diet Prey reported from western North Pacific includes various mesopelagic and epipelagic fishes (e.g. Myctophidae, Bathylagidae, Scombersocidae) and squids (e.g. Gonatidae, Enopleteuthidae) [388, 2626].

Habitat The short-beaked common dolphin occurs in pelagic waters of 18–26°C surface water temperature [1357]. In the eastern Pacific they most commonly occur in waters with complicated topography, with upwelling, or with higher chlorophyll concentration [348, 2281, 2876]. Groups with calves or neonates tend to use nearshore habitat in the Mediterranean and off Australia [348, 554]. They conduct seasonal movements [729] but no information is available for the North Pacific region.

Behavior Short-beaked common dolphins form schools of hundreds to thousands of individuals, but basic unit may be 20–30 animals. These groups are suggested to be not necessary kin-based [3779]. Segregation by sex and reproductive status was reported [1230]. Short-beaked common dolphins are known to be active at the surface and show various aerial behaviors.

Natural enemies Killer whales and larger sharks are possible predators, though there are no direct observations.

Parasites A few species of copepod, cirriped, and cyamid ectoparasites and various helminth endoparasites were reported from *Delphinus* spp. [279, 533, 1309, 1966, 2031, 2859, 2860]. A cestode *Diphyllobothrium stemmacephalum* was reported from Japan [1763].

5. Remarks

In recent times, short-beaked common dolphins have not been taken in Japan. Catches by drive and harpoon fisheries of the genus *Delphinus* were recorded in various areas, but the species is unknown. In the offshore western North Pacific, a large number of individuals belonging to this genus, possibly *D. delphis*, were taken incidentally by multinational high seas driftnets. There should be some incidental kills by driftnets off Japan but no information exists. Naval activities with intense mid-frequency sonars were suggested to have caused a mass stranding of this species in the U.K. [1233].

M. AMANO

Cetacea   DELPHINIDAE

153

Red list status: R (MSJ); DD (IUCN)

Feresa attenuata Gray, 1874

EN pygmy killer whale   JP ユメゴンドウ (yume gondou)   CM 小虎鯨   CT 小虎鯨
KR 들고양이고래   RS карликовая косатка

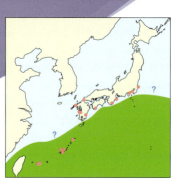

Pygmy killer whales swimming in an unusual location near a beach in northern Kyushu (Shimonoseki Marine Science Museum "Kaikyokan").

1. Distribution

Distributed in tropical and subtropical waters worldwide. Although sightings are not common, Japan has records of relatively frequent strandings many of which are from Okinawa, reflecting their warm-water habitats [3677]. The northernmost record is in Ibaraki Pref. Pygmy killer whales are relatively common in Hawaiian waters and the eastern tropical Pacific. Red solid circles in the map are known records in Japan.

2. Fossil record

Not reported in Japan.

3. General characteristics

Morphology   Round melon is projected forward against the tip of mouth. Dorsal fin is high and large. Flipper is also relatively large and round at its tip which is distinct from melon-headed whales. Body posterior to dorsal fin becomes abruptly thin in contrast to robust anterior part. Color is mostly dark gray and dorsal cape is narrower and more or less parallel to the body axis, whereas it extends downward at the dorsal fin in melon-head whales. Lip is often white. Ventral light patch from thoracic to genital area often extends laterally around the umbilicus and genital area. Measurement data are limited. Body length of sexually mature animals ranges from 220 to 227 cm ($n = 3$) for females and from 207 to 261 cm ($n = 3$) for males [475]. Skull is robust with broad rostrum less tapering anteriorly until a third length from the tip. Antorbital notches and preorbital processes are distinct. Skull measurements were reported [309, 358, 2906].

Dental formula   8–11/10–13 [2906].

Genetics   Chromosome: not reported but number of chromosomes presumed to be the same as for other delphinids, $2n = 44$. There is no information on intraspecific genetic variation. Pygmy killer whales show close genetic affinity to pilot whales, melon-headed whales, false killer whales, and Risso's dolphins [2034, 2042, 3778].

4. Ecology

Reproduction   Almost nothing is known of the life history of pygmy killer whales. They attain sexual maturity at body length about 200 cm and length at birth is about 80 cm [2906].

Lifespan   Not reported.

Feresa attenuata Gray, 1874

Diet Cephalpod beaks, fish otoliths and lens, and tunas were reported from their stomach contents [358, 2062, 2906, 3396]. Pygmy killer whales are assumed to feed on other small odontocetes (see Behavior).

Habitat Pygmy killer whales usually occur in offshore warm tropical and subtropical waters. In the Gulf of Mexico, they were sighted in the areas with 24.5–28.2°C surface water temperature [2035].

Behavior Reported mean school sizes are 20–30 animals; larger groups are rare [544, 2035]. Pygmy killer whales usually swim calmly, but porpoise, bow-ride, or leap on occasion [2906]. Aggressive, or chasing and attacking behavior toward other small dolphins were reported both in captivity and in the purse seine fishery operation, although there is no direct observation of predation on other small odontocetes [475].

Natural enemies No records of predation of pygmy killer whales were reported, but killer whales and larger sharks are possible predators.

Parasites Barnacles *Conchoderma* sp. and *Xenobalanus globicipitis* were reported [1309, 2860, 2906]. Various genera of helminth parasites including acanthocephalan, *Bolbosoma*; nematodes, *Anisakis*, *Filocapsularia*, *Stenurus*, *Terranova*, *Phaurus*; cestodes, *Tetrabothrius*, *Trigonocotyle*, *Monorygma*; and trematodes, *Nasitrema* were reported [2062, 2906].

Lip of pygmy killer whales is often white (Shimonoseki Marine Science Museum "Kaikyokan").

5. Remarks

In recent times, pygmy killer whales have not been taken directly or indirectly around Japan. Because of its rare occurrence, human effects on this species are thought to be limited in Japanese waters.

M. AMANO

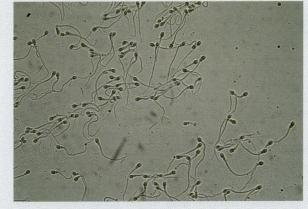

Sperms of a pygmy killer whale (M. Yoshioka).

The profile of a female pygmy killer whale stranded on a beach in Miyazaki Pref. (T. Kurita).

Cetacea   DELPHINIDAE

Red list status: R for the southern and northern forms, Common for the offshore population in the western North Pacific (MSJ); DD (IUCN); DC for the population north of Choshi to Hokkaido coasts, ND for south of Choshi (JFA)

Globicephala macrorhynchus Gray, 1846

EN short-finned pilot whale   JP コビレゴンドウ (kobire gondou)   CM 短肢领航鲸   CT 短肢領航鯨
KR 돌쇠고래   RS короткоплавниковая гринда, южная гринда

A group of short-finned pilot whales (northern form) off Muroran, Hokkaido (M. Amano).

■ Possible species range
▨ Range of northern form
▨ Range of southern form

1. Distribution

Occurs in tropical to temperate pelagic waters worldwide. In the western North Pacific, northern limit is off the Pacific coast of Hokkaido. Off the Pacific coast of Japan 2 distinct forms are segregated south and north of Kuroshio front near 35–37°N. Southern forms are found continuously as east as approximately 165°E, whereas northern forms do not occur west of 150°E [1397]. The range of southern forms extends south to around Philippines [367]. The form of short-finned pilot whales found in the Sea of Japan is unknown.

2. Fossil record

Not reported in Japan.

3. General characteristics

Morphology   Head is bulbous and melon is usually projected ahead of the snout. In the southern form, head of adult males becomes more squarish when seen from above, whereas this change is not obvious in northern-form males [1401]. Large and falcate dorsal fin is located posteriorly about 1/3 length of the body from the head. Adult males have dorsal fins with a wider base and rounder leading edge. Flippers are long and curved backward. Adult males have a massive hump on the back anterior to the dorsal fin. Body is mostly dark brownish gray to black with a pale anchor-shaped patch on the chest, saddle patch behind the dorsal fin and faint diagonal mark above the eye. The saddle patch and mark above the eye are distinct in the northern form but not obvious in the southern form. Body size shows drastic sexual and inter-form variation. Mean body length ($\pm$ SD) after cessation of growth is 3.64 ± 0.13 m (> 22 years of age, $n = 181$) for southern-form females, 4.74 ± 0.10 m (> 27 years of age, $n = 35$) for southern-form males, 4.67 ± 0.15 m (> 30 years of age, $n = 58$) for northern-form females, and 6.5 ± 0.33 m (> 30 years, $n = 11$) for northern-form males [1394, 1401]. The cranium is broad with well-developed preorbital processes and antorbital notches. Rostrum is short and wide and premaxillae widen anteriorly to cover maxillae. Mean condylobasal length of cranially

Globicephala macrorhynchus Gray, 1846

mature whales are: southern-form females, 57.4 ± 1.34 cm ($n = 15$); southern-form males, 64.1 ± 2.71 cm ($n = 10$); northern-form females, 62.9 ± 1.05 cm ($n = 9$); and northern-form males, 73.0 ± 2.77 cm ($n = 8$) [2159]. Other skull measurements were reported [2159]. Southern-form males show greater sexual dimorphism in skull morphology than northern-form males [2159].

Dental formula 5–10/7–10 in Japanese specimens [2159]. Teeth are conical in younger specimens. Their tips are worn to round shape with age.

Genetics Chromosome: $2n = 44$ [190]. There are significant genetic differences between northern and southern forms off Japan, but the differences are smaller than those observed between the short-finned and long-finned pilot whales (*G. melas*) [1256, 3802]. Based on the mtDNA analysis, the existence of 3 distinct populations off Japan is suggested and Japanese waters is considered to be a center of diversity of this species [2734]. Some gene flow between northern forms and eastern Pacific also is suggested [2734].

4. Ecology

Reproduction Average age and body length at attainment of sexual maturity of females is 9 years of age and 3.2 m for the southern form and 8.5 years of age and 3.9–4.0 m for the northern form [1392, 1401]. Average calving interval for the southern form is 7.8 years, but annual pregnancy rate decreases with age and females do not get pregnant after 35 years. Inversely, lactating period increases with age from about 1 year in females of 6–12 years to 6.6 years in those older than 36 years of age. All females older than 50 years of age are neither pregnant nor lactating and show a reproductive senescence or menopause [1392, 1957, 1958]. Gestation lasts about 15 months. Males attain sexual maturity at 17 years of age and 4.2 m in body length in the southern form and 16–18 years of age and 5.5–5.6 m in the northern form [1392, 1401].

Lifespan The oldest individuals known from the western North Pacific were a female of 62 years of age and males of 45 years of age [1394].

Diet Squids predominate in the diet of short-finned pilot whales with small amounts of fishes [2693]. It was suggested that the Japanese southern form preys mainly on epi- and meso-pelagic squids, whereas the northern form depends on more neritic squids and bentic octopods when their availability is high [1756].

Habitat Surface water temperature where the northern form is found ranges from 16° to 17°C in January–March and 19–23°C in July–September. Areas inhabited by the southern form ranges from 20° to 29°C in January–March and 24–30°C in July–September [1397].

Behavior Basic social unit (pod) might consist of 20–30 whales but aggregations of 2 or more pods are common [875, 1392, 2035]. Each pod usually contains 1–3 adult males [1392] who show a high association with each other [875]. Based on genetic study, pods are thought to be an aggregation of several matrilineages and several unrelated pods aggregate to make larger groups [102]. Mating may occur between whales from different pods [1256]. Old post-reproductive females are known to copulate with males. This may help mitigate male–male competition and/or make males stay in the pod for possible protection [1393].

Natural enemies Short-finned pilot whales were found in the stomach of killer whales [2495]. Larger sharks are also possible predators.

Parasites A trematode *Nasitrema gondo* was reported from Japan [1763]. Elsewhere, cestode *Phyllobothrium delphini*, nematodes *Anisakis simplex* and *A. typica*, and trematodes *Nasitrema globicephalae* and *Brachycladium pacificum* were reported as endoparasites [277, 428, 429, 2031]. Barnacle *Xenobalanus* and cyamid *Isocyamus* were reported as ectoparasites [277, 1309, 2860].

5. Remarks

In Japan, short-finned pilot whales were taken by small-type whaling (both forms), drive and crossbow-harpoon fisheries (southern form). The current total quota is 36 for the northern form and 232 for the southern form. Estimated abundance of the northern form is 5,300 (CV = 0.43) [1180] and that of the southern form is 14,000 (CV = 0.23) for coastal areas of Japan (30–37°N, west of 145°E), 20,900 (CV = 0.33) for eastern offshore areas (30–37°N, 145–165°E) and 18,700 (CV = 0.42) for southern offshore areas (25–30°N, 130–155°E) [2152]. Abundance of the northern form was estimated to decrease in the 1980s probably due to the commercial takes exceeded the Potential Biological Removal, and then slightly increase to a few thousands after the reduction of the quota [1304]. Although the northern form should consist of a single population, more than one population may be included within the southern form. Decrease in the catch of the southern form has been observed in each fishing area during recent decades. Short-finned pilot whales are kept in several aquaria in Japan.

M. AMANO

Cetacea   DELPHINIDAE

Red list status: K (MSJ); LC (IUCN); ND (JFA)

Grampus griseus (G. Cuvier, 1812)

EN Risso's dolphin   JP ハナゴンドウ (hana gondou)   CM 瑞氏海豚   CT 瑞氏海豚，花紋海豚
KR 큰머리돌고래   RS серый дельфин

Risso's dolphins are common in coastal waters off Kii Pen. (I. Kanda).

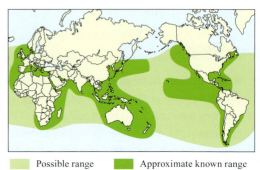

Possible range   Approximate known range

1. Distribution

Widely distributed in tropical and temperate waters worldwide. In North Pacific, the northern limit is at about Hokkaido in the western coast and at Gulf of Alaska in the eastern. Risso's dolphins are found off the coasts of Japan and are particularly common in warmer waters.

2. Fossil record

No reliable record is known in Japan.

3. General characteristics

Morphology   Risso's dolphins are robust anterior to the dorsal fin, which is tall and sharp in its tip. Head is bulbous and has a vertical cleft in the middle of melon. Flippers are relatively long. Body color changes with age. Infants are gray to brown dorsally and cream ventrally, then color becomes silvery gray to almost black. As individuals age, their body becomes lighter colored with scars mostly caused by conspecifics' teeth. Some old adults become almost completely white except for the dorsal fins and flippers. Whitish anchor-shaped thoracic patch is distinct and larger than that of pilot whales (*Globicephala*). Umbilical-genital area is also whitish and is connected with a fine white line to the thoracic patch. Although the largest reported specimen was 408.9 cm, most dolphins are shorter than 350 cm [1744]. Body lengths of 10 mature females taken in Iki Isl. ranged from 260–289 cm, and those of 9 males from 260–310 cm [1389]. Asymptotic length based on 49 females and 30 males captured at Taiji, Wakayama Pref. was 270.7 and 272.7 cm, respectively, though that of males should be an underestimate [114]. Size of Japanese Risso's dolphins may be smaller than those from other areas [114]. Rostrum of cranium attenuates anteriorly to show a triangular shape and lacks tooth. Antorbital notch is distinct and wide causing an indent in the edge of rostrum base. Average condylobasal length (± SD) of sexually mature specimens is 271.4 ± 9.4 mm ($n = 26$) for females and 282.0 ± 4.4 mm ($n = 4$) for males [2158].

Dental formula   0/2–7 [1744]. Teeth are conical with a pointed tip but wear down to round in old individuals.

Genetics   Chromosome: not reported but number of chromosomes presumed to be same as for other delphinids, $2n = 44$. Significant genetic differentiation between individuals in UK waters and the Mediterranean Sea and lower genetic diversity in the former area was reported [712]. There was a dispute on

Grampus griseus (G. Cuvier, 1812)

the phylogenetic position of *Grampus, i.e.* closer to dolphins, *Tursiops* in particular, or to pilot whales. Recent genetic studies show *Grampus* is in a clade together with pilot whales [1833, 2034, 2042, 3778]. Hybrids between *Grampus* and *Tursiops* were reported in captivity [3170, 3383] and possibly in the wild [582].

4. Ecology

Reproduction Age at maturity is estimated at 8–10 years of age for females and 10–12 years of age for males based on a school captured at Taiji, Wakayama Pref. [114]. Gestation lasts about 13–14 months and calving and mating is supposed to occur in summer to autumn off Japan [114]. Shortest calving interval is estimated at 2–3 years [114].

Lifespan The oldest individual known is 34.5 years of age from Taiji [114]. An individual survived over 40 years in an aquarium in Japan.

Diet The diet of Risso's dolphins consists almost completely of cephalopods [301, 1744, 2796, 3827]. In the Mediterranean Sea, Risso's dolphins have wider feeding habits than other squid-eating odontocetes [229].

Habitat Although Risso's dolphins are found in pelagic waters, they show a preference for continental shelf and slope waters throughout its range [268, 1231, 1609, 2854]. Risso's dolphins occur in waters of 10–28°C surface water temperature, although they have been found in warmer waters [1744].

Behavior Risso's dolphins are usually found in schools of 10–50 animals [242]. Mean size of 147 groups observed in the Gulf of Mexico was 10.2 (1–40) [2035], although much larger aggregations of over hundreds or thousands of animals also were reported [1357, 1744]. Schools have a fission-fusion nature with small groups showing high group fidelity [1745]. These groups would be those of mature females of similar reproductive condition and those of mature males, possibly roving among schools [114]. Risso's dolphins commonly form mixed-species associations with other cetaceans [1744]. They engage in various aerial behaviors but rarely porpoise or bow-ride [1744].

Natural enemies Killer whales and larger sharks are possible predators, although there are no direct observations.

Parasites Many kinds of internal (cestode *Phyllobothrium*, *Terabothrius*; nematode *Anisakis*, *Crassicauda*, *Stenurus*; trematode *Nasitrema*) and external (cyamid *Isocyamus* and barnacle *Xenobalanus*) parasites were reported [62, 233, 1744]. Nematode *A. simplex*, *A. physeteris*, *Pseudoterranova ceticola* were reported from Japan [1765]. Heavy infections of *Crassicauda* in the pterygoid and tympanic sinuses are common [4256]. *Nasitrema gondo* in typmanic sinuses are suspected to damage auditory nerve [2230]. Toxosoplasmosis also was reported [2879].

5. Remarks

In Japan, Risso's dolphins are taken commercially by small-type whaling, hand harpoon and drive fisheries. Recent annual quota is around 500 and recent actual catches are 350–500 dolphins. About 530 Risso's dolphins were killed to control the interference of dolphins with the fisheries in Iki-shima Isl., northern Kyushu [1381]. Risso's dolphins do well in captivity.

Estimated abundance of Risso's dolphins in the western Pacific is 31,012 (CV = 0.21) in coastal waters (north of 30°N, west of 145°E), 45,233 (CV = 0.27) in the eastern offshore area (30–42°N, 145–180°E), and 7,044 (CV = 0.79) in the southern offshore area (23–30°N, 127–180°E) of Japan [2152].

M. AMANO

A female Risso's dolphin (BL = 175.5 cm) stranded on a beach in Chiba Pref. (T. K. Yamada).

Cetacea DELPHINIDAE

156

Red list status: R (MSJ); LC (IUCN); ND (JFA)

Lagenodelphis hosei Fraser, 1956

EN Fraser's dolphin   JP サラワクイルカ (sarawaku iruka)   CM 弗氏海豚   CT 弗氏海豚
KR 삼각등지느러미돌고래   RS малазийский дельфин

Fraser's dolphins have a very short beak and a relatively small triangular dorsal fin (M. Amano).

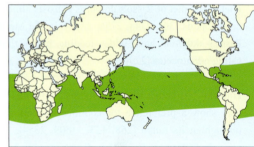

1. Distribution

Distributed in tropical pelagic waters worldwide. They are not common in Japanese waters with only 9 records of strandings [1113, 3578], and the capture of a school at Taiji, Wakayama Pref. [116]. There are unpublished sporadic sightings off the Kii Pen. and the Ogasawara Isls. Red solid circles in the map are known records in Japan.

2. Fossil record

Not reported in Japan.

3. General characteristics

Morphology Beak is short but distinct. Flippers and dorsal fin are relatively small. Dorsal fin becomes triangular from falcate shape in adult males. Postanal hump and deepened tail stock are also secondary sexual features in adult males [116]. Color is dark gray on the back and white on the belly, with a dark stripe that runs from tip of melon, through eye patch, to postanal region. Distinct flipper stripe extends from lower jaw to flipper. Both stripes become darker and wider with age, are very distinct in adult males and make a sharp contrast with the white belly and light gray flank. Mean body length of adult animals (> 10 years of age) is significantly greater in males (241.5 cm, $n = 28$) than females (234.0 cm, $n = 21$) [116]. In the skull, distinct palatal grooves, separated by a broad medial ridge, is present in the basal half of the wide rostrum. Left and right premaxillae come into contact in midlength of the rostrum in adult animals. Cranial measurements were reported [2825].

Dental formula 34–46 (mean = 41.0, $n = 122$)/37–43 (mean = 40.4, $n = 115$) [2825].

Genetics Chromosome: not reported, but number of chromosomes presumed to be as for other delphinids, $2n = 44$. Phylogenetic studies reported that Fraser's dolphin was included in a clade with *Sousa*, *Stenella*, *Delphinus*, and *Tursiops* [1833, 2034, 2042].

4. Ecology

Reproduction Information on life history is known only from a single school captured at Taiji, Japan. Females attain sexual maturity at 5–8 years of age and 210–220 cm in body length, males at 7–10 years and 220–240 cm [116]. Gestation lasts about 12–13 months and calving interval is approximately 2–2.3 years [116].

Lagenodelphis hosei Fraser, 1956

Lifespan The oldest individual in a school of 81 dolphins was 17.5 years of age [116]. Specimens of 16 years old and 19 years old were reported from France and Brazil, respectively [3269, 3767]. These suggest shorter longevity of Fraser's dolphin than for those of other pelagic dolphins.

Diet Stomach contents of Fraser's dolphins from eastern tropical Pacific, Philippins and Taiwan included mesopelagic fishes, crustaceans and cephalopods, suggesting they feed in a wide depth range and usually dive to over 500 m [473, 2894, 3827].

Habitat Fraser's dolphins inhabit deep waters over 500 m [472]. The range of water temperature where they were found was 16–30°C [116, 2035, 2166].

Behavior Fraser's dolphins form schools of tens to hundreds of individuals. Fraser's dolphins usually are found with many odontocete species and most frequently associated with melon-headed whales [472].

Natural enemies Predations by killer whales and tiger shark *Galeocerdo cuvier* were reported [481, 2783]. False killer whales and large sharks are other possible predators. Wounds possibly made by cookie cutter sharks *Isistius brasiliensis* were observed [2829].

Parasites Cestodes *Strobicephalus*, *Phyllobothrium*, *Monorygma*, and *Tetrabothrius*, tramatode *Campula*, nematodes *Anisakis* and *Stenurus*, and acanthocephalan *Bolbosoma* were reported [472].

5. Remarks

Fraser's dolphins currently are not allowed to be caught in Japan, but incidental catches by gill nets are suspected. In the eastern tropical Pacific and Philippines they are caught by purse seines [472].

M. AMANO

A Fraser's dolphin at Kamogawa Sea World, Chiba Pref. (M. Yoshioka). The head looks like that of *Lagenorhynchus* dolphins.

Fraser's dolphins off the coast of the Kii Pen. (M. Kanda).

Cetacea   DELPHINIDAE

Red list status: C-1 (MSJ); LC (IUCN); ND (JFA)

Lagenorhynchus obliquidens Gill, 1865

EN Pacific white-sided dolphin   JP カマイルカ (kama iruka)   CM 太平洋斑纹海豚   CT 太平洋斑紋海豚
KR 낫돌고래   RS тихоокеанский белобокий дельфин

A cow–calf pair of the Pacific white-sided dolphin in captivity at Kamogawa Sea World, Chiba Pref. (M. Yoshioka).

1. Distribution

Endemic to cool temperate waters in the North Pacific and adjacent seas including the Sea of Japan, the Sea of Okhotsk and the southern Bering Sea. Around Japan, the species is found from the southern Sea of Okhotsk off Hokkaido to Kumano-nada Sea off Wakayama Pref. and in the Sea of Japan into the East China Sea north of Taiwan [1174, 3313]. Funka-wan (Uchiura-wan or Volcano) Bay off Muroran in southern Hokkaido is one of the known calving areas [1358].

2. Fossil record

Not reported in Japan.

3. General characteristics

Morphology Pacific white-sided dolphins have a short and thick beak like all other members of *Lagenorhynchus*. Dorsal fin is large and falcate and hooked in older adult males [1368]. Body is dark gray on the back and white on the belly. Light-gray suspenders run from tail-stalk forward to near melon on the sides. Several color morphs are known, for example, anomalously-white Pacific white-sided dolphins are found in California and Hokkaido [1230, 3668]. The largest reported male and female specimens from the eastern North Pacific are 250 cm and 236 cm, respectively [3819]. Largest known specimens in the central and western North Pacific are smaller than in the eastern Pacific [1176, 1381]. Geographical differences in skull morphology and asymptotic length were reported between the populations of Pacific side and East China Sea–Sea of Japan–southern Okhotsk Sea side [2164, 2165].

Dental formula 23–36/23–36. Teeth are relatively fine and have a sharply-pointed tip [1230].

Genetics MtDNA and nuclear microsatellite DNA analyses suggest that the 2 populations located in offshore North Pacific and Japanese coastal waters are distinct and gene flow has been restricted severely [856]. However, no genetic differences have been reported among populations in Japanese coastal waters. Hybrids between Pacific white-sided dolphins and bottlenose dolphins (mother) were born in Japanese aquaria.

4. Ecology

Reproduction Males attain sexual maturity at about 10 years of age and 170–180 cm of body length and females at 8–11 years and 175–186 cm in the central North Pacific. In the Sea of Japan, males and females reach sexual maturity at 170–180 cm in body

Lagenorhynchus obliquidens Gill, 1865

length and 7–9 years of age [3463]. Sperm production is strictly seasonal in May to October in a captive individual in Japan [4187]. Gestation period is about 12 months. Calving season is from summer to autumn. Length at birth is 80–100 cm [1175]. Nursing period is about a year [3463].

Lifespan At least 40 years of age [3819].

Diet Pacific white-sided dolphins feed on lanternfish, anchovies, saury, horse mackerel, hake and cephalopods. However, the majority of their diet consists of mesopelagic and epipelagic fish [332].

Habitat The Pacific white-sided dolphin is an oceanic and deep water species in cool temperate waters. It is commonly found on the edge of the continental shelves and slopes.

Behavior Pacific white-sided dolphins are gregarious and often seen in large herds of hundreds or more. They are sometimes found with other species, such as, northern right whale dolphins, Risso's dolphins and short-beaked common dolphins. Behavior is highly acrobatic, and the dolphins like to ride bow-waves. Pacific white-sided dolphins have a fission–fusion society [3463].

Natural enemies Predation by killer whales was reported [1230].

Parasites Pacific white-sided dolphins are infected with various endoparasites [332]. Cestode *Tetrabothrius* sp. and Trematodes *Hadwenius seymouri* and *Nasitrema lagenorhynchus* were reported from Japan [1561, 1763]. A barnacle *Xenobalanus globicipitus* was found on this species [332].

5. Remarks

Pacific white-sided dolphins have been caught by dolphin fishery (drive fishery and harpoon fishery) in Japan. In 1970–80s, 4,800 dolphins were culled around Iki Isl. [2633]. Until 1992, squid gill-net fishery incidentally caught 4,000–6,000 Pacific white-sided dolphins per year [4104, 4105]. This species has been a target of Japanese dolphin fishery since 2007 with quota of 360, but only several to tens of dolphins were taken recently. Skeleton (not fossil) of the Pacific white-sided dolphins was found in the Mawaki Site, Noto Pen. on the Sea of Japan side from Holocene epoch [894]. As well as bottlenose dolphins, Pacific white-sided dolphins are one of the most common species in captivity in Japan. Newborn calves, conceived through artificial insemination were successfully born in captivity [2889].

M. YOSHIOKA

Pacific white-sided dolphins are commonly seen off northern Japan. (M. Yoshioka).

Cetacea   DELPHINIDAE

Red list status: C-1 (MSJ); LC (IUCN)

Lissodelphis borealis (Peale, 1848)

EN northern right whale dolphin   JP セミイルカ (semi iruka)   CM 北露脊海豚   CT 北露脊海豚
KR 고추돌고래   RS северный китовидный дельфин

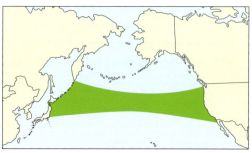

The nothern right whale dolphin is one of the 2 small cetacean species which are distributed around Japan and have no dorsal fin (M. Yoshioka).

1. Distribution

Endemically occurs in temperate offshore waters of the North Pacific. In the eastern North Pacific, the distribution range is 30–50°N, and in the western North Pacific, 35–51°N [1358]. Off Japan, the species is found north of Choshi (Chiba Pref.). Southernmost stranding was recorded in Sagami Bay [1113]. No sightings and strandings are recorded from the Sea of Japan or the Sea of Okhotsk.

2. Fossil record

Not reported in Japan.

3. General characteristics

Morphology   Northern right whale dolphins have a long-slender and vertically compressed body shape without dorsal fin. Coloration is mostly black with a white ventral band from the notch of tail flukes to throat, which widens in the area between the flippers. There is a small white patch just behind the tip of the lower jaw. Flippers and tail flukes are small. The largest specimen is 3.1 m long [1832], and males are larger than females [552].

Dental formula   43–53/46–54 [1226]. Teeth are small, slender and sharp.

Genetics   Chromosome: not reported but number of chromosomes presumed to be same as for other delphinids, $2n = 44$. Genetic analyses suggest northern right whale dolphins have a close relationship with a clade of *Lagenorhynchus* sp. and *Cephalorhynchus* sp. [1833, 2042].

4. Ecology

Reproduction   Information on life history is limited. Males attain sexual maturity at 9.9 years of age and average length of 214.7 cm [552]. Females attain sexual maturity at 9.7 years of age and average length of 199.8 cm [552]. Length at birth is about 80–100 cm. Birth peaks in summer. Female reproductive cycle is 2 years [552].

Lifespan   Known longest longevity is 23 years in males and 42 years in females from by-catch specimens [552].

Diet   Northern right whale dolphins primarily eat mesopelagic fishes (especially lanternfish, Myctophidae) and squids [1226].

Lissodelphis borealis (Peale, 1848)

Habitat Northern right whale dolphins are usually found in cold and deep offshore waters with 8–19°C [1358].

Behavior Northern right whale dolphins are highly gregarious and often found in large schools of a few hundred to a few thousand individuals. In the western North Pacific, mean school size was 85 with a maximum of 400 individuals [2148]. The schools occasionally are found in association with many other species, including Pacific white-sided dolphins, Dall's porpoises, pilot whales and bottlenose dolphins [1226, 2148]. They cleanly jump out of water when they swim fast. Little is known about their social behavior.

Natural enemies Poorly known, possibly killer whales are predators.

Parasites Cestodes *Phyllobothrium* and *Monorygma*, nematodes *Anisakis* and *Crassicauda*, trematode *Nasitrema* and protozoan *Sarcosporidia* were reported as endoparasites [1226]. A barnacle *Xenobalanus* and copepod *Pennella* were reported as ectoparasites [1226].

5. Remarks

Northern right whale dolphins were under threat of by-catches in the squid driftnet fishery until 1992 [916]. After the fishery was closed-down, there have been no serious threats for the species including direct catches and incidental takes by other fishing gear.

M. YOSHIOKA

The northern right whale dolphin has a very slender body (M. Amano).

Cetacea DELPHINIDAE

159

Red list status: R for the population in adjacent waters of Japan (MSJ); DD (IUCN); R for the population in adjacent waters of Japan (JFA)

Orcinus orca (Linnaeus, 1758)

EN killer whale, orca   JP シャチ (shachi), サカマタ (sakamata)   CM 虎鲸   CT 虎鯨
KR 범고래   RS косатка

A killer whale sighted in coastal waters around Japan (M. Amano).

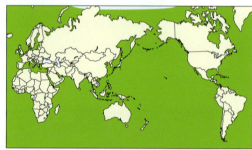

1. Distribution

The most widely distributed marine mammal species in the world. They are found in all oceans and seas from tropical to polar waters. Around Japan, they are found in coastal and offshore waters in the western North Pacific, the Sea of Okhotsk, the Sea of Japan and the East China Sea [1412].

2. Fossil record

No reliable record is known in Japan.

3. General characteristics

Morphology The black and white body is robust and a tall dorsal fin is characteristic. Pectoral fins are large and round in shape. Maximum body length is 9.0 m in males and 7.7 m in females [427]. Sexual dimorphism is clearly expressed in larger body size and pectoral fins, tail flukes with tips curling down and taller dorsal fin up to 1.8 m in males. Black (dorsal and lateral) and white (ventral) coloration are the most distinctive features of the species with white eye patches and white to gray saddle patches. White-colored killer whales have been sighted in the western and eastern North Pacific [2017, 2509]. Several distinct ecotypes, which differ in feeding habits, morphological and genetic traits from each other, are recognized. They may be subspecies or distinct species [575, 1230, 1835, 2234].

Dental formula 10–14/10–14. Upper and lower teeth interlock when the jaws are closed. In older individuals, tips of teeth are worn [427].

Genetics Chromosome: $2n = 44$ [190]. Low genetic variation and diversity is known among killer whales throughout the World [919–922]. DNA analyses indicate a similarity of school structure in the eastern North Pacific, a pod of killer whales entrapped by sea ice off northern Japan [3949] and a pod caught by the driving method off Wakayama Pref. [853].

4. Ecology

Reproduction Sexual maturity is achieved at 10–15 years of age for females and over 15 years for males. Ovulation (estrus) cycle is estimated at 42 days from hormonal profiles of captive mature females [2890]. Gestation period is estimated at 18 months from data of captive animals [2890]. Size at birth is 2.1–2.6 m in length. Calving occurs year around with a peak from October and March in the eastern North Pacific [1230]. Calving interval is > 5 years. The calf weans at about 2 years of age. As for pilot whales, reproductive senescence (post reproductive

380

period) is known in older females indicating that menopause and long post-reproductive lifespan are not a human specific phenomenon. Long term studies obtained support for menopause evolving to provide fitness benefits to old mothers through their son's reproductive success, and son's surviverhip may be increased by transferring mother's ecological knowledge [327, 579, 1250, 3831].

Lifespan Killer whales are long-lived animals with males reaching 50–60 years and females reaching 80–90 years [427].

Diet Killer whales are the only cetacean that routinely preys on marine mammals including over 35 species of baleen whales, dolphins and porpoises, seals and sea lions, and sea otters [1227]. They also attack penguins and sea turtles. In the eastern North Pacific, resident killer whales prey upon fish, whereas offshore transients prey upon marine mammals. Similar ecotypes, having specialized feeding habits, to those in the eastern North Pacific are reported in the Antarctic and Atlantic as well [574, 2846].

Habitat Killer whales are adaptable to both coastal and pelagic habitats and warm to cold temperatures. Surface water temperature when killer whales were found ranged from 3.6° to 26.7°C (average 13.2°C) in the western North Pacific in May to September [2017].

Behavior Killer whales are highly active. Breaching, spy-hopping, tail- and flipper-slaps occasionally are observed. Mean school (or pod) size is 6.2 individuals and more than 90% of sighted schools consisted of <10 animals in the western North Pacific [2017]. Killer whales are highly sociable and pods of residents off Vancouver Isl., Canada, represent one of the most stable societies. A pod of killer whales (11 or 12 individuals) stranded in the ice off Hokkaido in February 2005, included 7 females (3 lactating, 2 resting and 2 immature) and 2 males (1 mature and 1 immature) [119]. No resident populations have been known around Japan.

Natural enemies No natural enemies are known for killer whales.

Parasites Three species of cestode, *Diphyllobothrium fuhrmanni D. orcini, Trigonocotyle spassky* were reported from killer whales from Japan [1763, 3265]. In other areas, several endoparasites are known: Cestoda *Trigonocotyle spassky* and *Phyllobothrium* sp., Nematoda *Anasakis simplex* and *A. pacificus,* Trematoda *Fasciola skrjabini, Leucasiella subtilla* and *Oschmarinella albamarina,* and Acanthocephala *Bolbosoma physeteris* and *B. nipponicum* [1966, 3878]. A barnacle *Xenobalanus globicipitis* was found at the highest rate among cetaceans in the eastern tropical Pacific [1309]. Three species of Amphipod cyamids *Cyamus orcini, C. antarcticensis,* and *Isocyamus delphinii* were reported [3878].

5. Remarks

Historically, killer whales were the target of directed fisheries, culling, and persecution. In Japan, about 1,600 killer whales were caught by small-type whaling from 1940s to 1960s. Levels of pollutants (organohalogen and organotin compounds and heavy metals) and the mother-to-calf transfer of anthropogenic and natural organohalogens were reported in 9 individuals mass-stranded in Hokkaido [524, 525, 798, 802, 1275]. Killer whales are one of the most popular species in captivity, and relatively adaptable to captive environments. Newborn calves, conceived through artificial insemination were born successfully in captivity [2890], and a captive-born mother gave birth in an aquarium in Chiba Pref.

M. YOSHIOKA

A cow–calf pair of the killer whale at Kamogawa Sea World, Chiba Pref. (M. Yoshioka).

Cetacea DELPHINIDAE

160

Red list status: R (MSJ); LC (IUCN)

Peponocephala electra (Gray, 1846)

EN melon-headed whale   JP カズハゴンドウ (kazuha gondou)   CM 瓜头鲸   CT 瓜頭鯨
KR 고양이고래   RS бесклювый дельфин, широкомордый дельфин

A group of melon-headed whales off the Kii Pen. (M. Amano).

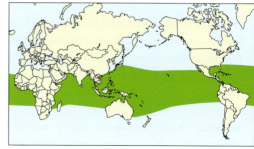

1. Distribution

Distributed in tropical and subtropical pelagic deep waters worldwide. In the western North Pacific, the northern limit is off southern Honshu, which is associated with the Kuroshio current. Although it is considered rare off Japan, occasionally it is sighted off the Kii Pen. At least 13 instances of mass-stranding have been reported on the coast of Japan; this species is one of the most frequently mass-stranded cetacean in Japan [334, 1054]. Although 9 of them occurred on the coasts of Chiba Pref. through Ibaraki Pref. in winter, it is unlikely to reflect its normal range in this season. Red solid circles in the map are known records in Japan.

2. Fossil record

Not reported in Japan.

3. General characteristics

Morphology Head is round but more pointed in the tip of mouth than that of the pygmy killer whale, which quite resembles this species. In older males, tip of the head becomes rounder and melon projects over tip of mouth. Postanal hump is observed in mature males. Dorsal fin is distinct and falcate. Tip of flipper is sharply pointed, which distinguishes it from pygmy killer whales. Body is mostly dark gray with darker dorsal cape. Lips are often white. There is light gray chevron shaped mark on the throat which extends posteriorly to the lighter genital–anus area. Eye patch is wide and extends to the melon. Mean body length (± SD) of adult animals (> 15.5 years of age) stranded at Aoshima, Miyazaki Pref. was 242.6 ± 7.6 cm ($n = 29$) for females and 253.1 ± 5.5 cm ($n = 24$) for males [2160]. Similar asymptotic body lengths are recorded in the mass-stranding in Tane-gashima Isl., Kagoshima Pref.; 246.0 cm for females and 252.1 cm for males, while asymptotic length of females from Ibaraki Pref. was slightly smaller; 236.3 cm [120]. Skull looks like that of other dolphins despite distinct differences in external shape, although a relatively wide rostrum, and the deep and round antorbital notches are shared with *Feresa* and *Pseudorca*. Skull measurements of specimens from a mass stranding at Aoshima, Miyazaki Pref. were reported [2160].

Dental formula 14–25/18–25 was reported for specimens from Aoshima [2160], but numbers < 20 would be anomalous. Teeth are small and conical.

Genetics Chromosome: not reported, but number of chromosomes presumed to be as for other delphinids, $2n = 44$. Lower genetic diversity in mtDNA, which is observed in some other matrilineal odontocetes, and some geographic differences among individuals involved in mass-strandings along the Pacific coast of Japan were reported [855].

4. Ecology

Reproduction Based on the specimens from several mass-strandings in Japan, females attain sexual maturity at approximately 10 years of age and 235 cm in body length, males about 15 years and 240 cm [120, 2160]. Annual ovulation rate is estimated at 0.28–0.37, which is somewhat lower than other pelagic small delphinids, suggesting longer calving intervals [120, 2160]. Mature males are thought to rove among groups according to reproductive condition of females.

Lifespan The oldest female reported was 45.5 years of age and the oldest male was 38.5 years of age [2160].

Diet Melon-headed whales feed mainly on mesopelagic squids, fishes and shrimps, and are thought to forage in deeper water columns [1224]. Cephalopods and fishes inhabit continental shelf seabed are found in the stomach from the Bay of Biscay [3299].

Habitat Melon-headed whales usually occur in warm tropical and subtropical pelagic waters. In the Gulf of Mexico, sightings were recorded in areas with surface water temperature 24.1–28.7°C [2035]. Melon-headed whales are considered to utilize edges of eddies as foraging grounds in Hawaiian waters [3890]. Occasional findings in temperate regions usually are associated with warm water currents [2833].

Behavior Melon-headed whales form large schools of hundreds of individuals [2035, 2304]. They often are associated with Fraser's dolphins, and sometimes with other oceanic dolphins such as spinner, common bottlenose, and rough-toothed dolphins [2833]. They are known as a species frequently mass-stranded [334].

Melon-headed whales mass-stranded on Hasaki Coast, Ibaraki Pref. in February 2002 (A. Hayano).

Natural enemies Observation of avoidance of killer whales suggests predation by this species [244]. Large sharks also could be predators, although there are no direct observations. Wounds caused by cookie-cutter sharks *Isistius brasiliensis* often were observed [317].

Parasites Infestations of a tremadode, *Nasitrema gondo*, in the cranial sinuses were found in mass-stranded specimens from Japan and damage to the auditory nerve by this species were suspected to be a cause of the stranding [2232]. Nematodes *Anisakis simplex* and *A. physeteris* were reported from Japan [1765]. Trematode *Nasitrema*, Cestodes *Monorygma*, *Phyllobothrium*, *Strobicephalus*, Nematodes *Halocercus*, *Stenurus*, *Anisakis*, and Acanthocephalans *Bolbosoma* were reported from other areas [2834, 3767].

5. Remarks

Not currently taken in Japan. They may be taken incidentally by offshore fisheries but status is unknown. Captured or stranded animals have been kept in aquaria, but none have lived long.

M. AMANO

The profile of a melon-headed whale stranded on a beach in Chiba Pref. (T. K. Yamada). This stranded whale was successfully returned to sea by local people.

A lateral view of a female melon-headed whale (NSMT-M 32782) (N. Kurihara).

Cetacea DELPHINIDAE

Red list status: LP for the population off the Pacific coast of Japan, C-2 for the other populations in the western North Pacific and adjacent seas (MSJ); DD (IUCN); DT for the population off the Pacific coast of Japan, DC for the populations off Iki-shima Isl. and in the East China Sea (JFA)

Pseudorca crassidens (Owen, 1846)

EN false killer whale   JP オキゴンドウ (oki gondou)   CM 伪虎鲸   CT 偽虎鯨
KR 흑범고래   RS малая косатка, чёрная косатка

False killer whales are one of the target species for whale watching off Kii Pen. (I. Kanda).

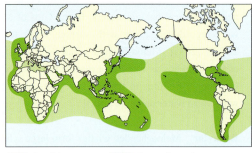

Possible range   Approximate known range

1. Distribution

Occurs in tropical to temperate pelagic waters worldwide. Off Japan, northern limit is at about Hokkaido off both the Pacific and Sea of Japan coasts, but they are commonly found south of central Honshu, off Kii Pen., Kyushu, and Okinawa in particular.

2. Fossil record

Not reported in Japan.

3. General characteristics

Morphology   Body is slender. Melon is projected against the tip of jaw, and gradually attenuated toward its tip. Dorsal fin is small and falcate. Flipper has a distinctive hump in the middle of its leading edge, making it appear to bend backwards. Color is mostly dark gray with a large lighter anchor-shaped patch in the thoracic area. Average asymptotic body length ($\pm$ SD) is 436.0 ± 21.7 cm ($n = 41$, older than 24 years) for females, and 520.1 ± 25.9 cm ($n = 14$, older than 29 years) for males based on specimens from Iki Isl. [1364]. Body lengths of whales from Japan are larger than those from South Africa [550]. Skull is robust with a broad and short rostrum. Temporal fossa is large and posterior margin is projected backwards in adults. Average condylobasal length of specimens with 10 or more growth layers in teeth was 611.4 cm ($n = 31$) for females and 613.0 cm ($n = 21$) for males (calculated from data of [2858]). Other skull measurements were reported [1627, 2571]. There is geographic variation in the skull morphology among specimens from Australia, Scotland, and South Africa. Individuals from Scotland develop larger skulls [1627].

Dental formula   7–11/8–12 [2571]. Teeth are large and pointed at the tip.

Genetics   Chromosome: not reported, but number of chromosomes presumed to be same as for other delphinids, $2n = 44$. Genetic differentiation was reported among populations in different oceans and among those in the North Pacific including 2 island-associated population off the Hawaiian Isls. [386, 1965]. Lower mtDNA nucleotide diversity is comparable to that of social

Pseudorca crassidens (Owen, 1846)

odontocetes such as sperm, killer and pilot whales [386]. False killer whales are phylogenetically closer to *Feresa*, *Globicephala*, *Grampus*, and *Peponocephala* than *Orcinus* [1833, 2034, 2042, 3778].

4. Ecology

Reproduction Based on specimens taken off Iki Isl., females attain sexual maturity at 8–11 years of age and 340–380 cm in body length, males 18.5 years and 430–439 cm [1364]. Gestation lasts about 15–16 months. Length at birth is about 175 cm and calving peak is in March–April [1364]. Proportion of pregnant females decreases with age, whereas that of resting females increases, suggesting occurrence of the post-reproductive period or menopause in females older than about 41 years [1364, 1958].

Lifespan The oldest animal among 133 captured off Iki Isl. is 62.5 years of age for females and 57.5 years for males [1364].

Diet False killer whales prey on relatively larger fishes and squids. Stomach contents of individuals taken off Japan included yellowtail (*Seriola quinqueradiata*), perch (*Lateolabrax japonicus*), and squids (Onchoteuthidae, Loliginidae, and Ommastrephidae) [1364, 2170, 3464, 3669]. False killer whales were observed to forage kahawai (*Arripis trutta*) in corporation with bottlenose dolphins off New Zealand [4242]. False killer whales were also observed to attack other cetaceans including small dolphins, and humpback and sperm whales [2797, 2835].

Habitat False killer whales usually are found in waters warmer than 17°C [1357]. In the Gulf of Mexico, false killer whales were sighted on the continental slope with surface temperature of 25–29°C [2035].

Behavior False killer whales are found in groups of several to hundreds of animals, but groups of tens are common [386, 544, 1375, 2035, 2147]. Associations with bottlenose dolphins (*Tursiops* sp.) are commonly observed off Japan [2147, 3669]. Strong and long lasting bonds among individuals were suggested [241]. Groups are thought to be matrilineal and some males leave their natal group at around puberty [550, 1364]. Strong social fidelity with some male-biased dispersal also was suggested based on genetics [1965]. A tagged false whale off Japan conducted deep foraging dives to 300–400 m more frequently in daytime than at night [2076].

Natural enemies There is a record that a false killer whale group was attacked and a calf was consumed by killer whales [3780].

Parasites Trematodes *Campula oblonga*, *Nasitrema dalli*, *N. gondo*, *Odhneriella gondo*, and acanthocephalan *Bolbosoma capitatum* were reported from Japan [1763, 2231, 2370]. Auditory nerve damage by trematode *N. gondo* was suggested as a cause of mass strandings, which are common in this species [2231]. From other waters, trematodes *N. attenuata*, *N. globicephalae*, nematodes *Anisakis simplex*, *A. typica*, *Stenurus globicephalus*, and acanthocephalan *B. capitatum* were reported [126, 2031, 2571]. Cyamid amphipods, *Isocyamus delphinii* and *Syncyamus pseudorcae*, and barnacle *Xenobalanus globicipitus* are known as ectoparasites [2571].

5. Remarks

False killer whales have been taken by drive, hand harpoon, crossbow fisheries, and small-type whaling in Japan [1608]. Current catch quota is 100 for drive and crossbow fisheries. More than 1,400 whales were killed for conflict control with yellowtail fishery in Iki-shima Isl. from 1945 to 1986. False killer whales are known to steal hooked fishes in the longline fishery and occasionally become entangled or hooked causing injury or death [243, 351]. Artificial mortality by other fisheries is unknown. False killer whales are kept in several aquaria in Japan.

Abundance off Japan is estimated at 2,029 (CV = 0.429) off the Pacific coast (north of 30°N, west of 145°E), 8,569 (CV = 0.338) in the eastern offshore area (30–42°N, 145–180°E), 6,070 (CV = 0.525) in the southern offshore area (23–30°N, 127–180°E), and 3,259 (CV = 0.808) in the eastern East China Sea including off Okinawa [2147, 2152].

M. AMANO

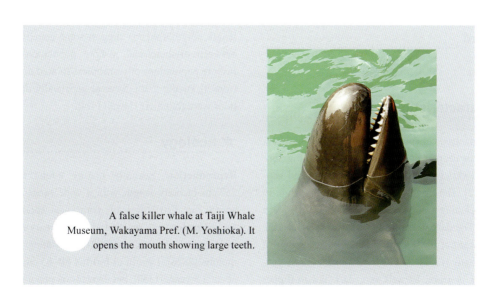

A false killer whale at Taiji Whale Museum, Wakayama Pref. (M. Yoshioka). It opens the mouth showing large teeth.

Cetacea   DELPHINIDAE

Red list status: K (MSJ); LC (IUCN); DT for population off Japan in the western North Pacific (JFA)

Stenella attenuata (Gray, 1846)

EN pantropical spotted dolphin   JP マダライルカ（madara iruka）   CM 热带斑海豚   CT 熱帶點斑原海豚
KR 점박이돌고래   RS пятнистый продельфин

Pantropical spotted dolphins occasionally bow ride, off the Kii Pen. (F. Sato).

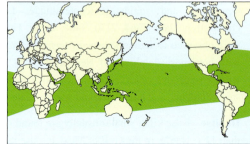

1. Distribution

Distributed worldwide in tropical and some subtropical waters, from approximately 30–40°N to 20–40°S [1225]. The pantropical spotted dolphin is one of the most abundant species of cetaceans on Earth [2873]. Around Japan, the species is found from south of Miyagi Pref. in the Pacific to the East China Sea and south of Akita Pref. in the Sea of Japan [1370].

2. Fossil record

Not reported in Japan.

3. General characteristics

Morphology   The pantropical spotted dolphin has a long beak, slim body with various degrees of spots (only in adults). Pigmentation and coloration varies greatly among individuals and geographical areas [2873]. The tip of the beak in adults is white in some populations including around Japan. The dorsal fin is very narrow and falcate. The dark dorsal cape that dips low onto the sides below and forward of the dorsal fin is the most distinctive color pattern of the body [1230]. Adults have a black mask, visible from the side or in front, as well as a dark jaw-to-flipper stripe. Physical maturity is attained at length of about 203 cm for males and 195 cm for females in the Pacific coast of Japan [1398]. Significant sexual dimorphism is known in the skulls of individuals in the western and eastern tropical Pacific populations [4064].

Dental formula   35–50/35–50 [2821]. The teeth are slender and sharply-pointed.

Genetics   MtDNA control region sequences and microsatellite length polymorphisms suggest some genetic isolation is present between offshore and inshore populations and among some inshore populations of pantropical spotted dolphins in the eastern tropical Pacific [530]. Several genetically differentiated populations were found off Hawaii [414].

4. Ecology

Reproduction   Sexually mature adults range from 166 to 257 cm in body length with a wide range of variation present across the distribution of pantropical spotted dolphins [2821]. Sexual maturity is reached at 12–15 years of age for males and 9–11 years for females [2821]. Gestation period is 11.2–11.5 months [2828]. Length at birth is 80–85 cm. Calving interval is about 3.3 years [1380], but depends on population status. Breeding

Stenella attenuata (Gray, 1846)

is seasonal, with multiple peaks in some regions. Off the Pacific coast of Japan mating occurs in February, March, July and November [1398].

Lifespan   Lifespan is 40–45 years. Shorter in males [1398].

Diet   Pantropical spotted dolphins prey on small pelagic fishes, squids and crustaceans [2893]. Flying fish (Exocoetidae) are a major diet item in some regions [2821]. This species utilizes prey near surface [3827].

Habitat   Pantropical spotted dolphins are oceanic. They inhabit tropical and subtropical waters, and warm temperate waters with the surface water temperature >22°C [1358].

Behavior   In the eastern North Pacific, pantropical spotted dolphins associate with other small cetaceans (e.g. spinner dolphins) and yellowfin tuna (*Thunnus albacares*). Evasive response of dolphins from the encirclement with the purse-seine net for tuna differs by area and periods and varies by longer-term learning [1842]. They also show a variety of aerial behaviors (jump, summersault and tail-slap etc.), and come to vessels to ride bow-waves. Pantropical spotted dolphins are one of the most gregarious cetaceans and school sizes are usually < 50–100 individuals in coastal waters, but offshore herds may number in the thousands [1230]. The groups can consist of mature females with calves, only juveniles, or only mature males and the membership can be fluid as with striped dolphins [1398]. In the pelagic waters of the eastern tropical Pacific, radio-tracked and tagged pantropical spotted dolphins changed group size and membership over the course of a day. At night, the dolphins traveled more slowly but dove deeper and longer, with more rapid ascents and descents than during daylight hour characterizing night time diving patterns [3156].

Tip of beak of pantropical spotted dolphins off Japan is white (M. Yoshioka).

Natural enemies   Killer whales and sharks, probably the pygmy killer whale, and possibly false killer whales and short-finned pilot whales are predators [2821].

Parasites   Fourteen species of helminth endoparasites were recorded [2828]. Parasites may cause direct or indirect mortality [2821]. Cyamid *Syncyamus* sp. and barnacles *Conchoderma auritum* and *Xenobalanus globicipitis* were reported [2828].

5. Remarks

Millions of pantropical spotted dolphins were taken incidentally together with spinner dolphins in the purse-seine fishery for tuna in the eastern North Pacific in the last 40 years. Populations were severely reduced and have not recovered [723, 3809]. Along the Pacific coast of Japan, drive and harpoon fisheries annually have taken about 1,000 pantropical spotted dolphins since the 1970s [1366]. Recent annual catch is about 100–150 dolphins to the quota of 400. Pantropical spotted dolphins are kept well in captivity in a few Japanese aquaria.

M. YOSHIOKA

Pantropical spotted dolphins are usually active near the surface of the sea (M. Amano).

Cetacea   DELPHINIDAE

Red list status: V for the populations off the Izu Pen. and Wakayama Pref. in Japanese waters, Common for the other populations in the western North Pacific (MSJ); LC (IUCN); R for the population in Pacific coasts of Japan, ND for populations in offshore North Pacific (JFA)

Stenella coeruleoalba (Meyen, 1833)

EN striped dolphin   JP スジイルカ（suji iruka）   CM 条纹海豚   CT 條紋海豚
KR 줄박이돌고래   RS полосатый продельфин

A striped dolphins clearly leaping out of water off Wakayama Pref., Kii Pen. (M. Yoshioka).

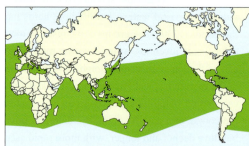

1. Distribution

A cosmopolitan species that occurs in warm-temperate to tropical areas in the western and eastern Pacific, Atlantic and Indian oceans as well as many adjacent seas including the Mediterranean Sea [1230]. Around Japan, the northern limit is south of Sanriku coast in the Pacific [1365]. This species is rare in the Sea of Japan and East China Sea [1385].

2. Fossil records

Not reported in Japan.

3. General characteristics

Morphology   Striped dolphins have a diagnostic color pattern of bluish gray and white. There are 2 narrow black stripes; one going from the eye to the flipper and the other from the eye to the anal area. A light-colored shoulder blaze is present along the lateral and dorsal sides of the robust body. Striped dolphins have a moderately long, gray beak and falcate dorsal fin. Maximal body length is 2.56 m. Geographical variation in skull and body size was reported [185].

Dental formula   39–53/39–55 [2830]. Teeth are small.

Genetics   Microsatellite DNA data suggest that the western Mediterranean population might be further subdivided [319]. No clear genetic difference suggesting differentiation of populations is found off Japan.

4. Ecology

Reproduction   Males attain sexual maturity at 7–15 years of age and females 5–13 years (length at 2.1–2.2 m) [184]. Density dependent changes in biological parameters caused by large catches by the driving fisheries have been reported for the western North Pacific population: the mean age at female sexual maturity declined from 9.7 to 7.2 years and the mean reproductive cycle from 4 to 2.8 years [1380, 1385]. The mating seasons are in May–June and November–December in the western North Pacific [1376]. Gestation period is about 12 months. Length at birth is about 100 cm [1376]. Females can have about 9 calves in a life [1365].

Lifespan Estimated maximum age for both sexes is 57.5 years [185].

Diet Striped dolphins have a fairly varied diet consisting of pelagic and benthopelagic fish and squid, e.g. Mycotphidae off Japan and cod in the Northeast Atlantic and the Mediterranean Sea [2829]. Diets of striped dolphins in the oceanic waters of the northeast Atlantic are composed primarily of fish (39% by mass) and cephalopods (56%) and secondarily of crustaceans (5%). The most significant fish family and squids are the lanternfish (Myctophidae) with *Notoscopelus kroeyeri* and *Lobianchia gemellarii* and the oceanic *Teuthowenia megalops* and *Histioteuthis* sp. [2887].

Habitat Striped dolphins most commonly occur in highly productive warm-temperate to tropical waters. They are seen mainly seaward of continental shelf and in deeper slope waters. Most records were reported from about 18–22°C in surface water temperature (range: 10–26°C) [185]. Around Japan, they occur in the waters of 18–26°C [1365].

Behavior Striped dolphins are fairly gregarious, social, and very acrobatic bow-riders. School size is from less than a hundred to a few thousand. In Japan, a school of about 3,000 individuals was taken in the fishery [1365]. School structure of striped dolphins in the western North Pacific is known; animals move among juvenile, adult, and mixed schools according to growth stage [2162]. Three mass strandings including more than 2 dolphins were reported in Yamaguchi (southern Sea of Japan), Ehime (Seto Inland Sea), and Kagoshima (eastern East China Sea) [1113, 2448].

Natural enemies Killer whales and sharks attack striped dolphins [185].

Parasites Various endo- and ectoparasites were reported for this species [2829]. Cestodes *Monorygma* sp. and *Phyllobothrium* sp. were reported from stranded specimens from Japan [1128].

5. Remarks

Striped dolphins have been caught by direct dolphin fishery from 1940s to present off the Pacific coast of Japan. The number of annual catch was about 10,000 in 1967–1983 and decreased to < 1,000 in early 1980s [1365, 1385, 1608]. Incidental takes have been reported in various fishing gears. High organochlorine concentrations in striped dolphins have been reported from the Mediterranean Sea, which led to the large die-off by an epizootic of striped dolphins in 1990–1992 [76]. In the liver of striped dolphins around Japan during 1977–1982, concentrations of Se, Sr, Ag, Cd, Cs, Ba, Hg, and Pb were higher in the tissues of adults than those of fetuses, whereas the opposite trend was observed for Cr and Tl [79].

M. YOSHIOKA

A striped dolphin with serious injuries stranded on a beach in Nagasaki Pref. (M. Amano).

Cetacea   DELPHINIDAE

164

Red list status: K (MSJ); DD (IUCN)

Stenella longirostris (Gray, 1828)

EN spinner dolphin   JP ハシナガイルカ（hashinaga iruka）   CM 长吻飞旋海豚   CT 長吻飛旋原海豚
KR 긴부리돌고래   RS длиннорылый продельфин

Spinner dolphins are found in sub-tropical areas in southern Japan including around Amami-ohshima Isl. and the Ogasawara Isls. (K. Oki).

1. Distribution

Pantropical, occurring in all tropical and most subtropical waters around the world [1225]. Four subspecies are recognized. Of these, *S. l. longirostris* is distributed in oceanic tropical waters worldwide [1230], including southwestern Japanese waters [1367]. Spinner dolphins are commonly found in the waters of the Ogasawara Isls. throughout the year [2203]. There are some opportunistic sightings and strandings in other Japanese waters [1370].

2. Fossil record

Not reported in Japan.

3. General characteristics

Morphology   The body of spinner dolphins is slender, but can be robust depending on the population. They have a long, narrow beak with a flat melon. The dorsal fin is located at the mid point on its back and is erect. It can be triangular and cants forward in adult males [2875]. The color pattern of *S. l. longirostris* is tripartite, consisting of a dark gray dorsal field or cape, lighter lateral field, and white or very light-gray ventral field [2819]. Adults reach 2.0 m in females and 2.35 m in males [1230]. Dwarf spinner dolphins (*S. l. roseiventris*) with 2/3–3/4 the length of other subspecies are known from waters of Southeast Asia and northern Australia [1230, 2831].

Dental formula   40–60 in each row [2822].

Genetics   Not reported in Japan.

4. Ecology

Reproduction   Sexually matured body length is 147–195 cm (n = 10) for females and 160–194 cm (n = 15) for males, and physically matured body length is 167–195 cm (n = 5) for females and 181–200 cm (n = 6) for males of *S. l. longirostris* from Philippines in the western Pacific [2826]. Females attain sexual maturity at 4–7 years of age and males at 7–10 years. Ovulation may be spontaneous and breeding is seasonal [3865]. Gestation is about 10 months and average length at birth is about 75–80 cm [2822]. Calving interval is about 3 years [2822].

Lifespan   Maximum age of spinner dolphins estimated from layers in teeth is about 20 years [2827].

Diet   Diets of spinner dolphins include primarily small (< 20 cm) mesopelagic fishes, squids and sergestid shrimps obtained in

Stenella longirostris (Gray, 1828)

dives to depths of at least 200–300 m [2827].

Habitat Habitats of spinner dolphins are oceanic, but depending on areas and behaviors, they use shallow and coastal waters, sandy bottomed bays of oceanic islands and atolls, and shallow reefs [1230].

Behavior Aerial behavior, such as spin jumps and tail slaps, is a trademark of spinner dolphins. These aerial behaviors decrease in the mid-day hours in the Ogasawara waters [2196]. School sizes range between 20 and 200 individuals (mean = 96.0, SD = 57.3) in the Ogasawara waters [2203]. Diurnal movement was observed in the waters of the Ogasawara Isls. They use coastal waters of these islands for resting in the daytime and move to offshore waters at night [2203]. Daily and seasonal rhythms also are reported in the behavior of spinner dolphins in the coastal area of the southwestern Atlantic [3271]. Spinner dolphins are gregarious and form mixed schools with other oceanic dolphins including pantropical spotted dolphins and pilot whales [1358].

Natural enemies Sharks, probably killer whales and possibly false killer whales, pygmy killer whales and short-finned pilot whales are predators [2822].

Parasites An acanthocephalan *Corynosoma* sp. was reported in Japan [3975]. Twelve endoparasite species were listed [2827]. Barnacle *Conchoderma auritum* and *Xenobalanus* sp. were reported [1309, 2827].

5. Remarks

There are no direct fisheries in Japan. Dolphin watching programs are commonly operated in the Ogasawara Isls. and sometimes in some other areas including Amami-ohshima Isl. in the East China Sea. Many spinner dolphins, along with pantropical spotted dolphins, were taken incidentally in the purse-seine fishery for tuna in the eastern North Pacific in the last 40 years. Populations were severely reduced and have not recovered [3809].

K. MORI

A spinner dolphin school consists of large number of animals (Ogasawara Whale Watching Association).

A ventral view of the skull of a spinner dolphin (NSMT-M 26472) (N. Kurihara).

Cetacea   DELPHINIDAE

Red list status: K (MSJ); LC (IUCN)

Steno bredanensis (G. Cuvier *in* Lesson, 1828)

EN rough-toothed dolphin   **JP** シワハイルカ (shiwaha iruka)   **CM** 糙齿海豚   **CT** 糙齒海豚
KR 뱀머리돌고래   **RS** крупнозубый дельфин, гребнезубый дельфин

Rough-toothed dolphins are surfacing off Miyake Isl., Tokyo (K. Kogi).

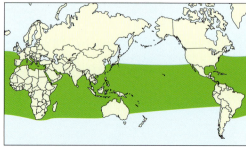

1. Distribution

Distributed in tropical, subtropical and warm-temperate waters of the world [1230]. Around Japan, they occur south of northern Honshu in the Pacific and the East China Sea, and also may occur in the Sea of Japan in summer [1370].

2. Fossil record

Not reported in Japan.

3. General characteristics

Morphology   Head is distinctive in that it lacks the crease usually separating the melon from the beak which usually is present in delphinids. The head tapers smoothly from blowhole to tip of the beak. The coloration is counter-shaded. Blackish brown dorsal cape is characteristic. Ventral side is pale gray. Dorsal fin is large and falcate. Beak is long and slender and occupies 60% of the condylobasal length. Lips are white. The largest record is 265 cm for males and 255 cm for females [2163]. Spots and scars by cookie-cutter shark (*Isistius brasiliensis*) occasionally are found on the body surface.

Dental formula   19–26/19–28 [2163]. Teeth of rough-toothed dolphins have fine longitudinal grooves, which leads to English vernacular name.

Genetics   Chromosome: $2n = 44$ [190]. Genetic analyses showed closer relationships with the tucuxi *Sotalia fluriatilis* than humpbacked dolphins *Sousa* sp., whereas those including nuclear genes indicated affiliation to Globicephalinae [74, 1833, 2042, 3778]. While rough tooth dolphins are generally pelagic, genetically differentiated small coastal populations are reported off tropical islands [2735]. Hybrid between *Steno* (mother) and *Tursiops* (father) was born in captivity [465].

4. Ecology

Reproduction   Little is known about the reproduction of rough-toothed dolphins. Mean length and age at sexual maturity are 225 cm and 14 years of age for males, and 210–220 cm and 10 years for females in the western North Pacific, respectively [2163]. Length at birth is about 100 cm.

Lifespan   Maximum age is estimated at 36 years of age [2157].

Diet   Prey items found in stomach contents were fish and cephalopods [2163]. Rough-toothed dolphins were observed preying on flying fish (Exocoetidae) while sliding beneath the water surface

Steno bredanensis (G. Cuvier *in* Lesson, 1828)

[1358] and sharing large mahi-mahi (dolphin fish, *Coryphaena*) in Hawaii [2874].

Habitat Rough-toothed dolphins usually occur in deep and offshore areas in warm areas with surface water temperatures > 25°C [1358].

Behavior Schools of a few tens of individuals usually are reported, but occasionally schools up to 300 individuals are found [2163]. Rough-toothed dolphins often are found with many other species including pilot whales, and bottlenose, spotted and spinner dolphins.

Natural enemies Not reported, but probably killer whales and large sharks are predators.

Parasites Trematoda *Odhneriella gondo* was reported from Japan [1763]. Cestodes *Tetrabothrius forsteri*, *Trigonocotyle prudhoei*, *Strobilocephalus triangularis*, acanthocephalan *Bolbosoma capitatum*, and nematode *Anisakis typica*, Trematoda *Nasitrema attenuata* and *N. globicephalae* were reported from other areas [487, 2031, 2163]. Cyamid ampipod *Isocyamus delphinii* was reported as an ectoparasite [2163].

5. Remarks

In Japan, small numbers of rough-toothed dolphins were hunted by drive and harpoon fishery [1370]. This species was one of the victims of incidental kills in gill and driftnets. Rough-toothed dolphins have been kept in captivity in Japanese aquaria.

M. YOSHIOKA

The rough-toothed dolphin lacks the crease separating the beak from melon (M. Yoshioka).

A rough-toothed dolphin at Okinawa Churaumi Aquarium, Okinawa Pref. (M. Yoshioka).

Cetacea   DELPHINIDAE

Red list status: DD (IUCN)

Tursiops aduncus (Ehrenberg, 1833)

EN Indo-Pacific bottlenose dolphin   JP ミナミハンドウイルカ (minami handou iruka), ミナミバンドウイルカ (minami bandou iruka)
CM 印太瓶鼻海豚   CT 印太洋瓶鼻海豚   KR 남방큰돌고래   RS индийская афалина

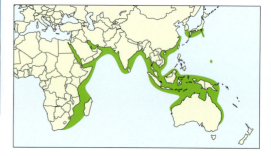

Two adult males. A resident population of Indo-Pacific bottlenose dolphins around Mikura Isl., Tokyo is one of the best-known targets of swim-with-dolphin programs in Japan (Mikurashima Tourist Association).

1. Distribution

Distributed in warm temperate to tropical waters in the Indian Ocean and western Pacific Ocean. In Japanese waters, they have a limited distribution in warm temperate waters including Mikura Isl., Ogasawara Isls. (Tokyo), Bungo Channel, Amakusa (western Kyushu), Kagoshima Bay and Amami-ohshima Isl. [1278, 2161, 2198, 2438, 3251, 3257, 3259, 3447]. A small number of Indo-Pacific bottlenose dolphins commonly are found off Noto-jima (Ishikawa Pref., the Sea of Japan side), and opportunistically in the areas of the Izu Isls. Chain and southern Izu and Boso Pens. [2197, 2237].

2. Fossil record

Not reported in Japan.

3. General characteristics

Morphology   Indo-Pacific bottlenose dolphins are very similar to common bottlenose dolphins *T. truncatus*, but smaller. They have a proportionately longer rostrum and robust (somewhat more slender than common bottlenose dolphins) and tall falcate (a little triangular and wider based) dorsal fin. They develop ventral spotting at about the time of sexual maturity [2905]. Coloration of the body is somewhat lighter than common bottlenose dolphins. The dark gray cape is distinct. Indo-Pacific bottlenose dolphins grow to lengths of about 2.7 m and 230 kg [1230].

Dental formula   21–29 in each half of the upper and lower jaws [1230].

Genetics   MtDNA and microsatellite analyses indicate differentiation among Amami, Amakusa, the Ogasawara Isls. and Mikura Isl. populations [856].

4. Ecology

Reproduction   Indo-Pacific bottlenose dolphins develop ventral spots at sexual maturity, when males are > 235 cm in length and females are > 230 cm [2905]. Sexual maturity may be attained at an older age for *T. aduncus* than for *T. truncatus*, with females producing their first calf at 12 years of age or older [399]. Calving interval averages 3.4 years. Neonates frequently are observed from April through October. The crude annual birth rate averages 0.07 around Mikura Isl. [1664].

Lifespan   Lifespan is probably similar to that of common bottlenose dolphins (< 40–50 years).

Tursiops aduncus (Ehrenberg, 1833)

Diet Prey includes a variety of fish and squids, including those in reefs or sandy bottoms and some inshore schooling fish [3867]. Stomach contents of individuals from around Mikura Isl. and Amakusa are occupied by epipelagic fish and cephalopods [1278], and *Chelon* sp., *Ilisha elongata*, conger eel (Congridae), filefishes (Mugilidae), mullet (Monacanthidae) and loliginid squid (*Loligo* sp.) [4045], respectively.

Habitat Indo-Pacific bottlenose dolphins occur over continental shelves, mostly in shallow costal and inshore waters. They also occur around oceanic islands [1230]. Habitats are limited and intermittent in Japan. Most of the dolphins off Amakusa, Kagoshima Bay, Mikura Isl. and Ogasawara Isls. are year-round residents [1664, 2202, 2438, 3257].

Behavior Indo-Pacific bottlenose dolphins form schools of variable size. Schools commonly consist of more than 100 individuals in Amakusa, 2–50 individuals (mean ± SD: 30.74 ± 13.90) in Kagoshima Bay, 1–30 (7.49 ± 6.23) around Ogasawara Isls. and 1–30 around Mikura Isl. [2067, 2198, 2438, 3257]. Geographic variation in the whistles were found among individuals along Amakusa, Mikura Isl. and Ogasawara Isls. [2236]. Dolphins in Shark Bay, west Australia are known to use marine sponges as foraging tools [1751].

Natural enemies Sharks are probably the most common predators.

Parasites *Phyllobothrium* sp. (Cestoda) and possibly *Crassicauda* sp. (Nematoda) were found in stranded specimens in Mikura Isl. [1278]. Barnacles *Xenobalanus* sp. is reported from Japan [2970].

5. Remarks

Until recently, the species was classified as *T. truncatus*. There is no direct fishery in Japan. Dolphin watching and swim-with-dolphin programs are operated around Mikura Isl., Toshima Isl., Ogasawara Isls., Noto-jima and Amakusa, western Kyushu.

K. MORI & M. YOSHIOKA

Indo-Pacific bottlenose dolphins appear in the Ogasawara shallow costal waters (Ogasawara Whale Watching Association).

Cetacea DELPHINIDAE

Red list status: C-2, LP for the population in offshore waters of the western North Pacific (MSJ); LC (IUCN); ND (JFA)

Tursiops truncatus (Montagu, 1821)

EN common bottlenose dolphin JP ハンドウイルカ (handou iruka), バンドウイルカ (bandou iruka)
CM 瓶鼻海豚 CT 瓶鼻海豚 KR 큰돌고래 RS афалина

Common bottlenose dolphins at Kamogawa Sea World, Chiba Pref. The common bottlenose dolphin is the most popular species in Japanese aquariums (M. Yoshioka).

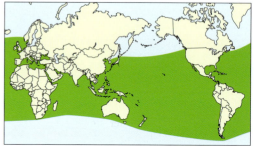

1. Distribution

Cosmopolitan: found in most of the world's warm temperate to tropical seas in coastal and offshore waters. Around Japan, they are found as far north as southern Hokkaido in the western North Pacific and in the Sea of Japan and the East China Sea [3465].

2. Fossil record

Not reported in Japan.

3. General characteristics

Morphology Common bottlenose dolphins have a mid-sized and robust body with falcate fins. The beak is short and thick. They have light- to dark-gray dorsal and lighter gray ventral coloration. Black stripes run from melon to eyes and blowhole and from eyes to flippers. A great deal of geographical variation in body size and morphology is known among populations, although some of them may be questionable because of small sample size and natural hybridization [1230]. Asymptotic lengths (over 20 years of age) are estimated at 305.3 cm for males and 288.0 cm for females [1369].

Dental formula 19–26/19–26. Teeth are stouter and fewer in number than in most of the smaller delphinids [3866].

Genetics Chromosome: $2n = 44$ [190]. Hybridization with other odontocete species is known to include Risso's dolphin, false killer whales, short-finned pilot whales, Pacific white-sided dolphins, rough-toothed dolphins and short-beaked common dolphins in captivity and the wild [465, 1230, 2505, 3170]. Two distinct populations were suggested around Japan from the comparison of biological parameters of life history [1369]; one is in the Pacific side (migrating off Izu–Taiji) and the other migrates around Iki Isl. in the Korean Str. to the Sea of Japan, but genetic evidence has not been obtained.

4. Ecology

Reproduction Males attain sexual maturity with 290–300 cm in body length and over 11 years of age, and females reach sexual maturity with length of 270–290 cm at 6–13 years of age in the Pacific population off Japan [1369]. Mean neonatal length is 128 cm and gestation is 370 days in Japanese populations. Parturition occurs during February to October with a peak in June. Mean calving interval is 3.07 years including 1.8-year lactation period [1369]. Simultaneously pregnant and lactating females have been noted on occasion. Ovulation (estrous) cycles were estimated to be 1 month from hormonal profiles in captive mature females [2891, 4191].

Lifespan Longevity is 45 years, but is less for males [1369].

Diet Generalist feeders. Their stomach contents consist mainly of small fish with occasional squid, crabs, shrimp, and other smaller animals [1230].

Habitat The species primarily inhabits most warm temperate and tropical shorelines, adapting to a variety of marine and estuarine habitats, even ranging into rivers. But, occasionally is found in offshore pelagic waters.

Behavior Common bottlenose dolphins normally live in small groups, containing 2–15 individuals. However, group size may be highly variable as they live in fission-fusion societies within which individuals associate in small groups that change in composition, often on a daily or hourly basis [3867]. In offshore waters, schools of more than a few hundred individuals were found. Mixed schools with pilot whales, false killer whales and Risso's dolphins occasionally have been observed. Common bottlenose dolphins are active in behavior. They frequently leap out of the water and ride bow-waves. The species is commonly known for its friendly character and curiosity towards humans immersed in or near water. However, dolphins are predators and frequently exhibit aggressive behaviors. This includes fights among males for social rank and access to estrous females, as well as aggression towards sharks, certain killer whales, and other smaller species of dolphins. In a population, off Scotland, infanticide was observed, and attacking and killing of harbor porpoises was filmed [482].

Natural enemies Large shark species prey on common bottlenose dolphins. Killer whales also may prey on them. As evidence of occasional encounters with sharks, many common bottlenose dolphins bear shark-bite scars on the body surface [3867].

A common bottlenose dolphin bearing shark-bite scars on the body surface is porpoising out of water off Kii Pen. (M. Kanda).

Parasites A cestode *Tetrabothrius delphini* and a trematode *Nasitrema dalli* were reported from Japan [1763]. A list of known endoparasites from other areas was reported [3866].

5. Remarks

Common bottlenose dolphins are the best known species of all cetaceans in captivity in the world including Japan. They also take advantage of human activities to facilitate prey capture in Africa and South American countries. In Japan, the species is one of the main targets in the driving fishery. Off Iki-shima Isl., about 6,600 dolphins were culled in mid 1960s to mid 1980s [1370]. Incidental catches also are known in various fishing gears including gillnets in countries around the World. Extremely high levels of chlorinated hydrocarbon residues have been detected in tissues of common bottlenose dolphins in many parts of the world including Japan [2533]. Births of calves by artificial insemination are successful in captivity, including in Japan [2891]. A skeleton (not fossil) of the bottlenose dolphin was found in the Mawaki Site, Noto Pen. on the Sea of Japan side from Holocene epoch [894].

M. YOSHIOKA

A cow–calf pair of the common bottlenose dolphin off the coast of Nagasaki Pref. (M. Amano).

Cetacea   PHOCOENIDAE

168

Red list status: En for the Omura Bay population as *N. phocaenoides*, R for the other populations off the Japanese coasts (MSJ); VU (IUCN); R as *N. phocaenoides* (JFA)

Neophocaena asiaeorientalis (Pilleri & Gihr, 1972)

EN narrow-ridged finless porpoise   JP スナメリ (sunameri)   CM 江豚，露脊鼠海豚   CT 江豚，露脊鼠海豚
KR 쇠물돼지   RS беспёрая морская свинья

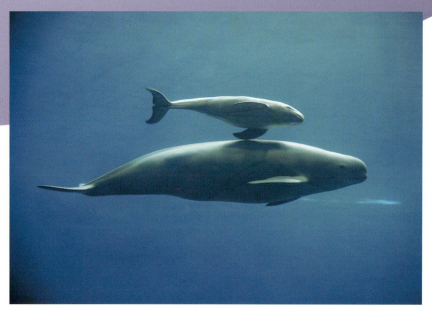

A cow–calf pair of the finless porpoise in captivity at Toba Aquarium, Mie Pref. (Toba Aquarium).

1. Distribution

Temperate coastal waters in the western Pacific Ocean from the Taiwan Str. to the waters of northern China, Korea and Japan, including the middle and lower reaches of the Yangtze River [1229]. The distribution within Japanese waters is discontinuous [3253]. The species is mainly distributed in 5 waters: the Ariake Sound–Tachibana Bay area, Omura Bay, the Seto Inland Sea–Hibiki Nada area, the Ise Bay–Mikawa Bay area, and coastal waters from Sendai Bay to Tokyo Bay. Reports of strandings and by-catches are few in other waters [1112, 1113]. No information on the occurrence of the porpoise remains at the Tsushima Isls., which are located between Korea and Japan [3253]. A porpoise discovered in the waters around Okinawa-jima Isl. is considered to be a stray from Chinese waters [4163].

2. Fossil record

Obtained from the Seto Inland Sea (Setonai kai) from the Late Pleistocene [1572].

3. General characteristics

Morphology   Narrow-ridged finless porpoises have a rounded head without an apparent beak, and no dorsal fin. Instead of a dorsal fin, there is a narrow dorsal ridge running from at or anterior to the middle of the back to the tail stock. The ridge is covered with 1–10 rows of tubercles [1229]. Adult coloration is dark or light gray [2171], but neonates are blackish [1391]. Maximum body length varies geographically ranging from 175 cm to 229 cm [688, 703, 1784, 1839, 2392, 2972, 3258]. Males are larger than females. Body weight at 160 cm is estimated to be 50–60 kg [1368]. Condylobasal length of adult size is 209–295 cm [1229]. Geographical variation in skull morphology is found among the Yangtze River, the Yellow Sea and Japanese waters [115, 704]. Variation has also been detected in Japanese waters, e.g. the porpoises in the Ise Bay–Mikawa Bay area have narrower skulls than other Japanese waters [4167]. Two subspecies are currently recognized: Yangtze finless porpoise, *N. a. asiaeorientalis* and East Asian finless porpoise, *N. a. sunameri* [1229].

Dental formula   16–21/15–20 [1223], 15–18/17–20 for the Ariake Sound–Tachibana Bay population in Japan [2171].

Neophocaena asiaeorientalis (Pilleri & Gihr, 1972)

Genetics Chromosome: $2n = 44$ [4248]. Differentiation is found in Chinese and Japanese populations and the genetic diversity of each population is low [4061, 4062]. In Japanese waters, the existence of 5 genetically isolated local populations and at least 2 distinct origins have been suggested [83, 4168]. The Japanese narrow-ridged finless porpoise has a relatively high MHC heterozygosity [863].

4. Ecology

Reproduction Information on the life history of narrow-ridged finless porpoises has been mainly obtained from by-caught and stranded specimens. In addition, the species has been held in captivity since the 1960s and various information on its life history has accumulated, including information on blood properties [84, 689, 1360, 2022, 2394, 3813]. Calves are born at approximately 80 cm [690, 816, 1391, 1402, 1784, 2972, 3258]. The peak of the calving season is in April–May in the Yangtze River and the Yellow Sea and most of Japanese waters but in November–December in the Ariake Sound–Tachibana Bay area [690, 702, 1391, 1784, 2972, 3254]. Serum testosterone concentrations of adult males show seasonal variation: high in March and April and low in October and December in the Yangtze River, suggesting that reproduction in the male is seasonal [783]. The porpoises attain sexual maturity at 4–6 years and at 125–155 cm in body length [368, 702, 1384, 1839, 3258]. Gestation lasts 10.6–11.2 months [1402] and a 2-year breeding cycle is common [1391]. The onset of ingestion of solid food begins at 4–6 months and lactation lasts for about 1 year in captivity [688, 3813].

Lifespan Maximum known age is 18–25 years [702, 1384, 3258]. One male survived for 28.83 years in the Toba Aquarium, Japan [688].

Diet Fish, cephalopods and shrimps are the main diet: fishes of the families Clupeidae, Engraulidae, Sciaenidae, Gobidae, Apogonidae, Leiognathidae, Embiotocidae, Pholidae, and Paralichthyidae; cephalopods of the families Octopodidae, Sepiidae, Sepiolidae, and Loliginidae; crustaceans of the families Penaeidae, Crangonidae, Pasiphaeidae, Palaemonidae, and Squillidae [2801, 3254]. At the Toba Aquarium, Japan, diurnal and seasonal variation of the food consumption has been reported and average daily food consumption was estimated to be 5.2–5.8% of whale's the body weight [1406].

Habitat In Japanese waters, the habitat must fulfill 2 conditions: shallow depth < 50 m and a non-rocky bottom [3253]. In the Seto Inland Sea, narrow-ridged finless porpoises frequent waters within 1.85 km off the coast [1391] and waters 10–20 m deep [3252]. The distribution is restricted by water depth in the coastal waters from Sendai Bay to Tokyo Bay [117]. In the Yellow Sea, the species is sighted at distances of 65–241 km from the coast, where the water depth is 30–55 m [2154]. The species shows seasonal movement. Seasonal density fluctuation may be linked to reproduction in the Seto Inland Sea and the Ariake Sound–Tachibana Bay area [1391, 3256]. Seasonal and diurnal occurrence patterns are reported in various waters and the relation with prey availability has been discussed [85, 86, 1570, 1709, 2329, 2483, 2939, 3395].

Behavior Group size is commonly small. Mean group size is reported to be below 2 individuals in Japanese waters [1368], but large groups sometimes form: 40 in the Ise Bay–Mikawa Bay area [3395], 21 and at least 30 in the Seto Inland Sea [1307, 2391], and 82 and 117 in the Ariake Sound–Tachibana Bay area [4165]. Such groups appear to form opportunistically for prey resources or due to the occurrence of sharks [1230, 3395]. Aerial behavior is rare. Narrow-ridged finless porpoises do not ride bow waves and avoid vessels by deep diving or alternating swimming direction, but they occasionally ride stern waves [1391]. The Yangtze narrow-ridged finless porpoise carries its calves on its back [2842, 3861]. Studies using electronic tags attached to the animals have revealed the acoustical and social behavior of narrow-ridged finless porpoises [87, 88, 1571, 2971]. Narrow-ridged finless porpoises can swim horizontally more than 90 km in a day.

Natural enemies The remnant of a finless porpoise was found in the stomach of a great white shark (*Carcharodon carcharias*) in the East China Sea off Okinawa [3677].

Parasites *Corynosoma* sp. (helminth) is reported in Ibaraki Pref. [1767].

5. Remarks

By-catch, habitat loss, boat traffic, and environmental contaminants are believed to threaten their survival [1368, 3824]. A significant decline in the density of the narrow-ridged finless porpoise has been reported in the Yangtze River and the Seto Inland Sea, and habitat fragmentation may be taking place there [1403, 2057, 3252, 4250]. In addition to organotin and organohalogen compounds, synthetic musk fragrances and hydroxylated polychlorinated biphenyls were identified in narrow-ridged finless porpoises in Japanese waters [1274, 2001, 2411, 2566]. A relationship between chemical contaminant level and parasitic infection has been suggested [1129, 2424]. In Japan, stranding networks have been established in several regions [816, 1054, 1819, 2447, 2972, 3544] and information on the status of local populations has accumulated by various methods, e.g. questionnaire to fishermen [1238, 3379, 3812]. Also, aerial sighting surveys for abundance estimation have been carried out. Estimates of abundance are as follows: 3,093 (CV = 0.16) and 2,963 (0.25) in the Ariake Sound–Tachibana Bay area, 187 (0.20) in Ohmura Bay, 7,572 (0.17) and 9,177 (0.20) in the Seto Inland Sea, 3,743 (0.24) and 2,961 (0.25) in the Ise Bay–Mikawa Bay area and 3,387 (0.33) and 7,745 (0.59) in coastal waters from Sendai Bay to Tokyo Bay [117, 2578, 3252, 4165, 4166, 4186]. Studies on by-catch magnitude and its risk assessment suggest the necessity of further investigations on by-catch situations [827, 828, 3255].

M. SHIRAKIHARA & M. YOSHIOKA

Cetacea PHOCOENIDAE

Red list status: R for population(s) inhabiting coasts of northern Japan, K for other populations in the Far East (MSJ); LC (IUCN); R (JFA)

Phocoena phocoena (Linnaeus, 1758)

EN harbor porpoise   JP ネズミイルカ (nezumi iruka)   CM 港湾鼠海豚   CT 港灣鼠海豚
KR 쇠돌고래   RS морская свинья

A harbor porpoise raising its head up out of water in captivity at Kamogawa Sea World, Chiba Pref., Japan (M. Yoshioka).

1. Distribution

Temperate to subpolar coastal waters of the Northern Hemisphere. The distribution range is separated into the North Atlantic, the North Pacific and the Black Sea including the Sea of Azov, and porpoises in each area are recognized as different nominal subspecies, *P. p. phocoena*, *P. p. vomeria*, and *P. p. relicata*, respectively [296]. Harbor porpoises in the western North Pacific are considered to be a fourth undescribed subspecies. In Japan, harbor porpoises are found off northern Honshu north of Chiba and Ishikawa Prefs., and they are common off the Hokkaido coasts [105, 1054].

2. Fossil record

Not reported in Japan.

3. General characteristics

Morphology   Harbor porpoises are short and robust without an apparent beak. A low and triangular dorsal fin is located in the middle of the back. There are small horny tubercles on the leading edge of the dorsal fin. Male's genital aperture is closer to the umbilicus than the anus, a trait shared with a few *Phocoena* species. Body is darker gray on the back and lighter on the belly, and the border is rather striking in the face and posterior half of the body. Clear dark flipper stripes run from around the gape to the anterior insertion of flipper. Lips, eye patches, flippers, the dorsal fin, and flukes are darker gray. There is considerable individual variation in coloration. Females become larger than males.

Based on the limited number of specimens, the maximum body length of Japanese harbor porpoise is about 170 cm for females and 150 cm for males [711]. Cranium has shared characteristics with other phocoenids, such as a short triangular rostrum with round tip, lumps of premaxillae in front of external nares, and a round occipital. No sexual dimorphism is observed in skull measurements [111]. Condylobasal length (mean ± SD) = 271.4 ± 13.1 mm (range = 246.2–303.2, n = 31) in Japan [111]. Other skull measurements of Japanese specimens have been reported [111]. There is morphological differentiation among skulls from the eastern North Atlantic, the western North Atlantic, the eastern North Pacific and the western North Pacific [111, 4240].

Dental formula   24–31/23–30 (Japan), 25–29/21–29 (eastern

Phocoena phocoena (Linnaeus, 1758)

North Pacific), 22–28/21–27 (eastern North Atlantic), 24–29/21–28 (western North Atlantic) [111]. There is a significant difference in the number of teeth between Atlantic and Pacific specimens [111, 4240]. Teeth are spatulate, like those of other phocoenids.

Genetics Chromosome: $2n = 44$ and $FN = 84$ [190]. Harbor porpoises are genetically distinct among 5 separated ranges; the Black Sea, the eastern North Atlantic, the western North Atlantic, the eastern North Pacific, and the western North Pacific [2898, 3825]. Distinct mtDNA differentiation has been reported between the north/northwestern and east/southeastern North Pacific, with the border near British Columbia [3393]. Differentiation among local populations has also been reported along both coasts of the North Atlantic [563, 2900, 3601]. Off British Columbia hybrids between harbor and Dall's porpoise (*Phocoenoides dalli*) are not rare [421, 3880].

4. Ecology

Reproduction Information on the life history of harbor porpoises around Japan is limited. The youngest mature animals are 3 years for both sexes [711]. Porpoises from the western North Atlantic also attain sexual maturity at 3 years [2869]. Mature females in the North Atlantic calve every year, while a 2-year calving interval is reported in California [925]. Lactation is thought to last for 8–12 months [2868]. Reproduction is quite seasonal; calving likely occurs in May–June in the western North Pacific [711] and in the western North Atlantic [2869], and mating follows about a month later. Male testis size and spermatogenetic activity shows seasonal variation [2455]. Large testis size over 2 kg and 4% of body weight suggests promiscuous mating with intensive sperm competition [564].

Lifespan Harbor porpoises are short-lived species and rarely live longer than 15 years. The oldest individual reported from the western North Atlantic was a 17 year old female but most specimens are younger than 12 years [2869]. The oldest porpoise from Japanese waters was 11 years [711].

Diet Stomach contents of five harbor porpoises taken off Japan included small squid beaks and otoliths tentatively identified as herring, anchovy and hake (likely misidentified) [711]. Prey reported in other areas include various small schooling fishes such as capelin, herring, and sprat as well as benthic fishes such as cusk-eel, hake, and rockfish [105, 2868]. Diet may vary depending on geographic range, season, year, reproductive condition, and age [2868].

Habitat Harbor porpoises are found in coastal waters of depth ranging 20–200 m [314, 352, 2864]. In the western North Atlantic, they are common in cold waters between 5–16°C but they also occur in colder waters. Off Japan, sightings are usually in waters of 7–16°C [1357].

Behavior Harbor porpoises are not gregarious and often found alone or in small schools of a few individuals. These groupings are temporary and large aggregations of tens to hundreds of porpoises are occasionally found around food concentrations [2868]. Harbor porpoises usually swim calmly and aerial behavior is rare. Distributional shifts in strandings suggest north–south seasonal migration off the coast of northern Japan [3394]. Offshore–inshore movements have also been observed in many areas. Radio-tagged porpoises in the Gulf of Maine traveled long distances in short periods of time (14–58 km/day) and estimated home ranges were as large as 50,000 km^2 [2870]. Harbor porpoises make short dives (up to 5 minutes) continuously but most dives are shallower than 20 m [2784, 2870]. Dives with a bottom part, which are likely for foraging, are deeper and longer and maximum recorded depths are 98.6 m off Japan and 226 m off east coast of Canada [2784, 3870].

Natural enemies Killer whales and white sharks (*Carcharodon carcharias*) are major predators of harbor porpoises [2868]. Off the northeast coast of Scotland, bottlenose dolphins have been observed attacking harbor porpoises and are suspected to be responsible for many dead strandings. This behavior is thought to be related to infanticide in bottlenose dolphins [2809, 2907]. In the North Sea, attacks of grey seals (*Halichoerus grypus*) on harbor porpoises have been reported [322, 1220, 3766].

Parasites Not reported in Japan, but heavy infection of parasites (nematodes and trematode) is common in other waters [2868].

5. Remarks

Because of their coastal habitat, harbor porpoises are threatened by incidental catches in the various fisheries throughout their range. In Japan, incidental catches of 24 individuals were reported during 2003–2007 and most of them were by trap nets in Hokkaido [1054], but the number of cases reported should be only a small part of the actual catches. Annual incidental catch and mortality in the 3 large set nets at Usujiri, Hokkaido were reported as 3.67/net/year and 0.17/net/year, respectively [1178]. Considering that the total trap net along the coast of Hokkaido is close to 5,000 [1181], substantial incidental catches are suspected. Harbor porpoises are also thought to be under threat of artificial persistent organic chemicals and heavy metals. High concentrations of these chemicals in the North Atlantic, and especially in the North Sea, are suspected to cause immune depression and to be responsible for mortality from infectious disease [77, 276, 1232, 1829].

M. AMANO

Cetacea   PHOCOENIDAE

Red list status: K for the 2 populations in Japanese waters (MSJ); LC (IUCN); ND (JFA)

Phocoenoides dalli (True, 1885)

EN Dall's porpoise   JP イシイルカ (ishi iruka)   CM 白腰鼠海豚   CT 白腰鼠海豚
KR 작은곱등어   RS белокрылая морская свинья

A Dall's porpoise (*truei*-type) swimming fast with a rooster-tail of spray (M. Amano).

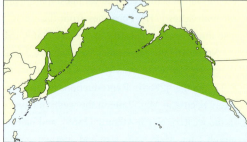

1. Distribution

Endemic to the northern North Pacific and adjacent seas (the Bering Sea, the Sea of Okhotsk and the Sea of Japan). Off Japan, this species is found from the north of Chiba Pref. on the Pacific side and the north of Shimane Pref. on the Sea of Japan side [2153].

2. Fossil record

Not reported in Japan.

3. General characteristics

Morphology   Dall's porpoises are robust, without an apparent beak, and with a wide-based and triangular dorsal fin located on the middle of the back. Body is black with a large white flank patch that differs in size among populations, and with white trailing edges on the dorsal fin and tail flukes. There are 2 color morphs: the *dalli*-type with a smaller white flank patch extending forward from the anus to the level of the dorsal fin and the *truei*-type with a larger white flank patch reaching to the front of the flippers. Also, the 2 *dalli*-type populations inhabiting Japanese coastal waters (the Sea of Japan population and the northwestern North Pacific population) have different size of white flank patches (smaller in the former) [110, 113]. The *truei*-type and the 2 forms of the *dalli*-type are considered to be 3 separate subspecies which can be discriminated by color pattern [107]. Geographical variation in body size and skull morphology has been reported [112, 4189]. Average lengths at physical maturity are 202.6 cm for male and 192.7 cm for the female *dalli*-type in the central North Pacific [551]. Maximum body lengths are 240 cm for males and 220 cm for females [1225].

Dental formula   21–28/21–28; tiny spade-shaped teeth in each side of the upper and lower jaws in most individuals [961].

Genetics   Chromosome: $2n = 44$ [190]. The *dalli*-type porpoises in the Sea of Japan are genetically differentiated from the *truei*-type and other *dalli*-type porpoises [854]. No clear difference in genetics has been found among the *dalli*-type populations in the Pacific although at least 11 populations are suggested to exist in the range of the species from the differences in pollutant levels, morphology, body size and calving grounds [1187]. In the eastern North Pacific, hybrids between Dall's and harbor porpoises are relatively common, and they may occur in other areas [421, 1230, 3880].

3. Ecology

Reproduction Information on the life history of Dall's porpoises around Japan has been accumulated by specimens collected from harpoon [1378, 1379] and driftnet fisheries [551, 2463]. Calves are born at an average length of 100 cm [551, 1378]. Males attain sexual maturity at 179.7 cm and ca 5 years and females 172 cm and ca 4 years for the *dalli*-type in the central North Pacific [551], 192 cm and 187 cm for the *dalli*-type in the Sea of Japan [109] and 196 cm and 7–8 years and 187 cm and 6–7 years for the *truei*-type population [1378], respectively. Most females give birth annually with an 11.4 month gestation period and 40 day nursing period. Reproduction is quite seasonal with a peak from spring to summer, and with an earlier peak in May in the Sea of Japan population [109]. Males having testes exceeding 40 g (one side) are identified as sexually mature [1390].

Lifespan Dall's porpoises are short-lived species, as with other Phocoenidae species, and rarely live longer than 20 years [2463].

Diet Dall's porpoises opportunistically take a variety of surface and midwater fish and squid. Diet may vary depending on geographic range, season and year [2625].

Habitat Dall's porpoises are found in deep cold temperate to subpolar coastal and offshore waters. They are frequent in cold waters between 3–9°C [961]. Off Japan, sightings are usually in waters of 3–20°C [1358].

Behavior Dall's porpoises swim fast and like to ride bow waves, especially younger animals [1390]. They seldom jump out of the water. When they swim fast, they produce a characteristic rooster-tail of spray. The species is not gregarious and often found alone or in small schools of a few to several individuals.

Natural enemies Killer whales and sharks are major predators of Dall's porpoises [1830].

A cow–calf pair of *dalli*-type Dall's porpoise in the Gulf of Alaska (M. Yoshioka).

Parasites *Corynosoma* spp. (helminth) are reported in Iwate Pref. [1763]. Heavy infection by internal parasites including *Campula oblonga* (helminth), is common in this species as well as in the harbor porpoise, but this situation does not usually cause health problems [1764].

5. Remarks

Large numbers of Dall's porpoises were incidentally taken by salmon- and squid-driftnet fisheries in the northern North Pacific and the Bering Sea in 1950–1990s. The handheld-harpoon fishery operated in northern Japan is a major threat for the species. This fishery was seriously damaged by the tsunami disaster in 2011, and currently hunts of about 1,300 porpoises per year occur, although the allowable quota is 13,500 [1607]. Recent estimates of population size are 174,000 (range: 115,000–262,000) for the *dalli*-type in the Sea of Japan–the western Sea of Okhotsk and 178,000 (114,000–279,000) for the *truei*-type in the northwestern North Pacific–the Sea of Okhotsk [2149].

M. YOSHIOKA

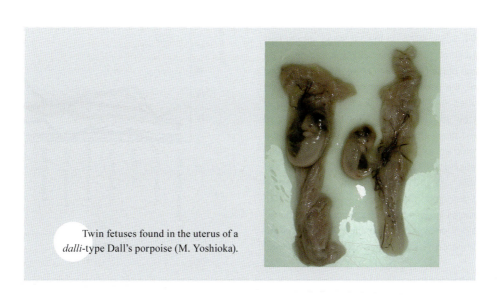

Twin fetuses found in the uterus of a *dalli*-type Dall's porpoise (M. Yoshioka).

RESEARCH TOPIC 11

History of mammalogy in Japan
Yukibumi KANEKO

The history of mammalogy in Japan can be divided into four stages. The 1st stage is the beginning of mammalogy led by Western researchers, who primarily described new species, subspecies, or new localities. During the 2nd stage, Japanese researchers studied by themselves, mainly revising taxonomy as well as adding new species, subspecies, or new records of localities. In the 3rd stage, following World War II, Japanese mammalogists studied mainly the taxonomy and ecology of small mammals. In the 4th stage, mammalogy in Japan diversified in research fields, methodology and techniques as well as in target mammals.

1. The outline of the fauna of Japanese Mammals

Currently, Japan is comprised of four main islands, which are Hokkaido, Honshu, Shikoku, and Kyushu, with neighboring small islands as well as the Ryukyu Isls. (Arch.). These Japanese islands form three long chains geographically. One chain is located at the East Asian coast from the Amur River which runs into the Sea of Okhotsk in eastern Siberia, to the southeast coast of China and Taiwan. The second one is the Kuril Isls., extending from the east of Hokkaido to the tip of the Kamchatka Pen. The third one is composed of the Ogasawara (Bonin) and Mariana Isls. in the Western Pacific Ocean, extending southward from southeastern Honshu.

The mammal fauna of Japan is very complex due to its geographical location in East Asia. The fauna of Japan is composed of the Palaearctic Region (Hokkaido; Honshu, Kyushu, and Shikoku; and Tsushima Isls.) and the Indo-Malayan Region (the Ryukyu Isls.) biogeographically. The demarcation line between the two regions is situated at the Tokara Str. between Akuseki-jima Isl. and Kodakara-jima Isl. in the Tokara Isls. of the Nansei Isls.

As of 2006, Japan had 112 native terrestrial mammals, 40 species of cetaceans, one species of sirenians, and 19 exotic species of mammals [33, 513, 1371]. We can follow changes in the number of species within a mammalian order in the current territory of Japan (Fig. 1) [24, 33, 1000, 1011, 1594, 1789, 1794, 3564, 3566, 3821]. In the 1st stage, Japan had 42 species of mammals [3564, 3566, 3821]; it had about 80 species in the 2nd stage [165, 1594, 1794]. After World War II (the 3rd stage), the number of species slightly decreased because it excluded the mammals of the Ryukyu Isls. due to political control by the U.S. [1000, 1789]. The number of species increased again after 1960, when Okinawa and the adjacent isls. returned to Japanese jurisdiction, especially due to the increasing number of species in Chiroptera and Soricomorpha [24, 33, 1011]. In the 4th stage, some Chiroptera and Soricomorpha species were divided into two or three species according to new taxonomic findings, based on karyology or molecular genetics. This increasing tendency in numbers of species has occurred worldwide [3882, 3883].

The number of exotic species of Japanese mammals increased from the 2nd to 4th stages (Fig. 2). Exotic mammals consisted of only *Mus musculus*, *Rattus norvegicus*, and *R. rattus* in the 2nd stage [165, 1794]. However, the number increased from 7–10 species listed in the 3rd stage [1000, 1011, 1789] to 19 species listed in 2005 [33].

2. The 1st stage

In the Edo (Tokugawa) period (17–19th century), natural history was popular among Japanese herbalists. They studied the Chinese literature to learn of medicinal uses for Japanese plants and animals. They identified Japanese plants and animals by their resemblance with drawings or descriptions in Chinese books in which plants and animals were not classified by the system of Linnean hierarchy or binominal nomenclature.

Scientific descriptions of mammals in Japan were first provided by Western mammalogists: John L. Bonhote, John E. Gray, Albert Günther, Alphone Milne-Edwards, Witmer Stone, Coenraad J. Temminck, Oldfield Thomas, Frederick W. True, etc. The pioneer in studies of Japanese mammals was Coenraad Temminck, Director of National Museum of

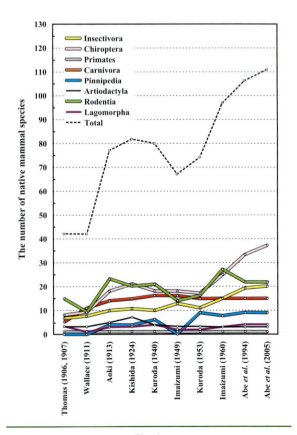

Fig. 1.

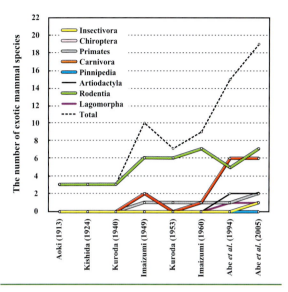

Fig. 2.

Natural History (RMNH), Leiden, the Netherlands. He studied many specimens collected by Phillip F. von Siebold during the Edo period (early 19th century), when Japan closed its doors to foreigners except from the Netherlands, Korea, and China. Siebold, a German, visited Japan as a Dutch medical doctor. He collected many scientific specimens that were gathered by his Japanese friends and students, with the aid of his assistant, Heinrich Bürger, because Siebold was prohibited from going outside the Dejima, Nagasaki Pref., Kyushu, controlled as a Dutch concession by the Edo government. Using his collections, Temminck studied terrestrial mammals and Hermann Schlegel studied marine mammals. They published "Fauna Japonica" with colored drawings. Among the new terrestrial species described by Temminck, specific names of 17 species are used currently [3547]: *Chimarrogale platycephalus, Crocidura dsinezumi, Urotrichus talpoides, Mogera wogura, Pteropus dasymallus, Rhinolophus cornutus, Myotis macrodactylus, Pipistrellus abramus, Cervus nippon, Capricornis crispus, Sciurus lis, Petaurista leucogenys, Pteromys momonga, Apodemus speciosus, A. argenteus,* and *Lepus brachyurus.*

Following Temminck, Oldfield Thomas, the curator of the British Museum (Natural History) (BM (NH)), London, England, was the next significant person for Japanese mammalogy. Thomas published a complete list of Japanese mammals excluding marine mammals at that time [3564, 3565]. Although the BM (NH) had received a very complete set of specimens from the RMNH, these specimens were recorded mostly as "Japan" and were degraded by exposure to light for 60–70 years before being transferred to the BM (NH). Thomas examined the specimens brought through the Zoological Expedition of the Duke of Bedford. The aim of the expedition was to collect specimens systematically throughout the Far East including Sakhalin, Korea, Inner Mongolia, and China, as well as Japan including the islands of Oki, Yaku-shima, and Tanegashima. Among these new species described by Thomas, specific names of nine species are still in use [3563–3565]: *Sorex gracillimus, S. shinto, Pipistrellus sturdeei, Nyctalus aviator, Tadarida latouchei, Eothenomys andersoni, E. smithii, Apodemus peninsulae,* and *Diplothrix legata.*

As a consequence of Thomas's study, the mammal fauna of Japan was characterized as a mixture of temperate and tropical forms [1000]. Of 42 terrestrial and volant mammals listed, 28 (66.7%) were endemic to Japan. Wallace stated that the majority of endemic mammals in Japan were related to those of the temperate or cold regions of the continent, either as the same or congeneric species [3821]. Wallace further perceived that several mammals which have no congeners in China are the antelope (*Capricornis crispus*), whose nearest related species are in Taiwan and Sumatra; a water-shrew (*Chimarrogale platycephalus*), whose nearest related species are in the Himalayas; and a unique shrew mole (*Urotrichus talpoides*).

3. The 2nd stage

After the Edo (Tokugawa) government reopened Japan to foreign intercourse at the end of the Edo period, the Meiji government was established in 1868. In the Meiji period, traditional Japanese natural history in the Edo period was excluded from the academic systems of the Faculty of Science in Japanese universities or colleges, and Western disciplines were imported as zoology or botany. Some American teachers of zoology such as Sylvester Morse and Charles O. Whitman taught at Tokyo Imperial College, the predecessor of the University of Tokyo.

Bunichiro Aoki at Tokyo Imperial College was the first Japanese mammalogist in the Meiji period. He listed and synonymized the scientific names of Japanese mammals based on 70 descriptions published up to that time [165]. Japan went to war against China (Qing dynasty) and Russia; the first Sino-Japanese War lasted from 1894 to 1895; the Russo-Japanese War lasted from 1904 to 1905. As a result, Japanese territory expanded to Taiwan and the southern part of Sakhalin. Aoki studied the mammals not only in current Japanese territory, but also in Taiwan, Korea, Sakhalin, and the Kurile Isls. Among his descriptions, 76 terrestrial and 23 marine mammals were listed in the current territory of Japan.

Following B. Aoki, Kyukichi Kishida, Nagamichi Kuroda, Yoshio Abe, Tetsuo Inukai, Mitoshi Tokuda, Tadao Kano, and Ryo Tanaka studied the mammals of Japan, Taiwan, Korea, or Northeastern China until the end of World War II (1945). During this time, the Japanese Empire Military Power expanded to Korea, Taiwan, North China, Philippines, and Indo-China. Except for Kano and Tanaka, they engaged mainly in taxonomy and reported new species and subspecies from new localities. However, they did not examine the type specimens housed in the RMNH and BM (NH). Therefore, several problems with regard to scientific names presented in publications by Temminck and Thomas remained unresolved until the 1960s–1970s: e.g. *Mus tanezumi, M. argenteus* (= *geisha*), *Vespertilio akokomuli, V. molossus, V. blepotis,* etc. Kishida published the "Monograph of Japanese mammals" [1594].

Kuroda published "A monograph of the Japanese mammals", which was richly illustrated with finely colored plates and included the geographic areas of not only Honshu, Kyushu, Shikoku, Hokkaido, and many small islands adjacent to them, but also Taiwan, Korea, the Bonin and the Mariana Isls. of the Western Pacific Ocean [1794]. Abe published a synopsis of the Japanese leporine and described a new spiny rat of *Tokudaia osimensis* [58, 59]. Tokuda published "A revised monograph of the Japanese and Manchou-Korean Muridae" including Taiwan [3592]. Inukai reported the occurrence of the pika and studied the brown bear in Hokkaido [1061, 1062]. Kano published zoogeographical studies of Mt. Tsugitake, Taiwan [1343]. Tanaka studied allometric growth for the small rodents of Taiwan [3518–3520]. Finally, the history of Japanese mammalogy during the 1st and 2nd stages was reviewed by Kuroda [1788, 1795].

In the 1930s, lectureships in taxonomy disappeared from most universities in Japan. Except for entomology and ichthyology, other disciplines (i.e. mammalogy, ornithology, etc.) of natural history were not developed due to the expansion of modern biological fields, such as cell biology, genetics, physiology, ecology, embryology, etc. Entomology and ichthyology were taught in the Faculty of Agriculture, because these subjects were applicable to agriculture, forestry, and fisheries. Furthermore, the Japanese government did not help to establish national museums of natural history except the National Science Museum, Tokyo, despite the demand for such facilities by many researchers [1323]. Most holotype specimens described by Japanese researchers before World War II were destroyed by fire during World War II [1330].

4. The 3rd stage

This stage includes the time span from the end of World War II (1945) to 1987. During this period most Japanese mammalogists belonged to the following two societies: the Nippon Honyuu-Dobutsu Gakkai (the former Mammalogical Society of Japan) and the Group of Mammalian Science. In addition to these two societies, the Institute of Cetacean Research and the Primate Research Institute, Kyoto University, were established in 1946, and 1967, respectively. These institutes vigorously promoted the international study of whales and primates.

After World War II, forested areas of Japan decreased extensively because the Ministry of Agriculture of Japan promoted the development of plantations of conifers for wood to build Japanese-style houses and produce furniture. As plantations expanded, afforested lands became plagued by feeding damage from Arvicolinae rodents, such as *Myodes rufocanus*, *Microtus montebelli*, and *Eothenomys smithii*. Subsequently, many applied studies (e.g. [2181, 3711, 3841]) were conducted to learn how to control these small rodents. Together with these applied studies, studies on taxonomy, ecology, physiology and morphology were developed for small mammals, such as moles, small rodents, and bats.

5. The 4th stage

Since 1987, Japanese mammalogists have been able to concentrate on their activities. This is because the current Mammalogical Society of Japan was established in that year by unifying the two societies previously mentioned. As of 1987, the Society had 566 members; as of 2007, membership was about 1,000. Following unification, research productivity increased because this society published two journals; "Mammal Study" in English and "Mammalian Science" in Japanese.

The number of papers focused on Japanese mammals written in Japanese or foreign languages has increased over time: from 150 (7.5%) in 1842–1950, to 100 (5%) in 1951–1960, to 300 (15%) in 1961–1970, 640 (32%) in 1971–1980, and to 810 (40.5%) in 1981–1991 [23]. The number of papers increased dramatically after the 1970s, owing to the increase in studies on chromosomes and biochemistry.

Now Japanese mammalogists investigate various fields of morphology, ecology, cytology, molecular genetics, ethology, etc. as well as wildlife conservation, management, or control. Their interests have diversified to include small mammals, and large mammals such as wild boars and deer and exotic species, because of increased interest in large native wild mammals and increasing numbers of exotic mammalian species in the 4th stage. They study in the field, museums and/or laboratories not only in Japan but also abroad.

Urotrichus talpoides Temminck, 1841

A portrait of Coenraad J. Temminck at National Museum of Natural History, Leiden, the Netherlands (photo by L. B. Holthuis).

References

1. Abe, C. (阿部近一), et al. (1989) 徳島県における哺乳類，両生類および爬虫類の生息状況. *徳島県立博物館開設準備調査報告* **4**: 1–53 (in Japanese).
2. Abe, G. (阿部豪) (2008) 北海道におけるアライグマ対策の地域生態学的研究～農作物被害防止に主眼をおいた外来種対策の実践と検証～. Sapporo: Hokkaido University. 152 pp. (in Japanese).
3. Abe, G., Ikeda, T., & Tatsuzawa, S. (2006) Differences in habitat use of the native raccoon dog (*Nyctereutes procyonoides albus*) and the invasive alien raccoon (*Procyon lotor*) in the Nopporo Natural Forest Park, Hokkaido, Japan. *in* Assessment and Control of Biological Invasion Risks (Koike, F., et al., Editors). Gland: IUCN, pp. 116–121.
4. Abe, H. Unpublished.
5. Abe, H. (1958) Individual and age variation in two species of genus *Sorex*, Insectivora in Hokkaido. *Memoirs of the Faculty of Agriculture, Hokkaido University* **3**: 201–209.
6. Abe, H. (1961) Two rare mammals obtained in Hokkaido. *The Journal of the Mammalogical Society of Japan* **2**: 3–7 (in Japanese with English abstract).
7. Abe, H. (1966) Habitats of wild mice in Hokkaido. *Japanese Journal of Applied Entomology and Zoology* **10**: 78–83 (in Japanese with English summary).
8. Abe, H. (1967) Classification and biology of Japanese Insectivora (Mammalia) I. Studies on variation and classification. *Journal of the Faculty of Agriculture, Hokkaido University* **55**: 191–269.
9. Abe, H. (1968) Classification and biology of Japanese Insectivora (Mammalia) II. Biological aspects. *Journal of the Faculty of Agriculture, Hokkaido University* **55**: 429–458.
10. Abe, H. (1968) Growth and development in two forms of *Clethrionomys*. 1. External characters, body weight, sexual maturity and behavior. *Bulletin of Hokkaido Forestry Experimental Station* **6**: 69–89 (in Japanese with English summary).
11. Abe, H. (1968) Notes on the ecology of *Sciurus vulgaris orientis* THOMAS. *The Journal of the Mammalogical Society of Japan* **3**: 118–124 (in Japanese).
12. Abe, H. (1971) Small mammals of central Nepal. *The Journal of Faculty of Agriculture, Hokkaido University* **56**: 367–426.
13. Abe, H. (阿部永) (1971) 日本の哺乳類 (10) げっ歯目 ヤチネズミ属 エゾヤチネズミ. *Mammalian Science* **11(1)**: 1–8 (in Japanese).
14. Abe, H. (1973) Growth and development in two forms of *Clethrionomys*: tooth characters, with special reference to phylogenetic relationships. *Journal of the Faculty of Agriculture, Hokkaido University* **57**: 229–253.
15. Abe, H. (1973) Growth and development in two forms of *Clethrionomys*: cranial characters, with special reference to phylogenetic relationships. *Journal of the Faculty of Agriculture, Hokkaido University* **57**: 255–274.
16. Abe, H. (1974) Change of the boundary-line of two mole's distributions in a period of 14 years. *The Journal of the Mammalogical Society of Japan* **6**: 13–23 (in Japanese with English summary).
17. Abe, H. (1975) Present status and problems on the natural history studies of Japanese small mammals. *Mammalian Science* **15(2)**: 59–80 (in Japanese).
18. Abe, H. (1975) Winter food of the red fox, *Vulpes vulpes schrencki* Kishida (Carnivora: Canidae), in Hokkaido, with special reference to vole populations. *Applied Entomology and Zoology* **10**: 40–51.
19. Abe, H. (阿部永) (1984) 北海道産ネズミ類の分類. *in* Study on Wild Murid Rodents in Hokkaido (北海道産野ネズミ類の研究) (Ota, K. (太田嘉四夫), Editor). Sapporo: Hokkaido University Press, pp. 1–20 (in Japanese).
20. Abe, H. (1985) Changing mole distributions in Japan. *in* Contemporary Mammalogy in China and Japan (Kawamichi, T., Editor). Tokyo: Mammalogical Society of Japan, pp. 108–112.
21. Abe, H. (1986) The range enlargement of *Echinococcus multilocularis* by red foxes. *Mammalian Science* **26(2)**: 1–9 (in Japanese).
22. Abe, H. (1986) Vertical space use of voles and mice in woods of Hokkaido, Japan. *The Journal of the Mammalogical Society of Japan* **11**: 93–106.
23. Abe, H. (1993) A review of the mammalogy in Japan. *Mammalian Science* **32**: 117–124 (in Japanese).
24. Abe, H. (阿部永), ed. (1994) A Pictorial Guide to the Mammals of Japan (日本の哺乳類). Tokyo: Tokai University Press, 195 pp. (in Japanese).
25. Abe, H. (1995) Revision of the Asian moles of the genus *Mogera*. *Journal of the Mammalogical Society of Japan* **20**: 51–68.
26. Abe, H. (1997) Habitat factors affecting the geographic size variation of Japanese moles. *Journal of the Mammalogical Society of Japan* **21**: 71–87.
27. Abe, H. (1998) Classification and morphology of Talpidae. *in* The Natural History of Insectivora (Mammalia) in Japan (Abe, H. & Yokohata, Y., Editors). Shobara: Hiba Society of Natural History, pp. 25–58.
28. Abe, H. (1999) Diversity and conservation of mammals of Japan. *in* Recent Advances in the Biology of Japanese Insectivora: Proceedings of the Symposium on the Biology of Insectivores in Japan and on the Wildlife Conservation (Yokohata, Y. & Nakamura, S., Editors). Shobara: Hiba Society of Natural History, pp. 89–104.
29. Abe, H. (阿部永) (2000) Illustrated Skulls of Japanese Mammals (日本産哺乳類頭骨図説). Sapporo: Hokkaido University Press, 279 pp. (in Japanese).
30. Abe, H. (2001) Isolated relic populations and their keeping mechanisms in moles. *Mammalian Science* **41**: 35–52 (in Japanese with English summary).
31. Abe, H. (2003) The habitation status of bats in Niijima, the Izu islands, Japan. *Bulletin of the Asian Bat Research Institute* **3**: 15–20 (in Japanese with English abstract).
32. Abe, H. (2003) Trapping, habitats, and activity of the Japanese water shrew, *Chimarrogale platycephala*. *Mammalian Science* **43**: 51–65 (in Japanese with English summary).
33. Abe, H., ed. (2005) A Guide to the Mammals of Japan. Revised ed. Hadano; Tokai University Press, 206 pp. (in Japanese and English).
34. Abe, H. (2005) Order Insectivora. *in* A Guide to the Mammals of Japan (Abe, H., Editor). Hadano: Tokai University Press, pp. 4–24 (in Japanese and English).
35. Abe, H. (阿部永) (2007) Illustrated Skulls of Japanese Mammals. Revised ed. (日本産哺乳類頭骨図説 増補版). Sapporo: Hokkaido University Press, 300 pp. (in Japanese).
36. Abe, H. (2009) A peculiar cusp on the fourth upper premolar of the Japanese water shrew *Chimarrogale platycephala*. *Mammal Study* **34**: 37–40.
37. Abe, H. (2011) Stomach contents of the Japanese water shrew, *Chimarrogale platycephala*. *Mammalian Science* **51**: 311–313 (in Japanese with English abstract).
38. Abe, H. (2012) The distribution of *Mogera etigo* in the Mitsuke City–Tochio, Nagaoka City Area and in the surrounding area of Gosen City–Niitsu, Niigata City in Niigata Prefecture in 2011. *Mammalian Science* **52**: 55–62 (in Japanese with English abstract).
39. Abe, H. (阿部永), et al. (1987) 野生動物分布等実態調査報告書：ヒグマ生態等調査報告書. Sapporo: Hokkaido Government, 75 pp. (in Japanese).
40. Abe, H. (阿部永) & Hayashi, Y. (早矢仕有子) (1990) 野生動物分布等実態調査報告書：シマフクロウ生態等調査報告書. Sapporo: Hokkaido Government, 63 pp. (in Japanese).
41. Abe, H., et al. (1970) Faunal survey of the Mt. Ishizuchi area, JIBP main area-II. Results of the small mammal survey on the Mt. Ishizuchi Area. *in* Studies on the Animal Communities in the Terrestrial Ecosystems and their Conservation; Annual Report of JIBP/CT-S for the Fiscal Year of 1969 (Kato, M., Editor). Sendai: JIBP/CT-S Section, pp. 7–14.
42. Abe, H., Maeda, K., & Kawabe, M. (1982) Mammals of the Tokachi-gawa Genryubu wilderness area. *in* 十勝川源流部原生自然環境保全地域調査報告書 (Reports of the Tokachi-gawa Genryubu Wilderness Area). Tokyo: Nature Conservation Society of Japan (日本自然保護協会), pp. 233–245 (in Japanese with English abstract).
43. Abe, H., et al. (2006) Dispersal of *Camellia japonica* seeds by *Apodemus speciosus* revealed by maternity analysis of plants and behavioral observation of animal vectors. *Ecological Research* **21**: 732–740.
44. Abe, H. (安倍博), et al. (1988) 野生ネズミのサーカディアンリズム—アカネズミ，ヒメネズミ，カゲネズミの比較. *Annual of Animal Psychology* **38**: 42 (in Japanese).
45. Abe, H., Shida, T., & Saitoh, T. (1989) Effects of reduced vertical space and arboreal food supply on densities of three forest rodents. *Journal of the Mammalogical Society of Japan* **14**: 43–52.
46. Abe, H., Shiraishi, S., & Arai, S. (1991) A new mole from Uotsuri-jima, the Ryukyu Islands. *Journal of the Mammalogical Society of Japan* **15**: 47–60.
47. Abe, M. Unpublished.
48. Abe, M., et al. (1994) Distribution, food habit and territory of *Pteropus pselaphon*. *Annual Report of Ogasawara Studies* **18**: 4–43 (in Japanese).
49. Abe, S. Unpublished.
50. Abe, S. (阿部慎太郎) (1992) マングースたちは奄美で何を食べているのか? (What kind of food do mongooses eat on Amami Island?) *Chirimosu* (チリモス) **3**: 1–18 (in Japanese).
51. Abe, S. (阿部慎太郎) (1994) 沖縄島の移入マングースの現状 (Present status of introduced mongooses on Okinawa Island). *Chirimosu* (チリモス) **5**: 34–43 (in Japanese).
52. Abe, S. (阿部慎太郎) (1995) 水晶体重量による奄美大島産マングースの齢査定 (Age determination of mongoose on Amami-oshima Island based on eye lens weight). *Chirimosu* (チリモス) **6**: 34–43 (in Japanese).
53. Abe, S., et al. (1997) Differential staining profiles of B-chromosomes in the East Asiatic wood mouse, *Apodemus peninsulae*. *Chromosome Science* **1**: 7–12.
54. Abe, S., et al. (1999) Food habits of feral mongoose (*Herpestes* sp.) on Amamioshima, Japan. *in* Problem Snake Management: the Habu and the Brown Treesnake (Rodda, G.H., et al., Editors). Ithaca: Comstock Pub. Associates, a division of Cornell University Press, pp. 372–383.
55. Abe, S., et al. (1991) Establishment in the wild of the mongoose (*Herpestes* sp.) on Amami-oshima Island. *Mammalian Science* **31**: 23–36 (in Japanese with English summary).

56. Abe, S., et al. (2006) Reproductive responses of the mongoose (*Herpestes javanicus*), to control operations on Amami-oshima Island, Japan. *in* Assessment and Control of Biological Invasion Risks (Koike, F., et al., Editors). Gland: IUCN, pp. 157–164.
57. Abe, T. (1974) An analysis of age structure and reproductive activity of *Microtus montebelli* population based on yearly trapping data. *Japanese Journal of Applied Entomology and Zoology* **18**: 21–27.
58. Abe, Y. (1931) A synopsis of the Leporine mammals of Japan. *Journal of Science of the Hiroshima University. Series B, Div. 1, Zoology* **1**: 45–63.
59. Abe, Y. (阿部余四男) (1933) アマミトゲネズミに就いて. *Botany and Zoology* **1**: 936–942 (in Japanese).
60. Abe, Y. & Maeda, K. (2004) Population dynamics of tube-nosed bats, *Murina leucogaster* Milne-Edwards, 1872 during year in Kawachi no kaza-ana cave, Taga-Cho, Shiga Prefecture. *Bulletin of Center for Natural Environmental Education, Nara University of Education* **6**: 19–23 (in Japanese with English abstract).
61. Abernethy, K. (1994) The establishment of a hybrid zone between red and sika deer (genus *Cervus*). *Molecular Ecology* **3**: 551–562.
62. Abollo, E., et al. (1998) Macroparasites in cetaceans stranded on the northwestern Spanish Atlantic coast. *Diseases of Aquatic Organisms* **32**: 227–231.
63. Abramov, A. & Wozencraft, C. (2008) *Martes zibellina. The IUCN Red List of Threatened Species. Version 2014.3*. Available from: http://www.iucnredlist.org.
64. Abramov, A.V. (1999) A taxonomic review of the genus *Mustela* (Mammalia, Carnivola). *Zoosystematica Rossica* **8**: 357–364.
65. Abramov, A.V. (2003) The head colour pattern of the Eurasian badgers (Mustelidae, *Meles*). *Small Carnivore Conservation* **29**: 5–7.
66. Abramov, A.V. & Baryshnikov, G.F. (2000) Geographic variation and intraspecific taxonomy of weasel *Mustela nivalis* (Carnivora, Mustelidae). *Zoosystematica Rossica* **8**: 365–402.
67. Abramov, A.V. & Puzachenko, A.Y. (2005) Sexual dimorphism of craniological charactors in Eurasian badgers, *Meles* spp. (Carnivora, Mustelidae). *Zoologischer Anzeiger* **244**: 11–29.
68. Abramson, N.I., Abramov, A.V., & Baranove, G.I. (2009) New species of red-backed vole (Mammalia: Rodentia: Cricetidae) in fauna of Russia: Molecular and morphological evidences. *Proceedings of the Zoological Institute RAS* **313**: 3–9.
69. Acebes, J.M.V., et al. (2005) Stranding of *Indopacetus pacificus* in Davao, Philippines. *in* Proceedings of the 16th Biennial Conference on the Biology of Marine Mammals. San Diego: Society for Marine Mammalogy, p. 67.
70. Adams, J.R., Leonard, J.A., & Waits, L.P. (2003) Widespread occurrence of a domestic dog mitochondrial DNA haplotype in southeastern US coyotes. *Molecular Ecology* **12**: 541–546.
71. Aduyanukosol, K. Personal communication.
72. Agetsuma, N. & Nakagawa, N. (1998) Effects of habitat differences on feeding behaviors of Japanese monkeys: Comparison between Yakushima and Kinkazan. *Primates* **39**: 275–289.
73. Agetsuma-Yanagihara, Y. (2004) Process of establishing an introduced raccoon (*Procyon lotor*) population in Aichi and Gifu Prefectures, Japan: policy for managing threats posed by introduced raccoons. *Mammalian Science* **44**: 147–160 (in Japanese with English abstract).
74. Agnarsson, I. & May-Collado, L. (2008) The phylogeny of Cetartiodactyla: The importance of dense taxon sampling, missing data, and the remarkable promise of cytochrome *b* to provide reliable species-level phylogenies. *Molecular Phylogenetics and Evolution* **48**: 964–985.
75. Aguilar, A. (2002) Fin Whale (*Balaenoptera physalus*). *in* Encyclopedia of Marine Mammals (Perrin, W.F., Würsig, B., & Thewissen, J.G.M., Editors). San Diego: Academic Press, pp. 435–438.
76. Aguilar, A. & Borrell, A. (1994) Abnormally high polychlorinated biphenyl levels in striped dolphins (*Stenella coeruleoalba*) affected by the 1990–1992 Mediterranean epizootic. *Science of the Total Environment* **154**: 237–247.
77. Aguilar, A. & Borrell, A. (1995) Pollution and harbour porpoises in the eastern North Atlantic: a review. *Report of the International Whaling Commission (Special Issue)* **16**: 231–242.
78. Aguilar, A., et al. (2008) The distribution of nuclear genetic variation and historical demography of sea otters. *Animal Conservation* **11**: 35–45.
79. Agusa, T., et al. (2008) Interelement relationships and age-related variation of trace element concentrations in liver of striped dolphins (*Stenella coeruleoalba*) from Japanese coastal waters. *Marine Pollution Bulletin* **57**: 807–815.
80. Aimi, M. (1980) A revised classification of the Japanese red-backed voles. *Memoirs of the Faculty of Science, Kyoto University. Series of Biology* **8**: 35–84.
81. Aimi, M. & Takahata, Y. (1994) Mammals of Japan. 18. *Macaca fuscata* (Japanese macaque). *Mammalian Science* **33**: 141–157 (in Japanese).
82. Aizawa, T. & Makino, S. (1938) On a vole, *Clethrionomys rufocanus bedfordiae* with an aberrant coat colour. *Transactions of the Sapporo Natural History Society* **15**: 187–190.
83. Aizu, M., et al. (2013) Genetic population structure of finless porpoise in Japanese coastal waters. (Paper SC/65a/SM24 presented to the IWC Scientific Committee). 12 pp.
84. Akagi, F. (2014) About the finless porpoise reproduction in the Miyajima Aquarium. *Aquabiology* **36**: 142–149 (in Japanese with English abstract).
85. Akamatsu, T., et al. (2010) Seasonal and diurnal presence of finless porpoises at a corridor to the ocean from their habitat. *Marine Biology* **157**: 1879–1887.
86. Akamatsu, T., et al. (2008) Evidence of nighttime movement of finless porpoises through Kanmon Strait monitored using a stationary acoustic recording device. *Fisheries Science* **74**: 970–975.
87. Akamatsu, T., et al. (2010) Scanning sonar of rolling porpoises during prey capture dives. *Journal of Experimental Biology* **213**: 146–152.
88. Akamatsu, T., et al. (2002) Diving behaviour of freshwater finless porpoises (*Neophocaena phocaenoides*) in an oxbow of the Yangtze River, China. *ICES Journal of Marine Science* **59**: 438–443.
89. Akasaka, T. (1977) Food habits and feeding behavior of Japanese serow in Nibetsu, Akita Prefecture. *Annual Reports of WWF-Japan* **1**: 67–80 (in Japanese with English abstract).
90. Akasaka, T. (1979) Social organization of Japanese serow in Nibetsu, Akita Pref.—Especially on the social units—. *in* Conservation Reports of Japanese Serow, Special Natural Monuments II (Nature Conservation Society of Japan, Editor). Tokyo: Nature Conservation Society of Japan, pp. 5–17 (in Japanese with English abstract).
91. Akasaka, T. & Maruyama, N. (1977) Social organization and habitat use of Japanese serow in Kasabori. *The Journal of the Mammalogical Society of Japan* **7**: 87–102.
92. Akasaka, T., Yanagawa, H., & Nakamura, F. (2007) Use of bridges as day roosts by bats in Obihiro. *Japanese Journal of Conservation Ecology* **12**: 87–93 (in Japanese with English abstract).
93. Akasaka, T., et al. (2004) Bat fauna of Hidaka and Tokachi districts, central and eastern Hokkaido. (2). Records of bats in Meto area, Ashoro, northeastern Tokachi. *Journal of the Japanese Wildlife Research Society* **30**: 9–14 (in Japanese with English summary).
94. Akazawa, T., et al. (1976) Report on the excavation at Kurosegawa Cave in Shirokawa-cho, Ehime Prefecture, Shikoku. *Memoirs of National Science Museum* **9**: 191–198 (in Japanese with English summary).
95. Aketa, K., et al. (2001) Digestibility of dugong feeding on *Zonstra marina* in captive. *Mammalian Science* **41**: 23–34 (in Japanese).
96. Akiba, Y. (秋保employee), Takahashi, O. (高橋修), & Takahashi, Y. (高橋雄一) (1996) 宮城県の野生哺乳動物. Sendai: 宮城野野生動物研究会, 101 pp. (in Japanese).
97. Akiba, Y., et al. (2013) Collection records of endo- and exoparasites obtained from small mammals in the University Tokyo Hokkaido Forest, Furano, Hokkaido, Japan. *Bulletin of the Biogeographical Society of Japan* **68**: 129–133 (in Japanese with English abstracts).
98. Akita, K. (1955) On Japanese tube-nosed bat collected in Mt. Mikasa, Nara Pref. *The Journal of the Mammalogical Society of Japan* **1**: 21–22 (in Japanese).
99. Akuzawa, M., et al. (2005) Clinical pathological examination of the Iriomote cat, *Felis iriomotensis*. *Island Studies in Okinawa* **13**: 7–15 (in Japanese with English abstract).
100. Allen, G. (1920) A bat new to the Japanese fauna. *Journal of Mammalogy* **1**: 139.
101. Allen, G.M. (1938) The Mammals of China and Mongolia. *in* Natural history of central Asia. Volume 11. Part 1 (Granger, W., Editor). New York: American Museum of Natural History, pp. 345–350.
102. Alves, F., et al. (2013) Population structure of short-finned pilot whales in the oceanic archipelago of Madeira based on photo-identification and genetic analyses: implications for conservation. *Aquatic Conservation: Marine and Freshwater Ecosystems* **23**: 758–776.
103. Amaha, A. (1994) Geographic variation of the common dolphin, *Delphinus delphis* (Odontoceti: Delphinidae). Doctoral dissertation. Tokyo: Tokyo University of Fisheries. 211 pp.
104. Amami Mammalogical Society (奄美哺乳類研究会) (2009) 奄美大島の野生化ヤギに関する基礎的研究. 2008 年度期 WWF ジャパン エコ・パートナーズ事業最終報告書. 25 pp. (in Japanese with English abstract).
105. Amano, M. (天野雅男) (1996) ネズミイルカ (Harbor porpoise). *in* 日本の希少な野生水生生物に関する基礎資料 (III). Tokyo: Japan Fisheries Resources Conservation Association (水産資源保護協会), pp. 340–348 (in Japanese).
106. Amano, N. (天野典子), Amano, Y. (天野洋祐), & Naruse, H. (成瀬裕昭) (2011) 伊豆大島におけるユビナガコウモリの初記録. *Bat Study and Conservation Report* **23**: 7–9 (in Japanese).

107. Amano, M. & Hayano, A. (2007) Intermingling of *dalli*-type Dall's porpoises into a wintering truei-type population off Japan: implication from color patterns. *Marine Mammal Science* **23**: 1–14.
108. Amano, M., et al. (2014) Differences in sperm whale coda between two waters off Japan: possible geographic separation of vocal clans. *Journal of Mammalogy* **95**: 169–175.
109. Amano, M. & Kuramochi, T. (1992) Segregative migration of Dall's porpoise (*Phocoenoides dalli*) in the Sea of Japan and Sea of Okhotsk. *Marine Mammal Science* **8**: 143–151.
110. Amano, M., et al. (2000) Re-evaluation of geographic variation in the white flank patch of *dalli*-type Dall's porpoises. *Marine Mammal Science* **16**: 631–636.
111. Amano, M. & Miyazaki, N. (1992) Geographic variation in skulls of the harbor porpoise, *Phocoena phocoena*. *Mammalia* **56**: 133–144.
112. Amano, M. & Miyazaki, N. (1992) Geographical variation and sexual dimorphism in the skull of Dall's porpoise, *Phocoenoides dalli*. *Marine Mammal Science* **8**: 240–261.
113. Amano, M. & Miyazaki, N. (1996) Geographic variation in external morphology of Dall's porpoise *Phocoenoides dalli*. *Aquatic Mammals* **22**: 167–174.
114. Amano, M. & Miyazaki, N. (2004) Composition of a school of Risso's dolphins, *Grampus griseus*. *Marine Mammal Science* **20**: 152–160.
115. Amano, M., Miyazaki, N., & Kureha, K. (1992) A morphological comparison of skulls of the finless porpoise *Neophocaena phocaenoides* from the Indian Ocean, Yangtze River and Japanese waters. *Journal of the Mammalogical Society of Japan* **17**: 59–69.
116. Amano, M., Miyazaki, N., & Yanagisawa, F. (1996) Life history of Fraser's dolphin, *Lagenodelphis hosei*, based on a school captured off the Pacific coast of Japan. *Marine Mammal Science* **12**: 199–214.
117. Amano, M., et al. (2003) Abundance estimate of finless porpoises off the Pacific coast of eastern Japan based on aerial surveys. *Mammal Study* **28**: 103–110.
118. Amano, M., Oi, T., & Hayano, A. (2004) Morphological differentiation between adjacent populations of Asiatic black bear, *Ursus thibetanus japonicus*, in northern Japan. *Journal of Mammalogy* **85**: 311–315.
119. Amano, M., et al. (2011) Age determination and reproductive traits of killer whales entrapped in ice off Aidomari, Hokkaido, Japan. *Journal of Mammalogy* **92**: 275–282.
120. Amano, M., et al. (2014) Life history and group composition of melon-headed whales based on mass strandings in Japan. *Marine Mammal Science* **30**: 480–493.
121. Amano, M., et al. (2010) Distribution and ecology of the Pallas's squirrel *Callosciurus erythraeus* in the Uto Peninsula, Kumamoto Prefecture, Japan. *Bulletin of the Kumamoto Wildlife Society* **6**: 13–22 (in Japanese).
122. Amano, M. & Yoshioka, M. (2003) Sperm whale diving behavior monitored using a suction-cup-attached TDR tag. *Marine Ecology Progress Series* **258**: 291–295.
123. Amaral, A.R., et al. (2007) New insights on population genetic structure of *Delphinus delphis* from the northeast Atlantic and phylogenetic relationships within the genus inferred from two mitochondrial markers. *Marine Biology* **151**: 1967–1976.
124. Ambrose, S.T., et al. (2013) Winter diet shift of long-beaked common dolphins (*Delphinus capensis*) feeding in the sardine run off KwaZulu-Natal, South Africa. *Marine Biology* **160**: 1543–1561.
125. Ames, J.A. & Morejohn, G.V. (1980) Evidence of white shark, *Carcharodon carcharias*, attacks on sea otters, *Enhydra lutris*. *Canadian Journal of Fish Game* **66**: 196–209.
126. Amin, O.M. & Margolis, L. (1998) Redescription of *Bolbosoma capitatum* (Acanthocephala: Polymorphidae) from false killer whale off Vancouver Island, with taxonomic reconsideration of the species and synonymy of *B. physeteris*. *Comparative Parasitology* **65**: 179–188.
127. Amori, G. (1999) *Tamias sibiricus* (Laxmann, 1769). *in* The Atlas of European Mammals (Mitchell-Jones, A.J., et al., Editors). London: T. & A. D. Poyser Ltd., pp. 194–195.
128. Anderson, J. (1878) Anatomical and Zoological Researches: Comprising an Account of the Zoological Results of the Two Expeditions to Western Yunnan in 1868 and 1875; and Monograph of the Two Cetacean Genera, *Platanista* and *Orcella*. London: Bernard Quaritch, 985 pp.
129. Anderson, P.K. (1981) The behaviour of the dugong (*Dugong dugon*) in relation to conservation and management. *Bulletin of Marine Science* **31**: 640–647.
130. Anderson, P.K. & Birtles, A. (1978) Behaviour and ecology of the dugong, *Dugong dugon* (Sirenia): Observations in Shoalwater and Cleveland bays, Queensland. *Australian Wildlife Research* **5**: 1–23.
131. Anderson, P.K. & Prince, R.I.T. (1985) Predation on dugongs: attacks by killer whales. *Journal of Mammalogy* **66**: 554–556.
132. Anderson, R.C., et al. (2006) Observations of Longman's beaked whale (*Indopacetus pacificus*) in the western Indian Ocean. *Aquatic Mammals* **32**: 223–231.
133. Anderson, S.S. & Fedak, M.A. (1985) Grey seal males: energetic and behavioural links between size and sexual success. *Animal Behaviour* **33**: 829–838.
134. Ando, M. (安藤元一) (2008) The Japanese Otter: Lessons from its Extinction (ニホンカワウソ：絶滅に学ぶ保全生物学). Tokyo: University of Tokyo Press, 235 pp. (in Japanese).
135. Ando, S. (安藤志郎) & Kajiura, K. (梶浦敬一) (1985) 岐阜県におけるアライグマの生息状況. *Bulletin of the Gifu Prefectural Museum* **6**: 23–30 (in Japanese).
136. Ando, A., Nigi, H., & Ohsawa, N. (1977) A rapid assay of urinary estrogens of the Japanese monkey for routine use. *Primates* **18**: 271–275.
137. Ando, A., et al. (1988) A karyological study of two intraspecific taxa in Japanese *Eothenomys* (Mammalia: Rodentia). *Journal of the Mammalogical Society of Japan* **13**: 93–104.
138. Ando, A., et al. (1991) Variation of the X chromosome in the Smith's red-backed vole *Eothenomys smithii*. *Journal of the Mammalogical Society of Japan* **15**: 83–90.
139. Ando, A., Shiraishi, S., & Uchida, T.A. (1988) Reproduction in a laboratory colony of the Smith's red-backed vole, *Eothenomys smithii*. *Journal of the Mammalogical Society of Japan* **13**: 11–20.
140. Ando, K., et al. (1994) Migration and development of the larvae of *Gnathostoma nipponicum* in the rat, second intermediate or paratenic host, and the weasel, definitive host. *Journal of Helminthology* **68**: 13–17.
141. Ando, K., et al. (1994) The occurrence of *Bertiella studeri* (Cestoda: Anolpocephalidae) in Japanese macaque, *Macaca fuscata*, from Mie Prefecture, Japan. *Japanese Journal of Parasitology* **43**: 211–213.
142. Ando, K., Tagawa, T., & Uchida, T.A. (1977) Considerations of karyotypic evolution within Vespertilionidae. *Experientia* **33**: 877–879.
143. Ando, K., Tagawa, T., & Uchida, T.A. (1980) C-banding pattern on the chromosomes of the Japanese house shrew, *Suncus murinus riukiuanus*, and its implication. *Experientia* **36**: 1040–1041.
144. Ando, K., Tagawa, T., & Uchida, T.A. (1980) Karyotypes of Taiwanese and Japanese bats belonging to the families Rhinolophidae and Hipposideridae. *Cytologia* **45**: 423–432.
145. Ando, K., Tagawa, T., & Uchida, T.A. (1987) A karyological study on five Japanese species of *Myotis* and *Pipisterellus*, with special attention to composition of their C-band materials. *Journal of the Mammalogical Society of Japan* **12**: 25–29.
146. Ando, K., et al. (1992) Life cycle of *Gnathostoma nipponicum* Yamaguti, 1941. *Journal of Helminthology* **66**: 53–61.
147. Ando, K. & Uchida, T.A. (1974) Karyotype analysis in Chiroptera II. Phylogenetic relationships in the genus *Rhinolophus*. *Science Bulletin of the Faculty of Agriculture, Kyushu University* **8**: 119–129 (in Japanese with English summary).
148. Ando, K., et al. (1983) Further study on the karyotypic evolution in the genus *Rhinolophus* (Mammalia: Chiroptera). *Caryologia* **36**: 101–111.
149. Ando, M. (1999) Changing association of Japanese people with the Japanese giant squirrel, *Petaurista leucogenys*. *Mammalian Science* **39**: 175–179 (in Japanese).
150. Ando, M. (2005) Improvement of nest box investigation techniques for study of arboreal rodents. *Mammalian Science* **45**: 165–176 (in Japanese with English abstract).
151. Ando, M., Funakoshi, K., & Shiraishi, S. (1983) Use patterns of nests by the Japanese giant flying squirrel, *Petaurista leucogenys*. *Science Bulletin of Faculty of Agriculture, Kyushu University* **38**: 27–43 (in Japanese with English abstract).
152. Ando, M. & Imaizumi, Y. (1982) Habitat utilization of the white-cheeked giant flying squirrel *Petaurista leucogenys* in a small shrine grove. *Journal of the Mammalogical Society of Japan* **9**: 70–81 (in Japanese with English abstract).
153. Ando, M. & Shiraishi, S. (1983) The nest and nest-building behavior of the Japanese giant flying squirrels, *Petaurista leucogenys*. *Science Bulletin of Faculty of Agriculture, Kyushu University* **38**: 59–69 (in Japanese with English abstract).
154. Ando, S. (1987) An essay on *Mustela itatsi* in Gifu Pref.—from comparison with *Mustela sibirica*—. *Bulletin of Gifu Prefectural Museum* **8**: 49–53 (in Japanese).
155. Ando, S. (1989) *Mustela sibirica* naturalized in Gifu Pref. *Bulletin of the Gifu Prefectural Museum* **10**: 15–16 (in Japanese).
156. Ando, Y., Nojima, N., & Yoshiyuki, M. (2004) Emergence patterns of the Japanese house bat (*Pipistrellus abramus*) from March to November in Atsugi city—focusing on the emergence points—. *ANIMATE* **5**: 29–35 (in Japanese).

157. Angeli, C.B. (2000) Death by an Asiatic black bear in Japan: predatory attack? *International Bear News* **9**: 10–11.
158. Angerbjörn, A. & Flux, J.E.C. (1995) *Lepus timidus*. *Mammalian Species* **495**: 1–11.
159. Aoi, T. (1985) Seasonal change in food habits of ezo brown bear (*Ursus arctos yesoensis* LYDEKKER) in northern Hokkaido. *Research Bulletin of Teshio Experimental Forest of Hokkaido University* **42**: 721–732.
160. Aoi, T. (1985) Two twenty-six year old, with pups, and a thirty-four year old wild females of the Yezo brown bear (*Ursus arctos yezoensis*) from Hokkaido. *Journal of Mammalogical Society of Japan* **10**: 165–167 (in Japanese).
161. Aoi, T. (1990) The effects of hunting and forest environmental change upon the population trends for brown bears (*Ursus arctos yesoensis* Lydekker) in northern Hokkaido. *Research Bulletin of Experimental Forest in Hokkaido University* **47**: 249–298 (in Japanese with English summary).
162. Aoki, Y. (青木由親) (2007) 倒木からコテングコウモリ～長野県下諏訪での事例～. *Bat Study and Conservation Report* **15**: 23 (in Japanese).
163. Aoki, Y. (青木雄司) (2007) 日中に飛翔するヤマコウモリの観察記録. *BINOS* **14**: 91–93 (in Japanese).
164. Aoki, Y. (青木雄司) (2014) Japanese Pipistrelle *Pipistrellus abramus* impaled by bull-headed shrike *Lanius bucephalus* (モズのはやにえになったアブラコウモリ). *BINOS* **21** (in Japanese).
165. Aoki, B. (1913) A hand-list of Japanese and Formosan mammals. *Annotationes Zoologicae Japonenses* **8**: 261–353.
166. Aoki, B. (1915) Japanese Muridae. Tokyo: Zoological Society of Tokyo, 88 pp. (in Japanese).
167. Aoki, K., et al. (2007) Diel diving behavior of sperm whales off Japan. *Marine Ecology Progress Series* **349**: 277–287.
168. Aoki, Y. (2002) Notes on a colony of Japanese large noctule *Nyctalus aviator* found in Sagamihara City. *Natural History Report of Kanagawa* **23**: 25–26 (in Japanese).
169. Aoki, Y. & Akiyama, S. (2006) Notes on Asian particolored bat *Vespertilio superans* looked after in a residential area of Sagamihara City, Kanagawa Prefecture. *Natural History Report of Kanagawa* **27**: 41–43 (in Japanese).
170. Aoyagi, A., et al. (2014) Observations of a gray whale, *Eschrichtius robustus*, appeared off the Ohkouzu diversion channel of the Shinano River, Niigata, Sea of Japan in the spring of 2014. *Japan Cetology* **24**: 15–22 (in Japanese).
171. Aplin, K.P., et al. (2011) Multiple geographic origins of commensalism and complex dispersal history of Black Rats. *PLoS ONE* **6**: e26357.
172. Appleton, B.R., McKenzie, J.A., & Christidis, L. (2004) Molecular systematics and biogeography of the bent-wing bat complex *Miniopterus schreibersii* (Kuhl, 1817) (Chiroptera: Vespertilionidae). *Molecular Phylogenetics and Evolution* **31**: 431–439.
173. Apps, P.J. (1983) Aspects of the ecology of feral cats on Dassen Island, South Africa. *South African Journal of Zoology* **18**: 393–399.
174. Arai, S. (荒井秋晴) (2002) チョウセンイタチ (*Mustela sibirica*). *in* Handbook of Alien Species in Japan (外来種ハンドブック) (Ecological Society of Japan, Editor). Tokyo: Chijin Shokan (地人書館), p. 73 (in Japanese).
175. Arai, K., Yamada, T.K., & Takano, Y. (2004) Age estimation of male Stejneger's beaked whales (*Mesoplodon stejnegeri*) based on counting of growth layers in tooth cementum. *Mammal Study* **29**: 125–136.
176. Arai, S., et al. (1985) A note on the Japanese water shrew, *Chimarrogale himalayica platycephala*, from Kyushu. *The Journal of the Mammalogical Society of Japan* **10**: 193–203.
177. Arai, S., et al. (2008) Molecular phylogeny of a newfound hantavirus in the Japanese shrew mole (*Urotrichus talpoides*). *Proceedings of the National Academy of Sciences of the United States of America* **105**: 16292–16301.
178. Arai, S., et al. (2005) Investigations for the distribution of Chiroptera at forests, caves and tunnels in Kumamoto Prefecture (I). *Bulletin of the Kumamoto Wildlife Society* **4**: 1–8 (in Japanese with English abstract).
179. Arai, S. & Shiraishi, S. (1978) Growth and development of the Charles' mouse, *Mus caroli caroli* Bonhote. *Zoological Magazine* **87**: 274–282 (in Japanese with English abstract).
180. Arakawa, Y., et al. (2002) X-chromosomal localization of mammalian Y-linked genes in two XO species of the Ryukyu spiny rat. *Cytogenetics and Genome Research* **99**: 303–309.
181. Araki, J. (荒木潤) (1999) Acanthocephalans in Japan (日本産鉤頭虫類). *in* Progress of Medical Parasitology in Japan. Volume 6 (日本における寄生虫学の研究 第 6 巻) (Otsuru, M. (大鶴正満), Kamegai, S. (亀谷了), & Hayashi, S. (林滋夫), Editors). Tokyo: Meguro Parasitological Museum, pp. 147–162 (in Japanese).
182. Aramilev, V.A. (2006) The conservation status of Asiatic black bears in the Russian Far East. *in* Understanding Asian Bears to Secure Their Future (Japan Bear Network, Compiler). Ibaraki: Japan Bear Network, pp. 86–89.
183. Aramilev, V.V. (2009) Sika deer in Russia. *in* Sika Deer—Biology and Management of Native and Introduced Populations (McCullough, D.R., Takatsuki, S., & Kaji, K., Editors). Tokyo: Springer, pp. 475–499.
184. Archer, F.I., II (2002) Striped dolphin (*Stenella coeruleoalba*). *in* Encyclopedia of Marine Mammals (Perrin, W.F., Würsig, B., & Thewissen, J.G.M., Editors). San Diego: Academic Press, pp. 1201–1203.
185. Archer, F.I., II & Perrin, W.F. (1999) *Stenella coeruleoalba*. *Mammalian Species* **603**: 1–9.
186. Arikawa, J. (1996) Hantavirus infection. *Virus* **46**: 119–129 (in Japanese).
187. Arikawa, J. & Hashimoto, N. (1986) Hemorrhagic fever with renal syndrome. *Virus* **36**: 233–251 (in Japanese).
188. Arima, H. (有馬浩史) (2006) 京都府芦生研究林におけるコテングコウモリ (*Murina ussuriensis*) の観察記録. *Nature Study* **52**: 4 (in Japanese).
189. Arimoto, I., et al. (2014) Behavior and habitat of Asiatic black bear (*Ursus thibetanus*) inhabiting near settlements. *Mammalian Science* **54**: 19–31 (in Japanese with English summary).
190. Árnason, Ú. (1974) Comparative chromosome studies in Cetacea. *Hereditas* **77**: 1–36.
191. Árnason, Ú. (1974) Comparative chromosome studies in Pinnipedia. *Hereditas* **76**: 179–226.
192. Árnason, Ú. (1977) The relationship between the four principal pinniped karyotypes. *Hereditas* **87**: 227–242.
193. Árnason, Ú. (1984) Karyotype stability in marine mammals. *Cytogenetics and Cell Genetics* **33**: 274–276.
194. Árnason, Ú., et al. (2002) Mammalian mitogenomic relationships and the root of the eutherian tree. *Evolution* **99**: 8151–8156.
195. Árnason, Ú., et al. (1977) Banded karyotypes of three whales: *Mesoplodon europaeus*, *M. carlhubbsi* and *Balaenoptera acutorostrata*. *Hereditas* **87**: 189–200.
196. Asada, M. (2011) Distribution and population estimation for Reeves's muntjac in 2010 in Chiba Prefecture, Japan. *Report of the Chiba Biodiversity Center* **3**: 36–43 (in Japanese).
197. Asada, M. & Ochiai, K. (1996) Conception dates of Sika deer on the Boso Peninsula, central Japan. *Mammal Study* **21**: 153–159.
198. Asada, M., Ochiai, K., & Hasegawa, K. (2000) Introduced Reeves' Muntjac in Boso Peninsula and Izu-Oshima, central Japan. *Journal of the Natural History Museum and Institute, Chiba* **6**: 87–94 (in Japanese with English abstract).
199. Asada, M., et al. (2014) Bayesian estimation of Reeves' muntjac (*Muntiacus reevesi*) population using state-space models. *Mammalian Science* **54**: 53–72 (in Japanese with English abstract).
200. Asada, M., et al. (2005) The first record of the common parti-coloured bat (*Vespertilio superans*) in Chiba Prefecture, Japan. *Journal of the Natural History Museum and Institute, Chiba* **8**: 49–51 (in Japanese with English abstract).
201. Asahi, M. (朝日稔) & Mori, M. (森美保子) (1980) タヌキの歯数異常. *Zoological Magazine* **89**: 61–64 (in Japanese).
202. Asahi, M. (1975) Stomach contents of wild boars (*Sus scrofa leucomystax*) in winter. *The Journal of the Mammalogical Society of Japan* **6**: 115–120 (in Japanese with English summary).
203. Asahi, M. (1985) Dispersion of mammals introduced to Japan. *in* Contemporary Mammalogy in China and Japan (Kawamichi, T., Editor). Tokyo: Mammalogical Society of Japan, pp. 124–145.
204. Asahi Shimbun (朝日新聞) (1973) 昭和 48 年 10 月 11 日発行 (Published on 11 October). Tokyo: Asahi Shimbun (in Japanese).
205. Asakawa, M. (浅川満彦) & Ikeda, T. (池田透) (2007) 北海道で野生化したアライグマの病原体疫学調査：外来種対策における感染症対策の一具体例として開始 12 年の総括. *Wildlife Forum* **12**: 25–31 (in Japanese).
206. Asakawa, M. (1989) Helminth fauna of the Japanese Microtidae and Muridae. *Mammalian Science* **29(1)**: 17–35 (in Japanese with English abstract).
207. Asakawa, M. (1995) A biogeographical study on the parasitic nematodes of Japanese Microtinae and Murinae with systematic and phylogenetic studies of the genera *Heligmosomoides* and *Heligmosomus* (Nematoda: Heligmosomidae). *Journal of Rakuno Gakuen University* **19**: 285–379 (in Japanese with English summary).
208. Asakawa, M. (1998) Host-parasite relationship between the red-backed voles and its parasitic nematodes in Japan. *Mammalian Science* **38**: 171–180 (in Japanese with English abstract).
209. Asakawa, M. (2005) Perspective of host-parasite relationships between rodents and nematodes in Japan. *Mammal Study* **30**: S95–S99.
210. Asakawa, M. & Harada, M. (1989) Faunal and zoogeographical study on the internal parasites of the Japanese red-backed vole, *Eothenomys* spp. *Bulletin of the Biogeographical Society of Japan* **44**: 199–210 (in Japanese).
211. Asakawa, M., Kamiya, H., & Ohbayashi, M. (1987) Studies on the parasites of Insectivora. II. Four new capillarid nematodes from the Japanese shrews, genera *Sorex* and *Crocidura*. *Journal of Rakuno Gakuen University* **12**:

335–347.

212. Asakawa, M. & Kitamura, K. (2003) Outbreaks of infectious diseases recorded in zoos and aquariums in Japan with overview of references cited in Journal of Japanese Association of Zoo and Aquarium. *Journal of Rakuno Gakuen University* **28**: 79–84 (in Japanese with English abstract).
213. Asakawa, M., Kurachi, T., & Wildlife Ecological Society (1999) Parasitic helminths of raccoons in Hokkaido, Japan. *Japanese Journal of Zoo and Wildlife Medicine* **4**: 101–103 (in Japanese with English abstract).
214. Asakawa, M., Mano, T., & Gardner, S.L. (2006) First record of *Ancylostoma malaynum* (Alessandrini, 1905) from brown bears (*Ursus arctos* L). *Comparative Parasitology* **73**: 282–284.
215. Asakawa, M., et al. (2003) Faunal studies of parasites in Nopporo Forest Park—Tube-nosed bat, *Murina ussuriensis*. *Nopporo Forest* **2**: 28–30 (in Japanese).
216. Asakawa, M., et al. (2000) Review of the parasitological state of feral raccoons captured in Nopporo National Park and its proximity, Hokkaido. *Journal of Rakuno Gakuen University* **25**: 1–8 (in Japanese with English abstract).
217. Asakawa, M., et al. (2012) Parastic helminthes of rodents on Habomai Is. *Memoires of the Nemuro City Museum of History and Nature* **24**: 45–48 (in Japanese with English summary).
218. Asakawa, M., et al. (2009) Further helminthological survey on alien rodents, coypu (*Myocastor coypus*: Myocastoridae) in Aichi and Hyogo Prefectures, Japan. *Journal of the College of Dairying. Natural Science* **33**: 291–292 (in Japanese with English summary).
219. Asano, M. (2003) Reproduction, growth and population dynamics of feral raccoon (*Procyon lotor*) in Hokkaido. Sapporo: Hokkaido University. 76 pp.
220. Asano, M., et al. (2003) Reproductive characteristics of the feral raccoon (*Procyon lotor*) in Hokkaido, Japan. *Journal of Veterinary Medical Science* **65**: 369–373.
221. Asano, R., Murasugi, E., & Yamamoto, Y. (1997) Detection of intestinal parasites from main-land raccoon dogs, *Nyctereutes procyonoides viverrinus*, in southeastern Kanagawa Prefecture. *Journal of the Japanese Association for Infectious Diseases* **71**: 664–667.
222. Asanuma, K. & Uchikawa, K. (1977) Studies on mesostigmatid mites parasitic on mammals and birds in Japan : IV. *Steatonyssus nakazimai* sp. nov. (Acarina, Macronyssidae), a specific parasite of *Glirulus japonicus* (Rodentia, Muscardinidae). *Bulletin of the National Science Museum. Series A, Zoology* **3**: 219–224.
223. Asanuma, Y. (1963) Records of a laelaptid mite, *Laelaps algericus* Hirst from *Mus musculus molossinus* in Tokyo and Urawa, Japan. *Medical Entomology and Zoology* **14**: 251–252 (in Japanese).
224. Asari, Y. (2004) Records of bats in Tashiro-machi, Akita Prefecture. *ANIMATE* **5**: 36–38 (in Japanese).
225. Asari, Y., Nakama, S., & Yanagawa, H. (2009) Investigation of tree cavities used by Siberian flying squirrels and their selectivity factors. *Journal of the Japanese Wildlife Research Society* **34**: 16–20 (in Japanese with English summary).
226. Asari, Y., Yamaguchi, Y., & Yanagawa, H. (2008) Field observations of the food items of the Siberian flying squirrel, *Pteromys volans orii*. *Journal of the Japanese Wildlife Research Society* **33**: 7–11 (in Japanese with English summary).
227. Asari, Y. & Yanagawa, H. (2008) Daily nest site use by the Siberian flying squirrel *Pteromys volans orii* in fragmented small woods. *Wildlife Conservation Japan* **12**: 7–10 (in Japanese with English abstract).
228. Association of Higashiyama Zoo (東山動物園友の会) (1986) トピックス. 東山動物園友の会 (Association of Higashiyama Zoo) **1**: 13 (in Japanese).
229. Astruc, G. & Beaubrun, P. (2005) Do Mediterranean cetaceans diets overlap for the same resources? *European Research on Cetaceans* **19**: 81.
230. Atoji, Y., Sugimura, M., & Suzuki, Y. (1989) *Ulex europaeus* agglutinin I binding in the apocrine glands of the interdigital gland and skin in the Japanese serow *Capricornis crispus*. *Japanese Journal of Veterinary Science* **51**: 194–196.
231. Atoji, Y., Suzuki, Y., & Sugimura, M. (1988) Lectin histochemistry of the interdigital gland in the Japanese serow (*Capricornis crispus*) in winter. *Journal of Anatomy* **161**: 159–170.
232. Atoji, Y., Suzuki, Y., & Sugimura, M. (1989) The preputial gland of the Japanese serow *Capricornis crispus*: Infrastructure and lectin histochemistry. *Acta Anatomica* **134**: 245–252.
233. Aznar, F.J., et al. (2007) Insight into the role of cetaceans in the life cycle of the terapyllideans (Platyhelminthes: Cestoda). *International Journal for Parasitology* **37**: 243–255.
234. Azzaroli, M.L. (1968) Second specimen of *Mesoplodon pacificus* the rarest living beaked whale. *Monitore Zoologico Italiano. N.S.* **2 Suppl.**: 67–79.
235. Baba, N. (馬場徳寿) (1997) キタオットセイ. in 日本の希少な野生水生生物に関する基礎資料 (IV). Tokyo: Japan Fisheries Resource Conservation Association (日本水産資源保護協会), pp. 349–355 (in Japanese).
236. Baba, H. (1979) A morphological study on the teeth of the Japanese monkeys, *Macaca fuscata*. *Journal of the Kyushu Dental Society* **32**: 741–768 (in Japanese with English summary).
237. Baba, M., Doi, T., & Ono, Y. (1982) Home range utilization and nocturnal activity of the giant flying squirrel, *Petaurista leucogenys*. *Japanese Journal of Ecology* **32**: 189–198.
238. Baba, N., Boltnev, A.I., & Stus, A.I. (2000) Winter migration of female northern fur seals *Callorhinus ursinus* from the Commander Islands. *Bulletin of National Research Institute of Far Seas Fishery* **37**: 39–44.
239. Baba, N., Nitto, H., & Nitta, A. (2000) Satellite tracking of young Steller sea lion off the coast of northern Hokkaido. *Fisheries Science* **66**: 180–181.
240. Baillie, J. & Groombridge, B., eds. (1996) 1996 IUCN Red List of Threatened Animals. Gland: IUCN, 341 pp.
241. Baird, R.W. (2009) False killer whale (*Pseudorca crassidens*). in Encyclopedia of Marine Mammals. 2nd ed. (Perrin, W.F., Würsig, B., & Thewissen, J.G.M., Editors). Amsterdam: Academic Press, pp. 405–406.
242. Baird, R.W. (2009) Risso's dolphin (*Grampus griseus*). in Encyclopedia of Marine Mammals. 2nd ed. (Perrin, W.F., Würsig, B., & Thewissen, J.G.M., Editors). Amsterdam: Academic Press, pp. 975–976.
243. Baird, R.W. & Gorgone, A.M. (2005) False killer whale dorsal fin disfigurements as a possible indicator of long-line fishery interactions in Hawaiian waters. *Pacific Science* **59**: 593–601.
244. Baird, R.W., et al. (2006) Killer whales in Hawaiian waters: information on population identity and feeding habits. *Pacific Science* **60**: 523–530.
245. Baird, R.W., et al. (2006) Diving behaviour of Cuvier's (*Ziphius cavirostris*) and Blainville's (*Mesoplodon densirostris*) beaked whales in Hawai'i. *Canadian Journal of Zoology* **84**: 1120–1128.
246. Baker, A.R., et al. (2005) Variation of mitochondrial control region sequences of Steller sea lions: the three-stock hypothesis. *Journal of Mammalogy* **86**: 1075–1084.
247. Baker, C.S., et al. (1990) Influence of seasonal migration of geographic distribution of mitochondrial DNA haplotypes in humpback whales. *Nature* **344**: 238–240.
248. Baker, J.D., et al. (1995) Natal site fidelity in northern fur seals, *Callorhinus ursinus*. *Animal Behaviour* **50**: 237–247.
249. Bakeyev, N.N. & Sinitsyn, A.A. (1994) Status and conservation of sables in the commonwealth of independent states. in Martens, Sables, and Fishers: Biology and Conservation (Buskirk, S.W., et al., Editors). New York: Comstock Publishing, pp. 246–254.
250. Balcomb, K.C. (1989) Baird's beaked whale *Berardius bairdii* Stejneger, 1883: Arnoux's beaked whale *Berardius arnuxii* Duvernoy, 1851. in Handbook of Marine Mammals. Volume 4. River Dolphins and the Larger Toothed Whales (Ridgway, S.H. & Harrison, R., Editors). London: Academic Press, pp. 261–288.
251. Balcomb, K.C., III & Claridge, D.E. (2001) A mass stranding of cetaceans caused by naval sonar in the Bahamas. *Bahamas Journal of Science* **8**: 2–12.
252. Bando, T., et al. (2013) Cruise report of the second phase of the Japanese Whale Research Program under Special Permit in the western North Pacific (JARPN II) in 2012 (part I)—offshore component—(Paper SC/65a/O3 presented to the IWC Scientific Committee). 33 pp.
253. Bannikov, A.G. (1964) Biologie du chien viverrin en U.R.S.S. *Mammalia* **28**: 1–39 (in French).
254. Bannikova, A.A., Lavrenchenko, L.A., & Kramerov, D.A. (2005) Phylogenetic relationships between Afrotropical and Palearctic *Crocidura* species inferred from Inter-SINE-PCR. *Biochemical Systematics and Ecology* **33**: 45–59.
255. Bannikova, A.A., et al. (2006) Phylogeny and systematics of the *Crocidura suaveolens* species group: corroboration and controversy between nuclear and mitochondrial DNA markers. *Mammalia* **70**: 106–119.
256. Banwell, D.B. (2009) The sika in New Zealand. in Sika Deer—Biology and Management of Native and Introduced Populations (McCullough, D.R., Takatsuki, S., & Kaji, K., Editors). Tokyo: Springer, pp. 643–656.
257. Barratt, E.M., et al. (1999) Genetic structure of fragmented populations of red squirrel (*Sciurus vulgaris*) in the UK. *Molecular Ecology* **8**: S55–S63.
258. Barrett-Hamilton, G.E.H. (1900) Exhibition of skins of the variable hare (*Lepus timidus* Linn.) showing colour-variations, and descriptions of subspecies and varieties of this species. *Proceedings of the General Meetings for Scientific Business of the Zoological Society of London* **1900**: 87–92.
259. Barrette, C. (1977) Scent-marking in captive muntjacs (*Muntiacus reevesi*). *Animal Behaviour* **25**: 536–541.
260. Barrette, C. (1977) The social behaviour of captive muntjacs (*Muntiacus reevesi* Ogilby 1839). *Zeitschrift für Tierpsychologie* **43**: 188–213.
261. Bartholomew, G.A. & Hoel, P.G. (1953) Reproductive behavior of the Alaska fur seal, *Callorhinus ursinus*. *Journal of Mammalogy* **34**: 417–436w.
262. Bartoš, L. (2009) Sika deer in continental Europe. in Sika Deer—Biology

and Management of Native and Introduced Populations (McCullough, D.R., Takatsuki, S., & Kaji, K., Editors). Tokyo: Springer, pp. 573–594.
263. Bartoš, L., Hyánek, J., & Žirovnický, J. (1981) Hybridization between red and sika deer. I. Craniological analysis. *Zoologischer Anzeiger, Jena* **207**: 260–270.
264. Barun, A., et al. (2011) A review of small Indian mongoose management and eradications on islands. *in* Island Invasives: Eradication and Management (Veitch, C.R., Clout, M.N., & Towns, D.R., Editors). Gland: IUCN, pp. 17–25.
265. Baryshnikov, G.F., Mano, T., & Masuda, R. (2004) Taxonomic differentiation of *Ursus arctos* (Carnivora, Ursidae) from south Okhotsk islands on the basis of morphometrical analysis of skull and teeth. *Russian Journal of Theriology* **3**: 77–88.
266. Baryshnikov, G.F., Puzachenko, A.Y., & Abramov, A.V. (2002) New analysis of variability of cheek teeth in Eurasian badgers (Carnivore, Mustelidae, *Meles*). *Russian Journal of Theriology* **1**: 133–149.
267. Bat Research Group of Centennial Woods Fan Club (2001) Bats in Mt. Yotei and Niseko range, Hokkaido, Japan, No.1. —Report on 1997~2000 faunal survey—. *Bulletin of the Otaru Museum* **14**: 127–132 (in Japanese with English abstract).
268. Baumgartner, M.F. (1997) The distribution of Risso's dolphin (*Grampus griseus*) with respect to the physiography of the northern Gulf of Mexico. *Marine Mammal Science* **13**: 614–638.
269. Bearzi, M. (2005) Habitat partitioning by three species of dolphins in Santa Monica Bay, California. *Bulletin of the Southern California Academy of Sciences* **104**: 113–124.
270. Beatson, E. (2007) The diet of pygmy sperm whales, *Kogia breviceps*, stranded in New Zealand: implications for conservation. *Reviews in Fish Biology and Fisheries* **17**: 295–303.
271. Beklemisheva, V.R., et al. (2011) Reconstruction of karyotype evolution in core Glires. I. The genome homology revealed by comparative chromosome painting. *Chromosome Research* **19**: 549–565.
272. Belkin, A.N. (1966) Summer distribution, abundance, possibility of harvest and biological characteristics of Steller sea lions in the Kuril Islands. *Izvestiia TINRO* **58**: 69–95 (in Russian. Translated into Japanese by Japan Fisheries Agency in 1994).
273. Bellemain, E., Swenson, J.E., & Taberlet, P. (2006) Mating strategies in relation to sexually selected infanticide in a non-social Carnivore: the brown bear. *Ethology* **112**: 238–246.
274. Benda, P., et al. (2008) Bats (Mammalia: Chiroptera) of the Eastern Mediterranean and Middle East. Part 6. Bats of Sinai (Egypt) with some taxonomic, ecological and echolocation data on that fauna. *Acta Societatis Zoologicae Bohemicae* **72**: 1–103.
275. Benda, P. & Tsytsulina, K.A. (2000) Taxonomic revision of *Myotis mystacinus* group (Mammalia: Chiroptera) in the western Palearctic. *Acta Socciettatis Zoologicae Bohemoslovenicae* **64**: 331–398.
276. Bennett, P.M., et al. (2001) Exposure to heavy metals and infectious disease mortality in harbour porpoises from England and Wales. *Environmental Pollution* **12**: 33–40.
277. Bernard, H.J. & Reilly, S.B. (1999) Pilot whales *Globicephala* Lesson, 1828. *in* Handbook of Marine Mammals. Volume 6. The Second Book of Dolphins and the Porpoises (Ridgway, S.H. & Harrison, R., Editors). London: Academic Press, pp. 245–279.
278. Bernays, E.A., Driver, G.C., & Bilgener, M. (1989) Herbivores and plant tannins. *Advances in Ecological Research* **19**: 263–302.
279. Berón-Vera, B., et al. (2007) Parasite communities of common dolphins (*Delphinus delphis*) from Patagonia: the relation with host distribution and diet and comparison with sympatric hosts. *Journal of Parasitology* **93**: 1056–1060.
280. Bérubé, M. (2002) Hybridism. *in* Encyclopedia of Marine Mammals (Perrin, W.F., Würsig, B., & Thewissen, J.G.M., Editors). San Diego: Academic Press, pp. 596–600.
281. Best, P.B. (1960) Further information on Bryde's whale (*Balaenoptera edeni* Anderson) from Saldanha Bay, South Africa. *Norsk Hvalfangst-tidende* **49**: 201–215.
282. Best, P.B. (1977) Two allopatric forms of Bryde's whales off South Africa. *Reports of the International Whaling Commission (Special Issue)* **1**: 10–35.
283. Best, P.B. (2007) Whales and Dolphins of the Southern African Subregion. Cape Town: Cambridge University Press, 352 pp.
284. Best, P.B., Butterworth, D.S., & Rickett, L.H. (1984) An assessment cruise for the South African inshore stock of Bryde's whale (*Balaenoptera edeni*). *Report of the International Whaling Commission* **34**: 403–423.
285. Best, P.B., Canham, P.A.S., & Macleod, M. (1984) Patterns of reproduction in sperm whales, *Physeter macrocephalus*. *Reports of the International Whaling Commission (Special Issue)* **6**: 51–79.
286. Bibikov, D.I., Filimonov, A.N., & Kudaktin, A.N. (1983) Territoriality and migration of the wolf in the USSR. *Acta Zoologica Fennica* **174**: 267–268.
287. Bickham, J., et al. (2004) Molecular phylogenetics, karyotypic diversity, and partition of the genus *Myotis* (Chiroptera: Vespertilionidae). *Molecular Phylogenetics and Evolution* **33**: 333–338.
288. Bickham, J.W., et al. (1998) Geographic variation in the mitochondrial DNA of Steller sea lions: Haplotype diversity and endemism in the Kuril Islands. *Biosphere Conservation* **1**: 107–117.
289. Bigg, M.A. (1981) Harbor seal *Phoca virtulina* and *P. largha*. *in* Handbook of Marine Mammals. Volume 2. Seals (Ridgway, S.H. & Harrison, R.J., Editors). London: Academic Press, pp. 1–28.
290. Bigg, M.A. (1990) Migration of northern fur seals (*Callorhinus ursinus*) off western North America. (Canadian Technical Report of Fisheries and Aquatic Sciences, Number 1764). Ottawa: Government of Canada, Fisheries and Oceans, 64 pp.
291. Biodiversity Center of Japan (2004) 自然環境保全基礎調査：種の多様性調査（奈良県）報告書. Fujiyoshida: Biodiversity Center of Japan, 37 pp. (in Japanese).
292. Biodiversity Center of Japan (2007) 平成18年度自然環境保全基礎調査：種の多様性調査（アライグマ生息情報収集）業務報告書. Fujiyoshida: Biodiversity Center of Japan, 130 pp. (in Japanese).
293. Biodiversity Information Standards (TDWG) (2007) Biodiversity Information Projects of the World: China Species Information Service/China Species Redlist. Available from: http://www.tdwg.org/biodiv-projects/projects-database/view-project/224/.
294. Birks, J.D.S. & Dunstone, N. (1991) Mink *Mustela vison*. *in* The Handbook of British Mammals (Corbet, G.B. & Harris, S., Editors). Oxford: Blackwell, pp. 406–415.
295. Birky, W.A. (2002) Mating Patterns and Social Structure in a Wild Group of Formosan Macaques (*Macaca cyclopis*). Doctoral dissertation. New Brunswick: Rutgers University.
296. Bjørge, A. & Tolley, K.A. (2009) Harbor porpoise, *Phocoena phocoena*. *in* Encyclopedia of Marine Mammals. 2nd ed. (Perinn, W.F., Würsig, B., & Thewissen, J.G.M., Editors). Amsterdam: Academic Press, pp. 530–533.
297. Bjørnstad, O.N., et al. (1996) Cyclicity and stability of grey-sided voles, *Clethrionomys rufocanus* of Hokkaido: evidence from spectral and principal components analyses. *Philosophical Transactions of the Royal Society of London. Series B-Biological Sciences* **351**: 867–875.
298. Bjørnstad, O.N., et al. (1998) Mapping the regional transition to cyclicity in *Clethrionomys rufocanus*: Spectral densities and functional data analysis. *Researches on Population Ecology* **40**: 77–84.
299. Blair, D. (1981) Helminth parasites of the dugong, their collection and preservation. *in* The Dugong: Proceedings of a seminar/workshop held at James Cook University 8–13 May 1979. Townsville, Queensland, Australia: James Cook University pp. 275–285.
300. Blanchard, B.M. & Knight, R.R. (1991) Movements of Yellowstone grizzly bears. *Biological Conservation* **58**: 41–67.
301. Blanco, C., Raduán, Á., & Raga, J.A. (2006) Diet of Risso's dolphin (*Grampus griseus*) in the western Mediterranean Sea. *Scientia Marina* **70**: 407–411.
302. Blasetti, A., et al. (1988) Activity budgets and use of enclosed space by wild boars (*Sus scrofa*) in captivity. *Zoo Biology* **7**: 69–79.
303. Blokhin, S.A., Maminov, M.K., & Kosygin, G.M. (1985) On the Korean–Okhotsk population of gray whales. *Report of the International Whaling Commission* **35**: 375–376.
304. Bloodworth, B.E. & Odell, D.K. (2008) *Kogia breviceps* (Cetacea: Kogiidae). *Mammalian Species* **819**: 1–12.
305. Blytt, H.J., Guscar, T.K., & Butler, L.G. (1988) Antinutritional effects and ecological significance of dietary condensed tannins may not be due to binding and inhibiting digestive enzymes. *Journal of Chemical Ecology* **14**: 1455–1465.
306. Bobrinskii, N.A., Kuznetsov, B.A., & Kuzyakin, A.P. (1944) Key to the Mammals of the USSR. Moscow: Sovetskaya Nauka, 440 pp. (in Russian).
307. Bodkin, J.L., et al. (1999) Population demographics and genetic diversity in remnant and translocated populations of sea otters. *Conservation Biology* **13**: 1378–1385.
308. Bodkin, J.L., Mulcahy, D., & Lensink, C.J. (1993) Age-specific reproduction in female sea otters (*Enhydra lutris*) from south-central Alaska: Analysis of reproductive tracts. *Canadian Journal of Zoology* **71**: 1811–1815.
309. Bolaños, J. & Villarroel-Marin, A. (2003) Three new records of cetacean species for Venezuelan waters. *Caribbean Journal of Science* **39**: 230–232.
310. Bonesi, L. & Palazon, S. (2004) The American mink in Europe: Status, impacts, and control. *Biological Conservation* **134**: 470–483.
311. Boness, D.J. (1991) Determinants of mating systems in the Otariidae (Pinnipedia). *in* Behavior of Pinnipeds (Renouf, D., Editor). London: Chapman and Hall, pp. 1–44.

312. Boness, D.J., Clapham, P.J., & Mesnick, S.L. (2002) Life history and reproductive strategies. in Marine Mammal Biology: An Evolutionary Approach (Hoelzel, A.S., Editor). Malden: Blackwell, pp. 278–324.
313. Bonhomme, F., et al. (1984) Biochemical diversity and evolution in the genus *Mus. Biochemical Genetics* **22**: 275–303.
314. Booth, C.G., et al. (2013) Habitat preference and distribution of the hourbour porpoise *Phocoena phocoena* west of Scotland. *Marine Ecology Progress Series* **478**: 273–285.
315. Borissenko, A.V. & Kruskop, S.V. (2003) Bats of Vietnam and Adjacent Territories: An Identification Manual. Moscow: Joint Russian-Vietnamese Science and Technological Tropical Centre and Zoological Museum of Moscow. M.V. Lomonosov State University, 212 pp.
316. Borkowski, J. & Furubayashi, K. (1998) Seasonal changes in number and habitat use of foraging sika deer at the high altitude of Tanzawa Mountains, Japan. *Acta Theriologica* **43**: 95–106.
317. Bossart, G.D., et al. (2007) Pathological findings in a rare mass stranding of melon-headed whales (*Peponocephala electra*) in Florida. *Aquatic Mammals* **33**: 235–240.
318. Bourne, G.H. (1975) Collected anatomical and physiological data from the rhesus monkey. in The Rhesus Monkey (Bourne, G.H., Editor). New York: Academic Press, pp. 1–63.
319. Bourret, V.J.R., Mace, M.R.J.M., & Crouau-Roy, B. (2007) Genetic variation and population structure of western Mediterranean and northern Atlantic *Stenella coeruleoalba* populations inferred from microsatellite data. *Journal of the Marine Biological Association of the United Kingdom* **87(1), Special Issue**: 265–269.
320. Boursot, P., et al. (1993) The evolution of house mice. *Annual Review of Ecology and Systematics* **24**: 119–152.
321. Boursot, P., et al. (1996) Origin and radiation of the house mouse: mitochondrial DNA phylogeny. *Journal of Evolutionary Biology* **9**: 391–415.
322. Bouveroux, T., et al. (2014) Direct evidence for gray seal (*Halicoerus grypus*) predation and scavenging on harobor porpoise (*Phocoena phocoena*). *Marine Mammal Science* **30**: 1542–1548.
323. Bowen, S.L. (1974) Probable extinction of the Korean stock of the gray whale (*Eschrichtius robustus*). *Journal of Mammalogy* **55**: 208–209.
324. Boyd, I.L., Lockyer, C., & Helene, M.D. (1999) Reproduction in marine mammals. in Biology of Marine Mammals (Reynolds, J.E., III & Rommel, S.A., Editors). Washington, D.C.: Smithsonian Institution Press, pp. 218–286.
325. Bradbury, J.W. (1977) Social organization and communication. in Biology of Bats. Volume 3 (Wimsatt, W.A., Editor). New York: Academic Press, pp. 1–72.
326. Brazil, M. & Sasaki, N. (2004) The bat fauna of Nopporo Forest Park, Hokkaido. *Mammal Study* **29**: 191–195.
327. Brent, L.J.N., et al. (2015) Ecological knowledge, leadership, and the evolution of menopause in killer whales. *Current Biology* **25**: 746–750.
328. Broker, M. (2002) Tick-borne encephalitis virus within and outside Japan: a cause for concern? *Japanese Journal of Infectious Diseases* **55**: 55–56.
329. Bromley, G.F. (Бромлей, Г.Ф.) (1965) Медведи юга дальнего Востока СССР (Bears of the South of the Far East of the USSR). Moscow-Leningrad: Nauka, 120 pp. (in Russian).
330. Brown, D.H. & Norris, K.S. (1956) Observations of captive and wild cetaceans. *Journal of Mammalogy* **37**: 311–326.
331. Brownell, R.L., Jr. & Chun, C. (1977) Probable existence of the Korean stock of the gray whale (*Eschrichtius robustus*). *Journal of Mammalogy* **58**: 237–239.
332. Brownell, R.L., Jr., Walker, W.A., & Forney, K.A. (1999) Pacific whitesided dolphin *Lagenorhynchus obliquidens* Gill, 1866. in Handbook of Marine Mammals. Volume 6. The Second Book of Dolphins and the Porpoises (Ridgway, S.H. & Harrison, R., Editors). London: Academic Press, pp. 57–84.
333. Brownell, R.L., Jr., et al. (2005) Mass strandings of Cuvier's beaked whales in Japan: US Naval acoustic link? (Paper SC/56/E37 presented to the IWC Scientific Committee) 10 pp.
334. Brownell, R.L., et al. (2006) Mass strandings of melon-headed whales, *Peponocephala electra*: a worldwide review. (Paper SC/58/SM8 presented to the IWC Scientific Committee). 12 pp.
335. Brunner, S. (2003) Fur seals and sea lions (Otariidae): identification of species and taxonomic review. *Systematics and Biodiversity* **1**: 339–439.
336. Buckton, K.E. & Cunningham, C. (1971) Variations of the chromosome number in the red fox (*Vulpes vulpes*). *Chromosoma* **33**: 268–272.
337. Budylenko, G.A. (1981) Distribution and some aspects of the biology of killer whales in the South Atlantic. *Report of the International Whaling Commission* **31**: 523–525.
338. Bunnell, F.L. & Tait, D.E.N. (1981) Population dynamics of bears—Implications. in Dynamics of Large Mammal Populations (Fowler, C.W. & Smith, T.D., Editors). New York: John Willey and Sons, pp. 75–98.
339. Burg, T.M., Andrew, W.T., & Smith, M.J. (1999) Mitochondrial and microsatellite DNA analyses of harbour seal population structure in the northeast Pacific Ocean. *Canadian Journal of Zoology* **77**: 930–943.
340. Burns, J.J. (1970) Remarks on the distribution and natural history of pagophilic pinnipeds in the Bering and Chukuchi Seas. *Journal of Mammalogy* **51**: 445–454.
341. Burns, J.J. (1981) Bearded seal, *Erignathus barbatus* Erxleben, 1777. in Handbook of Marine Mammals. Volume 2. Seals (Ridgway, S.H. & Harrison, R.J., Editors). London: Academic Press, pp. 145–170.
342. Burns, J.J. (1981) Ribbon seal, *Phoca fasciata* Erxleben,1777. in Handbook of Marine Mammals. Volume 2. Seals (Ridgway, S.H. & Harrison, R.J., Editors). London: Academic Press, pp. 89–109.
343. Burns, J.J. (2002) Harbor seal and spotted seal (*Phoca vitulina* and *P. largha*). in Encyclopedia of Marine Mammals (Perrin, W.F., Würsig, B., & Thewissen, J.G.M., Editors). San Diego: Academic Press, pp. 552–560.
344. Burns, J.J. & Fay, F.H. (1970) Comparative morphology oh the skull of the ribbon seal, *Histriophoca fasciata*, with remarks on systematics of Phocidae. *Journal of Zoology* **161**: 363–394.
345. Calambokidis, J., et al. (2008) SPLASH: Structure of Populations, Levels of Aboundance and Status of Humback Whales in the North Pacific (Final Report for Contract AB133F-03-RP-00078). Olympia: Cascadia Research, 57 pp. (http://www.cascadiaresearch.org/SPLASH/SPLASH-contract-Report-May08.pdf).
346. Caldwell, D.K. & Caldwell, M.C. (1989) Pygmy sperm whale *Kogia breviceps* (de Blainvillem, 1838): Dwarf sperm whale *Kogia simus* Owen, 1866. in Handbook of Marine Mammals. Volume 4. River Dolphins and the Larger Toothed Whales (Ridgway, S.H. & Harrison, R., Editors). London: Academic Press, pp. 235–260.
347. Calkins, D.G. & Schneider, K.B. (1984) Species Account: The Sea Otter (*Enhydra lutris*). Juneau: Alaska Department of Fish and Game, 14 pp.
348. Cañadas, A. & Hammond, P. (2008) Abundance and habitat preferences of the short-beaked common dolphin *Delphinus delphis* in the southwestern Mediterranean: implications for conservation. *Endangered Species Research* **4**: 309–331.
349. Carleton, M.D., Musser, G.G., & Pavlinov, L.Y. (2003) *Myodes* Pallas, 1811, is the valid name for the genus of red-backed voles. in Systematics, Phylogeny and Paleontology of Small Mammals (An International Conference Devoted to the 90th Anniversary of Prof. I. M. Gromov) (Averianov, A.O. & Abramson, N.I., Editors). Saint Petersburg: Proceedings of the Zoological Institute, pp. 96–98.
350. Carne, P. (2000) Deer of Britain and Ireland: their origins and distribution. Shrewsbury: Swan Hill Press, 240 pp.
351. Carretta, J.V., et al. (2007) U.S. Pacific Marine Mammal Stock Assessments: 2006. (U.S. Department of Commerce, NOAA, National Marine Fisheries Service, Southwest Fisheries Center, Technical Memorandum 398). 312 pp. (http://www.nmfs.noaa.gov/pr/pdfs/sars/po2006.pdf).
352. Carretta, J.V., Taylor, B.L., & Chivers, S.J. (2001) Abundance and depth distribution of harbor porpoise (*Phocoena phocoena*) in northern California determined from a 1995 ship survey. *Fishery Bulletin* **99**: 29–39.
353. Cavagna, P., Stone, G., & Stanyon, R. (2002) Black rat (*Rattus rattus*) genomic variability characterized by chromosome painting. *Mammalian Genome* **13**: 157–163.
354. Cerling, T.E., et al. (1997) Global vegetation change through the Miocene–Pliocene boundary. *Nature* **389**: 153.
355. Chabaud, A.G., Rausch, R.L., & Desset, M.C. (1963) Nématodes parasites de rongeurs et insectivores japonais. *Bulletin de la Société Zoologique de France* **138**: 489–512 (in French).
356. Chamura, S. (茶村真一郎) (1980) ハクビシンの飼育と繁殖. 広島市安佐動物公園飼育記録集 **9**: 15–21 (in Japanese).
357. Chanin, P. (1983) Observations on two populations of feral mink in Devon, U.K. *Mammalia* **47**: 463–476.
358. Chantrapornsyl, S. (1996) The first record of a pygmy killer whale (*Feresa attenuata*) from Thailand. *Phuket Marine Biological Center Research Bulletin* **61**: 29–37.
359. Chapman, D.I. & Chapman, N.G. (1982) The antler cycle of adult Reeves' muntjac. *Acta Theriologica* **27**: 107–114.
360. Chapman, D.I., Chapman, N.G., & Colles, C.M. (1985) Tooth eruption in Reeves' muntjac (*Muntiacus reevesi*) and its use as a method of age estimation (Mammalia: Cervidae). *Journal of Zoology. Series A* **205**: 205–221.
361. Chapman, D.I., Chapman, N.G., & Dansie, O. (1984) The periods of conception and parturition in feral Reeves' muntjac (*Muntiacus reevesi*) in southern England, based upon age of juvenile animals. *Journal of Zoology* **204**: 575–578.
362. Chapman, D.I. & Dansie, O. (1970) Reproduction and foetal development in female muntjac deer (*Muntiacus reevesi* Ogilby). *Mammalia* **34**: 303–319.

363. Chapman, N., Harris, S., & Stanford, A. (1994) Reeves' Muntjac *Muntiacus reevesi* in Britain: their history, spread, habitat selection, and the role of human intervention in accelerating their dispersal. *Mammal Review* **24**: 113–160.
364. Chapman, N.G. (1993) Reproductive performance of captive Reeves' muntjac. *in* Deer of China: Biology and Management (Ohtaishi, N. & Sheng, H.L., Editors). Amsterdam: Elsevier, pp. 199–203.
365. Chapman, N.G., et al. (1993) Sympatric populations of muntjac (*Muntiacus reevesi*) and roe deer (*Capreolus capreolus*): a comparative analysis of their ranging behaviour, social organization and activity. *Journal of Zoology* **229**: 623–640.
366. Chatani, K. (2003) Positional behavior of free-ranging Japanese macaques (*Macaca fuscata*). *Primates* **44**: 13–23.
367. Chen, I., et al. (2014) The "Southern form" of short-finned pilot whale (*Globicephala macrorhynchus*) in tropical west Pacific Ocean off Taiwan. *Ruffles Bulletin of Zoology* **62**: 188–199.
368. Chen, P., Liu, R., & Harrison, R.J. (1982) Reproduction and reproductive organs in *Neophocaena asiaeorientalis* from the Yangtze River. *Aquatic Mammals* **9**: 9–16.
369. Chen, Y., et al. (1981) Chromosome of the Primates of China. Beijing: Academic Press, 208 pp. (in Chinese).
370. Cheng, S.-C. & Wang, Y. (1993) The biological study of Formosan gem-faced civet, *Paguma larvata*. *Taipei Zoo Bulletin* **5**: 59–69 (in Chinese with English summary).
371. Chiarelli, B. (1962) Comparative morphometric analysis of the primate chromosomes: II. The chromosomes of the genera *Macaca*, *Papio*, *Theropithecus*, and *Cercocebus*. *Caryologia* **15**: 401–419.
372. Chiba, N. (千葉昇) (2001) 愛媛県立博物館所蔵ニホンカワウソ標本目録. *愛媛県立博物館研究報告 (Bulletin of the Ehime Prefectural Museum)* **15**: 1–12 (in Japanese).
373. Chiba, N. (千葉伸幸) (2000) テングコウモリのコロニー. *Bat Study and Conservation Report* **8**: 26 (in Japanese).
374. Chiba, N. (千葉伸幸) (2006) 福島の鍾乳洞におけるテングコウモリのコロニー観察報告. *Bat Study and Conservation Report* **14**: 15–16 (in Japanese).
375. Chiba, A., Onojima, M., & Kinoshita, T. (2005) Prey of the long-eared owl *Asio otus* in the suburbs of Niigata City, central Japan, as revealed by pellet analysis. *Ornithological Science* **4**: 169–172.
376. Chiba, H. & Yamaguchi, Y. (1975) The food habit of Japanese serow *Capricornis crispus crispus* (TEMMINCK), in the basin of River Takase, the Japan North Alps. *Bulletin of the Kanagawa Prefectural Museum (Natural Science)* **8**: 21–36 (in Japanese with English abstract).
377. Chiba Prefectural Government (2011) 千葉県の保護上重要な野生生物―千葉県レッドデータブック―動物編. Chiba: Chiba Prefectural Government, 542 pp. (in Japanese).
378. Chiba Prefectural Government (2013) 千葉県キョン防除実施計画. Chiba: Chiba Prefectural Government, 13 pp. (in Japanese).
379. Chikuni, K., et al. (1995) Molecular phylogeny based on the *k*-casein and cytochrome *b* sequences in the mammalian suborder Ruminantia. *Journal of Molecular Evolution* **41**: 859–866.
380. Chikuni, Y. (1957) *Plecotus auritus sacrimontis* Allen. *Collecting and Breeding* **19**: 136,138 (in Japanese).
381. Chilversa, B.L., et al. (2004) Diving behaviour of dugongs, *Dugong dugon*. *Journal of Experimental Marine Biology and Ecology* **304**: 203–224.
382. Chinen, A.A., et al. (2005) Preliminary genetic characterization of two lineages of black rats (*Rattus rattus* sensu lato) in Japan, with evidence for introgression at several localities. *Genes and Genetic Systems* **80**: 367–375.
383. Chinen, M.A., Aplin, K., & Suzuki, H. Unpublished.
384. Chishima, J. (千島淳) (1997) 1996 年ゼニガタアザラシ個体数調査 (センサス) 報告. *ワイルドライフ・フォーラム (Wildlife Forum)* **3** 113–118 (in Japanese).
385. Chittleborough, R.G. (1959) *Balaenoptera brydei* Olsen on the west coast of Australia. *Norwegian Whaling Gazette* **48**: 62–66.
386. Chivers, S.J., et al. (2007) Genetic variation and evidence for population structure in eastern North Pacific false killer whales (*Pseudorca crassidens*). *Canadian Journal of Zoology* **85**: 783–794.
387. Chivers, S.J., et al. (2005) Genetic variation in *Kogia* spp., with preliminary evidence for two species of *Kogia sima*. *Marine Mammal Science* **21**: 619–634.
388. Chou, L.-S., Bright, A.M., & Yeh, S.-Y. (1995) Stomach contents of dolphins (*Delphinus delphis* and *Lissodelphis borealis*) from North Pacific Ocean. *Zoological Studies* **34**: 206–210.
389. Chou, L.S., Lin, Y.S., & Mok, H.K. (1985) Study of the maintenance behavior of the red-bellied tree squirrel, *Callosciurus erythraeus*. *Bulletin of the Institute of Zoology, Academia Sinica* **24**: 39–50.
390. Chu, J.-H., Lin, Y.-S., & Wu, H.-Y. (2006) Evolution and dispersal of three closely related macaque species, *Macaca mulatta*, *M. cyclopis* and *M. fuscata*, in the eastern Asia. *Molecular Phylogenetics and Evolution* **43**: 418–429.
391. Chubu Regional Forest Office (中部森林管理局) (2007) 平成 18 年度南アルプスの保護林におけるシカ被害調査報告書 南アルプス北部の保護林内. Nagano: Chubu Regional Forest Office, 109 pp. (in Japanese).
392. Chung-MacCoubrey, A.L., Hagerman, A.E., & Kirkpatrick, R.L. (1997) Effect of tannins on digestion and detoxification activity in gray squirrels (*Sciurus carolinensis*). *Physiological Zoology* **70**: 270–277.
393. Churchfield, S., Nesterenko, V.A., & Shvarts, E.A. (1999) Food niche overlap and ecological separation amongst six species of coexisting forest shrews (Insectivora: Soricidae) in the Russian Far East. *Journal of Zoology, London* **248**: 349–359.
394. Clapham, P.J. & Mead, J.G. (1999) *Megaptera novaeangliae*. *Mammalian Species* **604**: 1–9.
395. Clark, E.L., et al., eds. (2006) Mongolian Red List of Mammals. Regional Red List Series. Volume 1. Zoological Society of London: London, 96 pp.
396. Clarke, M.R. (2003) Production and control of sound by the small sperm whales, *Kogia breviceps* and *K. sima* and their implications for other Cetacea. *Journal of the Marine Biological Association of the United Kingdom* **83**: 241–263.
397. Committee on Taxonomical names and Collections (日本哺乳類学会種名・標本検討委員会) (2001) 哺乳類標本の取り扱いに関するガイドライン (Guidelines for the procedure of obtaining mammal specimens as approved by the Mammalogical Society of Japan). *Mammalian Science* **41**: 215–233 (in Japanese).
398. Committee on Taxonomy, The Society for Marine Mammalogy (2014) List of Marine Mammal Species and Subspecies. Available from: https://www.marinemammalscience.org/species-information/list-of-marine-mammal-species-subspecies/.
399. Connor, R.C., et al. (1999) The bottlenose dolphin: Social relationships in a fission-fusion society. *in* Cetacean Societies: Field Studies of Dolphins and Whales (Mann, J., et al., Editors). Chicago: University of Chicago Press, pp. 91–126.
400. Cook, C.E., Wang, Y., & Sensabaugh, G. (1999) A mitochondrial control region and cytochrome *b* phylogeny of sika deer (*Cervus nippon*) and report of tandem repeats in the control region. *Molecular Phylogenetics and Evolution* **12**: 47–56.
401. Cooke, A.S. & Farrell, L. (2001) Impact of muntjac (*Muntiacus reevesi*) at Monks Wood National Nature Reserve, Cambridgeshire, eastern England. *Forestry* **74**: 241–250.
402. Cooke, A.S. & Lakhani, K.H. (1996) Damage to coppice regrowth by muntjac deer *Muntiacus reevesi* and protection with electric fencing. *Biological Conservation* **75**: 231–238.
403. Cooke, J.G., et al. (2007) Population assessment of western gray whales in 2007 (Paper SC/59/BRG41 presented to the IWC Scientific Committee). 10 pp.
404. Corbet, G.B. (1978) The Mammals of the Palaearctic Region: A Taxonomic Review. London: British Museum (Natural History), 314 pp.
405. Corbet, G.B. (1983) A review of classification in the Family Leporidae. *Acta Zoologica Fennica* **174**: 11–15.
406. Corbet, G.B. (1988) The family Erinaceidae: a synthesis of its taxonomy, phylogeny, ecology and zoogeography. *Mammal Review* **18**: 117–172.
407. Corbet, G.B. & Hill, J.E. (1980) A World List of Mammalian Species. London: British Museum (Natural History), 226 pp.
408. Corbet, G.B. & Hill, J.E. (1986) A World List of Mammalian Species. 2nd ed. London: British Museum (Natural History), 254 pp.
409. Corbet, G.B. & Hill, J.E. (1991) A World List of Mammalian Species. 3rd ed. London: British Museum (Natural History), 243 pp.
410. Corbet, G.B. & Hill, J.E. (1992) The Mammals of the Indomalayan Region: A Systematic Review. Oxford: Oxford University Press, 488 pp.
411. Costa, D.P. (1982) Energy, nitrogen, and electrolyte flux and sea-water drinking in the sea otter *Enhydra lutris*. *Physiological Zoology* **55**: 35–44.
412. Costa, D.P. (1991) Reproductive and foraging energetics of pinnipeds: implications for life history patterns. *in* The Behaviour of Pinnipeds (Renouf, D., Editor). London: Chapman & Hall, pp. 300–344.
413. Couch, L., Uni, S., & Duszynski, D.W. (2003) Eimeria species from serows (*Capricornis* spp.) in Japan with descriptions of two new species. *Journal of Parasitology* **89**: 580–584.
414. Courbis, S., et al. (2014) Multiple populations of pantropical spotted dolphins in Hawaiian waters. *Journal of Heredity* **105**: 627–641.
415. Cousse, S., et al. (1995) Temporal ontogeny in the wild boar (*Sus scrofa* L.): A systemic point of view. *Journal of Mountain Ecology* **3**: 122–125.
416. Cox, T.M., et al. (2006) Understanding the impacts of anthropogenic sound on beaked whales. *Journal of Cetacean Research and Management* **7**: 177–187.
417. Craighead, J.J., Summer, J.S., & Mitchell, J.A. (1995) The Grizzly Bears of Yellowstone. Their Ecology in the Yellowstone Ecosystem, 1959–1992.

Washington D.C.: Island Press, 535 pp.
418. Craighead, L., et al. (1995) Microsatellite analysis of paternity and reproduction in Arctic grizzly bears. *Journal of Heredity* **86**: 255–261.
419. Crespin, L., et al. (2002) Survival in fluctuating bank vole populations: Seasonal and yearly variations. *Oikos* **98**: 467–479.
420. Cronin, M.A., et al. (1996) Mitochondrial-DNA variation among subspecies and populations of sea otters (*Enhydra lutris*). *Journal of Mammalogy* **77**: 546–557.
421. Crossman, C.A., Barrett-Lennard, L.G., & Taylor, E.B. (2014) Population structure and intergeneric hybridization in harbour porpoises *Phocoena phocoena* in British Columbia, Canada. *Endangered Species Research* **26**: 1–12.
422. Csorba, G., et al. (2014) The reds and the yellows: a review of Asian *Chrysopteron* Jentink, 1910 (Chiroptera: Vespertilionidae: *Myotis*). *Journal of Mammalogy* **95**: 663–678.
423. Csorba, G., Ujhelyi, P., & Thomas, N. (2003) Horseshoe Bats of the World (Chiroptera: Rhinolophidae). Shropshire: Alana Books, 160 pp.
424. Cultural Property Center of Ichihara (市原市文化財センター) (1999) 千葉県市原市祇園原貝塚．上総国分寺台遺跡調査報告 V (財団法人市原市文化財センター調査報告書) **60**: 1–1136 (in Japanese).
425. Cummings, W.C. (1985) Right whales *Eubalaena glacialis* (Müller, 1776) and *Eubalaena australis* (Desmoulins, 1822). *in* Handbook of Marine Mammals. Volume 3. The Sirenians and Baleen Whales (Ridgway, S.H. & Harrison, R.J., Editors). London: Academic Press, pp. 275–304.
426. d'Huart, J.P. (1991) Habitat utilization of old world wild pigs. *in* Biology of Suidae (Barrett, R.H. & Spitz, F., Editors). Briancon: IRGM, pp. 30–48.
427. Dahlheim, M.E. & Heyning, J.E. (1999) Killer whale *Orcinus orca* (Linnaeus, 1758). *in* Handbook of Marine Mammals. Volume 6. The Second Book of Dolphins and the Porpoises (Ridgway, S.H. & Harrison, R., Editors). London: Academic Press, pp. 281–322.
428. Daily, M.D. (2007) A new species of *Digenea* (Trematoda: Brachycladiidae) from the Gervais' beaked whale, *Mesoplodon europaeus*, with comments on other cetacean liver flukes. *Comparative Parasitology* **74**: 229–232.
429. Daily, M.D. & Brownell, R.L., Jr. (1972) A checklist of marine mammal parasites. *in* Mammals of the Sea (Ridgway, S.H., Editor). Springfield: Charles C. Thomas, pp. 528–589.
430. Dalebout, M.L., et al. (2014) Resurrection of *Mesoplodon hotaula* Deraniyagala 1963: A new species of beaked whale in the tropical Indo-Pacific. *Marine Mammal Science* **30**: 1081–1108.
431. Dalebout, M.L., et al. (2003) Appearance, distribution, and genetic distinctiveness of Longman's beaked whale, *Indopacetus pacificus*. *Marine Mammal Science* **19**: 421–461.
432. Dalebout, M.L., Steel, D., & Baker, C.S. (2008) Phylogeny of beaked whale genus *Mesoplodon* (Ziphiidae: Cetacea) revealed by nuclear introns: Implications for the evolution of male tusks. *Systematic Biology* **57**: 857–875.
433. Danil, K. Unpublished.
434. Darling, J.D., Jones, M.E., & Nicklin, C.P. (2006) Humpback whale songs: Do they organize males during the breeding season? *Behavior* **143**: 1051–1101.
435. Darling, J.D. & Mori, K. (1993) Recent observations of humpback whales (*Megaptera novaeangliae*) in Japanese waters off Ogasawara and Okinawa. *Canadian Journal of Zoology* **71**: 322–333.
436. Date, I. (2007) A record of *Murina leucogaster* captured by the sweeping method over the trees. *Study Report on Bat Conservetin in Tohoku Region* **1**: 23 (in Japanese).
437. Dave, M.J., et al. (1965) Chromosome studies on the hare and the rabbit. *Proceedings of the Japan Academy. Series B* **41**: 244–248.
438. Davidson, M. (1973) Characteristics, liberation and dispersal of sika deer (*Cervus nippon*) in New Zealand. *New Zealand Journal of Forestry Science* **3**: 153–180.
439. Daxner, G. & Fejfar, O. (1967) Über die Gattungen *Alilepus* Dice, 1931 und *Pliopentalagus* Gureev, 1964 (Lagomorpha, Mammalia). *Annalen des Naturhistorischen Museums in Wien* **71**: 37–55.
440. de Guia, A.P., et al. (2007) Taxonomic status of the vole in Daikoku Island, Hokkaido, Japan: examination based on morphology and genetics. *Mammal Study* **32**: 33–44.
441. de Teilhard, C.P. & Young, C.C. (1931) Fossil mammals from the late Cenozoic of northern China. *Geological Survey of China Paleontology Sinica* **C9**: 1–88.
442. Deer Research Group on Boso (房総のシカ調査会) (2001) 千葉県イノシシ・キョン管理対策調査報告書 1. Chiba: Chiba Prefectural Government & Deer Research Group on Boso (千葉県環境生活部自然保護課 & 房総のシカ調査会), 95 pp. (in Japanese).
443. Deer Research Group on Boso (房総のシカ調査会) (2002) 千葉県イノシシ・キョン管理対策調査報告書 2. Chiba: Chiba Prefectural Government & Deer Research Group on Boso (千葉県環境生活部自然保護課 & 房総のシカ調査会), 97 pp. (in Japanese).
444. Deer Research Group on Boso (房総のシカ調査会) (2007) 平成 18 年度外来種緊急特別対策事業 (キョンの生息状況等調査) 報告書. Chiba: Chiba Prefectural Government & Deer Research Group on Boso (千葉県環境生活部自然保護課 & 房総のシカ調査会), 88 pp. (in Japanese).
445. Deer Research Group on Boso (房総のシカ調査会) (2008) 平成 19 年度外来種緊急特別対策事業 (キョンの生息状況等調査) 報告書. Chiba: Chiba Prefectural Government, Natural History Museum and Institute, Chiba & Deer Research Group on Boso (千葉県環境生活部自然保護課, 千葉県立中央博物館 & 房総のシカ調査会), 73 pp. (in Japanese).
446. Deguchi, Y., et al. (2000) Estimating the proportion of various crops in the diet of wild Japanese serow (*Capricornis crispus*) in Yamagata Prefecture. *Wildlife Conservation Japan* **5**: 13–20 (in Japanese with English abstract).
447. Deguchi, Y., et al. (2001) Feeding behavior of the wild Japanese serow (*Capricornis crispus*) in human cultivated fields. *Wildlife Conservation Japan* **7**: 49–62 (in Japanese with English abstract).
448. Dewa, H. (出羽寛) (2009) コウモリのねぐらと樹木. *Forest Protection* **313**: 4–6 (in Japanese).
449. Dewa, H. (1975) Seasonal variation of daily activity rhythms of the red-backed vole, *Clethrionomys rufocanus bedfordiae* (Thomas). 1. Daily rhythms in snow season. *Research Bulletin of College Experimental Forests Hokkaido University* **32**: 105–120 (in Japanese with English summary).
450. Dewa, H. (2001) Faunal survey of bats in Asahikawa area, Hokkaido, Japan II. *Annual Report of the Regional Research Institute Asahikawa University* **24**: 79–90 (in Japanese with English abstract).
451. Dewa, H. (2002) Distribution of bats and habitat utilization in the south area of northern Hokkaido. *Journal of Asahikawa University* **54**: 31–59 (in Japanese with English abstract).
452. Dewa, H. (2005) Faunal survey of bats in Asahikawa area, Hokkaido, Japan III. *Journal of Asahikawa University* **59**: 23–45 (in Japanese with English abstract).
453. Dewa, H. (2010) Faunal survey of bats in the river bed of Teshio River Basin (Otoineppu, Bifuka, Shimokawa and Shibetsu), in Northern Hokkaido, Japan. *Rishiri Studies* **29**: 25–33 (in Japanese with English abstract).
454. Dewa, H., et al. (2005) Faunal survey of bats in Tokachimitsumata, Kamishihoro, Hokkaido, Japan. *Bulletin of the Higashi Taisetsu Museum of Natural History* **27**: 21–26 (in Japanese with English abstract).
455. Dewa, H. & Kosuge, M. (2001) Faunal survey of bats in Asahikawa area, Hokkaido. *Bulletin of the Asahikawa Museum* **7**: 31–38 (in Japanese with English summary).
456. Dewa, H., et al. (2006) First records of four bat species from Shibetsu City. *Bulletin of the Shibetsu City Museum* **24**: 1–5 (in Japanese with English abstract).
457. Dewa, H., Shimizu, S., & Murayama, M. (2011) Daily and seasonal variation of the number caught by mist nets: Fraternal Myotis, *Myotis frater* compared with Ikonnikov's Myotis, *M. ikonnikovi*. *Annual Report of the Regional Institute Asahikawa* **33**: 15–32 (in Japanese).
458. Dewa, H., Shimizu, S., & Murayama, M. (2013) Drinking behavior of the fraternal Myotis, *Myotis frater* and Iknnnikov's Myotis, *M. ikonnikovi*, at water holes and predation by Ikonnikov's Myotis of large sized moths in an experimental net house. *Bulletin of the Asian Bat Research Institute* **9**: 20–26 (in Japanese with English abstract).
459. Dewa, H., et al. (2009) A maternity colony of Ikonnikov's whiskered bat in a barn. *Bulletin of the Asian Bat Research Institute* **8**: 1–8 (in Japanese with English abstract).
460. Dickerson, B.R., et al. (2010) Population structure as revealed by mtDNA and microsatellites in Northern fur seals, *Callorhinus ursinus*, throughout their range. *PLoS ONE* **5**: 1–9.
461. Din, W., et al. (1996) Origin and radiation of the house mouse: clues from nuclear genes. *Journal of Evolutionary Biology* **9**: 519–539.
462. Dobson, A.P. & Lyles, A.M. (1989) The population dynamics and conservation of primate populations. *Conservation Biology* **3**: 362–380.
463. Dobson, M. (1994) Patterns of distribution in Japanese land mammals. *Mammal Review* **24**: 91–111.
464. Dobson, M. & Kawamura, Y. (1998) Origin of the Japanese land mammal fauna: Allocation of extant species to historically-based categories. *The Quaternary Research* **37**: 385–395.
465. Dohl, T.P., Norris, K.S., & Kang, I. (1974) A porpoise hybrid: *Tursiops* × *Steno*. *Journal of Mammalogy* **55**: 217–221.
466. Doi, T. (土肥昭夫) & Izawa, M. (伊澤雅子) (1997) ツシマヤマネコの現在と未来 (Present status and future prospects of the Tsushima leopard cat). *Animals and Zoos* **49**: 288–294 (in Japanese).
467. Dokuchaev, N.E. (1990) Ecology of Shrews in North-east Asia. Moscow: Nauka, 160 pp. (in Russian with English summary).
468. Dokuchaev, N.E. (1994) Siberian shrew *Sorex minutissimus* found in

Alaska. *Zoologichesky Zhurnal* **73**: 254–256 (in Russian with English summary).
469. Dokuchaev, N.E. (1997) A new species of shrew (Soricidae, Insectivora) from Alaska. *Journal of Mammalogy* **78**: 811–817.
470. Dokuchaev, N.E., Kohno, N., & Ohdachi, S.D. (2010) Reexamination of fossil shrews (*Sorex* spp.) from the Middle Pleistocene of Honshu Island, Japan. *Mammal Study* **35**: 157–168.
471. Dokuchaev, N.E., Ohdachi, S., & Abe, H. (1999) Morphometric status of shrews of the *Sorex caecutiens/shinto* group in Japan. *Mammal Study* **24**: 67–78.
472. Dolar, M.L.L. (2009) Fraser's dolphin (*Lagenodelphis hosei*). *in* Encyclopedia of Marine Mammals. 2nd ed. (Perrin, W.F., Würsig, B., & Thewissen, J.G.M., Editors). Amsterdam: Academic Press, pp. 469–471.
473. Dolar, M.L.L., et al. (2003) Comparative feeding ecology of spinner dolphins (*Stenella longirostris*) and Fraser's dolphins (*Lagenodelphis hosei*) in the Sulu Sea. *Marine Mammal Science* **19**: 1–19.
474. Dolgov, V.A. (1985) Shrews of the Old World. Moscow: Moscow State University Press, 221 pp. (in Russian with English summary).
475. Donahue, M.A. & Perryman, W.L. (2009) Pygmy killer whale (*Feresa attenuata*). *in* Encyclopedia of Marine Mammals. 2nd ed. (Perrin, W.F., Würsig, B., & Thewissen, J.G.M., Editors). Amsterdam: Academic Press, pp. 938–939.
476. Doroff, A. & Burdin, A. (2013) *Enhydra lutris*. The IUCN Red List of Threatened Species. Version 2014.3. Available from: http://www.iucnredlist.org.
477. Dubey, S., et al. (2007) Mediterranean populations of the lesser white-toothed shrew (*Crocidura suaveolens* group): an unexpected puzzle of Pleistocene survivors and prehistoric introductions. *Molecular Ecology* **16**: 3438–3452.
478. Dubey, S., et al. (2007) Molecular phylogenetics of shrews (Mammalia: Soricidae) reveal timing of transcontinental colonizations. *Molecular Phylogenetics and Evolution* **44**: 126–137.
479. Dubey, S., et al. (2006) Pliocene and Pleistocene diversification and multiple refugia in a Eurasian shrew (*Crocidura suaveolens* group). *Molecular Phylogenetics and Evolution* **38**: 625–647.
480. Dulic, B. (1981) Chromosomes of three species of Indian Microchiroptera. *Myotis* **18/19**: 76–82.
481. Dunn, C. & Claridge, D. (2014) Killer whale (*Orcinus orca*) occurrence and predation in the Bahamas. *Journal of the Marine Biological Association of the United Kingdom* **94**: 1305–1309.
482. Dunn, D.G., et al. (2002) Evidence for infanticide in bottlenose dolphins of the Western North Atlantic Source. *Journal of Wildlife of Diseases* **38**: 505–510.
483. Dunphy-Daly, M.M., Heithaus, M.R., & Claridge, D.E. (2008) Temporal variation in dwarf sperm whale (*Kogia sima*) habitat use and group size off Great Abaco Island, Bahamase options. *Marine Mammal Science* **24**: 171–182.
484. Duszynski, D.W. & Wattam, A.R. (1988) Coccidian parasites (Apicomplexa: Eimeriidae) from insectivores. V. Ten forms from the moles of Japan (*Euroscaptor, Mogera* spp.). *Journal of Protozoology* **35**: 55–57.
485. Dutrillaux, B., et al. (1979) Comparison of the karyotypes of four Cercopithecidae: *Papio papio, P. anubis, Macaca mulatta*, and *M. fascicularis*. *Cytogenetics and Cell Genetics* **23**: 77–83.
486. Eager, J.L. (1990) Patterns of geographic variation in the skull of Nearctic ermine (*Mustela erminea*). *Canadian Journal of Zoology* **68**: 1241–1249.
487. Ebert, M.B. & Valente, A.L.S. (2013) New records of *Nasitrema atenuatta* and *Nasitrema globicephalae* (Tremadoda: Brachycladiidae) Neiland, Rice and Holden, 1970 in delphinids from South Atlantic. *Check List* **9**: 1538–1540.
488. Echenique-Díaz, L.M. (2005) Factor affecting genetic structure at different spatial and temporal scales in the bat *Hipposideros turpis* (Chiroptera, Hipposideridae). Doctral dissertation. Sendai: Tohoku University. 92 pp.
489. Echenique-Díaz, L.M., et al. (2002) Isolation and migration in island populations of *Hipposideros turpis turpis* (Chiroptera, Hipposideridae): History or current patterns? (A preliminary Report). *in* Proceedings of Two Symposia on Ecology and Evolution in VIII INTECOL, Seoul, Korea (Yoshimura, J., Nakagiri, N., & Shields, W.M., Editors). Otsu: Sangaku Publisher, pp. 26–33.
490. Echenique-Díaz, L.M., et al. (2002) Isolation and characterization of microsatellite loci in the Bang's leaf-nosed bat *Hipposideros turpis*. *Molecular Ecology Notes* **2**: 396–397.
491. Echenique-Díaz, L.M., et al. (2009) Genetic structure of island populations of the endangered bat *Hipposideros turpis turpis*: implications for conservation. *Population Ecology* **51**: 153–160.
492. Eduardo, S.I., Yaptinchay, A.A.P., & Lim, T.M.S. (1998) Some helminth parasites of sea cow (*Dugong dugon*, Müller, 1976) (Mammalia: Sirenia) in the Philippines. *Philippine Journal of Veterinary Medicine* **35**: 27–36.
493. Egi, H. (江木寿男) (2008) 岡山県におけるヒナコウモリの確認記録. *Bat Study and Conservation Report* **16**: 16–17 (in Japanese).
494. Eguchi, Y. (江口祐輔), Miura, S. (三浦慎吾), & Fujioka, M. (藤岡正博) (2002) 鳥獣害対策の手引き. Tokyo: Japan Plant Protection Association (日本植物防疫協会), 154 pp. (in Japanese).
495. Eguchi, K. & Nakazono, T. (1980) Activity studies of Japanese red foxes, *Vulpes vulpes japonica* GRAY. *Japanese Journal of Ecology* **30**: 9–17.
496. Eguchi, K., Nakazono, T., & Higashi, K. (1977) Tracking of the hondo fox (*Vulpes vulpes japonica* Gray) by radiotelemetry. *in* Studies on Methods of Estimating Population Density, Biomass and Productivity in Terrestrial Animals (Morishita, M., Editor). Tokyo: University Tokyo Press, pp. 216–225.
497. Eguchi, Y., Tanaka, T., & Yoshimoto, T. (2001) Some aspects of farrowing in Japanese wild boars, *Sus scrofa leucomystax*, under captive conditions. *Animal Science Journal* **72**: J49–J54 (in Japanese with English summary).
498. Ehime Prefectural Board of Education (愛媛県教育委員会) (1964) にっぽんかわうそ. 57 pp. (in Japanese).
499. Ellerman, J.R. & Morrison-Scott, T.C.S. (1951) Checklist of Palaearctic and Indian Mammals. 1758 to 1946. London: Trustees of the British Museum (Natural History), 810 pp.
500. Ellerman, J.R. & Morrison-Scott, T.C.S. (1966) Checklist of Palaearctic and Indian Mammals. 1758 to 1946. 2nd ed. London: Trustees of the British Museum (Natural History), 810 pp.
501. Elton, C.S. (1924) Periodic fluctuations in the numbers of animals: their causes and effects. *Journal of Experimental Biology* **2**: 119–163.
502. Elton, C.S. (1942) Voles, Mice and Lemmings: Problems in Population Dynamics. Oxford: Clarendon Press, 496 pp.
503. Enari, H. & Sakamaki-Enari, H. (2013) Influence of heavy snow on the feeding behavior of Japanese macaques (*Macaca fuscata*) in Northern Japan. *American Journal of Primatology* **75**: 534–544.
504. Endo, H. (遠藤秀紀) (2002) Evolution of the Mammals (哺乳類の進化). Tokyo: University of Tokyo Press, 383 pp. (in Japanese).
505. Endo, A. & Doi, T. (1996) Home range of female sika deer *Cervus nippon* on Nozaki Island, the Goto Archiperago, Japan. *Mammal Study* **21**: 27–35.
506. Endo, A. & Doi, T. (2001) Asynchronous estrus of female sika deer (*Cervus nippon*) during the rutting season. *Mammal Study* **26**: 69–72.
507. Endo, H. (1996) Scientific and Japanese names of artiodactyls of Japan. *Mammalian Science* **35**: 203–209 (in Japanese with English abstract).
508. Endo, H. (2008) Bibliography of the macaques of Vietnam, Japan and the other Southeast Asian Districts. *in* Checklist of Wild Mammal Species of Vietnam (Can, D.N., et al., Editors). Kyoto: Shoukadoh, pp. 359–368.
509. Endo, H., et al. (1995) Cardiac musculature of the intrapulmonary venous wall as an endocrine organ of atrial natriuretic polypeptide in Watase's shrew (*Crocidura watasei*) and musk shrew (*Suncus murinus*). *Journal of the Mammalogical Society of Japan* **20**: 109–116.
510. Endo, H., et al. (1997) Osteometrical and CT examination of the Japanese wolf skull. *Journal of Veterinary Medical Science* **59**: 531–538.
511. Endo, H., et al. (2004) Osteological examination and image analysis of a cranium of the Japanese wolf found at private house in Yamanashi Prefecture. *Japanese Journal Zoo and Wildlife Medicine* **9**: 109–114 (in Japanese with English summary).
512. Endo, H., et al. (1999) MRI examination of the masticatory muscles in the gray wolf (*Canis lupus*), with special reference to the *M. temporalis*. *Journal of Veterinary Medical Science* **61**: 581–586.
513. Endo, H. & Tsuchiya, K. (2006) A new species of Ryukyu spiny rat, *Tokudaia* (Muridae: Rodentia), from Tokunoshima Island, Kagoshima Prefecture, Japan. *Mammal Study* **31**: 47–57.
514. Endo, H., Xiaodi, Y., & Kogiku, H. (2000) Osteometrical study of the Japanese otter (*Lutra nippon*) from Ehime and Kochi Prefectures. *Memoirs of the Natural Science Museum (Tokyo)* **33**: 195–201.
515. Endo, K. (1957) Notes on *Murina ussuriensis* Ognev. *The Journal of the Mammalogical Society of Japan* **1**: 63–65 (in Japanese with English abstract).
516. Endo, K. (1958) *Murina hilgendolf* (Peters) collected in Iwate Pref., northen Honshu, Japan. *The Journal of the Mammalogical Society of Japan* **1**: 102–103 (in Japanese with English abstract).
517. Endo, K. (1961) The first record of propagation, including flight behavior on *Murina aurata ussuriensis* Ognev in Japan. *The Journal of the Mammalogical Society of Japan* **2**: 14–16 (in Japanese with English abstract).
518. Endo, K. (1963) *Mytos hosonoi* Imaizumi, recently obtained from Iwate Pref. Japan. *The Journal of the Mammalogical Society of Japan* **2**: 60–61 (in Japanese with English abstract).
519. Endo, K. (1963) Notes on pregnant females of *Nyctalus lasiopterus aviator* Thomas. *The Journal of the Mammalogical Society of Japan* **2**: 61–62 (in

Japanese).

520. Endo, K. (1967) Hunting flies of *Myotis frater kaguyae*. *The Journal of the Mammalogical Society of Japan* **3**: 64–67 (in Japanese with English abstract).
521. Endo, K. (1971) The oriental frosted bat, *Vespertilio orientalis*, living in forest in summer season. *The Journal of the Mammalogical Society of Japan* **5**: 149 (in Japanese).
522. Endo, K. (1976) Notes on some habits and habitat of *Myotis pruinosus*. *The Journal of the Mammalogical Society of Japan* **6**: 259–260 (in Japanese with English abstract).
523. Endo, K. (1977) Two species of bat obtained from high mountain in Hokkaido. *The Journal of the Mammalogical Society of Japan* **7**: 118–119 (in Japanese).
524. Endo, T., et al. (2007) Age-dependent accumulation of heavy metals in a pod of killer whales (*Orcinus orca*) stranded in the northern area of Japan. *Chemosphere* **67**: 51–59.
525. Endo, T., et al. (2006) Distribution of total mercury, methyl mercury and selenium in pod of killer whales (*Orcinus orca*) stranded in the northern area of Japan: Comparison of mature females with calves. *Environmental Pollution* **144**: 145–150.
526. Environment Agency (1993) 第4回自然環境保全基礎調査：動植物分布調査報告書 (哺乳類). Tokyo: Environment Agency, 208 pp. (in Japanese; http://www.biodic.go.jp/reports/4-02/P2_000_3.html).
527. Environment Agency (1999) 平成10年度島しょ地域の移入種駆除・制御モデル事業 (奄美大島：マングース) 調査報告書 (Report on the Investigation of the Mongoose for Eradication on Amami-oshima Island). Tokyo: Environment Agency, 51 pp. (in Japanese).
528. Environment Agency (2000) 平成11年度島しょ地域の移入種駆除・制御モデル事業 (奄美大島：マングース) 調査報告書 (Report on the Investigation of the Mongoose for Eradication on Amami-oshima Island). Tokyo: Environment Agency, 115 pp. (in Japanese).
529. Erftemeijer, P.L.A., Djunarlin, & Moka, W. (1993) Stomach content analysis of a Dugong (*Dugong dugon*) from South Sulawesi, Indonesia. *Australian Journal of Marine and Freshwater Research* **44**: 229–233.
530. Escorza-Trevino, S., et al. (2005) Genetic differentiation and intraspecific structure of Eastern Tropical Pacific spotted dolphins, *Stenella attenuata*, revealed by DNA analyses. *Conservation Genetics* **6**: 587–600.
531. Estes, J.A. (1980) *Enhydra lutris*. *Mammalian Species* **133**: 1–8.
532. Estes, J.A. (1996) The influence of large, mobile predators in aquatic food webs: examples from sea otters and kelp forests. *in* Aquatic Predators and Their Prey (Greenstreet, S.P.R. & Tasker, M.L., Editors). Oxford: Blackwell Scientific, pp. 65–72.
533. Evans, W.E. (1994) Common dolphin, white-bellied porpoise *Delphinus delphis* Linnaeus, 1758. *in* Handbook of Marine Mammals. Volume 5. The First Book of Dolphins (Ridgway, S.H. & Harrison, R., Editors). London: Academic Press, pp. 191–224.
534. Ewer, R.F. (1973) The Carnivores. London: Cornell University Press, 494 pp.
535. Fang, Y.-P., et al. (1997) Systematics of white-toothed shrews (*Crocidura*) (Mammalia: Insectivora: Soricidae) of Taiwan: karyological and morphological studies. *Journal of Zoology, London* **242**: 151–166.
536. Fedoseev, G. (2002) Ribbon seal (*Histriophoca fasciata*). *in* Encyclopedia of Marine Mammals (Perrin, W.F., Würsig, B., & Thewissen, J.G.M., Editors). San Diego: Academic Press, pp. 1027–1030.
537. Fedoseev, G.A. (1973) Morpho-ecological characteristics of ribbon seal populations and the grounds for protection of its stocks. *Izvestiya TINRO* **86**: 148–177 (in Russian).
538. Fedoseev, G.A. (1975) Principal population indicators of dynamics of numbers of seals of the family Phocidae. *Ekologiya* **5**: 62–70.
539. Fedoseev, G.A. (2000) Population Biology of Ice-Associated Forms of Seals and their Roles in the Northern Pacific Ecosystems. Moscow: Center for Russian Environmental Policy, 271 pp.
540. Feldhamer, G.A. (1980) *Cervus nippon*. *Mammalian Species* **128**: 1–7.
541. Feldhamer, G.A. (2002) Acorns and white-tailed deer: interrelationships in forest ecosystems. *in* Oak Forest Ecosystems (McShea, W.J. & Healy, W.M., Editors). Baltimore: Johns Hopkins University Press, pp. 215–223.
542. Feldhamer, G.A. & Demarais, S. (2009) Free-ranging and confined sika deer in North America: current status, biology, and management. *in* Sika Deer—Biology and Management of Native and Introduced Populations (McCullough, D.R., Takatsuki, S., & Kaji, K., Editors). Tokyo: Springer, pp. 615–641.
543. Fenner, F. & Ross, J. (1994) Myxomatosis. *in* The European Rabbit: The History and Biology of a Successful Colonizer (Thompson, H.V. & King, C.M., Editors). Oxford: Oxford University Press, pp. 205–239.
544. Ferguson, M.C., et al. (2006) Spatial models of delphinid (family Delphinidae) encounter rate and group size in the eastern tropical Pacific Ocean. *Ecological Modelling* **193**: 645–662.
545. Ferguson, W.W. (1981) The systematic position of *Canis aureus lupaster* (Carnivora: Canidae) and the occurrence of *Canis lupus* in North Africa, Egypt, and Sinai. *Mammalia* **45**: 459–465.
546. Fernández, A., et al. (2005) "Gas and fat embolic syndrome" involving a mass stranding of beaked whales (Family Ziphiidae) exposed to anthropogenic sonar signals. *Veterinary Pathology* **42**: 446–457.
547. Fernandez-Llario, P. & Carranza, J. (2000) Reproductive performance of the wild boar in a Mediterranean ecosystem under drought conditions. *Ethology Ecology and Evolution* **12**: 335–343.
548. Fernandez-Llario, P. & Mateos-Quesada, P. (1998) Body size and reproductive parameters in the wild boar *Sus scrofa*. *Acta Theriologica* **43**: 439–444.
549. Ferrand, N. (2008) Inferring the evolutionary history of the European rabbit (*Oryctolagus cuniculus*) from molecular markers. *in* Lagomorph Biology: Evolution, Ecology, and Conservation (Alves, P.C., Ferrand, N., & Hackländer, K., Editors). Berlin: Springer, pp. 47–63.
550. Ferreira, I.M., et al. (2014) False killer whales (*Pseudorca crassidens*) from Japan and South Africa: difference in growth and reproduction. *Marine Mammal Science* **30**: 64–84.
551. Ferrero, F.C. & Walker, W.A. (1999) Age, growth, and reproductive patterns of Dall's porpoise (*Phocoenoides dalli*) in the central North Pacific Ocean. *Marine Mammal Science* **15**: 273–313.
552. Ferrero, R.C. & Walker, W.A. (1993) Growth and reproduction of the northern right whale dolphin, *Lissodelphis borealis*, in the offshore waters of the North Pacific Ocean. *Canadian Journal of Zoology* **71**: 2335–2344.
553. Ferrero, R.C. & Walker, W.A. (1995) Growth and reproduction of the common dolphin, *Delphinus delphis* Linnaeus, in the offshore waters of the North Pacific Ocean. *Fishery Bulletin* **93**: 483–494.
554. Filby, N.E., et al. (2010) Distribution and population demographics of common dolphins (*Delphinus delphis*) in the Gulf St. Vincent, South Australia. *Aquatic Mammals* **36**: 33–45.
555. Fisher, E.M. (1939) Habits of the southern sea otter. *Journal of Mammalogy* **20**: 21–36.
556. Fisheries Agency, Ministry of Agriculture, Forestry and Fisheries (2014) トド管理基本方針 (in Japanese; http://www.jfa.maff.go.jp/j/press/sigen/pdf/140806-01.pdf).
557. Fisheries Research Agency (2014) 国際漁業資源の現況：55 トド 北太平洋沿岸・オホーツク海・ベーリング海 (Steller sea lion, *Eumetopias jubatus*). Available from: http://kokushi.job.affrc.go.jp/H25/H25_55.html (in Japanese).
558. Fitzgerald, B.M. (1988) Diet of domestic cats and their impact on prey populations. *in* The Domestic Cat: The Biology of its Behaviour (Turner, D.C. & Bateson, P., Editors). Cambridge: Cambridge University Press, pp. 123–147.
559. Flanders, J., et al. (2009) Phylogeography of the greater horseshoe bat, *Rhinolophus ferrumequinum*: contrasting results from mitochondrial and microsatellite data. *Molecular Ecology* **18**: 306–318.
560. Flux, J.E.C. (1994) World distribution. *in* The European Rabbit: The History and Biology of a Successful Colonizer (Thompson, H.V. & King, C.M., Editors). Oxford: Oxford University Press, pp. 8–21.
561. Flux, J.E.C. & Angermann, R. (1990) The hares and Jackrabbits. *in* Rabbits, Hares and Pikas: Status Survey and Conservation Action Plan (Chapman, J.A. & Flux, J.E.C., Editors). Gland: IUCN, pp. 61–94.
562. Flux, J.E.C. & Fullagar, P.J. (1992) World distribution of the rabbit *Oryctolagus cuniculus* on islands. *Mammalian Review* **22**: 151–205.
563. Fontaine, M.C., et al. (2007) Rise of oceanographic barriers in continuous populations of a cetacean: the genetic structure of harbour porpoises in Old World waters. *BMC Biology* **5**: 30.
564. Fontaine, P.M. & Barrette, C. (1997) Megatestes: anatomical evidence for sperm competition in the harbor porpoise. *Mammalia* **61**: 65–71.
565. Fooden, J. (1976) Provisional classification and key to living species of macaques (Primates: Macaca). *Folia Primatologica* **25**: 225–236.
566. Fooden, J. (1980) Classification and distribution of living macaques (*Macaca* Lacépède, 1799). *in* The Macaques: Studies in Ecology, Behavior and Evolution (Lindburg, D.G., Editor). New York: Van Nostrand Reinhold, pp. 1–9.
567. Fooden, J. (1982) Ecogeographic segregation of macaque species. *Primates* **23**: 574–579.
568. Fooden, J. (2000) Systematic Review of the Rhesus Macaque, *Macaca mulatta* (Zimmermann, 1780). Fieldiana Zoology, New Series. Vol. 96. Chicago: Field Museum of Natural History, 180 pp.
569. Fooden, J. & Aimi, M. (2003) Birth-season variation in Japanese macaques, *Macaca fuscata*. *Primates* **44**: 109–117.
570. Fooden, J. & Aimi, M. (2005) Systematic Review of Japanese Macaques, *Macaca fuscata* (Gray, 1870). Fieldiana Zoology, New Series. Vol. 104. Chicago: Field Museum of Natural History, 200 pp.
571. Fooden, J. & Albrecht, G.H. (1999) Tail-length evolution in *fascicularis*-group macaques (Cercopithecidae: *Macaca*). *International Journal of*

572. Fooden, J. & Lanyon, S.M. (1989) Blood-protein allele frequencies and phylogenetic relationships in *Macaca*: A review. *American Journal of Primatology* **17**: 209–241.
573. Fooden, J. & Wu, H.-Y. (2001) Systematic Review of the Taiwanese Macaques, *Macaca cyclopis* Swinhoe, 1863. Fieldiana Zoology, New Series. Vol. 98. Chicago: Field Museum of Natural History, 70 pp.
574. Foote, A.D. (2012) Investigating ecological speciation in non-model organisms: a case study on killer whale ecotypes. *Evolutionary Ecology Research* **14**: 447–465.
575. Foote, A.D., et al. (2009) Ecological, morphological and genetic divergence of sympatric North Atlantic killer whale populations. *Molecular Ecology* **18**: 5207–5217.
576. Formozov, N.A., Grigor'eva, T.V., & Surin, V.L. (2006) Molecular systematics of pikas of the subgenus *Pika* (*Ochotona*, Lagomorpha). *Zoologicheskii Zhurnal* **85**: 1465–1473 (in Russian with English abstract).
577. Formozov, N.A. & Nikolski, A.A. (1979) Status of the Ural pikas in the "alpina" group (a bioacoustical analysis). *in* Mammals of the Ural Mountains. Sverdlovsk: Akademii Nauk SSSR, Ural Science Center, pp. 80–82.
578. Forsyth, D.M. & Caley, P. (2006) Testing the irruptive paradigm of large-herbivore dynamics. *Ecology* **87**: 297–303.
579. Foster, E.A., et al. (2012) Adaptive prolonged postreproductive life span in killer whales. *Science* **337**: 1313.
580. Fowler, M.E. & Richards, W.P.C. (1965) Acorn poisoning in a cow and a sheep. *Journal of the American Veterinary Medical Association* **147**: 1215–1220.
581. Fraser, F.C. (1936) Vestigial teeth in specimens of Cuvier's whale (*Ziphius cavirostris*) stranded on the Scottish coast. *Scottish Naturalist* **1936**: 153–157.
582. Fraser, F.C. (1940) Three anomalous dolphins from Blacksod Bay, Ireland. *Proceedings of the Royal Irish Academy* **45B**: 413–455.
583. Fraser, F.C. (1946) Report on Cetacea stranded on the British coasts from 1933 to 1937. London: British Museum (Natural History), 56 pp. & 7 maps.
584. Fritts, S.H. & Mech, L.D. (1981) Dynamics, movements, and feeding ecology of a newly protected wolf population in northwestern Minnesota. *Wildlife Monographs* **80**: 1–79.
585. Frölich, K. & Lavazza, A. (2008) European brown hare syndrome. *in* Lagomorph Biology: Evolution, Ecology, and Conservation (Alves, P.C., Ferrand, N., & Hackländer, K., Editors). Berlin: Springer, pp. 253–261.
586. Frost, K.J. & Lowry, L.F. (1981) Ringed, Baikal and Caspian seals *Phoca hispida* Schreder, 1775, *Phoca sibirica* Gmelin, 1788 and *Phoca caspica* Gmelin, 1788. *in* Handbook of Marine Mammals. Volume 2. Seals (Ridgway, S.H. & Harrison, R.J., Editors). London: Academic Press, pp. 29–53.
587. Fujii, K. (藤井啓) & Nakagawa, E. (中川恵美子) (2002) 漁業被害とアザラシの混獲：襟裳地域. *in* アザラシ類の保護管理報告書. Sapporo: Hokkaido Government, pp. 133–140 (in Japanese).
588. Fujii, N., et al. (2002) Chloroplast DNA phylogeography of *Fagus crenata* (Fagaceae). *Plant Systematics and Evolution* **23**: 21–33.
589. Fujii, T., Maruyama, N., & Kanzaki, N. (1998) Seasonal changes in food habits of Japanese weasel in a middle stream of the Tamagawa River. *Mammalian Science* **38**: 1–8 (in Japanese with English abstract).
590. Fujii, T., et al. (2011) Wild bird and animal which use cavity of black woodpecker *Dryocopus martius*—Why is it that black woodpecker breeding was late? *Wildlife Conservation Japan* **13**: 37–40 (in Japanese with English abstract).
591. Fujimaki, Y. (藤巻裕蔵) (1970) 日本の哺乳類 (9) げっ歯目 アカネズミ属 ヒメネズミ. *Mammalian Science* **10(1)**: 1–11 (in Japanese).
592. Fujimaki, Y. (藤巻裕蔵) (1977) ノネズミ発生予察の変遷. 野ねずみ **142**: 74–75 (in Japanese).
593. Fujimaki, Y. (藤巻裕蔵) & Kuwahata, T. (桑畑勤) (1984) 生活史. *in* Study on Wild Murid Rodents in Hokkaido (北海道野ネズミ類の研究) (Ota, K. (太田嘉四夫), Editor). Sapporo: Hokkaido University Press, pp. 47–76 (in Japanese).
594. Fujimaki, Y. (1974) An albino red-backed vole, *Clethrionomys rufocanus bedofordiae*, from Hokkaido. *The Journal of the Mammalogical Society of Japan* **5**: 194.
595. Fujimaki, Y. & Takeuchi, Y. (1983) Small mammals of an agricultural land in the Tokachi district, Hokkaido. *The Journal of the Mammalogical Society of Japan* **9**: 162–167.
596. Fujimoto, T. (富士元寿彦) (1986) Blue Hare (野ウサギの四季). Tokyo: Heibonsha (平凡社), 71 pp. (in Japanese).
597. Fujimoto, A. & Iwasa, M.A. (2010) Intra- and interspecific nuclear ribosomal gene variation in the two Japanese *Eothenomys* species, *E. andersoni* and *E. smithii*. *Zoological Science* **27**: 907–911.
598. Fujimoto, K., Yamaguti, N., & Takahashi, M. (1986) Ecological studies on ixodid ticks. 1. Ixodid ticks on vegetations and wild animals at the low mountain zone lying south-western part of Saitama Prefecture. *Japanese Journal of Sanitary Zoology* **37**: 325–331 (in Japanese with English summary).
599. Fujimoto, R., et al. (2010) Rearing methods for the Japanese water shrew *Chimarrogale platycephala*. *Journal of agriculture science, Tokyo University of Agriculture* **55**: 219–226 (in Japanese with English summary).
600. Fujimoto, R., et al. (2011) Efficient capture methods for the Japanese water shrew *Chimarrogale platycepha*. *Journal of agriculture science, Tokyo University of Agriculture* **55**: 290–296 (in Japanese with English summary).
601. Fujimoto, T. (2001) The observation records of *Myotis daubentonii* in tree cavity. *Rishiri Studies* **20**: 35–37 (in Japanese with English abstract).
602. Fujino, K. (1960) Immunogenetic and marking approaches to identifying subpopulations of the North Pacific whales. *Scientific Reports of the Whales Research Institute* **15**: 85–142.
603. Fujinoki, M. (藤ノ木正美) (2007) 穴澤隧道（新潟県十日町）のコウモリ調査. *Kashiwazaki City Museum* **21**: 67–80 (in Japanese).
604. Fujinoki, M. (藤ノ木正美) (2009) 清津川流域で確認したカグヤコウモリ *Myotis frater* 保育コロニーと出産保育期の雄の行動. *Kashiwazaki City Museum* **23**: 85–92 (in Japanese).
605. Fujinoki, M. (藤ノ木正美) & Minowa, K. (箕輪一博) (2006) 清津川流域（新潟県）のコウモリ類について—ヒナコウモリ・カグヤコウモリの新たな記録—. *Kashiwazaki City Museum* **20**: 99–104 (in Japanese).
606. Fujisaki, K., et al. (1993) *Lipoptena fortisetosa* (Diptera, Hippoboscidae) collected from the Japanese deer, *Cervus nippon nippon*. *Medical Entomology and Zoology* **44**: 107–108 (in Japanese with English summary).
607. Fujise, Y. (藤瀬良弘), et al. (2004) イワシクジラとニタリクジラ. 鯨研叢書 (Geiken-sosyo) No.11. Tokyo: Institute of Cetacean Research, 168 pp. (in Japanese).
608. Fujise, Y., et al. (2002) Cruise report of the feasibility study of the Japanese Whales Research Program under Special Permit in the western North Pacific–Phase II (JARPN II) in 2001 (Paper SC/54/O16 presented to the IWC Scientific Committee). 51 pp.
609. Fujishita, A., Ohba, T., & Torii, H. (1998) The Japanese squirrels cause damage to the bed logs of shiitake mushrooms. *Forest Pests* **47**: 168–172 (in Japanese).
610. Fujita, H. (藤田博己) (2004) 野兎病 Tularemia. *Modern Media* **50**: 99–103 (in Japanese).
611. Fujita, H., et al. (1981) Ixodid ticks (Acarina: Ixodidae) parasitic on mammals and birds in Saitama and Gunma prefectures, central Japan. I. Host–tick relationships, geographical and vertical distributions, and medical problems. *Annual Report of Ohara General Hospital* **24**: 13–27 (in Japanese with English abstract).
612. Fujiwara, M. (1958) Mammal fauna of the northern mountainous part of Hiroshima prefecture. *Miscellaneus Report of the Hiwa Museum for Natural History* **1**: 1–13 (in Japanese).
613. Fujiwara, M. (1960) The seasonal change of testes in two species of Japanese moles (primary report). *Miscellaneous Report of the Hiwa Museum for Natural History* **3**: 18–23 (in Japanese).
614. Fujiwara, M. (1961) Miscellaneous notes of mammals and birds in the northern part of Hiroshima-Prefecture (1). *Miscellaneus Report of the Hiwa Museum for Natural History* **4**: 27 (in Japanese).
615. Fujiwara, M. (1962) Some examples of pregnant female and the breeding season of Japanese mole, *Mogera kobeae kobeae* Thomas, in Hiroshima Prefecture. *Miscellaneous Report of the Hiwa Museum for Natural History* **5**: 28–29 (in Japanese).
616. Fujiwara, S., et al. (2013) Direct observation of bear myrmecophagy: Relationship between bears' feeding habits and ant phenology. *Mammalian Biology* **78**: 34–40.
617. Fujizuka, H. (藤塚治義) & Minowa, K. (箕輪一博) (2012) 長岡市においてヒナコウモリのコロニーを発見. *Kashiwazaki City Museum* **26**: 59–64 (in Japanese).
618. Fukasawa, K., et al. (2013) Reconstruction and prediction of invasive mongoose population dynamics from history of introduction and management: a Bayesian state-space modeling approach. *Journal of Applied Ecology* **50**: 469–478.
619. Fukase, T., Chinone, S., & Itagaki, H. (1985) *Strongyloides planiceps* (Nematoda: Strongyloididae) in some wild carnivores. *Japanese Journal of Veterinary Science* **47**: 627–632.
620. Fukuda, D., Kamijo, T., & Yasui, S. (2006) Day roosts of parturient Ikonnikov's whiskered bat, *Myotis ikonnikovi* Ognev. *Mammalian Science* **46**: 177–180 (in Japanese with English abstract).
621. Fukue Municipal Board of Education (福江市教育委員会) (1980) 白浜貝塚 福江市文化財調査報告書 第2集. Nagasaki: Fukue Municipal Board of Education, 141 pp. (in Japanese).
622. Fukuhara, R., et al. (2010) Development and introduction of detection dogs

623. Fukui, D. & Agetsuma, N. (2010) Seasonal change in the diet composition of the Asian parti-coloured bat *Vespertilio sinensis*. *Mammal Study* **35**: 227–233.
624. Fukui, D., Agetsuma, N., & Hill, D.A. (2004) Acoustic identification of eight species of bat (Mammalia: Chiroptera) inhabiting forests of southern Hokkaido, Japan: Potential for conservation monitoring. *Zoological Science* **21**: 947–955.
625. Fukui, D., Agetsuma, N., & Hill, D.A. (2007) Bat fauna in the Nakagawa Experimental Forest, Hokkaido University. *Research Bulletin of the Hokkaido University Forests* **64**: 29–36 (in Japanese with English summary).
626. Fukui, D., et al. (2010) Bats in the Wakayama Experimental Forest, Hokkaido University. *Research Bulletin of the Hokkaido University Forests* **67**: 13–23 (in Japanese with English summary).
627. Fukui, D. & Bat Research Group of Centennial Woods Fan Club (2001) Bats in Mt. Yotei and Niseko range, Hokkaido, Japan, No.2. —Seasonal dynamics of Asian parti-colored bat *Vespertilio superans* in and around the Centennial Woods, Kutchan—. *Bulletin of the Otaru Museum* **14**: 133–138 (in Japanese with English abstract).
628. Fukui, D., et al. (2013) Bird predation by the birdlike noctule in Japan. *Journal of Mammalogy* **94**: 657–661.
629. Fukui, D., Hill, D.A., & Matsumura, S. (2012) Maternity roosts and behaviour of the Ussurian tube-nosed bat *Murina ussuriensis*. *Acta Chiropterologica* **14**: 93–104.
630. Fukui, D., et al. (2009) Recent records of bats from south-western Hokkaido. *Bulletin of the Asian Bat Research Institute* **8**: 2–27.
631. Fukui, D., et al. (2011) Effects of treefall gaps created by windthrow on bat assemblages in a temperate forest. *Forest Ecology and Management* **261**: 1546–1552.
632. Fukui, D., et al. (2001) Efficiency of harp trap for capturing bats in boreal broad-leaved forest in Japan. *Eurasian Journal of Forest Research* **3**: 23–26.
633. Fukui, D., et al. (2005) Bat fauna in southwestern Hokkaido, Japan. *Mammalian Science* **45**: 181–191 (in Japanese with English abstract).
634. Fukui, D., et al. (2005) Geographical variation in the cranial and external characters of the little tube-nosed bats, *Murina silvatica* in the Japanese archipelago. *Acta Theriologica* **50**: 309–322.
635. Fukui, D., et al. (2003) New record of the Japanese house-dwelling bat, *Pipistrellus abramus* from Hokkaido. *Mammalian Science* **43**: 39–43 (in Japanese with English abstract).
636. Fukui, D., et al. (2013) Roost and echolocation call structure of the Alashanian pipistrelle *Hypsugo alaschanicus*—First confirmation as a resident species in Japan. *Mammal Study* **38**: 61–66.
637. Fukui, D., Okazaki, K., & Maeda, K. (2009) Diet of three sympatric insectivorous bat species in Ishigaki Island, Japan. *Endangered Species Research* **8**: 117–128.
638. Fukui, D., et al. (2010) The effect of roost environments on roost selection by non-reproductive and dispersing Asian parti-coloured bats *Vespertilio sinensis*. *Mammal Study* **35**: 99–110.
639. Fukui, E., Koganezawa, M., & Yoshizawa, M. (2001) Genetic analysis of Japanese sika deer in Nikko National Park by random amplified polymorphic DNA method. *Animal Science Journal* **72**: 200-206 (in Japanese with English abstract).
640. Fukui, E., Koganezawa, M., & Yoshizawa, M. (2006) Determination of nucleotide sequence of SRY gene in sika deer (*Cervus nippon*). *Animal Science Journal* **77**: 250–252.
641. Fukui, E., Koganezawa, M., & Yoshizawa, M. (2006) Sexing of sika deer *Cervus nippon*, based on detection of SRY genes from DNA extracted from teeth. *Biosphere Conservation* **7**: 83–86.
642. Fukui, E., et al. (2001) Genetic variation in six loci of serum proteins and red blood cell enzymes in Japanese sika deer (*Cervus nippon*) in Nikko National Park. *Animal Science Journal* **72**: 295–302 (in Japanese with English abstract).
643. Fukui, E., Muramatsu, S., & Yoshizawa, M. (2000) Comparison of genetic variation of group-specific component, heomoglobin, plasminogen and transferrin between elk (*Cervus canadensis*) and Japanese sika deer (*Cervus nippon*). *Journal of Animal Genetics* **28**: 13–17 (in Japanese with English abstract).
644. Fukumoto, S., et al. (2000) First record of *Lipoptena fortisetosa* Maa, 1965 (Diptera: Hippoboscidae) from a *Cervus nippon yesoensis* in Hokkaido, northern Japan. *Medical Entomology and Zoology* **51**: 227–230 (in Japanese with English abstract).
645. Fukuoka, S. (福岡昇三) & Kasuya, T. (粕谷俊雄) (2014) 網走の小型捕鯨・一追鯨士の記録. Tokyo: Seibutsukenkyusha (生物研究社), 148 pp. (in Japanese).
646. Fukushi, D., et al. (2001) Molecular cytogenetic analysis of the highly repetitive DNA in the genome of *Apodemus argenteus*, with comments on the phylogenetic relationships in the genus *Apodemus*. *Cytogenetics and Cell Genetics* **92**: 254–263.
647. Fuller, T.K. & Keith, L.B. (1980) Wolf population dynamics and prey relationships in northwestern Alberta. *Journal of Wildlife Management* **44**: 583–601.
648. Funakoshi, K. (船越公威) (1984) テングコウモリ *Murina leucogaster hilgendorfi* (Peters). *Mogura* **11**: 90–91 (in Japanese).
649. Funakoshi, K. (船越公威) (1992) 人間との共存は可能か？イエコウモリ. 週刊朝日百科 動物たちの地球 **8**: 120–121 (in Japanese).
650. Funakoshi, K. (船越公威) (2000) 霧島山および山麓地域のコウモリ相. *Shizen Aigo (自然愛護)* **26**: 1–4 (in Japanese).
651. Funakoshi, K. (船越公威) (2004) 口永楽部島，中之島および宝島に生息する食虫性コウモリ類の新記録. *Shizen Aigo (自然愛護)* **30**: 5–7 (in Japanese).
652. Funakoshi, K. (船越公威) & Irie, T. (入江照雄) (1982) 九州におけるユビナガコウモリの個体群動態—特に大瀬洞を中心として—. *Mogura* **10**: 23–34 (in Japanese).
653. Funakoshi, K. (1977) Ecological studies on the bat fly, *Penicillidia jenynsii* (Diptera: Nycteribiidae), infested on the Japanese long-fingered bat, with special reference to the adaptability to their hosts from the viewpoint of life history. *Japanese Journal of Ecology* **27**: 125–140 (in Japanese with English synopsis).
654. Funakoshi, K. (1986) Maternal care and postnatal development in the Japanese long-fingered bat, *Miniopterus schreibersi fuliginosus*. *The Journal of the Mammalogical Society of Japan* **11**: 15–26.
655. Funakoshi, K. (1988) Habitat selection and population dynamics during the breeding season and in the Natterer's bat, *Myotis nattereri bombinus*. *Regional Studies* **16**: 137–147 (in Japanese with English abstract).
656. Funakoshi, K. (1990) Chiropteran fauna of the Tokara Islands. *Shizen Aigo (自然愛護)* **16**: 3–6 (in Japanese).
657. Funakoshi, K. (1991) Reproductive ecology and social dynamics in nursery colonies of the Natterer's bat, *Myotis nattereri bombinus*. *Mammal Study* **15**: 61–71.
658. Funakoshi, K. (1997) Bat fauna in Miyazaki Prefecture. *Bulletin of the Miyazaki Prefectural Museum of Nature and History* **20**: 17–24 (in Japanese with English abstract).
659. Funakoshi, K. (1998) Notes on the bats and shrews from the islands of Kuchinoerabujima, Yakushima and Tanegashima, Kagoshima Prefecture. *Mammalian Science* **38**: 293–298 (in Japanese with English abstract).
660. Funakoshi, K. (2010) Acoustic identification of thirteen insectivorous bat species from the Kyushu District, Japan. *Mammalian Science* **50**: 165–175 (in Japanese with English abstract).
661. Funakoshi, K., Arai, A., & Inoue, T. (2013) Development of sounds during postnatal growth of the eastern bent-winged bat *Miniopterus fuliginosus*. *Mammal Study* **38**: 49–56.
662. Funakoshi, K., et al. (2012) Food habits of the small Indian mongoose *Herpestes auropunctatus* in Kagoshima City, Japan. *Mammalian Science* **52**: 157–165 (in Japanese with English abstract).
663. Funakoshi, K. & Fukue, Y. (2001) Prenatal and postnatal growth in the greater horseshoe bat, *Rhinolophus ferrumequinum*, in the Kyushu District, Japan. *Mammalian Science* **41**: 171–186 (in Japanese with English abstract).
664. Funakoshi, K., Fukue, Y., & Tabata, S. (1992) Tooth development and replacement in the Japanese greater horseshoe bat, *Rhinolophus ferrumequinum nippon*. *Zoological Science* **9**: 445–450.
665. Funakoshi, K., et al. (in press) First record of the Japanese long-eared bat, *Plecotus sacrimontis*, in the Kyushu District, Japan, with special reference to the variations in their external character, skull, echolocation calls and mitochondrial cytochrome *b* sequences. *Mammalian Science* (in Japanese with English abstract).
666. Funakoshi, K., et al. (2013) Ecology of Ryukyu tube-nosed bat, *Murina ryukyuana*, on Tokunoshima island in Kagoshima Prefecture, Japan. *Nature of Kagoshima* **39**: 1–6 (in Japanese).
667. Funakoshi, K. & Kunisaki, T. (2000) On the validity of *Tadarida latouchei*, with reference to morphological divergence among *T. latouchei*, *T. insignis* and *T. teniotis* (Chiroptera, Molossidae). *Mammal Study* **25**: 115–123.
668. Funakoshi, K. & Kunisaki, T. (2003) Morphometric characters and systematics of the Erabu flying fox, *Pteropus dasymallus dasymallus*. in Erabu Flying Fox (*Pteropus dasymallus dasymallus*), a Report of the Urgent Survey on Natural Monuments of Kagoshima Prefecture. Kamiyaku: Kamiyaku-Town Board of Education (上屋久町教育委員会), pp. 7–11 (in Japanese).
669. Funakoshi, K. & Maeda, F. (2003) Foraging activity and night-roost usage in the Japanese greater horseshoe bat, *Rhinolophus ferrumequinum nippon*. *Mammal Study* **28**: 1–10.
670. Funakoshi, K., et al. (1999) Roost selection, population dynamics and activities of the Oriental free-tailed bat, *Tadarida insignis* on the Islet of

Biroujima, Miyazaki Prefecture. *Mammalian Science* **39**: 23–33 (in Japanese with English abstract).

671. Funakoshi, K., et al. (2009) Studies on the ecology and roosting behavior of Ussuri tube-nosed bat, *Murina ussuriensis*, based on the use of dead-leaf *Mollotus japonicus* trap. *Mammalian Science* **49**: 245–256 (in Japanese with English abstract).
672. Funakoshi, K., et al. (2010) Postnatal growth and vocalization development of the lesser horseshoe bat, *Rhinolophus cornutus*, in the Kyushu District, Japan. *Mammal Study* **35**: 65–78.
673. Funakoshi, K., Osawa, Y., & Osawa, K. (2006) Distribution of the Ryukyu flying-fox (*Pteropus dasymallus inopinatus*) on islands adjacent to Okinawa Island, with special reference to their ecology on Yoron Island. *Mammalian Science* **46**: 29–34 (in Japanese with English abstract).
674. Funakoshi, K., Osawa, Y., & Osawa, K. (2012) First record of the Orii's flying-fox, *Pteropus dasymallus inopinatus* on Okinoerabu-jima Island in Kagoshima Prefecture, Japan, with special reference to their inhabitation. *Mammalian Science* **52**: 179–184 (in Japanese with English abstract).
675. Funakoshi, K., et al. (2013) First record of the forested Myotis, *Myotis pruinosus*, in Kumamoto Prefecture, Japan, with special reference to variations in their external character, skull and mitochondrial cytochrome *b* sequences. *Mammalian Science* **53**: 351–357 (in Japanese with English abstract).
676. Funakoshi, K. & Takeda, Y. (1998) Food habits of sympatric insectivorous bats in southern Kyushu, Japan. *Mammal Study* **23**: 49–62.
677. Funakoshi, K. & Uchida, T.A. (1975) Studies on the physiological and ecological adaptation of temperate insectivorous bats. I. Feeding activities in the Japanese long-fingered bat, *Miniopterus schreibersi fuliginosus*. *Japanese Journal of Ecology* **25**: 217–234 (in Japanese with English synopsis).
678. Funakoshi, K. & Uchida, T.A. (1978) Studies on the physiological and ecological adaptation of temperate insectivorous bats III. Annual activity of the Japanese house-dwelling bat, *Pipistrellus abramus*. *Journal of the Faculty of Agriculture, Kyushu University* **23**: 95–115 (in Japanese).
679. Funakoshi, K. & Uchida, T.A. (1978) Studies on the physiological and ecological adaptation of temperate insectivorous bats. II. Hibernation and winter activity in some cave-dwelling bats. *Japanese Journal of Ecology* **28**: 237–261.
680. Funakoshi, K. & Uchida, T.A. (1980) Feeding activity of the Japanese lesser horseshoe bat, *Rhinolophus cornutus cornutus*, during the hibernation period. *Journal of Mammalogy* **61**: 119–121.
681. Funakoshi, K. & Uchida, T.A. (1981) Feeding activity during the breeding season and postnatal growth in the Namie's frosted bat, *Vespertilio superans superans*. *Japanese Journal of Ecology* **31**: 67–77.
682. Funakoshi, K. & Uchida, T.A. (1982) Age composition of summer colonies in the Japanese house-dwelling bat, *Pipistrellus abramus*. *Journal of the Faculty of Agriculture, Kyushu University* **27**: 55–64 (in Japanese).
683. Funakoshi, K. & Uchida, T.A. (1982) Annual cycles of body weight in the Namie's frosted bat, *Vespertilio superans superans*. *Journal of Zoology* **196**: 417–430.
684. Funakoshi, K., Watanabe, H., & Kunisaki, T. (1993) Feeding ecology of the northern Ryukyu fruit bat, *Pteropus dasymallus dasymallus*, in a warm-temperate region. *Journal of Zoology* **230**: 221–230.
685. Funakoshi, K. & Yamamoto, T. (2001) The first record of the roosts of the Oriental free-tailed bat, *Tadarida insignis* on the Islet of Biroujima, Kochi Prefecture. *Mammalian Science* **41**: 87–92 (in Japanese with English abstract).
686. Furubayashi, K., et al. (1980) Relationships between occurrence of bear damage and clearcutting in central Honshu, Japan. *International Conference on Bear Research and Management* **4**: 80–84.
687. Furukawa, Y. Unpublished.
688. Furuta, M. (古田正美) (2003) 水族館におけるスナメリの飼育と生態研究. *Kaiyo Monthly (月刊海洋)* **35**: 559–564 (in Japanese).
689. Furuta, M. (2014) History of finless porpoises breeding, especially at Toba Aquarium. *Aquabiology* **36**: 169–175 (in Japanese with English abstract).
690. Furuta, M., et al. (1989) Growth of the finless porpoise *Neophocaena phocaenoides* (G. Cuvier, 1829) from the Ise Bay, Central Japan. *Annual Report of Toba Aquarium* **1**: 89–102.
691. Furuta, Y. (2004) Reproduction and external measurements of the Japanese water shrew, *Chimarrogale platycephala*, in the western part of Hyogo Prefecture, Japan. *Bulletin of the Biological Society of Kagawa* **31**: 45–58 (in Japanese).
692. Furuya, Y. (古屋義男) (1974) カワウソの食性 in ニホンカワウソ生息調査報告. Kochi: Kochi Prefectural Board of Education (高知県教育委員会), pp. 24–31 (in Japanese).
693. Furuya, Y. & Yoshimura, N. (1989) Reduction of the range of otter (*Lutra lutra whiteleyi*) in Kochi Prefecture (1977–1987). *Bulletin of Kochi Women's University, Series of Natural Sciences* **37**: 1–11 (in Japanese with English abstract).
694. Fuziwara, M. (藤原仁) (1958) 広島県北部山地の哺乳類. *Miscellaneous Reports of the Hiwa Museum for Natural History* **1**: 1–13 (in Japanese).
695. Fuziwara, M. (1963) Record of a pregnant female of Japanese water shrew, *Chimarrogale platycephala* Temminck. *The Journal of the Mammalogical Society of Japan* **2**: 62–63 (in Japanese).
696. Galbreath, G.J., Hean, S., & Montgomery, S.M. (2000) A new color phase of *Ursus thibetanus* (Mammalia: Ursidae) from Southeast Asia. *Natural History Bulletin of the Siam Society* **49**: 107–111.
697. Gallo-Reynoso, J.P. & Figueroa-Carranza, A.L. (1995) Occurrence of bottlenose whales in the waters of Isla Guadalupe, Mexico. *Marine Mammal Science* **11**: 573–575.
698. Gambell, R. (1968) Seasonal cycles and reproduction in sei whales of the Southern Hemisphere. *Discovery Report* **35**: 31–134.
699. Gambell, R. (1985) Fin whale *Balaenoptera physalus* (Linnaeus, 1758). in Handbook of Marine Mammals. Volume 3. The Sirenians and Baleen Whales (Ridgway, S.H. & Harrison, R.J., Editors). London: Academic Press, pp. 171–192.
700. Gambell, R. (1985) Sei whale *Balaenoptera borealis* (Lesson, 1828). in Handbook of Marine Mammals. Volume 3. The Sirenians and Baleen Whales (Ridgway, S.H. & Harrison, R.J., Editors). London: Academic Press, pp. 155–170.
701. Gannon, W.L., Sikes, R.S., & The Animal Care and Use Committee of the American Society of Mammalogists (2007) Guidelines of the American Society of Mammalogists for the use of wild mammals in research. *Journal of Mammalogy* **88**: 809–823.
702. Gao, A. & Zhou, K. (1993) Growth and reproduction of three populations of finless porpoise, *Neophocaena phocaenoides*, in Chinese waters. *Aquatic Mammals* **19**: 3–12.
703. Gao, A. & Zhou, K. (1995) Geographical variation of external measurements and three subspecies of *Neophocaena phocaenoides* in Chinese waters. *Acta Theriologica Sinica* **15**: 81–92 (in Chinese with English abstract).
704. Gao, A. & Zhou, K. (1995) Geographical variations of skull among the populations of *Neophocaena* in Chinese Waters. *Acta Theriologica Sinica* **15**: 161–169 (in Chinese with English abstract).
705. Garcia, C., Huffman, M., & Shimizu, K. (2010) Seasonal and reproductive variation in body condition in captive female Japanese macaques (*Macaca fuscata*). *American Journal of Primatology* **72**: 277–286.
706. Garshelis, D.L. (1984) Movements and management of sea otters in Alaska. *Journal of Wildlife Management* **48**: 456–463.
707. Garshelis, D.L., Johnson, A.M., & Garshelis, J.A. (1984) Social organization of sea otters in Prince William Sound, Alaska. *Canadian Journal of Zoology* **62**: 2648–2658.
708. Garshelis, D.L. & Steinmetz, R. (2008) *Ursus thibetanus*. The IUCN Red List of Threatened Species. Version 2014.3. Available from: http://www.iucnredlist.org.
709. Gasaway, W.C., et al. (1992) The role of predation in living moose at low densities in Alaska and Yukon and implications for conservation. *Wildlife Monographs* **120**: 1–59.
710. Gaskin, D.E. (1982) The Ecology of Whales and Dolphins. London: Heinemann, 434 pp.
711. Gaskin, D.E., Yamamoto, S., & Kawamura, A. (1993) Harbor porpoise, *Phocoena phocoena* (L.), in the coastal waters of northern Japan. *Fishery Bulletin* **91**: 440–454.
712. Gaspari, S., Airoldi, S., & Hoelzel, A.R. (2007) Risso's dolphins (*Grampus griseus*) in UK waters are differentiated from a population in the Mediterranean Sea and genetically less diverse. *Conservation Genetics* **8**: 727–732.
713. Gatesy, J., et al. (1997) A cladistic analysis of mitochondrial ribosomal DNA from the Bovidae. *Molecular Phylogenetics and Evolution* **7**: 303–319.
714. Gehrt, S.D. (2003) Raccoon *Procyon lotor* and allies. in Wild Mammals of North America : Biology, Management, and Conservation (Feldhamer, G.A., Thompson, B.C., & Chapman, J.A., Editors). Baltimore: Johns Hopkins University Press, pp. 611–634.
715. Geist, V. (1987) On the evolution of the Caprinae. in The Biology and Management of *Capricornis* and Related Mountain Antelopes (Soma, H., Editor). London: Croom Helm, pp. 3–40.
716. Geist, V. (1998) Deer of the World. 1st ed. Mechanicsburg: Stackpole Books, 421 pp.
717. Gelatt, T. & Lowry, L. (2008) *Callorhinus ursinus*. The IUCN Red List of Threatened Species. Version 2014.3. Available from: http://www.iucnredlist.org.
718. Gentry, R.L. (1981) Northern fur seal *Callorhinus ursinus*. in Handbook of

718. ...Marine Mammals. Volume 1. The Walrus, Sea Lions, Fur Seals and Sea Otter (Ridgway, S.H. & Harrison, R.J., Editors). London: Academic Press, pp. 143–160.
719. Gentry, R.L. (1998) Behavior and Ecology of the Northern Fur Seal. New Jersey: Princeton University Press, 376 pp.
720. Gentry, R.L. & Holt, J.R. (1986) Attendance behavior of northern fur seals. *in* Fur Seals: Maternal Strategies on Land and at Sea (Gentry, R.L. & Kooyman, G.L., Editors). New Jersey: Princeton University Press, pp. 41–60.
721. Gentry, R.L., Kooyman, G.L., & Goebel, M.E. (1986) Feeding and diving behavior of northern fur seals. *in* Fur Seals: Maternal Strategies on Land and at Sea (Gentry, R.L. & Kooyman, G.L., Editors). New Jersey: Princeton University Press, pp. 66–78.
722. Geptner, V.G., et al. (1976) The Marine Mammals of the Soviet Union. Volume 2. Moscow: High School Publishers, 718 pp.
723. Gerrodette, T. & Forcada, J. (2005) Non–recovery of two spotted and spinner dolphin populations in the eastern tropical Pacific Ocean. *Marine Ecology Progress Series* **291**: 1–21.
724. Gibb, J.A. (1990) The European rabbit *Oryctolagus cuniculus*. *in* Rabbits, Hares and Pikas: Status Survey and Conservation Action Plan (Chapman, J.A. & Flux, J.E.C., Editors). Gland: IUCN, pp. 116–120.
725. Gilchrist, J.S., et al. (2009) Family Herpestidae. *in* Handbook of the Mammals of the World. Vol. 1. Carnivores (Wilson, D.E. & Mittermeier, R.A., Editors). Barcelona: Lynx Editions, pp. 262–328.
726. Glenn, L.P. & Miller, L.H. (1980) Seasonal movements of an Alaska Peninsula brown bear population. *Proceedings of the International Conference on Bear Research and Management* **4**: 307–312.
727. Gohar, H.A.F. (1957) The Red Sea Dugong, *Dugong dugon* (Erxleben). *Publication of the Marine Biological Station, Al Ghardaqa* **9**: 3–49.
728. Goodman, S.J., et al. (2001) Bottlenecks, drift and differentiation: the population structure and demographic history of Sika deer (*Cervus nippon*) in the Japanese archipelago. *Molecular Ecology* **10**: 1357–1370.
729. Goold, J.C. (1998) Acoustic assessment of populations of common dolphin off the west Wales coast, with perspectives from satellite infrared imagery. *Journal of the Marine Biological Association of the UK* **78**: 1353–1364.
730. Goto, M. (後藤睦夫) (2007) 遺伝的解析に基づくナガスクジラの海洋間の系統関係. 鯨研通信 *(Geiken Tsuushin)* **434**: 1–8 (in Japanese).
731. Goto, M. & Pastene, L.A. (1997) Population structure of the western North Pacific minke whale based on an RFLP analysis of the mtDNA control region. *Report of the International Whaling Commission* **47**: 531– 537.
732. Goto, Y. (1999) Feeding ecology of Steller sea lions. *in* Migratory Ecology and Conservation of Steller Sea Lions (Ohtaishi, N. & Wada, K., Editors). Tokyo: Tokai University Press, pp. 13–58 (in Japanese with English summary).
733. Goto, Y. (1999) Foraging ecology and nutritional condition of Steller sea lions, spotted seal, and ribbon seal. Doctoral dissertation. Hakodate: Hokkaido University. 185 pp. (in Japanese with English summary).
734. Goto, Y. & Shimazaki, K. (1998) Diet of Steller sea lions off the coast of Rausu, Hokkaido, Japan. *Biosphere Conservation* **1**: 141–148.
735. Granger, W., ed. (1938) Natural History of Central Asia. Volume 11. Part 1. New York: American Museum of Natural History, 620 pp.
736. Gray, J.E. (1870) Catalogue of Monkeys, Lemurs, and Fruit-eating Bats in the Collection of the British Museum. London: British Museum, 137 pp.
737. Groves, C.P. & Grubb, P. (1985) Reclassification of the serows and gorals (*Nemorhaedus*: Bovidae). *in* The Biology and Management of Mountain Ungulates (Lovari, S., Editor). London: Croom Helm, pp. 45–50.
738. Groves, P. & Shields, G.F. (1996) Phylogenetics of the Caprinae based on cytochrome *b* sequence. *Molecular Phylogenetics and Evolution* **5**: 467–476.
739. Grubb, P. (1993) Order Artiodactyla. *in* Mammal Species of the World: A Taxonomic and Geographic Reference (Wilson, D.E. & Reeder, D.M., Editors). Washington: Smithsonian Institution Press, pp. 377–414.
740. Grubb, P. (2005) Order Artiodactyla. *in* Mammal Species of the World: A Taxonomic and Geographic Reference (Wilson, D.E. & Reeder, D.M., Editors). Baltimore: Johns Hopkins University Press, pp. 637–722.
741. Grzimek, B. (1975) Grzimek's Animal Life Encyclopedia, Mammals I–IV. New York: Von Nostrand Reinhold.
742. Gulyaev, V.D., et al. (2009) Revised description of *Ecrinolepis ezoensis* (Sawada et Koyasu, 1995) comb. n. (Cestoda, Cyclophyllidea, Ditestolepidini) in shrews from Hokkaido, Japan. *Bulletin of the North-East Scientific Centre of the Far Eastern Branch of Russian Academy of Sciences* **4**: 54–57 (in Russian with English summary).
743. Gureev, A.A. (1964) Fauna of the USSR, mammals, Lagomorpha. *Nauk, Moscow, USSR* **3**: 1–276.
744. Gustavsson, I. (1971) Mitotic and meitic chromosomes of the variable hare (*Lepus timidus* L.), the common hare (*Lepus europaeus* Pall.) and their hybrids. *Hereditas* **67**: 27–34.
745. Ha, J.C., Robinette, R.L., & Sackett, G.P. (2000) Demographic analysis of the Washington Regional Primate Research Center pigtail macaque colony, 1967–1996. *American Journal of Primatology* **52**: 187–198.
746. Habe, S., Ashizawa, H., & Saito, T. (1977) On the lung flukes, *Paragonimus*, found in badgers, a dog and pigs in Japan. *Japanese Journal of Parasitology* **26**: 63–66 (in Japanese with English abstract).
747. Hafez, E.S.E. (1971) Comparative Reproduction of Non-human Primates. Springfield: Charles Thomas, 557 pp.
748. Haga, R. (芳賀良一), Fujimaki, Y. (藤巻裕蔵), & Onoyama, K. (小野山敬一) (1979) 哺乳類. *in* 日高山系自然生態系総合調査報告書 (動物篇). Sapporo: Hokkaido Government, pp. 3–56 (in Japanese).
749. Haga, R. (1960) Observations on the ecology of the Japanese pika. *Journal of Mammalogy* **41**: 200–212.
750. Haga, Y. & Iwasa, M.A. (2014) A note on the presence of B chromosome in the Japanese small field mouse, *Apodemus argenteus*. *Russian Journal of Genetics* **50**: 957–960.
751. Hagen, K.W. (1978) Raising management of rabbit colony. *in* 実験用ウサギの生物学: 繁殖, 疾病と飼育管理 (The Biology of the Laboratory Rabbit) (Weisbroth, S.H., Flatt, R.E., & Kraus, A.L., Editors). Tokyo: Buneido Publishing (文永堂), pp. 31–70 (in Japanese. Translated by 板垣博, 大倉永治, 伊藤昭吾 & 北村佐三郎. Originally published by Academic Press, New York in 1974).
752. Hagiwara, K., et al. (2003) Habitat status and genetic profile of the macaque populations containing alien species in the Bousou Peninsula. *Primate Research* **19**: 229–241 (in Japanese with English summary).
753. Hamada, K., et al. (2002) Changes in body temperature pattern in vertebrates do not influence the codon usages of α-globin genes. *Genes & Genetic Systems* **77**: 197–207.
754. Hamada, Y. (1982) Longitudinal somatometrical study on the growth patterns of newborn Japanese monleys. *Primates* **23**: 542–557.
755. Hamada, Y., et al. (1999) Adolescent growth and development in Japanese macaques (*Macaca fuscata*): Punctuated adolescent growth spurt by season. *Primates* **40**: 439–452.
756. Hamada, Y., et al. (2003) Seasonal variation in the body fat of Japanese macaques *Macaca fuscata*. *Mammal Study* **28**: 79–88.
757. Hamada, Y., Watanabe, T., & Iwamoto, M. (1992) Variation of body color within macaques, especially in the Japanese macaques. *Primate Research* **8**: 1–23 (in Japanese with English summary).
758. Hamada, Y., Watanabe, T., & Iwamoto, M. (1996) Morphological variations among local populations of Japanese macaque (*Macaca fuscata*). *in* Variation in the Asian Macaques (Shotake, T. & Wada, K., Editors). Tokyo: Tokai University Press, pp. 97–115.
759. Hamada, Y., Watanabe, T., & Iwamoto, M. (1996) Physique index for Japanese macaques (*Macaca fuscata*): Age change and regional variation. *Anthropological Science* **104**: 305–323.
760. Hamada, Y., et al. (2012) Variability of tail length in hybrids of the Japanese macaque (*Macaca fuscata*) and the Taiwanese macaque (*Macaca cyclopis*). *Primates* **53**: 397–411.
761. Hamaguchi, T. & Chokki, H. (1985) Notes on Oriental wrinkled-lipped bat *Tadarida insignis* found in Minamiashigara City. *Natural History Report of Kanagawa* **6**: 37–40 (in Japanese).
762. Hamajima, F. (浜島房則) (1969) 日本の哺乳類 (8) げっ歯目 ハツカネズミ属. *Mammalian Science* **9(2)**: 11–23 (in Japanese).
763. Hamajima, F. (1962) Studies on the parasites of the Japanese house mouse, *Mus musculus*. I. Helminth parasites in the mouse living in different environments. *Kyushu Journal of Medical Science* **13**: 319–323.
764. Hamajima, F. (1963) Studies on the parasites of the Japanese house mouse, *Mus musculus*. II. Intestinal protozoa in the mouse living in two different environments and seasons. *Kyushu Journal of Medical Science* **14**: 55–59.
765. Han, S.-H., et al. (2002) Molecular phylogeny of *Crocidura* shrews in northeastern Asia: a special reference to specimens on Cheju Island, South Korea. *Acta Theriologica* **47**: 369–379.
766. Han, S.-H., et al. (1996) Variation of the mitochondrial DNA and the nuclear ribosomal DNA in the striped field mouse *Apodemus agrarius* on the mainland and offshore islands of South Korea. *Mammal Study* **21**: 125–136.
767. Hanai, M. (花井正光) (1980) ツキノワグマの分布について. *in* 第2回自然環境保全基礎調査 動物分布調査報告書(哺乳類) 全国版 (Japan Wildlife Research Center, Editor). Tokyo: Japan Wildlife Research Center, pp. 69–86 (in Japanese).
768. Hanamura, H. (1985) Identification of the murid remains from Shimohieda Site. *in* Shimohieda Site. Yukuhashi: Yukuhashi City Board of Education, pp. 528–531 (in Japanese).
769. Hanamura, H., Ishikawa, A., & Namikawa, T. (1990) Molar size difference between two strains derived from the house musk shrew (*Suncus murinus*, Insectivora) in Bangladesh and Japan. *Journal of Growth* **29**: 227–238 (in

Japanese with English abstract).
770. Hanamura, H. & Kondo, S. (1993) Development of the pulpal floor for the upper first molar in *Suncus murinus* (Soricidae, Insectivora). *Japanese Journal of Oral Biology* **35**: 102–106.
771. Hanamura, H., et al. (1979) A morphological study on dentition of *Suncus murinus riukiuanus. Journal of Growth* **18**: 28–37 (in Japanese with English abstract).
772. Hanamura, H., Shigehara, N., & Oda, S. (1983) Calcification of molar and premolar of the house musk shrew (Insectivora). *Journal of Growth* **22**: 28–43 (in Japanese with English abstract).
773. Hanaoka, T. (1938) Notes on Muridae in the highlands of central Japan. *Zoological Magazine* **49**: 271–281 (in Japanese with English résumé).
774. Hanawa, S., et al. (1980) Ecological survey of Japanese serow, *Capricornis crispus* in Wakinosawa Village. *The Journal of the Mammalogical Society of Japan* **8**: 70–77 (in Japanese with English abstract).
775. Hanihara, K. (1991) Dual structure model for the population history of Japanese. *Japan Review* **2**: 1–33.
776. Hanski, I. (1989) Population biology of Eurasian shrews: towards a synthesis. *Annales Zoologici Fennici* **26**: 469–479.
777. Hanson, R.P. & Karstad, L. (1959) Feral swine in the southern United States. *Journal of Wildlife Management* **23**: 64–74.
778. Hansson, L. & Henttonen, H. (1988) Rodent dynamics as community processes. *Trends in Ecology and Evolution* **3**: 195–200.
779. Hanya, G. (2003) Age differences in food intake and dietary selection of wild Japanese macaques. *Primates* **44**: 333–339.
780. Hanya, G. (2004) Diet of a Japanese macaque troop in the coniferous forest of Yakushima. *International Journal of Primatology* **25**: 55–71.
781. Hanya, G., et al. (2004) Environmental determinants of the altitudinal variations in relative group densities of Japanese macaques on Yakushima. *Ecological Research* **19**: 485–493.
782. Hanya, G., et al. (2003) New method on census primate groups: Estimating group density of Japanese macaques by point census. *American Journal of Primatology* **60**: 43–56.
783. Hao, J., et al. (2007) Serum concentrations of gonadotropins and steroid hormones of *Neophocaena phocaenoides asiaeorientalis* in middle and lower regions of the Yangtze River. *Theriogenology* **67**: 673–680.
784. Harada, T. (原田知子) (2010) 聟島におけるオガサワラオオコウモリ (*Pteropus pselaphon*) の初観察記録. *Annual Report of Ogasawara Studies* **33**: 95–96 (in Japanese).
785. Harada, M. (1973) Chromosomes of nine Chiropteran species in Japan (Chiroptera). *La Kromosomo* **91**: 2885–2895 (in Japanese with English abstract).
786. Harada, M. (1983) Karyological study of the Bonin Flying fox (*Pteropus pselaphon*). *in* The Conservation Reports of the Minami-Iwojima Wilderness Area, Tokyo, Japan. Tokyo: Environment Agency, pp. 243–248 (in Japanese with English abstract).
787. Harada, M. (1988) Karyotypic evolution in the family Vespertilionidae. *Mammalian Science* **28(1)**: 69–83 (in Japanese).
788. Harada, M., et al. (2001) Geographical variations in chromosomes of the greater Japanese shrew-mole, *Urotrichus talpoides* (Mammalia: Insectivora). *Zoological Science* **18**: 433–442.
789. Harada, M., et al. (1987) A karyological study on two Japanese species of *Murina* (Mammalia: Chiroptera). *Journal of the Mammalogical Society of Japan* **12**: 15–23 (in Japanese with English abstract).
790. Harada, M., et al. (1987) Karyotypic evolution of two Japanese *Vespertilio* species and its taxonomic implications (Chiroptera: Mammalia). *Caryologia* **40**: 175–184.
791. Harada, M., Takagishi, N., & Okabana, K. (2006) A population study of three bat species from Shizushi limestone cave, Kyoto Prefecture, Japan. *ANIMATE* **Special Issue 1**: 92–94 (in Japanese).
792. Harada, M. & Torii, H. (1993) Karyological study of the masked palm civet *Paguma larvata* in Japan (Viverridae). *Journal of the Mammalogical Society of Japan* **18**: 39–42.
793. Harada, M. & Uchida, T.A. (1982) Karyotype of a rare species, *Myotis pruinosus* involving pericentirc inversion and duplicated translocation. *Cytologia* **47**: 539–543.
794. Harada, M., et al. (1982) Karyological studies of two Japanese noctule bats (Chiroptera). *Caryologia* **35**: 1–9.
795. Harada, M., et al. (1985) Karyotypes and chromosome banding patterns of a rare leporids species, *Pentalagus furnessi* (Lagomorpha, Leporidae). *Proceedings of the Japan Academy. Series B* **61**: 319–321.
796. Harada, M. & Yosida, T.H. (1978) Karyological study of four Japanese *Myotis* bats (Chiroptera: Mammalia). *Chromosoma* **65**: 283–291.
797. Harada, M., et al. (1985) Cytogenetical studies on Insectivora. III Karyotype comparison of two *Crocidura* species in Japan. *Proceedings of Japan Academy. Series B* **61**: 371–374.

798. Haraguchi, K., Hisamichi, Y., & Endo, T. (2009) Accumulation and mother-to-calf transfer of anthropogenic and natural organohalogens in killer whales (*Orcinus orca*) stranded on the Pacific coast of Japan. *Science of Total Environment* **407**: 2853–2859.
799. Harasawa, R., et al. (2006) Evidence for pestivirus infection in free-living Japanese serow, *Capricornis crispus*. *Microbiology and Immunology* **50**: 817–821.
800. Harashina, K., et al. (2013) Habitat use of the Japanese squirrel, *Sciurus lis*, in a patchy woodland landscape: a case study of Isawa Alluvial Fan, Iwate Prefecture. *Mammalian Science* **53**: 257–266 (in Japanese with English abstract).
801. Hareyama, S., Sato, T., & Shiratsuki, N. (1959) On mammalia in the Sandankyo Gorge and the Yawata Highland. *in* Scientific Researches of the Sandankyo Gorge and the Yawata Highland. Hiroshima: Hiroshima Prefectural Board of Education (広島県教育委員会), pp. 302–305 (in Japanese).
802. Harino, H., et al. (2008) Concentrations of organotin compounds in the stranded killer whales from Rausu, Hokkaido, Japan. *Archives of Environmental Contamination and Toxicology* **55**: 137–142.
803. Harrington, F.H. & Mech, L.D. (1979) Wolf howling and its role in territory maintenance. *Behaviour* **68**: 207–249.
804. Harrington, R. (1973) Hybridisation among deer and its implications for conservation. *Irish Forestry* **30**: 64–78.
805. Harunari, H. (2001) Relationship between the extinction of the big mammals and the human activities at the late Pleistocene in Japan. *Bulletin of the National Museum of Japanese History* **90**: 1–52.
806. Hasegawa, M. (長谷川雅美) (1986) Introduction of the weasel to Miyake-jima and its influences upon native biota. *Collecting and Breeding* **10**: 444–447 (in Japanese).
807. Hasegawa, Y. (長谷川善和) (1963) 日本の哺乳類化石に関する最近の知識. *Mammalian Science* **3(1)**: 11–16 (in Japanese).
808. Hasegawa, Y. (長谷川善和) (1980) 秋吉台の石灰洞と哺乳類化石. *in* 秋吉台の鍾乳洞：石灰洞の科学 (Kawano, M. (河野通弘), Editor). Yamaguchi: 河野通弘教授退官記念事業会, pp. 219–230 (in Japanese).
809. Hasegawa, Y. (長谷川善和) & Nohara, T. (野原朝秀) (1978) 石垣市石城山動物遺骸群集の概要. *in* 沖縄県文化財調査報告書第 15 集 石城山：緊急発掘調査概報. Naha: Okinawa Prefectural Board of Education (沖縄県教育委員会), pp. 49–78 (in Japanese).
810. Hasegawa, H. (1986) Presence of *Syphacia vandenbrueli* Bernard, 1961 (Nematoda: Oxyuridae) in Japan. *Japanese Journal of Parasitology* **35**: 265–267.
811. Hasegawa, H. (1988) *Pradujardinia halicoris* (Owen, 1833) (Nematoda: Ascarididae) collected from a dugong, *Dugong dugon*, of Okinawa, Japan. *Biological Magazine, Okinawa* **26**: 23–25 (in Japanese with English abstract).
812. Hasegawa, H. & Asakawa, M. (2003) Parasitic helminth fauna of terrestrial vertebrates in Japan. *Progress of Medical Parasitology in Japan* **7**: 129–145.
813. Hasegawa, H. & Otsuru, M. (1982) Helminth fauna of the Ryukyu archipelago, Japan. I. *Tenorastrongylus ryukyensis* n. sp. from *Mus caroli* on Okinawa Island (Nematoda: Heligmosomidae). *Zoological Magazine* **91**: 194–196.
814. Hasegawa, H., et al. (2012) Description of *Riouxgolvania kapapkamui* sp. n. (Nematoda: Muspiceoidea: Muspiceidae), a peculiar intradermal parasite of bats in Hokkaido, Japan. *Journal of Parasitology* **98**: 995–1000.
815. Hasegawa, M. (1992) Parasites of the Iriomote cat, *Felis iriomotensis* (III). *Island Studies in Okinawa* **10**: 1–24 (in Japanese with English abstract).
816. Hasegawa, S., et al. (2014) Stranded and by-catch finless porpoises in Ise and Mikawa Bays. *Aquabiology* **36**: 135–141 (in Japanese with English abstract).
817. Hasegawa, T. & Hiraiwa, M. (1980) Social interactions of orphans observed in a free-ranging troop of Japanese monkeys. *Folia Primatologica* **33**: 129–158.
818. Hasegawa, Y. (1964) Discovery of the common otter from Ojika-do (Limestone cave), Hiraodai (Karst plateau), Kyushu, Japan. *The Journal of the Mammalogical Society of Japan* **2**: 82–84 (in Japanese).
819. Hasegawa, Y. (1966) Quaternary smaller mammalian fauna from Japan. *Fossils* **11**: 31–40 (in Japanese).
820. Hasegawa, Y. (1966) Two fumerus remains of common otter from the Shiroyama shell mound of the early Jomon age in Yokosuka city, Japan. *Science Report of the Yokosuka City Museum* **12**: 9–13 (in Japanese).
821. Hasegawa, Y. (1972) The Naumann's elephant, *Palaeoloxodon naumanni* (Makiyama) from the Late Pleistocene off Shakagahana, Shodoshima Is. in Seto Inland Sea, Japan. Doctoral dissertation. Tokyo: The University of Tokyo.
822. Hasegawa, Y. (1980) Notes on vertebrate fossils from Late Pleistocene to Holocene of Ryukyu Islands, Japan. *The Quaternary Research* **18**: 263–267

(in Japanese with English abstract).

823. Hasegawa, Y. & Nohara, T. (1982) Two large tusks of Dugong from Okinawa and Iriomote Islands, Ryukyu Islands. *Science Reports of the Yokohama National University, Section II* **No. 29**: 29–31 with a plate.

824. Hasegawa, Y., et al. (1978) Mammalian fossil assemblage from Gohezy Cave. *in* Survey of Gohezu Cave, Ie Island, Okinawa Prefecture (Second Preliminary Report). Ie: Educational Board of Ie-mura, pp. 8–17, 23–26, pls. 3–12 (in Japanese).

825. Hasegawa, Y., et al. (1988) Quaternary vertebrates from Shiriya area, Shimokita Penninsula, northeastern Japan. *Memoirs of the National Science Museum* **21**: 18–36 (in Japanese with English summary).

826. Hasegawa, Y., Yamauchi, H., & Okafuji, G. (1968) A fossil assemblage of *Macaca* and *Homo* from Ojikado-cave of Hiraodai Karst plateau, Northern Kyusyu, Japan. *Transactions and Proceedings of the Palaeontological Society of Japan, New Series* **69**: 218–229.

827. Hashimoto, M., Shirakihara, K., & Shirakihara, M. (2015) Bycatch effects on the population viability of the narrow-ridged finless porpoises in Ariake Sound and Tachibana Bay, Japan. *Endangered Species Research* **27**: 87–94.

828. Hashimoto, M., et al. (2013) Estimating the rate of increase for the finless porpoise with special attention to predictions for the Inland Sea population in Japan. *Population Ecology* **55**: 441–449.

829. Hashimoto, T. & Abe, M. (2001) Body size and reproductive schedules in two parapatric moles, *Mogera tokudae* and *Mogera imaizumii*, in the Etigo Plain. *Journal of the Mammalogical Society of Japan* **26**: 35–44.

830. Hashimoto, Y. (2003) An ecological study of the Asiatic black bear in the Chichibu Mountains with special reference to food habits and habitat conservation. Doctoral dissertation. University of Tokyo. 97 pp.

831. Hashimoto, Y., et al. (2003) Five-year study on the autumn food habits of the Asiatic black bear in relation to nut production. *Ecological Research* **18**: 485–492.

832. Hashimoto, Y. & Takatsuki, S. (1997) Food habits of Japanese black bears: A review. *Mammalian Science* **37**: 1–19 (in Japanese with English summary).

833. Hassanin, A., Pasquet, E., & Vigne, J.D. (1998) Molecular systematics of the subfamily Caprinae (Artiodactyla, Bovidae) as determined from cytochrome *b* sequences. *Journal of Mammalian Evolution* **5**: 217–236.

834. Hatanaka, H. & Miyashita, T. (1997) On the feeding migration of the Okhotsk Sea–West Pacific stock of minke whales, estimates based on length composition data. *Report of the International Whaling Commission* **47**: 557–567.

835. Hatase, J. (2005) Exhibition and conservation of Oriental free-tailed bat. *Mammalian Science* **45**: 69–72 (in Japanese).

836. Hattori, K. (服部薫) (2008) 漁業被害問題—トドの回遊と消長. *in* Mammalogy in Japan. 3. Marine Mammals (日本の哺乳類学 第 3 巻 水生哺乳類) (Kato, H. (加藤秀弘), Editor). Tokyo: University of Tokyo Press, pp. 254–280 (in Japanese).

837. Hattori, S. (服部正策) (1991) 南西諸島のトガリネズミ科. *Chirimosu (チリモス)* **2**: 17–26 (in Japanese).

838. Hattori, K. Unpublished.

839. Hattori, K. (1966) The insectivorous bat in Hokkaido. *Report of the Hokkaido Institute of Public Health* **16**: 69–77 (in Japanese with English abstract).

840. Hattori, K. (1971) Studies on the Chiroptera in Hokkaido I. Historical review, habitats and species of the Chiroptera in Hokkaido. *Report of the Hokkaido Institute of Public Health* **21**: 68–101 (in Japanese with English abstract).

841. Hattori, K., et al. (2003) Sex determination in the sea otter (*Enhydra lutris*) from tissue and dental pulp using PCR amplification. *Canadian Journal of Zoology* **81**: 52–56.

842. Hattori, K., et al. (2013) Wintering ecology and population management of Steller sea lions (*Eumetopias jubatus*) in Hokkaido. *in* Ecosystem and its Conservation in the Sea of Okhotsk (Sakurai, Y., Ohshima, K.I., & Ohtaishi, N., Editors). Sapporo: Hokkaido University Press, pp. 223–228 (in Japanese with English summary).

843. Hattori, K., et al. (2009) The distribution of Steller sea lions (*Eumetopias jubatus*) in the Sea of Japan off Hokkaido, Japan: A preliminary report. *Marine Mammal Science* **25**: 949–954.

844. Hattori, K., et al. (2005) History and status of sea otters, *Enhydra lutris* along the coast of Hokkaido, Japan. *Mammal Study* **30**: 41–51.

845. Hattori, S., Noboru, Y., & Yamanouchi, K. (1986) Domestication of the Watase's shrew, *Crocidura horsfieldi watasei*, for a laboratory animal. *Japanese Journal of Experimental Medicine* **56**: 75–79.

846. Hattori, S. & Yamanouchi, H. (1984) Gross anatomy of Watase's shrew, *Crocidura horsfieldi watasei*. *Experimental Animals* **33**: 519–524.

847. Hattori, S., et al. (1990) Geographical and ecological distributions and morphological variations of *Crocidura watasei* Kuroda, 1924 from the Amami Islands. *Memoirs of the National Science Museum* **23**: 167–172 (in Japanese with English abstract).

848. Haukisalmi, V. & Hanski, I.K. (2007) Contrasting seasonal dynamics in fleas of the Siberian flying squirrel (*Pteromys volans*) in Finland. *Ecological Entomology* **32**: 333–337.

849. Hauser, N. Personal communication.

850. Hayaishi, S. & Kawamoto, Y. (2006) Low genetic diversity and biased distribution of mitochondrial DNA haplotypes in the Japanese macaque (*Macaca fuscata yakui*) on Yakushima island. *Primates* **47**: 158–164.

851. Hayama, S. (羽山伸一), Uno, H. (宇野裕之), & Wada, K. (和田一雄) (1986) ゼニガタアザラシの回遊様式 (A migration model based on sex and age composition of Kuril seals captured along the coast of Nemuro Peninsula, Hokkaido). *in* ゼニガタアザラシの生態と保護 (Wada, K. (和田一雄), et al., Editors). Tokyo: Tokai University Press, pp. 140–157 (in Japanese).

852. Hayama, H., Kaneda, M., & Tabata, M. (2006) Rapid range expansion of the feral raccoon (*Procyon lotor*) in Kanagawa Prefecture, Japan and its impact on native organisms. *in* Assessment and Control of Biological Invasion Risks (Koike, F., et al., Editors). Gland: IUCN, pp. 196–199.

853. Hayano, A. (早野あづさ) (2009) 採捕したシャチに関する DNA 解析結果. *in* シャチの現状と繁殖研究にむけて—2007 シンポジウムプロシーディングス (Killer Whales in Japan: Proceedings of the Symposium on their Status and Breeding Research in Captivity, November 23, 2007, Tokyo) 鯨研叢書 (Geiken-sosyo) No. 14. Tokyo: Institute of Cetacean Research, pp. 60–64 (in Japanese).

854. Hayano, A., Amano, M., & Miyazaki, N. (2003) Phylogeography and population structure of the Dall's porpoise, *Phocoenoides dalli*, in Japanese waters revealed by mitochondrial DNA. *Genes & Genetic Systems* **78**: 81–91.

855. Hayano, A., et al. (2007) Genetic composition of mass-stranded melon-headed whales (*Peponocephala electra*) on the Japanese coast. *in* 17th Biennial Conference on the Biology of Marine Mammals, Abstracts. Cape Town: Society for Marine Mammalogy, Abstract #291 (CD-ROM).

856. Hayano, A., et al. (2004) Population differentiation in the Pacific white-sided dolphin *Lagenorhynchus obliquuidens* inferred from mitochondrial DNA and microsatellite analysis. *Zoological Science* **21**: 989–999.

857. Hayasaka, K., Fujii, K., & Horai, S. (1996) Molecular phylogeny of macaques: Implications of nucleotide sequences from an 896-base pair region of mitochondrial DNA. *Molecular Biology and Evolution* **13**: 1044–1053.

858. Hayasaka, K., et al. (1988) Phylogenetic relationships among Japanese, rhesus, Formosan, and crab-eating monkeys, inferred from restriction-enzyme analysis of mitochondrial DNAs. *Molecular Biology and Evolution* **5**: 270–281.

859. Hayasaka, K., et al. (1987) Population genetical study of Japanese macaques, *Macaca fuscata*, in Shimokita A1 troop, with special reference to genetic variability and relationships to Japanese macaques in other troops. *Primates* **28**: 507–516.

860. Hayasaki, M. & Oishi, I. (1982) Incidence of canine heartworm, *Dirofilaria immitis*, in wild raccoon dogs in the central area of Japan. *Japanese Journal of Parasitology* **31**: 175–183 (in Japanese with English summary).

861. Hayashi, C. (林知己夫) (1990) 無人島は語る：自然生態観察. Tokyo: Kyoritsu Shuppan (共立出版), 232 pp. (in Japanese).

862. Hayashi, C., Komazawa, T., & Hayashi, F. (1979) A new statistical method to estimate the size of animal population—Estimation of population size of hare—. *Annuals of the Institute of Statistical Mathematics* **31**: 325–348.

863. Hayashi, K., et al. (2006) Genetic variation of the MHC *DQB* locus in the finless porpoise (*Neophocaena phocaenoides*). *Zoological Science* **23**: 147–153.

864. Hayashi, S., Shibata, S., & Kawamichi, T. (2008) Uniparental care and activity of nursing females of *Apodemus argenteus* during the lactation period on Mt. Asama, central Japan. *Mammal Study* **33**: 111–114.

865. Hayashi, Y., Nishida, T., & Mochizuki, K. (1977) Sex and age determination of the Japanese wild boar (*Sus scrofa leucomystax*) by the lower teeth. *Japanese Journal of Veterinary Science* **39**: 165–174 (in Japanese with English summary).

866. Hayashi, Y. & Suzuki, H. (1977) Mammals. *in* Animals of Medical Importance in the Nansei Islands in Japan (Sasa, M., Editor). Tokyo: Shinjuku Shobo, pp. 9–28.

867. Hayashida, M. (1988) The influence of social interactions on the pattern of scatterhoarding in red squirrels. *Bulletin of Hokkaido University, Faculty of Agriculture* **45**: 267–278 (in Japanese).

868. Hayata, I. (1973) Chromosomal polymorphism caused by supernumerary chromosomes in the field mouse, *Apodemus giliacus*. *Chromosoma* **42**: 403–414.

869. Hayata, I. & Shimba, H. (1969) A note on the somatic chromosomes of the Japanese pika, *Ochotona hyperborea yesoensis* Kishida. *Journal of Faculty*

of Science of Hokkaido University, Series VI, Zoology **17**: 393–396.
870. Hayata, I., et al. (1970) Preliminary accounts on the chromosomal polymorphism in the field mouse, *Apodemus giliacus*, a new form from Hokkaido. *Proceedings of Japan Academy* **46**: 567–571.
871. Haydon, D., et al. (2003) Spatio-temporal dynamics of the grey-sided vole in Hokkaido: Identifying coupling using state-based Markov-chain modeling. *Proceedings of the Royal Society of London. Series B* **270**: 435–445.
872. Hays, W.S. & Conant, S. (2007) Biology and impacts of Pacific island invasive species. 1. A Worldwide review of effects of the small Indian mongoose, *Herpestes javanicus* (Carnivora: Herpesitidae). *Pacific Science* **61**: 3–16.
873. Hazumi, T. (1999) Status and management of the Asiatic black bears in Japan. in Bears. Status Survey and Conservation Action Plan (Servheen, K., Herrero, S., & Peyton, B., Compilers). Gland and Cambridge: IUCN/SSC Bear and Polar Bear Specialist Groups, pp. 207–211.
874. Heaney, L.R., et al. (1998) A synopsis of the mammalian fauna of the Philippine Islands. *Fieldiana Zoology, New Series* **88**: 1–61.
875. Heimlich-Boran, J.R. (1993) Social organization of the short-finned pilot whale, *Globicephala macrorhynchus*, with special reference to the comparative social ecology of delphinids. Doctoral dissertation. Cambridge: University of Cambridge. 134 pp.
876. Heinsohn, G.E. & Birch, W.R. (1972) Food and feeding habits of the dugong, *Dugong dugon* (Erxleben), in northern Queensland, Australia. *Mammalia* **36**: 414–422.
877. Heinsohn, G.E., Marsh, H., & Anderson, P.K. (1979) Australian dugong. *Oceans* **12**: 48–52.
878. Helgen, K.M. & Wilson, D.E. (2002) The bats of Flores, Indonesia, with remarks on Asian *Tadarida*. *Breviora* **511**: 1–12.
879. Hemami, M.R., Watkinson, A.R., & Dolman, P.M. (2004) Habitat selection by sympatric muntjac (*Muntiacus reevesi*) and roe deer (*Capreolus capreolus*) in a lowland commercial pine forest. *Forest Ecology and Management* **194**: 49–60.
880. Hemami, M.R., Watkinson, A.R., & Dolman, P.M. (2005) Population densities and habitat associations of introduced muntjac *Muntiacus reevesi* and native roe deer *Capreolus capreolus* in a lowland pine forest. *Forest Ecology and Management* **215**: 224–238.
881. Heptner, V.G. (Гептнер, В.Г.), et al. (1967) Млекопитающие Советского Союза. Том 2. Часть 1. Морские коровы и хищные. Moscow: Vysshaya Shkola (Высшая школа), 1004 pp. (in Russian).
882. Heptner, V.G. & Sludskii, A.A. (1992) Mammals of the Soviet Union. Volume II, part 2. CARNIVORA (Hyenas and Cats). New Delhi: Amerind Publishing, 784 pp.
883. Heuvelmans, B. (1941) Note sur la dentition des Siréniens. III. La dentition du dugong. *Bulletin du Musée Royal d'Histoire Naturelle de Belgique* **17**: 1–14.
884. Heyning, J.E. & Perrin, W.F. (1994) Evidence for two species of common dolphins (genus *Delphinus*) from the eastern North Pacific. *Contribution in Science, Natural History Museum of Los Angeles County* **442**: 1–35.
885. Hidaka, Y., et al. (1995) Pulmonary arterial lesions of *Dirofilaria immitis* in the raccoon dog (*Nyctereutes procyonoides*). *Advances in Animal Cardiology* **28**: 11–16 (in Japanese with English summary).
886. Higa, G. (比嘉源和) (1997) イリオモテヤマネコ「ケイ太」飼育日誌 (Ikehara, S. (池原貞雄), Editor). Ginowan: Okinawa Shuppan (沖縄出版), 125 pp. (in Japanese).
887. Higa, G. (比嘉源和), Katekari, R. (嘉手苅林俊), & Asato, S. (安里巽) (1981) イリオモテヤマネコの飼育経過 (1). *majaa* **1**: 6–12 (in Japanese).
888. Higashi, H. (東英生) (1988) 多摩川河川敷におけるイタチの生息状況の把握並びに行動圏の調査 (ラジオテレメトリー法による): とうきゅう環境浄化財団学術研究助成報告書研究報告 115. Tokyo: Tokyu Foundation for Better Environment (とうきゅう環境浄化財団), 50 pp. (in Japanese).
889. Higashi, N. (東直人) (2001) ザトウクジラ調査. in 海の王者ザトウクジラ II: 日本近海の鯨類基礎調査 1991～2000 (Kings of the Sea. Humpback Whale II) (Uchida, S. (内田詮三), Editor). Nagoya: Tokai Foundation (東海財団), pp. 8–15 (in Japanese).
890. Hikida, T. & Murakami, O. (1980) Age determination of the Japanese wood mouse, *Apodemus speciosus*. *Japanese Journal of Ecology* **30**: 109–116 (in Japanese with English synopsis).
891. Hill, B.D., Fraser, I.R., & Prior, H.C. (1997) Cryptosporidium infection in a dugong (*Dugong dugon*). *Australian Veterinary Journal* **75**: 670–671.
892. Hill, J.E. (1963) A revision of the genus *Hipposideros*. *Bulletin of the British Museum (Natural History)* **2**: 1–129.
893. Hill, J.E. & Yoshiyuki, M. (1980) A new species of *Rhinolophus* (Chiroptera: Rhinolophidae) from Iriomote Island, Ryukyu Islands, with notes on the Asiatic members of the *Rhinolophus pusillus* group. *Bulletin of the National Science Museum, Tokyo, Series A Zoology* **6**: 179–189.
894. Hiraguchi, T. (1988) A dolphin scapula stuck with a stone tool from the Mawaki Site, Noto Peninsula. *Journal of the Anthropological Society of Nippon* **96**: 209.
895. Hirai, T. & Kimura, S. (2004) Diet composition of the common bat *Pipistrellus abramus* (Chiroptera; Vespertilionidae), revealed by fecal analysis. *Japanese Journal of Ecology* **54**: 159–163 (in Japanese with English abstract).
896. Hiraiwa, Y. (平岩馨邦) & Hamajima, F. (浜島房則) (1960) 野棲ハツカネズミの生活史 V. 性的成熟期の到来と生殖機能の衰退および寿命. *Science Bulletin of the Faculty of Agriculture, Kyushu University* **18**: 181–186 (in Japanese).
897. Hirakawa, H. (平川浩文) (2007) 1997年札幌市羊ヶ丘におけるクロテン (*Martes zibellina*) の記録とその意味. *Northern Forestry, Japan* **59**: 101–104 (in Japanese).
898. Hirakawa, H. (平川浩文) (2008) 野幌森林公園におけるクロテン (*Martes zibellina*) の初記録. *Northern Forestry, Japan* **60**: 79–81 (in Japanese).
899. Hirakawa, H. (2001) Coprophagy in leporids and other mammalian herbivores. *Mammalian Review* **31**: 61–80.
900. Hirakawa, H. (2007) Summer roost use of Ussurian tube-nosed bats (*Murina ussuriensis*). *Bulletin of the Asian Bat Research Institute* **6**: 1–7 (in Japanese with English abstract).
901. Hirakawa, H. & Fukui, D. (2008) Predation by a rat snake (*Elaphe climacophora*) on a foliage-roosting bat (*Murina ussuriensis*) in Japan. *Bat Research News* **49**: 37–39.
902. Hirakawa, H. & Fukui, D. (2009) Roost use and activity of an Ussurian tube-nosed bat (*Murina ussuriensis*) during late autumin in Hokkaido. *Bulletin of the Asian Bat Research Institute* **8**: 45–51.
903. Hirakawa, H. & Kawai, K. (2006) Hiding low in the thicket: roost use by Ussurian tube-nosed bats (*Murina ussuriensis*). *Acta Chiropterologica* **8**: 263–269.
904. Hirakawa, H. & Kosaka, I. (2009) A record of an Ussurian tube-nosed bat (*Murina ussuriensis*) found in the snow in early winter and its implication. *Bulletin of Forestry and Forest Products Research Institute* **8**: 175–178 (in Japanese with English abstract).
905. Hirakawa, H., et al. (2010) Can we distinguish between the sable, which is native to Hokkaido, and the Japanese marten, which was introduced to Hokkaido, in photographs? *Mammalian Science* **50**: 145–155 (in Japanese with English summary).
906. Hirakawa, H., et al. (1992) Insular variation of the Japanese hare (*Lepus brachyurus*) on the Oki Island, Japan. *Journal of Mammalogy* **73**: 672–679.
907. Hirama, K., et al. (2004) Phylogenetic analysis of the hemagglutinin (H) gene of canine distemper viruses isolated from wild masked palm civet (*Paguma larvata*). *Journal of Veterinary Medical Science* **66**: 1575–1578.
908. Hirasawa, M., Kanda, E., & Takatsuki, S. (2006) Seasonal food habits of the raccoon dog at a western suburb of Tokyo. *Mammal Study* **31**: 9–14.
909. Hirasawa, S., Nakagawa, R., & Numanami, S. (2008) A mandible of *Hipposideros* (leaf-nosed bat) from Nakabari Cave on Miyako Island, Okinawa Prefecture, Japan. *Journal of the Speleological Society of Japan* **33**: 43–51 (in Japanese with English abstract).
910. Hirata, K. (平田和彦) (2004) コシアカツバメの巣内におけるアブラコウモリの繁殖記録. *Bat Study and Conservation Report* **12**: 2–3 (in Japanese).
911. Hirata, D., et al. (2014) Mitochondrial DNA haplogrouping of the brown bear, *Ursus arctos* (Carnvora: Ursidae) in Asia, based on a newly developed APLP analysis. *Molecular Biology and Evolution* **30**: 1644–1652.
912. Hirata, D., et al. (2013) Molecular phylogeography of the brown bear (*Ursus arctos*) in northeastern Asia based on analysis of complete mitochondrial DNA sequences. *Molecular Biology and Evolution* **30**: 1644–1652.
913. Hirose, K. (広瀬憲也) & Ohashi, N. (大橋直哉) (2008) 東京都墨田区のマンションでヒナコウモリを保護. *Bat Study and Conservation Report* **16**: 13–15 (in Japanese).
914. Hirose, A. (1987) Come across of *Myotis frater kaguyae*. 東北の自然 (*The Nature of Tōhoku*) **26**: 2–3 (in Japanese).
915. Hisamatsu, M. (久松問孝) & Masuda, Y. (増田勇次郎) (1891) 蝙蝠通信. *Zoological Magazine* **3**: 507–508 (in Japanese).
916. Hobbs, R.C. & Jone, L.L. (1993) Impact of high seas driftnet fisheries on marine mammal population in the North Pacific. *Bulletin of the International North Pacific Fishery Commission* **53**: 409–434.
917. Hobbs, R.P. & Samuel, W.M. (1974) Coccidia (Protozoa, Eimeriidae) of the pikas *Ochotona collaris*, *O. princeps*, and *O. hyperborea yesoensis*. *Canadian Journal of Zoology* **53**: 1079–1085.
918. Hoekstra, H.E. & Fagan, W.F. (1998) Body size, dispersal ability and compositional disharmony: the carnivore-dominated fauna of the Kuril Islands. *Diversity and Distributions* **4**: 135–149.
919. Hoelzel, A.R. (1998) Genetic structure of cetacean populations in sympatry, parapatry, and mixed assemblages: Implications for conservation policy. *Journal of Heredity* **89**: 451–458.
920. Hoelzel, A.R., Dahlheim, M., & Stern, S.J. (1998) Low genetic variation among killer whales (*Orcinus orca*) in the eastern North Pacific and genetic

differentiation between foraging specialist. *Journal of Heredity* **89**: 121–128.

921. Hoelzel, A.R., et al. (2007) Evolution of population structure in a highly social top predator, the killer whale. *Molecular Evolution and Biology* **24**: 1407–1415.
922. Hoelzel, A.R., et al. (2002) Low worldwide genetic diversity in the killer whale (*Orcinus orca*): implications for demographic history. *Proceeding of the Royal Society of London B* **269**: 1467–1473.
923. Hoffmann, R.S. (1996) Noteworthy shrews and voles from the Xizang-Qinghai Plateau. *in* Contributions in Mammalogy: A Memorial Volume Honoring Dr. J. Knox Jones, Jr. (Genoways, H.H. & Baker, R.J., Editors). Lubbock, Texas: Museum of Texas Tech University, pp. 155–168.
924. Hoffmann, R.S. & Smith, A.T. (2005) Order Lagomorpha. *in* Mammal Species of the World: A Taxonomic and Geographic Reference (Wilson, D.E. & Reeder, D.M., Editors). Baltimore: Johns Hopkins University Press, pp. 185–193.
925. Hohn, A.A. & Brownell, R.L., Jr. (1990) Harbor porpoise in central Californian waters: life history and incidental catches (Paper SC/42/SM47 presented to the IWC Scientific Committee). 21 pp.
926. Hokkaido Government (1985) 野生動物分布等実態調査報告書 (野生化ミンク). Sapporo: Hokkaido Government, 62 pp. (in Japanese).
927. Hokkaido Government (2001) 北海道の希少野生生物：北海道レッドデータブック 2001. Sapporo: Hokkaido Government, 309 pp. (in Japanese).
928. Hokkaido Government (2015) ヒグマ捕獲数及び被害状況の推移 (in Japanese; http://www.pref.hokkaido.lg.jp/ks/skn/grp/01/higuma/capture-damage 150216.pdf).
929. Hokkaido Institute of Environmental Sciences (1996) ヒグマ・エゾシカ生息実態調査報告書 II. 野生動物分布等実態調査 (ヒグマ：1991～1995年度). Sapporo: Hokkaido Institute of Environmental Sciences, 85 pp. (in Japanese).
930. Hokkaido Institute of Environmental Sciences (2000) ヒグマ・エゾシカ生息実態調査報告書 IV. 野生動物分布等実態調査 (ヒグマ：1991～1998年度). Sapporo: Hokkaido Institute of Environmental Sciences, 118 pp. (in Japanese).
931. Hokkaido Institute of Environmental Sciences (2004) 渡島半島地域ヒグマ対策推進事業調査研究報告書 (1999～2003年度). Sapporo: Hokkaido Institute of Environmental Sciences, 77 pp. (in Japanese).
932. Hokkaido Shimbun Press (北海道新聞社), ed. (2000) 検証 士幌高原道路と時のアセス. Sapporo: Hokkaido Shimbun Press, 248 pp. (in Japanese).
933. Holmala, K. & Kauhala, K. (2006) Ecology of wildlife rabies in Europe. *Mammal Review* **36**: 17–36.
934. Honda, N. (本多宣仁) (2002) コテングコウモリの休息場所. *Bat Study and Conservation Report* **10**: 5 (in Japanese).
935. Honda, M. & Satô, M. (2003) A record of *Murina ussuriensis* from an urban area in Horonobe. *Rishiri Studies* **22**: 9–10 (in Japanese).
936. Honda, T., Suzuki, H., & Itoh, M. (1977) An unusual sex chromosome constitution found in the Amami spinous country-rat, *Tokudaia osimensis osimensis*. *Japanese Journal of Genetics* **52**: 247–249.
937. Honda, T., et al. (1978) Karyotypical differences of the Amami spinous country-rats, *Tokudaia osimensis osimensis* obtained from two neighbouring islands. *Japanese Journal of Genetics* **53**: 297–299.
938. Honma, H. (本間浩昭) (2014) ラッコ：おなじみ…「母親おなかの上に子」根室で初確認. *in* 毎日新聞 (Mainichi Simbun). Tokyo: Mainichi Newspapers (in Japanese).
939. Hope, A.G., et al. (2010) High-latitude diversification within Eurasian least shrews and Alaska tiny shrews (Soricidae). *Journal of Mammalogy* **91**: 1041–1057.
940. Horáček, I. (1997) Status of *Vesperus sinensis* Peters, 1880 and remarks on the genus *Vespertilio*. *Vespertilio* **2**: 59–72.
941. Horáček, I., Hanák, V., & Gaisler, J. (2000) Bats of the Palearctic Region: A taxonomic and biogeographic review. *Proceedings of the VIIIth EBRS. Volume 1, Approaches to Biogeography and Ecology of Bats* **1**: 11–157.
942. Horai, S., et al. (2006) Accumulation of Hg and other heavy metals in the Javan mongoose (*Herpestes javanicus*) captured on Amamioshima Island, Japan. *Chemosphere* **65**: 657–665.
943. Hori, S. (堀繁久) & Matoba, Y. (的場洋平) (2001) 移入種アライグマが捕食していた節足動物. *Bulletin of the Historical Museum of Hokkaido* **29**: 67–76 (in Japanese).
944. Hori, S. (堀繁久) & Mizushima, M. (水島未記) (2002) 野幌森林公園の両生類について. *Bulletin of the Historical Museum of Hokkaido* **30**: 21–26 (in Japanese).
945. Horie, M., et al. (1981) Studies on *Strongyloides* sp. isolated from a cat and raccoon dog. *Japanese Journal of Parasitology* **30**: 215–223 (in Japanese with English summary).
946. Horimoto, T., et al. (2012) Distribution and stranding records of the northern fur seal *Callorhinus ursinus* near Tsugaru Strait during winter and spring 2009. *Nippon Suisan Gakkaishi* **78**: 256–258 (in Japanese).
947. Horwood, J. (1987) The Sei Whale: Population Biology, Ecology & Management. London: Croom Helm, 375 pp.
948. Hoshina, H. & Minowa, T. (2005) Distributional notes of the cavernicolous bats in Fukui Pref., Honshu, Japan. *Bulletin of the Fukui City Museum of Natural History* **52**: 75–82 (in Japanese with English abstract).
949. Hoshino, H. (星野広志) (2004) トドの来遊状況 (The status of the Steller sea lions migrating to Hokkaido). *in* Management of Marine Mammals along the Coast of Hokkaido, Japan (北海道の海生哺乳類管理：シンポジウム「人と獣の生きる海」報告書) (Kobayashi, M. (小林万里), Isono, T. (磯野岳志), & Hattori, K. (服部薫), Editors). Sapporo: Marine Wildlife Center of Japan (北の海の動物センター), pp. 2–5 (in Japanese).
950. Hoshino, H., et al. (2006) Distribution of the Steller sea lion *Eumetopias jubatus* during winter in the northern Sea of Japan, along the west coast of Hokkaido, based on aerial and land sighting surveys. *Fisheries Science* **72**: 922–931.
951. Hoshizaki, K. (星崎和彦) (2009) トチノキ *Aesculus turbinata* Blume. *in* 日本樹木誌 (Watanabe, S. (渡邊定元) & Osumi, K. (大住克博), Editors). Tokyo: J-FIC (日本林業調査会), pp. 497–527 (in Japanese).
952. Hoshizaki, K. & Miguchi, H. (2005) Influence of forest composition on tree seed predation and rodent responses: A comparison of monodominant and mixed temperate forests in Japan. *in* Seed Fate: Predation, Dispersal, and Seedling Establishment (Forget, P.-M., et al., Editors). Wallingford: CABI Publishing, pp. 253–267.
953. Hoslett, S.A. & Imaizumi, Y.-H. (1966) Age structure of a Japanese mole population. *The Journal of the Mammalogical Society of Japan* **2**: 151–156.
954. Hoslett, S.A. & Imaizumi, Y.-H. (1967) Seasonal weight changes in a Japanese mole population. *The Journal of the Mammalogical Society of Japan* **3**: 53–36.
955. Hosoda, T. (細田徹治) & Tatara, M. (鑪雅哉) (1996) テンとエゾクロテン. *in* The Encyclopaedia of Animals in Japan. 1. Mammals I (日本動物大百科 1 哺乳類 I) (Kawamichi, T. (川道武男), Editor). Tokyo: Heibonsha (平凡社), pp. 136–139 (in Japanese).
956. Hosoda, T. & Oshima, K. (1993) Color variation of the fur of Japanese marten (*Martes melampus melampus* WAGNER) in Japan. *Nanki Seibutu* **35**: 19–23 (in Japanese with English abstract).
957. Hosoda, T., et al. (2005) Independent nonframeshift deletions in the *MC1R* gene are not associated with melanistic coat coloration in three mustelid lineages. *Journal of Heredity* **96**: 607–613.
958. Hosoda, T., et al. (2000) Evolutionary trends of the mitochondrial lineage differentiation in species of genera *Martes* and *Mustela*. *Genes & Genetic Systems* **75**: 259–267.
959. Hosoda, T., et al. (1999) Genetic relationships within and between the Japanese marten *Martes melampus* and the sable *M. zibellina*, based on variation of mitochondrial DNA and nuclear ribosomal DNA. *Mammal Study* **24**: 25–33.
960. Hosokawa, S. (2013) Population dynamics of bats during year in Koyo abandoned mine, Gojo-city, Nara Prefecture. *Wild Animals of Kii Peninsula* **10**: 46–50 (in Japanese).
961. Houck, W.J. & Jefferson, T.A. (1999) Dall's porpoise *Phocoenoides dalli* (TRUE, 1885). *in* Handbook of Marine Mammals. Volume 6. The Second Book of Dolphins and the Porpoises (Ridgway, S.H. & Harrison, R., Editors). London: Academic Press, pp. 443–472.
962. Hsu, M.-J., Agoramoorthy, G., & Lin, J.-F. (2001) Birth seasonality and interbirth intervals in free-ranging Formosan macaques, *Macaca cyclopis*, at Mt. Longevity, Taiwan. *Primates* **42**: 15–25.
963. Hsu, M.-J. & Lin, J. (2001) Troop size and structure in free ranging Formosan macaques (*Macaca cyclopis*) at Mt. Longevity, Taiwan. *Zoological Studies* **40**: 49–60.
964. Hsu, M.-J., et al. (2000) High incidence of supernumerary nipples and twins in Formosan macaques (*Macaca cyclopis*) at Mt. Longevity, Taiwan. *American Journal of Primatology* **52**: 199–205.
965. Hsu, M.J. & Agoramoorthy, G. (1997) Wildlife conservation in Taiwan. *Conservation Biology* **11**: 834–836.
966. Hsu, T.C. & Benirschke, K. (1967) An Atlas of Mammalian Chromosomes. Volume 1, Folios 8. Berlin: Springer.
967. Huckstadt, L.A.T. (2001) An observation of parturition in a stranded *Kogia breviceps*. *Marine Mammal Science* **17**: 362–365.
968. Hutterer, R. (2005) Order Soricomorpha. *in* Mammal Species of the World: A Taxonomic and Geographic Reference. (Wilson, D.E. & Reeder, D.M., Editors). Baltimore: Johns Hopkins University Press, pp. 220–311.
969. Hutterer, R. & Zaitsev, M.V. (2004) Cases of homonymy in some Palaearctic and Nearctic taxa of the genus *Sorex* L. (Mammalia: Soricidae). *Mammal Study* **29**: 89–91.
970. Huygens, O.C. & Hayashi, H. (2001) Use of stone pine seeds and oak acorns by Asiatic black bears in central Japan. *Ursus* **12**: 47–50.
971. Huygens, O.C., et al. (2003) Diet and feeding habits of Asiatic black bears in the Northern Japanese Alps. *Ursus* **14**: 236–245.

972. Hwang, M.H. & Garshelis, D.L. (2007) Activity patterns of Asiatic black bears (*Ursus thibetanus*) in the Central Mountains of Taiwan. *Journal of Zoology* **271**: 203–209.
973. Ichihara, T. (1957) An application of liner discriminant function to external measurement of fin whale. *Scientific Reports of the Whales Research Institute* **12**: 127–190.
974. Ichishima, H., et al. (2006) The oldest record of Eschrichtiidae (Cetacea: Mysticeti) from the late Pliocene, Hokkaido, Japan. *Journal of Paleontology* **80**: 367–379.
975. Ichiyanagi, H. (一柳英隆) (2009) けものが川で生きるということ―カワネズミの生活史 (特集 川と生態系 (2) 本州). *in* 季刊河川レビュー (Review), pp. 12–18 (in Japanese).
976. Igarashi, Y. (1980) Studies on the population fluctuation of the Smith's red backed vole, *Eothenomys smithi* (Thomas), in young plantations of Sugi and Hinoki in the central highlands of Shikoku. *Bulletin of the Government Forestry Experiment Station* **311**: 45–64 (in Japanese with English abstract).
977. Igota, H., et al. (2004) Seasonal migration patterns of female sika deer in eastern Hokkaido, Japan. *Ecological Research* **19**: 169–178.
978. Ihara, S., et al. (2013) Estimated density of American mink, *Neovison vison*, in two tributaries of the Abukuma Rivers. *Wildlife Conservation Japan* **14**: 9–13 (in Japanese with English abstract).
979. Iida, S. (1996) Quantitative analysis of acorn transportation by rodents using magnetic locator. *Vegetatio* **124**: 39–43.
980. Iida, T. (1999) Predation of Japanese macaque *Macaca fuscata* by mountain hawk eagle *Spizaetus nipalensis*. *Japanese Journal of Ornithology* **47**: 125–127.
981. Ikeda, K. (池田浩一) (1983) 福岡県上陽町におけるムササビの造林被害. *森林防除 (Forest Pests)* **32**: 86–90 (in Japanese).
982. Ikeda, T. (池田透) (1999) 野幌森林公園におけるアライグマ問題について. *Forest Protection* **272**: 28–29 (in Japanese).
983. Ikeda, T. (池田透) (2000) 移入アライグマをめぐる諸問題. *遺伝 (The Heredity)* **54**: 59–63 (in Japanese).
984. Ikeda, T. (池田透) (2008) 外来種問題―アライグマを中心に (Invasive alien species issues). *in* Mammalogy in Japan. 2. Middle-, and Large-Sized Mammals Including Primates (日本の哺乳類学 第2巻 中大型哺乳類・霊長類) (Takatsuki, S. (高槻成紀) & Yamagiwa, J. (山極寿一), Editors). Tokyo: University of Tokyo Press, pp. 369–400 (in Japanese).
985. Ikeda, T. (池田透) & Yamada, F. (山田文雄) (2011) 海外の外来哺乳類対策先進国に学ぶ. *in* Invasive Alien Mammals in Japan: Biology of Control Strategy and Conservation (日本の外来哺乳類：管理戦略と生態系保全) (Yamada, F. (山田文雄), Ikeda, T. (池田透), & Ogura, G. (小倉剛), Editors). Tokyo: University of Tokyo Press, pp. 59–101 (in Japanese).
986. Ikeda, H. (1982) Socio-ecological study on the raccoon dog, *Nyctereutes procyonoides viverrinus*, with reference to the habitat utilization pattern. Doctoral dissertation. Fukuoka: Kyushu University. 76 pp.
987. Ikeda, H. (1983) Development of young and parental care of raccoon dog, *Nyctereutes procyonoides viverrinus* Temminck, in captivity. *The Journal of the Mammalogical Society of Japan* **9**: 229–236.
988. Ikeda, H. (1984) Raccoon dog scent marking by scats and its significance in social behaviour. *Journal of Ethology* **2**: 77–84.
989. Ikeda, H. (1988) Parapox infection on the Japanese serow. *in* A Fundamental Research on Ecology and Conservation of the Japanese Serow (A Report for the Research Fund by Japan Society for the Promotion of Science) (Ono, Y., Editor). Tokyo: Japan Society for the Promotion of Science, pp. 91–99.
990. Ikeda, H., Eguchi, K., & Ono, Y. (1979) Home range utilization of a raccoon dog, *Nyctereutes procyonoides viverrinus* Temminck, in a small islet in western Kyushu. *Japanese Journal of Ecology* **29**: 35–48.
991. Ikeda, H., et al. (2011) Origin of *Callosciurus erythraeus* introduced into the Uto Peninsula, Kumamoto, Japan, inferred from mitochondrial DNA analysis. *Mammal Study* **36**: 61–65.
992. Ikeda, T. Unpublished.
993. Ikeda, T. (1997) Some aspects of introduced mammals and related problems in Japan: Introduced animals as a factor influencing ecosystem. *Annual Reports on Cultural Science, Hokkaido University* **46**: 195–215 (in Japanese).
994. Ikeda, T. (1999) Progress of naturalization of raccoons and related problems in Hokkaido. *Annual Reports on Cultural Science, Hokkaido University* **47**: 149–175 (in Japanese).
995. Ikeda, T. (2000) Toward fundamental management of invasive raccoons. *Japanese Journal of Conservation Ecology* **5**: 159–170 (in Japanese).
996. Ikeda, T. (2006) Problems controlling the invasive raccoon in Japan. *Mammalian Science* **46**: 95–97 (in Japanese).
997. Ikeda, T., et al. (2004) Present status of invasive alien raccoon and its impact in Japan. *Global Environmental Research* **8**: 125–131.
998. Ikehara, S. (池原貞雄) (1973) 大東島の陸産脊椎動物. *in* 大東島天然記念物特別調査報告. Tokyo: Agency for Cultural Affairs, pp. 52–63 (in Japanese).
999. Imaizumi, Y. (今泉吉晴) & Takashima, Y. (高島幸男) (1974) ニホンカワウソの衰退を辿る―主に四国のカワウソについて. *Biological Science (生物科学)* **26**: 24–29 (in Japanese).
1000. Imaizumi, Y. (今泉吉典) (1949) 日本哺乳動物図説：分類と生態 (The Natural History of Japanese Mammals). Tokyo: Yoyo-shobo (洋々書房), 348 pp. (in Japanese).
1001. Imaizumi, Y. (今泉吉典) (1963) ヤマコウモリの採集と飼育. *The Journal of the Mammalogical Society of Japan* **2**: 64 (in Japanese).
1002. Imaizumi, Y. (今泉吉典) (1973) カワウソ最後の生息地を探る. *アニマ (Anima)* **2**: 5–17 (in Japanese).
1003. Imaizumi, Y. (1953) A new species of *Eptesicus* from Japan (Mammalia; Chiroptera). *Bulletin of the National Science Museum* **33**: 91–95.
1004. Imaizumi, Y. (1954) New species and subspecies of *Sorex* from Japan (Mammalia: Insectivora). *Bulletin of National Science Museum, Tokyo* **1**: 42–48.
1005. Imaizumi, Y. (1954) Taxonomic studies on Japanese *Myotis* with descriptions of three new forms (Mammalia: Chiroptera). *Bulletin of the National Science Museum* **1**: 40–62.
1006. Imaizumi, Y. (1955) On the characters distinguishing *Eptesicus japonensis* from *E. parvus*. *The Journal of the Mammalogical Society of Japan* **1**: 27–28 (in Japanese with English abstract).
1007. Imaizumi, Y. (1955) Systematic notes on the Korean and Japanese bats of *Pipistrellus savii* group. *Bulletin of the National Science Museum* **2**: 55–63.
1008. Imaizumi, Y. (1956) A new species of *Myotis* from Japan (Chiroptera). *Bulletin of the National Science Museum* **3**: 42–46.
1009. Imaizumi, Y. (1957) Taxonomic studies on the red-backed vole from Japan. Part I. Major divisions of the vole and descriptions of *Eothenomys* with a new species. *Bulletin of the National Science Museum* **3**: 195-216.
1010. Imaizumi, Y. (1959) A new bat of the "*Pipistrellus javanicus*" group from Japan. *Bulletin of the National Science Museum* **4**: 363-371.
1011. Imaizumi, Y. (今泉吉典) (1960) Coloured Illustrations of the Mammals of Japan (原色日本哺乳類図鑑). Osaka: Hoikusha (保育社), 196 pp. (in Japanese).
1012. Imaizumi, Y. (1967) A new genus and species of cat from Iriomote, Ryukyu Islands. *The Journal of the Mammalogical Society of Japan* **3**: 75-105.
1013. Imaizumi, Y. (1968) Taxonomic status of the Japanese lesser noctule, *Nyctalus noctula motoyoshii*. *The Journal of the Mammalogical Society of Japan* **4**: 35-39 (in Japanese with English abstract).
1014. Imaizumi, Y. (今泉吉典) (1970) The Handbook of Japanese Land Mammals. Volume 1 (日本哺乳動物図説・上巻). Tokyo: Shin-shicho-sha (新思潮社), 350 pp. (in Japanese with English summary).
1015. Imaizumi, Y. (1970) Systematic status of the extinct Japanese wolf, *Canis hodophilax*. 1. Identification of specimens. *The Journal of the Mammalogical Society of Japan* **5**: 27–32 (in Japanese with English abstract).
1016. Imaizumi, Y. (1970) Systematic status of the extinct Japanese wolf, *Canis hodophilax*. 2. Similarity relationship of *hodophilax* among species of the genus *Canis*. *The Journal of the Mammalogical Society of Japan* **5**: 62–66 (in Japanese with English abstract).
1017. Imaizumi, Y. (1971) A new vole of the *Clethrionomys rufocanus* group from Rishiri Island, Japan. *The Journal of the Mammalogical Society of Japan* **5**: 99–103.
1018. Imaizumi, Y. (1972) Land mammals of the Hidaka mountains, Hokkaido, Japan, with special reference to the origin of an endemic species of the genus *Clethrionomys*. *Bulletin of National Science Museum* **5**: 131–149 (in Japanese with English summary).
1019. Imaizumi, Y. (1979) Taxonomy of muroid rodents from Japan. *Animals and Nature* **9**: 2–6 (in Japanese).
1020. Imaizumi, Y. & Endo, K. (1959) *Barbastella leucomelas darjelingensis*, recently obtained from Iwate Pref., Japan. *The Journal of the Mammalogical Society of Japan* **1**: 127–132 (in Japanese).
1021. Imaizumi, Y. & Yoshiyuki, M. (1965) Taxonomic studies on *Tadarida insignis* from Japan. *The Journal of the Mammalogical Society of Japan* **2**: 105–108 (in Japanese with English resume).
1022. Imaizumi, Y. & Yoshiyuki, M. (1968) A new species of insectivorous bat of the genus *Nyctalus* from Japan. *Bulletin of the National Science Museum* **11**: 123–134.
1023. Imaizumi, Y. & Yoshiyuki, M. (1989) Taxonomic status of the Japanese otter (Carnivora, Mustelidae), with a description of a new species. *Bulletin of the National Science Museum, Tokyo, Series A* **15**: 177–188.
1024. Imaizumi, Y.-H. (1969) Reproduction in the Japanese shrew mole in Niigata Pref., Honshu. *The Journal of the Mammalogical Society of Japan* **4**: 81–86 (in Japanese with English summary).
1025. Imaizumi, Y.-H. (1978) Hunting methods in relation to hunting situations in Japanese shrew-mole, *Urotrichus talpoides*. *Annotations Zoologicae Japonensis* **51**: 245–253.

1026. Imaizumi, Y.-H. (1979) Hunting methods in relation to hunting situations in Japanese shrew-mole, *Urotrichus talpoides*. II. Detection of the earthworm "Head". *Annotations Zoologicae Japonensis* **52**: 212–224.

1027. Imaizumi, Y.-H. & Imaizumi, T. (1970) Interspecific relationship in two mole species in the plains of Niigata, Honshu. I. Geographic distribution. *The Journal of the Mammalogical Society of Japan* **5**: 15–18 (in Japanese with English summary).

1028. Imaizumi, Y.-H. & Imaizumi, T. (1972) Habitat segregation between two species of Japanese shrew-moles. *Zoological Magazine* **81**: 49–55 (in Japanese with English summary).

1029. Imaizumi, Y.-H. & Kubota, K. (1978) Numerical identification of teeth in Japanese shrew-moles, *Urotrichus talpoides* and *Dymecodon pilirostoris*. *Bulletin of Tokyo Medical and Dental University* **25**: 91–99.

1030. Inaba, M. (稲葉慎) (1999) オガサワラオオコウモリの生態 (Ecology of Bonin flying-fox). *in* 天然記念物緊急調査 (オガサワラオオコウモリ). Tokyo: Ogasawara Village Educational Commission (小笠原村教育委員会), pp. 29–40 (in Japanese).

1031. Inaba, M., et al. (2005) Food habits of Bonin flying foxes, *Pteropus pselaphon*, Layard 1829 on the Ogasawara (Bonin) Islands, Japan. *Ogasawara Research* **30**: 15–23.

1032. Inaba, M., et al. (2002) An urgent appeal for conservation of the Bonin flying fox, *Pteropus pselaphon* Layard, an endangered species. *Japanese Journal of Conservation Ecology* **7**: 51–61 (in Japanese with English abstract).

1033. Inada, S. (2007) A case of *Myotis yanbarensis* preyed by *Nehila pilipes* (Febricius 1793). *Study Report on Bat Conservetin in Tohoku Region* **1**: 20 (in Japanese).

1034. Inada, T. & Kawamura, Y. (2004) Middle Pleistocene cave sediments and their mammalian fossil assemblage discovered at Tarumi, Niimi, Okayama Prefecture, western Japan. *The Quaternary Research* **43**: 331–344 (in Japanese with English abstract).

1035. Inagaki, H. (1985) A preliminary study on hair length in the Japanese monkey (*Macaca fuscata fuscata*). *Primates* **26**: 334–338.

1036. Inagaki, H. (1986) Morphological characteristics of the hair of Japanese monkeys (*Macaca fuscata fuscata*): Length, diameter and shape in cross-section, and rearrangement of the medulla. *Primates* **27**: 115–123.

1037. Inokuma, H., et al. (2002) Tick infestation of sika deer (*Cervus nippon*) in the western part of Yamaguchi Prefecture, Japan. *Journal of Veterinary Medical Science* **64**: 615–617.

1038. Inoshima, Y. (2013) Parapoxvirus infection in wild Japanese serows (*Capricornis crispus*). *Journal of the Japan Veterinary Medical Association* **66**: 557–563 (in Japanese).

1039. Inoue, T. (井上忠行) (1990) オオアシトガリネズミ (*Sorex unguiculatus* Dobson) における社会構造と生活史に関する研究: 特に若齢個体における出生地からの移動分散の性差について. Doctoral dissertation. Sapporo: Hokkaido University. 134 pp. (in Japanese).

1040. Inoue, I. (1989) *Eimeria capricornis* n. sp., *E. nihonis* n. sp., *E. naganoensis* n. sp., and *E. kamoshika* n. sp. (Protozoa: Eimeriidae) from the Japanese serow, *Capricornis crispus*. *Japanese Journal of Veterinary Science* **51**: 163–168.

1041. Inoue, I., et al. (1990) Prevalence of *Sarcocytis* (Protozoa, Apicomplexa) in voles in Japan. *Japanese Journal of Parasitology* **39**: 415–417.

1042. Inoue, M., et al. (1991) Male mating behavior and paternity discrimination by DNA fingerprinting in a Japanese macaque group. *Folia Primatologica* **56**: 202–210.

1043. Inoue, T. (1972) The food habit of Tsushima leopard cat, *Felis bengalensis* ssp., analysed from their scats. *The Journal of the Mammalogical Society of Japan* **5**: 155–169 (in Japanese with English summary).

1044. Inoue, T. (1988) Territory establishment of young big-clawed shrew, *Sorex unguiculatus* (Dobson) (Insectivora, Soricidae). *Researches on Population Ecology* **30**: 83–93.

1045. Inoue, T. (1991) Sex difference in spatial distribution of the big-clawed shrew *Sorex unguiculatus*. *Acta Theriologica* **36**: 229–237.

1046. Inoué, T. (1995) Approaches to reconstruct Japanese sea lions (11), Japanese sea lions in Take-shima Island located in the Japan Sea—Part2 Numerical transition of captured animals. *Aquabiology* **17**: 41–46 (in Japanese with English abstract).

1047. Inoue, T., et al. (2012) Genetic population structure of the masked palm civet *Paguma larvata*, (Carnivora: Veverridae) in Japan, revealed from analysis of newly identified compound microsatellites. *Conservation Genetics* **13**: 1095–1107.

1048. Inoue, T. & Maekawa, K. (1990) Difference in diets between two species of soricine shrews, *Sorex unguiculatus* and *S. caecutiens*. *Acta Theriologica* **35**: 253–260.

1049. Inoue, T., et al. (2010) Mitochondrial DNA control region variations in the sable *Martes zibellina* of Hokkaido Island and the Eurasian Continent, compared with the Japanese marten *M. melampus*. *Mammal Study* **35**: 145–155.

1050. Inoue, T., et al. (2007) Mitochondrial DNA phylogeography of the red fox (*Vulpes vulpes*) in northern Japan. *Zoological Science* **24**: 1178–1186.

1051. Institute of Boninology (小笠原自然文化研究所) (2012) オガサワラオオコウモリ保全調査委託報告書 2012 年 (Ogasawara Islands Branch Office (東京都小笠原支庁), Editor). Ogasawara: Ogasawara Islands Branch Office, 199 pp. (in Japanese).

1052. Institute of Boninology (小笠原自然文化研究所) (2013) オガサワラオオコウモリ保全調査委託報告書 2013 年 (Ogasawara Islands Branch Office (東京都小笠原支庁), Editor). Ogasawara: Ogasawara Islands Branch Office, 264 pp. (in Japanese).

1053. Institute of Boninology (小笠原自然文化研究所) (2014) オガサワラオオコウモリ保全調査委託報告書 2014 年 (Ogasawara Islands Branch Office (東京都小笠原支庁), Editor). Osawara: Ogasawara Islands Branch Office, 211 pp. (in Japanese).

1054. Institute of Cetacean Research (2011) 座礁した鯨等の情報 (ストランディングレコード). Available from: http://icrwhale.org/zasho.htm (in Japanese).

1055. Inukai, T. (犬飼哲夫) (1933) 北海道産狼とその滅亡経路. *Botany and Zoology* **1**: 1091–1098 (in Japanese).

1056. Inukai, T. (犬飼哲夫) (1934) 貂の北海道内侵入径路とその利用. *Plants and Animals* **2**: 1309–1317 (in Japanese).

1057. Inukai, T. (犬飼哲夫) (1974) わが動物記. Tokyo: Kurashi-no-techo (暮しの手帖社), 353 pp. (in Japanese).

1058. Inukai, T. (1931) A food-hoard of *Ochotona* from Taisetsuzan, the central mountains of Hokkaido. *Transactions of Sapporo Natural History Society* **11**: 210–214.

1059. Inukai, T. (1949) The result of introduction of the Japanese mink into the adjacent islands of Hokkaido with the purpose of rat control. *Transaction of Sapporo Natural History Society* **18**: 56–59 (in Japanese with English abstract).

1060. Inukai, T. & Kadosaki, M. (1979) Observation on the hibernation den of the Yeso brown bear (*Ursus arctos yesoensis*). *Journal of the Mammalogical Society of Japan* **7**: 280–299 (in Japanese with English abstract).

1061. Inukai, T. & Mukasa, K. (1934) Über den Zhanweschsel des Yezo Braunbaren, *Ursus arctos yesoensis* Lyd. *Folia Anatomica Japonica* **112**: 291–300 (in German).

1062. Inukai, T. & Shimakura, K. (1930) On *Ochotona*, a new rodent unrecorded from Hokkaido. *Transactions of Sapporo Natural History Society* **11**: 115–118.

1063. Invasive Species Specialist Group (2000) 100 of the World's Worst Invasive Alien Species: A Selection from the Global Invasive Species Database. Auckland: Invasive Species Specialist Group, 10 pp. (http://www.issg.org/database/species/reference_files/100English.pdf).

1064. Irie, T. (入江照雄) (1982) 九州中・南部におけるコウモリ類の動態調査 (II) —新地の穴のコウモリ—. 宇土半島 自然と文化 **2**: 105–112 (in Japanese).

1065. Iriomote Wildlife Conservation Center. Personal communication.

1066. Iriomote-Island Natural History Research Association. Unpublished.

1067. Ishibashi, O., et al. (2006) Distribution of *Leptospira* spp. on the small Asian mongoose and the roof rat inhabiting the northern part of Okinawa Island. *Japanese Journal of Zoo and Wildlife Medicine* **11**: 35–41 (in Japanese with English abstract).

1068. Ishibashi, O., et al. (2009) Survey of parasitic Ixodid ticks on small Asian mongoose on Okinawajima island, Japan. *Japanese Journal of Zoo and Wildlife Medicine* **14**: 51–57 (in Japanese with English abstract).

1069. Ishibashi, O., et al. (2009) Survey of parasitic fleas on small Asian mongoose on Okinawajima island, Japan. *Japanese Journal of Zoo and Wildlife Medicine* **14**: 67–72 (in Japanese with English abstract).

1070. Ishibashi, O., et al. (2007) Salmonella research for livestock and feral mammals on Okinawa Island. *Science Bulletin of the Faculty of Agriculture, University of the Ryukyus* **54**: 47–52 (in Japanese with English abstract).

1071. Ishibashi, Y. & Saitoh, T. (2004) Phylogenetic relationships among fragmented Asian black bear (*Ursus thibetanus*) populations in western Japan. *Conservation Genetics* **5**: 311–323.

1072. Ishibashi, Y. & Saitoh, T. (2008) Effect of local male density on the occurrence of multi-male mating in the gray-sided vole (*Myodes rufocanus*). *Journal of Mammalogy* **89**: 388–397.

1073. Ishibashi, Y. & Saitoh, T. (2008) Role of male-biased dispersal in inbreeding avoidance in the grey-sided vole (*Myodes rufocanus*). *Molecular Ecology* **17**: 4887–4896.

1074. Ishibashi, Y., et al. (1995) Polymorphic microsatellite DNA markers in the grey red-backed vole *Clethrionomys rufocanus bedfordiae*. *Molecular Ecology* **4**: 127–128.

1075. Ishibashi, Y., et al. (1997) Cross-species amplification of microsatellite DNA in Old World microtine rodents with PCR primers for the grey-sided vole *Clethrionomys rufocanus*. *Mammal Study* **22**: 5–10.

1076. Ishibashi, Y., et al. (1997) Sex-related spatial kin structure in a spring popu-

REFERENCES

lation of grey-sided voles *Clethrionomys rufocanus* as revealed by mitochondrial and microsatellite DNA analyses. *Molecular Ecology* **6**: 63–71.

1077. Ishibashi, Y., et al. (1998) Kin-related social organization in a winter population of the vole *Clethrionomys rufocanus*. *Researches on Population Ecology* **40**: 51–59.

1078. Ishibashi, Y., Saitoh, T., & Kawata, M. (1998) Social organization of the vole *Clethrionomys rufocanus* and its demographic and genetic consequences: A review. *Researches on Population Ecology* **40**: 39–50.

1079. Ishibashi, Y., Zenitani, J., & Saitoh, T. (2013) Male-biased dispersal causes intersexual differences in the subpopulation structure of the gray-sided vole. *Journal of Heredity* **104**: 718–724.

1080. Ishida, K., Miyashita, T., & Yamada, F. (2003) Ecosystem management considering community dynamics, an example of Amami-oshima Island. *Japanese Journal of Conservation Ecology* **8**: 159–168 (in Japanese with English abstract).

1081. Ishida, K., et al. (2013) Evolutionary history of the sable (*Martes zibellina brachyura*) on Hokkaido inferred from mitochondrial *Cytb* and nuclear *Mc1r* and *Tcf25* gene sequences. *Acta Theriologica* **58**: 13–24.

1082. Ishida, K., Tani, S., & Itagaki, H. (1982) Studies on *Pygidiopsis summa* Onji et Nishio, 1916 in Akita Prefecture. *Bulletin of Azabu University Veterinary Medicine* **3**: 77–84.

1083. Ishida, M. (2012) Ecological studies of the Hilgendorf's tube-nosed bat, *Murina hilgendorfi*, and the greater horseshoe bat, *Rhinolophus ferrumequinum*, in the Akiyoshi-dai Plateau, Yamaguchi Prefecture, Japan. Doctoral dissertation. Yamaguchi: Yamaguchi University. 104 pp. (in Japanese with English summary).

1084. Ishida, M., et al. (2014) Echolocation calls of bats inhabiting Rishiri Island, Hokkaido, Japan. *Rishiri Studies* **33**: 77–81 (in Japanese with English abstract).

1085. Ishida, M., et al. (2010) Usage frequency of night roosts and diet preference of the greater horseshoe bat *Rhinolophus ferrumequinum* in the Akiyoshi-dai Plateau, Yamaguchi, Japan. *Journal of the Speleological Society of Japan* **35**: 11–17.

1086. Ishida, M., et al. (2012) Population dynamics and long-term survival of Hilgendorf's tube-nosed bat *Murina hilgendorfi* in the Akiyoshi-dai karst area, Yamaguchi, Japan. *Mammal Study* **37**: 249–253.

1087. Ishiguro, N. (石黒直隆) & Horiuchi, M. (堀内広) (2003) 絶滅した日本のオオカミの遺伝的特徴と系統解析 (The genetic characters and the molecular phylogenetic analysis of the extinct wolf in Japan), *in* 第135回日本獣医学会学術集会 (The 135th Meeting of the Japanese Association of Veterinary Science). p. 55 (in Japanese).

1088. Ishiguro, F., et al. (1994) Diversity of *Borrelia* isolates found in rodent-tick relationship in Fukui Prefecture and some additional areas. *Japanese Journal of Sanitary Zoology* **45**: 141–145 (in Japanese with English summary).

1089. Ishiguro, F. & Takada, Y. (1996) Lyme *Borrelia* from *Ixodes persulcatus* and small rodents from northern to central parts of mainland Japan. *Medical Entomology and Zoology* **47**: 183–185.

1090. Ishiguro, N., et al. (2010) Osteological and genetic analysis of the extinct Ezo wolf (*Canis lupus hattai*) from Hokkaido Island, Japan. *Zoological Science* **27**: 320–324.

1091. Ishiguro, N., et al. (2002) A genetic method to distinguish crossbred Inobuta from Japanese wild boars. *Zoological Science* **19**: 1313–1319.

1092. Ishii, K. (石井健太), Yamagawa, H. (柳川久), & Nakajima, H. (中島宏章) (2008) コウモリ類にとっての防風林の有用性について. 「野生生物と交通」研究発表会講演論文集 **7**: 61–72 (in Japanese).

1093. Ishii, N. (石井信夫) (1983) 南硫黄島の哺乳類. *in* 南硫黄島の自然 (Ministry of the Environment, Editor). Tokyo: Japan Wildlife Research Center, pp. 225–242 (in Japanese).

1094. Ishii, K. & Yanagawa, H. (2012) Foraging habitats of northern bats (*Eptesicus nilssonii*) in the agricultural areas of the Tokachi district, Hokkaido, Japan. *Journal of the Japanese Wildlife Research Society* **37**: 17–26 (in Japanese with English summary).

1095. Ishii, N. Unpublished.

1096. Ishii, N. (1982) Reproductive activity of Japanese shrew-mole, *Urotrichus talpoides* Temminck. *The Journal of the Mammalogical Society of Japan* **9**: 25–36.

1097. Ishii, N. (1993) Size and distribution of home ranges of the Japanese shrew-mole *Urotrichus talpoides*. *Journal of the Mammalogical Society of Japan* **18**: 87–98.

1098. Ishii, N. (2003) Controlling mongooses introduced to Amami-Oshima Island: A population estimate and program evaluation. *Japanese Journal of Conservation Ecology* **8**: 73–82 (in Japanese with English abstract).

1099. Ishii, N. (2005) Amur hedgehog. *in* A Guide to the Mammals of Japan (Abe, H., Editor). Hadano: Tokai University Press, p. 4 (in Japanese and English).

1100. Ishii, N. (2005) Primates. *in* A Guide to the Mammals of Japan (Abe, H., Editor). Hadano: Tokai University Press, pp. 65–70.

1101. Ishii, N. (2005) Eurasian red squirrel *Sciurus vulgaris* Linnaeus, 1758. *in* A Guide to the Mammals of Japan (Abe, H., Editor). Hadano: Tokai University Press, p. 118.

1102. Ishii, N. (2005) Eurasian red squirrel *Sciurus lis* Temminck, 1844. *in* A Guide to the Mammals of Japan (Abe, H., Editor). Hadano: Tokai University Press, p. 119.

1103. Ishii, N. (2005) Pallas's squirrel *Callosciurus erythraeus* (Pallas, 1778). *in* A Guide to the Mammals of Japan (Abe, H., Editor). Hadano: Tokai University Press, p. 120.

1104. Ishii, N. (2005) Siberian chipmunk *Tamias sibiricus* (Laxmann, 1769). *in* A Guide to the Mammals of Japan (Abe, H., Editor). Hadano: Tokai University Press, p. 121.

1105. Ishii, N. (2005) Japanese giant flying squirrel *Petaurista leucogenys* (Temminck, 1827). *in* A Guide to the Mammals of Japan (Abe, H., Editor). Hadano: Tokai University Press, p. 122.

1106. Ishii, N. (2005) Japanese flying squirrel *Pteromys momonga* Temminck, 1844. *in* A Guide to the Mammals of Japan (Abe, H., Editor). Hadano: Tokai University Press, p. 123.

1107. Ishii, N. (2005) Japanese flying squirrel *Pteromys volans* (Linnaeus, 1758). *in* A Guide to the Mammals of Japan (Abe, H., Editor). Hadano: Tokai University Press, p. 124.

1108. Ishii, N. (2005) Northern Pika *Ochotona hyperborea* (Pallas, 1811). *in* A Guide to the Mammals of Japan (Abe, H., Editor). Hadano: Tokai University Press, p. 148 (in Japanese and English).

1109. Ishii, N. (2005) Japanese hare. *in* A Guide to the Mammals of Japan (Abe, H., Editor). Hadano: Tokai University Press, p. 151 (in Japanese and English).

1110. Ishikawa, H. (石川創) (1994) 日本沿岸のストランディングレコード (1901～1993) 鯨研叢書 (Geiken-sosyo) No.6 (Stranding Records in the Coast of Japan, 1901–1993). Tokyo: Institute of Cetacean Research, 94 pp. (in Japanese).

1111. Ishikawa, H. (石川創) (1998) 1997年度北西北太平洋鯨類捕獲調査航海記. 鯨研通信 (Geiken Tsuushin) **399**: 6–15 (in Japanese).

1112. Ishikawa, H. (石川創) (2014) ストランディングレコード (2013年収集) (Stranding Record in Japan 2013). *in* 下関鯨類研究室報告 No.2 (Shimonoseki Marine Science Report No.2). Shimonoseki: Whale Laboratory (下関鯨類研究室), pp. 21–43 (in Japanese).

1113. Ishikawa, H. (石川創), Goto, M. (後藤睦夫), & Mogoe, T. (茂addy敏弘) (2013) 日本沿岸のストランディングレコード：1901–2012 (Stranding Record in Japan 1901–2012). 下関鯨類研究室報告 No.1 (Shimonoseki Marine Science Report No.1). Shimonoseki: Whale Laboratory (下関鯨類研究室), 314 pp. (in Japanese; http://whalelab.org/stranding.html).

1114. Ishikawa, S. (石川慎也) (2004) 北海道襟裳岬におけるラッコ (*Enhydra lutris*) の生息について. *Bulletin of the Erimo Town Museum* **1**: 15–19 (in Japanese).

1115. Ishikawa, A. & Namikawa, T. (1987) Postnatal growth and development in laboratory strains of large and small musk shrews (*Suncus murinus*). *Journal of Mammalogy* **68**: 766–774.

1116. Ishikawa, A. & Namikawa, T. (1991) The inheritance of body weight in two strains of large and small musk shrews (*Suncus murinus*): an estimate of the minimum number of genes for the weight difference. *Journal of Growth* **30**: 197–206.

1117. Ishikawa, A. & Namikawa, T. (1991) Postanatal growth pattern of F1 hybrids of a cross between two strains of large and small musk shrews, *Suncus murinus*. *Experimental Animal* **40**: 223–230.

1118. Ishikawa, A., Yamagata, T., & Namikawa, A. (1991) An attempt at reciprocal crosses between laboratory strains of large and small musk shrews (*Suncus murinus*)—influence of body-weight difference between sexes on mating success—. *Experimental Animal* **40**: 145–152.

1119. Ishikawa, A., Yamagata, T., & Namikawa, A. (1995) Relationships between morphometric and mitochondrial DNA differentiation in laboratory strains of musk shrews (*Suncus murinus*). *Japanese Journal of Genetics* **70**: 57–74.

1120. Ishikawa, J. (1958) The lime cave and their faunae in the Kyushu District. *Bulletin of Kochi Women's University. Series of Natural Sciences* **2**: 7–22 (in Japanese with English abstract).

1121. Ishinazaka, T. (石名坂豪) (2000) 知床のトド・アザラシ (Steller sea lions and seals of Shiretoko). *in* 知床のほ乳類 I (Shiretoko Museum (斜里町立知床博物館), Editor). Sapporo: Hokkaido Shimbun Press, pp. 164–205 (in Japanese).

1122. Ishinazaka, T. Unpublished.

1123. Ishinazaka, T. (1999) Reproductive aspects and population dynamics of Steller sea lions. *in* Migratory Ecology and Conservation of Steller Sea Lions (Ohtaishi, N. & Wada, K., Editors). Tokyo: Tokai University Press, pp. 59–77 (in Japanese with English summary).

1124. Ishinazaka, T. (2002) The physiology and life-history parameters of reproduction in Steller sea lions, spotted seals, and ribbon seals from the coastal waters of Hokkaido. Doctoral dissertation. Sapporo: Hokkaido University. 82 pp. (in Japanese with English summary).

1125. Ishinazaka, T. & Endo, T. (1999) The reproductive status of Steller sea lions in the Nemuro Strait, Hokkaido, Japan. *Biosphere Conservation* **2**: 11–19.

1126. Ishinazaka, T., et al. (2009) Wintering of Steller sea lions *Eumetopias jubatus* at the resting sites along the coast of Rausu Town, Shiretoko Peninsula, Japan in 2006–07 and 2007–08. *Bulletin of the Shiretoko Museum* **30**: 27–53 (in Japanese with English abstract).

1127. Isobe, Y. (磯部ゆう), Masuda, A. (益田敦子), & Oishi, T. (大石正) (1998) 坑道時間秩序を決める生物時計. *The Heredity (遺伝)* **52**: 15–20 (in Japanese).

1128. Isobe, T., et al. (2008) Organohalogen contaminants in striped dolphins (*Stenella coeruleoalba*) from Japan: present contamination status, body distribution and temporal trends (1978–2003). *Marine Pollution Bulletin* **58**: 396–401.

1129. Isobe, T., et al. (2011) Contamination status of POPs and BFRs and relationship with parasitic infection in finless porpoises (*Neophocaena phocaenoides*) from Seto Inland Sea and Omura Bay, Japan. *Marine Pollution Bulletin* **63**: 564–571.

1130. Isono, T. (磯野岳臣) (2000) Comparative morphology in growth pattern and geographical variation of Steller sea lions (トド *Eumetopias jubatus* の成長様式, 成長量の経年変化および地理的変異に関する比較形態学的研究). Doctoral dissertation. Hakodate: Hokkaido University. 123 pp. (in Japanese).

1131. Isono, T., et al. (2010) Resightings of branded Steller sea lions at wintering haul-out sites in Hokkaido, Japan 2003–2006. *Marine Mammal Science* **26**: 698–706.

1132. Isono, T., et al. (2004) Changes in abundance and sightings of marked Steller sea lions in Hokkaido. *in* 22nd Lowell Wakefield Fisheries Symposium "Sea Lions of the World: Conservation and Research in the 21st Century". September 30–October 3, 2004, Anchorage, Alaska, USA., p. 99.

1133. Isono, T. & Wada, K. (1999) Migration of Steller sea lions *Eumetopias jubatus* around Hokkaido Island, Japan. *in* Migratory Ecology and Conservation of Steller Sea Lions (Ohtaishi, N. & Wada, K., Editors). Tokyo: Tokai University Press, pp. 229–247 (in Japanese with English summary).

1134. Itabashi, M., et al. (2007) Notes on hibernation and breeding of captive Asian particolored bat relieved at Zama city. *Natural History Report of Kanagawa* **28**: 51–53 (in Japanese).

1135. Itagaki, H., Uchida, A., & Udagawa, T. (1979) Helminth fauna of the Amami-Islands, Japan (8) A nematode of the family Heligmosomatidae from a feral rat *Diplothrix legata*. *Japanese Journal of Parasitology* **28 (Suppl.)**: 100.

1136. Itani, J. (伊谷純一郎) (1954) 高崎山のサル. Tokyo: Kodansha (講談社), 284 pp. (in Japanese).

1137. Ito, K. (伊藤圭子), et al. (2004) 奄美大島に生息する外来種ジャワマングースのトキソプラズマ抗体保有. *in* Abstracts of the Annual Meeting of the Japanese Society of Zoo and Wildlife Medicine. p. 84 (in Japanese).

1138. Ito, T. (伊東拓也) & Takahashi, K. (高橋健一) (2006) エゾシカ寄生マダニ類の生態. *in* エゾシカの保全と管理 (Kaji, K. (梶光一), Miyaki, M. (宮木雅美), & Uno, H. (宇野裕之), Editors). Sapporo: Hokkaido University Press, pp. 165–181 (in Japanese).

1139. Ito, M. & Itagaki, T. (2003) Survey on wild rodents for endoparasites in Iwate Prefecture, Japan. *Journal Veterinary Medical Science* **65**: 1151–1153.

1140. Ito, T. (1971) On the oestrouse cycle and gestation period of the Japanese serow, *Capricornis crispus*. *The Journal of the Mammalogical Society of Japan* **5**: 104–108 (in Japanese with English abstract).

1141. Ito, T. & Takatsuki, S. (1987) The distribution and seasonal movements of sika deer (*Cervus nippon*) in the Mt. Goyo area, Iwate Prefecture. *Bulletin of the Yamagata University Natural Science* **11**: 411–430 (in Japanese with English summary).

1142. Itoh, Y. (伊藤勇樹), et al. (2001) 大雪山におけるエゾヒグマ (*Ursus arctos yesoensis*) の食性の再構築. *Bears Japan* **2**: 20–24 (in Japanese).

1143. Itoh, K., et al. (1988) Helminth parasites of the Japanese monkey, *Macaca fuscata fuscata* in Ehime Prefecture, Japan. *Japanese Journal of Veterinary Research* **36**: 235–247.

1144. Itoh, T., et al. (2012) Effective dispersal of brown bears (*Ursus arctos*) in eastern Hokkaido, Inferred from analyses of mitochondrial DNA and microsatellites. *Mammal Study* **37**: 29–41.

1145. Itoh, T., et al. (2013) Estimating the population structure of brown bears in eastern Hokkaido based on microsatellite analysis. *Acta Theriologica* **58**: 127–138.

1146. Itoigawa, N., et al. (1992) Demography and reproductive parameters of a free-ranging group of Japanese macaques (*Macaca fuscata*) at Katsuyama. *Primates* **33**: 49–68.

1147. Itoo, T. (伊藤徹魯) (1998) ニホンアシカ. *in* 日本の希少な野生水生生物に関するデータブック (Data Book of Rare Aquatic Animals and Plants of Japan) (Fishery Agency, Editor). Tokyo: Japan Fisheries Resource Conservation Association (日本水産資源保護協会), pp. 256–257 (in Japanese).

1148. Itoo, T. (1985) New cranial materials of the Japanese sea lion, *Zalophus californianus japonicus* (Peters, 1866). *The Journal of the Mammalogical Society of Japan* **10**: 135–148.

1149. Itoo, T. & Nakamura, K. (1994) Approaches to reconstruct Japanese sea lions (9). An attempt to reconstruct the previous distribution of the Japanese sea lion. *Aquabiology* **94**: 373–393 (in Japanese with English abstract).

1150. Itoo, T. & T. Inoué (1993) Reconstruction of adult body size in the Japanese sea lion. *Journal of Growth* **32**: 89–97.

1151. IUCN (2009) Western Gray Whales: status, threats and the potential for recovery. Report of the Western Gray Whales Rangewide Workshop, Tokyo, September 2008, 45 pp. (Unpublished).

1152. IUCN/SSC Seal Specialist Group (1993) Seals, Fur Seals, Sea Lions, and Walrus: Status Survey and Conservation Action Plan. Gland: IUCN, 88 pp. (http://data.iucn.org/dbtw-wpd/edocs/1993-034.pdf).

1153. Ivanitskaya, E.Y. (1994) Comparative cytogenetics and systematics of *Sorex*: A cladistic approach. *in* Advances in the Biology of Shrews (Merritt, J.F., Kirkland, G.L., Jr., & Rose, R.K., Editors). Pittsburgh: Carnegie Museum of Natural History, pp. 313–324.

1154. Iwaki, T., et al. (1993) Survey on larval *Echinococcus multilocularis* and other hepatic helminths in rodents and insectivores in Hokkaido, Japan, from 1985 to 1992. *Japanese Journal of Parasitology* **42**: 502–506.

1155. Iwamoto, M. & Hasegawa, Y. (1972) Two macaque fossil teeth from the Japanese Pleistocene. *Primates* **13**: 77–81.

1156. Iwamoto, M. & Hasegawa, Y. (1991) Fossil macaques from Fujisawa and Kisarazu, Kanto district, Japan. *Primate Research* **7**: 96–102 (in Japanese with English summary).

1157. Iwamoto, T. (1974) A bioeconomic study on a provisioned troop of Japanese monkeys (*Macaca fuscata*) at Koshima Islet, Miyazaki. *Primates* **15**: 241–262.

1158. Iwamoto, T. (1982) Food and nutritional condition of free-ranging Japanese monkeys on Koshima Islet during winter. *Primates* **23**: 153–170.

1159. Iwasa, M.A. Unpublished.

1160. Iwasa, M.A. (2000) Variation of skull characteristics in the Anderson's red-backed vole, *Eothenomys andersoni*, from Nagano City, central Honshu, Japan. *Mammal Study* **25**: 125–139.

1161. Iwasa, M.A. (2004) A note on aberrant pelage colors in a wild population of the gray red-backed vole *Clethrionomys rufocanus bedfordiae* in Hokkaido. *Mammal Study* **29**: 93–95.

1162. Iwasa, M.A. & Abe, H. (2006) Colonization history of the Japanese water shrew *Chimarrogale platycephala*, in the Japanese Islands. *Acta Theriologica* **51**: 29–38.

1163. Iwasa, M.A. & Hosoda, T. (2002) A note on karyotype of the sable, *Martes zibellina brachyura*, in Hokkaido, Japan. *Mammal Study* **27**: 83–86.

1164. Iwasa, M.A., et al. (2002) Local differentiation of *Clethrionomys rutilus* in northeastern Asia inferred from mitochondrial gene sequences. *Mammalian Biology* **67**: 157–166.

1165. Iwasa, M.A., et al. (2006) Intraspecific differentiation in the lesser Japanese mole in eastern Honshu, Japan, indicated by nuclear and mitochondrial gene analyses. *Zoological Science* **23**: 955–961.

1166. Iwasa, M.A. & Nakata, K. (2011) A note on the genetic status of the dark red-backed vole, *Myodes rex*, in Hokkaido, Japan. *Mammal Study* **36**: 99–103.

1167. Iwasa, M.A., et al. (2001) Karyotype and RFLP of the nuclear rDNA of *Crocidura* sp on Cheju Island, South Korea (Mammalia, Insectivora). *Mammalia* **65**: 451–459.

1168. Iwasa, M.A., Serizawa, K., & Satoh, M. (2001) Taxonomic problems of the dark red-backed vole, *Clethrionomys rex*. *Rishiri Studies* **20**: 43–53 (in Japanese with English summary).

1169. Iwasa, M.A. & Suzuki, H. (2002) Evolutionary networks of maternal and paternal gene lineages in voles (*Eothenomys*) endemic to Japan. *Journal of Mammalogy* **83**: 852–865.

1170. Iwasa, M.A. & Suzuki, H. (2002) Evolutionary significance of chromosome changes in northeastern Asiatic red-backed voles inferred from the aid of intron 1 sequences of the *G6pd* gene. *Chromosome Research* **10**: 419–428.

1171. Iwasa, M.A. & Suzuki, H. (2003) Intra- and interspecific genetic complexities of two *Eothenomys* species in Honshu, Japan. *Zoological Science* **20**: 1305–1313.

1172. Iwasa, M.A. & Tsuchiya, K. (2000) Karyological analysis of the *Eothenomys* sp. from Nagano City, central Honshu, Japan. *Chromosome Science* **4**: 31–38.

1173. Iwasa, M.A., et al. (2000) Geographic patterns of cytochrome *b* and *Sry* gene lineages in the gray red-backed vole *Clethrionomys rufocanus* from Far East Asia including Sakhalin and Hokkaido. *Zoological Science* **17**: 477–484.

1174. Iwasaki, T. (岩﨑俊秀) (1996) カマイルカ. *in* The Encyclopaedia of Animals in Japan. 2. Mammals II (日本動物大百科2 哺乳類II) (Izawa, K. (伊澤紘生), Kasuya, T. (粕谷俊雄), & Kawamichi, T. (川道武男), Editors). Tokyo: Heibonsha (平凡社), pp. 70–71 (in Japanese).

1175. Iwasaki, T. (岩﨑俊秀) (1997) カマイルカ (Pacific white-sided dolphin). *in* 日

本の希少な野生水生生物に関する基礎資料. Tokyo: Japan Fisheries Resources Conservation Association (水産資源保護協会), pp. 410–413 (in Japanese).

1176. Iwasaki, T. (1991) Preliminary analysis of life history parameters of Pacific white-sided dolphins and northern right whale dolphins (Document submitted to "Scientific Review of the North Pacific high seas driftnet fisheries", Sidney, B.C., June 11–14). 13 pp.

1177. Iwasaki, T. & Kubo, N. (2001) Northbound migration of humpback whale *Megaptera novaeangliae* along the Pacific coast of Japan. *Mammal Study* **26**: 77–82.

1178. Iwata, T., et al. (2003) Incidental catch of harbor porpoises in set nets in the coastal waters of southern Hokkaido, Japan. *Fisheries Science* **69**: 657–659.

1179. IWC (1990) Report of the special meeting of the scientific committee on the assessment of gray whales (Paper IWC/42/4A presented to the IWC Scientific Committee). 29 pp.

1180. IWC (1992) Report of the sub-committee on small cetaceans. *Report of the International Whaling Commission* **42**: 178–234.

1181. IWC (1994) Report of the workshop on mortality of cetaceans in passive fishing nets and traps. *Report of the International Whaling Commission (Special Issue)* **15**: 1–70.

1182. IWC (1998) Report of the intersessional working group to review data and results from special permit research on minke whales in the Antarctic, Tokyo, 12–16 May 1997. *Report of the International Whaling Commission* **48**: 377–390.

1183. IWC (2000) Report of the Scientific Committee. *Journal of Cetacean Research and Management* **2 (Suppl.)**: 1–318.

1184. IWC (2001) Report of the working group on nomenclature. *Journal of Cetacean Research and Management* **3 (Suppl.)**: 363–367.

1185. IWC (2001) Report of the workshop on the comprehensive assessment of right whales: a worldwide comparison. *in* Journal of Cetacean Research and Management. Special Issue 2: Right Whales: Worldwide Status (Best, P.B., et al., Editors). Cambridge: IWC, pp. 1–60.

1186. IWC (2001) Report of the Workshop to review the Japanese Whale Research Programme under Special Permit for North Pacific minke whales (JARPN), Tokyo, 7–10 February 2000. *Journal of Cetacean Research and Management* **3 (Suppl.)**: 375–413.

1187. IWC (2002) Report of the standing sub-committe on small cetaceans. *Journal of Cetacean Research and Management* **4 (Suppl.)**: 325–338.

1188. IWC (2004) Report of the Workshop on the Western Gray Whale: Research and Monitoring Needs. *Journal of Cetacean Research and Management* **6 (Suppl.)**: 487–500.

1189. IWC (2005) Annex Q. Progress reports. *Journal of Cetacean Research and Management* **7 (Suppl.)**: 355–384.

1190. IWC (2010) Report of the Scientific Committee. *Journal of Cetacean Research and Management* **11 (Suppl. 2)**: 1–98.

1191. IWC (2013) Report of the Scientific Committee. *Journal of Cetacean Research and Management* **14 (Suppl.)**: 1–86.

1192. Izawa, K. (伊沢紘生) (1982) ニホンザルの生態：豪雪の白山に野生を問う. Tokyo: Dobutsusha (どうぶつ社), 418 pp. (in Japanese).

1193. Izawa, M. (伊澤雅子) (2002) ノネコ (*Felis catus*). *in* Handbook of Alien Species in Japan (外来種ハンドブック) (Ecological Society of Japan, Editor). Tokyo: Chijin Shokan (地人書館), p. 76 (in Japanese).

1194. Izawa, M. (伊澤雅子) & Doi, T. (土肥昭夫) (1997) イエネコからのウイルス感染—ツシマヤマネコは生き残れるか？ (New threat to the survival of the Tsushima leopard cat: transmission of infectious disease from domestic cat). *Kagaku* (科学) **67**: 705–707 (in Japanese).

1195. Izawa, M. (伊澤雅子), Doi, T. (土肥昭夫), & Ono, Y. (小野勇一) (1985) 行動圏と日周期活動 (Home range and daily activity). *in* イリオモテヤマネコ生息環境等保全対策調査報告書 (Environment Agency, Editor). Tokyo: Environment Agency, pp. 81–128 (in Japanese).

1196. Izawa, M. & Doi, T. (1993) Flexibility of the social system of the feral cat, *Felis catus*. *in* Animal Societies: Individuals, Interactions, and Organisation (Jarman, P.J. & Rossiter, A., Editors). Kyoto: Kyoto University Press, pp. 237–247.

1197. Izawa, M., et al. (2009) Ecology and conservation of two endangered subspecies of the leopard cat (*Prionailurus bengalensis*) on Japanese islands. *Biological Conservation* **142**: 1884–1890.

1198. Izawa, M., Doi, T., & Ono, Y. (1982) Grouping patterns of feral cat (*Fells catus*) living on a small island in Japan. *Japanese Journal of Ecology* **32**: 373–382.

1199. Izawa, M., Doi, T., & Ono, Y. (1990) Notes on the spacing pattern of the feral cats at high density. *Bulletin of the Kitakyushu Museum of Natural History* **10**: 109–113.

1200. Izawa, M., et al. Unpublished.

1201. Izawa, M., Kinjo, K., & Nakamoto, A. (2003) Utilization of Mangrove Forests as Day-Roost Sites by the Daito Flying-Fox, *Pteropus dasymallus daitoensis*. *in* 平成14年度内閣府委託事業：マングローブに関する調査研究報告書. Naha: Research Institute for Subtropics (亜熱帯総合研究所), pp. 51–56 (in Japanese with English abstract).

1202. Izawa, M. & Maeda, K. (1998) Record of *Pipistrellus abramus* from Iriomote Island, southern Ryukyus, Japan. *Island Studies in Okinawa* **16**: 17–18 (in Japanese).

1203. Izumi, I., et al. (2011) Preliminary survey of habitat use by *Sciurus vulgaris orientis* in a natural forest of Hokkaido Island, Japan. *Mammal Study* **36**: 109–112.

1204. Izumimaya, S. & Shiraishi, T. (2004) Seasonal changes in elevation and habitat use of the Asiatic black bear (*Ursus thibetanus*) in the Northern Japan Alps. *Mammal Study* **29**: 1–8.

1205. Izumiyama, S., Mochizuki, T., & Shiraishi, T. (2003) Troop size, home range area and seasonal range use of the Japanese macaque in the Northern Japan Alps. *Ecological Research* **18**: 465–474.

1206. Jadejaroen, J., et al. (2015) Use of photogrammetry as a means to assess hybrids of rhesus (*Macaca mulatta*) and long-tailed (*M. fascicularis*) macaques. *Primates* **56**: 77–88.

1207. Jaenson, T.G.T. (1991) The epidemiology of Lyme borreliosis. *Parasitology Today* **7**: 39–45.

1208. Jameson, E.W., Jr. (1961) Relationships of the red-backed voles of Japan. *Pacific Science* **15**: 594–604.

1209. Jameson, R.J. & Johnson, A.M. (1993) Reproductive characteristics of female sea otters. *Marine Mammal Science* **9**: 156–167.

1210. Japan Bear Network, ed. (2014) ツキノワグマおよびヒグマの分布域拡縮の現況把握と軋轢抑止および危機個体群回復のための支援事業報告書. Ibaraki: Japan Bear Network, 172 pp. (in Japanese).

1211. Japan Bear Network (compiler) (2006) Understanding Asian Bears to Secure Their Future. Ibaraki: Japan Bear Network, 145 pp.

1212. Japan Wildlife Research Center, ed. (2004) The National Survey on the Natural Environment: Report of the Distributional Survey of Japanese Animals (Mammals). 213 pp. (in Japanese; http://www.biodic.go.jp/reports2/6th/6_mammal/6_mammal.pdf).

1213. Japan Wildlife Research Center (2005) Research Report on Widespread Appearance of Asiatic Black Bears. 115 pp.

1214. Japan Wildlife Research Center (1979) 第2回自然環境保全基礎調査：動物分布調査報告書 (哺乳類) 全国版. 91 pp. (in Japanese with English summary; http://www.biodic.go.jp/reports/2-5/ac000.html).

1215. Japan Wildlife Research Center (1988) ツシマヤマネコ生息環境等調査報告 (Research Report on Environmental Habitat of the Tsushima Leopard Cat). 106 pp. (in Japanese).

1216. Japan Wildlife Research Center (1992) 小笠原諸島における山羊の異常繁殖による動植物への被害緊急調査報告書. 147 pp. (in Japanese).

1217. Japan Wildlife Research Center (2005) 平成16年度移入種 (ほ乳類) 生息状況等報告書—長崎県委託調査. 32 pp. (in Japanese).

1218. Japan Wildlife Research Center, ed. (2011) 平成22年度小笠原地域自然再生事業外来ほ乳類対策調査業務報告書. 216 pp. (in Japanese).

1219. Jarman, P.J. (1966) The status of the dugong (*Dugong dugon* Müller); Keneya, 1961. *East African Wildlife Journal* **4**: 82–88.

1220. Jauniaux, T., et al. (2014) Bite injuries of grey seals (*Halichoerus grypus*) on harbour porpoises (*Phocoena phocoena*). *PLoS ONE* **9**: e108993.

1221. Jayasekara, P. & Takatsuki, S. (2000) Seasonal food habits of a sika deer population in the warm temperate forest of the westernmost part of Honshu, Japan. *Ecological Research* **15**: 153–157.

1222. Jefferson, T.A. (トマス・A・ジェファソン), Leatherwood, S. (スティーブン・レザウッド), & Webber, M.A. (マーク・A・ウェバー) (1999) 海の哺乳類FAO種同定ガイド (FAO Species Identification Guide: Marine Mammals of the World). Tokyo: NTT Publishing (NTT出版), 336 pp. (in Japanese. Translated by 山田格. Originally published by FAO, Rome in 1993).

1223. Jefferson, T.A. (2002) Preliminary analysis of geographic variation in cranial morphometrics of the finless porpoise (*Neophocaena phocaenoides*). *Raffles Bulletin of Zoology Supplement.* **10**: 3–14.

1224. Jefferson, T.A. & Barros, N.B. (1997) *Peponocephala electra*. *Mammalian Species* **553**: 1–6.

1225. Jefferson, T.A., Leatherwood, S., & Webber, M.A. (1993) FAO Species Identification Guide. Marine Mammals of the World. Rome: Food and Agriculture Organization of the United Nations, 320 pp.

1226. Jefferson, T.A., et al. (1994) Right whale dolphins *Lissodelphis borealis* (Peale, 1848) and *Lissodelphis peronii* (Lacepede, 1804). *in* Handbook of Marine Mammals. Volume 5. The First Book of Dolphins (Ridgway, S.H. & Harrison, R., Editors). London: Academic Press, pp. 335–362.

1227. Jefferson, T.A., Stacey, P.F., & Baird, R.W. (1991) A review of killer whale interactions with other marine mammals: Predation to co-existence. *Mammalian Review* **21**: 151–180.

1228. Jefferson, T.A. & Van Waerebeek, K. (2002) The taxonomic status of the nominal dolphin species *Delphinus tropicalis* van Bree, 1971. *Marine Mammal Science* **18**: 787–818.

1229. Jefferson, T.A. & Wang, J.Y. (2011) Revision of the taxonomy of finless porpoises (genus *Neophocaena*): The existence of two species. *Journal of Marine Animals and Their Ecology* **4**: 3–16.

1230. Jefferson, T.A., Webber, M.A., & Pitman, R.L. (2008) Marine Mammals of the World: A Comprehensive Guide to their Identification. London: Academic Press, 573 pp.

1231. Jefferson, T.A., et al. (2013) Global distribution of Risso's dolphin *Grampus griseus*: a review and critical evaluation. *Mammal Review* **44**: 56–68.

1232. Jepson, P.D., et al. (2005) Relationships between polychlorinated biphenyls and health status in harbor porpoises (*Phocoena phocoena*) stranded in the United Kingdom. *Environmental Toxicology and Chemistry* **24**: 238–248.

1233. Jepson, P.D., et al. (2013) What caused the UK's largest common dolphin (*Delphinus delphis*) mass stranding event? *PloS ONE* **8**: e60953.

1234. Jezierski, W. (1977) Longevity and mortality rate in a population of wild boar. *Acta Theriologica* **22**: 337–348.

1235. Jia, Z., et al. (2002) Copulatory plugs in masked palm civets: prevention of semen leakage, sperm storage, or chastity enhancement? *Journal of Mammalogy* **83**: 1035–1038.

1236. Jiang, X.-L. & Hoffmann, R.S. (2001) A revision of the white-toothed shrews (*Crocidura*) of southern China. *Journal of Mammalogy* **82**: 1059–1079.

1237. Jiang, Z. (1998) Feeding Ecology and Digestive Systems of Ruminants: A Case Study of the Mongolian Gazelle (*Procapra gutturosa*) and the Japanese Serow (*Capricornis crispus*). Tokyo: University of Tokyo. 100 pp.

1238. Jinguu, J. (2014) The finless porpoise of Sendai Bay. *Aquabiology* **36**: 8–13 (in Japanese with English abstract).

1239. Jinnai, M., et al. (2010) Molecular evidence of the multiple genotype infection of a wild Hokkaido brown bear (*Ursus arctos yesoensis*) by *Babesia* sp. UR1. *Veterinary Parasitology* **173**: 128–133.

1240. Jogahara, T., et al. (2007) Quest for the cause of oligodontia in *Suncus murinus* (Soricomorpha, Soricidae): morphological re-examination. *Archives of Oral Biology* **52**: 836–843.

1241. Jogahara, T., et al. (2005) The isolation and seroprevalence of antibodies against *Leptospira* in *Mus caroli* and *M. musculus yonakunii* on Okinawa Island. *Japanese Journal of Zoo and Wildlife Medicine* **10**: 85–90 (in Japanese with English abstract).

1242. Jogahara, T., et al. (2003) Food habits of cats (*Felis catus*) in forests and villages and their impacts on native animals in the Yaeyame area, northern part of Okinawa Island, Japan. *Mammalian Science* **43**: 29–37 (in Japanese with English abstract).

1243. Johnson, D.H. (1946) The spiny rat of the Riu Kiu islands. *Proceedings of the Biological Society of Washington* **59**: 169–172.

1244. Johnson, J.H. & Wolman, A.A. (1984) The humpback whale, *Megaptera novaeangliae*. *Marine Fisheries Review* **46**: 30–37.

1245. Johnson, M., et al. (2006) Foraging Blainville's beaked whales (*Mesoplodon densirostris*) produce distinct click types matched to different phases of echolocation. *Journal of Experimental Biology* **209**: 5038–5050.

1246. Johnson, M.L., et al. (1966) Marine mammals. *in* Environment of the Cape Thompson Region, Alaska (Wilimovsky, N.J. & Wolfe, J.N., Editors). Oak Ridge: U.S. Atomic Energy Commission, pp. 897–924.

1247. Johnstone, I.M. & Hudson, B.E.T. (1981) The dugong diet: mouth sample analysis. *Bulletin of Marine Science* **31**: 681–690.

1248. Jones, M.L. & Swartz, S.L. (2002) Gray whale (*Eschrichtius robustus*). *in* Encyclopedia of Marine Mammals. 2nd ed. (Perrin, W.F., Würsig, B., & Thewissen, J.G.M., Editors). San Diego: Academic Press, pp. 806–813.

1249. Jones, M.L. & Swartz, S.L. (2009) Gray whale (*Eschrichtius robustus*). *in* Encyclopedia of Marine Mammals. 2nd ed. (Perrin, W.F., Würsig, B., & Thewissen, J.G.M., Editors). Amsterdam: Academic Press, pp. 503–511.

1250. Jonstone, R.A. & Cant, M.A. (2010) The evolution of menopause in cetaceans and humans: the role of demography. *Proceedings of the Royal Society B: Biological Science* **277**: 3765–3771.

1251. Junge, G.C.A. (1950) On a specimen of the rare fin whale, *Balaenoptera edeni* Anderson, stranded on Pulu Sugi Near Singapore. *Zoologische Verhandelingen* **9**: 1–26, 29 pls.

1252. Kadosaki, M., Inage, M., & Kudo, A. (2000) Criteria for identification of *Sorex minutissimus*. *The Journal of the Japanese Wildlife Research Society* **30**: 3–8 (in Japanese with English abstract).

1253. Kadosaki, M., et al. (1991) Deer attacked by bears. *Bulletin of Higashi Taisetsu Museum of Natural History* **13**: 57–62 (in Japanese with English abstract).

1254. Kadosaki, M., Odazima, M., & Yamashita, S. (2004) *Sorex minutissimus* captured in Daisetsuzan area of central Hokkaido. *The Journal of the Japanese Wildlife Research Society* **30**: 3–8 (in Japanese with English summary).

1255. Kadoya, N., et al. (2010) A preliminary survey on nest cavity use by Siberian flying squirrels, *Pteromys volans orii*, in forests of Hokkaido Island, Japan. *Russian Journal of Theriology* **9**: 27–32.

1256. Kage, T. (景崇洋) (1999) DNA 多型によるコビレゴンドウ (*Globicephala macrorhynchus*) の群構造の解析に関する研究. Doctoral dissertation. Tsu: Mie University. 141 pp. (in Japanese).

1257. Kagei, N. & Sawada, I. (1973) Helminth fauna of bats in Japan XIII. *Annotationes Zoologicae Japonenses* **46**: 49–52.

1258. Kagei, N. & Sawada, I. (1973) Helminth fauna of bats in Japan XIV. *Annotationes Zoologicae Japonenses* **46**: 53–56.

1259. Kagei, N. & Sawada, I. (1977) Helminth fauna of bats in Japan XVII. *Annotationes Zoologicae Japonenses* **50**: 178–181.

1260. Kagei, N. & Sawada, I. (1977) Helminth fauna of bats in Japan XVIII. *Annotationes Zoologicae Japonenses* **50**: 245–248.

1261. Kagei, N. & Sawada, I. (1984) Helminth fauna of bats in Japan XXXI. *Proceeding of the Japanese Society of Systematic Zoology* **28**: 13–18.

1262. Kagei, N., Sawada, I., & Kifune, T. (1979) Helminth fauna of bats in Japan XX. *Annotationes Zoologicae Japonenses* **52**: 54–62.

1263. Kagei, N., et al. (1983) On the helminths from raccoon dogs in Japan. *Japanese Journal of Parasitology* **32**: 367–369.

1264. Kagoshima Prefectural Government (2014) 鹿児島県マングース防除実施計画. Kagoshima: Kagoshima Prefectural Government, 8 pp. (in Japanese).

1265. Kahn, B., Wawandono, N.B., & Sublijanto, J. (2003) Protecting the Cetaceans of Komodo National Park, Indonesia: Positive Identification of the Rare Pygmy Bryde's Whale (*Balaenoptera edeni*) with the Assistance of Genetic Profiling. Cairns & Bali: Apex Environmental & The Nature Conservancy Coastal and Marine Program, 10 pp.

1266. Kaji, K. Personal communication.

1267. Kaji, K. (1997) Analysis of female sika deer population during the 1994–96 hunting seasons. *Annual report of Hokkaido Institute of Environmental Sciences* **24**: 53–59 (in Japanese with English abstract).

1268. Kaji, K., Koizumi, T., & Ohtaishi, N. (1988) Effects of resource limitation on the physical and reproductive condition of Sika deer on Nakanoshima Island, Hokkaido. *Acta Theriologica* **33**: 187–208.

1269. Kaji, K., et al. (2000) Spatial distribution of an expanding sika deer population in Hokkaido, Japan. *Wildlife Society Bulletin* **28**: 699–707.

1270. Kaji, K., et al. (2006) The Shiretoko deer herd: Management policy and natural regulation. *in* Wildlife in Shiretoko and Yellowstone National Parks (McCullough, D.R., Kaji, K., & Yamanaka, M., Editors). Shari: Shiretoko Nature Foundation, pp. 43–55 (in Japanese with English abstract).

1271. Kaji, K., et al. (2004) Irruption of a colonizing sika deer population. *Journal of Wildlife Management* **68**: 889–899.

1272. Kaji, K., et al. (2010) Adaptive management of sika deer populations in Hokkaido, Japan: Theory and practice. *Population Ecology* **52**: 373–387.

1273. Kajimura, H. (1985) Opportunistic feeding by the northern fur seal, *Callorhinus ursinus*. *in* Marine Mammals and Fisheries (Beddington, J.R., Beverton, R.J.H., & Lavigine, D.M., Editors). London: George Allen & Unwin, pp. 300–318.

1274. Kajiwara, N., et al. (2006) Geographical distribution of polybrominated diphenyl ethers (PBDEs) and organochlorines in small cetaceans from Asian waters. *Chemosphere* **64**: 287–295.

1275. Kajiwara, N., et al. (2006) Organohalogen and organotin compounds in killer whales mass-stranded in the Shiretoko Peninsula, Hokkaido, Japan. *Marine Pollution Bulletin* **52**: 1066–1076.

1276. Kakogawa, M., et al. (2013) Case reports on two species of parasitic insects obtained from sika deer *Cervus nippon yesoensis* in Shiretoko Peninsula and its proximity, Hokkaido, Japan. *Bulletin of the Shiretoko Museum* **35**: 11–14 (in Japanese with English abstract).

1277. Kakuda, H., Shiraishi, S., & Uchida, T. (1989) Ticks from wild mammals in the Kyushu district including Okinawa Prefecture, Japan. *Journal of Faculty of Agriculture, Kyushu University* **33**: 267–273.

1278. Kakuda, T., et al. (2002) On the resident "bottlenose dolphins" from Mikura water. *Memories of the National Science Museum* **38**: 255–272.

1279. Kakuda, T. & Yamada, T.K. (2003) Generic variability of Stejneger's beaked whale (*Mesoplodon stejnegeri*) stranded on the shore of Sea of Japan based on mitochondrial DNA sequences. *Mammalian Science* **Suppl. 3**: 93–96 (in Japanese with English abstract).

1280. Kamada, S., et al. (2012) Genetic distinctness and variation in the Tsushima Islands population of the Japanese marten, *Martes melampus* (Carnivora: Mustelidae), revealed by microsatellite analysis. *Zoological Science* **29**: 827–833.

1281. Kamada, S., Murakami, T., & Masuda, R. (2013) Multiple origins of the Japanese marten *Martes melampus* introduced into Hokkaido Island, Japan, revealed by microsatellite analysis. *Mammal Study* **38**: 261–267.

1282. Kambe, Y., et al. (2012) Genetic characterization of Okinawan black rats showing coat color polymorphisms of white spotting and melanism. *Genes & Genetic Systems* **87**: 29–38.

1283. Kambe, Y., et al. (2013) Introgressive hybridization of two major lineages of invasive black rats, *Rattus rattus* and *R. tanezumi* on the Japanese Islands inferred from *Mc1r* sequences. *Mammalian Science* **53**: 28–299 (in Japanese with English abstract).

1284. Kambe, Y., et al. (2011) Origin of agouti-melanistic polymorphism in wild black rats (*Rattus rattus*) inferred from *Mc1r* gene sequences. *Zoological Science* **28**: 560–567.

1285. Kamegai, S. & Ichihara, A. (1973) A check list of the helminths from Japan and adjacent areas. Part 2. Parasites of amphibia, reptiles, birds and mammals reported by S. Yamaguchi. *Research Bulletin of the Meguro Parasitological Museum* **7**: 33–64.

1286. Kameyama, Y., et al. (2014) Sexing of long-clawed shrew (*Sorex unguiculatus*) by using genomic DNA from the hair. *Journal of Agriculture Science, Tokyo University of Agriculture* **59**: 163–167 (in Japanese with English summary).

1287. Kamijo, T. & Endo, Y. (2001) A record of the brown long-eared bat, *Plecotus auritus* in Tsukuba University Forest at Yatsugatake, Nagano Prefecture, Japan. *Bulletin of Tsukuba University Forests* **17**: 85–86 (in Japanese).

1288. Kamiya, M. (神谷正男), et al. (1987) 特別天然記念物アマミノクロウサギ *Pentalagus furnessi*— その捕獲・飼育・寄生虫 (I). *Journal of the Hokkaido Veterinary Medical Association* **31**: 221–228 (in Japanese).

1289. Kamiya, H. & Ishigaki, K. (1972) Helminths of Mustelidae in Hokkaido. *Japanese Journal of Veterinary Research* **20**: 117–128.

1290. Kamiya, H., et al. (1977) An epidemiological survey of multilocular echinococcosis in small mammals of Eastern Hokkaido, Japan. *Japanese Journal of Parasitology* **26**: 148–156 (in Japanese).

1291. Kamiya, H. & Suzuki, Y. (1975) Parasites of the Japanese badger, *Meles meles anakuma* Temminck, especially on *Isthmiophora mels* (Schrank, 1788) Luhe,1909. *Japanese Journal of Veterinary Research* **23**: 125–130.

1292. Kamiya, M., et al. (2006) Current control strategies targeting sources of echinococcosis in Japan. *Revue Scientifique et Technique de l'Office International des Epizooties* **25**: 1055–1066.

1293. Kamiya, M., Suzuki, H., & Hayashi, Y. (1977) *Macracanthorhynchus hirudinaceus* (Pallas, 1781) from a wild boar, *Sus scrofa riukiuanus* on Amami Island, southern Japan. *Japanese Journal of Parasitology* **26**: 260–264 (in Japanese with English summary).

1294. Kamiya, Y. & Ohbayashi, M. (1975) Some helminthes of the red fox, *Vulpes vulpes schrencki* Kishida, in Hokkaido, Japan, with a description of a new trematode, *Massaliatrema yamashitai* n. sp. *Japanese Journal of Parasitology* **23**: 60–68.

1295. Kamiyama, N. (1979) The growth and development of the gonad in Japanese house shrew (*Suncus murinus riukiuanus*). *The Journal of the Mammalogical Society of Japan* **7**: 274–279 (in Japanese with English abstract).

1296. Kamiyama, T. (2004) Plaque. *Journal of Clinical and Experimental Medicine* **208**: 57–62 (in Japanese).

1297. Kamo, H., Kawashima, K., & Nishimura, K. (1957) Notes on the trombiculid mites infesting bats in Kyushu, Japan, including descriptions of two new species (Acarina: Trombiculidae). *Kyushu Journal of Medical Science* **8**: 209–216.

1298. Kamo, H., Maejima, J., & Hatsushika, R. (1980) First record of *Diphyllobothrium macroovatum* Jurachno, 1973 from minke whale, *Balaenoptera acutorostrata* Lacepede, 1804 (Cestoda: Diphyllobothriidae) in Japan. *Japanese Journal of Parasitology* **29**: 499–505 (in Japanese with English summary).

1299. Kamogawa Sea World (鴨川シーワールド) (1992) 海獣類の漂着記録. *in* 鴨川シーワールド報告 I 業績集 1970–1991 (Kamogawa Sea World Report. I. Collected Reprints 1970–1991). Kamogawa: Kamogawa Sea World, pp. 347–349 (in Japanese).

1300. Kanai, Y., et al. (2007) Epizootiological survey of *Trichinella* spp. infection in carnivores, rodents and insectivores in Hokkaido, Japan. *Japanese Journal of Veterinary Research* **54**: 175–82.

1301. Kanaizuka, T. (金井塚務) (1992) コテングコウモリの雪中越冬. モンキータイムズ **25**: 3–5 (in Japanese).

1302. Kanaizuka, T. (金井塚務) (1993) 宮島町におけるヒナコウモリ (*Vespertilio superans*). おおの自然観察の森研究報告 **3**: 36 (in Japanese).

1303. Kanaizuka, T., Shiramizu, T., & Hoshino, K. (1991) Some notes on Japanese mountain moles (*Euroscaptor mizura*) in Hiroshima. *Miscellaneous Report of the Hiwa Museum for Natural History* **29**: 47–51 (in Japanese with English summary).

1304. Kanaji, Y., Okamura, H., & Miyashita, T. (2011) Long-term abundance trends of the northern form of the short-finned pilot whale (*Globicephala macrorhynchus*) along the Pacific coast of Japan. *Marine Mammal Science* **27**: 477–492.

1305. Kanamori, M. & Tanaka, R. (1968) Studies on population ecology of the vole, *Microtus montebelli*, in mountain grasslands of Sugadaira and its adjacent areas. I. Results of research on five populations in 1966–1967. *Bulletin of the Sugadaira Biological Laboratory* **2**: 17–39 (in Japanese with English summary).

1306. Kanazawa, Y. & Nishikata, S. (1976) Disappearance of acorns from the floor in *Quercus crispula* forests. *Journal of the Japanese Forest Society* **58**: 52–56.

1307. Kanda, I., et al. (2007) Study on habitat of finless porpoise *Neophocaena phocaenoides* around Kansai International Airport. *Bulletin of the Osaka Prefectural Fisheries Experimental Station* **17**: 27–34 (in Japanese).

1308. Kanda, N., Goto, M., & Pastene, L.A. (2006) Genetic characteristics of western North Pacific sei whales, *Balaenoptera borealis*, as revealed by microsatellites. *Marine Biotechnology* **8**: 86–93.

1309. Kane, E.A., et al. (2008) Prevalence of the commensal barnacle *Xenobalanus globicipitis* on cetacean species in the eastern tropical Pacific Ocean, and a review of global occurrence. *Fishery Bulletin* **106**: 395–404.

1310. Kaneko, H. (金子浩昌), Koyanagi, M. (小柳美登里), & Ushizawa, Y. (牛沢百合子) (1973) 飯富山野貝塚出土の脊椎動物遺存体. *in* 袖ヶ浦山野貝塚 (千葉県都市公社, Editor). Tokyo: Tokyo Electric Power Company (東京電力), pp. 221–229 (in Japanese).

1311. Kaneko, K. (金子清俊) & Hattori, K. (服部睢作) (1968) コウモリの内部寄生ダニ類について. *Japanese Journal of Sanitary Zoology* **19**: 128–129 (in Japanese).

1312. Kaneko, Y. (金子之史) (1994) ネズミ目 Rodentia. *in* A Pictorial Guide to the Mammals of Japan (日本の哺乳類) (Abe, H. (阿部永), Editor). Tokyo: Tokai University Press, pp. 81–110 & 168–182 (in Japanese).

1313. Kaneko, Y. (金子之史) (2007) エゾヤチネズミ, ムクゲネズミ, およびミカドネズミの属名は *Myodes* に変更. *Forest Protection* **308**: 25–27 (in Japanese).

1314. Kaneko, Y. (金子弥生) (2001) 東京都日の出町におけるアナグマの生態学的研究 (An Ecological Study of the Badger in Hinode-town, Tokyo). Doctoral dissertation. Tokyo: Tokyo University of Agriculture and Technology. 119 pp. (in Japanese).

1315. Kaneko, K. (1955) Studies on the murine lice in Japan (Part I). A revision of the 9 species of Japanese murine-lice. *Japanese Journal of Sanitary Zoology* **6**: 104–110 (in Japanese with English summary).

1316. Kaneko, Y. Unpublished.

1317. Kaneko, Y. (1975) Mammals of Japan (12): order Rodentia, genus *Microtus*. *Mammalian Science* **15(1)**: 3–26 (in Japanese).

1318. Kaneko, Y. (1985) Examinations of diagnostic characters (mammae and bacula) between *Eothenomys smithi* and *E. kageus*. *The Journal of the Mammalogical Society of Japan* **10**: 221–229 (in Japanese with English abstract).

1319. Kaneko, Y. (1988) Relationships of skull dimensions with latitude in the Japanese field vole. *Acta Theriologica* **33**: 35–46.

1320. Kaneko, Y. (1989) Seasonal changes of the number collected and reproduction in *Eothenomys smithii* at the foot of a lower mountain, Minoura, Kagawa Prefecture, Japan. *Kagawa Seibutsu* **15/16**: 67–74 (in Japanese with English abstract).

1321. Kaneko, Y. (1992) Mammals of Japan. 17. *Eothenomys smithii* (Smith's red-backed vole). *Mammalian Science* **32**: 39–54 (in Japanese).

1322. Kaneko, Y. (1996) Age variation of the third upper molar in *Eothenomys smithii*. *Mammal Study* **21**: 1–13.

1323. Kaneko, Y. (1998) Taxonomy (哺乳類の生物学 1 分類). Tokyo: University of Tokyo Press, 148 pp. (in Japanese).

1324. Kaneko, Y. (2001) Life cycle of the Japanese badger (*Meles meles anakuma*) in Hinode Town, Tokyo. *Mammalian Science* **41**: 53–64 (in Japanese with English abstract).

1325. Kaneko, Y. (2001) Morphological discrimination of the Ryukyu spiny rat (genus *Tokudaia*) between the islands of Okinawa and Amami Oshima, in the Ryukyu Islands, southern Japan. *Mammal Study* **26**: 17–33.

1326. Kaneko, Y. (2002) Inner structure of badger home range in Hinode-town. *Japanese Journal of Ecology* **52**: 243–252 (in Japanese).

1327. Kaneko, Y. (2005) Rodentia. *in* A Guide to the Mammals of Japan (Abe, H., Editor). Hadano: Tokai University Press, pp. 115–146 (in Japanese and English).

1328. Kaneko, Y. (2006) The Taxonomy of Japanese Muridae (Mammalia, Rodentia): From the View Point of Biogeography. Tokyo: University of Tokyo Press, 302 pp. (in Japanese).

1329. Kaneko, Y., et al. (2014) The socio-spatial dynamics of the Japanese badger (*Meles anakuma*). *Journal of Mammalogy* **95**: 290–300.

1330. Kaneko, Y. & Maeda, K. (2002) A list of scientific names and the types of mammals published by Japanese researchers. *Mammalian Science* **42**: 1–21 (in Japanese with English abstract).

1331. Kaneko, Y. & Maruyama, N. (2005) Changes in Japanese badger (*Meles meles anakuma*) body weight and condition caused by the provision of food by local people in a Tokyo suburb. *Mammalian Science* **45**: 157–164 (in Japanese with English abstract).

1332. Kaneko, Y., Maruyama, N., & Kanzaki, N. (1996) Growth and seasonal changes in body weight and size of Japanese badger in Hinodecho, suburb of Tokyo. *Journal of Wildlife Research* **1**: 42–46.

1333. Kaneko, Y., Maruyama, N., & Macdonald, D.W. (2006) Food habits and habitat selection of suburban badgers (*Meles meles*) in Japan. *Journal of Zoology (Peking)* **270**: 78–89.

1334. Kaneko, Y., Nakashima, T., & Kimura, Y. (1992) Identification and vertical distribution of two species of *Eothenomys* on Ryo-Hakusan Mountains, central Honshu, Japan. *Bulletin of Gifu Prefectural Museum* **13**: 23–34 (in Japanese with English abstract).

1335. Kaneko, Y., et al. (1998) The biology of the vole *Clethrionomys rufocanus*: a review. *Researches on Population Ecology* **40**: 21–37.

1336. Kaneko, Y. & Sato, M. (1993) Identification and distribution of red-backed voles from Is. Rishiri, Hokkaido (Preliminary study). *Rishiri Studies* **12**: 37–47 (in Japanese).

1337. Kaneko, Y., Suzuki, T., & Atoda, O. (2009) Latrine use in a low density Japanese badger (*Meles anakuma*) population determined by a continuous tracking system. *Mammal Study* **34**: 179–186.

1338. Kanemitsu, H., Fujii, T., & Kannan, Y. (1988) Breeding season, gestation period and litter size of the wild boar *Sus scrofa leucomystax*, in captivity. *Journal of Japanese Association of Zoological Gardens and Aquariums* **30**: 6–8 (in Japanese with English abstract).

1339. Kaneshiro, Y., et al. (2009) Bat fly, *Penicillidia jenynsii*, collected from Saijo city, Ehime Prefecture. *Bulletin of the Shikoku Institute of Natural History* **5**: 19–20 (in Japanese).

1340. Kaneshiro, Y., et al. (2009) Two bat species captured from Saijo city, Ehime Prefecture. *Bulletin of the Shikoku Institute of Natural History* **5**: 15–18 (in Japanese).

1341. Kanno, M., et al. (2004) Geographical distribution of two haplotypes of chloroplast DNA in four oak species (*Quercus*) in Japan. *Journal of Plant Research* **117**: 311–317.

1342. Kano, R. (加納六郎) & Shinonaga, S. (篠永哲) (1997) 日本の有害節足動物 生態と環境変化に伴う変遷 (Venomous and Noxious Arthropods of Japan). Hadano: Tokai University Press, 402 pp. (in Japanese).

1343. Kano, T. (1940) Zoogeographical Studies of the Tsugitake Mountains of Formosa. Tokyo: Shibusawa Institute for Ethnographical Researches, 145 pp.

1344. Kanzaki, N. (神崎伸夫) (2002) イノシシ・イノブタ (*Sus scrofa*). *in* Handbook of Alien Species in Japan (外来種ハンドブック) (Ecological Society of Japan, Editor). Tokyo: Chijin Shokan (地人書館), pp. 77 (in Japanese).

1345. Kanzaki, N. (1995) A study on population dynamics, hunting, and trade of Japanese wild boar (*Sus scrofa leucomystax*). Tokyo: Tokyo University of Agriculture and Technology. 158 pp. (in Japanese).

1346. Kanzaki, N. & Ohtuka, E. (1991) Winter diet and reproduction of Japanese wild boars. *in* Wildlife Conservation: Present Trends and Perspectives for the 21st Century (Proceedings of the International Symposium on Wildlife Conservation in Tsukuba and Yokohama, Japan, August 21–25, 1990: Intecol '90) (Maruyama, N., et al., Editors). Tokyo: Japan Wildlife Research Center, pp. 217–219.

1347. Kanzaki, N. & Perzanowski, K. (1998) A note for a wild boar study in the Bieszczady mountains, Poland. *Willife Forum* **3**: 139–143 (in Japanese).

1348. Kartavtseva, I.V., et al. (2000) The B-chromosome system of the Korean field mouse *Apodemus peninsulae* in the Russian Far East. *Chromosome Science* **4**: 21–29.

1349. Kartavtseva, I.V. & Roslik, G.V. (2004) A complex of B chromosomes in Korean field mouse *Apodemus peninsulae*. *Cytogenetics and Genome Research* **106**: 271–278.

1350. Kasahara, H. (笠原昊) (1950) 日本近海の捕鯨業とその資源. 日本水産株式会社研究所報告 第 4 号. Odawara: Nippon Suisan Co., Ltd, 103 pp. (in Japanese).

1351. Kasahi, T. (重昆達也) & Nagaoka, H. (長岡浩子) (2005) 東京都町田市で保護されたヒナコウモリ. *Bat Study and Conservation Report* **13**: 5–7 (in Japanese).

1352. Kasahi, T., et al. (2014) Attempted eradication of the Pallas's squirrel, an invasive alien species, at the early stage of establishment in Iruma City, Saitama Prefecture, Central Japan. *Bulletin of the Saitama Museum of Natural History* **8**: 19–32 (in Japanese with English summary).

1353. Kasahi, T., et al. (2013) A newly discovered maternity colony and the winter population of the Asian parti-colored bat *Vespertilio sinensis* in the elevated railways of the Joetsu-shinkansen in Gunma Prefecture. *Bulletin of Gunma Museum of Natural History* **17**: 131–146 (in Japanese with English abstract).

1354. Kasahi, T., et al. (2006) Spring roosts of the Endo's pipistrelle, *Pipistrellus endoi*, in the Okutama region, central Japan. *ANIMATE* **6**: 19–26 (in Japanese).

1355. Kasahi, T., et al. (2006) Spring roosts of the Asian parti-coloured bat, *Vespertilio sinensis*, in the Okutama region, central Japan. *ANIMATE* **6**: 27–32 (in Japanese).

1356. Kasahi, T., Urano, M., & Takamizu, Y. (2014) Records of bats in Okutama-machi, Nishitama-gun, Tokyo, central Japan. *ANIMATE* **11**: 36–41 (in Japanese).

1357. Kasamatsu, F. (笠松不二男) & Miyashita, T. (宮下富夫) (1991) 鯨とイルカのフィールドガイド (Field Guide of Whales and dolphins). Tokyo: University of Tokyo Press, 148 pp. (in Japanese).

1358. Kasamatsu, F. (笠松不二男), Miyashita, T. (宮下富夫), & Yoshioka, M. (吉岡基) (2009) Field Guide to Whales, Dolphins and Porpoises in the Western North Pacific and Adjacent Waters. New Edition (新版 鯨とイルカのフィールドガイド). Tokyo: University of Tokyo Press, 148 pp. (in Japanese).

1359. Kasamatsu, F. & Hata., T. (1985) Notes on minke whales in the Okhotsk Sea–West Pacific area. *Report of the International Whaling Commission* **35**: 299–304.

1360. Kasamatsu, M., et al. (2012) Hematology and serum biochemistry values in five captive finless porpoises (*Neophocaena phocaenoides*). *Journal of Veterinary Medical Science* **74**: 1319–1322.

1361. Kashif, M.S. (2006) The status and conservation of bears in Pakistan. *in* Understanding Asian Bears to Secure Their Future (Japan Bear Network, Compiler). Ibaraki: Japan Bear Network, pp. 1–6.

1362. Kashiramoto, A. (頭本昭夫) & Hirotani, H. (広谷浩子) (2002) ニホンザルがムササビを襲う. Available from: http://nh.kanagawa-museum.jp/tobira/8-1/sub3.html (in Japanese).

1363. Kashiwabara, S. & Onoyama, K. (1988) Karyotypes and G-banding patterns of the red-backed voles, *Clethrionomys montanus* and *C. rufocanus bedfordiae* (Rodentia, Microtinae). *Journal of the Mammalogical Society of Japan* **13**: 33–41.

1364. Kasuya, T. (粕谷俊雄) (1986) 生活史的特性値 オキゴンドウ. *in* 漁業公害 (有害生物駆除) 対策調査委託事業調査報告書 (昭和 56–60 年度) (Tamura, T. (田村保), Ohsumi, S. (大隅清治), & Arai, S. (荒井修亮), Editors). Tokyo: Fisheries Agency, pp. 178–187 (in Japanese).

1365. Kasuya, T. (粕谷俊雄) (1996) スジイルカ. *in* The Encyclopaedia of Animals in Japan. 2. Mammals II (日本動物大百科 2 哺乳類 II) (Izawa, K. (伊澤紘生), Kasuya, T. (粕谷俊雄), & Kawamichi, T. (川道武男), Editors). Tokyo: Heibonsha (平凡社), pp. 74–76 (in Japanese).

1366. Kasuya, T. (粕谷俊雄) (1996) マダライルカ. *in* The Encyclopaedia of Animals in Japan. 2. Mammals II (日本動物大百科 2 哺乳類 II) (Izawa, K. (伊澤紘生), Kasuya, T. (粕谷俊雄), & Kawamichi, T. (川道武男), Editors). Tokyo: Heibonsha (平凡社), pp. 76–78 (in Japanese).

1367. Kasuya, T. (粕谷俊雄) (1996) ハシナガイルカ. *in* The Encyclopaedia of Animals in Japan. 2. Mammals II (日本動物大百科 2 哺乳類 II) (Izawa, K. (伊澤紘生), Kasuya, T. (粕谷俊雄), & Kawamichi, T. (川道武男), Editors). Tokyo: Heibonsha (平凡社), pp. 78–79 (in Japanese).

1368. Kasuya, T. (粕谷俊雄) (2011) Conservation Biology of Small Cetaceans around Japan (イルカ: 小型鯨類の保全生物学). Tokyo: University of Tokyo Press, 640 pp. (in Japanese).

1369. Kasuya, T. (粕谷俊雄), et al. (1997) 日本近海産ハンドウイルカの生活史特性値 (Life history parameters of bottlenose dolphins off Japan). *国際海洋生物研究所報告 (IBI Reports)* **7**: 71–107 (in Japanese with English summary).

1370. Kasuya, T. (粕谷俊雄) & Miyazaki, N. (宮崎信之) (1997) 鯨類. *in* レッドデータ 日本の哺乳類 (Mammalogical Society of Japan, Editor). Tokyo: Bun-ichi (文一総合出版), pp. 139–185 (in Japanese).

1371. Kasuya, T. (粕谷俊雄) & Yamada, T. (山田格) (1995) 日本鯨類目録. 鯨研叢書 (Geiken-sosyo) No.7. Tokyo: Institute of Cetacean Research, 90 pp. (in Japanese).

1372. Kasuya, K., et al. (2000) Status of dugong and its protection in Japan. *Report of PRO NATURA FUND* **9**: 29–36 (in Janapese with English summary).

1373. Kasuya, K., et al. (1999) Status of dugong and its protection in Japan. *Report of PRO NATURA FUND* **8**: 6–13 (in Japanese with English summary).

1374. Kasuya, T. (1966) Karyotype of a sei whale. *Scientific Reports of the Whales Research Institute* **20**: 83–88.

1375. Kasuya, T. (1971) Consideration of distribution and migration of toothed whales off the Pacific coast of Japan based upon aerial sighting record. *Scientific Reports of the Whales Research Institute, Tokyo* **23**: 37–60.

1376. Kasuya, T. (1972) Growth and reproduction of *Stenella coeruleoalba* based on the age determination by means of dentinal growth layers. *Scientific Reports of the Whales Research Institute, Tokyo* **24**: 57–79.

1377. Kasuya, T. (1977) Age determination and growth of the Baird's beaked whale with a comment on the fetal growth rate. *Scientific Reports of the Whales Research Institute* **29**: 1–20.

1378. Kasuya, T. (1978) The life history of Dall's porpoise with special reference to the stock off the Pacific coast of Japan. *Scientific Reports of the Whales Research Institute, Tokyo* **30**: 1–63.

1379. Kasuya, T. (1983) Preliminary report of the biology, catch and populations of Phocoenoides in the western North Pacific. *in* Mammals in the Seas:

Small Cetaceans, Seals, Sirenians and Otters. Rome: Food and Agriculture Organization of the United Nations, pp. 3–19.

1380. Kasuya, T. (1985) Effect of exploitation on reproductive parameters of the spotted and striped dolphins off the Pacific coast of Japan. *Scientific Reports of the Whales Research Institute, Tokyo* **36**: 107–138.

1381. Kasuya, T. (1985) Fishery-dolphin conflict in the Iki Island area of Japan. *in* Marine Mammal and Fisheries (Beddington, J.R., Beverton, R.J.H., & Lavigne, D.M., Editors). London: George Allen & Unwin, pp. 253–272.

1382. Kasuya, T. (1986) Distribution and behavior of Baird's beaked whales off the Pacific coast of Japan. *Scientific Reports of the Whales Research Institute* **37**: 61–83.

1383. Kasuya, T. (1991) Density dependent growth in North Pacific sperm whales. *Marine Mammal Science* **7**: 230–257.

1384. Kasuya, T. (1999) Finless porpoise *Neophocaena phocaenoides* (G. Cuvier, 1829). *in* Handbook of Marine Mammals. Volume 6. The Second Book of Dolphins and the Porpoises (Ridgway, S.H. & Harrison, R., Editors). London: Academic Press, pp. 411–442.

1385. Kasuya, T. (1999) Review of biology and exploitation of striped dolphins off Japan. *Journal of Cetacean Research and Management* **1**: 81–100.

1386. Kasuya, T. (2002) Giant beaked whales (*Berardius bairdii* and *B. arnuxii*). *in* Encyclopedia of Marine Mammals (Perrin, W.F., Würsig, B., & Thewissen, J.G.M., Editors). San Diego: Academic Press, pp. 519–522.

1387. Kasuya, T. (2009) Japanese whaling. *in* Encyclopedia of Marine Mammals. 2nd ed. (Perrin, W.F., Würsig, B., & Thewissen, J.G.M., Editors). Amsterdam: Academic Press, pp. 643–649.

1388. Kasuya, T., Brownell, R.L., Jr., & Balcomb, K.C. (1997) Life history of Baird's beaked whales off the Pacific Coast of Japan. *Reports of the International Whaling Commission* **47**: 969–979.

1389. Kasuya, T. & Izumizawa, Y. (1981) The fishery-dolphin conflict in the Iki Island area of Japan (Report No.MMC-80/02 prepared for U.S. Marine Mammal Commission, Washington, D.C.). 31 pp.

1390. Kasuya, T. & Jones, L. (1984) Behavior and segregation of the Dall's porpoise in the northwestern North Pacific Ocean. *Scientific Reports of the Whales Research Institute, Tokyo* **35**: 107–128.

1391. Kasuya, T. & Kureha, K. (1979) The population of finless porpoise in the Inland Sea of Japan. *Scientific Reports of the Whales Research Institute, Tokyo* **31**: 1–44.

1392. Kasuya, T. & Marsh, H. (1984) Life history and reproductive biology of the short-finned pilot whale, *Globicephala macrorhynchus*, off the Pacific coast of Japan. *Reports of the International Whaling Commission (Special Issue)* **6**: 259–310.

1393. Kasuya, T., Marsh, H., & Amino, A. (1993) Non-reproductive mating in short-finned pilot whales. *Reports of the International Whaling Commission (Special Issue)* **14**: 425–437.

1394. Kasuya, T. & Matsui, S. (1984) Age determination and growth of the short-finned pilot whale off the Pacific coast of Japan. *Scientific Reports of the Whales Research Institute, Tokyo* **35**: 57–91.

1395. Kasuya, T. & Miyashita, T. (1988) Distribution of sperm whale stocks in the North Pacific. *Scientific Reports of the Whales Research Institute, Tokyo* **39**: 31–75.

1396. Kasuya, T. & Miyashita, T. (1997) Distribution of Baird's beaked whales off Japan. *Reports of the International Whaling Commission* **47**: 963–968.

1397. Kasuya, T., Miyashita, T., & Kasamatsu, F. (1988) Segregation of two forms of short-finned pilot whales off the Pacific coast of Japan. *Scientific Reports of the Whales Research Institute, Tokyo* **39**: 77–90.

1398. Kasuya, T., Miyazaki, N., & Dawbin, W.H. (1974) Growth and reproduction of *Stenella attenuata* in the Pacific coast of Japan. *Scientific Reports of the Whales Research Institute, Tokyo* **26**: 157–226.

1399. Kasuya, T. & Ohsumi, S. (1966) A secondary sexual character of the sperm whale. *Scientific Reports of the Whales Research Institute, Tokyo* **20**: 89–93.

1400. Kasuya, T. & Rice, D. (1970) Notes on baleen plates and on arrangement of parasitic barnacles of gray whale. *Scientific Reports of the Whales Research Institute* **22**: 39–43.

1401. Kasuya, T. & Tai, S. (1993) Life history of short-finned pilot whale stocks off Japan. *Reports of the International Whaling Commission (Special Issue)* **14**: 439–473.

1402. Kasuya, T., et al. (1986) Perinatal growth of Delphinoids: information from aquarium reared bottlenose dolphins and finless porpoises. *Scientific Reports of the Whales Research Institute, Tokyo* **37**: 85–97.

1403. Kasuya, T., Yamamoto, Y., & Iwatsuki, T. (2002) Abundance decline in the finless porpoise population in the Inland Sea of Japan. *Raffles Bulletin of Zoology Supplement* **10**: 57–65.

1404. Kataoka, T., et al. (2010) Home range and population dynamics of the Japanese squirrel in natural red pine forests. *Mammal Study* **35**: 79–84.

1405. Kataoka, T. & M., W. (2005) Uses of underground dens by the Japanese squirrel, *Sciurus lis*. *in* Abstracts of the 9th International Mammalogical Congress, Sapporo, Japan, P303.

1406. Kataoka, T., et al. (1967) On the food quantity of the finless black porpoises (*Neophocaena phocaenoides*) in captivity. *Journal of Japanese Association of Zoological Gardens and Aquariums* **9**: 46–50 (in Japanese with English summary).

1407. Kataoka, T. & Tamura, N. (2005) Effects of habitat fragmentation on the presence of Japanese squirrels, *Sciurus lis*, in suburban forests. *Mammal Study* **30**: 131–137.

1408. Katayama, A., et al. (1996) Reproductive evaluation of Japanese black bears (*Selenarctos thibetanus japonicus*) by observation of the ovary and uterus. *Japanese Journal of Zoo and Wildlife Medicine* **1**: 26–32 (in Japanese with English summary).

1409. Kato, H. (加藤秀弘) (1994) ミンククジラ. *in* 日本の希少な野生水生生物に関する基礎資料 (I). Tokyo: Japan Fisheries Resources Conservation Association (日本水産資源保護協会), pp. 601–615 (in Japanese).

1410. Kato, H. (加藤秀弘) (1995) セミクジラ. *in* 日本の希少な野生水生生物に関する基礎資料 (II) Tokyo: Japan Fisheries Resource Conservation Association (日本水産資源保護協会), pp. 507–512 (in Japanese).

1411. Kato, H. (加藤秀弘) (1996) セミクジラ. *in* The Encyclopaedia of Animals in Japan. Volume 2: Mammals II (日本動物大百科 2 哺乳類 II) (Izawa, K. (伊沢紘生), Kasuya, T. (粕谷俊雄), & Kawamichi, T. (川道武男), Editors). Tokyo: Heibonsha (平凡社), pp. 38–39 (in Japanese).

1412. Kato, H. (加藤秀弘) & Yoshioka, M. (吉岡基), eds. (2009) シャチの現状と繁殖研究にむけて (Killer whales in Japan: Proceedings of the Symposium of Their Status and Breeding Research in Captivity, November 23, 2007, Tokyo) 鯨研叢書 (Geiken-sosyo) No.14. Tokyo: Institute of Cetacean Research, 96 pp. (in Japanese).

1413. Kato, H. (加藤博) & Iwamoto, M. (岩本雅郎) (1967) ヒナコウモリの採集例. *The Journal of the Mammalogical Society of Japan* **3**: 128 (in Japanese).

1414. Kato, C. (2011) Records on rescued wild animals at Kanagawa Prefecture Natural Environment Conservation Center (2008.4–2010.3). *Bulletin of the Kanagawa Prefecture Natural Environment Conservation Center* **8**: 65–84 (in Japanese).

1415. Kato, H. (1992) Body length, reproduction and stock separation of minke whales off northern Japan. *Report of the International Whaling Commission* **42**: 443–453.

1416. Kato, H. (1995) The Natural History of Sperm Whales (マッコウクジラの自然誌). Tokyo: Heibonsha (平凡社), 317 pp. (in Japanese).

1417. Kato, H. & Kasuya, T. (2002) Some analyses on the modern whaling catch history of the western North Pacific stock of gray whales (*Eschrichtius robustus*), with special reference to the Ulsan whaling ground. *Journal of Cetacean Research and Management* **4**: 277–282.

1418. Kato, H. & Perrin, W.F. (2009) Bryde's whale (*Balaenoptera edeni/B. brydei*). *in* Encyclopedia of Marine Mammals. 2nd ed. (Perrin, W.F., Würsig, B., & Thewissen, J.G.M., Editors). Amsterdam: Academic Press, pp. 158–163.

1419. Kato, H. & Yoshioka, M. (1995) Biological parameters and morphology of Bryde's whales in the western North Pacific, with special reference to stock identification (Paper SC/47/NP11 presented to the IWC Scientific Committee). 18 pp.

1420. Kato, J. (1985) Food and hoarding behavior of the Japanese squirrel. *Japanese Journal of Ecology* **35**: 13–20.

1421. Kato, T. & Ohata, J. (1994) The catalogue of animals of the Oki Island in Mr. Y. Kimura's collection (2). 隠岐の文化財 **11**: 1–10 (in Japanese with English abstract).

1422. Katsuta, S. (勝田節子) & Sato, A. (佐藤顕義) (2010) コウモリに外部寄生するハエ目及びノミ目. 寄せ蛾記 **137**: 33–38 (in Japanese).

1423. Katsuta, S., et al. (2014) Food habits of the birdlike noctule (*Nyctalus aviator*) in Saitama Prefecture, the analysis of feces collected at Kojima, Kumagaya in 2012. *Bulletin of the Saitama Museum of Natural History* **8**: 45–48 (in Japanese).

1424. Kaufmann, J.H. (1982) Raccoon and Allies. *in* Wild Mammals of North America: Biology, Management, and Economics (Chapman, J.A. & Feldhamer, G.A., Editors). Baltimore: Johns Hopkins University Press, pp. 567–585.

1425. Kauhala, K. & Helle, E. (1994) Home ranges and monogamy of the raccoon dog in southern Finland. *Suomen Riista* **40**: 32–41.

1426. Kauhala, K., Helle, E., & Pietila, H. (1998) Time allocation of male and female raccoon dogs to pup rearing at the den. *Acta Theriologica* **43**: 301–310.

1427. Kauhala, K., Helle, E., & Taskinen, K. (1993) Home range of the raccoon dog (*Nyctereutes procyonoides*) in southern Finland. *Journal of Zoology* **231**: 95–106.

1428. Kauhala, K. & Saeki, M. (2004) Raccoon dogs. *in* The Biology and Conservation of Wild Canids (Macdonald, D.W. & Sillero-Zubiri, C., Editors). Oxford: Oxford University Press, pp. 217–226.

1429. Kauhala, K. & Saeki., M. (2004) Raccoon dog. in Canids: Foxes, Wolves, Jackals and Dogs. Status Survey and Conservation Action Plan (Sillero-Zubiri, C., Hoffmann, M., & Macdonald, D.W., Editors). Gland: IUCN, pp. 136–142.

1430. Kawabe, M. (2008) The distribution of Ochotona hyperborea yesoensis in Hokkaido. Bulletin of the Higashi Taisetsu Museum of Natural History 30: 1–20 (in Japanese with English summary).

1431. Kawabe, M. (2014) Ochotona hyperborea yesoensis Kishida, 1930. in Red Data Book 2014: Threatened Wildlife of Japan. 1 Mammalia (Ministry of the Environment, Editor). Tokyo: Gyosei Corporation, p. 82 (in Japanese).

1432. Kawabe, M., Shimizu, C., & Sawada, Y. (2009) The habitat of Ochotona hyperborea yesoensis in the Yubari Mountains. Bulletin of the Higashi Taisetsu Museum of Natural History 31: 23–34 (in Japanese with English summary).

1433. Kawabuchi, T., et al. (2005) Babesia microti-like parasites detected in feral raccoons (Procyon lotor) captured in Hokkaido, Japan. Journal of Veterinary Medical Science 67: 825–827.

1434. Kawada, S. (2006) Recent advance in the taxonomy of Japanese moles, genus Mogera; a new perspective based on karyological research. Taxa, Proceedings of the Japanese Society of Systematic Zoology 20: 41–50 (in Japanese).

1435. Kawada, S. (川田伸一郎), et al. (2001) ロシアデスマン (Desmana moschata) の歯式の再検討とその歯式決定における諸問題. Special Publication of Nagoya Society of Mammalogists 3: 41–48 (in Japanese).

1436. Kawada, S. Unpublished.

1437. Kawada, S., et al. (2001) Karyosystematic analysis of Japanese talpine moles in the genera Euroscaptor and Mogera (Insectivora, Talpidae). Zoological Science 18: 1003–1010.

1438. Kawada, S. & Obara, Y. (1999) Reconsideration of the karyological relationship between two Japanese species of shrew-moles Dymecodon pilirostoris and Urotrichus talpoides. Zoological Science 16: 167–174.

1439. Kawaguchi, Y. (川口嘉夫) (1998) 七尾市鵜浦町中浦の更新世後期中位海成段丘層動物化石群集について (Fossil fauna from the late Pleistcene middle marine terrace deposits of Nakaura, Unoura, Nana, Noto, central Japan). Report of the Nanao Children Science Museum 2: 29–36 (in Japanese).

1440. Kawaguchi, S. (2004) Distribution of Apodemus speciosus and Mogera wogura on islands in the inland sea. Mammalian Science 43: 121–126 (in Japanese with English abstract).

1441. Kawaguchi, S., Kaneko, Y., & Hasegawa, Y. (2009) A new species of the fossil murine rodent from Pinza-Abu Cave, the Miyako Island of the Ryukyu Archipelago, Japan. Bulletin of Gunma Museum of Natural History 13: 15–28.

1442. Kawahara, A. (2005) Sorex minutissimus in Hamanaka-cho, Akkeshi-gun, Hokkaido (1). Habitats. Journal of the Japanese Wildlife Research Society 31: 11–18 (in Japanese with English abstract).

1443. Kawahara, A. (2005) Sorex minutissimus in Hamanaka-cho, Akkeshi-gun, Hokkaido (2) Habitat on Kenbokki Island and an observation of behavior. Journal of the Japanese Wildlife Research Society 31: 19–24 (in Japanese with English abstract).

1444. Kawahara, A. (河原淳) (2013) チビトガリネズミ Sorex minutissimus の生態学的研究. Doctoral dissertation. Ebetsu: Rakuno Gakuen University (in Japanese with English summary).

1445. Kawahara, A., Morishita, T., & Yanagawa, H. (2003) A collection of Chiroptera from the western area of Hidaka, central Hokkaido. Journal of the Japanese Wildlife Research Society 29: 12–18 (in Japanese with English summary).

1446. Kawahara, A., et al. (2004) Records of bats in Akan and Shiranuka, eastern Hokkaido. Journal of the Japanese Wildlife Research Society 30: 15–20 (in Japanese with English summary).

1447. Kawai, K. (2000) Faunal survey of bats in Nukabira, Tokachi area, Hokkaido. Bulletin of the Higashi Taisetsu Museum of Natural History 22: 1–4 (in Japanese with English abstract).

1448. Kawai, K. (2000) Notes on the specimens of bats preserved in the Higashi Taisetu Museum of Natural History. Bulletin of the Higashi Taisetsu Museum of Natural History 22: 5–7 (in Japanese with English abstract).

1449. Kawai, K. (2006) Faunal survey on bats in Tokachi Station of the National Livestock Breeding Center (NLBC). Bulletin of the Asian Bat Research Institute 5: 1–8 (in Japanese with English abstract).

1450. Kawai, K. & Akasaka, T. (2008) Day roost character of bats in Tokachi Station of National Livestock Breeding Center (NLBC), Hokkaido. Bulletin of the Asian Bat Research Institute 7: 9–16 (in Japanese with English abstract).

1451. Kawai, K., et al. (2015) Insights into the natural hisotry of Pipistrellus endoi Imaizumi, 1959 from survey records in Miyagi prefecture. Research Bulletin of Environmental Education Center, Miyagi University of Education 17: 49–53.

1452. Kawai, K., et al. (2007) Bat fauna in Tsushima, Nagasaki, Japan. Mammalian Science 47: 239–253 (in Japanese with English abstract).

1453. Kawai, K., et al. (2010) Vespertilio murinus Linnaeus, 1758 confirmed in Japan from morphology and mitochondrial DNA. Acta Chiropterologica 12: 463–470.

1454. Kawai, K., et al. (2013) Refugia in glacial ages led to the current discontinuous distribution patterns of the dark red-backed vole Myodes rex on Hokkaido, Japan. Zoological Science 30: 642–650.

1455. Kawai, K. & Kondo, N. (2007) Survey of bats in Oketo, Hokkaido. Bulletin of the Asian Bat Research Institute 6: 11–15 (in Japanese with English abstract).

1456. Kawai, K., et al. (2011) Faunal survey of bats in Kunashir Island. Memoirs of the Nemuro City Museum of History and Nature 23: 63–68 (in Japanese with English abstract).

1457. Kawai, K., et al. (2006) Distinguishing between cryptic species Myotis ikonnikovi and M. brandtii gracilis in Hokkaido, Japan: Evaluation of a novel diagnostic morphological feature using molecular methods. Acta Chiropterologica 8: 95–102.

1458. Kawai, K., et al. (2003) The status of the Japanese and East Asian bats of the genus Myotis (Vespertilionidae) based on mitochondrial sequences. Molecular Phylogenetics and Evolution 28: 297–307.

1459. Kawai, K., et al. (2002) Intra- and interfamily relationships of Vespertilionidae inferred by various molecular markers including SINE insertion data. Journal of Molecular Evolution 55: 284–301.

1460. Kawai, K., et al. (2014) Bats from Kunashir and Iturup Islands. Biodiversity and Biogeography of the Kuril Islands and Sakhalin 4: 74–81.

1461. Kawai, K., et al. (in press) First record of the parti-coloured bat Vespertilio murinus (Chiroptera: Vespertilionidae) from Ishikawa prefecture provides insights into the migration of bats to Japan. Mammal Study.

1462. Kawai, K. & Yokohata, Y. Unpublished.

1463. Kawai, M., Azuma, S., & Yoshiba, K. (1967) Ecological studies of Japanese monkeys (Macaca fuscata): I. Problems of the birth season. Primates 9: 1–12.

1464. Kawakami, T. (1980) A review of sperm whale food. Scientific Reports of the Whales Research Institute, Tokyo 32: 199–218.

1465. Kawamichi, M. (川道美枝子) (2000) シマリス (Siberian chipmunk). in Hibernation in Mammals (冬眠する哺乳類) (Kawamichi, T. (川道武男), Kondo, N. (近藤宣昭), & Morita, T. (森田哲夫), Editors). Tokyo: University of Tokyo Press, pp. 143–161 (in Japanese).

1466. Kawamichi, M. (1980) Food, food hoarding and seasonal changes of Siberian chipmunks. Japanese Journal of Ecology 30: 211–220.

1467. Kawamichi, M. (1989) Nest structure dynamics and seasonal use of nests by Siberian chipmunks (Eutamias sibiricus). Journal of Mammalogy 70: 44–57.

1468. Kawamichi, M. (1996) Ecological factors affecting annual variation in commencement of hibernation in wild chipmunks (Tamias sibiricus). Journal of Mammalogy 77: 731–744.

1469. Kawamichi, M. & Kawamichi, T. (1984) Maternal care and independence of young Siberian chipmunk. Mammalian Science 48: 3–17 (in Japanese).

1470. Kawamichi, T. (1969) Behavior and daily activities of the Japanese pika, Ochotona hyperborea yesoensis. Journal of the Faculty of Science Hokkaido University Series VI, Zoology 17: 127–151.

1471. Kawamichi, T. (1970) Social pattern of the Japanese pika Ochotona hyperborea yesoensis, preliminary report. Journal of the Faculty of Science Hokkaido University Series VI, Zoology 17: 462–473.

1472. Kawamichi, T. (1971) Annual cycle of behavior and social pattern of the Japanese pika Ochotona hyperborea yesoensis. Journal of the Faculty of Science Hokkaido University Series VI, Zoology 18: 173–185.

1473. Kawamichi, T. (1981) Vocalizations of Ochotona as a taxonomic character. in Proceedings of the World Lagomorph Conference (Myers, K. & MacInners, C.D., Editors). Guelph, Ontario: Guelph University Press, pp. 324–339.

1474. Kawamichi, T. (1997) Seasonal changes in the diet of Japanese giant flying squirrels in relation to reproduction. Journal of Mammalogy 78: 204–212.

1475. Kawamichi, T. (1998) Seasonal change in the testis size of the Japanese giant flying squirrel, Petaurista leucogenys. Mammal Study 23: 79–82.

1476. Kawamichi, T. (2010) Biannual reproductive cycles in the Japanese giant flying squirrel (Petaurista leucogenys). Journal of Mammalogy 91: 905–913.

1477. Kawamichi, T. & Kawamichi, M. (1993) Gestation period and litter size of Siberian chipmunk Eutamias sibiricus lineatus in Hokkaido, northern Japan. Journal of Mammalogical Society of Japan 18: 105–109.

1478. Kawamichi, T., Kawamichi, M., & Kishimoto, R. (1987) Social organizations of solitary mammals. in Animal Societies: Theories and Facts (Ito, Y., Brown, J.L., & Kikkawa, J., Editors). Tokyo: Japan Scientific Societies Press, pp. 173–188.

1479. Kawamichi, T. & Yamada, F. (1995) Present taxonomic status of Japanese lagomorphs. *Mammalian Science* **35**: 193–202 (in Japanese).
1480. Kawamoto, Y., Hagihara, K., & Aizawa, K. (2004) Finding of hybrid individuals between native Japanese macaques and introduced rhesus macaques in the Bousou Peninsula, Chiba, Japan. *Primate Research* **20** 89–95 (in Japanese with English summary).
1481. Kawamoto, Y., Kawamoto, S., & Kawai, S. (2005) Hybridization of introduced Taiwanese macaques with native Japanese macaques in Shimokita Peninsula, Aomori, Japan. *Primate Research* **21**: 11–18 (in Japanese with English summary).
1482. Kawamoto, Y., et al. (2001) Genetic assessment of a hybrid population between Japanese and Taiwan macaques in Wakayama Prefecture. *Primate Research* **17**: 13–24 (in Japanese with English summary).
1483. Kawamoto, Y., et al. (1999) A case of hybridization between the Japanese and Taiwan macaques found in Wakayama Prefecture. *Primate Research* **15**: 53–60 (in Japanese with English summary).
1484. Kawamoto, Y., et al. (2007) Postglacial population expansion of Japanese macaques (*Macaca fuscata*) inferred from mitochondrial DNA phylogeography. *Primates* **48**: 27–40.
1485. Kawamoto, Y., et al. (2008) Genetics of the Shimokita macaque population suggest an ancient bottleneck. *Primates* **49**: 32–40.
1486. Kawamura, A. (河村章人) (1996) イワシクジラ. in The Encyclopaedia of Animals in Japan. 2. Mammals II (日本動物大百科 2 哺乳類 II) (Izawa, K. (伊沢紘生), Kasuya, T. (粕谷俊雄), & Kawamichi, T. (川道武男), Editors). Tokyo: Heibonsha (平凡社), pp. 40–41 (in Japanese).
1487. Kawamura, A. (河村愛) & Kawamura, Y. (河村善也) (2013) 白保竿根田原洞穴遺跡の後期更新世と完新世の小型哺乳類遺体. in 沖縄県立埋蔵文化財センター調査報告書第65集 白保竿根田原洞穴遺跡—新石垣空港建設工事に伴う緊急発掘調査報告書. Nishihara: Okinawa Prefectural Archaeological Center (沖縄県埋蔵文化財センター), pp. 154–175 (in Japanese).
1488. Kawamura, R. (川村麟也) & Ikeda, K. (池田嘉平) (1935) 恙虫病発生原野に於けるハタネズミ *Microtus montebelli* の生態的観察. *Zoological Magazine* **47**: 90–101 (in Japanese).
1489. Kawamura, Y. (河村善也) (1977) ウルム氷期の小哺乳類—岐阜県熊石洞の小型哺乳動物化石—. *Journal of Fossil Research* **14**: 5–10 (in Japanese).
1490. Kawamura, Y. (河村善也) (1991) ナウマンゾウと共存した哺乳類 (Mammals co-distributed with Naumann's elephant). in 日本の長鼻類化石 (Japanese Proboscidean Fossils) (Kamei, T. 亀井節夫, Editor). Tokyo: Tsukiji Shokan (築地書館), pp. 164–170 (in Japanese).
1491. Kawamura, Y. (河村善也) (2003) アバクチ洞穴の完新世小型哺乳類遺体. in 北上山地に日本更新世人類化石を探る—岩手県大迫町アバクチ・風穴洞穴遺跡の発掘— (Dodo, Y. (百々幸雄), Takigawa, W. (瀧川渉), & Sawada, J. (澤田純明), Editors). Sendai: Tohoku University Press, pp. 156–184 (in Japanese).
1492. Kawamura, Y. (河村善也) (2003) アバクチ洞穴の後期更新世脊椎動物遺体. in 北上山地に日本更新世人類化石を探る—岩手県大迫町アバクチ・風穴洞穴遺跡の発掘— (Dodo, Y. (百々幸雄), Takigawa, W. (瀧川渉), & Sawada, J. (澤田純明), Editors). Sendai: Tohoku University Press, pp. 185–200 (in Japanese).
1493. Kawamura, Y. (河村善也) (2003) 風穴洞穴の完新世および後期更新世の哺乳類遺体. in 北上山地に日本更新世人類化石を探る—岩手県大迫町アバクチ・風穴洞穴遺跡の発掘— (Dodo, Y. (百々幸雄), Takigawa, W. (瀧川渉), & Sawada, J. (澤田純明), Editors). Sendai: Tohoku University Press, pp. 284–386 (in Japanese).
1494. Kawamura, A. (1977) On the food of Bryde's whales caught in the south Pacific and Indian Oceans. *Scientific Reports of the Whales Research Institute* **29**: 49–58.
1495. Kawamura, A. (1980) A review of food of Balaenopterid whales. *Scientific Reports of the Whales Research Institute* **32**: 155–197.
1496. Kawamura, A. (1982) Food habits and prey distribution of three rorqual species in the North Pacific Ocean. *Scientific Reports of the Whales Research Institute* **34**: 59–61.
1497. Kawamura, A. & Kawamura, Y. (2013) Quaternary mammal fossil newly collected from Tanabaru Cave on Miyako Island, Okinawa prefecture, Japan. *Journal of the Speleological Society of Japan* **38**: 20–36.
1498. Kawamura, A. & Satake, Y. (1976) Preliminary report on the geographical distribution of the Bryde's whale in the North pacific with special reference to the structure of filtering apparatus. *Scientific Reports of the Whales Research Institute* **28**: 1–35.
1499. Kawamura, Y. (1977) Micro-mammals in the Würm Glacial Time. Micromammalian fossils from Kumaishi-do Cave, Gifu Prefecture, Central Japan. *Fossil Club Bulletin* **14**: 5–10 (in Japanese).
1500. Kawamura, Y. (1978) Small mammalian remains of Taishaku-Kannondo Cave Site, Part 1. *Annual Bulletin of Hiroshima University Taishaku-kyo Sites Research Centre* **1**: 55–67 (in Japanese).
1501. Kawamura, Y. (1979) Small mammalian remains of Taishaku-Kannondo Cave Site, Part 2. *Annual Bulletin of Hiroshima University Taishaku-kyo Sites Research Centre* **2**: 45–55 (in Japanese).
1502. Kawamura, Y. (1980) Mammalian remains of the Pre-Jomon Period from Taishaku-Kannondo Cave Site (Part 1). Mammalian remains obtained by the excavation of 1975. *Annual Bulletin of Hiroshima University Taishaku-kyo Sites Research Centre* **3**: 61–74 (in Japanese).
1503. Kawamura, Y. (1981) Mammalian remains of the Pre-Jomon Period from Taishaku-Kannondo Cave Site (Part 2). Mammalian remains obtained by the excavation of 1976. *Annual Bulletin of Hiroshima University Taishaku-kyo Sites Research Centre* **4**: 67–88 (in Japanese).
1504. Kawamura, Y. (1982) Biogeographical aspects of the Quaternary mammals of Japan. *Mammalian Science* **22(1&2)**: 99–130 (in Japanese).
1505. Kawamura, Y. (1982) Mammalian remains of the Pre-Jomon Period from Taishaku-Kannondo Cave Site (Part 3). Mammalian remains obtained by the excavation of 1978. *Annual Bulletin of Hiroshima University Taishaku-kyo Sites Research Centre* **5**: 57–70 (in Japanese).
1506. Kawamura, Y. (1983) Holocene vertebrate remains from Taishaku-Anagami Rockshelter Site. *Annual Bulletin of Hiroshima University Taishaku-kyo Sites Research Centre* **6**: 53–64 (in Japanese).
1507. Kawamura, Y. (1988) Quaternary rodent faunas in the Japanese Islands (Part 1). *Memoirs of the Faculty of Science, Kyoto University, Series of Geology and Mineralogy* **53**: 31–348.
1508. Kawamura, Y. (1989) Quaternary rodent faunas in the Japanese Islands (Part 2). *Memoirs of the Faculty of Science, Kyoto University, Series of Geology & Mineralogy* **54**: 1–235.
1509. Kawamura, Y. (1991) Quaternary mammalian faunas in the Japanese Islands. *The Quaternary Research* **30**: 213–220 (in Japanese with English abstract).
1510. Kawamura, Y. (1994) Late Pleistocene to Holocene mammalian faunal succession in the Japanese Islands, with comments on the Late Quaternary extinctions. *Archeozoologia* **6**: 7–22.
1511. Kawamura, Y. (2009) Fossil record of sika deer in Japan. in Sika Deer–Biology and Management of Native and Introduced Populations (McCullough, D.R., Takatsuki, S., & Kaji, K., Editors). Tokyo: Springer, pp. 11–25.
1512. Kawamura, Y. & Ishida, S. (1976) Preliminary report on the Late Pleistocene micro-mammalian fossils from Kumaishi-do Cave, Gifu Prefecture, central Japan. *Journal of the Speleological Society of Japan* **1**: 28–34 (in Japanese with English abstract).
1513. Kawamura, Y. & Kajiura, K. (1980) Mammalian fossils from Sugi-ana Cave, Gifu Prefecture, Central Japan. *Journal of the Speleological Society of Japan* **5**: 50–65 (in Japanese with English abstract).
1514. Kawamura, Y., Kamei, T., & Taruno, H. (1989) Middle and Late Pleistocene mammalian faunas in Japan. *The Quaternary Research* **28**: 317–326 (in Japanese with English abstract).
1515. Kawamura, Y. & Matsuhashi, Y. (1989) Late Pleistocene fissure sediments and their mammalian fauna at Site 5 of Yage Quarry, Inasa, Shizuoka Prefecture, central Japan. *The Quaternary Research* **28**: 95–102 (in Japanese with English).
1516. Kawamura, Y., Matsuhashi, Y., & Matsuura, S. (1990) Late Quaternary mammalian faunas at Suse Quarry, Toyohashi, central Japan, and their implications for the reconstruction of the faunal succession. *The Quaternary Research* **29**: 307–317 (in Japanese with English).
1517. Kawamura, Y. & Nakagoshi, T. (1997) Late Quaternary mammalian extinctions in central and western Honshu, and related problems. *Annual Bulletin of Hiroshima University Teishaku-kyo Sites Research Center* **XII**: 155–168 (in Japanese).
1518. Kawamura, Y. & Sotsuka, T. (1984) Preliminary report on the Quaternary mammalian remains from several caves on the Hiraodai Plateau, Fukuoka Prefecture, Northern Kyushu, Japan. *Bulletin of the Kitakyushu Museum of Natural History* **5**: 163–188 (in Japanese with English abstract).
1519. Kawamura, Y. & Tamiya, S. (1980) Report of the first to the third excavations of Tanuki-ana cave in the Akiyoshi-dai Plateau, Yamaguchi Prefecture, western Japan. *Bulletin of the Akiyoshi-dai Museum of Natural History* **15**: 15–46.
1520. Kawamura, Y. & Nishioka, Y. (2011) Significance of vole remains of the genus *Microtus* from Shikoku Island, Japan. in Abstracts with Programs, Regular Meeting of the Palaeontological Society of Japan (in Japanese).
1521. Kawamura, Y., Yamada, Y., & Ando, Y. (1986) Late Pleistocene micromammals from Taishaku-Kannondo Cave Site (first preliminary report). *Annual Bulletin of Hiroshima University Taishaku-kyo Sites Research Centre* **9**: 67–85 (in Japanese).
1522. Kawano, M. (河野通弘), Fujii, A. (藤井厚志), & Cave Research Club of Yamaguchi University (山口大学洞穴研究会) (1985) 秋吉台雲出原の洞窟 (Limestone caves in Kumode-hara area on the Akiyoshi Plateau). in 秋吉台雲出原—中規模観光レクリエーション地区学術調査報告 (中規模観光レクリエーション地区学術調査団, Editor). Shuho: Shuho Town Office (秋芳町),

pp. 20–33 (in Japanese with English summary).
1523. Kawasaki, S. Personal communication.
1524. Kawata, M. (1985) Mating system and reproductive success in a spring population of the red-backed vole, *Clethrionomys rufocanus bedfordiae*. *Oikos* **45**: 181–190.
1525. Kawata, M. (1985) Sex differences in the spatial distribution of genotypes in the red-backed vole, *Clethrionomys rufocanus bedfordiae*. *Journal of Mammalogy* **66**: 384–387.
1526. Kawata, M. (1987) Pregnancy failure and suppression by female–female interaction in enclosed populations of the red-backed vole, *Clethrionomys rufocanus bedfordiae*. *Behavioral Ecology and Sociobiology* **20**: 89–97.
1527. Kawata, M. (1988) Mating success, spatial organization, and male characteristics in experimental field populations of the red-backed vole, *Clethrionomys rufocanus bedfordiae*. *Journal of Animal Ecology* **57**: 217–235.
1528. Kazacos, K.R. (1986) Raccoon ascarids as a cause of larva migrans. *Parasitology Today* **2**: 253–255.
1529. Kazacos, K.R. & Boyce, W.M. (1989) *Baylisasaris* larva migrans. *Journal of the American Veterinary Medical Association* **195**: 894–903.
1530. Kazacos, K.R., Vestre, W.A., & Kazacos, E.A. (1984) Raccoon ascarid larvae (*Baylisasaris procyonis*) as a cause of ocular larva migrans. *Investigative Ophthalmology and Visual Science* **25**: 1177–1183.
1531. Kazacos, K.R., et al. (1981) Raccoon ascarid larvae as a cause of fetal central nervous system disease in subhuman primates. *Journal of the American Veterinary Medical Association* **179**: 1089–1094.
1532. Kelly, B.P. (1988) Ringed seal, *Phoca hispida*. *in* Selected Marine Mammals of Alaska (Lentfe, J.W., Editor). Washington, D.C.: Marine Mammal Commission, pp. 57–76.
1533. Kelly, B.P. (1988) Bearded seal, *Erignathis barbatus*. *in* Selected Marine Mammals of Alaska (Lentfe, J.W., Editor). Washington, D.C.: Marine Mammal Commission, pp. 77–94.
1534. Kelly, B.P. (1988) Ribbon seal, *Phoca fasciata*. *in* Selected Marine Mammals of Alaska (Lentfe, J.W., Editor). Washington, D.C.: Marine Mammal Commission, pp. 95–106.
1535. Kenney, R.D. (2002) North Atlantic, North Pacific, and southern right whales *Eubalaena glacialis*, *E. japonica*, and *E. australis*. *in* Encyclopedia of Marine Mammals (Perrin, W.F., Würsig, B., & Thewissen, J.G.M., Editors). San Diego: Academic Press, pp. 806–813.
1536. Kenyon, K.W. (1969) The sea otters in the eastern Pacific Ocean. *North American Fauna* **68**: 1–352.
1537. Kenyon, K.W. & Scheffer, V.B. (1954) A Population Study of the Alaska Fur Seal Herd. (U.S. Fish and Wildlife Service. Special Scientific Report: Wildlife. No.12). Washington: U.S. Department of the Interior, 77 pp.
1538. Kenyon, K.W. & Wilke, F. (1953) Migration of the northern fur seal *Callorhinus ursinus*. *Journal of Mammalogy* **34**: 86–98.
1539. Kershaw, F., et al. (2013) Population differentiation of 2 forms of Bryde's whales in the Indian and Pacific Oceans. *Journal of Heredity* **104**: 755–764.
1540. Keyes, M.C. (1965) Pathology of the northern fur seal. *Journal of American Veterinary Medical Association* **147**: 1091–1095.
1541. Kifune, T. (1980) Description of *Duboisitrema sawadai* gen. et. sp. nov. from some Japanese Chiroptera (Trematoda: Lecithodendriidae). *Japanese Journal of Parasitology* **29**: 393–397.
1542. Kifune, T. (1980) Records of the trematode parasites of bats from Honshu and Kyushu with descriptions of two new species of the genus *Prosthodendrium*. *Medical Bulletin of Fukuoka University* **7**: 387–394.
1543. Kifune, T. (1998) A tentative checklist of trematode parasites of Japanese Insectivora. *Transactions of Nagasaki Biological Society* **49**: 23–32 (in Japanese).
1544. Kifune, T., Iwata, K., & Satô, M. (2004) Records of trematode parasites of several bats in Hokkaido. *Rishiri Studies* **23**: 15–17 (in Japanese with English abstract).
1545. Kifune, T. & Saito, T. (1996) Records of three trematode parasites in two Japanese shrew moles. *Medical Bulletin of Fukuoka University* **23**: 241–243.
1546. Kifune, T. & Sawada, I. (1979) Helminth fauna of bats in Japan XXI. *Medical Bulletin of Fukuoka University* **6**: 291–301.
1547. Kifune, T. & Sawada, I. (1980) Helminth fauna of bats in Japan XXIII. *Medical Bulletin of Fukuoka University* **7**: 169–181.
1548. Kifune, T. & Sawada, I. (1980) Helminth fauna of bats in Japan XXIV. *Medical Bulletin of Fukuoka University* **7**: 263–272.
1549. Kifune, T. & Sawada, I. (1982) Helminth fauna of bats in Japan XXVI. *Medical Bulletin of Fukuoka University* **9**: 101–113.
1550. Kifune, T. & Sawada, I. (1984) Helminth fauna of bats in Japan XXX. *Medical Bulletin of Fukuoka University* **11**: 95–111.
1551. Kifune, T. & Sawada, I. (1986) Helminth fauna of bats in Japan XXXV. *Medical Bulletin of Fukuoka University* **13**: 197–208.
1552. Kifune, T. & Sawada, I. (1989) Two new species of the genus *Acanthatrium* and records of coexisting trematodes in Japanese bats. *Medical Bulletin of Fukuoka University* **16**: 353–358.
1553. Kifune, T. & Sawada, I. (1993) Helminth fauna of bats in Japan XLIV. *Medical Bulletin of Fukuoka University* **20**: 1–10.
1554. Kifune, T. & Sawada, I. (1993) Helminth fauna of bats in Japan XLV. *Medical Bulletin of Fukuoka University* **20**: 67–78.
1555. Kifune, T. & Sawada, I. (1993) Helminth fauna of bats in Japan XLVI. *Medical Bulletin of Fukuoka University* **20**: 221–231.
1556. Kifune, T. & Sawada, I. (1997) Helminth fauna of bats in Japan L. *Medical Bulletin of Fukuoka University* **24**: 91–100.
1557. Kifune, T., Sawada, I., & Harada, M. (1992) Trematode parasites of the Japanese shrew moles (*Urotrichus* and *Dymecodon*) with descriptions of two new species. *Medical Bulletin of Fukuoka University* **19**: 101–108.
1558. Kifune, T., Sawada, I., & Harada, M. (1994) Trematode parasites of some Japanese insectivorous mammals. *Medical Bulletin of Fukuoka University* **21**: 61–64.
1559. Kifune, T., Sawada, I., & Harada, M. (2001) Helminth fauna of bats in Japan LIV. *Medical Bulletin of Fukuoka University* **28**: 1–9.
1560. Kikuchi, S., et al. (1995) Morphology of *Crassicauda giliakiana* (Nematoda; Spirurida) from a Cuvier's beaked whale *Ziphius cavirostris*. *Japanese Journal of Parasitology* **44**: 228–237.
1561. Kikuchi, S., Okuyama, Y., & Nakajima, M. (1987) *Nasitrema lagenorhynchus* n. sp. from the larynx and lungs of a pacific striped dolphin (Nasitrematidae, Trematoda). *Japanese Journal of Parasitology* **36**: 42–48.
1562. Kikuzawa, K. (1988) Dispersal of *Quercus mongolica* acorns in a broad-leaved deciduous forest 1. Disappearance. *Forest Ecology and Management* **25**: 1–8.
1563. Kimura, Y. (木村吉幸), Kaneko, Y. (金子之史), & Yoshida, T. (吉田忠義) (1994) 安達太良山系の小哺乳類—特にビロードネズミ属について—. *Fukushima Seibutsu (福島生物)* **37**: 13–19 (in Japanese).
1564. Kimura, Y. (木村吉幸), et al. (2002) 福島県の翼手類 III. *Fukushima Seibutsu (福島生物)* **45**: 15–18 (in Japanese).
1565. Kimura, Y. (木村吉幸), Togashi, Y. (富樫祐美子), & Sato, M. (佐藤正幸) (2003) 福島県の翼手類 IV. *Fukushima Seibutsu (福島生物)* **46**: 29–35 (in Japanese).
1566. Kimura, K., Takeda, A., & Uchida, T.A. (1987) Changes in progesterone concentrations in the Japanese long-fingered bat, *Miniopterus schreibersii fuliginosus*. *Journal of Reproduction and Fertility* **80**: 59–63.
1567. Kimura, K. & Uchida, T.A. (1983) Ultrastructural observations of delayed implantation in the Japanese long-fingered bat, *Miniopterus schreibersii fuliginosus*. *Journal of Reproduction and Fertility* **69**: 187–193.
1568. Kimura, M. (2000) Paleogeography of the Ryukyu Islands. *Tropics* **10**: 5–24.
1569. Kimura, R., et al. (2014) Mutations in the testis-specific enhancer of *SOX9* in the *SRY* independent sex-determining mechanism in the genus *Tokudaia*. *PLoS ONE* **9**: e108779.
1570. Kimura, S., et al. (2012) Seasonal changes in the local distribution of Yangtze finless porpoises related to fish presence. *Marine Mammal Science* **28**: 308–324.
1571. Kimura, S., et al. (2013) Variation in the production rate of biosonar signals in freshwater porpoises. *Journal of the Acoustical Society of America* **133**: 3128–3134.
1572. Kimura, T. & Hasegawa, Y. (2005) Fossil finless porpoise, *Neophocaena phocaenoides*, from the Seto Inland Sea, Japan. *Bulletin of Gunma Museum of Natural History* **9**: 65–72 (in Japanese with English summary).
1573. Kimura, T., et al. (2007) A new species of *Eubalaena* (Cetacea: Mysticeti: Balaenidae) from the Gonda Formation (latest Miocene–Early Pliocene) of Japan. *Bulletin of Gunma Museum of Natural History* **11**: 15–27 (in Japanese).
1574. Kimura, Y. (1974) Notes on the bats from Fukushima Prefecture I. *Fukushima Seibutsu* **17**: 16–18 (in Japanese).
1575. Kimura, Y. (2001) Notes on the bats from Fukushima Prefecture II. *ANIMATE* **2**: 19–21 (in Japanese).
1576. Kimura, Y., Kaneko, Y., & Konno, M. (2001) The shift in the altitudinal distributions of *Dymecodon pilirostris* and *Urotrichus talpoides* in the Mt. Bandai area, Fukushima Prefecture, Japan. *Mammalian Science* **41**: 71–82 (in Japanese with English summary).
1577. Kimura, Y., et al. (2002) Bat fauna in Fukushima Prefecture, Japan. *Mammalian Science* **42**: 71–77 (in Japanese with English abstract).
1578. King, C.M. (1983) *Mustela erminea*. *Mammalian Species* **195**: 1–8.
1579. King, C.M. (1990) Natural History of Weasels and Stoats. New York: Comstock Publishing Associates, 253 pp.
1580. King, J.E. (1964) Seals of the World. London: British Museum (Natural History), 240 pp.
1581. King, J.E. (1983) Seals of the World. 2nd ed. London: British Museum

(Natural History), 240 pp.
1582. Kingston, S.E. & Rosel, P.E. (2004) Genetic differentiation among recently diverged delphinid taxa determined using AFLP markers. *Journal of Heredity* **95**: 1–10.
1583. Kinjo, K. Unpublished.
1584. Kinjo, K. & Maeda, K. (1999) A record of the Japanese large noctule *Nyctalus aviator* from Okinawa Island. *Biological Magazine Okinawa* **37**: 61–64 (in Japanese).
1585. Kinjo, K. & Nakamoto, A. Unpublished.
1586. Kinjo, K., Nakamoto, A., & Izawa, M. Unpublished.
1587. Kinjo, T. & Minamoto, N. (1987) Serologic survey for selected microbial pathogens in Japanese serow (*Capricornis crispus*) in Gifu Prefecture, Japan. in The Biology and Management of Capricornis and Related Mountain Antelopes (Soma, H., Editor). London: Croom Helm, pp. 299–311.
1588. Kinoshita, E. (木下栄次郎) (1928) 野鼠ノ森林保護學的研究. *Research Bulletins of the College Experiment Forests, Hokkaido Imperial University* **5**: 167–282 (in Japanese).
1589. Kinoshita, G., et al. (2012) Ancient colonization and within-island vicariance revealed by mitochondrial DNA phylogeography of the mountain hare (*Lepus timidus*) in Hokkaido, Japan. *Zoological Science* **29**: 776–785.
1590. Kinoshita, Y., et al. (2014) Prey pursuit strategy of Japanese horseshoe bats during an in-flight target-selection task. *Journal of Comparative Physiology A* **200**: 799–809.
1591. Kinugasa, J., et al. (2012) Records of three bat species captured in Mt. Hyonosen of Ohya-cho, Yabu City, Hyogo. *Humans and Nature* **23**: 95–99 (in Japanese).
1592. Kirihara, T., et al. (2013) Spatial and temporal aspects of occurrence of *Mogera* species in the Japanese islands inferred from mitochondrial and nuclear gene sequences. *Zoological Science* **30**: 267–281.
1593. Kirkpatrick, R.L. & Pekins, P.J. (2002) Nutritional value of acorns for wildlife. in Oak Forest Ecosystems (McShea, W.J. & Healy, W.M., Editors). Baltimore: Johns Hopkins University Press, pp. 173–181.
1594. Kishida, K. (岸田久吉) (1924) Monograph of Japanese Mammals (哺乳動物図解). (Ministry of Agriculture (農商務省農務局), Editor) Tokyo: Ornithological Society of Japan, 429 pp. (in Japanese).
1595. Kishida, K. (岸田久吉) (1928) カウモリ雑記. *Zoological Magazine* **40**: 28–29 (in Japanese).
1596. Kishida, K. (岸田久吉) (1931) 渡瀬先生とマングース輸入. *Zoological Magazine* **43**: 70–78 (in Japanese).
1597. Kishida, K. (1930) Notes on Japanese ticks of the genus *Ixodes* Latreille, 1795. *Lansania* **2**: 1–5 (in Japanese).
1598. Kishida, K. (1934) The mammal fauna of the great city of Tokyo. *Lansania* **6**: 17–30 (in Japanese).
1599. Kishimoto, R. (1987) Family break-up in Japanese serow *Capricornis crispus*. in The Biology and Management of *Capricornis* and Related Mountain Antelopes (Soma, H., Editor). London: Croom Helm, pp. 104–109.
1600. Kishimoto, R. (1988) Age and sex determination of the Japanese serows, *Capricornis crispus*. *Journal of the Mammalogical Society of Japan* **13**: 51–58.
1601. Kishimoto, R. (1989) Early mother and kid behavior of a typical "follower", Japanese serow *Capricornis crispus*. *Mammalia* **53**: 165–176.
1602. Kishimoto, R. (1989) Social organization of a solitary ungulate, Japanese serow *Capricornis crispus*. Osaka: Osaka City University. 155 pp.
1603. Kishimoto, R. (2003) Social monogamy and social polygyny in a solitary ungulate, the Japanese serow (*Capricornis crispus*). in Monogamy: Mating Strategies and Partnerships in Birds, Humans and Other Mammals (Reichard, U.H. & Boesch, C., Editors). Cambridge: Cambridge University Press, pp. 147–158.
1604. Kishimoto, R. (2005) Invasion of an alien species, American mink (*Mustela vison*), into the upper area of Chikuma River. *Bulletin of Nagano Environmental Conservation Research Institute* **1**: 65–68 (in Japanese).
1605. Kishimoto, R. (2005) Scent-marking and territoriality of the Japanese serow, *Capricornis crispus*. in Abstract of Plenary, Symposium, Poster and Oral Papers at the 9th International Mammalogical Congress, IMC9. Sapporo: Science Council of Japan & Mammalogical Society of Japan, p. 266.
1606. Kishimoto, R. & Kawamichi, T. (1996) Territoriality and monogamous pairs in a solitary ungulate, the Japanese serow, *Capricornis crispus*. *Animal Behaviour* **52**: 673–682.
1607. Kishiro, T. (木白俊哉) (2014) 日本の小型鯨類調査研究についての進捗報告：2013年4月から2014年3月. Tokyo: Fisheries Agency, 12 pp. (in Japanese).
1608. Kishiro, T. & Kasuya, T. (1993) Review of Japanese dolphin drive fisheries and their status. *Report of the International Whaling Commission* **43**: 439–452.
1609. Kiszka, J., et al. (2007) Distribution, encounter rates, and habitat characteristics of toothed cetaceans in the Bay of Biscay and adjacent waters from platform-of-opportunity data. *ICES Journal of Marine Science* **64**: 1033–1043.
1610. Kita, I., et al. (1983) Reproduction of female Japanese serows, *Capricornis crispus*, based on pregnancy and macroscopical ovarian findings. *Research bulletin of the Faculty College of Agriculture, Gifu University* **48**: 137–146 (in Japanese with English summary).
1611. Kita, I., et al. (1987) Reproduction of female Japanese serow based on the morphology of ovaries and fetuses. in The Biology and Management of Capricornis and Related Mountain Antelopes (Soma, H., Editor). London: Croom Helm, pp. 321–331.
1612. Kitagaki, K. (北垣憲仁) (1996) カワネズミの谷. Tokyo: Froebel-kan (フレーベル館), 55 pp. (in Japanese).
1613. Kitagaki, K. (北垣憲仁) (2000) カワネズミ. in 日本の希少な野生水生生物に関するデータブック (Fisheries Agency, Editor). Tokyo: Japan Fisheries Resource Conservation Association (日本水産資源保護協会), pp. 250–251 (in Japanese).
1614. Kitahara, E. (北原英治) (1988) ワカヤマヤチネズミについて. *Forest Pests* **37**: 12–15 (in Japanese).
1615. Kitahara, E. (1995) Reproductive traits of captive Anderson's red-backed voles *Eothenomys andersoni* from the Kii Peninsula. *Journal of the Mammalogical Society of Japan* **20**: 95–108.
1616. Kitahara, E. (1995) Taxonomic status of Anderson's red-backed vole on the Kii Peninsula, Japan, based on skull and dental characters. *Journal of the Mammalogical Society of Japan* **20**: 9–28.
1617. Kitahara, E. & Harada, M. (1996) Karyological identity of Anderson's red-backed voles from the Kii Peninsula and central Honshu in Japan. *Bulletin of the Forestry and Forest Products Research Institute* **370**: 21–30.
1618. Kitahara, E. & Kimura, Y. (1995) Taxonomic reexamination among three local populations of Anderson's red-backed voles from crossbreeding experiments. *Journal of the Mammalogical Society of Japan* **20**: 43–49.
1619. Kitamura, E., et al. (1997) Metazoan parasites of sika deer from east Hokkaido, Japan and ecological analyses of their abomasal nematodes. *Journal of Wildlife Diseases* **33**: 278–284.
1620. Kitamura, E. (北村榮次) (1934) 實際養狸. Tokyo: Sougou Kagaku Shuppan Kyokai (総合科学出版協会), 125 pp. (in Japanese).
1621. Kitamura, S., et al. (2013) Two genetically distinct stocks in Baird's beaked whale (Cetacea: Ziphiidae). *Marine Mammal Science* **29**: 755–766.
1622. Kitamura, Y. & Machida, M. (1973) *Cryptocotyle lingua* (Creplin) from the red fox *Vulpes vulpes schrencki* Kishida. *Research Bulletin of the Meguro Parasitological Museum* **7**: 15–16.
1623. Kitao, N., et al. (2009) Overwintering strategy of wild free-ranging and enclosure-housed Japanese raccoon dogs (*Nyctereutes procyonoides albus*). *International Journal of Biometeorology* **53**: 159–165.
1624. Kitaoka, S. (北岡茂男) (1989) 大型野生動物とマダニ. *Kashiwazaki City Museum* **4**: 44–50 (in Japanese).
1625. Kitaoka, S. & Suzuki, H. (1974) Reports of medico-zoological investigations in the Nansei Islands. Part II. Ticks and their seasonal prevalences in southern Amami-oshima. *Medical Entomology and Zoology* **25**: 21–26.
1626. Kitchener, A. (1991) The Natural History of the Wild Cats. London: Christopher Helm Publishers, 280 pp.
1627. Kitchener, D.J., Ross, G.J.B., & Caputi, N. (1990) Variation in skull and external morphology in the false killer whale, *Pseudorca crassidens*, from Australia, Scotland and South Africa. *Mammalia* **54**: 119–135.
1628. Kiuchi, M. (木内正敏), et al. (1978) 朝日連峰・朝日川流域におけるニホンカモシカ. in 特別天然記念物カモシカに関する調査研究報告書 (NACS-J (日本自然保護協会), Editor). Tokyo: NACS-J, pp. 27–93 (in Japanese).
1629. Kiuchi, M. (木内正敏), et al. (1979) 朝日連峰・朝日川流域におけるニホンカモシカ(第二報). in 特別天然記念物カモシカに関する調査研究報告書 (NACS-J (日本自然保護協会), Editor). Tokyo: NACS-J, pp. 19–72 (in Japanese).
1630. Kiuchi, M. (木内盛郷) & Yoshida, M. (吉田正隆) (1969) 徳島県の洞窟動物相. 徳島県博物館紀要 **1**: 41–63 (in Japanese).
1631. Kiyota, M. & Baba, N. (1999) Records of northern fur seals and other pinnipeds stranded or taken incidentally by coastal fisheries in Japan, 1977–1998. *Bulletin of National Research Institute of Far Seas Fishery* **36**: 9–16.
1632. Kiyota, M. & Baba, N. (2001) Entanglement in marine debris among adult female northern fur seals at St. Paul Island, Alaska in 1991–1999. *Bulletin of National Research Institute of Far Seas Fisheries* **38**: 13–20.
1633. Kiyota, M., Insley, S.J., & Lance, S.L. (2008) Effectiveness of territorial polygyny and alternative mating strategies in northern fur seals, *Callorhinus ursinus*. *Behavioral Ecology and Sociobiology* **62**: 739–746.
1634. Kiyota, M., Ohashi, E., & Kobayashi, A. (2000) Non-radioacitve oligonucleotide probes for DNA fingerprinting in Northern fur seals. *Bulletin of National Research Institute of Far Seas Fishery* **37**: 19–26.
1635. Klimpel, S., Kellermanns, E., & Palm, H.W. (2008) The role of pelagic swarm fish (Myctophidae: Teleostei) in the oceanic life cycle of *Anisakis* sibling species at the Mid-Atlantic Ridge, Central Atlantic. *Parasitology*

Research **104**: 43–53.

1636. Kobayashi, H. (小林恒明) (1972) 道内で発見されたカラフトアカネズミ (2). *野ねずみ* **108**: 6–8 (in Japanese).

1637. Kobayashi, M. (小林万里) (2008) 世界自然遺産知床半島の海獣類―アザラシ類の実態. in Mammalogy in Japan. 3. Marine Mammals (日本の哺乳類学 第3巻 水生哺乳類) (Kato, H. (加藤秀弘), Editor). Tokyo: University of Tokyo Press, pp. 75–100 (in Japanese).

1638. Kobayashi, M. (小林万里) & Kakumoto, C. (角本千治) (2002) 漁業被害とアザラシの混獲：納沙布地域. in アザラシ類の保護管理報告書. Sapporo: Hokkaido Government, pp. 147–153 (in Japanese).

1639. Kobayashi, Y. (小林由美) (2002) アザラシ類の生態1 ゼニガタアザラシ. in アザラシ類の保護管理報告書. Sapporo: Hokkaido Government, pp. 6–11 (in Japanese).

1640. Kobayashi, A., et al. (2009) Habitat selection and the selection of Japanese white pine cones (*Pinus parviflora*) by Japanese squirrels (*Sciurus lis*) in the subalpine zone of Mt. Fuji. *Mammalian Science* **49**: 13–24 (in Japanese with English abstract).

1641. Kobayashi, K., Sato, Y., & Kaji, K. (2012) Increased brown bear predation on sika deer fawns following a deer population irruption in eastern Hokkaido, Japan. *Ecological Research* **27**: 849–855.

1642. Kobayashi, M. Unpublished.

1643. Kobayashi, T., Abe, H., & Maeda, K. (1970) Faunal survey of the Mt. Odaigahara area, JIBP supplementary area-IV. Results of the small mammal fauna of the Mt. Odaigahara Area, Kii Peninsula. in Studies on the Animal Communities in the Terrestrial Ecosystems and their Conservation; Annual Report of JIBP/CT-S for the Fiscal Year of 1969 (Kato, M., Editor). Sendai: JIBP/CT-S Section, pp. 317–320.

1644. Kobayashi, T. & Hayata, I. (1971) Revision of the genus *Apodemus* in Hokkaido. *Annual Zoology of Japan* **44**: 236–240.

1645. Kobayashi, T., et al. (2007) Epidemiology, histpathology, and muscle distribution of *Trichinella* T9 in feral raccoons (*Procyon lotor*) and wildlife of Japan. *Parasitology Research* **100**: 1287–1291.

1646. Kobayashi, T., et al. (2007) Exceptional minute sex-specific region in the XO mammal, Ryukyu spiny rat. *Chromosome Research* **15**: 175–187.

1647. Kobayashi, T., et al. (2008) Centromere repositioning in the X chromosome of XO/XO mammals, Ryukyu spiny rat. *Chromosome Research* **16**: 587–593.

1648. Kobayashi, Y., et al. (2011) The occurrence of Pinnipeds and fishery conflict records in the Oshima Peninsula, Hokkaido, Sea of Japan. *Bulletin of Fisheries Sciences, Hokkaido University* **61**: 75–82 (in Japanese with English abstract).

1649. Kobuchi, Y. (小渕幸輝) (2005) 高尾山でのテングコウモリの記録. *Bat Study and Conservation Report* **13**: 7 (in Japanese).

1650. Kock, D. (1999) *Tadarida* (*Tadarida*) *latouchei*, a separate species recorded from Thailand with remarks on related Asia taxa (Mammalia, Chiroptera, Molossidae). *Senckenbergiana Biologica* **78**: 237–240.

1651. Kodama, S., et al. (2015) Ancient onset of geographic divergence, interpopulation genetic exchange, and natural selection on the *Mc1r* coat-color gene in the house mouse (*Mus musculus*). *Biological Journal of the Linnean Society* **114**: 778–794.

1652. Kodera, Y. (小寺祐二) & Kanzaki, N. (神崎伸夫) (2001) 島根県石見地方におけるニホンイノシシの食性および栄養状態の季節的変化. *Wildlife Conservation Japan* **6**: 109–117 (in Japanese).

1653. Kodera, S., Suzuki, S., & Sugimura, M. (1982) Postnatal development and histology of the infraorbital glands in the Japanese serow, *Capricornis crispus*. *Japanese Journal of Veterinary Science* **44**: 839–843.

1654. Kodera, Y. & Kanzaki, N. (2001) Food habits and nutritional condition of Japanese wild boar in Iwami district, Shimane prefecture, western Japan. *Wildlife Conservation Japan* **6**: 109–117 (in Japanese with English abstract).

1655. Kodera, Y., et al. (2013) Food habits of wild boar (*Sus scrofa*) inhabiting Iwami District, Shimane Prefecture, western Japan. *Mammalian Science* **53**: 279–287 (in Japanese with English summary).

1656. Kodera, Y., et al. (2001) Habitat selection of Japanese wild boar in Iwami district, Shimane prefecture, western Japan. *Wildlife Conservation Japan* **6**: 119–129 (in Japanese with English summary).

1657. Kodera, Y., et al. (2010) How dose spreading maize on fields influence behavior of wild boars (*Sus scrofa*)? *Mammalian Science* **50**: 137–244 (in Japanese with English summary).

1658. Kodera, Y., et al. (2012) The estimation of birth periods in wild boar by detailed aging. *Mammalian Science* **52**: 185–191 (in Japanese with English summary).

1659. Kodutsumi, T. (小堤知行), et al. (1988) 宮城県の野生動物とくにタヌキにおける疥癬の流行とその発生被害. *Journal of the Veterinary Medicine* **803**: 28–30 (in Japanese).

1660. Koford, C.B., Farber, P.A., & Windle, W.F. (1966) Twins and teratisms in rhesus monkeys. *Folia Primatologica* **4**: 221–226.

1661. Koganezawa, M. (1995) Extraction and isolated degree of the local population of Japanese monkeys using geographic information system. *Primate Research* **11**: 59–66 (in Japanese with English summary).

1662. Koganezawa, M. (1999) Changes in the population dynamics of Japanese serow and sika deer as a result of competitive interactions in the Ashio Mountains, central Japan. *Biosphere Conservation* **2**: 35–44.

1663. Koganezawa, M. & Kurokawa, M. (1983) Altitudinal distribution of middle-sized mammals in Nikko area, Tochigi Prefecture (II) Focused on the winter distribution and the food-habits of the red fox (*Vulpes vulpes*). *Memoirs of Tochigi Prefecture Museum* **1**: 67–82 (in Japanese with English summary).

1664. Kogi, K., et al. (2004) Demographic parameters of Indo-Pacific bottlenose dolphins (*Tursiops aduncus*) around Mikura Island, Japan. *Marine Mammal Science* **20**: 510–526.

1665. Koh, H.S., et al. (2013) A preliminary study on genetic divergence of the Asian lesser white-toothed shrew *Crocidura shantungensis* (Mammalia: Soricomorpha) in mainland Korea, adjacent islands and continental East Asia: cytochrome *b* sequence analysis. *Russian Journal of Theriology* **12**: 71–77.

1666. Koh, H.S., et al. (2006) Mitochondrial DNA variation in the red squirrel (*Sciurus vulgaris mantchuricus*) from Korea and northeast China. *Acta Theriologica Sinica* **26**: 1–7.

1667. Kohira, M., Okada, H., & Yamanaka, M. (2006) Controlled exposure. Demographic trends, dispersal patterns, and management of brown bear in Shiretoko National Park. in Wildlife in Shiretoko and Yellowstone National Parks. Lessons in Wildlife Conservation from Two World Heritage Sites (McCullough, D.R., Kaji, K., & Yamanaka, M., Editors). Shari: Shiretoko Nature Foundation, pp. 238–242.

1668. Kohno, N. & Yanagisawa, Y. (1997) The first record of the Pliocene Gilmore fur seal in the western North Pacific Ocean. *Bulletin of the National Science Museum. Series C, Geology & Paleontology* **23**: 119–130.

1669. Koide, T., et al. (1998) A new inbred strain JF1 established from Japanese fancy mouse carrying the classic piebald allele. *Mammalian Genome* **9**: 15–19.

1670. Koike, S., ed. (2006) Historical status of Asiatic black bear, *Ursus thibetanus*, in Japan. Proceedings of the 17th International Conference on Bear Research and Management. Ibaraki: International Association for Bear Research and Management, 402 pp.

1671. Koike, S., et al. (2008) Fruit phenology of *Prunus jamasakura* and the feeding habit of the Asiatic black bear as seed disperser. *Ecological Research* **23**: 385–392.

1672. Koike, S., et al. (2012) Effect of hard mast production on foraging and sex-specific behavior of the Asiatic black bear (*Ursus thibetanus*). *Mammal Study* **37**: 21–28.

1673. Koike, S. & Masaki, T. (2008) Frugivory of canivora in central and southern parts of Japan analyzed by literature search. *Journal of the Japanese Forestry Society* **90**: 26–35 (in Japanese with English summary).

1674. Koike, S., et al. (2011) Estimate of the seed shadow created by the Asiatic black bear (*Ursus thibetanus*) and its characteristics as a seed disperser in Japanese cool-temperate forest. *Oikos* **120**: 280–290.

1675. Koizumi, T. (1992) Reproductive characteristics of female sika deer, *Cervus nippon*, in Hyogo Prefecture, Japan. in Proceedings of the International Symposium "Ongulés/Ungulates 91", pp. 561–563.

1676. Koizumi, T., Nogami, S., & Yokohata, Y. (2011) Gastrointestinal helminth fauna of the lesser Japanese mole (*Mogera imaizumii*) in Kanagawa Prefecture, Japan, and analyses on infection status of two parasitic nematode species of the host. *Japanese Journal of Zoo and Wildlife Medicine* **16**: 121–126.

1677. Kojima, K. (児島恭子) (1994) ラッコ皮と蝦夷錦の道. in 歴史の道・再発見. 第1巻. Osaka: Forum-A (フォーラム・A), pp. 71–99 (in Japanese).

1678. Kojima, N., Onoyama, K., & Kawamichi, T. (2006) Year-long stability and individual differences of male long calls in Japanese pikas *Ochotona hyperborea yesoensis* (Lagomorpha). *Mammalia* **70**: 80–85.

1679. Komatsu, M. & Sasaki, Y. (1995) Reproduction and brood care in the Japanese marten. *Animals and Zoos* **47**: 120–123 (in Japanese with English abstract).

1680. Komatsu, T., et al. (1994) Puberty and stem cell for the initiation and resumption spermatogenesis in the male Japanese black bear (*Selenarctos thibetanus japonicus*). *Journal of Reproduction and Development* **40**: 65–71 (in Japanese with English summary).

1681. Komiya, T. (1987) Reproduction of native Japanese hares at Tama Zoo. *Animals and Zoos* **39**: 4–7 (in Japanese with English summary).

1682. Komiya, T. & Yamada, F. Unpublished.

1683. Komori, A. (小森厚) (1975) ニホンカモシカの繁殖についての一考察. *Journal of Japanese Association of Zoological Gardens and Aquariums* **17**: 53–61 (in Japanese).

1684. Komura, H., et al. (2002) Composition of Stejneger's beaked whale

(*Mesoplodon stejnegeri*) milk and characteristics of milk fat. *Milk Science* **51**: 133–135.
1685. Kon, N. & Yamaga, Y. (2003) The activity pattern of the attracted bats and insects to a street lamp, in Bihoro town. *Bulletin of the Bihoro Museum* **10**: 69–78 (in Japanese with English abstract).
1686. Kondo, K. (近藤敬治) (2013) 日本産哺乳動物毛図鑑：走査電子顕微鏡で見る毛の形態 (Hair of Japanese Mammals: Observations with a Scanning Electoron Miciroscope). Sapporo: Hokkaido Univeristy Press, 244 pp. (in Japanese).
1687. Kondo, N. (近藤憲久) (2004) 翼手目. *in* 野付半島野生生物生息状況調査 (Wildlife Preservation Bureau of Hokkaido Corporation (北海道野生生物保護公社), Editor). Kushiro: Wildlife Preservation Bureau of Hokkaido Corporation, pp. 71–81 (in Japanese).
1688. Kondo, A. & Shiraki, S. (2012) Preferences for specific food species of the red fox *Vulpes vulpes* in Abashiri, eastern Hokkaido. *Mammal Study* **37**: 43–46.
1689. Kondo, N. (1980) Seasonal fluctuation of population size, activity and activity area of *Apodemus speciosus ainu* (Thomas) in a small stand. *The Journal of the Mammalogical Society of Japan* **8**: 129–138 (in Japanese with English abstract).
1690. Kondo, N. (2011) Roosts of Ussuri whiskered bat (*Myotis gracilis* Ognev, 1927) in Nemuro and Kushiro districts, Hokkaido. *Memoirs of the Nemuro City Museum of History and Nature* **23**: 57–62 (in Japanese).
1691. Kondo, N. & Abe, H. (1978) Reproductive activity of *Apodemus speciosus ainu*. *Memoirs of Faculty of Agriculture, Hokkaido University* **11**: 159–165 (in Japanese with English summary).
1692. Kondo, N., et al. (2012) A maternity colony of *Vespertilio murinus* in Ozora, Abashiri District, Hokkaido. *Mammalian Science* **52**: 63–70 (in Japanese with English abstract).
1693. Kondo, N. & Hattori, K. (1999) Two sea otters captured in Nemuro. *Memoirs of the Preparative Office of Nemuro Municipal Museum* **13**: 71–75 (in Japanese with English abstract).
1694. Kondo, N., et al. (2013) Japanese large-footed bat *Myotis macrodactylus* (Temminck, 1840) inhabiting the two Nikishoro sea caves of Kunashir Island. *Memoirs of the Nemuro City Museum of History and Nature* **25**: 1–7 (in Japanese).
1695. Kondo, N., Kawai, K., & Murano, N. (2011) New record of *Hypsugo alaschanicus* (Bobrinskii, 1926) from Sapporo. *Mammalian Science* **51**: 39–45 (in Japanese with English abstract).
1696. Kondo, N. & Kondo, J. (1996) Structural aspects of complex of hibernation-specific proteins. *in* Adaptation to The Cold (Geiser, F., Hulbelt, A.J., & Nicol, S.C., Editors). Armidale, Australia: University of New England Press, pp. 351–355.
1697. Kondo, N., et al. (2013) The first records of Japanese little horseshoe bats, *Rhinolophus cornutus* from eastern Hokkaido. *Bulletin of the Bihoro Museum* **20**: 3–6 (in Japanese with English summary).
1698. Kondo, N., et al. (1988) Age determination, growth and sexual dimorphism of the feral mink (*Mustela vison*) in Hokkaido. *Journal of the Mammalogical Society of Japan* **13**: 69–75.
1699. Kondo, N. & Sasaki, N. (2005) An external taxonomic character suitable for separating live *Myotis ikonnikovi* and *M. mystacinus*. *Mammal Study* **30**: 29–32.
1700. Kondo, N. & Sasaki, N. (2006) Faunal survey of bats in Nakashibetsu. *in* 中標津の格子状防風林保存事業報告書. Nakashibetsu: 中標津町文化的景観検討委員会, pp. 110–118 (in Japanese with English abstract).
1701. Kondo, N., et al. (2006) Mammal survey in forest habitat around lake Chimikeppu, Hokkaido. Bat survey. *Hokkaido Institute of Environmental Sciences* **32**: 94–99 (in Japanese with English abstract).
1702. Kondo, N., et al. (2006) Circannual control of hibernation by HP complex in the brain. *Cell* **125**: 161–172.
1703. Kondo, N. & Serizawa, Y. (2007) Summer roost of *Myotis nattereri* (Kuhl, 1817) in eastern Hokkaido. *Bulletin of the Asian Bat Research Institute* **6**: 16–19 (in Japanese).
1704. Kondo, N., Serizawa, Y., & Sasaki, N. (2005) Bat survey in Hamanaka town, Hokkaido. *Bulletin of the Asian Bat Research Institute* **4**: 1–6 (in Japanese with English abstract).
1705. Kondo, N., Takahashi, K., & Yagi, K. (1986) Winter food of the red fox, *Vulpes vulpes schrencki* Kishida, in the endemic area of multilocular echinococcosis. *Memoirs of the Preparative Office of Nemuro Municipal Museum* **1**: 23–31 (in Japanese with English summary).
1706. Kondo, N., et al. (2003) Faunal survey of bats in Akkeshi town, Hokkaido. *Bulletin of the Asian Bat Research Institute* **3**: 1–9 (in Japanese with English abstract).
1707. Kondo, S. (1985) Morphological study on the development of cheek tooth-germs of *Suncus murinus* fetus (Soricidae, Insectivora). *Aichi-Gakuin Journal of Dental Science* **23**: 697–730 (in Japanese with English abstract).
1708. Kondo, S., et al. (1988) A morphological study of the dental roots in house shrew, *Suncus murinus* (Soricidae, Insectivora). *Japanese Journal of Oral Biology* **30**: 794–806 (in Japanese with English abstract).
1709. Kondo, S., et al. (2010) Distribution and density of finless porpoise (*Neophocaena phocaenoides*) in Osaka Bay, Japan. *Mammalian Science* **50**: 13–20 (in Japanese with English abstract).
1710. Kondo, T. (1977) Social behavior of the Japanese wood mouse, *Apodemus speciosus* (Temminck et Schlegel), in the field. *Japanese Journal of Ecology* **27**: 301–310.
1711. Konishi, K. (1999) Parasites of pinnipeds. *in* Migratory Ecology and Conservation of Steller Sea Lions (Ohtaishi, N. & Wada, K., Editors). Tokyo: Tokai University Press, pp. 123–162 (in Japanese with English summary).
1712. Koopman, K.F. (1993) Order Chiroptera. *in* Mammal Species of the World. A Taxonomic and Geographic Reference (Wilson, D.E. & Reeder, D.M., Editors). Washington: Smithsonian Institution Press, pp. 137–241.
1713. Koopman, K.F. (1994) Chiroptera: Systematics. *in* Handbook of Zoology, Volume VIII: Mammalia (Niethammer, J., Schliemann, H., & Starck, D., Editors). Berlin: Walter de Gruyter, p. 217.
1714. Kooyman, G.L., Gentry, R.L., & Urquhart, D.L. (1976) Northern fur seal diving behavior; a new approach to its study. *Science* **193**: 411–412.
1715. Korablev, V.P., Yakimenko, L.V., & Tiunov, M.P. (1989) Karyotypes of bats in the Far East. *in* Present-day Approaches to Studies of Variability. Vladivostok: The Far East Branch, Academy of Sciences of the USSR, pp. 95–98 (in Russian with English abstract).
1716. Kornev, S.I. & Korneva, S.M. (2004) Historical trends in sea otters populations of the Kuril Islands and South Kamchatka. *in* Alaska Sea Otter Research Workshop: Addressing the Decline of the Southwestern Alaska Sea Otter Population. Alaska Sea Grant College Program (Maldini, D., et al., Editors). University of Alaska, Fairbanks, pp. 21–23.
1717. Kornev, S.I. & Korneva, S.M. (2006) Some criteria for assessing the state and dynamics of sea otter (*Enhydra lutris*) populations in the Russian part of the species range. *Russian Journal of Ecology* **37**: 172–179.
1718. Korosue, T. (1999) A record on *Mustela itatsi* from the Shodoshima, Kagawa Prefecture, Japan. *Bulletin of the Biological Society of Kagawa* **26**: 27–30 (in Japanese).
1719. Kostenko, V.A. (2000) Rodents (Rodentia) of the Russian Far East. Vladivostok: Dalnauka, 208 pp. (in Russian).
1720. Kostenko, V.A., Nesterenko, V.A., & Trukhin, A.M. (2004) Mammals of the Kuril Archipelago. Vladivostok: Dalnauka, 186 pp. (in Russian with English abstract).
1721. Kotaka, N. & Matsuoka, S. (2002) Secondary users of great spotted woodpecker (*Dendrocopos major*) nest cavities in urban and suburban forests in Sapporo City, northern Japan. *Ornithological Science* **1**: 117–122.
1722. Kotaka, N., et al. (2008) A record of Java mongoose *Herpestes javanicus*, by camera trapping on Mt. Nishime, northern part of Yambaru, in Okinawa prefecture, Japan. *Kyushu Journal of Forest Research* **60**: 104–105 (In Japanese).
1723. Koumorinokai Chosa Group (コウモリの会調査グループ) (1998) 山形県朝日鉱泉周辺のコウモリ類. *Bat Study and Conservation Report* **6**: 11–15 (in Japanese).
1724. Kovacs, K.M. (2002) Bearded seal (*Erignathus barbatus*). *in* Encyclopedia of Marine Mammals (Perrin, W.F., Würsig, B., & Thewissen, J.G.M., Editors). San Diego: Academic Press, pp. 84–87.
1725. Kowalski, K. & Hasegawa, Y. (1976) Quaternary rodents from Japan. *Bulletin of the National Science Museum, Series C (Geology and Paleontology)* **2**: 31–66.
1726. Koyama, H. Unpublished.
1727. Koyama, N., et al. (1992) Reproductive parameters of female Japanese macaques: Thirty years data from the Arashiyama troops, Japan. *Primates* **33**: 33–47.
1728. Koyama, T. (1970) Changes in dominance rank and division of a wild Japanese monkey troop in Arashiyama. *Primates* **11**: 335–390.
1729. Koyanagi, H. (小柳秀章) (1995) オリイジネズミの死体発見と胃内容. *Chirimosu* (チリモス) **6**: 29–30 (in Japanese).
1730. Koyanagi, K. Unpublished.
1731. Koyanagi, K., et al. (2013) Seasonal roost distribution of cave-dwelling bats (*Hipposideros turpis* Bangs, 1901; *Rhinolophus perditus* Anderson, 1918; and *Miniopterus fuscus* Bonhote, 1902) on Ishigaki Island, Okinawa Prefecture, Japan. *Bulletin of the Asian Bat Research Institute* **9**: 1–19 (in Japanese with English summary).
1732. Koyanagi, K. & Tsuji, A. (2006) A new record of lesser tube-nosed bat (*Murina ussuriensis* Ognev, 1913) in Kumamoto Prefecture: roost characteristics and roostig conditions at winter. *Bulletin of the Asian Bat Research Institute* **5**: 20–23 (In Japanese with English abstract).
1733. Koyanagi, K., Tsuji, A., & Yamamoto, T. (2002) Fauna of bats in Fujihashi

Village, Ibi-Country, Gifu Prefecture. *Bulletin of the Asian Bat Research Institute* **2**: 8–12 (in Japanese with English abstract).

1734. Koyanagi, K., et al. (2005) Distribution records of bats in Kuriyama village, Shioya-country, Tochigi Prefecture. *Bulletin of the Asian Bat Research Institute* **4**: 7–14 (in Japanese with English abstract).

1735. Koyanagi, K., et al. (2011) Bat fauna of Yunishigawa area in Nikko, Tochigi Prefecture. *Bulletin of Tochigi Prefectural Museum* **28**: 35–43 (in Japanese with English abstract).

1736. Koyanagi, K., et al. (2007) The first record of the Japanese northern bat, *Eptesicus japonensis* Imaizumi, from Tochigi Prefecture, Japan. *Bulletin of the Tochigi Prefectural Museum* **24**: 1–4 (in Japanese with English abstract).

1737. Koyasu, K. (子安和弘) (1993) 足跡図鑑. Tokyo: Nikkei Science (日経サイエンス社), 178 pp. (in Japanese).

1738. Koyasu, K. (子安和弘) (1998) 日本産トガリネズミ亜科の自然史. in 食虫類の自然史 (Abe, H. (阿部永) & Yokohata, Y. (横畑泰志), Editors). Shobara: Hiba Society of Natural History, pp. 201–267 (in Japanese).

1739. Koyasu, K. (1999) Some morphological characters of the Japanese water shrew, *Chimarrogale platycephala*. in Recent advances in the biology of Japanese Insectivora. Proceedings of the symposium on the biology of Insectivora in Japan and on the wildlife conservation (Yokohata, Y. & Nakamura, S., Editors). Shiobara: Hiba Society of Natural History, pp. 29–31.

1740. Koyasu, K., et al. (2005) Dental anomalies in *Suncus murinus*. in Advances in the Biology of Shrews II (Merritt, J.F., et al., Editors). New York: International Society of Shrew Biologists, pp. 405–411.

1741. Koyasu, K., et al. (1995) Breeding activity of the Sado shrew, *Sorex sadonis*. *Annals of The Research Institute of Environmental Medicine, Nagoya University* **46**: 192–193 (in Japanese).

1742. Kozakai, C., et al. (2011) Effect of mast production on home range use of Japanese black bears. *Journal of Wildlife Management* **75**: 867–875.

1743. Kozakai, K., et al. (2013) Fluctuation of daily activity time budgets of Japanese black bears: relationship to sex, reproductive status, and hardmast availability. *Journal of Mammalogy* **94**: 351–360.

1744. Kruse, S., Caldwell, D.K., & Caldwell, M.C. (1999) Risso's dolphin *Grampus griseus* G. Cuvier, 1812. in Handbook of Marine Mammals. Volume 6. The Second Book of Dolphins and the Porpoises (Ridgway, S.H. & Harrison, R., Editors). London: Academic Press, pp. 183–212.

1745. Kruse, S.L. Unpublished.

1746. Kruskop, S.V. (2004) Subspecific structure of *Myotis daubentonii* (Chiroptera, Vespertilionidae) and composition of the "daubentonii" species group. *Mammalia* **68**: 299–306.

1747. Kruskop, S.V. (2005) Towards the taxonomy of the Russian *Murina* (Vespertilionidae, Chiroptera). *Russian Journal of Theriology* **4**: 91–99 (in English with Russian abstruct).

1748. Kruskop, S.V. (2012) Chiroptera. in The Mammals of Russia: A Taxonomic and Geographic Reference (Pavlinov, I.Y. & Lissovsky, A.A., Editors). Moscow: MK Scientific Press, pp. 73–126 (in English and Russian).

1749. Kruskop, S.V., et al. (2012) Genetic diversity of northeastern Palaearctic bats as revealed by DNA barcodes. *Acta Chiropterologica* **14**: 1–14.

1750. Krüssmann, G. (1986) Manual of Cultivated Broad-leaved Trees and Shrubs. Vol. III. London: B T Batsford Ltd., 510 pp.

1751. Krützen, M., et al. (2005) Cultural transmission of tool use in bottlenose dolphins. *Proceedings of the National Academy of Sciences of the United States of America* **102**: 8939–8943.

1752. Kubo, M., et al. (2010) *Hepatozoon* sp. infection in Hokkaido brown bears (*Ursus arctos yesoensis*). *Japanese Journal of Zoo and Wildlife Medicine* **15**: 111–113 (in Japanese with English abstract).

1753. Kubo, M., Miyoshi, N., & Yasuda, N. (2006) Hepatozoonosis in two species of Japanese wildcat. *Journal of Veterinary Medical Science* **68**: 833–837.

1754. Kubo, M., et al. (2013) Histopathological examination of spontaneous lesions in Amami rabbits (*Pentalagus furnessi*): a preliminary study using formalin-fixed archival specimens. *Japanese Journal of Zoo and Wildlife Medicine* **18**: 65–70.

1755. Kubo, M.O. & Takatsuki, S. (2015) Geographical body size clines in sika deer: path analysis to discern amongst environmental influences. *Evolutionary Biology* **42**: 115–127.

1756. Kubodera, T. & Miyazaki, N. (1993) Cephalopods eaten by short-finned pilot whales, *Globicephala macrorhynchus*, caught off Ayukawa, Ojika Peninsula, Japan, in 1982 and 1983. in Recent Acvances in Cephalopod Fisheries Biology (Okutani, T., O'Dor, R.K., & Kubodera, T., Editors). Tokyo: Tokai University Press, pp. 215–227.

1757. Kucherenko, S.P. (クチェレンコ, S.P.) (1993) Mammals in the southern area of far-eastern Russia. *Science Report of the Kushiro City Museum* **343**: 51–56 (in Japanese. Translated from Russian by 藤巻裕蔵. Originally published in 1973).

1758. Kugi, G. (1983) Pathological changes in hosts infected with *Mesocestoides paucitesticulus*. *Journal of the Japanese Veterinary Medical Association* **36**: 665–668 (in Japanese with English summary).

1759. Kumada, N., et al. (1987) *Pipistrellus abramus* (Temminck) as a nuisance in the Tokai district, Japan, and arthropods associated with this bat. *Japanese Journal of Sanitary Zoology* **29**: 261–263 (in Japanese with English abstract).

1760. Kumazawa, H. & Yachimori, S. (2005) Ectoparasites found on wild mammals in Kochi Prefecture. *Bulletin of the Shikoku Institute of Natural History* **2**: 45–50 (in Japanese).

1761. Kuntz, R. & Myers, B. (1969) A check-list of parasites and commensals reported for the Taiwan macaque. *Primates* **10**: 71–80.

1762. Kuo, P.C. (1985) Silvicultural studies on damage to forest plantations by the Formosan red-bellied tree squirrel and its control procedures. *Technical Bulletin of National Taiwan University* **159**: 1–106 (in Chinese).

1763. Kuramochi, T. (倉持利明) (1999) 海産哺乳類の寄生虫相. in Progress of Medical Parasitology in Japan. Volume 6 (日本における寄生虫学の研究 第 6 巻) (Otsuru, M. (大鶴正満), Kamegai, S. (亀谷了), & Hayashi, S. (林滋生), Editors). Tokyo: Meguro Parasitological Museum, pp. 121–128 (in Japanese).

1764. Kuramochi, T. (倉持利明) (2004) イルカ・クジラの寄生虫を追って. in Aquaparasitology in the field in Japan (フィールドの寄生虫学：水族寄生虫学の最前線) (Nagasawa, K. (長澤和也), Editor). Hadano: Tokai University Press, pp. 243–255 (in Japanese).

1765. Kuramochi, T. (2001) Stomach nematodes of the family Anisakidae collected from the cetaceans stranded on or incidentally caught off the coasts of the Kanto districts and adjoining areas. *Memoirs of the National Science Museum, Tokyo* **37**: 177–192.

1766. Kuramochi, T., et al. (2008) Helminth parasites collected from the gray whale entangled in a set net off Tomiyama-cho, Chiba, on the Pacific coast of Japan (Document RW2008-4 submitted to the IUCN workshop of Western Gray Whales: Status, Threats and the Potential for Recovery, Tokyo. September 21–24, 2008). 5 pp. (unpublished).

1767. Kuramochi, T., et al. (2000) Parasitic helminth and epizoit fauna of finless porpoise in the Inland Sea of Japan and the Western North Pacific with a preliminary note on faunal difference by host's local population. *Memoirs of National Science Museum, Tokyo* **33**: 83–95.

1768. Kuramochi, T., et al. (1996) Minke whales (*Balaenoptera acutorostrata*) are one of the major final hosts of *Anisakis simplex* (Nematoda: Anisakidae) in the northwestern North Pacific Ocean. *Report of the International Whaling Commission* **46**: 415–419.

1769. Kuramoto, T. (1964) On the Japanese tube-nosed bat, *Murina leucogaster hilgendorfi* (PETERS) caught in the caves at Akiyosidai Plateau. *Bulletin of the Akiyoshi-dai Science Museum* **3**: 35–37 (in Japanese with English abstract).

1770. Kuramoto, T. (1972) Studies on bats at the Akiyoshi-dai Plateau, with special reference to the ecological and phylogenic aspects. *Bulletin of the Akiyoshi-dai Science Museum* **8**: 7–119 (in Japanese with English abstract).

1771. Kuramoto, T. (1977) Mammals of Japan (15): order Chiroptera, genus *Rhinolophus*. *Mammalian Science* **17(2)**: 31–57 (in Japanese).

1772. Kuramoto, T. (1979) Nursery colony of the Japanese horseshoe bat, *Rhinolophus ferrumequinum nippon*. *Bulletin of the Akiyoshi-dai Museum of Natural History* **14**: 27–44 (in Japanese with English abstract).

1773. Kuramoto, T., Nakamura, H., & Uchida, T.A. (1978) Habitat selection, mode of social life and population dynamics in *Myotis macrodactylus*. *Bulletin of the Akiyoshi-dai Museum of Natural History* **13**: 35–54 (in Japanese with English abstract).

1774. Kuramoto, T., Nakamura, H., & Uchida, T.A. (1985) A survey of bat-banding on the Akiyoshi-dai Plateau. IV. Results from April 1975 to March 1983. *Bulletin of the Akiyoshi-dai Museum of Natural History* **20**: 25–44 (in Japanese with English abstract).

1775. Kuramoto, T., Nakamura, H., & Uchida, T.A. (1988) A survey of bat-banding on the Akiyoshi-dai Plateau. V. Results from April 1983 to March 1987. *Bulletin of the Akiyoshi-dai Museum of Natural History* **23**: 39–54 (in Japanese with English abstract).

1776. Kuramoto, T., Nakamura, H., & Uchida, T.A. (1995) A survey of bat-banding on the Akiyoshi-dai Plateau. VI. Results from April 1987 to March 1993. *Bulletin of the Akiyoshi-dai Museum of Natural History* **30**: 37–49 (in Japanese with English abstract).

1777. Kuramoto, T., Nakamura, H., & Uchida, T.A. (1998) A survey of bat-banding on the Akiyoshi-dai Plateau. VII. Results from April 1993 to March 1997. *Bulletin of the Akiyoshi-dai Museum of Natural History* **33**: 31–43 (in Japanese with English abstract).

1778. Kuramoto, T., et al. (1973) A survey of bat-banding on the Akiyoshi-dai Plateau. II. Results from April 1967 to March 1972. *Bulletin of the Akiyoshi-dai Science Museum* **9**: 1–18 (in Japanese with English abstract).

1779. Kuramoto, T., et al. (1975) A survey of bat-banding on the Akiyoshi-dai

Plateau. III. Results from April 1972 to March 1975. *Bulletin of the Akiyoshi-dai Museum of Natural History* **11**: 29–47 (in Japanese with English abstract).
1780. Kuramoto, T. & Uchida, T.A. (1981) Growth of newborn young in the Japanese tube-nosed bat, *Murina leucogaster hilgendolfi* (Peters). *Bulletin of the Akiyoshi-dai Museum of Natural History* **16**: 55–69 (in Japanese with English abstract).
1781. Kuramoto, T. & Uchida, T.A. (1991) Life table for the Japanese long-fingered bat, *Miniopterus schreibersii fuliginosus*, on the Akiyoshi-dai Plateau. *Bulletin of the Akiyoshi-dai Museum of Natural History* **26**: 53–64 (in Japanese with English abstract).
1782. Kuramoto, T., Uchida, T.A., & Nakamura, H. (1969) Weasel and rat snake as predators on bats. *Bulletin of the Akiyoshi-dai Science Museum* **6**: 27–33 (in Japanese with English abstract).
1783. Kuramoto, T., et al. (1969) Further studies on the dense mixed colony consisting of different species in cave bats. *Bulletin of the Akiyoshi-dai Science Museum* **6**: 47–58 (in Japanese with English abstract).
1784. Kurihara, N., et al. (2013) Records of stranded or by-caught finless porpoises in Mikawa Bay, Japan. *Mammalian Science* **53**: 99–106 (in Japanese with English abstract).
1785. Kurita, H., Shimomura, T., & Fujita, T. (2002) Temporal variation in Japanese macaque bodily mass. *International Journal of Primatology* **23**: 411–428.
1786. Kuroda, N. (黒田長禮) (1931) ヤチネズミの新産地と其学名. *Zoological Magazine* **43**: 661–666 (in Japanese).
1787. Kuroda, N. (黒田長禮) (1935) 甲州産のコテングカウモリ. *Botany and Zoology* **3**: 467 (in Japanese).
1788. Kuroda, N. (黒田長禮) (1942) 日本哺乳動物學の濫觴並に沿革. *Zoological Magazine* **54**: 442–453 (in Japanese).
1789. Kuroda, N. (黒田長禮) (1953) 日本獣類図説. Tokyo: Sogensya (創元社), 177 pp. (in Japanese).
1790. Kuroda, N. (1920) On a collection of Japanese and Formosan mammals. *Annotationes Zoologicae Japonenses* **9**: 599–611.
1791. Kuroda, N. (1921) On three new mammals from Japan. *Journal of Mammalogy* **2**: 208–211.
1792. Kuroda, N. (1922) Notes on the mammal fauna of Tsushima and Iki Islands. *Journal of Mammalogy* **3**: 42–45.
1793. Kuroda, N. (1939) Distribution of mammals in the Japanese empire. *Journal of Mammalogy* **20**: 37–50.
1794. Kuroda, N. (黒田長禮) (1940) A Monograph of the Japanese Mammals, Exclusive of Sirenia and Cetacea (原色日本哺乳類図説). Tokyo & Osaka: Sanseido (三省堂), 311 pp. (in Japanese).
1795. Kuroda, N. (1952) Mammalogical history of Formosa, with zoogeography and bibliography. *Quarterly Journal of the Taiwan Museum* **5**: 262–304.
1796. Kuroda, N. (1957) A new name for the lesser Japanese mole. *The Journal of the Mammalogical Society of Japan* **1**: 74 (in Japanese with English summary).
1797. Kuroda, N. (1964) A new locality for *Pipistrellus endoi*. *The Journal of the Mammalogical Society of Japan* **2**: 84–85 (in Japanese).
1798. Kuroda, N. (1965) A rare white-cuffed bat of the genus *Pipistrellus*. *The Journal of the Mammalogical Society of Japan* **2**: 143–144 (in Japanese with English abstract).
1799. Kuroda, N. (1969) On the discovery of the hibernating place of *Murina aurata ussuriensis* in Hokkaido. *The Journal of the Mammalogical Society of Japan* **4**: 125–126 (in Japanese with English abstract).
1800. Kuroda, S. (1992) Geographic variation of nonmetrical traits of Japanese monkey (*Macaca fuscata*) skulls and their stability against environmental change. *Primate Research* **8**: 75–82 (in Japanese with English summary).
1801. Kuroe, M., et al. (2007) Nest-site selection by the harvest mouse *Micromys minutus* in seasonally changing environments. *Acta Theriologica* **52**: 355–360.
1802. Kurohmaru, M., et al. (1982) Morphology of the intestine of the Watase's shrew, *Crocidura horsfieldi watasei*. *Japanese Journal of Veterinary Science* **44**: 795–799.
1803. Kuroiwa, A. Unpublished.
1804. Kuroiwa, A., et al. (2011) Additional copies of *CBX2* in the genomes of males of mammals lacking *SRY*, the Amami spiny rat (*Tokudaia osimensis*) and the Tokunoshima spiny rat (*Tokudaia tokunoshimensis*). *Chromosome Research* **19**: 635–644.
1805. Kuroiwa, A., et al. (2010) The process of a Y-loss event in an XO/XO mammal, the Ryukyu spiny rat. *Chromosoma* **119**: 519–526.
1806. Kuroko, H. (黒子浩) (1958) 彦山の動物 翼手目. *in* 英彦山 (田川郷土研究会, Editor). Fukuoka: Ashishobo (葦書房), pp. 668–671 (in Japanese).
1807. Kurosawa, Y. (黒澤弥悦) (2001) イノシシとブター人との関わりを通して. *in* イノシシと人間：共に生きる (Takahashi, S. (高橋春成), Editor). Tokyo: Kokon (古今書院), pp. 2–44 (in Japanese).

1808. Kurose, N., Abramov, A.V., & Masuda, R. (2000) Intrageneric diversity of the cytochrome *b* gene and phylogeny of Eurasian species of the genus *Mustela* (Mustelidae, Carnivora). *Zoological Science* **17**: 673–679.
1809. Kurose, N., Abramov, A.V., & Masuda, R. (2005) Comparative phylogeography between the ermine *Mustela erminea* and the least weasel *M. nivalis* of Palaearctic and Nearctic regions, based on analysis of mitochondrial DNA control region sequences. *Zoological Science* **22**: 1069–1078.
1810. Kurose, N., et al. (2001) Low genetic diversity in Japanese populations of the Eurasian badger *Meles meles* (Mustelidae, Carnivora) revealed by mitochondrial cytochrome *b* gene sequences. *Zoological Science* **18**: 1145–1151.
1811. Kurose, N., et al. (2000) Karyological differentiation between two closely related mustelids, the Japanese weasel *Mustela itatsi* and the Siberian weasel *Mustela sibirica*. *Caryologia* **53**: 269–275.
1812. Kurose, N., et al. (1999) Intraspecific variation of mitochondrial cytochrome *b* gene sequences of the Japanese marten *Martes melampus* and the sable *Martes zibellina* (Mustelidae, Carnivora, Mammalia) in Japan. *Zoological Science* **16**: 693–700.
1813. Kurose, N., Masuda, R., & Tatara, M. (2005) Fecal DNA analysis for identifying species and sex of sympatric carnivores: a noninvasive method for conservation on the Tsushima Islands, Japan. *Journal of Heredity* **96**: 688–697.
1814. Kurose, N., Masuda, R., & Yoshida, M.C. (1999) Phylogeographic variation in two mustelines, the least weasel *Mustela nivalis* and the ermine *Mustela erminea* of Japan, based on mitochondrial DNA control region sequences. *Zoological Science* **16**: 971–977.
1815. Kuru, D.D. (1972) Evolution and cytogenetics. *in* Mammals of the Sea: Biology and Medicine (Ridgway, S.H., Editor). Springfield: Charles C. Thomas, pp. 503–527.
1816. Kusuda, S., et al. (2010) Induced estrus in female small Asian mongooses (*Herpestes javanicus*) for the purpose of controlling invasive alien species in Okinawa Island. *Mammal Study* **35**: 217–219.
1817. Kusumoto, K. & Saitoh, T. (2008) Effects of cold stress on immune function in the grey-sided vole, *Clethrionomys rufocanus*. *Mammal Study* **33**: 11–18.
1818. Kuwabara, K. & Okuda, M. (2002) Ecological study on cavernicolous bat in Mizuho-cho, Shimane Prefecture. *Natural History of Nishi-Chugoku Mountains* **7**: 59–83 (in Japanese).
1819. Kuwano, R. (2014) Stranding survey of finless porpoise in Oita Prefecture and educational usage with skeletal specimen. *Aquabiology* **36**: 156–162 (in Japanese with English abstract).
1820. Kyoya, T., Obara, Y., & Nakata, A. (2008) Chromosomal aberrations in Japanese grass voles in and around an illegal dumpsite at the Aomori-Iwate prefectural boundary. *Zoological Science* **26**: 307–312.
1821. Kyuzaki, H. (1951) Note on the *Plecotus auritus sacrimontis* and *Dinodon orientale* found in Kanagawa-Pref. *Collecting and Breeding* **252** (in Japanese).
1822. Lai, J. & Sheng, H.L. (1993) A comparative study on scent-marking behavior of captive forest musk deer and Reeves' muntjac. *in* Deer of China: Biology and Management (Ohtaishi, N. & Sheng, H.L., Editors). Amsterdam: Elsevier, pp. 204–208.
1823. Lambersten, R.H., et al. (1988) Cytogenetic determination of sex among individually identified humpback whales (*Megaptera novaeangliae*). *Canadian Journal of Zoology* **66**: 1243–1248.
1824. Lambertsen, R.H. (1986) Disease of the common fin whale (*Balaenoptera physalus*): crassicaudiosis of the urinary system. *Journal of Mammalogy* **67**: 353–366.
1825. Lambertsen, R.H. (1992) Crassicaudosis: a parasitic disease threatening the health and population recovery of large baleen whales. *Scientific and Technical Review of the Office International des Epizooties* **11**: 1131–1141.
1826. Lander, R.H. & Kajimura, H. (1982) Status of northern fur seals. *in* Mammals in the Seas (FAO Fisheries Reports), pp. 319–345.
1827. Larson, S., et al. (2002) Microsatellite DNA and mitochondrial DNA variation in remnant and translocated sea otter (*Enhydra lutris*) populations. *Journal of Mammalogy* **83**: 893–906.
1828. Lavazza, A. & Capucci, L. (2008) How many caliciviruses are there in rabbits? A review on RHDV and correlated viruses. *in* Lagomorph Biology: Evolution, Ecology, and Conservation (Alves, P.C., Ferrand, N., & Hackländer, K., Editors). Berlin: Springer, pp. 263–278.
1829. Law, R.J., et al. (1998) Organotin compounds in liver tissue of harbour porpoises (*Phocoena phocoena*) and grey seals (*Halichoerus grypus*) from the coastal waters of England and Wales. *Marine Pollution Bulletin* **36**: 241–247.
1830. Leatherwood, S., et al. (1982) Whales, Dolphins, and Porpoises of the Eastern North Pacific and Adjacent Waters: A Guide to Their Identification. NOAA Technical Report, NMFS Circular, 444. Seattle: U.S. Department of Commerce, National Oceanic and Atmospheric Administration, National Marine Fisheries Service, 245 pp.

1831. Leatherwood, S., et al., eds. (1988) Whales, Dolphins and Porpoises of the Eastern North Pacific and Adjacent Arctic Waters. A Guide to their Identification. National Marine Fisheries Service. Dover Publications: New York, 245 pp.
1832. Leatherwood, S. & Walker, W.A. (1979) The northern right whale dolphin *Lissodelphis borealis* Peale in the eastern North Pacific. in Behavior of Marine Animals (Winn, H.E. & Olla, B.L., Editors). New York: Plenum Press, pp. 85–141.
1833. LeDuc, C.A., Perrin, W.F., & Dizon, A. (1999) Phylogenetic relationships among the delphinid cetaceans based on full cytochrome *b* sequences. *Marine Mammal Science* **15**: 619–648.
1834. LeDuc, R.G. & Dizon, A.E. (2002) Reconstructing the rorqual phylogeny, with comments on the use of molecular and morphological data for systematic study. in Molecular and Cell Biology of Marine Mammals (Pfeiffer, C.J., Editor). Malabar: Krieger Publishing, pp. 100–110.
1835. LeDuc, R.G., Robertson, K.M., & Pitman, R.L. (2008) Mitochondrial sequence divergence among Antarctic killer whale ecotypes is consistent with multiple species. *Biology Letters* **4**: 426–429.
1836. Lee, L.-L. & Lin, Y.-S. (1990) Status of Formosan macaques in Taiwan. *Primate Conservation* **11**: 18–20.
1837. Lee, T.H. (1999) Habitat use, food habits and mating ecology of the Eurasian red squirrel (*Sciurus vulgaris* L.). Doctoral dissertation. Sapporo: Hokkaido University. 72 pp.
1838. Lee, T.H. & Fukuda, H. (1999) The distribution and habitat use of the Eurasian red squirrel *Sciurus vulgaris* L. during summer, in Nopporo Forest Park, Hokkaido. *Mammal Study* **24**: 7–15.
1839. Lee, Y.R., et al. (2013) Age and reproduction of the finless porpoises, *Neophocaena asiaeorientalis*, in the Yellow Sea, Korea. *Animal Cells and Systems* **17**: 366–373.
1840. Lekagul, B. & McNeely, J.A. (1977) Mammals of Thailand. Bangkok: Association for the Conservation of Wildlife, 758 pp.
1841. Lekagul, B. & McNeely, J.A. (1988) Rhesus macaque. in Mammals of Thailand. Bangkok: Saha Karn Bhaet, Co., pp. 288–290.
1842. Lennert-Cody, C.E. & Scott, M.D. (2005) Spotted dolphin evasive response in relation to fishing effort. *Marine Mammal Science* **21**: 13–28.
1843. Levan, G. (1974) Nomenclature for G-bands in rat chromosomes. *Hereditas* **77**: 37–52.
1844. Lever, C. (1985) Naturalised Mammals of the World. London: Longman, 487 pp.
1845. Li, G., et al. (2006) Phylogenetics of small horseshoe bats from East Asia based on mitochondrial DNA sequence variation. *Journal of Mammalogy* **87**: 1234–1240.
1846. Li, T.-C., et al. (2005) Hepatitis E virus transmission from wild boar meat. *Emerging Infectious Diseases* **11**: 1958–1960.
1847. Liberg, O. & Sandell, M. (1988) Spatial organization and reproductive tactics in the domestic cat and other felids. in The Domestic Cat: The Biology of its Behaviour (Turner, D.C. & Bateson, P., Editors). Cambridge: Cambridge University Press, pp. 83–98.
1848. Lin, L.K. (林良恭), Lee, L.L. (李玲玲), & Cheng, H.C. (鄭錫奇) (2004) Bats of Taiwan (台灣的蝙蝠). 2nd ed. Taichung: National Museum of Natural Science, 177 pp. (in Chinese).
1849. Lin, L.-K. & Pei, K. (1999) On the current status of field population of Formosan fruit bat (*Pteropus dasymallus formosus*). *Endemic Species Research* **1**: 12–19 (in Chinese with English abstract).
1850. Lin, L.K., Motokawa, M., & Harada, M. (2002) Karyological study of the house bat *Pipistrellus abramus* (Mammalis: Chiroptera) from Taiwan with comments on its taxonomic status. *Raffles Bulletin of Zoology* **50**: 507–510.
1851. Lin, L.K., Motokawa, M., & Harada, M. (2002) Karyology of ten vespertilionid bats (Chiroptera: Vespertilionidae) from Taiwan. *Zoological Studies* **41**: 347–354.
1852. Lindburg, D.G. (1971) The rhesus monkey in north India: an ecological and behavioral study. *Primate Behavior* **2**: 1–106.
1853. Lissovsky, A.A., Ivanova, N.V., & Borisenko, A.V. (2007) Molecular phylogenetics and taxonomy of the subgenus *Pika* (*Ochotona*, Lagomorpha). *Journal of Mammalogy* **88**: 1195–1204.
1854. Lissovsky, A.A. & Serdyuk, N.V. (2004) Identification of Late Pleistocene pikas (*Ochotona*, Lagomorpha, Mammalia) of the Alpine-Hyperborea group from Denisova Cave (Altai) on the basis of the anterior lower premolar (P3). *Paleontologicheskii Zhurnal* **6**: 89–95 (in Russian with English abstract).
1855. Liu, W., et al. (2013) Genetic variation and phylogenetic relationship between three serow species of the genus *Capricornis* based on the complete mitochondrial DNA control region sequences. *Molecular Biology Reports* **40**: 6793–6802.
1856. Lloyd, H.G. (1980) The Red Fox. London: Batsford, 320 pp.
1857. Lockley, R.M. (1973) The Private Life of the Rabbit (アナウサギの生活). Tokyo: Sisakusya (思索社), 265 pp. (in Japanese. Translated by 立川賢一. Originally published by MacMillan & Co., London in 1964).
1858. Lockyer, C. (1981) Growth and energy budgets of large baleen whales from the Southern Hemisphere. in Mammals in the Seas. Volume III. Rome: Food and Agriculture Organization of the United Nations, pp. 379–487.
1859. Long, J.L. (2003) Introduced Mammals of the World: Their History, Distribution and Influence. Collingwood: CSIRO Publishing, 589 pp.
1860. Long, J.L. (2003) Raccoon. in Introduced Mammals of the World (Long, J.L., Editor). Collingwood: CSIRO Publishing, pp. 265–269.
1861. Longman, H.A. (1926) New records of Cetacea, with a list of Queensland species. *Memoirs of Queensland Museum* **8**: 266–278.
1862. Lönneberg, E. (1931) The skeleton of *Balaenoptera brydei*. *Arkiv für Zoologi* **23A**: 1–23.
1863. Lopez-Martinez, N. (2008) The lagomorph fossil record and the origin of the European rabbit. in Lagomorph Biology: Evolution, Ecology, and Conservation (Alves, P.C., Ferrand, N., & Hackländer, K., Editors). Berlin: Springer, pp. 27–46.
1864. Loreille, O., et al. (2001) Ancient DNA analysis reveals divergence of the cave bear, *Ursus spelaeus*, and brown bear, *Ursus arctos*, lineages. *Current Biology* **11**: 200–203.
1865. Lotze, J. & Anderson, S. (1979) *Procyon lotor*. *Mammalian Species* **119**: 1–8.
1866. Loughlin, T.R. (1985) *Mesoplodon stejnegeri*. *Mammalian Species* **250**: 1–6.
1867. Loughlin, T.R. (1997) Using the phylogeographic method to identify Steller sea lion stocks. in Molecular Genetics of Marine Mammals. Special Publication Number 3 (Dizon, A., Chivers, S.J., & Perrin, W.F., Editors). Lawrence: Society for Marine Mammalogy, pp. 159–171.
1868. Loughlin, T.R. (2002) Steller's sea lion. in Encyclopedia of Marine Mammals (Perrin, W.F., Würsig, B., & Thewissen, J.G.M., Editors). San Diego: Academic Press, pp. 1181–1185.
1869. Loughlin, T.R., Ames, J.A., & Vandevere, J.E. (1981) Annual reproduction, dependency period, and apparent gestation period in two California sea otters, *Enhydra lutris*. *Fisheries Bulletin* **79**: 347–349.
1870. Loughlin, T.R., et al. (1982) Observations of *Mesoplodon stejnegeri* (Ziphiidae) in the central Aleutian Islands, Alaska. *Journal of Mammalogy* **63**: 697–700.
1871. Loy, A., et al. (2005) Origin and evolution of Western European moles (genus *Talpa*, Insectivora): a multidisciplinary approach. *Mammal Study* **30 (Suppl. 1)**: S13–S17
1872. Lu, J.-F., Lin, Y.-S., & Lee, L.-L. (1991) Troop composition, activity pattern and habitat utilization of Formosan macaque (*Macaca cyclopis*) at Nanshi logging road in Yushan National Park. in Primatology Today (Ehara, A., et al., Editors). Amsterdam: Elsevier Science Publishers, pp. 93–96.
1873. Luca, M., et al. (2009) Population structure of short-beaked common dolphins (*Delphinus delphis*) in the North Atlantic Ocean as revealed by mitochondrial and nuclear genetic markers. *Marine Biology* **156**: 821–834.
1874. Lukoschus, F.S., Kroos, A., & Uchikawa, K. (1977) *Lophioglyphus japonensis* sp. nov. (Acarina, Glycyphagidae) from *Apodemus speciosus* (Rodentia, Muridae). *Bulletin of National Science Museum, Series A* **3**: 9–17.
1875. Macchi, E., et al. (1995) Cytogenetic variability in the wild boar (*Sus scrofa scrofa*) in Piedmont (Italy): Preliminary data. *Journal of Mountain Ecology* **3**: 17–18.
1876. Macdonald, D.W. (1979) 'Helpers' in fox society. *Nature* **282**: 69–71.
1877. Macdonald, D.W. (1980) Rabies and Wildlife. Oxford: Oxford University Press, 151 pp.
1878. Macdonald, D.W. (1987) Running with the fox. London: Unwin Hyman, 224 pp.
1879. Macdonald, D.W. & Reynolds, J.C. (2004) Red fox. in Canids: Foxes, Wolves, Jackals and Dogs. Status Survey and Conservation Action Plan (Sillero-Zubiri, C., Hoffmann, M., & Macdonald, D.W., Editors). Gland: IUCN, pp. 129–136.
1880. Machida, K., et al. (1986) Bat fauna in Osorezan district, Aomori prefecture. *The Journal of the Mammalogical Society of Japan* **11**: 173–181 (in Japanese with English abstract).
1881. Machida, K. & Sasaki, M. (1987) Bats collected from Rishiri Island, Hokkaido, Japan. *Bulletin of the Saitama Museum of Natural History* **5**: 1–6 (in Japanese with English abstract).
1882. Machida, M., Kitamura, M., & Kamiya, H. (1975) Occurrence of *Alaria alata* (Diplostomidae: Digenea) from the red fox in Hokkaido, Japan. *Japanese Journal of Parasitology* **24**: 144–147 (in Japanese with English summary).
1883. Machida, M. & Mikuria, M. (1968) Studies on skrjabingyliasis in Japan. *Bulletin of National Science Museum Tokyo* **11**: 153–156.
1884. Machida, M. & Uchida, A. (1982) Some helminth parasites of the Japanese shrew mole from the Izu Peninsula. *Memoirs of the National Science*

1884. *Museum, Tokyo* **15**: 149–154.
1885. Machida, N., et al. (1992) Canine distemper virus infection in a masked palm civet (*Paguma larvata*). *Journal of Comparative Pathology* **107**: 439–443.
1886. Machida, N., et al. (1993) Pathology and epidemiology of canine distemper in raccoon dogs (*Nyctereutes procyonoides*). *Journal of Comparative Pathology* **108**: 383–392.
1887. MacPhee, R.D.E. & Flemming, C. (1999) Requiem æternam: the last five hundred years of mammalian extinctions, extinctions in near time. *in* Extinctions in Near Time: Causes, Contexts, and Consequences (MacPhee, R.D.E., Editor). New York: Kluwer Academic/Plenum Publishers, pp. 333–371.
1888. Maeda, K. (前田喜四雄) (1973) 日本の哺乳類 (11) 翼手目 ヤマコウモリ属. *Mammalian Science* **13(2)**: 1–28 (in Japanese).
1889. Maeda, K. (前田喜四雄) (1986) 岐阜県のコウモリ類 1. 文献にあらわれた記録. 岐阜ふるさとと動物通信 **12**: 132–135 (in Japanese).
1890. Maeda, K. (前田喜四雄) (1987) 岐阜県下のコウモリ類 2. 昭和 61 年の白山国立公園における調査結果. 岐阜ふるさとと動物通信 **17**: 238–240 (in Japanese).
1891. Maeda, K. (前田喜四雄) (1991) 岐阜県下のコウモリ類 (15) 尾上郷国有林 (1). 岐阜ふるさとと動物通信 **39**: 629 (in Japanese).
1892. Maeda, K. (前田喜四雄) (1993) 奈良県のコウモリ類 (1) 奈良県からのクロホオヒゲコウモリ、アブラコウモリとコテングコウモリの記録. *Wild Animals of Kii Peninsula* **1**: 19–20 (in Japanese).
1893. Maeda, K. (前田喜四雄) (1994) コウモリ目 Chiroptera. *in* A Pictorial Guide to the Mammals of Japan (日本の哺乳類) (Abe, H. (阿部永), Editor). Tokyo: Tokai University Press, pp. 37–70 & 158–167 (in Japanese).
1894. Maeda, K. (前田喜四雄) (1994) 奈良県のコウモリ類 (2) ウサギコウモリ. *Wild Animals of Kii Peninsula* **2**: 12 (in Japanese).
1895. Maeda, K. (前田喜四雄) (2001) Natural History of Japanese Bats (日本コウモリ研究誌：翼手類の自然史). Tokyo: University of Tokyo Press, 203 pp. (in Japanese).
1896. Maeda, K. (前田喜四雄) (2002) コウモリ. *in* 新名寄市史 (名寄市史編纂委員会, Editor). Nayoro: Nayoro City Office, pp. 65–67 (in Japanese).
1897. Maeda, K. (前田喜四雄) & Kojo, H. (校条博光) (1987) 岐阜県のコウモリ類 3. チチブコウモリとモリアブラコウモリ. 岐阜ふるさとと動物通信 **17**: 240–241 (in Japanese).
1898. Maeda, K. (前田喜四雄) & Matsumura, S. (松村澄夫) (1997) 翼手目 (Chiroptera). *in* レッドデータ 日本の哺乳類 (Mammalogical Society of Japan, Editor). Tokyo: Bun-ichi (文一総合出版), pp. 31–55 (in Japanese).
1899. Maeda, M. (前田満) (1969) ヒメネズミによるトドマツ稚苗 (芽) の食害. 野ねずみ **92**: 5 (in Japanese).
1900. Maeda, M. (前田満), Igarashi, B. (五十嵐文吉), & Mizushima, S. (水島俊一) (1984) 食物. *in* Study on Wild Murid Rodents in Hokkaido (北海道産野ネズミ類の研究) (Ota, K. (太田嘉四夫), Editor). Sapporo: Hokkaido University Press, pp. 119–138 (in Japanese).
1901. Maeda, N. (前田菜穂子) (2006) ヒグマの飼育からわかること. *in* ヒグマ学入門 (Amano, T. (天野哲也), Masuda, R. (増田隆一), & Mano, T. (間野勉), Editors). Sapporo: Hokkaido University Press, pp. 22–34 (in Japanese).
1902. Maeda, K. (1972) Growth and development of large noctule, *Nyctalus lasiopterus* Schreber. *Mammalia* **36**: 269–278.
1903. Maeda, K. (1974) Eco-ethologie de la grande noctule, *Nyctalus lasiopterus*, a Sapporo Japon. *Mammalia* **38**: 461–487 (in French with English abstract).
1904. Maeda, K. (1976) Growth and development of the Japanese large-footed bat, *Myotis macrodactylus* (Temminck, 1840). I. External characters and breeding habits. *Journal of Growth* **15**: 29–40 (in Japanese with English).
1905. Maeda, K. (1978) Baculum of the large noctule, *Nyctalus Lasoipterus aviator* Thomas, 1911. *Acta Anatomica Nipponica* **53**: 447–453 (in Japanese with English abstract).
1906. Maeda, K. (1978) Variation in bent-winged bats, *Miniopterus schreibersi* Kuhl, and least horseshoe bats, *Rhinolophus cornutus* Temminck, in the Japanese Islands I. External characters. *in* Proceeding of the Fourth International Bat Research Conference. Nairobi: Kenya National Academy for Advancement of Arts and Sciences, pp. 177–187.
1907. Maeda, K. (1979) Mammals of Japan (16): order Chiroptera, genus *Murina*, *Murina ussuriensis*. *Mammalian Science* **19(1)**: 1–16 (in Japanese).
1908. Maeda, K. (1980) Review on the classification of little tube-nosed bats, *Murina aurata* group. *Mammalia* **44**: 531–551.
1909. Maeda, K. (1982) Studies on the classification of *Miniopterus* in Eurasia, Australia and Melanesia. *Mammalian Science* **Supplement 1**: 1–179.
1910. Maeda, K. (1983) Classificatory study of the Japanese large noctule, *Nyctalus lasiopterus aviator* Thomas, 1911. *Zoological Magazine* **92**: 21–36.
1911. Maeda, K. (1984) Collected records of Chiroptera in Japan (I). *Mammalian Science* **24(2)**: 55–78 (in Japanese).
1912. Maeda, K. (1985) New records of the eastern Daubenton's bats, *Myotis daubentoni ussuriensis* Ognev, 1927, in Hokkaido and variations in external and skull dimensions. *The Journal of the Mammalogical Society of Japan* **10**: 159–164 (in Japanese with English abstract).
1913. Maeda, K. (1988) Age and sexual variations of the cranial characters in the least horseshoe bat, *Rhinolophus cornutus* Temminck. *Journal of the Mammalogical Society of Japan* **13**: 43–50.
1914. Maeda, K. (1993) *Eptesicus nilssonii parvus* (Chiroptera, Mammalia) from Rishiri Island, northern Hokkaido. *Rishiri Studies* **12**: 11–13 (in Japanese).
1915. Maeda, K. (1996) Review and coments on the classification of Japanese bats (Chiroptera). *Mammalian Science* **36**: 1–23 (in Japanese).
1916. Maeda, K. (2005) Chiroptera. *in* A Guide to the Mammals of Japan (Abe, H., Editor). Hadano: Tokai University Press, pp. 25–64 (in Japanese and English).
1917. Maeda, K. (2005) Key to the species of Japanese Chiroptera. *in* A Guide to the Mammals of Japan (Abe, H., Editor). Hadano: Tokai University Press, pp. 159–169 (in Japanese and English).
1918. Maeda, K. & Akazawa, Y. (1999) An attempt to estimate the population size of passing bats. *Mammalian Science* **39**: 221–228 (in Japanese with English abstract).
1919. Maeda, K., Akazawa, Y., & Matsumura, S. (2001) Current conditions of inhabiting bats and new bat records in Tokunoshima Island, Nansei Archipelago. *Bulletin of the Asian Bat Research Institute* **1**: 1–9 (in Japanese with English abstract).
1920. Maeda, K. & Aoi, T. (2001) New record of tube-nosed bat, *Murina leucogaster* Milne-Edwards, 1872 from Wakayama Prefecture, Kii Peninsula. *Wild Animals of Kii Peninsula* **6**: 13–14 (in Japanese).
1921. Maeda, K., Arimoto, S., & Wakabayashi, R. (1985) First records of common long-eared bats in Wakayama Prefecture. *The Nature and Animals* **15**: 26–28 (in Japanese).
1922. Maeda, K. & Dewa, H. (1982) Breeding habits of Japanese long-legged whiskered bats, *Myotis frater kaguyae* in Asahikawa, Japan. *The Journal of the Mammalogical Society of Japan* **9**: 82–87 (in Japanese with English abstract).
1923. Maeda, K. & Hashimoto, H. (2002) Foraging environments of three microchiropteran bats in Iriomote Is., Okinawa Prefecture (1) Differences between so-called open area and covered area. *Bulletin of the Asian Bat Research Institute* **2**: 18–20 (in Japanese with English abstract).
1924. Maeda, K., et al. (2008) Isolation of novel Adenovirus from fruit bat (*Pteropus dasymallus yayeyamae*). *Emerging Infectious Diseases* **14**: 347–349.
1925. Maeda, K. & Ito, F. (2007) Note record of *Murina leucogaster* Milne-Edwards, 1872 in Ouda, Uta City, Nara Prefecture. *Wild Animals of Kii Peninsula* **9**: 17–18 (in Japanese).
1926. Maeda, K. & Kawamiti, M. (1991) Report of the survey on hollow-tree dwelling bats in Shari-cho. *Bulletin of the Shiretoko Museum* **12**: 55–58 (in Japanese).
1927. Maeda, K., Kawamiti, M., & Segawa, Y. (1993) Report of the survey on hollow-tree dwelling bats in Shari-cho (II). *Bulletin of the Shiretoko Museum* **14**: 9–15 (in Japanese).
1928. Maeda, K. & Matsumoto, M. (2004) Population changes of the Bang's leaf-nosed bats (*Hipposideros turpis* Bangs, 1901) for ten years in Otomi-Daiichi Cave, Iriomote Island, Nansei Archipelago. *Biological Magazine Okinawa* **42**: 57–60 (in Japanese with English abstract).
1929. Maeda, K. & Matsumura, S. (1998) Two new species of vespertilionid bats, *Myotis* and *Murina* (Vespertilionidae: Chiroptera) fromYanbaru, Okinawa Island, Okinawa Prefecture, Japan. *Zoological Science* **15**: 301–307.
1930. Maeda, K., Mizutani, T., & Taguchi, F. (2011) Viruses isolated from bats and their importance as emerging infectious diseases. *Journal of Veterinary Epidemiology* **15**: 88–93 (in Japanese with English summary).
1931. Maeda, K., Nagaoka, H., & Tamura, H. (2003) Extinct microchiropteran bats and forest size in Nansei Archipelago. *Bulletin of the Asian Bat Research Institute* **3**: 21–23 (in Japanese with English abstract).
1932. Maeda, K., Nishii, K., & Oguri, T. (2002) New records of *Myotis yanbarensis* and *Murina ryukyuana* in Amami-Oshima, Kagoshima Prefecture. *Bulletin of the Asian Bat Research Institute* **2**: 16–17 (in Japanese with English abstract).
1933. Maeda, K., et al. (2013) First record of *Myotis pruinosus* in Shiga-Prefecture. *Bulletin of the Asian Bat Research Institute* **9**: 27–29 (in Japanese with English abstract).
1934. Maeda, K., et al. (2007) New records of *Myotis pruinosus* Yoshiyuki, 1971 in Kii Peninsula. *Wild Animals of Kii Peninsula* **9**: 15 (in Japanese).
1935. Maeda, K. & Satô, M. (1995) Distribution of bats in Rishiri Island, Northern Hokkaido. *Rishiri Studies* **15**: 45–48 (in Japanese).
1936. Maeda, K. & Uno, H. (1997) Faunal survey of bats in Bihoro, Hokkaido (1). *Bulletin of the Bihoro Museum* **4**: 33–40 (in Japanese).
1937. Maeda, N. & Ohdachi, S. (1994) Growth and body measurement of Hokkaido brown bear in captivity. *Proceedings of the Second East Asiatic*

Bear Conference: 68–76.
1938. Maeji, I., et al. (2000) Home range of sika deer (*Cervus nippon*) on Mt. Ohdaigahara, central Japan. *Nagoya University Forest Science* **19**: 1–10 (in Japanese with English abstract).
1939. Maekawa, K. Unpublished.
1940. Maekawa, K., Yoneda, M., & Togashi, H. (1980) A preliminary study of the age structure of the red fox in eastern Hokkaido. *Japanese Journal of Ecology* **30**: 103–108.
1941. Mäkinen, A. (1974) Exceptional karyotype in a raccoon dog. *Hereditas* **78**: 150–152.
1942. Mäkinen, A. (1975) G-band karyotype and aneuploid cell lines of the European raccoon dogs, *Nyctereutes procyonoides. in* Proceedings of the 2nd Europäisches Kolloquium über Zytogenetik (Chromosomenpathologie) in Veterinärmedizin, Tierzucht und Säugetierkunde, Giessen. pp. 53–58.
1943. Mäkinen, A., Kuokkanen, M.T., & Valtonen, M. (1986) A chromosome-banding study in the Finnish and the Japanese raccoon dog. *Hereditas* **105**: 97–105.
1944. Makino, T. (牧野敬) & Fukui, H. (福井秀雄) (2002) 神奈川県自然環境保全センターに搬送されたアライグマの記録. 神奈川県自然環境保全センター自然情報 **1**: 1–6 (in Japanese).
1945. Makita, Y. (1980) Notes on *Mogera kobeae* (4). *The Journal of the Mammalogical Society of Japan* **8**: 113–116 (in Japanese with English summary).
1946. Mano, T. (1987) Population characteristics of brown bears on the Oshima Peninsula, Hokkaido. *Proceedings of the International Conference on Bear Research and Management* **7**: 69–73.
1947. Mano, T. (1995) Sex and age characteristics of harvested brown bears in the Oshima Peninsula, Japan. *Journal of Wildlife Management* **59**: 199–204.
1948. Mano, T. & Moll, J. (1999) Status and management of the Hokkaido brown bear in Japan. *in* Bears. Status Survey and Conservation Action Plan (Servheen, C., Herrero, S., & Peyton, B., Compilers). Gland and Cambridge: IUCN/SSC Bear and Polar Bear Specialist Groups, pp. 128–130.
1949. Mano, T. & Tsubota, T. (2002) Reproductive characteristics of brown bears on the Oshima Peninsula, Hokkaido, Japan. *Journal of Mammalogy* **83**: 1026–1034.
1950. Margolis, L. & Pike, G.C. (1955) Some helminth parasites of Canadian Pacific whales. *Journal of the Fisheries Research Board of Canada* **12**: 97–120.
1951. Marmi, J., et al. (2004) Radiation and phylogeography in the Japanese macaque, *Macaca fuscata. Molecular Phylogenetics and Evolution* **30**: 676–685.
1952. Marmi, J., et al. (2006) Mitochondrial DNA reveals a strong phylogeographic structure in the badger across Eurasia. *Molecular Ecology* **15**: 1007–1020.
1953. Marsh, H. (1980) Age determination of the dugong (*Dugong dugon* (Müller)) in northern Australia and its biological implications. *Reports of the International Whaling Commission (Special Issue)* **3**: 181–201.
1954. Marsh, H. (1995) The life history, pattern of breeding, and population dynamics of the dugong. *in* Information and Technology Report 1: Population Biology of the Florida Manatee (O'Shea, T., Ackerman, B.B., & Percival, H.F., Editors). Washington, D.C.: U.S. Department of the Interior National Biological Service, pp. 75–83.
1955. Marsh, H., et al. (1982) Analysis of stomach contents of dugongs from Queensland. *Australian Wildlife Research* **9**: 55–67.
1956. Marsh, H., Heinsohn, G.E., & Marsh, L.M. (1984) Breeding cycle, life history and population dynamics of the dugong, (*Dugong dugon*) (Sirenia: Dugongidae). *Australian Journal of Zoology* **32**: 767–788.
1957. Marsh, H. & Kasuya, T. (1984) Changes in the ovaries of the short-finned pilot whale, *Globicephala macrorhynchus*, with age and reproductive activity. *Reports of the International Whaling Commission (Special Issue)* **6**: 311–335.
1958. Marsh, H. & Kasuya, T. (1986) Evidence for reproductive senescence in female cetaceans. *Reports of the International Whaling Commission (Special Issue)* **8**: 57–74.
1959. Marsh, H., et al., eds. (2002) Dugong: Status Reports and Action Plans for Countries and Territories. Early Warning and Assessment Report Series (UNEP/DEWA/RS.02-1). Nairobi: United Nations Environment Programme, 162 pp.
1960. Marsh, H. & Rathbun, G.B. (1990) Development and application of conventional and satellite radiotracking techniques for studying dugong movements and habitat usage. *Australian Wildlife Research* **17**: 83–100.
1961. Marshal, J.T., Jr. (1977) A synopsis of Asian species of *Mus* (Rodentia, Muridae) *Bulletin of the American Museum of Natural History* **158**: 175–220.
1962. Marshal, J.T., Jr. (1998) Identification and scientific names of Eurasian house mice and European allies, subgenus *Mus* (Rodentia: Muridae). Privately published at Springfield, Virginia, 80 pp.
1963. Marshal, J.T., Jr. & Sage, R.D. (1981) Taxonomy of the house mouse. *Symposia of the Zoological Society of London* **47**: 15–25.
1964. Marshall, J.T. (1981) Taxonomy. *in* The Mouse in Biomedical Research (Foster, H.L., Small, J.D., & Fox, J.G., Editors). New York: Academic Press, pp. 17–26
1965. Martien, K.K., et al. (2014) Nuclear and mitochondrial patterns of population structure in North Pacific false killer whales (*Pseudorca crassidens*). *Journal of Heredity* **105**: 611–626.
1966. Martínez, R., et al. (2008) Occurrence of the etoparasite *Isocyamus deltobranchium* (Amphipoda: Cyamidae) on cetaceans from Atlantic waters. *Journal of Parasitology* **94**: 1239–1242.
1967. Martinoli, A., et al. (2006) Recapture of ringed *Eptesicus nilssonii* (Chiroptera, Vespertilionidae) after 12 years: an example of high site fidelity. *Mammalia* **70**: 331–332.
1968. Maruhashi, T., Sprague, D.S., & Matsubayashi, K. (1992) What future for Japanese monkeys? *Asian Primates* **2**: 1–4.
1969. Maruyama, K. (丸山勝彦) (2005) オキナワコキクガシラコウモリ. *in* Threatened Wildlife in Okinawa (Animals): Red Data Okinawa (Okinawa Prefectural Government, Editor). Naha: Okinawa Prefectural Government, pp. 29–30 (in Japanese).
1970. Maruyama, N. (丸山直樹) & Furubayashi, K. (古林賢恒) (1980) ニホンカモシカの分布域・生息密度・生息頭数の推定について. Tokyo: Environment Agency, 48 pp. (in Japanese).
1971. Maruyama, K. (1992) New records of Vespertilionidae (Chiroptera) from Okinawa Prefecture. *Biological Magazine Okinawa* **30**: 55–57 (in Japanese).
1972. Maruyama, K. (1999) Seasonal changes in forearm length, body weight and size of testes of cave-dwelling bat species at Haneji, Okinawa Island. *Biological Magazine Okinawa* **37**: 15–19 (in Japanese with English abstract).
1973. Maruyama, N. (1981) A study of the seasonal movements and aggregation patterns of sika deer. *Bulletin of Faculty of Agriculture Tokyo, University of Agriculture and Technology* **23**: 1–85 (in Japanese with English summary).
1974. Maruyama, N., Suzuki, Y., & Sugimura, M. (1988) Widespread of Parapoxvirus infection and nutritional conditions in Japanese serows in Gifu Prefecture in the winter of 1984–1985. *in* A Fundamental Research on Ecology and Conservation of the Japanese Serow (A Report for the Research Fund by Japan Society for the Promotion of Science) (Ono, Y., Editor). Tokyo: Japan Society for the Promotion of Science, pp. 81–90 (in Japanese with English abstract).
1975. Maruyama, N., Totake, Y., & Okabayashi, R. (1976) Seasonal movements of sika in Omote-Nikko, Tochigi Prefecture. *The Journal of the Mammalogical Society of Japan* **6**: 187–198.
1976. Maruyama, T. (2006) Some analysis of Japanese black bears captured in Tochigi Prefecture. *Bulletin on Wildlife in Tochigi Prefecture* **32**: 1–15 (in Japanese).
1977. Masaki, Y. (1976) Biological studies on the North Pacific sei whale. *Bulletin of the Far Seas Fisheries Research Laboratory* **14**: 1–104.
1978. Massei, G., et al. (1997) Factors influencing home range and activity of wild boar (*Sus scrofa*) in a Mediterranean coastal area. *Journal of Zoology* **242**: 411–423.
1979. Masuda, M. (増田戻樹) (1996) オコジョ. *in* The Encyclopaedia of Animals in Japan. 1. Mammals I (日本動物大百科1哺乳類I) (Kawamichi, T. (川道武男), Editor). Tokyo: Heibonsha (平凡社), pp. 132–135 (in Japanese).
1980. Masuda, R., Amano, T., & Ono, H. (2001) Ancient DNA analysis of brown bear (*Ursus arctos*) remains from the archeological site of Rebun Island, Hokkaido, Japan. *Zoological Science* **18**: 741–751.
1981. Masuda, R., et al. (2008) Genetic variations of the masked palm civet *Paguma larvata* inferred from mitochondrial cytochrome *b* sequences. *Mammal Study* **33**: 19–24.
1982. Masuda, R., et al. (2012) Molecular phylogeography of the Japanese weasel, *Mustela itatsi* (Carnivora: Mustelidae), endemic to the Japanese islands, revealed by mitochondrial DNA analysis. *Biological Journal of the Linnean Society* **107**: 307–321.
1983. Masuda, R., et al. (2010) Origins and founder effects on the Japanese masked palm civet *Paguma larvata* (Viverridae, Carnivora), revealed from a comparison with its molecular phylogeography in Taiwan. *Zoological Science* **27**: 499–505.
1984. Masuda, R., Tamura, T., & Takahashi, O. (2006) Ancient DNA analysis of brown bear skulls from a ritual rock shelter site of the Ainu culture at Bihue, central Hokkaido, Japan. *Anthropological Science* **114**: 211–215.
1985. Masuda, R. & Yoshida, M.C. (1994) A molecular phylogeny of the family Mustelidae (Mammalia, Carnivora), based on comparison of mitochondrial cytochrome *b* nucleotide sequences. *Zoological Science* **11**: 605–612.
1986. Masuda, R. & Yoshida, M.C. (1994) Nucleotide sequence variation of cytochrome *b* genes in three species of weasels *Mustela itatsi*, *Mustela sibirica*,

and *Mustela nivalis*, detected by improved PCR product-direct sequencing technique. *Journal of the Mammalogical Society of Japan* **19**: 33–43.

1987. Masuda, R. & Yoshida, M.C. (1995) Two Japanese wildcats, the Tsushima cat and the Iriomote cat, show the same mitochondrial DNA linage as the leopard cat *Felis bengalensis*. *Zoological Science* **12**: 655–659.

1988. Masuda, R., et al. (1994) Molecular phylogenetic status of the Iriomote cat *Felis iriomotensis*, inferred from mitochondrial DNA sequence analysis. *Zoological Science* **11**: 597–604.

1989. Masuda, Y. (2003) Notes on nesting trees of the flying squirrel (*Pteromys volans orii*). *Bulletin of the Shiretoko Museum* **24**: 67–70 (in Japanese).

1990. Masui, M. (1987) Social behaviour of Japanese serow, *Capricornis crispus crispus*. in The Biology and Management of Capricornis and Related Mountain Antelopes (Soma, H., Editor). London: Croom Helm, pp. 134–144.

1991. Matoba, Y., et al. (2002) First records of the genera *Eimeria* and *Isospora* (Protozoa : Eimeriidae) obtained from feral raccoons (*Procyon lotor*) alien species in Japan and prevalence of serum antibodies to *Toxoplasma gondii* among the raccoons. *Japanese Journal of Zoo and Wildlife Medicine* **7**: 87–90 (in Japanese with English abstract).

1992. Matoba, Y., et al. (2003) Detection of a taeniid species *Taenia taeniaeformis* from a feral raccoon *Procyon lotor* and its epidemiological significance. *Mammal Study* **28**: 157–160.

1993. Matoba, Y., Sakata, K., & Asakawa, M. (2002) A helminthological survey of raccoon dogs captured in Sado Island, Japan. *Bulletin of the Biogeographical Society of Japan* **57**: 31–36.

1994. Matoba, Y., et al. (2006) Parasitic helminths from feral raccoons (*Procyon lotor*) in Japan. *Helminthologia* **43**: 139–146.

1995. Matsubara, K., et al. (2001) A new primer set for sex identification in the genus *Sorex* (Soricidae, Insectivora). *Molecular Ecology Notes* **1**: 241–242.

1996. Matsubara, K., et al. (2007) Japanese serow (*Capricornis crispus*) inhabits in a lone forest: the ecology and wild animal medical research. *Japanese Journal of Zoo and Wildlife Medicine* **12**: 27–34 (in Japanese with English abstract).

1997. Matsubayashi, M., et al. (2004) First record of *Cryptosporidium* infection in a raccoon dog (*Nyctereutes procyonoides viverrinus*). *Veterinary Parasitology* **120**: 171–175.

1998. Matsuda, H., et al. (1999) A management policy for sika deer based on sex-specific hunting. *Researches on Population Ecology* **41**: 139–149.

1999. Matsuda, H., et al. (2002) Harvest-based estimation of population size for Sika deer on Hokkaido Island, Japan. *Wildlife Society Bulletin* **30**: 1160–1171.

2000. Matsuda, Y., et al. (1994) Chromosomal mapping of mouse 5S rRNA genes by direct R-banding fluorescence in situ hybridization. *Cytogenetics and Cell Genetics* **66**: 246–249.

2001. Matsudaira, C., et al. (2007) Contamination of organotin compounds in finless porpoises (*Neophocaena phocaenoides*) stranded along coastal waters of Japan and Hong Kong. *Frontier Science Series* **48**: 183–186.

2002. Matsudate, H., et al. (2003) A survey of the parasitic helminths of alien rodents (belly-banded squirrel *Callosciurus erythraeus* and nutria *Myocastor coypus*) in Japan. *Japanese Journal of Zoo and Wildlife Medicine* **8**: 63–67 (in Japanese with English abstract).

2003. Matsuhashi, T., et al. (1999) Microevolution of the mitochondrial DNA control region in the Japanese brown bear (*Ursus arctos*) population. *Molecular Biology and Evolution* **16**: 676–684.

2004. Matsui, A. (松井章), et al. (2001) 野生のブタ？飼育されたイノシシ？—考古学から見るイノシシとブタ. in イノシシと人間：共に生きる (Takahashi, S. (高橋春成), Editor). Tokyo: Kokon (古今書院), pp. 45–78 (in Japanese).

2005. Matsui, H. Unpublished.

2006. Matsui, M., Takahashi, K., & Goto, S. (2005) Scatter-hoarding of a walnut seed (*Juglans ailanthifolia*) by squirrels (*Sciurus vulgaris orientis*) and its seedling establishment. *Transactions of the Meeting in Hokkaido Branch of the Japanese Forest Society* **53**: 27–29 (in Japanese).

2007. Matsumaru, O., Watanabe, S., & Hosoda, T. (1989) Variation in the skull of Japanese matens (*Martes melampus melampus* WAGNER, 1841) from Honshu, Japan. *Nanki Seibutu* **31**: 93–98 (in Japanese).

2008. Matsumoto, H. (1930) Report of the mammalian remains obtained from the sites at Aoshima and Hibiku, Province of Rikuzen. *Science Reports of Tohoku Imperial University. Second Series (Geology)* **13**: 59–93 with plates.

2009. Matsumoto, H., et al. (1959) On the discovery of the Upper Pliocene fossiliferous and culture-bearing bed at Kanamori, Hanaizumi Town, Province of Rikuchu. *Bulletin of the National Science Museum, Tokyo* **4**: 287–324 with plates.

2010. Matsumura, S. (松村澄子) (1988) コウモリの生活戦略序論 (An Introduction to the Life Strategies in Bats). 動物その適応戦略と社会. Tokyo: Tokai University Press, 192 pp. (in Japanese).

2011. Matsumura, S. (松村澄子) (2005) 小コウモリ類超音波音声の地理的変異. in 動物地理の自然史：分布と多様性の進化学 (Masuda, R. (増田隆一) & (Abe, H. (阿部永), Editors). Sapporo: Hokkaido University Press, pp. 225–241 (in Japanese).

2012. Matsumura, S. (松村澄子), et al. (2005) 山口県内の洞窟におけるテングコウモリの生息状況. 山口ケイビングクラブ会報 **40**: 5–6 (in Japanese).

2013. Matsumura, S. Unpublished.

2014. Matsumura, S. (1979) Mother-infant communication in a horseshoe bat (*Rhinolophus ferrumequinum nippon*): development of vocalization. *Journal of Mammalogy* **60**: 76–84.

2015. Matsumura, S. (1981) Mother-infant communication in a horseshoe bat (*Rhinolophus ferrumequinum nippon*): vocal communication in three-week-old infant. *Journal of Mammalogy* **62**: 20–28.

2016. Matsumura, S., Inoshima, Y., & Ishiguro, N. (2014) Reconstructing the colonization history of lost wolf lineages by the analysis of the mitochondrial genome. *Molecular Phylogenetics and Evolution* **80**: 105–112.

2017. Matsuoka, K. (松岡耕二) (2009) 北西太平洋鯨類捕獲調査で発見されたシャチ. in シャチの現状と繁殖研究にむけて—2007シンポジウムプロシーディングス (Killer Whales in Japan: Proceedings of the Symposium on their Status and Breeding Research in Captivity, November 23, 2007, Tokyo) 鯨研叢書 (Geiken-sosyo) No.14. Tokyo: Institute of Cetacean Research, pp. 65–79 (in Japanese).

2018. Matsuoka, K., et al. (2009) Distribution of blue (*Balaenoptera musculus*), fin (*B. physalus*), humpback (*Megaptera novaeangliae*) and north pacific right (*Eubalaena japonica*) whales in the western North Pacific based on JARPN and JARPN II sighting surveys (1994 to 2007). (Paper SC/J09/JR35 presented to the IWC expert workshop to review the ongoing JARPN II Programme). 12 pp.

2019. Matsuoka, S. (1974) Prey taken by long-eared owl *Asio otus* in the breeding season in Hokkaido. *Journal of the Yamashina Institute for Ornithology* **7**: 324–329 (in Japanese with English summary).

2020. Matsuoka, S. (1977) Winter food habits of the Ural owl *Strix uralensis* Pallas in the Tomakomai experiment forest of Hokkaido University. *Research Bulletins of the College Experiment Forests, Faculty of Agriculture, Hokkaido University (Sapporo)* **34**: 161–174 (in Japanese with English summary).

2021. Matsuoka, S. (2008) Use of artifical roosts by Ussuri tube-nosed bats *Murina ussuriensis*. *Bulletin of Forestry and Forest Products Research Institute* **17**: 9–12 (in Japanese with English abstract).

2022. Matsuoka, T. (2014) Pregnancy, delivery and growth records of a finless porpoise. *Aquabiology* **36**: 44–49 (in Japanese with English abstract).

2023. Matsuura, Y. (松浦義雄) (1935) 日本近海に於ける白長須鯨の分布及び習性に就いて. *Zoological Magazine* **47**: 742–759 (in Japanese).

2024. Matsuura, Y., et al. (2004) The effects of age, body weight and reproductive status on conception dates and gestation periods in captive sika deer. *Mammal Study* **29**: 15–20.

2025. Matsuyama, T. (松山利夫) (1982) 木の実．ものと人間の文化誌. Tokyo: Hosei University Press, 371 pp. (in Japanese).

2026. Matsuyama, Y. & Ueno, H. (1955) Faunastic survey of the ectoparasite of the wood mice, *Apodemus speciosus*: 1. Chigger mites (Trombiculidae). *Medical Entomology and Zoology* **6**: 158–163.

2027. Matsuzaki, T., Suzuki, H., & Kamiya, M. (1989) Laboratory rearing of the Amami rabbits (*Pentalagus furnessi* Stone, 1900) in captivity. *Experimental Animals* **38**: 65–69 (in Japanese with English summary).

2028. Matsuzaki, Y., Mukaida, M., & Imai, T. (2005) Analysis of MHC gene in raccoons (*Procyon lotor*). *DNA Polymorphism* **13**: 120–125 (in Japanese).

2029. Matsuzaki, Y., et al. (2004) Geographical distribution of Raccoon (*Procyon lotor*) mitochondrial DNA polymorphism. *DNA Polymorphism* **12**: 39–44 (in Japanese).

2030. Matthee, C.A., et al. (2004) A molecular supermatrix of the rabbits and hares (Leporidae) allows for the identification of five intercontinental exchanges during the Miocene. *Systematic Biology* **53**: 433–447.

2031. Mattiucci, S. & Nascetti, G. (2006) Molecular systematics, phylogeny and ecology of anisakid nematodes of the Genus *Anisakis* Dujardin, 1845: an update. *Parasite* **13**: 99–113.

2032. Matveev, V.A., Kruskop, S.V., & Kramerov, D.A. (2005) Revalidation of *Myotis petax* Hollister, 1912 and its new status in connection with *M. daubentonii* (Kuhl, 1817) (Vespertilionidae, Chiroptera). *Acta Chiropterologica* **7**: 23–37.

2033. Mauget, R. (1982) Seasonality of reproduction in the wild boar. in Control of Pig Reproduction (Cole, D.J.A. & Foxcroft, G.R., Editors). London: Butterworth Scientific, pp. 509–526.

2034. May-Collado, L. & Agnarsson, I. (2006) Cytochrome *b* and Bayesian inference of whale phylogeny. *Molecular Phylogenetics and Evolution* **38**: 344–354.

2035. Maze-Foley, K. & Mullin, K.D. (2006) Cetaceans of the oceanic northern Gulf of Mexico: Distributions, group size and interspecific associations.

Journal of Cetacean Research and Management **8**: 203–213.
2036. McAlpine, D.F. (2002) Pygmy and dwarf sperm whales *Kogia breviceps* and *K. sima*. *in* Encyclopedia of Marine Mammals (Perrin, W.F., Würsig, B., & Thewissen, J.G.M., Editors). San Diego: Academic Press, pp. 1007–1009.
2037. McCullough, D.R. (2009) Sika deer in Korea and Vietnam. *in* Sika Deer—Biology and Management of Native and Introduced Populations (McCullough, D.R., Takatsuki, S., & Kaji, K., Editors). Tokyo: Springer, pp. 541–548.
2038. McCullough, D.R. (2009) Sika deer in Taiwan. *in* Sika Deer—Biology and Management of Native and Introduced Populations (McCullough, D.R., Takatsuki, S., & Kaji, K., Editors). Tokyo: Springer, pp. 549–560.
2039. McCullough, D.R., Jiang, Z.-G., & Li, C.-W. (2009) Sika deer in mainland China. *in* Sika Deer—Biology and Management of Native and Introduced Populations (McCullough, D.R., Takatsuki, S., & Kaji, K., Editors). Tokyo: Springer, pp. 521–539.
2040. McCullough, D.R., Pei, K.C.J., & Wang, Y. (2000) Home range, activity patterns, and habitat relations of Reeves' muntjac in Taiwan. *Journal of Wildlife Management* **64**: 430–441.
2041. McDonald, M. (2005) Population genetics of dugongs around Australia: Implications of gene flow and migration. Doctoral dissertation. Townsville: James Cook University of North Queensland. 184 pp.
2042. McGowen, M.R. (2011) Toward the resolution of an explosive radiation—A multilocus phylogeny of oceanic dolphins (Delphinidae). *Molecular Phylogenetics and Evolution* **60**: 345–357.
2043. McLaren, I.A. (1958) The biology of the ringed seal (*Phoca hispida* Schreber) in the eastern Canadian arctic. *Bulletin (Fisheries Research Board of Canada)* **118**: 1–97.
2044. McLaren, I.A. (1993) Growth in pinnipeds. *Biological Review* **68**: 1–79.
2045. McShea, W.J. (2000) The influence of acorn crops on annual variation in rodent and bird populations. *Ecology* **81**: 228–238.
2046. McShea, W.J. & Healy, W.M. (2002) Oaks and acorns as a foundation for ecosystem management. *in* Oak Forest Ecosystems (McShea, W.J. & Healy, W.M., Editors). Baltimore: Johns Hopkins University Press, pp. 1–12.
2047. Mead, J.G. (1989) Beaked whales of the genus *Mesoplodon*. *in* Handbook of Marine Mammals. Volume 4. River Dolphins and the Larger Toothed Whales (Ridgway, S.H. & Harrison, R., Editors). London: Academic press, pp. 349–430.
2048. Mead, J.G., Walker, W.A., & Houck, W.J. (1982) Biological observations on *Mesoplodon carlhubbsi* (Cetacea: Ziphiidae). *Smithsonian Contribution to Zoology* **344**: 1–25.
2049. Measures, L.N. (1993) Annotated list of Metazoan parasites reported from the blue whale *Balaenoptera musculus*. *Journal of the Helminthological Society of Washington* **60**: 62–66.
2050. Mech, L.D. (1970) The Wolf: the Ecology and Behavior of an Endangered Species. New York: Natural History Press, 384 pp.
2051. Mech, L.D. (1974) *Canis lupus*. *Mammalian Species* **37**: 1–6.
2052. Mech, L.D. (1974) A new profile for the wolf. *Natural History* **83**: 26–31.
2053. Mech, L.D. (1977) Productivity, mortality , and population trends of wolves in northeastern Minnesota. *Journal of Mammalogy* **58**: 559–574.
2054. Mech, L.D. & Nelson, M.E. (1989) Polygyny in a wild wolf pack. *Journal of Mammalogy* **70**: 675–676.
2055. Medway, L. (1978) The Wild Mammals of Malaya (Peninsular Malaysia) and Singapore. 2nd ed. Oxford: Oxford University Press, 127 pp.
2056. Meguro Parasitological Museum (2013) Check List of Parasitic Helminths of Mammals in Japan. Tokyo: Meguro Parasitological Museum, 198 pp. (in Japanese with English abstract; http://homepage2.nifty.com/mpm-hp/archive/Parasitic helminths of mammals in Japan.pdf).
2057. Mei, Z., et al. (2014) The Yangtze finless porpoise: On an accelerating path to extinction? *Biological Conservation* **172**: 117–123.
2058. Melis, C., et al. (2007) Raccoon dogs in Norway—Potential expansion rate, distribution area and management implications. *NTNU Vitenskapsmuseet Rapport Zoologisk Serie 2007* **3**: 1–49.
2059. Melnick, D.J., et al. (1993) MtDNA diversity in rhesus monkeys reveals overestimates of divergence time and paraphyly with neighboring species. *Molecular Biology and Evolution* **10**: 282–295.
2060. Mercer, J.M. & Roth, V.L. (2003) The effects of Cenozoic global change on squirrel phylogeny. *Science* **299**: 1568–1572.
2061. Mesnick, S.L. & Ralls, K. (2002) Mating systems. *in* Encyclopedia of Marine Mammals (Perrin, W.F., Würsig, B., & Thewissen, J.G.M., Editors). San Diego: Academic Press, pp. 726–733.
2062. Mignucci-Giannoni, A.A., et al. (1999) Mass stranding of pygmy killer whales (*Feresa attenuata*) in the British Virgin Islands. *Journal of the Marine Biological Association of the United Kingdom* **80**: 759–760.
2063. Miguchi, H. (1988) Two years of community dynamics of murid rodents after a beechnut mast year. *Journal of the Japanese Forest Society* **70**: 472–480 (in Japanese with English abstract).
2064. Miguchi, H. (1994) Role of wood mice on the regeneration of cool temperate forest. *in* Proceedings of NAFRO Seminar on Sustainable Forestry and its Biological Environment (Kobayashi, S., Editor). Tokyo: Japan Society of Forest Planning Press, pp. 115–121.
2065. Miguchi, H. (1996) Dynamics of beech forest from the view point of rodent ecology—Ecological interactions of the regeneration characteristics of *Fagus crenata* and rodents. *Japanese Journal of Ecology* **46**: 185–189.
2066. Mikasa, A. (三笠暁子) (1996) 戸隠森林植物園のコウモリ調査報告. *Bat Study and Conservation Report* **4**: 3–4 (in Japanese).
2067. Mikura-jima Bottlenose Dolphin Research Group (御蔵島バンドウイルカ研究会) (2005) 御蔵島周辺のミナミハンドウイルカ個体識別調査報告書 (1994 ~ 2003 年). Tokyo: Mikura-jima Bottlenose Dolphin Research Group (御蔵島バンドウイルカ研究会), 160 pp. (in Japanese).
2068. Mikuriya, M. (御厨正治), ed. (1969) いたち. Utsunomiya: Forest Agency, Utsunomiya Forestry Office (宇都宮営林署), 70 pp. (in Japanese).
2069. Mikuriya, M. (御厨正治) (1980) 有益獣増殖事業 20 年のあしあと. Utsunomiya: Forest Agency, Utsunomiya Forestry Office (宇都宮営林署), 146 pp. (in Japanese).
2070. Mikuriya, M. & Yoshiyuki, M. (1972) New record of three species of bats from Nikko, Tochigi Pref. *The Journal of the Mammalogical Society of Japan* **5**: 196 (in Japanese).
2071. Miller, P.J.O., et al. (2008) Stereotypical resting behavior of the sperm whale. *Current Biology* **18**: R21–R23.
2072. Miller, P.J.O., Johnson, M.P., & Tyack, P.L. (2004) Sperm whale behaviour indicates the use of rapid echolocation click buzzes "creaks" in prey capture. *Proceedings of the Royal Society of London. Series B* **271**: 2239–2247.
2073. Miller-Butterworth, C.M., et al. (2007) A family matter: conclusive resolution of the taxonomic position of the long-fingered bats, *Miniopterus*. *Molecular Biology and Evolution* **24**: 1553–1561.
2074. Min, M.S., et al. (2004) Molecular phylogenetic status of the Korean goral and Japanese serow based on partial sequences of the mitochondrial cytochrome *b* gene. *Molecules and Cells* **17**: 365–372.
2075. Minamikawa, S., Iwasaki, T., & Kishiro, T. (2007) Diving behaviour of a Baird's beaked whale, *Berardius bairdii*, in the slope water region of the western North Pacific: first dive records using a data logger. *Fisheries Oceanography* **16**: 573–577.
2076. Minamikawa, S., Watanabe, H., & Iwasaki, T. (2013) Diving behavior of a false killer whale, Pseudorca crassidens, in the Kuroshio-Oyashio transition region and the Kuroshio front region of the western North Pacific. *Marine Mammal Science* **29**: 177–185.
2077. Minasian, S.M., Balcomb, III, K. C., & Foster, L. (1984) The World's Whales: The Complete Illustrated Guide. Washington, D.C.: Smithsonian Books, 224 pp.
2078. Minato, S. (湊秋作) (2000) ヤマネって知ってる？：ヤマネおもしろ観察記. Tokyo: Tsukiji-shokan (築地書館), 126 pp. (in Japanese).
2079. Minezawa, M., Moriwaki, K., & Kondo, K. (1979) Geographical distribution of Hbb^p allele in the Japanese wild mice, *Mus musculus molossinus*. *Japanese Journal of Genetics* **54**: 165–173.
2080. Ministry of Agriculture, Forestry and Fisheries (2006) 耕作放棄地対策推進の手引き. Tokyo: Ministry of Agriculture, Forestry and Fisheries (農林水産省農村振興局), 153 pp. (in Japanese).
2081. Ministry of the Environment, Personal communication.
2082. Ministry of the Environment (1992) Wildlife Protection and Hunting Law (Law No.32 1918). Available from: http://www.env.go.jp/en/nature/biodiv/law.html.
2083. Ministry of the Environment (2002) Threatened Wildlife of Japan—Red Data Book 2nd ed.—Volume 1, Mammalia (改訂・日本の絶滅のおそれのある野生生物：レッドデータブック 1 哺乳類). Tokyo: Japan Wildlife Research Center, 177 pp. (in Japanese).
2084. Ministry of the Environment (2004) ジュゴンと藻場の広域的調査：平成 13 ~ 15 年度 結果概要. 31 pp. (in Japanese; http://www.env.go.jp/nature/yasei/jugon/h13-h15.pdf).
2085. Ministry of the Environment (2005) Web Page on Invasive Alien Species Act. Available from: https://www.env.go.jp/en/press/2005/0203a.html.
2086. Ministry of the Environment (2007) Japanese Red List on Mammals (レッドリスト 哺乳類). Tokyo: Ministry of the Environment, 3 pp. (in Japanese; https://www.env.go.jp/press/files/jp/9941.pdf).
2087. Ministry of the Environment (2009) 野生鳥獣の保護管理に係る計画制度：特定鳥獣保護管理計画の概要. Available from: http://www.env.go.jp/press/press.php?serial=2002 (in Japanese).
2088. Ministry of the Environment (2014) アライグマ防除の手引き (計画的な防除の進め方) 改訂版. Tokyo: Ministry of Environment (in Japanese; https://www.env.go.jp/nature/intro/4control/files/manual_racoon.pdf).
2089. Ministry of the Environment (2014) 鳥獣関係統計. Available from: https://www.env.go.jp/nature/choju/docs/docs2.html (in Japanese).
2090. Ministry of the Environment & Ministry of Agriculture, Forestry and

Fisheries (2015) 我が国の生態系等に被害を及ぼすおそれのある外来種リスト (生態系被害防止外来種リスト) (Invasive Alien Species Lists of Japan). 90 pp. (in Japanese; http://www.env.go.jp/nature/intro/1outline/list/list.pdf).

2091. Ministry of the Environment, Ministry of Agriculture, Forestry and Fisheries, & Ministry of Land, Infrastructure, Transport and Tourism (2015) 外来種被害防止行動計画〜生物多様性条約・愛知目標の達成に向けて〜 (Invasive Alien Species Control Action Plan). 116 pp. (in Japanese; http://www.env.go.jp/nature/intro/1outline/actionplan/actionplan.pdf).

2092. Mio, H. (1951) The state of hibernation of the Japanese long-eared bat. *Collecting and Breeding* **13**: 310 (in Japanese).

2093. Misawa, E. (三沢英一) (1992) 狩猟統計が語る狐の皮算用. in 下中科学研究助成金30周年記念論文集. Tokyo: Shimonaka Memorial Foundation (下中記念財団), pp. 35–42 (in Japanese).

2094. Misawa, E. (1979) Change in the food habits of the red fox, *Vulpes vulpes schrencki* Kishida according to habitat conditions. *The Journal of the Mammalogical Society of Japan* **7**: 311–320 (in Japanese with English summary).

2095. Misawa, E., Abe, H., & Ota, K. (1987) The ecological study of red fox (*Vulpes vulpes schrencki* Kishida) in the Tomakomai experiment forest—The home range and land usage of red fox—. *Research Bulletins Experiment Forest Faculty of Agriculture Hokkaido University* **44**: 675–687 (in Japanese with English summary).

2096. Mitani, N., et al. (2014) An evaluation of detection dogs for alien mongoose control on Amami-oshima Island. *Wildlife and Human Society* **2**: 11–22 (in Japanese with English abstract).

2097. Mitchell, E. & Houck, W.J. (1967) Cuvier's beaked whale (*Ziphius cavirostris*) stranded in northern California. *Journal of Fisheries Research Board of Canada* **24**: 2503–2513.

2098. Mitchell, J. (1973) Determination of relative age in the dugong *Dugong dugon* (Müller) from a study of skulls and teeth. *Zoological Journal of the Linnean Society* **53**: 1–23.

2099. Mitchell-Jones, A.J., et al., eds. (1999) The Atlas of European Mammals. London: T & AD Poyster, 484 pp.

2100. Mitsuki, M. (光木将意) & Tanabe, T. (田辺) (1960) 広島県帝釈峡のコウモリ. *Journal of the Hiba Society of Natural History* **8**: 12–17 (in Japanese).

2101. Miura, S. (三浦慎吾) (1991) 日本産偶蹄類の生活史戦略とその保護管理―標本個体群の検討から―. in 現代の哺乳類学 (Asahi, M. (朝日稔) & Kawamichi, T. (川道武男), Editors). Tokyo: Asakura Publishing (朝倉書店), pp. 244–273 (in Japanese).

2102. Miura, S. (三浦慎吾) (1994) ニホンジカ. in A Pictorial Guide to the Mammals of Japan (日本の哺乳類) (Abe, H. (阿部永), Editor). Tokyo: Tokai University Press, pp. 148–149 (in Japanese).

2103. Miura, S. (三浦慎吾) (1994) ヌートリア (Nutoria). in 日本の希少な野生水生生物に関する基礎資料 I. Tokyo: Japan Fisheries Resource Conservation Association (日本水産資源保護協会), pp. 539–541 (in Japanese).

2104. Miura, S. (三浦慎吾) (1999) 野生動物の生態と農林業被害―共存の論理を求めて. Tokyo: Zenrinkyou (全国林業改良普及協会), 174 pp. (in Japanese).

2105. Miura, S. (三浦慎吾) (2002) 日本は野生動物とどのように向き合ってきたか. *Kagaku* (科学) **72**: 95–101 (in Japanese).

2106. Miura, S. (1976) Dispersal of nutoria in Okayama Prefecture. *The Journal of the Mammalogical Society of Japan* **6**: 231–237 (in Japanese with English abstract).

2107. Miura, S. (1977) Social studies on sika deer in Nara Park with reference to spatial structure and behavior. *Annual Report of Nara Deer Research Association* **3**: 3–41 (in Japanese with English summary).

2108. Miura, S. (1981) Behavior and social structure of captive muntjacs (1). *Animals and Zoos* **33**: 80–86 (in Japanese with English abstract).

2109. Miura, S. (1981) Behavior and social structure of captive muntjacs (2). *Animals and Zoos* **33**: 120–125 (in Japanese with English abstract).

2110. Miura, S. (1983) Grouping behavior of male sika deer in Nara park, Japan. *The Journal of the Mammalogical Society of Japan* **9**: 279–284.

2111. Miura, S. (1984) Annual cycles of coat changes, antler regrowth, and reproductive behavior of sika deer in Nara Park, Japan. *The Journal of the Mammalogical Society of Japan* **10**: 1–7.

2112. Miura, S. (1984) Dominance hierarchy and space use pattern in male captive muntjac, *Muntiacus reeves*. *Journal of Ethology* **2**: 69–75.

2113. Miura, S. (1984) Social behavior and territoriality in male sika deer (*Cervus nippon* Temminck 1838) during the rut. *Zeitschrift für Tierpsychologie* **64**: 33–73.

2114. Miura, S. (1985) Horn and cementum annulation as age criteria in Japanese serow. *Journal of Wildlife Management* **49**: 152–156.

2115. Miura, S. (1986) Body and horn growth patterns in the Japanese serow, *Capricornis crispus*. *The Journal of the Mammalogical Society of Japan* **11**: 1–13.

2116. Miura, S. (1986) Preliminary report on age composition, sex ratio, pregnancy rate, and life table of Japanese serow from Gifu and Nagano prefecture population in 1979, 1980, and 1981. in Report on Analysis of Shooting Samples of Japanese Serow in Gifu and Nagano Prefectures. Tokyo Agency for Cultural Affairs, pp. 58–69 (in Japanese with English abstract).

2117. Miura, S. (2005) Goat. in A Guide to the Mammals of Japan (Abe, H., Editor). Hadano: Tokai University Press, p. 114 (in Japanese and English).

2118. Miura, S., Kita, I., & Sugimura, M. (1987) Horn growth and reproductive history in female Japanese serow. *Journal of Mammalogy* **68**: 826–836.

2119. Miura, S. & Maruyama, N. (1986) Winter weight loss in Japanese serow. *Journal of Wildlife Management* **50**: 336–338.

2120. Miura, T. & Kitaoka, M. (1977) Viruses isolated from bats in Japan. *Archives of Virology* **53**: 281–286.

2121. Miyagi, K., Shiraishi, S., & Uchida, T. (1983) Age determination in the yellow weasel *Mustela sibirica coreana*. *Journal of Faculty of Agriculture, Kyushu University* **27**: 109–114.

2122. Miyake, T. (三宅隆), et al. (2008) 特定外来生物静岡県ハリネズミ生息実態調査報告書 (Report on the Status of Alien Hedgehog in Shizuoka Prefecture). Shizuoka: Natural History Museum Network of Shizuoka Prefecture (静岡県自然誌博物館ネットワーク), 77 pp. (in Japanese).

2123. Miyake, T. Personal communication.

2124. Miyaki, M. & Kaji, K. (2004) Summer forage biomass and the importance of litterfall for a high-density sika deer population. *Ecological Research* **19**: 405–409.

2125. Miyaki, M. & Kikuzawa, K. (1988) Dispersal of *Quercus mongolica* acorns in a broadleaved deciduous forest. 2. Scatterhoarding by mice. *Forest Ecology and Management* **25**: 9–16.

2126. Miyamoto, F. (1991) On the skull of Japanese wolf (*Canis lupus hodophiliax* Temminck) taken out from the mounted specimen preserved in Wakayama University. *Bulletin of Faculty of Education, Wakayama University, Natural Science* **39**: 55–60 (in Japanese with English abstract).

2127. Miyamoto, F. & Maki, I. (1983) On the repaired specimen of Japanese wolf (*Canis lupus hodophiliax* Temminck) and its skull newly taken out. *Bulletin of Faculty of Education, Wakayama University, Natural Science* **32**: 9–16 (in Japanese with English abstract).

2128. Miyamoto, K. (1985) Studies on zoonoses in Hokkaido. 7. Survey of natural definitive hosts of *Metagonimus yokogawai*. *Japanese Journal of Parasitology* **34**: 371–376 (in Japanese with English summary).

2129. Miyamoto, K. (2002) On an epidemiological study of Lyme disease in Hokkaido, Japan. *Medical Entomology and Zoology* **53**: 1–6 (in Japanese with English abstract).

2130. Miyamoto, K., Nakao, M., & Inaoka, T. (1983) Studies on the zoonoses in Hokkaido, Japan. 5. On the epidemiological survey of *Echinostoma hortense* Asada, 1926. *Japanese Journal of Parasitology* **30**: 261–269 (in Japanese with English abstract).

2131. Miyamoto, K., et al. (1992) Prevalence of Lyme borreliosis spilochetes in ixodid ticks of Japan, with special reference to a new potential vector, *Ixodes ovatus* (Acari: Ixodidae). *Journal of Medical Entomology* **29**: 216–220.

2132. Miyamoto, T. & Furukawa, M. (2011) A record of Ussuri tube-nosed bat *Murina ussuriensis* in the Onigajo Mountain Range. *Bulletin of the Shikoku Institute of Natural History* **6**: 15–17 (in Japanese).

2133. Miyao, T. (宮尾嶽雄), et al. (1984) 哺乳類. in 愛知の動物：愛知文化シリーズ (3) (Sato, M. (佐ండ正孝) & Ando, H. (安藤尚), Editors). Nagoya: 愛知県郷土資料刊行会, pp. 286–325 (in Japanese).

2134. Miyao, T. (宮尾嶽雄) & Morozumi, M. (両角源美) (1972) コウモリの胎仔数ならびに産仔数. in 日本哺乳類雑記第1集 (Miyao, T. (宮尾嶽雄), Editor). Matsumoto: 信州哺乳類研究会, pp. 15–17 (in Japanese).

2135. Miyao, T. (宮尾嶽雄), Morozumi, M. (両角徹郎), & Morozumi, M. (両角源美) (1972) 食虫類およびネズミ類の1腹の胎児数. in 日本哺乳類雑記第1集 (Miyao, T. (宮尾嶽雄), Editor). Matsumoto: 信州哺乳類研究会, pp. 87–94 (in Japanese).

2136. Miyao, T. (1961) Variation of the form of the third upper molar in Microtinae. *Japanese Journal of Applied Entomology and Zoology* **5**: 212–214 (in Japanese).

2137. Miyao, T. (1973) Supernumerary and missing teeth in nine species of the suborder Microchiroptera. *The Journal of the Mammalogical Society of Japan* **5**: 230–233 (in Japanese with English abstract).

2138. Miyao, T. (1976) Stomach contents of Japanese serow indigenous to southern parts of the Japan North Alps. *The Journal of the Mammalogical Society of Japan* **6**: 199–209 (in Japanese with English abstract).

2139. Miyao, T. (1984) Animal remains from Miwa Cave. in Report on the Excavation of the Cave and Rockshelter in Miwa-cho, Gifu Prefecture. The Academic Year 1982. Inuyama: Primate Research Institute of Kyoto University, pp. 13–16 (in Japanese).

2140. Miyao, T., et al. (1983) Mammals of the southern part of Awajishima lying in the Setouchi Sea, Japan. *The Journal of the Mammalogical Society of Japan* **9**: 128–140 (in Japanese with English abstract).

2141. Miyao, T. & Morozumi, M. (1969) Notes on the embryo-size in Japanese native bats (I). *The Journal of the Mammalogical Society of Japan* **4**: 87–89 (in Japanese with English abstract).

2142. Miyao, T., Morozumi, M., & Morozumi, T. (1966) Small mammals of Mt. Yatsugatake in Honshu VI. Seasonal variation in sex ratio, body weight and reproduction in the vole, *Microtus montebelli*. *Zoological Magazine* **95**: 98–102.
2143. Miyao, T., et al. (1965) Small mammals of Mt. Yatsugatake in Honshu IV. Seasonal differences in body weight and reproduction of *Dymecodon pilirostris* and *Sorex shinto* in the subalpine forest zone on Mt. Yatsugatake. *Zoological Magazine* **74**: 76–81.
2144. Miyao, T., et al. (1964) Small mammals on Mt. Yatsugatake in Honshu. III. Smith's red-backed vole (*Eothenomys smithi*) in the subalpine forest zone on Mt. Yatsugatake. *Zoological Magazine* **73**: 189–195 (in Japanese with English summary).
2145. Miyao, T., Nishizawa, T., & Suzuki, S. (1980) Mammalian remains of the Earliest Jomon Period at the Rockshelter site of Tochibara, Nagano Pref., Japan. II. Size differences of the lower molars between the earliest Jomon Period and the living specimens of *Apodemus speciosus* (Muridae, Rodentia). *Japanese Journal of Oral Biology* **23**: 141–146 (in Japanese with English abstract).
2146. Miyashita, K. (宮下和喜) (1963) 帰化動物 (5). *Nature (自然)* **18**: 69–75 (in Japanese).
2147. Miyashita, T. (宮下富夫) (1986) IV. 資源量の推定 2. 調査船. *in* 漁業公害 (有害生物駆除) 対策調査委託事業調査報告書 (昭和 56–60 年度) (Tamura, T. (田村保), Ohsumi, S. (大隅清治), & Arai, S. (荒井ené亮), Editors). Tokyo: Fisheries Agency, pp. 178–187 (in Japanese).
2148. Miyashita, T. (宮下富夫) (1996) セミイルカ. *in* The Encyclopaedia of Animals in Japan. 2. Mammals II (日本動物大百科2哺乳類II) (Izawa, K. (伊澤紘生), Kasuya, T. (粕谷俊雄), & Kawamichi, T. (川道武男), Editors). Tokyo: Heibonsha (平凡社), p. 81 (in Japanese).
2149. Miyashita, T. (宮下富夫), Iwasaki, T. (岩崎俊秀), & Moronuki, H. (諸貫秀樹) (2007) 北西太平洋におけるイシイルカの資源量推定. *in* Abstracts of the Annual Meeting of the Japanese Society of Fisheries Science. p. 164 (in Japanese).
2150. Miyashita, M. (1993) Prevalence of *Baylisascaris procyonis* in raccoons in Japan and experimental infections of the worm in laboratory animals. *Journal of Urban Living and Health Association* **37**: 137–151 (in Japanese with English abstract).
2151. Miyashita, M. (1993) Study of raccoon roundworm (*Baylisascaris procyonis*) larva migrans. *Journal of Urban Living and Health Association* **37**: 137–151 (in Japanese).
2152. Miyashita, T. (1993) Abundance of dolphin stocks in the western North Pacific taken by the Japanese drive fishery. *Report of the International Whaling Commission* **43**: 417–437.
2153. Miyashita, T. & Kasuya, T. (1988) Distribution and abundance of Dall's porpoises off Japan. *Scientific Reports of the Whales Research Institute* **39**: 121–150.
2154. Miyashita, T., et al. (1995) Report of the Japan/China joint whale sighting cruise in the Yellow Sea and the East China Sea in 1994 summer (Paper SC/47/NP17 presented to the IWC Scientific Committee). 12 pp.
2155. Miyatsu, T., Konno, M., & Nitta, S. (1988) An albino red-backed vole, *Clethrionomys rufocanus bedofordiae*. *Forest Protection* **207**: 40 (in Japanese).
2156. Miyazaki, N. (宮崎信之) & Hiraguchi, T. (平口哲夫) (1986) 動物遺体. *in* 石川県能都町真脇遺跡 (石川県能都町教育委員会 & 真脇遺跡発掘調査団, Editors), pp. 346–400 (in Japanese).
2157. Miyazaki, N. (1980) Preliminary note on age determination and growth of the rough-toothed dolphin *Steno bredanensis*, off the Pacific coast of Japan. *Reports of the International Whaling Commission (Special Issue)* **3**: 171–179.
2158. Miyazaki, N. & Amano, M. Unpublished.
2159. Miyazaki, N. & Amano, M. (1994) Skull morphology of two forms of short-finned pilot whales off the Pacific coast of Japan. *Report of the International Whaling Commission* **44**: 499–508.
2160. Miyazaki, N., Fujise, Y., & Iwata, K. (1998) Biological analysis of a mass stranding of melon-headed whales (*Peponocephala electra*) at Aoshima, Japan. *Bulletin of the National Science Museum, Series A* **24**: 31–60.
2161. Miyazaki, N. & Nakayama, K. (1989) Records of cetaceans in the waters of the Amami Islands. *Memories of the National Science Museum* **22**: 235–249.
2162. Miyazaki, N. & Nishiwaki, M. (1978) School structure of the striped dolphin off the Pacific coast of Japan. *Scientific Reports of the Whales Research Institute, Tokyo* **30**: 65–115.
2163. Miyazaki, N. & Perrin, W.F. (1994) Rough-toothed dolphin *Steno bredanensis* (Lesson, 1828). *in* Handbook of Marine Mammals. Volume 5. The First Book of Dolphins (Ridgway, S.H. & Harrison, R., Editors). London: Academic Press, pp. 1–21.
2164. Miyazaki, N. & Shikano, C. (1997) Comparison of growth and skull morphology of Pacific white-sided dolphin, *Lagenorhynchus obliquidens*, between the coastal waters of Iki Island and the oceanic waters of the western North Pacific. *Mammalia* **61**: 561–572.
2165. Miyazaki, N. & Shikano, C. (1997) Preliminary study on comparative skull morphology and vertebral formula among the six species of the genus *Lagenorhynchus* (Cetacea: Delphinidae). *Mammalia* **61**: 573–587.
2166. Miyazaki, N. & Wada, S. (1978) Fraser's dolphin, *Lagenodelphis hosei* in the western north Pacific. *Scientific Reports of the Whales Research Institute, Tokyo* **30**: 231–244.
2167. Miyoshi, K. (2006) Home range, habitat use and food habits of the Japanese sable *Martes zibellina brachyura* in a cool-temperate mixed forest. Sapporo: Hokkaido University. 58 pp.
2168. Miyoshi, K. & Higashi, S. (2005) Home range and habitat use by the sable *Martes zibellina brachyura* in a Japanese cool-temperate mixed forest. *Ecological Research* **20**: 95–101.
2169. Mizoguchi, N., et al. (1996) Effect of yearly fluctuations in beechnut production on food habits of Japanese black bear. *Mammalian Science* **36**: 33–44 (in Japanese with English abstract).
2170. Mizue, K. & Yoshida, K. (1961) Studies on the little toothed whales in the west sea area of Kyusyu-VII. About *Pseudorca crassidens* caught at Arikawa in Goto Is., Nagasaki Pref. *Bulletin of the Faculty of Fisheries, Nagasaki University* **11**: 39–48 (in Japanese with English summary).
2171. Mizue, K., Yoshida, K., & Masaki, Y. (1965) Studies on the little toothed whales in the west sea area of Kyushu-XII. *Neomeris phocaenoides*, so-called Japanese "SUNAMERI", caught in the coast of Tachibana Bay, Nagasaki Pref. *Bulletin of Faculty of Fisheries, Nagasaki University* **18**: 7–29.
2172. Mizukami, R., et al. (2005) Estimation of feeding history of measuring carbon and nitrogen stable isotope ratios in hair of Asiatic black bear. *Ursus* **16**: 93–101.
2173. Mizukoshi, A., Johnson, K.P., & Yoshizawa, K. (2012) Co-phylogeography and morphological evolution of sika deer lice (*Damalinia sika*) with their hosts (*Cervus nippon*). *Parasitology* **139**: 1614–1629.
2174. Mizuno, Y. (水野裕子), et al. (1993) 道東のエゾシカ *Cervus nippon yesoensis* の第四胃内線虫症の季節的変動. 北海道家畜寄生虫研究会報 **9**: 8 (in Japanese).
2175. Mizuno, A. (1996) Infant mortality of northern fur seals, *Callorhinus Ursinus*, in Bering Island, Russia. *in* Migratory Ecology and Conservation of Steller Sea Lions (Ohtaishi, N. & Wada, K., Editors). Tokyo: Tokai University Press, pp. 351–363 (in Japanese with English summary).
2176. Mizuno, A.W., et al. (2003) Population genetic structure of the spotted seal *Phoca largha* along the coast of Hokkaido, based on mitochondrial DNA sequences. *Zoological Science* **20**: 783–788.
2177. Mizuno, A.W., et al. (2002) Distribution and abundance of spotted seals *Phoca largha* and ribbon seals *Phoca fasciata* in the southern Sea of Okhotsk. *Ecological Research* **17**: 79–96.
2178. Mizuno, T. (1970) Note on *Myotis hosonoi* from Nagano. *The Journal of the Mammalogical Society of Japan* **5**: 70 (in Japanese).
2179. Mizuno, T. (1973) Notes on some mammals in Kiso, Nagano Pref. *The Journal of the Mammalogical Society of Japan* **5**: 205 (in Japanese).
2180. Mizushima, S. & Yamada, E. (1974) On the distribution and food habits of the murid rodents in agrosystems in Hokkaido. *Japanese Journal of Applied Entomology and Zoology* **18**: 81–88 (in Japanese with English abstract).
2181. Mochizuki, M. (1962) Ecological studies of the paddy mouse, *Apodemus speciosus*, and the vole, *Microtus montebelli*, in cultivated fields. *Bulletin of the Toyama Agricultural Experiment Station* **4**: 1–135 (in Japanese).
2182. Modi, W.S. & Gamperl, R. (1989) Chromosomal banding comparisons among American and European red-backed mice, genus *Clethrionomys*. *Zeitschrift für Säugetierkunde* **54**: 141–152.
2183. Mogi, M. (茂木幹義) (1976) 対馬のコウモリ寄生バエ. *in* 対馬の生物 (Nagasaki Biological Society (長崎県生物学会), Editor). Nagasaki: Nagasaki Biological Society, pp. 343–348 (in Japanese).
2184. Mogi, M. (1975) A new species of *Lipoptena* (Diptera, Hippoboscidae) from the Japanese deer. *Kontyu* **43**: 387–392.
2185. Mogi, M. (1976) A new species of *Brachytarsina* (Diptera, Streblidae) from the Ryukyu Islands. *Kontyû* **44**: 323–326.
2186. Mogi, M. (1979) Two species of batflies (Diptera, Nycteribiidae) new to Japan with description of a new subspecies. *Tropical Medicine* **21**: 145–151.
2187. Mogi, M., Mano, T., & Sawada, I. (2002) Records of Hippoboscidae, Nycteribiidae and Streblidae (Diptera) from Japan. *Medical Entomology and Zoology* **53**: 141–165.
2188. Mohnot, S.M. (1978) The conservation of non-human primates in India. *in* Recent Advances in Primatology. II. Conservation (Chivers, D.J. & Lane-Petter, W., Editors). London: Academic Press, pp. 47–53.
2189. Möller, L., et al. (2010) Fine-scale genetic structure in short-beaked com-

REFERENCES

2189. mon dolphins (*Delphinus delphis*) along the East Australian Current. *Marine Biology* **158**: 113–126.
2190. Momosaki, T. (2009) Artificial caves in Fukui Prefecture and inhabiting species (bats and trechid beetles). *Bulletin of the Fukui City Museum of Natural History* **56** (in Japanese).
2191. Moore, J.C. (1968) Relationships among the living genera of beaked whales, with classification, diagnoses and keys. *Fieldiana Zoology* **53**: 206–298.
2192. Moore, J.C. & Gilmore, R.M. (1965) A beaked whale new to the Western Hemisphere. *Nature* **205**: 1239–1240.
2193. Moors, P.J. (1980) Sexual dimorphism in the body size of mustelids (Carnivora) : the roles of food habits and breeding systems. *Oikos* **34**: 147–158.
2194. Morgan, U.M., et al. (2000) Detection of the *Cryptosporidium parvum* ''Human'' Genotype in a Dugong (*Dugong dugon*). *Journal of Parasitology* **86**: 1352–1354.
2195. Mori, K. (森恭一) (1994) 小笠原諸島周辺海域におけるザトウクジラの分布・回遊と系群に関する研究 (Distribution, migration and population structure of humpback whale in the waters of the Ogasawara Islands). Doctoral dissertation. Shizuoka: Tokai University. 129 pp. (in Japanese).
2196. Mori, K. (森恭一), et al. (2002) 小笠原諸島父島列島周辺海域におけるハシナガイルカの日周行動 (Diurnal behavior of spinner dolphins in the waters of the Ogasawara Islands). in Abstracts of the Annual Meeting of the Mammalogical Society of Japan. p. 205 (in Japanese).
2197. Mori, K. (森恭一) (2013) 能登島のミナミハンドウイルカ. *Kaiyo Monthly (月刊海洋)* **45**: 253–256 (in Japanese).
2198. Mori, K. (森恭一) & Okamoto, R. (岡本亮介) (2013) 小笠原のハンドウイルカ. *Kaiyo Monthly (月刊海洋)* **45**: 226–231 (in Japanese).
2199. Mori, A. (1979) Analysis of population changes by measurement of body weight in the Koshima troop of Japanese monkeys. *Primates* **20**: 371–397.
2200. Mori, A. & Moriguchi, H. (1988) Food habits of the snakes in Japan: A critical review. *The Snake* **20**: 98–113.
2201. Mori, K., et al. (1999) School structure, distribution and food habitats of sperm whales near the Ogasawara Islands, Japan. in 13th Biennial Conference on the Biology of Marine Mammals, Abstracts. Maui: Society for Marine Mammalogy, p. 130.
2202. Mori, K., et al. (2005) Distribution and residency of Indo-Pacific bottlenose dolphins (*Tursiops aduncus*) in the waters of the Ogasawara (Bonin) Islands, Japan. in 16th Biennial Conference on the Biology of Marine Mammals, Abstracts. San Diego: Society for Marine Mammalogy, p. 199.
2203. Mori, K., et al. (2001) Distribution and diurnal movement of spinner dolphins in the waters of the Ogasawara Islands. in 14th Biennial Conference on the Biology of Marine Mammals, Abstracts. Vancouver: Society for Marine Mammalogy, p. 149.
2204. Mori, K., et al. (1998) Distribution, migration and local movements of humpback whale (*Megaptera novaeangliae*) in the adjacent waters of the Ogasawara (Bonin) Islands, Japan. *Journal of School of Marine Science and Technology, Tokai University* **45**: 197–213.
2205. Mori, K., Tsunoda, T., & Fujimagari, M. (1995) Ixodid ticks on Sika deer *Cervus nippon* Temminck in Chiba Prefecture. *Medical Entomology and Zoology* **46**: 313–316 (in Japanese with English abstract).
2206. Mori, T., et al. (1991) Ultrastructural observations on spermatozoa of the Soricidae, with special attention to a subfamily revision of the Japanese water shrew *Chimarrogale himalayica*. *Journal of the Mammalogical Society of Japan* **16**: 1–12.
2207. Mōri, T., Oh, Y.K., & Uchida, T.A. (1982) Sperm storage in the oviduct of the Japanese greater horseshoe bat, *Rhinolophus ferrumequinum nippon*. *Journal of the Faculty of Agriculture, Kyushu University* **27**: 47–53.
2208. Mōri, T. & Uchida, T.A. (1981) Ultrastructural observations of fertilization in the Japanese long-fingered bat, *Miniopterus schreibersii fuliginosus*. *Journal of Reproduction and Fertility* **63**: 231–235.
2209. Mōri, T. & Uchida, T.A. (1981) Ultrastructural observations of ovulation in the Japanese long-fingered bat, *Miniopterus schreibersii fuliginosus*. *Journal of Reproduction and Fertility* **63**: 391–395.
2210. Moribe, J. Unpublished.
2211. Moribe, J., et al. (2007) The karyotype of the Azumi shrew *Sorex hosonoi*. *Acta Theriologica* **52** 69–74.
2212. Morii, R. (森井隆三) (1991) コウモリ (ほ乳類). in 龍河洞の自然. Kami: 龍河洞保存会, pp. 39–49 (in Japanese).
2213. Morii, R. (森井隆三) (2005) コウモリとともに. Takamatsu: Bikohsha (美巧社), 122 pp. (in Japanese).
2214. Morii, M. (1976) Biological study of the Japanese house bat, *Pipistrellus abramus* (Temminck, 1840) in Kagawa Prefecture. Part 1. External, cranial and dental characters of embryos and litters. *The Journal of the Mammalogical Society of Japan* **6**: 248–258 (in English with Japanese abstruct).
2215. Morii, M. (1980) Postnatal development of external charcters and behavior in young *Pipistrellus abramus*. *The Journal of the Mammalogical Society of Japan* **8**: 117–121 (in English with Japanese abstruct).
2216. Morii, M. (1981) 岡山県内同一採集地点, 8月の2時期において採集したアブラコウモリ (*Pipistrellus abramus*) 同一集団の外部形態と個体群構成の比較. 香川県高等学校教育研究会理化部会・生地部会会誌 **17**: 31–35 (in Japanese).
2217. Morii, R. (1979) The first record of two species of *Rhinolophus* (CHIROPTERA) from Yakushima Island, Japan. *The Journal of the Mammalogical Society of Japan* **8**: 12–13.
2218. Morii, R. (1982) Seasonal changes of emergence time in the Japanese house bat, *Pipistrellus abramus*, from 1974 to 1980 in Kan-onji City, Kagawa Prefecture, Japan. *Bulletin of the Biological Society of Kagawa* **10**: 97–104 (in Japanese).
2219. Morii, R. (1986) Seasonal variation of Trematode in *Pipistrellus abramus* in Kagawa Prefecture. *Bulletin of the Biological Society of Kagawa* **14**: 9–15 (in Japanese).
2220. Morii, R. (1989) Observation of a colony and note of measurements of *Plecotus auritus* from Mt. Tsurugi, Tokushima prefecture and new collecting record of *Vespertilio orientalis* from Mt. Hiei, Shiga prefecture, Japan. *Bulletin of the Biological Society of Kagawa* **15/16**: 23–25 (in Japanese).
2221. Morii, R. (1992) Horizontal distribution of Chiroptera in Shikoku, Japan. *Bulletin of the Biological Society of Kagawa* **19**: 21–36 (in Japanese).
2222. Morii, R. (1993) 香川県産アブラコウモリ *Pipistrellus abramus* の水平分布. *Bulletin of the Biological Society of Kagawa* **20**: 1–5 (in Japanese).
2223. Morii, R. (1996) Growth of external dimensions in *Pipistrellus abramus*. *Bulletin of the Biological Society of Kagawa* **23**: 1–13 (in Japanese).
2224. Morii, R. (2000) Seasonal changes of body weight in *Pipistrellus abramus* in Kagawa Prefecture. *Bulletin of the Biological Society of Kagawa* **27**: 33–42 (in Japanese).
2225. Morii, R. (2001) Seasonal changes of emergence numebr, sex ratio and age composition in the same colony of *Pipistrellus abramus* in Kagawa Prefecture, Japan. *Bulletin of the Biological Society of Kagawa* **28**: 37–44 (in Japanese).
2226. Morii, R. (2007) Relationship between monthly emergence frequencies against times and annual life cycle in *Pipistrellus abramus*, Kagawa Prefecture, Japan. *Bulletin of the Biological Society of Kagawa* **34**: 107–116 (in Japanese).
2227. Morii, R. & Shioiri, T. (1996) On the pellet contents of the Ural owl, *Strix uralensis hondoensis*. *Bulletin of the Biological Society of Kagawa* **23**: 15–20 (in Japanese with English abstract).
2228. Morii, R., et al. (1998) On some bats in 17 caves on Shikoku karst, Ehime Prefecture. *Bulletin of the Biological Society of Kagawa* **25**: 25–29 (in Japanese).
2229. Morikawa, K. (1981) Japanese otter of the special natural monument, and their decline and fall. *The Nature and Animals* **11**: 18–23 (in Japanese).
2230. Morimitsu, T., et al. (1992) Histopathology of eighth cranial nerve of mass stranded dolphins at Goto Islands, Japan. *Journal of Wildlife Diseases* **28**: 656–658.
2231. Morimitsu, T., et al. (1987) Mass stranding of odontoceti caused by parasitogenic eighth cranial neuropathy. *Journal of Wildlife Diseases* **23**: 586–590.
2232. Morimitsu, T., et al. (1986) Parasitogenic octavius neuropathy as a cause of mass stranding of Odontoceti. *Journal of Parasitology* **72**: 469–472.
2233. Morimoto, I. (2004) Preliminary survey of excavated dugong bones. *Journal of Okinawa Prefectural Archaeological Center* **2**: 23–42 (in Japanese).
2234. Morin, P.A., et al. (2010) Complete mitochondrial genome phylogeographic analysis of killer whales (*Orcinus orca*) indicates multiple species. *Genome Research* **20**: 908–916.
2235. Morioka, T. (森岡照明), et al. (1995) 図鑑 日本のワシタカ類 (The Birds of Prey in Japan). 2nd ed. Tokyo: Bun-ichi Sogo Shyuppan (文一総合出版), 631 pp. (in Japanese with English abstract).
2236. Morisaka, T., et al. (2005) Geographic variations in the whistles among three Indo-Pacific bottlenose dolphin *Tursiops aduncus* population in Japan. *Fisheries Science* **71**: 568–576.
2237. Morisaka, T. (森坂匡通), et al. (2013) 伊豆鳥島のミナミハンドウイルカ. *Kaiyo Monthly (月刊海洋)* **45**: 232–238 (in Japanese).
2238. Morishima, Y., et al. (1999) Coproantigen survey for *Echinococcus multilocularis* prevalence of red foxes in Hokkaido, Japan. *Parasitology International* **48**: 121–134.
2239. Morita, C. (森田忠義) (1977) 名瀬市と周辺地域の動物調査―陸生脊椎動物及び淡水産魚について―. 南日本文化 **10**: 251–263 (in Japanese).
2240. Morita, C., et al. (1991) Isolation of lymphocytic choriomeningitis virus from wild house mice (*Mus musculus*) in Osaka Port, Japan. *Journal of Veterinary Medical Science* **53**: 889–892.
2241. Morita, S. (1964) On the breeeding season, litter size and gestation period of the Riukiu musk shrew, *Suncus murinus riukiuanus* Kuroda (Insectivora).

2241. *Zoological Magazine* **73**: 196–201 (in Japanese with English abstract).

2242. Morita, S. (1964) Reproduction of the Riukiu musk shrew, *Suncus murinus riukiuanus* Kuroda. 1. On the breeding season, size of litter, embryonic mortality, transference of ovum and duration of gestation. *Scientific Bulletin of Faculty of Liberal Arts and Education, Nagasaki University* **15**: 17–40.

2243. Morita, S. (1964) Reproduction of the Riukiu musk shrew, *Suncus murinus riukiuanus* Kuroda. 2. Growth and development of the ovaian follicle. *Scientific Bulletin of Faculty of Liberal Arts and Education, Nagasaki University* **15**: 41–61.

2244. Morita, S. (1968) On the copulation and ovulation of the Riukiu musk shrew, *Suncus murinus riukiuanus* Kuroda. *Scientific Bulletin of Faculty of Liberal Arts and Education, Nagasaki University* **19**: 85–95 (in Japanese with English abstract).

2245. Morita, S. (1971) Seasonal changes in the body weight and the reproduction of males of *Suncus murinus temminckii* Fitzinger in Nagasaki. *Scientific Bulletin of Faculty of Liberal Arts and Education, Nagasaki University* **22**: 65–75 (in Japanese with English abstract).

2246. Morita, S. & Jinno, N. (1970) Electron microscopic observations on the so-called prostate gland of *Suncus murinus temminckii* Fitzinger. *Scientific Bulletin of Faculty of Liberal Arts and Education, Nagasaki University* **21**: 53–67 (in Japanese with English abstract).

2247. Moriwaki, K. (1994) Wild mouse from a geneticist's viewpoint. *in* Genetics in Wild Mice: Its Application to Biomedical Research (Moriwaki, K., Shiroishi, T., & Yonekawa, H., Editors). Basel: S Karger, pp. xii–xxv.

2248. Moriwaki, K., et al. (2009) Unique inbred strain MSM/Ms established from the Japanese wild mouse. *Experimental Animals* **58**: 123–134.

2249. Moriwaki, K., et al. (1986) Genetic features of major geographic isolates of *Mus musculus*. *in* The Wild Mouse in Immunology (Potter, M., Nadeau, J.H., & Cancro, M.P., Editors). Berlin: Springer-Verlag, pp. 255–267.

2250. Moriwaki, K., et al. (1979) Frequency distribution of histocompatibility-2 antigenic specificities in the Japanese wild mouse genetically remote from the European subspecies. *European Journal of Immunogenetics* **6**: 99–113.

2251. Moriwaki, K., Shiroishi, T., & Yonekawa, H. (1994) Genetics in Wild Mice: Its Application to Biomedical Research. Tokyo: Japan Scientific Societies Press, 333 pp.

2252. Moriwaki, K., et al. (1984) Implications of the genetic divergence between European wile mice with Robertsonian translocations from the viewpoint of mitochondrial DNA. *Genetical Research* **43**: 277–287.

2253. Morozumi, M. (両角源美) (1972) コウモリの冬眠. 日本哺乳類雑記 **1**: 5–7 (in Japanese).

2254. Mörzer Bruyns, W.F.J. (1971) Field Guide of Whales and Dolphins. Amsterdam: C.A.Mees, 258 pp.

2255. Motokawa, M. (本川雅治) (1995) 奄美諸島および沖縄島におけるハツカネズミ属の外部形態と分布状況. *Chirimosu* (チリモス) **6**: 10–14 (in Japanese).

2256. Motokawa, M. (本川雅治) (1998) 日本産ジネズミ亜科の自然史. *in* 食虫類の自然史 (Abe, H. (阿部永) & Yokohata, Y. (横畑泰志), Editors). Shobara: Hiba Society of Natural History, pp. 275–349 (in Japanese).

2257. Motokawa, M. Unpublished.

2258. Motokawa, M. (1995) Juvenile tooth of the Japanese white-toothed shrew, *Crocidura dsinezumi* (Mammalia: Insectivora). *Journal of Growth (Nagoya)* **34**: 49–52 (in Japanese with English abstract).

2259. Motokawa, M. (1997) New locality records and a case of dental abnormality of *Crocidura watasei* Kuroda, 1924 (Insectivora: Soricidae) in the Okinawa Group of the Ryukyu Archipelago, Japan. *Biological Magazine Okinawa* **35**: 43–46 (in Japanese with English abstract).

2260. Motokawa, M. (1998) Reevaluation of the Orii's shrew, *Crocidura dsinezumi orii* Kuroda, 1924 (Insectivora, Soricidae) in the Ryukyu Archipelago, Japan. *Mammalia* **62**: 259–267.

2261. Motokawa, M. (1999) Taxonomic history of the genus *Crocidura* (Insectivora: Soricidae) from Japan. *in* Recent Advances in the Biology of Japanese Insectivora: Proceedings of the Symposium on the Biology of Insectivore in Japan and on the Wildlife Conservation (Yokohata, Y. & Nakamura, S., Editors). Shobara: Hiba Society of Natural History, pp. 63–71.

2262. Motokawa, M. (2000) Biogeography of living mammals in the Ryukyu Islands. *Tropics* **10**: 63–71.

2263. Motokawa, M. (2003) Geographic variation in the Japanese white-toothed shrew *Crocidura dsinezumi*. *Acta Theriologica* **48**: 145–156.

2264. Motokawa, M. (2004) Phylogenetic relationships within the family Talpidae (Mammalia: Insectivora). *Journal of Zoology* **263**: 147–157.

2265. Motokawa, M. (2007) Distribution and new localities of small mammals in the central part of the Ryukyu Archipelago, Japan. *Bulletin of the Biogeographical Society of Japan* **62**: 3–9 (in Japanese with English abstract).

2266. Motokawa, M. (2008) Taxonomic status of *Neoaschizomys sikotanensis* Tokuda, 1935 (Rodentia, Muridae) after re-examination of type specimens. *Mammal Study* **33**: 71–75.

2267. Motokawa, M. & Abe, H. (1996) On the specific names of the Japanese moles of the genus *Mogera*. *Mammal Study* **21**: 115–123.

2268. Motokawa, M., et al. (2006) Taxonomic study of the water shrew *Chimarrogale himalayica* and *C. platycephala*. *Acta Theriologica* **51**: 215–223.

2269. Motokawa, M., Harada, M., & Lin, L.-K. (2004) Variation in the Y chromosome of *Crocidura tadae kurodai* (Insectivora, Soricidae). *Mammalian Biology* **69**: 273–276.

2270. Motokawa, M., Hasegawa, M., & Mori, A. (2000) Additional record of *Crocidura dsinezumi* from Nii-jima and Toshima, Izu Islands. *Journal of the Natural History Museum and Institute, Chiba* **6**: 95–96 (in Japanese with English abstract).

2271. Motokawa, M., et al. (1996) Geographic variation in the Watase's shrew *Crocidura watasei* (Insectivora, Soricidae) from the Ryukyu archipelago, Japan. *Mammalia* **60**: 243–254.

2272. Motokawa, M., et al. (2001) Taxonomic status of the Senkaku mole, *Nesoscaptor uchidai*, with special reference to variation in *Mogera insularis* from Taiwan (Mammalia: Insectivora). *Zoological Science* **18**: 733–740.

2273. Motokawa, M., et al. (2003) Morphometric geographic variation in the Asian lesser white-toothed shrew *Crocidura shantungensis* (Mammalia, Insectivora) in East Asia. *Zoological Science* **20**: 789–795.

2274. Motokawa, M., Lin, L.-K., & Motokawa, J. (2003) Morphological comparison of Ryukyu mouse *Mus caroli* (Rodentia: Muridae) populations from Okinawajima and Taiwan. *Zoological Studies* **42**: 258–267.

2275. Motokawa, M., et al. (2001) New records of the Polynesian rat *Rattus exulans* (Mammalia: Rodentia) from Taiwan and the Ryukyus. *Zoological Studies* **40**: 299–304.

2276. Motokawa, M. & Maeda, K. (2002) Preserved paratypes of Kuroda's (1924) Ryukyu mammals. *Mammal Study* **27**: 145–147.

2277. Motokawa, M., et al. (2000) Phylogenetic relationships among East Asian species of *Crocidura* (Mammalia, Insectivora) inferred from mitochondrial cytochrome *b* gene sequences. *Zoological Science* **17**: 497–504.

2278. Motokawa, M., Yu, H.-T., & Harada, M. (2005) Diversification of the white-toothed shrews of the genus *Crocidura* (Insectivora: Soricidae) in East and Southeast Asia. *Mammal Study* **30 (Suppl.)**: S53–S64.

2279. Motoki, T. & Yoshida, T. (2000) New breeding method of the water shrew, *Chimarrogale platycephala* and the diurnal activity by using the new breeding apparatus. *The Annals of Environmental Science, Shinshu University* **22**: 37–40 (in Japanese).

2280. Moura, A.E., et al. (2012) Atypical panmixia in a European dolphin species (*Delphinus delphis*): implications for the evolution of diversity across oceanic boundaries. *Journal of Evolutionary Biology* **26**: 63–75.

2281. Moura, A.E., Sillero, N., & Rodrigues, A. (2012) Common dolphin (*Delphinus delphis*) habitat preferences using data from two platforms of opportunity. *Acta Oecologica* **38**: 24–32.

2282. Mouri, T. (1996) Multivariate cranial ontogenetic allometries in crab-eating, rhesus, and Japanese Macaques. *Anthropological Science* **104**: 281–303.

2283. Mouri, T. & Nishimura, T. (2002) Spatial variation in Japanese macaques—Evidence from cranial metrics of adult females. *Asian Paleoprimatology* **2**: 45–54 (in Japanese).

2284. Mukai, H., et al. (2000) Dugong grazing on *Halophila* beds in Haad Chao Mai National park, Thailand. *Biologia Marina Mediterranea* **7**: 268–270.

2285. Mukohyama, M. (向山満) (1978) 天間館神社のトウヨウヒナコウモリ（その保護と現状）. 三戸高校研究紀要 **8**: 25–31 (in Japanese).

2286. Mukohyama, M. (向山満) (1985) 青森県のコウモリ類―その保護と生態研究―. 東北の自然 (*The Nature of Tōhoku*) **6**: 4–9 (in Japanese).

2287. Mukohyama, M. (向山満) (1988) 郷土の生物写真集 18 の解説―本州北部の翼手類―. *Paulownia* **20**: 119–121 (in Japanese).

2288. Mukohyama, M. (向山満) (1988) 青森県産翼手目の分布. 青森県立三戸高等学校研究紀要 **18**: 20–27 (in Japanese).

2289. Mukohyama, M. (向山満) (1989) 青森県産翼手目の検討. *Journal of Aomori-ken Biological Society* **26**: 36 (in Japanese).

2290. Mukohyama, M. (向山満) (1989) 青森県初記録の翼手目 2 種. *Paulownia* **21**: 112 (in Japanese).

2291. Mukohyama, M. (向山満) (1990) 自然の備忘録；1989 年の霞網調査. *Paulownia* **22**: 77 (in Japanese).

2292. Mukohyama, M. (向山満) (1990) 自然の備忘録；岩手県中山峠のテングコウモリ. *Paulownia* **22**: 78 (in Japanese).

2293. Mukohyama, M. (向山満) (1993) 岩手県一戸町西岳動物相調査報告. *Paulownia* **25**: 64–74 (in Japanese).

2294. Mukohyama, M. (向山満) (1996) 青森県 2 頭目のクロオオアブラコウモリ. *Journal of the Natural History of Aomori* **1**: 34 (in Japanese).

2295. Mukohyama, M. (向山満) (1997) 標識ヒナコウモリの青森県外における再捕獲 3 例. *Journal of the Natural History of Aomori* **2**: 22 (in Japanese).

2296. Mukohyama, M. (向山満) (2000) ヒナコウモリ M13705 の秋田市における回収報告. *Bat Study and Conservation Report* **8**: 3 (in Japanese).
2297. Mukohyama, M. Unpublished.
2298. Mukohyama, M. (1985) On the Oriental frosted bat, *Vespertilio orientalis*, of the Temmadate Shrine, with some biological notes. *The Nature and Animals* **15**: 22–26 (in Japanese).
2299. Mukohyama, M. (1987) An attempt to compulsively change the roost of the bat, *Vespertilio orientalis*, and some aspects of its ecology. *Collecting and Breeding* **49**: 444–449 (in Japanese).
2300. Mukohyama, M. (1987) Biology of bats in Aomori-ken 1. Confirmation of breeding. *Journal of Aomori-ken Biological Society* **24**: 31–34 (in Japanese).
2301. Mukohyama, M. (1995) A status of animals (Amphibia, Reptilia, and Chiroptera) in Shirakami-sanchi. *in* 平成 6 年度特定地域自然林総合調査報告書 (白神山地自然環境保全地域総合調査報告書) (National Parks Association of Japan, Editor). Tokyo: Environment Agency, pp. 325–366 (in Japanese with English abstract).
2302. Mukohyama, M. (1996) Notes on breeding colonies of *Vespertilio superans* Thomas, 1899 in Aomori prefecture, Japan. *Journal of the Natural History of Aomori* **1**: 9–12 (in Japanese).
2303. Mukohyama, M. (2007) A recapture case of *Myotis frater kaguyae* four years after the release. *Study Report on Bat Conservation in Tohoku Region* **1**: 28–29 (in Japanese).
2304. Mullin, K.D., Hoggard, W., & Hansen, L.J. (2004) Abundance and seasonal occurrence of Cetaceans in outer continental shelf and slope waters of the north-central and northwestern Gulf of Mexico. *Gulf of Mexico Science* **22**: 62–73.
2305. Murai, H. & Anada, S. (1993) A faunal survey of artificial cave-dwelling Chiroptera in Toyama Prefecture, central Japan. *Bulletin of the Toyama Biological Society* **32**: 24–29 (in Japanese with English abstract).
2306. Murai, H., et al. (2004) Records of mammals in Toyama Prefecture (2003). *Bulletin of the Toyama Biological Society* **43**: 1–8 (in Japanese).
2307. Murai, H., et al. (2003) Records of mammals in Toyama Prefecture (2002). *Bulletin of the Toyama Biological Society* **42**: 27–37 (in Japanese).
2308. Murakami, O. (村上興正) (1980) アカネズミの生態. *The Heredity (遺伝)* **34**: 75–81 (in Japanese).
2309. Murakami, O. (村上興正) (2002) ヌートリア (*Myocastor coypus*). *in* Handbook of Alien Species in Japan (外来種ハンドブック) (Ecological Society of Japan, Editor). Tokyo: Chijin Shokan (地人書館), p. 69 (in Japanese).
2310. Murakami, T. (村上隆広) (2001) エゾクロテン (Japanese sable). *in* 知床のほ乳類 II (Shiretoko Museum (斜里町立知床博物館), Editor). Sapporo: Hokkaido Shimbun Press, pp. 138–161 (in Japanese).
2311. Murakami, T. (2010) Morphological characteristics of mutelids in Hokkaido. *Bulletin of the Shiretoko Museum* **31**: 35–40 (in Japanese with English summary).
2312. Murakami, T. (2015) Seasonal change in testicular and scrotal sizes of the *Martes zibellina brachyura*. *Bulletin of the Shiretoko Museum* **37**: 41–44 (in Japanese with English abstract).
2313. Murakami, O. (1974) Growth and development of the Japanese wood mouse (*Apodemus speciosus*) I. The breeding season in the field. *Japanese Journal of Ecology* **24**: 194–206 (in Japanese with English synopsis).
2314. Murakami, T. (2003) Food habits of the Japanese sable *Martes zibellina brachyura* in eastern Hokkaido, Japan. *Mammal Study* **28**: 129–134.
2315. Murakami, T. (2008) Habitat use of Japanese sable *Martes zibellina brachyura* in and around the Shiretoko 100m^2 area. *Bulletin of the Shiretoko Museum* **29**: 31–39 (in Japanese with English abstract).
2316. Murakami, T., Asano, M., & Ohtaishi, N. (2004) Mitochondrial DNA variation in the Japanese marten *M. melampus* and Japanese sable *M. zibellina*. *Japanese Journal of Veterinary Research* **51**: 135–142.
2317. Murakami, T., Ashizawa, H., & Saito, I. (1976) Incidence of *Concinnum ten* in some wild carnivorous animals inhabiting in Miyazaki Prefecture. *Bulletin of the Faculty of Agriculture, Miyazaki University* **23**: 461–464.
2318. Murakami, T., et al. (1970) Infection of raccoon-dogs captured in Miyazaki prefecture with *Concinnum ten* (Yamaguti, 1939). *Bulletin of the Faculty of Agriculture, Miyazaki University* **17**: 96–103 (in Japanese with English abstract).
2319. Murakami, T. & Ohtaishi, N. (2000) Current distribution of the sable and introduced Japanese marten in Hokkaido. *Mammal Study* **25**: 149–152.
2320. Murakawa, M. (1986) A capturing record of a white colored *Clethrionomys rufocanus bedofordiae*. *Forest Protection* **196**: 47 (in Japanese).
2321. Muraki, N. & Yanagawa, H. (2006) Seasonal change in the utilization of tree cavities by wildlife in Obihiro City. *Tree and Forest Health* **10** (in Japanese with English abstract).
2322. Murano, N., et al. (2012) Movement between roosts of *Myotis macrodactylus* and *M. petax* in and around Ebetsu, Northern Japan. *Journal of Rakuno Gakuen University* **36**: 377–391 (in Japanese).
2323. Murase, M., et al. (2014) An ectoparasite and epizoite from a western gray whale (*Eschrichtius robustus*) stranded on Tomakomai, Hokkaido, Japan. *Journal of Rakuno Gakuen University* **38**: 149–152.
2324. Murashima, Y. (村島祐希) (2007) ヒナコウモリを保護しました. *Nature Study* **53**: 3 (in Japanese).
2325. Murasugi, E., et al. (1996) Anti-toxoplasma antibody levels in the main-land raccoon dog, *Nyctereutes procyonoides viverrinus*, in southeastern Kanagawa prefecture. *Journal of the Japanese Association for Infectious Diseases* **70**: 1068–1071 (in Japanese with English summary).
2326. Murata, C., et al. (2010) Multiple copies of SRY on the large Y chromosome of the Okinawa spiny rat, *Tokudaia muenninki*. *Chromosome Research* **18**: 623–634.
2327. Murata, C., et al. (2012) The Y chromosome of the Okinawa spiny rat, *Tokudaia muenninki*, was rescued through fusion with an autosome. *Chromosome Research* **20**: 111–125.
2328. Murata, K., Taki, M., & Masuda, R. (1996) PCR-amplification of SRY gene for sexing a newborn sea otter, *Enhydra lutris*, using the placenta and umbilical cord. *Journal of Japanese Association of Zoological Gardens and Aquarium,* **38**: 48–51 (in Japanese with English abstract).
2329. Murayama, T., et al. (2014) Movement of finless porpoises in Kamogawa Bay, Chiba. *Aquabiology* **36**: 22–28 (in Japanese with English abstract).
2330. Muroyama, Y. (室山泰之) (2003) 里のサルとつきあうには―野生動物の被害管理. Kyoto: Kyoto University Press, 245 pp. (in Japanese).
2331. Muroyama, Y. (2005) Damage management of Japanese macaques—from the viewpoint of feeding ecology. *Mammalian Science* **45**: 99–103 (in Japanese with English abstract).
2332. Muroyama, Y. & Eudey, A.A. (2004) Do macaque species have a future? *in* Macaque Societies: A Model for the Study of Social Organizations (Thierry, B., Singh, M., & Kaumanns, W., Editors). Cambridge: Cambridge University Press, pp. 328–332.
2333. Murphy, S., Collet, A., & Rogan, E. (2005) Mating strategy in the male common dolphin (*Delphinus delphis*): what gonadal analysis tells us. *Journal of Mammalogy* **86**: 1247–1258.
2334. Musser, G.G. & Carleton, M.D. (2005) Superfamily Muroidea. *in* Mammal Species of the World: A Taxonomic and Geographic Reference (Wilson, D.E. & Reeder, D.M., Editors). Baltimore: Johns Hopkins University Press, pp. 894–1531.
2335. Mustonen, A.-M., et al. (2007) Seasonal rhythms of body temperature in the free-ranging raccoon dogs (*Nyctereutes procyonoides*) with special emphasis on winter sleep. *Chronobiology International* **24**: 1095–1107.
2336. Muya, A. (撫養明美) (1979) カモシカ社会. *in* ニホンカモシカ生息環境調査研究報告書 (Japan Forest Technology Association (日本林業技術協会), Editor). Nagano: Forestry Agency, Nagano Regional Forest Office (長野営林局), pp. 13–80 (in Japanese).
2337. Nabata, D., et al. (2007) Genetic structure changes of expanding sika deer (*Cervus nippon*) populations in central and western Hokkaido, revealed by mitochondrial DNA analysis. *Mammal Study* **32**: 17–22.
2338. Nabata, D., et al. (2004) Bottleneck effects on the sika deer *Cervus nippon* population in Hokkaido, revealed by ancient DNA analysis. *Zoological Science* **21**: 473–481.
2339. Nadler, C.F., Hoffmann, R.S., & Lay, D.M. (1969) Chromosome of the Asian chipmunk *Eutamias sibiricus* Laxmann (Rodentia: Sciuridae). *Experientia* **25**: 868–869.
2340. Nagahama, M., et al. (1977) Does the nutria, *Myocaster coypus*, serve as the reservoir of liver fluke, *Clonorchis sinensis*? *Japanese Journal of Parasitology* **26**: 41–45 (in Japanese with English abstract).
2341. Nagahama, M., et al. (1984) Epidemiological studies on clonorchiasis in Okayama Prefecture. (V) Survey on the distribution of the reservoir host. *Japanese Journal of Parasitology* **33**: 1–6 (in Japanese with English abstract).
2342. Nagai, T., Murakami, T., & Masuda, R. (2012) Genetic variation and population structure of the sable *Martes zibellina* on Hokkaido Island, Japan, revealed by microsatellite analysis. *Mammal Study* **37**: 323–330.
2343. Nagai, Y., Fukuda, D., & Yanagawa, H. (2009) Bat fauna in Yubari, central Hokkaido, Japan. *Journal of the Japanese Wildlife Research Society* **34**: 21–30 (in Japanese with English summary).
2344. Nagasawa, K. (1999) Parasites of pinnipeds (Mammalia: Carnivora) in Japan: Checklist and bibliography. *Bulletin. National Research Institute of Far Seas Fisheries* **36**: 27–32 (in Japanese).
2345. Nagasawa, K. & Mitani, Y. (2004) A humpback whale, *Megaptera novaeangliae* (Boriwski, 1781), from the Pleistocence Kioroshi Formation of Inba-numa, Chiba Prefecture, central Japan. *Paleontological Research* **8**: 155–165.
2346. Nagata, J. (2009) Two genetically distinct lineages of the Japanese sika deer based on mitochondrial control regions. *in* Sika Deer—Biology and Management of Native and Introduced Populations (McCullough, D.R., Takatsuki, S., & Kaji, K., Editors). Tokyo: Springer, pp. 27–41.

2347. Nagata, J., et al. (2006) Genetic characteristics of the wild boars in Tochigi prefecture and neighboring prefectures. *Bulletin on Wildlife in Tochigi Prefecture* **32**: 58–62 (in Japanese).

2348. Nagata, J., et al. (1998) Genetic variation and population structure of the Japanese sika deer (*Cervus nippon*) in Hokkaido Island, based on mitochondrial D-loop sequences. *Molecular Ecology* **7**: 871–877.

2349. Nagata, J., et al. (1998) Microsatellite DNA variations of the sika deer, *Cervus nippon*, in Hokkaido and Chiba. *Mammal Study* **23**: 95–101.

2350. Nagata, J., et al. (1999) Two genetically distinct lineages of the Sika deer, *Cervus nippon*, in Japanese Islands: comparison of mitochondrial D-loop region sequences. *Molecular Phylogenetics and Evolution* **13**: 511–519.

2351. Nagata, J., Masuda, R., & Yoshida, M.C. (1995) Nucleotide sequences of the cytochrome b and 12S rRNA genes in the Japanese sika deer *Cervus nippon*. *Journal of the Mammalogical Society of Japan* **20**: 1–8.

2352. Nagatomi, N. (永富直子) (1992) 茅野市内の神社に生息するコウモリの生態 (1). 茅野市八ヶ岳総合博物館紀要 **2**: 23–29 (in Japanese).

2353. Nagorsen, D. (1985) *Kogia simus*. *Mammalian Species* **239**: 1–6.

2354. Naito, Y. (1976) The occurrence of the phocid seals along the coast of Japan and possible dispersal of pups. *Scientific Reports of the Whales Research Institute* **28**: 175–185.

2355. Naito, Y. & Konno, S. (1979) The post-breeding distribution of ice-breeding harbour seal (*Phoca largha*) and ribbon seal (*Phoca fasciata*) in the southern sea of Okhotsk. *Scientific Reports of the Whales Research Institute* **31**: 105–119.

2356. Naitoh, Y., et al. (2005) Restriction fragment length polymorphism of nuclear rDNA in *Sorex caecutiens/shinto* group (Eulipotyphla, Soricidae). *Mammal Study* **30**: 101–107.

2357. Naitoh, Y. & Ohdachi, S.D. (2005) Population genetic structure of *Sorex unguiculatus* and *Sorex caecutiens* (Soricidae, Mammalia) in Hokkaido, based on microsatellite DNA polymorphism. *Ecological Research* **21**: 586–596.

2358. Nakagaki, H. (中垣和英) & Suzuki, T. (鈴木隆史) (1994) タヌキ個体群動態に与えるフィラリア感染の影響. *Journal of the Veterinary Medicine* **47**: 29–32 (in Japanese).

2359. Nakagawa, H. (1991) Notes on the specimens of bats deposited in Shiretoko Museum. *Bulletin of the Shiretoko Museum* **12**: 53–84 (in Japanese with English abstract).

2360. Nakagawa, R., Kawamura, Y., & Fujita, M. (1997) Quaternary mammalian remains from Fudô-dô Cave on the Hirao-dai Karst Plateau, Fukuoka Prefecture, Japan. *Journal of the Speleological Society of Japan* **22**: 43–54 (in Japanese with English abstract).

2361. Nakagawa, R., et al. (2012) A new OIS 2 and OIS 3 terrestrial mammal assemblage on Miyako Island (Ryukyus), Japan. *in* Environmental Changes and Human Occupation in East Asia during OIS3 and OIS2 (Ono, A. & Izuho, M., Editors). Oxford: British Archaeological Reports, pp. 55–64.

2362. Nakahara, Y. (中原ゆうじ) (2006) 長野県入笠山でのコテングコウモリの確認. *Bat Study and Conservation Report* **14**: 14 (in Japanese).

2363. Nakai, M. (2001) Vertebral age changes in Japanese Macaques. *American Journal of Physical Anthropology* **116**: 59–65.

2364. Nakajima, F. (中島福男) (1993) 森の珍獣ヤマネ—冬眠の謎を探る. Nagano: Shinano Mainichi Shimbun, 191 pp. (in Japanese).

2365. Nakajima, F. (中島福男) (2006) 日本のヤマネ 改訂版. Nagano: Shinano Mainichi Shimbun, 179 pp. (in Japanese).

2366. Nakajima, H. (中島宏章) (2008) 北海道札幌市手稲山におけるコテングコウモリ (*Murina ussuriensis*) の残雪上での発見. *Northern Forestry, Japan* **60**: 127–129 (in Japanese).

2367. Nakajima, H. (中島宏章), Serizawa, Y. (芹澤裕二), & Yamaguchi, Y. (山口裕司) (2009) 北海道全域におけるコテングコウモリ・テングコウモリの枯葉ねぐら利用. *Northern Forestry, Japan* **61**: 73–75 (in Japanese).

2368. Nakajima, A., et al. (2012) Spatial and elevational variation in fruiting phenology of a deciduous oak (*Quercus crispula*) and its effect on foraging behavior of the Asiatic black bear (*Ursus thibetanus*). *Ecological Research* **27**: 529–538.

2369. Nakajima, H. & Ishii, K. (2005) Records of *Myotis daubentonii* from Sapporo, Ishikari and Tobetsu Town, Hokkaido. *Journal of the Japanese Wildlife Research Society* **31**: 42–47 (in Japanese with English abstract).

2370. Nakajima, M., Kohyama, K., & Kikuchi, S. (1995) Parasitism and pathobiological study on Acanthocephala, *Balbosoma capitatum*, from false killer whales, *Pseudorca crassidens*. *Journal of Japanese Association of Zoo and Aquarium* **36**: 71–79 (in Japanese).

2371. Nakajima, M., et al. (2005) Four stranding records in two species of beaked whales, Genus *Mesoplodon*. *Journal of Japanese Association of Zoos and Aquariums* **46**: 97–109.

2372. Nakajima, S., Adachi, M., & Furui, S. (1998) Seasonal occurrences of larval trombiculid mites found on wild rodents in Nodagawa River basin, Kyoto Prefecture. *Journal of the Acarological Society of Japan* **7**: 39–45 (in Japanese with English abstract).

2373. Nakajima, Z. (1958) On the occurrence of the Quaternary mammalian fauna from the limestone fissures near Shiriya-zaki, Shimokita Peninsula, Aomori Prefecture, Japan (No.2). *Miscellaneous Reports of the Research Institute for Natural Resources* **46&47**: 37–39 (in Japanese with English abstract).

2374. Nakajima, Z. & Kuwano, Y. (1957) Mammalian fauna from the lemonstone fissures near Shiriya-zaki, Shimokita Peninsula, Aomori Prefecture. *Miscellaneous Reports of the Research Institute for Natural Resources* **43&44**: 153–159 (in Japanese with English abstract).

2375. Nakajima, Z. & Kuwano, Y. (1957) On the occurrence of the Quaternary mammalian fauna from the limestone fissures near Shiriya-zaki, Shimokita Peninsula, Aomori prefecture, Japan. *Miscellaneous Reports of the Research Institute for Natural Resources* **43&44**: 153–159 (in Japanese with English summary).

2376. Nakama, H. (中間弘) & Komizo, K. (小溝勝己) (2009) On the mongoose in Kiire-sesekushi-cho Kagoshima City, Kagoshima Prefecture. *Bulletin of the Kagoshima Prefectural Museum* **28**: 103–104 (in Japanese).

2377. Nakamatsu, M., Goto, M., & Morita, M. (1966) *Concinnum ten* (Yamaguti, 1939) from a racoon-dog *Nyctereutes procyonoides viverrinus* Temminck et Schlegel. *Japanese Journal of Parasitology* **15**: 528–532 (in Japanese with English summary).

2378. Nakamichi, M., et al. (1995) Interactions among adult males and females before and after the death of the alpha male in a free-ranging troop of Japanese macaques. *Primates* **36**: 385–396.

2379. Nakamoto, A. (2008) Foraging Ecology of the Orii's Flying-Fox. Nishihara: University of the Ryukyus. 123 pp.

2380. Nakamoto, A., et al. (2011) Geographical distribution pattern and interisland movements of Orii's flying fox in Okinawa Islands, the Ryukyu Archipelago, Japan. *Population Ecology* **53**: 241–252.

2381. Nakamoto, A., Kinjo, K., & Izawa, M. (2007) Diet of the Ryukyu flying-fox (*Pteropus dasymallus*). *Biological Magazine Okinawa* **45**: 61–77.

2382. Nakamoto, A., Kinjo, K., & Izawa, M. (2007) Food habits of Orii's flying-fox, *Pteropus dasymallus inopinatus*, in relation to food availability in an urban area of Okinawa-jima Island, the Ryukyu Archipelago, Japan. *Acta Chiropterologica* **9**: 237–249.

2383. Nakamoto, A., Kinjo, K., & Izawa, M. (2009) The role of Orii's flying-fox (*Pteropus dasymallus inopinatus*) as a pollinator and a seed disperser on Okinawa-jima Island, the Ryukyu Archipelago, Japan. *Ecological Research* **24**: 405–414.

2384. Nakamoto, A., Kinjo, K., & Izawa, M. (2012) Ranging patterns and habitat use of a solitary flying fox (*Pteropus dasymallus*) on Okinawa-jima Island, Japan. *Acta Chiropterologica* **14**: 387–399.

2385. Nakamoto, A., Kinjo, K., & Izawa, M. (2015) A capturing record of old aged individuals of Ryukyu flying foxes. *Biological Magazine Okinawa* **53**: 37–44 (in Japanese with English abstract).

2386. Nakamoto, A. & Nakanishi, N. (2013) Home range, habitat selection, and activity of male Asian house shrews, *Suncus murinus*, on Okinawa-jima Island. *Mammal Study* **38**: 147–153.

2387. Nakamoto, A., et al. (2009) Distribution and abundance of the Orii's flying fox in Okinawa Islands, the Ryukyu Archipelago, Japan. *Mammalian Science* **49**: 53–60 (in Japanese with English abstract).

2388. Nakamoto, A., et al. (2011) Population growth of Orii's flying fox, *Pteropus dasymallus inopinatus*, on Okinawa-jima Island. *Japanese Journal of Conservation Ecology* **16**: 45–53 (in Japanese with English summary).

2389. Nakamura, T. (中村禎里) (1990) 狸とその世界. Tokyo: Asahi Shimbun Publications, 254 pp. (in Japanese).

2390. Nakamura, Y. (中村豊) & Morita, J. (森田純一) (2003) 鹿児島県におけるカワネズミ *Chimarrogale platycephala* の記録. *ANIMATE* **4**: 67–68 (in Japanese).

2391. Nakamura, K. (2014) Finless porpoise researches in the eastern Inland Sea of Japan. *Aquabiology* **36**: 66–70 (in Japanese with English abstract).

2392. Nakamura, K., et al. (2003) Notes on strandings and incidental catch of finless porpoise *Neophocaena phocaenoides* in coastal waters of Yamaguchi and Fukuoka Prefectures, western Japan. *Nihonkai Cetology* **13**: 13–18 (in Japanese).

2393. Nakamura, K., et al. (1994) A preliminary list of the marine mammals recorded from the coast of Kanagawa prefecture. *Natural History Report of Kanagawa* **16**: 1–9.

2394. Nakamura, M. & Fujimaru, K. (2014) Reproductive studies of finless porpoises in captivity: Ultrasonographic evaluation of testis and epididymis, and collection and storage of post-mortem spermatozoa. *Aquabiology* **36**: 163–168 (in Japanese with English abstract).

2395. Nakamura, T., et al. (2008) Pollen recovered from the faeces of the Bonin flying fox (*Pteropus pselaphon* Layard, 1829) on Minami-Iwo-To and Chichi-jima Islands. *Japanese Journal of Palynology* **54**: 53–60 (in Japanese with English abstract).

2396. Nakamura, T. & Hirose, A. (1986) 宮城県のコウモリ. 東北の自然 *(The*

Nature of Tōhoku) **17**: 2–7 (in Japanese).
2397. Nakamura, T., Kanzaki, M., & Maruyama, N. (2001) Seasonal changes in food habits of Japanese martens in Hinode-cho and Akiruno-shi, Tokyo. *Wildlife Conservation Japan* **6**: 15–24 (in Japanese with English abstract).
2398. Nakamura, T., et al. (2007) Comparative chromosome painting map between two Ryukyu spiny rat species, *Tokudaia osimensis* and *Tokudaia tokunoshimensis* (Muridae, Rodentia). *Chromosome Research* **15**: 799–806.
2399. Nakanishi, N. (中西希) & Izawa, M. (伊澤雅子) (2014) イリオモテヤマネコの山地部における繁殖情報 (Breeding records of the Iriomote cat in moutainous areas). *Biological Magazine Okinawa* **52**: 45–51 (in Japanese).
2400. Nakanishi, S. (中西せつ子) & Hayama, S. (羽山伸一) (1996) 飼育下におけるハクビシンの成長, 性成熟, 出産. *in* 静岡県ハクビシン調査報告書. Shizuoka: Shizuoka Prefectural Government, pp. 33–39 (in Japanese).
2401. Nakanishi, N., et al. (2009) Age determination of the Iriomote cat, *Prionailurus bengalensis iriomotensis*, by using cementum annuli. *Journal of Zoology* **279**: 338–348.
2402. Nakanishi, N., et al. (2005) The effect of habitat on home range size in the Iriomote cat *Prionailurus bengalensis iriomotensis*. *Mammal Study* **30**: 1–10.
2403. Nakanishi, N., et al. Unpublished.
2404. Nakao, M. (1995) *Ixodes persulcatus* and Lyme disease spirochaete: Comparison of development between infected and noninfected ticks. *Japanese Journal of Sanitary Zoology* **46**: 241–247.
2405. Nakao, M. & Miyamoto, K. (1993) Long-tailed shrew, *Sorex unguiculatus*, as a potential reservoir of the spirochetes transmitted by *Ixodes ovatus* in Hokkaido, Japan. *Japanese Journal of Sanitary Zoology* **44**: 237–245 (in Japanese with English summary).
2406. Nakao, M. & Miyamoto, K. (1993) Reservoir competence of the wood mouse, *Apodemus speciosus ainu*, for the Lyme disease spirochete, *Borrelia burgdorferi*, in Hokkaido, Japan. *Japanese Journal of Sanitary Zoology* **44**: 237–245.
2407. Nakao, M., et al. (1992) Comparison of *Borrelia burgdorferi* isolated from humans and ixodid ticks in Hokkaido, Japan. *Microbiology and Immunology* **36**: 1189–1193.
2408. Nakaoka, T. (中岡利泰), et al. (1986) ゼニガタアザラシとゴマフアザラシの食性 (Food and feeding habits of Kuril and Spotted seals captured at the Nemuro Peninsula) *in* ゼニガタアザラシの生態と保護 (Wada, K. (和田一雄), et al., Editors). Tokyo: Tokai University Press, pp. 103–125 (in Japanese).
2409. Nakashita, R., et al. (2007) Carbon and nitrogen stable isotope evidence for the involvement of a captured Asiatic black bear in damages at a rainbow trout farm. *Mammal Study* **47**: 19–23 (in Japanese with English summary).
2410. Nakata, K. (中田圭亮) (1998) 野ネズミの予察調査と防除の手引 (Handbook for Vole Census Methods and Control). 2nd ed. Sapporo: Hokkaido Forest Conservation Association (北海道森林保全協会), 71 pp. (in Japanese).
2411. Nakata, H. (2005) Occurrence of synthetic musk fragrances in marine mammals and sharks from Japanese coastal waters. *Environmental Science & Technology* **39**: 3430–3434.
2412. Nakata, K. Unpublished.
2413. Nakata, K. (1978) A new record and notes on *Clethrionomys rex* Imaizumi, 1971 at low altitudes in Daisetsu area. *The Journal of the Mammalogical Society of Japan* **7**: 231–232 (in Japanese).
2414. Nakata, K. (1986) Litter size of *Apodemus argenteus* in relation to the population cycle. *The Journal of the Mammalogical Society of Japan* **11**: 117–125.
2415. Nakata, K. (1989) Regulation of reproduction rate in a cyclic population of the red-backed vole, *Clethrionomys rufocanus bedfordiae*. *Researches on Population Ecology* **31**: 185–209.
2416. Nakata, K. (1990) Feeding experiment on trees and twigs by *Clethrionomys rex*. *Forest Protection* **220**: 44–45 (in Japanese).
2417. Nakata, K. (1995) Microhabitat selection in two sympatric species of voles, *Clethrionomys rex* and *Clethrioonomys rufocanus bedfordiae*. *Journal of the Mammalogical Society of Japan* **20**: 135–142.
2418. Nakata, K. (1998) Regulation of reproduction in a natural population of the small Japanese field mouse, *Apodemus argenteus*. *Mammal Study* **23**: 19–30.
2419. Nakata, K. (2000) Distribution and habitat of the dark red-backed vole *Clethrionomys rex* in Japan. *Mammal Study* **25**: 87–94.
2420. Nakata, K., Fukumura, S., & Satoh, H. (1989) A list of pelage-color variants in the red-backed vole, *Clethrionomys rufocanus bedfordiae*, with records of two new phenotypes. *Mammalian Science* **29(2)**: 29–32 (in Japanese with English abstract).
2421. Nakatani, M. & Sasaki, N. (2005) The relation between bat and insect I. Ectoparasites 1. *Sylvicola* **23**: 15–16 (in Japanese).
2422. Nakatsu, A. & Maeda, M. (1979) A note on individuals of *Clethrionomys rufocanus bedfordiae* with black colored pelage captured from Nakajima Islet of Lake Toya in Hokkaido. 野ねずみ **154**: 30–31 (in Japanese).
2423. Nakayama, F., et al. (2008) The record of bats in Gassan Mountain, Nishikawa-cho, Yamagata. *Sagae River Basin Natural History Studies* **2**: 3–7 (in Japanese with English abstract).
2424. Nakayama, K., et al. (2009) Temporal and spatial trends of organotin contamination in the livers of finless porpoises (*Neophocaena phocaenoides*) and their association with parasitic infection status. *Science of the Total Environment* **407**: 6173–6178.
2425. Nakayama, T., et al. (2009) Home range and day roost of an Ussurian tube-nosed bat *Murina ussuriensis* in northern Hokkaido. *Bulletin of the Asian Bat Research Institute* **28**: 83–85 (in Japanese with English abstract).
2426. Nakazawa, A., et al. (2004) Analysis of gene structures and promoter activities of the chipmunk α^1-antitrypsin-like genes. *Genes & Genetic Systems* **329**: 71–79.
2427. Nakazono, T. (1970) Note on the burrows of *Vulpes vulpes japonica* in Kyushu, Japan. *The Journal of the Mammalogical Society of Japan* **5**: 45–49 (in Japanese with English abstract).
2428. Nakazono, T. (1970) Researches of burrows of *Vulpes vulpes japonica*, in Kyushu, Japan 1. Examples of the burrows and the distributions. *The Journal of the Mammalogical Society of Japan* **5**: 1–7 (in Japanese with English abstract).
2429. Nakazono, T. (1994) A study on the social system and habitat utilization of the Japanese red foxes, *Vulpes vulpes japonica*. Doctoral dissertation. Fukuoka: Kyushu University. 73 pp.
2430. Namba, T. (2013) Field experiments of top-down and bottom-up effects on the structure of soil animal community in a cool-temperate forest of Japan. Doctoral dissertation. Sapporo: Hokkaidou University. 69 pp.
2431. Namba, T., Iwasa, M.A., & Murata, K. (2007) A new method for the identification of *Martes melampus* in Honshu by a multiplex PCR for fecal DNAs. *Mammal Study* **32**: 129–133.
2432. Namba, T. & Ohdachi, S.D. (2009) Diets of the Eurasian least shrew (*Sorex minutissimus*) from various localities in Hokkaido, Japan. *Mammal Study* **34**: 219–221.
2433. Nambu, H., Ishikawa, H., & Yamada, T.K. (2010) Records of the western gray whale *Eschrichtius robustus*: its distribution and migration. *Japan Cetology* **20**: 21–29 (in Japanese).
2434. Namie, M. (波江元吉) (1889) 日本に栖息する蝙蝠の話 三十七回. *Zoological Magazine* **1**: 510–511 (in Japanese).
2435. Namie, M. (波江元吉) (1889) 日本に栖息する蝙蝠の話 (第卅二版). *Zoological Magazine* **1**: 418–419 (in Japanese).
2436. Namie, M. (波江元吉) (1889) 日本に栖息する蝙蝠の話 第五 (第二十一版). *Zoological Magazine* **1**: 212–213 (in Japanese).
2437. Namie, M. (波江元吉) (1889) 日本に栖息する蝙蝠の話 (続). 第廿三版 *Zoological Magazine* **1**: 256–257 (in Japanese).
2438. Nanbu, Y., et al. (2006) Location and number of individuals of Indo-Pacific bottlenose dolphins (*Tursiops aduncus*) in Kagoshima Bay. *Memoirs of Faculty of Fisheries Kagoshima University* **55**: 51–60 (in Japanese with English abstract).
2439. Naora, N. (直良信夫) (1944) 日本哺乳動物史 (History of Japanese Mammals). Tenri: Yotokusha (養徳社), 265 pp. (in Japanese).
2440. Naora, N. (直良信夫) (1954) 日本旧石器時代の研究 (早稲田大学考古学研究室報告第 2 冊) (Old Stone Age in Japan). Tokyo: Neiraku-shobo (寧楽書房), 298 pp. (in Japanese).
2441. Naora, N. (直良信夫) (1972) 古代遺跡発掘の脊椎動物遺体 (Vertebrate Remains Excavated from the Sites of Ancient Times). Tokyo: Azekura Shobo (校倉書房), 198 pp. (in Japanese).
2442. Naruse, I., Oda, S., & Kameyama, Y. (1978) Behavioral development and postnatal growth of the musk shrew, *Suncus murinus*. *Annual Report of Institute of Environmental Medicine, Nagoya University* **29**: 200–202 (in Japanese).
2443. Nash, W.G. (2006) Family Canidae. *in* Atlas of Mammalian Chromosomes (O'Brien, S.J., Menninger, J.C., & Nash, W.G., Editors). Hoboken: John Wiley & Sons, pp. 446–447.
2444. Nash, W.G. & O'Brien, S.J. (1987) A comparative chromosome banding analysis of the Ursidae and their relationship to other carnivores. *Cytogenetics and Cell Genetics* **45**: 206–212.
2445. Nash, W.G., et al. (1998) Comparative genomics: Tracking chromosome evolution in the family Ursidae using reciprocal chromosome painting. *Cytogenetics and Cell Genetics* **83**: 182–192.
2446. National Institute of Infectious Diseases (2014) Severe fever with thrombocytopenia syndrome (SFTS) in Japan, 2013. *Infectious Agents Surveillance Report* **35**: 31–32 (in Japanese and English).
2447. National Museum of Nature and Science (2009) 海棲哺乳類情報データベース (Marine Mammal Information Database). Available from: http://svrsh1.kahaku.go.jp/marmam/ (in Japanese).
2448. National Museum of Nature and Science (2015) 海棲哺乳類ストランディングデータベース (The Marine Mammals Stranding Database). Available from:

http://svrsh2.kahaku.go.jp/drift/ (in Japanese).

2449. Natoli, A., et al. (2005) Phylogeography and alpha taxonomy of the common dolphin (*Delphinus* sp). *Journal of Evolutionary Biology* **19**: 943–954.

2450. Natori, M. & Shigehara, N. (1993) A craniometrical analysis of geographical difference between house shrews (*Suncus murinus*) on the Okinawa and Taiwan Islands. *Journal of Growth* **32**: 47–51.

2451. Natori, M. & Shigehara, N. (1997) Loss of the third upper premolar in *Suncus murinus* and its relation to jaw size. *Acta Theriologica* **42**: 99–104.

2452. Nawa, A. (1991) An ecological study of Japanese serow on Mt. Ryozen, in the Suzuka mountains, central Japan. *in* Landscape and Environment of Shiga (Scientific Studies of Shiga Prefecture, Japan). Shiga: Foundation of Nature Conservation in Shiga Prefecture, pp. 1459–1472 (in Japanese with English summary).

2453. Neagari, Y., et al. (1998) Incidence of antibodies in raccoon dogs and deer inhabiting suburban areas. *Journal of the Japanese Association for Infectious Diseases* **72**: 331–334 (in Japanese with English summary).

2454. Neal, E. & Cheeseman, C. (1996) Badgers. London: Poyster Natural History, 271 pp.

2455. Neimanis, A.S., et al. (2000) Seasonal regression in testicular size and histology in harbour porpoises (*Phocoena phocoena* L.) from the Bay of Fundy and Gulf of Maine. *Journal of Zoology, London* **250**: 221–229.

2456. Nellis, D.W. (1989) *Herpestes auropunctatus*. *Mammalian Species* **342**: 1–6.

2457. Nemoto, T. (1959) Food of baleen whales with reference to whale movements. *Scientific Reports of the Whales Research Institute* **14**: 149–290.

2458. Nemuro City Board of Education (根室市教育委員会) (2001) 根室半島コウモリ類調査報告書. 52 pp. (in Japanese).

2459. Nerini, M. (1984) A review of gray whale feeding ecology. *in* The Gray Whale *Eschrichtius robustus* (Jones, M.L., Swartz, S.L., & Leatherwood, S., Editors). Orlando: Academic Press, pp. 423–450.

2460. Nesterenko, V. & Ohdachi, S.D. (2001) Postnatal growth and development in *Sorex unguiculatus*. *Mammal Study* **26**: 145–148.

2461. Nesterenko, V., Sheremetyev, I.S., & Alexeeva, E.V. (2002) Dynamics of the shrew taxocene structure in the late Quaternary of the southern Far East. *Paleontological Journal* **36**: 535–540.

2462. Nesterenko, V.A. (1999) Insectivores of the south Far East and their communities. Vladivostok: Dalnauka, 172 pp. (in Russian with English summary).

2463. Newby, T.C. (1982) Life history of Dall porpoise (*Phocoenoides dalli*, TRUE 1885) incidentally taken by the Japanese high seas salmon mothership fishery in the northwestern North Pacific and western Bering Sea, 1978 to 1980. Doctoral dissertation. Seattle: University of Washington. 155 pp.

2464. Newman, C., Buesching, C.D., & Wolff, J.O. (2005) The function of facial masks in "midguild" carnivores. *Oikos* **108**: 623–633.

2465. Nigi, H. (1977) Laparoscopic observations of ovaries before and after ovulation in the Japanese monkey (*Macaca fuscata*). *Primates* **18**: 243–259.

2466. Nigi, H., Hayama, S.-I., & Torii, R. (1989) Rise in age of sexual maturation in male Japanese monkeys at Takasakiyama in relation to nutritional condition. *Primates* **30**: 571–576.

2467. Niida, S., et al. (1995) Development of the chondrocranial base of the musk shrew, *Suncus murinus* (Insectivora). *Experimental Animal* **44**: 79–86.

2468. Niida, S., et al. (1994) Occipital roof development in the Japanese musk shrew, *Suncus murinus*. *Journal of Anatomy* **185**: 433–437.

2469. Niioka, T. (新岡武彦) (1977) 樺太・北海道の古文化. 北方歴史文化叢書 2. Sapporo: Hokkaido Shuppan-kikaku Center (北海道出版企画センター), 256 pp. (in Japanese).

2470. Niizuma, A. (新妻昭夫) (1986) ゼニガタアザラシの社会生態と繁殖戦略 (Socio-ecology and reproductive strategy of the Kuril seal) *in* ゼニガタアザラシの生態と保護 (Wada, K. (和田一雄), et al., Editors). Tokyo: Tokai University Press, pp. 59–102 (in Japanese).

2471. Nikaido, M., et al. (2000) Monophyletic origin of the order Chiroptera and its phylogenetic position among Mammalia, as inferred from the complete sequence of the mitochondrial DNA of a Japanese megabat, the Ryukyu flying fox (*Pteropus dasymallus*). *Journal of Molecular Evolution* **51**: 318–328.

2472. Nikaido, M., et al. (2001) Maximum likelihood analysis of the complete mitochondrial genomes of eutherians and a reevaluation of the phylogeny of bats and insectivores. *Journal of Molecular Evolution* **53**: 508–516.

2473. Nikoh, N., et al. (2011) Phylogenetic comparison between nycteribiid bat flies and their host bats. *Medical Entomology and Zoology* **62**: 185–194.

2474. Niño-Torres, C.A., et al. (2006) Isotopic analysis of $\partial^{13}C$, $\partial^{15}N$, and $\partial^{34}S$ "A feeding tale" in teeth of the longbeaked common dolphin *Delphinus capensis*. *Marine Mammal Science* **22**: 831–846.

2475. Ninomiya, H., Ogata, M., & Makino, M. (2003) Notoedric mange in free-ranging masked palm civets (*Paguma larvata*) in Japan. *Veterinary Dermatology* **14**: 339–344.

2476. Nishi, C., Deguchi, Y., & Aoi, T. (2011) Home range size and its overlap in Japanese squirrels in relation to the walnut resource level in a suburban forest in Morioka City. *Mammalian Science* **51**: 277–285 (in Japanese with English abstract).

2477. Nishi, C., Deguchi, Y., & Aoi, T. (2014) Nest use of Japanese squirrels in a suburban forest in Morioka city. *Mammalian Science* **54**: 11–18 (in Japanese with English abstract).

2478. Nishigaki, M. (西垣正男) & Kawamichi, T. (川道武男) (1996) ニホンリス (*Sciurus lis*). *in* The Encyclopaedia of Animals in Japan. 1. Mammals I (日本動物大百科 1 哺乳類 I) (Kawamichi, T. (川道武男), Editor). Tokyo: Heibonsha (平凡社), pp. 70–73 (in Japanese).

2479. Nishigaki, M. (2003) Reproductive biology of the Japanese squirrel (*Sciurus lis*). *in* Workshop and Symposium for Japanese Squirrels at Inokashira, Tokyo, pp. 7–8 (in Japanese).

2480. Nishikata, S. (1979) Ecological studies on the population of *Apodemus argenteus argenteus* on Mt. Kiyosumi, Chiba Pref. (I). A life cycle and fluctuation of population size. *The Journal of the Mammalogical Society of Japan* **7**: 240–253 (in Japanese with English abstract).

2481. Nishikata, S. (1981) Habitat preference of *Apodemus speciosus* and *A. argenteus*. *Journal of the Japanese Forest Society* **63**: 151–155.

2482. Nishikata, S. (1982) Ecological studies on the population of *Apodemus argenteus* on Mt. Kiyosumi, Chiba Prefecture (II) Social structure and its role. *Journal of the Japanese Forest Society* **64**: 249–256 (in Japanese with English summary).

2483. Nishimoto, S., Ishikawa, M., & Chimoto, K. (2014) About the research activity of the finless porpoise in the Gulf of Osaka. *Aquabiology* **36**: 50–57 (in Japanese with English abstract).

2484. Nishimura, S. (1970) Recent records of Baird's beaked whale in the Sea of Japan. *Publications of the Seto Marine Biological Laboratory* **18**: 61–68.

2485. Nishimura, T., et al. (2007) Serum protein polymorphism of Japanese serow (*Capricornis crispus*) by polyacrylamide gel electrophoresis: analysis of albumin and transferring. *Japanese Journal of Zoo and Wildlife Medicine* **12**: 105–109 (in Japanese with English abstract).

2486. Nishimura, T., et al. (2011) Development of microsatellite marker assay for individual identification in Japanese serows (*Capricornis crisps*). *Japanese Journal of Zoo and Wildlife Medicine* **16**: 75–78.

2487. Nishimura, T., et al. (2010) Sex determination of the Japanese serow (*Capricornis crispus*) by fecal DNA anallysis. *Japanese Journal of Zoo and Wildlife Medicine* **15**: 73–78.

2488. Nishimura, Y., et al. (1999) Interspecies transmission of feline immunodeficiency virus from the domestic cat to the Tsushima cat (*Felis bengalensis euprilura*) in the wild. *Journal of Virology* **73**: 7916–7921.

2489. Nishinakagawa, H., et al. (1994) Mammal from archaeological sites of the Jomon period in Kagoshima Prefecture. *Journal of the Mammalogical Society of Japan* **19**: 57–66.

2490. Nishiwaki, M. (西脇昌治) (1965) 鯨類・鰭脚類. Tokyo: University of Tokyo Press, 439 pp. (in Japanese).

2491. Nishiwaki, M. (1959) Humpback whales in Rukyuan waters. *Scientific Reports of the Whales Research Institute* **14**: 49–87.

2492. Nishiwaki, M. (1962) *Mesoplodon bowdoini* stranded at Akita Beach, Sea of Japan. *Scientific Reports of the Whales Research Institute* **16**: 61–77.

2493. Nishiwaki, M. (1967) Distribution and migration of marine mammals in the North Pacific area. *Bulletin of the Ocean Research Institute University of Tokyo* **1**: 1–64.

2494. Nishiwaki, M. (1972) General biology. *in* Mammals of the Sea: Biology and Medicine (Ridgway, S.H., Editor). Springfield: Charles C. Thomas Publisher, pp. 3–204.

2495. Nishiwaki, M. & Handa, C. (1958) Killer whales caught in the coastal waters off Japan for recent 10 years. *Scientific Reports of the Whales Research Institute, Tokyo* **13**: 85–96.

2496. Nishiwaki, M. & Hasegawa, Y. (1969) The discovery of the right whale skull in the Kisagata shell bed. *Scientific Reports of the Whales Research Institute* **21**: 79–84.

2497. Nishiwaki, M., Hibiya, T., & Kimura, S. (1954) On the sexual maturity of the sei whale of the Bonin Island waters. *Scientific Reports of the Whales Research Institute* **9**: 165–177.

2498. Nishiwaki, M. & Kamiya, T. (1958) A beaked whale *Mesoplodon* stranded at Oiso beach, Japan. *Scientific Reports of the Whales Research Institute* **13**: 53–83.

2499. Nishiwaki, M. & Kamiya, T. (1959) *Mesoplodon stejnegeri* from the coast of Japan. *Scientific Reports of the Whales Research Institute* **14**: 25–48.

2500. Nishiwaki, M., et al. (1972) Further comments on *Mesoplodon ginkgodens*. *Scientific Reports of the Whales Research Institute* **24**: 43–56.

2501. Nishiwaki, M., et al. (1979) Present distribution of the dugong in the world. *Scientific Reports of the Whales Research Institute* **31**: 133–141.

2502. Nishiwaki, M. & Oguro, N. (1971) Baird's beaked whales caught on the coast of Japan in recent 10 years. *Scientific Reports of the Whales Research*

Institute **23**: 111–122, pls 1&2.
2503. Nishiwaki, M. & Oguro, N. (1972) Catch of the Cuvier's beaked whales off Japan in recent years. *Scientific Reports of the Whales Research Institute* **24**: 35–41.
2504. Nishiwaki, M., Ohsumi, S., & Maeda, Y. (1963) Change of form in the sperm whale accompanied with growth. *Scientific Reports of the Whales Research Institute, Tokyo* **17**: 1–14.
2505. Nishiwaki, M. & Tobayama, T. (1982) Morphological study on the hybrid between *Tursiops* and *Pseudorca*. *Scientific Reports of the Whales Research Institute, Tokyo* **34**: 109–121.
2506. Nitta, N. Unpublished.
2507. Niu, Y., et al. (2004) Phylogeny of pikas (Lagomorpha, *Ochotona*) inferred from mitochondrial cytochrome *b* sequences. *Folia Zoologica* **53**: 141–155.
2508. Niwa, H. (1936) On the period of gestation, number of litter and morphology of new born young in *Pipistrellus abramus* (Temminck). *Botany and Zoology* **4**: 67–76 (in Japanese).
2509. NOAA Fisheries (2008) Rare white killer whale spotted in Alaskan waters from NOAA ship Oscar Dyson (News Release, March 6, 2008). Available from: http://alaskafisheries.noaa.gov/newsreleases/2008/whitewhale030608.htm.
2510. Noda, R. & Kugi, G. (1980) On hookworms from raccoon dogs and foxes, with a note on some related species. *Bulletin of the University of Osaka Prefecture. Series B, Agriculture and Biology* **32**: 63–68.
2511. Nojima, Y., Tsutsuki, K., & Oshida, T. (2013) Effect of different soil horizons on distribution of *Sorex* species in Hokkaido, Japan. *Acta Zoologica Academiae Scientiarum Hungaricae* **59**: 297–304.
2512. Nojiriko Excavation Research Group (1975) Lake Nojiri excavation 1962–1973. Tokyo: Kyoritsu Shuppan, 278 pp. (in Japanese with English abstract).
2513. Nomenclature, I.C.o.Z. (1999) International Code of Zoological Nomenclature. 4th ed. London: International Trust for Zoological Nomenclature, 306 pp.
2514. Noro, T. (2014) First record of Oriental free-tailed bat *Tadarida insignis* (Blyth, 1861) in Nagoya, Aichi Prefecture, Japan. *Bulletin of Nagoya Biodiversity Center* **1**: 65–69 (in Japanese with English abstract).
2515. Nowak, R.M. (1991) Walker's Mammals of the World. 5th ed. Baltimore: Johns Hopkins University Press, 2 volumes.
2516. Nowak, R.M. (1999) Genus *Canis*. *in* Walker's Mammals of the World (Nowak, R.M., Editor). Baltimore: Johns Hopkins University Press, pp. 655–672.
2517. Nowak, R.M. (1999) Macaques. *in* Walker's Mammals of the World. Baltimore and London: Johns Hopkins University Press, pp. 580–588.
2518. Nowak, R.M. (1999) Walker's Mammals of the World. 6th ed. Baltimore: Johns Hopkins University Press, 1936 pp.
2519. Nowell, K. & Jackson, P., eds. (1996) Wild Cats: Status Survey and Conservation Action Plan. Gland: IUCN, 383 pp.
2520. Nozaki, E. (野崎英吉) (2002) 石川県七ツ島大島におけるカイウサギ対策とその成果 (Rabbit eradication programs in Nanatsujima Island in Ishikawa Prefecture). *in* Handbook of Alien Species in Japan (外来種ハンドブック) (Ecological Society of Japan, Editor). Tokyo: Chijin Shokan (地人書館), pp. 82–83 (in Japanese).
2521. Nozaki, T. (野崎達也) & Takahashi, Y. (高橋良明) (2011) 志河川ダムにおけるコウモリ類の保全対策事例. 水と土 **162**: 67–73 (in Japanese).
2522. Nozaki, E. & Mizuno, A. (1986) On the age related sizes of upper canine teeth and skulls of Japanese black bear, *Selenarctos thibetanus japonicus*, in Ishikawa Prefecture. *Annual Report of Hakusan Nature Conservation Center, Ishikawa* **13**: 49–64 (in Japanese with English summary).
2523. Nozaki, M., Mori, Y., & Oshima, K. (1992) Environmental and internal factors affecting seasonal breeding of Japanese monkeys (*Macaca fuscata*). *in* Topics in Primatology: Volume 3. Evolutionary Biology, Reproductive Endocrinology, and Virology (Matano, S., et al., Editors). Tokyo: University of Tokyo Press, pp. 301–317.
2524. Nozaki, M., et al. (1990) Changes in circulating inhibin levels during pregnancy and early lactation in the Japanese monkey. *Biology of Reproduction* **43**: 444–449.
2525. Nozawa, S. (野沢俊次郎) (1892) 北海道ノかわほり. *Zoological Magazine* **4**: 294 (in Japanese).
2526. Nozawa, K., et al. (1996) Population genetic study of the Japanese macaque, *Macaca fuscata*. *in* Variation in the Asian Macaques (Shotake, T. & Wada, K., Editors). Tokyo: Tokai University Press, pp. 1–36.
2527. Nozawa, K., et al. (1991) Population genetics of Japanese monkeys: III. Ancestry and differentiation of local populations. *Primates* **32**: 411–435.
2528. Nozawa, K., et al. (1977) Genetic variations within and between species of Asian macaques. *Japanese Journal of Genetics* **52**: 15–30.
2529. Nunome, M., et al. Unpublished.
2530. Nunome, M., et al. (2014) Lack of association between winter coat colour and genetic population structure in the Japanese hare, *Lepus brachyurus* (Lagomorpha: Leporidae). *Biological Journal of the Linnean Society* **111**: 761–776.
2531. Nunome, M., et al. (2007) Phylogenetic relationships and divergence times among dormice (Rodentia, Gliridae) based on three nuclear genes. *Zoological Scripta* **36**: 537–546
2532. Nunome, M., et al. (2010) The influence of Pleistocene refugia on the evolutionary history of the Japanese hare, *Lepus brachyurus*. *Zoological Science* **27**: 746–754.
2533. O'shea, T.L. (1999) Environmental contaminants and marine mammals. *in* Biology of Marine Mammals (Reynolds, J.E., III & Rommel, S.A., Editors). Washington, D.C.: Smithsonian Institution Press, pp. 485–563.
2534. Obara, I. (小原巌) (1968) 北海道産と思われるチチブコウモリ. *The Journal of the Mammalogical Society of Japan* **4**: 43 (in Japanese).
2535. Obara, Y. (小原良孝) (1991) 進化と核型. *in* 現代の哺乳類学 (Asahi, M. (朝日稔) & Kawamichi, T. (川道武男), Editors). Tokyo: Asakura Publishing (朝倉書店), pp. 23–44 (in Japanese).
2536. Obara, I. (1990) A fore paw of Japanese wolf, *Canis hodophilax*, preserved in Kiyokawa-mura, Kanagawa Prefecture. *Natural History Report of Kanagawa* **11**: 67–69 (in Japanese).
2537. Obara, I. (1990) Skulls of Japanese wolf, *Canis hodophilax* preserved as old private houses in Atsugi-shi and Kiyokawamura, Kanagawa Prefecture. *Natural History Report of Kanagawa* **11**: 53–65 (in Japanese).
2538. Obara, I. & Nakamura, K. (1992) Notes on a skull of socalled Yama-Inu or wild canine preserved in the Minamiashigara Municipal Folklore Museum. *Bulletin of Kanagawa Prefectural Museum. Natural Science* **21**: 105–110 (in Japanese with English abstract).
2539. Obara, Y. (1982) Comparative analysis of karyotypes in the Japanese mustelids, *Mustela nivalis namiyei* and *M. erminea nippon*. *The Journal of the Mammalogical Society of Japan* **9**: 59–69.
2540. Obara, Y. (1985) G-band homology and C-band variation in the Japanese mustelids, *Mustela erminea nippon* and *M. sibirica itatsi*. *Genetica* **68**: 59–64.
2541. Obara, Y. (1987) Karyological kinship of two species of mustelids, *Mustela erminea nippon* and *Martes melampus melampus*. *Proceedings of the Japan Academy* **63B**: 197–200.
2542. Obara, Y. (1991) Karyosystematics of the mustelid carnivores of Japan. *Mammalian Science* **30**: 197–220 (in Japanese with English abstract).
2543. Obara, Y., et al. (1995) Revised karyotypes of the Japanese northern red-backed vole, *Clethrionomys rutilus mikado*. *Journal of the Mammalogical Society of Japan* **20**: 125–133.
2544. Obara, Y., et al. (2009) Genotoxic assessment of small mammals at an illegal dumpsite at the Aomori-Imate prefectual boundary. *Zoological Science* **26**: 139–144.
2545. Obara, Y. & Miyai, T. (1981) A preliminary study on the sex chromosome variation of the Ryukyu house shrew, *Suncus murinus riukiuanus*. *Japanese Journal of Genetics* **56**: 365–371.
2546. Obara, Y. & Saitoh, K. (1977) Chromosome studies in the Japanese vespertilionid bats: IV. Karyotypes and C-banding pattern of *Vespertilio orientalis*. *Japanese Journal of Genetics* **52**: 159–161.
2547. Obara, Y. & Sasaki, S. (1997) Fluorescent approaches on the origin of B chromosomes of *Apodemus argenteus hokkaidi*. *Chromosome Science* **1**: 1–5.
2548. Obara, Y., Sasaki, S., & Igarashi, Y. (1997) Delayed response of QM- and DA/DAPI-fluorescence in C-heterochromatin of the small Japanese field mouse, *Apodemus argenteus*. *Zoological Science* **14**: 57–64.
2549. Obara, Y., Sasamori, K., & Mukohyama, M. (1997) Inhabitation and distribution of the least weasel in Aomori Prefecture. *Mammalian Science* **37**: 81–85 (in Japanese with English abstract).
2550. Obara, Y. & Tada, T. (1985) Karyotypes and chromosome banding patterns of the Japanese water shrew *Chimarrogale platycephala platycephala*. *Proceedings of the Japan Academy. Series B* **61**: 20–23.
2551. Obara, Y., Tomiyasu, T., & Saitoh, K. (1976) Chromosome studies in the Japanese vespertilionid bats: I. Karyotypic variation in *Myotis macrodactylus* Temminck. *Japanese Journal of Genetics* **51**: 201–206.
2552. Obolenskaya, E.V., et al. (2009) Diversity of Palaearctic chipmunks (*Tamias*, Sciuridae). *Mammalia* **73**: 281–298.
2553. Ochiai, K. (落合けいこ) (1996) ヤマコウモリの死体をひろった. *Bat Study and Conservation Report* **4**: 6 (in Japanese).
2554. Ochiai, K. (落合啓二) (1992) カモシカの生活誌. Tokyo: Doubutsu-sha (どうぶつ社), 226 pp. (in Japanese).
2555. Ochiai, K. (落合啓二) & Asada, M. (浅田正彦) (2007) 同所的に生息する外来種キョンと在来種ニホンジカの食性・食物成分の比較. *in* 日本哺乳類学会 2007 年度大会. p. 110 (in Japanese).
2556. Ochiai, K. Unpublished.
2557. Ochiai, K. (1983) Pair-bond and mother-offspring relationships of Japanese serow in Kusoudomari, Wakinosawa Village. *The Journal of the Mammalogical Society of Japan* **9**: 192–203 (in Japanese with English abstract).

2558. Ochiai, K. (1983) Territorial behavior of the Japanese serow in Kusoudomari, Wakinosawa Village. *The Journal of the Mammalogical Society of Japan* **9**: 253–259 (in Japanese with English abstract).

2559. Ochiai, K. (1993) Dynamics of population density and social interrelation in the Japanese serow *Capricornis crispus*. Fukuoka: Kyushu University. 72 pp.

2560. Ochiai, K. (1999) Diet of the Japanese serow (*Capricornis crispus*) on the Shimokita Peninsula, northern Japan, in reference to variations with a 16-year interval. *Mammal Study* **24**: 91–102.

2561. Ochiai, K., Ishii, M., & Hiraoka, T. (2003) Status of sarcoptic mange among raccoon dog and other wild mammals in Chiba Prefecture, Central Japan. *Journal of Natural History Museum Institute, Chiba* **7**: 89–103 (in Japanese with English summary).

2562. Ochiai, K., et al. (1993) Population dynamics of Japanese serow in relation to social organization and habitat conditions. I. Stability of Japanese serow density in stable habitat conditions. *Ecological Research* **8**: 11–18.

2563. Ochiai, K., et al. (1993) Population dynamics of Japanese serow in relation to social organization and habitat conditions. II. Effects of clear-cutting and planted tree growth on Japanese serow population. *Ecological Research* **8**: 19–25.

2564. Ochiai, K. & Susaki, K. (2002) Effects of territoriality on population density in the Japanese serow (*Capricornis crispus*). *Journal of Mammalogy* **83**: 964–972.

2565. Ochiai, K. & Susaki, K. (2007) Causes of natal dispersal in a monogamous ungulate, the Japanese serow, *Capricornis crispus*. *Animal Behaviour* **73**: 125–131.

2566. Ochiai, M., et al. (2013) Accumulation of hydroxylated polychlorinated biphenyls (OH-PCBs) and implications for PCBs metabolic capacities in three porpoise species. *Chemosphere* **92**: 803–810.

2567. Oda, S., et al., eds. (1985) *Suncus murinus*: Biology of the Laboratory Shrew. Tokyo: Japan Scientific Societies Press, 535 pp. (in Japanese with English abstract).

2568. Oda, S. & Kondo, K. (1976) Progress in domestication of *Suncus murinus riukiuanus* for experiments. *Mammalian Science* **16(2)**: 13–30 (in Japanese).

2569. Oda, S. & Shigehara, N. (1978) Capture of the Riukiu house musk shrew, *Suncus murinus riukiuanus* in the suburbs of Naha, Okinawa. *The Journal of the Mammalogical Society of Japan* **7**: 150–154 (in Japanese).

2570. Oda, S., Tohya, K., & Miyaki, T., eds. (2011) *Suncus murinus* (Biology of Suncus スンクスの生物学). Tokyo: Japan Scientific Societies Press, 449 pp. (in Japanese with English abstract).

2571. Odell, D.K. & McClune, K.M. (1999) False killer whale *Pseudorca crassidens* Owen, 1846. in Handbook of Marine Mammals. Volume 6. The Second Book of Dolphins and the Porpoises (Ridgway, S.H. & Harrison, R., Editors). London: Academic Press, pp. 213–243.

2572. Ogasawara Marine Center (小笠原海洋センター), ed. (2000) 事典「クジラの尾ビレ」―小笠原・沖縄― (Humpback Whales in Ogasawara and Okinawa). 139 pp. (in Japanese).

2573. Ogata, M., Itagaki, H., & Wakuri, H. (1977) A case of Chorioptes mite infestation of a Japanese serow *Capricornis crispus* (Temminck) in Morioka, Iwate Prefectures, Japan. *Bulletin of Azabu Veterinary College* **2**: 223–225.

2574. Ogawa, T. (小川鼎三) (1936) 本邦の歯鯨目録に加ふべき 4 属 (抄). *Zoological Magazine* **48**: 175 (in Japanese).

2575. Ogawa, T. (小川鼎三) (1937) 本邦の歯鯨類に関する研究 (XII). *Botany and Zoology* **5**: 25–34 (in Japanese).

2576. Ogawa, T. (小川鼎三) (1937) 本邦の歯鯨類に関する研究 (XIII). *Botany and Zoology* **5**: 409–416 (in Japanese).

2577. Ogawa, F., Yagihashi, T., & Tanaka, N. (2002) Discovery of *Murina ussuirensis* Ognev on remaining snow in Mt. Naeba, central Japan. *Bulletin of the Asian Bat Research Institute* **2**: 13–15 (in Japanese with English abstract).

2578. Ogawa, N. & Yoshida, H. (2014) Abundance estimation of finless porpoises in Japan. *Aquabiology* **36**: 182–190 (in Japanese with English abstract).

2579. Ogiwara, Y. (荻原弥生), et al. (2002) エゾシカの処理実態および疾病状況調査. *Journal of Hokkaido Veterinary Medical Association* **46**: 35–39 (in Japanese).

2580. Ognev, S.I. (1940) Mammals of the USSR and Adjacent Countries. IV. Rodents. Moscow: Izvestii Akademii Nauk USSR, 429 pp.

2581. Ogura, G. (小倉剛) & Yamada, F. (山田文雄) (2011) フイリマングース 日本の最優先対策種. in Invasive Alien Mammals in Japan: Biology of Control Strategy and Conservation (日本の外来哺乳類：管理戦略と生態系保全) (Yamada, F. (山田文雄), Ikeda, S. (池田透), & Ogura, G. (小倉剛), Editors). Tokyo: University of Tokyo Press, pp. 105–137 (in Japanese).

2582. Ogura, G., et al. (2005) Investigation of the northern limit of dugong habitat in the Tokara Islands and Amami-Oshima in the Ryukyu Archipelago, Japan. *Wildlife Conservation Japan* **9**: 49–58 (in Japanese with English abstract).

2583. Ogura, G., et al. (2003) A case of accidental death of a Kerama deer (*Cervus nippon keramae*) in Tokashiki Island. *Japanese Journal of Zoo and Wildlife Medicine* **8**: 55–62 (in Japanese with English abstract).

2584. Ogura, G., et al. (2000) Postnatal growth in the Javan mongoose, *Herpestes javanicus auropunctatus*, raised in captivity on Okinawa. *Japanese Journal of Zoo and Wildlife Medicine* **5**: 77–85.

2585. Ogura, G., Kawashima, Y., & Oda, S. (2003) Analysis of captured small Asian mongooses, and present situation of countermeasures and problems. *Journal of the Veterinary Medicine* **56**: 295–301 (in Japanese).

2586. Ogura, G., et al. (2001) Relationship between body size and sexual maturation, and seasonal change of reproductive activities in female feral small Asian mongoose (*Herpestes javanicus*) on Okinawa Island. *Japanese Journal of Zoo and Wildlife Medicine* **6**: 7–14.

2587. Ogura, G., Sakashita, M., & Kawashima, Y. (1998) External morphology and classification of mongoose on Okinawa Island. *Mammalian Science* **38**: 259–270 (in Japanese with English abstract).

2588. Ogura, G., et al. (2002) Food habits of the feral small Asian mongoose (*Herpestes javanicus*) and impacts on native species in the northern part of Okinawa Island. *Mammalian Science* **42**: 53–62 (in Japanese with English abstract).

2589. Oh, H.S. & Mori, T. (1998) Growth, development and reproduction in captive of the large Japanese field mouse, *Apodemus speciosus* (Rodentia, Muridae). *Journal of Faculty of Agriculture, Kyushu University* **43**: 397–408.

2590. Oh, Y.K., Mōri, T., & Uchida, T.A. (1985) Prolonged survival of the Graafian follicle and fertilization in the Japanese greater horseshoe bat, *Rhinolophus ferrumequinum nippon*. *Journal of Reproduction and Fertility* **73**: 121–126.

2591. Oh, Y.K., Mōri, T., & Uchida, T.A. (1985) Spermiogenesis in the Japanese greater horseshoe bat, *Rhinolophus ferrumequinum nippon*. *Journal of the Faculty of Agriculture, Kyushu University* **29**: 203–209.

2592. Ohata, J. (2007) Bats in Ohkubo-mabu tunnel of "Iwami-ginzan" old silver mine, Ohda city, Shimane prefecture. *Bulllletin of the Shimane Nature Museum of Mt. Sanbe (Sahimel)* **5**: 15–24 (in Japanese).

2593. Ohbayashi, M. (大林正士) (1988) 寄生虫病. in 新編 獣医ハンドブック (Nakamura, R. (中村良一), et al., Editors). Tokyo: Yokendo (養賢堂), pp. 219–267 (in Japanese).

2594. Ohbayashi, M. (1975) Incidence of helminth parasites in shrew moles. *Japanese Journal of Veterinary Research* **23**: 101–102.

2595. Ohbayashi, M. (1985) Helminth parasites of the Soricidae. in *Suncus murinus*. Biology of the laboratory shrew (Oda, S., et al., Editors). Tokyo: Japan Scientific Press, pp. 88–93 (in Japanese with English abstract).

2596. Ohbayashi, M. & Kitamura, Y. (1959) *Sarcocystis clethrionomysi* n. sp. from *Clethrionomys rufocanus bedfordiae* Thomas. *Japanese Journal of Veterinary Research* **7**: 115–119.

2597. Ohbayashi, M. & Masegi, T. (1972) Some nematodes of the Japanese shrew mole, *Urotrichus talpoides* Temminck. *Japanese Journal of Veterinary Research* **20**: 111–116.

2598. Ohbayashi, M., Masegi, T., & Kubota, K. (1972) Parasites of the Japanese shrew mole, *Urotrichus talpoides* Temminck. *Japanese Journal of Veterinary Research* **20**: 50–61.

2599. Ohbayashi, M., Masegi, T., & Kubota, K. (1973) Further observations on parasites of the Japanese shrew mole, *Urotrichus talpoides* Temminck. *Japanese Journal of Veterinary Research* **21**: 15–25.

2600. Ohdachi, S. (大舘智氏) (2005) DNA より示唆される北海道産トガリネズミ群集の成立過程. in 動物地理の自然史：分布と多様性の進化学 (Masuda, R. (増田隆一) & Abe, H. (阿部永), Editors). Sapporo: Hokkaido University Press, pp. 15–31 (in Japanese).

2601. Ohdachi, S. (1992) Female reproduction in three species of *Sorex* in Hokkaido, Japan. *Journal of Mammalogy* **73**: 455–457.

2602. Ohdachi, S. (1992) Home ranges of sympatric soricine shrews in Hokkaido, Japan. *Acta Theriologica* **37**: 91–101.

2603. Ohdachi, S. (1994) Total activity rhythms of three soricine species in Hokkaido. *Journal of the Mammalogical Society of Japan* **19**: 89–99.

2604. Ohdachi, S. (1995) Burrowing habits and earthworm preferences of three species of *Sorex* in Hokkaido. *Journal of the Mammalogical Society of Japan* **20**: 85–88.

2605. Ohdachi, S. (1995) Diets and abundances of three sympatric shrew species in northern Hokkaido. *Journal of the Mammalogical Society of Japan* **20**: 69–83.

2606. Ohdachi, S. (1996) Longevity of captive shrews in Hokkaido. *Mammal Study* **21**: 65–69.

2607. Ohdachi, S. (1997) Laboratory experiments on spatial use and aggression in three sympatric species of shrew in Hokkaido, Japan. *Mammal Study* **22**: 11–26.

2608. Ohdachi, S. (1999) Some problems about Blakiston's line in insectivores: a special reference to shrews. *Mammalian Science* **39**: 329–336 (in Japanese).

2609. Ohdachi, S. & Aoi, T. (1987) Food habits of brown bears in Hokkaido, Japan. *Proceedings of the International Conference on Bear Research and Management* **7**: 215–220.

2610. Ohdachi, S., et al. (1992) Growth, sexual dimorphism, and geographical variation of skull dimensions of the brown bear *Ursus arctos* in Hokkaido. *Journal of the Mammalogical Society of Japan* **17**: 27–47.

2611. Ohdachi, S., et al. (2001) Intraspecific phylogeny and geographical variation of six species of northeastern Asiatic *Sorex* shrews based on the mitochondrial cytochrome *b* sequences. *Molecular Ecology* **10**: 2199–2213.

2612. Ohdachi, S. & Maekawa, K. (1990) Geographic distribution and relative abundance of four species of soricine shrews in Hokkaido, Japan. *Acta Theriologica* **35**: 261–267.

2613. Ohdachi, S. & Maekawa, K. (1990) Relative age, body weight, and reproductive condition in three species of *Sorex* (Soricidae; Mammalia) in Hokkaido. *Research Bulletins of the College Experiment Forests, Faculty of Agriculture, Hokkaido University* **47**: 535–546.

2614. Ohdachi, S., et al. (1997) Phylogeny of Eurasian soricine shrews (Insectivora, Mammalia) inferred from the mitochondrial cytochrome *b* gene sequences. *Zoological Science* **14**: 527–532.

2615. Ohdachi, S. & Takahashi, M. (1985) Ticks of the wild Yezo brown bears (*Ursus arctos yesoensis*). 新ひぐま通信 **11**: 24–25 (in Japanese).

2616. Ohdachi, S.D. Unpublished.

2617. Ohdachi, S.D., Abe, H., & Han, S.-H. (2003) Phylogenetical Positions of *Sorex* sp. (Insectivora, Mammalia) from Cheju Island and *S. caecutiens* from the Korean Peninsula, inferred from mitochondrial cytochrome *b* gene sequences. *Zoological Science* **20**: 91–95.

2618. Ohdachi, S.D., et al. (2005) Morphological relationships among populations in the *Sorex caecutiens/shinto* group (Eulipotyphla, Soricidae) in East Asia, with description of a new subspecies from Cheju Island, Korea. *Mammalian Biology* **70**: 345–358.

2619. Ohdachi, S.D., et al. (2006) Molecular phylogenetics of soricid shrews (Mammalia) based on mitochondrial cytochrome *b* gene sequences: with special reference to the Soricinae. *Journal of Zoology* **270**: 177–191.

2620. Ohdachi, S.D., et al. (2004) Molecular phylogenetics of *Crocidura* shrews (Insectivora) in East and Central Asia. *Journal of Mammalogy* **85**: 396–403.

2621. Ohdachi, S.D. & Kawahara, A. (2015) The first record of the Eurasian common shrew (*Sorex caecutiens*) with completely white fur, captured in Hokkaido, Japan. *Special Publication of Nagoya Society of Mammalogists (マンモ・ス)* **16**: 14–17 (in Japanese with English summary).

2622. Ohdachi, S.D. & Seo, Y. (2004) Small mammals and a frog found in the stomach of a Sakhalin taimen *Hucho perryi* (Brevoort) in Hokkaido. *Mammal Study* **29**: 85–87.

2623. Ohdachi, S.D., et al. (2012) Intraspecific phylogeny and nucleotide diversity of the least shrews the *Sorex minutissimus–S. yukonicus* complex based on nucleotide sequences of the mitochondrial cytochrome *b* gene and the control region. *Mammal Study* **37**: 281–297.

2624. Ohizumi, H., et al. (2003) Feeding habits of Baird's beaked whale *Berardius bairdii*, in the western North Pacific and Sea of Okhotsk off Japan. *Fisheries Science* **69**: 11–20.

2625. Ohizumi, H., et al. (2003) Feeding habits of Dall's porpoises (*Phocoenoides dalli*) in the subarctic North Pacific and the Bering Sea basin and the impact of predation on mesopelagic micronecton. *Deep-Sea Research Part I* **50**: 593–610.

2626. Ohizumi, H., et al. (1998) Stomach contents of common dolphins (*Delphinus delphis*) in the pelagic western North Pacific. *Marine Mammal Science* **14**: 835–844.

2627. Ohkawa, Y. (大河康隆), Tajima, H. (田島裕志) , & Tanada, E. (棚田英治) (1979) エゾヒグマ (*Ursus arctos yesoensis* LYDEKKER) の冬眠穴に関する研究. 新ひぐま通信 **6**: 37–51 (in Japanese).

2628. Ohnishi, N., Saitoh, T., & Ishibashi, Y. (2000) Spatial genetic relationships in a population of the Japanese wood mouse *Apodemus argenteus*. *Ecological Research* **15**: 285–292.

2629. Ohnishi, N., Saitoh, T., & Ishibashi, Y. (2007) Low genetic diversities in isolated populations of the Asian black bear (*Ursus thibetanus*) in Japan, in comparison with large stable populations. *Conservation Genetics* **8**: 1331–1337.

2630. Ohnishi, N., et al. (2009) The influence of climatic oscillations during the Quaternary Era on the genetic structure of Asian black bears in Japan. *Heredity* **102**: 579–589.

2631. Ohno, K., et al. (1992) The domestication of *Crocidura dsinezumi* as a new laboratory animal. *Experimental Animals* **41**: 449–454.

2632. Ohshima, K. (1990) The history of straits around the Japanese Islands in the Late-Quaternary. *The Quaternary Research* **29**: 193–208 (in Japanese with English abstract).

2633. Ohsumi, S. (大隅清治) (1986) 間引き可能量. *in* 漁業公害 (有害生物駆除) 対策調査委託事業調査報告書 (昭和56–60年度) (Tamura, T. (田村保), Ohsumi, S. (大隅清治), & Arai, S. (荒井修亮), Editors). Tokyo: Fisheries Agency, pp. 221–227 (in Japanese).

2634. Ohsumi, S. (大隅清治) (1994) シロナガスクジラ. *in* 日本の希少な野生水生生物に関する基礎資料 (I). Tokyo: Japan Fisheries Resource Conservation Association (日本水産資源保護協会), pp. 592–600 (in Japanese).

2635. Ohsumi, S. (大隅清治) (1996) シロナガスクジラ. *in* The Encyclopaedia of Animals in Japan. 2. Mammals II (日本動物大百科2哺乳類II) (Izawa, K. (伊沢紘生), Kasuya, T. (粕谷俊雄), & Kawamichi, T. (川道武男), Editors). Tokyo: Heibonsha (平凡社), pp. 36–37 (in Japanese).

2636. Ohsumi, S. (大隅清治) (1996) ナガスクジラ. *in* The Encyclopaedia of Animals in Japan. 2. Mammals II (日本動物大百科2哺乳類II) (Izawa, K. (伊沢紘生), Kasuya, T. (粕谷俊雄), & Kawamichi, T. (川道武男), Editors). Tokyo: Heibonsha (平凡社), pp. 38–39 (in Japanese).

2637. Ohsumi, S. (大隅清治) (1997) イワシクジラ. *in* 日本の希少な野生水生生物に関する基礎資料 (IV). Tokyo: Japan Fisheries Resource Conservation Association (日本水産資源保護協会), pp. 364–369 (in Japanese).

2638. Ohsumi, S. (大隅清治) (1998) シロナガスクジラ. *in* 日本の希少な野生生物に関するデータブック (Fisheries Agency, Editor). Tokyo: Japan Fisheries Resource Conservation Association (日本水産資源保護協会), pp. 286–287 (in Japanese).

2639. Ohsumi, S. (大隅清治) (1998) ナガスクジラ. *in* 日本の希少な野生水生生物に関するデータブック (Fisheries Agency, Editor). Tokyo: Japan Fisheries Resource Conservation Association (日本水産資源保護協会), pp. 288–289 (in Japanese).

2640. Ohsumi, S. (1965) Reproduction of the sperm whale in the north-west Pacific. *Scientific Reports of the Whales Research Institute, Tokyo* **19**: 1–35.

2641. Ohsumi, S. (1977) Bryde's whales in the pelagic whaling ground of the North Pacific. *Reports of the International Whaling Commission (Special Issue)* **1**: 140–150.

2642. Ohsumi, S. (1978) Provisional report on the Bryde's whales in the Southern Hemisphere under scientific permit in the three seasons, 1976/77–1978/79. *Report of the International Whaling Commission* **28**: 281–287.

2643. Ohsumi, S. (1979) Interspecific relationships among some biological parameters in cetaceans and estimation of the Southern Hemisphere minke whale. *Report of the International Whaling Commission* **29**: 397–406.

2644. Ohsumi, S. (1980) Population study of the Bryde's whales in the Southern Hemisphere under scientific permit in the three seasons, 1976/77–1978/79. *Report of the International Whaling Commission* **30**: 319–331.

2645. Ohsumi, S. & Wada, S. (1972) Stock assessment of blue whales in the North Pacific (Paper SC/24/13 presented to the IWC Scientific Committee). 20 pp.

2646. Ohta, K. Unpublished.

2647. Ohtaishi, N. (大泰司紀之) & Honma, H. (本間浩昭) (1998) エゾシカを食卓へ (Yezo Deer to the Table). Tokyo: Maruzen Planet (丸善プラネット), 215 pp. (in Japanese).

2648. Ohtaishi, N. (大泰司紀之), Ibe, M. (井部真理子), & Masuda, Y. (増田泰), eds. (1998) 野生動物の交通事故対策—エコロード事始め. Sapporo: Hokkaido University Press, 210 pp. (in Japanese).

2649. Ohtaishi, N. (大泰司紀之) & Nakagawa, H. (中川元) (1988) 鰭脚類. *in* 知床の動物. Sapporo: Hokkaido University Press, pp. 225–248 (in Japanese).

2650. Ohtaishi, N. (1976) Life table for Japanese deer at Nara Park and its characteristic. *Annual Report of Nara Deer Research Association* **3**: 83–95 (in Japanese with English summary).

2651. Ohtaishi, N. (1986) Preliminary memorandum of classification, distribution and geographic variation on sika deer. *Mammalian Science* **26(2)**: 13–17 (in Japanese).

2652. Ohtaishi, N. & Gao, Y. (1990) A review of the distribution of all species of deer (Tragulidae, Moschidae and Cervidae) in China. *Mammal Review* **20**: 125–144.

2653. Ohtaishi, N., Hachiya, N., & Shibata, Y. (1976) Age determination of the hare from annual layers in the mandibular bone. *Acta Theriologica* **21**: 168–171.

2654. Ohtaishi, N. & Yoneda, M. (1981) A thirty four years old male Kuril seal from Shiretoko Pen., Hokkaido. *Scientific Reports of the Whales Research Institute* **33**: 131–135.

2655. Oi, T. (1988) Sociological study on the troop fission of wild Japanese monkeys (*Macaca fuscata yakui*) on Yakushima Island. *Primates* **29**: 1–19.

2656. Oi, T. (2003) For better management plans of wild Japanese macaques: A case study of the plan by Shiga Prefecture. *Mammalian Science* **Suppl. 3**: 31–34 (in Japanese).

2657. Oi, T., et al. (2012) Characteristics of food habits of Asiatic black bears in the Nishi-Chugoku Mountains, southern Japan. *Mammalian Science* **53**: 1–13 (in Japanese with English summary).

2658. Oi, T. & Yamazaki, K. (2006) The Status of Asiatic Black Bears in Japan. *in* Understanding Asian Bears to Secure Their Future (Japan Bear Network,

Compiler). Ibaraki: Japan Bear Network, pp. 122–133.
2659. Oikawa, E., et al. (2012) First case of *Echinococcus multilocularis* infection in a zoo-housed flying squirrel (*Pteromys volans orii*). *Journal of Veterinary Medical Science* **75**: 659–661.
2660. Oikawa, N., Matsui, M., & Hirakawa, H. (2013) Notably high nighttime activity in the autumn: Daily rhythm of activity in the northern pika (*Ochotona hyperborea*) in a stunted subalpine forest in central Hokkaido, Japan. *Mammalian Science* **53**: 79–87 (in Japanese with English abstract).
2661. Oishi, M. & Hasegawa, Y. (1994) A list of fossil cetaceans in Japan. *The Island Arc* **3**: 493–505.
2662. Oka, T. (1992) Home range and mating system of two sympatric field mouse species, *Apodemus speciosus* and *Apodemus agenteus*. *Ecological Research* **7**: 163–169.
2663. Oka, T., et al. (2010) Mitochondrial DNA polymorphisms of alien hedgehogs in Ito and Odawara, Japan. *Journal of Agricultural Science, Tokyo Nogyo Daigaku* **55**: 158–162 (in Japanese with English summary).
2664. Oka, T., et al. (2004) Relationship between changes in beechnut production and Asiatic black bears in northern Japan. *Journal of Wildlife Management* **68**: 979–986.
2665. Okabe, K., et al. (1968) Natural infection of the Japanese vole with *Schistosoma japonicum* in the Chikugo Riverbed in Kurume City. XX. *Kurume Medical Journal* **31**: 827–837 (in Japanese with English abstract).
2666. Okada, A., Miura, S., & Murakami, O. (1998) Changes in distribution of nutria (*Myocastor coypus*) in Gifu Prefecture. *Kansai Organization for Nature Conservation* **20**: 77–81 (in Japanese).
2667. Okada, A., et al. (2014) Genetic structure and cryptic genealogy of the Bonin flying fox *Pteropus pselaphon* revealed by mitochondrial DNA and microsatellite markers. *Acta Chiropterologica* **16**: 15–26.
2668. Okada, A. & Tamate, H.B. (2000) Pedigree analysis of the sika deer (*Cervus nippon*) using microsatellite markers. *Zoological Science* **17**: 335–340.
2669. Okada, A., et al. (2005) Use of microsatellite markers to assess the spatial genetic structure of a population of sika deer *Cervus nippon* on Kinkazan island, Japan. *Acta Theriologica* **50**: 227–240.
2670. Okada, H. (1956) On two species of Vespertilionidae from Aomori prefecture. *The Journal of the Mammalogical Society of Japan* **1**: 29–31 (in Japanese with English abstract).
2671. Okada, M., Kuroda, T., & Katsuno, T. (2007) Distribution and habitat of the Japanese weasel (*Mustela itatsi*) at several watersheds in Kanagawa Prefecture. *Natural History Report of Kanagawa* **28**: 55–58.
2672. Okada, T. & Okada, S. (2006) First record of *Vespertilio superans* (Mammalia: Chiroptera, Vespertilionidae) from Mt. Hyonosen, Tottori Prefecture, Honshu. *Natural History Research of San'in* **2**: 39–40 (in Japanese).
2673. Okada, T., Okada, S., & Ichisawa, K. (2008) First records of two species of *Murina* and second record of *Vespertilio sinensis* in Tottori prefecture, Honshu, Japan (Mammalia: Chiroptera: Vespertilionidae). *Bulletin of the Tottori Prefectural Museum* **45**: 7–9 (in Japanese).
2674. Okamoto, S. (岡本省吾) & Kitamura, S. (北村四郎) (1978) Coloured Illustrations of Trees and Shrubs of Japan (原色日本樹木図鑑). Osaka: Hoikusha (保育社), 306 pp. (in Japanese).
2675. Okamoto, M. (1999) Phylogeny of Japanese moles inferred from mitochondrial CO I gene sequences. *in* Recent Advances in the Biology of Japanese Insectivora: Proceedings on the Symposium on the Biology of Insectivores in Japan and on the Wildlife Conservation (Yokohata, Y. & Nakamura, S., Editors). Shobara & Hiwa: Hiba Society of Natural History & Hiwa Museum for Natural History, pp. 21–27.
2676. Okamura, M., et al. (2000) Annual reproductive cycle of the Iriomote cat *Felis iriomotensis*. *Mammal Study* **25**: 75–85.
2677. Okawara, Y., et al. (2014) Food habits of the urban Japanese weasels *Mustela itatsi* revealed by faecal DNA analysis. *Mammal Study* **39**: 155–161.
2678. Okayama Prefectural Government (2009) 岡山県版レッドデータブック—絶滅のおそれのある野生生物—. Okayama: Okayama Prefectural Government, 542 pp. (in Japanese).
2679. Okazaki, H. Personal communication.
2680. Okimoto, K., et al. Unpublished.
2681. Okinawa Board of Education (沖縄県教育委員会) (1981) ケナガネズミ実態調査報告書. Naha: Okinawa Board of Education (沖縄県教育委員会), 65 pp. (in Japanese).
2682. Oku, Y., et al. (2001) Japan. *in* Helminths of Wildlife (Chowdhury, N. & Aguirre, A.A., Editors). Enfield: Science Publishers, pp. 255–283.
2683. Okubo, M., Hobo, T., & Tamura, N. (2005) Vegetation types selected by alien Formosan squirrel in Kanagawa Prefecture. *Natural History Report of Kanagawa* **26**: 53–56 (in Japanese).
2684. Okubo, M., Tamura, N., & Katsuki, T. (2005) Nest site and materials of the alien squirrel, *Callosciurus erythraeus*, in Kanagawa Prefecture. *Journal of the Japanese Wildlife Research Society* **31**: 5–10 (in Japanese with English summary).
2685. Okumura, H. (奥村栄朗), Ito, T. (伊藤健雄), & Miura, S. (三浦慎悟) (1996) 山形市滝山地区におけるカモシカの行動圏と移動分散—ラジオ・テレメトリー法による行動解析—. *in* 西蔵王のカモシカ—特別天然記念物カモシカ保護地域管理技術策定調査報告書 (特別天然記念物カモシカ保護地域管理技術策定調査会, Editor). Yamagata: Yamagata Prefectural Board of Education (山形県教育委員会), pp. 19–55 (in Japanese).
2686. Okumura, H. (2004) Complete sequence of mitochondrial DNA control region of the Japanese serow *Capricornis crispus* (Bovidae: Caprinae). *Mammal Study* **29**: 137–145.
2687. Okumura, K., et al. (1982) Latest Pleistocene mammalian assemblage of Kumaishi-do Cave, Gifu Prefecture and the significance of its ^{14}C age. *Earth Science* **36**: 214–218 (in Japanese with English).
2688. Okumura, K., et al. (1982) Latest Pleistocene mammalian assemblage of Kumaishi-do Cave, Gifu Prefecture, Central Japan. *Bulletin of Osaka Museum of Natural History* **31**: 13–24 (in Japanese with English abstract).
2689. Okumura, N., et al. (1996) Geographic population structure and sequence divergence in the mitochondrial DNA control region of the Japanese wild boar (*Sus scrofa leucomystax*), with reference to those of domestic pigs. *Biochemical Genetics* **34**: 179–189.
2690. Okumura, N., et al. (2001) Genetic relationship amongst the major non-coding regions of mitochondrial DNAs in wild boars and several breeds of domesticated pigs. *Animal Genetics* **32**: 139–147.
2691. Okuzaki, M. (1979) Reproduction of raccoon dogs, *Nyctereutes procyonoides viverrinus* Temminck, in captivity. *Journal of Kagawa Nutrition College* **10**: 99–103 (in Japanese).
2692. Olsen, Ø. (1913) On the external characters and biology of Bryde's whale (*Balaenoptera brydei*), a new rorqual from the coast of South Africa. *Proceedings of the Zoological Society of London* **1913**: 1073–1090.
2693. Olson, P.A. & Reilley, S.B. (2002) Pilot whales (*Globicephala melas* and *G. macrorhynchus*. *in* Encyclopedia of Marine Mammals (Perrin, W.F., Würsig, B., & Thewissen, J.G.M., Editors). San Diego: Academic Press, pp. 898–903.
2694. Omata, Y., et al. (2005) Antibodies to *Toxoplasma gondii* in free-ranging wild boar (*Sus scrofa leucomystax*) in Shikoku, Japan. *Japanese Journal of Zoo and Wildlife Medicine* **10**: 99–102.
2695. Omura, H. (1950) Whales in the adjacent water of Japan. *Scientific Reports of the Whales Research Institute* **4**: 27–113.
2696. Omura, H. (1959) Bryde's whale from the coast of Japan. *Scientific Reports of the Whales Research Institute* **14**: 1–33.
2697. Omura, H. (1962) Further information on Bryde's whales from the coast of Japan. *Scientific Reports of the Whales Research Institute* **16**: 7–18.
2698. Omura, H. (1974) Possible migration route of the gray whale on the coast of Japan. *Scientific Reports of the Whales Research Institute* **26**: 1–14.
2699. Omura, H. (1984) History of Gray Whales in Japan. *in* The Gray Whale *Eschrichtius robustus* (Jones, M.L., Swartz, S.L., & Leatherwood, S., Editors). Orlando: Academic Press, pp. 57–77.
2700. Omura, H. (1986) History of right whale catches in the waters around Japan. *Reports of the International Whaling Commission (Special Issue)* **10**: 35–41.
2701. Omura, H. (1988) Distribution and migration of the western Pacific stock of the gray whale. *Scientific Reports of the Whales Research Institute* **39**: 1–9.
2702. Omura, H., Fujino, K., & Kimura, S. (1955) Beaked whale *Berardius bairdi* of Japan, with notes on *Ziphius cavirostris*. *Scientific Reports of the Whales Research Institute* **10**: 89–132.
2703. Omura, H., Nishimoto, S., & Fujino, K. (1952) Sei Whales (*Balaenoptera borealis*) in the Adjacent Waters of Japan. Tokyo: Fisheries Agency, 80 pp.
2704. Omura, H., et al. (1969) Black right whales in the North Pacific. *Scientific Report of the Whales Research Institute* **21**: 1–78.
2705. Omura, Y., Fukumoto, Y., & Ohtaki, K. (1983) Chromosome polymorphism in Japanese sika, *Cervus* (*Sika*) *nippon*. *Japanese Journal of Veterinary Science* **45**: 23–30.
2706. Ondo, Y. & Fukuda, K. (1961) Ecological studies on the Temminck's mole, *Talpa micrura kobeae*. (1) An outline of ecological informations. *Liberal Arts Journal. Natural Science* **12**: 50–60 (in Japanese with English summary).
2707. Ono, Y. (小野勇一) (2000) ニホンカモシカのたどった道：野生動物との共生を探る. Tokyo: Chuokoron-shinsha (中央公論新社), 184 pp. (in Japanese).
2708. Ono, Z. (大野善右衛門) (1967) 北海道のコウモリダニ. *Japanese Journal of Sanitary Zoology* **18**: 17 (in Japanese).
2709. Ono, M., et al. (2003) Comparative study of HP-27 gene promoter activities between the chipmunk and tree squirrel. *Gene* **302**: 193–199.
2710. Ono, T. & Obara, Y. (1989) Karyotype and Ag-NORs Asian free-tailed bat, *Tadarida insignis* (Molossidae: Chiroptera). *Chromosome Information Service* **46**: 17–19.
2711. Ono, T. & Obara, Y. (1994) Karyotypes and Ag-NOR variations in Japanese

vespertilionid bats (Mammalia: Chiroptera). *Zoological Science* **11**: 473–484.

2712. Ono, T. & Yoshida, M.C. (1995) Banded karyotype of *Eptesicus nilssonii parvus* (Mammalia: Chiroptera). *Chromosome Information Service* **59**: 19–21.

2713. Ono, T. & Yoshida, M.C. (1997) Differences in the chromosomal distribution of telomeric (TTAGGG)n sequences in two species of the vespertilionid bats. *Chromosome Research* **5**: 203–205.

2714. Ono, Y., et al. (2013) Parasitic nematodes obtained from small mammals in Shiretoko Peninsula, Japan. *Bulletin of the Shiretoko Museum* **35**: 5–10 (in Japanese with Engilsh abstract).

2715. Ono, Z. (1965) Supplemental notes of the studies on Japanese fleas Part (6). On four bat-fleas from Hokkaido, Japan. *Japanese Journal of Sanitary Zoology* **16**: 99–103 (in Japanese with English abstract).

2716. Ono, Z. (1966) Studies on ticks parasitic on small rodents of the family Muridae in Hokkaido, Japan. *Report of the Hokkaido Institute of Public Health* **16**: 62–68 (in Japanese with English abstract).

2717. Ono, Z. (1968) Murine lice of Hokkaido. *Report of the Hokkaido Institute of Public Health* **18**: 74–78 (in Japanese with English abstract).

2718. Ono, Z. (1968) Two myobiid mite species (Acarina: Myobiidae) from shrews of Hokkaido, Japan. *Sanitary Zoology* **19**: 191–193 (in Japanese with English summary).

2719. Ono, Z. (1970) Fleas collected from shrews in Hokkaido (Supplemental notes of the studies on Japanese fleas Part 9). *Report of the Hokkaido Institute of Public Health* **20**: 81–86 (in Japanese with English summary).

2720. Ono, Z. (1971) Ectoparasites of small mammals caught on the southern slope of Mt. Daisetsuzan. *in* Studies on the Animal Communities in the Terrestrial Ecosystems and their Conservation (Annual Report of JIBP/CT-S for the Fiscal Year of 1970) (Kato, M., Editor). Sendai: JIBP/CT-S Section, pp. 225–235 (in Japanese).

2721. Ono, Z. (1971) Faunal survey of the Mt. Daisetsu area, JIBP main area. XVI. Ectoparasites of small mammals caught on the southern slope of Mt. Daisezuzan. *in* Studies on the Animal Communities in the Terrestrial Ecosystems and their Conservation (Annual Report of JIBP/CT-S for the Fiscal Year of 1970) (Kato, M., Editor). Sendai: JIBP/CT-S Section, pp. 225–235 (in Japanese with English summary).

2722. Ono, Z. (1973) Ectoparasites from small mammals of the Nopporo Forest Park near Sapporo, Hokkaido, Japan. *Report of the Hokkaido Institute of Public Health* **23**: 49–56 (in Japanese with English summary).

2723. Ono, Z. & Takada, N. (1973) Murine lice of northern part of Honshu, Japan. *Report of the Hokkaido Institute of Public Health* **23**: 45–48 (in Japanese with English abstract).

2724. Ono, Z. & Takada, N. (1973) Siphonaptera from small mammals in northern Honshu, Japan. *Research Bulletin of the Meguro Parasitological Museum* **7**: 28–31.

2725. Ono, Z. & Yoshida, H. (1973) Fleas found on field rodents in northern Kyushu, Japan. *Japanese Journal of Sanitary Zoology* **23**: 275 (in Japanese).

2726. Onoyama, K. (1989) Allometric study of internal organs in three species of *Clethrionomys* of Hokkaido (Rodentia, Microtinae). *Research Bulletin of Obihiro Zootechnical University, Series I* **16**: 141–148 (in Japanese).

2727. Onoyama, K. (1989) Small mammals in a coniferous forest at high altitudes of the Daisetsu Mountains, Hokkaido, with special reference to the home range and microhabitat of *Clethrionomys rutilus mikado*. *Research Bulletin of Obihiro Zootechnical University, Series I* **16**: 131–139.

2728. Oohata, J. (2011) Bats in the abandoned mine at Kawahira in Gohtsu City, Shimane Prefecture. *Bulletin of the Shimane Nature Museum of Mt. Sanbe* **9**: 89–98 (in Japanese).

2729. Oohata, J., Inoue, M., & Mishima, H. (2011) Bats in Ohkubo-mabu tunnel of 'Iwami-ginzan' old silver mine in Ohda city, Shimane prefecture (Part II): Conservation of hibernating bat's colonies in the World Cultural Heritage. *Bulletin of the Shimane Nature Museum of Mt. Sanbe* **9**: 77–87 (in Japanese).

2730. Ooi, H., et al. (1984) Diplostomulum of *Pharyngostomum cordatum* in the muscle of a raccoon dog *Nyctereutes procyonoides*. *Journal of Veterinary Medical Science* **46**: 409–412.

2731. Oosawa, Y. (2007) Observation records of bats mainly inhabiting abandoned mines in Yamagata. *Study Report on Bat Conservation in Tohoku Region* **1**: 11–13 (in Japanese).

2732. Ootsu, M. (1972) Winter food habit of the Japanese marten. *Japanese Journal of Applied Entomology and Zoology* **16**: 75–78 (in Japanese with English abstract).

2733. Ooya, T. (大宅利之) & Hisatomi, S. (久冨伸一郎) (2013) 佐賀県で初確認されたオヒキコウモリについて. 佐賀県立博物館・美術館調査研究書 **37**: 5–10 (in Japanese).

2734. Oremus, M., et al. (2009) Worldwide mitochoncrial DNA diversity and phylogeography of pilot whales (*Globicephala* spp.). *Biological Journal of the Linnean Society* **98**: 729–744.

2735. Oremus, M., et al. (2010) Pelagic or insular? Genetic differentiation of rough-toothed dolphins in the Society Islands, French Polynesia. *Journal of Experimental Marine Biology and Ecology* **432&433**: 37–46.

2736. Osada, H. (長田英己) (1994) 南部千島おける海獣類の生息状況と今後の課題—1992 年南部千島海獣調査を中心として. *Rise* **4**: 134–135 (in Japanese).

2737. Osada, N., et al. (2010) Ancient genome-wide admixture beyond the current hybrid zone between *Macaca fascicularis* and *M. mulatta*. *Molecular Ecology* **19**: 2884–2895.

2738. Osaka Center for Cultural Heritage (大阪府文化財センター) (1984) 亀井遺跡 II. 361 pp. (in Japanese).

2739. Osaka Museum of Natural History (2015) 大阪平野の地下から見つかったクジラ類の化石リスト (List of cetacean fossils discovered in the Osaka Plain). Available from: http://www.mus-nh.city.osaka.jp/dai3ki_zoo/whale.html (in Japanese).

2740. Osawa, K. (大沢啓子), et al. (2013) 埼玉県立川の博物館におけるヒナコウモリ *Vespertilio sinensis* の越冬期間中の活動状況. *Bulletins and Reports* **13**: 1–12 (In Japanese).

2741. Osawa, K. & Osawa, Y. (2013) Distribution of the Yaeyama flying-fox *Pteropus dasymallus yayeyamae* Kuroda, 1933 (Chiroptera: Pteropodidae) on the Miyako Islands. *Fauna Ryukyuana* **4**: 1–3 (in Japanese with English abstract).

2742. Osawa, K. & Osawa, Y. (2013) A new record of the Ryukyu flying-fox *Pteropus dasymallus* Temminck, 1825 (Chiroptera: Pteropodidae) from Kume-jima Island. *Fauna Ryukyuana* **4**: 9–11 (in Japanese with English abstract).

2743. Osawa, K., et al. (2013) Seasonal population dynamics of *Vespertilio sinensis* at Kojima, Kumagaya, Saitama, Japan. *Bulletin of the Saitama Museum of Natural History* **7**: 95–100 (in Japanese).

2744. Osawa, K., et al. (2013) A new record of the Ryukyu flying-fox *Pteropus dasymallus* Temminck, 1825 (Chiroptera: Pteropodidae) from Aguni-jima Island. *Fauna Ryukyuana* **4**: 5–7 (in Japanese with English abstract).

2745. Osawa, Y., et al. (2012) New wintering spots of *Vespertilio sinensis* in Saitama, Japan. *Bulletin of the Saitama Museum of Natural History* **6**: 53–58 (in Japanese).

2746. Oshida, T. (押田龍夫) & Yanagawa, H. (柳川久) (2002) 外来リス類 (Alien squirrels). *in* Handbook of Alien Species in Japan (外来種ハンドブック) (Ecological Society of Japan, Editor). Tokyo: Chijin Shokan (地人書館), p. 67 (in Japanese).

2747. Oshida, T. Unpublished.

2748. Oshida, T. (2014) *Tamias sibiricus lineatus* (Siebold, 1824). *in* Red Data Book 2014: Threatened Wildlife of Japan. 1 Mammalia (Ministry of the Environment, Editor). Tokyo: Gyosei Corporation, p. 89 (in Japanese).

2749. Oshida, T., et al. (2005) Phylogeography of the Russian flying squirrel (*Pteromys volans*): implication of refugia theory in arboreal small mammal of Eurasia. *Molecular Ecology* **14**: 1191–1196.

2750. Oshida, T., et al. (2001) Phylogeography of the Japanese giant flying squirrel, *Petaurista leucogenys*, based on mitochondrial DNA control region sequences. *Zoological Science* **18**: 107–114.

2751. Oshida, T., Itoya, M., & Yoshida, M.C. (1996) Q-banded karyotype of a male Japanese squirrel, *Sciurus lis*. *Chromosome Information Service* **61**: 22–24.

2752. Oshida, T., et al. (2006) Phylogeography of the Pallas's squirrel (*Callosciurus erythraeus*) in Taiwan: geographical isolation in arboreal small mammals. *Journal of Mammalogy* **87**: 247–254.

2753. Oshida, T., et al. (2000) Phylogenetic relationships among Asian species of *Petaurista* (Rodentia, Sciuridae), inferred from mitochondrial cytochrome *b* gene sequences. *Zoological Science* **17**: 123–128.

2754. Oshida, T. & Masuda, R. (2000) Phylogeny and zoogeography of six squirrel species of the genus *Sciurus* (Mammalia, Rodentia), inferred from cytochrome *b* gene sequences. *Zoological Science* **17**: 405–409.

2755. Oshida, T., Masuda, R., & Ikeda, K. (2009) Phylogeography of the Japanese giant flying squirrel, *Petaurista leucogenys* (Rodentia: Sciuridae): implication of glacial refugia in an arboreal small mammal in the Japanese Islands. *Biological Journal of the Linnean Society* **98**: 47–60.

2756. Oshida, T., Masuda, R., & Yoshida, M.C. (1996) Phylogenetic relationships among Japanese species of the family Sciuridae (Mammalia, Rodentia), inferred from nucleotide sequences of mitochondrial 12S ribosomal RNA genes. *Zoological Science* **13**: 615–620.

2757. Oshida, T., Matsushima, M., & Yoshida, M.C. (1993) Chromosome banding patterns of the Eurasian squirrel, *Sciurus vulgaris orientis*, Thomas. *Chromosome Information Service* **55**: 10–12.

2758. Oshida, T. & Obara, Y. (1991) Karyotypes and chromosome banding patterns of a male Japanese giant flying squirrel, *Petaurista leucogenys* Temminck. *Chromosome Information Service* **50**: 26–28.

2759. Oshida, T. & Obara, Y. (1993) C-band variation in the chromosomes of the

Japanese giant flying squirrel, *Petaurista leucogenys*. *Journal of the Mammalogical Society of Japan* **18**: 61–67.

2760. Oshida, T., et al. (2005) A note on karyotypes of *Sorex caecutiens* (Mammalia, Insectivora) from Cheju Island, Korea. *Caryologia* **58**: 52–55.

2761. Oshida, T., et al. (2004) A preliminary study on molecular phylogeny of giant flying squirrels, genus *Petaurista* (Rodentia, Sciuridae) based on mitochondrial cytochrome *b* gene sequences. *Russian Journal of Theriology* **3**: 15–24.

2762. Oshida, T., et al. (2007) A preliminary study on origin of *Callosciurus* squirrels introduced into Japan. *Mammal Study* **32**: 75–82.

2763. Oshida, T., et al. (2000) Comparisons of the banded karyotypes between the small Japanese flying squirrel, *Pteromys momonga* and the Russian flying squirrel, *Pteromys volans* (Rodentia, Sciuridae). *Caryologia* **53**: 133–140.

2764. Oshida, T., Yanagawa, H., & Yoshida, M.C. (1996) Chromosome banding patterns of a male Pallas squirrel, *Callosciurus erythraeus*. *Chromosome Information Service* **60**: 7–9.

2765. Oshida, T., Yanagawa, H., & Yoshida, M.C. (1997) Chromosomal characterization of the Japanese dormouse *Glirulus japonicus* Schinz (Rodentia, Muscardinidae). *Chromosome Science* **1**: 13–16.

2766. Oshida, T., et al. (2001) Molecular phylogeny of five squirrels of the genus *Callosciurus* (Mammalia, Rodentia) inferred from cytochrome *b* gene sequences. *Mammalia* **65**: 473–482.

2767. Oshida, T. & Yoshida, M.C. (1994) Banded karyotype of Asiatic chipmunk, *Tamias sibiricus lineatus* Siebold. *Chromosome Information Service* **57**: 27–28.

2768. Oshida, T. & Yoshida, M.C. (1996) Banded karyotypes and the localization of ribosomal RNA genes of Eurasian flying squirrel, *Pteromys volans orii* (Mammalia, Rodentia). *Caryologia* **49**: 219–225.

2769. Oshida, T. & Yoshida, M.C. (1997) Comparison of banded karyotypes between the Eurasian red squirrel *Sciurus vulgaris* and the Japanese squirrel *Sciurus lis*. *Chromosome Science* **1**: 17–20.

2770. Oshida, T. & Yoshida, M.C. (1999) Chromosomal localization of nucleolus organizer regions in eight Asian squirrel species. *Chromosome Science* **3**: 55–58.

2771. Oshima, K., Hayashi, M., & Matsubayashi, K. (1977) Progesterone levels in the Japanese monkey (*Macaca fuscata*) during breeding and non-breeding season and pregnancy. *Journal of Medical Primatology* **6**: 99–107.

2772. Oshiro, I. (1994) Terrestrial vertebrate fossils from the Futenma-gu site, Ginowan City, Okinawa, Japan. *Journal of Geography* **103**: 49–63 (in Japanese with English abstract).

2773. Ota, H. (太田英利), Miyagi, K. (宮城邦治), & Sakaguchi, N. (阪口法明) (1992) ダイトウオオコウモリの測定形質の検討 (Preliminary analyses of morphometric characters of the Daito flying fox, *Pteropus dasymallus daitoensis*, from Kita-Daitojima Island, Ryukyu Archipelago). in 沖縄県天然記念物調査シリーズ第31集 ダイトウオオコウモリ：保護対策緊急調査報告書 (Daito-Okomori (Daito Flying Fox), Survey Reports on Natural Monuments of Okinawa Prefecture, No.31). Naha: Okinawa Prefectural Board of Education (沖縄県教育委員会), pp. 99–104 (in Japanese).

2774. Ota, K. (太田嘉四夫) (1984) 生態的分布. in Study on Wild Murid Rodents in Hokkaido (北海道産野ネズミ類の研究) (Ota, K. (太田嘉四夫), Editor). Sapporo: Hokkaido University Press, pp. 313–354 (in Japanese).

2775. Ota, K. (太田嘉四夫) & Abe, H. (阿部永) (1984) すみ場所と巣. in Study on Wild Murid Rodents in Hokkaido (北海道産野ネズミ類の研究) (Ota, K. (太田嘉四夫), Editor). Sapporo: Hokkaido University Press, pp. 119–138 (in Japanese).

2776. Ota, K. (太田嘉四夫) & Higuchi, S. (樋口輔三郎) (1984) 行動. in Study on Wild Murid Rodents in Hokkaido (北海道産野ネズミ類の研究) (Ota, K. (太田嘉四夫), Editor). Sapporo: Hokkaido University Press, pp. 77–118 (in Japanese).

2777. Ota, H. (1998) Geographic patterns of endemism and speciation in amphibians and reptiles of the Ryukyu Archipelago, Japan, with special reference to their paleogeographical implications. *Researches on Population Ecology* **40**: 189–204.

2778. Ota, H., et al. (1984) On the specimen of *Rhinolophus cornutus* collected in Yakushima Island of the Ohsumi Group. *The Journal of the Mammalogical Society of Japan* **10**: 67–68 (in Japanese).

2779. Ota, K. (1956) The Muridae of the islands adjacent to Hokkaido. *Memoirs of the Faculty of Agriculture, Hokkaido University* **2**: 123–136 (in Japanese with English summary).

2780. Ota, K., et al. (1973) Faunal survey of Oketo area, JIBP supplementary area—I. Report of the survey on bird and mammal communities in a subarctic coniferous ecosystem in Oketo, Hokkaido. in Studies on the Animal Communities in the Terrestrial Ecosystems and their Conservation; Annual Report of JIBP/CT-S for the Fiscal Year of 1972 (Kato, M., Editor). Sendai: JIBP/CT-S Section, pp. 208–235 (in Japanese with English abstract).

2781. Ota, K. & Jameson, E.W., Jr. (1961) Ecological relationships and economic inportance of Japanese microtinae. *Ecology* **42**: 184–186.

2782. Ôta, K., et al. (1991) Lactation in the Japanese monkey (*Macaca fuscata*): Yield and composition of milk and nipple preference of young. *Primates* **32**: 35–48.

2783. Ota, Y. (2010) A record of the Fraser's dolphin *Lagenodelphis hosei* Fraser, 1956 (Odontoceti: Delphinidae) from a stomach of the tiger shark *Galeocerdo cuvier* (Peron and LeSueur, in LeSueur, 1822). *Biological Magazine of Okinawa* **48**: 95–99 (in Japanese).

2784. Otani, S., et al. (1998) Diving behavior and performance of harbor porpoise, *Phocoena phocoena*, in Funka Bay, Hokkaido, Japan. *Marine Mammal Science* **14**: 209–220.

2785. Otani, T. (2002) Seed dispersal by Japanese marten *Martes melampus* in the subalpine shrubland of northern Japan. *Ecological Research* **17**: 29–38.

2786. Otani, Y., et al. (2013) Density of Japanese macaque (*Macaca fuscata yakui*) males ranging alone: Seasonal and regional variation in male cohesiveness with the group. *Mammal Study* **37**: 105–115.

2787. Otsu, S. (1971) On the food habit of Japanese mink, *Mustela itatsi itatsi* Temminck, in winter and its protection. *Japanese Journal of Applied Entomology and Zoology* **15**: 87–88.

2788. Otsu, S. (1974) The ecology and control of the Tohoku hare *Lepus brachyurus angustidens* Hollister. *Bulletin of the Yamagata Prefecture Forest Experiment Station* **5**: 1–94 (in Japanese with English summary).

2789. Otsubo, M. & Hayashi, D. (2003) Neotectonics in Southern Ryukyu arc by means of paleostress analysis. *Bulletin of the Faculty of Science, University of the Ryukyus* **76**: 1–73.

2790. Otsuka, T. Unpublished.

2791. Ou, W., et al. (2014) Temporal change in the spatial genetic structure of a sika deer population with an expanding distribution range over a 15-year period. *Population Ecology* **56**: 311–325.

2792. Oyamada, T., et al. (1995) The first record of *Gnathostoma nipponicum* in Aomori Prefecture. *Japanese Journal of Parasitology* **44**: 128–132.

2793. Ozaki, K. (1986) Food and feeding behaviour of the Formosan squirrel, *Callosciurus* sp. *Journal of the Mammalogical Society of Japan* **11**: 165–172 (in Japanese).

2794. Ozaki, M., et al. (2007) Correlations between feeding type and mandibular morphology in the sika deer. *Journal of Zoology* **272**: 244–257.

2795. Ozawa, T. (小澤智生) (2009) 古脊椎動物群の変遷からみた琉球列島の固有動物相の起源と成立プロセス (Origin and formative process of endemic fauna of the Ryukyu Archipelago, viewed from temporal changes of Pleistocene vertebrate fauna). in 日本古生物学会 第158回例会 学会講演予稿集 (Abstracts with Programs: The 158th Regular Meeting, Palaeontological Society of Japan. January 30–February 1, 2009, Okinawa Prefecture). Naha & Nishihara: Paleontological Society of Japan, p. 5 (in Japanese).

2796. Öztürk, B., et al. (2007) Cephalopod remains in the diet of striped dolphins (*Stenella coeruleoalba*) and Risso's dolphins (*Grampus griseus*) in the eastern Mediterranean Sea. *Life and Environment* **57**: 53–59.

2797. Palacios, D.M. & Mate, B.R. (1996) Attack by false killer whales (*Pseudorca crassidens*) on sperm whales (*Physeter macrocephalus*) in the Galápagos Islands. *Marine Mammal Science* **12**: 582–587.

2798. Palsbøll, P., et al. (1995) Distribution of mtDNA haplotypes in North Atlantic humpback whales: the influence of behaviour on population structure. *Marine Ecology Progress Series* **116**: 1–10.

2799. Pan, Y., et al. (2013) Scatter hoarding and hippocampal cell proliferation in Siberian chipmunks. *Neuroscience* **255**: 76–85.

2800. Panin, K.I. & Panina, G.K. (1968) Fur seal ecology and migration to the Sea of Japan during winter and spring. in Pinnipeds of the North Pacific (Arsen'ev, V.A. & Panin, K.I., Editors). Jerusalem: Israel Program for Scientific Translations, pp. 66–77 (Translated from Russian in 1971).

2801. Park, K.J., et al. (2011) Feeding habits and consumption by finless porpoises (*Neophocaena asiaeorientalis*) in the Yellow Sea. *Korean Journal of Fisheries and Aquatic Sciences* **44**: 78–84 (in Korean with English abstract).

2802. Park, S.R. & Won, P.O. (1978) Chromosomes of the Korean bats. *The Journal of the Mammalogical Society of Japan* **7**: 199–203.

2803. Parker, H.G., et al. (2004) Genetic structure of the purebred domestic dog. *Science* **304**: 1160–1164.

2804. Parker, S., et al. (2003) Mammal. London: Dorling Kindersley, 216 pp.

2805. Parkes, J.P. (2005) Feral goat. in The Handbook of New Zealand Mammals (King, C.M., Editor). South Melbourne: Oxford University Press, pp. 374–392.

2806. Pastene, L.A., Kanda, N., & Hatanaka, H. (2012) Discription and summary of evidence supporting stock structure hypotheses I and II for western North Pacific common minke whales. *Journal of Cetacean Research and Management* **13 (Suppl.)**: 435–439.

2807. Paterson, R. (1990) Effects of long term anti-shark measures on target and non-target species in Queensland, Australia. *Biological Conservation* **52**: 147–159.

2808. Patou, M.L., et al. (2009) Molecular phylogeny of the Herpestidae (Mammalia, Carnivora) with a special emphasis on the Asian *Herpestes*. *Molecular Phylogenetics and Evolution* **53**: 69–80.
2809. Patterson, I.A.P., et al. (1998) Evidence for infanticide in bottlenose dolphins: an explanation for violent interactions with harbour porpoises? *Proceedings of the Royal Society of London. Series B* **265**: 1167–1170.
2810. Pavelka, M.S. & Fedigan, L.M. (1999) Reproductive termination in female Japanese monkeys: A comparative life history perspective. *American Journal of Physical Anthropology* **109**: 455–464.
2811. Pawlinin, W.N. (1966) Der Zobel (*Martes zibellina* L.). Wittenberg: A. Ziemsen, 102 pp. (in German).
2812. Payne, R. (1986) Long term behavioral studies of the southern right whale (*Eubalaena australis*). *Reports of the International Whaling Commission (Special Issue)* **10**: 161–176.
2813. Pei, K.J.-C. (2009) The present status of the re-introduced sika seer in Kenting National Park, Southern Taiwan. *in* Sika Deer—Biology and Management of Native and Introduced Populations (McCullough, D.R., Takatsuki, S., & Kaji, K., Editors). Tokyo: Springer, pp. 561–570.
2814. Peng, M.-T., et al. (1973) Reproductive parameters of the Taiwan monkey (*Macaca cyclopis*). *Primates* **14**: 201–213.
2815. Peng, Y.-J., et al. (2009) Description of a new record species of whales from Chinese coastal waters. *Journal of Marine Sciences* **27**: 117–120.
2816. Penry, G.S. (2010) The Biology of South African Bryde's Whales. Doctoral dissertation. United Kingdom: University of St. Andrews. 173 pp.
2817. Perkins, S.E., Cattadori, I., & Hudson, P.J. (2005) The role of mammals in emerging zoonoses. *Japanese Journal of Veterinary Parasitology* **30**: S67–S71.
2818. Perlov, A.S. (1971) The onset of sexual maturity in sea lions. *All Union Institute of Marine Fisheries and Oceanography (VNIRO)* **82**: 174–189 (in Russian with English summary).
2819. Perrin, W.F. (1998) *Stenella longirostris*. *Mammalian Species* **599**: 1–7.
2820. Perrin, W.F. (2009) Common dolphins (*Delphinus delphis* and *D. capensis*). *in* Encyclopedia of Marine Mammals. 2nd ed. (Perrin, W.F., Würsig, B., & Thewissen, J.G.M., Editors). Amsterdam: Academic Press, pp. 255–259.
2821. Perrin, W.F. (2002) Pantropical spotted dolphin (*Stenella attenuata*). *in* Encyclopedia of Marine Mammals (Perrin, W.F., Würsig, B., & Thewissen, J.G.M., Editors). San Diego: Academic Press, pp. 865–867.
2822. Perrin, W.F. (2002) Spinner Dolphin (*Stenella longirostris*). *in* Encyclopedia of Marine Mammals (Perrin, W.F., Würsig, B., & Thewissen, J.G.M., Editors). San Diego: Academic Press, pp. 1174–1178.
2823. Perrin, W.F. & Brownell, R.L. (2002) Minke whales (*Balaenoptera acutorostrata* and *B. bonaerensis*). *in* Encyclopedia of Marine Mammals (Perrin, W.F., Würsig, B., & Thewissen, J.G.M., Editors). San Diego: Academic Press, pp. 750–754.
2824. Perrin, W.F., Dolar, M.L., & Ortega, E. (1996) Osteological comparison of Bryde's whales from the Philippines with specimens from other regions. *Report of the International Whaling Commission* **46**: 409–413.
2825. Perrin, W.F., et al. (2003) Cranial sexual dimorphism and geographic variation in Fraser's dolphin, *Lagenodelphis hosei*. *Marine Mammal Science* **19**: 484–501.
2826. Perrin, W.F., Dolar, M.L.L., & Robineau, D. (1999) Spinner dolphins (*Stenella longirostris*) of the western Pacific and southeast Asia: pelagic and shallow-water forms. *Marine Mammal Science* **15**: 1029–1053.
2827. Perrin, W.F. & Gilpatrick, J.W., Jr. (1994) Spinner Dolphin. *in* Handbook of Marine Mammals. Volume 5. The First Book of Dolphins (Ridgway, S.H. & Harrison, R., Editors). London: Academic Press, pp. 99–128.
2828. Perrin, W.F. & Hohn, A.A. (1994) Pantropical spotted dolphin *Stenella attenuata*. *in* Handbook of Marine Mammals. Volume 5. The First Book of Dolphins (Ridgway, S.H. & Harrison, R., Editors). London: Academic Press, pp. 71–98.
2829. Perrin, W.F., Leatherwood, S., & Collet, A. (1994) Fraser's dolphin *Lagenodelphis hosei* Fraser, 1956. *in* Handbook of Marine Mammals, Volume 5: The First Book of Dolphins (Ridgway, S.H. & Harrison, R., Editors). London: Academic Press, pp. 225–240.
2830. Perrin, W.F., et al. (1981) *Stenella clymene*, a rediscovered tropical dolphins of the Atlantic. *Journal of Mammalogy* **62**: 583–598.
2831. Perrin, W.F., Miyazaki, N., & Kasuya, T. (1989) A dwarf form of the spinner dolphin (*Stenella longirostris*) from Thailand. *Marine Mammal Science* **5**: 213–227.
2832. Perrin, W.F. & Reilly, S.B. (1984) Reproductive parameters of dolphins and small whales of the family Delphinidae. *in* Reproduction in whales, dolphins and porpoises (Perrin, W.F., Brownell, R.L., Jr., & DeMaster, D.P., Editors). Cambridge: IWC, pp. 97–133.
2833. Perryman, W.L. (2009) Melon-headed whale (*Peponocephala electra*). *in* Encyclopedia of Marine Mammals. 2nd ed. (Perrin, W.F., Würsig, B., & Thewissen, J.G.M., Editors). Amsterdam: Academic Press, pp. 719–721.
2834. Perryman, W.L., et al. (1994) Melon-headed whale *Peponocephala electra* Gray, 1846. *in* Handbook of Marine Mammals, Volume 5: The First Book of Dolphins (Ridgway, S.H. & Harrison, R., Editors). London: Academic Press, pp. 363–386.
2835. Perryman, W.L. & Foster, T.C. (1980) Preliminary report of predation by small whales, mainly the false killer whales, *Pseudorca crassidens*, on dolphins (*Stenella* spp. and *Delphinus delphis*) in the eastern tropical Pacific (NOAA National Marine Fisheries Service, Southwest Fisheries Center, Administrative Report LJ-80-05). 9 pp.
2836. Peters, W. (1880) Mittheilungen uber die von Hrn. Dr. F. Hilgendorf in Japan gesammelten Chiropteren. *Monatsberichte der K. Preussischen Akademie der Wissenschaften zu Berlin* **1880**: 3–25.
2837. Petersen, J.M. & Schriefer, M.E. (2005) Tularemia: Emergence/re-emergence. *Veterinary Research* **36**: 455–467.
2838. Peterson, R.S. (1968) Social behavior in pinnipeds with particular reference to the northern fur seal. *in* The Behavior and Physiology of Pinnipeds (Harrison, R.J., et al., Editors). New York: Appleton-Century-Crofts, pp. 3–53.
2839. Phillips, C.D., et al. (2009) Systematics of Steller sea lions (*Eumetopias jubatus*): Subspecies recognition based on concordance of genetics and morphometrics. *Occasional Papers: The Museum of Texas Tech University* **283**: 1–15.
2840. Phillips, C.D., et al. (2009) Assessing substitution patterns, rates and homoplasy at HVRI of Steller sea lions, *Eumetopias jubatus*. *Molecular Ecology* **18**: 3379–3393.
2841. Piaggio, A.J. & Spicer, G.S. (2001) Molecular phylogeny of the chipmunks inferred from mitochondrial cytochrome *b* and cytochrome oxidase II gene sequences. *Molecular Phylogenetics and Evolution* **20**: 335–350.
2842. Pilleri, G. & Chen, P. (1979) How the finless porpoise (*Neophocaena asiaeorientalis*) carries its calves on its back, and the function of the denticulate area of skin, as observed in the Changjiang River, China. *Investigations on Cetacea* **10**: 105–108.
2843. Pimlott, D.H., Shannon, J.A., & Kolenosky, G.B. (1969) The Ecology of the Timber Wolf in Algonquin Provincial Park. Toronto: Ontario Department of Lands and Forests, 92 pp.
2844. Pitcher, K.W. & Calkins, D.G. (1981) Reproductive biology of Steller sea lions in the Gulf of Alaska. *Journal of Mammalogy* **62**: 599–605.
2845. Pitman, R.L. (2002) Alive and whale. A missing cetacean resurfaces in the tropics. *Natural History* **2002**: 32–36.
2846. Pitman, R.L. & Ensor, P. (2003) Three forms of killer whales (*Orcinus orca*) in Antarctic waters. *Journal of Cetacean Research and Management* **5**: 131–139.
2847. Pitman, R.L., et al. (1999) Sightings and possible identity of a bottlenose whale in the tropical Indo-Pacific: *Indopacetus pacificus*? *Marine Mammal Science* **15**: 531–549.
2848. Pocock, R.I. (1932) The black and brown bears in Europe and Asia Part2. *Journal of the Bombay Natural History Society* **36**: 101–138.
2849. Pocock, R.I. (1935) The races of *Canis lupus*. *Proceedings of the Zoological Society of London* **105**: 647–686.
2850. Pocock, R.I. (1939) The fauna of British India, including Ceylon and Burma. Mammalia. Volume 1. Primates and Carnivora (in part). London: Taylor and Francis, 463 pp.
2851. Poirier, F.E. (1982) Taiwan macaques: ecology and conservation needs. *in* Advanced View of Primate Biology (Chiarelli, A.B. & Corruccini, R.S., Editors). Berlin: Springer, pp. 138–142.
2852. Poirier, F.E. & Davidson, D.M. (1979) A preliminary study of the Taiwan macaque (*Macaca cyclopis*). *Quarterly Journal of the Taiwan Museum* **32**: 123–191.
2853. Pollard, E. & Cooke, A.S. (1994) Impact of muntjac deer *Muntiacus reevesi* on egg-laying site of the white admiral butterfly *Ladoga camilla* in a Cambridgeshire wood. *Biological Conservation* **70**: 189–191.
2854. Praca, E. & Gannier, A. (2008) Ecological niches of three teuthophageous odontocetes in the northwestern Mediterranean Sea. *Ocean Science* **4**: 49–59.
2855. Prater, S.H. (1971) The Book of Indian Animals. 3rd ed. Bombay: Bombay Natural History Society, 324 pp.
2856. Preen, A. (2001) Dugongs, boats, dolphins and turtles in the Townsville–Cardwell region and recommendations for a boat-traffic management plan for the Hinchinbrook Dugong Protection Area (Great Barrier Reef Marine Park Authority Research Publication 67). Great Barrier Reef Marine Park Authority Research Publication 67. Townsville: Great Barrier Reef Marine Park Authority, 88 pp.
2857. Preen, A. & Marsh, H. (1995) Response of dugongs to large-scale loss of seagrass from Hervey Bay, Queensland, Australia. *Wildlife Research* **22**: 507–519.
2858. Purves, P.E. & Pilleri, G. (1978) The functional anatomy and general biol-

ogy of *Pseudorca crassidens* (Owen) with a review of the hydrodynamics and acoustics in Cetacea. *Investigations on Cetacea* **9**: 67–227.

2859. Quiñones, R., et al. (2013) Intestinal helminth fauna of bottlenose dolphin *Tursiops truncatus* and common dolphin *Delphinus delphis* from the western Mediterranean. *Journal of Parasitology* **99**: 576–579.

2860. Rajaguru, A. & Shantha, G. (1992) Association between the sessile barnacle *Xenobalanus globicipitis* (Coronulidae) and the bottlenose dolphin *Tursiops truncatus* (Delphinidae) from the Bay of Bengal, India, with a summary of previous records from cetaceans. *Fishery Bulletin* **90**: 197–202.

2861. Ralls, K. & Harvey, P.H. (1985) Geographic variation in size and sexual dimorphism of North American weasels. *Biological Journal of the Linnean Society* **25**: 119–167.

2862. Randi, E. (1995) Conservation genetics of the genus *Sus*. *Journal of Mountain Ecology* **3**: 6–12.

2863. Ratajszczak, R., Adler, J., & Smielowski, J. (1993) The Vietnamese sika *Cervus nippon pseudaxis* conservation project. *International Zoo Yearbook* **32**: 56–60.

2864. Raum-Suryan, K.L. & Harvey, J.T. (1998) Distribution and abundance of and habitat use by harbor porpoise, *Phocoena phocoena*, off the northern San Juan Islands, Washington. *Fishery Bulletin* **96**: 808–822.

2865. Rausch, V.R. & Rausch, R.L. (1979) Karyotype of the Red Fox, *Vulpes vulpes* L., in Alaska. *Northwest Science* **53**: 54–57.

2866. Rausch, V.R. & Rausch, R.L. (1982) The karyotype of the Eurasian flying squirrel, *Pteromys volans* (L.), with a consideration of karyotypic and other distinctions in *Glaucomys* spp. (Rodentia: Sciuridae). *Proceedings of Biological Society of Washington* **95**: 58–66.

2867. Raven, H.C. (1942) On the structure of *Mesoplodon densirostris*, a rare beaked whale. *Bulletin American Museum of Natural History* **80**: 23–50.

2868. Read, A.J. (1999) Harbour porpoise *Phocoena phocoena* (Linnaeus, 1758). *in* Handbook of Marine Mammals. Volume 6. The Second Book of Dolphins and the Porpoises (Ridgway, S.H. & Harrison, R., Editors). London: Academic Press, pp. 323–355.

2869. Read, A.J. & Hohn, A.A. (1995) Life in the fast lane: The life history of harbour porpoises from the Gulf of Maine. *Marine Mammal Science* **11**: 423–440.

2870. Read, A.J. & Westgate, A.J. (1997) Monitoring the movements of harbour porpoises (*Phocoena phocoena*) with satellite telemetry. *Marine Biology* **130**: 315–322.

2871. Ream, R.R., Sterling, J.T., & Loughlin, T.R. (2005) Oceanographic features related to northern fur seal migratory movements. *Deep-Sea Research II* **52**: 823–843.

2872. Reeves, R.R., et al. (2002) National Audubon Society Guide to Marine Mammals of the World. New York: Knopf, 528 pp.

2873. Reeves, R.R., et al. (2002) Pantropical spotted dolphin. *in* Guide to Marine Mammals of the World. New York: Alfred A. Knoph, pp. 366–369.

2874. Reeves, R.R., et al. (2002) Rough-toothed dolphin. *in* Guide to Marine Mammals of the World. New York: Alfred A. Knoph, pp. 346–349.

2875. Reeves, R.R., et al. (2002) Spinner dolphin. *in* Guide to Marine Mammals of the World. New York: Alfred A. Knoph, pp. 374–377.

2876. Reilly, S.B. & Fiedler, P.C. (1994) Interannual variability of dolphin habitats in the eastern tropical Pacific. I: research vessel surveys, 1986–1990. *Fishery Bulletin* **92**: 434–450.

2877. Rendell, L.E. & Whitehead, H. (2003) Vocal clans in sperm whales (*Physeter macrocephalus*). *Proceedings of the Royal Society of London. Series B* **270**: 225–231.

2878. Repenning, C.A. (1967) Subfamilies and genera of the Soricidae. *Geological Survey Professional Paper* **565**: 1–74.

2879. Resendes, A.R., et al. (2002) Disseminated toxoplasmosis in a Mediterranean pregnant Risso's dolphin (*Grampus griseus*) with transplacental fetal infection. *Journal of Parasitology* **88**: 1029–1032.

2880. Rice, D.W. (1978) The humpback whale in the North Pacific: distribution, exploitation, and numbers. *in* Report on a Workshop on Problems Related to Humpback Whales (*Megaptera novaeangliae*) in Hawaii (Norris, K.S. & Reeves, R., Editors). Washington, D.C.: U.S. Marine Mammal Commission, pp. 29–44.

2881. Rice, D.W. (1989) Sperm whale. *in* Handbook of Marine Mammals. Volume 4. River Dolphins and the Larger Toothed Whales (Ridgway, S.H. & Harrison, R., Editors). London: Academic Press, pp. 177–233.

2882. Rice, D.W. (1998) Marine Mammals of the World: Systematics and Distribution. Special Publication Number 4, the Society for Marine Mammalogy. Lawrence: Society for Marine Mammalogy, 231 pp.

2883. Rice, D.W. & Wolman, A.A. (1971) The Life History and Ecology of the Gray Whale (*Eschrichtius robustus*). Special Publication Number 3 of the American Society of Mammalogists. Lawrence: American Society of Mammalogists, 142 pp.

2884. Riedman, M. (1990) The Pinnipeds: Seals, Sea Lions, and Walruses. Berkeley: University of California Press, 439 pp.

2885. Riedman, M.L. & Estes, J.A. (1988) Predation on seabirds by sea otters. *Canadian Journal of Zoology* **66**: 1396–1402.

2886. Riedman, M.L. & Estes, J.A. (1990) The Sea Otter (*Enhydra lutris*): Behavior, Ecology and Natural History. U.S. Fish and Wildlife Service, Biological Report 90(14). Washington, D.C.: U.S. Department of the Interior, Fish and Wildlife Service, 126 pp.

2887. Ringelstein, J., et al. (2006) Food and feeding ecology of the striped dolphin, *Stenella coeruleoalba*, in the oceanic waters of the north-east Atlantic. *Journal of the Marine Biological Association of the United Kingdom* **86**: 909–918.

2888. Robbins, C.T. (1983) Wildlife Feeding and Nutrition. Animal feeding and nutrition. Vol. 1 (Cunha, T.J., Editor). Orlando: Academic Press, 343 pp.

2889. Robeck, T.R., et al. (2003) Artificial insemination using frozen-thawed semen in the Pacific white-sided dolphin (*Lagenorhynchus obliquidens*). *in* Proceedings for the 34th Annual Conference of the International Association for Aquatic Animal Medicine, May 9–14, 2003, Waikoloa, Hawaii, pp. 50–52.

2890. Robeck, T.R., et al. (2004) Reproductive physiology and development of artificial insemination technology in killer whales (*Orcinus orca*). *Biology of Reproduction* **71**: 650–660.

2891. Robeck, T.R., et al. (2005) Estrous cycle characterization and artificial insemination using frozen-thawed spermatozoa in the bottlenose dolphin (*Tursiops truncatus*). *Reproduction* **129**: 659–674.

2892. Roberts, T.J. (1977) The Mammals of Pakistan. Oxford: Oxford University Press, 525 pp.

2893. Robertson, K.M. & Chivers, S. (1997) Prey occurrence in pantropical spotted dolphins, *Stenella attenuata*, from the eastern North Pacific. *Fishery Bulletin* **95**: 334–348.

2894. Robinson, B.H. & Craddock, J.E. (1983) Mesopelagic fishes eaten by Fraser's dolphin, *Lagenodelphis hosei*. *Fishery Bulletin* **81**: 283–289.

2895. Rogers, P.M., Arthur, C.P., & Soriguer, R.C. (1994) The rabbit in continental Europe. *in* The European Rabbit: The History and Biology of a Successful Colonizer (Thompson, H.V. & King, C.M., Editors). Oxford: Oxford University Press, pp. 22–63.

2896. Rommel, S.A., et al. (2006) Elements of beaked whale anatomy and diving physiology and some hypothetical causes of sonar-related stranding. *Journal of Cetacean Research and Management* **7**: 189–209.

2897. Roonwal, M.L. & Mohnot, S.M. (1977) Primates of South Asia: Ecology, Sociobiology, and Behavior. Cambridge: Harvard University Press, 421 pp.

2898. Rosel, P.E., Dizon, A.E., & Haygood, M.G. (1995) Variability of the mitochondrial control region in populations of the harbour porpoise, *Phocoena phocoena*, on interoceanic and regional scales. *Canadian Journal of Fisheries and Aquatic Sciences* **52**: 1210–1219.

2899. Rosel, P.E., Dizon, A.E., & Heyning, J.E. (1994) Genetic analysis of sympatric morphotypes of common dolphins (genus *Delphinus*). *Marine Biology* **119**: 159–167.

2900. Rosel, P.E., et al. (1999) Genetic structure of harbour porpoise *Phocoena phocoena* populations in the northwest Atlantic based on mitochondrial and nuclear markers. *Molecular Ecology* **8**: S41–S54.

2901. Rosel, P.E. & Wilcox, L.A. (2014) Genetic evidence reveals a unique lineage of Bryde's whales in the northern Gulf of Mexico. *Endangered Species Research* **25**: 19–34.

2902. Roslik, G.V., Kartavtseva, I.V., & Iwasa, M. (2003) Variability and stability of B chromosome number in the Korean field mouse, *Apodemus peninsulae* (Rodentia, Muridae) from continental and insular populations. *in* Problems of Evolution 5. Vladivostok: Dalnauka, pp. 136–149 (in Russian).

2903. Ross, G.J.B. (1979) Records of pygmy and dwarf sperm whales, genus *Kogia*, from southern Africa, with biological notes and some comparisons. *Annals of the Cape Provincial Museums (Natural History)* **15**: 259–327.

2904. Ross, G.J.B. (1984) The smaller cetaceans of the south east coast of southern Africa. *Annals of the Cape Provincial Museums. Natural History* **15**: 173–410.

2905. Ross, G.J.B. & Cockcroft, V.G. (1990) Comments on Australian bottlenose dolphins and taxonomic status of *Tursiops aduncus* (Ehrenberg, 1832). *in* The Bottlenose Dolphin (Leatherwood, S. & Reeves, R.R., Editors). San Diego: Academic Press, pp. 101–128.

2906. Ross, G.J.B. & Leatherwood, S. (1994) Pygmy killer whale *Feresa attenuata* Gray, 1874. *in* Handbook of Marine Mammals, Volume 5: The First Book of Dolphins (Ridgway, S.H. & Harrison, R., Editors). London: Academic Press, pp. 387–404.

2907. Ross, H.M. & Wilson, B. (1986) Violent interactions between bottlenose dolphins and harbour porpoise. *Proceedings of the Royal Society of London. Series B, Biological Sciences* **263**: 283–286.

2908. Rossiter, S.J., et al. (2007) Rangewide phylogeography in the greater horseshoe bat inferred from microsatellites: Implications for population history,

taxonomy and conservation. *Molecular Ecology* **16**: 4699–4717.
2909. Rowe, K.C., Heske, E.J., & Paige, K.N. (2006) Comparative phylogeography of eastern chipmunks and white-footed mice in relation to the individualistic nature of species. *Molecular Ecology* **15**: 4003–4020.
2910. Rowe, N. (1996) The Pictorial Guide to the Living Primates. New York: Pogonias Press, 263 pp.
2911. Royama, T. (1992) Analytical Population Dynamics. London: Chapman and Hall, 392 pp.
2912. Russo, L., Genov, P., & Massei, G. (1995) Preliminary data of activity patterns of wild boar (*Sus scrofa*) in the Maremma national park (Italy). *Journal of Mountain Ecology* **3**: 126–127.
2913. Rydell, J. (1993) *Eptesicus nilssonii*. *Mammalian Species* **430**: 1–7.
2914. Ryu, S.H., Kwak, M.J., & Hwang, U.W. (2013) Complete mitochondrial genome of the Eurasian flying squirrel *Pteromys volans* (Sciuromorpha, Sciuridae) and revision of rodent phylogeny. *Molecular Biological Reports* **40**: 1917–1926.
2915. Rzebik-Kowalska, B. (1998) Fossil history of shrews in Europe. *in* Evolution of Shrews (Wójcik, J.M. & Wolsan, M., Editors). Białowieża: Mammalian Research Institute, Polish Academy of Sciences, pp. 23–92.
2916. Rzebik-Kowalska, B. & Hasegawa, Y. (1976) New materials to the knowledge of the genus *Shikamainosorex* Hasegawa 1957 (Insectivora, Mammalia). *Acta Zoologica Cracoviensia* **2**: 341–35.
2917. Saeki, M. (佐伯緑) (2008) 里山の動物の生態―ホンドタヌキ (Ecology of wildlife living in satoyama: the raccoon dog). *in* Mammalogy in Japan. 2. Middle-, and Large-Sized Mammals Including Primates (日本の哺乳類学 第2巻 中大型哺乳類・霊長類) (Takatsuki, S. (高槻成紀) & Yamagiwa, J. (山極寿一), Editors). Tokyo: University of Tokyo Press, pp. 321–345 (in Japanese).
2918. Saeki, M. (佐伯緑) & Takeuchi, M. (竹内正彦) (2008) タヌキによる農作物被害の現状とその対策 (1)―タヌキの生態と農作物被害の現状―. *Agriculture and Horticulture* **83**: 657–665 (in Japanese).
2919. Saeki, M. Unpublished.
2920. Saeki, M. (2001) Ecology and Conservation of the Raccoon dog (*Nyctereutes procyonoides*) in Japan. Doctoral dissertation. Oxford: University of Oxford. 294 pp.
2921. Saeki, M., Johnson, P., & Macdonald, D.W. (2007) Movements and habitat selection of raccoon dogs (*Nyctereutes procyonoides*) in a mosaic landscape. *Journal of Mammalogy* **88**: 1098–1111.
2922. Saeki, M. & Macdonald, D.W. (2004) The Effects of traffic on the raccoon dog (*Nyctereutes procyonoides viverrinus*) and other mammals in Japan. *Biological Conservation* **118**: 559–571.
2923. Safronov, V.M. & Anikin, R.K. (2000) Ecology of the sable, *Martes zibellina* (Carnivora, Mustelidae), in northeastern Yakutia. *Zoologicheskii Zhurnal* **79**: 471–479 (in Russian).
2924. Sagara, N. Personal communication.
2925. Sagara, N. (1995) Association of ectomycorrhizal fungi with decomposed animal wastes in forest habitats: a cleaning symbiosis? *Canadian Journal of Botany* **73**, Suppl. 1: S1423–S1433.
2926. Sagara, N. & Abe, H. (1993) A case of late breeding in the mole *Mogera kobeae* and its nest. *Journal of the Mammalogical Society of Japan* **18**: 53–59.
2927. Sagara, N., Abe, H., & Okabe, H. (1993) The persistence of moles in nesting at the same site as indicated by mushroom fruiting and nest reconstruction. *Canadian Journal of Zoology* **71**: 1690–1693.
2928. Sagara, N., et al. (1989) Finding *Euroscaptor mizura* (Mammalia: Insectivora) and its nest from under *Hebeloma radicosum* (Fungi: Agaricales) in Ashiu, Kyoto, with data of possible contiguous occurrences of three talpine species in this region. *Contributions from the Biological Laboratory, Kyoto University* **27**: 261–272.
2929. Saheki, M. (1966) Morphological Studies of *Macaca fuscata*. IV. Dentition. *Primates* **7**: 407–422.
2930. Saheki, M., Hayama, S., & Tashiro, K. (1961) Morphological Studies on dentition of the macaque. I. Morphology of molars. *Journal of Stomatological Society, Japan* **28**: 81–100 (in Japanese).
2931. Saitama Museum of Natural History (1987) 哺乳類 (1). *in* 埼玉県立自然史博物館収蔵資料目録 第1集. Nagatoro: Saitama Museum of Natural History, pp. 1–7 (in Japanese).
2932. Saitama Museum of Natural History, ed. (1990) 埼玉の希少動物：天然記念物基礎調査報告書. Urawa: Saitama Prefectural Boards of Education (埼玉県教育委員会), 42 pp. (in Japanese).
2933. Saito, M. (斉藤正寛) (1987) 江戸川河口域のマスクラット. *日本の生物* **1**: 60–61 (in Japanese).
2934. Saito, S. (齋藤幸子) (2002) 漁業被害とアザラシの混獲：厚岸地域. *in* アザラシ類の保護管理報告書. Sapporo: Hokkaido Government, pp. 141–147 (in Japanese).
2935. Saito, S. (齊藤幸子) & Watanabe, Y. (渡邊有希子) (2004) ゼニガタアザラシの概要と問題点. *in* Management of Marine Mammals along the Coast of Hokkaido, Japan (北海道の海生哺乳類管理：シンポジウム「人と獣の生きる海」報告書) (Kobayashi, M. (小林万里), Isono, T. (磯野岳臣), & Hattori, K. (服部薫), Editors). Sapporo: Marine Wildlife Center of Japan (北の海の動物センター), pp. 23–28 (in Japanese).
2936. Saito, A. (2003) Babesiosis. *Japanese Journal of Veterinary Parasitology* **2**: 21–26 (in Japanese).
2937. Saito, C., et al. (1998) Aggressive intergroup encounters in two populations of Japanese macaques (*Macaca fuscata*). *Primates* **39**: 303–312.
2938. Saito, S. & Yamaguchi, T. (1985) *Trichinella spiralis* in a raccoon dog *Nyctereutes procyonoides viverrinus* from Yamagata Prefecture, Honshu, Japan. *Japanese Journal of Parasitology* **34**: 311–314.
2939. Saito, T., Douzaki, M., & Soichi, M. (2014) Research on finless porpoises in the port of Nagoya. *Aquabiology* **36**: 29–35 (in Japanese with English abstract).
2940. Saito, Y. (1955) On five species of Ectoparasites on bats (*Pipistrellus abramus*) collected in Niigata City, Japan. *Niigata Medical Journal* **69**: 7–12.
2941. Saito, Y., Watanabe, T., & Yamashita, T. (1982) Two new intestinal trematodes, *Macroorchis chimarrogalus* n. sp. and *Macroorchis elongatus* n. sp. from Japanese water shrew and others, with a description of *Macroorchis chimarrogalus* n. sp. (Trematoda: Nanophyetidae). *Acta Medica et Biologica* **30**: 47–56.
2942. Saitoh, T. (齊藤隆) (1985) 野生化ミンクの生息調査報告 (その1)―野生化の過程と食性―. *Journal of Fur Research* **2**: 7–17 (in Japanese).
2943. Saitoh, T. (齊藤隆), Kawaji, N. (川路則友), & Kudo, T. (工藤琢磨) (2000) ドングリを持ち去るのは誰か？―「分散貯蔵」されたミズナラ堅果の消失. *Northern Forestry, Japan* **52**: 215–218 (in Japanese).
2944. Saitoh, H., Kazama, K., & Hino, T. (2013) Environmental factors affecting capture rate of the Japanese water shrew, *Chimarrogale platycephala*. *Mammalian Science* **53**: 117–121 (in Japanese with English abstract).
2945. Saitoh, M. & Obara, Y. (1986) Chromosome banding patterns in five intraspecific taxa of the large Japanese field mouse, *Apodemus speciosus*. *Zoological Science* **3**: 785–792.
2946. Saitoh, M. & Obara, Y. (1988) Meiotic studies of interracial hybrids from the wild population of the large Japanese field mouse, *Apodemus speciosus speciosus*. *Zoological Science* **5**: 815–822.
2947. Saitoh, T. (1981) Control of female maturation in high density populations of the red-backed vole, *Clethrionomys rufocanus bedfordiae*. *Journal of Animal Ecology* **50**: 79–87.
2948. Saitoh, T. (1985) Practical definition of territory and its application to the spatial distribution of voles. *Journal of Ethology* **3**: 143–149.
2949. Saitoh, T. (1989) Communal nesting and spatial structure in an early spring population of the grey red-backed vole, *Clethrionomys rufocanus bedfordiae*. *Journal of the Mammalogical Society of Japan* **14**: 27–41.
2950. Saitoh, T. (1990) Lifetime reproductive success in reproductively suppressed female voles. *Researches on Population Ecology* **32**: 391–406.
2951. Saitoh, T. (1991) The effects and limits of territoriality on population regulation in grey red-backed voles, *Clethrionomys rufocanus bedfordiae*. *Researches on Population Ecology* **33**: 367–386.
2952. Saitoh, T. (1995) Sexual differences in natal dispersal and philopatry of the grey-sided vole. *Researches on Population Ecology* **37**: 49–57.
2953. Saitoh, T., Bjørnstad, O.N., & Stenseth, N.C. (1999) Density-dependence in voles and mice: a comparative study. *Ecology* **80**: 638–650.
2954. Saitoh, T., et al. (2008) Taxonomic, genetic and ecological status of the Daikoku vole. *in* Proceedings of International Symposium "The Origin and Evolution of Natural Diversity", 1–5 October 2007, Sapporo. Sapporo pp. 145–150.
2955. Saitoh, T., et al. (2001) Genetic status of fragmented populations of the Asian black bear *Ursus thibetanus* in western Japan. *Population Ecology* **43**: 221–227.
2956. Saitoh, T., et al. (2007) Effects of acorn masting on population dynamics of three forest-dwelling rodent species in Hokkaido, Japan. *Population Ecology* **49**: 249–256.
2957. Saitoh, T., Stenseth, N.C., & Bjørnstad, O.N. (1997) Density dependence in fluctuating grey-sided vole populations. *Journal of Animal Ecology* **66**: 14–24.
2958. Saitoh, T., Stenseth, N.C., & Bjørnstad, O.N. (1998) The population dynamics of the vole *Clethrionomys rufocanus* in Hokkaido, Japan. *Researches on Population Ecology* **40**: 61–76.
2959. Saitoh, T. & Takahashi, K. (1998) The role of vole populations in prevalence of the parasites (*Echinococcus multilocularis*) in foxes. *Researches on Population Ecology* **40**: 97–105.
2960. Saitoh, T., et al. (2008) Effects of acorn abundance on density dependence in a Japanese wood mouse (*Apodemus speciosus*) population. *Population Ecology* **50**: 159–167.

2961. Sakagami, S., Mori, H., & Kikuchi, H. (1956) Miscellaneous observations on a pika, *Ochotona* sp. inhabiting near the lake Shikaribetsu, Taisetsuzan National Park, Hokkaido. *Japanese Journal of Applied Zoology* **21**: 1–9.

2962. Sakaguchi, N. (阪口法明) & Toyama, M. (当山昌直) (1992) 北大東島におけるダイトウオオコウモリの行動圏及び場所利用 (Home range and habitat utilization of the Daito flying fox, *Pteropus dasymallus daitoensis*, in Kita-Daitojima Island, Ryukyu Archipelago). in 沖縄県天然記念物調査シリーズ第31集 ダイトウオオコウモリ: 保護対策緊急調査報告書 (Daito-Okomori (Daito Flying Fox), Survey Reports on Natural Monuments of Okinawa Prefecture, No.31). Naha: Okinawa Prefectural Board of Education (沖縄県教育委員会), pp. 105–141 (in Japanese).

2963. Sakaguchi, K. (1962) A Monograph of the Siphonoptera of Japan. Osaka: Nippon Printing and Publishing, 255 pp. (in Japanese).

2964. Sakaguchi, N. & Ono, Y. (1994) Seasonal change in the food habits of the Iriomote cat *Felis iriomotensis*. *Ecological Research* **9**: 167–174.

2965. Sakaguchi, S., et al. (2004) Reemerging murine typhus, Japan. *Emerging Infectious Diseases* **10**: 963–964.

2966. Sakaguti, K. (1957) A species of bat flea, *Nycteridopsylla galba* Dampf, 1910 unreported in Japan, with description of the female sex. *Japanese Journal of Sanitary Zoology* **8**: 171–173 (in Japanese with English abstract).

2967. Sakaguti, K. (1962) A Monograph of the Shiphonaptera of Japan. Osaka: Nippon Printing and Publishing Co. Ltd., 255 pp.

2968. Sakaguti, K. & Jameson, E.W., Jr. (1962) The Siphonaptera of Japan. *Pacific Insects Monograph* **3**: 1–169.

2969. Sakahira, F. & Niimi, M. (2007) Ancient DNA analysis of the Japanese sea lion (*Zalophus californianus japonics* Peters, 1866): Preliminary results using mitochondrial control-region sequences. *Zoological Science* **24**: 81–85.

2970. Sakai, M., et al. (2006) Flipper rubbing behaviors in wild bottlenose dolphins (*Tursiops aduncus*). *Marine Mammal Science* **22**: 966–978.

2971. Sakai, M., et al. (2010) Do porpoises choose their associates? A new method for analyzing social relationships among cetaceans. *PLoS ONE* **6**: e28836.

2972. Sakai, T. (2014) Stranded and by-catch finless porpoises in Ibaraki. *Aquabiology* **36**: 14–21 (in Japanese with English abstract).

2973. Sakai, T. & Hanamura, H. (1969) A morphological study on the dentition of Insectivora I. Soricidae. *Aichi-Gakuin Journal of Dental Science* **7**: 1–26 (in Japanese with English abstract).

2974. Sakai, T., et al. (2003) Molecular phylogeny of Japanese Rhinolophidae based on variations in the complete sequence of the mitochondrial cytochrome b gene. *Genes & Genetic Systems* **78**: 179–189.

2975. Sakai, Y., et al. (2013) Rearing method to induce natural mating of the large Japanese field mouse, *Apodemus speciosus*. *Mammalian Science* **53**: 57–65 (in Japanese with English abstract).

2976. Sakamoto, S.H., et al. (2012) Seasonal habitat partitioning between sympatric terrestrial and semi-arboreal Japanese wood mice, *Apodemus speciosus* and *A. argenteus* in spatially heterogeneous environment. *Mammal Study* **37**: 261–272.

2977. Sakata, T. (坂田拓司), et al. (1991) 牛深市大島における野生化したカイウサギの生態研究. *Bulletin of the Kumamoto Wildlife Society* **1**: 27–33 (in Japanese).

2978. Sakata, K. & Asakawa, M. (2003) Parasitic nematodes of Sado moles (*Mogera tokudae*) with the first geographical record of *Tricholinstowia talpae* (Morgan, 1928) from Sado I. and a brief description of the species. *Journal of Rakuno Gakuen University* **27**: 211–214 (in Japanese with English summary).

2979. Sakata, K., et al. (2006) The first report of parasitic nematodes obtained from the large Japanese field mouse, *Apodemus speciosus* (Muridae, Rodentia), collected on the Izu Islands, Japan, with a brief zoogeographical comment for its nematode fauna. *Bulletin of the Biogeographical Society of Japan* **61**: 135–139 (in Japanese with English abstract).

2980. Sako, T., Uchimura, M., & Koreeda, Y. (1991) Keeping and breeding of the Amami rabbit in the Kagoshima Hirakawa Zoo. *Animals and Zoos* **43**: 272–274 (in Japanese).

2981. Sakuragi, M., et al. (2003) Seasonal habitat selection of an expanding sika deer *Cervus nippon* population in eastern Hokkaido, Japan. *Wildlife Biology* **9**: 141–153.

2982. Sakuragi, M., et al. (2004) Female sika deer fidelity to migration route and seasonal ranges in eastern Hokkaido, Japan. *Mammal Study* **29**: 113–118.

2983. Sakurai, Y. (桜井泰憲), et al. (2004) なぜトドは越冬来遊するのか (Why do Steller sea lions come and winter around Hokkaido?). in Management of Marine Mammals along the Coast of Hokkaido, Japan (北海道の海生哺乳類管理：シンポジウム「人と獣の生きる海」報告書) (Kobayashi, M. (小林万里), Isono, T. (磯野岳臣), & Hattori, K. (服部薫), Editors). Sapporo: Marine Wildlife Center of Japan (北の海の動物センター), pp. 69–87 (in Japanese).

2984. Sakurai, M. (1981) Socio-ecological study of the Japanese serow *Capricornis crispus* (Temminck) (Mammalia; Bovidae) with special reference to the flexibility of its social structure. *Physiology and Ecology Japan* **18**: 163–212.

2985. Sakuyama, M., Gotoh, J., & Mukohyama, M. (2007) The distribution of nursery colonies of *Vespertilio superans* in inland Iwate. *Study Report on Bat Conservation in Tohoku Region* **1**: 14–19 (in Japanese).

2986. Sakuyama, M. & Shiraishi, A. (2007) Records of *Nyctalus aviator* flying in the daytime in Morioka City, Iwate. *Study Report on Bat Conservation in Tohoku Region* **1**: 25–27 (in Japanese).

2987. Salgueiro, P., et al. (2007) Genetic divergence and phylogeography in the genus *Nyctalus* (Mammalia, Chiroptera): implications for population history of the insular bat *Nyctalus azoreum*. *Genetica* **130**: 169–181.

2988. Sanderson, G.C. (1987) Raccoon. in Wild Furbearer Management and Conservation in North America (Novak, M., et al., Editors). North Bay: Ontario Trappers Association, pp. 487–499.

2989. Sano, A. (佐野明) & Sano, J. (佐野順子) (1997) 平成8年度アブラコウモリ生息実態調査報告書. Tsu: Mie Prefectural Government, 23 pp. (in Japanese).

2990. Sano, A. Unpublished.

2991. Sano, A. (2000) Distribution of four cave-dwelling bat species in Ishikawa Prefecture, with references to utilization of roosts. *Mammalian Science* **40**: 167–173 (in Japanese with English abstract).

2992. Sano, A. (2000) Postnatal growth and development of thermoregulative ability in the Japanese greater horseshoe bat, *Rhinolophus ferrumequinum nippon*, related to maternal care. *Mammal Study* **25**: 1–15.

2993. Sano, A. (2000) Regulation of the creche size by intercolonial migrations in the Japanese greater horseshoe bat, *Rhinolophus ferrumequinum nippon*. *Mammal Study* **25**: 95–105.

2994. Sano, A. (2001) New record of long-eared bat, *Plecotus auritus* from Mie Prefecture. *Wild Animals of Kii Peninsula* **6**: 15 (in Japanese).

2995. Sano, A. (2001) A population study of the Japanese greater horseshoe bat, *Rhinolophus ferrumequinum*, in the Izumo mines, Ishikawa Prefecture, Japan. *Bulletin of the Mie Prefectural Science and Technology Promotion Center (Forestry)* **13**: 1–68 (in Japanese with English summary).

2996. Sano, A. (2003) A first record of the Natterer's bat, *Myotis nattereri*, from Mie Prefecture. *Wild Animals of Kii Peninsula* **7**: 20 (in Japanese).

2997. Sano, A. (2004) New distributional records for the greater tube-nosed bat, *Murina leucogaster*, in Mie Prefecture (II). *Mie Natural History* **8/9/10**: 19 (in Japanese).

2998. Sano, A. (2006) Impact of predation by a cave-dwelling bat, *Rhinolophus ferrumequinum*, on the diapausing population of a troglophilic moth, *Goniocraspidum pryeri*. *Ecological Research* **21**: 321–324.

2999. Sano, A. (2013) First record of the Asian parti-colored bat, *Vespertilio sinensis*, from Hokusei region, northern Mie Prefecture. *Bulletin of the Fujiwara-dake Nature Museum* **35**: 1–3 (in Japanese).

3000. Sano, A. (2014) First record of the birdlike noctule bats, *Nyctalus aviator*, and their day roost from Mie Prefecture, Japan. *Mie Natural History* **14**: 29–30 (in Japanese).

3001. Sano, A. & Akita, K. (2001) Burrowing behavior of the Japanese large-footed bat, *Myotis macrodactylus*. *Mammalian Science* **41**: 83–86 (in Japanese with English abstract).

3002. Sano, A. & Ohnishi, K. (2006) First record of the Namie's frosted bat, *Vespertilio superans*, from Mie Prefecture. *Wild Animals of Kii Peninsula* **8**: 1–2 (in Japanese).

3003. Sano, A. & Sano, J. (2000) New distributional records for the greater tube-nosed bat, *Murina leucogaster*, in Mie Prefecture. *Mie Natural History* **6**: 67–38 (in Japanese).

3004. Sano, A., et al. (2006) The first record for the day roosts of the Oriental free-tailed bat, *Tadarida insignis*, from Kii Peninsula. *Wild Animals of Kii Peninsula* **8**: 3–5 (in Japanese).

3005. Sano, A. & Ueuma, Y. (1981) Notes on bats in the Hakusan region. *Annual Report of the Hakusan Nature Conservation Center* **7**: 23–29 (in Japanese with English abstract).

3006. Santos, M.B., et al. (2006) Pygmy sperm whales *Kogia breviceps* in the Northeast Atlantic: New information on stomach contents and strandings. *Marine Mammal Science* **22**: 600–616.

3007. Sasa, M. (1965) Mites; An Introduction to Classification, Bionomics and Control Acarina. Tokyo: Tokyo University Press, 486 pp.

3008. Sasagawa, M. (1984) Growth of the skull and eruption sequences of permanent teeth in red fox, *Vulpes vulpes*. *Japanese Journal of Oral Biology* **26**: 1210–1227 (in Japanese with English summary).

3009. Sasaki, C., Sato, T., & Kozawa, Y. (2001) Apoptosis in regressive deciduous tooth germs of *Suncus murinus* evaluated by the TUNEL method and electron microscopy. *Archives of Oral Biology* **46**: 649–660.

3010. Sasaki, H. Unpublished.

3011. Sasaki, H. (1994) Ecological study of the Siberian weasel *Mustela sibirica coreana* related to habitat preference and spacing pattern. Doctoral dissertation. Fukuoka: Kyushu University. 92 pp.

3012. Sasaki, H. (1995) History of river otters in Japan. *in* Proceedings (Abstracts) of Korea–Japan Otter Symposium 1995. Kouchi & Masan, pp. 16–17.
3013. Sasaki, H. & Kawabata, M. (1994) Food habits of the raccoon dog *Nyctereutes procyonoides viverrinus* in a mountainous area of Japan. *Journal of the Mammalogical Society of Japan* **19**: 1–8.
3014. Sasaki, H., et al. (2014) Factors affecting the distribution of the Japanese weasel *Mustela itatsi* and the Siberian weasel *M. sibirica* in Japan. *Mammal Study* **39**: 133–139.
3015. Sasaki, H. & Ono, Y. (1994) Habitat use and selection of the Siberian weasel *Mustela sibirica coreana* during the non-mating season. *Journal of the Mammalogical Society of Japan* **19**: 21–32.
3016. Sasaki, M., Shimba, H., & Itoh, M. (1968) Note on the somatic chromosomes of two species of Asiatic squirrels. *Chromosome Information Service* **9**: 6–8.
3017. Sasaki, N., Kondo, N., & Serizawa, Y. (2006) Faunal survey of bats in around Kushiro Marsh, Hokkaido. *Bulletin of the Shibecha-cho Folk Museum* **18**: 99–115 (in Japanese with English abstract).
3018. Sasaki, N., et al. (2012) Bats in Minakami, Gunma Prefecture. *Bulletin of the Gunma Museum of Natural History* **16**: 131–144 (in Japanese with English abstract).
3019. Sasaki, S., et al. (2006) Chemical composition of Blainville's whale (*Mesoplodon densirostris*) milk. *Milk Science* **55**: 37–41.
3020. Sasaki, S. & Ito, K. (2003) The Japanese squirrel exhibit at Inokashira Park Zoo. *in* Workshop and Symposium for Japanese Squirrels at Inokashira, Tokyo, pp. 39–49 (in Japanese with English abstract).
3021. Sasaki, S., et al. (2004) Composition and characteristics of Ginkgo-toothed beaked whale (*Mesoplodon ginkgodens*) milk. *Milk Science* **53**: 15–18.
3022. Sasaki, T., et al. (2006) *Balaenoptera omurai* is a newly discovered baleen whale that represents an ancient evolutionary lineage. *Molecular Phylogenetics and Evolution* **41**: 40–52.
3023. Sato, A. (佐藤顕義) & Katsuta, S. (勝田節子) (2006) 天竜川上流域で越冬していたチチブコウモリとヒナコウモリ. *Bat Study and Conservation Report* **14**: 5–9 (in Japanese).
3024. Sato, A. (佐藤顕義) & Katsuta, S. (勝田節子) (2007) 天竜川水系で確認したテングコウモリ *Murina leucogaster* の繁殖と周年動態. *Bat Study and Conservation Report* **15**: 2–5 (in Japanese).
3025. Sato, A. (佐藤顕義), et al. (2014) 埼玉県で2例目となるコウモリトコジラミをヤマコウモリから採集. *寄せ蛾記* **154**: 50–52 (in Japanese).
3026. Sato, A. (佐藤顕義), et al. (2011) 静岡県伊東市城ケ崎海岸燕黒岩におけるオヒキコウモリの集団ねぐら. *Bat Study and Conser-vation Report* **23**: 2–4 (in Japanese).
3027. Sato, F. (佐藤文彦), et al. (1998) 小笠原諸島海域におけるザトウクジラの生態調査. *Kaiyo Monthly (月刊海洋)* **339**: 572–576 (in Japanese).
3028. Satô, M. (佐藤雅彦) (2012) 北海道におけるコウモリ寄生性トコジラミ. *Forest Protection* **328**: 28–31 (in Japanese).
3029. Sato, M. (佐藤美穂子) (2004) ヒナコウモリ. 北海道苫前郡羽幌町での初記録. *Bat Study and Conservation Report* **12**: 3–4 (in Japanese).
3030. Sato, Y. (佐藤喜和) (2013) 浦幌町おけるヒグマ駆除数の推移 II. *Bulletin of the Historical Museum of Urahoro* **13**: 15–23 (in Japanese).
3031. Sato, Y. (佐藤喜和), et al. (2004) 浦幌町におけるヒグマ捕獲・計測記録. *Bulletin of the Historical Museum of Urahoro* **4**: 17–19 (in Japanese).
3032. Sato, Y. (佐藤喜和), et al. (2015) 北海道東部におけるヒグマの分散. *Bulletin of the Historical Museum of Urahoro* **15**: 9–13 (in Japanese).
3033. Sato, A. Unpublished.
3034. Sato, A., Katsuta, S., & Yamamoto, T. (2010) The distribution and roost use of the Ussurian tube-nosed bat, *Murina ussuriensis*, in the Minami-Alps area. *Journal of the Japanese Wildlife Research Society* **35**: 33–41 (in Japanese with English summary).
3035. Sato, A., Katsuta, S., & Yamamoto, T. (2011) The distribution and roost use of the Ikkonikov's Myotis, *Myotis ikoonnikovi*, in the Minami-Alps area. *Journal of the Japanese Wildlife Research Society* **36**: 1–7 (in Japanese with English summary).
3036. Sato, A., et al. (2013) Winter ecology of the birdlike noctule (*Nyctalus aviator*) in Saitama Prefecture 1. Distribution along the Joetsu Shinkansen and seasonal migration. *Bulletin of the Saitama Museum of Natural History* **7**: 101–108 (in Japanese).
3037. Sato, H. (1978) The role of small mammals as predators of sewfly coccons in northern Honshu, Japan. *Bulletin of Iwate Prefectural Forestry Experimental Station* **2**: 1–26 (in Japanese with English summary).
3038. Sato, H. (2005) Zoonotic roundworm infection, with special reference to *Baylisascaris procyonis* larva migrans. *Modern Media* **51**: 177–186 (in Japanese).
3039. Sato, H., Furuoka, H., & Kamiya, H. (2002) First outbreak of *Baylisascaris procyonis* larva migrans in rabbits in Japan. *Parasitology International* **51**: 105–108.
3040. Sato, H., et al. (1999) Parasitological survey on wild carnivora in north-western Tohoku, Japan. *Journal of Veterinary Medical Science* **61**: 1023–1026.
3041. Sato, H., et al. (1999) Helminth fauna of carnivores distributed in north-western Tohoku, Japan, with special reference to *Mesocestoides paucitesticulus* and *Brachylaima tokudai*. *Journal of Veterinary Medical Science* **61**: 1339–1342.
3042. Sato, H., Kamiya, H., & Furuoka, H. (2003) Epidemiological aspects of the first outbreak of *Baylisascaris procyonis* larva migrans in rabbits in Japan. *Journal of Veterinary Medical Science* **65**: 453–457.
3043. Sato, H. & Suzuki, K. (2006) Gastrointestinal helminthes of feral raccoons (*Procyon lotor*) in Wakayama Prefecture, Japan. *Journal of Veterinary Medical Science* **68**: 311–318.
3044. Sato, H., Suzuki, K., & Aoki, M. (2006) Juvenile bird acanthocephalans recovered incidentally from raccoon dogs (*Nyctereutes procyonoides viverrinus*) on Yakushima Island, Japan. *Journal of Veterinary Medical Science* **68**: 689–692.
3045. Sato, H., Suzuki, K., & Aoki, M. (2007) Nematodes from raccoon dogs (*Nyctereutes procyonoides viverrinus*) introduced recently on Yakushima Island, Japan. *Journal of Veterinary Medical Science* **68**: 693–700.
3046. Sato, H., et al. (2006) *Paragonimus westermani* and some rare intestinal trematodes recovered from raccoon dogs (*Nyctereutes procyonoides viverrinus*) introduced recently on Yakushima Island, Japan. *Journal of Veterinary Medical Science* **68**: 681–687.
3047. Sato, H., et al. (2006) Identification and characterization of the threadworm, *Strongyloides procyonis*, from feral raccoons (*Procyon lotor*) in Japan. *Journal of Parasitology* **92**: 63–68.
3048. Sato, J.J., et al. (2011) Genetic diversity of the sable (*Martes zibellina*, Mustelidae) in Russian Far East and Hokkaido inferred from mitochondrial NADH dehydrogenase subunit 2 gene sequences. *Mamal Study* **36**: 209–222.
3049. Sato, J.J., et al. (2004) Molecular phylogeny of arctoids (Mammalia: Carnivora) with emphasis on phylogenetic and taxonomic positions of the ferret-badgers and skunks. *Zoological Science* **21**: 111–118.
3050. Sato, J.J., et al. (2003) Phylogenetic relationships and divergence times among mustelids (Mammalia: Carnivora) based on nucleotide sequences of the nuclear interphotoreceptor retinoid binding protein and mitochondrial cytochrome *b* genes. *Zoological Science* **20**: 243–264.
3051. Sato, J.J., et al. (2014) A few decades of habitat fragmentation has reduced population genetic diversity: a case study of landscape genetics of the large Japanese field mouse, *Apodemus speciosus*. *Mammal Study* **39**: 1–10.
3052. Sato, J.J. & Suzuki, H. (2004) Phylogenetic relationships and divergence times of the genus *Tokudaia* within Murinae (Muridae; Rodentia) inferred from the nucleotide sequences encoding the Cyt*b* gene, RAG 1, and IRBP. *Canadian Journal of Zoology* **82**: 1343–1351.
3053. Sato, K., et al. (2006) Microhabitat use of wood mice ranging from a reserved belt with evergreen broad-leaved trees to a coniferous plantation. *Journal of Forest Research* **11**: 275–280.
3054. Satô, M. (2000) Some records of bat flies (Diptera: Nycteribidae) collected in northern Hokkaido. *Rishiri Studies* **19**: 15–17 (in Japanese with English abstract).
3055. Satô, M. (2002) The first record of *Murina ussuriensis* from Wakkanai and Toyotomi, northern Hokkaido. *Rishiri Studies* **21**: 1–2 (in Japanese with English abstract).
3056. Satô, M. (2010) Record of the northern bat captured at the Fishing Port of Shinminato, western Rishiri Island, northern Hokkaido. *Rishiri Studies* **29**: 35–36 (in Japanese with English abstract).
3057. Satô, M. (2011) A long-legged Whiskered bat, *Myotis frater*, with reduced pigmentation, from Oshidomari, Rishiri Island. *Rishiri Studies* **30**: 3–5 (in Japanese with English abstract).
3058. Satô, M. (2012) Catalogue of bat specimens deposited in Rishiri Town Museum. *Rishiri Studies* **31**: 1–6 (in Japanese with English abstract).
3059. Satô, M., et al. (2013) The northernmost record of a maternal colony of Ussuri whiskered bat in Japan. *Rishiri Studies* **32**: 53–60 (in Japanese with English abstract).
3060. Satô, M. & Kosugi, K. (1994) First record of *Murina ussuriensis* (Chiroptera, Mammalia) from Rishiri Island, northern Hokkaido. *Rishiri Studies* **13**: 1–2 (in Japanese).
3061. Satô, M. & Maeda, K. (1999) Distribution of bats in Rebun and Esashi, northern Hokkaido. *Rishiri Studies* **18**: 37–42 (in Japanese with English abstract).
3062. Satô, M. & Maeda, K. (2003) First record of *Vespertilio murinus* Linnaeus, 1758 (Vespertilionidae, Chiroptera) from Japan. *Bulletin of the Asian Bat Research Institute* **3**: 10–14.
3063. Satô, M., Maeda, K., & Akazawa, Y. (2001) Distribution of bats in Toyotomi and Horonobe, Northern Hokkaido. *Rishiri Studies* **20**: 23–28 (In Japanese with English abstract).

3064. Satô, M., et al. (2000) Distribution of bats in Hamatonbetsu, Northern Hokkaido. *Rishiri Studies* **19**: 23–26 (In Japanese with English abstract).

3065. Satô, M., et al. (2002) Determination of an unknown bat flying around streetlights in Northern Hokkaido. *Rishiri Studies* **21**: 65–73 (In Japanese with English abstract).

3066. Satô, M., et al. (2014) Utilization of three old railway tunnels by bats in Esashi, nothern Hokkaido: Reports of monitoring and banding for eleven years. *Rishiri Studies* **33**: 35–51 (in Japanese with English abstrast).

3067. Satô, M., et al. (2003) Distiribution of bats in Wakkanai, northern Hokkaido. *Rishiri Studies* **22**: 13–22 (in Japanese with English abstract).

3068. Satô, M. & Mogi, M. (2008) First descriptions of the males of *Ornithomya candia* Maa and *Nycteribia pleuralis* Maa (Diptera: Hippoboscidae and Nycteribiidae). *Medical Entomology and Zoology* **59**: 19–23.

3069. Satô, M. & Mogi, M. (2008) Records of some blood-sucking flies from birds and bats of Japan (Diptera: Hippodoscidae, Nycteribiidae and Streblidae). *Rishiri Studies* **27**: 41–48 (in Japanese with English abstract).

3070. Satô, M., Murayama, R., & Maeda, K. (2004) Current conditions of inhabiting bats in tunnels at Esashi & Utanobori towns, northern Hokkaido. *Rishiri Studies* **23**: 25–32 (In Japanese with English abstract).

3071. Satô, M., Murayama, R., & Maeda, K. (2004) Distribution of bats in Utanobori, northern Hokkaido. *Rishiri Studies* **23**: 33–43 (In Japanese with English abstract).

3072. Satô, M., Murayama, R., & Maeda, K. (2006) Distribution of bats in Sarufutsu, northern Hokkaido. *Rishiri Studies* **25**: 37–45 (In Japanese with English abstract).

3073. Satô, M., et al. (2008) Distiribution of bats in Bifuka, Northern Hokkaido. *Rishiri Studies* **27**: 27–32 (in Japanese with English abstract).

3074. Satô, M., et al. (2007) Distiribution of bats in Teshio and Enbetsu, northern Hokkaido. *Rishiri Studies* **26**: 39–44 (in Japanese with English abstract).

3075. Satô, M., et al. (2011) Distribution of bats in Otoineppu, northern Hokkaido. *Rishiri Studies* **30**: 35–44 (in Japanese with English abstract).

3076. Satô, M., Murayama, Y., & Maeda, K. (2004) Bat fauna research in the limestone cave 'Nakatonbetsu-shonyu-do', northern Hokkaido. *Rishiri Studies* **23**: 9–14 (in Japanese with English abstract).

3077. Satô, M., Murayama, Y., & Maeda, K. (2005) Distribution of bats in Nakatonbetsu, northern Hokkaido. *Rishiri Studies* **24**: 19–27 (in Japanese with English abstract).

3078. Satô, M., et al. (2009) Distribution of bats in Oumu, northeast Hokkaido. *Rishiri Studies* **28**: 33–42 (in Japanese with English abstract).

3079. Satô, M., Murayama, Y., & Sato, R. (2012) Distribution of bats in Tomamae, northern Hokkaido. *Rishiri Studies* **31**: 19–26 (in Japanese with English abstract).

3080. Satô, M., Murayama, Y., & Sato, R. (2013) Distribution of bats in Obira, Northeast Hokkaido. *Rishiri Studies* **32**: 29–35 (in Japanese with English abstract).

3081. Satô, M., Murayama, Y., & Sato, R. (2014) Distribution of bats in Rumoi, northeast Hokkaido. *Rishiri Studies* **33**: 27–33 (in Japanese with English abstract).

3082. Satô, M., Murayama, Y., & Sato, R. (2015) Distribution of bats in Mashike, northern Hokkaido. *Rishiri Studies* **34**: 19–26 (in Japanese with English abstract).

3083. Sato, M., et al. (2000) Changes in serum progesterone, estradiol-17β, luteinizing hormone and prolactin in lactating and non-lactating Japanese black bears (*Ursus thibetanus japonicus*). *Journal of Reproduction and Development* **46**: 301–308.

3084. Satô, M., Nakatani, M., & Mogi, M. (2007) Four bat fly species collected by N. Sasaki from eastern Hokkaido. *Sylvicola* **25**: 59–62 (in Japanese).

3085. Satô, M., Sato, M., & Maeda, K. (2002) Distribution of bats in Haboro and Shosanbetsu, northern Hokkaido. *Rishiri Studies* **21**: 55–64 (in Japanese with English abstract).

3086. Satô, M., et al. (2010) Distribution of bats in Horokanai, northern Hokkaido. *Rishiri Studies* **29**: 13–23 (in Japanese with English abstract).

3087. Satô, M. & Takahashi, M. (2014) Records of bat fleas from Hokkaido. *Rishiri Studies* **33**: 73–75 (in Japanese with English abstract).

3088. Satô, S. & Yachimori, Y. (2007) A record of the tube-nosed bat *Murina hilgendorfi* from Ino Town. *Shikoku Institute of Natural History* **4**: 30–33 (in Japanese with English abstract).

3089. Sato, Y. Unpublished.

3090. Sato, Y., Aoi, T., & Takatsuki, S. (2004) Temporal changes in the population density and diet of brown bears in eastern Hokkaido, Japan. *Mammal Study* **29**: 47–53.

3091. Sato, Y. & Endo, M. (2006) Relationship between crop use by brown bears and *Quercus crispula* acorn production in Furano, central Hokkaido, Japan. *Mammal Study* **31**: 93–104.

3092. Sato, Y., et al. (2011) Dispersal of male bears into peripheral habitats inferred from mtDNA haplotypes. *Ursus* **22**: 120–132.

3093. Sato, Y., et al. (2014) Selection of rub trees by brown bears (*Ursus arctos*) in Hokkaido, Japan. *Acta Theriologica* **59**: 127–137.

3094. Sato, Y., et al. (2008) Home range and habitat use of female brown bear (*Ursus arctos*) in Urahoro, eastern Hokkaido, Japan. *Mammal Study* **33**: 99–108.

3095. Sato, Y., Mano, T., & Takatsuki, S. (2005) Stomach contents of brown bears *Ursus arctos* in Hokkaido, Japan. *Wildlife Biology* **11**: 133–144.

3096. Sato, Y., et al. (2011) The white-colored brown bears of the Southern Kurils. *Ursus* **22**: 84–90.

3097. Sato, Y., et al. (1980) A survey of *Angiostrongylus cantonensis* in the Amami Islands. 1. The occurrence of *A. cantonensis* in snails and rodents in Yoron-jima. *Japanese Journal of Parasitology* **27**: 143–150 (in Japanese with English abstract).

3098. Satoh, A., et al. (2012) Bat fauna in middle-western Shizuoka Prefecture, Japan. *Natural History of the Tokai District* **5**: 51–68 (in Japanese).

3099. Satoh, A., et al. (2014) Recapture of ringed *Myotis ikonnikovi* after 5 years. *Natural History of the Tokai District* **7**: 31–33 (in Japanese).

3100. Satoyoshi, A., et al. (2004) A preliminary report of parasitological and microbiological survey of freeranging Japanese macaques (*Macaca fuscata* (Blyth)) in Boso Peninsula, Japan. *Japanese Jourrnal of Zoo and Wildlife Medicine* **9**: 79–83 (in Japanese with English absataract).

3101. Savolainen, P., et al. (2004) A detailed picture of the origin of the Australian dingo, obtained from the study of mitochondrial DNA. *Proceedings of the National Academy of Science, U. S. A.* **101**: 12387–12390.

3102. Savolainen, P., et al. (2002) Genetic evidence for an East Asian origin of domestic dogs. *Science* **298**: 1610–1613.

3103. Sawada, I. (沢田勇), et al. (1987) 北摂・丹波地方のコウモリ. *Nature Study* **33**: 3–4 (in Japanese).

3104. Sawada, Y. (澤田佳長) (1972) 四国西南地区におけるニッポンカワウソの棲息 (野生動物調査報告) 第一報. 高知県立中村高等学校研究紀要 **17**: 3–22 (in Japanese).

3105. Sawada, I. (1967) Helminth fauna of bats in Japan I. *Annotationes Zoologicae Japonenses* **4**: 61–66.

3106. Sawada, I. (1967) Helminth fauna of bats in Japan II. *Japanese Journal of Parasitology* **16**: 103–106.

3107. Sawada, I. (1976) Notes on the distribution of Japanese Rhinolophidae bats from the standpoint of their tapeworm fauna. *Zoological Magazine* **85**: 140–155 (in Japanese with English abstract).

3108. Sawada, I. (1978) Cestode fauna of cave bats in the Ryukyu Islands. *Proceeding of the Japanese Society of Systematic Zoology* **14**: 5–9 (in Japanese with English abstract).

3109. Sawada, I. (1978) Helminth fauna of bats in Japan XIX. *Annotationes Zoologicae Japonenses* **51**: 155–163.

3110. Sawada, I. (1980) Helminth fauna of bats in Japan XXII. *Annotationes Zoologicae Japonenses* **53**: 194–201.

3111. Sawada, I. (1982) Helminth fauna of bats in Japan XXVII. *Bulletin of Nara University of Education, Natural Science* **31**: 39–46.

3112. Sawada, I. (1983) Helminth fauna of bats in Japan XXIX. *Annotationes Zoologicae Japonenses* **56**: 209–220.

3113. Sawada, I. (1983) Helminth fauna of cave bats in Wakayama Prefecture. *Bulletin of Nara University of Education* **32**: 37–47 (in Japanese with English summary).

3114. Sawada, I. (1985) On the distribution of cave bats and the cestode fauna in Fukui Prefecture. *Bulletin of Nara Sangyo University* **1**: 123–128 (in Japanese with English abstract).

3115. Sawada, I. (1986) On the distribution of bats and the endoparasite fauna on the Oki Islands. *Bulletin of Nara Sangyo University* **2**: 145–151 (in Japanese with English abstract).

3116. Sawada, I. (1986) Two new hymenolepidid tapeworms, *Vampirolepis kawasakiensis* and *Insectivolepis mukooyamai*, with records of known tapeworms from bats of Japan. *Zoological Science* **3**: 707–713.

3117. Sawada, I. (1987) On the distribution of bats and their endoparasite fauna in Toyama Prefecture. *Bulletin of Nara Sangyo University* **3**: 197–204 (in Japanese with English summary).

3118. Sawada, I. (1988) On the distribution of Japanese Rhinolophidae bats with special reference to the cestode fauna. *Bulletin of Nara Sangyo University* **4**: 169–207 (in Japanese with English summary).

3119. Sawada, I. (1988) A survey on cestodes in Japanese bats, with descriptions of five new species of the genus *Vampirolepis* (Cestoda: Hymenolepididae). *Japanese Journal of Parasitology* **37**: 156–168.

3120. Sawada, I. (1989) On the distribution of Japanese Vespertilionidae bats with special reference to the cestode fauna. *Bulletin of Nara Sangyo University* **5**: 161–178 (in Japanese with English summary).

3121. Sawada, I. (1989) *Vampirolepis urawaensis* sp. n. (Cestoda: Hymenolepididae), with records of known cestodes, from Japanese bats. *Japanese Journal of Parasitology* **38**: 226–231.

3122. Sawada, I. (1990) *Vampirolepis ezoensis* sp. nov. (Cestoda: Hymenolepididae) from the Japanese northern bat, *Epitesicus nilssoni parvus* Kishida, a list of known species of the genus *Vampirolepis* Spassky from bats. *Japanese Journal of Parasitology* **39**: 176–185.

3123. Sawada, I. (1992) A bat tapeworm, *Vampirolepis ozensis* Sawada (Cestoda: Hymenolepididae), with records of known Cestodes from Japanese bats. *Bulletin of the Biogeographical Society of Japan* **47**: 51–55.

3124. Sawada, I. (1994) A list of caves of bat habitation in Japan. *Journal of the Natural History of Japan* **(2/3/4)**: 53–80 (in Japanese).

3125. Sawada, I. (1997) Check-list of new cestode species recorded by Sawada. *Nara Sangyo University, the Journal of Industrial Economics* **11**: 111–127.

3126. Sawada, I. (1998) A bat cestode, *Vampirolepis ozensis* Sawada (Cestoda: Hymenolepididae) of the Japanese long-eared bats, *Plecotus auritus sacrimontis* from Ehime Prefecture. *Bulletin of Nara Sangyo University* **14**: 35–36.

3127. Sawada, I. (1998) Helminth fauna of bats in Japan LI. *Nara Sangyo University, the Journal of Industrial Economics* **12**: 95–99.

3128. Sawada, I. (1999) A checklist of cestode species from Asian Insectivora. *Nara Sangyo University, the Journal of Industrial Economics* **13**: 109–157.

3129. Sawada, I. (2002) A list of nematodes parasitic on the Japanese bats. *Bulletin of Nara Sangyo University, the Journal of Industrial Economics* **17**: 81–85 (in Japanese).

3130. Sawada, I. (2008) The endoparasites parasitic on *Vampirolepis superans superans* Thomas, 1898. *Transactions of the Nagasaki Biological Society* **64**: 67–70 (in Japanese).

3131. Sawada, I. & Harada, M. (1990) Cestodes of field micromammalians (Insectivora) from central Honshu, Japan. *Zoological Science* **7**: 469–475.

3132. Sawada, I. & Harada, M. (1990) A new *Hymenolepis* species (Cestoda: Hymenolepididae) from the lesser Japanese shrew-mole *Dymecodon pilirostoris* of Nagano Prefecture, Japan. *Proceedings of the Japanese Society of Systematic Zoology* **42**: 10–13.

3133. Sawada, I. & Harada, M. (1991) A new species of the genus *Pseudohymenolepis* (Cestoda: Hymenolepididae) from Insectivora of central Japan, with a record of the known cestoda species. *Proceedings of the Japanese Society of Systematic Zoology* **44**: 8–14.

3134. Sawada, I. & Harada, M. (1994) Helminth fauna of bats in Japan XLVIII. *Nara Sangyo University, the Journal of Industrial Economics* **8**: 319–324.

3135. Sawada, I. & Harada, M. (1998) Redescription of *Vampirolepis multihamata* Sawada (Cestoda: Hymenolepididae) from the noctule bat, *Nyctalus aviator*. *Bulletin of the Biogeographical Society of Japan* **53**: 49–51.

3136. Sawada, I. & Inoue, R. (1985) The distribution and the endoparasite fauna of cave bats in Nara Prefecture. *Proceeding of the Japanese Society of Systematic Zoology* **30**: 11–17 (in Japanese with English synopsis).

3137. Sawada, I. & Koyasu, K. (1990) Further studies on cestodes from Japanese shrews. *Bulletin of Nara Sangyo University* **6**: 187–202.

3138. Sawada, I. & Kugi, G. (1973) A new cestode, *Mesocestoides paucitesticulus*, from a badger, *Nyctereutes procyonoides*, in Japan. *Japanese Journal of Parasitology* **22**: 45–47.

3139. Sawada, I. & Kugi, G. (1979) Studies on the helminth fauna of Kyushu: Part 5, Cestode parasites of wild mammals and birds. *Annotationes Zoologicae Japonenses* **52**: 133–141.

3140. Sawada, I., Uchikawa, K., & Harada, M. (1987) Effect of artificial destruction of forest upon existence of bats: Case studies on Nansei Islands and Taiwan. in Research Projects in Review 1987. Tokyo: Nissan Science Foundation, pp. 229–242 (in Japanese with English).

3141. Sawahata, T. (澤畠拓夫) (2007) 新潟県十日町市のコテングコウモリ情報. *Bat Study and Conservation Report* **15**: 24 (in Japanese).

3142. Sawahata, T. (澤畠拓夫) (2008) 新潟県におけるヒナコウモリの分布情報. *Bat Study and Conservation Report* **16**: 18 (in Japanese).

3143. Sawaya, H., et al. Unpublished.

3144. Scheffer, V.B. (1951) Measurements of sea otters from western Alaska. *Journal of Mammalogy* **32**: 10–14.

3145. Scheffer, V.B. (1958) Seals, Sea Lions and Walruses: A Review of the Pinnipedia. Stanford: Stanford University Press, 179 pp.

3146. Scheffer, V.B. & Kraus, B.S. (1964) Dentition of the northern fur seal. *Fishery Bulletin* **63**: 293–342.

3147. Scheffer, V.B. & Wilke, F. (1953) Relative growth in the northern fur seal. *Growth* **17**: 129–145.

3148. Schmidt, K., et al. (2003) Movements and use of home range in the Iriomote cat (*Prionailurus bengalensis iriomotensis*). *Journal of Zoology, London* **261**: 273–283.

3149. Schneider, K.B. (1973) Reproduction in the Female Sea Otter. Project Progress Report, Federal Aid in Wildlife Restoration Project W-17-4 and W-17-5. Anchorage: Alaska Department of Fish and Game, 23 pp.

3150. Schneider, K.B. (1978) Sex and Age Segregation of Sea Otters. Project Progress Report, Federal Aid in Wildlife Restoration Project W-17-4 and W-17-5. Anchorage: Alaska Department of Fish and Game, 45 pp.

3151. Schorr, G.S., et al. (2014) First long-term behavioral records from Cuvier's beaked whales (*Ziphius cavirostris*) reveal record-breaking dives. *PLoS ONE* **9**: e92633.

3152. Schulz, T., et al. (2008) Overlapping and matching of codas in vocal interactions between sperm whales: insights into communication function. *Animal Behaviour* **76**: 1977–1988.

3153. Schwartz, M., et al. (1992) Stomach contents of beach cast cetaceans collected along the San Diego County coast of California, 1972–1991 (NOAA National Marine Fisheries Service, Southwest Fisheries Center, Administrative Report LJ-92-18). 33 pp.

3154. Schwarz, E. & Schwarz, H. (1943) The wild and commensal stocks of the house mouse, *Mus musculus linnaeus*. *Journal of Mammalogy* **24**: 59–72.

3155. Sclater, P.L. (1862) Note on the Japanese black bear. *Proceeding of the Scientific Meeting of the Zoological Society of London* **1862**: 261.

3156. Scott, M.D. & Chivers, S.J. (2009) Movements and diving behavior of pelagic spotted dolphins. *Marine Mammal Science* **25**: 137–160.

3157. Scott, W.E. (1951) Wisconsin's first prairie spotted skunk, and other notes. *Journal of Mammalogy* **32**: 363.

3158. Scribner, K.T., et al. (1997) Population genetic studies of the sea otter (*Enhydra lutris*): a review and interpretation of available data. in Molecular Genetics of Marine Mammals. Special Publication No. 3 of The Society for Marine Mammalogy (Dizon, A.E., Chivers, S.J., & Perrin, W.F., Editors). Lawrence: Society for Marine Mammalogy, pp. 197–208.

3159. Scribner, K.T., et al. (2005) Verification of sex from harvested sea otters using DNA testing. *Wildlife Society Bulletin* **33**: 1027–1032.

3160. Sears, R. (2002) Blue whale (*Balaenoptera musculus*). in Encyclopedia of Marine Mammals (Perrin, W.F., Würsig, B., & Thewissen, J.G.M., Editors). San Diego: Academic Press, pp. 112–116.

3161. Sekiguchi, K. (1994) Studies on feeding habits and dietary analytical methods for smaller odontocete species along the southern African coast. Doctoral dissertation. Pretoria: University of Pretoria. 259 pp.

3162. Sekiguchi, K., et al. (2001) Genealogical relationship between introduced mongooses in Okinawa and Amamiohsima Islands, Ryukyu Archipelago, inferred from sequences of mtDNA cytochrome *b* gene. *Mammalian Science* **41**: 65–70 (in Japanese with English summary).

3163. Sekiguchi, K., et al. (2002) Food habits of introduced Japanese weasels (*Mustela itatsi*) and impacts on native species on Zamami Island. *Mammalian Science* **42**: 153–160 (in Japanese with English abstract).

3164. Sekijima, T. & Sone, K. (1994) Role of interspecific competition in the coexistence of *Apodemus argenteus* and *A. speciosus* (Rodentia: Muridae). *Ecological Research* **9**: 237–244.

3165. Serizawa, K., Suzuki, H., & Tsuchiya, K. (2000) A phylogenetic view on species radiation in *Apodemus* inferred from variation of nuclear and mitochondrial genes. *Biochemical Genetics* **38**: 27–40.

3166. Serizawa, Y. (2006) Bat survey in the town of Kushiro, Hokkaido. *Bulletin of the Asian Bat Research Institute* **5**: 9–18 (in Japanese with English abstract).

3167. Servheen, C. (1990) The status and conservation of the bears of the world. *International Conference on Bear Research and Management Monograph Series* **2**: 1–32.

3168. Servheen, C., Herrero, S., & Peyton, B. (compilers) (1999) Bears. Status Survey and Conservation Action Plan. Gland and Cambridge: IUCN/SSC Bear and Polar Bear Specialist Groups, 309 pp.

3169. Setoguchi, M. (1981) Utilization of holes and home ranges in the Japanese long-tailed mice (*Apodemus argenteus*). *Japanese Journal of Ecology* **31**: 385–394 (in Japanese with English summary).

3170. Sezaki, K., et al. (1984) Electrophoretic characters of the hybrids between two dolphins *Tursiops truncatus* and *Grampus griseus*. *Bulletin of the Japanese Society of Scientific Fisheries* **50**: 1771–1776.

3171. Shaughnessy, P.D. & Fay, F.H. (1977) A review of the taxonomy and nomenclature of North Pacific harbour seals. *Journal of Zoology* **182**: 385–419.

3172. Sheffield, S.R. & King, C.M. (1994) *Mustela nivalis*. *Mammalian Species* **454**: 1–10.

3173. Shenbrot, G.I. & Krasnov, B.R. (2005) An Atlas of the Geographic Distribution of the Arvicoline Rodents of the World (Rodentia, Muridae: Arvicolinae). Moscow: Pensoft, 366 pp.

3174. Sheng, H. (1979) The reproduction of the Siberian weasel *Mustela sibirica*. *Journal of Zoology (Peking)* **4**: 36–39 (in Chinese).

3175. Sheng, H. (1987) Sexual dimorphism and geographical variation in the body size of the yellow weasel *Mustela sibirica*. *Acta Theriologica Sinica* **7**: 92–95 (in Chinese with English abstract).

3176. Sheng, H., Ohtaishi, N., & Lu, H. (2000) Chinese Wild Mammals. Shenzhen: Chinese Forestry Publishing, 298 pp.

3177. Sheng, H.L. (1992) Reeves' Muntjac *Muntiacus reevesi*. in The Deer in China (Sheng, H.-l., Editor). Shanghai: East China Normal University Press,

pp. 126–144 (in Chinese with English summary).
3178. Sheng, H.L. (1992) Sika deer *Cervus nippon*. *in* The Deer in China (Sheng, H.L., Editor). Shanghai: East China Normal University Press, pp. 202–212 (in Chinese with English summary).
3179. Sheppard, J., et al. (2006) Movement heterogeneity of dugongs, *Dugong dugon* (Müller) over large spatial scales. *Journal of Experimental Marine Biology and Ecology* **334**: 64–83.
3180. Sherrod, S.K., Estes, J.A., & White, C.M. (1975) Depredation of sea otter pups by bald eagles at Amchitka Island, Alaska. *Journal of Mammalogy* **56**: 701–703.
3181. Shi, L., Ye, Y., & Duan, X. (1980) Comparative cytogenetic studies on the red muntjac, Chinese muntjac, and their F_1 hybrids. *Cytogenetics and Cell Genetics* **26**: 22–27.
3182. Shibahara, T. & Nishida, H. (1985) An epidemiological survey of the lung fluke, *Paragonimus* spp. in wild mammals of the northern part of Hyogo Prefecture Japan. *Japanese Journal of Veterinary Science* **47**: 911–919.
3183. Shibata, F. (芝田史仁) (2000) ヤマネ. *in* 冬眠する哺乳類 (Hibernation in Mammals) (Kawamichi, T. (川道武男), Kondo, N. (近藤宣昭), & Morita, T. (森田哲夫), Editors). Tokyo: University of Tokyo Press, pp. 162–186 (in Japanese).
3184. Shibata, A., et al. (2003) Chorioptic mange in a wild Japanese serow. *Journal of Wildlife Diseases* **39**: 437–440.
3185. Shibata, F. & Kawamichi, T. (1999) Decline of raccoon dog populations resulting from sarcoptic mange epizootics. *Mammalia* **63**: 281–290.
3186. Shibata, F. & Kawamichi, T. (2009) Female-biased sex allocation of offspring by an *Apodemus* mouse in an unstable environment. *Behavioral Ecology and Sociobiology* **63**: 1307–1317.
3187. Shibata, Y. (1981) Age and population dynamics of hares. *Journal of the Japanese Society for Hares* **8**: 21–24 (in Japanese).
3188. Shibata, Y. (1983) Rabbits on Toshima Ohshima. *Journal of the Japanese Society for Hares* **10**: 35–40 (in Japanese).
3189. Shibata, Y. (1987) The Ezo varying or blue hare. *Animals and Zoos* **39**: 8–11 (in Japanese).
3190. Shibata, Y. & Yamamoto, T. (1980) A study on the population dynamics of *Lepus timidus ainu* Barrett-Hamilton (I) Age distributions and life table. *Bulletin of the Forestry and Forest Products Research Institute* **309**: 13–21 (in Japanese with English summary).
3191. Shibecha Town Folk Museum (標茶町郷土館), ed. (1992) 標茶町郷土館収蔵・展示資料目録 (3) 哺乳類・鳥類・は虫類・両生類・魚類. 33 pp. (in Japanese).
3192. Shida, T. (1987) *Betula*'s catkins gathered by *Apodemus argenteus*. *Forest Protection* **197**: 4–6 (in Japanese).
3193. Shiga Prefectural Board of Education (滋賀県教育委員会) & Shiga Institute for Cultural Heritage Protection (滋賀県文化財保護協会) (1997) 粟津湖底遺跡第 3 貝塚 (Awazu No.3 Shell Midden). 453 pp. (in Japanese with English summary).
3194. Shigehara, N. (茂原信生), Hongo, H. (本郷一美), & Amitani, K. (網谷克彦) (1991) Mammal remains from the 1985 excavation of the Torihama shell mound. *Bulletin of the National Museum of Japanese History* **29**: 329–342 with 15 plates (in Japanese with English abstract).
3195. Shigehara, N. (1980) Epiphyseal union and tooth eruption of the Ryukyu house shrew, *Suncus murinus*, in captivity. *The Journal of the Mammalogical Society of Japan* **8**: 151–159.
3196. Shigematsu, H. (重松雄), Ochiai, K. (落合啓二), & Asada, M. (浅田正彦) (1994) 電波発信機による個体追跡. *in* 千葉県房総半島におけるニホンジカの保護管理に関する調査報告. Chiba: Chiba Prefectural Government & Deer Research Group on Boso (千葉県環境生活部自然保護課・房総のシカ調査会), pp. 27–32 (in Japanese).
3197. Shigeta, M., Fujii, Y., & Shigeta, Y. (2014) Damages of camellia trees by an alien squirrel *Callosciurus erythraeus* on Izu-ohshima Island. *Forest Pests* **63**: 8–18 (in Japanese).
3198. Shigeta, M., Oshida, T., & Okazaki, H. (2000) Alien Eurasian red squirrel in Sayama Hills. *Sciurid Information* (リスとムササビ) **7**: 6–9 (in Japanese).
3199. Shigeta, M., Shigeta, Y., & Endo, H. (2006) Day roosts and foraging areas of the common Japanese pipistrelle bat in the Imperial Palace, Tokyo, Japan. *Memoirs of the National Science Museum* **43**: 21–28 (in Japanese with English abstract).
3200. Shigeta, M., et al. (2005) Population fluctuation of large colonies of the long-fingered bats (*Miniopterus fuliginosus*) in Chiba Prefecture, Japan. *Journal of the Natural History Museum and Institute, Chiba* **8**: 33–40 (in Japanese with English abstract).
3201. Shikama, T. (1937) Geological study of the Kuzuü Formation (fissure deposits). *Contributions from the Institute of Geology and Paleontology Tohoku Imperial University* **27**: 1–34, pls. 1–18 (in Japanese).
3202. Shikama, T. (1937) Short notes on the excavation of the ossiferous fissures and caves in Kuzuu during the years 1931 to 1936. *Journal of the Geological Society of Japan* **44**: 405–420 (in Japanese with English resume).
3203. Shikama, T. (1949) The Kuzuu ossuaries: Geological and palaeontological studies of the limestone fissure deposits, in Kuzuu, Totigi Prefecture. *The Science Reports of the Tohoku University. Second Series, Geology* **23**: 1–201.
3204. Shikama, T. & Hasegawa, Y. (1962) Discovery of the fossil giant salamander (*Megalobatrachus*) in Japan. *Transactions and Proceedings of the Palaeontological Society of Japan, New Series* **45**: 197–200, pl. 29.
3205. Shikama, T. & Okafuji, G. (1958) Quaternary cave and fissure deposits and their fossils in Akiyoshi District, Yamaguti Prefecture. *Science Reports of the Yokohama National University. Section II, Biological and Geological Sciences* **7**: 43–103.
3206. Shikama, T., et al. (1952) Same Cave, Siga Prefecture (Japanese speleological works No.1). 趣味の地学 *(Syumi-no-tigaku)* **5**: 351–359 (in Japanese).
3207. Shikama, T. & Takayasu, T. (1971) Fossil mammals from the Shibikawa Formation in Oga Peninsula, Akita Prefecture. *Science Reports of the Yokohama National University, Section II* **18**: 43–47.
3208. Shikano, K. (鹿野一厚) (1996) 野生化ヤギ. *in* The Encyclopaedia of Animals in Japan. 2. Mammals II (日本動物大百科 2 哺乳類 II) (Izawa, K. (伊沢紘生), Kasuya, T. (粕谷俊雄), & Kawamichi, T. (川道武男), Editors). Tokyo: Heibonsha (平凡社), pp. 126–130 (in Japanese).
3209. Shimada, H. (島田裕之) (1995) 日本海におけるオオギハクジラ属の夏期分布. *in* Abstracts of the Annual Meeting of the Mammalogical Society of Japan. p. 109 (in Japanese).
3210. Shimada, T. (2001) Hoarding behaviors of two wood mouse species: different preference for acons of two Fagaceae species. *Ecological Research* **16**: 127–133.
3211. Shimada, T., et al. (2007) Rediscovery of *Mus nitidulus* Blyth, 1859 (Rodentia, Muridae), an endemic murine rodent of the central basin of Myanmar. *Zootaxa* **1498**: 45–68.
3212. Shimada, T., et al. (2007) Complex phylogeographic structuring in a continental small mammal from East Asia, the rice field mouse, *Mus caroli* (Rodentia, Muridae). *Mammal Study* **32**: 49–62.
3213. Shimada, T. & Saitoh, T. (2003) Negative effects of acorns on the wood mouse *Apodemus speciosus*. *Population Ecology* **45**: 7–17.
3214. Shimada, T. & Saitoh, T. (2006) Re-evaluation of the relationship between rodent populations and acorn masting: a review from the aspect of nutrients and defensive chemicals in acorns. *Population Ecology* **48**: 341–352.
3215. Shimada, T., et al. (2006) The role of tannin-binding salivary proteins and tannase-producing bacteria in acclimation of the Japanese wood mouse to acorn tannins. *Journal of Chemical Ecology* **32**: 1165–1180.
3216. Shimada, T., et al. (2006) Role of tannin-binding salivary proteins and tannase-producing bacteria in the acclimation of the Japanese wood mouse to acorn tannins. *Journal of Chemical Ecology* **32**: 1165–1180.
3217. Shimatani, Y., et al. (2010) Genetic variation and population structure of the feral American mink (*Neovison vison*) in Nagano, Japan, revealed by microsatellite analysis. *Mammal Study* **35**: 1–7.
3218. Shimatani, Y., et al. (2008) Genetic identification of mammalian carnivore species in the Kushiro Wetland, eastern Hokkaido, Japan, by analysis of fecal DNA. *Zoological Science* **25**: 714–720.
3219. Shimatani, Y., et al. (2010) Sex determination and individual identification of American minks (*Neovison vison*) on Hokkaido, northern Japan, by fecal DNA analysis. *Zoological Science* **27**: 243–247.
3220. Shimizu, E. (清水栄盛) (1975) ニッポンカワウソ物語. Matsuyama: Ehime Shimbun, 160 pp. (in Japanese).
3221. Shimizu, T. (清水孝頼) (2000) 群馬県でのコウモリ観察記録. *Bat Study and Conservation Report* **8**: 7 (in Japanese).
3222. Shimizu, E. (1959) On the skull of the Japanese otter. *The Journal of the Mammalogical Society of Japan* **1**: 137–138 (in Japanese).
3223. Shimizu, K. (1988) Ultrasonic assessment of pregnancy and fetal development in three species of macaque monkeys. *Journal of Medical Primatology* **17**: 247–256.
3224. Shimizu, M. & Miyao, T. (1966) A study of bats. Skeletal development and physical proportion. *Journal of Growth* **5**: 4–21.
3225. Shimizu, R. & Shimano, K. (2010) Food and habitat selection of *Lepus brachyurus lyoni* Kishida, a near-threatened species on Sado Island, Japan. *Mammal Study* **35**: 169–177.
3226. Shimizu, Z. (1987) Distribution of *Eothenomys andersoni* and *E. smithii* on the Kii Peninsula. *Nanki Seibutsu* **29**: 89–95 (in Japanese with English summary).
3227. Shimizu, Z. (2007) First record of Ikonnikov's whiskered bat *Myotis ikonnikovi* in Mie Prefecture. *Mie Natural History* **11**: 127 (in Japanese).
3228. Shimizu, Z. (2009) Mammalian fauna of the Mt. Odaigahara area, Kii Peninsula. *Mie Natural History* **12**: 2–21 (in Japanese).
3229. Shimizu, Z., et al. (2004) Caves of bat habitation and distributional records of bats in Mie Prefecture. *Mie Netural History* **8/9/10**: 77–90 (in Japanese).
3230. Shimokita Peninsula Serow Research Group (下北半島ニホンカモシカ調査

会) (1980) 下北半島のニホンカモシカ. Sendai: Shimokita Peninsula Serow Research Group, 166 pp. (in Japanese).

3231. Shinohara, A., Campbell, K.L., & Suzuki, H. (2003) Molecular phylogenetic relationships of moles, shrew moles, and desmans from the New and Old Worlds. *Molecular Phylogenetics and Evolution* **27**: 247–258.

3232. Shinohara, A., Campbell, K.L., & Suzuki, H. (2005) An evolutionary view on the Japanese talpids based on nucleotide sequences. *Mammal Study* **30**: S19–S24.

3233. Shinohara, A., et al. (2008) Phylogenetic relationships of the short-faced mole, *Scaptochirus moschatus* (Mammalia: Eulipotyphla), among Eurasian fossorial moles. As inferred from mitochondrial and nuclear gene sequences. *Mammal Study* **33**: 77–82.

3234. Shinohara, A., et al. (2014) Molecular phylogeny of East and Southeast Asian fossorial moles (Lipotyphla, Talpidae). *Journal of Mammalog* **95**: 455–466.

3235. Shinohara, A., et al. (2004) Evolution and biogeography of talpid moles from continental East Asia and the Japanese Islands inferred from mitochondrial and nuclear gene sequences. *Zoological Science* **21**: 1177–1185.

3236. Shinozaki, Y., et al. (2004) Ectoparasites of the Pallas squirrel, *Callosciurus erythraeus*, introduced to Japan. *Medical and Veterinary Entomology* **18**: 61–63.

3237. Shinozaki, Y., et al. (2004) The first record of sucking louse, *Neohaematopinus callosciuri*, infesting Pallas squirrels in Japan. *Journal of Veterinary Medical Science* **66**: 333–335.

3238. Shintaku, Y. (新宅勇太), Wu, Y. (呉毅), & Harada, M. (原田正史) (2010) 京都府におけるモモジロコウモリの初記録. *Nature Study* **56**: 9–10 (in Japanese).

3239. Shiota, T., et al. (1983) Studies on *Babesia* first found in murines in Japan. (1) Epidemiology and morphology. *Japanese Journal of Parasitology* **32**: 165–175 (in Japanese with English abstract).

3240. Shioya, K., Shiraishi, S., & Uchida, T.A. (1990) Microhabitat segregation between *Apodemus argenteus* and *A. speciosus* in northern Kyushu. *Journal of the Mammalogical Society of Japan* **14**: 105–118.

3241. Shirai, M. & Muramatsu, S. (2006) Robertsonian translocations in the wild population of Japanese Sika deer (*Cervus nippon*). *Chromosome Science* **9**: 89–93.

3242. Shiraishi, H. (白石浩隆) (1998) 河口湖周辺のコウモリ. *Bat Study and Conservation Report* **6**: 6–9 (in Japanese).

3243. Shiraishi, S. (白石哲) (1965) 日本の哺乳類　2　齧歯目カヤネズミ属. *Mammalian Science* **5(1)**: 1–13 (in Japanese).

3244. Shiraishi, S. (白石哲) (1968) 日本住血吸虫症と野ネズミ. *Mammalian Science* **8(2)**: 27–36 (in Japanese).

3245. Shiraishi, S. (白石哲) (1975) ハタネズミの飼育実験. *Zoological Magazine* **84**: 462 (in Japanese).

3246. Shiraishi, S. (白石哲) & Arai, A. (荒井秋晴) (1980) 陸上動物調査 (2) 主に哺乳動物. *in* 尖閣諸島調査報告書 (学術調査編). Tokyo: Okinawa Development Agency, pp. 47–86 (in Japanese).

3247. Shiraishi, S. (1964) Size of six external characters of adult harvest mouse in Kyushu, Japan, with special reference to statistical examination of its subspecific characters. *Science bulletin of the Faculty of Agriculture, Kyushu University* **21**: 97–109 (in Japanese with English abstract).

3248. Shiraishi, S. (1967) Some observation on the Japanese vole, *Microtus montebelli*, in the Chikugo River-bed, Kurume City. *The Journal of the Mammalogical Society of Japan* **3**: 57–63.

3249. Shiraishi, S. (1982) Rats control by using weasels and progression. *Collecting and Breeding* **44**: 414–419 (in Japanese).

3250. Shiraishi, S. (1982) The weasel as a rat-control agent. *Collecting and Breeding* **44**: 414–419 (in Japanese).

3251. Shirakihara, M. (白木原美紀), Shirakihara, K. (白木原国雄), & Nishiyama, M. (西山雅人) (2004) 豊後水道に出現するハンドウイルカ 2 種. *in* Abstracts of the Annual Meeting of the Japanese Society of Fisheries Science. p. 35 (in Japanese).

3252. Shirakihara, K., Shirakihara, M., & Yamamoto, Y. (2007) Distribution and abundance of finless porpoise in the Inland Sea of Japan. *Marine Biology* **150**: 1025–1032.

3253. Shirakihara, K., et al. (1992) A questionnaire survey on the distribution of the finless porpoise, *Neophocaena phocaenoides*, in Japanese waters. *Marine Mammal Science* **8**: 160–164.

3254. Shirakihara, M., et al. (2008) Food habits of finless porpoises *Neophocaena phocaenoides* in western Kyushu, Japan. *Journal of Mammalogy* **89**: 1248–1256.

3255. Shirakihara, M. & Shirakihara, K. (2013) Finless porpoise bycatch in Ariake Sound and Tachibana Bay, Japan. *Endangered Species Research* **21**: 255–262.

3256. Shirakihara, M., Shirakihara, K., & Takemura, A. (1994) Distribution and seasonal density of the finless porpoise *Neophocaena phocaenoides* in the coastal waters of western Kyushu, Japan. *Fisheries Science* **60**: 41–46.

3257. Shirakihara, M., et al. (2002) A resident population of Indo-Pacific bottlenose dolphins (*Tursiops aduncus*) in Amakusa, western Kyusyu, Japan. *Marine Mammal Science* **18**: 30–41.

3258. Shirakihara, M., Takemura, A., & Shirakihara, K. (1993) Age, growth, and reproduction of the finless porpoise, *Neophocaena phocaenoides*, in the coastal waters of western Kyushu, Japan. *Marine Mammal Science* **9**: 392–406.

3259. Shirakihara, M., Yoshida, H., & Shirakihara, K. (2003) Indo-Pacific bottlenose dolphins *Tursiops aduncus* in Amakusa, western Kyushu, Japan. *Fisheries Science* **69**: 654–656.

3260. Shirato, K., et al. (2012) Detection of bat coronaviruses from *Miniopterus fuliginosus* in Japan. *Virus Genes* **44**: 40–44.

3261. Shiratsuki, N., Asahi, M., & Yoshida, H. (1973) Food habit of Japanese marten *Martes melampus mepampus*, with a consideration on the home range. *Bulletin of Education of Mukogawa Woman's University* **20&21**: 45–56 (in Japanese with English abstract).

3262. Shiroma, T. (2000) Morphological survey on the Kerama deer, *Cervus nippon keramae*. *Midoriishi* (みどりいし) **11**: 8–11 (in Japanese).

3263. Shirota, M., et al. (1990) Anatomy on the nose of the oriental frosted bat (*Vespertilio orientalis*). *Bulletin of the College of Agriculture and Veterinary Medicine, Nihon University* **47**: 97–106 (in Japanese with English abstract).

3264. Shirotani, Y. (城谷義則) (1985) 福井県の翼手目 (コウモリ類). *Bulletin of the Fukui City Museum of Natural History* **31**: 85–93 (in Japanese).

3265. Shirouzu, H., Hatsushika, R., & Okino, T. (1999) Morphological studies on the diphyllobothriid tapeworms in killer whale, *Orcinus orca* (Linnaeus, 1758) and bottlenose dolphin, *Tursiops truncatus* (Montague, 1821). *Japanese Journal of Zoo and Wildlife Medicine* **4**: 53–60.

3266. Shizuokaken Shizenkankyou Chosaiinkai Honyuurui Bukai (静岡県自然環境調査委員会哺乳類部会), ed. (2005) 静岡県の哺乳類. 149 pp. (in Japanese).

3267. Short, R.V. (1984) Hopping mad. *in* Frontiers in Physiological Research (Garlick, D.G. & Komer, P.I., Editors). Cambridge: Cambridge University Press, pp. 371–386.

3268. Shou, T. (2005) Predator confirmed around breeding place of *Eptesicus nilssonii* in Bihoro. *Bulletin of the Bihoro Museum* **12**: 39–46 (in Japanese with English abstract).

3269. Siciliano, S., et al. (2007) Age and growth of some delphinids in southeastern Brazil. *Journal of the Marine Biological Association of the United Kingdom* **87**: 293–303.

3270. Silber, G.K., Newcomer, M.W., & Pérez-Cortés, M.H. (1990) Killer whales (*Orcinus orca*) attack and kill a Bryde's whale (*Balaenoptera edeni*). *Canadian Journal of Zoology* **68**: 1603–1606.

3271. Silva, F.J.D. & Da Silva, J.M. (2009) Circadian and seasonal rhythms in the behavior of spinner dolphins (*Stenella longirostris*). *Marine Mammal Science* **25**: 176–186.

3272. Simmons, N.B. (2005) Order Chiroptera. *in* Mammal Species of the World: A Taxonomic and Geographic Reference (Wilson, D.E. & Reeder, D.M., Editors). Baltimore: Johns Hopkins University Press, pp. 312–529.

3273. Simonds, P.E. (1965) The bonnet macaque in south India. *in* Primate Behavior: Field Studies of Monkeys and Apes (De Vore, I., Editor). New York: Holt, Reinhart & Winston, pp. 175–196.

3274. Simpfendorfer, C.A., Goodreid, A.B., & McAuley, R.B. (2001) Size, sex and geographic variation in the diet of the tiger shark, *Galeocerdo cuvier*, from western Australian waters. *Environmental Biology of Fishes* **61**: 37–46.

3275. Sinha, A.A., Conaway, C.H., & Kenyo, K.W. (1966) Reproduction in the female sea otter. *Journal of Wildlife Management* **30**: 121–130.

3276. Slate, J., et al. (1998) Bovine microsatellite loci are highly conserved in red deer (*Cervus elaphus*), sika deer (*Cervus nippon*) and Soay sheep (*Ovis aries*). *Animal Genetics* **29**: 307–315.

3277. Smith, A.T., et al. (1990) The pikas. *in* Rabbits, Hares and Pikas: Status Survey and Conservation Action Plan (Chapman, J.A. & Flux, J.E.C., Editors). Gland: IUCN, pp. 14–60.

3278. Smith, A.T. & Xie, Y., eds. (2008) A Guide to the Mammals of China. Princeton: Princeton University Press, 544 pp.

3279. Smith, D.G. & McDonough, J. (2005) Mitochondiral DNA variation in Chinese and Indian rhesus macaques (*Macaca mulatta*). *American Journal of Primatology* **65**: 1–25.

3280. Smith, T.G. (1973) Population dynamics of the ringed seal in the Canadian eastern Arctic. *Bulletin (Fisheries Research Board of Canada)* **181**: 1–55.

3281. Smith, T.G. (1987) The ringed seal, *Phoca hispida*, of the Canadian western arctic. *Bulletin (Fisheries Research Board of Canada)* **216**: 1–81.

3282. Smith-Jones, C. (2004) Muntjac—Managing an Alien Species—. Wiltshire: Cromwell Press, 216 pp.

3283. Snow, H.J. (1980) In Forbidden Seas: Recollections of Sea-otter Hunting in the Krils (千島列島黎明記). Tokyo: Kodansha (講談社), 363 pp. (in Japanese. Translated by 馬場脩 & 大久保義昭. Originally published by Edward

Arnold, London in 1910).

3284. Society for the Protection of Nature in Kagoshima Prefecture (鹿児島県自然愛護協会) (1981) ヤクシカの生息・分布に関する緊急調査報告書. *鹿児島県自然愛護協会調査報告* **5**: 1–34 (in Japanese).

3285. Society for Wildlife and Nature (2006) Formosan flying fox found on Green Island again. *International Conservation Newsletter* **14**: 3–4.

3286. Society for Wildlife and Nature (2010) Flying foxes spotted on Turtle Island. *International Conservation Newsletter* **18**: 3–4.

3287. Sohn, P.K. (1984) The palaeoenvironment of Middle and Upper Pleistocene Korea. in The Evolution of the East Asian Environment (White, R.O., Editor). Hong Kong: Center of Asian Studies, University of Hong Kong, pp. 877–893.

3288. Sokolov, V.E. & Arsen'ev, V.A. (2006) Baleen Whales: Mammals of Russia and Adjacent Regions (Mead, J. & Hoffman, R.S., Editors). Enfield: Science Publishers, 317 pp.

3289. Soma, H., Kada, H., & Matayoshi, K. (1987) Evolutionary pathways of karyotypes of the tribe Rupicaprini. in The Biology and Management of *Capricornis* and Related Mountain Antelopes (Soma, H., Editor). London: Croom Helm, pp. 62–71.

3290. Son, S.W., et al. (1988) Reproduction of two rare *Pipistrellus* species, with special attention to the fate of spermatozoa in their female genial tracts. *Journal of the Mammalogical Society of Japan* **13**: 77–91.

3291. Sone, K., et al. (1999) Biomass of food plants and density of Japanese serow, *Capricornis crispus. Memoirs of the Faculty of Agriculture, Kagoshima University* **35**: 7–16.

3292. Song, Y., Lan, Z., & Kohn, M.H. (2014) Mitochondrial DNA phylogeography of the Norway rat. *PLoS ONE* **9**: e88425.

3293. Sonoda, Y. & Kuramoto, N. (2004) Environmental selectivity to the lower-layer vegetation structure by raccoon dogs at Tama Hill. *Selected Papers of Environmental Systems Research* **32**: 335–342 (in Japanese with English summary).

3294. Soot-Ryen, T. (1961) On a Bryde's whale stranded on Curaçao. *Norsk Hvalfangst-tidende* **1961**: 323–332.

3295. Soullier, S., et al. (1998) Male sex determination in the spiny rat *Tokudaia osimensis* (Rodentia: Muridae) is not *Sry* dependent. *Mammalian Genome* **9**: 590–592.

3296. Southwick, C.H., Beg, M.A., & Sidiqqi, M.F. (1965) Rhesus monkey in north India. in Primate Behavior: Field Studies of Monkeys and Apes (De Vore, I., Editor). New York: Holt, Reinhart & Winston, pp. 111–159.

3297. Southwick, C.H., et al. (1974) Xenophobia among free-ranging rhesus group in India. in Primate Aggression, Territoriality, and Xenophobia: A Comparative Perspective (Holloway, R.L., Editor). New York: Academic Press, pp. 185–209.

3298. Spitz, F. (1992) General model of the spatial and social organization of the wild boar (*Sus scrofa* L.). in Ongulés/Ungulates 91 (Spitz, F., et al., Editors). Toulouse: Institut de Recherche sur les Grands Mammiferes, pp. 385–389.

3299. Spitz, J., et al. (2011) Prey preferences among the community of deep-diving odontocetes from the Bay of Biscay, Northeast Atlantic. *Deep Sea Research Part I* **58**: 273–282.

3300. Spitzenberger, F., et al. (2006) A preliminary revision of the genus *Plecotus* (Chiroptera, Vespertilionidae) based on genetic and morphological results. *Zoologica Scripta* **35**: 187–230.

3301. Stafford, B.J., Thorington Jr, R.W., & Kawamichi, T. (2002) Gliding behavior of Japanese giant flying squirrels (*Petaurista leucogenys*). *Journal of Mammalogy* **83**: 553–562.

3302. Stains, H.J. (1975) Distribution and taxonomy of the Canidae. in The Wild Canids: Their Systematics, Behavioural Ecology and Evolution (Fox, M.W., Editor). New York: Van Nostrand Reinhold, pp. 3–26.

3303. Stanyon, R., et al. (1983) The banded karyotypes of *Macaca fuscata* compared with *Cercocebus aterrimus. Folia Primatologica* **41**: 137–146.

3304. Staudinger, M.D., et al. (2014) Foraging ecology and niche overlap in pygmy (*Kogia breviceps*) and dwarf (*Kogia sima*) sperm whales from waters of the U.S. mid-Atlantic coast. *Marine Mammal Science* **30**: 626–655.

3305. Stenseth, N.C. & Saitoh, T. (eds.) (1998) The population ecology of the vole *Clethrionomys rufocanus. Researches on Population Ecology* **40**: 1–120.

3306. Steppan, S.J., Storz, B.L., & Hoffmann, R.S. (2004) Nuclear DNA phylogeny of the squirrels (Mammalia: Rodentia) and the evolution of arboreality from c-myc and RAG1. *Molecular Phylogenetics and Evolution* **30**: 703–719.

3307. Stewart, B.S. & Leatherwood, S.L. (1985) Minke whale *Balaenoptera acutorostrata* Lacépède, 1804. in Handbook of Marine Mammals. Volume 3. The Sirenians and Baleen Whales (Ridgway, S.H. & Harrison, R.J., Editors). London: Academic Press, pp. 91–136.

3308. Steyaert, S.M.J.G., et al. (2012) The mating system of the brown bear *Ursus arctos. Mammal Review* **42**: 12–34.

3309. Stirling, I., ed. (1993) Bears Majestic Creature of the World. Pennsylvania: Rodale Press, 240 pp.

3310. Stock, A.D. & Hsu, T.-C. (1973) Evolutionary conservation in arrangement of genetic material. *Chromosoma* **43**: 211–224.

3311. Stone, W. (1900) Descriptions of a new rabbit from the Liu kiu Islands and a new flying squirrel from Borneo. *Proceedings of the Academy of Natural Sciences of Philadelphia* **52**: 460–463.

3312. Su, H. & Lee, L. (2001) Food habits of Formosan rock macaques (*Macaca cyclopis*) in Jentse, northeastern Taiwan, assessed by fecal analysis and behavioral observation. *International Journal of Primatology* **22**: 359–377.

3313. Sudo, R., et al. (2008) Sighting survey of cetaceans in the Tsugaru Strait, Japan. *Fisheries Science* **74**: 211–213.

3314. Suématu, S. (1950) On the third stuffed specimen of *Canis lupus japonicus*, a scientifically valuable specimen. *Bulletin of Liberal Arts College, Wakayama University. Natural Science* **1**: 85–88 (in Japanese with English abstract).

3315. Suga, M., Nakamura, S., & Abe, S. (1974) On the *Oestromyia leporina* in pikas, *Ochotona hyperborea*, kept in Obihiro Zoo. *Journal of Japanese Association of Zoos and Aquariums* **17**: 66–68 (in Japanese).

3316. Sugai, S., et al. (2011) Bat fauna in three river basins originating in Mt. Mokoto, Hokkaido. *Journal of Agricultural Science of Tokyo University of Agriculture* **56**: 155–161 (in Japanese with English summary).

3317. Sugawara, R. (菅原竜二) (1972) 栄養はなかなか豊富. *科学朝日* **1972**: 44–48 (in Japanese).

3318. Sugimura, K., et al. (2000) Distribution and abundance of the Amami rabbit *Pentalagus furnessi* in the Amami and Tokuno Islands, Japan. *Oryx* **34**: 198–206.

3319. Sugimura, K. & Yamada, F. (2004) Estimating population size of the Amami rabbit *Pentalagus furnessi* based on fecal pellet counts on Amami Island, Japan. *Acta Zoologica Sinica* **50**: 519–526.

3320. Sugimura, M., et al. (1981) Reproduction and prenatal growth in the wild Japanese serow, *Capricornis crispus. Japanese Journal of Veterinary Science* **43**: 553–555.

3321. Sugimura, M., et al. (1983) Prenatal development of Japanese serow, *Capricornis crispus*, and reproduction in females. *Journal of Mammalogy* **64**: 302–304.

3322. Sugimura, K., et al. (2014) Monitoring the effects of forest clear-cutting and mongoose *Herpestes auropunctatus* invasion on wildlife diversity on Amami Island, Japan. *Oryx* **48**: 241–249.

3323. Suginome, H. (杉野目斉) (2004) 鳥の標識調査中に誤捕獲したコウモリについて. *Bat Study and Conservation Report* **12**: 11 (in Japanese).

3324. Sugita, N., Inaba, M., & Ueda, K. (2009) Roosting pattern and reproductive cycle of Bonin flying foxes (*Pteropus pselaphon*). *Journal of Mammalogy* **90**: 195–202.

3325. Sugita, N., et al. (2013) Possible spore dispersal of a bird-nest fern *Asplenium setoi* by Bonin flying foxes *Pteropus pselaphon. Mammal Study* **38**: 225–229.

3326. Sugita, N. & Ueda, K. (2014) Sexual size dimorphism in Bonin flying foxes *Pteropus pselaphon* on Chichijima, Ogasawara Islands. *Mammal Study* **39**: 185–189.

3327. Sugiyama, H., et al. (2003) Raccoon roundworm, *Baylisascaris procyonis*, as a cause of larva migrans. *Japanese Journal of Veterinary Parasitology* **2**: 13–19 (in Japanese).

3328. Sugiyama, Y., Minamoto, N., & Kinjo, T. (1998) Serological surveillance of Lyme borreliosis in wild Japanese serows (*Capricornis crispus*). *Journal of Veterinary Medical Science* **60**: 745–747.

3329. Sugiyama, Y. & Ohsawa, H. (1982) Population dynamics of Japanese monkeys with special reference to the effect of artificial feeding. *Folia Primatologica* **39**: 238–263.

3330. Sugiyama, Y. & Ohsawa, H. (1988) Population dynamics and management of baited Japanese monkeys at Takasakiyama. *Primate Research* **4**: 33–41 (in Japanese with English summary).

3331. Sumino, T., et al. (2007) Parasitic helminths obtained from small intestines of free-ranging brown bears (*Ursus arctos*) in Hokkaido, Japan, with special reference to geographical distribution of *Ancylostoma malayanum* (Nematoda). in Abstracts of the Annual Meeting of the Japan Society of Zoo and Wildlife Medicine. p. 93 (in Japanese).

3332. Surridge, A.K., et al. (1999) Population structure and genetic variation of European wild rabbits (*Oryctolagus cuniculus*) in East Anglia. *Heredity* **82**: 479–487.

3333. Sutou, S., Mitsui, Y., & Tsuchiya, K. (2001) Sex determination without the Y chromosome in two Japanese rodents *Tokudaia osimensis osimensis* and *Tokudaia osimensis* spp. *Mammalian Genome* **12**: 17–21.

3334. Suzuki, K. (鈴木欣司) (2012) 外来どうぶつミニ図鑑. Tokyo: Zennokyo (全国農村教育協会), 191 pp. (in Japanese).

3335. Suzuki, K. (鈴木和男) (2005) 捕獲個体から見えるアライグマの生物学. in 田辺市におけるアライグマ調査報告書—平成16年度農作物鳥獣害防止対策

事業 (田辺鳥獣害対策協議会, Editor). Tanabe: Tanabe City Office, pp. 15–32 (in Japanese).
3336. Suzuki, K. (鈴木和男) (2007) アライグマの繁殖情報. in 田辺鳥獣害調査研究報告書 (田辺鳥獣害対策協議会, Editor). Tanabe: Tanabe City Office, pp. 62–67 (in Japanese).
3337. Suzuki, M. (鈴木正嗣) & Yamashita, T. (山下忠幸) (1986) ゼニガタアザラシの性成熟と発育段階区分 (Sexual maturity and developmental stages of Kuril seals). in ゼニガタアザラシの生態と保護 (Wada, K. (和田一雄), et al., Editors). Tokyo: Tokai University Press, pp. 179–194 (in Japanese).
3338. Suzuki, T. (鈴木敏秀) (2006) 福島県裏磐梯でのコテングコウモリの確認. *Bat Study and Conservation Report* **14**: 17 (in Japanese).
3339. Suzuki, A. (1965) An ecological study of wild Japanese monkeys in snowy areas: Focused on their food habits. *Primates* **6**: 31–72.
3340. Suzuki, A. (1983) Keeping and exhibiting long-eared bats. *Animals and Zoos* **35**: 8–13 (in Japanese with English abstract).
3341. Suzuki, A., et al. (1975) Population dynamics and group movement of Japanese monkeys in Yokoyugawa Valley, Shiga Heights. *Physiological Ecology Japan* **16**: 15–23 (in Japanese).
3342. Suzuki, H. (1975) Reports of medico-zoological investigations in the Nansei Islands. Part III. Descriptions of two new species of *Walchia* from southern Amami Island (Prostigmata, Trombiculidae). *Japanese Journal of Experimental Medicine* **45**: 235–239.
3343. Suzuki, H. (1980) Trombiculid fauna in Nansei Islands and their characteristics (Prostigmata, Trombiculidae). *Tropical Medicine* **22**: 137–159.
3344. Suzuki, H. & Aplin, K.P. (2012) Phylogeny and biogeography of the genus *Mus* in Eurasia. in Evolution of the house mouse (Cambridge studies in morphology and molecules) (Macholán, M., et al., Editors). Cambridge: Cambridge University Press, pp. 35–64.
3345. Suzuki, H., et al. (2008) A biogeographic view of *Apodemus* in Asia and Europe inferred from nuclear and mitochondrial gene sequences. *Biochemical Genetics* **46**: 329–346.
3346. Suzuki, H., et al. (1994) Phylogenetic relationship between the Iriomote cat and the leopard cat, *Felis bengalensis*, based on the ribosomal DNA. *Japanese Journal of Genetics* **69**: 397–406.
3347. Suzuki, H. & Inaba, M. (2010) Future of the flying guardian of forest and the islands. *The Heredity (遺伝)* **64**: 61–67 (in Japanese).
3348. Suzuki, H., et al. (1999) Molecular phylogeny of red-backed voles in Far East Asia based on variation in ribosomal and mitochondrial DNA. *Journal of Mammalogy* **80**: 512–521.
3349. Suzuki, H., et al. (1999) The genetic status of two insular populations of the endemic spiny rat *Tokudaia osimensis* (Rodentia, Muridae) of the Ryukyu Islands, Japan. *Mammal Study* **24**: 43–50.
3350. Suzuki, H., Kawakami, K., & Fujita, T. (2008) Flying fox of Minami-Iwo-To Island, Volcano Isls, the Bonin Islands. *Ogasawara Research* **33**: 89–104 (in Japanese with English summary).
3351. Suzuki, H., et al. (1997) Phylogenetic position and geographic differentiation of the Japanese dormouse, *Glirulus japonicus*, revealed by variations among rDNA, mtDNA and the *Sry* gene. *Zoological Science* **14**: 167–173.
3352. Suzuki, H., et al. (2003) Molecular phylogeny of wood mice (*Apodemus*, Muridae) in East Asia. *Biological Journal of the Linnean Society* **80**: 469–481.
3353. Suzuki, H., et al. (2004) Temporal, spatial, and ecological modes of evolution of Eurasian *Mus* based on mitochondrial and nuclear gene sequences. *Molecular Phylogenetics and Evolution* **33**: 626–646.
3354. Suzuki, H. & Takai, F. (1959) Entdeckung eines pleistozänen hominiden Humerus in Zentral-Japan. *Anthropologischer Anzeiger* **23**: 224–235.
3355. Suzuki, H., et al. (1994) Evolution of restriction sites of ribosomal DNA in natural populations of the field mouse, *Apodemus speciosus*. *Journal of Molecular Evolution* **38**: 107–112.
3356. Suzuki, H., Tsuchiya, K., & Takezaki, N. (2000) A molecular phylogenetic framework for the Ryukyu endemic rodents *Tokudaia osimensis* and *Diplothrix legata*. *Molecular Phylogenetics and Evolution* **15**: 15–24.
3357. Suzuki, H., et al. Unpublished.
3358. Suzuki, H., et al. (2004) Differential geographic patterns of mitochondrial DNA variation in two sympatric species of Japanese wood mice, *Apodemus speciosus* and *A. argenteus*. *Genes & Genetic Systems* **79**: 165–176.
3359. Suzuki, K. (1978) Mammals of Saitama. in 埼玉県動物誌 (埼玉県動物誌編集委員会, Editor). Urawa: Saitama Prefectural Board of Education (埼玉県教育委員会), pp. 31–44 (in Japanese).
3360. Suzuki, K., Asari, Y., & Yanagawa, H. (2012) Gliding locomotion of Siberian flying squirrels in low-canopy forces: the role of energy-inefficient short-distance gliders. *Acta Theriologica* **57**: 131–135.
3361. Suzuki, K., et al. (2011) Nest site characteristics of *Pteromys momonga* in the Tanzawa Mountains. *Mammalian Science* **51**: 65–69 (in Japanese with English abstract).

3362. Suzuki, K. & Takatsuki, S. (1986) Winter foods habits and sexual monomorphism in Japanese serow. *Proceedings of the Biennial Symposium of the Northern Wild Sheep & Goat Council* **5**: 396–402.
3363. Suzuki, K. & Yanagawa, H. (2011) Small mammals preyed on by long-eared owls, *Asio otus*, wintering in an urban riparian zone. *Mammalian Science* **51**: 315–319 (in Japanese with English abstract).
3364. Suzuki, K. & Yoshinaga, S. (1999) Prevention and control of wildlife damage in fruit production. *Bulletin of the National Institute of Fruit Tree Science* **32**: 39–64 (in Japanese with English abstract).
3365. Suzuki, M., et al. (1996) Gestational age determination, variation of conception date, and external fetal development of sika deer (*Cervus nippon yesoensis* Heude, 1884) in eastern Hokkaido. *Journal of Veterinary Medical Science* **58**: 505–509.
3366. Suzuki, M., et al. (2011) Preliminary estimation of population density of the Siberian flying squirrel (*Pteromys volans orii*) in natural forest of Hokkaido, Japan. *Mammal Study* **36**: 155–158.
3367. Suzuki, M., Koizumi, T., & Kobayashi, M. (1992) Reproductive characteristics and occurrence of accessory corpora lutea in sika deer *Cervus nippon centralis* in Hyogo Prefecture, Japan. *Journal of the Mammalogical Society of Japan* **17**: 11–18.
3368. Suzuki, M. & Ohtaishi, N. (1993) Reproduction of female sika deer (*Cervus nippon yesoensis* HEUDE, 1884) in Ashoro District, Hokkaido. *Journal of Veterinary Medical Science* **55**: 833–836.
3369. Suzuki, M., Ohtaishi, N., & Nakane, F. (1990) Supernumerary postcanine teeth in the Kuril seal (*Phoca vitulina stejnegeri*), the Larga seal (*Phoca largha*) and the ribbon seal (*Phoca fasciata*). *Japanese Journal of Oral Biology* **32**: 323–329.
3370. Suzuki, N. & Ikeda, T. (1985) Echinococcosis in Hokkaido. *Mammalian Science* **Suppl. 2**: 34 (in Japanese).
3371. Suzuki, S., et al. (1990) Frog- and lizard-eating behaviour of wild Japanese macaques in Yakushima, Japan. *Primates* **31**: 421–426.
3372. Suzuki, T., et al. (2006) Bats fauna of Hidaka, Tokachi districts, central and eastern Hokkaido. 5. Records of bats in Kobukariishi, Urahoro, southern Tokachi. *Bulletin of the Higashi Taisetsu Museum of Natural History* **28**: 1–4 (in Japanese with English abstract).
3373. Suzuki, T., et al. (2013) Food habits of the Ural owl (*Strix uralensis*) during the breeding season in central Japan. *Journal of Raptor Research* **47**: 304–310.
3374. Suzuki, T., Yuasa, H., & Machida, Y. (1996) Phylogenetic position of the Japanese river otter *Lutra nippon* inferred from the nucleotide sequence of 224 bp of the mitochondrial cytochrome *b* gene. *Zoological Science* **13**: 621–626.
3375. Suzuki, T.A. & Iwasa, M.A. (2013) A cross-experimental analysis of coat color variations and morphological characteristics of the Japanese wild mouse, *Mus musculus*. *Experimental Animals* **62**: 25–34.
3376. Suzuki, Y., Sugimura, M., & Atoji, Y. (1987) Pathological studies on Japanese serow (*Capricornis crispus*). in The Biology and Management of *Capricornis* and Related Mountain Antelopes (Soma, H., Editor). London: Croom Helm, pp. 283–298.
3377. Suzuki, Y., et al. (1986) Widespread of parapox infection in wild Japanese serow, *Capricornis crispus*. *Japanese Journal of Veterinary Science* **48**: 1279–1282.
3378. Suzuki, Z. (1958) Epidemiological studies on paragonimiasis in south Izu District, Shizuoka Prefecture, Japan. *Japanese Journal of Parasitology* **7**: 112–114 (in Japanese with English summary).
3379. Suzuno, M., Shirakihara, M., & Furota, T. (2010) Occurrence of finless porpoise (*Neophocaena phocaenoides*) in the coastal waters of Chiba, Central Japan. *Bulletin of the Biological Society of Chiba* **60**: 11–16 (in Japanese with English abstract).
3380. Swanson, G.M. & Putman, R. (2009) Sika deer in the British Isles. in Sika Deer—Biology and Management of Native and Introduced Populations (McCullough, D.R., Takatsuki, S., & Kaji, K., Editors). Tokyo: Springer, pp. 595–614.
3381. Swenson, J.E., et al. (1997) Infanticide caused by hunting of male bears. *Nature* **386**: 450–451.
3382. Switonski, M., et al. (2003) Chromosome polymorphism and karyotype evolution of four canids: the dog, red fox, arctic fox and raccoon dog. *Caryologia* **56**: 375–385.
3383. Sylvestre, J. & Tasaka, S. (1985) On the intergeneric hybrids in cetaceans. *Aquatic Mammals* **11**: 101–108.
3384. T'sui, W.H., Lin, F.Y., & Huang, C.C. (1982) The reproductive biology of the red-bellied tree squirrel, *Callosciurus erythraeus*, at Ping-Lin, Taipei Hsien. *Proceeding of National Science Council R.O.C.* **6**: 443–451.
3385. Tabata, M. & Iwasa, M.A. (2013) Environmental factors for the occurrence of the Smith's red-backed vole, *Eothenomys smithii*, in rocky terrains at the foot of Mt. Fuji in central Japan. *Mammal Study* **38**: 243–250.

3386. Tachi, C., et al. (2002) Successful molecular cloning and nucleotide sequences determination of partial amelogenin (AMELX) exon DNA fragment recovered from a mounted taxidermic pelt specimen tentatively identified as an extinct wolf species, *Canis lupus hodophilax* Temminck, the Japanese wolf and stocked at School of Agriculture and Life Sciences, the University of Tokyo. *Journal of Reproduction and Development* **48**: 633–638.

3387. Tachibana, T. (立花隆) (1991) サル学の現在 (The Frontiers of Primatology). Tokyo: Heibonsha (平凡社), 714 pp. (in Japanese).

3388. Tachibana, K., Ishii, N., & Kato, S. (1988) Food habits of small mammals during a sawfly (*Pristifora erichsoni* Hartig) outbreak. *Journal of the Japanese Forest Society* **70**: 525–528 (in Japanese).

3389. Tachibana, S. (1971) Notes on mammals observed in Kahoku-machi, Miyagi Pref. *The Journal of the Mammalogical Society of Japan* **5**: 120–121 (in Japanese).

3390. Tachibana, T., et al. (1970) Diurnal activity of *Lepus brachyurus brachyurus*, under semi-natural conditions. *The Journal of the Mammalogical Society of Japan* **5**: 50–57 (in Japanese with English summary).

3391. Tada, T. & Obara, Y. (1986) Karyological relationship between the Japanese house shrew, *Suncus murinus riukiuanus* and the Japanese white-toothed shrew, *Crocidura dsinezumi chisai*. *Proceeding of Japan Academy. Series B* **62**: 125–128.

3392. Tada, T. & Obara, Y. (1988) Karyological relationships among four species and subspecies of *Sorex* revealed by differential staining techniques. *Journal of the Mammalogical Society of Japan* **13**: 21–31.

3393. Taguchi, M., et al. (2010) Mitochondrial DNA phylogeography of the harbour porpoise *Phocoena phocoena* in the North Pacific. *Marine Biology* **157**: 1489–1498.

3394. Taguchi, M., Ishikawa, H., & Matsuishi, T. (2010) Seasonal distribution of harbour porpoise (*Phocoena phocoena*) in Japanese waters inferred from stranding and bycatch records. *Mammal Study* **35**: 133–138.

3395. Taguchi, M., Yoshioka, M., & Kashiwagi, M. (2007) Seasonal changes in density of the finless porpoise (*Neophocaena phocaenoides*) around the mouth of Mikawa Bay, Japan. *Mammalian Science* **47**: 11–17 (in Japanese with English summary).

3396. Tajima, Y. (2001) A report on the investigations of marine mammals stranded on the coasts of the sea of Japan. Jan. 1, 1999–Apr. 30, 2000. *Nihonkai Cetology* **11**: 1–5.

3397. Tajima, Y., Maeda, K., & Yamada, T.K. (2015) Pathological findings and probable causes of the death of Stejneger's beaked whales (*Mesoplodon stejnegeri*) stranded in Japan from 1999 and 2011. *Journal of Veterinary Medical Science* **77**: 45–51.

3398. Takada, N. (高田伸弘), Fujita, H. (藤田博巳), & Nara, N. (奈良典明) (1978) トウヨウヒナコウモリの寄生虫相の調査 (1) 外部寄生虫について. *Japanese Journal of Sanitary Zoology* **29**: 48 (in Japanese).

3399. Takada, Y. (高田雄三), Matobe, Y. (的場洋平), & Asakawa, M. (浅川満彦) (2007) アライグマMHCの地理的分布. *MHC* **14**: 163–175 (in Japanese).

3400. Takada, N. (1970) On the distribution of *Microti*-group and *Acomatacarus yosanoi* Fukuzumi et Obata, 1953, trombiculid mites in the Northern area of Honshu. *Japanese Journal of Sanitary Zoology* **21**: 222–223 (in Japanese).

3401. Takada, N. (1979) Descriptions of four new species of chiggers from bats in northern Japan (Acarina; Trombiculidae). *Japanese Journal of Sanitary Zoology* **30**: 99–106.

3402. Takada, N. (1995) Recent findings on vector acari for rickettia and spirochaete in Japan. *Japanese Journal of Sanitary Zoology* **46**: 91–108.

3403. Takada, N., Fujita, H., & Takahashi, M. (1977) The haemogamasid mites from rodents in Japan, with some new subspecies, additional records and a key to the females (Acarina; Haemogamasidae). *Hirosaki Medical Journal* **29**: 733–747.

3404. Takada, N., et al. (2001) First records of tick-borne pathogens *Borrelia* and spotted fever group rickettsiase in Okinawajima Island, Japan. *Microbiology and Immunology* **45**: 163–165.

3405. Takada, N., et al. (1994) Prevalence of Lyme *Borrelia* in ticks, especially *Ixodes persulcatus* (Acari: Ixodidae), in central and western Japan. *Journal of Medical Entomology* **31**: 474–478.

3406. Takada, N., et al. (2001) Prevalence of Lyme disease *Borrelia* in ticks and rodents in northern Kyushu Japan. *Medical Entomology and Zoology* **52**: 117–123.

3407. Takada, N. & Yamaguchi, T. (1971) Studies on laelaptid mites found in wild rodents in Tohoku district, Honshu, Japan. *Hirosaki Medical Journal* **23**: 363–373 (in Japanese with English summary).

3408. Takada, N. & Yamaguchi, T. (1974) Studies on ixodid fauna in the northern part of Honshu, Japan. 1. Ixodid ticks (Ixodidae) parasitic on wild mammals and some cases of human infestation. *Japanese Journal of Sanitary Zoology* **25**: 35–40 (in Japanese with English summary).

3409. Takada, Y. (1986) Foods of the feral house mouse, *Mus muculus molossinus*. *The Journal of the Mammalogical Society of Japan* **11**: 71–75 (in Japanese).

3410. Takada, Y., et al. (2006) Mitochondrial DNA polymorphisms in the raccoon (*Procyon lotor*) population. *Journal of Analytical Bio-Science* **29**: 167–170.

3411. Takada, Y., et al. (1999) Distribution of small mammals on the Izu Islands. *Bulletin of the Biogeographical Society of Japan* **54**: 9–19 (in Japanese with English abstract).

3412. Takada, Y., et al. (2012) Collection and distribution of rodents and insectivores on islands of the Inland Sea and Kyushu. *Bulletin of the Biogeographical Society of Japan* **67**: 81–92 (in Japanese with English abstract).

3413. Takada, Y., et al. (2013) Morphological variation among island populations of Japanese white-toothed shrews (*Crocidura dsinezumi*) in central and western Japan. *Mammalian Science* **53**: 67–77 (in Japanese with English abstract).

3414. Takagi, M., Personal communication.

3415. Takagi, M. & Higuchi, H. (1992) Habitat preference of the Izu Islands Thrush *Turdus celaenops* and the effect on weasel introduction on the population of the thrush on Miyake Island. *Strix* **11**: 47–57 (in Japanese with English abstract).

3416. Takahashi, K. (高橋啓一), et al. (2003) 佐浜ナウマンゾウ発掘調査で産出した脊椎動物化石について. *Shizuoka Chigaku* **87**: 15–21 (in Japanese).

3417. Takahashi, S. (高橋春成) (1995) 野生動物と野生化家畜. Tokyo: Daimeido (大明堂), 309 pp. (in Japanese).

3418. Takahashi, S. (高橋春成) (1998) ノヤギ. *in* 野生化哺乳類実態調査報告書 (Japan Wildlife Research Center, Editor). Tokyo: Japan Wildlife Research Center, pp. 113–123 (in Japanese).

3419. Takahashi, E., et al. (2014) Adaptive changes in echolocation sounds by *Pipistrellus abramus* in response to artificial jamming sounds. *Journal of Experimental Biology* **217**: 2885–2891.

3420. Takahashi, H. (1997) Huddling relationships in night sleeping groups among wild Japanese macaques in Kinkazan Island during winter. *Primates* **38**: 57–68.

3421. Takahashi, H. & Kaji, K. (2001) Fallen leaves and unpalatable plants as alternative foods for sika deer under food limitation. *Ecological Research* **16**: 257–262.

3422. Takahashi, K. (1988) Capturing examples of white-colored individuals of *Sorex unguiculatus* and *Clethrionomys rufocanus bedfordiae* in Hokkaido. *Forest Protection* **203**: 6 (in Japanese).

3423. Takahashi, K. & Ito, T. (1995) Ixodid ticks parasitic on sika deer (*Cervus nippon yesoensis*). *Bulletin of Bihoro Museum* **3**: 1–5 (in Japanese with English summary).

3424. Takahashi, K. & Nakata, K. (1995) Note on the first occurrence of larval *Echinococcus multilocularis* in *Clethrionomys rex* in Hokkaido, Japan. *Journal of Helminthology* **69**: 265–266.

3425. Takahashi, K. & Uraguchi, K. (2001) Sarcoptic mange among wildlife in Japan: Prevalence among red foxes in Hokkaido. *Infectious Agents Surveillance Report* **22**: 247–248 (in Japanese).

3426. Takahashi, K., Uraguchi, K., & Kudo, S. (2005) The epidemiological status of *Echinococcus multilocularis* in animals in Hokkaido, Japan. *Mammal Study* **30 (Suppl.)**: S101–S105.

3427. Takahashi, M. (1989) Ecological study of *Leptotrombidium* (*Leptotrombidium*) *pallidum*, 6. Morphological observations and ecological features of each developmental stage of *Leptotrombidium* (*Leptotrombidium*) *pallidum*, and its' life history observed in the containers placed in the field. *Bulletin of the Saitama Museum of Natural History* **7**: 7–24.

3428. Takahashi, M., et al. (1998) Sexing carcass remains of the sika deer (*Cervus nippon*) using PCR amplification of the *Sry* gene. *Journal of Veterinary Medical Science* **60**: 713–716.

3429. Takahashi, M., et al. (2001) Mixed infestation of sarcoptic and chorioptic mange mites in Japanese serow, *Capricornis crispus* Temminck, 1845 in Japan, with a description of *Chorioptes japonensis* sp. nov. (Acari: Psoroptidia). *Medical Entomology and Zoology* **52**: 297–306.

3430. Takahashi, M., et al. (2001) Mange caused by *Sarcoptes scabiei* (Acari: Sarcoptidae) in wild raccoon dogs, *Nyctereutes procyonoides*, in Kanagawa Prefecture, Japan. *Journal of Veterinary Medical Science* **63**: 457–460.

3431. Takahashi, M., et al. (1988) Studies on the life history of chigger mites in the laboratory 3. Observations of the spermatophores deposition of the chigger mites *Leptotrombidium* (*Leptotrombidium*) *miyazakii* and *L*. (*L*.) *intermedium*. *Bulletin of the Saitama Museum of Natural History* **6**: 1–10 (in Japanese with English abstract).

3432. Takahashi, S. (1980) Recent changes in the distribution of wild boars and the trade of their flesh in Japan. *Geographical Sciences* **34**: 24–31 (in Japanese).

3433. Takahashi, S. & Tisdell, C. (1989) The trade in wild pig meat: Australia and Japan. *Australian Geographer* **20**: 88–94.

REFERENCES

3434. Takahashi, T. (1986) Food habits and interspecific relationships in two red-backed voles, *Clethrionomys* in Hokkaido. *The Journal of the Mammalogical Society of Japan* **11**: 107–115.
3435. Takahashi, Y. & Kimura, K. (1993) A myological study of the musk shrew (*Suncus murinus riukiuanus*) I. gluteal and thigh muscles. *Acta Anatomica Nippon* **68**: 58–66.
3436. Takahata, C., et al. (2013) An evaluation of habitat selection of Asiatic black bears in a season of prevalent conflicts. *Ursus* **24**: 16–26.
3437. Takahata, Y. (1980) The reproductive biology of free-ranging troop of Japanese monkeys. *Primates* **21**: 303–329.
3438. Takahata, Y., et al. (1995) Are daughters more costly to produce for Japanese macaque mothers?: Sex of the offspring and subsequent interbirth intervals. *Primates* **36**: 571–574.
3439. Takahata, Y., Koyama, N., & Suzuki, S. (1995) Do the old aged females experience a long post-reproductive life span?: The cases of Japanese macaques and chimpanzees. *Primates* **36**: 169–180.
3440. Takahata, Y., et al. (1998) Reproduction of wild Japanese macaque females of Yakushima and Kinkazan islands: a preliminary report. *Primates* **39**: 339–349.
3441. Takahira, K. (高平兼司) (1982) 西表島大富洞における翼手類3種の日周期活動. *Majaa* **2**: 1–7 (in Japanese).
3442. Takai, F. (1962) Vertebrate fossils from the Tadaki Formation. *Journal of Anthropological Society of Nippon* **70**: 36–40 (in Japanese with English abstract).
3443. Takai, F. & Hasegawa, Y. (1971) On the fossil vertebrates of Ryukyu Islands, Japan. *in* Geological Problems of Kyushu and its Adjacent Seas (Annual Meeting of the Geological Society of Japan). pp. 107–109 (in Japanese).
3444. Takakuwa, Y. (高桒祐二) (2004) 青森県上北郡下田町錦ヶ丘の中部更新統から産出したニホンジカ化石. *in* 下田町錦ヶ丘産出のニホンジカ化石調査報告書 (Educational Board of Shimoda Town, Aomori Pref. (青森県下田町教育委員会), Editor). Shimoda: Educational Board of Shimoda Town, Aomori Pref., pp. 3–23 (in Japanese).
3445. Takakuwa, Y. (2006) Cervid Fossils from the Kiyokawa Formation of Shimosa Group, Sodegaura, Chiba Prefecture, Japan. *The Quaternary Research* **45**: 197–206 (in Japanese with English abstract).
3446. Takakuwa, Y., Anezaki, T., & Kimura, T. (2007) Fossil of the brown bear from Fuji-do cave, Ueno Village, Gunma Prefecture, Japan. *Bulletin of Gunma Museum of Natural History* **11**: 63–72 (in Japanese with English abstract).
3447. Takanawa, N. (高輪奈々) & Jiromaru, M. (次郎丸光帆) (2012) 利島に移動したと思われるミナミハンドウイルカの個体識別調査報告. *Mikurensis* **1**: 37–47 (in Japanese).
3448. Takara, T. (1954) Fauna of Senkaku Islands, Ryukyus. *Science bulletin of the Faculty of Agriculture, University of the Ryukyus* **1**: 57–74 (in Japanese).
3449. Takaragawa, N. (宝川範久) (1996) エゾリス (*Sciurus vulgaris orientis*). *in* The Encyclopaedia of Animals in Japan. 1. Mammals I (日本動物大百科1 哺乳類I) (Kawamichi, T. (川道武男), Editor). Tokyo: Heibonsha (平凡社), pp. 68–69 (in Japanese).
3450. Takasaki, H. (1981) Troop size, habitat quality, and home range area in Japanese macaques. *Behavioral Ecology and Sociobiology* **9**: 277–281.
3451. Takasaki, H. & Masui, K. (1984) Troop composition data of wild Japanese macaques reviewed by multivariate methods. *Primates* **25**: 308–318.
3452. Takatsuki, S. (高槻成紀) (1991) 草食獣の採食生態—シカを中心に— (Feeding ecology of ungulates with reference to cervids). *in* 現代の哺乳類学 (Asahi, M. (朝日稔) & Kawamichi, T. (川道武男), Editors). Tokyo: Asakura Publishing (朝倉書店), pp. 119–144 (in Japanese).
3453. Takatsuki, S. (1992) Foot morphology and distribution of Sika deer in relation to snow depth in Japan. *Ecological Research* **7**: 19–23.
3454. Takatsuki, S. & Ikeda, S. (1993) Botanical and chemical composition of rumen contents of sika deer on Mt. Goyo, northern Japan. *Ecological Research* **8**: 57–64.
3455. Takatsuki, S., Osugi, N., & Ito, T. (1988) A note on the food habits of the Japanese serow at the western foothill of Mt. Zao, northern Japan. *Journal of the Mammalogical Society of Japan* **13**: 139–142.
3456. Takatsuki, S. & Suzuki, K. (1984) Status and food habits of Japanese serow. *Proceedings of the Biennial Symposium of the Northern Wild Sheep & Goat Council* **4**: 231–240.
3457. Takatsuki, S., Suzuki, K., & Higashi, H. (2000) Seasonal elevational movements of sika deer on Mt. Goyo, northern Japan. *Mammal Study* **25**: 107–114.
3458. Takayama, N. (高山直秀) (2000) ヒトの狂犬病：忘れられた死の病. Tokyo: Jiku Publishers (時空出版), 130 pp. (in Japanese).
3459. Takayanagi, A. (2007) Relocation and monitoring for control and conservation of bears. *Mammalian Science* **47**: 143–144 (in Japanese).
3460. Takazawa, Y., et al. (1990) Reproductive biology of the female oriental frosted bat (*Vespertilio orientalis*) in northern Japan. I. The ovary. *Bulletin of the College of Agriculture and Veterinary Medicine, Nihon University* **47**: 145–149 (in Japanese with English abstract).
3461. Takeda, H., Oh-ueno, Y., & Nakata, K. (1999) A note on an individual of *Clethrionomys rufocanus bedofordiae* with black colored coat captured in Hokkaido. *Forest Protection* **273**: 40 (in Japanese).
3462. Takeda, M. & Ogino, M. (2005) Record of a whale louse, *Cyamus scammoni* Dall (Crustacea: Amphipoda: Cyamidae), from the gray whale strayed into Tokyo Bay, the Pacific coast of Japan. *Bulletin of the National Science Museum, Series A* **31**: 151–156.
3463. Takemura, A. (竹村暘) (1986) イルカ類の生物学的特性値. *in* 漁業公害 (有害生物駆除) 対策調査委託事業調査報告書 (昭和56–60年度) (Tamura, T. (田村保), Ohsumi, S. (大隅清治), & Arai, S. (荒井修亮), Editors). Tokyo: Fisheries Agency, pp. 161–187 (in Japanese).
3464. Takemura, A. (竹村暘) (1986) 食性と生態系における地位. *in* 漁業公害 (有害生物駆除) 対策調査委託事業調査報告書 (昭和56–60年度) (Tamura, T. (田村保), Ohsumi, S. (大隅清治), & Arai, S. (荒井修亮), Editors). Tokyo: Fisheries Agency, pp. 187–195 (in Japanese).
3465. Takemura, A. (竹村暘) (1996) ハンドウイルカ. *in* The Encyclopaedia of Animals in Japan. 2. Mammals II (日本動物大百科2 哺乳類II) (Izawa, K. (伊沢紘生), Kasuya, T. (粕谷俊雄), & Kawamichi, T. (川道武男), Editors). Tokyo: Heibonsha (平凡社), pp. 68–70 (in Japanese).
3466. Taketazu, M. (竹田津実) (1996) イイズナ. *in* The Encyclopaedia of Animals in Japan. 1. Mammals I (日本動物大百科1 哺乳類I) (Kawamichi, T. (川道武男), Editor). Tokyo: Heibonsha (平凡社), p. 135 (in Japanese).
3467. Takeuchi, M. (1995) Morphological and ecological study of the red fox (*Vulpes vulpes*) in Tochigi, central Japan. Doctoral dissertation. Kanazawa: Kanazawa University. 139 pp.
3468. Takeuchi, M. & Koganezawa, M. (1992) Home range and habitat utilization of the red fox *Vulpes vulpes* in the Ashio Mountains, central Japan. *Journal of the Mammalogical Society of Japan* **17**: 95–110.
3469. Takeyama, K., Kondo, N., & Asakawa, M. (2013) Ecto- and endoparasites obtained from bats (Chiroptera) captured on Hokkaido, Japan. *Memoirs of the Nemuro City Museum of History and Nature* **25**: 21–29 (in Japanese with English summary).
3470. Takiguchi, H., et al. (2012) Genetic variation and population structure of the Japanese sika deer (*Cervus nippon*) in the Tohoku District based on mitochondrial D-loop sequences. *Zoological Science* **29**: 433–436.
3471. Takiguchi, M. Unpublished.
3472. Takiguchi, M. Unpublished (Hearing investigation in Minamisatsuma City).
3473. Takiguchi, M. Unpublished (Hearing investigation in Mishima Village).
3474. Takikawa Environment Forum (2014) 北海道滝川市内の農機具庫を利用するコウモリたち. カグヤコウモリ *Myotis frater* の出産・哺育集団を中心に. Takikawa: Takikawa Environment Forum, 64 pp. (in Japanese).
3475. Talbot, S.L. & Shields, G.F. (1996) A phylogeny of the bears (Ursidae) inferred from complete sequences of three mitochondrial genes. *Molecular Phylogenetics and Evolution* **5**: 567–575.
3476. Tamada, T., et al. (2008) Molecular diversity and phylogeography of the Asian leopard cat, *Felis bengalensis*, inferred from mitochondrial and Y-chromosomal DNA sequences. *Zoological Science* **25**: 154–163.
3477. Tamate, H. (2013) Genetic diversity in the mammalian fauna of Japan: past, present and future. *Chikyu Kankyo* (地球環境) **17**: 159–166 (in Japanese).
3478. Tamate, H.B. (2009) Evolutionary significance of admixture and fragmentation of sika deer populations in Japan. *in* Sika Deer—Biology and Management of Native and Introduced Populations (McCullough, D.R., Takatsuki, S., & Kaji, K., Editors). Tokyo: Springer, pp. 43–60.
3479. Tamate, H.B., et al. (2000) Genetic differentiation among subspecies of the sika deer (*Cervus nippon*), with special reference to the phylogeny of *C. n. keramae* in the Kerama island group. *Tropics* **10**: 73–78.
3480. Tamate, H.B., et al. (2000) Genetic variations revealed by microsatellite markers in a small population of the sika deer (*Cervus nippon*) on Kinkazan island, northern Japan *Zoological Science* **17**: 47–53.
3481. Tamate, H.B., et al. (1995) Assessment of genetic variations within populations of sika deer in Japan by analysis of randomly amplified polymorphic DNA (RAPD). *Zoological Science* **12**: 669–673.
3482. Tamate, H.B., et al. (1998) Mitochondrial DNA variations in local populations of the Japanese Sika deer, *Cervus nippon*. *Journal of Mammalogy* **79**: 1396–1403.
3483. Tamate, H.B. & Tsuchiya, T. (1995) Mitochondrial DNA polymorphism in subspecies of the Japanese sika deer, *Cervus nippon*. *Journal of Heredity* **86**: 211–215.
3484. Tamiya, T., ed. (1962) Recent advances in studies of Tsutsugamushi disease in Japan. Tokyo: Medical Culture, 309 pp.
3485. Tamura, N. (田村典子) (2002) タイワンリス (*Callosciurus erythraeus thaiwanensis*). *in* Handbook of Alien Species in Japan (外来種ハンドブック) (Ecological Society of Japan, Editor). Tokyo: Chijin Shokan (地人書館), p. 66 (in Japanese).

3486. Tamura, N. (田村典子) (2005) ニホンリスとタイワンリス. 森林科学 **44**: 37–41 (in Japanese).
3487. Tamura, N. Unpublished.
3488. Tamura, N. (1989) Sociobiological studies on the Formosan squirrel, *Callosciurus erythraeus thaiwanensis* (Bonhote). Doctoral dissertation. Tokyo: Tokyo Metropolitan University. 138 pp.
3489. Tamura, N. (1995) Postcopulatory mate guarding by vocalization in the Formosan squirrel. *Behavioral Ecology and Sociobiology* **36**: 377–386.
3490. Tamura, N. (1998) Forest type selection by the Japanese squirrel, *Sciurus lis*. *Japanese Journal of Ecology* **48**: 123–127 (in Japanese).
3491. Tamura, N. (1999) Seasonal change in reproductive states of the Formosan squirrel on Izu-Oshima Island, Japan. *Mammal Study* **24**: 121–124.
3492. Tamura, N. (2004) Effects of habitat mosaic on home range size of the Japanese squirrel, *Sciurus lis*. *Mammal Study* **29**: 9–14.
3493. Tamura, N. (2008) Pine wilt disease and squirrel distribution in Yamanashi Prefecture. *Journal of the Japanese Wildlife Reserach Society* **33**: 20–24 (in Japanese).
3494. Tamura, N., Aikyo, C., & Kataoka, T. (2007) Habitat evaluation of red pine forest as a habitat of Japanese squirrels, *Sciurus lis*. *Journal of the Japanese Forest Society* **89**: 71–75 (in Japanese with English abstract).
3495. Tamura, N., Hashimoto, Y., & Hayashi, F. (1999) Optimal distances for squirrels to transport and hoard walnuts. *Animal Behaviour* **58**: 635–642.
3496. Tamura, N. & Hayashi, F. Unpublished.
3497. Tamura, N. & Hayashi, F. (2007) Five-year study of the genetic structure and demography of two subpopulations of the Japanese squirrel (*Sciurus lis*) in a continuous forest and an isolated woodlot. *Ecological Reserach* **22**: 261–267.
3498. Tamura, N., Hayashi, F., & Miyashita, K. (1988) Dominance hierarchy and mating behavior of the Formosan squirrel, *Callosciurus erythraeus thaiwanensis*. *Journal of Mammalogy* **69**: 320–331.
3499. Tamura, N., Hayashi, F., & Miyashita, K. (1989) Spacing and kinship in the Formosan squirrel living in different habitats. *Oecologia* **79**: 344–352.
3500. Tamura, N., Katsuki, T., & Hayashi, F. (2005) Walnut seed dispersal: Mixed effects of tree squirrels and field mice with different hoarding ability. *in* Seed Fate: Predation, Dispersal, and Seedling Establishment (Forget, P.-M., et al., Editors). Wallingford: CABI Publishing, pp. 241–252.
3501. Tamura, N., et al. (2007) Current distribution of *Sciurus lis* in the Chugoku district, Japan. *Mammalian Science* **47**: 231–237 (in Japanese).
3502. Tamura, N., et al. (2004) Environmental factors affecting distribution of the alien Formosan squirrel in fragmented landscape. *Ecology and Civil Engineering* **6**: 211–218 (in Japanese with English summary).
3503. Tamura, N. & Miyashita, K. (1984) Diurnal activity of the Formosan squirrel, *Callosciurus erythraeus thaiwanensis*, and its seasonal change with feeding. *Journal of the Mammalogical Society of Japan* **10**: 37–40.
3504. Tamura, N. & Ohara, S. (2005) Chemical components of hardwood barks stripped by the alien squirrel *Callosciurus erythraeus* in Japane. *Journal of Forest Research* **10**: 429–433.
3505. Tamura, N. & Shibasaki, E. (1996) Fate of walnut seeds, *Juglans ailanthifolia*, hoarded by Japanese squirrels, *Sciurus lis*. *Journal of Forest Research* **1**: 219–222.
3506. Tamura, N., Takahashi, B., & Satou, N. (2006) Habitat variables of the Japanese squirrel identified by regression tree model. *Mammal Study* **31**: 1–8.
3507. Tamura, N. & Terauchi, M. (1994) Variation in body weight among three populations of the Formosan squirrel *Callosciurus erythraeus thaiwanensis*. *Journal of the Mammalogical Society of Japan* **19**: 101–111.
3508. Tamura, T. & Fujise, Y. (2000) Geographical and seasonal changes of prey species in the western North Pacific minke whale (Paper SC/F2K/J22 presented to the IWC workshop). 26 pp.
3509. Tamura, T., et al. (2007) Cruise report of the second phase of the Japanese Whale Research Program under Special Permit in the western North Pacific (JARPN II) in 2006 (part 1)—offshore component—(Paper SC/59/O5 presented to the IWC Scientific Committee). 26 pp.
3510. Tamura, T., et al. (2006) Cruise report of the second phase of the Japanese Whale Research Program under Special Permit in the western North Pacific (JARPN II) in 2005—Offshore component (Paper SC/58/O8 presented to the IWC Scientific Committee). 52 pp.
3511. Tanaka, H. (田中浩) (2002) ニホンアナグマの生態と社会システム (Ecology and social system of the Japanese badger, *Meles meles anakuma* (Carnivora; Mustelidae) in Yamaguchi, Japan. Yamaguchi: Yamaguchi University. 117 pp. (in Japanese).
3512. Tanaka, H. (2005) Seasonal and daily activity patterns of Japanese badgers (*Meles meles anakuma*) in Western Honshu, Japan. *Mammal study* **30**: 11–17.
3513. Tanaka, H. (2006) Winter hibernation and body temperature fluctuation Japanese badger, *Mele meles anakuma*. *Zoological Science* **23**: 991–997.
3514. Tanaka, H., Yamanaka, A., & Endo, K. (2002) Spatial distribution and sett use by the Japanese badger, *Meles meles anakuma*. *Mammal Study* **27**: 15–22.
3515. Tanaka, I. (1982) Rodents and their Control. Pest Control Series No.1. Kawasaki: Japan Environmental Sanitation Center, 115 pp. (in Japanese).
3516. Tanaka, M., et al. (2004) Mitochondrial genome variation in Eastern Asia and the peopling of Japan. *Genome Research* **14**: 1832–1850.
3517. Tanaka, M. & Shibata, E. (2006) Abundance and microhabitat preference of small rodents in man-made forests. *Nagoya University Forest Science* **25**: 1–6 (in Japanese with English summary).
3518. Tanaka, R. (1940) A biostatistical analysis of *Apodemus agraius* (Pallas) from Formosa with special reference to its systematic characters. *Memoirs of the Faculty of Science and Agriculture, Taihoku Imperial University* **23**: 211–285.
3519. Tanaka, R. (1944) Comparative analysis of two species of *Mus* from Formosa from the standpoint of variation-statistics. *Memoirs of the Faculty of Science and Agriculture, Taihoku Imperial University* **1**: 1–64.
3520. Tanaka, R. (1944) Comparative study on Formosan rats and mice of four species belonging to the genera *Rattus*, *Apodemus*, and *Mus* from the standpoint of variation-statistics. *Memoirs of the Faculty of Science and Agriculture, Taihoku Imperial University* **2**: 65–118 (in Japanese with English abstract).
3521. Tanaka, R. (1953) Home range and territories in *Clethrionomys* population on a peat-bog grassland in Hokkaido. *Bulletin of Kochi Women's College* **2**: 10–20.
3522. Tanaka, R. (1962) A population ecology of rodent hosts of the scrub-typhus vector of Shikoku district with special reference to their true range in Japan. *Japanese Journal of Zoology* **13**: 395–402.
3523. Tanaka, R. (1964) Population dynamics of the Smith's red-backed vole in highland of Shikoku. *Researches on Population Ecology* **6**: 54–66.
3524. Tani, S. (1976) Studies on *Echinostoma hortense* (Asada, 1926) (2) The intermediate and final hosts in Akita Prefecture. *Japanese Journal of Parasitology* **25**: 461–467 (in Japanese with English abstract).
3525. Taniguchi, A. (谷口明) (1986) 鹿児島県におけるノウサギによる造林木の被害とその個体群生態に関する研究. *Research Report of Kagoshima Prefectural Forest Experiment Station* **2**: 1–38 (in Japanese).
3526. Taniguchi, H. (1983) Comparative morphological study of molar teeth of *Macaca* Genus of primates. *Kyushu Shika Gakkai Zasshi (Journal of the Kyushu Dental Society)* **37**: 979–1003 (in Japanese with English summary).
3527. Taniguchi, I. (1985) Echolocation sounds and hearing of the greater Japanese horseshoe bat (*Rhinolophus ferrumequinum nippon*). *Journal of Comparative Physiology, A* **156**: 185–188.
3528. Tanizaki, M. (谷崎美由記), et al. (2009) 北海道帯広市のコウモリ用ボックスカルバートのモニタリング (続報). 「野生生物と交通」研究発表会講演論文集 **8**: 95–102 (in Japanese).
3529. Tano, N., et al. (1994) Ecological study of Japanese serow (*Capricornis crispus*) at subalpine zone (I) Home range. *Transactions of the Japanese Forestry Society* **105**: 543–546 (in Japanese).
3530. Taruno, H. (樽野博幸) & Ishii, M. (石井みき子) (1978) 森の宮遺跡出土の動物遺体 (第3次調査) *in* 森の宮遺跡第3・4次発掘調査報告書 (Naniwanoato Kenshokai (難波宮址顕彰会), Editor). Osaka: Naniwanoato Kenshokai, pp. 160–165 (in Japanese).
3531. Tashima, S., et al. (2011) Identification and molecular variations of CAN-SINEs from the *ZFY* gene final intron of the Eurasian badgers (genus *Meles*). *Mammal Study* **36**: 41–48.
3532. Tashima, S., et al. (2011) Phylogeographic sympatry and isolation of the Eurasian badgers (*Meles*, Mustelidae, Carnivora): implication for an alternative analysis using maternally as well as paternally inherited genes. *Zoological Science* **28**: 293–303.
3533. Tashima, S., et al. (2010) Genetic diversity among the Japanese badger (*Meles anakuma*) populations, revealed by microsatellite analysis. *Mammal Study* **35**: 221–226.
3534. Tashiro, M. (田代道彌) (1989) 動物. *in* 南足柄市史 (1)：資料編. 自然. Minamiashigara: Minamiashigara City Office, pp. 191–242 (in Japanese).
3535. Tasumi, M. (田隅本生) (1991) ニホンオオカミの実態を頭骨から探る. *The Bone* **6**: 119–128 (in Japanese).
3536. Tatara, M. (1994) Ecology and conservation status of the Tsushima marten. *in* Martens, Sables, and Fishers: Biology and Conservation (Buskirk, S.W., et al., Editors). New York: Comstock Publishing, pp. 272–280.
3537. Tatara, M. (1994) Notes on the breeding ecology and behavior of Japanese martens on Tsushima Islands, Japan. *Journal of the Mammalogical Society of Japan* **19**: 67–74.
3538. Tatara, M. (1994) Social system and habitat ecology of the Japanese marten *Martes mepampus tsuensis* (Carnivora; Mustelidae) on the islands of Tsushima. Doctoral dissertation. Fukuoka: Kyushu University. 79 pp.
3539. Tatara, M. & Doi, T. (1994) Comparative analyses on food habits of

Japanese marten, Siberian weasel and leopard cat in the Tsushima Islands, Japan. *Ecological Research* **9**: 99–107.

3540. Tate, G.H.H. (1942) Results of the Archbold expeditions. No.47. Review of the vespertilionid bats, with special attention to genera and species of the Archbold collections. *Bulletin of the American Museum of Natural History* **80**: 221–297.

3541. Tateishi, T. (2006) Reproductive activity of the small Japanese field mouse (*Apodemus argenteus*) in the Oze district of northeastern Honshu, Japan. *Mammalian Science* **46**: 161–167 (in Japanese with English abstract).

3542. Tatsugami, M., et al. (2006) Bat fauna of Hidaka and Tokachi districts, central and eastern Hokkaido. (7). Capture record of bats in the Inada and Kawanishi areas of Obihiro City. *Journal of the Japanese Wildlife Research Society* **32**: 11–15 (in Japanese with English summary).

3543. Tatsukawa, K. & Murakami, O. (1976) On the food utilization of the Japanese wood mouse *Apodemus speciosus* (Mammalia: Murideae). *Physiology and Ecology Japan* **17**: 133–144 (in Japanese with English synopsis).

3544. Tatsukawa, T. & Ishibashi, T. (2014) Live strandings and rescue operations of finless porpoises at the coats of the western areas of Inland Sea. *Aquabiology* **36**: 150–155 (in Japanese with English abstract).

3545. Taylor, K.R., et al. (2013) Differential tick burdens may explain differential *Borrelia afzelii* and *Borrelia garinii* rates among four, wild, rodent species in Hokkaido, Japan. *Journal of Veterinary Medical Science* **75**: 785–790.

3546. Teduka, M. (手塚牧人) (1992) ニホンコテングコウモリの捕獲記録. *FIELD NOTE* **36**: 34 (in Japanese).

3547. Temminck, C.J. (1842–1844) Aperçu général et spécifique sur les mammifères qui habitent le Japon et les îles qui en dépendent. *in* Fauna Japonica (de Siebold, P.F., Temminck, C.J., & Schlegel, H., Editors). Lugduni Batavorum: A. Arnz et Socios, pp. 1–59, pls. 1–20 (in French).

3548. Terada, C., Tatsuzawa, S., & Saitoh, T. (2012) Ecological correlates and determinants in the geographical variation of deer morphology. *Oecologia* **169**: 981–994.

3549. Terada, C., et al. (2013) New mtDNA haplotypes of the sika deer (*Cervus nippon*) found in Hokkaido, Japan suggest human-mediated immigration. *Mammal Study* **38**: 123–129.

3550. Teranishi, T. (寺西敏夫) (1994) 三重県北西地方の洞穴棲コウモリ. *Bat Study and Conservation Report* **2**: 10–12 (in Japanese).

3551. Teranishi, T. (寺西敏夫) (2008) 篠立の風穴とコウモリ (Life of bat in the Shinodachi-no-kazaana limestone-cave Mie Pref.) *in* 第二次篠立の風穴自然科学調査報告書 (A Report on the Second Investigation of 'Shinodachi-no-Kaza-ana' Limestone-cave, a Natural Monument Designated by Mie prefecture). Inabe: Group of Natural Science Investigators to the Second Investigation of Shinodachi-no-Kaza-ana (第二次篠立の風穴自然科学調査会), pp. 45–64 (in Japanese).

3552. Teranishi, T. (2002) Records of bats excluding *Pipistrellus abramus*, Aichi Prefecture. *Special Publication of Nagoya Society of Mammalogists* (マンモ・ス) **4**: 3–13 (in Japanese).

3553. Teranishi, T. (2003) Distributional records for bats in Shiga Prefecture: Additional records in 1999~2001. *Bulletin of the Shiga Society of Naturalists* **5**: 11–20 (in Japanese).

3554. Teranishi, T. (2008) The least horseshoe bat (*Rhinolophus cornutus*) from Ohtsudou, Mie Prefecture, Japan: the records of its longevity and migration. *Special Publication of Nagoya Society of Mammalogists* **10**: 9–14 (in Japanese).

3555. Teranishi, T., Harada, I., & Harada, C. (2009) Record of Ussurian tube-nosed bat, *Murina ussuriensis* in Higashi-Mikawa area, Aichi, Japan. *Special Publication of Nagoya Society of Mammalogists* **11**: 35–38 (in Japanese).

3556. Terashima, M., et al. (2006) Phylogeographic origin of Hokkaido house mice (*Mus musculus*) as indicated by genetic markers with maternal, paternal and biparental inheritance. *Hereditas* **96**: 123–138.

3557. Terashima, M., et al. (2003) Geographic variation of *Mus caroli* from East and Southeast Asia based on mitochondrial cytochrome *b* gene sequences. *Mammal Study* **28**: 67–72.

3558. Tezuka, H. (1958) Notes on thigmotaxis of *Talpa wogura wogura*. *The Journal of the Mammalogical Society of Japan* **1**: 84–86 (in Japanese with English summary).

3559. Theberge, J.B. (1991) Ecological classification, status, and management of the gray wolf, *Canis lupus*, in Canada. *Canadian Field-Naturalist* **105**: 459–463.

3560. Thomas, J., et al. (1980) Observations of a newborn Ross seal pup (*Ommatophoca rossi*) near the Antarctic Peninsula. *Canadian Journal of Zoology* **58**: 2156–2158.

3561. Thomas, N. (1997) A systematic review of selected Afro-Asiatic Rhinolophidae (Mammalia: Chiroptera): an evaluation of taxonomic methodologies. Doctoral dissertation. Sevenoaks: University of Aberdeen/Harrison Zoological Museum. 211 pp.

3562. Thomas, O. (1905) Abstract. *Proceedings of Zoological Society of London* No. 23: 18–19.

3563. Thomas, O. (1905) On some new Japanese mammals presented to the British Museum by Mr. R. Gordon Smith. *Annals and Magazine of Natural History: Series 7* **15**: 487–495.

3564. Thomas, O. (1906) The Duke of Bedford's zoological exploration in Eastern Asia. I. List of mammals obtained by Mr. M. P. Anderson in Japan. *Proceedings of the Zoological Society of London* **1905**: 331–363.

3565. Thomas, O. (1906) On a second species of *Lenothrix* from the Liu Kiu Islands. *Annals and Magazine of Natural History, including Zoology, Botany and Geology. Series 7* **17**: 88–89.

3566. Thomas, O. (1907) The Duke of Bedford's zoological exploration in Eastern Asia. IV. List of mammals from the islands of Saghalin and Hokkaido. *Proceedings of the Zoological Society of London* **1907**: 404–414.

3567. Thomas, O. (1915) On bats of the genera *Nyctalus*, *Tylonycteris*, and *Pipistrellus*. *Proceedings of the Zoological Society of London* **15**: 225–232.

3568. Thomas, O. (1920) Two new Asiatic bats of the genera *Tadarida* and *Dyacopterus*. *Annals Magazine Natural History* **5**: 283–285.

3569. Thong, V.D., et al. (2012) Systematics of the *Hipposideros turpis* complex and a description of a new subspecies from Vietnam. *Mammal Review* **42**: 166–192.

3570. Thorington, R.W., Jr. & Hoffmann, R.S. (2005) Family Sciuridae. *in* Mammal Species of the World: A Taxonomic and Geographic Reference (Wilson, D.E. & Reeder, D.M., Editors). Baltimore: Johns Hopkins University Press, pp. 754–818.

3571. Thorsteinson, F.V. & Lensink, C.J. (1962) Biological observations of Steller sea lions taken during an experimental harvest. *Journal of Wildlife Management* **26**: 353–359.

3572. Thulin, C.G., et al. (2006) Genetic divergence in the small Indian mongoose (*Herpestes auropunctatus*), a widely distributed invasive species. *Molecular Ecology* **15**: 3947–3956.

3573. Tian, L., et al. (2004) Molecular studies on the classification of *Miniopterus schreibersii* (Chiroptera: Vespertilionidae) inferred from mitochondrial cytochrome *b* sequences. *Folia Zoologica* **53**: 303–311.

3574. Tiba, T., et al. (1988) An annual rhythm in reproductive activities and sexual maturation in male Japanese serows (*Capricornis crispus*). *Zeitschrift für Säugetierkunde* **53**: 178–187.

3575. Tikel, D. (1997) Using a Genetic Approach to Optimise Dugong (*Dugong dugon*) Conservation Management. Doctoral dissertation. Townsville: James Cook University of North Queensland. 277 pp.

3576. Tikhomirov, E.A. (1961) Distribution and migration of seals in waters of the Far East. *Trudi Soveshchaniy Ikhtiologicheskoi Komissii Akademii Nauk SSSR* **12**: 199–210 (in Russian).

3577. Tinker, S.W. (1988) Whales of the World. Leiden: E. J. Brill, 310 pp.

3578. Tobayama, T., Nishiwaki, M., & Yang, H.C. (1973) Records of the Fraser's Sarawak dolphin (*Lagenodelphis hosei*) in the western north Pacific. *Scientific Reports of the Whales Research Institute, Tokyo* **25**: 251–263.

3579. Todd, N.B., Mulvaney, D.A., & Pressman, S.R. (1969) Chromosomal polymorphism in the raccoon dog (*Nyctereutes* sp.). *Carnivore Genetic Newsletter* **1**: 167–168.

3580. Tojima, S. (2013) Tail length estimation from sacrocaudal skeletal morphology in catarrhines. *Anthropological Science* **121**: 13–24.

3581. Tokashiki, S. (渡嘉敷綏宝) (1984) 沖縄の山羊. Okinawa: Nahashuppan-sha (那覇出版社), 184 pp. (in Japanese).

3582. Tokida, K. (常田邦彦) (2002) ヤギ (ノヤギ) *Capra hircus*. *in* Handbook of Alien Species in Japan (外来種ハンドブック) (Ecological Society of Japan, Editor). Tokyo: Chijin Shokan (地人書館), pp. 80–81 (in Japanese).

3583. Tokida, K. (常田邦彦) (2006) 自然保護公園におけるシカ問題：人とシカのかかわりの歴史をふまえて. *in* 世界遺産をシカが喰う：シカと森の生態学 (Yumoto, T. (湯本貴和) & Matsuda, H. (松田裕之), Editors). Tokyo: Bun-ichi Sogo Shyuppan (文一総合出版), pp. 20–27 (in Japanese).

3584. Tokida, K. (2006) The feral goat eradication program on the Ogasawara Islands. *Mammalian Science* **46**: 93–94 (in Japanese with English abstract).

3585. Tokida, K. & Ikeda, H. (1988) Present status of Japanese serow (*Capricornis crispus crispus*): distribution and density. *in* A Fundamental Research on Ecology and Conservation of the Japanese Serow (A Report for the Research Fund by Japan Society for the Promotion of Science) (Ono, Y., Editor). Tokyo: Japan Society for the Promotion of Science, pp. 1–10.

3586. Tokida, K. & Miura, S. (1988) Mortality and life table of a Japanese serow, (*Capricornis crispus*) population in Iwate Prefecture, Japan. *Journal of the Mammalogical Society of Japan* **13**: 119–126.

3587. Tokida, K., et al. (2004) A deer management approach to promote ecosystem management in National Parks: A case study of sika deer in Shiretoko, Hokkaido Island, Japan. *Journal of Conservation Ecology* **9**: 193–202 (in Japanese with English abstract).

3588. Tokita, K. (時田克夫), Yokoyama, K. (横山恵一), & Mukohyama, M. (向山満) (2001) 哺乳類. in 平成12年度早池峰地区自然環境調査報告書. Morioka: Iwate Prefectural Government, pp. 280–293 (in Japanese).

3589. Tokuda, M. (徳田御稔) (1951) イタチの棲み分け. 科学朝日 (Scientific Asahi) **11**: 38–39 (in Japanese).

3590. Tokuda, M. (徳田御稔) (1969) 生物地理学. Tokyo: Tsukiji-shokan (築地書館), 199 pp. (in Japanese).

3591. Tokuda, M. (1935) *Neoaschizomys*, a new genus of Microtinae from Sikotan, a south Kurile island. *Memoirs of the College of Science, Kyoto Imperial University, Series B* **X-3**: 241–250.

3592. Tokuda, M. (1941) A revised monograph of the Japanese and Manchou–Korean Muridae. Tokyo: Biogeographical Society of Japan, 155 pp.

3593. Tokyo Metropolitan Government (1974) 東京における野生哺乳動物の生息地のうつり変わり. in 自然環境保全に関する基礎調査報告書 (1) 野生哺乳動物. Tokyo: Tokyo Metropolitan Government, pp. 15–16 (in Japanese).

3594. Tokyo Metropolitan Government (2007) 平成18年度特定外来生物 (キョン) 生息実態調査報告書. 60 pp. (in Japanese).

3595. Tokyo Metropolitan Government (2011) 特定外来生物 (キョン) 生息状況等調査委託報告書. 100 pp. (in Japanese).

3596. Tokyo Metropolitan Government, Ogasawara Islands Branch Office & Japan Wildlife Research Center (2005) 小笠原国立公園兄島植生回復調査委託報告書. Tokyo: Japan Wildlife Research Center, 80 pp. (in Japanese).

3597. Tokyo Metropolitan Government, Ogasawara Islands Branch Office & Japan Wildlife Research Center (2006) 小笠原国立公園兄島植生回復調査委託報告書. Tokyo: Japan Wildlife Research Center, 104 pp. (in Japanese).

3598. Tokyo Metropolitan Government, Ogasawara Islands Branch Office & Japan Wildlife Research Center (2006) 小笠原国立公園兄島植生回復調査委託報告書. Tokyo: Japan Wildlife Research Center, 174 pp. (in Japanese).

3599. Tokyo Metropolitan Government, Ogasawara Islands Branch Office & Japan Wildlife Research Center (2007) 小笠原国立公園兄島植生回復調査委託報告書. Tokyo: Japan Wildlife Research Center, 105 pp. (in Japanese).

3600. Tokyo Metropolitan Government, Ogasawara Islands Branch Office & Japan Wildlife Research Center (2009) 小笠原国立公園兄島・弟島植生回復調査委託報告書. Tokyo: Japan Wildlife Research Center, 173 pp. (in Japanese).

3601. Tolley, K.A. & Rosel, P.E. (2006) Population structure and historical demography of eastern North Atlantic harbour porpoises inferred through mtDNA sequences. *Marine Ecology* **327**: 297–308.

3602. Tomida, S. (1978) On the Quaternary cave and fissure deposits and vertebrate fossils from Yage Quarry, near Lake Hamana, central Japan. *Bulletin of the Mizunami Fossil Museum* **5**: 113–141 with plates (in Japanese with English abstract).

3603. Tomida, Y. (1979) Mammalian fauna of Mie Prefecture. *Bulletin of the Mie Prefectural Museum, Natural Science* **1**: 5–68 (in Japanese with English summary).

3604. Tomida, Y. & Jin, C. (2002) Morphological evolution of the genus *Pliopentalagus* based on the fossil material from Anhui Province, China: a preliminary study. Proceedings of the Third and Fourth Symposia on Collection Building and Natural History Studies in Asia and the Pacific Rim. *National Science Museum Monographs* **22**: 97–107.

3605. Tomida, Y. & Otsuka, H. (1993) First discovery of fossil Amami rabbit (*Pentalagus furnessi*) from Tokunoshima, Southern Japan. *Bulletin of the National Science Museum, Tokyo. Series C* **19**: 73–79.

3606. Tomilin, A.G. (1957) Mammals of the U.S.S.R. and Adjacent Countries. Vol. 9. Cetacea (Heptner, V.G., Editor). Moscow: Izdatel'stvo Akademi Nauk SSSR, 717 pp. (Translated from Russian by the Israel Program for Scientific Translations, Jerusalem, 1967).

3607. Tomisawa, A. (富沢章) (1990) コウモリの捕食した蛾類2 (List of moths fallen prey to bats. 2). *Journal of Research on Moths* **120**: 65–68 (in Japanese).

3608. Tomiyama, K. (冨山清升) (1984) 宇治向島採集記 (その一). *Reports of the Kyushu Mollusc Study Society* **23**: 25–34 (in Japanese).

3609. Tomozawa, M. & Suzuki, H. (2008) A trend of central versus peripheral structuring in mitochondrial and nuclear gene sequences of the Japanese wood mouse, *Apodemus speciosus*. *Zoological Science* **25**: 273–285.

3610. Topal, G. (1993) Taxonomic status of *Hipposideros larvatus alongensis* Bourret, 1942 and occurrence of *H. turpis* Bangs, 1901 in Vietnam (Mammalia, Chiroptera). *Acta Zoologica Hungarica* **39**: 267–288.

3611. Torii, H. (鳥居春己) (1989) 静岡県の哺乳類. 静岡県の自然環境シリーズ. Tokyo: Dai-ichi Hoki (第一法規出版), 231 pp. (in Japanese).

3612. Torii, H. (鳥居春己) (1991) 静岡県におけるハクビシンの分布 (Distribution of the masked palm civet, *Paguma larvata*, in Shizuoka prefecture). *Bulletin of Shizuoka Prefecture Forestry and Forest Products Research Institute* **19**: 35–42 (in Japanese).

3613. Torii, H. (鳥居春己) (1996) 静岡県内市町村別のハクビシンの分布 (Distribution of the masked palm civet in Shizuoka Prefecture by municipality). in 静岡県ハクビシン調査報告書. Shizuoka: Shizuoka Prefectural Government, pp. 1–7 (in Japanese).

3614. Torii, H. (鳥居春己) & Ohba, T. (大場孝裕) (1996) ハクビシンの行動域について (The home range of the masked palm civet). in 静岡県ハクビシン調査報告書. Shizuoka: Shizuoka Prefectural Government, pp. 13–28 (in Japanese).

3615. Torii, H. (鳥居春己) & Tezuka, M. (手塚牧人) (1996) ハクビシンの糞内容物分析 (The food of the masked palm civet by droppings contents analysis). in 静岡県ハクビシン調査報告書. Shizuoka: Shizuoka Prefectural Government, pp. 33–39 (in Japanese).

3616. Torii, A. & Nigi, H. (1998) Successful artificial insemination for indoor breeding in the Japanese monkey (*Macaca fuscata*) and cynomolgus monkey (*Macaca fascicularis*). *Primates* **39**: 399–406.

3617. Torii, H. Unpublished.

3618. Torii, H. (1986) Food habits of the masked palm civet, *Paguma larvata* Hamilton-Smith. *The Journal of the Mammalogical Society of Japan* **11**: 39–43.

3619. Torii, H. (1990) Survey of hare tracks on the snow. *Journal of the Japanese Society for Hares* **17**: 21–28 (in Japanese with English abstract).

3620. Torii, H. (1993) Hinoki damage caused by Formosan gray-headed squirrels. *Bulletin of Shizuoka Prefecture Forestry and Forest Product Research Institute* **21**: 1–7 (in Japanese).

3621. Torii, H. & Miyake, T. (1986) Litter size and sex ratio of the masked palm civet, *Paguma larvata*, in Japan. *The Journal of the Mammalogical Society of Japan* **11**: 35–38.

3622. Toriumi, H. (鳥海隼夫) (1967) ウサギコウモリの捕獲. *The Journal of the Mammalogical Society of Japan* **3**: 67 (in Japanese).

3623. Toriumi, H. (鳥海隼夫), Hirose, A. (広瀬章裕), & Kusakari, K. (草刈広一) (1988) 山形県のコウモリ類. 東北の自然 (*The Nature of Tōhoku*) **39**: 2–12 (in Japanese).

3624. Tosi, A.J., Morales, J.C., & Melnick, D.J. (2000) Comparisons of Y chromosome and mtDNA phylogenies leads to unique inferences of macaque evolutionary history. *Molecular Phylogenetics and Evolution* **17**: 133–144.

3625. Tosi, A.J., Morales, J.C., & Melnick, D.J. (2002) Y chromosome and mitochondrial markers in *Macaca fascicularis* indicate introgression with Indochinese *M. mulatta* and a biogeographic barrier in the Isthmus of Kra. *International Journal of Primatology* **23**: 161–178.

3626. Tosuji, T. & Shibata, E. (2003) Seasonal activity of common Japanese pipistrelle bat (*Pipistrellus abramus*) in urban area in relation to feeding-site selection. *Mammalian Science* **43**: 113–120 (in Japanese with English abstract).

3627. Toyama, C., et al. (2012) Feeding behavior of the Orii's flying-fox, *Pteropus dasymallus inopinatus*, on *Mucuna macrocarpa* and related explosive opening of petals, on Okinawajima Island in the Ryukyu Archipelago, Japan. *Mammal Study* **37**: 205–212.

3628. Trewhella, W.J., Harris, S., & McAllister, F.E. (1988) Dispersal distance, home-range size and population density in the red fox (*Vulpes vulpes*): a quantitative analysis. *Journal of Applied Ecology* **25**: 423–434.

3629. Tsubota, T., et al. (1985) Observation of sexual behavior under captive conditions in Hokkaido brown bears (*Ursus arctos yesoensis*). *Japanese Journal of Animal Reproduction* **31**: 203–210.

3630. Tsubota, T., Mizoguchi, N., & Kita, I. (1998) Ecological and physiological studies of the Japanese black bear, *Ursus thibetanus japonicus*. *Japanese Journal of Zoo and Wildlife Medicine* **3**: 17–24 (in Japanese with English summary).

3631. Tsubota, T., Takahashi, Y., & Kanagawa, H. (1987) Changes in serum progesterone levels and growth of fetuses in Hokkaido brown bears. *Proceedings of the International Conference on Bear Research and Management* **7**: 355–358.

3632. Tsuchiya, K. (土屋公幸) (1985) 食虫類染色体の数と形態. in スンクス. 実験動物としての食虫目トガリネズミ科動物の生物学 (*Suncus murinus*: Biology of the Laboratory Shrew) (Oda, S. (織田銑一), et al., Editors). Tokyo: Japan Scientific Societies Press, pp. 51–67 (in Japanese).

3633. Tsuchiya, K. (1974) Cytological and biochemical studies of *Apodemus speciosus* group in Japan. *The Journal of the Mammalogical Society of Japan* **6**: 67–87 (in Japanese with English abstract).

3634. Tsuchiya, K. (1976) Karyological investigations of Japanese mammals. *Mammalian Science* **16(2)**: 53–59 (in Japanese).

3635. Tsuchiya, K. (1979) A contribution to the chromosome study in Japanese mammals. *Proceedings of the Japan Academy, Series B* **55**: 191–195.

3636. Tsuchiya, K. (1981) On the chromosome variations in Japanese cricetid and murid rodents. *Mammalian Science* **21(1)**: 51–58 (in Japanese).

3637. Tsuchiya, K. (1985) The chromosomes of Insectivora. in *Suncus murinus*. Biology of the Laboratory Shrews (Oda, S., et al., Editors). Tokyo: Japanese Scientific Societies Press, pp. 51–67 (in Japanese with English abstract).

3638. Tsuchiya, K. (1985) Notes on wood mouse groups for laboratory animal. *Association Report of Kyushu Branch, Japanese Association of Laboratory Animals and Technology* **8**: 4–12 (in Japanese).

3639. Tsuchiya, K. (1986) Wild rodents in Japan—Candidates for experimental animals. *Laboratory Animals* **3**: 43–46 (in Japanese).
3640. Tsuchiya, K. (1987) Cytological and biochemical studies of Insectivora in Tsushima Island. *in* Nature of Tsushima (Nagasaki Prefectural Government, Editor). Nagasaki: Nagasaki Prefectural Government, pp. 111–124 (in Japanese with English abstract).
3641. Tsuchiya, K. (1988) Cytotaxonomic studies of the family Talpidae from Japan. *Mammalian Science* **28(1)**: 49–61 (in Japanese).
3642. Tsuchiya, K., Moriwaki, K., & Yosida, T.H. (1973) Cytogenetical survey in wild populations of Japanese wood mouse, *Apodemus speciosus* and its breeding. *Experimental Animals* **22**: 221–229.
3643. Tsuchiya, K., et al. (2000) Molecular phylogeny of East Asian moles inferred from the sequence variation of the mitochondrial cytochrome *b* gene. *Genes & Genetic Systems* **75**: 17–24.
3644. Tsuchiya, K., et al. (1989) Taxonomic study of *Tokudaia* (Rodentia: Muridae): I. Genetic differentiation. *Memoirs of the National Science Museum, Tokyo* **22**: 227–234 (in Japanese with English summary).
3645. Tsuchiya, K. & Yosida, T.H. (1971) Chromosome survey of small mammals in Japan. *in* Annual Report / National Institute of Genetics No. 21. Mishima: National Institute of Genetics, pp. 54–55.
3646. Tsuda, K., et al. (1977) Extensive interbreeding occurred among multiple matriarchal ancestors during the domestication of dogs: Evidence from inter- and intraspecies polymorphisms in the D-loop region of mitochondrial DNA between dogs and wolves. *Genes & Genetic Systems* **72**: 229–238.
3647. Tsuda, K., et al. (2007) Risk of accidental invasion and expansion of allochthonous mice in Tokyo metropolitan coastal areas in Japan. *Genes and Genetic Systems* **82**: 421–428.
3648. Tsuji, A. & Koyanagi, K. (2001) A record of the black whiskered bat, *Myotis pruinosus* Yoshiyuki, 1971, in Niigata Prefecture. *Bulletin of the Asian Bat Research Institute* **1**: 10–12 (In Japanese with English abstract).
3649. Tsuji, H., et al. (2007) Molecular biological position of *Baylisascaris transfuga* in Family Ascarididae and pathological study on its pathogenicity. *in* Abstracts of the Annual Meeting of the Japan Society of Zoo and Wildlife Medicine. p. 90 (in Japanese).
3650. Tsuji, K. & Ishikawa, T. (1982) Life mode and behavior of the Ryukyu musk shrew (*Suncus murinus* var *riukiuanus*). *The Journal of the Mammalogical Society of Japan* **9**: 96–103 (in Japanese with English abstract).
3651. Tsuji, K. & Ishikawa, T. (1984) Some observations of the caravaning behavior in the musk shrew (*Suncus murinus*). *Behaviour* **90**: 167–183.
3652. Tsuji, K., Matsuo, T., & Ishikawa, T. (1986) Developmental changes in the caravaning behaviour of the house musk shrew (*Suncus murinus*). *Behaviour* **99**: 117–138.
3653. Tsuji, M., et al. (2001) Human babesiosis in Japan: epizootiologic survey of rodent reservoir and isolation of new type of *Babesia microti*-like parasite. *Journal of Clinical Microbiology* **39**: 4316–4322.
3654. Tsuji, M., et al. (2006) *Babesia microti*-like parasites detected in Eurasian red squirrels (*Sciurus vulgaris orientis*) in Hokkaido, Japan. *Journal of Veterinary Medical Science* **68**: 643–646.
3655. Tsuji, O., et al. (2004) Estimating of Russian flying squirrel habitat using GIS. *Journal of the Japanese Society of Irrigation, Drainage and Reclamation Engineering* **72**: 37–40 (in Japanese).
3656. Tsuji, T., et al. (2013) Estimation of the fertility rates of Japanese wild boars (*Sus scrofa leucomystax*) using fetuses and corpora albicans. *Acta Theriologica* **58**: 315–323.
3657. Tsuji, Y. (2011) Sleeping-site preferences of wild Japanese macaques (*Macaca fuscata*): the importance of non-predatory factors. *Journal of Mammalogy* **92**: 1261–1269.
3658. Tsuji, Y. & Takatsuki, S. (2012) Interannual variation in nut abundance is related to agonistic interactions of foraging female Japanese macaques (*Macaca fuscata*). *International Journal of Primatology* **33**: 489–512.
3659. Tsukada, H. (塚田英晴) (2000) キタキツネ. *in* 知床のほ乳類 I (Shiretoko Museum (斜里町立知床博物館), Editor). Sapporo: Hokkaido Shimbun Press, pp. 74–129 (in Japanese).
3660. Tsukada, H. (1997) A division between foraging range and territory related to food distribution in the red fox. *Journal of Ethology* **15**: 27–37.
3661. Tsukada, H. (1997) External measurements, breeding season, litter size, survival rate, and food habits of red foxes (*Vulpes vulpes schrencki*) in the Shiretoko National Park. *Bulletin of the Shiretoko Museum* **18**: 35–44 (in Japanese).
3662. Tsukada, H., et al. (1999) The spreading of Sarcoptic mange among red foxes *Vulpes vulpes* and its impact on the local fox population in Shiretoko Peninsula, Hokkaido, Japan. *Mammalian Science* **39**: 247–256.
3663. Tsumura, Y., et al. (2007) Genome scan to detect genetic structure and adaptive genes on natural populations of *Cryptomeria japonica*. *Genetics* **176**: 2393–2403.
3664. Tsunoda, T. (2012) The tick fauna parasitizing middle- and large-sized mammals in Chiba Prefecture, central Japan. *Journal of the Natural History Museum and Institute, Chiba* **12**: 33–42 (in Japanese with English abstract).
3665. Tsunoda, T., Chinone, S., & Yamazaki, K. (2001) Ticks of the Asiatic black bear, *Ursus thibetanus*, in the Okutama Mts., central Japan. *Bulletin of Ibaraki Nature Museum* **4**: 101–102.
3666. Tsushima, T. (1994) Development and growth of the baculum of *Pipistrellus abramus* (Temminck, 1840). *Bulletin of the Biological Society of Kagawa* **21**: 39–50 (in Japanese).
3667. Tsushima Wildlife Conservation Center. Personal communication.
3668. Tsutsui, S., et al. (2001) Pacific white-sided dolphins (*Lagenorhynchus obliquidens*) with anomalous colour patterns in Volcano Bay, Hokkaido, Japan. *Aquatic Mammals* **27**: 172–182.
3669. Tsutsumi, T., Kamimura, Z., & Mizue, K. (1961) Studies on the little toothed whales in the west sea area of Kyusyu - V. About the food of the little toothed whales. *Bulletin of the Faculty of Fisheries, Nagasaki University* **11**: 19–28 (in Japanese with English summary).
3670. Tsytsulina, K. (2001) *Myotis ikonnikovi* (Chiroptera, Vespertilionidae) and its relationships with similar species. *Acta Chiropterologica* **3**: 11–19 (in Japanese with English abstract).
3671. Tsytsulina, K. (2004) On the taxonomical status of *Myotis abei* Yoshikura, 1944 (Chiroptera, Vespertilionidae). *Zoological Science* **21**: 963–966.
3672. Tul'skaya, O.L., Derenko, M.V., & Malyarchuk, B.A. (1999) Low level of mitochondrial DNA variation in sea otter populations from Kamchatka and Commander Islands. *Russian Journal of Genetics* **35**: 11–15.
3673. Tuomi, P. (2001) Sea otters. *in* CRC Handbook of Marine Mammal Medicine (Dierauf, L.A. & Gulland, F.M.D., Editors). Boca Raton: CRC Press, pp. 961–987.
3674. Tyack, P.L., et al. (2006) Extreme diving of beaked whales. *Journal of Experimental Biology* **209**: 4238–4253.
3675. Uchida, S. (内田清之助) (1920) 鹿児島県奄美大島ノ動物ニ関スルモノ (史跡名勝天然記念物調査報告 第23号). Vol. 23. Tokyo: Department of the Interior (内務省), 24 pp. (in Japanese).
3676. Uchida, S. (内田清之助) & Imaizumi, Y. (今泉吉典) (1939) ヘビ類の食性に関する調査成績. *Ornithological and Mammalogical Report* **9**: 143–208 (in Japanese).
3677. Uchida, S. (内田詮三) (1994) 沖縄近海の鯨類. *in* ピトゥと名護人―沖縄県名護のイルカ漁―. Nago: Nago Museum, pp. 75–118 (in Japanese).
3678. Uchida, T.A. (内田照章) (1966) 日本の哺乳類 (4) 翼手目 イエコウモリ属. *Mammalian Science* **6(2)**: 5–23 (in Japanese).
3679. Uchida, A., Itagaki, H., & Kugi, G. (1976) *Concinnum ten* (Yamaguti, 1939) from carnivorous mammals in Japan. *Japanese Journal of Parasitology* **25**: 319–323 (in Japanese).
3680. Uchida, A., Uchida, K., & Murata, Y. (2000) New-host records of *Metagonimus yokogawai* and *Spirometra erinaceieuropaei* from the masked palm civet (*Paguma larvata*). *Journal of the Japan Veterinary Medical Association* **53**: 232–234.
3681. Uchida, T., et al. (1992) *Rickettsia japonica* sp. nov., the etiological agent of spotted fever group rickettsiosis in Japan. *International Journal of Systematic Bacteriology* **42**: 303–305.
3682. Uchida, T. & Yoshida, H. (1968) On the distribution and morphological characters of *Dymecodon pilirostris* True from Kyusyu. *Mammalian Science* **8(2)**: 17–26 (in Japanese).
3683. Uchida, T.A. (1950) Studies on the embryology of the Japanese house bat, *Pipistrellus tralatitus abramus* (Temminck). I. On the period of gestation and the number of litter. *Journal of the Faculty of Agriculture, Kyushu University* **12**: 11–14 (in Japanese with English abstract).
3684. Uchida, T.A. (1953) Studies on the embryology of the Japanese house bat, *Pipistrellus tralatitus abramus* (Temminck). II. From the maturation of the ova to the fertilization, at the period of fertilization. *Journal of the Faculty of Agriculture, Kyushu University* **14**: 153–168 (in Japanese with English abstract).
3685. Uchida, T.A. (1956) Japanese tube-nosed bat, *Murina leucogaster hilgendorfi* (Peters) found in Kyushu. *The Journal of the Mammalogical Society of Japan* **1**: 32–34 (in Japanese with English abstract).
3686. Uchida, T.A. & Ando, K. (1972) Karyotype analysis in Chiroptera. I. Karyotype of the Eastern barbastelle, *Barbastella leucomelas darjelingensis* and comments on its phylogenetic position. *Science Bulletin of the Faculty of Agriculture, Kyushu University* **26**: 393–398 (in Japanese with English abstract).
3687. Uchida, T.A. & Kuramoto, T. (1968) Modes of life in the colony of cave bats, with special remarks on dense mixed colony consisting of different species. *Mammalian Science* **8(2)**: 3–15 (in Japanese).
3688. Uchida, T.A. & Mōri, T. (1972) Electron microscope studies on the fine structure of germ cells in Chiroptera. I. Spermiogenesis in some bats and notes on its phylogenetic significance *Journal of the Faculty of Agriculture,*

Kyushu University **26**: 399–418 (in Japanese with English summary).

3689. Uchida, T.A. & Mōri, T. (1987) Prolonged storage of spermatozoa in hibernating bats. *in* Recent Advances in the Study of Bats (Fenton, M.B., Racey, P., & Rayner, J.M.V., Editors). Cambridge: Cambridge University Press, pp. 351–365.

3690. Uchikawa, K. (内川公人) (1974) 長野県下の哺乳類および鳥類の外部寄生虫相について (II). *Japanese Journal of Sanitary Zoology* **24**: 292 (in Japanese).

3691. Uchikawa, K. (内川公人) (1976) 日本産のコウモリ寄生ダニ類—対馬における調査成績—. *in* 対馬の生物 (Nagasaki Biological Society (長崎県生物学会), Editor). Nagasaki: Nagasaki Biological Society, pp. 839–844 (in Japanese).

3692. Uchikawa, K. (内川公人) (1977) 日本産food虫類および齧歯類に寄生するケモチダニ二類 (Notes on the myobiid mites parasitic on Insectivora and Rodentia in Japan). *in* ダニ学の進歩 (Contributions to Acarology) (Sasa, M. (佐々学) & Aoki, J. (青木淳一), Editors). Tokyo: Hokuryukan (北隆館), pp. 415–431 (in Japanese).

3693. Uchikawa, K. (内川公人) (1984) Insecta and Arachnida. *in* 大町市の動物. 大町市史 (Catalog of Animals in Ohmachi City. Ohmachi City History) Vol. 1. Ohmachi: Ohmachi City Office, pp. 156–160 (in Japanese).

3694. Uchikawa, K. (1967) An ecological study on the fleas in the Yatsugatake range: (I) On the fleas found on the small mammals trapped in the subalpine forest. *Japanese Journal of Ecology* **17**: 43–49 (in Japanese with English summary).

3695. Uchikawa, K. (1969) An ecological study on the fleas in the Yatsugatake range: (II) On the fleas found on the small mammals trapped in the mountain zone. *Japanese Journal of Ecology* **19**: 48–52 (in Japanese with English summary).

3696. Uchikawa, K. (1969) Faunal survey of the Mt. Ontake, JIBP main area III. Ectoparasites of small mammals caught on the western slope of Mt. Ontake. *in* Studies on the Animal Communities in the Terrestrial Ecosystems and their Conservation (Annual Report of JIBP/CT-S for the Fiscal Year of 1968) (Kato, M., Editor). Sendai: JIBP/CT-S Section, pp. 22–32.

3697. Uchikawa, K. (1971) Ectoparasite fauna of small mammals on Mt. Fuji. *in* Results of the co-operative scientific survey of Mt. Fuji. Tokyo: Fuji Kyuko, pp. 848–855 (in Japanese with English summary).

3698. Uchikawa, K. (1978) Myobiid mites (Acarina, Myobiidae) parasitic on bats in Japan VI. Genus *Pteracarus* Jameson et Chow, 1952 (Part I). *Annotationes Zoologicae Japonenses* **51**: 107–110.

3699. Uchikawa, K. (1985) Ectoparasites *sens latum* of the Soricidae, with special reference to those of *Suncus* spp. *in* Suncus murinus. Biology of the Laboratory Shrew (Oda, S., et al., Editors). Tokyo: Japan Scientific Press, pp. 77–87 (in Japanese with English abstract).

3700. Uchikawa, K. (1986) *Chimarrogalobia yoshiyukiae* gen. n. sp. n. (Acarina, Myobiidae) parasitic on *Chimarrogale* (Insectivora, Soricidae). *Zoological Science* **3**: 187–192.

3701. Uchikawa, K. (1998) Phylogeny of *Clethrionomys* voles deduced from their myobiid mites. *Mammalian Science* **38**: 159–170 (in Japanese with English abstract).

3702. Uchikawa, K. & Asanuma, K. (1974) Studies on mesostimatid mites parasitic on mammals and birds in Japan. II. *Androlaelaps himizu* (Jameson, 1966) and *Androlaelaps himehimizu* sp. n. parasitic on the shrew-moles (Acarina: Laelapidae). *Japanese Journal of Sanitary Zoology* **25**: 65–77 (in Japanese with English summary).

3703. Uchikawa, K. & Kumada, N. (1977) Studies on mesostigmatid mites parasitic on mammals and birds in Japan: VI. Bat mites of the genus *Steatonyssus* Kolenati, with redescription of *Steatonyssus longispinosus* Wang, 1963. *Japanese Journal of Sanitary Zoology* **28**: 423–429.

3704. Uchikawa, K., Nakata, K., & Takahashi, K. (1997) *Radfordia* (*Microtimyobia*) (Acari, Myobiidae) associated with arvicoline voles (Rodentia, Muridae) in Japan. *Zoological Science* **14**: 671–682.

3705. Uchikawa, K. & Suzuki, H. (1984) Medio-zoological studies in Tokara Archipelago, Kagoshima Prefecture, Japan- Records of acari and flea found on an *Apodemus* mouse on Nakanoshima Island comprised in the Archopelago. *Tropical Medicine* **26**: 17–25.

3706. Uchikawa, K. & Wada, Y. (1979) Studies on mesostigmatid mites parasitic on mammals and birds in Japan IX. Bat mites of the genus *Spinturnix* von Heyden 1829 (Part I) (Spinturnicidae). *Japanese Journal of Sanitary Zoology* **30**: 121–125.

3707. Uchikawa, K. & Yoshiyuki, M. (1978) Myobiid mites (Acarina, Myobiidae) parasitic on bats in Japan. V. Genus *Binuncus* Radford, 1954. *The Journal of the Mammalogical Society of Japan* **7**: 204–205.

3708. Uchiyama, J. (内山純蔵) (2002) 低湿地立地の遺跡にみる縄文時代本州西部地域の生業活動：縄文時代早期末から中期初頭における若狭湾沿岸と琵琶湖周辺地域の動物考古学的考察. Doctoral dissertation. Kanagawa: Graduate University for Advanced Studies. 169 pp. (in Japanese).

3709. Ueda, H. & Takatsuki, S. (2005) Sexual dimorphism of *Apodemus specious* in wild populations. *Mammal Study* **30**: 65–68.

3710. Ueda, M. (1972) Problems about the black colored pelage variants in *Clethrionomys rufocanus bedfordiae*. 森林防疫 (*Forest Pests*) **21**: 10–12 (in Japanese).

3711. Ueda, M., et al. (1966) Historical review of studies on the Bedford's red-backed vole, *Clethrionomys rufocanus bedfordiae* (Thomas). *Bulletin of the Government Forestry Experiment Station* **191**: 1–100 (in Japanese).

3712. Ueda, M., et al. (1970) The black type of *Clethrionomys rufocanus bedfordiae* found on the isle in Lake Toya. *Northern Forestry, Japan* **22**: 309–310 (in Japanese).

3713. Uehara, S. (1975) The importance of the temperate forest elements among woody food plants utilized by Japanese monkeys and its possible historical meaning for the establishment of the monkey's range: A preliminary report. *in* Contemporary Primatology; Proceedings of the Fifth International Congress of Primatology (Kondo, S., Kawai, M., & Ehara, A., Editors). Basel: Karger, pp. 392–400.

3714. Ueno, Y. (上野吉雄) (1993) 西中国山地におけるモリアブラコウモリの記録. おおのの自然観察の森研究報告 **3**: 35 (in Japanese).

3715. Ueno, Y., et al. (1996) The mammals in Geihoku-cho, Hiroshima Prefecture. *Natural history of Nishi-Chugoku Mountains* **1**: 395–441 (in Japanese with English abstract).

3716. Ueno, Y., et al. (2002) Bat fauna in Nishi-Chugoku Mountains I. *Natural History of Nishi-Chugoku Mountains* **7**: 85–97 (in Japanese with English abstract).

3717. Uesugi, T., Maruyama, N., & Kanzaki, N. (1998) Public attitude toward introduced Japanese weasels in Miyake-jima Island, Tokyo. *Wildlife Conservation Japan* **3**: 85–94 (in Japanese with English abstract).

3718. Ueuma, Y. & Mihara, Y. (1995) The first record of *Vespertilio superans* in the Mt. Hakusan area, Ishikawa prefecture. *Annual Report of the Hakusan Nature Conservation Center* **22**: 17–18 (in Japanese).

3719. Ueuma, Y. & Minami, T. (1984) Notes on the hibernation of the Japanese large noctule, *Nyctalus aviator* Thomas, 1911 in Kanazawa city. *Annual Report of the Hakusan Nature Conservation Center* **11**: 85–86 (in Japanese).

3720. Ueuma, Y., Tokuno, C., & Tsuji, M. (2005) Food of red fox (*Vulpes vulpes japonica*), Japanese Marten (*Mustelmelanpus mepanpus*) and Hondo stoat (*Mustela erminea nippon*) analyzed dropping contents on the trails in Mt. Hakusan. *Annual Report of the Hakusan Nature Conservation Center, Ishikawa Prefecture* **32**: 31–36.

3721. Ueyama, T., Hayashida, M., & Mukouyama, M. (2007) First records of *Myotis nattereri* utilizing tree hollows. *Study Report on Bat Conservation in Tohoku Region* **1**: 2–4 (in Japanese).

3722. Ullrey, D.E., et al. (1984) Blue-green color and composition of Stejneger's beaked whales (*Mesoplodon stejnegeri*) milk. *Comparative Biochemistry and Physiology. Part B* **79**: 349–352.

3723. Umeda, Y. Unpublished.

3724. Uni, Y. (宇仁義和) (2001) 北海道沿岸の近代海獣猟業の統計と関連資料. *Bulletin of the Shiretoko Museum* **22**: 81–92 (in Japanese).

3725. Uni, S. (1983) Filarial parasites from the black bear of Japan. *Annales de Parasitologie Humaine et Comparee* **58**: 71–84.

3726. Uni, S. (1984) Note on *Dipetalonema* (*Chenofilaria*) *japonica* Uni, 1983 from Japanese black bear supplementary description. *Annales de Parasitologie Humaine et Comparee* **59**: 531–534.

3727. Uni, S., et al. (2006) New filarial nematode from Japanese serows (*Naemorhedus crispus*: Bovidae) close to parasites from elephants. *Parasite* **13**: 193–200.

3728. Uni, S., et al. (2013) Infective larvae of *Cercopithifilaria* spp. (Nematoda: Onchocercidae) from hard ticks (Ixodidae) recovered from the Japanese serow (Bovidae). *Parasite* **20**: 1–7.

3729. Uni, S., et al. (2001) Coexistence of five *Cercopithifilaria* species in the Japanese rupicaprine bovid, *Capricornis crispus*. *Parasite* **8**: 197–213.

3730. Uno, H. (宇野裕之) (1986) ゼニガタアザラシ (*Phoca virtulina stejnegeri*) およびゴマフアザラシ (*Phoca largha*) の頭骨の成長・発育に関する比較 (Comparison of cranial growth and development between Kuril and Largha seals). *in* ゼニガタアザラシの生態と保護 (Wada, K. (和田一雄), et al., Editors). Tokyo: Tokai University Press, pp. 158–178 (in Japanese).

3731. Uno, H. (宇野裕之) & Yamanaka, M. (山中正美) (1988) 鰭脚類 (Marine mammals (pinnipeds)). *in* 知床の動物：原生的自然環境下の脊椎動物群集とその保護 (Animals of Shiretoko) (Ohtaishi, N. (大泰司紀之) & Nakagawa, H. (中川元), Editors). Sapporo: Hokkaido University Press, pp. 225–248 (in Japanese with English summary).

3732. Uno, H., Chiba, M., & Yamaki, M. (1997) The collected records of Chiroptera in Bihoro, Hokkaido. *Bulletin of the Bihoro Museum* **4**: 29–31 (in Japanese).

3733. Uno, H. & Kaji, K. (2000) Seasonal movements of female sika deer in eastern Hokkaido, Japan. *Mammal Study* **25**: 49–57.

3734. Uno, H., et al. (2007) Population management and monitoring for sika deer on Hokkaido Island, Japan. *Mammalian Science* **47**: 133–138 (in Japanese).

3735. Uno, H., et al. (2006) Evaluation of relative density indices for sika deer in eastern Hokkaido, Japan. *Ecological Research* **21**: 624–632.
3736. Uno, H., Maeda, K., & Yamaki, M. (1998) Faunal survey of bats in Bihoro, Hokkaido (2). *Bulletin of the Bihoro Museum* **5**: 27–36 (in Japanese with English summary).
3737. Uno, H., et al. (2007) Current status of and perspectives on conservation and management for sika deer populations in Japan. *Mammalian Science* **47**: 25–38 (in Japanese with English abstract).
3738. Uraguchi, K. (浦口宏二) (1996) ミンク (Mink). *in* The Encyclopaedia of Animals in Japan, Vol.2, Mammal II (日本動物大百科2哺乳類 II) (Izawa, K. (伊沢紘生), Kasuya, T. (粕谷俊雄), & Kawamichi, T. (川道武男), Editors). Tokyo: Heibonsya (平凡社), p. 139 (in Japanese).
3739. Uraguchi, K. Unpublished.
3740. Uraguchi, K. (1988) A preliminary study on the range size and activity pattern of the feral mink *Mustela vison* in Hokkaido, Japan. *Journal of Fur Research* **5**: 9–11.
3741. Uraguchi, K., et al. (1987) Food habits of the feral mink (*Mustela vison* Schreber) in Hokkaido. *Journal of the Mammalogical Society of Japan* **12**: 57–67.
3742. Uraguchi, K. & Takahashi, K. (1998) Den site selection and utilization by the red fox in Hokkaido, Japan. *Mammal Study* **23**: 31–40.
3743. Uraguchi, K., Takahashi, K., & Maekawa, K. (1991) The age structure of the red fox population in Hokkaido, Japan. *in* Wildlife Conservation: Present Trends and Perspectives for the 21st Century (Maruyama, N., et al., Editors). Tokyo: Japan Wildlife Research Center, pp. 228–230.
3744. Uraguchi, K., et al. (2014) Demographic analyses of a fox population suffering from sarcoptic mange. *Journal of Wildlife Management* **78**: 1356–1371.
3745. Uramoto, M. (浦本昌紀) (1963) ウサギコウモリの採集. *The Journal of the Mammalogical Society of Japan* **2**: 64 (in Japanese).
3746. Urano, M. (浦野守雄) (2000) 富士山5合目で発見したコウモリ. *Bat Study and Conservation Report* **8**: 18 (in Japanese).
3747. Urano, N. (浦野信孝) (2002) 北摂の洞穴性のコウモリについて. *Bat Study and Conservation Report* **10**: 1–4 (in Japanese).
3748. Urano, N. (浦野信孝) (2003) 人工洞穴のコウモリ. *Nature Study* **49**: 11 (in Japanese).
3749. Urano, N. (浦野信孝) (2003) 大阪府で発見されたヒナコウモリの繁殖コロニー. *Bat Study and Conservation Report* **11**: 11–12 (in Japanese).
3750. Urano, N. (浦野信孝) (2008) 隧道に生息するコウモリ. *Nature Study* **54**: 2 (in Japanese).
3751. Urano, N. (浦野信孝), Nagai, E. (永井英司), & Minobe, N. (美濃部直久) (2011) 京都府の人口洞に生息するコウモリ. *Nature Study* **57**: 6–8 (in Japanese).
3752. Urano, N. (浦野信孝) & Nishida, Y. (西田安則) (2003) 高野山でウサギコウモリ発見. *Nature Study* **49**: 16 (in Japanese).
3753. Urano, N. (浦野信孝), et al. (2006) 兵庫県佐用郡佐用町のコウモリ. *Nature Study* **52**: 6–7 (in Japanese).
3754. Urano, N. (浦野信孝) & Takeda, S. (武田忍) (2008) 宝塚市でヒナコウモリ発見. *Nature Study* **54**: 8 (in Japanese).
3755. Urano, M. (1998) 西多摩郡檜原村で確認されたモリアブラコウモリ *Pipistrellus endoi*. 東京都の自然 **24**: 22 (in Japanese).
3756. Urano, M., Kasahi, T., & Takamizu, Y. (2002) Distribution records of bats in Okutama region, Tokyo Prefecture. (1) Notes of bats in Akiruno-shi, Ome-shi and Hinohara-mura. *Science Report of the Takao Museum of Natural History* **21**: 13–20 (in Japanese with English abstract).
3757. Urano, N. (2007) Mammals using artificial caves in north-western region of Osaka Prefecture. *Bulletin of Kansai Organization for Nature Conservation* **29**: 57–64 (in Japanese).
3758. Urata, A. (浦田明夫) & Yamaguchi, T. (山口鉄男) (1976) 対馬の哺乳類. *in* 対馬の生物 (Nagasaki Biological Society (長崎県生物学会), Editor). Nagasaki: Nagasaki Biological Society, pp. 155–166 (in Japanese).
3759. Urban-R., J., Ramirez-S., S., & Salinas-V., J.C. (1994) First record of bottlenose whales, *Hyperoodon* sp., in the Gulf of California. *Marine Mammal Science* **10**: 471–473.
3760. Uryudo Site Research Group (瓜生堂遺跡調査会) (1980) 恩智遺跡：一級河川恩智川改修工事に伴う恩智遺跡発掘調査報告書. 254 pp. (in Japanese).
3761. Usinger, R.L. (1966) Monograph of Cimicidae (Hemiptera: Heteroptera). The Thomas Say Foundation Vol. VII. Lanham: Entomological Society of America, 585 pp.
3762. Usuki, H. (臼杵秀明) (1965) コテングコウモリを入手. *The Journal of the Mammalogical Society of Japan* **2**: 145 (in Japanese).
3763. Vaisfeld, M.A. & Chestin, I.E. (1993) Bears—Brown Bear, Polar Bear and Asian Black Bear—Distribution, Ecology, Use and Protection. Moscow: Nauka, 519 pp.
3764. van Ballenberghe, V., Erickson, A.W., & Byman, D. (1975) Ecology of the Timber Wolf in Northeastern Minnesota. *Wildlife Monographs* **43**: 43.
3765. van Ballenberghe, V. & Mech, L.D. (1975) Weights, growth, and survival of timber wolf pups in Minnesota. *Journal of Mammalogy* **56**: 44–63.
3766. van Bleijswijk, J.D.L., et al. (2014) Detection of grey seal *Halichoerus grypus* DNA in attack wounds on stranded harbour porpoises *Phocoena phocoena*. *Marine Ecology Progress Series* **513**: 277–281.
3767. van Bree, P.J.H., et al. (1986) Le dauphin de Fraser, *Lagenodelphis hosei* (Cetacea, Odontoceti), espece nouvelle pour la faune d'Europe. *Mammalia* **50**: 57–86 (in French with English summary).
3768. Van Bressen, M., et al. (2006) Diseases, lesions and malformations in the long-beaked common dolphin *Delphinus capensis* from the southeast Pacific. *Diseases of Aquatic Organisms* **68**: 149–165.
3769. Van Wagene, R.F., Foster, M.S., & Burns, F. (1981) Sea otter predation on birds near Monterey, California. *Journal of Mammalogy* **62**: 433–434.
3770. van Wagenen, G. (1972) Vital statistics from a breeding colony: Reproduction and pregnancy outcome in *Macaca mulatta*. *Journal of Medical Primatology* **1**: 3–28.
3771. van Zyll de Jong, C.G. (1987) A phylogenetic study of the Lutrinae (Carnivora; Mustelidae) using morphological data. *Canadian Journal of Zoology* **65**: 2536–2544.
3772. VanBlaricom, G.R. & Estes, J.A. (1988) The Community Ecology of Sea Otters. Berlin: Springer, 247 pp.
3773. Vander Wall, S.B. (2001) The evolutionary ecology of nut dispersal. *Botanical Review* **67**: 74–117.
3774. Vaughan, M.A. (2002) Oak trees, acorns, and bears. *in* Oak Forest Ecosystems (McShea, W.J. & Healy, W.M., Editors). Baltimore: Johns Hopkins University Press, pp. 224–240.
3775. Veron, G., et al. (2007) Systematic status and biogeography of the Javan and small Indian mongooses (Herpestidae, Carnivora). *Zoologica Scripta* **36**: 1–10.
3776. Vilà, C., et al. (1997) Multiple and ancient origins of the domestic dog. *Science* **276**: 1687–1689.
3777. Vilà, C., et al. (2003) Combined use of maternal, paternal and bi-parental genetic markers for the identification of wolf-dog hybrids. *Heredity* **90**: 17–24.
3778. Vilstrup, J.T., et al. (2011) Mitogenomic phylogenetic analyses of the Delphinidae with an emphasis on the Globicephalinae. *BMC Evolutionary Biology* **11**: 65.
3779. Viricel, A., et al. (2008) Insights on common dolphin (*Delphinus delphis*) social organization from genetic analysis of a mass-stranded pod. *Behavioral Ecology and Sociobiology* **63**: 173–185.
3780. Visser, I.N., et al. (2010) First Record of Predation on False Killer Whales (*Pseudorca crassidens*) by Killer Whales (*Orcinus orca*). *Aquatic Mammals* **36**: 195–204.
3781. Volleth, M. (1985) Chromosomal homologies of the genera *Vespertilio*, *Plecotus* and *Barbastella* (Chiroptera: Vespertilionidae). *Genetica* **66**: 231–236.
3782. Volleth, M., et al. (2001) Karyotype comparison and phylogenetic relationships of *Pipistrellus*-like bats (Vespertilionidae; Chiroptera; Mammalia). *Chromosome Research* **9**: 25–46.
3783. Voloshina, I.V. & Myslenkov, A.I. (2009) Sika deer distribution changes at the northern extent of their range in the Sikhote-Alin Mountains of the Russian Far East. *in* Sika Deer—Biology and Management of Native and Intro-duced Populations (McCullough, D.R., Takatsuki, S., & Kaji, K., Editors). Tokyo: Springer, pp. 501–519.
3784. Voronov, V.G. (Воронов, В.Г.) & Voronov, G.A. (Воронов, Г.А.) (1977) О находке пятнистого оленя на Малой Курильской гряде. *in* Редкие виды млекопитающих и их охрана Moscow: Soviet Academy of Sciences (Академия Наук СССР), pp. 195–196 (in Russian).
3785. Vorontsov, N.N. & Ivanitskaya, E.Y. (1973) Comparative karyology of North Palaearctic pikas (*Ochotona*, Ochotonidae, Lagomorpha). *Caryologia* **26**: 213–223.
3786. Wada, H. (和田久), Sasamori, K. (笹森耕二), & Seki, T. (関哲朗) (1991) 赤石川流域の自然. 青森県郷土館調査報告 **28**: 55-69 (in Japanese).
3787. Wada, Y. (和田芳武) & Sasa, M. (佐々学) (1967) 日本産コウモリに寄生するダニについて. *Japanese Journal of Sanitary Zoology* **18**: 156–157 (in Japanese).
3788. Wada, H., Sasamori, K., & Seki, T. (1993) Nature section; Preliminary report of the scientific survey of the Shirakami mountain range, Aomori prefecture, Japan (3) Mammal. *Annual Report of the Aomori Prefectural Museum* **17**: 30–33 (in Japanese).
3789. Wada, H., Sasamori, K., & Seki, T. (1994) Nature section; Preliminary report of the scientific survey of the Shirakami mountain range, Aomori prefecture, Japan (4) Mammal. *Annual Report of the Aomori Prefectural Museum* **18**: 12–14 (in Japanese).
3790. Wada, K. (1969) Migration of northern fur seals along the coast of Sanriku. *Bulletin of Tokai Regional Fisheries Research Laboratory* **58**: 19–82 (in Japanese with English summary).

3791. Wada, K. (1971) Food and feeding habit of northern fur seals along the coast of Sanriku. *Bulletin of Tokai Regional Fisheries Research Laboratory* **64**: 1–37 (in Japanese with English summary).
3792. Wada, K. (1971) Some comments on the migration of Northern fur seals. *Bulletin of Tokai Regional Fisheries Research Laboratory* **67**: 47–80 (in Japanese with English summary).
3793. Wada, K. (1975) Ecology of wintering among Japanese monkeys in Shiga Heights and its adaptive significance. *Seiri Seitai (Physiology and Ecology)* **16**: 9–14 (in Japanese with English summary).
3794. Wada, K. (1997) The establishment and succession of the sea otter and northern fur seal sealing industry to resource management (2). *Wildlife Conservation Japan* **2**: 141–163 (in Japanese with English abstract).
3795. Wada, K., Hayakawa, C., & Yokohama, M. (2003) Determination of the mitochondrial DNA sequences in the Yeso Sika deer (*Cervus nippon yesoensis*)—12S rRNA gene, COX II gene and D-loop region. *Journal of Agricultural Science, Tokyo University of Agriculture* **47**: 298–312 (in Japanese with English abstract).
3796. Wada, K., Nishibori, M., & Yokohama, M. (2007) The complete nucleotide sequence of mitochondrial genome in the Japanese Sika deer (*Cervus nippon*), and a phylogenetic analysis between Cervidae and Bovidae. *Small Ruminant Research* **69**: 46–54.
3797. Wada, K. & Yokohama, M. (2004) Analysis of mitochondrial DNA protein-coding region in the Yeso Sika deer (*Cervus nippon yesoensis*). *Animal Science Journal* **75**: 295–302.
3798. Wada, M.Y. & Imai, H.T. (1991) On the Robertsonian polymorphism found in the Japanese raccoon dog (*Nyctereutes procyonoides viverrinus*). *Japanese Journal of Genetics* **66**: 1–11.
3799. Wada, M.Y., Lim, Y., & Wurster-Hill, D.H. (1991) Banded karyotype of a wild-caught male Korean raccoon dog *Nyctereutes procyonoides koreensis*. *Genome* **34**: 302–306.
3800. Wada, M.Y., Suzuki, T., & Tsuchiya, K. (1998) Re-examination of the chromosome homology between two subspecies of Japanese raccoon dogs (*Nyctereutes procyonoides albus* and *N. p. viverrinus*). *Caryologia* **51**: 13–18.
3801. Wada, N. (1993) Dwarf bamboos affect the regeneration of zoochorous trees by providing habitats to acorn-feeding rodents. *Oecologia* **94**: 403–407.
3802. Wada, S. (1988) Genetic differentiation between two forms of short-finned pilot whales off the Pacific coast of Japan. *Scientific Report of the Whales Research Institute, Tokyo* **39**: 91–101.
3803. Wada, S. (1989) Latitudinal Segregation of the Okhotsk Sea–West Pacific Stock of Minke Whales. *Report of the International Whaling Commission* **39**: 229–233.
3804. Wada, S. & Numachi, K.-I. (1991) Allozyme analyses of genetic differentiation among the populations and species of the *Balaenoptera*. *Reports of the International Whaling Commission (Special Issue)* **13**: 125–154.
3805. Wada, S., Oishi, M., & Yamada, T.K. (2003) A newly discovered species of living baleen whale. *Nature* **426**: 278–281.
3806. Wada, Y. (1967) Studies on mites associated with bats in Japan. I. Description of *Alabidocarpus fujii* n. sp. (Acarina: Listrophoridae). *Japanese Journal of Sanitary Zoology* **18**: 1–3 (in Japanese).
3807. Wada, Y. & Sasa, M. (1966) A few mites parasiting on bats. *Japanese Journal of Sanitary Zoology* **17**: 140 (in Japanese).
3808. Wade, P.R. & Gerrodette, T. (1993) Estimation of cetacean abundance and distribution in the eastern tropical Pacific. *Reports of the International Whaling Commission* **43**: 477–493.
3809. Wade, P.R., et al. (2007) Depletion of spotted and spinner dolphins in the eastern tropical Pacific: modeling hypotheses for their lack of recovery. *Marine Ecology Progress Series* **343**: 1–14.
3810. Wainio, W.W. & Forbes, E.B. (1941) The chemical composition of forest fruits and nuts from Pennsylvania. *Journal of Agricultural Research* **62**: 627–635.
3811. Waits, L.P., et al. (1999) Rapid radiation events in the family ursidae indicated by likelihood phylogenetic estimation from multiple fragments of mtDNA. *Molecular Phylogenetics and Evolution* **13**: 82–92.
3812. Wakabayashi, I. & Nakamura, M. (2007) The habitat conditions of finless porpoise *Neophocaena phocaenoides* in Ise Bay surveyed by a questionnaire to fishermen. *Mie Natural History* **11**: 117–122 (in Japanese).
3813. Wakabayashi, I., et al. (2014) Reproduction and artificial nursing of finless porpoises at Toba Aquarium. *Aquabiology* **36**: 36–43 (in Japanese with English abstract).
3814. Wakabayashi, I., Tsukada, O., & Kubota, T. (1998) Records of the boreal marine mammals from Suruga Bay, Enshu-Nada and Kumano-Nada, Central Japan. *Bulletin of Institute of Oceanic Research and Development, Tokai University* **19**: 105–112 (in Japanese with English summary).
3815. Wakabayashi, M. & Yamaga, Y. (2004) The use style near the breeding site of the bat *Eptesicus nilssonii* in Bihoro. *Bulletin of the Bihoro Museum* **11**: 55–62 (in Japanese with English abstract).
3816. Wakana, S., et al. (1996) Phylogenetic implications of variations in rDNA and mtDNA in red-backed voles collected in Hokkaido, Japan, and in Korea. *Mammal Study* **21**: 15–25.
3817. Waku, D., et al. (2014) A systematic study of the extinct Japanese otter. *in* Abstracts of XII IUCN OSG International Otter Congress in Rio de Janeiro, p. 60.
3818. Walker, W.A. & Hanson, B.M. (1999) Biological observations on Stejneger's beaked whale, *Mesoplodon stejnegeri*, from strandings on Adak Island, Alaska. *Marine Mammal Science* **15**: 1314–1329.
3819. Walker, W.A., et al. (1986) Geographical variation and biology of the Pacific white-sided dolphin, *Lagenorhynchus obliquidens*, in the north-eastern Pacific. *in* Research on Dolphins (Bryden, M.M. & Harrison, R., Editors). Oxford: Clarendon Press, pp. 441–465.
3820. Walker, W.A., Mead, J.G., & Brownell, R.L., Jr. (2002) Diets of Baird's beaked whales, *Berardius bairdii*, in the Southern Sea of Okhotsk and off the Pacific coast of Honshu, Japan. *Marine Mammal Science* **18**: 902–919.
3821. Wallace, A.R. (1911) Island Life or the Phenomena and Causes of Insular Fauna and Floras Including a Revision and Attempted Solution of the Problem of Geological Climates. 3rd and revised ed. London: Macmillan and Co., 563 pp. (Reprinted by AMS Press).
3822. Wallin, L. (1962) Notes on *Vespertilio namiyei* (Chiroptera). *Zoologiska bidrag från Uppsala* **35**: 397–416.
3823. Wallin, L. (1969) The Japanese bat fauna. *Zoologiska bidrag från Uppsala* **37**: 223–440.
3824. Wang, J. & Reeves, R. (2012) *Neophocaena asiaeorientalis*. The IUCN Red List of Threatened Species. Version 2014.3. Available from: http://www.iucnredlist.org.
3825. Wang, J.Y. & Berggren, P. (1997) Mitochondrial DNA analysis of harbour porpoises (*Phocoena phocoena*) in the Baltic Sea, the Kattegat–Skagerrak Seas and off the west coast of Norway. *Marine Biology* **127**: 531–537.
3826. Wang, J.Y. & Yang, S.C. (2006) Unusual cetacean stranding events of Taiwan in 2004 and 2005. *Journal of Cetacean Research and Management* **8**: 283–92.
3827. Wang, M.C., et al. (2012) Food partitioning among three sympatric odontocetes (*Grampus griseus*, *Lagenodelphis hosei*, and *Stenella attenuata*). *Marine Mammal Science* **28**: E143–E157.
3828. Wang, P. (1984) Distribution of the gray whale (*Eschrichtius robustus*) off the coast of China. *Acta Theriologica Sinica* **4**: 21–26 (in Chinese with English abstract).
3829. Wang, S., et al. (1962) On the small mammals from Southwestern Kwangsi, China. *Acta Zoologica Sinica* **14**: 555–570.
3830. Wang, Z., et al. (1984) Karyotypes of three species of Carnivora. *Acta Zoologica Sinica* **30**: 188–195.
3831. Ward, E.J., et al. (2009) The role of menopause and reproductive senescence in a long-lived social mammal. *Frontiers in Zoology* **6**: Article No.4.
3832. Ward, O.G., et al. (1987) Comparative cytogenetics of Chinese and Japanese raccoon dogs, *Nyctereutes procyonoides*. *Cytogenetics and Cell Genetics* **45**: 177–186.
3833. Waseda, K. Unpublished.
3834. Waseda, K. & Kameyama, A. (2006) Public education. *in* Understanding Asian Bears to Secure Their Future (Japan Bear Network, Compiler). Ibaraki: Japan Bear Network, pp. 116–117.
3835. Watanabe, S. (渡辺茂樹) & Harada, M. (原田正史) (2007) 大阪府北部におけるイタチ類2種の分布について—1980年代半ばより2006年4月までの調査結果より—. 成安紀要 **14**: 63–86 (in Japanese).
3836. Watanabe, S. (渡辺茂樹), Tanigaki, T. (谷垣岳人), & Yoshihiro, S. (好廣眞一) (2007) 滋賀県南部におけるイタチ類2種の分布について—2006年の調査より—. *in* 里山から見える世界: 龍谷大学里山学・地域共生学オープンリサーチセンター2006年度年次報告書. Kyoto: Ryukoku University, pp. 168–180 (in Japanese).
3837. Watanabe, A. (1959) Studies on trematode parasites of bats in Hiroshima Prefecture, with some reference to the function of Laurer's canal. *Japanese Journal of Parasitology* **8**: 849–857 (in Japanese with English abstract).
3838. Watanabe, A. (2004) Q fever caused by *Coxiella burnetii*. *Journal of Clinical and Experimental Medicine* **208**: 48–52 (in Japanese).
3839. Watanabe, H. (1980) Damage to conifers by the Japanese black bear. *International Conference on Bear Research and Management* **4**: 67–70.
3840. Watanabe, H. (2003) The total mobilization and wildlife: The introduction of nutorias in Japan. *Journal of History of Science, Japan* **42**: 129–139 (in Japanese with English résumé).
3841. Watanabe, K. (1962) On the taxonomical and ecological studies of voles, rats and mice, from the point of view of the plant protection of agriculture. *Technical Bulletin of Miyagi Prefectural Agricultural Experiment Station* **31**: 1–106 (in Japanese with English summary).
3842. Watanabe, K. (1989) Fish: A new addition to the diet of Japanese macaques

3843. Watanabe, K. (1993) Bibliography: Field Research on Japanese monkeys, 1975–1992—With some remarks on the history of researches. *Primate Research* **9**: 33–60 (in Japanese).
3844. Watanabe, K., Mori, A., & Kawai, M. (1992) Characteristics features of the reproduction of Koshima monkeys, *Macaca fuscata fuscata*: A summary of thirty-four years of observation. *Primates* **33**: 1–32.
3845. Watanabe, S. Unpublished.
3846. Watanabe, S., Nakanishi, N., & Izawa, M. (2003) Habitat and prey resource overlap between the Iriomote cat *Prionailurus iriomotensis* and introduced feral cat *Felis catus* based on assessment of scat content and distribution. *Mammal Study* **28**: 47–56.
3847. Watanabe, Y. Unpublished.
3848. Watanobe, T., Ishiguro, N., & Nakano, M. (2003) Phylogeography and population structure of the Japanese wild boar *Sus scrofa leucomystax*: mitochondrial DNA variation. *Zoological Science* **20**: 1477–1489.
3849. Watanobe, T., et al. (2001) Ancient mitochondrial DNA reveals the origin of *Sus scrofa* from Rebun Island, Japan. *Journal of Molecular Evolution* **52**: 281–289.
3850. Watanobe, T., et al. (1999) Genetic relationship and distribution of the Japanese wild boar (*Sus scrofa leucomystax*) and Ryukyu wild boar (*Sus scrofa riukiuanus*) analysed by mitochondrial DNA. *Molecular Ecology* **8**: 1509–1512.
3851. Watanuki, Y. & Nakayama, Y. (1993) Age differences in activity pattern of Japanese monkeys: Effects of temperature, snow and diet. *Primates* **34**: 419–430.
3852. Watari, Y. & Funakoshi, K. (2013) Use of dead-leaf foliage as day-roosts by the Ryukyu tube-nosed bat, *Murina ryukyuana*. *Mammalian Science* **53**: 331–334 (in Japanese with English abstract).
3853. Watari, Y., Nagata, J., & Funakoshi, K. (2010) New detection of a 30-year-old population of introduced mongoose *Herpestes auropunctatus* on Kyushu Island, Japan. *Biological Invasions* **13**: 269–276.
3854. Watari, Y., et al. (2013) Evaluating the "recovery-level" of endangered species without prior information before alien invasion. *Ecology and Evolution* **3**: 4711–4721.
3855. Watari, Y., Takatsuki, S., & Miyashita, T. (2008) Effects of exotic mongoose (*Herpestes javanicus*) on the native fauna of Amami-Oshima Island, southern Japan, estimated by distribution patterns along the historical gradient of mongoose invasion. *Biological Invasion* **10**: 7–17.
3856. Waterman, P.G. & Mole, S. (1994) Analysis of Phenolic Plant Metabolites. Methods in ecology. Vol. 1 (Lawton, J.H. & Likens, G.E., Editors). Oxford: Blackwell Scientific Publications, 238 pp.
3857. Watkins, W.A. (1981) Activities and underwater sounds of fin whales. *Scientific Reports of the Whales Research Institute* **33**: 83–117.
3858. Watwood, S.L., et al. (2006) Deep-diving foraging behaviour of sperm whales (*Physeter macrocephalus*). *Journal of Animal Ecology* **75**: 814–825.
3859. Wayne, R.K., et al. (1991) Conservation genetics of the endangered Isle Royale gray wolf. *Conservation Biology* **5**: 41–51.
3860. Weerasinghe, U.R. & Takatsuki, S. (1999) A record of acorn eating by sika deer in western Japan. *Ecological Research* **14**: 205–209.
3861. Wei, Z., et al. (2002) Observations on behavior and ecology of the Yangtze finless porpoise (*Neophocaena phocaenoides asiaeorientalis*) group at Tian-e-Zhou Oxbow of the Yangtze River. *Raffles Bulletin of Zoology Supplement* **10**: 97–103.
3862. Weigl, R. (2005) Longevity of Mammals in Captivity; from the Living Collections of the World. Stuttgart: E.Schweizerbart, Borntraeger and Cramer Science Publishers, 214 pp.
3863. Weller, D.W., et al. (2002) The western gray whale: a review of past exploitation, current status and potential threats. *Journal of Cetacean Research and Management* **4**: 7–12.
3864. Weller, D.W., et al. (1999) Gray whales *Eschrichtius robustus* off Sakhalin Island, Russia: seasonal and annual patterns of occurrence. *Marine Mammal Science* **15**: 1208–1227.
3865. Wells, R.S. (1984) Reproductive behavior and hormonal correlates in Hawaiian spinner dolphins, *Stenella longirostris*. *Reports of the International Whaling Commission (Special Issue)* **6**: 465–474.
3866. Wells, R.S. & Scott, M.D. (1999) Bottlenose dolphin *Tursiops truncatus* (Montagu, 1821). in Handbook of Marine Mammals, Volume 6: The Second Book of Dolphins and the Porpoises (Ridgway, S.H. & Harrison, R., Editors). London: Academic Press, pp. 137–182.
3867. Wells, R.S. & Scott, M.D. (2002) Bottlenose dolphins (*Tursiops truncatus* and *T. aduncus*). in Encyclopedia of Marine Mammals (Perrin, W.F., Würsig, B., & Thewissen, J.G.M., Editors). San Diego: Academic Press, pp. 122–128.
3868. West, K.L., et al. (2012) A Longman's beaked whale (*Indopacetus pacificus*) strands in Maui, Hawaii, with first case of morbillivirus in the central Pacific. *Marine Mammal Science* **29**: 767–776.
3869. Westgate, A.J. & Read, A.J. (2006) Reproduction in short-beaked common dolphins (*Delphinus delphis*) from the western North Atlantic. *Marine Biology* **150**: 1011–1024.
3870. Westgate, A.J., et al. (1995) Diving behaviour of habour porpoises, *Phocoena phocoena*. *Canadian Journal of Fisheries and Aquatic Sciences* **52**: 1064–1073.
3871. Westlake, R.L. & O'Corry-Crowe, G.M. (2002) Macrogeographic structure and patterns of genetic diversity in harbor seals (*Phoca vitulina*) from Alaska to Japan. *Journal of Mammalogy* **83**: 1111–1126.
3872. Whitehead, G.K. (1993) The Whitehead Encyclopedia of Deer. Shrewsbury: Swan Hill Press, 597 pp.
3873. Whitehead, H. (1993) The behaviour of mature male sperm whales on the Galapagos breeding grounds. *Canadian Journal of Zoology* **71**: 689–699.
3874. Whitehead, H. (2009) Sperm whale. in Encyclopedia of Marine Mammals. 2nd ed. (Perrin, W.F., Würsig, B., & Thewissen, J.G.M., Editors). Amsterdam: Academic Press, pp. 1091–1097.
3875. Whitehead, H. (2003) Sperm whales: Social Evolution in the Ocean. Chicago: University of Chicago Press, 431 pp.
3876. Wildlife Management Office (野生動物保護管理事務所) (1998) 里地性の獣類に関する緊急疫学調査報告書 (Commission Report of the Environment Agency). 64 pp. (in Japanese).
3877. Wildlife Management Office (野生動物保護管理事務所) (2007) 平成18年度関東地域アライグマ防除モデル事業調査報告書. 49 pp. (in Japanese).
3878. Wiles, G.J. (2004) Washington State Status Report for the Killer Whales. Olympia: Washington Department Fish and Wildlife, 106 pp.
3879. Williams, K., et al. (1995) Managing Vertebrate Pests: Rabbits. Canberra: Australian Government Publishing Service, 284 pp.
3880. Willis, P.M., et al. (2004) Natural hybridization between Dall's porpoises (*Phocoenoides dalli*) and harbour porpoises (*Phocoena phocoena*). *Canadian Journal of Zoology* **82**: 828–834.
3881. Wilson, D.E., et al. (1991) Geographic variation in sea otters, *Enhydra lutris*. *Journal of Mammalogy* **72**: 22–36.
3882. Wilson, D.E. & Reeder, D.M., eds. (1993) Mammal Species of the World. A Taxonomic and Geographical Reference. 2nd ed. Washington: Smithsonian Institute Press, 1206 pp.
3883. Wilson, D.E. & Reeder, D.M., eds. (2005) Mammal Species of the World. A Taxonomic and Geographical Reference. 3rd ed. Baltimore: Johns Hopkins University Press, 2142 pp.
3884. Winn, H.E. & Perkins, P.J. (1976) Distribution and sounds of the minke whale, with a review of mysticete sounds. *Cetology* **19**: 1–12.
3885. Winn, H.E. & Reichley, N.E. (1985) Humpback whale *Megaptera novaeangliae* (Browski, 1781). in Handbook of Marine Mammals. Volume 3. The Sirenians and Baleen Whales (Ridgway, S.H. & Harrison, R.J., Editors). London: Academic Press, pp. 241–273.
3886. Witteveen, B.H., et al. (2004) Abundance and mtDNA differentiation of humpback whales (*Megaptera novaeangliae*) in the Shumagin Islands, Alaska. *Canadian Journal of Zoology* **82**: 1352–1359.
3887. Wolman, A.A. (1985) Gray whale *Eschrichtius robustus* (Lilljeborg, 1861). in Handbook of Marine Mammals. Volume 3. The Sirenians and Baleen Whales (Ridgway, S.H. & Harrison, R.J., Editors). London: Academic Press, pp. 67–90.
3888. Won, C. & Smith, K.G. (1999) History and current status of mammals of the Korean Peninsula. *Mammal Review* **29**: 3–33.
3889. Wood, N.A. (1922) The mammals of Washtenaw County, Michigan. *University of Michigan Museum of Zoology Occasional Papers* **123**: 1–23.
3890. Woodworth, P.A., et al. (2012) Eddies as offshore foraging grounds for melon-headed whales (*Peponocephala electra*). *Marine Mammal Science* **28**: 638–647.
3891. Working Group for Damage on Agriculture, Forestry and Fishery by Wild Animals (鳥獣による農林水産業被害対策に関する検討会) (2007) 鳥獣による農林水産業被害対策に関する検討会報告書. Tokyo: Ministry of Agriculture, Forestry and Fisheries, 24 pp. (in Japanese).
3892. World Wide Fund for Nature (1978) World Wildlife Yearbook 1978–79. Gland: WWF, 100 pp.
3893. Wozencraft, W.C. (1993) Order Carnivora. in Mammal Species of the World: A Taxonomic and Geographic Reference (Wilson, D.E. & Reeder, D.M., Editors). Washington: Smithsonian Institution Press, pp. 279–348.
3894. Wozencraft, W.C. (2005) Order Carnivora. in Mammal Species of the World: A Taxonomic and Geographic Reference (Wilson, D.E. & Reeder, D.M., Editors). Baltimore: Johns Hopkins University Press, pp. 532–628.
3895. Wu, C.-H., et al. (2005) Molecular phylogenetics and biogeography of *Lepus* in Eastern Asia based on mitochondrial DNA sequences. *Molecular Phylogenetics and Evolution* **37**: 45–61.
3896. Wu, H.-Y. & Lin, Y.-S. (1992) Life history variables of wild troops of Formosan macaques (*Macaca cyclopis*) in Kenting, Taiwan. *Primates* **33**:

85–97.

3897. Wu, J., et al. (in press) Phylogeographic and demographic analysis of Asian black bear (*Ursus thibetanus*) based on the mitochondrial DNA. *PLoS ONE*.
3898. Wurster, D.H. & Benirschke, K. (1967) Chromosome studies in some deer, the springbok, and the pronghorn, with notes on placentation in deer. *Cytologia* **32**: 273–285.
3899. Wurster, D.H. & Benischke, K. (1968) Comparative cytogenetic studies in the order Carnivore. *Chromosoma* **24**: 336–382.
3900. Wurster-Hill, D.H., et al. (1987) Banded chromosome study of the Iriomote cat. *Journal of Heredity* **78**: 105–107.
3901. Wurster-Hill, D.H., et al. (1986) Banded chromosome studies and B chromosomes in wild-caught Japanese raccoon dogs, *Nyctereutes procyonoides viverrinus*. *Cytogenetics and Cell Genetics* **42**: 85–93.
3902. Wynen, L.P., et al. (2001) Phylogenetic relationships within the eared seals (Otariidae: Carnivora): Implications for the historical biogeography of the family. *Molecular Phylogenetics and Evolution* **21**: 270–284.
3903. Xu, H., et al. (2005) Migration of young bent-winged bats, *Miniopterus fuliginosus* born in Shirahama, Wakayama Prefecture (1) Records from the years 2003 and 2004. *Bulletin of Center for Natural Environment Education, Nara University of Education* **7**: 31–37 (in Japanese with English summary).
3904. Xu, H., et al. (2009) Fauna of bats in the Oku-Yoshino Forest for Practical Exercises, Center for Natural Environment Education, Nara University of Education. *Bulletin of Natural Environment Education, Nara University of Education* **9**: 10–12 (in Japanese).
3905. Xu, H.-F., et al. (1998) Status and current distribution of south China sika deer. *Chinese Biodiversity* **6**: 87–91 (in Chinese with English abstract).
3906. Yabe, T. (矢部恒晶) & Koizumi, T. (小泉透) (2003) 九州の生息地におけるニホンジカの行動. 九州の森と林業 **65**: 1–3 (in Japanese).
3907. Yabe, T. (矢部辰男) (2008) 日本の家ねずみ問題—これだけは知っておきたい (Control of Commensal Rodents in Japan). Tokyo: Chijin Shokan (地人書館), 169 pp. (in Japanese).
3908. Yabe, T. (1995) A fundamental study on habitat management for wildlife: habitat use of sika deer and a change in the vegetation on Shiretoko Peninsula, Hokkaido. *Research Bulletin of The Hokkaido University Forests* **52**: 115–180 (in Japanese with English abstract).
3909. Yabe, T. & Arakawa, O. (2009) Eye-lens weight curve for estimation of age in the Japanese grass vole, *Microtus montebelli* Milne-Edwards (Rodentia: Muridae). *Applied Entomology and Zoology* **44**: 501–504.
3910. Yachimori, S. (谷地森秀二) (1997) 野生ホンドタヌキにおける家族関係の推定と家族を構成する個体間の行動変化. Doctoral dissertation. Fujisawa: Nihon University. 148 pp. (in Japanese).
3911. Yachimori, S. (2000) A case of an assault on hoggery by an Asiatic black bear, *Ursus thibetanus*, in Fujiwara Town, Tochigi Prefecture. *Bulletin of Tochigi Prefectural Museum* **17**: 113–118 (in Japanese with English summary).
3912. Yachimori, S. (2007) Records of bats from Kochi Prefecture: Forest-dwelling bats caught by mist nets (II). *Bulletin of the Shikoku Institute of Natural History* **4**: 10–17 (in Japanese with English abstract).
3913. Yachimori, S. & Yamasaki, K. (2004) Record of bats from Kochi Prefecture: Woodland bats caught by mist nets. *Bulletin of the Shikoku Institute of Natural History* **1**: 43–49 (in Japanese).
3914. Yachimori, S. & Yasui, S. (2002) Mammals of the Nasu Imperial Villa in the northern part of Tochigi Prefecture, Central Japan. in Flora and Fauna of the Nasu Imperial Villa (Tochigi Prefectural Museum, Editor). Utsunomiya: Tochigi Prefectural Museum, pp. 21–27 (in Japanese with English abstract).
3915. Yagi, N. (八木誠政) (1934) アカモズの早贄の追加. *Botany and Zoology* **2**: 1888 (in Japanese).
3916. Yagi, S. (八木繁一) (1971) カワウソの糞と10年. in 愛媛の自然 13(4). Matsuyama: Ehime Prefectural Museum, pp. 8–9 (in Japanese).
3917. Yagisawa, M. (1978) Studies on zoonotic helminthes from mammals in northern Honshu, Japan. *Hirosaki Medical Journal* **30**: 239–284 (in Japanese with English summary).
3918. Yahara, T. (矢原徹一) (2006) シカの増加と野生植物の絶滅リスク. in 世界遺産をシカが喰う：シカと森の生態学 (Yumoto, T. (湯本貴和) & Matsuda, H. (松田裕之), Editors). Tokyo: Bun-ichi Sogo Shyuppan (文一総合出版), pp. 168–187 (in Japanese).
3919. Yahner, R.H. (1979) Temporal patterns in the male mating behavior of captive Reeves' muntjac (*Muntiacus reevesi*). *Journal of Mammalogy* **60**: 560–567.
3920. Yajima, K., et al. (2002) Seasonal changes in home range of female sika deer (*Cervus nippon*) on Mt. Ohdaigahara, central Japan. *Nagoya University Forest Science* **21**: 1–7 (in Japanese with English abstract).
3921. Yamada, F. (山田文雄) (1996) カイウサギ (Feral rabbits). in The Encyclopaedia of Animals in Japan. 2. Mammals II (日本動物大百科2哺乳類II) (Izawa, K. (伊沢紘生), Kasuya, T. (粕谷俊雄), & Kawamichi, T. (川道武男), Editors). Tokyo: Heibonsha (平凡社), p. 131 (in Japanese).
3922. Yamada, F. (山田文雄) (2002) マングース (*Herpestes javanicus*). in Handbook of Alien Species in Japan (外来種ハンドブック) (Ecological Society of Japan, Editor). Tokyo: Chijin Shokan (地人書館), p. 75 (in Japanese).
3923. Yamada, F. Unpublished.
3924. Yamada, F. (1987) Reproductive behavior in the *Lepus* and its properties. *Journal of the Japanese Society for Hares* **14**: 17–21 (in Japanese).
3925. Yamada, F. (1990) Habitat selection and feeding habits of the Japanese hare and its damage to seedlings. in Wildlife Conservation, Present Trends and Perspectives for the 21st Century (Maruyama, N., Editor). Tokyo: Japan Wildlife Research Center, pp. 111–113.
3926. Yamada, F. (1991) Feral rabbits on Japanese islands. *Lagomorph Newsletter* **14**: 9–11.
3927. Yamada, F. (2002) Impacts and control of introduced small Indian mongoose on Amami Island, Japan. in Turning the Tide: The Eradication of Invasive Species: Proceedings of the International Conference on Eradication of Island Invasives (Occasional Paper of the IUCN Species Survival Commission No. 27) (Veitch, C.R. & Clout, M.N., Editors). Gland: IUCN, pp. 389–392 (http://www.hear.org/articles/turningthetide/turningthetide.pdf).
3928. Yamada, F. (2008) A review of the biology and conservation of the Amami rabbit (*Pentalagus furnessi*). in Lagomorph Biology, Evolution, Ecology and Conservation (Alves, P.C., Ferrand, N., & Hackländer, K., Editors). Berlin: Springer-Verlag, pp. 369–378.
3929. Yamada, F. (2014) *Lepus brachyurus lyoni* Kishida, 1937. in Red Data Book 2014: Threatened Wildlife of Japan. 1 Mammalia (Ministry of the Environment, Editor). Tokyo: Gyosei Corporation, p. 82 (in Japanese).
3930. Yamada, F. & Cervantes, F.A. (2005) *Pentalagus furnessi*. *Mammalian Species* **782**: 1–5.
3931. Yamada, F., et al. (2010) Rediscovery after thirty years since the last capture of the critically endangered Okinawa spiny rat *Tokudaia muenninki* in the northern part of Okinawa Island. *Mammal Study* **35**: 243–255.
3932. Yamada, F., et al. (1989) Follicular growth and timing of ovulation after coitus in the Japanese hare, *Lepus brachyurus brachyurus*. *Journal of the Mammalogical Society of Japan* **14**: 1–9.
3933. Yamada, F., et al. (1990) Growth, development and age determination of the Japanese hare, *Lepus brachyurus*. *Journal of the Mammalogical Society of Japan* **14**: 65–77.
3934. Yamada, F., Shiraishi, S., & Uchida, T.A. (1988) Parturition and nursing behaviours of the Japanese hare, *Lepus brachyurus*. *Journal of the Mammalogical Society of Japan* **13**: 59–68.
3935. Yamada, F., Sugimura, K., & Abe, S. (1999) Present situation and problems of countermeasures against invasive alien mongooses in Amami Island. *Bulletin of Kansai Organization for Nature Conservation* **21**: 31–41 (in Japanese).
3936. Yamada, F., et al. (2000) Present status and conservation of the endangered Amami rabbit *Pentalagus furnessi*. *Tropics* **10**: 87–92.
3937. Yamada, F., Takaki, M., & Suzuki, H. (2002) Molecular phylogeny of Japanese Leporidae, the Amami rabbit *Pentalagus furnessi*, the Japanese hare *Lepus brachyurus*, and the mountain hare *Lepus timidus*, inferred from mitochondrial DNA sequences. *Genes & Genetic Systems* **77**: 107–116.
3938. Yamada, M. (1954) Some remarks on the pygmy sperm whale, *Kogia*. *Scientific Reports of the Whales Research Institute, Tokyo* **9**: 37–58.
3939. Yamada, M. & Egi, H. (2011) Records of *Murina ussuriensis* (Chiroptera, Vespertilionidae) in Okayama Prefecture. *Bulletin of Okayama Prefectural Nature Conservation Center* **18**: 81–87 (in Japanese with English abstract).
3940. Yamada, M., et al. (2007) Phylogenetic relationship of the southern Japan lineages of the sika deer (*Cervus nippon*) in Shikoku and Kyushu Islands, Japan. *Mammal Study* **32**: 121–127.
3941. Yamada, M., et al. (2006) Distribution of two distinct lineages of sika deer (*Cervus nippon*) on Shikoku Island revealed by mitochondrial DNA analysis. *Mammal Study* **31**: 23–28.
3942. Yamada, M., Shibuya, Y., & Matsuzaki, R. (2012) Records of *Nyctalus aviator* Thomas, 1911 (Chiroptera, Vespertilionidae) in Okayama Prefecture. *Bulletin of Okayama Prefectural Nature Conservation Center* **19**: 1–6 (in Japanese).
3943. Yamada, T.K. Unpublished.
3944. Yamada, T.K., et al. (2006) Middle sized balaenopterid whale specimens (Cetacea: Balaenopteridae) preserved at several institutions in Taiwan, Thailand, and India. *Memoirs of the National Science Museum, Tokyo* **44**: 1–10.
3945. Yamada, T.K., Kakuda, T., & Tajima, Y. (2008) Middle sized balaenopterid whale specimens in the Philippines and Indonesia. *Memoirs of the National Museum of Nature and Science, Tokyo* **45**: 75–83.
3946. Yamada, T.K., et al. (2006) Marine mammal collections in Australia. in Proceedings of the 7th and 8th Symposia on Collection Building and

Natural History Studies in Asia and the Pacific Rim. National Science Museum Monographs 34 (Tomida, Y., Editor). Tokyo: National Science Museum, pp. 117–126.

3947. Yamada, T.K., et al. (2004) On a Longman's beaked whale stranded in Kagoshima. in 15th Congress of the Sea of Japan Cetology Research Group, Abstracts (Kanazawa), p. 12 (in Japanese).

3948. Yamada, T.K., et al. (1995) Stomach contents of *Mesoplodon stejnegeri* (ZIPHIIDAE). *Nihonkai Cetology* 5: 31–36 (in Japanese).

3949. Yamada, T.K., et al. (2007) Biological indices obtained from a pod of killer whales entrapped by sea ice off northern Japan (Paper SC/59/SM12 presented to the IWC Scientific Committee). 15 pp.

3950. Yamada, T.K., Uni, Y., & Ishikawa, H. (2002) Recent gray whale strandings and sightings around Japan (Paper SC/02/WGW08 presented to the IWC Scientific Committee, Western Gray Whale Workshop, October 2002, Ulsan, Korea). 5 pp.

3951. Yamaga, Y. (2006) Notes of bats in Bihoro, Hokkaido, 2005. *Bulletin of the Bihoro Museum* 13: 87–90 (in Japanese with English abstract).

3952. Yamaga, Y. & Inoue, J. (2005) Notes on bats in Bihoro, Hokkaido, 2004. *Bulletin of the Bihoro Museum* 12: 75–80 (in Japanese with English abstract).

3953. Yamaga, Y., Kon, N., & Yamaki, M. (2002) Faunal survey of bats in Bihoro, Hokkaido, 2001. *Bulletin of the Bihoro Museum* 9: 103–108 (in Japanese with English summary).

3954. Yamaga, Y., Saito, O., & Yamaki, M. (2000) Faunal survey of bats in Bihoro, Hokkaido (3). *Bulletin of the Bihoro Museum* 7: 61–70 (in Japanese with English summary).

3955. Yamaga, Y. & Saitoh, O. (2003) Growth record of a bat *Eptesicus nilssonii* in Bihoro. *Bulletin of the Bihoro Museum* 10: 79–86 (in Japanese with English abstract).

3956. Yamaga, Y., Sekino, H., & Yamaki, M. (2001) Faunal survey of bats in Bihoro, Hokkaido, 2000. *Bulletin of the Bihoro Museum* 8: 87–90 (in Japanese with English summary).

3957. Yamaga, Y. & Shou, T. (2004) Faunal survey of bats in Bihoro, Hokkaido, 2002, 2003. *Bulletin of the Bihoro Museum* 11: 63–70 (in Japanese with English summary).

3958. Yamagata, T., et al. (1987) Genetic differentiation between laboratory lines of the musk shrew (*Suncus murinus*, Insectivora) based on restriction endonuclease cleavage patterns of mitochondrial DNA. *Biochemical Genetics* 25: 429–446.

3959. Yamagata, T., et al. (1995) Genetic variation and geographic distribution on the mitochondrial DNA in local populations of the musk shrew, *Suncus murinus*. *Japanese Journal of Genetics* 70: 321–337.

3960. Yamagata, T., et al. (1990) Genetic relationship among the musk shrews, *Suncus murinus* Insectivora, inhabiting islands and the continent based on mitochondrial DNA types. *Biochemical Genetics* 28: 185–195.

3961. Yamagishi, M., Matsubara, K., & Sakaizumi, M. (2012) Molecular cytogenetic identification and characterization of Robertsonian chromosomes in the large Japanese field mouse (*Apodemus speciosus*) using FISH. *Zoological Science* 29: 709–713.

3962. Yamaguchi, K. (山口敬治), Kamiya, H. (神谷晴夫), & Kudo, N. (工藤規雄) (1977) エゾシカ *Cervus nippon yesoensis* (HEUDE) の寄生蠕虫について. *Journal of the Hokkaido Veterinary Medical Association* 21: 167–170 (in Japanese).

3963. Yamaguchi, T. (山口鉄男) & Urata, A. (浦田明夫) (1976) ツシマヤマネコ. in 対馬の生物 (Nagasaki Biological Society (長崎県生物学会), Editor). Nagasaki: Nagasaki Biological Society, pp. 167–180 (in Japanese).

3964. Yamaguchi, Y. (山口佳秀) & Takahashi, H. (高橋秀男) (1979) 胃内容物からみたニホンカモシカの食性について. in 鳥獣害性調査報告書. Tokyo: Environment Agency, pp. 29–51 (in Japanese).

3965. Yamaguchi, M., Torii, H., & Higuchi, S. (2008) Habitat utilization of the Japanese hare (*Lepus brachyurus*) in hilly areas in the southern part of Kyoto Prefecture during the winter season. *Journal of the Japanese Wildlife Research Society* 33: 12–19 (in Japanese with English summary).

3966. Yamaguchi, T., et al. (1998) Prevalence of infectious agents, drug resistant- *Escherichia coli* and residual organochlorine in wild animals inhabiting the mountainous areas of central Japan. *Japanese Journal of Zoo and Wildlife Medicine* 3: 1–7 (in Japanese with English abstract).

3967. Yamaguchi, Y. (2006) Seasonal change of bats using tunnels as rosst in Kurokura Basin of the Tanzawa Mountains, Kanagawa Prefecture. *Natural History Report of Kanagawa* 27: 45–49 (in Japanese).

3968. Yamaguchi, Y., Sone, M., & Aimoto, D. (2002) Tracking of the Tube-nosed bat *Murina leucogaster* using the radio-transmitter. *Natural History Report of Kanagawa* 23: 15–18 (in Japanese).

3969. Yamaguchi, Y., et al. (2002) Distribution of bats in Tanzawa Mountains, Kanagawa Prefecture. *Natural History Report of Kanagawa* 23: 19–24 (in Japanese).

3970. Yamaguchi, Y., et al. (2005) Notes on hibernating colonies of *Vespertilio superans* found in Kanagawa Prefecture. *Natural History Report of Kanagawa* 26: 49–51 (in Japanese).

3971. Yamaguchi, Y. & Yanagawa, H. (1995) Field observation on circadian activity of the flying squirrel, *Pteromys volans orii*. *Mammalian Science* 34: 139–149 (in Japanese with English abstract).

3972. Yamaguchi, Y. & Yanagawa, H. (2010) The ecology of the red squirrel, *Sciurus vulgaris orientis* on the campus of Obihiro University. 1. Nest and nest tree selection. *Research Bulletin of Obihiro University* 31: 34–39 (in Japanese).

3973. Yamaguchi, N., et al. (1971) Ticks of Japan, Korea, and the Ryukyu Islands. *Brigham Young University Science Bulletin, Biological Series* 15: 1–226.

3974. Yamaguti, S. (1935) Studies on the helminth fauna of Japan. Part 7. Cestodes of mammals and snakes. *Japanese Journal of Zoology* 6: 233–246.

3975. Yamaguti, S. (1935) Studies on the helminth fauna of Japan. Part 8. Acanthocephala, I. *Japanese Journal of Zoology* 6: 247–278.

3976. Yamaguti, S. (1939) Studies on the helminth fauna of Japan. Part 29. Acanthocephala, II. *Japanese Journal of Zoology* 8: 318–351, pls. I–XLIX.

3977. Yamaguti, S. (1941) Studies on the helminth fauna of Japan. Part 35. Mammalian nematodes. II. *Japanese Journal of Zoology* 9: 409–438.

3978. Yamaguti, S. (1954) Helminth fauna of Mt. Ontake. Part 2. Trematoda and cestoda. *Acta Medica Okayama* 8: 393–405.

3979. Yamakage, K., et al. (1985) G-, C- and N-banding patterns on the chromosomes of the Japanese grass vole, *Microtus montebelli montebelli*, with special attention to the karyotypic comparison with the root vole, *M. oeconomus*. *The Journal of the Mammalogical Society of Japan* 10: 209–220 (in Japanese with English abstract).

3980. Yamamoto, E. (山本栄治) & Doi, M. (土居雅恵) (2000) 小田深山およびその周辺の哺乳動物. in 小田深山の自然 I (山本森林生物研究所 & 小田深山の自然編集委員会, Editors). Oda: Oda Town Office, pp. 293–377 (in Japanese).

3981. Yamamoto, I. (山本五男) & Shimazawa, T. (嶋沢匡寿) (1992) 一戸町西岳のコウモリ類とカワネズミ調査. *Paulownia* 24: 36 (in Japanese).

3982. Yamamoto, S. (山本純郎) (1999) シマフクロウ. Sapporo: Hokkaido Shimbun Press, 192 pp. (in Japanese).

3983. Yamamoto, S., Takahashi, M., & Nogami, S. (1998) Scabies in wild raccoon dogs, *Nyctereutes procyonoides* at the Tomioka–Kanra district in Gunma Prefecture, Japan. *Medical Entomology and Zoology* 49: 217–222 (in Japanese with English abstract).

3984. Yamamoto, T. (山本輝正) (1990) 石川県のコウモリ. in 石川の生物 (石川の生物編集委員会, Editor). Kanazawa: 石川県高等学校研究会生物部会, pp. 137–142 (in Japanese).

3985. Yamamoto, T. (山本輝正) (1991) 八百津町哺乳類 (2). 岐阜ふるさとと動物通信 42: 676 (in Japanese).

3986. Yamamoto, T. (山本輝正) (1994) クビワコウモリの生態調査と保護 (WWFJ自然保護事業報告書 事業番号：9304). Zushi: コウモリの会クビワコウモリ保護研究部会, 11 pp. (in Japanese).

3987. Yamamoto, T. (山本輝正) (1994) 白山石川県側のコウモリ 1. 岐阜ふるさとと動物通信 60: 967 (in Japanese).

3988. Yamamoto, T. (山本輝正) (1994) 白川村のコウモリ (1). 岐阜ふるさとと動物通信 60: 980 (in Japanese).

3989. Yamamoto, T. (山本輝正) (1995) 白山石川県側のコウモリ 2. 岐阜ふるさとと動物通信 62: 996 (in Japanese).

3990. Yamamoto, T. (山本輝正) (1998) 白山地域のコウモリ類. in 環境庁委託業務報告書：平成9年度生態系多様性地域調査（白山地区）報告書. Gifu & Kanazawa: Gifu & Ishikawa prefectural governments, pp. 227–235 (in Japanese).

3991. Yamamoto, T. (山本輝正) (2004) 岐阜県白川村大窪池周辺のコウモリ相. *Journal of the Biological Education* 49: 27–31 (in Japanese).

3992. Yamamoto, T. (山本輝正) (2006) テングコウモリとコテングコウモリの秋期ねぐら. *Bat Study and Conservation Report* 14: 13 (in Japanese).

3993. Yamamoto, T. (山本輝正), Hashimoto, H. (橋本肇), & Ueki, Y. (植木康徳) (1998) 乗鞍高原のコウモリ. *Journal of the Biological Education* 42: 12–18 (in Japanese).

3994. Yamamoto, A. (2007) Geographic Variations of Dental and Skeletal Morphology in Japanese Macaques (*Macaca fuscata*), Doctoral dissertation. Kyoto: Kyoto University.

3995. Yamamoto, A. & Kunimatsu, Y. (2006) Ontogenetic change and geographical variation of atlas bridging in Japanese macaques (*Macaca fuscata*). *Anthropological Science* 114: 153–160.

3996. Yamamoto, I. (1984) Latrine utilization and feces recognition in the raccoon dog, *Nyctereutes procyonoides*. *Journal of Ethology* 2: 47–54.

3997. Yamamoto, I. (1987) Male parental care in the raccoon dog *Nyctereutes procyonoides* during the early rearing period. in Animal Societies: Theories and Facts (Ito, Y., Brown, J.L., & Kikkawa, J., Editors). Tokyo: Japan Scientific Society Press, pp. 185–195.

3998. Yamamoto, I. & Hidaka, T. (1984) Utilisation of latrines in the raccoon dog, *Nyctereutes procyonoides*. *Acta Zoologica Fennica* **171**: 241–242.

3999. Yamamoto, K., Ohdachi, S.D., & Kasahara, Y. (2010) Detection of effects of a high trophic level predator, *Sorex unguiculatus* (Soricidae, Mammalia), on a soil microbial community in a cool temperate forest in Hokkaido, using the ARISA method. *Microbes and Environments* **25**: 197–203.

4000. Yamamoto, K., Tsubota, T., & Kita, I. (1998) Observation of sexual behavior of captive Japanese black bear, *Ursus thibetanus japonicus*. *Journal of Reproduction and Development* **44**: 13–18 (in Japanese with English summary).

4001. Yamamoto, S. & Noda, S. (1995) Collecting records of trombiculid mites in Kagoshima Prefecture in 1993 and 1994. *Journal of the Acarological Society of Japan* **4**: 123–127 (in Japanese with English abstract).

4002. Yamamoto, T., et al. (2004) List of Chiroptera in Ehime Prefecture, Shikoku, Japan. *Bulletin of the Ehime Prefectural Science Museum* **9**: 1–9 (in Japanese).

4003. Yamamoto, T., Kajiura, K., & Kondo, M. (2008) Fauna of Chiroptera around the Nomugi pass, Gifu. *Bulletin of the Gifu Prefectural Museum* **29**: 45–48 (in Japanese).

4004. Yamamoto, T. & Nozaki, E. (2002) Fauna of Chiroptera in Hakusan, Ishikawa Prefecture. *Annual Report of the Hakusan Nature Conservation Center* **29**: 73–76 (in Japanese).

4005. Yamamoto, T., Sato, A., & Katsuta, S. (2008) New distributional record of the Japanese lesser noctule bat, *Nyctalus furvus*, and the Japanese northern bat, *Eptesicus japonensis* from Nagano Prefecture, central Japan. *Mammalian Science* **48**: 277–280 (In Japanese with English abstract).

4006. Yamamoto, T., Ueuma, Y., & Nozaki, E. (2005) Fauna of Chiroptera in Mt. Hakusan, Ishikawa Prefecture—ecological survey from 1998 to 2005. *Annual Report of the Hakusan Nature Conservation Center* **32**: 25–30 (in Japanese).

4007. Yamamoto, Y. (1986) Researches of setts of *Meles meles anakuma* in Mt. Nyugasa, Nagano. *Journal of the Hiraoka Environmental Science Laboratory* **2**: 131–139 (in Japanese with English abstract).

4008. Yamamoto, Y. (1991) Diet and distribution of raccoon dog, *Nyctereutes procyonoides viverrinus*, in Kawasaki. *Report of the Kawasaki Municipal Natural Environment* **2**: 185–194 (in Japanese).

4009. Yamamoto, Y. (1991) Food habit of the Japanese badger (*Meles meles anakuma*) in Mt. Nyugasa, Nagano prefecture. *Natural Environmental Science Research* **4**: 73–83 (in Japanese with English abstract).

4010. Yamamoto, Y. (1994) Comparative analyses on food habits of Japanese marten, red fox, badger and raccoon dog in the Mt. Nyugasa, Nagano Prefecture, Japan. *Natural Environmental Scientific Research* **7**: 45–52 (in Japanese with English summary).

4011. Yamamoto, Y. (1995) Home range and habitat selection of the Japanese badger (*Meles meles anakuma*) in Mt. Nyugasa, Nagano prefecture. *Natural Environmental Science Research* **8**: 51–65 (in Japanese with English abstract).

4012. Yamamoto, Y. & Kinosita, A. (1994) Food composition of the raccoon dog *Nyctereutes procyonoides viverrinus* in Kawasaki. *Bulletin of the Kawasaki Municipal Science Museum for Youth* **5**: 29–34 (in Japanese).

4013. Yamamoto, Y., et al. (1994) Home range and dispersal of the raccoon dog (*Nyctereutes procyonoides viverrinus*) in the Mt. Nyugasa, Nagano Prefecture, Japan. *Natural Environmental Scientific Research* **7**: 53–61 (in Japanese with English summary).

4014. Yamamura, K., et al. (2008) Harvest-based Bayesian estimation of sika deer populations using state-space models. *Population Ecology* **50**: 131–144.

4015. Yamanaka, M. (山中正実), et al. (1995) 知床半島におけるヒグマの生息環境とその規模に関する研究 (Research on living environment and scale of habitat of brown bear in Shiretoko Peninsula). in 自然度の高い生態系の保全を考慮した流域管理に関するランドスケープエコロジー的研究 (平成6年度 科学技術振興調整費による生活・地域流動研究. N19961909). Sapporo: Forestry Technology Center, Hokkaido (北海道森林技術センター), pp. 122–130 (in Japanese).

4016. Yamanaka, M. & Aoi, T. (1988) Brown bears. in Animals of Shiretoko (Ohtaishi, N. & Nakagawa, H., Editors). Sapporo: Hokkaido University Press, pp. 181–223 (in Japanese with English summary).

4017. Yamanaka, M., Yasue, K., & Ohtaishi, N. (1985) Food habits, habitat use, and population trends of brown bear (*Ursus arctos yesoensis*) in the Onnebetsu-dake Wilderness Area and the surrounding areas, Shiretoko Peninsula, Hokkaido. in Conservation Reports of the Onnebetu-dake Wilderness Area Hokkaido Japan (Nature Conservation Bureau Environment Agency Japan, Editor). Nature Conservation Bureau, Environment Agency, Japan, pp. 333–357 (in Japanese with English summary).

4018. Yamanaka, S., Akasaka, T., & Nakamura, F. (2011) Geographic variation in the echolocation call of the Japanese large-footed bat, *Myotis macrodactylus* in Hokkaido, northern Japan. *Mammalian Science* **51**: 265–275 (in Japanese with English abstract).

4019. Yamasaki, M. & Matsumura, S. (2004) Development of sounds and mother–infant communication in the Natterer's bat *Myotis nattereri bombinus*. *Bulletin of the Akiyoshi-dai Museum of Natural History* **39**: 23–36 (in Japanese with English abstract).

4020. Yamasaki, M. & Matsumura, S. (2007) The greater tube-nosed bat, *Murina leucogaster*, captured in the west foot of the Akiyoshi-dai Plateau, Yamaguchi Prefecture, Japan. *Bulletin of the Akiyoshi-dai Museum of Natural History* **42**: 65–69 (in Japanese with English abstract).

4021. Yamasaki, M. & Matsumura, S. (2008) The tube-nosed bat *Murina leucogaster*, captured at a house in the northern Shu-ho town, Yamaguchi pref., Japan. *Bulletin of the Akiyoshi-dai Museum of Natural History* **43**: 51–54 (in Japanese with English abstract).

4022. Yamasaki, M., Sakamoto, Y., & Matsumura, S. (2006) Postnatal growth and development of sounds in the greater tube-nosed bat, *Murina leucogaster*. *Bulletin of the Akiyoshi-dai Museum of Natural History* **41**: 33–44.

4023. Yamashita, J. (1978) *Echinococcus*. Sapporo: Hokkaido University Press, 246 pp. (in Japanese).

4024. Yamashita, J. & Haga, R. (1954) On some ectoparasites of the bat, *Nyctalus maximus aviator* Thomas. *Japanese Journal of Sanitary Zoology* **4**: 217–223 (in Japanese with English abstract).

4025. Yamashita, J. & Mori, H. (1953) On some of the endoparasites of bats, *Nyctalus maximus aviator* Thomas. *Memoirs of the Faculty of Agriculture, Hokkaido University* **1**: 499–503 (in Japanese with English abstract).

4026. Yamashita, J., Ohbayashi, M., & Konno, S. (1957) On daughter cysts of *Coenurus serialis* Gervais, 1847. *Japanese Journal of Veterinary Research* **5**: 14–18.

4027. Yamashita, T., et al. (1978) A survey of *Angiostrongylus cantonensis* on Yoron-jima, Amami Islands, Japan. *Japanese Journal of Parasitology* **27**: 143–150 (in Japanese with English abstract).

4028. Yamauchi, K., et al. (2000) Sex determination based on fecal DNA analysis of the amelogenin gene in sika deer (*Cervus nippon*). *Journal of Veterinary Medical Science* **62**: 669–671.

4029. Yamauchi, T. (2005) A bibliographical survey on ixodid fauna of Shimane Prefecture, Japan (Acari: Ixodoidea). *Bulletin of the Hoshizaki Green Foundation* **8**: 289–301 (in Japanese with English abstract).

4030. Yamauchi, T. & Egusa, S. (2005) Fleas (Siphonaptera) parasitic on medium-sized mammals and birds from Hiroshima Prefecture, Japan. *Japanese Journal of Entomology* **8**: 37–42 (in Japanese with English abstract).

4031. Yamauchi, T. & Funakoshi, K. (2000) Ticks (Acari: Ixodoidea) from Chiroptera (Mammalia) of the Kyushu mainland, Japan. *Journal of the Acarological Society of Japan* **9**: 51–54.

4032. Yamauchi, T. & Nakayama, H. (2006) Two species of deer keds (Diptera: Hippoboscidae) in Miyajima, Hiroshima Prefecture, Japan. *Medical Entomology and Zoology* **57**: 55–58.

4033. Yamauchi, T., et al. (2013) *Lipoptena fortisetosa* (Diptera: Hippoboscidae) collected from Hokkaido sika deer *Cervus nippon yesoensis* in southern Hokkaido, Japan. *Bulletin of the Biogeographical Society of Japan* **68**: 103–105 (in Japanese with English abstract).

4034. Yamauchi, T., Tsurumi, M., & Kataoka, N. (2009) Distributional records of *Lipoptena* species (Diptera: Hippoboscidae) in Japan and Jeju-do, Korea. *Medical Entomology and Zoology* **60**: 131–133.

4035. Yamazaki, K. (山崎晃司), et al. (1996) 多摩川集水域におけるツキノワグマの生態に関する研究 (Ecological study on Japanese black bears in the catchment of Tama River). Tokyo: Tokyu Foundation for Better Environment (とうきゅう環境浄化財団), 67 pp. (in Japanese).

4036. Yamazaki, K. Unpublished.

4037. Yamazaki, K. (2003) Effects of pruning and brush cleaning on debarking within damaged conifer stands by Japanese black bears. *Ursus* **14**: 94–98.

4038. Yamazaki, K. (2004) Recent bear-human conflicts in Japan. *International Bear News* **13**: 16–17.

4039. Yamazaki, K., Koyanagi, K., & Tsuji, A. (2001) List of mammals found in Ibaraki Prefecture, central part of Japan. *Bulletin of the Ibaraki Natural History Museum* **4**: 103–108 (in Japanese with English abstract).

4040. Yamazaki, K., et al. (2008) A preliminary evaluation of activity sensing GPS collars for estimating daily activity patterns of Japanese black bears. *Ursus* **19**: 154–161.

4041. Yamazaki, K., et al. (2012) Myrmecophagy of Japanese black bear in the grasslands of the Ashio area, Nikko National Park, Japan. *Ursus* **23**: 52–64.

4042. Yamazaki, K. & Sato, Y. (2014) Country-wide range mapping of Asiatic black bears reveals increasing range in Japan. *International Bear News* **23**: 18–19.

4043. Yamazaki, K., Yasui, S., & Hirose, M. (2008) A new record of Asian particolored bat in Ibaraki prefecture, central Japan. *Bulletin of Ibaraki Nature Museum* **11**: 27–28 (in Japanese).

4044. Yamazaki, T. (2011) The old written records about introduction and distri-

bution of the Asian musk shrew (*Suncus murinus*) in Japan. *in Suncus murinus* (Biology of Suncus スンクスの生物学) (Oda, S., Tohya, K., & Miyaki, T., Editors). Tokyo: Japan Scientific Societies Press, pp. 73–77 (in Japanese with English summary).
4045. Yamazaki, T., Oda, S., & Shirakihara, M. (2008) Stomach contents of an Indo-Pacific bottlenose dolphin stranded in Amakusa, western Kyushu, Japan. *Fisheries Science* **74**: 1195–1197.
4046. Yanagawa, H. (柳川久) (1994) 小鳥用巣箱を用いたエゾモモンガの野外研究. *Forest Protection* **241**: 20–22 (in Japanese).
4047. Yanagawa, H. (柳川久), Akisawa, M. (秋沢成江), & Tsutsubuchi, M. (筒渕美幸) (2003) 北海道十勝地方におけるコウモリ類の交通事故. *Bat Study and Conservation Report* **11**: 9–10 (in Japanese).
4048. Yanagawa, H. (1998) Traffic accidents involving the red squirrel and measures to prevent such accidents: Actions taken by Obihiro City. *Sciurid Information (リスとムササビ)* **3**: 7–8 (in Japanese).
4049. Yanagawa, H. (1999) Ecological notes on the Russian flying squirrel (*Pteromys volans orii*) with a video camera. *Mammalian Science* **39**: 181–183 (in Japanese).
4050. Yanagawa, H. (2005) Traffic accidents involving the red squirrel and measures to prevent such accidents in Obihiro City, Hokkaido, Japan. *Research Bulletin of Obihiro University* **26**: 35–37.
4051. Yanagawa, H., et al. (2003) The bat fauna of Kitafushiko, Memuro, central Hokkaido (II). *Journal of the Japanese Wildlife Research Society* **29**: 19–24 (in Japanese with English summary).
4052. Yanagawa, H., et al. (1996) Notes on the Japanese flying squirrel, *Pteromys momonga* captured in Fukui Prefecture. *Journal of Japanese Wildlife Research Society* **22**: 8–16 (in Japanese).
4053. Yanagawa, H., Sasaki, Y., & Kataoka, K. (2001) Catching and classifying the Chiroptera in central Hokkaido. *Journal of the Japanese Wildlife Research Society* **27**: 20–26 (In Japanese with English abstract).
4054. Yanagawa, H., Sasaki, Y., & Takimoto, I. (2006) Bat fauna in Hidaka and Tokachi districts, central and eastern Hokkaido. (6). Capture record of bats in windbreak forests located in the agricultural areas of Obihiro City. *Journal of the Japanese Wildlife Research Society* **32**: 5–10 (in Japanese with English summary).
4055. Yanagawa, H., Sato, T., & Sugawara, M. (2005) Bats fauna of Hidaka and Tokachi districts, central and eastern Hokkaido. (4). Records of bats in Shihoro River and Osarushinai River, in central Tokachi. *Journal of the Japanese Wildlife Research Society* **31**: 37–41 (in Japanese with English summary).
4056. Yanagawa, H., Takimoto, I., & Sasaki, Y. (2009) Bat fauna in Hidaka and Tokachi districts, central and eastern Hokkaido (8) Capture records of bats in windbreak forests being located in the agricultural areas of Nakasatsunai Village. *Journal of the Japanese Wildlife Research Society* **34**: 1–6 (in Japanese with English summary).
4057. Yanagawa, H., et al. (1991) Annual and daily activities of the flying squirrel, *Pteromys volans orii*, in captivity. *Mammalian Science* **30**: 157–165 (in Japanese with English abstract).
4058. Yanagawa, H., et al. (2004) Bats fauna of Hidaka and Tokachi districts, central and eastern Hokkaido. (3). Records of bats in Saruru River, Erimo, southern Hidaka. *Journal of the Japanese Wildlife Research Society* **30**: 21–27 (in Japanese with English summary).
4059. Yanbaru Branch of the Wild Bird Society of Japan (1997) 沖縄本島北部 (やんばる) における貴重鳥獣の生息調査およびその保護 (Status and Effect of Introduced Animals on Endangered Animals in Yanbaru, Northern Part of Okinawajima Island). 86 pp. (in Japanese).
4060. Yang, F., Ma, C., & Shi, L. (1991) Studies on the karyotypes of *Erinaceus amurensis* and *Hemiechinus auritus*. *Zoological Research* **12**: 393–398 (in Chinese with English abstract).
4061. Yang, G., et al. (2008) Mitochondrial phylogeography and population history of finless porpoises in Sino-Japanese waters. *Biological Journal of the Linnean Society* **95**: 193–204.
4062. Yang, G., et al. (2002) Population genetic structure of finless porpoises, *Neophocaena phocaenoides*, in Chinese waters, inferred from mitochondrial control region sequences. *Marine Mammal Science* **18**: 336–347.
4063. Yang, W.-C., et al. (2008) Unusual cetacean mortality event in Taiwan, possibly linked to naval activities. *Veterinary Record* **162**: 184–186.
4064. Yao, C.J., et al. (2008) Cranial variation in the pantropical spotted dolphin, *Stenella attenuata*, in the Pacific Ocean. *Zoological Science* **25**: 1234–1246.
4065. Yashiki, H. (野紫木洋) (1995) オコジョの不思議. Tokyo: Doubutsu-sha (どうぶつ社), 135 pp. (in Japanese).
4066. Yashiki, H. (1987) Ecological study on the raccoon dog, *Nyctereutes procyonoides viverrinus*, in Shiga Heights. *Bulletin of Institution of Nature Education, Shiga Heights, Shinshu University* **24**: 43–53 (in Japanese with English summary).
4067. Yasuda, M. (2007) Threatened arboreal squirrels in Kyushu, southwestern Japan. *Mammalian Science* **47**: 195–206 (in Japanese).
4068. Yasuda, M. (2013) The Pallas's squirrel has likely established a new feral population on Mt. Kirishima, on the border between Miyazaki and Kagoshima Prefectures, Kyushu, Southwestern Japan. *Sciurid Information (リスとムササビ)* **30**: 15 (in Japanese).
4069. Yasuda, M. (2014) Environmental evaluation and control of invasive alien Pallas's squirrel *Callosciurus erythraeus* in Japan. *Sciurid Information* **32**: 11–14 (in Japanese).
4070. Yasuda, M., et al. (2012) Reproduction of the Pallas's squirrel *Callosciurus erythraeus* introduced into the Uto Peninsula, Kumamoto, Japan, from March 2010 to February 2012. *Bulletin of the Kumamoto Wildlife Society* **7**: 13–16 (in Japanese).
4071. Yasuda, M. & Matsuo, K. (2012) Alien squirrels found in the Shimabara Peninsula, Nagasaki Prefecture, Kyushu Island, Japan. *Sciurid Information (リスとムササビ)* **28**: 18–19 (in Japanese).
4072. Yasuda, N., et al. (1992) Experimental infection to domestic cats with *Arthrostoma hunanensis* derived from the feces of Tsushima leopard cats (*Felis bengalensis euptilura*). *Japanese Journal of Parasitology* **41**: 498–504.
4073. Yasuda, N., et al. (1993) Helminths of the Tsushima leopard cat (*Felis bengalensis euptilura*). *Journal of Wildlife Diseases* **29**: 153–155.
4074. Yasuda, N., et al. (1994) Helminth survey of wildcats in Japan. *Journal of Veterinary Medical Science* **56**: 1069–1073.
4075. Yasuda, S.P. Unpublished.
4076. Yasuda, S.P., et al. (2012) Spatial framework of nine distinct local populations of the Japanese dormouse *Glirulus japonicus* based on matrilineal cytochrome *b* and patrilineal *SRY* gene sequences. *Zoological Science* **29**: 111–120.
4077. Yasuda, S.P., et al. (2007) Onset of cryptic vicariance in the Japanese dormouse *Glirulus japonicus* (Mammalia, Rodentia) in the Late Tertiary, inferred from mitochondrial and nuclear DNA analysis. *Journal of Zoological Systematics and Evolutionary Research* **45**: 155–162.
4078. Yasuda, S.P., et al. (2005) Phylogeographic patterning of mtDNA in the widely distributed harvest mouse (*Micromys minutus*) suggests dramatic cycles of range contraction and expansion during the mid- to late Pleistocene. *Canadian Journal of Zoology* **83**: 1411–1420.
4079. Yasui, K. (1992) Embryonic development of the house shrew (*Suncus murinus*) I. Embryos at stages 9 and 10 with 1 to 12 pairs of somites. *Anatomy and Embryology* **186**: 49–65.
4080. Yasui, K. (1993) Embryonic development of the house shrew (*Suncus murinus*) II. Embryos at stages 11 and 12 with 13 to 29 pairs of somites, showing limb bud formation and closed cephalic neural tube. *Anatomy and Embryology* **187**: 45–65.
4081. Yasui, S. (2007) Mammals in the northwestern area of Ibaraki prefecture. *in* Nature in prefectural northwest district of Ibaraki including the Yamizo Mts. and the Kuji River: The 4th General Research Report of the Ibaraki Nature Museum. Bando: Ibaraki Nature Museum, pp. 255–260 (in Japanese).
4082. Yasui, S. (2010) Sex and reproductive class composition of solitary bats and roosting groups in *Pipistrellus abramus*. *Mammalian Science* **50**: 49–54 (in Japanese with English abstract).
4083. Yasui, S. & Hitomi, H. (1996) Distribution records of bats in the Nasu region, Tochigi Prefecture (1): Winter roosts in irrigation tunnels. *Bulletin of the Tochigi Prefectural Museum* **13**: 1–4 (in Japanese with English abstract).
4084. Yasui, S. & Kamijo, T. (1999) The first record of the black whiskered bat, *Myotis pruinosus* Yoshiyuki and the long-legged whiskered bat, *Myotis frater* Allen, from Tochigi Prefecture, Japan. *Bulletin of the Tochigi Prefectural Museum* **16**: 77–80 (in Japanese with English abstract).
4085. Yasui, S. & Kamijo, T. (1999) A record of the Endo's pipistrelle bat, *Pipistrellus endoi* Imaizumi, and the Natterer's bat, *Myotis nattereri* (Kuhl), from Tochigi Prefecture, Japan. *Bulltein of the Tochigi Prefectural Museum* **16**: 81 (in Japanese with English abstract).
4086. Yasui, S., et al. (2002) Summer roosts of the Ikonnikov's whiskered Bat, *Myotis ikonnikovi*, in Nikko, Japan. *Bulletin of the Asian Bat Research Institute* **2**: 1–7 (in Japanese with English abstract).
4087. Yasui, S., et al. (2004) Day roosts and roost-site selection of Ikonnikov's whiskered bat, *Myotis ikonnikovi*, in Nikko, Japan. *Mammals Study* **29**: 155–161.
4088. Yasui, S., et al. (2000) Distribution of the Ikonnikov's whiskered bat, *Myotis ikonnikovi* Ognev and its relationship to the habitat type in Tochigi Prefecture, Japan. *Mammalian Science* **40**: 155–165 (in Japanese with English abstract).
4089. Yasui, S., Maruyama, N., & Kanzaki, N. (1997) Roost site selection and colony size of the common Japanese pipistrelle (*Pipistrellus abramus*) in Fuchu, Tokyo. *Wildlife Conservation Japan* **2**: 51–59.
4090. Yasui, S., et al. (2012) Underground site gating for protection of a large colony of *Miniopterus fuliginosus*. *Japanese Journal of Conservation Ecology*

17: 73–80 (in Japanese with English abstract).

4091. Yasui, S. & Saito, O. (2010) 茨城県のコウモリ類. in Report of Comprehensive Surveys of Plants, Animals and Geology in Ibaraki Prefecture by the Ibaraki Nature Museum: Vertebrate fauna from around the Southwest District, Ibaraki Prefecture. Bando: Ibaraki Nature Museum, pp. 3–9 (in Japanese).

4092. Yasui, S., et al. (1997) Distribution records of bats in the Nasu region, Tochigi Prefecture (2) : A survey for bats in flight using mist nets. *Bulletin of the Tochigi Prefectural Science Museum* **14**: 33–37 (in Japanese with English abstract).

4093. Yasui, S., et al. (2001) Mammals of the Nasu area, the northern part of Tochigi Prefecture, central Japan. *Bulletin of the Tochigi Prefectural Museum* **18**: 1–21 (in Japanese with English abstract).

4094. Yasui, S. & Yamazaki, K. (2013) A new record of *Miniopterus fuliginosus* in Ibaraki Prefecture, central Japan. *Bulletin of Ibaraki Nature Museum* **16**: 63–67 (in Japanese).

4095. Yasukochi, Y., et al. (2009) Genetic structure of the Asiatic black bear in Japan using mitochondrial DNA analysis. *Journal of Heredity* **100**: 297–308.

4096. Yasumoto, M., et al. (1997) Seroepidemiological study of *Coxiella burnetii* in *Cervus nippon* in northern Japan. *Japanese Journal of Zoo and Wildlife Medicine* **2**: 101–106.

4097. Yatabe, A., et al. (2007) Stomach contents of Longman's beaked whale (*Indopacetus pacificus*) stranded in Kagoshima prefecture. *in* Abstract of the Annual Meeting of the Cetology Study Group of Japan. pp. 22–23 (in Japanese).

4098. Yatake, H. (2010) Nest and nesting habit of Japanese squirrels (*Sciurus lis*). *Journal of the Japanese Wildlife Research Society* **35**: 7–12 (in Japanese with English abstract).

4099. Yatake, H. (2014) Reproductive season of Japanese squirrels (*Sciurus lis*) in the northern part of Chiba Prefecture. *Mammalian Science* **54**: 265–268 (in Japanese with English abstract).

4100. Yatake, H., Akita, T., & Abe, M. (1999) Space used by introduced Japanese squirrels (*Sciurus lis* Temminck) in Shimizu Park. *Mammalian Science* **39**: 9–22 (in Japanese).

4101. Yatake, H., et al. (2005) Distribution of Japanese squirrel (*Sciurus lis*) in Chiba Prefecture, Central Japan. *Journal of Natural History Museum and Institute, Chiba* **8**: 41–48 (in Japanese).

4102. Yatake, H., et al. (2003) Density estimation of Japanese hare *Lepus brachyurus* by fecal pellet count and INTGEP in Akita-komagatake mountains area. *Mammalian Science* **43**: 99–111 (in Japanese with English abstract).

4103. Yatake, H. & Tamura, N. (2001) To make guidelines for the Japanese squirrel conservation. III. Ecological studies to conserving the Japanese squirrel. *Mammalian Science* **41**: 149–157 (in Japanese).

4104. Yatsu, A., Hiramatsu, K., & Hayase, S. (1993) Outline of the Japanese squid driftnet fishery with notes on by-catch. *International North Pacific Fisheries Commission Bulletin* **53**: 5–24.

4105. Yatsu, A., Hiramatsu, K., & Hayase, S. (1994) A review of the Japanese squid driftnet fishery with notes on the cetacean by-catch. *Reports of the International Whaling Commission (Special Issue)* **15**: 365–379.

4106. Yawata, I. (八幡一郎) (1973) 貝の花貝塚 (松戸市文化財調査報告 第 4 集). Matsudo: Board of Education of Matsudo City (松戸市教育委員会), 587 pp. (in Japanese with English abstract).

4107. Yimam, A.E., et al. (2002) Prevalence and intensity of *Echinococcus multilocularis* in red foxes (*Vulpes vulpes schrencki*) and raccoon dogs (*Nyctereutes procyonoides albus*) in Otaru City, Hokkaido, Japan. *Japanese Journal of Veterinary Research* **49**: 287–96.

4108. Yochem, P.K. & Leatherwood, S. (1985) Blue whale *Balaenoptera musculus* (Linnaeus, 1758). *in* Handbook of Marine Mammals. Volume 3. The Sirenians and Baleen Whales (Ridgway, S.H. & Harrison, R.J., Editors). London: Academic Press, pp. 193–240.

4109. Yokohama, M., et al. (1994) Karyotype analysis of the Yeso sika (*Cervus nippon yesoensis*) and its related species. *Journal of Agricultural Science, Tokyo University of Agriculture* **39**: 170–176 (in Japanese).

4110. Yokohata, Y. (横畑泰志) (1998) モグラ科動物の生態.in 食虫類の自然史 (Abe, H. (阿部永) & Yokohata, Y. (横畑泰志), Editors). Shobara: Hiba Society of Natural History, pp. 67–187 (in Japanese).

4111. Yokohata, Y. (1994) Age determination of *Mogera robusta* (Mammalia; Talpidae) and age structure of its population in Hiwa, Hiroshima Prefecture, Japan. *Memoirs of the Faculty of Education, Toyama University, Series B* **45**: 63–74 (in Japanese with English summary).

4112. Yokohata, Y. (1997) Analysis of the population structure of large Japanese moles, *Mogera wogura* (Mammalia; Talpidae) in Hiwa, Hiroshima Prefecture, using cohort analysis. *Memoirs of the Faculty of Education, Toyama University, Series B* **49**: 47–54 (in Japanese with English summary).

4113. Yokohata, Y. (1999) A case of estimation of species of a fossil of the genus *Mogera* from Japan using discriminant functions. *Memoirs of the Faculty of Education, Toyama University* **53**: 37–44 (in Japanese with English summary).

4114. Yokohata, Y. (2000) Age structure of the lesser Japanese moles *Mogera imaizumii* from Sagamihara Golf Club in Kanagawa Prefecture, Japan. *Memoirs of the Faculty of Education, Toyama University* **54**: 161–169 (in Japanese with English summary).

4115. Yokohata, Y. (2002) Parasitic Helminths of alien mammals in Japan. *Japanese Journal of Zoo and Wildlife Medicine* **7**: 91–102 (in Japanese with English abstract).

4116. Yokohata, Y. (2003) The problem of feral goats on Uotsuri-jima in Senkaku Islands and appeals for countermeasures to resolve the problem. *Japanese Journal of Conservation Ecology* **8**: 87–96 (in Japanese with English summary).

4117. Yokohata, Y. (2005) A brief review of the biology on moles in Japan. *Mammal Study* **30 (Suppl.)**: S25–S30.

4118. Yokohata, Y. & Abe, H. (1989) Two new spirurid nematodes in Japanese moles, *Mogera* spp. *Japanese Journal of Parasitology* **38**: 92–99.

4119. Yokohata, Y., et al. (1989) Gastrointestinal helminth fauna of Japanese moles, *Mogera* spp. *Japanese Journal of Veterinary Research* **37**: 1–13.

4120. Yokohata, Y., Abe, H., & Kamiya, M. (1988) Redescription and multivariate morphometrics of *Moguranema nipponicum* Yamaguti. *Japanese Journal of Veterinary Research* **36**: 223– 233.

4121. Yokohata, Y., et al. (1990) Parasites from Asiatic black bear (*Ursus thibetanus*) on Kyushu Island, Japan. *Journal of Wildlife Diseases* **26**: 137–138.

4122. Yokohata, Y., et al. (2003) The effects of introduced goats on the ecosystem of Uotsuri-Jima, Senkaku Islands, Japan, as assessed by remote-sensing techniques. *Biosphere Conservation* **5**: 39–46.

4123. Yokohata, Y., et al. (1988) Pseudoparasitism by thelastomatid nematodes in moles, *Mogera* spp., in Japan. *Japanese Journal of Veterinary Research* **36**: 53–67.

4124. Yokohata, Y. & Kamiya, M. (2004) Analyses of regional environmental factors on the prevalence of *Echinococcus multilocularis* in foxes in Hokkaido, Japan. *Japanese Journal of Zoo and Wildlife Medicine* **9**: 91–96.

4125. Yokohata, Y., et al. (1987) Histology and lipid analysis of the infraorbital glands of Japanese serow, and functional considerations. *in* The Biology and Management of *Capricornis* and Related Mountain Antelopes (Soma, H., Editor). London: Croom Helm, pp. 243–256.

4126. Yokohata, Y. & Onodera, A. (1997) Small mammals and parasitic nematode fauna in their alimentary tracts in Nakaikemi-marsh, Tsuruga City, Hukui Prefecture, Japan. *Memoirs of the Faculty of Education, Toyama University. Series B* **50**: 41–46 (in Japanese with English summary).

4127. Yokohata, Y. & Sagara, N. (1995) Some parasitic nematodes of the Japanese mountain mole, *Euroscaptor mizura*. *Memoirs of the Faculty of Education, Toyama University. Series B* **47**: 19–25.

4128. Yokohata, Y. & Suzuki, Y. (1993) The gullet nematode, *Gongylonema pulchrum* from sika deer, *Cervus nippon* in Hyogo Prefecture, Japan. *Japanese Journal of Parasitology* **42**: 440–444.

4129. Yokohata, Y. & Yokota, M. (2000) The problem of introduced goats on Uotsuri-Jima in the Senkaku Islands. *Wildlife Conservation Japan* **5**: 1–12 (in Japanese with English summary).

4130. Yokoyama, K. (横山恵一) (1992) 長い耳の秘密. 週刊朝日百科. 動物たちの地球 **40**: 118–119 (in Japanese).

4131. Yokoyama, K. (横山恵一) (1993) 岩手県から. *Bat Study and Conservation Report* **1**: 4 (in Japanese).

4132. Yokoyama, K. (横山恵一) (1997) モリアブラコウモリ (*Pipistrellus endoi*), 図書室で休眠. *Bat Study and Conservation Report* **5**: 1–2 (in Japanese).

4133. Yokoyama, K. (1996) Notes concerning the Fuji whiskered bat (*Myotis fujiensis*). *Annual of the Speleological Research Institute of Japan* **14**: 13–18.

4134. Yokoyama, K., Ohtsu, R., & Uchida, T.A. (1979) Growth and LDH isozyme patterns in the pectoral and cardiac muscles of the Japanese lesser horseshoe bat, *Rhinolophus cornutus cornutus* from the standpoint of adaptation for flight. *Journal of Zoology* **187**: 85–96.

4135. Yokoyama, K. & Uchida, T.A. (1979) Functional morphology of wings from the standpoint of adaptation for flight in Chiroptera II. Growth and changes in mode of life during the young period in *Rhinolophus cornutus cornutus*. *Journal of the Faculty of Agriculture, Kyushu University* **23**: 185–198.

4136. Yokoyama, K., Uchida, T.A., & Shiraishi, S. (1975) Functional morphology of wings from the standpoint of adaptation for flight in Chiroptera I. Relative growth and ossification in forelimb, wing loading and aspect ratio. *Zoological Magazine* **84**: 233–247 (in Japanese with English abstract).

4137. Yoneda, M. (米田政明) (1978) ノネズミ類の個体群動態とその捕食者に関する研究—自然植生地を含む農生態系における哺乳動物群集— Doctoral dissertation. Sapporo: Hokkaido University. 186 pp. (in Japanese with English summary).

4138. Yoneda, M. (米田政明) (1994) ハクビシン. *in* A Pictorial Guide to the Mammals of Japan (日本の哺乳類) (Abe, H. (阿部永), Editor). Tokyo: Tokai University Press, p. 129 (in Japanese).

4139. Yoneda, M. (米田政明) (2007) ツキノワグマ保護管理の課題 教訓を活かす. *in* JBN 緊急クマシンポジウム＆ワークショップ報告書—2006 年ツキノワグマ大量出没の総括と JBN からの提言— (Japan Bear Network (日本クマネットワーク), Editor). Gifu: Japan Bear Network, pp. 8–15 (in Japanese).

4140. Yoneda, M. (米田政明) & Nakata, K. (中田圭亮) (1984) 天敵. *in* Study on Wild Murid Rodents in Hokkaido(北海道産野ネズミ類の研究) (Ota, K. (太田嘉四夫), Editor). Sapporo: Hokkaido University Press, pp. 159–185 (in Japanese).

4141. Yoneda, H. (2012) The new record of the Ryukyu flying-fox (*Pteropus dasymallus*) on Tokashikijima Island. *Biological Magazine Okinawa* **50**: 99 (in Japanese).

4142. Yoneda, M. (1979) Influence of red fox predation on a local population of small rodents II. Food habits of the red fox. *Applied Entomology and Zoology* **17**: 308–318.

4143. Yoneda, M. (1979) Prey preference of the red fox, *Vulpes vulpes schrencki* Kishida (Carnivora: Canidae), on small rodents. *Applied Entomology and Zoology* **14**: 28–35.

4144. Yoneda, M. (1983) Influence of red fox predation upon a local population of small rodents. III. Seasonal changes in predation pressure, prey preference and predation effect. *Applied Entomology and Zoology* **18**: 1–10.

4145. Yoneda, M. (2005) Gray Wolf. *in* A Guide to the Mammals of Japan (Abe, H., Editor). Hadano: Tokai University Press, p. 75 (in Japanese and English).

4146. Yoneda, M. & Abe, H. (1976) Sexual dimorphism and geographic variation in the skull of the Ezo brown bear (*Ursus arctos yesoensis*). *Memoirs of the Faculty of Agriculture Hokkaido University* **9**: 265–276 (in Japanese with English summary).

4147. Yoneda, M., Abe, H., & Nakao, H. (1979) Winter food habits of the Yezo Ural Owl *Strix uralensis japonica* in a wind shelter-belt. *Journal of the Yamashina Institute for Ornithology* **11**: 49–53 (in Japanese with English summary).

4148. Yoneda, M. & Maekawa, K. (1982) Effects of hunting on age structure and survival rates of red fox in eastern Hokkaido. *Journal of Wildlife Management* **46**: 781–786.

4149. Yonekawa, H. (米川洋) (1992) 北海道の集約農業地域におけるチゴハヤブサの食性. *Bulletin of the Higashi Taisetsu Museum of Natural History* **14**: 63–74 (in Japanese).

4150. Yonekawa, H., et al. (1981) Evolutionary relationships among five subspecies *Mus musculus* based on restriction enzyme cleavage patterns of mitochondrial DNA. *Genetics* **98**: 801–816.

4151. Yonekawa, H., et al. (1988) Hybrid origin of Japanese mice "*Mus musculus molossinus*": evidence from restriction analysis of mitochondrial DNA. *Molecular Biology and Evolution* **5**: 63–78.

4152. Yonekawa, H., et al. (2012) Origin and genetic status of *Mus musculus molossinus*: a typical example for reticulate evolution in the genus *Mus*. *in* Evolution of the House Mouse (Macholán, M., et al., Editors). Cambridge: Cambridge University Press, pp. 94–113.

4153. Yonekawa, H., et al. (2003) Genetic diversity, geographic distribution and evolutionary relationships of *Mus musculus* subspecies based on polymorphisms of mitochondrial DNA. *in* Problems of Evolution (Kryukov, A.P. & Yakimenko, L.V., Editors). Vladivostok: Dalnauka, pp. 90–108.

4154. Yonezaki, S., Kiyota, M., & Baba, N. (2008) Decadal changes in the diet of northern fur seal (*Callorchinus ursinus*) migrating the Pacific coast of northeastern Japan. *Fisheries Oceanography* **17**: 231–238.

4155. Yoo, D.H. & Yoon, M.H. (1992) A karyotypic study on six Korean vespertilionid bats. *Korean Journal of Zoology* **35**: 489–496.

4156. Yoon, M.H., Andoo, K., & Uchida, T.A. (1990) Taxonomic validity of scientific names in Japanese *Vespertilio* species by ontogenetic evidence of the penile pseudobaculum. *Journal of the Mammalogical Society of Japan* **14**: 119–128.

4157. Yoon, M.H., Kuramoto, T., & Uchida, T.A. (1981) Studies on taxonomy and phylogeny of bats' fossils from the Akiyoshi-dai Plateau. I. *Plecotus auritus* and *Barbastella leucomelas darjelingensis* belonging to the Tribe Plecotini. *Bulletin of the Akiyoshi-dai Museum of Natural History* **16**: 35–53 (in Japanese with English abstract).

4158. Yoon, M.H., Kuramoto, T., & Uchida, T.A. (1984) Studies on Late Pleistocene bats including two new extinct *Myotis* species from the Akiyoshi-dai Plateau, with reference to the Japanese microchiropteran faunal succession. *Bulletin of the Akiyoshi-dai Museum of Natural History* **19**: 1–14.

4159. Yoon, M.H., Kuramoto, T., & Uchida, T.A. (1984) Studies on Middle Pleistocene bats including two new extinct *Pleistmyotis* gen. et sp. nov. and two new extinct *Myotis* species from the Akiyoshi-dai Plateau. *Bulletin of the Akiyoshi-dai Museum of Natural History* **19**: 15–26.

4160. Yoshida, H. (吉田宏) (2008) 富山県薬師峠キャンプ指定地管理小屋でクビワコウモリを発見. *Bat Study and Conservation Report* **16**: 22 (in Japanese).

4161. Yoshida, H. (1973) Small mammals of Mt. Kiyomizu, Fukuoka Pref. 5. Reproduction in the Smith's red-backed vole, *Anteliomys smithi*. *The Journal of the Mammalogical Society of Japan* **5**: 206–212 (in Japanese with English abstract).

4162. Yoshida, H. (1985) A note on the morphology of the Smith's red-backed vole, *Eothenomys smithi*, collected in the mountain districts in Kyushu. *Seibutsu Fukuoka* **25**: 9–14 (in Japanese).

4163. Yoshida, H., et al. (2010) Finless porpoise (*Neophocaena phocaenoides*) discovered at Okinawa Island, Japan, with the source population inferred from mitochondorial DNA. *Aquatic Mammals* **36**: 278–283.

4164. Yoshida, H. & Kato, H. (1999) Phylogenetic relationships of Bryde's whales in the western north Pacific and adjacent waters inferred from mitochondrial DNA sequences. *Marine Mammal Science* **15**: 1269–1286.

4165. Yoshida, H., et al. (1997) A population size estimate of the finless porpoise, *Neophocaena phocaenoides*, from aerial sighting surveys in Ariake Sound and Tachibana Bay, Japan. *Researches on Population Ecology* **39**: 239–247.

4166. Yoshida, H., et al. (1998) Finless porpoise abundance in Omura Bay, Japan: Estimation from aerial sighting surveys. *Journal of Wildlife Management* **62**: 286–291.

4167. Yoshida, H., et al. (1995) Geographic variation in the skull morphology of the finless porpoise *Neophocaena phocaenoides* in Japanese waters. *Fisheries Science* **61**: 555–558.

4168. Yoshida, H., et al. (2001) Population structure of finless porpoises (*Neophocaena phocaenoides*) in coastal waters of Japan based on mitochondrial DNA sequences. *Journal of Mammalogy* **82**: 123–130.

4169. Yoshida, I., Obara, Y., & Matsuoka, N. (1989) Phylogenetic relationships among seven taxa of the Japanese microtine voles revealed by karyological and biochemical techniques. *Zoological Science* **6**: 409–420 (in Japanese).

4170. Yoshida, M., et al. (1982) Chromosomal analysis of the Japanese raccoon dog based on the G- and C-banding techniques. *Japanese Journal of Veterinary Research* **30**: 68–77.

4171. Yoshida, T., et al. (1999) Identification of the Japanese wolf skull in the 3D morphological method. *Mammalian Science* **39**: 239–246 (in Japanese with English abstract).

4172. Yoshida, Y. & Arizona, N. (1976) *Arthrostoma miyazakiense* (Nagayosi 1955) Comb. n., a parasite of the raccoon-like dog, *Nyctereutes procyonoides*, with a key to the genus *Arthrostoma* (Nematoda: Ancylostomatidae). *Journal of Parasitology* **62**: 766–770.

4173. Yoshida, Y., et al. (2002) Analysis of causes of bark stripping by the Japanese black bear (*Ursus thibetanus japonicus*). *Mammal Science* **42**: 35–43 (in Japanese with English summary).

4174. Yoshihara, M. (吉原正人) (1999) 動物園のオガサワラオオコウモリ. *in* 天然記念物緊急調査 (オガサワラオオコウモリ). Ogasawara: Ogasawara Village Educational Commission (小笠原村教育委員会), pp. 71–74 (in Japanese).

4175. Yoshihara, M. & Miura, S. (1984) Birth records of captive Reeves' muntjac. *The Journal of the Mammalogical Society of Japan* **10**: 35–36.

4176. Yoshikawa, T. & Saijo, M. (2014) Genomic phylogenetic analysis of SFTSV isolated from SFTS patients in Japan. *Infectious Agents Surveillance Report* **35**: 35–37 (in Japanese).

4177. Yoshikura, S., et al. (2009) Structure of nursery colonies and reproductive traits of the Japanese long-eared bat, *Plecotus auritus sacrimonts*. *Mammalian Science* **49**: 225–235 (in Japanese with English abstract).

4178. Yoshikura, S., Watanabe, M., & Yasui, S. (2011) New distribution record and echolocation call structure of the Japanese lesser noctule bat, *Nyctalus furvus*, from Tochigi Prefecture, central Japan. *Bulletin of Tochigi Prefectural Museum* **28**: 45–49 (in Japanese).

4179. Yoshimura, K., et al. (2013) Characteristic features of acorn hoarding of the wood mouse, *Apodemus argenteus* Temminck (Rodentia: Muridae). *Research Bulletin of the Kagoshima University Forests* **40**: 9–15 (in Japanese with English summary).

4180. Yoshino, H. & Abe, H. (1984) Comparative study on the foraging habits of two species of soricine shrews. *Acta Theriologica* **29**: 35–43.

4181. Yoshino, H., et al. (2008) Genetic and acoustic population structuring in the Okinawa least horseshoe bat: are intercolony acoustic differences maintained by vertical maternal transmission? *Molecular Ecology* **17**: 4978–4991.

4182. Yoshino, H., Armstrong, K.N., & Tamura, H. (2009) Cave-dwelling bat surveys on Kume-jima, Tokashiki-jima and Iheya-jima Islands in Okinawa Prefecture. *Bulletin of the Asian Bat Research Institute* **8**: 28–32 (in Japanese with English abstract).

4183. Yoshino, H., et al. (2006) Geographical variation in echolocation call and body size of the Okinawan least horseshoe bat, *Rhinolophus pumilus* (Mammalia: Rhinolophidae), on Okinawa-jima Island, Ryukyu Archipelago, Japan. *Zoological Science* **23**: 661–667.

4184. Yoshino, H. & Tamura, H. (2009) Habitat selection of the Okinawa least horseshoe bat (*Rhinolophus pumilus*) in the southern part of Okinawa-jima Island. *Bulletin of the Asian Bat Research Institute* **8**: 33–36 (in Japanese with English abstract).

4185. Yoshio, M., et al. (2008) Spatially heterogeneous distribution of mtDNA haplotypes in a sika deer (*Cervus nippon*) population on the Boso Peninsula, central Japan. *Mammal Study* **33**: 59–69.

4186. Yoshioka, M. (吉岡基) (2002) 伊勢湾・三河湾調査 (Aerial survey in Ise Bay–Mikawa Bay). *in* 海域自然環境保全基礎調査 海棲動物動物 (スナメリ生息調査) 報告書 Fujiyoshida: Biodiversity Center of Japan, pp. 27–52 (in Japanese).

4187. Yoshioka, M. (吉岡基) (2008) 人工繁殖の現状と将来—飼育下の小型鯨類 (Status and future of artificial breeding in small cetaceans in captivity). *in* Mammalogy in Japan. 3. Marine Mammals (日本の哺乳類学 第3巻 水生哺乳類) (Kato, H. (加藤秀弘), Editor). Tokyo: University of Tokyo Press, pp. 123–146 (in Japanese).

4188. Yoshioka, M. Unpublished.

4189. Yoshioka, M., Kasuya, T., & Aoki, M. (1990) Identity of *dalli*-type Dall's porpoise stocks in the northern North Pacific and adjacent seas (Paper SC/42/SM31 presented to the IWC Scientific Committee). 20 pp.

4190. Yoshioka, M., et al. (1990) Seasonal changes in serum levels of testosterone and progesterone on the Japanese raccoon dog, *Nyctereutes procyonoides viverrinus*. *Proceedings of the Japanese Society For Comparative Endocrinology* **5**: 17.

4191. Yoshioka, M., et al. (1986) Annual changes in serum reproductive hormone levels in the captive female bottle-nosed dolphins. *Bulletin of the Japanese Society of Scientific Fisheries* **52**: 1939–1946.

4192. Yoshioka, M. & Takekawa, Y. (2008) Adult sperm whale found in Ise Bay. *Bulletin of the Graduate School of Bioresources, Mie University* **35**: 79–84 (in Japanese).

4193. Yoshiyuki, M. (吉行瑞子) (1974) 四国ではじめて採集されたクロホオヒゲコウモリ. *Japan Caving* **6**: 43–45 (in Japanese).

4194. Yoshiyuki, M. (吉行瑞子) (1974) 尾瀬の翼手類. *in* 福島県文化財調査報告書 Volume 42. Fukushima: Fukushima Prefectural Board of Education (福島県教育委員会), pp. 1–4 (in Japanese).

4195. Yoshiyuki, M. (1965) On the subspecific character of *Myotis frater kaguyae*. *The Journal of the Mammalogical Society of Japan* **2**: 136–141 (in Japanese with English abstract).

4196. Yoshiyuki, M. (1968) Notes on some bats from Mt. Goyo and Mt. Hayachine. *Memoirs of the National Science Museum* **1**: 92–95 (in Japanese with English abstract).

4197. Yoshiyuki, M. (1968) Notes on the milk dentition of *Vespertilio superans*. *The Journal of the Mammalogical Society of Japan* **4**: 48–50 (in Japanese).

4198. Yoshiyuki, M. (1970) A new species of insectivorous bat of the genus *Murina* from Japan. *Bulletin of the National Science Museum, Tokyo* **13**: 195–198 (in Japanese with English abstract).

4199. Yoshiyuki, M. (1970) Notes on some bats from Tsushima Islands. *Memoirs of the National Science Museum* **3**: 177–184 (In Japanese with English abstract).

4200. Yoshiyuki, M. (1971) Insectivorous bats of Mt. Fuji. *in* 富士山：富士山総合学術調査報告書 (National Parks Association of Japan (国立公園協会), Editor). Fujiyoshida: Fuji Kyuko (富士急行), pp. 829–833 (in Japanese with English abstract).

4201. Yoshiyuki, M. (1971) A new bats of the Leuconoe Group in the genus *Myotis* from Honshu, Japan. *Bulletin of the National Science Museum* **14**: 305–310.

4202. Yoshiyuki, M. (1973) On the taxonomic status of *Miniopterus* from Ryukyu Islands 1. Variations of the forearm and the greatest length of skull. *The Journal of the Mammalogical Society of Japan* **5**: 234–239 (in Japanese with English abstract).

4203. Yoshiyuki, M. (1979) Insect remains of a Japanese long-eared bat, *Plecotus auritus sacrimontis*. *The Journal of the Mammalogical Society of Japan* **7**: 321–323 (in Japanese with English abstract).

4204. Yoshiyuki, M. (1980) Notes on some bats collected from Oze district, Nikko National Park, Honshu, Japan. *The Journal of the Mammalogical Society of Japan* **8**: 89–196 (in Japanese with English abstract).

4205. Yoshiyuki, M. (1983) A new species of *Murina* from Japan (Chiroptera, Vespertilionidae). *Bulletin of the National Science Museum, Tokyo, Series A Zoology* **9**: 141–150.

4206. Yoshiyuki, M. (1984) A new species of *Myotis* (Chiroptera, Vespertilionidae) from Hokkaido, Japan. *Bulletin of the National Science Museum, Tokyo, Series A Zoology* **10**: 153–159.

4207. Yoshiyuki, M. (1986) The phylogenetic status of *Mogera tokudae* Kuroda, 1940 on the basis of body skeletons. *Memoirs of the National Science Museum* **19**: 203–213 (in Japanese with English summary).

4208. Yoshiyuki, M. (1989) A Systematic Study of the Japanese Chiroptera. Tokyo: National Science Museum, 242 pp.

4209. Yoshiyuki, M. (1990) Notes on the genus *Nyctalus* from Japan (1). 日本の生物 **4(6)**: 74–78 (in Japanese).

4210. Yoshiyuki, M. (1990) Notes on the genus *Pipistrellus* from Japan. 日本の生物 **4(7)**: 74–77 (in Japanese).

4211. Yoshiyuki, M. & Endo, K. (1972) The bats from the Hidaka mountains, Hokkaido. *Memoirs of the National Science Museum* **5**: 123–131 (in Japanese with English abstract).

4212. Yoshiyuki, M., Hattori, S., & Tsuchiya, K. (1989) Taxonomic analysis of two rare bats from Amami Islands (Chiroptera, Molossidae and Rhinolophidae). *Memoirs of the National Science Museum, Tokyo* **22**: 215–225.

4213. Yoshiyuki, M., Iijima, M., & Ogawara, Y. (1970) The embryo-size in a population of *Pipistrellus abramus* in Saitama, Japan. *The Journal of the Mammalogical Society of Japan* **5**: 74–75 (in Japanese with English abstract).

4214. Yoshiyuki, M. & Imaizumi, Y. (1986) A new species of *Sorex* (Insectivora, Soricidae) from Sado Island, Japan. *Bulletins of the National Science Museum, Tokyo. Series A* **12**: 185–193.

4215. Yoshiyuki, M. & Imaizumi, Y. (1991) Taxonomic status of the large mole from the Echigo Plain, central Japan, with description of a new species (Mammalia, Insectivora, Talpidae). *Bulletin of the National Science Museum. Series A, Zoology* **17**: 101–110.

4216. Yoshiyuki, M. & Karube, H. (2002) New habitat of Japanese little tube-nosed bat, *Murina silvatica* Yoshiyuki, 1983. *ANIMATE* **3**: 15–16 (in Japanese).

4217. Yoshiyuki, M. & Kimura, T. (1975) New record of *Pipistrellus savii velox* Ognev, 1927. *The Journal of the Mammalogical Society of Japan* **6**: 138, 142–143 (in Japanese).

4218. Yoshiyuki, M. & Kinoshita, A. (1986) Notes on a hibernating colony of *Nyctalus aviator* Thomas, 1911 found in Kawasaki City, Kanagawa Prefecture. *Natural History Report of Kanagawa* **7**: 43–48 (in Japanese).

4219. Yoshiyuki, M. & Morita, C. (2003) The specific status of *Tadarida latouchei* Thomas, 1920 from Yoron Island, Nansei Islands, Kyushu, Japan. *ANIMATE* **4**: 33–37.

4220. Yoshiyuki, M., et al. (1999) Distribution of greater horseshoe bat in Oshima Island, Izu Is., Japan. *ANIMATE* **1**: 3–10 (in Japanese).

4221. Yoshiyuki, M. & Suzuki, K. (1970) The hibernating sites of *Barbastella leucomelas darjelingensis* in Honshu, Japan. *The Journal of the Mammalogical Society of Japan* **5**: 73 (in Japanese).

4222. Yosida, T.H. (1980) Cytogenetics of the Black Rat—Karyotype Evolution and Species Differentiation. Tokyo: University of Tokyo Press, 256 pp.

4223. Yosida, T.H. (1982) Cytogenetical studies on Insectivora. II. Geographical variation of chromosomes in the house shrew, *Suncus murinus* (Soricidae), in east, southeast and southwest Asia, with a note on the karyotype evolution and distribution. *Japanese Journal of Genetics* **57**: 101–111.

4224. Yosida, T.H., Moriguchi, Y., & Sonoda, J. (1968) Karyological studies of three species of Insectivora collected in Japan. *Annual Report of National Institute of Genetics* **18**: 24–25.

4225. Yosida, T.H. & Shi, L. (1986) Cytogenetical studies on the Japanese raccoon dog XIII. A preliminary note on the karyotype of a Chinese specimen with 57 chromosomes collected in Yunnan. *Proceedings of the Japan Academy. Series B* **62**: 49–52.

4226. Yosida, T.H., Wada, M.Y., & Ward, O.G. (1983) Karyotype of a Japanese raccoon dog with 40 chromosomes including two supernumeraries. *Proceedings of the Japan Academy, Series B* **59**: 267–270.

4227. Yosida, T.H., et al. (1984) Further studies on the Japanese raccoon dog karyotypes, with a special regard to somatic variation of B-chromosomes. *Proceedings of the Japan Academy, Series B* **60**: 17–20.

4228. Young, D.D. & Cockcroft, V.G. (1995) Stomach contents of stranded common dolphins *Delphinus delphis* from the south-east of Southern Africa. *Zeitschrift für Säugetierkunde* **60**: 343–351.

4229. Yuan, S.L., et al. (2013) A mitochondrial phylogeny and biogeographical scenario for Asiatic water shrews of the genus *Chimarrogale*: implications for taxonomy and low-latitude migration routes. *PLoS ONE* **8**: e77156.

4230. Yuasa, T., et al. (2007) The impact of habitat fragmentation on genetic structure of the Japanese sika deer (*Cervus nippon*) in southern Kantoh, revealed by mitochondrial D-loop sequences. *Ecological Research* **22**: 97–106.

4231. Yukawa, M. (湯川仁) (1968) カワネズミの巣について. *Miscellaneous Reports of the Hiwa Museum for Natural History* **11**: 31–32 (in Japanese).

4232. Yukawa, M. (湯川仁) (1977) 広島県比和町の哺乳類. *in* 比和の自然：比和を中心とした中国山地の総合学術調査報告 (Hiwa Museum for Natural History, Editor). Hiwa: 比和町郷土史研究会, pp. 157–180 (in Japanese).

4233. Yukawa, M. (1965) The breeding habits of the Japanese shrew mole (*Urotrichus talpoides talpoides* Thomas) in Hiroshima Pref. *Miscellaneous Report of the Hiwa Museum for Natural History* **8**: 1–3 (in Japanese).

4234. Yukawa, M. (1966) The breeding habits of the Smith's red-backed vole (*Eothenomys smithi* Thomas) in Hiroshima-Pref. *Miscellaneous Reports of*

Hiwa Museum for Natural History 9: 2–4 (in Japanese).
4235. Yukawa, M. (1966) Notes on some habits of lesser tube-nosed bat (*Murina aurata ussuriensis* OGNEV). *Miscellaneous Reports of the Hiwa Museum for Natural History* 10: 11–13 (in Japanese).
4236. Yukawa, M. (1971) Reproduction of the Smith red-backed vole (*Eothenomys smithi* Thomas) in Hiroshima-Pref. *Miscellaneous Reports of Hiwa Museum for Natural History* 14: 1–5 (in Japanese).
4237. Yukawa, M. (1976) The reproduction of the Smith red-backed vole (*Eothenomys smithi* Thomas) in Hiroshima-Pref. (II). *Miscellaneous Reports of Hiwa Museum for Natural History* 19: 9–15 (in Japanese).
4238. Yukawa, M. (1977) Mammals in Hiwa-cho, Hiroshima Prefecture. *in* Hiwa no Shizen (Hiwa Museum for Natural Science, Editor). Hiwa: 比和町郷土史研究会, pp. 157–180 (in Japanese).
4239. Yumoto, T., Noma, N., & Maruhashi, T. (1998) Cheek-pouch dispersal of seeds by Japanese monkeys (*Macaca fuscata yakui*) on Yakushima Island, Japan. *Primates* 39: 325–338.
4240. Yurick, D.B. & Gaskin, D.E. (1987) Morphometric and meristic comparisons of skulls of harbour porpoise *Phocoena phocoena* (L.) from the North Atlantic and North Pacific. *Ophelia* 27: 52–75.
4241. Yutani, S. & Ikeda, T. Unpublished.
4242. Zaeschmar, J.R., Dwyer, S.L., & Stockin, K.A. (2013) Rare observations of false killer whales (*Pseudorca crassidens*) cooperatively feeding with common bottlenose dolphins (*Tursiops truncatus*) in the Hauraki Gulf, New Zealand. *Marine Mammal Science* 29: 555–562.
4243. Zamoto, A., et al. (2004) U.S.-type *Babesia microti* isolated from small wild mammals in eastern Hokkaido, Japan. *Journal of Veterinary Medical Science* 66: 919–926.
4244. Zamoto, A., et al. (2004) Epizootiologic survey for *Babesia microti* among small wild mammals in northeastern Eurasia and a geographic deversity in the β-tubulin gene sequences. *Journal of Veterinary Medical Science* 66: 785–792.
4245. Zeveloff, S.I. (2002) Raccoons: A Natural History. Washington: Smithsonian Institution Press, 200 pp.
4246. Zhang, J.S., et al. (2007) A new species of *Barbastella* (Chiroptera: Vespertilionidae) from north China. *Journal of Mammalogy* 88: 1393–1403.
4247. Zhang, W., et al. (1995) Chromosome polymorphism found in the Japanese sika deer, *Cervus nippon*. *Japanese Journal of Zootechnical Science* 66: 462–464 (in Japanese).
4248. Zhang, X., Chen, Y., & Zhou, K. (1989) Banded karyotypes of the finless porpoise, *Neophocaena phocaenoides*, from the Yangtze River. *Acta Theriologica Sinica* 9: 281–284 (in Chinese with English summary).
4249. Zhang, Y., et al. (1997) Distribution of Mammalian Species in China. Beijing: China Forestry Publishing House, 280 pp. (in Chinese and English).
4250. Zhao, X., et al. (2008) Abundance and conservation status of the Yangtze finless porpoise in the Yangtze River, China. *Biological Conservation* 141: 3006–3018.
4251. Ziegler, A.C. (1971) Dental homologies and possible relationships of recent Talpidae. *Journal of Mammalogy* 52: 50–68.
4252. Zima, J. & Kral, B. (1984) Karyotypes of European mammals, part III. *Acta Scientiarium Naturalium, Brno* 18: 1–15.
4253. Zima, J., Lukácová, L., & Macholán, M. (1998) Chromosomal evolution in shrews. *in* Evolution of Shrews (Wójcik, J.M. & Wolsan, M., Editors). Białowieża: Polish Academy of Science, pp. 173–218.
4254. Zimen, E. (1975) Social dynamics of the wolf pack. *in* The Wild Canids: Their Systematics, Behavioural Ecology and Evolution (Fox, M.W., Editor). New York: Van Nostrand Reinhold, pp. 336–362.
4255. Zornetzer, H.R. & Duffield, D.A. (2003) Captive-born bottlenose dolphin x common dolphin (*Tursiops truncatus* x *Delphinus capensis*) intergeneric hybrids. *Canadian Journal of Zoology* 81: 1755–1762.
4256. Zucca, P., et al. (2004) Use of Computer tomography for imaging of *Crassicauda grampicola* in a Risso's dophin (*Grampus griseus*). *Journal of Zoo and Wildlife Medicine* 35: 391–394.
4257. Zuckerman, S. (1953) The breeding season of mammals in captivity. *Proceedings of the Zoological Society of London* 133: 827–950.

Supplements

4258. Kato, Y. (加藤嘉太郎) (1979) 家畜比較解剖図説 第二次増改訂版. Tokyo: Youkendo (養賢堂). 661 pp. (in Japanese).
4259. Kawai, K., Fujinoki, M. and Dewa, H. (in press) Bats in Kiyotsu-kyo: new distributional records of *Myotis nattereri* and *M. ikonnikovi* from Niigata Prefecture. *Kashiwazaki City Museum*.
4260. Ogasawara Village. Unpublished.
4261. Ruedas, L., Heaney, L. & Molur, S. 2008. *Rattus exulans*. The IUCN Red List of Threatened Species. Version 2015.1. Available from: http://www.iucnredlist.org.
4262. Sato, A. (佐藤顕義), et al. (2012) 埼玉県におけるコウモリトコジラミの初記録. 寄せ蛾記 147: 76–77 (in Japanese).
4263. Satô, M. & Satô, R. (2013) Distribution of bats in Wakkanai (2). *Rishiri Studies* 32: 11–14 (in Japanese with English abstract).
4264. Shibata, M. & Yasui, S. (2006) Accidental capture of *Murina ussuriensis* Ognev at canopy layer in Ogawa forest reserve, Ibaraki prefecture. *Bulletin of the Asian Bat Research Institute* 5: 27–29.
4265. Smith, A.T. & Boyer, A.F. (2008) *Oryctolagus cuniculus*. The IUCN Red List of Threatened Species. Version 2015.1. Available from: http://www.iucnredlist.org.
4266. Tomisawa, A. (富沢章) (1984) コウモリの捕食した蛾類. *Amica* 28: 59–60 (in Japanese).

Indices

❖Common names in English

— A —

Alashanian pipistrelle	92
Amami rabbit	212
Amami spiny rat	165
American mink	256
Amur hedgehog	48
Anderson's red-backed vole	158
Asian black bear	243
Asian lesser white-toothed shrew	19
Asian parti-colored bat	96
Asiatic black bear	243
Azumi shrew	4

— B —

Baird's beaked whale	350
bearded seal	290
birdlike noctule	76
black rat	181
Blainville's beaked whale	356
blue whale	336
Bonin flying fox	56
brown bear	240
Bryde's whale	333

— C —

Chinese muntjac	307
common bottlenose dolphin	396
common minke whale	328
Cuvier's beaked whale	362

— D —

Dall's porpoise	402
dark red-backed vole	154
Daubenton's bat	112
dsinezumi shrew	23
dugong	300
dwarf sperm whale	348

— E —

East Asian field mouse	173
East-Asian little bent-winged bat	128
eastern barbastelle	88
eastern bent-winged bat	126
eastern water bat	112
Echigo mole	38
Eden's whale	333
Endo's pipistrelle	85
ermine	254
Eurasian common shrew	8
Eurasian least shrew	2
Eurasian otter	262
Eurasian red squirrel	190
European rabbit	218

— F —

false killer whale	384
Far Eastern Myotis	110
feral goat	318
fin whale	340
Fraser's dolphin	374
fraternal Myotis	100
frosted Myotis	114

— G —

ginkgo-toothed beaked whale	358
gloomy tube-nosed bat	122
goose-beaked whale	362
gray red-backed vole	150
gray whale	324
gray wolf	226
greater horseshoe bat	58
greater Japanese shrew-mole	30

— H —

harbor porpoise	400
harbour seal	280
harvest mouse	170
Hilgendorf's tube-nosed bat	117
house mouse	185
house shrew	26
Hubbs' beaked whale	354
humpback whale	342

— I —

Ikonnikov's Myotis	104
Indo-Pacific bottlenose dolphin	394
Iriomote cat	236

— J —

Japanese badger	266
Japanese black bear	243
Japanese dormouse	148
Japanese eastern mole	34
Japanese field vole	163
Japanese flying squirrel	202
Japanese giant flying squirrel	200
Japanese hare	216
Japanese large-footed bat	107
Japanese little horseshoe bat	61
Japanese long-eared bat	90
Japanese macaque	134
Japanese marten	258
Japanese monkey	134
Japanese mountain mole	32
Japanese noctule	80
Japanese northern bat	72
Japanese pipistrelle	82
Japanese sable	260
Japanese sea lion	297
Japanese serow	314
Japanese shrew-mole	30
Japanese squirrel	192
Japanese water shrew	16
Japanese weasel	248
Japanese western mole	36
Japanese white-toothed shrew	23

— K —

killer whale	380
(Kuril) harbor seal	280

— L —

larga seal	282
large Japanese field mouse	175
large Japanese mole	36
Laxmann's shrew	8
least weasel	252
lesser Japanese mole	34
lesser Japanese shrew-mole	28
lesser leaf-nosed bat	68
long-beaked common dolphin	364
long-clawed shrew	11
Longman's beaked whale	352

— M —

masked palm civet	275
melon-headed whale	382
moon bear	243
mountain hare	214
musk shrew	26
muskrat	162

— N —

narrow-ridged finless porpoise	398
North American raccoon	232
North Pacific right whale	322
northern bat	74
northern fur seal	295
northern pika	210
northern red-backed vole	156
northern right whale dolphin	378
northern sea lion	292
Norway rat	180
nutria	188

— O —

Okinawa flying fox	54
Okinawa little horseshoe bat	63
Okinawa spiny rat	168
Omura's whale	338
orca	380
Oriental free-tailed bat	130
Oriental little free-tailed bat	132
Orii's shrew	25

— P —

Pacific white-sided dolphin	376
Pallas's squirrel	196
pantropical spotted dolphin	386
parti-colored bat	94
Polynesian rat	182
pygmy killer whale	368
pygmy sperm whale	346

— R —

raccoon	232
raccoon dog	224
red and black Myotis	99
red fox	222
Reeves's muntjac	307
rhesus macaque	139
rhesus monkey	139
ribbon seal	288
ringed seal	286
Risso's dolphin	372
river otter	262
rough-toothed dolphin	392
Russian flying squirrel	204
Ryukyu flying fox	52
Ryukyu long-furred rat	183
Ryukyu mouse	184
Ryukyu tube-nosed bat	120

— S —

sable	260
Sado mole	40
sea otter	264
sei whale	331
Senkaku mole	42
shinto shrew	6
short-beaked common dolphin	366
short-finned pilot whale	370
Siberian chipmunk	198
Siberian flying squirrel	204
Siberian weasel	250
sika deer	304
slender shrew	14
small Indian mongoose	272
small Japanese field mouse	178
Smith's red-backed vole	160
sperm whale	344
spinner dolphin	390
spotted seal	282
Stejneger's beaked whale	360
Steller sea lion	292
Steller's sea lion	292
stoat	254
striped dolphin	388
striped field mouse	172
Sturdee's pipistrelle	87

— T —

Taiwan macaque	137
Taiwanese macaque	137
Tokuda's mole	40
Tokunoshima spiny rat	169
True's shrew-mole	28
Tsushima leopard cat	234

— U —

Ussuri whiskered bat	102
Ussurian tube-nosed bat	123

— W —

Watase's shrew	21
wild boar	312

— Y —

Yaeyama little horseshoe bat	65
Yanbaru Myotis	116

❖Common names in Japanese ("Kana" order)

— ア —

アカギツネ	222
アカゲザル	139
アカネズミ	175
アカボウクジラ	362
アゴヒゲアザラシ	290
アジアコジネズミ	19
アズマモグラ	34
アズミトガリネズミ	4
アナウサギ	218
アブラコウモリ	82
アマミトゲネズミ	165
アマミノクロウサギ	212
アムールハリネズミ	48
アライグマ	232

— イ —

イイズナ	252
イシイルカ	402
イチョウハクジラ	358
イノシシ	312
イリオモテヤマネコ	236
イワシクジラ	331

— ウ —

ウスリホオヒゲコウモリ	102

— エ —

エゾクロテン	260
エゾシマリス	198
エゾトガリネズミ	8
エゾモモンガ	204
エゾリス	190
エチゴモグラ	38

— オ —

オウギハクジラ	360
オオアシトガリネズミ	11
オオカミ	226
オガサワラアブラコウモリ	87
オガサワラオオコウモリ	56
オガワコマッコウ	348
オキゴンドウ	384
オキナワオオコウモリ	54
オキナワコキクガシラコウモリ	63
オキナワトゲネズミ	168
オキナワハツカネズミ	184
オコジョ	254
オットセイ	295
オヒキコウモリ	130
オリイジネズミ	25

— カ —

カグヤコウモリ	100
カグラコウモリ	68
カズハゴンドウ	382
カツオクジラ	333
カマイルカ	376
カモシカ	314
カヤネズミ	170
カワウソ	262
カワネズミ	16

— キ —

キクガシラコウモリ	58
キタオットセイ	295
キタクビワコウモリ	74
キタナキウサギ	210
キタリス	190
キョン	307

— ク —

クチバテングコウモリ	122
クビワオオコウモリ	52
クビワコウモリ	72
クマネズミ	181
クラカケアザラシ	288
クリハラリス	196
クロアカコウモリ	99
クロオオアブラコウモリ	92
クロテン	260
クロホオヒゲコウモリ	114

— ケ —

ケナガネズミ	183

— コ —

コイワシクジラ	328
コウベモグラ	36
コキクガシラコウモリ	61
コククジラ	324
コテングコウモリ	123
コビレゴンドウ	370
コブハクジラ	356
コマッコウ	346
コヤマコウモリ	80
ゴマフアザラシ	282

— サ —

サカマタ	380
サドモグラ	40
サラワクイルカ	374
ザトウクジラ	342

— シ —

シベリアイタチ	250
シマリス	198
シャチ	380
シロナガスクジラ	336
シワハイルカ	392
シントウトガリネズミ	6
ジャコウネズミ	26
ジュゴン	300

— ス —

スジイルカ	388
スナメリ	398
スミイロオヒキコウモリ	132
スミスネズミ	160

— セ —

セスジネズミ	172
セミイルカ	378
セミクジラ	322
センカクモグラ	42
ゼニガタアザラシ	280

— タ —

タイヘイヨウアカボウモドキ	352
タイリクモモンガ	204
タイリクヤチネズミ	150
タイワンザル	137
タヌキ	224

― チ ―

チチブコウモリ	88
チビトガリネズミ	2

― ツ ―

ツキノワグマ	243
ツシマヤマネコ	234
ツチクジラ	350
ツノシマクジラ	338

― テ ―

テングコウモリ	117

― ト ―

トクノシマトゲネズミ	169
トド	292
ドーベントンコウモリ	112
ドブネズミ	180

― ナ ―

ナガスクジラ	340

― ニ ―

ニタリクジラ	333
ニホンアシカ	297
ニホンアナグマ	266
ニホンイタチ	248
ニホンウサギコウモリ	90
ニホンカモシカ	314
ニホンザル	134
ニホンシカ	304
ニホンジカ	304
ニホンジネズミ	23
ニホンテン	258
ニホンノウサギ	216
ニホンモモンガ	202
ニホンリス	192

― ヌ ―

ヌートリア	188

― ネ ―

ネズミイルカ	400

― ノ ―

ノヤギ	318
ノレンコウモリ	110

― ハ ―

ハクビシン	275
ハシナガイルカ	390
ハセイルカ	364
ハタネズミ	163
ハップスオウギハクジラ	354
ハツカネズミ	185
ハナゴンドウ	372
ハントウアカネズミ	173
ハンドウイルカ	396
バイカルトガリネズミ	8
バンドウイルカ	396

― ヒ ―

ヒグマ	240
ヒナコウモリ	96
ヒミズ	30
ヒメトガリネズミ	14
ヒメネズミ	178
ヒメヒナコウモリ	94
ヒメヒミズ	28
ヒメホオヒゲコウモリ	104
ヒメヤチネズミ	156

― フ ―

フイリマングース	272

― ホ ―

ホオジロムササビ	200
ポリネシアネズミ	182

― マ ―

マイルカ	366
マスクラット	162
マダライルカ	386
マッコウクジラ	344

― ミ ―

ミズラモグラ	32
ミナミハンドウイルカ	394
ミナミバンドウイルカ	394
ミンク	256
ミンククジラ	328

― ム ―

ムクゲネズミ	154
ムササビ	200

― モ ―

モモジロコウモリ	107
モリアブラコウモリ	85

― ヤ ―

ヤエヤマコキクガシラコウモリ	65
ヤギ	318
ヤチネズミ	158
ヤマコウモリ	76
ヤマネ	148
ヤンバルホオヒゲコウモリ	116

― ユ ―

ユキウサギ	214
ユビナガコウモリ	126
ユメゴンドウ	368

― ラ ―

ラッコ	264

― リ ―

リュウキュウテングコウモリ	120
リュウキュウユビナガコウモリ	128

― ワ ―

ワタセジネズミ	21
ワモンアザラシ	286

❖ Scientific names

— A —

Alces alces	144
Alilepus	220
Ailuropoda melanoleuca	241
Anteliomys	161
Apodemus	151, 207, 269, 271, 310
Apodemus agrarius	xxiii, 44, **172**, 271
Apodemus argenteus	xxiii, 45, 143, 173, 174, 176, **178**, 208, 228, 230, 235, 270, 310, 405
Apodemus peninsulae	xxiii, **173**, 176, 178, 179, 405
Apodemus peninsulae giliacus	xxiii, 173, 174
Apodemus speciosus	xxiii, 13, 45, 143, 173, 174, **175**, 178, 179, 208, 229, 230, 231, 235, 270, 271, 310, 405
Apodemus spp.	229, 261
Aschizomys	158, 159
Axis	279

— B —

Balaenoptera acutorostrata	xxiv, **328**
Balaenoptera acutorostrata acutorostrata	330
Balaenoptera acutorostrata scammoni	330
Balaenoptera bonaerensis	328, 329
Balaenoptera borealis	xxiv, **331**, 335
Balaenoptera brydei	xxv, **333**
Balaenoptera edeni	xxv, **333**
Balaenoptera edeni brydei	333
Balaenoptera edeni edeni	333
Balaenoptera musculus	xxv, **336**
Balaenoptera musculus brevicauda	336
Balaenoptera musculus intermedia	336
Balaenoptera musculus musculus	xxv, **336**
Balaenoptera omurai	xxv, 335, **338**
Balaenoptera physalus	xxv, **340**
Balaenoptera physalus physalus	340
Balaenoptera physalus quoyi	340
Barbastella darjelingensis	xxii, 88
Barbastella leucomelas	xxii, 88, 89
Barbastella leucomelas darjelingensis	89
Berardius bairdii	xxv, **350**
Bison priscus	144
Bos taurus	278

— C —

Callorhinus gilmorei	295
Callorhinus ursinus	xxiv, **295**
Callosciurus	207
Callosciurus erythraeus	xxiii, **196**, 277, 278, 279
Callosciurus erythraeus thaiwanensis	xxiii, 196
Callosciurus finlaysonii	197, 279
Canis familiaris	227, 278
Canis hodophilax	226
Canis latrans	227
Canis lupus	xxiv, 215, **226**, 242, 313
Canis lupus arabs	226
Canis lupus hattai	226
Canis lupus hodophilax	226
Capra hircus	xxiv, 43, 277, 278, **318**
Capreolus capreolus	308
Capricornis crispus	xxiv, **314**, 405
Capricornis milneedwardsii	316
Capricornis rubidus	316
Capricornis sp.	314
Capricornis sumatraensis	316
Capricornis swinhoei	316
Capricornis thar	316
Cephalorhynchus sp.	378
Cervus elaphus	306
Cervus mariannus	278
Cervus nippon	xxiv, 45, 142, 143, 145, 207, 227, 229, 231, 242, 244, 246, 278, **304**, 308, 316, 405
Cervus nippon centralis	279, 306
Cervus nippon grassianus	306
Cervus nippon hortulorum	306
Cervus nippon keramae	278, 279, 306
Cervus nippon kopschi	306
Cervus nippon mageshimae	278, 279, 306
Cervus nippon manchuricus	306
Cervus nippon mandarinus	306
Cervus nippon nippon	279, 306
Cervus nippon pseudaxis	306
Cervus nippon pulchellus	279
Cervus nippon sichuanicus	306
Cervus nippon taiouanus	278, 306
Cervus nippon yakushimae	144, 279, 306
Cervus nippon yesoensis	144, 279, 306
Chimarrogale himalayica	16, 17
Chimarrogale phaeura	17
Chimarrogale platycephalus	xxii, **16**, 45, 405
Chimarrogale platycephala	xxii, 16, 18
Clethrionomys	151, 155, 159, 161, 207, 270, 271, 309
Clethrionomys montanus	155
Clethrionomys rex	xxiii, 154, 155
Clethrionomys rufocanus bedfordiae	xxiii, 150
Clethrionomys rutilus mikado	xxiii, 156
Condylura	269
Craseomys	159
Crocidura dsinezumi	xxii, 21, **23**, 25, 26, 27, 45, 143, 230, 278, 405
Crocidura dsinezumi dsinezumi	xxii, 23
Crocidura dsinezumi quelpartis	24
Crocidura dsinezumi umbrina	xxii, 23, 24
Crocidura horsfieldii	21, 22
Crocidura horsfieldii watasei	xxii, 21
Crocidura kurodai	24
Crocidura lasiura	24
Crocidura orii	21, **25**, 26
Crocidura shantungensis	xxii, **19**, 24, 46, 235
Crocidura sibirica	19
Crocidura suaveolens	19, 20
Crocidura suaveolens shantungensis	xxii, 19
Crocidura watasei	xxii, **21**, 24, 25, 26, 277

Chrysopteron	99
Cynomys sp.	278

— **D** —

Dama	279
Delphinus	374
Delphinus capensis	xxv, **364**, 366
Delphinus capensis tropicalis	364
tropicalis-form	364
Delphinus delphis	xxv, 364, **366**
Delphinus spp.	367
Desmana	269
Dicrocerus sp.	307
Diplothrix	44
Diplothrix legata	xxiii, **183**, 405
Dugong dugon	xxiv, **300**
Dugong sp.	300
Dymecodon	45, 269
Dymecodon pilirostris	xxii, **28**, 31, 270

— **E** —

Elaphurus davidianus	279
Enhydra lutris	xxiv, **264**
Enhydra lutris kenyoni	264
Enhydra lutris lutris	264
Enhydra lutris nereis	264
Eothenomys	151, 163, 164
Eothenomys andersoni	xxiii, **158**, 160, 161, 163, 164, 230, 405
Eothenomys imaizumii	159
Eothenomys kageus	160, 161
Eothenomys niigatae	159
Eothenomys smithii	xxiii, 158, 159, **160**, 163, 164, 230, 405, 406
Eothenomys sp.	160
Eptesicus japonensis	xxii, **72**, 75
Eptesicus nilssonii	xxii, 73, **74**
Eptesicus nilssonii parvus	73, 75
Equus caballus	228, 278
Erignathus barbatus	xxiv, **290**
Erignathus barbatus nauticus	xxiv, 290
Erinaceus	279
Erinaceus amurensis	xxii, **48**, 277, 278
Erinaceus europaeus	48
Erinaceus sp.	48
Erinaceus spp.	48
Eschrichtius robustus	xxiv, **324**
Eschrichtius sp.	324
Eubalaena australis	323
Eubalaena glacialis	xxiv, 322, 323
Eubalaena japonica	xxiv, **322**
Eubalaena shinshuensis	322
Eumetopias jubatus	xxiv, 281, **292**
Eumetopias jubatus jubatus	292, 293
Eumetopias jubatus monteriensis	293
Euroscaptor	269, 270, 271
Euroscaptor mizura	xxii, **32**, 35, 38, 40, 270
Evotomys	159, 161

— **F** —

Felis bengalensis	197
Felis bengalensis euptilura	xxiv, 37, 234
Felis catus	13, 238, 277, 278
Felis cf. *microtis*	234, 236
Felis iriomotensis	xxiv, 236
Felis silvestris	215
Felis silvestris lybica	238
Feresa	382, 385
Feresa attenuata	xxv, **368**

— **G** —

Glirulus	45
Glirulus japonicus	xxiii, 45, 143, **148**, 271
Globicephala	372, 385
Globicephala macrorhynchus	xxv, **370**
Globicephala melas	371
Grampus	385
Grampus griseus	xxv, **372**

— **H** —

Halichoerus grypus	401
Herpestes auropunctatus	xxiv, 213, **272**, 277, 278, 279
Herpestes javanicus	xxiv, 22, 272, 274, 279
Hipposideros terasensis	68
Hipposideros turpis	xxii, **68**
Histriophoca fasciata	xxiv, **288**
Hydrochoerus hydrochaeris	278
Hyperoodon planifrons	352
Hypsugo alaschanicus	xxii, **92**
Hypsugo (Pipistrellus)	93
Hypsugo (Pipistrellus) savii	93

— **I** —

Indopacetus pacificus	xxv, **352**
Indopacetus sp.	353

— **K** —

Kogia breviceps	xxv, **346**
Kogia sima	xxv, **346**, 348
Kogia simus	xxv, 348

— **L** —

Lagenodelphis hosei	xxv, **374**
Lagenorhynchus obliquidens	xxv, **376**
Lagenorhynchus sp.	378
Lepus	218, 269
Lepus brachyurus	xxiv, 143, 214, **216**, 269, 405
Lepus brachyurus angustidens	217
Lepus brachyurus brachyurus	217
Lepus brachyurus lyoni	xxiv, 216, 217
Lepus brachyurus okiensis	217
Lepus europaeus	214, 215
Lepus spp.	229
Lepus timidus	xxiii, **214**, 269
Lepus timidus ainu	xxiii, 214, 215, 269
Lissodelphis borealis	xxv, **378**
Lutra lutra	xxiv, **262**
Lutra nippon	xxiv, 262, 263
Lutra sp.	262

— **M** —

Macaca cyclopis	xxiii, 135, **137**, 277, 278, 279

Macaca fascicularis	135, 137, 139, 278, 279	*Mus*	269, 271
Macaca fuscata	xxiii, 45, **134**, 137, 142, 143, 207, 246, 277, 278, 279	*Mus argenteus* (= *geisha*)	405
		Mus caroli	xxiii, 44, **184**, 230, 271
Macaca fuscata fuscata	xxiii, 134, 135	*Mus castaneus castaneus*	185
Macaca fuscata yakui	xxiii, 134, 135, 278	*Mus molossinus*	185
Macaca mulatta	xxiii, 135, 137, **139**, 277, 278, 279	*Mus musculus*	xxiii, 44, 166, 170, 171, 178, 184, **185**, 186, 187, 228, 277, 278, 404
Macaca nemestrina	136		
Martes americana	260	*castaneus* group	186
Martes flavigula	197	*domesticus* group	186
Martes martes	215	*musculus* group	186
Martes melampus	xxiv, 5, 7, 31, 194, 217, 244, **258**, 260, 261, 269, 278	*Mus musculus domesticus*	185
		Mus muscolus manchu	185
Martes melampus melampus	xxiv, 258	*Mus musculus molossinus*	185, 186
Martes melampus tsuensis	xxiv, 235, 258	*Mus tanezumi*	405
Martes sp.	258	*Mustela*	257, 258
Martes zibellina	xxiv, 10, 13, 15, 191, 211, **260**, 271	*Mustela erminea*	xxiv, 5, 7, 215, 252, 253, **254**
Martes zibellina brachyura	xxiv, 260	*Mustela erminea nippon*	xxiv, 254
Megaptera novaeangliae	xxv, **342**	*Mustela erminea orientalis*	xxiv, 254
Meles anakuma	xxiv, 31, 37, **266**	*Mustela furo*	278
Meles leucurus	266, 268	*Mustela itatsi*	xxiv, 18, 22, 62, 217, **248**, 250, 256, 257, 269, 277, 278
Meles meles	266, 268		
Meles meles anakuma	xxiv, 268	*Mustela namiyei*	253
Mesoplodon	352	*Mustela nivalis*	xxiv, **252**, 260
Mesoplodon carlhubbsi	xxv, **354**	*Mustela nivalis namiyei*	xxiv, 252, 253
Mesoplodon densirostris	xxv, **356**	*Mustela nivalis nivalis*	xxiv, 252, 253
Mesoplodon ginkgodens	xxv, **358**	*Mustela sibirica*	xxiv, 46, 235, 248, 249, **250**, 277, 278
Mesoplodon hotaula	359	*Mustela sibirica coreana*	xxiv, 250
Mesoplodon stejnegeri	xxv, 357, **360**	*Mustela* sp.	248
Micromys	269	*Mustela* spp.	229
Micromys minutus	xxiii, **170**, 235, 270	*Myocastor coypus*	xxiii, **188**, 277, 278, 279
Microtus	151, 310	*Myodes*	159, 161, 164, 176, 207, 310
Microtus montebelli	xxiii, 158, 160, **163**, 229, 230, 406	*Myodes glareolus*	156, 158, 161
Miniopterus	44	*Myodes japonicus*	150
Miniopterus fuliginosus	xxiii, 108, 109, **126**, 128	(*Myodes*) *microtinus*	155
Miniopterus fuscus	xxiii, **128**	*Myodes montanus*	155
Miniopterus schreibersi	127	*Myodes rex*	xxiii, 151, **154**, 228, 270, 271
Miniopterus schreibersi fuliginosus	127	*Myodes rufocanus*	xxiii, 13, **150**, 154, 155, 156, 157, 158, 161, 208, 228, 229, 230, 271, 309, 310, 406
Mogera	32, 269, 270, 271		
Mogera etigo	xxii, 35, **38**, 40, 41	*Myodes rufocanus bedfordiae*	151
Mogera imaizumii	xxii, 32, **34**, 36, 37, 38, 39, 40, 45, 270	*Myodes rutilus*	xxiii, 151, **156**, 228, 230
Mogera insularis	42	*Myodes rutilus mikado*	157
Mogera kobeae	37	*Myodes sikotanensis*	155
Mogera minor	35	*Myodes* spp.	229, 230, 261
Mogera robusta	36, 37	*Myotis abei*	113
Mogera spp.	33, 35, 37, 41, 43	*Myotis annectans*	115
Mogera tokudae	xxii, 38, 39, **40**, 270	*Myotis bombinus*	xxiii, 108, 109, **110**
Mogera uchidai	xxii, **42**, 44	*Myotis brandtii*	103
Mogera wogura	xxii, 32, 34, 35, **36**, 38, 39, 40, 270, 405	*Myotis brandtii gracilis*	xxiii, 102, 103
Mungos mungo	279	*Myotis daubentonii*	xxiii, 112, 113
Muntiacus reevesi	xxiv, 277, 278, 279, **307**	*Myotis daubentonii petax*	113
Muntiacus reevesi reevesi	307	*Myotis daubentonii ussuriensis*	112, 113
Murina aurata	124, 125	*Myotis formosus*	xxii, 99
Murina aurata ussuriensis	124	*Myotis frater*	xxii, **100**
Murina hilgendorfi	xxiii, 109, **117**	*Myotis frater bucharensis*	101
Murina leucogaster	xxii, 117, 118, 119	*Myotis frater frater*	101
Murina leucogaster hilgendorfi	119	*Myotis frater longicaudatus*	101
Murina silvatica	122, 124, 125	*Myotis frater kaguyae*	101
Murina ryukyuana	xxiii, **120**	*Myotis fujiensis*	106
Murina tenebrosa	xxiii, **122**	*Myotis gracilis*	xxiii, 102, 104, 106
Murina ussuriensis	xxiii, 122, **123**	*Myotis hosonoi*	105, 106

Myotis ikonnikovi	xxiii, 102, 103, **104**, 114
Myotis ikonnikovi hosonoi	xxiii, 104
Myotis kaguyae	101
Myotis macrodactylus	xxiii, **107**, 113, 405
Myotis montivaus	115
Myotis mystacinus	xxiii, 102, 103, 106
Myotis nattereri	xxiii, 107, 110, 111
Myotis nattereri bombinus	xxiii, 107, 110, 111
Myotis ozensis	106
Myotis petax	xxiii, 108, **112**
Myotis petax chosanensis	113
Myotis petax loukashkini	113
Myotis petax ussuriensis	113
Myotis pruinosus	xxiii, 104, **114**, 116
Myotis rufoniger	xxii, **99**
Myotis sibiricus	103
Myotis sp.	103
Myotis yanbarensis	xxiii, 115, **116**
Myotis yesoensis	103, 106

— N —

Nemorhaedus (*Naemorhaedus*)	316
Neoaschizomys sikotanensis	155
Neophocaena asiaeorientalis	xxv, **398**
Neophocaena asiaeorientalis asiaeorientalis	398
Neophocaena asiaeorientalis sunameri	398
Neophocaena phocaenoides	xxv, 398
Neovison vison	13, **256**, 277, 278, 279
Nesoscaptor	43
Neurotrichus	269
Nyctalus aviator	xxii, **76**, 405
Nyctalus furvus	xxii, **80**
Nyctalus lasiopterus	78
Nyctalus lasiopterus aviator	78
Nyctalus noctula	81
Nyctalus noctula motoyoshii	81
Nyctalus plancyi	81
Nyctereutes procyonoides	xxiv, 13, 37, **224**, 233, 244, 278
Nyctereutes procyonoides albus	xxiv, 224, 225
Nyctereutes procyonoides viverrinus	xxiv, 224, 267
Nyctereutes viverrinus genitor	224
Nyctereutes viverrinus nipponicus	224
Nyctereutes viverrinus okuensis	224

— O —

Ochotona	213
Ochotona hyperborea	xxiii, **210**
Ochotona hyperborea yesoensis	xxiii, 210
Odocoileus virginianus	208
Ondatra zibethicus	xxiii, **162**, 278, 279
Orcinus	385
Orcinus orca	xxv, 281, 291, 293, 332, **380**
Oryctolagus cuniculus	xxiv, 218, 277, 278

— P —

Paguma larvata	xxiv, **275**, 278
Palaeoloxodon naumanni	144
Pentalagus	44
Pentalagus furnessi	xxiii, **212**, 269, 273, 277
Pentalagus spp.	212
Peponocephala	385
Peponocephala electra	xxv, **382**
Petaurista	207
Petaurista leucogenys	xxiii, **200**, 405
Phaulomys	150, 158, 159, 161
Phaulomys cf. *smithii*	160
Phoca cf. *largha*	282
Phoca cf. *vitulina*	280
Phoca fasciata	xxiv, 288
Phoca hispida botnica	286
Phoca hispida hispida	286
Phoca hispida ochotensis	xxiv, 286
Phoca largha	xxiv, **282**
Phoca vitulina	xxiv, **280**
Phoca vitulina concolor	281
Phoca vitulina mellonae	281
Phoca vitulina richardii	281
Phoca vitulina stejnegeri	xxiv, 280, 281
Phoca vitulina vitulina	281
Phocoena phocoena	xxv, **400**
Phocoena phocoena phocoena	400
Phocoena phocoena relicata	400
Phocoena phocoena vomeria	400
Phocoenoides dalli	xxv, 401, **402**
truei-type	402, 403
Physeter macrocephalus	xxv, **344**
Pipistrellus abramus	xxii, **82**, 85, 86, 87, 278, 405
Pipistrellus coreensis	93
Pipistrellus coromandra	87
Pipistrellus endoi	xxii, 82, **85**
Pipistrellus javanicus	84, 87
Pipistrellus savii	xxii, 92, 93
Pipistrellus sturdeei	xxii, **87**, 405
Plecotus auritus	xxii, 90, 91
Plecotus auritus sacrimontis	91
Plecotus sacrimontis	xxii, **90**
Pliopentalagus	212
Praekogia cedrosensis	346, 348
Prionailurus bengalensis	xxiv, 31, 44, 46, 234, 236, 237
Prionailurus bengalensis euptilurus	xxiv, 37, **234**, 238
Prionailurus bengalensis iriomotensis	xxiv, 27, 44, 53, 234, **236**, 238
Prionailurus iriomotensis	237
Procapra gutturosa	315
Procyon cancrivorus	233, 279
Procyon lotor	xxiv, 13, 224, 231, **232**, 268, 277, 278, 279
Pseudorca	382
Pseudorca crassidens	xxv, **384**
Pteromys momonga	xxiii, **202**, 405
Pteromys volans	xxiii, **204**, 279
Pteromys volans orii	xxiii, 204, 279
Pteropus dasymallus	xxii, 44, **52**, 56, 405
Pteropus dasymallus daitoensis	xxii, 52, 53, 238, 311
Pteropus dasymallus dasymallus	xxii, 52, 53
Pteropus dasymallus formosus	52
Pteropus dasymallus inopinatus	xxii, 52, 53
Pteropus dasymallus yayeyamae	xxii, 52, 53, 237
Pteropus loochoensis	xxii, **54**
Pteropus mariannus	54
Pteropus pselaphon	xxii, **56**
Pusa hispida	xxiv, **286**

Pusa hispida botanica	287
Pusa hispida hispida	287
Pusa hispida ladogensis	287
Pusa hispida ochotensis	287
Pusa hispida saimensis	287

— R —

Rattus	269
Rattus exulans	xxiii, 44, **182**, 278
Rattus norvegicus	xxiii, 44, **180**, 181, 228, 229, 277, 278, 404
Rattus sp.	180, 181
Rattus spp.	229
Rattus rattus	xxiii, 44, 180, **181**, 182, 229, 237, 278, 279, 404
Rattus tanezumi	181
Rhinolophus cornutus	xxii, **61**, 63, 64, 65, 66, 108, 109, 405
Rhinolophus cornutus orii	xxii, 61, 62
Rhinolophus ferrumequinum	xxii, **58**, 61, 62, 108, 109
Rhinolophus ferrumequinum nippon	60
Rhinolophus imaizumii	66
Rhinolophus monoceros	62
Rhinolophus perditus	xxii, 62, 63, 64, **65**
Rhinolophus pusillus	62
Rhinolophus pumilus	xxii, 62, **63**, 65, 66
Rhinolophus pumilus miyakonis	xxii, 63, 64
Rhinolophus pumilus pumilus	xxii, 63, 64

— S —

Saimiri sciureus	278
Scalopus	269
Sciurus	207
Sciurus carolinensis	279
Sciurus lis	xxiii, **192**, 405
Sciurus vulgaris	xxiii, **190**, 192, 278, 279
Sciurus vulgaris orientis	xxiii, 190, 278, 279
Selenarctos thibetanus	xxiv, 243
Sorex caecutiens	xxii, 4, 5, 6, 7, **8**, 11, 13, 14, 15, 23, 228, 230
Sorex caecutiens kunashiriensis	10
Sorex caecutiens saevus	xxii, 10
Sorex caecutiens shikokensis	xxii, 6
Sorex caecutiens shinto	xxii, 6
Sorex chouei	7
Sorex gracillimus	xxii, 2, 3, 8, 10, 13, **14**, 46, 405
Sorex hosonoi	xxii, 2, 3, **4**, 7, 14
Sorex hosonoi shiroumanus	4
Sorex isodon	11, 12
Sorex minutissimus	xxii, **2**, 4, 14, 15
Sorex minutissimus hawkeri	xxii, 2
Sorex platycephalus	18
Sorex sadonis	xxii, 6
Sorex shinto	xxii, 4, 5, **6**, 8, 9, 23, 230, 405
Sorex shinto sadonis	6, 7
Sorex shinto shikokensis	6, 7
Sorex shinto shinto	6
Sorex spp.	229
Sorex unguiculatus	xxii, 3, 4, 8, 10, **11**, 15, 230
Sorex yukonicus	2, 3
Sotalia fluriatilis	392
Sousa	374
Sousa sp.	392
Stenella	374
Stenella attenuata	xxv, 366, **386**
Stenella coeruleoalba	xxv, 366, **388**
Stenella longirostris	xxv, **390**
Stenella longirostris longirostris	390
Stenella longirostris roseiventris	390
Steno bredanensis	xxv, **392**
Suncus montanus	27
Suncus murinus	xxii, 24, **26**, 44, 229, 278
Suncus murinus temmincki	xxii, 26
Sus scrofa	xxiv, 31, 44, 207, 228, 229, 231, 246, 277, 278, **312**
Sus scrofa domesticus	312
Sus scrofa leucomystax	xxiv, 312
Sus scrofa riukiuanus	xxiv, 312

— T —

Tadarida insignis	xxiii, **130**, 132
Tadarida latouchei	xxiii, **132**, 405
Tadarida teniotis	131
Tadarida teniotis insignis	132
Talpa	270
Talpa europaea var. *minor*	35
Tamias	207
Tamias sibiricus	xxiii, **198**, 261, 278
Tamias sibiricus lineatus	xxiii, 198
Tokudaia	44, 271
Tokudaia muenninki	xxiii, 44, 165, 166, **168**, 238
Tokudaia osimensis	xxiii, 44, **165**, 168, 169, 277, 406
Tokudaia osimensis muenninki	xxiii, 168
Tokudaia osimensis osimensis	xxiii, 165
Tokudaia sp.	165
Tokudaia tokunoshimensis	xxiii, 44, 166, 167, **169**
Tremarctos ornatus	241
Trichosurus vulpecula	279
Trischizolagus	220
Tursiops	373, 374, 392
Tursiops aduncus	xxv, **394**
Tursiops sp.	385
Tursiops truncatus	xxv, 364, 394, 395, **396**

— U —

Uropsilus	269
Urotrichus	45, 269
Urotrichus talpoides	xxii, 28, 29, **30**, 35, 43, 45, 177, 270, 405
Urotrichus talpoides adversus	30
Ursus americanus	208, 244
Ursus arctos	xxiv, 46, 207, **240**, 244, 246, 306
Ursus malayanus	244
Ursus maritimus	241, 287, 291
Ursus spelaeus	241
Ursus spp.	229
Ursus tanakai	243
Ursus thibetanus	xxiv, 142, 143, 207, **243**, 246, 316
Ursus thibetanus formosanus	243
Ursus thibetanus gedrosianus	243
Ursus thibetanus japonicus	243

— V —

Vespertilio akokomuli	405
Vespertilio blepotis	405
Vespertilio molossus	405

Vespertilio murinus	xxii, 53, **94**, 96
Vespertilio orientalis	98
Vespertilio sinensis	xxii, 81, 94, **96**
Vespertilio superans	xxii, 81, 96, 98
Vulpes vulpes	xxiv, 5, 7, 10, 13, 37, 191, 194, 215, 217, **222**, 228, 233, 249, 253, 255, 257, 259, 267, 277
Vulpes vulpes japonica	xxiv, 222, 223
Vulpes vulpes schrencki	xxiv, 222, 223, 278

— Z —

Zalophus japonicus	xxiv, **297**
Zalophus californianus	297
Zalophus californianus californianus	298
Zalophus californianus japonicus	xxiv, 297, 298
Zalophus californianus wollebaeki	298
Zalophus wollebaeki	297
Ziphius cavirostris	xxv, **362**

❖Research Topics

— A —

abandoned or escaped species	277
Abe	405, 406
acorn crop	206, 207, 208
alien mammal	231, 277, 278, 279
allopatric	144
animals transported and released into the food supply	277
Aoki	405

— B —

beechnut	206, 208, 246, 310
binominal nomenclature	404
biodiversity hotspot	269
Blakiston's line	45
Bonhote	404
Borrelia	228, 229, 230, 231
British Museum (Natural History) (BM (NH))	405
Bürger	405

— C —

castaneus group	186
cestode	228, 231
chestnut	208
colonization	44, 270, 271, 311
commensal	44, 186, 231
commensalism	186
coproantigen	228
countermeasures	228, 231, 277, 279
cyclical population fluctuation	309
cyclicity	309, 310

— D —

defensive chemical	206, 207
density	207, 208, 228, 246, 247, 279, 310
density dependence	310
divergence	45, 46, 144, 186, 269, 270, 271
domestication	238
domesticus group	186
Duke of Bedford	405

— E —

eastern vs. western	142
echinococcosis	228, 229, 231, 277
Edo era (or period)	186, 277, 404, 405
endemicity	270, 271
endemism	271
energy content	206
enforcement of hunting regulations	246
environmental change	46, 270
erythema	228
extinction	238, 246, 247, 271, 277, 309, 310, 311

— F —

Fauna Japonica	405
feedback management	247, 310, 311
Fennoscandia	310
feral cat	238, 239
flexibility	238
fluctuation	206, 207, 208, 309, 310

— G —

garbage	238, 239
genetic diversity	142, 145, 271
geographical boundary	142, 143
geomorphology (-gical)	45, 46
glacial period	45, 46, 144
Gray	404
Group of Mammalian Science	406
Günther	404

— H —

heterozygosity	145
hunting pressure	145, 247
hybridization	238, 277

— I —

immigration	143, 144
impacts of alien mammals	277, 279
intentional introduction as a natural enemy	277
Inukai	405, 406
irreplaceable niche	142

— J —

Japanese wild mouse	186

— K —

Kano	405, 406
Kerama Gap	44
Kishida	405
Korean Pen.	45, 46, 144, 186, 187
Kuroda	405, 406

— L —

land bridge	45, 46, 144
legal protection	246
Linnean hierarchy	404
Lyme disease	228, 229, 230

— M —

Mammalogical Society of Japan	406
masting	206, 207, 208, 310
microevolutionary	270, 271
microsatellite	145
migrans	228, 231
Milne-Edwards	404
Miocene	45, 269, 270
mitochondrial DNA (mtDNA)	142, 143, 144, 145, 186, 187, 271
Morse	405
multiple-colonization	143, 144
Mus musculus molossinus	186
musculus group	186

— N —

National Museum of Natural History (RMNH)	404, 406
national park	246, 247, 310
National Science Museum, Tokyo	406
Natural Park law	247
northern vs. southern	142, 143, 144
nutrition	206, 207, 246

— O —

overwinter	207

— P —

Palearctic element	44
paleontology (-ical)	143, 144, 269
pathogenicity	231
phylogenetic	142, 143, 186, 187, 269, 270, 271
phylogeography (-phical)	142, 143, 144, 145
Pleistocene	45, 46, 144, 269, 270, 271
Pliocene	44, 269, 270
polyphenolics	206
population dynamics	142, 206, 207, 208, 309, 310, 311
population size	144, 145, 238, 247, 311
predation	228, 238
predator	231, 238, 310, 311
primatology	310

— Q —

Quercus	144, 206, 207, 208, 310

— R —

radiation	269, 270
Red Data Book	246
refugia	143, 144
roundworm	231, 277
routes of introduction	277

— S —

Sakhalin	45, 46, 186, 187, 270, 405
Schlegel	405
single nucleotide polymorphism (SNP)	187
Soya Str.	46
Specified Wildlife Conservation and Management Plan (SWCMP)	246, 247
spirochete	228
Stone	404
synchronism	310
synchrony	206

— T —

taeniid egg	228
Tanaka	405, 406
tannin	206, 207, 208, 310
Temminck	404, 405, 406
Thomas	404, 405
tick	228, 229, 230, 231
Tokara Tectonic Str.	44, 45
Tokuda	405, 406
tolerance	208, 231, 310
transmission	229, 231, 238, 277
True	404
Tsugaru Str.	45
Tsushima & Korean Str.	45

— U —

unintentional introduction	277

— V —

vegetation	144, 246, 277
vicariance	144
von Siebold	405

— W —

Wallace	405
Watase's line	44
Whitman	405
wildlife management	246, 247
Wildlife Protection and Hunting Law (WPHL)	246, 247
World War II	277, 310, 404, 405, 406
World's Worst Invasive Alien Species	238

— Z —

zoogeography (-ical, -ic)	44, 45, 46, 269, 271, 406
zoonosis (-ses)	228, 229, 231

❖ Authors

— A —

Hisashi ABE	16
Shintaro ABE	272
Masao AMANO	344, 364, 366, 368, 370, 372, 374, 382, 384, 400
Kyle N. ARMSTRONG	61, 63, 65

— E —

Lazaro M. ECHENIQUE-DIAZ	68
Hideki ENDO	134, 137, 139, 226

— F —

Dai FUKUI	72, 74, 76, 80, 88, 90, 94, 96

— H —

Kaoru HATTORI	264, 295
Osamu HOSON	300

— I —

Tohru IKEDA	232, 277
Takao INOUÉ	297
Yasuyuki ISHIBASHI	150
Mari ISHIDA	107
Nobuo ISHII	28, 30, 48
Hajime ISHIKAWA	322, 324, 328, 331, 333, 336, 340
Tsuyoshi ISHINAZAKA	292
Takeomi ISONO	297
Masahiro A. IWASA	148, 150, 154, 156, 158, 160, 162, 163, 165, 168, 169, 170, 172, 173, 175, 178, 180, 181, 182, 183, 184, 185, 188
Masako IZAWA	56, 234, 236, 238

— K —

Koichi KAJI	246
Yayoi KANEKO	266
Yukibumi KANEKO	160, 404
Shin-ichiro KAWADA	32, 34, 36, 38, 40
Atsushi KAWAHARA	2
Kuniko KAWAI	82, 85, 87, 92, 99, 100, 102, 104, 112, 114, 116, 117, 120, 122, 123
Kazumitsu KINJO	52, 54, 56
Mari KOBAYASHI	280, 282, 286, 288, 290
Yuuji KODERA	312
Asato KUROIWA	165

— M —

Ryuichi MASUDA	248, 252, 254, 258
Kyoichi MORI	342, 344, 390, 394
Junji MORIBE	4, 6
Kazuo MORIWAKI	186
Masaharu MOTOKAWA	16, 19, 21, 23, 25, 26, 44
Takahiro MURAKAMI	260

— N —

Junco NAGATA	304
Atsushi NAKAMOTO	52
Nozomi NAKANISHI	234, 236
Keisuke NAKATA	150, 154, 156, 173, 175, 178
Mitsuo NUNOME	186

— O —

Keiji OCHIAI	307, 314
Go OGURA	272, 300
Satoshi D. OHDACHI	2, 8, 11, 14
Tatsuo OSHIDA	198, 200, 202, 204, 210

— S —

Midori SAEKI	224
Hiroaki SAITO	16
Takashi SAITOH	150, 175, 178, 309
Akira SANO	58, 61, 63, 65, 68, 107, 110, 126, 128, 130, 132
Hiroshi SASAKI	250, 262
Yoshikazu SATO	240
Takuya SHIMADA	206
Miki SHIRAKIHARA	398
Hitoshi SUZUKI	186, 269

— T —

Masaaki TAKIGUCHI	318
Hidetoshi TAMATE	142
Noriko TAMURA	190, 192, 196
Harumi TORII	275

— U —

Kohji URAGUCHI	222, 256

— W —

Shigeki WATANABE	248

— Y —

Fumio YAMADA	212, 214, 216, 218, 272
Tadasu K. YAMADA	333, 338, 350, 352, 354, 356, 358, 360, 362
Koji YAMAZAKI	243
Yasushi YOKOHATA	32, 34, 36, 38, 40, 42, 228
Motoi YOSHIOKA	344, 346, 348, 376, 378, 380, 386, 388, 392, 394, 396, 398, 402

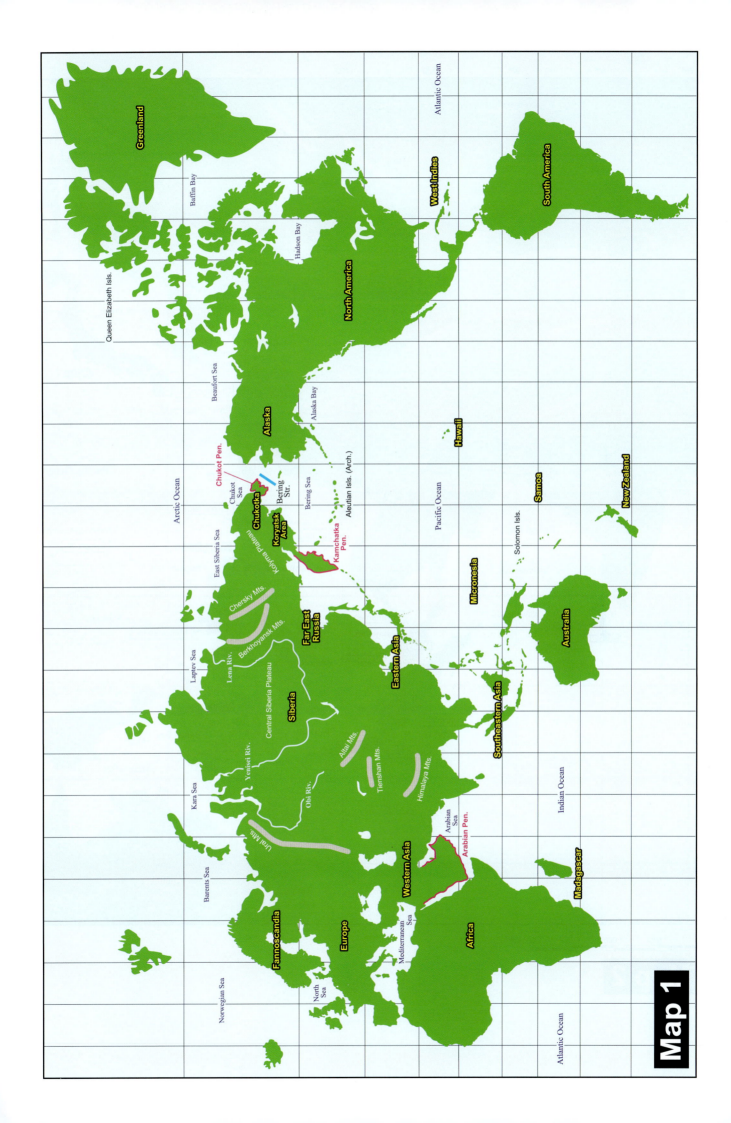

Map 1

Map 2

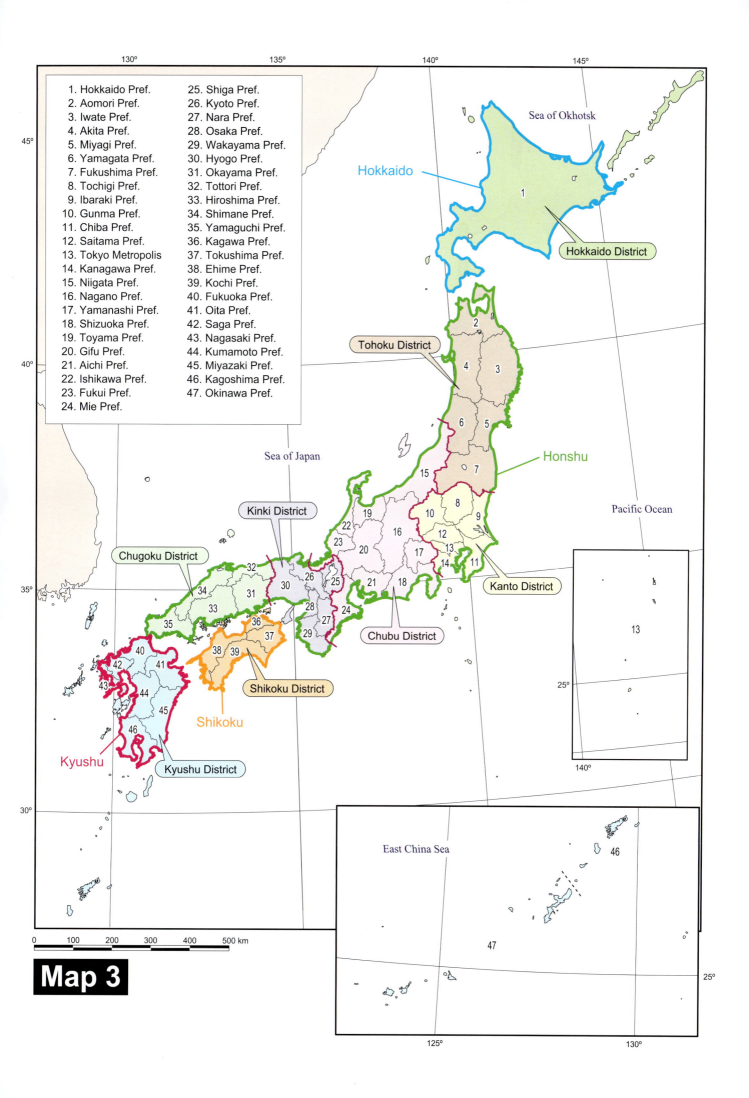

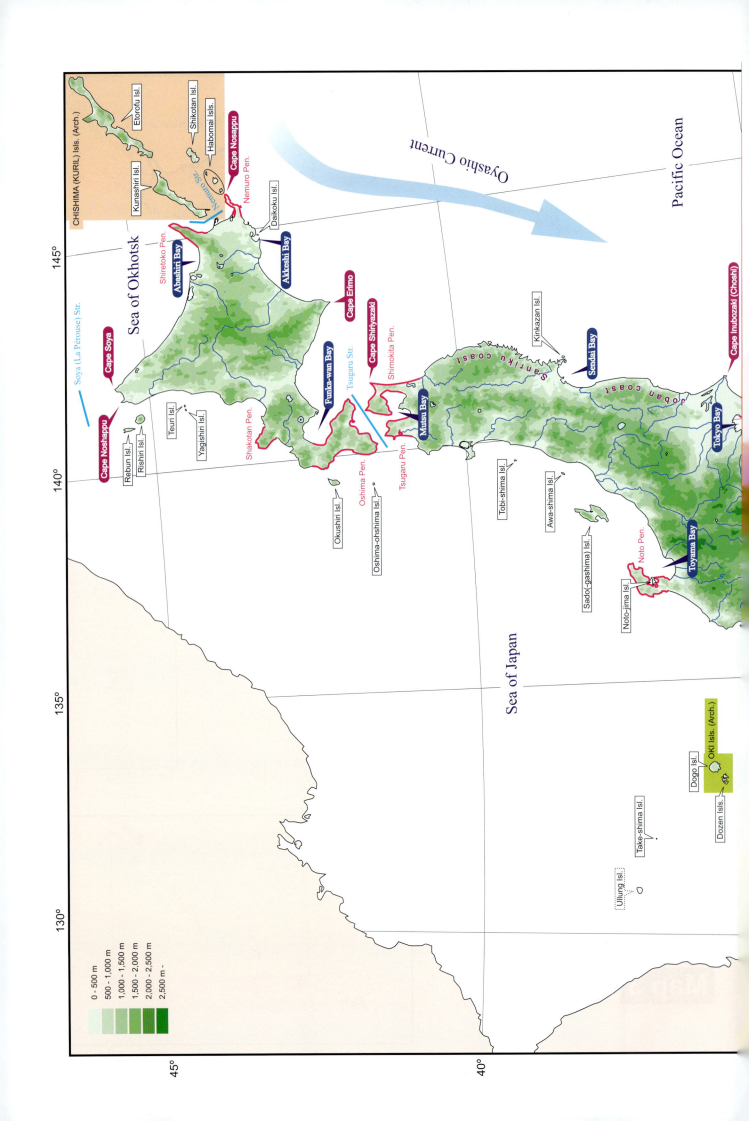

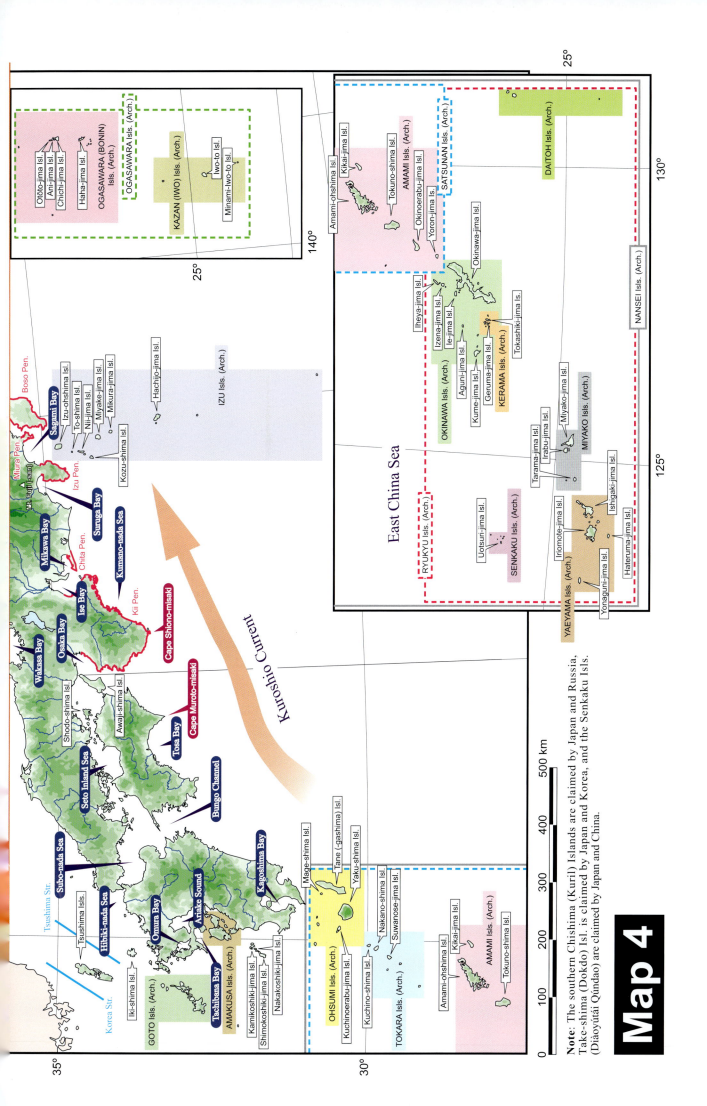